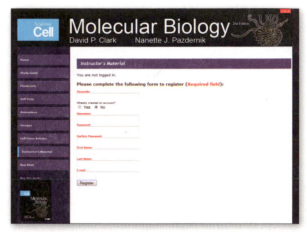

The website includes access to the study guide, self-quizzes, flashcards, Cell Press articles, animations and images

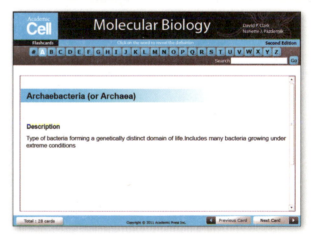

In order to register for access to the site, use the code on the facing page to create your personal username and password. The access code is only needed once—to set up the account.

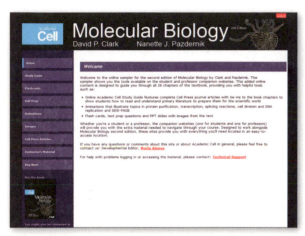

Quizzes are available for every chapter to help students review and prepare for tests.

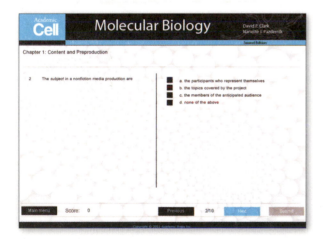

Flashcards are available for all glossary terms and are organized alphabetically, and by chapter.

Biology

Second Edition

David P. Clark

Nanette J. Pazdernik

AMSTERDAM • BOSTON • HEIDELBERG • LONDON
NEW YORK • OXFORD • PARIS • SAN DIEGO
SAN FRANCISCO • SINGAPORE • SYDNEY • TOKYO

Academic Press is an imprint of Elsevier

Academic Press is an imprint of Elsevier
225 Wyman Street, Waltham, MA 02451, USA
The Boulevard, Langford Lane, Kidlington, Oxford, OX5 1GB, UK

Notices
Knowledge and best practice in this field are constantly changing. As new research and experience
broaden our understanding, changes in research methods, professional practices, or medical treatment
may become necessary.

Practitioners and researchers must always rely on their own experience and knowledge in evaluating
and using any information, methods, compounds, or experiments described herein. In using such
information or methods they should be mindful of their own safety and the safety of others, including
parties for whom they have a professional responsibility.

To the fullest extent of the law, neither the Publisher nor the authors, contributors, or editors, assume
any liability for any injury and/or damage to persons or property as a matter of products liability, neg-
ligence or otherwise, or from any use or operation of any methods, products, instructions, or ideas contained in the material herein.

Library of Congress Cataloging-in-Publication Data
Clark, David P.
Molecular biology / David Clark, Nan Pazdernik. – 2nd ed.
   p. cm.
Includes bibliographical references and index.
ISBN 978-0-12-378594-7 (alk. paper)
1. Molecular biology. 2. Molecular genetics. I. Pazdernik, Nanette Jean. II. Title.
QH506.C533 2013
572.8–dc23   2011038218

British Library Cataloguing-in-Publication Data
A catalogue record for this book is available from the British Library

For information on all Academic Press publications
visit our website at www.elsevierdirect.com

Printed in the United States of America
12 13 14 15 10 9 8 7 6 5 4 3 2 1

Dedication

This book is dedicated to Lonnie Russell who was to have been my coauthor on the first edition. A few months after we started this project together, in early July 2001, Lonnie drowned in the Atlantic Ocean off the coast of Brazil in a tragic accident.

DPC

To my family, especially my husband and my three children. They have given me the gift of time, courage, and strength. Time to actually write, courage to continue even when I was tired, and the strength to do my very best work no matter the circumstances.

NJP

Preface to second edition

The last quarter of the 20th century saw major scientific revolutions in genetics and computer technology. Indeed, handling the vast amounts of genetic information generated nowadays depends on advanced computer technology. This book reflects this massive surge in our understanding of the molecular foundations of genetics. Today, we now know that genes are much more than the abstract entities proposed over a century ago by Mendel. Genes are segments of DNA molecules, carrying encoded information. Indeed, genes have now become chemical reagents to be manipulated in the test tube in ever more complex ways. Over the next half century our understanding of how living organisms function at the molecular level, together with our ability to intervene, will expand in ways we are only just beginning to perceive.

A full understanding of how living organisms function includes an appreciation of how cells operate at the molecular level. This is of vital importance to all of us as it becomes ever more clear that molecular factors underlie many health problems and diseases. While cancer is the "classic" case of a disease that only became understandable when its genetic basis was revealed, it is not the only one by any means. Today, the molecular aspects of medicine are expanding rapidly and it is becoming possible to tailor clinical treatment personally by taking into account the genetic make-up of individual patients, an area known as personal genomics.

This book is intended as a survey-oriented textbook for upper-division students in a variety of biological subdisciplines. In particular, it is aimed at final-year undergraduates and beginning graduate students. This book does not attempt to be exhaustive in its coverage, even as a textbook. There is a second book in this series, entitled "Biotechnology," which emphasizes the more practical applications of modern genetics. We hope that both books together effectively survey the foundations and applications of modern molecular genetics.

Some of the students using this book will be well-versed in the basics of modern molecular biology, having taken courses in genetics, biochemistry, and cell biology. However, others will not be so well-prepared, in part due to the continuing influx of students into molecular biology from biology programs that are not oriented in a molecular direction. For them we have tried to create a book whose early chapters cover the basics before launching out into the depths. The first unit of five chapters covers basic cell structure and genetics. This is followed by a survey of DNA, RNA, and proteins and how they interact to provide the cell with genetic information.

Because of the continuing interest in applying molecular biology to an ever-widening array of topics, we have tried to avoid overdoing detail (depth) in favor of breadth. Molecular biology is applicable to more than just human medicine and health. The genetic revolution has also greatly impacted other important areas such as agriculture, veterinary medicine, animal behavior, evolution, and microbiology. Students of these, and related disciplines, will all benefit from an improved understanding of molecular biology.

Changes in the Second Edition

This edition includes significant changes in comparison to the first edition of *Molecular Biology*.

The flood of new sequence information has necessitated updates to many areas of the book, especially Genomics and Systems Biology (Ch. 9), Proteomics (Ch. 15) Bacterial Genetics (Ch. 25), and Molecular Evolution (Ch. 26). Perhaps the most rapidly-changing area of molecular biology at present is the ever-expanding role of RNA. Although scattered items occur throughout the book, particularly in Unit 4, most of the major novel RNA topics, such as CRISPR and long-non-coding RNA, are grouped together in Chapter 18, Regulation at the RNA level.

In this second edition we have re-ordered some of the chapters in a more logical order. In particular, chapters on DNA technology and genomics that were toward the end of the book have been moved far forward. This reflects the much greater role that sequence analysis and genomics have come to play in the last two or three years.

We have also divided the book into modules, each of several related chapters. The first module contains introductory material that experienced students can either skip or skim through rapidly for the reasons cited above. Sections within chapters have been numbered to aid in cross-reference. Review questions and conceptual problems are now provided at the end of each chapter.

A new text element, "Focus on Relevant Research," now appears throughout the book. These feature discussions of recent papers in the field published by Cell Press. The content focuses on helping the student learn how to read and understand primary literature in hopes of preparing them for the scientific world. The complete articles are also provided on the accompanying website for easy reference by students and instructors.

The website also includes access to the Focus on Relevant Research Case Studies that discuss the main topics of each chapter and build case studies around the content to help students understand the basics of primary literature while allowing them to make the appropriate connections to the text.

Other online materials to supplement the text include flashcards, animations, quizzes to prepare for tests, and PowerPoint® slides with images for note-taking. Students also have access to online references as they can then be directly linked to Internet databases, such as PubMed® or ScienceDirect®.

Instructors also have access to the images from the book and test banks based on the text and accompanying journal articles.

We look forward to hearing about your experiences, whether you use our book for teaching or studying. Please send your comments, criticisms, and advice to MolecularBiologyAC2@elsevier.com. Thank you!

David Clark and Nan Pazdernik,
Carbondale, Illinois, April 2011

Acknowledgements

We would like to thank the following individuals for their help in providing information, suggestions for improvement and encouragement: Malikah Abdullah-Israel, Laurie Achenbach, Steven Ackerman, Rubina Ahsan, Kasirajan Ayyanathan, Marilyn Baguinon, Joan Betz, Blake Bextine, Gail Breen, Douglas Burks, Mehmet Candas, Jung-ren Chen, Helen Cronenberger, Phil Cunningham, Dennis Deluca, Linda DeVeaux, Elizabeth De Stasio, Justin DiAngelo, Susan DiBartolomeis, Brian Downes, Ioannis Eleftherianos, Robert Farrell, Elizabeth Blinstrup Good, Joyce Hardy, David L. Herrin, Walter M. Holmes, Karen Jackson, Mark Kainz, Nemat Keyhani, Rebecca Landsberg, Richard LeBaron, Richard Londraville, Larry Lowe, Charles Mallery, Boriana Marintcheva, Stu Maxwell, Michelle McGehee, Ana Medrano, Thomas Mennella, Donna Mueller, Khalil Nezhad , Dan Nickrent, Monica Oblinger, Rekha Patel, Marianna Patrauchan, Neena Philips, Wanda Reygaert, Veronica Riha, Phillip Ryals, Donald Seto, Dan Simmons, Joan Slonczewski, Malgosia Wilk-Blaszczak, Hongzhuan Wu and Ding Xue.

Contents

CHAPTER 7 *Cloning Genes for Analysis*   194

CHAPTER 8 *DNA Sequencing*   227

CHAPTER 9 *Genomics & Systems Biology*   248

UNIT 3 THE CENTRAL DOGMA OF MOLECULAR BIOLOGY   273

CHAPTER 10 *Cell Division and DNA Replication*   274

Basic Chemical and Biological Principles

Chapter 1

Cells and Organisms

Before tackling the complex details of biology at the molecular level, we need to get familiar with the subject of our investigation—the living world. First, we will consider what it means to be alive and then we shall survey a range of cells and organisms that are often studied by molecular biologists.

Life is impossible to define exactly, but a general idea is sufficient here. Living things consist of cells—some consist of a single cell, whereas others are made from assemblies of many million cells. Whatever the situation, living cells must grow, divide, and pass on their characteristics to their offspring. Molecular biology focuses on the details of growth and division. In particular, we are interested in how division is arranged so that each descendent can inherit their parents characteristics.

Scientists have devoted much effort in investigating certain favored organisms. In some cases this is a matter of convenience—bacteria, yeast, and other single-celled microorganisms are relatively easy to investigate. In other cases it is due to self-interest. Mice—and some other animals—reveal much about humans, plants provide our food, and viruses make us sick.

1. What Is Life?

Although there is no definition of life that suits all people, everyone has an idea of what being alive means. Generally, it is accepted that something is alive if it can *grow* and *reproduce*, at least during some stage of its existence. Thus, we still regard adults who are no longer growing and those individuals beyond reproductive age as being alive. We also regard sterile individuals, such as mules or worker bees as being alive, even though they lack the ability to reproduce. Part of the difficulty in defining life is the complication introduced by multicellular organisms. Although a multicellular organism as a whole may not grow or reproduce some of its cells may still retain these abilities.

No satisfactory technical definition of life exists. Despite this we understand what life entails. In particular, life involves a dynamic balance between duplication and alteration.

The basic ingredients needed to sustain life include the following:

- *Genetic information* Biological information is carried by the **nucleic acid** molecules, **deoxyribonucleic acid** (**DNA**) and **ribonucleic acid** (**RNA**). The units of genetic information are known as **genes**, and each consists physically of a segment of a nucleic acid molecule. DNA is used for long-term storage of large amounts of genetic information (except by some viruses—see Ch. 21). Whenever genetic information is actually used, working copies of the genes are carried on RNA. The total genetic information possessed by an organism is known as its **genome.** DNA genomes are maintained by **replication**, a process where a copy of the DNA is produced by enzymes and then passed on to the daughter cells.

- *Mechanism for energy generation* By itself, information is useless. Energy is needed to put the genetic information to use. Living creatures must all obtain energy for growth and reproduction. **Metabolism** is the set of processes in which energy is acquired, liberated, and used for biosynthesis of cell components, then catabolized and recycled. Living organisms use raw material from the environment to grow and reproduce.

- *Machinery for making more living matter* Synthesis of new cell components requires chemical machinery. In particular, the **ribosomes** are needed for making proteins, the **macromolecules** that make up the bulk of all living tissue. These tiny subcellular machines allow the organism to grow and maintain itself.

- *A characteristic outward physical form* Living creatures all have a material body that is characteristic for each type of life-form. This structure contains all the metabolic and biosynthetic machinery for generating energy and making new living matter. It also contains the DNA molecules that carry the genome. The form has levels of organization from single cells to tissues, organs, and organ systems of multicellular life-forms.

- *Identity or Self* All living organisms have what one might call an identity. The term self-replication implies that an organism knows to make a copy of itself— not merely to assemble random organic material. This concept of "self" versus "non-self" is very evident in the immune systems that protect higher animals against disease. But even primitive creatures attempt to preserve their own existence.

deoxyribonucleic acid (DNA)   The nucleic acid polymer of which the genes are made
gene   A unit of genetic information
genome   The entire genetic information from an individual
macromolecule   Large polymeric molecule; in living cells especially DNA, RNA, protein, or polysaccharide
metabolism   The processes by which nutrient molecules are transported and transformed within the cell to release energy and to provide new cell material
nucleic acid   Polymer made of nucleotides that carries genetic information
replication   Duplication of DNA prior to cell division
ribonucleic acid (RNA)   Nucleic acid that differs from DNA in having ribose in place of deoxyribose and having uracil in place of thymine
ribosome   The cell's machinery for making proteins

- *Ability to reproduce*   The organism uses energy and raw materials to make itself, and then uses the same materials to produce offspring. Some organisms simply reproduce with asexual reproduction (making offspring without creating gametes) and other organisms use sexual reproduction (two gametes fuse to form a new organism).
- *Adaptation*   The most important characteristic of a living creature is the ability to adapt to its current environment. This concept also encompasses evolution or the adaptations that get passed from generation to generation.

2. Living Creatures Are Made of Cells

Looking around at the living creatures that inhabit this planet, one is first struck by their immense variety: squids, seagulls, sequoias, sharks, sloths, snakes, snails, spiders, strawberries, soybeans, *Saccharomyces cerevisiae*, and so forth. Although highly diverse to the eye, the biodiversity represented by these creatures is actually somewhat superficial. The most fascinating thing about life is not its superficial diversity but its *fundamental unity*. All of these creatures, together with microscopic organisms too small to see with the naked eye, are made up of **cells,** structural units or compartments that have more or less the same components.

The idea that living cells are the structural units of life was first proposed by Schleiden and Schwann in the 1830s. Cells are microscopic structures that vary considerably in shape. Many are spherical, cylindrical, or roughly cuboidal, but many other shapes are found, such as the long-branched filaments of nerve cells. Many microscopic life-forms consist of a single cell, whereas creatures large enough to see usually contain thousands of millions. Each cell is enclosed by a cell membrane composed of **proteins** and **phospholipids** and contains a complete copy of the genome (at least at the start of its life). Living cells possess the machinery to carry out metabolic reactions and generate energy and are usually able to grow and divide. Moreover, living cells always result from the division of pre-existing cells, and they are never assembled from their component parts. This implies that living organisms too can only arise from pre-existing organisms. In the 1860s, Louis Pasteur confirmed experimentally that life cannot arise spontaneously from organic matter. Sterilized nutrient broth did not "spoil" or "go bad" unless it was exposed to microorganisms in the air.

In most multicellular organisms, the cells are specialized in a variety of ways (Fig. 1.01). The development of specialized roles by particular cells or whole tissues is referred to as **differentiation**. For example, the red blood cells of mammals lose their nucleus and the enclosed DNA during development. Once these cells are fully differentiated, they can perform only their specialized role as oxygen carriers and can no longer grow and divide. Some specialized cells remain functional for the life span of the individual organism, whereas others have limited life spans, sometimes lasting only a few days or hours. For multicellular organisms to grow and reproduce, some cells clearly need to keep a complete copy of the genome and retain the ability to become another organism. The other differentiated cells perform their function, but do not retain the ability to create an entirely new organism. In single-celled organisms, such as **bacteria** or protozoa, each individual cell has a complete genome and can grow and reproduce; hence, every cell is essentially the same.

> Matter is divided into atoms. Genetic information is divided into genes. Living organisms are divided into cells.

---

bacteria   Primitive, relatively simple, single-celled organisms that lack a cell nucleus
cell   The cell is the basic unit of life. Each cell is surrounded by a membrane and usually has a full set of genes that provide it with the genetic information necessary to operate
differentiation   Progressive changes in the structure and gene expression of cells belonging to a single organism that leads to the formation of different types of cell
phospholipid   A hydrophobic molecule found making up cell membranes and consisting of a soluble head group and two fatty acids both linked to glycerol phosphate
protein   Polymer made from amino acids that does most of the work in the cell

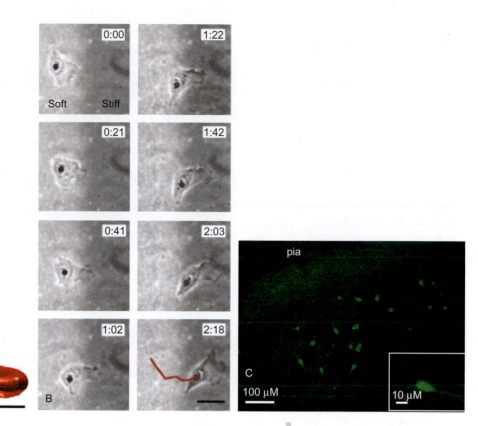

FIGURE 1.01
Some Cells Differentiate

Cells differentiate into all different shapes and sizes to provide specialized functions in a multicellular organism. In this figure, red blood cells (A) are specialized to exchange carbon dioxide and oxygen in the tissues of humans; fibroblasts (B) provide support to various organs; and neurons (C) transmit signals from the environment to the brain to elicit a response. *(Credit: A) Esposito, et al. (2010) Biophysical J 99(3): 953–960. B) Dokukina and Gracheva (2010) Biophysical J 98(12): 2794–2803. C) Fino and Yuste (2011) Neuron 69(6): 1188–1203.)*

2.1. Essential Properties of a Living Cell

At least in the case of unicellular organisms, each cell must possess the characteristics of life as discussed above. Each living cell must generate its own energy and synthesize its own macromolecules. Each must have a genome, a set of genes carried on molecules of DNA. (Partial exceptions occur in the case of multicellular organisms, where responsibilities may be distributed among specialized cells and some cells may lack a complete genome.)

A cell must also have a surrounding **membrane** that separates the cell interior, the **cytoplasm**, from the outside world. The cell membrane, or cytoplasmic membrane, is made from a double layer of phospholipids together with proteins (Fig. 1.02). Phospholipid molecules consist of a water-soluble head group, including phosphate, found at the surface of the membrane, and a lipid portion consisting of two hydrophobic chains that form the body of the membrane (Fig. 1.03). The phospholipids form a hydrophobic layer that greatly retards the entry and exit of water-soluble molecules. For the cell to grow, it must take up nutrients. For this, transport proteins that transverse the membranes are necessary. Many of the metabolic reactions involved in the breakdown of nutrients to release energy are catalyzed by soluble enzymes located in the cytoplasm. Other energy-yielding series of reactions, such as the respiratory chain or the photosynthetic system, are located in membranes. The proteins may be within or attached to the membrane surfaces.

The cytoplasmic membrane is physically weak and flexible. Many cells therefore have a tough structural layer, the cell wall, outside the cell membrane. Most bacterial and plant cells have hard cell walls, though animal cells usually do not. Thus, a cell wall is not an essential part of a living cell.

Membranes do not merely separate living tissue from the non-living exterior. They are also the site of many biosynthetic and energy-yielding reactions.

---

cytoplasm The portion of a cell that is inside the cell membrane but outside the nucleus
membrane A thin flexible structural layer made of protein and phospholipid that is found surrounding all living cells

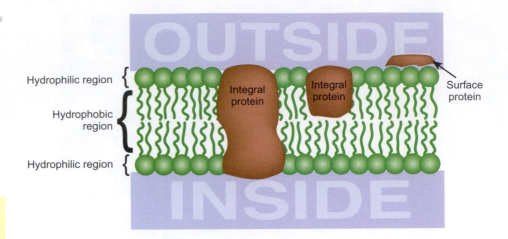

FIGURE 1.02
A Biological Membrane

A biological membrane is formed by phospholipid and protein. The phospholipid layers are oriented with their hydrophobic tails inward and their hydrophilic heads outward. Proteins may be within the membrane (integral) or lying on the membrane surfaces.

Some single-celled protozoa, such as *Paramecium*, have multiple nuclei within each single cell. In addition, in certain tissues of some multicellular organisms several nuclei may share the same cytoplasm and be surrounded by only a single cytoplasmic membrane. Such an arrangement is known as a syncytium when it is derived from multiple fused cells.

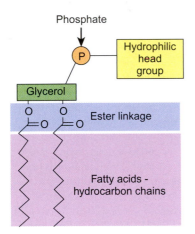

FIGURE 1.03
Phospholipid Molecules

Phospholipid molecules of the kind found in membranes have a hydrophilic head group attached via a phosphate group to glycerol. Two fatty acids are also attached to the glycerol via ester linkages.

Based on differences in compartmentalization, living cells may be divided into two types, the simpler **prokaryotic** cell and the more complex **eukaryotic** cell. By definition, prokaryotes are those organisms whose cells are not subdivided by membranes into a separate **nucleus** and cytoplasm. All prokaryote cell components are located together in the same compartment. In contrast, the larger and more complicated cells of higher organisms (animals, fungi, plants, and protists) are subdivided into separate compartments and are called eukaryotic cells. Figure 1.04 compares the design of prokaryotic and eukaryotic cells.

Another property of living cells are soluble enzymes located in the cytoplasm. Cellular enzymes catalyze biosynthesis of the low molecular weight precursors to protein and nucleic acids. Protein synthesis requires a special organelle, the ribosome. This is a subcellular machine that consists of several molecules of RNA and around 50 proteins. It uses information that is carried from the genome by special RNA molecules, known as **messenger RNA**. The ribosome decodes the nucleic acid-encoded genetic information on the messenger RNA to make protein molecules.

2.2. Prokaryotic Cells Lack a Nucleus

Bacteria (singular, bacterium) are the simplest living cells and are classified as prokaryotes. Bacterial cells (Fig. 1.05) are always surrounded by a membrane (the cell or cytoplasmic membrane) and usually also by a cell wall. Like all cells, they contain all the essential chemical and structural components necessary for the type of life they lead. Typically, each bacterial cell has a single **chromosome** carrying a full set of genes providing it with the genetic information necessary to operate as a living organism. Most bacteria have 3,000–4,000 genes, although some have as few as 500.

The minimum number of genes needed to allow the survival of a living cell is uncertain. *Mycoplasma genitalium* has the smallest genome of any cultured bacterium. Its 485 genes have been systematically disrupted and around 100 are dispensable. Although this suggests around 385 genes are needed, some are duplicated in function. In addition, *M. genitalium* relies on a host organism to survive, and therefore, it lacks genes for making certain essential components. The entire genome of *M. genitalium* has recently been successfully synthesized and re-introduced into a cell in order to continue this analysis.

chromosome   Structure containing the genes of a cell and made of a single molecule of DNA
eukaryote   Higher organism with advanced cells, which have more than one chromosome within a compartment called the nucleus
messenger RNA (mRNA)   The class of RNA molecule that carries genetic information from the genes to the rest of the cell
nucleus   An internal compartment surrounded by the nuclear membrane and containing the chromosomes. Only the cells of higher organisms have nuclei
prokaryote   Lower organism, such as a bacterium, with a primitive type of cell containing a single chromosome and having no nucleus

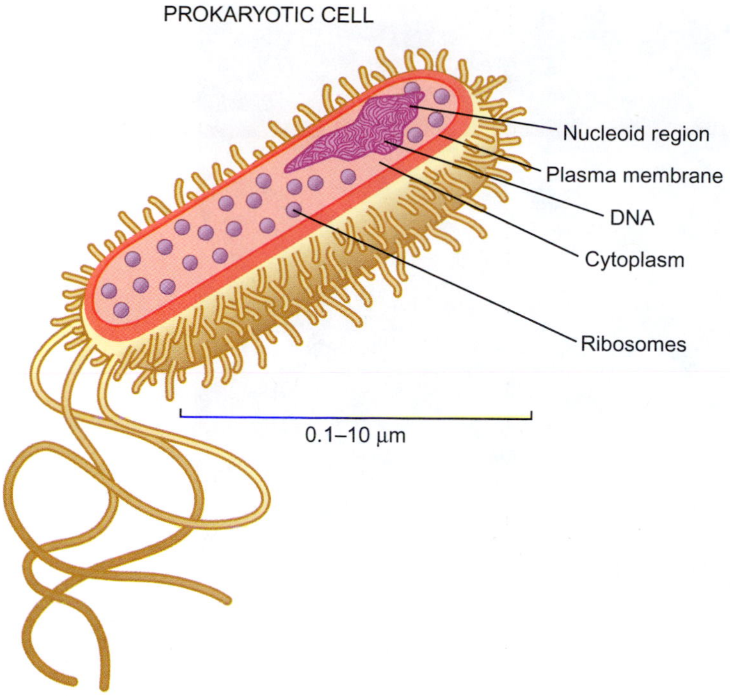

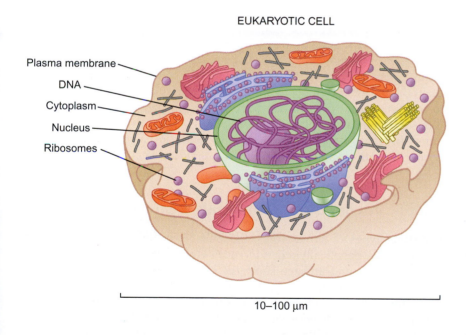

PROKARYOTIC CELL

Nucleoid region

Plasma membrane

DNA

Cytoplasm

Ribosomes

0.1–10 μm

EUKARYOTIC CELL

Plasma membrane

DNA

Cytoplasm

Nucleus

Ribosomes

10–100 μm

■ **FIGURE 1.04**
Typical Prokaryotic Cell

The components of a typical prokaryote, a bacterium, are depicted. There is no nucleus and the DNA is free in the cytoplasm where it is compacted into the nucleoid.

■ **FIGURE 1.05**
Typical Eukaryotic Cell

A typical eukaryotic cell showing a separate compartment called the nucleus that contains the DNA.

A typical bacterial cell, such as ***Escherichia coli***, is rod shaped and about two or three microns long and one micron wide. Bacteria are not limited to a rod shape (Fig. 1.06); spherical, filamentous, or spirally-twisted bacteria are also found. Occasional giant bacteria occur, such as *Epulopiscium fishelsoni*, which inhabits the

Escherichia coli   A bacterium commonly used in molecular biology

FIGURE 1.06
Bacillus subtilis *Cells Visualized by Scanning Electron Microscopy*

Scanning electron microscopy of *B. subtilis* shows tube shaped connections between bacterial cells. (Credit: Dubey, et al. (2011) Cell 144(4): 590–600.)

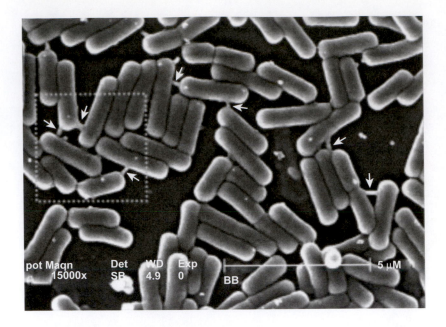

A micrometer (μm), also known as a micron, is a millionth of a meter (i.e., 10^{-6} meter).

Cells are separated from their environments by membranes. In the more complex cells of eukaryotes, the genome is separated from the rest of the cell by another set of membranes.

If higher organisms disappeared from the Earth, the prokaryotes would survive and evolve. They do not need us although we need them.

surgeonfish and measures a colossal 50 microns by 500 microns—an organism visible to the naked eye. In contrast, typical eukaryotic cells are 10 to 100 microns in diameter.

A smaller cell has a larger surface-to-volume ratio. Smaller cells transport nutrients relatively faster, per unit mass of cytoplasm (i.e., cell contents), and so can grow more rapidly than larger cells. Because bacteria are less structurally complex than animals and plants, they are often referred to as "lower organisms." However, it is important to remember that present-day bacteria are at least as well adapted to modern conditions as animals and plants and are just as highly evolved as so-called "higher organisms." In many ways, bacteria are not so much "primitive" as specialized for growing more efficiently in many environments than larger and more complex organisms.

3. Eubacteria and Archaea Are Genetically Distinct

There are two distinct types of prokaryotes, the **eubacteria** and **archaebacteria** or **archaea**, which are no more genetically related to each other than either group is to the eukaryotes. Both eubacteria and archaea show the typical prokaryotic structure—in other words, they both lack a nucleus and other internal membranes. Thus, cell structure is of little use for distinguishing these two groups. The eubacteria include most well-known bacteria, including all those that cause disease. When first discovered, the archaea were regarded as strange and primitive. This was largely because most are found in extreme environments (Fig. 1.07) and/or possessed unusual metabolic pathways. Some grow at very high temperatures; others in very acidic conditions; and others in very high salt. The only major group of archaea found under "normal" conditions

archaebacteria (or archaea)   Type of bacteria forming a genetically-distinct domain of life. Includes many bacteria growing under extreme conditions
eubacteria   Bacteria of the normal kind as opposed to the genetically-distinct *Archaebacteria*

■ **FIGURE 1.07**
Hot springs in Ethiopia

Hot springs are good sites to find archaea. These springs are in the Dallol area of the Danakil Depression, 120 meters below sea level. The Danakil Depression of Ethiopia is part of the East African Rift Valley. Hot water flows from underground to form these pools. The water is heated by volcanic activity and is at high pressure, causing minerals in the rock to dissolve in the water. The minerals precipitate out as the water cools at the surface, forming the deposits seen here. *(Credit: Bernhard Edmaier, Science Photo Library.)*

is the methane bacteria, which, however, have a very strange metabolism. They contain unique enzymes and co-factors that allow the formation of methane by a pathway found in no other group of organisms. Despite this, the **transcription** and **translation** machinery of archaea resembles that of eukaryotes, so they turned out to be neither fundamentally strange nor truly primitive when further analyzed.

Biochemically, there are major differences between the eubacterial and archaeal cells. In all cells, the cell membrane is made of phospholipids, but the nature and linkage of the lipid portion is quite different in the eubacteria and archaea (Fig. 1.08). The cell wall of eubacteria is always made of **peptidoglycan**, a molecule unique to this group of organisms. Archaea often have cell walls, but these are made of a variety of materials in different **species**, and peptidoglycan is never present. Thus, the only real cellular structures possessed by prokaryotes, the cell membrane and cell wall, are in fact chemically different in these two groups of prokaryotes. The genetic differences will be discussed later when molecular evolution is considered (see Ch. 26).

At a fundamental level, three domains of life, eubacteria, archaea, and eukaryotes, have replaced the old-fashioned division of animal and vegetable.

4. Eukaryotic Cells Are Subdivided into Compartments

A eukaryotic cell has its genome inside a separate compartment, the nucleus. In fact, eukaryotic cells have multiple internal cell compartments surrounded by membranes. The nucleus itself is surrounded by a double membrane, the **nuclear envelope**, which separates the nucleus from the cytoplasm, but allows some communication with the

nuclear envelope   Envelope consisting of two concentric membranes that surrounds the nucleus of eukaryotic cells
peptidoglycan   Mixed polymer of carbohydrate and amino acids that comprises the structural layer of bacterial cell walls
species   A group of closely-related organisms with a relatively recent common ancestor. Among animals, species are populations that breed among themselves but not with individuals of other populations. No satisfactory definition exists for bacteria or other organisms that do not practice sexual reproduction
transcription   Process by which information from DNA is converted into its RNA equivalent
translation   Making a protein using the information provided by messenger RNA

·FIGURE 1.08
Lipids of Archaea

In eubacteria and eukaryotes, the fatty acids of phospholipids are esterified to the glycerol. In archaea, the lipid portion consists of branched isoprenoid hydrocarbon chains joined to the glycerol by ether linkages (as shown here). Such lipids are much more resistant to extremes of pH, temperature, and ionic composition.

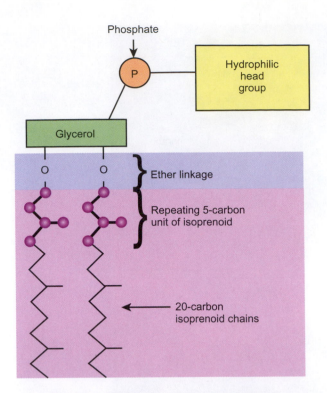

FIGURE 1.09
A Eukaryote Has Multiple Cell Compartments

False color transmission electron micrograph of a plasma cell from bone marrow. Multiple compartments surrounded by membranes, including a nucleus, are found in eukaryotic cells. Characteristic of plasma cells is the arrangement of heterochromatin (orange) in the nucleus, where it adheres to the inner nuclear membrane. Also typical is the network of rough endoplasmic reticulum (yellow dotted lines) in the cytoplasm. The oval or rounded crimson structures in the cytoplasm are mitochondria. Magnification ×4,500. *(Credit: Dr. Gopal Murti, Science Photo Library.)*

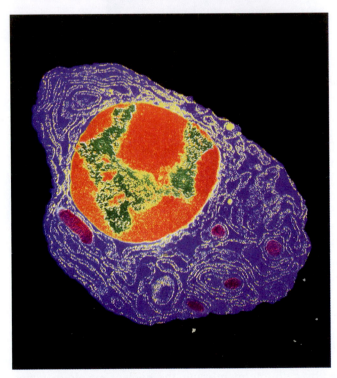

cytoplasm via **nuclear pores** (Fig. 1.09). The genome of eukaryotes consists of 10,000–50,000 genes carried on several chromosomes. Eukaryotic chromosomes are linear, unlike the circular chromosomes of bacteria. Most eukaryotes are diploid, with two copies of each chromosome. Consequently, they possess at least two copies of each gene. In addition, eukaryotic cells often have multiple copies of certain genes as the result of gene duplication.

nuclear pore   Pore in the nuclear membrane through which the nucleus communicates with the cytoplasm

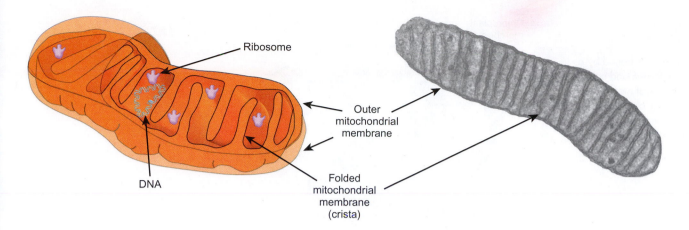

Ribosome

Outer
mitochondrial
membrane

Folded
mitochondrial
membrane
(crista)

DNA

FIGURE 1.10
Mitochondrion

A mitochondrion is surrounded by
two concentric membranes. The
inner membrane is folded inward
to form **cristae**. These are the
site of the respiratory chain that
generates energy for the cell.

Eukaryotes possess a variety of other **organelles**. These are subcellular structures that carry out specific tasks. Some are separated from the rest of the cell by membranes (so-called **membrane-bound organelles**) but others (e.g., the ribosome) are not. The **endoplasmic reticulum** is a membrane system that is continuous with the nuclear envelope and permeates the cytoplasm. The **Golgi apparatus** is a stack of flattened membrane sacs and associated vesicles that is involved in secretion of proteins, or other materials, to the outside of the cell. **Lysosomes** are membrane-bound structures containing degradative enzymes and specialized for digestion.

All except a very few eukaryotes contain **mitochondria** (singular, mitochondrion; Fig. 1.10). These are generally rod-shaped organelles, bounded by a double membrane. They resemble bacteria in their overall size and shape. As will be discussed in more detail (see Ch. 4), it is thought that mitochondria are indeed evolved from bacteria that took up residence in the primeval ancestor of eukaryotic cells. Like bacteria, mitochondria each contain a circular molecule of DNA. The mitochondrial genome is similar to a bacterial chromosome, though much smaller. The mitochondrial DNA has some genes needed for mitochondrial function.

Mitochondria are specialized for generating energy by respiration and are found in all eukaryotes. (A few eukaryotes are known that cannot respire; nonetheless, these retain remnant mitochondrial organelles—see below.) In eukaryotes, the enzymes of respiration are located on the inner mitochondrial membrane, which has numerous infoldings to create more membrane area. This contrasts with bacteria, where the respiratory chain is located in the cytoplasmic membrane, as no mitochondria are present.

Chloroplasts are membrane-bound organelles specialized for photosynthesis (Fig. 1.11). They are found only in plants and some single-celled eukaryotes. They are oval- to rod-shaped and contain complex stacks of internal membranes that contain the green, light-absorbing pigment **chlorophyll** and other components needed for trapping light energy. Like mitochondria, chloroplasts contain a circular DNA molecule and are thought to have evolved from a photosynthetic bacterium.

Eukaryotic cells have extensive intracellular architecture to maintain their shape and move materials and organelles around the cells. The **cytoskeleton** is a complex

chlorophyll   Green pigment that absorbs light during photosynthesis
crista (plural cristae)   Infolding of the respiratory membranes of mitochondria
cytoskeleton   Internal structural elements in eukaryotic cells that keep the cellular shape and provide structures to move intracellular materials and organelles from one location to another
endoplasmic reticulum   Internal system of membranes found in eukaryotic cells
Golgi apparatus   A membrane-bound organelle that takes part in export of materials from eukaryotic cells
lysosome   A membrane-bound organelle of eukaryotic cells that contains degradative enzymes
membrane-bound organelles   Organelles that are separated from the rest of the cytoplasm by membranes
mitochondrion   Membrane-bound organelle found in eukaryotic cells that produces energy by respiration
organelle   Subcellular structure that carries out a specific task. Membrane-bound organelles are separated from the rest of the cytoplasm by membranes, but other organelles such as the ribosome are not.

FIGURE 1.11
Chloroplast

The chloroplast is bound by a double membrane and contains infolded stacks of membrane specialized for photosynthesis. The chloroplast also contains ribosomes and DNA.

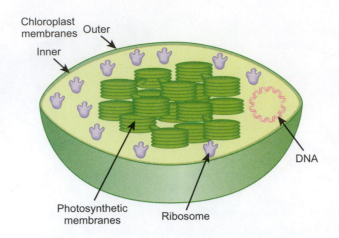

Chloroplast membranes   Outer
Inner
DNA
Photosynthetic membranes   Ribosome

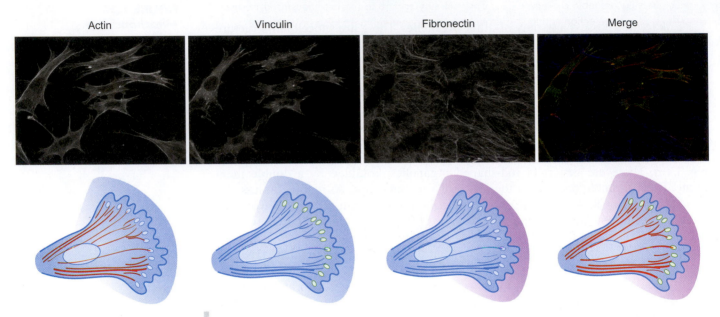

Actin   Vinculin   Fibronectin   Merge

FIGURE 1.12
Cytoskeleton

Actin, vinculin, and fibronectin are three cytoskeletal proteins that give this cell a flattened edge. This edge has adhesions that connect the cell to the dish *in vitro*, but function to keep the cell attached to other cells within the organs of a multicellular organism. *(Credit: Byron, et al. (2010) Curr Biol 20(24): R1063–R1067)*

Life is modular. Complex organisms are subdivided into organs. Large and complex cells are divided into organelles.

Eukaryotes have many membrane-bound organelles to perform functions like respiration (mitochondria), enzyme degradation (lysosomes), and protein processing and secretion (Golgi apparatus and endoplasmic reticulum).

Eukaryotic cells have internal structural elements called a cytoskeleton.

network of filaments made of proteins like **actin**, vinculin, and fibronectin (Fig. 1.12). Besides maintaining cell shape, the cytoskeleton is important for cellular transport. For example, cytoskeletal fibers run through the long axons of neurons, and vesicles filled with neurotransmitters travel up and down the axon to facilitate the communication between the nucleus and the nerve fibers. The cytoskeleton also initiates cellular movements. By increasing the length of fibers on one side of the cell and decreasing their length on the opposite side, the cell can physically move. This is especially true for smaller single-cell eukaryotes and for movements during a multicellular organism's development. Finally, these cytoskeletal movements are important to processes like cell division, since the very same fibers make up the spindle.

actin   A long filament of small subunits that is a component of cellular cytoskeleton

5. The Diversity of Eukaryotes

Unlike prokaryotes that fall into two distinct genetic lineages (the eubacteria and archaebacteria), all eukaryotes are genetically related, in the sense of being ultimately derived from the same ancestor. Perhaps this is not surprising since all eukaryotes share many advanced features that the prokaryotes lack. When it is said that all eukaryotes are genetically related, this refers to the nuclear part of the eukaryotic genome, not the mitochondrial or chloroplast DNA molecules that have become part of the modern eukaryotic cell.

Box 1.01 Anaerobic Eukaryote

Some single-celled eukaryotes lack true respiratory mitochondria and must grow by fermentation. For example, **Entamoeba** *histolytica* invades and destroys the tissues of the intestines, causing amoebic dysentery (Fig. 1.13). It may spread to the liver causing abscesses to develop. This infection is acquired from flies, or by contaminated food or water.

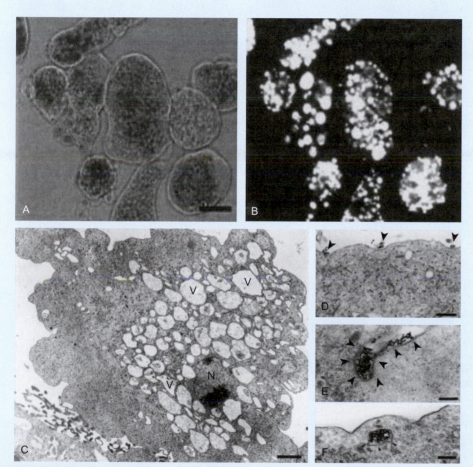

FIGURE 1.13
Entamoeba: An Anaerobic Eukaryote

Endocytosis in Entamoeba histolytica—Phase contrast (a) and acridine orange fluorescence (b) images showed many acidic vesicles and vacuoles in the cytoplasm of *Entamoeba*. Ultrathin section (c) of a trophozoite confirmed that the cytoplasm is filled with vacuoles (v); gold-labeled lactoferrin could be found bound to parasite surface (arrowheads in d), inside peripheral tubules (arrowheads in e) and vesicles (arrowheads in e). Bars: 17 μm (a, b); 1.6 μm (c); 400 nm (d); 500 nm (e); 170 nm (f). (Credit: de Souza, et al. (2009) Progress in Histochem Cytochem 44(2): 67–124.)

Entamoeba   A very primitive single-celled eukaryote that lacks mitochondria

A wide variety of eukaryotes live as microscopic single cells. However, the most visible eukaryotes are larger multicellular organisms that are visible to the naked eye. Traditionally, these higher organisms have been divided into the plant, fungus, and animal kingdoms. This classification must be modified to include several new groups to account for the single-celled eukaryotes. Some single-celled eukaryotes may be viewed as plants, fungi, or animals. Others are intermediate or possess a mixture of properties and need their own miniature kingdoms.

6. Haploidy, Diploidy, and the Eukaryote Cell Cycle

Most bacteria are **haploid**, having only one copy of each gene. Eukaryotes are normally **diploid**, having two copies of each gene carried on pairs of homologous chromosomes. While this is true of the majority of multicellular animals and many single-celled eukaryotes, there are significant exceptions. Many plants are **polyploid**, especially angiosperms (flowering plants). About half of the present-day angiosperms are thought to be polyploid, especially tetraploid or hexaploid. For example, coffee (ancestral haploid number = 11) exists as variants with 22, 44, 66, or 88 chromosomes (i.e., 2n, 4n, 6n, and 8n). Polyploid plants have larger cells, and the plants themselves are often larger. In particular, polyploids have often been selected among domesticated crop plants, since they tend to give bigger plants with higher yields (Table 1.01).

Polyploidy is unusual in animals, being found in occasional insects and reptiles. So far the only polyploid mammal known is a rat from Argentina that was discovered to be tetraploid in 1999. It actually has only 102 chromosomes, having lost several from the original tetraploid set of 4n = 112. The tetraploid rat has larger cells than its diploid relatives. The only haploid animal known is an arthropod, a mite, *Brevipalpus phoenicis*, which was discovered in 2001. Infection of these mites by an endosymbiotic bacterium causes feminization of the males. The genetic females of this species reproduce by parthenogenesis (i.e., development of unfertilized eggs into new individuals).

In most animals, only the gametes, the egg and sperm cells, are haploid. After mating, two haploid gametes fuse to give a diploid zygote that develops into a new animal. However, in plants and fungi, haploid cells often grow and divide for several generations before producing the actual gametes. It seems likely that in the ancestral eukaryote a phase consisting of haploid cells alternated with a diploid phase. In yeasts, both haploid and diploid cells may be found and both types grow and divide in essentially the same manner (see above). In lower plants, such as mosses and liverworts, the haploid phase, or **gametophyte**, may even form a distinct multicellular plant body.

TABLE 1.01	Polyploidy in Crop Plants		
Plant	Ancestral Haploid Number	Chromosome Number	Ploidy Level
wheat	7	42	6n
domestic oat	7	42	6n
peanut	10	40	4n
sugar cane	10	80	8n
white potato	12	48	4n
tobacco	12	48	4n
cotton	13	52	4n

diploid   Possessing two copies of each gene
gametophyte   Haploid phase of a plant, especially of lower plants such as mosses and liverworts, where it forms a distinct multicellular body
haploid   Possessing only a single copy of each gene
polyploidy   Possessing more than two copies of each gene

During animal development, there is an early division into **germline** and **somatic** cells. Only cells from the germline can form gametes and contribute to the next generation of animals. Somatic cells have no long-term future but grow and divide only as long as the individual animal continues to live. Hence, genetic defects arising in somatic cells cannot be passed on through the gametes to the next generation of animals. However, they may be passed on to other somatic cells. Such somatic inheritance is of great importance as it provides the mechanism for cancer. In plants and fungi there is no rigid division into germline and somatic cells. The cells of many higher plants are **totipotent**. In other words, a single cell from any part of the plant has the potential to develop into a complete new plant, which can develop reproductive tissues and produce gametes. This is not normally possible for animal cells. (The experimental cloning of animals such as Dolly the sheep is an artificial exception to this rule.)

> Many eukaryotes alternate between haploid and diploid phases. However, the properties and relative importance of the two phases varies greatly with the organism.

> The concept of germline versus somatic cells applies to animals but not to other higher organisms.

7. Organisms Are Classified

Living organisms are referred to by two names, both printed in italics; for example, *Escherichia coli* or *Saccharomyces cerevisiae*. The first name refers to the **genus** (plural, genera), a group of closely-related species. After its first use in a publication, the genus name is often abbreviated to a single letter, as in "*E. coli.*" After the genus, the scientific name contains the species, or individual, name. The genus and species are the smallest subdivision of the system of biological classification. Classification of living organisms facilitates the understanding of their origins and the relationships of their structure and function. In order to classify organisms, they are first assigned to one of the three **domains** of life, which are eubacteria, archaea, and eukaryotes. Next, the domains are divided into **kingdoms**. Within domain Eukarya, there are four kingdoms:

- *Protista*—An artificial accumulation of primitive, mostly single-celled eukaryotes often referred to as protists that don't belong to the other three main kingdoms. There are several groups that are distinct enough that some scientists would elevate them in rank to miniature kingdoms.
- *Plants*—Possess both mitochondria and chloroplasts and are photosynthetic. Typically, they are non-mobile and have rigid cell walls made of cellulose.
- *Fungi*—Possess mitochondria but lack chloroplasts. Once thought to be plants that had lost their chloroplasts, it is now thought they never had them. Their nourishment comes from decaying biomatter. Like plants, fungi are non-mobile, but they lack cellulose and their cell walls are made of chitin. They are genetically more closely related to animals than plants.
- *Animals*—Lack chloroplasts but possess mitochondria. Differ from fungi and plants in lacking a rigid cell wall. Many animals are mobile.

After an organism is classified into a kingdom, then each kingdom is divided into **phyla** (singular, phylum). For example, the animal kingdom is divided into 20–30 different phyla, including Platyhelminthes (flatworms), Nematoda (roundworms), Arthropoda (insects), and Mollusca (snails, squids, etc.). These divisions are then narrowed even further:

---

domain (of life)   Highest ranking group into which living creatures are divided, based on the most fundamental genetic properties
genus   A group of closely-related species
germline cells   Reproductive cells producing eggs or sperm that take part in forming the next generation
kingdom   Major subdivision of eukaryotic organisms; in particular the plant, fungus, and animal kingdoms
phylum (plural phyla)   Major groups into which animals are divided, roughly equivalent in rank to the divisions of plants or bacteria
somatic cells   Cells making up the body but which are not part of the germ cell line
totipotent   Capable of giving rise to a complete multicellular organism

Phyla are divided into classes, such as mammals.

Classes are divided into orders, such as primates.

Orders are divided into families, such as hominids.

Families are divided into genera, such as *Homo*.

Genera are divided into species, such as *Homo sapiens*

Biological classification attempts to impose a convenient filing system upon organisms related by continuous evolutionary branching. The current tree of life can be investigated at http://www.tolweb.org.

Box 1.02 Classification Summary

Domain

     Kingdom

         Phylum

            Class

               Order

                  Family

                     Genus

                        species

8. Some Widely-Studied Organisms Serve as Models

With so much biological diversity, biologists concentrate their attention on certain living organisms, either because they are convenient to study or are of practical importance, and they are termed "model organisms." Although scientists like to think that the model organism is representative of all bacteria, humans, or plants, model organisms are atypical in some respects. For example, few bacteria grow as fast as *E. coli* and few mammals breed as fast as mice. Nonetheless, information discovered in such model systems is assumed to apply also to related organisms. In practice this often proves to be true, at least to a first approximation. Model organisms are useful to the discovery of basic principles of biology. They help researchers gain knowledge that advances medicine and agriculture. But since they do have limitations; ultimately, human cells and agriculturally-useful animals and plants have to be studied directly.

8.1. Bacteria Are Used for Fundamental Studies of Cell Function

Most of the early experiments providing the basis for modern-day molecular biology were performed using bacteria such as *E. coli* (see below), because they are relatively simple to analyze. Some advantages of using bacteria to study cell function are:

1. Bacteria are single-celled microorganisms. Furthermore, a bacterial culture consists of many identical cells due to lack of sexual recombination during cell division. In contrast, in multicellular organisms, even an individual tissue or organ contains many different cell types. All the cells in a bacterial culture respond in a reasonably similar way, whereas those from a higher organism will give a variety of responses, making analysis much more difficult.

2. The most commonly used bacteria have about 4,000 genes as opposed to higher organisms, which have up to 50,000. Furthermore, different selections of genes are expressed in the different cell types of a single multicellular organism.

3. Bacteria are haploid, having only a single copy of most genes, whereas higher organisms are diploid, possessing at least two copies of each gene. Analyzing genes present in a single copy is far easier than trying to analyze two different alleles of the same gene simultaneously.

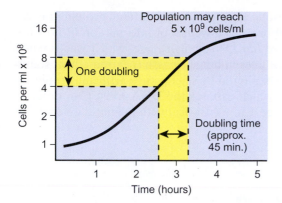

FIGURE 1.14
Graph of Exponential Growth of Bacterial Culture

The number of bacteria in this culture is doubling approximately every 45 minutes. This is typical for fast growing bacteria such as *E. coli*, which are widely used in laboratory research. The bacterial population may reach 5×10^9 cells per ml or more in only a few hours under ideal conditions.

4. In addition to their main chromosome, some bacteria harbor small rings of DNA called **plasmids**, which naturally carry extra genes. Plasmids have been used extensively by scientists as **vectors** to carry different genes from different organisms. Analysis of recombinant DNA in bacteria is one of the essential techniques used by molecular biologists to ascertain the function of different proteins without using the organism from which they originated.

5. Bacteria can be grown under strictly controlled conditions and many will grow in a chemically-defined culture medium containing mineral salts and a simple organic nutrient such as glucose.

6. Bacteria grow fast and may divide in as little as 20 minutes, whereas higher organisms often take days or years for each generation (Fig. 1.14).

7. A bacterial culture contains around 10^9 cells per ml. Consequently, genetic experiments that need to analyze large numbers of cells can be done conveniently.

8. Bacteria can be conveniently stored for short periods (a couple of weeks) by placing them in the refrigerator and for longer periods (20 years or more) in low temperature freezers at $-70°C$. Upon thawing, the bacteria resume growth. Thus, it is not necessary to keep hundreds of cultures of bacterial mutants constantly growing just to keep them alive.

In practice, bacteria are usually cultured by growing them as a suspension in liquid inside tubes, flasks, or bottles. They can also be grown as colonies (visible clusters of cells) on the surface of an agar layer in flat dishes known as Petri dishes. Agar is a carbohydrate polymer extracted from seaweed that sets, or solidifies, like gelatin.

It should be noted that the convenient properties noted above apply to commonly-grown laboratory bacteria. In contrast, many bacterial species found in the wild are difficult or, by present techniques impossible, to culture in the laboratory. Many others have specialized growth requirements and most rarely grow to the density observed with the bacteria favored by laboratory researchers.

The biological diversity of bacteria is immense since the total number of bacteria on our planet is estimated at an unbelievable 5×10^{30}. Over 90% are in the soil and subsurface layers below the oceans. The total amount of bacterial carbon is 5×10^{17} grams, nearly equal to the total amount of carbon found in plants. Probably over half of the living matter on Earth is microbial, yet we are still unable to identify the majority of these species. The bacteria that live in extreme environments, such as boiling hot water in the deep sea thermal vents, salty seas like the Dead Sea, and Antarctic lakes that thaw for only a few months each year, have very unique adaptations to the

Biologists have always been pulled in two directions. Studying simple creatures allows basic principles to be investigated more easily. And yet we also want to know about ourselves.

plasmid Self-replicating genetic elements that are sometimes found in both prokaryotic and eukaryotic cells. They are not chromosomes nor part of the host cell's permanent genome. Most plasmids are circular molecules of double-stranded DNA, although rare linear plasmids and RNA plasmids are known

vector (a) In molecular biology a vector is a molecule of DNA that can replicate and is used to carry cloned genes or DNA fragments; (b) in general biology a vector is an organism (such as a mosquito) that carries and distributes a disease-causing microorganism (such as yellow fever or malaria)

basic proteins that drive DNA replication, gene transcription, and protein translation. The research on these bacteria will provide a wealth of information of how life formed, exists, and persists. For example, *Taq* DNA polymerase, which was isolated from *Thermus aquaticus*, a bacterium that lives in hot springs, has extreme heat stability. This characteristic of the enzyme was exploited by scientists to artificially replicate DNA by the **polymerase chain reaction**, which exposes the DNA to 94°C in order to denature the helix into single strands (see Ch. 6). Basically, the genetic diversity of bacteria has barely been studied.

Box 1.03 Bacteria Have Alter Egos

E. coli is normally harmless, although occasional rogue strains occur. Even these few **pathogenic** *E. coli* strains mostly just cause diarrhea by secreting a mild form of a toxin related to that found in cholera and dysentery bacteria. However, the notorious *E. coli* O157:H7 carries two extra toxins and causes bloody diarrhea that may be fatal, especially in children or the elderly. In outbreaks of *E. coli* O157:H7, the bacteria typically contaminate ground meat used in making hamburgers. Several massive recalls of frozen meat harboring *E. coli* O157:H7 occurred in the late 1990s. For example, in 1997 the Hudson Foods plant in Columbus, Nebraska was forced to shut down and 25 million pounds of ground beef were recalled.

Box 1.04 Antibiotics Kill Bacteria

Patients are usually given **antibiotics** to treat bacterial infections. These are chemical substances capable of killing most bacteria by inhibiting specific biochemical processes, but which are relatively harmless to people. The most commonly used antibiotics, the penicillins and cephalosporins, are synthesized by a kind of fungus known as mold (Fig. 1.15). However, many antibiotics are made by one kind of bacteria in order to kill other types of bacteria. The *Streptomyces* group of soil bacteria produces a wide range of antibiotics including streptomycin, kanamycin, and neomycin. Some antibiotics, like chloramphenicol, were originally made by molds, but nowadays can be chemically synthesized. Finally, some antibiotics, such as sulfonamides, are entirely artificial and are only synthesized by chemical corporations.

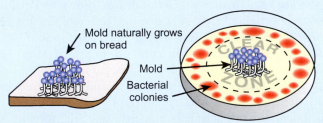

Mold naturally grows on bread

Mold

Bacterial colonies

FIGURE 1.15
Bacterial Growth Is Suppressed by Bread Mold

The blue mold that often grows on bread makes **penicillin**. When penicillin is produced by molds grown on agar in a Petri dish, it will diffuse outwards and suppress the growth of bacteria in a circle around it.

---

antibiotics   Chemical substances that inhibit specific biochemical processes and thereby stop bacterial growth selectively; that is, without killing the patient too

pathogenic   Disease-causing

penicillin   An antibiotic made by a mold called *Penicillium*, which grows on bread producing a blue layer of fungus

polymerase chain reaction (PCR)   Amplification of a DNA sequence by repeated cycles of strand separation and replication

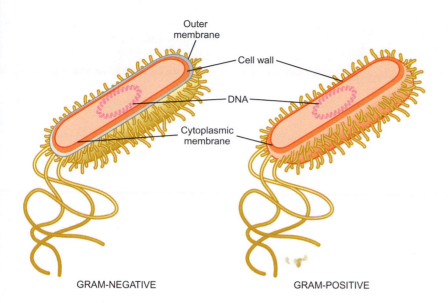

Outer
membrane

Cell wall

DNA

Cytoplasmic
membrane

GRAM-NEGATIVE

GRAM-POSITIVE

FIGURE 1.16
*Gram-Negative and
Gram-Positive Bacteria*

Gram-negative bacteria have an
extra membrane surrounding the
cell wall.

8.2. *E. coli* Is a Model Bacterium

Although many different types of bacteria are used in laboratory investigations, the bacterium used most often in molecular biology research is *E. coli*, a rod-shaped bacterium of approximately 1 by 2.5 microns. Its natural habitat is the colon (hence "coli"), the lower part of the large intestine of mammals, including humans. The knowledge derived by examining *E. coli* has been used to untangle the genetic operation of other organisms. In addition, bacteria, together with their viruses and plasmids, have been used experimentally during the genetic analysis of higher organisms.

E. coli is a **gram-negative bacterium**, which means that it possesses two membranes. Outside the cytoplasmic membrane possessed by all cells are the cell wall and a second, outer membrane (Fig. 1.16). (Although gram-negative bacteria do have two compartments, they are nonetheless genuine prokaryotes, as their chromosome is in the same compartment as the ribosomes and other metabolic machinery. They do not have a nucleus, the key characteristic of a eukaryote.) The presence of an outer membrane provides an extra layer of protection to the bacteria. However, it can be inconvenient to the biotechnologist who wishes to manufacture genetically-engineered proteins from genes cloned into *E. coli*. The outer membrane hinders protein secretion. Consequently, there has been a recent upsurge of interest in **gram-positive bacteria**, such as *Bacillus*, which lack the outer membrane.

Box 1.05 Bacteria Can Have Sex

The famous K-12 laboratory strain of *E. coli* was chosen as a research tool because of its fertility. In 1946, Joshua Lederberg was attempting to carry out genetic crosses with bacteria. Until then, no mechanisms for gene transfer had been demonstrated in bacteria, and genetic crosses were therefore thought to be restricted to higher organisms. Lederberg was lucky, as most bacterial strains, including most strains of *E. coli*, do not mate. But among those he tested was one strain (K-12) of *E. coli* that happened to give positive results. Mating in *E. coli* K-12 is actually due to a plasmid, an extra circular molecule of DNA within the bacterium that is separate from the chromosome. Because the plasmid carries the genes for fertility, it was named the **F-plasmid** (see Ch. 25).

F-plasmid A particular plasmid that confers ability to mate on its bacterial host, *Escherichia coli*
gram-negative bacterium Type of bacterium that has both an inner (cytoplasmic) membrane plus an outer membrane that is located outside the cell wall
gram-positive bacterium Type of bacterium that has only an inner (cytoplasmic) membrane and lacks an outer membrane

FIGURE 1.17
The E. coli Chromosome

The circular *E. coli* chromosome has been divided into 100 map units. Starting with zero at *thrABC*, the units are numbered clockwise from 0 to 100. Various genes are indicated with numbers corresponding to their position on the map. The replication origin (*oriC*) and termini (*ter*) of replication are also indicated. Note that chromosome replication does not start at zero map units—the zero point is an arbitrary designation.

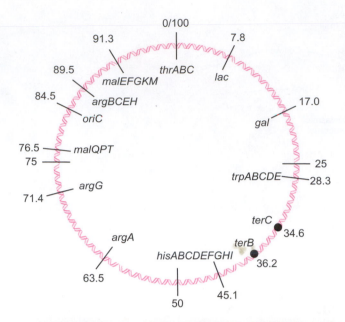

FIGURE 1.18
Yeast Cells

Colored scanning electron micrograph (SEM) of budding yeast cells (*Saccharomyces cerevisiae*). The smaller daughter cells are budding from the larger mother cells. Magnification: × 4,000. (Credit: Andrew Syred, Science Photo Library.)

According to Jacques Monod who discovered the operon (see Ch. 25): "What applies to *E. coli* applies to *E. lephant.*"

The *E. coli* chromosome was mapped before the advent of DNA sequencing by using conjugation experiments (see Ch. 25). The chromosome is divided into map units where the genes for *thrABC* are arbitrarily assigned the zero position (Fig. 1.17).

8.3. Yeast Is a Widely-Studied Single-Celled Eukaryote

Yeast is widely used in molecular biology for many of the same reasons as bacteria. It is the eukaryote about which the most is known and the first whose genome was sequenced—in 1996. Yeasts are members of the fungus kingdom and are slightly more related to animals than plants. A variety of yeasts are found in nature, but the one normally used in the laboratory is brewer's yeast, *Saccharomyces cerevisiae* (Fig. 1.18). This is a single-celled eukaryote that is easy to grow in culture. Even before the age of molecular biology, yeast was widely used as a source of material for biochemical analysis. The first enzymatic reactions were characterized in extracts of yeast, and the word enzyme is derived from the Greek for "in yeast."

Although it is a "higher organism," yeast measures up quite well to the list of useful properties that make bacteria easy to study. In addition, it is less complex genetically than many other eukaryotes. Some of its most useful attributes include:

1. Yeast is a single-celled microorganism. Like bacteria, a yeast culture consists of many identical cells. Although larger than bacteria, yeast cells are only about a tenth the size of the cells of higher animals.

2. Yeast has a haploid genome of about 12 Mb of DNA with about 6,000 genes, as compared to *E. coli*, which has 4,000 genes, and humans, which have approximately 25,000.

3. The natural life cycle of yeast alternates between a diploid phase and a haploid phase. Thus, it is possible to grow haploid cultures of yeast, which, like bacteria, have only a single copy of each gene, making research interpretations easy.

4. Unlike many higher organisms, yeast has relatively few of its genes—about 5%—interrupted by intervening sequences, or introns.

5. Yeast can be grown under controlled conditions in a chemically-defined culture medium and forms colonies on agar like bacteria.

6. Yeast grows fast, though not as fast as bacteria. The cell cycle takes approximately 90 minutes (compared to around 20 minutes for fast-growing bacteria).

7. Yeast cultures can contain around 10^9 cells per ml of culture media, like bacteria.

8. Yeast can be readily stored at low temperatures.

9. Genetic analysis using recombination is much more powerful in yeast than in higher eukaryotes. Furthermore, collections of yeast strains that each have one yeast gene deleted are available.

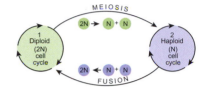

FIGURE 1.19
Yeast Life Cycle

The yeast cell alternates between haploid and diploid phases and is capable of growth and cell division in either phase.

Yeast may grow as diploid or haploid cells (Fig. 1.19). Both haploid and diploid yeast cells grow by **budding**, rather than symmetrical cell division. In budding, a bulge, referred to as a bud, forms on the side of the mother cell. The bud gets larger and one of the nuclei resulting from nuclear division moves into the bud. Finally, the cross wall develops and the new cell buds off from the mother. Especially under conditions of nutritional deprivation, diploid yeast cells may divide by meiosis to form haploid cells, each with a different genetic constitution. This process is analogous to the formation of egg and sperm cells in higher eukaryotes. However, in yeast, the haploid cells appear identical, and there is no way to tell the sexes apart so we refer to mating types. In contrast to the haploid gametes of animals and plants, the haploid cells of yeast may grow and divide indefinitely in culture. Two haploid cells, of opposite mating types, may fuse to form a zygote.

In its haploid phase, *Saccharomyces cerevisiae* has 16 chromosomes and nearly three times as much DNA as *E. coli*. Despite this, it only has 1.5 times as many genes as *E. coli*. It is easier to use the haploid phase of yeast for isolating mutations and analyzing their effects. Nonetheless, the diploid phase is also useful for studying how two alleles of the same gene interact in the same cell. Thus, yeast can be used as a model to study the diploid state and yet take advantage of its haploid phase for most of the genetic analysis.

Biotechnology is a new word but not a new occupation. Brewing and baking both use genetically-modified yeast that have been modified and selected for superior taste throughout history.

Yeast illustrates the genetic characteristics of higher organisms in a simplified manner.

Selker, E (2011) **Neurospora** Curr. Biol. 21(4): R139–140.

The list of model organisms presented in this chapter is only some of the actual model organisms used in laboratories across the world. Another eukaryotic model organism used in genetic research is *Neurospora*, a group of filamentous fungi. The associated article describes the attributes of *Neurospora* that make it an excellent model organism. First and foremost, *Neurospora* grows easily and quickly in the lab. The fungus has a haploid state and a well-defined sexual cycle. The genome is sequenced and thousands of different mutations have been isolated or created by mutagenesis. The most striking characteristic that makes *Neurospora* an interesting model organism is

FOCUS ON RELEVANT RESEARCH

the ability to see the actual cells after meiosis. The ascospores stay in the same order in which they divided, so it is easy to follow how genes segregate during meiosis.

budding   Type of cell division seen in yeasts in which a new cell forms as a bulge on the mother cell, enlarges, and finally separates

8.4. A Roundworm and a Fly Are Model Multicellular Animals

"If all the matter in the universe except the nematodes were swept away, our world would still be dimly recognizable…"

—N.A. Cobb, 1914

Ultimately, researchers have to study multicellular creatures. The most primitive of these that is widely used is the roundworm, *Caenorhabditis elegans* (Fig. 1.20). Nematodes, or roundworms, are best known as parasites both of animals and plants. Although it is related to the "eelworms"—nematodes that attack the roots of crop plants—*C. elegans* is a free-living and harmless soil inhabitant that lives by eating bacteria. A single acre of soil in arable land may contain as many as 3,000 million nematodes belonging to dozens of different species.

The haploid genome of *C. elegans* consists of 97 Mb of DNA carried on six chromosomes. This is about seven times as much total DNA as in a typical yeast genome. *C. elegans* has an estimated 20,000 genes and so contains a much greater proportion of non-coding DNA than lower eukaryotes such as yeast. Its genes contain an average of four intervening sequences each.

The adult *C. elegans* is about 1 mm long and has 959 cells, and the lineage of each has been completely traced from the fertilized egg (i.e., the zygote). It is thus a useful model for the study of animal development. In particular, **apoptosis**, or programmed cell death, was first discovered and has since been analyzed genetically using *C. elegans*. Although very convenient in the special case of *C. elegans*, such a fixed number of cells in an adult multicellular animal is extremely rare. *C. elegans*, which lives about 2–3 weeks, is also used to study life span and the aging process. RNA interference, a gene-silencing technique that relies on double-stranded RNA, was discovered in *C. elegans* in 1998 and is now used to study gene function during development in worms and other higher animals. RNA interference is discussed in Chapter 18.

The fruit fly, *Drosophila melanogaster* (usually called *Drosophila*), was chosen for genetic analysis in the early part of the twentieth century (Fig. 1.21). Fruit flies live on rotten fruit and have a 2-week life cycle, during which the female lays several hundred eggs. The adults are about 3 mm long and the eggs about 0.5 mm. Once molecular biology came into vogue it became worthwhile to investigate *Drosophila* at the molecular level in order to take advantage of the wealth of genetic information

> Nematodes that live in oceanic mud or inland soils may all look the same; nonetheless, they harbor colossal genetic diversity.

FIGURE 1.20
C. elegans

The soil-dwelling nematode *C. elegans* shown with low-magnification phase contrast microscopy. *C. elegans* is convenient for genetic analysis because it is a hermaphrodite; that is, it makes sperm and eggs. It takes only three days to reach maturity and thousands of worms can be kept on a culture plate. *(Credit: Jill Bettinger, Virginia Commonwealth University, Richmond, VA.)*

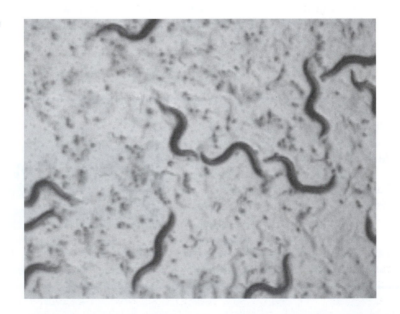

apoptosis   Programmed suicide of unwanted cells

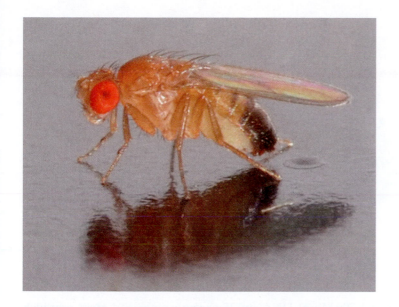

■ **FIGURE 1.21**
Drosophila melanogaster,
the Fruit Fly

Photo of a male *Drosophila* adult fly. (*Credit: André Karwath, Creative Commons Attribution-Share Alike 2.5 Generic license, Wikipedia*)

■ **FIGURE 1.22**
Danio rerio

Danio rerio, the zebrafish, has recently been adopted as a model for the genetic study of embryonic development in higher animals. (*Credit: James King-Holmes, SPL, Photo Researchers, Inc.*)

already available. The haploid genome has 180 Mb of DNA carried on four chromosomes. Although we normally think of *Drosophila* as more advanced than a primitive roundworm, it has an estimated 14,000 genes—6,000 fewer than the roundworm, *C. elegans*. Research on *Drosophila* has concentrated on cell differentiation, development, signal transduction, and behavior.

8.5. Zebrafish and Xenopus are used to Study Vertebrate Development

Danio rerio (previously *Brachydanio rerio*), the zebrafish, is increasingly being used as a model for studying genetic effects in vertebrate development. Zebrafish are native to the slow freshwater streams and rice paddies of East India and Burma, including the Ganges River. They are small, hardy fish, about an inch long that have been bred for many years by fish hobbyists in home aquariums where they may survive for about five years. The standard "wild-type" is clear-colored with black stripes that run lengthwise down its body (Fig. 1.22). Its eggs are laid in clutches of about 200. They are clear and develop outside the mother's body, so it is possible to watch a zebrafish egg grow into a newly-formed fish under a microscope. Development from egg to adult takes about three months. Zebrafish are unusual in being nearly transparent so it is possible to observe the development of the internal organs.

FIGURE 1.23
***Xenopus laevis*, African Clawed Frog**

Adult *Xenopus laevis* is shown (Credit: Michael Linnenback; Wikipedia Commons.)

Zebrafish have about 1,700 Mb of DNA on 25 chromosomes and most zebrafish genes have been found to be similar to human genes. The genome is now sequenced completely, which offers scientists another key element to study the organism. Genetic tagging is relatively easy and microinjecting the egg with DNA is straightforward. Consequently, the zebrafish has become a favorite model organism for studying the molecular genetics of embryonic development. In addition, zebrafish are able to regenerate their heart, nervous tissues, retina, hearing tissues, and fins, which can offer a glimpse into genes that control the growth of these tissues in humans also.

The use for zebrafish in molecular biology is growing. Zebrafish are now used in initial screening for new drugs. The fish and embryos absorb small molecules from the water, so to screen a drug for toxicity is fairly easy. Some screens use robotic microscopes to visualize embryos as they are exposed to new drugs. The microscopes can monitor thousands of embryos, exposed to thousands of new compounds. Compounds giving positive results are then used in mouse models and finally humans. Zebrafish are also able to grow human tumors. The tumor cells can be transplanted into a fish and each step of the tumor formation can be visualized because of the transparent nature of zebrafish.

Xenopus laevis, or the African clawed frog, is another key model organism for the understanding of vertebrate development (Fig. 1.23). These frogs live in any kind of water and are very easy to grow in the laboratory. Like zebrafish, *Xenopus* tadpoles develop outside the mother and are easily visualized throughout the entire developmental process. The size of the eggs allows researchers to inject different genes or chemical substances directly into the eggs. As the egg develops into a tadpole, the effect of this alteration can be visually determined.

Box 1.06 Brainbow Fish

Late in 2003, zebrafish became the first commercially-available, genetically-engineered pets. Fluorescent red zebrafish are marketed in the United States by Yorktown Technologies as GloFish™. They fluoresce red when illuminated with white light, or better, black light (i.e., near UV) due to the presence of a gene for a red fluorescent protein taken from a sea coral. The principle is similar to that of the widely-used green fluorescent protein taken from jellyfish (see Ch. 19 for use of GFP in genetic analysis). The price of about $5 per fish makes GloFish™ about five times as expensive as normal zebrafish. The fish were developed at the National University of Singapore by researcher Zhiyuan Gong with the ultimate objective of monitoring pollution. A second generation of more specialized red fluorescent zebrafish will fluoresce in response to toxins or pollutants in the environment.

Continued

Box 1.06 Continued

The ability to color zebrafish has now advanced. Recent work by Lichtman and Smith (2008) allows each individual cell to be labeled a different fluorescent color. The cells of the brain have been treated with over 90 different fluorescent colors, and each of the neurons is a different color, creating what scientists have called a "brainbow" (Fig. 1.24). The labeling provides researchers with the ability to visualize each and every neuron from the cell body all the way to its terminal branches. The different colors are crucial to follow the long axons of the neurons through their twists and turns in development. Using the color technology will enable scientists to trace each cell of the zebrafish through development, and create a lineage map similar to what has already been established in *C. elegans*.

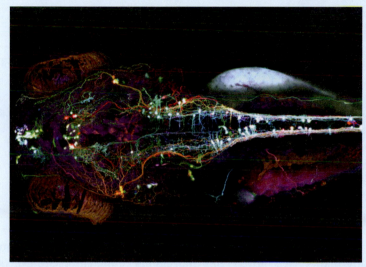

FIGURE 1.24
Brainbow image of 5-day-old zebrafish embryo

This beautiful image of the brain of a 5-day-old zebrafish larva, which was created by Albert Pan of Harvard University, won 4th place in the 2008 Olympus BioScapes Digital Imaging competition.

8.6. Mouse and Man

The ultimate aim of molecular medicine is to understand human physiology at the molecular level and to apply this knowledge in curing disease. The complete sequence of the human genome is now known, but researchers have little idea of what the products of most of these genes actually do. Since direct experimentation with humans is greatly restricted, animal models are necessary. Although a range of animals has been used to investigate various topics, the rat and the mouse are the most widespread laboratory animals. Rats were favored in the early days of biochemistry when metabolic reactions were being characterized. Mice are smaller and breed faster than rats, and are easier to modify genetically. Consequently, the mouse is used more often for experiments involving genetics and molecular biology. Mice live from 1 to 3 years and become sexually mature after about 4 weeks. Pregnancy lasts about three weeks and may result in up to 10 offspring per birth.

Humans have two copies each of approximately 20,000–25,000 genes scattered over 23 pairs of chromosomes. Mice have a similar genome, of 2,600 Mb of DNA carried on 20 pairs of chromosomes. Less than 1% of mouse genes lack a homolog in the human genome. The average mouse (or human) gene extends over 40 kilobases of DNA that consists mostly of non-coding introns (approximately seven per gene). Nowadays there are many strains of mutant mice in which one or more particular genes have been altered or disrupted. These are used to investigate gene function (Fig. 1.25).

FIGURE 1.25
Transgenic Mice

The larger mouse contains an artificially introduced human gene, which causes a difference in growth. Mice with the human growth hormone gene grow larger than normal mice. (Credit Palmiter, et al. Nature 300: 611–615.)

Intact humans cannot be used for routine experiments for ethical reasons. However, it is possible to grow cells from both humans and other mammals in culture. Many cell lines from humans and monkeys are now available. Such cells are much more difficult to culture than genuine single-celled organisms. Cell lines from multicellular organisms allow fundamental investigations into the genome and other cell components. Historically, the most commonly-used cell lines (e.g., HeLa cells) are actually cancer cells. Unlike cells that retain normal growth regulation, cancer cells are "immortalized," that is they are not limited to a fixed number of generations. In addition, cancer cell lines can often divide in culture in the absence of the complex growth factors needed to permit the division of normal cells.

> Only a few animals have been investigated intensively. The rest are assumed to be similar except for minor details.

Box 1.07 The first cell line

A recent account of the short life of Henrietta Lacks (*The Immortal Life of Henrietta Lacks* by Rebecca Skloot) was recently published. The book follows the life and times of Henrietta Lacks who was diagnosed and eventually died from cervical cancer. Prior to her death, the surgeons took a small portion of her tumor and tried to grow the cells in culture. Unlike all other human cells that were grown in culture, these cells survived beyond 24 hours. In fact, the descendants of these very cells are growing today for research purposes and are code named HeLa.

8.7. Arabidopsis Serves as a Model for Plants

Historically, the molecular biology of plants has lagged behind other groups of organisms. Ironically, plants have more genes than higher animals (40,000–50,000 genes for rice—about 20,000–25,000 more than humans). If our criterion for superiority is gene number, then it is the plants that represent the height of evolution, not mammals. Why do plants have so many genes? One suggestion is that because plants are immobile they cannot avoid danger by moving. Instead they must stand and face it like a man—or rather like a vegetable. This means that plants have accumulated many genes involved in defense against predators and pests as well as for adapting to changing environmental conditions. One of the most active areas in biotechnology today is the further genetic improvement of crop plants. Genetic manipulation of plants is not hindered by the ethical considerations that apply to research on animals or humans. Moreover, crop farming is big business.

Arabidopsis thaliana, the mouse-ear cress, has become the model for the molecular genetics of higher plants (Fig. 1.26). It is structurally simple and also has the smallest genome of any flowering plant, 125 Mb of DNA—just over 10 times as much DNA as yeast, yet carried on only five pairs of chromosomes. *Arabidopsis* has an estimated 25,000 genes with an average of four intervening sequences per gene. *Arabidopsis* can be grown indoors and takes about 6–10 weeks to produce several thousand offspring from a single original plant. Though slow by bacterial standards, this is much faster than waiting a year for a new crop of peas, corn, or soybeans, for example.

Arabidopsis shares with yeast the ability to grow in the haploid state, which greatly facilitates genetic analysis. Pollen grains are the male germline cells of plants and are therefore haploid. When pollen from some plants, including *Arabidopsis*, is grown in tissue culture, the haploid cells grow and divide and may eventually develop into normal-looking plants. These are haploid, and therefore sterile. Diploid plants may be reconstituted by fusion of cells from two haploid cell lines. Alternatively, diploidy can be artificially induced by agents such as colchicine that interfere with mitosis to cause a doubling of the chromosome number. In the latter case, the new diploid line will be homozygous for all genes.

FIGURE 1.26
Arabidopsis thaliana, *the Mouse-ear Cress*

The plant most heavily used as a model for molecular biology research is *Arabidopsis thaliana*, a member of mustard family (Brassicaceae). Common names include mouse-ear cress, thale cress, and mustard weed. *(Credit: Dr. Jeremy Burgess, Science Photo Library.)*

> Flowering plants have more genes than animals. The function of most of these genes is still a mystery.

TABLE 1.02	What Makes an Organism Amenable to Molecular Biology Research?

Location in the tree of life: Many lineages are not well represented in research, so new model organisms should lie in uncharacterized regions of the tree of life.

Ability to obtain candidate genes via either sequencing or cloning of the organism's DNA. Sequenced genomes also facilitate studies of quantitative traits that are controlled by multiple genes.

Ability to determine gene expression patterns at different stages of the organism's development.

Ability to grow or obtain large numbers of organisms, either in a laboratory or in the wild.

Ability to knockout or functionally inactivate genes to ascertain the role the gene plays in the organism's growth, development, or metabolism. In addition, *in vitro* or *in vivo* assays to determine the effect of genetic changes on various traits.

Ability to overexpress a gene of interest to ascertain what happens when too much of one protein is expressed.

Availability of close relatives that are genetically different than the model organism allows the study of trait evolution.

Comprehensive collection of the organism's genes into a gene library.

A genetic map of the genome to study the location of genes and investigate their phenotypes.

Ability to make genetic crosses to study inheritance of candidate genes.

9. Basic Characteristics of a Model Organism

In a recent article by Abzhanov, et al. (2008), the authors identify key features that are necessary to make an organism a good candidate for laboratory analysis. These are summarized in Table 1.02.

10. Purifying DNA from Model Organisms

One of the most important techniques for a molecular biologist is to purify DNA from the organism of interest. The purification of DNA is slightly different for each model organism because each has slightly different cellular structures. For each of the organisms, the first step to purifying DNA involves breaking open the cells, and for eukaryotic organisms, the nucleus too. For bacteria, the enzyme **lysozyme** is used to digest the peptidoglycan layer of the cell wall, and a **detergent** is used to dissolve the lipids in the cytoplasmic membrane. Chelating agents such as **EDTA (ethylene diamine tetraacetate)** are used to chelate or remove the metal ions that bind components of the outer membrane together in gram-negative bacteria. Yeast has a very tough cell wall, and this must be broken. Small glass beads are often mixed with the yeast cells and then vortexed at high speeds. The beads crash into the yeast and break open the tough outer wall. *C. elegans* does not have a tough outer coat, but isolating its DNA is best done after a freeze/thaw procedure. The freezing causes ice crystals to form inside their cells, which puncture the membranes, and then when the worm thaws, DNA is released from the nuclei. *Drosophila* are best ground into a fine powder with a mortar and pestle, or frozen in liquid nitrogen and then ground to break the fly apart into small pieces. Extracting DNA from humans or rodents only requires a small amount of blood, cheek cells, a hair follicle, or a small piece of tissue. For rodents, the tip of the tail can be used to isolate DNA. A small piece of the tail is first incubated with an enzyme that digests protein and with detergent to solubilize the cytoplasmic and nuclear membranes. The strong cellulose cell wall of plants also

> Modern technology has led to a steady decrease in the amount of DNA needed for analysis.

> Cell walls and membranes must be broken down to liberate the DNA from the cell.

detergent   Molecule that is hydrophobic at one end and highly hydrophilic at the other and which is used to dissolve lipids or grease
EDTA (ethylene diamine tetraacetate)   A widely-used chelating agent that binds di-positive ions such as Ca^{2+} and Mg^{2+}
lysozyme   An enzyme found in many bodily fluids that degrades the peptidoglycan of bacterial cell walls

FIGURE 1.27
*Phenol Extraction Removes
Proteins from Nucleic Acids*

Proteins can be removed from
a solution of DNA and RNA by
adding an equal volume of phenol.
Since phenol is very dense, it forms
a separate layer at the bottom of
the tube. When the two solutions
are shaken, the proteins dissolve
into the phenol. The two layers
separate again after a brief spin
in the centrifuge. The top phase,
which contains just DNA and RNA,
can then be isolated.

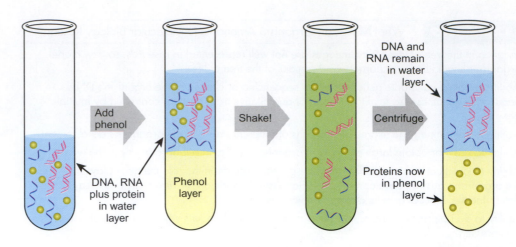

needs special treatment before DNA can be isolated. The cell wall is broken up by
grinding with a mortar and pestle or by homogenizing the material in a blender.

No matter how the cells are broken, the remaining steps of DNA isolation are
similar. The cytoplasmic membrane is usually dissolved by adding some sort of detergent. The proteins are digested by adding non-specific enzymes such as proteinase K
or even papain, which is found in meat tenderizer. In order to separate the DNA and
protein into separate fractions, **phenol extraction** dissolves the proteins from the sample. Phenol, also known as carbolic acid, is very corrosive and extremely dangerous
because it dissolves and denatures the proteins that make up 60–70% of all living
matter. When phenol is added to water, the two liquids do not mix to form a single
solution; instead, the denser phenol forms a separate layer below the water. When
shaken, the two layers mix temporarily, and the proteins dissolve in the phenol. When
the shaking stops, the two layers separate, where the water phase containing the DNA
is above the phenol phase containing the proteins (Fig. 1.27). To ensure that no phenol is trapped with the DNA, the sample is centrifuged briefly. Then the water containing the DNA and RNA is removed and kept. Generally, several successive phenol
extractions are performed to purify away the proteins from DNA.

A variety of newer techniques have been developed that avoid phenol extraction. Most of these involve purifying DNA by passing it through a column containing
a resin that binds DNA but not other cell components. The two main choices are silica
and anion exchange resins. Silica resins bind nucleic acids rapidly and specifically at
low pH and high salt concentrations. The nucleic acids are released at higher pH and
low salt concentration. Anion exchange resins, such as diethylaminoethyl-cellulose,
are positively-charged and bind DNA via its negatively-charged phosphate groups. In
this case, binding occurs at low salt concentration and the nucleic acids are eluted by
high concentrations of salt, which disrupt the ionic bonding.

Next, unwanted RNA is removed using an enzyme called **ribonuclease**, which
degrades RNA into short oligonucleotides but leaves DNA unchanged. Once the RNA
is digested, the large DNA fragments are isolated by adding an equal volume of ethanol.
The DNA precipitates out of the liquid, but the small pieces of RNA remain dissolved.
The solution is centrifuged at high speed, so the DNA pellets to the bottom, and the RNA
remains inside the liquid phase. This supernatant (liquid portion) containing the RNA
fragments is discarded (Fig. 1.28). The tiny pellet of DNA left at the bottom of the tube
is often scarcely visible. Nonetheless, it contains billions of DNA molecules, sufficient for
most investigations. This DNA is dissolved into buffered water and is now ready for use.

> Phenol dissolves proteins and
> separates them from DNA.

> Nucleic acids may be purified
> on columns containing resins
> that bind DNA and RNA.

> Unwanted RNA is often
> removed by degrading it
> enzymatically.

> Macromolecules such as DNA
> may be separated from smaller
> molecules by **centrifugation**.

---

centrifugation   Process in which samples are spun at high speed and the centrifugal force causes the larger or heavier components to sediment to
the bottom
phenol extraction   Technique for removing protein from nucleic acids by dissolving the protein in phenol
ribonuclease   An enzyme that degrades RNA

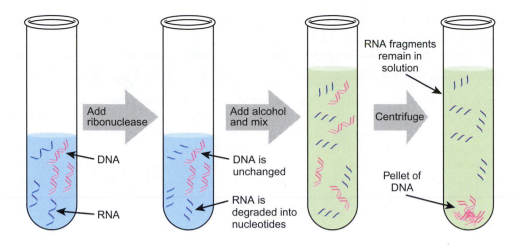

▸**FIGURE 1.28**
Removal of RNA by Ribonuclease

A mixture of RNA and DNA is incubated with ribonuclease, which digests all the RNA into small fragments and leaves the DNA unaltered. An equal volume of alcohol is added, and the larger pieces of DNA are precipitated out of solution. The solution is centrifuged, and the large insoluble pieces of DNA form a small pellet at the bottom of the tube. The RNA fragments remain in solution.

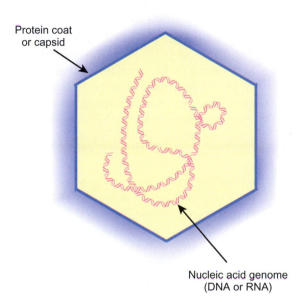

Nucleic acid genome
(DNA or RNA)

▸**FIGURE 1.29**
Structural Components of a Virus

A virus is composed of a protein coat and nucleic acid. Note that there are no ribosomes or cytoplasmic membranes and only one type of nucleic acid is present.

11. Viruses Are Not Living Cells

The characteristics of living cells were outlined early in this chapter. A common, but somewhat technical, definition of a living cell is as follows: Living cells contain both DNA and RNA and can use the genetic information encoded in these to synthesize proteins by using energy that they generate themselves. This definition is designed not so much to explain, positively, how a cell works as to exclude **viruses** from the realm of living cells. The essential features of a virus are shown in Figure 1.29. Viruses are packages of genes in protein coats and are usually much smaller than bacteria. Viruses are obligate **parasites** that must infect a host cell in order to replicate themselves. Whether viruses are alive or not is a matter of opinion; however, viruses are certainly not living cells. Virus particles (**virions**) do contain genetic information in the form of DNA or RNA but are incapable of growth or division by themselves. A virus may

---

parasite   An organism or genetic entity that replicates at the expense of another creature
virion   A virus particle
virus   Subcellular parasite with genes of DNA or RNA that replicates inside the host cell upon which it relies for energy and protein synthesis. In addition, it has an extracellular form in which the virus genes are contained inside a protective coat

have its genome made of DNA or RNA, but only one type of nucleic acid is present in the virion of any given type of virus.

Viruses lack the machinery to generate their own energy or to synthesize protein. After invading a host cell, the virus does not grow and divide like a cell itself. The virion disassembles and the viral genes are expressed using the machinery of the host cell. In particular, viral proteins are made by the host cell ribosomes, using virus genetic information. In many cases, only the virus DNA or RNA enters the host cell and the other components are abandoned outside. After infection, virus components are manufactured by the infected cell, as directed by the virus, and are assembled into new virus particles. Usually the host cell is killed and disintegrates. Typically, several hundred viruses may be released from a single infected cell. The viruses then abandon the cell and look for another host. (Note that some viruses cause "chronic" or "persistent" infections where virus particles are made slowly and released intermittently rather than as a single burst. In this case the host cell may survive for a long time despite infection. In addition, many viruses may persist inside the host cell for a long time in a latent, non-replicating state and only change to replicative mode under certain conditions—see Ch. 21.)

Some scientists regard viruses as being alive based on the viral possession of genetic information. The majority, however, do not accept that viruses are truly alive, since viruses are unable to generate energy or to synthesize protein. Viruses are thus on the borderline between living and non-living. Virus particles are in suspended animation, waiting for a genuine living cell to come along so they can infect it and replicate themselves. Nonetheless, a host cell whose life processes have been subverted by a virus does duplicate the viral genetic information and produces more virus particles. Thus, viruses possess some of the properties of living creatures. Viruses are very important from a practical viewpoint. Firstly, many serious diseases are due to virus infection. Secondly, many manipulations that are now used in genetic engineering are carried out using viruses.

> Viruses are packages of genes that are not alive by themselves but may take over living cells. Once in control the virus uses the cell's resources to manufacture more viruses.

Merely being a parasite does not prevent an organism from being a living cell. For example, **rickettsias** are degenerate bacteria that cause typhus fever and related diseases. They cannot grow and divide unless they infect a suitable host cell. However, rickettsias can generate energy and make their own proteins, provided they obtain sufficient complex nutrients from the animal cell they invade. Furthermore, rickettsias reproduce by growing and dividing like other bacteria. Viruses are subcellular parasites and totally dependent on other life-forms for their energy, materials, and even the equipment to manufacture their own components.

12. Bacterial Viruses Infect Bacteria

Even bacteria can get sick, usually as the result of infection by a virus. Bacterial viruses are sometimes referred to as **bacteriophages**, or phages for short. Phage comes from a Greek word meaning to eat. When bacteria catch a virus, they do not merely get a mild infection, like a cold, as humans usually do. They are doomed. The bacteriophage takes over the bacterial cell and fills it up by manufacturing more bacteriophages, as shown in Figure 1.30. Then the bacterial cell bursts and liberates the new crop of bacteriophages to infect more bacteria. This takes only about an hour or so. In a matter of hours, a bacteriophage epidemic could wipe out a culture of bacteria numbering several times the Earth's human population.

Bacterial viruses infect only bacteria. Some have relatively broad host ranges, whereas others infect only a single species or even just a few particular strains of bacteria. Generally speaking, any particular disease, whether caused by bacteria or by viruses, infects only a closely-related group of organisms.

bacteriophage A virus that infects bacteria
rickettsia Type of degenerate bacterium that is an obligate parasite and infects the cells of higher organisms

BACTERIAL VIRUS ATTACHES
TO A BACTERIAL CELL

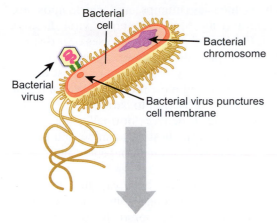

Bacterial
cell

Bacterial
chromosome

Bacterial
virus

Bacterial virus punctures
cell membrane

VIRAL GENOME IS REPLICATED

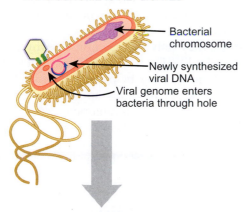

Bacterial
chromosome

Newly synthesized
viral DNA

Viral genome enters
bacteria through hole

VIRAL PROTEINS ARE SYNTHESIZED
AND ASSEMBLED INTO VIRAL PARTICLES

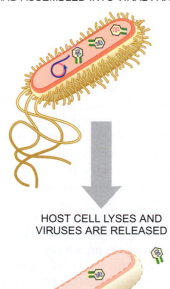

HOST CELL LYSES AND
VIRUSES ARE RELEASED

▸ **FIGURE 1.30**

Virus Entry into a Cell

Components of a new virus are synthesized under the direction of viral DNA but using the synthetic machinery of the host cell. First, a virus binds to the host cell and then inserts its nucleic acid into the host cell. The synthetic machinery of the host cell then manufactures the viral proteins and nucleic acids according to the genetic information carried by the viral DNA. Finally, the virus causes the cell to burst, releasing the newly synthesized viruses that seek a new host. The host cell dies as the result of the viral infection.

13. Human Viral Diseases Are Common

Many common childhood diseases such as measles, mumps, and chickenpox are caused by viruses, as are the common cold and flu. More dangerous viral diseases include polio, smallpox, herpes, Lassa fever, Ebola, and AIDS. Do viruses ever do anything useful? Yes; infection by a mild virus can provide resistance against a related but more dangerous virus (see Ch. 21). Viruses may carry genes from one host organism to another, in a process known as transduction, and have thereby played a major role in molecular evolution (see Ch. 26). The ability of viruses to carry genes between organisms is also used by genetic engineers hoping to deliver "normal" genes into a person who has a genetic disease. All the same, about the best that can be said for the natural role of viruses is that most of them do relatively little damage and only a few cause highly virulent diseases.

Viral diseases usually cannot be cured once they have been caught. Either the victim's body fights off the infection or it does not, although some antiviral drugs can help the host in the fight. However, viral diseases can often be prevented by **immunization** if a potential victim is **vaccinated** before catching the virus. In this case, the invading virus will be killed by the immune system, which has been put on alert by the vaccine, and the disease will be prevented.

Antibiotics are of no use against viruses; they only kill bacteria. So why do doctors often prescribe antibiotics for viral diseases like flu or colds? There are two main reasons. The valid reason is that giving antibiotics may help combat secondary or opportunistic infections caused by bacteria, especially in virally-infected patients who are in poor health. However, massive over-prescription of antibiotics occurs because many patients would be upset if faced with the truth. They would rather be given medicine, even if it is of no use, than face the fact that there is no cure. This abuse has in turn contributed to the spread of antibiotic resistance among many infectious bacteria (see Ch. 20)—thus creating a major health problem.

> An immense variety of viruses exists (see Ch. 21 for more details). Viruses infect every other life-form, from bacteria to eukaryotes, including humans.

14. A Variety of Subcellular Genetic Entities Exist

A whole range of entities exist that have genetic information, but do not themselves possess the machinery of life and cannot exist without a host cell to parasitize (Fig. 1.31). Viruses are the most complex of these subcellular genetic elements. In this book, these elements will sometimes be collectively referred to as "gene creatures" to emphasize that they possess genetic information, but have no cell structure or metabolism of their own. Gene creatures may be thought of as inhabiting cells, much as living cells live in their own, larger-scale environment. The term, gene creatures, is intended to focus attention on the properties of these genetic elements in contrast to the traditional viewpoint, which regards them merely as parasites or accessories to "real cells." These assorted genetic elements will be dealt with in subsequent chapters. Here, they will just be introduced to give some idea of the range of gene creatures that share the biosphere with the more traditional life forms (Fig. 1.32):

1. Viruses carry their genes inside a protective shell of protein. **DNA viruses** have their genes in the form of DNA, and **RNA viruses** contain genes as RNA. **Retroviruses** have RNA copies of their genes inside the virus particle, but once inside the host cell, they make a DNA copy of their genome (see Ch. 21).

DNA virus   A virus whose genome consists of DNA
immunization   Process of preparing the immune system for future infection by treating the patient with weak or killed versions of an infectious agent
retrovirus   Type of virus which has its genes as RNA in the virus particle but converts this to a DNA copy inside the host cell by using reverse transcriptase
RNA virus   A virus whose genome consists of RNA
vaccination   Artificial induction of the immune response by injecting foreign proteins or other antigens

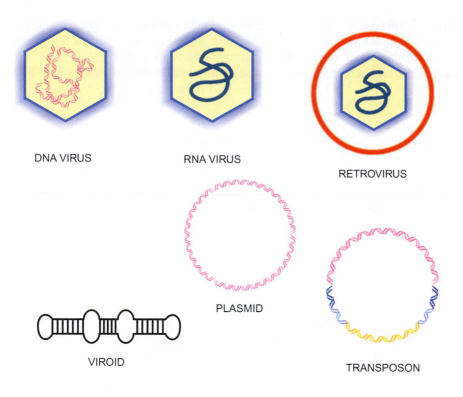

FIGURE 1.31

The Variety of Subcellular Genetic Elements—"Gene Creatures"

These structures possess some of the characteristics of life. However, they use their host's machinery to replicate. The plasmid and the viroid lack a protein shell. The transposon is merely a segment of DNA (yellow) with special ends (blue) inserted into another DNA molecule.

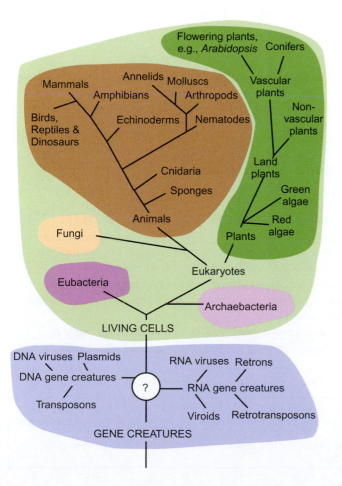

FIGURE 1.32

The Molecular Biologist's "Tree of Life"

This tree of life includes both the traditional living creatures, such as plants and animals, as well as the two genetically-distinct types of prokaryotic cell (eubacteria and archaebacteria). At the bottom, a variety of gene creatures are shown whose relationships are still mostly uncertain.

2. Viroids and plasmids are self-replicating molecules of nucleic acid that lack the protein coat characteristic of a virus. Viroids are naked molecules of RNA that infect plants and trick the infected plant cell into replicating more viroid RNA (see Ch. 21). Like a virus, they are released into the environment and must find a new cell to infect. Unlike a virus, their extracellular phase lacks a protective protein shell.

3. Plasmids are self-replicating molecules of DNA that live permanently inside host cells (see Ch. 20). Although some plasmids can be transferred from one host cell to another, they have no extracellular phase and so unlike viruses or viroids, they do not destroy their host cell. Plasmids are widely used to carry genes during many genetic engineering procedures.

4. Transposable elements, or **transposons**, are simpler still. They are nucleic acid molecules, usually DNA, which lack the ability to self-replicate. In order to get replicated, they must insert themselves into other molecules of DNA that are capable of replicating themselves. Thus, transposable elements require a host DNA molecule, such as the chromosome of a cell, a virus genome, or a plasmid. Transposable refers to the fact that these elements possess the ability to jump from one host DNA molecule to another, a property that is essential for their survival and distribution (see Ch. 22).

5. Prions are infectious protein molecules, the ultimate parasites. They contain no nucleic acid and possess genetic information only in the sense of being gene products. Prions infect cells in the nervous systems of animals and cause diseases, the most famous of which is bovine spongiform encephalopathy, better known as mad cow disease. The prion protein is actually a misfolded version of a normal protein found in nerve cells, especially in the brain. When the prion infects a nerve cell, it promotes the misfolding of the corresponding normal proteins, which causes the cell to die. The prion protein is actually encoded by a gene belonging to the host animal that it infects.

> An amazing variety of quasi-independent genetic elements are widespread in the biosphere. They range from those causing major diseases of cellular organisms to those whose existence is scarcely noticeable without sophisticated molecular analysis.

Key Concepts

- Although life is difficult to define, there are six key ingredients: genetic information (DNA or RNA); mechanism of energy production; machinery to make more living matter; an outward physical form; the ability to reproduce; and the ability to adapt.
- Organisms are made of discrete subunits called cells.
- Cells have a membrane layer that separates the inside portion or cytoplasm from the external environment.
- Cells have soluble enzymes in the cytoplasm that translate messenger RNA into proteins and other enzymes.
- Prokaryotic cells have a cell wall, cytoplasmic membrane, soluble cytoplasmic enzymes, and a nucleoid region that holds a single chromosome.
- There are three domains of life: eukaryotes, eubacteria, and archaea.
- Eubacteria are the most familiar prokaryotes since the members of this domain tend to cause human diseases.
- Archaea and eubacteria are both considered prokaryotes since they lack a nucleus surrounding their chromosome(s). Other cellular components of

---

prion   Distorted, disease-causing form of a normal brain protein that can transmit an infection
transposable element or transposon   Segment of DNA that can move as a unit from one location to another, but which always remains part of another DNA molecule
viroid   Naked single-stranded circular RNA that forms a stable highly base-paired rod-like structure and replicates inside infected plant cells. Viroids do not encode any proteins but possess self-cleaving ribozyme activity.

archaea, including the cell wall, enzymes that synthesize proteins, and metabolic enzymes, are very different from eubacteria, and in some cases resemble eukaryotes.

- Eukaryotes have nuclear envelopes to surround their chromosomes, a cytoskeleton to give the cells shape, and organelles such as endoplasmic reticulum, Golgi apparatus, lysosomes, mitochondria, and chloroplasts.
- Eukaryotes include a great variety of species that are classified into: kingdoms, phyla, class, order, family, followed by genus and species. The organism's scientific name is printed in text using the following format: *Genus species*.
- Model organisms are used to investigate how life develops, exists, and reproduces. Some model organisms include bacteria, yeast, *C. elegans*, *Drosophila*, zebrafish, *Xenopus*, and mice. In the plant world, *Arabidopsis* serves as the main model organism.
- Model organisms can be grown easily and reproduce fast, have their genomes completely sequenced, can be studied in each stage of their development, and are amenable to genetic manipulations.
- DNA isolation is a key technique used in molecular biology. The method involves removing the cellular proteins and RNA, leaving behind just the DNA.
- Besides model organisms, a variety of gene creatures are studied in molecular biology. These include viruses, bacteriophage, viroids, plasmids, transposable elements, and prions. Although these have genetic material, they do not possess the ability to make their own proteins or exist without a host organism.

Review Questions

1. Describe the controversy surrounding the question "what is life?" Describe the basic ingredients needed to sustain life.
2. What is a cell? What is the term used to describe the development of specialized cells?
3. What are the functions of a cell membrane? What is the structure of this membrane? What reactions are carried out on the cell membrane?
4. What organelle translates messenger RNA (mRNA) into proteins? Where is this organelle located?
5. Compare and contrast the structure of prokaryotic and eukaryotic cells.
6. Compare and contrast the two subgroups of prokaryotes.
7. What are the advantages of using bacteria to study cell function? What is the name of the model bacterium and where is it naturally found?
8. What surrounds the nucleus of eukaryotic cells? What allows communication to occur between the nucleus and cytoplasm?
9. What are organelles? What are the functions of the endoplasmic reticulum, Golgi apparatus, lysosomes, mitochondria, and chloroplasts? Where are each located?
10. What is the entire classification for modern-day humans? Start with domain and end with species.
11. What is *Saccharomyces cerevisiae* and why is it useful? How are the haploid and diploid stages useful?
12. What are the model organisms used for studying multicellular animals?
13. What has the zebrafish been useful for studying?
14. In terms of genetics, what characteristics about the mouse make using this organism beneficial for molecular medicine research?
15. What is one theory used to explain why plants have more genes than humans?
16. What is the model organism for studying plant genetics?
17. Can genetic defects in somatic cells be passed on to the next generation? Why or why not?
18. What is totipotent? Under natural circumstances, which organisms, plants or animals, have totipotent cells? Give an exception to this rule.

19. Are viruses living cells? Why or why not?
20. What are bacteriophages?
21. How can viral diseases be prevented? What is the principle behind this prevention?
22. How do viroids and plasmids differ from and compare to viruses?
23. How do transposable elements, or transposons, differ from and compare to plasmids?
24. What are prions and how do they act?

Conceptual Questions

1. A brand new unicellular organism was recently discovered by leaving some food outside for too long. Scientists have identified the following characteristic using electron microscopy. In the inside of the organism, there is an oval-shaped structure surrounding patches of dark staining material. The dark staining matter has no visible structures. There are visible structures in the rest of the cell, and each of these seems to have membranes. In addition, the outer edge of the organism has a cell membrane surrounded by another layer or wall. Classify this organism as prokaryote, eukaryote, or archaea, based on your knowledge of the structure of these organisms.

2. A researcher was growing two *E. coli* cultures in order to feed their organism, *C. elegans*. The *E. coli* strain normally grows in plain broth with nutrients. The researcher added a small colony of *E. coli* to each flask of nutrient broth and let these grow overnight at 37°C. In the morning, one flask was very cloudy due to the bacteria, but the other culture had far fewer bacteria and almost looked clear. The researcher isolated the *E. coli* from each culture and looked at the bacteria in the microscope. The cloudy culture had lots of rod-shaped bacteria moving around because of their flagella. In contrast, the clear culture had a few *E. coli*, but there were also lots of particulate matter. What could have happened to the culture with just a few *E. coli*?

3. Many different model organisms are used in molecular biology. List some of the traits that make a model organism useful for research. Specifically explain why *Drosophila*, *C. elegans*, and zebrafish are useful for molecular biology.

4. A male was born with one blue eye and one brown eye. No living relative has this trait, and he married someone with brown eyes. Would you expect any of his children to be born with two different colored eyes? Why?

5. Using this table of bacterial cell growth data, plot a graph (time on the X-axis; number of cells on the Y-axis) and determine the approximate doubling time for each growth condition.

Bacterial Cell Growth (n $\times$ 10^8 cells/ml)

Time (mins)	Minimal Media	Normal Media	Rich Media
10	1	1.3	1.3
20	1.3	1.7	2
30	1.6	2.5	3
40	1.8	3.7	4.4
50	2.4	4.5	7.8
60	3	6	12
70	3.9	8	18
80	4.8	12	19
90	6	15.8	19.1
100	7.7	17.2	19.4
110	9.5	17.4	19.5
120	12	17.5	19.3
130	13.6	17.5	19.1
140	14.3	17.6	18.8
150	14.5	17.4	18.9
160	15	17.3	18.9

Basic Genetics

Genetics is the study of biological inheritance. Before we move on to consider the chemical nature of genetic information in detail, we need to review the fundamentals of genetics. Modern genetics was established by Gregor Mendel who focused on discrete unambiguous characters due to single genes and discovered the basic laws of heredity. Today we know that inheritance is due to genes, most of which encode proteins. Consequently, differences in many observable genetic characters are due to alterations in individual proteins that make up biochemical pathways or that form cellular structures. Other characters, which are more difficult to study, result from the effects of multiple genes. Genes are carried on long DNA molecules known as chromosomes; consequently, nearby genes tend to be inherited together. Living organisms contain anywhere from several hundred to several thousand different genes carried on one or more chromosomes.

1. Gregor Mendel, The Father of Classical Genetics

From very ancient times, people have vaguely realized the basic premise of heredity. It was always a presumption that children looked like their fathers and mothers, and that the offspring of animals and plants generally resemble their ancestors. During the nineteenth century, there was great interest in how closely offspring resembled their parents. Some early investigators measured such quantitative characters as height, weight, or crop yield and analyzed the data statistically. However, they failed to produce any clear-cut theory of inheritance. It is now known that certain properties of higher organisms, such as height or skin color, are due to the combined action of many genes. Consequently, there is a gradation or quantitative variation in such properties. Such multigene characteristics caused much confusion for the early geneticists, and they are still difficult to analyze, especially if more than two or three genes are involved.

The birth of modern genetics was due to the discoveries of **Gregor Mendel** (1823–1884), an Augustinian monk who taught natural science to high school students in the town of Brno in Moravia (now part of the Czech Republic). Mendel's greatest insight was to focus on discrete, clear-cut characters rather than measuring continuously variable properties, such as height or weight. Mendel used pea plants and studied characteristics such as whether the seeds were smooth or wrinkled, whether the flowers were red or white, and whether the pods were yellow or green, etc. When asked if any particular individual inherited these characteristics from its parents, Mendel could respond with a simple "yes" or "no," rather than "maybe" or "partly." Such clear-cut, discrete characteristics are known as **Mendelian characters** (Fig. 2.01).

Today, scientists would attribute each of the characteristics examined by Mendel to a single **gene**. Genes are units of genetic information and each gene provides the instructions for some property of the organism in question. In addition to those genes that affect the characteristics of the organism more or less directly, there are also many regulatory genes. These control other genes; hence their effects on the organism are less direct and more complex. Each gene may exist in alternative forms known as **alleles**, which code for different versions of a particular inherited character (such as red versus white flower color). The different alleles of the same gene are closely related but have minor chemical variations in their DNA that may produce significantly different outcomes. Any segment of DNA, whether coding or not, can be referred to as a **locus** (plural, loci); that is, a location on a chromosome (or other molecule of DNA). Since any DNA sequence can occur in alternative versions, the term allele is used for these even if the DNA in question is non-coding.

The overall nature of an organism is due to the sum of the effects of all of its genes as expressed in a particular environment. The total genetic make-up of an organism is referred to as its **genome**. Among free-living bacteria, the genome may consist of approximately 2,000 to 6,000 genes, whereas in higher organisms such as plants and animals, there may be up to 50,000 genes.

> A century before the discovery of the DNA double helix, Mendel realized that inheritance was quantized into discrete units we now call genes.

Box 2.01 Etymological Note

Mendel did not use the word "gene." This term entered the English language in 1911 and was derived from the German "Gen," short for "Pangen." This in turn came via French and Latin from the original ancient Greek "genos," which means birth. "Gene" is related to such modern words as <u>gen</u>us, ori<u>gin</u>, <u>gen</u>erate, and <u>gen</u>esis. In Roman times, a "gen-ius" was a spirit representing the *inborn power* of individuals.

---

allele   One particular version of a gene
gene   A unit of genetic information
genome   The entire genetic information of an individual organism
Gregor Mendel   Discovered the basic laws of genetics by crossing pea plants
locus (plural, loci)   A place or location on a chromosome; it may be a genuine gene or just any site with variations in the DNA sequence that can be detected, like RFLPs or VNTRs
Mendelian character   Trait that is clear cut and discrete and can be unambiguously assigned to one category or another

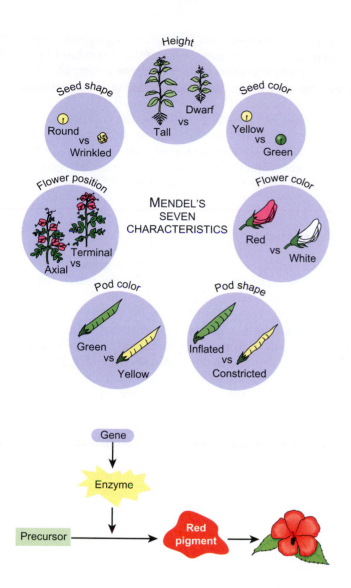

Mendelian Characters in Peas

Mendel chose specific characteristics, such as those shown.

► **FIGURE 2.02**
One Gene—One Enzyme

A single gene determines the presence of an enzyme, which, in turn, results in a biological characteristic such as a red flower.

2. Genes Determine Each Step in Biochemical Pathways

Mendelian genetics was a rather abstract subject, since no one knew what genes were actually made of, or how they operated. The first great leap forward came when biochemists demonstrated that each step in a biochemical pathway was determined by a single gene. Each biosynthetic reaction is carried out by a specific **protein** known as an **enzyme**. Each enzyme has the ability to mediate one particular chemical reaction and so the *one gene—one enzyme* model of genetics (Fig. 2.02) was put forward by G.W. Beadle and E. L. Tatum who won a Nobel prize for this scheme in 1958. Since then, a variety of exceptions to this simple scheme have been found. For example, some complex enzymes consist of multiple subunits, each of which requires a separate gene.

A gene determining whether flowers are red or white would be responsible for a step in the biosynthetic pathway for red pigment. If this gene were defective, no red pigment would be made and the flowers would take the default coloration—white. It is easy to visualize characters such as the color of flowers, pea pods, or seeds in terms of a biosynthetic pathway that makes a pigment. But what about tall versus dwarf plants and round versus wrinkled seeds? It is difficult to interpret these in terms of a single pathway and gene product. Indeed, these properties are affected by the action

enzyme   A protein that carries out a chemical reaction
protein   A polymer made from amino acids; proteins make up most of the structures in the cell and also do most of the work

of many proteins. However, as will be discussed in detail later, certain proteins control the expression of genes rather than acting as enzymes. Some of these **regulatory proteins** control just one or a few genes, whereas others control large numbers of genes. Thus, a defective regulatory protein may affect the levels of many other proteins. Modern analysis has shown that some types of dwarfism are due to defects in a single regulatory protein that controls many genes affecting growth. If the concept of "one gene—one enzyme" is broadened to "one gene—one protein," it still applies in most cases. (There are of course exceptions. Perhaps the most important are that in higher organisms multiple related proteins may sometimes be made from the same gene by alternative patterns of splicing at the RNA. In addition, non-coding RNA is not translated into protein, as discussed in Ch. 12.)

> Beadle and Tatum linked genes to biochemistry by proposing there was one gene for each enzyme.

3. Mutants Result from Alterations in Genes

Consider a simple pathway in which red pigment is made from its precursor in a single step. When everything is working properly, the flowers shown in Figure 2.02 will be red and will match thousands of other red flowers growing in the wild. If the gene for flower color is altered so as to prevent the gene from functioning properly, one may find a plant with white flowers. Such genetic alterations are known as **mutations**. The white version of the flower color gene is defective and is a mutant allele. The properly functioning red version of this gene is referred to as the **wild-type** allele (Fig. 2.03). As the name implies, the wild-type is supposedly the original version as found in the wild, before domestication and/or mutation altered the beauties of nature. In fact, there are frequent genetic variants in wild populations and it is not always obvious which version of a gene should be regarded as the true wild-type. Generally, the wild-type is taken as the form that is common and shows adaptation to the environment.

Geneticists often refer to the red allele as "R" and the white allele as "r" (not "W"). Although this may seem a strange way to designate the color white, the idea is that the r-allele is merely a defective version of the gene for red pigment. The r-allele is NOT a separate gene for making white color. In our hypothetical example, there is no enzyme that makes white pigment; there is simply a failure to make red pigment.

FIGURE 2.03
Wild-type and Mutant Genes

If red flowers are found normally in the wild, the "red" version of the gene is called the wild-type allele. Mutation of the wild-type gene may alter the function of the enzyme so ultimately affecting a visible characteristic. Here, no pigment is made and the flower is no longer red.

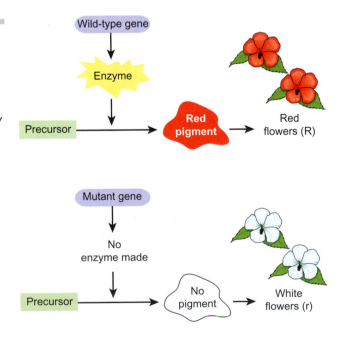

mutation   An alteration in the genetic information carried by a gene
regulatory protein   A protein that regulates the expression of a gene or the activity of another protein
wild-type   The original or "natural" version of a gene or organism

Originally, it was thought that each enzyme was either present or absent; that is, there were two alleles corresponding to Mendel's "yes" and "no" situations. In fact, things are often more complicated. An enzyme may be only partially active or even be hyperactive or have an altered activity and genes may actually have dozens of alleles, matters to be discussed later. A mutant allele that results in the complete absence of the protein is known as a **null allele**. (More strictly, a null allele is one that results in complete absence of the gene product. This includes the absence of RNA (rather than protein) in the case of those genes where RNA is the final gene product (e.g., ribosomal RNA, transfer RNA, etc.)—see Ch. 3).

> Determining the "wild-type" gene is arbitrary because it must be based on its occurrence within a specific population.

4. Phenotypes and Genotypes

In real life, most biochemical pathways have several steps, not just one. To illustrate this, extend the pathway that makes red pigment so it has three steps and three genes, called A, B, and C. If any of these three genes is defective, the corresponding enzyme will be missing, the red pigment will not be made, and the flowers will be white. Thus, mutations in any of the three genes will have the same effect on the outward appearance of the flowers. Only if all three genes are intact will the pathway succeed in making its final product (Fig. 2.04).

Outward characteristics—the flower color—are referred to as the **phenotype** and the genetic make-up as the **genotype**. Obviously, the phenotype "white flowers" may be due to several possible genotypes, including defects in gene A, B, or C, or in genes not mentioned here that are responsible for producing precursor P in the first place. If white flowers are seen, only further analysis will show which gene or genes are defective. This might involve assaying the biochemical reactions, measuring the build-up of pathway intermediates (such as P or Q in the example) or mapping the genetic defects to locate them in a particular gene(s).

If gene A is defective, it no longer matters whether gene B or gene C are functional or not (at least as far as production of our red pigment is concerned; some genes affect multiple pathways, a possibility not considered in this analysis). A defect near the beginning of a pathway will make the later reactions irrelevant. This is known in genetic terminology as **epistasis**. Gene A is epistatic to gene B and gene C; that is, it masks the effects of these genes. Similarly, gene B is epistatic to gene C. From a practical viewpoint, this means that a researcher cannot tell if genes B or C are defective or not, when there is already a defect in gene A.

> Remember that phenotypes refer to physical trait, and genotypes refer to the genes that confer the trait.

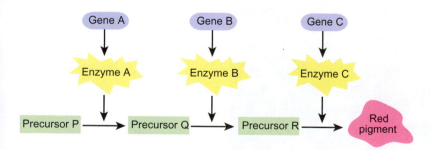

FIGURE 2.04
Three-Step Biochemical Pathway

In this scenario, genes A, B, and C are all needed to make the red pigment required to produce a red flower. If any precursor is missing due to a defective gene, the pigment will not be made and the flower will be white.

---

epistasis   When a mutation in one gene masks the effect of alterations in another gene
genotype   The genetic make-up of an organism
null allele   Mutant version of a gene that completely lacks any activity
phenotype   The visible or measurable effect of the genotype

5. Chromosomes Are Long, Thin Molecules That Carry Genes

Genes are aligned along very long, string-like molecules called **chromosomes** (Fig. 2.05). Organisms such as **bacteria** usually fit all their genes onto a single circular chromosome (Fig. 2.06), whereas higher, eukaryotic organisms have several chromosomes that accommodate their much greater number of genes. Genes are often drawn on a bar representing a chromosome (or a section of one), as shown in Figure 2.05.

One entire chromosome strand consists of a molecule of deoxyribonucleic acid, called simply DNA (see Ch. 3). The genes of living cells are made of DNA, as are the regions of the chromosome between the genes. In bacteria, the genes are closely packed together, but in higher organisms such as plants and animals, the DNA between genes comprises up to 96% of the chromosome and the functional genes only make up around 4 to 5% of the length. (Viruses also contain genetic information

FIGURE 2.05
Genes Arranged Along a Chromosome

Although a chromosome is a complex three-dimensional structure, the genes on a chromosome are in linear order and can be represented by segments of a bar, as shown here. Genes are often given alphabetical designations in genetic diagrams.

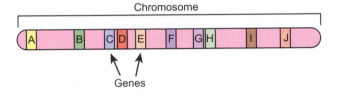

FIGURE 2.06
Circular DNA from a Bacterium

Hand-tinted transmission electron micrograph (TEM) of circular bacterial DNA. This figure actually shows a small plasmid, rather than a full-size chromosome. The double-stranded DNA is yellow. An individual gene has been mapped by using an RNA copy of the gene. The RNA forms base pairs to one strand of the DNA forming a DNA/RNA hybrid (red). The other strand of the DNA forms a single-stranded loop, known as an "R-loop" (blue). Magnification: ×28,600. *(Credit: P. A. McTurk and David Parker, Science Photo Library.)*

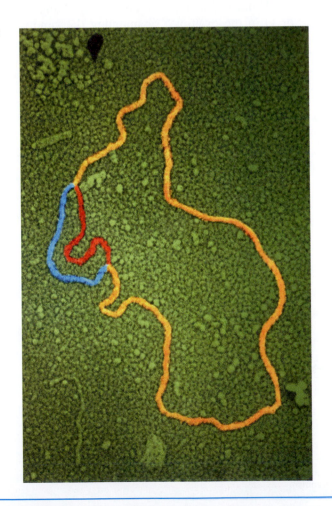

---

bacteria   Primitive single-celled organisms without a nucleus and with one copy of each gene
chromosome   Structure containing the genes of a cell and made of a single molecule of DNA

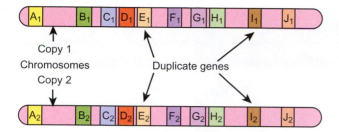

Copy 1
Chromosomes   Duplicate genes
Copy 2

■ **FIGURE 2.07**

Genes Match on Each of a Pair of Homologous Chromosomes

Higher organisms possess two copies of each gene arranged on pairs of homologous chromosomes. The genes of the paired chromosomes are matched along their length. Although corresponding genes match, there may be molecular variation between the two members of each pair of genes.

> Genes are not mere abstractions. They are segments of DNA molecules carrying encoded information.

and some have genes made of DNA. Other viruses have genes made of the related molecule, **ribonucleic acid, RNA**.)

In addition to the DNA, the genetic material itself, chromosomes carry a variety of proteins that are bound to the DNA. This is especially true of the larger chromosomes of higher organisms where histone proteins are important in maintaining chromosome structure (see Ch. 4). (Bacteria also have histone-like proteins. However, these differ significantly in both structure and function from the true histones of higher organisms.)

5.1. Different Organisms May Have Different Numbers of Chromosomes

The cells of higher organisms usually contain two copies of each chromosome. Each pair of identical chromosomes possesses copies of the same genes, arranged in the same linear order. In Figure 2.07, identical capital letters indicate sites where alleles of the same gene can be located on a pair of chromosomes. In fact, identical chromosomes are not usually truly identical, as the two members of the pair often carry different alleles of the same gene. The term **homologous chromosome** refers to chromosomes that carry the same set of genes in the same sequence, although they may not necessarily carry identical alleles of each gene.

A cell or organism that possesses two homologous copies of each of its chromosomes is said to be **diploid** (or "2n," where "n" refers to the number of chromosomes in one complete set). Those that possess only a single copy of each chromosome are haploid (or "n"). Thus, humans have 2×23 chromosomes (n = 23 and 2n = 46). Although the X and Y sex chromosomes of animals form a pair they are not actually identical (see below). Thus, strictly, a male mammal is not fully diploid. Even in a diploid organism, the reproductive cells, known as gametes, possess only a single copy of each chromosome and are thus haploid. Such a single, though complete, set of chromosomes carrying one copy of each gene from a normally diploid organism is known as its "**haploid genome**."

Bacteria possess only one copy of each chromosome and are therefore **haploid**. (In fact, most bacteria have only a single copy of a single chromosome, so that n = 1.) If one of the genes of a haploid organism is defective, the organism may be seriously endangered since the damaged gene no longer contains the correct information that the cell needs. Higher organisms generally avoid this predicament by being diploid and having duplicate copies of each chromosome and therefore of each gene. If one copy of the gene is defective, the other copy may produce the correct product required by the cell. Another advantage of diploidy is that it allows recombination between two copies of the same gene (see Ch. 24). Recombination is important in promoting the genetic variation needed for evolution.

---

diploid   Having two copies of each gene
haploid   Having one copy of each gene
haploid genome   A complete set containing a single copy of all the genes (generally used to describe organisms that have two or more sets of each gene)
homologous chromosomes   two chromosomes are homologous when they carry the same sequence of genes in the same linear order
ribonucleic acid (RNA)   Nucleic acid that differs from DNA in having ribose in place of deoxyribose and having uracil in place of thymine

FIGURE 2.08
Diploid, Tetraploid, and Hexaploid Wheats

The origin of modern hexaploid bread wheat is illustrated. Einkorn wheat hybridized with goat grass to give tetraploid wheat. This in turn hybridized with the weed *Triticum tauschii* to give hexaploid bread wheat. The increase in grain yield is obvious. (Credit: Dr. Wolfgang Schuchert Max-Planck Institute for Plant Breeding Research, Köln, Germany.)

Note that haploid cells may contain more than a single copy of certain genes. For example, the single chromosome of *E. coli* carries two copies of the gene for the elongation factor EF-Tu and seven copies of the genes for ribosomal RNA. In haploid cells of the yeast *Saccharomyces cerevisiae* as many as 40% of the genes are duplicate copies. Strictly speaking, duplicate copies of genes are only regarded as genuine alleles if they occupy the same location on **homologous chromosomes**. Thus, these other duplicate copies do not count as true alleles.

Occasionally, living cells with more than two copies of each chromosome can be found. **Triploid** means possessing three copies, **tetraploid** means having four copies, and so on. Animal and plant geneticists refer to the "**ploidy**" of an organism, whereas bacterial geneticists tend to use the term "**copy number**." Many modern crop plants are polyploids, often derived from hybridization between multiple ancestors. Such polyploids are often larger and give better yields. The ancestral varieties of wheat originally grown in the ancient Middle East were diploid. These were then displaced by tetraploids, which in turn gave way to modern bread wheat (*Triticum aestivum*), which is hexaploid (6n = 42) (Fig. 2.08). Hexaploid bread wheat is actually a hybrid that contains four sets of genes from emmer wheat and two sets from the wild weed, *Triticum tauschii* (= *Aegilops squarrosa*). Emmer wheat is a tetraploid (4n = 28) derived from two diploid ancestors—einkorn wheat (*Triticum monococcum*) and a weed similar to modern goat grass (*Triticum speltoides* = *Aegilops speltoides*). A small amount of tetraploid wheat (*Triticum turgidum* and relatives) is still grown for specialized uses, such as making pasta.

Cases are known where there are fewer or more copies of just a single chromosome. Cells that have irregular numbers of chromosomes are said to be **aneuploid**. In higher animals, aneuploidy is often lethal for the organism as a whole, although certain aneuploid cells may survive in culture under some conditions. Although aneuploidy is usually lethal in animals, it is tolerated to a greater extent in plants. Nonetheless, in rare cases, aneuploid animals may survive. Thus, partial triploidy is the cause of certain human conditions such as Down syndrome, where individuals have an extra copy of chromosome #21. The presence of three copies of one particular chromosome is known as **trisomy**.

> Different organisms differ greatly in the number of genes, the number of copies of each gene, and the arrangement of the genes on DNA.

---

aneuploid   Having irregular numbers of different chromosomes
copy number   The number of copies of a gene that is present
homologous   Related in sequence to an extent that implies common genetic ancestry
ploidy   The number of sets of chromosomes possessed by an organism
tetraploid   Having four copies of each gene
triploid   Having three copies of each gene
trisomy   Having three copies of a particular chromosome

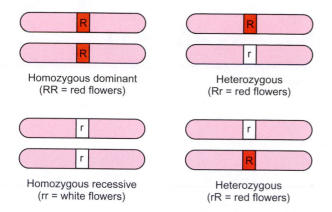

Homozygous dominant
(RR = red flowers)

Heterozygous
(Rr = red flowers)

Homozygous recessive
(rr = white flowers)

Heterozygous
(rR = red flowers)

■ FIGURE 2.09
Two Different Alleles
Produce Four Genotypes

The genotypes R and r can be combined in four ways.

6. Dominant and Recessive Alleles

Consider a diploid plant that has two copies of a gene involved in making red pigment for flowers. From a genetic viewpoint, there are four possible types of individual plant; that is, there are four possible genotypes: RR, Rr, rR, and rr. The genotypes Rr and rR differ only depending on which of the pair of chromosomes carries r or R (see Fig. 2.09). When two identical alleles are present the organism is said to be **homozygous** for that gene (either RR or rr), but if two different alleles are present the organism is **heterozygous** (Rr or rR). Apart from a few exceptional cases there is no phenotypic difference between rR and Rr individuals, as it does not usually matter which of a pair of homologous chromosomes carries the r allele and which carries the R allele.

If both copies of the gene are wild-type, R-alleles (genotype, RR), then the flowers will be red. If both copies are mutant r-alleles (genotype, rr), then the flowers will be white. But what if the flower is heterozygous, with one copy "red" and the other copy "white" (genotype, Rr or rR)? The enzyme model presented above predicts that one copy of the gene produces enzyme and the other does not. Overall, there should be half as much of the enzyme, so red flowers will still be the result. Most of the time one gene produces enough enzymes, since many enzymes are present in levels that exceed minimum requirements. (In addition, many genes are regulated by complex feedback mechanisms. These may increase or decrease gene expression so that the same final level of enzyme is made whether there are two functional alleles or only one.)

From the outside, a flower that is Rr will therefore look red, just like the RR version. When two different alleles are present, one may dominate the situation and is then known as the **dominant** allele. The other one, whose properties are masked (or perhaps just function at a lower level), is the **recessive** allele. In this case, the R allele is dominant and the r allele recessive. Overall, three of the genotypes, RR (homozygous dominant), Rr (heterozygous), and rR (heterozygous), share the same phenotype and have red flowers, while only rr (homozygous recessive) plants have white flowers.

Genes and their alleles may interact with each other in a variety of ways. Sometimes one copy of a gene may predominate. In other cases, both copies share influence.

6.1. Partial Dominance, Co-Dominance, Penetrance, and Modifier Genes

The assumption thus far is that one wild-type allele of the flower color gene will produce sufficient red pigment to give red flowers; in other words, the R-allele is dominant. Although one good copy of a gene is usually sufficient, this is not always the

dominant allele   Allele whose properties are expressed in the phenotype whether present as a single or double copy
heterozygous   Having two different alleles of the same gene
homozygous   Having two identical alleles of the same gene
recessive allele   The allele whose properties are not observed because they are masked by the dominant allele

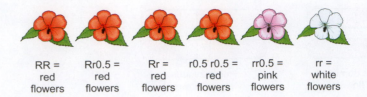

FIGURE 2.10
The Possible Phenotypes from Three Different Alleles

There are six possible pairs of three different alleles. Here, the r0.5 allele is a partly functional allele that makes only 50% of the normal pigment level. R is wild-type and r is null. The RR, Rr0.5, Rr, and r0.5 r0.5 combinations will all make 100% or more of the wild-type level of red pigment and so are red. The rr0.5 combination will make 50% as much pigment and so has pink flowers. The rr combination makes no pigment and so has white flowers.

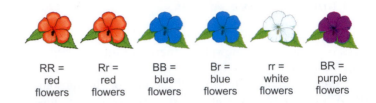

FIGURE 2.11
Phenotypes Resulting from Co-dominance

Here, the B allele makes an altered, blue, pigment. R is wild-type and r is null. The RR and Rr combinations will make red pigment. The Br combination will make only blue pigment and the RB combination makes both red and blue pigments so has purple flowers.

case. For example, the possession of only one functional copy of a gene for red pigment may result in half the normal amount of pigment being produced. The result may then be pale red or pink flowers. The phenotype resulting from Rr is then not the same as that seen with RR. This sort of situation, where a single good copy of a gene gives results that are recognizable but not the same as for two good copies, is known as **partial dominance**.

As indicated above, there may be more than two alleles. In addition to the wild-type and null alleles, there may be alleles with partial function. Assume that a single gene dosage of enzyme is sufficient to make enough red pigment to give red flowers. Suppose there is an allele that is 50% functional, or "r0.5." Any combination of alleles that gives a total of 100% (= one gene dosage) or greater will yield red flowers. If there are three alleles, R = wild-type, r = null, and r0.5 = 50% active, then the genotypes and resulting phenotypes shown in Fig. 2.10 are possible. In such a scenario, there are three different phenotypes resulting from six possible allele combinations.

Another possibility is alleles with an altered function. For example, there may be a mutant allele that gives rise to an altered protein that still makes pigment but that carries out a slightly altered biochemical reaction. Instead of making red pigment, the altered protein could produce a pigment whose altered chemical structure results in a different color, say blue. Let's name this allele "B." Both R and B are able to make pigment and so both are dominant over r (absence of pigment). The combination of R with B gives both red and blue pigment in the same flower, which will look purple, and so they are said to be **co-dominant**. There are six possible genotypes and four possible phenotypes (colors, in this case) of flowers, as shown in Figure 2.11.

co-dominance   When two different alleles both contribute to the observed properties
partial dominance   When a functional allele only partly masks a defective allele

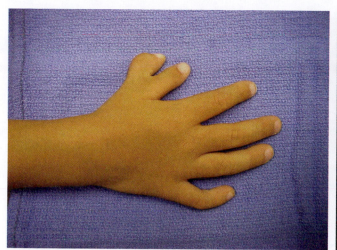 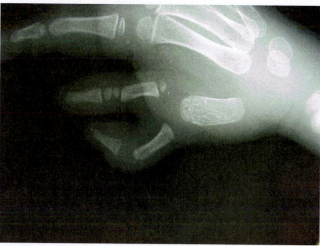

FIGURE 2.12
Polydactyly

A dominant mutation may cause the appearance of extra fingers and/or toes. (*Credit: Charles Eaton, MD. Used by permission.*)

As the above example shows, mutant alleles need not be recessive. There are even cases where the wild-type is recessive to a dominant mutation. Note also that a characteristic that is due to a dominant allele in one organism may be due, in another organism, to an allele that is recessive. For example, the allele for black fur is dominant in guinea pigs, but recessive in sheep. Note that a dominant allele receives a capital letter, even if it is a mutant rather than a wild-type allele. Sometimes a "+" is used for the wild-type allele, irrespective of whether the wild-type allele is dominant or recessive. A "-" is frequently used to designate a defective or mutant allele.

Does any particular allele always behave the same in each individual that carries it? Usually it does, but not always. Certain alleles show major effects in some individuals and only minor or undetectable effects in others. The term **penetrance** refers to the relative extent to which an allele affects the phenotypic in a particular individual. Penetrance effects are often due to variation in other genes in the population under study. In humans, there is a dominant mutation (allele = P) that causes polydactyly, a condition in which extra fingers and toes appear on the hands and feet (Fig. 2.12). This may well be the oldest human genetic defect to be noticed as the Bible mentions Philistine warriors with six fingers on each hand and six toes on each foot (II Samuel, Chapter 21, verse 20). About 1 in 500 newborn American babies shows this trait, although nowadays the extra fingers or toes are usually removed surgically, leaving little trace. Detailed investigation has shown that heterozygotes (Pp or P+) carrying this dominant allele do not always show the trait. Furthermore, the extra digits may be fully formed or only partially developed. The P allele is thus said to have variable penetrance.

Such variation in the expression of one gene is often due to its interaction with other genes. For example, the presence of white spots on the coat of mice is due to a recessive mutation, and in this case, the homozygote with two such recessive alleles is expected to show white spots. However, the size of the spots varies enormously, depending on the state of several other genes. These are consequently termed **modifier genes**. Variation in the modifier genes among different individuals will result in variation in expression of the major gene for a particular character. Environmental effects may also affect penetrance. In the fruit fly, *Drosophila*, alterations in temperature may change the penetrance of many alleles. The number of eye facets is one example where the environment effects gene expression. *Drosophila* flies grown at 15°C, have many more eye facets than flies that are born and raised in warm temperatures such as 30°C.

A complex and largely unresolved issue of genetics is that the same allele of certain genes may behave differently in different individuals. Some variation is due to the environment, and some is due to the effects of other genes.

---

modifier gene   Gene that modifies the expression of another gene
penetrance   Variability in the phenotypic expression of an allele

FIGURE 2.13
Meiosis—the Principle

Diploid organisms distribute their chromosomes among their gametes by the process of meiosis. Chromosome reduction means that the gametes contain only half of the genetic material of the diploid parental cell (i.e., each gamete has one complete haploid set of genes). Each chromosome of a pair has a 50% chance of appearing in any one gamete, a phenomenon known as random segregation. While only sperm are shown here, the same process occurs during the production of ova.

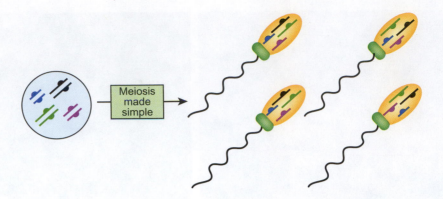

7. Genes from Both Parents Are Mixed by Sexual Reproduction

How are alleles distributed at mating? If both copies of both parents' genes were passed on to all their descendants, the offspring would have four copies of each gene, two from their mother and two from their father. The next generation would end up with eight copies and so on. Clearly, a mechanism is needed to ensure that the number of copies of each gene remains stable from generation to generation!

How does nature ensure that the correct copy number of genes is transferred? When diploid organisms such as animals or plants reproduce sexually, the parents both make **germ cells**, or **gametes**. These are specialized cells that pass on genetic information to the next generation of organisms, as opposed to the **somatic cells**, which make up the body. Female gametes are known as eggs or ova (singular = ovum) and male gametes as sperm. When a male gamete combines with a female gamete at fertilization, they form a **zygote**, the first cell of a new individual (Fig. 2.13). Although the somatic cells are diploid, the egg and sperm cells only have a single copy of each gene and are haploid. During the formation of the gametes, the diploid set of chromosomes must be halved to give only a single set of chromosomes. Reduction of chromosome number is achieved by a process known as **meiosis**. Figure 2.13 bypasses the technical details of meiosis, which are presented below, and just illustrates its genetic consequences. In addition to reducing the number of chromosomes to one of each kind, meiosis randomly distributes the members of each pair. Thus, different gametes from the same parent contain different assortments of chromosomes.

Because egg and sperm cells only have a single copy of each chromosome, each parent passes on a single allele of each gene to any particular descendent. Which of the original pair of alleles gets passed to any particular descendent is purely a matter of chance. For example, when crossing an RR parent with an rr parent, each offspring gets a single R-allele from the first parent and a single r-allele from the second parent. The offspring will therefore all be Rr (Fig. 2.14). Thus, by crossing a plant that has red flowers with a plant that has white flowers, the result is offspring that all have red flowers. Note that the offspring, while phenotypically similar, are not genetically identical to either parent; they are heterozygous. The parents are regarded as generation zero and the offspring are the first, or F_1, generation. Successive generations of descendants are labeled F_2, F_3, F_4, etc.; the "F" stands for **filial generation**.

---

filial generations   Successive generations of descendants from a genetic cross, which are numbered F1, F2, F3, etc., to keep track of them
gametes   Cells specialized for sexual reproduction that are haploid (have one set of genes)
germ cells   Cell specialized to pass genetic information to the next generation of organisms; see gametes
meiosis   Formation of haploid gametes from diploid parent cells
somatic cell   Cell making up the body, as opposed to the germline
zygote   Cell formed by union of sperm and egg which develops into a new individual

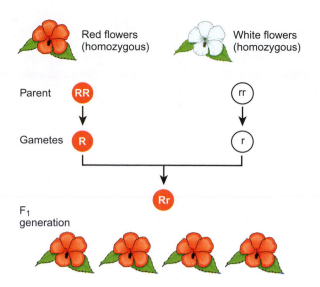

Cross between Homozygous Dominant and Recessive for Red Flower Color

When individuals with the genotypes RR and rr are crossed, all the progeny of the cross, known as the F1 generation, are red.

Extending the ideas presented above, Figure 2.16 shows the result of a cross between two Rr plants. Each parent randomly contributes one copy of the gene, which may be an R or an r allele, to its gametes. Sexual reproduction ensures that the offspring get one copy from each parent. The relative numbers of each type of progeny as depicted in Figure 2.16 are often referred to as **Mendelian ratios**. The Mendelian ratio in the F_2 generation is 3 red: 1 white. Note that white flowers have reappeared after skipping a generation. This is because the parents were both heterozygous for the r allele, which is recessive and so was masked by the R allele.

A similar situation exists with human eye color. In this case the allele for blue eyes (b) is recessive to brown (B). This explains how two heterozygous parents (Bb) who both have brown eyes can produce a child who has blue eyes (bb, homozygous recessive). The same scenario also explains why inherited diseases do not afflict all members of a family and often skip a generation.

To geneticists, sex is merely a mechanism for re-shuffling genes to promote evolution. From the gene's perspective, an organism is just a machine for making more copies of itself.

Box 2.02 Checkerboard Diagrams or Punnett Squares

First-generation or F_1 mating

Gametes	R	R
r	Rr	Rr
r	Rr	Rr

4 Rr = 4 red
F_1 offspring

Second-generation mating

Gametes	R	r
R	RR	Rr
r	Rr	rr

1 RR = red
2 Rr = red
1 rr = white

3 red: 1 white
F_2 offspring

FIGURE 2.15
Checkerboard Determination of Genotype Ratios

Checkerboard diagrams (also known as Punnett squares) are often used to determine the possible genotypes and their ratios that result from a genetic cross with two or more alleles. To construct a checkerboard diagram, place the possible alleles from one parent on the horizontal row and those from the other parent on the vertical row. Fill in the boxes with the combinations determined from the intersection of the vertical and horizontal rows. Then list the various phenotypes and add up the similar phenotypes. When adding similar phenotypes, Rr and rR, although genetically dissimilar, are equivalent phenotypically.

Mendelian ratios   Whole number ratios of inherited characters found as a result of a genetic cross

FIGURE 2.16
The Rr × Rr Cross: Checkerboard Determination of Phenotypes

The parents for this cross are the F1 generation from the mating shown in Figure 2.15. At the top of the figure the possible gametes are shown for each parent flower. The arrows demonstrate how the genes distribute to give a 3:1 ratio of red to white flowers in the F2 generation. At the bottom of the figure the F2 mating is analyzed by a checkerboard to yield the same 3:1 ratio. Note that although no white flowers were present among the parents of this second mating, they are found in their F2 offspring.

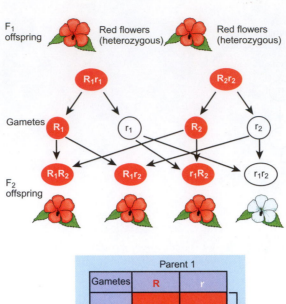

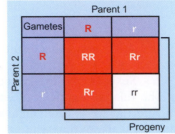

FIGURE 2.17
Checkerboard Diagram for Sex Determination

The male parent contributes an X- or a Y-chromosome to each gamete. The female contributes one X-chromosome to each gamete. When these gametes fuse, an equal proportion of males and females are created in each generation.

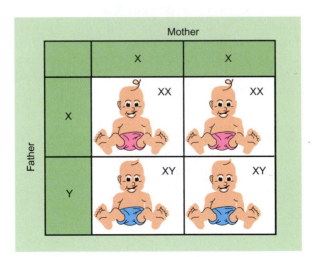

7.1. Sex Determination and Sex-Linked Characteristics

The genetic sex of many diploid organisms, including mammals and insects, is determined by which sex chromosomes they possess. Among mammals, possession of two **X-chromosomes** makes the organism a genetic female, whereas possession of one X-chromosome and one **Y-chromosome** makes the organism a genetic male. (The term "genetic" male or female is used because occasional individuals are found whose phenotypic sex does not match their genetic sex, due to a variety of complicating factors.) The checkerboard diagram for sex determination is shown in Figure 2.17.

Genes that have nothing to do with sex are also carried on the sex chromosomes. Which allele of these genes an individual will inherit correlates with the individual's

---

X-chromosome   Female sex chromosome; possession of two X-chromosomes creates a female gender in mammals
Y-chromosome   Male sex chromosome; possession of a Y-chromosome plus an X-chromosome results in male gender in mammals

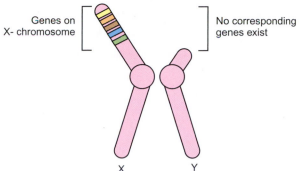

Genes on
X- chromosome

No corresponding
genes exist

X   Y

FIGURE 2.18
***The X- and Y-Chromosomes
Are Not of Equal Length***

The X- and Y-chromosome are not
of equal length. The Y-chromosome
lacks many genes corresponding
to those on the X-chromosome.
Therefore males have only one copy
of these genes.

sex and so they are called **sex-linked genes**. Most sex-linked genes are present in two copies in females but only one copy in males. This is because although the X- and Y-chromosomes constitute a pair, the Y-chromosome is much shorter. Thus many genes present on the X-chromosome do not have a corresponding partner on the Y-chromosome (Fig. 2.18). Conversely, a few genes, mostly involved in male fertility, are present on the Y-chromosome but missing from the X-chromosome.

If the single copy of a sex-linked gene present in a male is defective, there is no back-up copy and severe genetic consequences may result. In contrast, females with just one defective copy will usually have no problems because they usually have a good copy of the gene. However, they will be carriers and half of their male children will suffer the genetic consequences. The result is a pattern of inheritance in which the male members of a family often inherit the disease, but the females are carriers and suffer no symptoms. Figure 2.20 shows a family tree with several occurrences of an X-linked recessive disease. Males have only one X-chromosome and their Y-chromosome has no corresponding copy of the gene (symbolized by −). So any male who gets one copy of the defective allele ("a") will get the disease.

A well-known example of sex-linked inheritance is red-green color blindness in humans. About 8% of men are color blind, whereas less than 1% of women show the defect. Many genes are involved in the synthesis of the three pigments for color vision, which are sensitive to red, green and blue. About 75% of color-blind people carry a sex-linked recessive mutation in the gene for the green-sensitive pigment, which is located on the X-chromosome (but absent from the Y-chromosome). A variety of other hereditary diseases show sex linkage and their detrimental effects are therefore more commonly observed in males than females.

Sex determination complicates the inheritance of a variety of other characters in many animals. Among mammals, males are more likely to suffer from certain genetic defects.

Box 2.03 Standard Symbols for a Family Tree

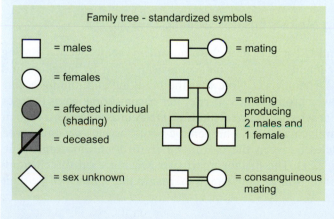

Family tree - standardized symbols

☐ = males

○ = females

● = affected individual
(shading)

▨ = deceased

◇ = sex unknown

☐—○ = mating

= mating
producing
2 males and
1 female

☐═○ = consanguineous
mating

FIGURE 2.19
***Symbols Used for
Pedigree Analysis***

In studying pedigrees, these
symbols are commonly used
to denote sex of the family
member; matings between
different people; and who
was affected by the particular
disease. Pedigree analysis is
used extensively to determine
whether or not a disease is
found on a particular sex
chromosome, and whether or
not a disease is inherited as a
dominant or recessive allele.

sex-linked   A gene is sex-linked when it is carried on one of the sex chromosomes

FIGURE 2.20
Inheritance of a Sex-linked Gene

This family tree shows the inheritance of the wild-type ("A") and deleterious ("a") alleles of a gene that is carried on the X-chromosome. Since males have only one X-chromosome, they have only a single allele of this gene. The symbol "-" is used to indicate the absence of a gene. When the defective allele "a" is passed on to males, they will suffer its deleterious effects.

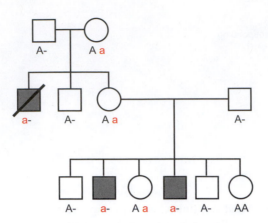

8. Neighboring Genes Are Linked During Inheritance Unless the DNA Recombines

Although there is a random distribution of strands of DNA (chromosomes) during sexual reproduction, there is not always a random distribution of alleles. To illustrate this point we must remember that higher organisms have tens of thousands of genes carried on multiple pairs of homologous chromosomes. Consider just a few of these genes—call them A, B, C, D, E, etc.—which have corresponding mutant alleles—a, b, c, d, e, etc. These genes may be on the same chromosome or they may be on different chromosomes. Let's assume that genes A, B, and C are on one pair of homologous chromosomes and D and E are on a separate pair. Organisms that are heterozygous for all of these genes will have the genotype Aa, Bb, Cc, Dd, Ee. Consequently, A, B, and C will be on one of a pair of homologous chromosomes and a, b, and c will be on the other member of the pair. A similar situation applies to D and E and d and e. Alleles carried on different chromosomes are distributed at random among the offspring of a mating. For example, there is as much chance of allele d accompanying allele A during inheritance as allele D. Because this segregation occurs independently, the likelihood that the two alleles are found in the same progeny after a mating is 50% or 0.5. In contrast, when genes are carried on the same chromosome, their alleles will not be distributed at random among the offspring. For example, because the three alleles A, B, and C are on the same chromosome, that is, the same molecule of DNA, they will tend to stay together. The same applies to a, b, and c. Such genes are said to be linked, and the phenomenon is known as **linkage**.

> When genes are on the same chromosome they are linked to each other.

8.1. Recombination During Meiosis Ensures Genetic Diversity

However, the alleles A, B, and C (or a, b, and c) do not *always* stay together during reproduction. Swapping of segments of the chromosomes can occur by breaking and rejoining of the neighboring DNA strands. Note that the breaking and joining occurs in equivalent regions of the two chromosomes and neither chromosome gains or loses any genes overall. The point at which the two strands of DNA cross over and recombine is called a **chiasma** (plural, chiasmata). The genetic result of such **crossing over**, the shuffling of different alleles between the two members of a chromosomal pair, is called **recombination** (Fig. 2.21). The farther apart two genes are on the chromosome, the more likely a crossover will form between them and the higher will be their

---

crossing over   When two different strands of DNA are broken and are then rejoined to one another
chiasma (pl. chiasmata)   Point at which two homologous chromosomes break and rejoin opposite strands
linkage   Two alleles are linked when they are inherited together more often than would be expected by chance, usually this is because they reside on the same DNA molecule (that is, on the same chromosome)
recombination   Mixing of genetic information from two chromosomes as a result of crossing over

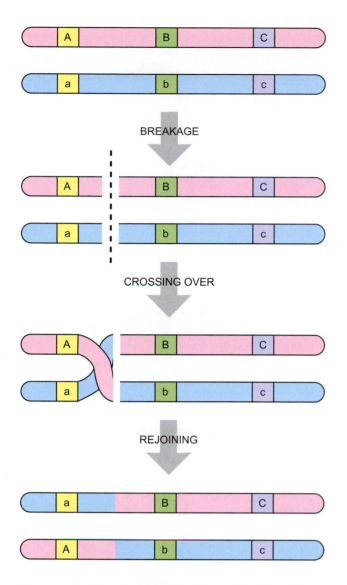

► **FIGURE 2.21**
Linkage of Genes and Recombination During Meiosis

At the top, the two members of a chromosome pair are shown, each carrying different alleles. Because the three alleles A, B, and C are on the same molecule of DNA, they will tend to stay together. So if the offspring inherits allele A from one parent, it will usually get alleles B and C, rather than b and c. If recombination occurs during meiosis, the DNA breaks and the chromosomes rejoin such that part of one chromosome is exchanged with the homologous partner. Now, the offspring can receive allele A with alleles b and c from one parent.

frequency of recombination. Recombination frequency is an important value for a geneticist, and the values range between 0% or 0, which means that the two genes are so close together, they are always found in the same progeny after a mating, to 50% or 0.5, which means that the two genes are so far apart that they appear to be on separate chromosomes.

This type of recombination occurs during meiosis, the process that reduces the genome from diploid to haploid. The process of meiosis is divided into two parts, meiosis I and meiosis II (Fig. 2.22). Table 2.01 describes the events that occur at each stage of meiosis. In a typical diploid organism, there are two homologues for each chromosome inside the normal cell. After the first replication of meiosis, there are four different copies of each chromosome, two copies of homologue 1 and two copies of homologue 2. These are attached at their centromeres and form what is called a **tetrad**.

When cells are in the substages of meiosis I, the genetic information on each of the four copies of the chromosomes aligns perfectly, matching gene for gene along the entire length of each chromosome. When the chromosomes are in this state, genetic information is exchanged with the other copies, forming new genetic combinations. The alignment stage, called **synapsis**, occurs due to a set of conserved proteins that

synapsis   The process in which homologous paternal and maternal chromosomes align so each gene is in the same location
tetrad   A structure found in prophase I of meiosis in which the two duplicated sister chromatids align to create a complex of four homologous chromosomes

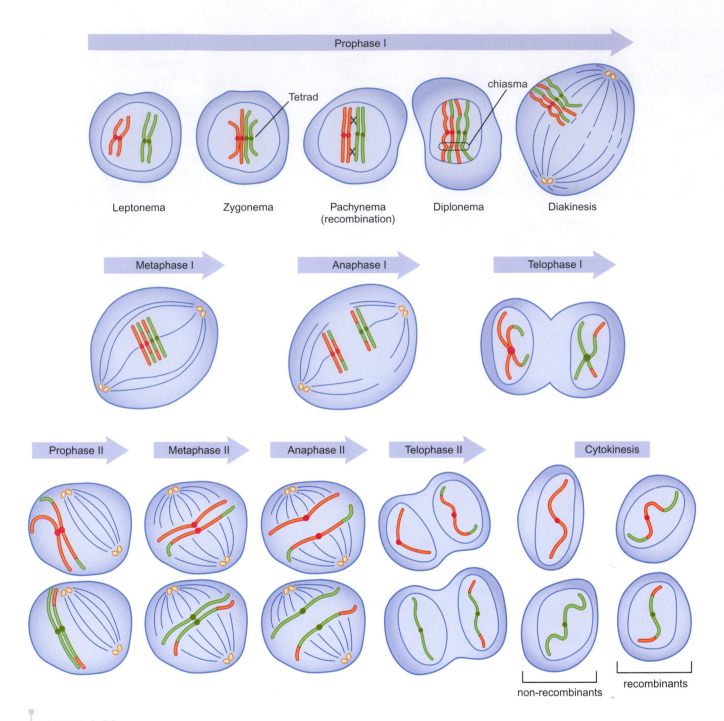

FIGURE 2.22
Meiosis Forms Haploid Gametes

This figure demonstrates how the diploid cell forms four haploid gametes during a special cell division called meiosis. Only one homologous chromosome (red and green) is shown for clarity, but it has undergone DNA replications to create two copies of each homolog.

link the chromosomes. These proteins form a structure where the DNA of each pair of homologous chromosomes is linked together with a zipper-like structure consisting of lateral elements and central elements connected by transverse fibers (Fig. 2.23).

Genetic linkage is often defined, from a molecular viewpoint, as the tendency of alleles carried by the same DNA molecule to be inherited together. However, if two genes are very far apart on a very long DNA molecule, linkage may not be observed in practice. In this example, consider a long chromosome, carrying all five genes, A, B, C, D, and E. It can be observed that A is linked to B and C, and that C and D are linked to E, but that no linkage is observed between A and E (Fig. 2.24). Given that A is on the same DNA molecule as B and that B is on the same DNA molecule as C, etc., it can be deduced that A, B, C, D, and E must all be on the same chromosome.

TABLE 2.01		Meiosis	
Division	**Stage of Meiosis**	**Substage of Prophase I**	**Chromosome Structure**
MEIOSIS I	Prophase I	Leptonema	Tetrads begin to condense
		Zygonema	Homologous chromosomes begin to pair up
		Pachynema	Homologous chromosomes are fully paired; recombination occurs
		Diplonema	Homologous chromosomes separate (except at the centromere); chiasmata are visible
		Diakinesis	Paired chromosomes condense further and attach to spindle fibers
	Metaphase I		Paired tetrads align at the middle of the cell
	Anaphase I		Homologous chromosomes split so that two copies move to each half of the cell
	Telophase I		Two new nuclei form, each containing a set of two sister chromatids
MEIOSIS II	Prophase II		Each chromosome condenses once again and start to attach to spindle fibers
	Metaphase II		Chromosomes align in the center of each new cell
	Anaphase II		Each of the sister chromatids separates and moves to each side of the new cell
	Telophase II		Chromosomes decondense and two new nuclei form
	Cytokinesis		Each of the two cells divides, completely forming four new cells, each containing one copy of each chromosome (a haploid genome)

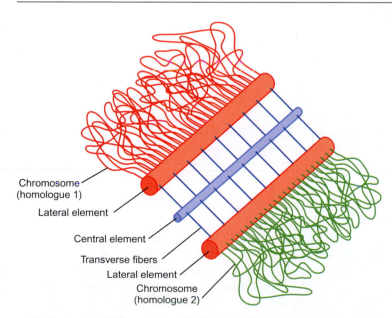

Chromosome (homologue 1)

Lateral element

Central element

Transverse fibers

Lateral element

Chromosome (homologue 2)

FIGURE 2.23
Synaptonemal Complex

The synaptonemal complex is a set of proteins that link the two homologous chromosomes during the zygotema stage of meiosis I. Only one chromosome pair is shown for clarity. The red and green chromosomes form a homologous pair.

In genetic terminology, it is said that A, B, C, D, and E are all in the same **linkage group**. Even though the most distant members of a linkage group may not directly show linkage to each other, their relationship can be deduced from their mutual linkage to intervening genes.

The exchange of alleles between homologous chromosomes and independent assortment of chromosomes during anaphase provide all the new combinations of genes to make each person unique.

linkage group   A group of alleles carried on the same DNA molecule (that is, on the same chromosome)

FIGURE 2.24
Linkage Groups

In this example chromosome, genes A and E are linked even though the recombination frequencies suggest they are not linked. After this parent mates, the percentage of progeny that have a recombination event between the labeled genes is indicated above the chromosome. When the progeny are assayed for the presence of both the A and C allele, there are only 30% that have this combination of genes. When progeny are assayed for the presence of both the C and E allele, about 25% of the progeny have this combination. Since A is linked to C, and C is linked to E, it can be deduced that A and E are on the same linkage group.

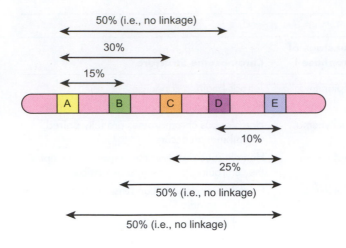

FOCUS ON RELEVANT RESEARCH

Hochwagen A, Marais GAB (2010) Meiosis: A PRDM9 guide to the hotspots of recombination. Curr Biol 20(6): R271-R274.

Meiosis is a unique process with two different methods of creating new genetic combinations. First, in metaphases I and II, the chromosomes align along the metaphase plate randomly. Therefore, the resulting cells contain a combination of different chromosomes, some from the paternal side and others from the maternal side. Second, during prophase I, recombination creates new genetic combinations within a single chromosome. Although crossovers can theoretically occur at any location along a chromosome, in reality, there are certain regions of chromosomes that are more likely to have recombination, and these are called **hotspots**. One of the areas of research on meiosis focuses on determining what types of proteins and/or DNA structures define the recombination site.

One potential marker for a recombination hotspot is the post-translational modification of histone H3. Eukaryotic genomes are very densely packed in order to fit inside the nucleus, and a group of proteins called histones help keep the DNA strands compacted. Eight histone proteins form a solid sphere, which the strand of DNA wraps around twice. When histone H3 has three methyl groups ($-CH_3$), groups added onto the amino acid lysine, then the DNA wrapped around this group of histones is more likely to recombine. However, this modification is found in areas that do not recombine, therefore, other recombination factors must exist. This article reviews the evidence that a protein called PRDM9 is a potential determinant of recombination hotspot location. In addition, the article discusses the changes in PRDM9 from species to species and how this may impact genetic diversity.

Box 2.04 Meiosis in Human Females

In human females, meiosis occurs in three different stages, separated by long periods of arrested development. During fetal development, meiosis of the female eggs begins at 11–12 weeks gestation. During this period, the eggs enter prophase, undergo synapsis, and finally, recombination. The cells then arrest in diplonema and enter a state of hibernation or arrest called dictyate. During the first arrest, the primordial follicle cells, which will ultimately provide protection and the proper environment for the egg, begin to continue development. These primordial follicle cells are fully formed at birth and then remain quiet until the female reaches sexual maturity in adolescence.

Once the correct hormones are produced by the pituitary gland in the brain, the follicle primordial cells begin to develop into a mature follicle surrounding one egg cell. This growth and development takes about 85 days. When the egg then receives a surge of luteinizing hormone at the mid-point of the menstrual cycle, meiosis resumes, continuing

Continued

Box 2.04 Continued

through metaphase II. During meiosis I, the chromosome number is reduced from four copies of each chromosome to only two copies. The other half of the chromosomes are released as a polar body rather than forming another egg. At this point the egg is released from the ovary for fertilization or degeneration. If fertilization occurs, the sperm triggers the last stages of meiosis in the egg. As before, after this final cell division, the two copies of homologous chromosomes are reduced to a single haploid genome, and the other half is released as a second polar body. After meiosis II completes, the egg haploid genome becomes packaged into a nucleus and then fuses with the sperm pronucleus to form the zygote, and eventually a new person. If fertilization does not occur, the metaphase II arrested egg is shed with the menstrual flow approximately 2 weeks after its release from the ovary.

9. Identifying Genes that Cause Human Diseases

Linkage is the primary technique used to identify the genes associated with inherited human diseases. In order to identify the gene responsible for a particular disease, each person from a family afflicted with the disorder is examined for the presence of genetic markers along each of their chromosomes. These genetic markers can be specific nucleotide sequences or specific genes or a combination of both. Since the human genome is so large and recombination tends to occur only at certain hotspots, certain combinations of genetic markers almost always stay together during meiosis and are called **haplotypes**. Haplotypes tend to stay together throughout a family pedigree. Genetic researchers try to identify whether or not the disease that they are studying is linked to one of these haplotypes. To refine the location of the potential gene that causes the disease, researchers then try to determine if there has been recombination among any of these haplotypes. If recombination is discovered, then the family pedigree is analyzed to determine if the recombination is only found in the members afflicted with the disease. If this is true, the recombination neighborhood may contain the mutation that causes the disease being studied.

Linkage of haplotypes is calculated in a similar manner as linkage of genes (see Fig. 2.24 above). Basically, the percentage of times a haplotype is linked to the disease is calculated. In Figure 2.22, there was recombination between the red and green chromosomes at two places during pachynema. The two gametes that inherited the hybrid (chimeric) chromosomes are called **recombinants**. The normal order of alleles and/or haplotypes along the red chromosome is disrupted because a segment of the green chromosome is present. The number of recombinant progeny in comparison to the total number of progeny is called the recombination frequency or recombination fraction. When the haplotype and disease is linked, the recombination frequency is low. When the haplotype and disease is unlinked, the recombination frequency is 50% because of independent assortment of each single chromatid during meiosis. The recombination frequency or recombination fraction is calculated with the following formula:

$$\text{Recombination Frequency} (\theta) = \frac{\text{Number of Recombinants}}{\text{Total Number of Progeny Analyzed}}$$

The tendency for chromosomes to have recombination hotspots and multiple chiasmata skew the values of recombination frequency, therefore, recombination frequency is only accurate when the value for θ is less than 10%.

---

haplotypes   A combination of alleles or genetic markers that are inherited as one unit during meiosis
recombinants   Gametes in which genetic recombination occurred

To examine a human pedigree for recombination is difficult because there are too few family members to truly determine a recombination frequency. There are simply not enough recombinants and non-recombinants. Therefore, a statistical method for examining linkage, called the **LOD score (logarithm of the odds) (Z)**, is used. LOD score compares the likelihood that two haplotypes are linked and the likelihood that two haplotypes are unlinked. Therefore, in order to calculate a LOD score, the probability that two haplotypes are linked through the birth order of the family and the probability that they are not linked is calculated. The actual LOD score is the log of the ratio of these two values. But since we do not know the actual recombination frequency, LOD scores are calculated for each recombination frequency between 0 and 0.5, where 0 is complete linkage and 0.5 is completely unlinked. A LOD score is calculated using the following formula:

$$Z(\theta) = \log_{10} \frac{\text{Likelihood of linkage}}{\text{Likelihood of no linkage}}$$

where θ represents the chosen recombination frequency and Z is the LOD score. Although the mathematical calculations are important for those entering the field of human genetics, for this text, interpretation of the final values is sufficient knowledge. When the LOD score is 3 or greater, then the odds are in favor of linkage, and the odds that the linkage is due to chance are 1000 to 1. This is considered the arbitrary cutoff for human genetic analysis of linkage. All of these values are calculated by computer, so the process can accommodate the large number of haplotypes found in the human genome.

> Identifying genes that cause disease is a complex process and relies upon statistical analysis of the probability for a particular genetic marker to be linked to the disease.

Key Concepts

- Some traits are complex and are controlled by a variety of genes, whereas other traits are due to the expression of one gene. Some traits, such as plant height and wrinkled seeds, are controlled by a single gene whose function is to regulate the expression of multiple genes.
- A gene can have different changes or mutations that affect the function of the gene product (either protein or RNA). Each different form of the gene is called an allele. Null alleles produce no detectable gene product.
- The total combination of genes and alleles are called genotypes. The outward physical appearance is called the phenotype.
- When multiple genes control a biochemical pathway, a defect in the beginning of the pathway will change the original phenotype whether or not the genes that control the other steps in the pathway are defective or wild-type. This effect is called epistasis.
- Genes are segments of DNA that are found on chromosomes. Most mammals are diploid; that is, they have two sets of homologous chromosomes, whereas bacteria have only one set. In contrast, plants tend to have multiple sets, also called polyploidy.
- Since mammals have two copies of each gene, they can have two different alleles. If one allele expresses a normal protein, which carries out the biochemical function, this allele is dominant and can mask or hide the presence of the other allele, which is called recessive.
- Some genetic mutations are not strictly dominant or recessive. Some mutations only affect the enzyme function partially and therefore are partially dominant alleles. Other mutations can create new enzyme functions that are

---

LOD score (logarithm of the odds) (Z)   Statistical estimate of whether two loci are found near each other along a chromosomes

dominant over a recessive allele, but co-dominant with the wild-type allele. Other mutations show complete dominance over the wild-type allele. In still other cases, some mutations will have different penetrance; that is, the mutation will be expressed differently in different individuals or environments.

- During meiosis, the diploid genome of each parent is split into germ cells (eggs or sperm) so that only one copy of each gene is present; that is, each germ cell contains a haploid genome. After fusing of gametes, the new organism will receive one copy of its genome from the egg and the other copy from the sperm to return to a diploid state.
- During meiosis, homologous chromosomes align so that each gene is side-by-side with its other copy, a process called synapsis. Recombination occurs when certain areas of the chromosome break and switch locations with its homolog. This is one key process that causes genetic diversity in sexually-reproducing organisms.
- When genes are far apart during a recombination event, they often end up on different homologous chromosomes, even though they started on the same. In contrast, genes that are close together in a recombination event will stay together. In human meiosis, certain combinations of genetic markers tend to stay together, which is called a haplotype.
- The percentage of times two genes or haplotypes stay together after meiosis is used to determine the relative distance from each other. In studying human genetic diseases, scientists determine if various recombination events with a known haplotype are only found in the affected individuals and not the normal individuals within the family.

Review Questions

1. What are genes? How do regulatory genes and alleles complicate the issue of genetic traits?
2. What is meant by the term "Mendelian characters"?
3. What is the difference between an enzyme and a protein? Are all proteins also enzymes?
4. What are regulatory proteins? How can a defect in a regulatory protein affect an organism?
5. What is meant by "wild-type"?
6. In terms of red versus white flowers, why is the designation for white flowers termed "r" as opposed to "R," for red flowers? Which would be the wild-type allele?
7. What is the difference between "phenotype" and "genotype"? Can a genotype be readily determined based on a specific phenotype? Why or why not?
8. What is the difference between haploid and diploid? What are the advantages of being diploid over haploid?
9. Why can male mammals be regarded as partially diploid? What term, diploid or haploid, would apply to the gametes of a diploid organism?
10. Why are duplicated genes on a haploid chromosome not regarded as genuine alleles?
11. Define "ploidy" and "copy number." What, if any, is the difference?
12. What is aneuploidy?
13. What is the condition in humans caused by trisomy?
14. What is the term used to describe a diploid organism carrying an RR or rr genotype? What is the term for Rr or rR?
15. What do the terms "dominant" and "recessive" mean?
16. In terms of R and r, write homozygous dominant, heterozygous, and homozygous recessive for diploid organisms.

17. If R represents the gene for red flowers, what would be the color of the flowers of an Rr or rR plant? How could partial dominance alter the phenotype of an Rr or rR genotype?
18. If B is a dominant allele for blue flowers and R is a dominant allele for red flowers, what color of flower would produce from co-dominance of these two alleles (i.e., BR)?
19. Give an example of a dominant allele in one organism that is recessive in another organism.
20. What is meant by penetrance? Give an example of a human case of penetrance. What other factors can affect penetrance?
21. What are modifier genes and how do they affect the phenotype of an organism?
22. What is the purpose of sexual reproduction?
23. What are the differences between gametes and somatic cells? How many sets of chromosomes does each have? What process produces the set of chromosomes for gametes?
24. What is a zygote and how is it formed?
25. What are filial generations and how are they labeled?
26. Write out a Punnett square or checkerboard diagram for the mating of homozygous dominant to a homozygous recessive organism for red flowers. What are the possible progeny from this mating? Assuming no partial dominance, what would be the phenotypes of these progeny?
27. What is the function of a sex chromosome? What genetic sex is given to an XX organism? What about XY? Is it possible to have YY organisms? Why or why not?
28. What are the genes called that are carried on the sex chromosomes but have nothing to do with the sex of an organism?
29. Why are males more likely to suffer genetic defects? Describe the situation that would cause females to be carriers of genetic defects. Describe color blindness in terms of sex-linked genes.
30. What is the term used to describe two genes that are inherited more often together than just by chance?
31. What is "crossing over"? How is this advantageous to an organism?
32. What would cause a higher frequency of recombination for genes?

Conceptual Questions

1. a. Determine the potential phenotype(s) for the offspring of a mother with dominant red flowers (RR) and purple stem (PP) and a father with recessive white flowers (rr) and all green stems (pp). b. Using the checkerboard diagram or Punnett square, determine all the potential genotypes for the F2 generation if the F1 was to self-cross.
2. Red-green color blindness in humans is caused by a recessive gene on the X-chromosome. The following pedigree was determined for a family in the United States. What is the probability that the daughter (III-1) will be a carrier? What is the probability that III-2 and III-3 will be color blind?

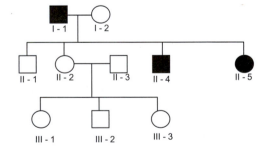

3. Three precursor substances (X, Y, and Z) for an essential vitamin were added to minimal growth media for a wild-type and three different mutant strains of *E. coli*. Wild-type *E. coli* can synthesize this vitamin and all three precursor substances, but mutations in the biosynthetic enzymes for any of the three precursors cause a vitamin deficiency that is lethal. The growth was assessed and tabulated in the following table (+ means growth, − means no growth). Which of the strains is wild-type?

E. coli strain	Min. Media (no vitamin) (MM)	MM + X	MM + Y	MM + Z
HB101	−	−	+	+
BL21	+	+	+	+
EB898	−	−	+	−
CJ765	−	+	+	+

4. In the plant *Arabidopsis thaliana*, there are three different genes on chromosome 4: CPK27 (ck), PUX3 (px), and IMMUTANS (im). To determine the order of the genes along the chromosome, the following parents were crossed and the following progeny recovered:
   ck^+ px im/ck px^+ im^+ × ck px im/ck px im

ck^+	px	im	207
ck	px^+	im^+	215
ck^+	px	im^+	167
ck	px^+	im	188
ck^+	px^+	im	71
ck	px	im^+	73
ck^+	px^+	im^+	35
ck	px	im	44

   What is the gene order for these three genes and what are the recombination frequencies between them? What genes are closer together on the chromosome?

5. Researchers are trying to find a gene that is responsible for absolute pitch, which is the ability to identify tones with the correct musical note without any help from a reference note. Many people do not think absolute pitch is a genetic trait, and instead believe that absolute pitch is actually a learned skill. DNA samples were collected from 73 families with at least two people that have absolute pitch. Linkage analysis was performed for the various families and the LOD scores were analyzed to determine whether or not a particular haplotype was associated with absolute pitch. One haplotype on chromosome 8 was found to have a LOD score of 3.231. Does the linkage analysis support the hypothesis that absolute pitch is a genetic trait? Why or why not?

Chapter 3

DNA, RNA, and Protein

Early geneticists studied the transmission of characters from parents to offspring, but they did not know the underlying chemical mechanisms. In many ways, molecular biology is the merger of biochemistry with genetics. The revelation that genes are made of DNA—in other words, that biological information is carried by the nucleic acids, DNA and RNA—has transformed biology. Understanding the chemical nature of DNA has provided a mechanistic basis not just for heredity but also for a variety of other phenomena from cell growth and division to cancer. In this chapter we discuss how DNA was discovered to be the genetic material. We then review the chemical nature of DNA and RNA and explain how these molecules provide a physical mechanism for storing biological information, distributing it inside a growing cell, and finally passing it from one generation to the next. Finally, we introduce the molecules that are encoded by the genes and carry out most of the day-to-day operations of the cell—the proteins.

1. History of DNA as the Genetic Material

Until early in the nineteenth century, it was believed that living matter was quite different from inanimate matter and was not subject to the normal laws of chemistry. In other words, organisms were thought to be made from chemical components unique to living creatures. Furthermore, there was supposedly a special vital force that mysteriously energized living creatures. Then, in 1828, Friedrich Wohler demonstrated the conversion in a test tube of ammonium cyanate, a

Molecular Biology.

laboratory chemical, to urea, a "living" molecule also generated by animals. This was the first demonstration that there was nothing magical about the chemistry of living matter.

> Despite their complexity, living organisms obey the laws of chemistry.

Further experiments showed that the molecules found in living organisms were often very large and complex. Consequently, their complete chemical analysis was time consuming and is indeed, still continuing today. The demystification of life chemistry reached its peak in the 1930s when the Russian biochemist Alexander Oparin wrote a book outlining his proposal for the chemical origin of life. Although the nature of the genetic material was still unknown, Oparin put forward the idea that life, with its complex molecular composition, evolved from small molecules in the primeval ocean as a result of standard physical and chemical forces (see Ch. 26).

Until the time of World War II, the chemical nature of the *inherited genetic information* remained very vague and elusive. **DNA** was actually discovered in 1869 by Frederich Miescher who extracted it from the pus from infected wounds! However, it was nearly a century before its true significance was revealed by Oswald Avery. In 1944, Avery found that the virulent nature of some strains of bacteria that caused pneumonia could be transmitted to related harmless strains by a chemical extract. Avery purified the essential molecule and demonstrated that it was DNA, although he did not use the name "DNA," since its structure was then uncharacterized. When DNA from virulent strains was added to harmless strains, some took up the DNA and were "transformed" into virulent strains. Avery concluded that the genes were made of DNA and that somehow genetic information was encoded in this molecule.

> Avery found that purified DNA could carry genetic information from one strain of bacterium to another. This revealed that DNA was the genetic material.

2. Nucleic Acid Molecules Carry Genetic Information

Chapter 2 discussed how the fundamentals of modern genetics were laid when Mendel found that hereditary information consists of discrete fundamental units now called genes. Each gene is responsible for a single inherited property or characteristic of the organism. The realization that genes are made up of DNA molecules opened the way both to a deeper understanding of life and to its artificial alteration by genetic engineering.

Genetic information is encoded by molecules named **nucleic acids** because they were originally isolated from the nucleus of eukaryotic cells. There are two related types of nucleic acid, **deoxyribonucleic acid** (**DNA**) and **ribonucleic acid** (**RNA**). The master copy of each cell's genome is stored on long molecules of DNA, which may each contain many thousands of genes. Each gene is thus a linear segment of a long DNA molecule. In contrast, RNA molecules are much shorter, are used to transmit the genetic information to the cell machinery, and carry only one or a few genes. (Certain viruses use RNA to encode their genomes as well as transmit genetic information to the cell machinery. These RNA viruses have short genomes, rarely more than a dozen genes, as opposed to the hundreds or thousands of genes carried on the DNA genomes of cells.)

> Genetic information is carried on long linear polymers, the nucleic acids. Two classes of nucleic acid, DNA and RNA, divide up the responsibility of storing and deploying the genetic information.

3. Chemical Structure of Nucleic Acids

DNA and RNA are linear polymers made of subunits known as **nucleotides**. The information in each gene is determined by the order of the different nucleotides, just as the information in this sentence is due to the order of the 26 possible letters of the alphabet. There are four different nucleotides in each type of nucleic acid and their order determines the genetic information (Fig. 3.01).

---

deoxyribonucleic acid (DNA) Nucleic acid polymer of which the genes are made
DNA Deoxyribonucleic acid, nucleic acid polymer of which the genes are made
nucleic acid Class of polymer molecule consisting of nucleotides that carries genetic information
nucleotide Monomer or subunit of a nucleic acid, consisting of a pentose sugar plus a base plus a phosphate group
ribonucleic acid (RNA) Nucleic acid that differs from DNA in having ribose in place of deoxyribose

FIGURE 3.01
The Order of the Nucleotides Encodes the Genetic Information

Nucleotides are ordered along a string of DNA or RNA. It is the ordering of the different nucleotides that dictates the nature of the information within the nucleic acid.

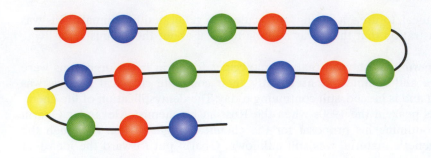

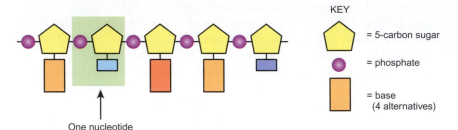

KEY

⬠ = 5-carbon sugar

● = phosphate

▮ = base (4 alternatives)

One nucleotide

FIGURE 3.02
Three Views of a Nucleotide

The three components of a nucleotide are shown to the left. The structures on the right show the pentose sugar (deoxyribose) connected to the phosphate and the base.

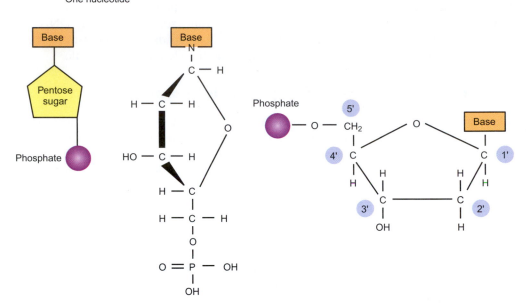

Each nucleotide has three components: a **phosphate group**, a five-carbon sugar, and a nitrogen-containing **base** (Fig. 3.02). The phosphate groups and the sugars form the backbone of each strand of DNA or RNA. The bases are joined to the sugars and stick out sideways.

In DNA, the sugar is always **deoxyribose**; whereas, in RNA, it is **ribose**. Both sugars are **pentoses**, or five-carbon sugars. Deoxyribose has one less oxygen than ribose (Fig. 3.03). It is this chemical difference that gave rise to the names deoxyribonucleic acid and ribonucleic acid. Both sugars have five-membered rings consisting of four carbon atoms and one oxygen atom. The fifth carbon forms a side chain to the ring. The five carbon atoms of the sugar are numbered 1', 2', 3', 4', and 5' as shown in Figure 3.02. By convention, in nucleic acids, numbers with prime marks refer to the sugars and numbers without prime marks refer to the positions around the rings of the bases.

---

base   Alkaline chemical substance, in molecular biology especially refers to the cyclic nitrogen compounds found in DNA and RNA
deoxyribose   The sugar with five carbon atoms that is found in DNA
pentose   A five-carbon sugar, such as ribose or deoxyribose
phosphate group   Group of four oxygen atoms surrounding a central phosphorus atom found in the backbone of DNA and RNA
ribose   The five-carbon sugar found in RNA

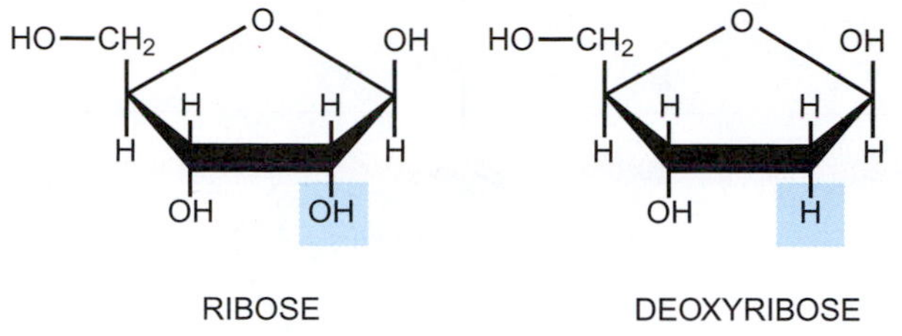

FIGURE 3.03
The Sugars Composing RNA and DNA

Ribose is the five-carbon sugar (pentose) found in RNA. Deoxyribose is the pentose of DNA, and it has one less oxygen than ribose as it has hydrogen in place of the hydroxyl group on position 2' of the ribose ring.

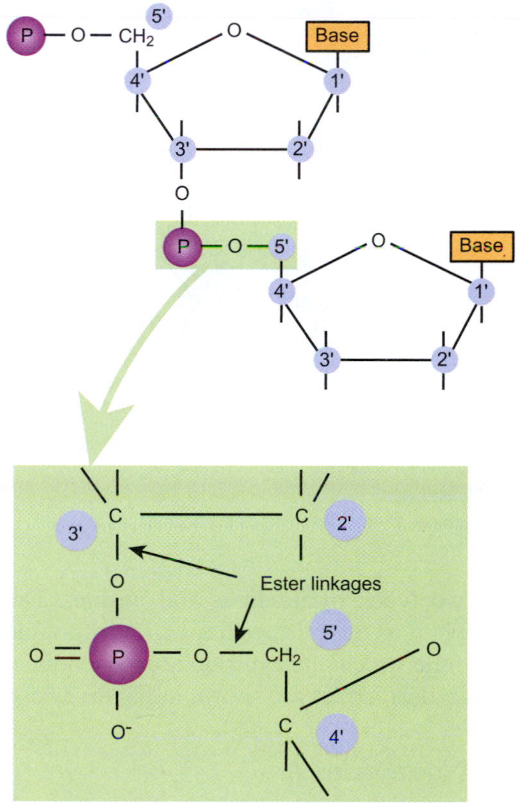

FIGURE 3.04
Nucleotides Are Joined by Phosphodiester Linkages

The nucleotides that form the backbone of DNA and RNA are joined together by linkages involving their phosphate groups. One nucleotide is linked via its 5'-carbon to the oxygen of the phosphate group and another nucleotide is linked via its 3'-carbon to the other side of the central phosphate. These linkages are termed phosphodiester groups.

Nucleotides are joined by linking the phosphate on the 5'-carbon of the (deoxy) ribose of one nucleotide to the 3'-hydroxyl of the next as shown in Figure 3.04. The phosphate group is joined to the sugar on either side by ester linkages, and the overall structure is therefore a **phosphodiester** linkage. The phosphate group linking the sugars has a negative charge.

Two nucleotides are linked together via a phosphodiester bond between the 5'-carbon of the sugar in one nucleotide to the 3'-hydroxyl of the next nucleotide.

3.1. DNA and RNA Each Have Four Bases

There are five different types of nitrogenous bases associated with nucleotides. DNA contains the bases **adenine**, **guanine**, **cytosine**, and **thymine**. These are often abbreviated to A, G, C, and T, respectively. RNA contains A, G, and C, but T is replaced by **uracil** (U). From the viewpoint of genetic information, T in DNA, and U in RNA are equivalent.

---

adenine (A) A purine base that pairs with thymine, found in DNA or RNA
cytosine (C) One of the pyrimidine bases found in DNA or RNA and which pairs with guanine
guanine (G) A purine base found in DNA or RNA that pairs with cytosine
phosphodiester The linkage between nucleotides in a nucleic acid that consists of a central phosphate group esterified to sugar hydroxyl groups on either side
thymine (T) A pyrimidine base found in DNA that pairs with adenine
uracil (U) A pyrimidine base found in RNA that may pair with adenine

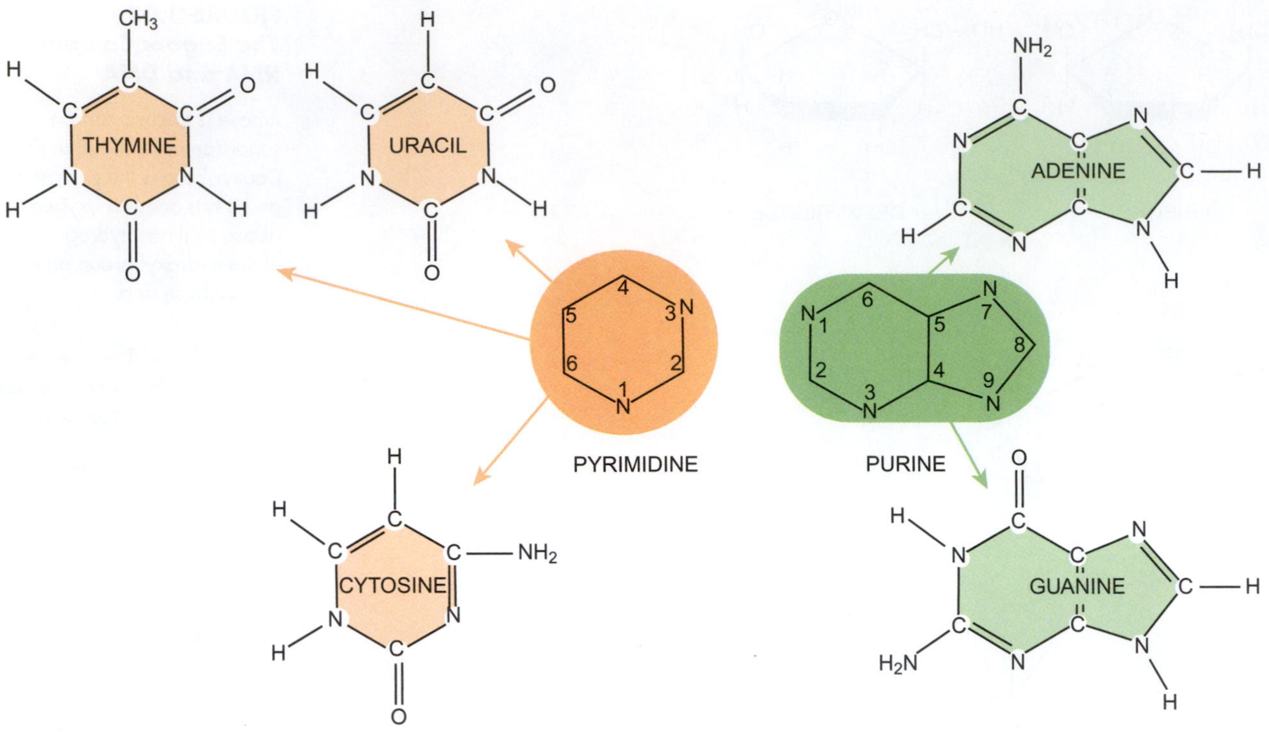

FIGURE 3.05
The Bases of the Nucleic Acids

The four bases of DNA are adenine, guanine, cytosine, and thymine. In RNA, uracil replaces thymine. Pyrimidine bases contain one-ring structures, whereas purine bases contain two-ring structures.

Purines have two rings and pyrimidines have a single ring.

The bases found in nucleic acids are of two types, **pyrimidines** and **purines**. The smaller pyrimidine bases contain a single ring whereas the purines have a fused double ring. Adenine and guanine are purines; and thymine, uracil and cytosine are pyrimidines. The purine and pyrimidine ring systems and their derivatives are shown in Figure 3.05.

3.2. Nucleosides Are Bases Plus Sugars; Nucleotides Are Nucleosides Plus Phosphate

A base plus a sugar is known as a **nucleoside**. A base plus a sugar plus phosphate is known as a nucleotide. If necessary, one may distinguish between **deoxynucleosides** or **deoxynucleotides** where the sugar is deoxyribose, and **ribonucleosides** or **ribonucleotides** that contain ribose. The names of the nucleosides are similar to the names of the corresponding bases (see Table 3.01). The nucleotides do not have names of their own but are referred to as phosphate derivatives of the corresponding nucleoside. For example, the nucleotide of adenine is **adenosine monophosphate**, or **AMP**.

Nucleotides have the sugar base attached to a phosphate and nitrogenous base. Nucleosides are missing the phosphate group.

Three-letter abbreviations for the bases such as ade, gua, etc., are sometimes used when writing biochemical pathways or for the names of genes involved in nucleotide metabolism. When writing the sequence of a nucleic acid, the single letter abbreviations are used (A, T, G, and C for DNA or A, U, G, and C for RNA). The letter N is often used to refer to an unspecified base.

adenosine monophosphate (AMP)   The nucleotide consisting of adenine, (deoxy)ribose, and one phosphate
deoxynucleoside   A nucleoside containing deoxyribose as the sugar
deoxynucleotide   A nucleotide containing deoxyribose as the sugar
nucleoside   The union of a purine or pyrimidine base with a pentose sugar
purine   Type of nitrogenous base with a double ring found in DNA and RNA
pyrimidine   Type of nitrogenous base with a single ring found in DNA and RNA
ribonucleoside   A nucleoside whose sugar is ribose (not deoxyribose)
ribonucleotide   A nucleotide whose sugar is ribose (not deoxyribose)

4. Double-Stranded DNA Forms a Double Helix

A strand of nucleic acid may be represented in various ways, either in full or abbreviated to illustrate the linkages (Fig. 3.06). As illustrated above, nucleotides are linked by joining the 5'-phosphate of one to the 3'-hydroxyl group of the next. Typically,

TABLE 3.01	Naming Bases, Nucleosides, and Nucleotides		
Base	**Abbreviations**	**Nucleoside**	**Nucleotide**
Adenine	ade A	adenosine	adenosine monophosphate (AMP)
Guanine	gua G	guanosine	guanosine monophosphate (GMP)
Cytosine	cyt C	cytidine	cytidine monophosphate (CMP)
Thymine	thy T	thymidine	thymidine monophosphate (TMP)
Uracil	ura U	uridine	uridine monophosphate (UMP)

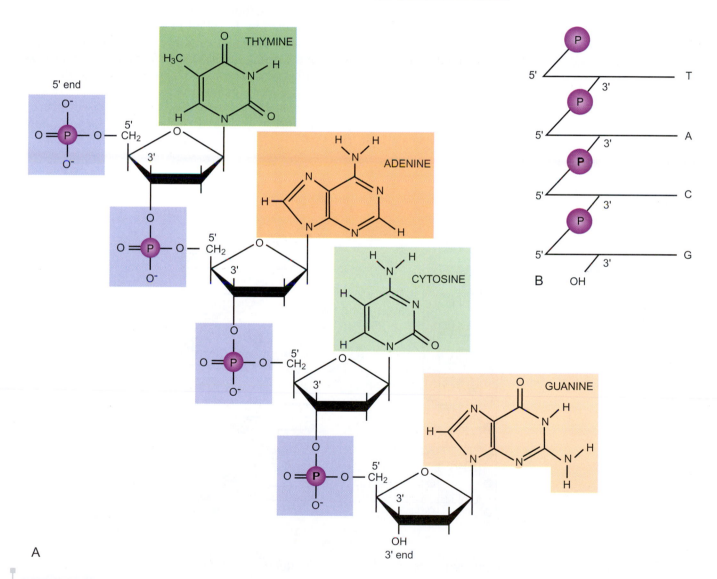

FIGURE 3.06
Some Variations in the Ways Nucleic Acids are Represented

A) More elaborate drawings show the chemical structures of the nucleic acid components, including the pentose sugar, phosphate groups, and bases. B) Simple line drawings may be used to summarize the linkage of sugars by the 5' and 3' phosphodiester bonds. Here, the protruding bases have been abbreviated to a single letter.

there is a free phosphate group at the 5'-end of the chain and a free hydroxyl group at the 3'-end of a nucleic acid strand. Consequently, a strand of nucleic acid has polarity and it matters in which direction the bases are read off. The 5'-end is regarded as the beginning of a DNA or RNA strand. This is because genetic information is read starting at the 5'-end. (In addition, when DNA is replicated, nucleic acids are synthesized starting at the 5'-end as described in Ch. 9.)

RNA is normally found as a single-stranded molecule, whereas DNA is double-stranded. Note that the two strands of a DNA molecule are **antiparallel**, as they point in opposite directions. This means that the 5'-end of one strand is opposite the 3'-end of the other strand (Fig. 3.07). Not only is DNA double-stranded, but the two separate strands are wound around each other in a helical arrangement. This is the famous **double helix** first proposed by Francis Crick and James Watson in 1953 (Fig. 3.08). The DNA double helix is stabilized both by hydrogen bonds between the bases (see below) and by stacking of the aromatic rings of the bases in the center of the helix.

In order to determine the molecular arrangement of the phosphates, sugars, and bases, Watson and Crick interpreted X-ray crystallographic data. In 1950 Maurice Wilkins and his assistant Raymond Gosling took the first images of DNA using X-ray diffraction. Gosling's work was continued by Rosalind Franklin who joined Wilkins' group the following year. Watson and Crick used an X-ray diffraction picture taken by Rosalind Franklin and Raymond Gosling in 1952 as the basis for their structural model. Rosalind Franklin died in 1958 of cancer aged 37, probably due to the effects of the X-rays. Unraveling the chemical basis for inheritance won Watson, Crick, and Wilkins the Nobel Prize in Physiology or Medicine for 1962 "for their discoveries concerning the molecular structure of nucleic acids and its significance for information transfer in living material" (Fig. 3.09).

> The complementary structure of the DNA double helix is critical to replication of the genes, as described in more detail in Chapter 10.

FIGURE 3.07
Representations of Double-Stranded DNA

On the left DNA is represented as a double line consisting of two complementary strands. Actually, DNA forms a double helix, as shown to the right.

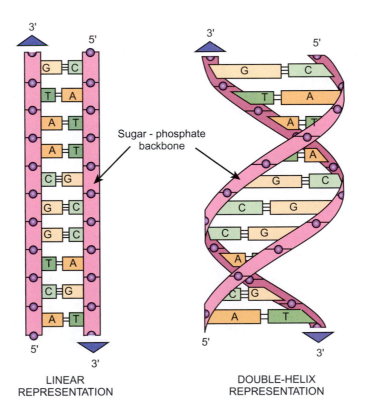

Sugar - phosphate backbone

LINEAR REPRESENTATION

DOUBLE-HELIX REPRESENTATION

antiparallel   Parallel, but running in opposite directions
double helix   Structure formed by twisting two strands of DNA spirally around each other

NATURE

No. 4356   **April 25, 1953**

MOLECULAR STRUCTURE OF NUCLEIC ACIDS

A structure for Deoxyribose Nucleic Acid

We wish to suggest a structure for the salt of deoxyribose nucleic acid (D.N.A.). This structure has novel features which are of considerable biological interest.

A structure for nucleic acid has already been proposed by Pauling and Corey[1]. They kindly made their manuscript available to us in advance of publication. Their model consists of three intertwined chains, with the phosphates near the fibre axis, and the bases on the outside. In our opinion, this structure is unsatisfactory for two reasons: (1) We believe that the material which gives the X-ray diagrams is the salt, not the free acid. Without the acidic hydrogen atoms it is not clear what forces would hold the structure together, especially as the negatively charged phosphates near the axis will repel each other. (2) Some of the van der Waals distances appear to be too small.

Another three-chain structure has also been suggested by Fraser (in the press). In his model the phosphates are on the outside and the bases on the inside, linked together by hydrogen bonds. This structure as described is rather ill-defined, and for this reason we shall not comment on it.

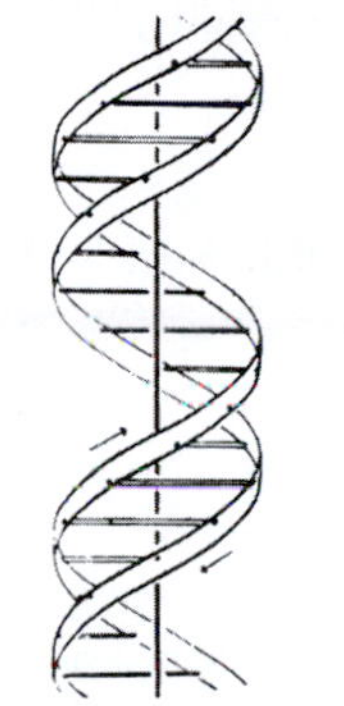

We wish to put forward a radically different structure for the salt of deoxyribose nucleic acid. This structure has two helical chains each coiled round the same axis (see diagram). We have made the usual chemical assumptions, namely, that each chain consists of phosphate diester groups joining β-D-deoxyribofuranose residues with 3', 5' linkages. The two chains (but not their bases) are related by a dyad perpendicular to the fibre axis. Both chains follow right-handed gelices, but owing to the dyad the sequences of the atoms in the two chains run in opposite directions. Each chain loosely resembles Furberg's[2] model No. 1; that is the bases are on the inside of the helix and the phosphates on the outside. The configuration of the sugar and the atoms near it is close to Furberg's 'standard configuration', the sugar being roughly perpendicular to the attached base.

This figure is purely diagrammatic. The Two ribbons symbolize the two phosphate—sugar chains, and the horizontal rods the pairs of bases holding the chains together. The vertical line marks the fibre axis

There is a residue on each chain every 3·4. A. in the z-direction. We have assumed an angle of 36° between adjacent residues in the same chain, so that the structure repeats after 10 residues on each chain, that is, after 34 A. The distance of a phosphorus atom from the fibre axis is 10 A. As the phosphates are on the outside, cations have easy access to them.

The structure is an open one, and its water content is rather high. At lower water contents we would expect the bases to tilt so that the structure could become more compact.

The novel feature of the structure is the manner in which the two chains are held together by the purine and pyrimidine bases. The planes of the bases are perpendicular to the fibre axis. They are joined together in pairs, a single base from one chain being hydrogen-bonded to a single base from the other chain, so that the two lie side by side with identical z-co-ordinates. One of the pair must be a purine and the other a pyrimidine for bonding to occur. The hydrogen bonds are made as follows: purine position 1 to pyrimidine position 1; purine position 6 to pyrimidine position 6.

If it is assumed that the bases only occur in the structure in the most plausible tautomeric forms (that is, with the keto rather than the enol configurations) it is found that only specific pairs of bases can bond together. These pairs are: adenine (purine) with thymine (pyrimidine), and guanine (purine) with cytosine (pyrimidine).

In other words, if an adenine forms one member of a pair, on either chain, then on these assumptions the other member must be thymine; similarly for guanine and cytosine. The sequence of bases on a single chain does not appear to be restricted in any way. However, if only specific pairs of bases can be formed, it follows that if the sequence of bases on one chain is given, then the sequence on the other chain is automatically determined.

It has been found experimentally[3,4] that the ratio of the amounts of adenine to thymine, and the ratio of guanine to cytosine, are always very close to unity for deoxyribose nucleic acid.

It is probably impossible to build this structure with a ribose sugar in place of the deoxyribose, as the extra oxygen atom would make too close a van der Waals contact.

The previously published X-ray data[5,6] on deoxyribose nucleic acid are insufficient for a rigorous test of our structure. So far as we can tell, it is roughly compatible with the experimental data, but it must be regarded as unproved until it has been checked against more exact results. Some of these are given in the following communications. We were not aware of the details of the results presented there when we devised our structure, which rests mainly though not entirely on published experimental data and stereochemical arguments.

It has not escaped our notice that the specific pairing we have postulated immediately suggests a possible copying mechanism for the genetic material.

Full details of the structure, including the conditions assumed in building it, together with a set of co-ordinates for the atoms, will be published elsewhere.

We are much indebted to Dr. Jerry Donohue for constant advice and criticism, especially on inter-atomic distances. We have also been stimulated by a knowledge of the general nature of the unpublished experimental results and ideas of Dr. M. H. F. Wilkins, Dr. R. E. Franklin and their co-workers at King's College, London. One of us (J. D. W.) has been aided by a fellowship from the National Foundation for Infantile Paralysis.

J. D. Watson
F. H. C. Crick

Medical Research Council Unit for the
Study of the Molecular Structure of
Biological Systems,
Cavendish Laboratory, Cambridge.
April 2.

[1] Pauling, L., and Corey, R. B., *Nature*, 171, 346 (1953); *Proc. U.S. Nat. Acad. Sci.*, 39, 84 (1953).
[2] Furberg, S., *Acta Chem. Scand.*, 6, 634 (1952).
[3] Chargaff, E., for references see Zamenhof, S., Brawerman, G., and Chargaff, E., *Biochim. et Biophys. Acta*, 9, 402 (1952).
[4] Wyatt, G. R., *J. Gen. Physiol.*, 36, 201 (1952).
[5] Astbury, W. T., Symp. Soc. Exp. Biol. 1, Nucleic Acid, 66 (Camb. Univ. Press, 1947).
[6] Wilkins, M. H. F., and Randall, J. T., *Biochim. et Biophys. Acta*, 10, 192 (1953).

FIGURE 3.08
DNA is a Double Helix

This one-page paper published in *Nature* described the now-famous double helix. J.D. Watson & F.H.C. Crick (1953) Molecular Structure of Nucleic Acids, A Structure for Deoxyribose Nucleic Acid. *Nature* 171: 737.

FIGURE 3.09
Watson and Crick in the 1950s

James Watson (b. 1928) at left and Francis Crick (b. 1916) are shown with their model of part of a DNA molecule in 1953. (Credit: A. Barrington Brown, Science Photo Library.)

X-ray diffraction showed that two strands of DNA are twisted together forming a double helix.

DNA forms a **right-handed double helix**. To tell a right-handed helix from a left-handed helix, the observer must look down the helix axis (in either direction). In a right-handed helix, each strand turns clockwise as it moves away from the observer (in a left-handed helix it would turn counterclockwise).

Box 3.01 *The Double Helix* **by James D. Watson Published in 1968 by Atheneum, New York**

This book gives a personal account of the greatest biological advance of the twentieth century—the unraveling of the structure of the DNA double helix by James Watson and Francis Crick. Like the bases of DNA, Watson and Crick formed a complementary pair. Crick, a physicist with an annoying laugh, was supposed to be working towards a Ph.D. on protein X-ray crystallography. Watson was a homeless American biologist, wandering around Europe with a postdoctoral fellowship, looking for something to do.

Despite spending much time carousing, the intrepid heroes, Crick and Watson, beat their elders to the finish line. Watson describes with relish how the great American chemist, Linus Pauling, placed the phosphate backbone of DNA down the middle, so failing to solve the structure. The data proving the phosphate backbone was on the outside of the double helix came from Rosalind Franklin, an X-ray crystallographer at London University.

The Director of the Cavendish Laboratory at Cambridge was Sir William Bragg, the august inventor of X-ray crystallography. Despite being depicted as a stuffy has-been who nearly threw Crick out for loud-mouthed insubordination, Bragg wrote the foreword to the book. After all, when younger scientists under your direction make the greatest discovery of the century, it is no time to bear a grudge!

The biographies of great scientists are usually exceedingly dull. Who cares, after all, what Darwin liked for breakfast or what size shoes Mendel wore? It is their discoveries and how they changed the world that is fascinating. "The Double Helix" is different. Biographers are generally minor figures, understandably hesitant to criticize major achievers. Watson, himself a big name, happily lacks such respect, and cheerfully castigates other top scientists. It is this honest portrayal of the flaws and fantasies of those involved in unraveling the DNA double helix that keeps the reader's attention.

If your stomach can't stand any more sagas about caring investigators who work on into the early hours hoping that their discoveries will help sick children, this book is for you. Like most candid scientists, Watson and Crick did not work for the betterment of mankind; they did it for fun.

---

right-handed double helix   In a right-handed helix, as the observer looks down the helix axis (in either direction), each strand turns <u>clockwise</u> as it moves away from the observer

Box 3.02 50 Years After Determining the Structure of DNA

In 2003 the double helix celebrated its 50th anniversary. In Great Britain, the Royal Mail issued a set of five commemorative stamps illustrating the double helix together with some of the technological advances that followed, such as comparative genomics and genetic engineering. In addition, the Royal Mint issued a £2 coin depicting the DNA double helix itself (Fig. 3.10).

FIGURE 3.10
Double Helix—50th Anniversary Coin

A £2 coin commemorating the discovery of the double helix was issued in 2003 by Great Britain.

4.1. Base Pairs are Held Together by Hydrogen Bonds

In double-stranded DNA, the bases on each strand protrude into the center of the double helix where they are paired with the bases in the other strand by means of **hydrogen bonds**. Adenine (A) in one strand is always paired with thymine (T) in the other, and guanine (G) is always paired with cytosine (C) (Fig. 3.11). Consequently, the number of adenines in DNA is equal to the number of thymines, and similarly the numbers of guanine and cytosine are equal. This is called **Chargaff's rule** in honor of Edwin Chargaff who determined that there were equimolar amounts of C and G and equimolar amounts of A and T in DNA. Notice that the nucleic acid bases have amino or oxygen side-groups attached to the ring. It is these chemical groups, along with the nitrogen atoms, that are part of the rings themselves, which allow the formation of hydrogen bonds. The hydrogen bonding in DNA **base pairs** involves either oxygen or nitrogen as the atoms that carry the hydrogen, giving three alternative arrangements: O–H–O, N–H–N, and O–H–N.

Each base pair consists of one larger purine base paired with a smaller pyrimidine base. So, although the bases themselves differ in size, all of the allowed base pairs are the same width, providing for a uniform width of the helix. The A-T base pair has two hydrogen bonds and the G-C base pair is held together by three, as shown in Figure 3.11. Before the hydrogen bonds form and the bases pair off, the shared hydrogen atom is found attached to one or the other of the two bases (shown by the complete lines in Fig. 3.11). During base pairing, this hydrogen also bonds to an atom of the second base (shown by the dashed lines).

> Guanine always pairs with cytosine; adenine pairs with thymine in DNA and uracil in RNA. There is always a 1:1 ratio of G to C and A to T in DNA.

---

base pair   A pair of two complementary bases (A with T or G with C) held together by hydrogen bonds
Chargaff's rule   For each strand of DNA the ratio of purines to pyrimidines is always 1:1 because A always pairs with T and G always pairs with C.
hydrogen bond   Bond resulting from the attraction of a positive hydrogen atom to both of two other atoms with negative charges

Although RNA is normally single-stranded, many RNA molecules fold up, giving double-stranded regions. In addition, a strand of RNA may be found paired with one of DNA under some circumstances. Furthermore, the genome of certain viruses consists of double-stranded RNA (see Ch. 21). In all of these cases, the uracil in RNA will base pair with adenine. Thus, the base-pairing properties of the uracil found in RNA are identical to those of the thymine of DNA.

4.2. Complementary Strands Reveal the Secret of Heredity

If one of the bases in a base pair of double-stranded DNA is known, then the other can be deduced. If one strand has an A, then the other will have a T, and vice versa. Similarly, G is always paired with C. This is termed complementary base pairing. The significance is that if the base sequence of either one of the strands of a DNA molecule is known, the sequence of the other strand can be deduced. Such mutually deducible sequences are known as **complementary sequences**. It is this complementary nature of a DNA double helix that allows genetic information to be inherited. Upon cell division, each daughter cell must receive a copy of the parental genome. This requires accurate duplication or replication of the DNA (Fig. 3.12). This is achieved by separating the two strands of DNA and using complementary base pairing to make a new partner for each original strand (see Ch. 10 for details).

> Due to the rules for base pairing, the sequence of a DNA strand can be deduced if the sequence of its partner is known.

4.3. Melting Separates DNA Strands; Cooling Anneals Them

Hydrogen bonds are rather weak, but since a molecule of DNA usually contains millions of base pairs, the added effect of millions of weak bonds is strong enough to

---

complementary sequences Two nucleic acid sequences whose bases pair with each other because A, T, G, C in one sequence correspond to T, A, C, G, respectively, in the other

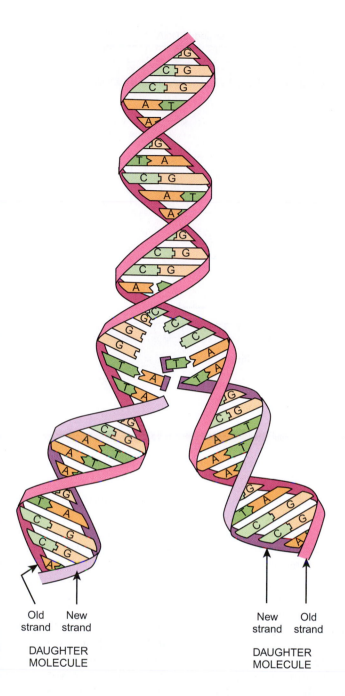

■ **FIGURE 3.12**
***Complementary Strands
Allow Duplication***

Because DNA strands are
complementary, double-stranded
DNA can be split into single
strands, each carrying sufficient
information to recreate the original
molecule. Complementary base
pairing allows the synthesis of two
new strands so restoring double-
stranded DNA.

Old   New   New   Old
strand strand   strand strand

DAUGHTER   DAUGHTER
MOLECULE   MOLECULE

keep the two strands together (Fig. 3.13). When DNA is heated, the hydrogen bonds
begin to break and the two strands will eventually separate if the temperature rises
high enough. This is referred to as "**melting**" or **denaturation**, and each DNA molecule has a **melting temperature** (T_m) that depends on its base composition. The
melting temperature of a DNA molecule is defined strictly as the temperature at the
halfway point on the melting curve, as this is more accurate than trying to guess where
exactly melting is complete.

The melting temperature is affected by the pH and salt concentration of the solution, so these must be standardized if comparisons are to be made. Extremes of pH
disrupt hydrogen bonds. A highly alkaline pH deprotonates the bases, which abolishes

> Heating breaks hydrogen
> bonds and eventually causes
> the two strands of a DNA
> double helix to separate—the
> DNA "melts."

denaturation   In reference to DNA, the breaking apart of a double strand of DNA into two single strands; when used of proteins, refers to the
  loss of correct 3D structure
melting   When used of DNA, refers to its separation into two strands as a result of heating
melting temperature (T_m)   The temperature at which the two strands of a DNA molecule are half unpaired

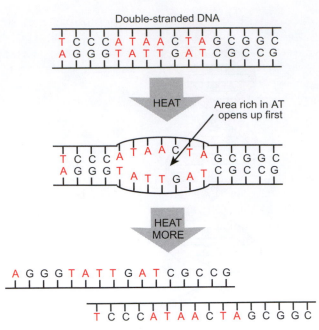

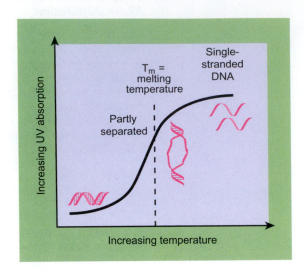

FIGURE 3.13
Melting of DNA

DNA strands separate, or "melt," with increasing temperature. This curve shows the measurement of DNA separation by ultraviolet absorption. As the temperature increases more UV is absorbed by the individual strands. The T_m or melting temperature is the point at which half of the double-stranded DNA has separated. During the melting process regions rich in A/T base pairs melt first since these base pairs have only two hydrogen bonds.

> Melted DNA absorbs more UV light than double-helical DNA.

> The more GC base pairs (with three hydrogen bonds) the higher the melting temperature for DNA.

their ability to form hydrogen bonds, and at a pH of 11.3 DNA becomes fully denatured. Conversely, a very low pH causes excessive protonation, which also prevents hydrogen bonding. When DNA is deliberately denatured by pH, alkaline treatment is used because unlike acid, this does not affect the glycosidic bonds between bases and deoxyribose. DNA is relatively more stable at higher ionic concentrations. This is because ions suppress the electrostatic repulsion between the negatively-charged phosphate groups on the backbone and hence exert a stabilizing effect. In pure water, DNA will melt even at room temperature.

A spectrophotometer detects the amount absorbed when light is passed through a solution containing DNA. This is compared with the light absorbed by a solution containing no DNA to determine the amount absorbed by the DNA itself. Melting is followed by measuring the absorption of ultraviolet (UV) light at a wavelength of 260 nm (the wavelength of maximum absorption), since disordered DNA absorbs more UV light than a double helix.

Overall, the higher the proportion of GC base pairs, the higher the melting temperature of a DNA molecule. This is because A/T base pairs are weaker, as they have only two hydrogen bonds, as opposed to G/C pairs, which have three. In addition, the stacking of G/C base pairs with their neighbors is also more favorable than for A/T base pairs. In the early days of molecular biology, melting temperatures were used to estimate the percentage of G/C versus A/T in samples of DNA. DNA base compositions are often cited as the **G/C ratio**. The G/C content (%G+C) is calculated from the fractional composition of bases as follows:

$$\%GC = \frac{(G + C)}{(A + T + G + C)} \times 100$$

GC ratio   The amount of G plus C divided by the total of all four bases in a sample of DNA. The GC ratio is usually expressed as a percentage

G/C contents for the DNA from different bacterial species vary from 20% to 80%, with *E. coli* having a ratio of 50%. Despite this, there is no correlation between G/C content and optimum growth temperature. Apparently, this is because the genomes of bacteria are circular DNA molecules with no free ends, and this greatly hinders unraveling at elevated temperatures. In fact, small circular DNA molecules, like plasmids, may remain base paired up to 110–120°C. In contrast, the range of G/C contents for animals (which have linear chromosomes) is much narrower, from approximately 35–45%, with humans having 40.3% G/C.

As a DNA molecule melts, regions with a high local concentration of A/T pairs will melt earlier and G/C-rich regions will stay double-stranded longer. When DNA is replicated, the two strands must first be pulled apart at a region known as the origin of replication (see Ch. 9). The DNA double helix must also be opened up when genes are transcribed to make mRNA molecules. In both cases, AT-rich tracts are found where the DNA double helix will be opened up more readily.

If the single strands of a melted DNA molecule are cooled, the single DNA strands will recognize their partners by base pairing and the double-stranded DNA will re-form. This is referred to as **annealing** or **renaturation**. For proper annealing, the DNA must be cooled slowly to allow the single strands time to find the correct partners. Furthermore, the temperature should remain moderately high to disrupt random H-bond formation over regions of just one or a few bases. A temperature 20–25°C below the T_m is suitable. If DNA from two different, but related, sources is melted and reannealed, **hybrid DNA** molecules may be obtained (Fig. 3.14).

> Upon cooling, the bases in the separated strands of DNA can pair up again and the double helix can re-form.

In the days before direct sequencing of DNA became routine, **hybridization** of DNA and/or RNA was originally used to estimate the relatedness of different organisms, especially bacteria where the amount of DNA is relatively small. Other uses for hybridization include detection of specific gene sequences and gene cloning. Several extremely useful techniques used in molecular biology rely upon hybridization, including *in situ* hybridization, which determines the location of different molecules within an organism, bacteria, or cell.

> Hybrid DNA molecules may be formed by heating and cooling a mixture of two different, but related, DNA molecules.

5. Constituents of Chromosomes

Genes are segments of large DNA molecules known as **chromosomes** (Fig. 3.15). Each chromosome is thus an exceedingly long single molecule of DNA. In addition to the DNA, which comprises the genes themselves, the chromosome has some accessory protein molecules, which help maintain its structure. The term **chromatin** refers to this mixture of DNA and protein, especially as observed with the microscope in the nuclei of eukaryotic cells. The genes are arranged in linear order. In front of each gene is a **regulatory region** of DNA involved in switching the gene on or off. In prokaryotes, groups of genes may be clustered close together with no **intergenic regions**. Such clusters are called **operons** and each is under the control of a single regulatory region. Operons are transcribed to give single mRNA molecules, each consisting of several genes.

> Genetic information includes both the genes themselves and regions of DNA involved in controlling gene expression.

Chromosomes from bacteria are circular molecules of double-stranded DNA. Since bacteria generally have only around 3,000–4,000 genes, and the intergenic regions are very short, one chromosome is sufficient to accommodate all of their

---

annealing   The re-pairing of separated single strands of DNA to form a double helix
chromatin   Complex of DNA plus protein which constitutes eukaryotic chromosomes
chromosome   Structure containing the genes of a cell and made of a single molecule of DNA
hybrid DNA   Artificial double-stranded DNA molecule made by pairing two single strands from two different sources
hybridization   Pairing of single strands of DNA or RNA from two different (but related) sources to give a hybrid double helix
intergenic region   DNA sequence between genes
operon   A cluster of prokaryotic genes that are transcribed together to give a single mRNA (i.e., polycistronic mRNA)
regulatory region   DNA sequence in front of a gene, used for regulation rather than to encode a protein
renaturation   Re-annealing of single-stranded DNA or refolding of a denatured protein to give the original natural 3D structure

FIGURE 3.14
Annealing of DNA

When pure DNA is heated and cooled the two strands pair up again (i.e., they re-anneal). When DNA from two related sources is heated and re-annealed, hybrid strands may form. The likelihood of hybridization depends on percentage of identical sequence between the two related DNAs.

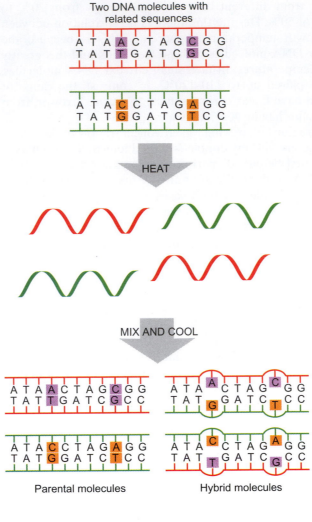

FIGURE 3.15
The General Pattern of Information on a Chromosome

Genes are normally preceded by regions of DNA involved in regulation. Between the genes are regions of DNA that apparently do not carry useful genetic information. These are called intergenic regions and vary greatly in size.

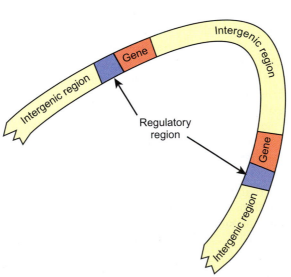

genes. When bacteria divide, the chromosome opens up at the origin of replication and replication proceeds around the circle in both directions (Fig. 3.16).

Chromosomes from higher organisms such as animals and plants are linear molecules of double-stranded DNA. They have a **centromere**, usually located more

centromere   Region of eukaryotic chromosome, usually more or less central, where the microtubules attach during mitosis and meiosis

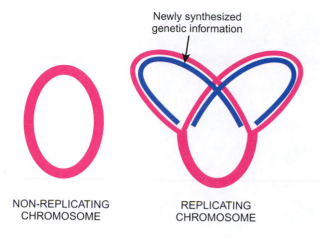

Newly synthesized genetic information

NON-REPLICATING
CHROMOSOME

REPLICATING
CHROMOSOME

■ FIGURE 3.16
The Circular Bacterial Chromosome and Its Replication

The bacterial chromosome is circular and not linear. When the double-stranded DNA is duplicated, the chromosome is opened forming loops that allow replication of each DNA strand.

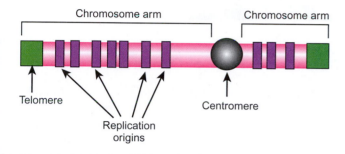

Chromosome arm   Chromosome arm

Telomere

Replication origins

Centromere

■ FIGURE 3.17
Structural Components of a Eukaryotic Chromosome

The eukaryotic chromosome is a linear molecule with specific DNA sequences called telomeres at each end. More or less in the center is an organized region called the centromere that is involved in chromosome division. Along the chromosome are multiple regions where replication is initiated.

or less in the middle, and structures known as **telomeres** at the two ends (Fig. 3.17). Both centromeres and telomeres contain special repetitive DNA sequences allowing their recognition by particular proteins. (One exception to this rule is that the yeast, *Saccharomyces*, lacks repetitive sequences at its centromere. However, this is not a general property of fungi, as other fungi do have repetitive centromere sequences.) The centromere is used at cell division when the chromosomes replicate. The newly-divided daughter chromosomes are pulled apart by spindle fibers (or microtubules) attached to the centromeres via protein structures known as **kinetochores**.

Telomeres are critically important to maintain chromosome stability. Due to the mechanism of initiating DNA replication by an RNA primer (see Ch. 10), the far ends of linear DNA molecules are shortened by a few bases each round of replication. In those cells that are permitted to continue growing and dividing, the end sequences are restored by the enzyme **telomerase**. If telomeres become too short, cells commit suicide in order to prevent problems in cell differentiation, cancer, and aging.

Eukaryotic chromosomes are only visible under the light microscope during cell division and it is only then that a complete set of chromosomes can be visualized (Fig. 3.18). The complete set of chromosomes found in the cells of a particular individual is known as the **karyotype**. Chromosomes and specific regions of chromosomes may be identified by their staining patterns after using specific stains that emphasize regions lacking genes, known as **non-coding DNA**. This **chromosome banding technique** has been used to identify major chromosome abnormalities (Fig. 3.19).

Details of replication mechanism and structure vary between the linear chromosomes of eukaryotes and the circular chromosomes of most bacteria.

Humans have vast amounts of DNA making up 46 linear chromosomes. However, most of this sequence is non-coding DNA as discussed further in Chapter 4.

chromosome banding technique   Visualization of chromosome bands by using specific stains that emphasize regions lacking genes
karyotype   The complete set of chromosomes found in the cells of a particular individual
kinetochore   Protein structure that attaches to the DNA of the centromere during cell division and also binds the microtubules
non-coding DNA   DNA sequences that do not code for proteins or functional RNA molecules
telomerase   Enzyme that adds DNA to the end, or telomere, of a chromosome
telomere   Specific sequence of DNA found at the end of linear eukaryotic chromosomes

FIGURE 3.18
A Set of Human Chromosomes

A human karyotype is a complete set of chromosomes containing 22 pairs plus one "X" and one "Y" chromosome (lower right) if the individual is male (as shown here). Females possess two "X" chromosomes. *(Credit: Alfred Pasieka, Science Photo Library.)*

> Under normal circumstances, genetic information flows from DNA to RNA to protein. As a result, proteins are often referred to as "gene products." Some RNA molecules are also "gene products" as they act without being translated into protein.

6. The Central Dogma Outlines the Flow of Genetic Information

Genetic information flows from DNA to RNA to protein during cell growth. In addition, all living cells must replicate their DNA when they divide. The **central dogma** of molecular biology is a scheme showing the flow of genetic information during both the growth and division of a living cell (Fig. 3.20). During cell division each daughter cell receives a copy of the genome of the parent cell. As the genome is present in the form of DNA, cell division involves the duplication of this DNA. **Replication** is the process by which two identical copies of DNA are made from an original molecule of DNA. Replication occurs prior to cell division. An important point is that information does not flow from protein to RNA or DNA. However, information flow from RNA "backwards" to DNA is possible in certain special circumstances due to the operation of reverse transcriptase. In addition, replication of RNA occurs in viruses with an RNA genome (neither complication is shown in Fig. 3.20).

The genetic information stored as DNA is not used directly to make protein. During cell growth and metabolism, temporary, working copies of the genes known as **messenger RNA (mRNA)** are used. These are RNA copies of genetic information stored by the DNA and are made by a process called **transcription**. The mRNA molecules carry information from the genome to the cytoplasm, where the information is used by the **ribosomes** to synthesize **proteins**. In eukaryotes, mRNA is not made directly. Instead, transcription yields precursor RNA molecules (pre-mRNA) that must be processed to produce the actual mRNA as detailed in Chapter 12.

The DNA that carries the primary copy of the genes is present as gigantic molecules, each carrying hundreds or thousands of genes. In contrast, any individual mRNA molecule carries only one or a few genes' worth of information. Thus, in practice, multiple short segments of DNA are transcribed simultaneously to give many different mRNA molecules. In eukaryotes, each mRNA normally carries only a single gene, whereas in prokaryotes, anywhere from one to a dozen genes may be transcribed as a block to give an mRNA molecule carrying several genes, usually with related functions (Fig. 3.21).

---

central dogma   Basic plan of genetic information flow in living cells which relates genes (DNA), message (RNA), and proteins
messenger RNA (mRNA)   The molecule that carries genetic information from the genes to the rest of the cell
protein   Polymer made from amino acids; may consist of several polypeptide chains
replication   Duplication of DNA prior to cell division
ribosome   The cell's machinery for making proteins
transcription   Conversion of information from DNA into its RNA equivalent

FIGURE 3.19
Banding Patterns of Human Chromosomes

Representation of the banding patterns seen in metaphase chromosomes during meiosis is shown. The bands are originally visualized by dyes. The relative distances between these bands are the same for an individual chromosome, so this is a useful way of identifying a particular chromosome. (Credit: Dept. of Clinical Cytogenetics, Addenbrookes Hospital, Cambridge, UK, Science Photo Library.)

FIGURE 3.20
The Central Dogma (Simple Version)

The information flow in cells begins with DNA, which may either be replicated, giving a duplicate molecule of DNA, or be transcribed to give RNA. The RNA is read (translated) as a protein is built.

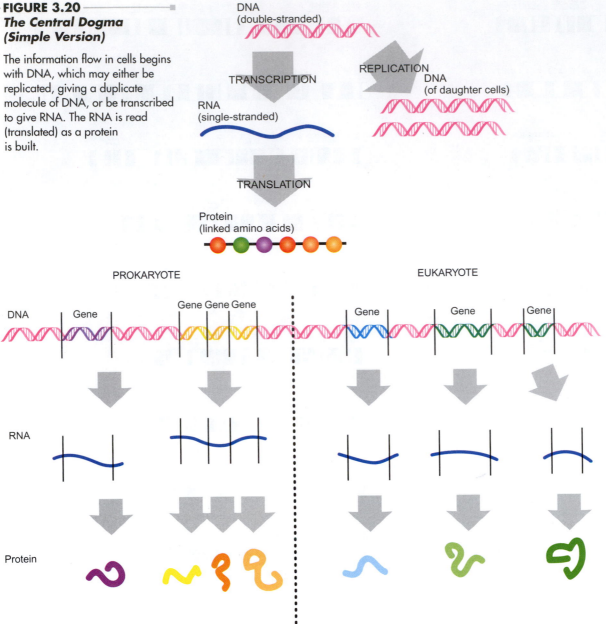

FIGURE 3.21
Differing Patterns of Transcription

In eukaryotes, each gene is transcribed to give a separate mRNA that encodes only a single protein. In prokaryotes, an mRNA molecule may carry information from a single gene or from several genes that are next to each other on the chromosome.

Translation is the synthesis of proteins using genetic information carried by mRNA. Proteins consist of one or more polymer chains known as **polypeptides**. These are made from subunits called **amino acids**. Translation thus involves transfer of information from nucleic acids to an entirely different type of macromolecule. This decoding process is carried out by ribosomes. These submicroscopic machines read the mRNA and use the information to make a polypeptide chain. Proteins, which make up about two-thirds of the organic matter in a typical cell, are directly responsible for

---

amino acid   Monomer from which the polypeptide chains of proteins are built
polypedtide chain   A polymer that consists of amino acids
translation   Making a protein using the information provided by mRNA

most of the processes of metabolism. Proteins perform most of the enzyme reactions and transport functions of the cell. They also provide many structural components and some act as regulatory molecules, as described below.

7. Ribosomes Read the Genetic Code

This introductory section will summarize protein synthesis as it occurs in bacteria. Although the overall process is similar, the details of protein synthesis differ between bacteria and higher organisms (see Ch. 13). The bacterial ribosome consists of two subunits, small (30S) and large (50S). **S-values** indicate how fast particles sediment through a particular solution in an ultracentrifuge. They give a rough indication of size but are not linearly related to molecular weight. A complete ribosome with a 30S and a 50S subunit has an S-value of 70S (not 80S).

By weight, the ribosome itself consists of about two-thirds **ribosomal RNA (rRNA)** and one-third protein. In bacteria, the large subunit has two rRNA molecules, 5S rRNA and 23S rRNA, and the small subunit has just the one 16S rRNA. In addition to the rRNA, there are 52 different proteins, 31 in the large subunit and the other 21 in the small subunit (Fig. 3.22). The rRNA molecules are NOT themselves translated into protein; instead, they form part of the machinery of the ribosome that translates the mRNA.

> Proteins are made by a subcellular machine, the ribosome that uses the genetic code to read information encoded by nucleic acids.

7.1. The Genetic Code Dictates the Amino Acid Sequence of Proteins

There are 20 amino acids in proteins but only four different bases in the mRNA. So nature cannot simply use one base of a nucleic acid to code for a single amino acid when making a protein. During translation, the bases of mRNA are read off in groups of three, which are known as **codons**. Each codon represents a particular amino acid. Since there are four different bases, there are 64 possible groups of three bases; that is,

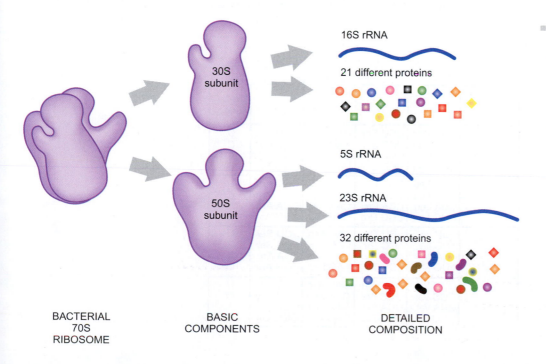

**BACTERIAL
70S
RIBOSOME**

**BASIC
COMPONENTS**

**DETAILED
COMPOSITION**

30S subunit

16S rRNA

21 different proteins

50S subunit

5S rRNA

23S rRNA

32 different proteins

■ **FIGURE 3.22**

Structural Components of a Ribosome

The bacterial ribosome can be broken down into two smaller subunits and finally into RNA molecules and proteins.

codon   Group of three RNA or DNA bases that encodes a single amino acid
ribosomal RNA (rRNA)   Class of RNA molecule that makes up part of the structure of a ribosome
S-value   The sedimentation coefficient is the velocity of sedimentation divided by the centrifugal field. It is dependent on mass and is measured in Svedberg units

64 different codons in the **genetic code**. However, there are only 20 different amino acids making up proteins, so some amino acids are encoded by more than one codon. Three of the codons are used for punctuation to stop the growing chain of amino acids (Fig. 3.23). In addition, the codon, AUG, encoding methionine, acts as a start codon. Thus, newly made polypeptide chains start with the amino acid methionine. (Much less often, GUG-encoding valine may also act as the start codon. However, even if the start codon is GUG the first amino acid of the newly made protein is still methionine (not valine).)

To read the codons a set of adapter molecules is needed. These molecules, known as **transfer RNA (tRNA)**, recognize the codon on the mRNA at one end and carry the corresponding amino acid attached to their other end (Fig. 3.24). These adapters represent a third class of RNA and were named tRNA since they transport amino acids to the ribosome in addition to recognizing the codons of mRNA. Since there are numerous codons, there are many different tRNAs. (Actually, there are fewer different tRNA molecules than codons as some tRNA molecules can read multiple codons—see Ch. 13 for details.) At one end, the tRNA has an **anticodon** consisting of three bases that are complementary to the three bases of the codon on the mRNA. The codon and anticodon recognize each other by base pairing and are held together by hydrogen bonds. At its other end, each tRNA carries the amino acid corresponding to the codon it recognizes.

The small (30S) subunit binds the mRNA, and the large (50S) subunit is responsible for making the new polypeptide chain. Figure 3.25 shows the relationship between the mRNA and the tRNAs in a stylized way. In practice, only two tRNA molecules are base paired to the mRNA at any given time. After binding to the mRNA, the ribosome moves along it, adding a new amino acid to the growing polypeptide chain each time it reads a codon from the message (Fig. 3.26).

8. Various Classes of RNA Have Different Functions

Originally, genes were regarded as units of heredity and alleles were defined as alternative versions of a gene. However, these concepts have been broadened as knowledge of

> The bases of DNA or RNA are grouped in threes for decoding.

FIGURE 3.23
The Genetic Code

A codon consisting of three base pairs determines each amino acid to be added to a growing polypeptide chain. The codon table shows the 64 different codons, in RNA language, alongside the amino acids they encode. Three of the codons act as stop signals. AUG (methionine) and GUG (valine) act as start codons.

1st base	2nd (middle) base				3rd base
	U	C	A	G	
U	UUU Phe UUC Phe UUA Leu UUG Leu	UCU Ser UCC Ser UCA Ser UCG Ser	UAU Tyr UAC Tyr UAA stop UAG stop	UGU Cys UGC Cys UGA stop UGG Trp	U C A G
C	CUU Leu CUC Leu CUA Leu CUG Leu	CCU Pro CCC Pro CCA Pro CCG Pro	CAU His CAC His CAA Gln CAG Gln	CGU Arg CGC Arg CGA Arg CGG Arg	U C A G
A	AUU Ile AUC Ile AUA Ile AUG Met	ACU Thr ACC Thr ACA Thr ACG Thr	AAU Asn AAC Asn AAA Lys AAG Lys	AGU Ser AGC Ser AGA Arg AGG Arg	U C A G
G	GUU Val GUC Val GUA Val GUG Val	GCU Ala GCC Ala GCA Ala GCG Ala	GAU Asp GAC Asp GAA Glu GAG Glu	GGU Gly GGC Gly GGA Gly GGG Gly	U C A G

---

anticodon Group of three complementary bases on tRNA that recognizes and binds to a codon on the mRNA
genetic code The code for converting the base sequence in nucleic acids, read in groups of three, into the sequence of a polypeptide chain
transfer RNA (tRNA) RNA molecules that carry amino acids to a ribosome

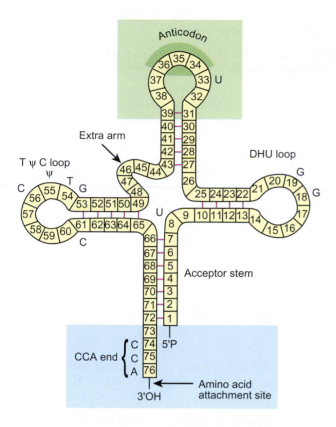

■ **FIGURE 3.24**
Transfer RNA Contains the Anticodon

Each tRNA molecule has an anticodon that is complementary to the codon carried on the mRNA. The codon and anticodon bind together by base pairing. At the far end of the tRNA is the acceptor stem ending in the bases CCA (cytosine, cytosine, and adenine). Here, the amino acid that corresponds to the codon is attached.

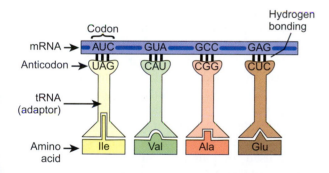

■ **FIGURE 3.25**
Stylized Relationship of Charged tRNA to mRNA

This figure shows a stylized relationship between the mRNA and tRNA molecules. In protein translation, there are at most two tRNAs attached to an mRNA. These are held together with the 30S subunit of the ribosome (not shown).

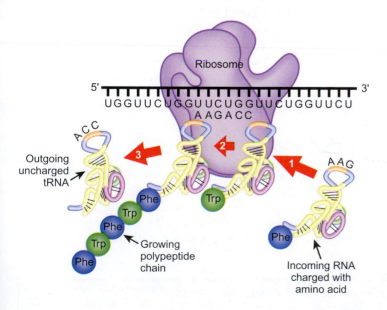

■ **FIGURE 3.26**
Ribosome Elongating a Polypeptide Chain

A new amino acid is added to the polypeptide chain each time a new tRNA arrives at the ribosome, bringing its attached amino acid. The anticodon of the tRNA binds to the mRNA. The large subunit crosslinks the incoming amino acid to the growing chain, such that the incoming tRNA ends up carrying the growing polypeptide chain. The 30S subunit of the ribosome then moves one step along the mRNA. This results in ejection of the left-most tRNA and readies the mRNA to accept the next incoming tRNA. The polypeptide chain continues to grow until a "stop codon" is reached.

TABLE 3.02	Major Classes of Non-Translated RNA
Ribosomal RNA (rRNA)	comprises major portion of ribosome and is involved in synthesis of polypeptide chains
Transfer RNA (tRNA)	carries amino acids to ribosome and recognizes codons on mRNA
Small nuclear RNA (snRNA)	involved in the processing of mRNA molecules in the nucleus of eukaryotic cells
microRNA (miRNA)	small RNAs encoded by the genome and used to regulate gene expression
Short interfering RNA (siRNA)	short RNA created by enzymatic cleavage of a larger double-stranded RNA and used in defense against viruses
Guide RNA	involved in processing of RNA or DNA in some organisms
Regulatory RNA	functions in the regulation of gene expression by binding to proteins or DNA or to other RNA molecules
Antisense RNA	functions in regulating gene expression by base pairing to mRNA
Recognition RNA	part of a few enzymes (e.g., telomerase); enables them to recognize certain short DNA sequences
Ribozymes	enzymatically-active RNA molecules

RNA is not so simple after all. Several classes of RNA exist that carry out a variety of roles in addition to carrying information for protein synthesis. (See especially Ch. 18 for novel insights into the role of RNA in regulation.)

genome structure has increased. Molecular insights led first to the view of genes as segments of DNA encoding proteins—the one gene—one enzyme model of Beadle and Tatum. In this case, mRNA acts as an intermediary between the DNA, which is used for storage of genetic information, and the protein, which functions in running the cell. The concept of a gene was then further extended to include segments of DNA that encode RNA molecules that are not translated into protein but function as RNA. The most common examples are the ribosomal RNA and tRNA involved in protein synthesis. The term "gene products" therefore includes such non-translated RNA molecules as well as proteins. For convenience, the major classes of non-translated RNA are summarized in Table 3.02.

In addition to the chemical differences discussed above (ribose instead of deoxyribose and uracil instead of thymine), RNA differs from DNA in several respects. RNA is usually single-stranded, although most RNA molecules do fold up, thus producing double-stranded regions. RNA molecules are usually much shorter than DNA and only carry the information for one or a few genes. Moreover, RNA is usually much shorter-lived than DNA, which is used for long-term storage of the genome. Some classes of RNA molecules, especially tRNA, contain unusual, chemically-modified bases that are never found in DNA (see Ch. 12).

The above differences in function between RNA and DNA apply to living cells. However, certain viruses carry their genomes as either single- or double-stranded RNA. In such cases, multiple genes will obviously be present on these RNA genomes. Furthermore, double-stranded viral RNA can form a double helix, similar though not identical in structure to that of DNA. The properties of viruses and the novel aspects of their genomes are discussed more fully in Chapter 21.

9. Proteins Carry Out Many Cell Functions

Typically, about 60% of the organic matter in a cell is protein. Most of the cell's activities and many of its structures depend on its proteins.

Proteins are made from a linear chain of monomers, known as amino acids, and are folded into a variety of complex 3D shapes. A chain of amino acids is called a **polypeptide chain**. There are 20 different amino acids used in making proteins. All have a central carbon atom, the **alpha carbon**, surrounded by a hydrogen atom, an amino group (—NH₂), a carboxyl group (—COOH), and a variable side chain, the

alpha carbon   The central carbon atom of an amino acid to which the amino, carboxyl, and R-groups are attached
polypeptide chain   A polymer that consists of amino acids

R-group (Fig. 3.27). Amino acids are joined together by **peptide bonds** (Fig. 3.28). The first amino acid in the chain retains its free amino group, and this end is often called the **amino-** or **N-terminus** of the polypeptide chain. The last amino acid to be added is left with a free carboxyl group, and this end is often called the **carboxy-** or **C-terminus**.

Some proteins consist of a single polypeptide chain; others contain more than one. To function properly, many proteins need extra components, called **cofactors** or **prosthetic groups**, which are not made of amino acids. Many proteins use single metal atoms as cofactors; others need more complex organic molecules.

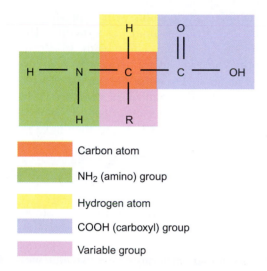

	Carbon atom
	NH₂ (amino) group
	Hydrogen atom
	COOH (carboxyl) group
	Variable group

■ FIGURE 3.27
General Features of Amino Acids

Almost all amino acids found in proteins have an amino group, hydrogen atom, and carboxyl group surrounding a central carbon. In addition, the central carbon has an R-group that varies from one amino acid to the next. In glycine, the simplest amino acid, the R group is a single hydrogen atom. In proline, the R group consists of a ring structure that bonds to the nitrogen atom shown. This therefore only has a single attached hydrogen and becomes an imino group (—NH—).

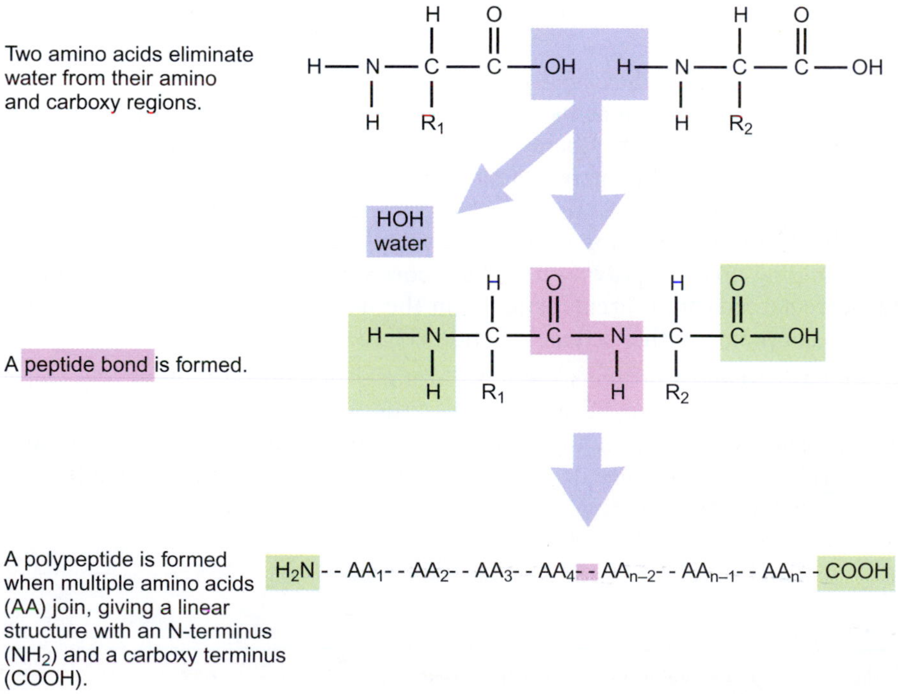

■ FIGURE 3.28
Formation of a Polypeptide Chain

A polypeptide chain is formed as amino and carboxyl groups on two neighboring amino acids combine and eliminate water. The linkage formed is known as a peptide bond. No matter how many amino acids are added, the growing chain always has an N- or amino-terminus and a C- or carboxy-terminus.

Two amino acids eliminate water from their amino and carboxy regions.

HOH
water

A peptide bond is formed.

A polypeptide is formed when multiple amino acids (AA) join, giving a linear structure with an N-terminus (NH₂) and a carboxy terminus (COOH).

H_2N - - AA_1 - - AA_2 - - AA_3 - - AA_4 - - AA_{n-2} - - AA_{n-1} - - AA_n - - COOH

amino- or N-terminus   The end of a polypeptide chain that is made first and that has a free amino group
carboxy- or C-terminus   The end of a polypeptide chain that is made last and has a free carboxy group
cofactor   Extra chemical group non-covalently attached to a protein that is not part of the polypeptide chain
peptide bond   Type of chemical linkage holding amino acids together in a protein molecule
prosthetic group   Extra chemical group covalently attached to a protein that is not part of the polypeptide chain
R-group   Chemical group forming side chain of amino acid

9.1. The Structure of Proteins Has Four Levels of Organization

For a protein to be functional, the polypeptide chains must be folded into their correct 3D structures. The structures of biological polymers, both protein and nucleic acid, are often divided into levels of organization (Fig. 3.29). The first level, or **primary structure**, is the linear order of the monomers—that is, the sequence of the amino acids for a protein. **Secondary structure** is the folding or coiling of the original polymer chains by means of hydrogen bonding. In proteins, hydrogen bonding between peptide groups results in several possible helical or wrinkled sheet-like structures (see Ch. 14 for details).

The next level is the **tertiary structure**. The polypeptide chain, with its pre-formed regions of secondary structure, is then folded to give the final 3D structure. This level of folding depends on the side chains of the individual amino acids. In certain cases, proteins known as chaperonins help other proteins to fold correctly. As there are 20 different amino acids, a great variety of final 3D conformations is possible. Nonetheless, many proteins are roughly spherical. Lastly, **quaternary structure** is the assembly of several individual polypeptide chains to give the final structure. Not all proteins have more than one polypeptide chain; some just have one, so they have no quaternary structure.

> Proteins have four different levels of structure, from the primary sequence of amino acids to the assembly of multiple polypeptide chains.

9.2. Proteins Vary in Their Biological Roles

Functionally, proteins may be divided into four main categories: **structural proteins, enzymes, regulatory proteins,** and **transport proteins**:

1. Structural proteins make up many subcellular structures. The flagella with which bacteria swim around, the microtubules used to control traffic flow inside cells of higher organisms, the fibers involved in contractions of a muscle cell, and the outer coats of viruses are examples of structures constructed using proteins.

2. Enzymes are proteins that catalyze chemical reactions. An enzyme first binds another molecule, known as its **substrate**, and then performs chemical operations with it. Some enzymes bind only a single substrate molecule; others may bind two or more and combine them to make the final product. In any case, the enzyme needs an **active site**, a pocket or cleft in the protein, where the substrate binds and the reaction occurs. The active site of the protein is produced by the folding up of its polypeptide chain correctly so that amino acid residues that were spread out at great distances in the linear chain now come together and can cooperate in the enzyme reaction (Fig. 3.30).

3. Although regulatory proteins are not enzymes, they do bind other molecules and so they also need active sites to accommodate these. Regulatory proteins vary enormously. Many of them can bind both small signal molecules and DNA. The presence or absence of the signal molecule determines whether or not a gene is switched on (Fig. 3.31).

active site   Special site or pocket on a protein where the substrate binds and the enzyme reaction occurs
enzyme   A protein or RNA molecule that catalyzes a chemical reaction
primary structure   The linear order in which the subunits of a polymer are arranged
quaternary structure   Aggregation of more than one polymer chain in the final structure
regulatory protein   A protein that regulates the expression of a gene or the activity of another protein
secondary structure   Initial folding up of a polymer due to hydrogen bonding
structural protein   A protein that forms part of a cellular structure
substrate   The molecule altered by the action of an enzyme
transport protein   A protein that carries other molecules across membranes or around the body
tertiary structure   Final 3D folding of a polymer chain

(a) Primary structure

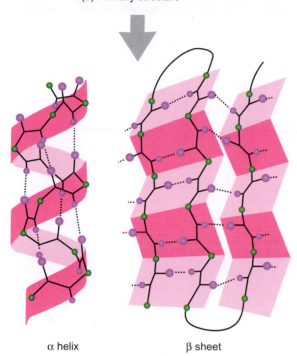

α helix   β sheet

(b) Secondary structures

(c) Tertiary structure

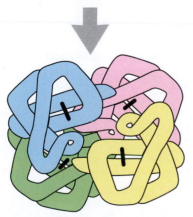

(d) Quaternary structure

■ **FIGURE 3.29**
Four Levels of Protein Structure

The final protein structure is best understood by following the folding process from simple to complex. The primary structure is the specific order of the amino acids (a). The secondary structure is due to regular folding of the polypeptide chain due to hydrogen bonding (b). The tertiary structure results from further folding of the polypeptide due to interactions between the amino acid side chains (c). Finally, the quaternary structure is the assembly of multiple polypeptide chains (d).

FIGURE 3.30
Polypeptide Forms an Active Site After Folding

Folding of the protein brings together several regions of the polypeptide chain that are needed to perform its biological role. The active site forms a pocket for binding the substrate. Some of the amino acid residues at the active site are also involved in chemical reactions with the substrate.

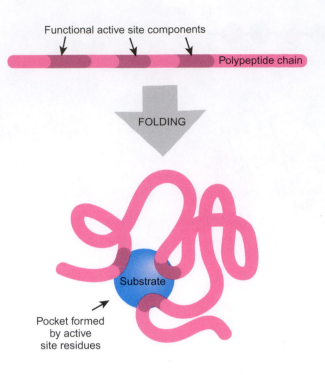

FIGURE 3.31
Regulatory Protein

Regulatory proteins usually exist in two conformations. Receiving a signal promotes a change in shape. The regulatory protein may then bind to DNA and alter the expression of a gene.

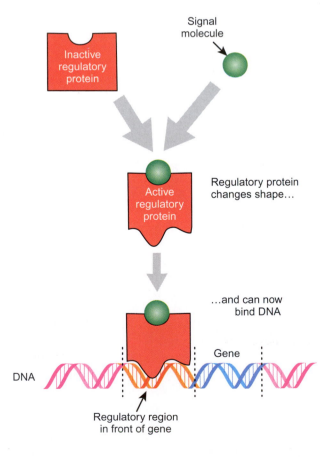

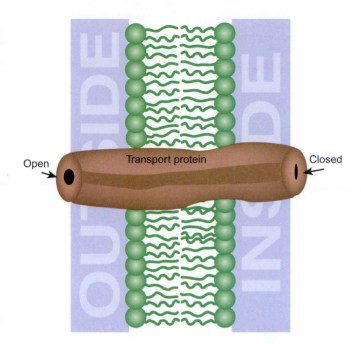

FIGURE 3.32
Transport Proteins

Transport proteins are often found in cell membranes where they are responsible for the entry of nutrients or the export of waste products.

4. Transport proteins are found mostly in biological membranes, as shown in Figure 3.32, where they carry material from one side to the other. Nutrients, such as sugars, must be transported into cells of all organisms, whereas waste products are deported. Multicellular organisms also have transport proteins to carry materials around the body. An example is hemoglobin, which carries oxygen in blood.

Since proteins are so abundant within the cell, they assume many roles, including providing structure, regulating other proteins or genes, catalyzing reactions, and importing and exporting various solutes or molecules.

9.3. Protein Structure is Elucidated by X-Ray Crystallography

Figuring out the exact 3D structure of a protein provides a lot of insight into its function and in the case of enzymes, the transition states of its reactions. The structure of many proteins has been elucidated using X-ray crystallography, the same technique used to elucidate the structure of DNA. The first step, and by far the most challenging, is to consolidate the pure protein into a regularly ordered crystal. The key to creating a good protein crystal is to slowly condense the protein by removing water from the protein solution. In the hanging drop method, a drop of low concentration protein solution is placed on a glass or acrylic cover slip (Fig. 3.33). This is inverted over a well containing the same protein sample at a higher concentration. The drop will become more concentrated as the water vapor moves toward equilibrium with the solution in the well. As the drop becomes more concentrated, the proteins begin to order themselves in crystals. Other methods can also be used to create crystals, such as the sitting drop method or the microdialysis methods, but each of these follow the principle of the hanging drop method.

To determine the structure of a protein, the soluble protein must be first concentrated into orderly crystals, and then X-rays are passed through the crystals. The pattern of X-ray diffraction is then interpreted and used to create a model for the actual protein structure.

 The second step of X-ray crystallography is to pass X-rays through the protein crystal. The X-rays are scattered into regular repeating patterns as they pass through the atoms of the protein. The patterns of scattered X-rays are evaluated by computer programs and interpreted into a set of data that describes the location and orientation of the amino acids within the protein (Fig. 3.34).

FIGURE 3.33
Hanging Drop Method

The more concentrated protein solution is at the bottom of the chamber, and the drop contains a lower concentration. Water molecules move to an equilibrium, meaning that the proteins in the drop become more and more concentrated. As they concentrate, the proteins order themselves into a regular repeating array—a crystal.

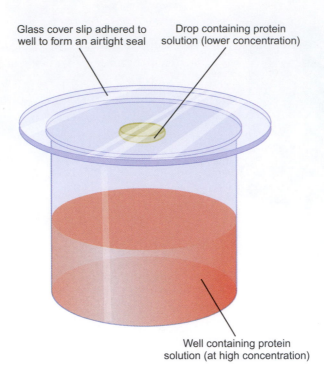

Glass cover slip adhered to well to form an airtight seal

Drop containing protein solution (lower concentration)

Well containing protein solution (at high concentration)

FIGURE 3.34
X-ray Crystal Structure of NUP120 and ACE1

The figure shows the structure of two different proteins that form the nuclear pore complex in yeast. These proteins provide the structure of the pore and lie adjacent to the lipid membrane. A. Overall structure. B. Surface model shown with subunits together and pulled apart. (Credit: Leksa, et al. (2009) Structure 17(8): 1082–1091.

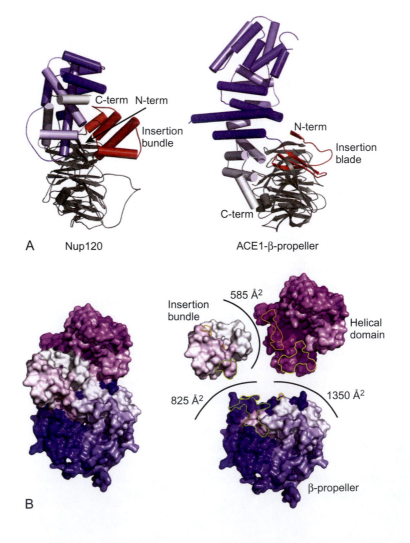

C-term   N-term
Insertion bundle

N-term
Insertion blade
C-term

A   Nup120   ACE1-β-propeller

585 Å²
Insertion bundle
Helical domain
825 Å²
1350 Å²
β-propeller

B

Rodnina MV, Wintermeyer W. (2010) The ribosome goes Nobel. Trends Biochem Sci 35(1): 1–5.

FOCUS ON RELEVANT RESEARCH

As defined above, the central dogma of molecular biology describes the key cellular processes that convert the information stored in DNA into proteins that perform most of the functions in the cell. One of the key molecules in the central dogma is the ribosome. The actual structure of the ribosome was elusive for many years simply because it is a large complex of protein and RNA. Such a large complex was simply too difficult to crystallize into an ordered manner for X-ray diffraction. Although a formidable project, the structure of the 50S or the 30S subunits was finally crystallized by three independent groups within the same year. A discovery so grand that it earned the three primary investigators, Thomas A. Steitz, Ada E. Yonath, and Venkatraman Ramakrishnan, a Nobel prize in Chemistry. This associated article describes how the discovery of the ribosome structure has advanced our understanding of how the catalytic activity and the structure of the ribosome are intimately linked for proper protein production.

Key Concepts

- Genetic material can be classified as DNA or RNA based upon the chemical structure of the molecule. Both DNA and RNA are long polymers made of nucleotides composed of a phosphate group, a five-carbon sugar, and a nitrogen-containing base.
- DNA nucleotides have deoxyribose attached to a phosphate group and one nitrogenous base (guanine, cytosine, adenine, or thymine). RNA nucleotides have ribose which has an extra hydroxyl group attached to the 2'-carbon instead of deoxyribose, and uracil instead of thymine. Nucleotides are connected via phosphodiester bonds. The bases of a nucleotide polymer can connect via hydrogen bonds such that adenine always pairs to thymine (or uracil in RNA), and guanine always attaches to cytosine to form antiparallel double-stranded DNA or RNA.
- Two DNA strands with complementary bases are normally found connected in an antiparallel direction, where the 5'-phosphate of one strand is opposite the 3'-hydroxyl group of the other strand. The double strands spontaneously twist into a helix.
- The complementary nature of DNA is the basis for heredity.
- Two DNA strands can be melted or denatured by heat and then re-annealed by cooling. The temperature that is halfway up the melting curve is called the T_m or melting temperature. Annealing two related DNA strands is called hybridization.
- Long DNA strands form chromosomes that have different structures depending upon the organism of origin. Most bacteria have circular chromosomes, whereas humans have linear chromosomes with centromeres and telomeres.
- The central dogma of molecular biology shows the flow of genetic material during the growth and division of living cells. Replication is when one DNA molecule is copied and then transferred to the daughter cell after cell division. Transcription creates temporary copies of a gene in the form called mRNA. Translation creates the proteins used to carry out vital cellular functions by the decoding of the mRNA sequence into amino acid chains via the ribosome and tRNA.
- Ribosomes are large molecular machines consisting of both protein and RNA (ribosomal RNA—rRNA). Interestingly, the RNA components are responsible for the catalytic activity. Other types of RNA are also important to cellular

function, including transfer RNA (tRNA), microRNA (miRNA), short interfering RNA (siRNA), **antisense RNA**, and **ribozymes**.
- The genetic code in DNA and mRNA is based on three nucleotides read together as a codon.
- Proteins carry out many of the cellular functions. Structural proteins provide structure, enzymes catalyze chemical reactions, regulatory proteins regulate other molecules, and transport proteins move other molecules in and out of the cell.
- Protein structure is controlled by the interaction of the amino acid side chains with each other and the environment. Proteins have primary structure, the linear order of amino acids, secondary structure, the folding or coiling of the amino acids due to hydrogen bonding, tertiary structure, where the coils or sheets fold into a 3D shape, and in some proteins, quaternary structure where multiple polypeptide chains combine into one complex. Protein structure can be elucidated via X-ray diffraction patterns.

Review Questions

1. What are nucleic acids? Name the two types of nucleic acids.
2. What are the three components of a nucleotide?
3. Name the bond that links the nucleotides.
4. Name the four bases in DNA. Name the four bases in RNA.
5. What is the difference between: a) a purine and a pyrimidine; b) a nucleotide and a nucleoside?
6. What is at the beginning and the end of a DNA or RNA molecule?
7. Name the bond that holds together the double strands of DNA.
8. Describe the Watson-Crick model of a DNA molecule.
9. What is the difference between A/T pairing and G/C pairing?
10. What are chromosomes? What is the difference between the prokaryotic and eukaryotic chromosomes?
11. How are genes arranged on a chromosome?
12. How are DNA strands duplicated? What is this process called?
13. What is a centromere? What is its function?
14. What are telomeres? What is the function of telomerase? What is the importance of telomeres?
15. What is a karyotype? Give an example. When can it be visualized?
16. Describe a technique used to identify chromosome abnormalities.
17. What is the central dogma of molecular biology?
18. Define replication, transcription, and translation.
19. What are the differences between prokaryotic and eukaryotic mRNA?
20. What are the structural components of a bacterial ribosome?
21. What is the function of ribosomal RNA?
22. What is a codon? How many codons are possible in the genetic code?
23. What is the function of a start codon? What is the first amino acid in a newly made protein?
24. What are tRNAs? What are their functions?
25. Name the three RNAs involved in the formation of protein.
26. List the major differences between DNA and RNA.
27. What is the general structure of an amino acid? How many different amino acids are used to make proteins?
28. What are proteins? How are they formed?
29. What are the N-terminus and C-terminus of a protein?
30. What are the four levels of folding that makes the protein active?
31. List the four main functional roles of proteins.

antisense RNA   RNA complementary in sequence to mRNA and which, therefore, base pairs with it and prevents translation
ribozyme   RNA molecule that acts as an enzyme

Conceptual Questions

1. a. Draw the complementary sequence for the following single strand of DNA:

    5' CTATCGATTCAACGAAATTCGCAAGGCATT 3'

    b. Transcribe the double-stranded DNA from question 1a into a single-stranded mRNA using the top strand as the template.
    c. Translate the mRNA from question 2b into protein using the codon chart.
2. Explain the central dogma of molecular biology.
3. A scientist was given a solution containing ribosomes and was told to isolate the different subunits from the mixture. Present an experimental technique to isolate ribosomal subunits.
4. A researcher studying the gene for curly hair in mice has found that one nucleotide is different between the gene for hair shape enzyme in the curly-haired mouse and straight-haired mouse. This mutation alters the gene so that a different amino acid is added at the location. Explain the results of this research based on your knowledge of protein functions.
5. Calculate the percent content of each of the four bases for the following organisms. Deduce the information based on the provided data.

Organism	Guanine (G)	Cytosine (C)	Thymine (T)	Adenine (A)
Streptomyces coelicolor		38%		
Saccharomyces cerevisiae			19%	
Arabidopsis thaliana	18%			
Plasmodium falciparum				30%

Chapter 4

Genomes and DNA

The term genome refers to the total genetic information that an organism (whether a living cell or a virus) possesses. The number of genes in the genome can range from a mere handful in certain viruses to many thousand in complex higher organisms. Not surprisingly, parasitic organisms that rely on others to provide the essentials of life often have relatively smaller genomes than corresponding free-living organisms. Genes are carried on DNA molecules known as chromosomes. The number, structure, and arrangement of the chromosomes vary considerably between different life forms. Bacteria generallwy have a single circular chromosome, whereas eukaryotes have multiple linear chromosomes. Since the chromosomes are longer than the cells that contain them, they are compressed by a variety of mechanisms to fit in the space available. In addition to the genes themselves, chromosomes include many other stretches of DNA. Some of these segments are regulatory in nature; others appear to have no clear function. Especially in higher organisms there are many repeated sequences that, in extreme cases, may comprise a major portion of the DNA.

1. Genome Organization

Genomes vary from organism to organism, and this section introduces the basic pattern for viruses, prokaryotes, organelles, and eukaryotes. As the field of genomics advances, more and more understanding of how the order of nucleotides in the DNA relates to the expression of

Molecular Biology.

genes into their functional products will become clearer. As for now, research into the structural organization of the genome is still a burgeoning field of study with exciting breakthroughs in the interphase structure of DNA and how this relates to the expression and interactions of genes.

1.1. Genome Organization of Viruses and Prokaryotes

Now that molecular biology has firmly entered the age of genomics, over 1000 different bacterial species have had their entire genome sequenced. Bacteria typically carry all of their genes on a single circular chromosome, although some species of bacteria have two or more different chromosomes. A few bacteria even contain linear chromosomes such as *Borrelia burgdorferi*, the bacterium responsible for Lyme's disease. The circular genomes are packed with genes with very little space between them. In addition, many genes are grouped based upon function, so that only one set of

TABLE 4.01	Genome Sizes		
Organism	**Number of Genes**	**Amount of DNA (bp)**	**Number of Chromosomes**
Viruses			
Bacteriophage MS2	4	3,600	1 (ssRNA)*
Tobacco Mosaic Virus	4	6,400	1 (ssRNA)*
ΦX174 bacteriophage	11	5,387	1 (ssDNA)
Influenza	12	13,500	8 (ssRNA)
T4 bacteriophage	200	165,000	1
Poxvirus	300	187,000	1
Bacteriophage G	680	498,000	1
Prokaryotes			
Mitochondrion (human)	37	16,569	1
Mitochondrion (*Arabidopsis*)	57	366,923	1
Chloroplast (*Arabidopsis*)	128	154,478	1
Nanoarchaeum equitans	550	490,000	1
Mycoplasma genitalium	480	580,000	1
Methanococcus	1,500	1.7 Mbp	1
Escherichia coli	4,000	4.6 Mbp	1
Myxococcus	9,000	9.5 Mbp	1
Eukaryotes (haploid genome)			
Encephalitozoon	2,000	2.5 Mbp	11
Saccharomyces	5,700	12.5 Mbp	16
Caenorhabditis	19,000	97 Mbp	6
Drosophila	12,000	180 Mbp	5
Mus musculus	25,000	2,600 Mbp	20
Homo sapiens	25,000	3,300 Mbp	23
Arabidopsis	25,000	115 Mbp	5
Oryza sativa (Rice)	45,000	430 Mbp	12

*ssRNA = single-stranded RNA; ssDNA = single-stranded DNA; all other genomes consist of double-stranded DNA.

promoter elements controls the expression of multiple genes. This structure is called an **operon**.

To an even greater extreme, the genome of most viruses is more compact than bacteria. In some viruses, the number of genes can be as few as four, as seen in bacteriophage MS2. They can survive with so few genes because viruses can rely on the host cell for almost all of their functions. The viral genome is only necessary to trick the host into making more viral genomes and the production of the viral coat (see Table 4.01). Despite this, the largest known viruses have around 1000 genes, most of unknown function.

The smallest genome sizes are found in symbiotic bacteria and viruses.

Box 4.01: Smallest Prokaryotic Genome

The smallest prokaryotic genomes belong to symbiotic bacteria since these bacteria can afford to give up certain biosynthetic pathways. These bacteria can simply rely on their host to provide the nutrient, metabolite, or replication/transcription/translation machinery instead of creating it themselves. In 2006, a group of researchers from the USA and Japan discovered that a symbiotic bacterium called *Carsonella ruddii* had a genome that was only 159,662 base pairs in length and only had 182 genes. The bacterium resides inside a small bug called a psyllid, which feeds on the sap of various plants (see Fig. 4.01). The psyllid's diet is rich in sugars, but lacking in amino acids, so it was originally thought that *C. ruddii* was providing the missing amino acids. The bacterial genome was analyzed for genes that provide essential living functions for itself and genes that were needed to provide its host with essential amino acids. The genome is missing key genes for replication and protein synthesis, so it must rely upon its host for these functions. In addition, the genome is missing genes for making histidine, phenylalanine, and tryptophan. These are essential amino acids for psyllids, so the psyllid must be getting these amino acids from a different source, which is probably *Wolbachia*, another symbiont—many insects possess multiple symbionts. *C. ruddii* is thought to be an ancient symbiont that is now providing only some functions for its host.

FIGURE 4.01
Psyllid

Adult hackberry petiole gall psyllid, *Pachypsylla venusta*, contain an endosymbiotic bacteria, *Carsonella ruddii*, that have the smallest cellular genome known. (*Credit: Jerry F. Butler, University of Florida*)

operon   A group of genes controlled by one promoter or regulatory region

1.2. Genome Organization of Organelles

In eukaryotes, mitochondria and chloroplasts contain their own genomes, and like prokaryotes, the genomes are typically small circular molecules of DNA. These circular genomes are very compact and appear to be similar to prokaryotic genomes in organization and content, although much smaller. The main difference is that these organelles often contain multiple copies of their genome. These genomes contain genes for respiration (mitochondria), photosynthesis (chloroplasts), and the construction of their own ribosomes (both organelles). All these genes appear to be more similar to prokaryotes. Because organelle genomes do not contain enough genes for the organelle to survive outside the cell, organelles rely on proteins and enzymes that are encoded by nuclear genes to provide all the necessary components for their function.

Neither the amount of DNA nor the numbers of organelle genes correlate with complexity of the organism. In humans, the mitochondrial genome encodes 37 genes on 16,569 base pairs of DNA, whereas the yeast mitochondrion has 87,779 base pairs of DNA and 43 genes.

Since these genomes and the genes that they encode are so similar to bacteria, the prevailing theory is that these organelles arose by **symbiosis**. The **symbiotic theory** proposes that the complex eukaryotic cell arose by a series of symbiotic events in which organisms of different lineages merged. Throughout time, the symbionts lost the ability to survive on their own, and became specialized to provide one specific function for the host. The organelles of higher organisms are thus remnants of ancient bacteria (Fig. 4.02). The nuclear genes of a eukaryotic cell are sometimes referred to

> Mitochondria are derived from ancestral bacteria that specialized in respiration whereas chloroplasts are descended from ancestral photosynthetic bacteria.

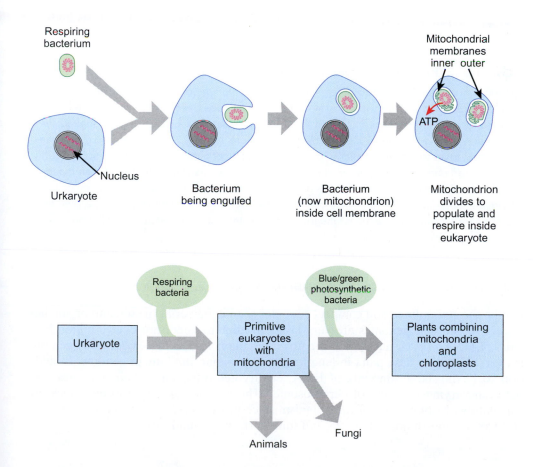

FIGURE 4.02

Symbiosis with Respiring Bacteria Gives Rise to the Primitive Eukaryote

The ancestor to the eukaryote, or "urkaryote," engulfs a respiring bacterium by surrounding it with an infolding of the cell membrane. Consequently, there is now a double membrane around the newly enveloped bacterium. The symbiont, now called a "mitochondrion," divides by fission like a bacterium and provides energy for the primitive eukaryote. The mitochondrion develops infoldings of the inner membrane that increase its energy producing capacity.

symbiosis   Association of two living organisms that interact
symbiotic theory   Theory that the organelles of eukaryotic cells are derived from symbiotic prokaryotes

as derived from the "**urkaryote**." The urkaryote is the hypothetical ancestor that provided the genetic information found in the present day eukaryotic nucleus.

According to the symbiotic theory, mitochondria are descended from bacteria that were trapped long ago by the ancestors of modern eukaryotic cells. These bacteria received shelter and nutrients, and in return, devoted themselves to generating energy by respiration. During the eons following their capture, these bacteria became narrowly specialized for energy production, lost the ability to survive on their own, and evolved into mitochondria. The term **endosymbiosis** is sometimes used to indicate those symbiotic associations where one partner is physically inside the other (endo is from the Greek word for inside), as in the present case. The symbiotic bacterium *Carsonella ruddii* is a modern example of such a symbiont. The bacteria has lost so many genes that it may be evolving into an organelle.

Plant cells contain chloroplasts that perform photosynthesis with light harvesting pigments such as chlorophyll. The rRNA from chloroplasts matches rRNA from photosynthetic bacteria better than rRNA from the plant cell nucleus. Thus, chloroplasts probably descended from photosynthetic bacteria trapped by the ancestors of modern-day plants. Some plants have lost the ability to photosynthesize but still contain defective chloroplasts. The term **plastid** refers to all organelles that are genetically equivalent to chloroplasts, whether functional or not. Since fungi do not contain chlorophyll, the green light-absorbing pigment of plants, it was once thought that fungi were degenerate plants that had lost their chlorophyll through evolution. However, fungi contain no trace of a plastid genome and rRNA analysis implies that the ancestral fungus was never photosynthetic, but split off from the ancestors of green plants before the capture of the chloroplast. If anything, rRNA sequencing implies that fungi are more closely related to animals than plants.

Box 4.02: The Symbiotic Theory of Organelle Origins

Certain primitive single-celled eukaryotes, such as *Entamoeba* and *Giardia*, lack the ability to respire and instead live by fermentation. It was once believed that they lacked mitochondria and had branched off from the ancestral eukaryote before it had captured the bacterium that gave rise to the mitochondrion. More recently, it was suggested that the ancestors to these organisms did originally possess mitochondria, but lost them secondarily during the course of evolution. In addition, recent work has shown that even *Entamoeba* and *Giardia* retain small remnant organelles ("mitosomes") corresponding to mitochondria. Although the capability for respiration has indeed been completely lost, the remnant organelles function in assembling the iron sulfur clusters found in several essential proteins.

1.3. Genome Organization of Eukaryotes

Unlike the streamlined and compact genomes seen in bacteria, viruses, and organelles, eukaryotic nuclear genomes are very large. In humans, there are over 6 billion base pairs of DNA in each nucleus. In the flowering plant *Fritillaria assyriaca*, the genome is over 120 billion base pairs in length, 20× larger than the human genome. In addition to having large amounts of DNA, eukaryotes often have their genomes split between varying numbers of chromosomes. The number of chromosomes does not correlate with complexity of the organism since yeast (*Saccharomyces cerevisiae*) has 16 chromosomes, humans have 23, and some ferns have hundreds.

endosymbiosis Form of symbiosis where one organism lives inside the other
plastid Any organelle that is genetically equivalent to a chloroplast, whether functional in photosynthesis or not
urkaryote Hypothetical ancestor that provided the genetic information of the eukaryotic nucleus

One of the most interesting observations about genome structure is that the amount of DNA does not correlate with the number of genes. This phenomenon is called the C-value paradox, a name that was coined in the early 1970s because the complexity of an organism does not relate to the amount of DNA in its genome. This discrepancy is due to **non-coding DNA** in eukaryotes. This, as its name indicates, is DNA whose base sequence consists of non-coding regions that may not have any function.

Although bacteria have relatively little non-coding DNA, eukaryotes have significant amounts. Even a relatively primitive eukaryote, such as yeast, has nearly 50% non-coding DNA. Thus, yeast has about three times as much DNA as *E. coli* but only 1.5 times as many genes. Higher eukaryotes have even greater proportions of non-coding DNA. Mammals such as mice and men have an estimated 20,000 to 30,000 genes carried on a total of 300 **Mbp** of DNA. This means that just over 95% is non-coding. However, flowering plants, which are estimated to have roughly as many genes as mammals, possess 100-fold more DNA. Some amphibians, such as frogs and newts, also possess almost as much DNA.

The large amount of non-coding DNA found in eukaryotes can be categorized based on location and sequence. In prokaryotes, almost all the non-coding DNA is found between genes as **intergenic DNA**. In the human genome, the amount of intergenic DNA is highly variable. In some areas of the human genome, large areas of intergenic DNA are found, and in other areas genes are closer together. Unlike the human genome, species like *Drosophila* and *C. elegans* have genes spaced evenly throughout their genome.

Not only is non-coding DNA scattered throughout the eukaryotic chromosomes between the genes, but the actual genes themselves are often interrupted with non-coding DNA. These **intervening sequences** are known as **introns**, whereas the regions of the DNA that contain coding information are known as **exons**. Most eukaryotic genes consist of exons alternating with introns (Fig. 4.03). In lower, single-celled eukaryotes, such as yeast, introns are relatively rare and often quite short. In contrast, in higher eukaryotes, most genes have introns and they are often longer than the exons. In some genes, the introns may occupy 90% or more of the DNA. For example, the CFTR gene, whose mutation causes cystic fibrosis, was found to occupy 250,000 base pairs and have 24 exons, which encoded a protein of 1,480 amino acids. Since 1,480 amino acids need only 4,440 base pairs to encode them, this means that scarcely 2% of the cystic fibrosis gene is actually coding DNA. The rest consists of intervening sequences—23 introns.

Although rare, eukaryotic genes can be clustered together, and there is evidence that some eukaryotic genes are actually transcribed as an operon. In *Leishmania major*, a small unicellular eukaryote commonly called a trypanosome, the genome has many different clusters of genes of varying sizes. Each cluster is transcribed as

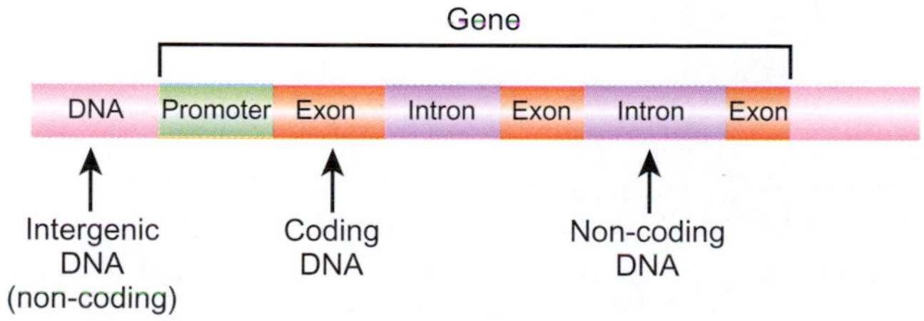

> Non-coding DNA accounts for the majority of the DNA in higher animals and plants.

FIGURE 4.03
Intervening Sequences Interrupt Eukaryotic Genes

Regions of non-coding DNA between genes are called intergenic DNA. Non-coding regions that interrupt the coding regions of genes are called introns.

exon   Segment of a gene that codes for protein and that is still present in the messenger RNA after processing is complete
intergenic DNA   Non-coding DNA that lies between genes
intervening sequence   An alternative name for an intron
intron   Segment of a gene that does not code for protein but is transcribed and forms part of the primary transcript
Mbp   Megabase pairs or million base pairs
non-coding DNA   DNA that does not code for proteins or functional RNA molecules

one unit. In *C. elegans*, about 15% of its genes are present in operons. These operons do not contain genes that are functionally related, but may be genes that are regulated by the same factors. In other eukaryotes, such as *Arabidopsis*, there is evidence that some small non-coding RNA genes are polycistronic. In humans, genes for micro-RNAs (see Ch. 18) are found in clusters and function in similar pathways during development, although there is no evidence as of now that these are transcribed as one message.

2. Repeated Sequences Are a Feature of Eukaryotic DNA

> Repeated sequences are frequently found in the DNA of higher organisms.

Unique sequences usually refer to the sequences that encode proteins, which occur only one time in the genome. Unique sequences account for almost all bacterial DNA. However, in higher organisms, unique sequences may comprise as little as 20% of the total DNA. In humans, only 2% of the entire genome actually encodes proteins. About 50% of the human genome is **repeated sequences** (or **repetitive sequences**) of one kind or another. Repeated sequences are what their name suggests, DNA sequences that are repeated multiple times throughout the genome. In some cases, the repeat sequences follow each other directly—tandem repeats (see below)— whereas others are spread separately around the genome—interspersed sequences. Some repeated sequences are genuine genes, but the majority consists of non-coding DNA.

> Consensus sequences are used to describe many different DNA motifs, including transcription factor binding sites, RNA polymerase binding sites, enhancer elements, silencer elements, etc. Consensus sequences can also be used to describe conserved protein domains, but instead of nucleotides, a protein consensus sequence is described by the most common amino acid at each position.

Individual members of a family of repeated sequences are rarely identical in every base. Nonetheless, one may imagine an ideal, so-called **consensus sequence**, from which they are all derived by only minor alterations (Fig. 4.04). Such a consensus sequence is deduced in practice by examining many related individual sequences and including those bases most often found at each position. In other words, consensus sequences are found by comparing many sequences and taking the average.

> Genes for ribosomal RNA are usually found in multiple copies. In higher organisms there may be thousands of copies.

One small category of repeats, called **pseudogenes,** is found in eukaryotic cells. Some of these are defective duplicate copies of genuine genes whose defects prevent them from being expressed. Other pseudogenes are expressed, but their mRNA regulates expression of other genes rather than coding for proteins. Pseudogenes are present in only one or two copies and may be next to the original, functional version of the gene or may be far away, even on a different chromosome. In quantitative terms, pseudogenes account for only a tiny fraction of the DNA. However, they are

---

FIGURE 4.04
Deduction of Consensus Sequence

The frequency of base appearances is used to derive a consensus sequence that is most representative of the series of related sequences shown.

Actual sequence observed:

(bases that differ from consensus are shown in lower case)

```
A t C C G T A T G T
A G C a t T A T G T
A G g C G T t T G T
c G C C G c A T G a
A a t C G T A T c T
A G C g a g A T G T
A G C C G T A T G T
g G C C a T A g t T
A G a C G c A a G T
A G t C G T A T a T
```

Number of times most common base appears at each position:

8 8 6 8 7 7 9 8 7 9

Derived *consensus sequence:* A G C C G T A T G T

---

consensus sequence   Idealized base sequence consisting of the bases most often found at each position
pseudogene   Defective copy of a genuine gene
repeated sequences   DNA sequences that exist in multiple copies
repetitive sequences   Same as repeated sequences

believed to be of great importance in molecular evolution as the precursors to new genes (see Ch. 26). Sometimes both copies of a duplicated gene remain functional and repeated duplication may even give families of related genes. The multiple copies gradually diverge to a greater or lesser extent as they adapt to carry out similar but related roles. Thus, the repeated sequences due to a gene family are closely related but not absolutely identical. One example of this is the HOX gene cluster, which encodes transcription factors that establish body patterns during development. These genes are organized such that the first gene is expressed earliest in development, and the last gene in the cluster is expressed last.

Sequences present in hundreds or thousands of copies are referred to as **moderately repetitive sequences**, and in the human genome, 25% of the total DNA falls into this category. This includes multiple copies of highly used genes, like those for ribosomal RNA, as well as non-functional stretches of DNA that are repeated many times. Since each prokaryotic cell contains 10,000 or more ribosomes, it is not surprising that their DNA usually contains half a dozen copies of the genes for rRNA. As you might expect, in the much larger and more complex eukaryotic cell, there are hundreds or thousands of copies of the rRNA genes. In every life form studied to date, rRNA genes are arranged in linear clusters in the genome. These are expressed as polycistronic RNA and then processed into separate rRNAs.

Much of the moderately repetitive non-coding DNA is formed of **LINEs**, which are "**Long INterspersed Elements.**" They are thought to be derived from retrovirus-like ancestors. (See Ch. 21 for information about retroviruses). In mammalian genomes, there are 20,000–50,000 copies of the LINE-1 (L1) family (Fig. 4.05). A complete L1 element is around 7,000 bp and contains two coding sequences. However, most individual L1 elements are shorter and many contain sequence rearrangements that disrupt the coding sequences, rendering them non-functional.

In addition, 10% of human DNA consists of sequences present in hundreds of thousands to millions of copies. Much of this **highly repetitive DNA** consists of **SINEs**, or **Short INterspersed Elements**. These sequences are almost all non-functional as far as is known. The best known SINE is the 300 base pair **Alu element**. It is named after the restriction enzyme *Alu I*, which cuts it at a single site 170 bp from the front of the sequence. (See Ch. 5 for information about restriction enzymes.) From 300,000 to 500,000 copies (per haploid genome) of the Alu element are scattered throughout

> About 7% of human DNA consists of repeats of the 300 bp Alu element.

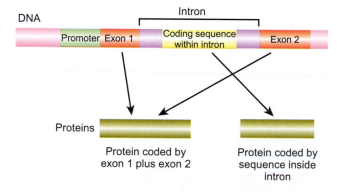

■-**FIGURE 4.05**
Structure of the LINE-1 Element

An example of a LINE-1 or L1 element is shown. L1 contains blocks of DNA that show homology with the *pol* and LTR sequences of retroviruses, as well as two coding sequences or open reading frames (ORF1 and ORF2) involved in its own replication.

Alu element   An example of a SINE, a particular short DNA sequence found in many copies on the chromosomes of humans and other primates
bp   Abbreviation for base pair(s)
highly repetitive DNA   DNA sequences that exist in hundreds of thousands of copies
LINE   Long interspersed element
long interspersed element (LINE)   Long sequence found in multiple copies that makes up much of the moderately repetitive DNA of mammals
moderately repetitive sequence   DNA sequences that exist in thousands of copies (but less than a hundred thousand)
short interspersed element (SINE)   Short sequence found in multiple copies that makes up much of the highly or moderately repetitive DNA of mammals
SINE   Short interspersed element

human DNA. They make up 6–8% of a human's genetic information, and recent studies suggest they bind to RNA polymerase II to repress gene transcription.

Most mammals contain SINEs topographically related to the Alu element. However, the original sequence, as found in mice, hamsters, etc., is only 130 bp long. For example, the mouse contains about 50,000 copies of the Alu-related **B1 element**. The human Alu element possesses two tandem repeats of this ancestral 130 bp B1 sequence plus an extra, unrelated 31 bp insertion of obscure origin. The mouse has less than 100,000 copies of the B1 element, so it would be classified as moderately repetitive DNA, whereas humans have more than 100,000 copies of the related Alu element, which is therefore classified as highly repetitive DNA. Clearly, the division into "moderately" repetitive and "highly" repetitive DNA is somewhat arbitrary.

Box 4.03: Purifying DNA Using a Cesium Chloride Density Gradient

DNA that consists of very large numbers of tandem repeats may well have a base composition different from that of the genome as a whole. If so, the satellite DNA will have a different buoyant density from the rest of the DNA, since this property depends on the base composition. DNA may be fractionated according to density by ultracentrifugation in a gradient of the heavy metal salt, cesium chloride (CsCl). A small tube containing a solution of CsCl and DNA is centrifuged at high speeds (450,000 x **g**) overnight. As the centrifugal forces push on the Cs + ions and the DNA molecules, the DNA settles in a band where its density is equal to the density of Cs + ions, a point called **neutral buoyancy**. If the %GC varies by 5% or more, separate bands are obtained. When mouse DNA is separated on a CsCl density gradient, two DNA bands are seen (Fig. 4.06). One contains 92% of the DNA with a density of 1.701 gm/cm^3 and the smaller, satellite band contains 8% of the DNA with a density of 1.690 gm/cm^3. Satellite DNA was originally defined by this density separation. However, in cases where the average satellite DNA base composition is close to that of the genome as a whole, the satellite DNA cannot be physically separated using a density gradient.

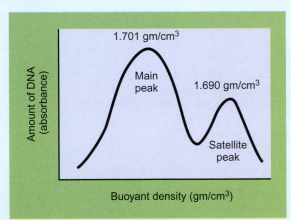

FIGURE 4.06
Density Gradient Centrifugation and Satellite Bands

A cesium chloride gradient will reveal two (or more) bands of fragmented DNA if these differ in density. In this case, the lighter DNA contains sequences that are primarily satellite DNA.

B1 element   An example of a SINE found in mice; the precursor sequence from which the human Alu element evolved
neutral buoyancy   Point at which the density of the substance is the same as the solution in which it is floating

2.1. Satellite DNA Is Non-Coding DNA in the Form of Tandem Repeats

Unlike the LINEs and SINEs, which by definition are scattered throughout the genome, a significant amount of highly repetitive DNA in eukaryotic cells is found as long clusters of **tandem repeats**. This is also known as **satellite DNA** because the repeated DNA forms lighter bands when genomic DNA is isolated by cesium chloride density gradient centrifugation (see Box 4.03). Tandem means that the repeated sequences are next to each other in the DNA without gaps between. The amount of satellite DNA is highly variable. In mammals such as the mouse, satellite DNA accounts for about 8% of the DNA, whereas in the fruit fly, *Drosophila*, it comprises nearly 50%.

Long series of tandem repeats tend to misalign when pairs of chromosomes line up for recombination during meiosis. **Unequal crossing over** will then produce one shorter and one longer segment of repetitive DNA (Fig. 4.07). Thus, the exact number of tandem repeats varies from individual to individual within the same population. (See Ch. 24 for information about crossing over).

In insects, the repeating sequences of satellite DNA are very short and consist of only one or very few different sequences. Thus, in *Drosophila virilis* a 7 bp repeat with a consensus sequence of ACAAACT accounts for almost all of the satellite DNA. About half of the repeats have the consensus itself and the rest differ by one or, rarely, two bases. Satellite sequences vary considerably from one organism to another. The more commonly used *Drosophila melanogaster* has more complex satellite DNA that includes the 7 bp sequence just described as well as other 5, 10, and 12 bp repeats. In mammals, the satellite sequences are relatively complex. Although there is an overall 9 bp consensus in the mouse, there is much more variation among the repeats (Fig. 4.08).

Satellite DNA is inert and is permanently coiled tightly into what is known as **heterochromatin**. A large proportion of satellite DNA, and therefore heterochromatin, is located around the **centromeres** of the chromosomes in humans, suggesting that it serves some structural role. These repeats are called **alpha DNA**, and in humans

> Tandem repeats cluster together forming regions of satellite DNA.

> The opposite of heterochromatin is called euchromatin, which is a looser and more accessible form of DNA in eukaryotes. Expressed genes are generally found within euchromatin.

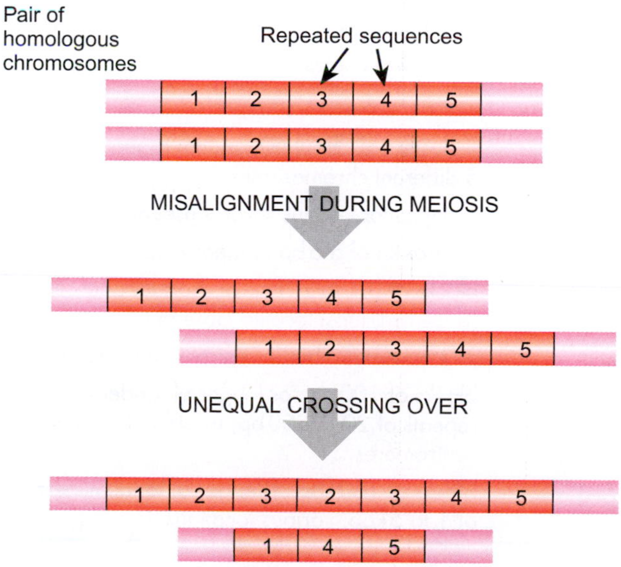

Pair of homologous chromosomes — Repeated sequences

MISALIGNMENT DURING MEIOSIS

UNEQUAL CROSSING OVER

FIGURE 4.07

Unequal Crossover due to Misalignment

A pair of homologous chromosomes contains repeated elements. Since repeated elements may be readily misaligned during meiosis, crossing over will sometimes occur in regions that are not comparable in each chromosome. The result is one longer and one shorter DNA fragment.

---

alpha DNA   Tandem DNA repeats found around the centromere in human DNA
centromere   Structure found on a chromosome and used to build and organize microtubules during mitosis
heterochromatin   Highly condensed form of chromatin that is genetically inert
satellite DNA   Highly repetitive DNA of eukaryotic cells that is found as long clusters of tandem repeats and is permanently coiled tightly into heterochromatin
tandem repeats   Repeated sequences of DNA (or RNA) that lie next to each other
unequal crossing over   Crossing over in which the two segments that cross over are of different lengths; often due to misalignment during pairing of DNA strands

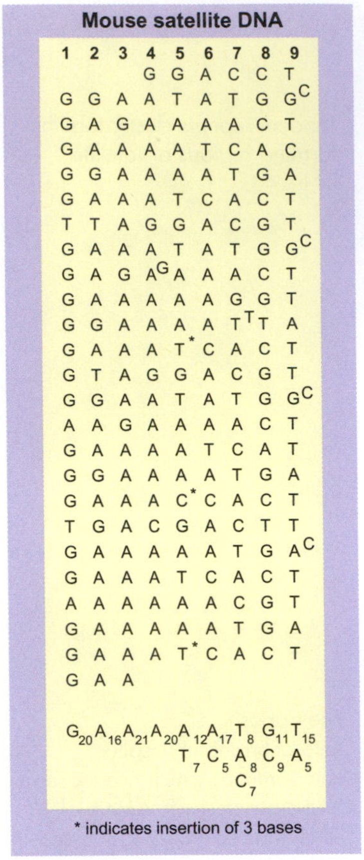

Mouse satellite DNA

	1	2	3	4	5	6	7	8	9
			G	G	A	C	C	T	
	G	G	A	A	T	A	T	G	G^C
	G	A	G	A	A	A	A	C	T
	G	A	A	A	A	T	C	A	C
	G	G	A	A	A	A	T	G	A
	G	A	A	A	T	C	A	C	T
	T	T	A	G	G	A	C	G	T
	G	A	A	A	A	T	A	T	G G^C
	G	A	G	A^G	A	A	A	C	T
	G	A	A	A	A	A	G	G	T
	G	G	A	A	A	A	T^T	T	A
	G	A	A	A	T*	C	A	C	T
	G	T	A	G	G	A	C	G	T
	G	G	A	A	T	A	T	G	G^C
	A	A	G	A	A	A	A	C	T
	G	A	A	A	A	T	C	A	T
	G	G	A	A	A	A	T	G	A
	G	A	A	A	C*	C	A	C	T
	T	G	A	C	G	A	C	T	T
	G	A	A	A	A	A	T	G	A^C
	G	A	A	A	T	C	A	C	T
	A	A	A	A	A	A	C	G	T
	G	A	A	A	A	A	T	G	A
	G	A	A	A	T*	C	A	C	T
	G	A	A						

$G_{20} A_{16} A_{21} A_{20} A_{12} A_{17} T_8 G_{11} T_{15}$
$T_7\ C_5\ A_8\ C_9\ A_5$
C_7

* indicates insertion of 3 bases

FIGURE 4.08
Repeating Motifs in Mouse Satellite DNA

Variations in the consensus 9 bp satellite DNA sequence GAAAAATGT are shown.

Person-to-person variation in the overall length of short tandem repeats allows individual identification and is used in forensic analysis.

Regulatory proteins often bind to DNA at inverted repeat sequences.

TABLE 4.02	Distribution of a 64 bp Human VNTR
% of population	**# of repeats**
7	18
11	16
43	14
36	13
4	10

TABLE 4.03	Components of the Eukaryotic Genome*

Unique sequences

Protein encoding genes—comprising upstream regulatory region, exons, and introns

Genes encoding non-translated RNA (snRNA, snoRNA, 7SL RNA, telomerase RNA, Xist RNA, a variety of small regulatory RNAs)

Non-repetitive intragenic non-coding DNA

Interspersed Repetitive DNA

Pseudogenes	
Short Interspersed Elements (SINEs)	
Alu element (300 bp)	~1,000,000 copies
MIR families (average ~130 bp)	~400,000 copies
(mammalian-wide interspersed repeat)	
Long Interspersed Elements (LINEs)	
LINE-1 family (average ~800 bp)	~200,000–500,000 copies
LINE-2 family (average ~250 bp)	~270,000 copies
Retrovirus-like elements (500–1300 bp)	~250,000 copies
DNA transposons (variable; average ~250 bp)	~200,000 copies

Tandem Repetitive DNA

Ribosomal RNA genes	5 clusters of about 50 tandem repeats on 5 different chromosomes
Transfer RNA genes	multiple copies plus several pseudogenes
Telomere sequences	several kb of a 6 bp tandem repeat
Mini-satellites (= VNTRs)	blocks of 0.1 to 20 kbp of short tandem repeats (5–50 bp), most located close to telomeres
Centromere sequence (α-satellite DNA)	171 bp repeat, binds centromere proteins
Satellite DNA	blocks of 100 kbp or longer of tandem repeats of 20 to 200 bp, most located close to centromeres
Mega-satellite DNA	blocks of 100 kbp or longer of tandem repeats of 1 to 5 kbp, various locations

*Numbers of copies given is for the human genome.

there is a 171 bp repeat that is situated in a head to tail fashion throughout the centromere region. Note, however, that these satellite DNA sequences are quite distinct from the **centromere sequences**, which are needed for attachment of the spindle fibers during cell division.

centromere sequence (CEN)   A recognition sequence found at the centromere and needed for attachment of the spindle fibers

2.2. Minisatellites and VNTRs

Segments of DNA consisting of short tandem repeats, but in much fewer copies than satellites, are known as **VNTRs (Variable Number of Tandem Repeats)**. These repeats are sometimes categorized as either **minisatellite** DNA if the repeat is around 25 base pairs in length, or **microsatellites** if the repeat is less than 13 nucleotides in length. In mammals, VNTRs are common and are scattered over the genome. One of the most common minisatellites is the repeat found in eukaryotic telomeres.

Due to unequal crossing over, the number of repeats in a given VNTR varies among individuals. Although VNTRs are non-coding DNA and not true genes, nonetheless the different versions are referred to as **alleles**. For example, Table 4.02 shows the distribution of one human VNTR of 64 bp among the population.

Some hyper-variable VNTRs may have as many as 1,000 different alleles and give unique patterns for almost every individual. This quantitative variation may be used for the identification of individuals by **DNA fingerprinting**.

3. Palindromes, Inverted Repeats, and Stem and Loop Structures

Palindromes are words or phrases that read the same backwards as forwards. In the case of DNA, which is double stranded, two types of palindromes are theoretically possible. **Mirror-like palindromes** are like those of ordinary text, but involve two strands of DNA. However, in practice, the **inverted repeat** type of palindrome is much more common and of major biological significance. In an inverted repeat, the sequence reads the same forwards on one strand as it reads backwards on the complementary strand (Fig. 4.09).

Inverted repeats are extremely important as recognition sites on the DNA for the binding of a variety of proteins. Many regulatory proteins recognize inverted repeats, as do most restriction and modification enzymes (see Ch. 5). In such cases, the inverted repeat usually remains as normal double helical DNA and does not need to be distorted by supercoiling. The term "direct repeat" refers to the situation where the repeated sequences point in the same direction and are on the same strand.

Consider just a single strand of the inverted repeat sequence. Note that the right and left halves of the sequence on each single strand must be complementary to each other. Thus, such a sequence (e.g., GGATATCC) can be folded into a **hairpin** whose two halves are held together by base pairing (Fig. 4.10). The U-turn at the top of the hairpin is possible, but energetically unfavorable. In practice, normally a few unpaired bases (shown as N = any base in the diagram below) are found forming a loop at the top of the base paired stem—the so-called **stem and loop** motif. Such a stem and loop can form from one strand of any inverted repeat that has a few extra bases in the middle (Fig. 4.11).

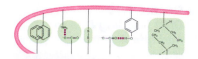

FIGURE 4.09
Palindromes and Inverted Repeats

A mirror-like palindrome and an inverted repeat are shown. Similar colors indicate palindromic or inverted sequences.

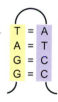

FIGURE 4.10
A Hairpin

If a single strand of DNA containing inverted repeats is folded back upon itself, base pairing occurs forming a hairpin structure.

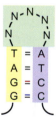

FIGURE 4.11
Stem and Loop Motif

If inverted repeats are separated by a few bases, a stem and loop structure results. The loop contains unpaired bases (NNN).

---

allele One particular version of a gene, or more broadly, a particular version of any locus on a molecule of DNA

DNA fingerprint Individually unique pattern due to multiple bands of DNA produced using restriction enzymes, separated by electrophoresis and usually visualized by Southern blotting

hairpin A double-stranded base-paired structure formed by folding a single strand of DNA or RNA back upon itself

inverted repeat Sequence of DNA that is the same when read forwards as when read backwards, but on the other complementary strand; one type of palindrome

microsatellite Another term for a VNTR (variable number tandem repeats) with repeats around 13 base pairs in length

minisatellite Another term for a VNTR (variable number tandem repeats) with repeats around 25 base pairs in length

mirror-like palindrome Sequence of DNA that is the same when read forwards and backwards on the same strand; one type of palindrome

palindrome A sequence that reads the same backwards as forwards

stem and loop Structure made by folding an inverted repeat sequence

variable number tandem repeats (VNTR) Cluster of tandemly-repeated sequences in the DNA whose number of repeats differs from one individual to another

VNTR See variable number tandem repeats

FIGURE 4.12
DNA Bending Due to Multiple A-tracts

A) Bending of DNA occurs to the 3'-side of A-tracts. B) Such bending decreases the speed at which DNA travels during electrophoresis. Indeed, the mobility of a DNA molecule of a given length varies depending on the location of bent regions within the molecule. Bends in the middle have greater effect than those close to the ends.

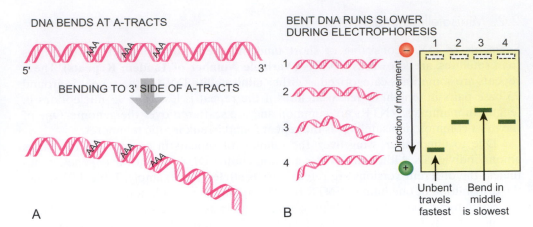

DNA BENDS AT A-TRACTS

BENDING TO 3' SIDE OF A-TRACTS

A

BENT DNA RUNS SLOWER DURING ELECTROPHORESIS

Direction of movement

Unbent travels fastest

Bend in middle is slowest

B

4. Multiple A-Tracts Cause DNA to Bend

A DNA sequence that consists of several runs of adenine (A) residues (three to five nucleotides long) separated from each other by 10 base pairs forms bends in the helix. Note that the spacing of the A-tracts corresponds to one turn of the double helix. Bending occurs at the 3'-end of the adenines (Fig. 4.12). **Bent DNA** moves more slowly during gel electrophoresis than unbent DNA of the same length because the agarose beads catch this structure and prevent it from migrating as far (see Section 6 of this chapter for information about electrophoresis).

Bent DNA is found at the origins of replication of some viruses and of yeast chromosomes. It is thought to help the binding of the proteins that initiate DNA replication (see Ch. 10). In addition to "naturally" bent DNA, certain regulatory proteins also bend DNA into U-turns when activating transcription (see Ch. 11).

5. Supercoiling Is Necessary for Packaging of Bacterial DNA

Bacterial DNA is 1000 times longer than the cell that contains it. The DNA must be **supercoiled** in order to fit into the cell. The length of the single DNA molecule needed to carry the 4,000 or so genes of a bacterial cell is about 1.5 millimeters, and this structure is compacted into 0.2 cubic micrometers! Thus, a stretched out bacterial chromosome is a thousand times longer than a bacterial cell. In order to create supercoils, the DNA, which is already a double helix, is twisted again, as shown in Figure 4.13. The original double helix has a right-handed twist, but the supercoils twist in the opposite sense; that is, they are left-handed or **"negative" supercoils**. There is roughly one supercoil for every 200 nucleotides in typical bacterial DNA. Negative (rather than positive) supercoiling helps promote the unwinding and strand separation necessary during replication and transcription. Bacterial chromosomes and plasmids are double-stranded circular DNA molecules and are often referred to as **covalently closed circular DNA**, or **cccDNA**. If one strand of a double-stranded circle is nicked, the supercoiling can unravel. Such a molecule is known as an **open circle**.

Although supercoiling reduces the size of the DNA, the chromosome still will not fit into the bacterial cell. The second level of compaction occurs when approximately 50 giant loops of supercoiled DNA are arranged around a protein scaffold. In Figure 4.14, the protein scaffold is created when DNA binding proteins bind at regular

DNA gyrase puts negative supercoils into a bacterial chromosome.

bent DNA Double helical DNA that is bent due to several runs of A's
covalently closed circular DNA (cccDNA) Circular DNA with no nicks in either strand
negative supercoiling Supercoiling with a left-handed or counterclockwise twist
open circle Circular DNA with one strand nicked and hence with no supercoiling
supercoiling Higher-level coiling of DNA that is already a double helix

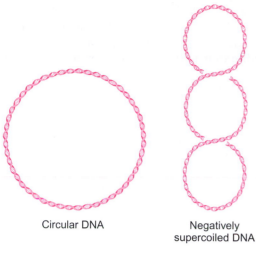

Supercoiling of DNA

Bacterial DNA is negatively supercoiled in addition to the twisting imposed by the double helix.

Circular DNA   Negatively supercoiled DNA

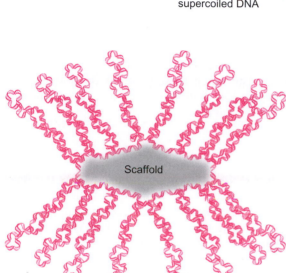

Scaffold

Bacterial Chromosomes Loop from a Protein Scaffold

Supercoiling of bacterial DNA results in giant loops of supercoiled DNA extending from a central scaffold.

intervals along the bacterial chromosome. These proteins then come together in one mass, and the unbound DNA loops out from the proteins.

These two levels of compaction are essential and actually keep the bacterial chromosomal DNA compacted into a small area called the **nucleoid**. This structure contains the chromosome and its associated proteins. When visualized using electron microscopy, the nucleoid is a densely staining area offset from the rest of the cytoplasm, but not compartmentalized by a membrane. Several different proteins are responsible for DNA compaction into the nucleoid. These include HU and IHF, proteins that compact the DNA; H-NS, a protein that forms filaments with two different DNA duplexes; and Fis, a DNA binding and bending protein.

Although the nucleoid seems randomly generated, the chromosome is highly organized. The location of a gene on the circle correlates with a specific location within a nucleoid region. Recent research has identified some proteins that help organize the bacterial chromosome in bacteria. The organization occurs after the DNA is copied by replication. Bacterial chromosomes initiate the replication process by opening the double helix at the origin of replication, and then adding new complementary nucleotides as DNA polymerase travels around the ring. To save time, two DNA polymerase complexes travel in opposite directions and meet opposite the origin.

The bacterial chromosome consists of about 50 giant supercoiled loops of DNA.

nucleoid   Area within a bacterial cell in which the chromosome is usually found; not surrounded by membranes

The condensation of the newly synthesized DNAs begins at the origin of replication, shortly after the replication enzymes complete synthesis. The first step to proper positioning of the chromosome within the nucleoid is to get each new copy into the proper cell half. This process is called **partitioning**, and is mediated by a complex of ParA, ParB, and specific DNA sequences near the origin of replication called *parS* sites. First, the DNA binding protein, ParB, binds to the *parS* sites and spreads along the DNA to cover adjacent nucleotides. The spreading creates a nucleoprotein filament that acts somewhat like the eukaryotic centromere. This structure then binds ParA, an ATPase that creates long polymers that help partition the new chromosome into each half of the cell.

The ParB protein also attracts a protein called SMC (structural maintenance of chromosomes) to the origin. SMC has a unique shape. It has an ATPase domain attached to a long coiled coil of amino acids. Two of these structures attach to each other at the opposite end of the ATPase domain with a hinge, resulting in a V-shaped protein. The V is able to connect via the ATPase domains either within the same protein or to other SMC proteins. Thus, this protein can act like a tether to hold strands of DNA together like a pair of chopsticks or like a diamond shaped cage. The SMC proteins then diffuse along the DNA, and hold the various domains in the correct location within the nucleoid. Although the picture on how these proteins function is becoming clearer, much is still unknown. But when these proteins do not function properly, bacteria are unable to properly express their genes and die, suggesting that the organization is essential for the bacteria to survive.

5.1. Topoisomerases and DNA Gyrase

The total amount of twisting present in a DNA molecule is referred to as the **linking number (L)**. This is the sum of the contributions due to the double helix plus the supercoiling. (The number of double helical turns is sometimes known as the **twist**, **T**, and the number of superhelical turns as the **writhe** or **writhing number**, **W**. In this terminology, the linking number, L, is the sum of the twist plus the writhe (L = T + W).)

The same circular DNA molecule can have different numbers of supercoils. These forms are known as topological isomers, or **topoisomers**. The enzymes that insert or remove supercoils are therefore named **topoisomerases**. **Type I topoisomerases** break only one strand of DNA, which changes the linking number in steps of one. In contrast, **type II topoisomerases** (including **DNA gyrases**) break both strands of the DNA and pass another part of the double helix through the gap. This changes the linking number in steps of two (Fig. 4.15).

DNA gyrase, a type II topoisomerase, introduces negative supercoils into closed circular molecules of DNA, such as plasmids or the bacterial chromosome. Gyrase works by cutting both strands of the DNA, twisting the DNA strands, and then rejoining the cut ends. Gyrase can generate 1,000 supercoils per minute. As each supertwist is introduced, gyrase changes conformation to an inactive form. Reactivation requires energy, provided by breakdown of ATP. DNA gyrase can also remove negative supercoils (but not positive ones) without using ATP, but this occurs 10 times more slowly. The steady-state level of supercoiling in *E. coli* is maintained by a balance between topoisomerase IV, acting in concert with topoisomerase I, to remove excess negative

> Enzymes known as topoisomerases change the level of supercoiling.

> Ciprofloxacin kills bacteria by inhibiting DNA gyrase. It is harmless to animals as they do not use DNA gyrase for compacting their DNA.

DNA gyrase An enzyme that introduces negative supercoils into DNA, a member of the type II topoisomerase family
linking number (L) The sum of the superhelical turns (the writhe, W) plus the double helical turns (the twist, T)
partitioning Movement of each replicated copy of the chromosome into each respective daughter cell during cell division
topoisomerase Enzyme that alters the level of supercoiling or catenation of DNA (i.e., changes the topological conformation)
topoisomers Isomeric forms that differ in topology (i.e., their level of supercoiling or catenation)
twist, T The number of double helical turns in a molecule of DNA (or double-stranded RNA)
type I topoisomerase Topoisomerase that cuts a single strand of DNA and therefore changes the linking number by one
type II topoisomerase Topoisomerase that cuts both strands of DNA and therefore changes the linking number by two
writhing number, W The number of supercoils in a molecule of DNA (or double-stranded RNA)
writhe Same as writhing number, W

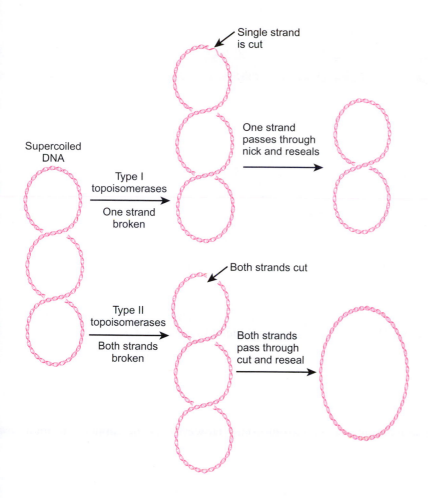

Single strand
is cut

One strand
passes through
nick and reseals

Supercoiled
DNA

Type I
topoisomerases

One strand
broken

Both strands cut

Type II
topoisomerases

Both strands
broken

Both strands
pass through
cut and reseal

▶ FIGURE 4.15
Mechanism of Type I and II Topoisomerases

The difference in action between topoisomerases of Type I and Type II is in the breakage of strands. Type I breaks only one strand, while Type II breaks both strands. When one strand is broken, the other strand is passed through the break to undo one supercoil. When two strands are broken, double-stranded DNA is passed through the break and the supercoiling is reduced by two. After uncoiling, the breaks are rejoined.

supercoils and thus acting in opposition to DNA gyrase. In the absence of topoisomerase I and topoisomerase IV, the DNA becomes negatively hyper-supercoiled.

DNA gyrase is a tetramer of two different subunits. The GyrA subunit cuts and rejoins the DNA and the GyrB subunit is responsible for providing energy by ATP hydrolysis. DNA gyrase is inhibited by **quinolone antibiotics**, such as **nalidixic acid** and their fluorinated derivatives such as **norfloxacin** and **ciprofloxacin**, which bind to the GyrA protein. An inactive complex is formed in which GyrA protein is inserted into the DNA double helix and covalently attached to the 5'-ends of both broken DNA strands. **Novobiocin** also inhibits gyrase by binding to the GyrB protein and preventing it from binding ATP.

5.2. Catenated and Knotted DNA Must Be Corrected

Circular molecules of DNA may become interlocked during replication or recombination. Such structures are called **catenanes**. The circles may be liberated by certain type II topoisomerases, such as topoisomerase IV (Fig. 4.16) of *E. coli* and related enzymes. Circular DNA molecules may also form knots. Type II topoisomerases can both create and untie knots. Like DNA gyrases, these enzymes are tetramers of two

catenane   Structure in which two or more circles of DNA are interlocked
ciprofloxacin   A fluoroquinolone antibiotic that inhibits DNA gyrase
norfloxacin   A fluoroquinolone antibiotic that inhibits DNA gyrase
nalidixic acid   A quinolone antibiotic that inhibits DNA gyrase
novobiocin   An antibiotic that inhibits type II topoisomerases, especially DNA gyrase, by binding to the B subunit
quinolone antibiotics   A family of antibiotics, including nalidixic acid, norfloxacin, and ciprofloxacin that inhibit DNA gyrase and
other type II topoisomerases by binding to the A subunit

Unlinking of Catenanes by Topoisomerase

Topoisomerases may uncoil, unknot, or unlink DNA as well as carry out the coiling, knotting, or interlinking of DNA. Topoisomerases act (at the locations shaded blue) by cutting both strands of the DNA at one location and passing another region of the DNA through the gap.

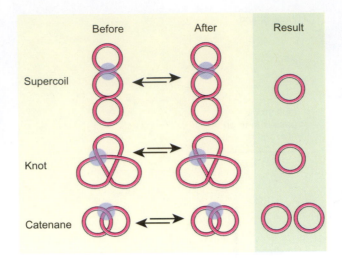

FIGURE 4.17
Cruciform Structure Formed from an Inverted Repeat

Because the DNA is palindromic, the strands can separate and base pair with themselves to form lateral cruciform extensions.

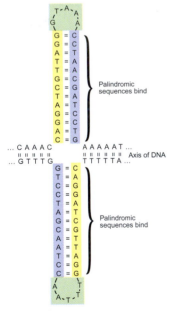

different subunits, one for cutting the DNA and the other for energy coupling. Like gyrase, topoisomerase IV is inhibited by quinolone antibiotics.

5.3. Local Supercoiling

Whether the DNA is found in a prokaryote or eukaryote, when DNA is replicated or when genes are expressed, the double helix must first be unwound. This is aided by the negative supercoiling of the chromosome. However, as the replication apparatus proceeds along a double helix of DNA, it creates positive supercoiling ahead of itself. Similarly, during transcription, when RNA polymerase proceeds along a DNA molecule, it also creates positive supercoiling ahead of itself. For replication and transcription to proceed more than a short distance, DNA gyrase must insert negative supercoils to cancel out the positive ones. Behind the moving replication and transcription apparatus, a corresponding wave of negative supercoiling is generated. Excess negative supercoils are removed by topoisomerase I.

As a result, at any given instant, the extent of supercoiling varies greatly in any particular region of the chromosome. It has been suggested that supercoiling might regulate gene expression. However, only rare examples are known; thus transcription of the gene for DNA gyrase in *E. coli* is regulated by supercoiling. More often, the opposite is the case. Local supercoiling depends largely on the balance between transcription and the restoration of normal supercoiling by gyrase and topoisomerase.

5.4. Supercoiling Affects DNA Structure

Supercoiling places DNA under physical strain. This may lead to the appearance of alterations in DNA structure that serve to relieve the strain. Three of these alternate forms are **cruciform structures**; the left-handed double helix, or **Z-DNA**; and the triple helix, or **H-DNA**. All of these structures depend on certain special characteristics within the DNA sequence, as well as supercoiling stress.

The cruciform (cross-like) structures are formed when the strands in a double-stranded DNA palindrome are separated and formed into two stem and loop structures opposite each other (Fig. 4.17). The probability that cruciform structures will

---

cruciform structure   Cross-shaped structure in double-stranded DNA (or RNA) formed from an inverted repeat
H-DNA   A form of DNA consisting of a triple helix; its formation is promoted by acid conditions and by runs of purine bases
Z-DNA   An alternative form of DNA double helix with left-handed turns and 12 base pairs per turn

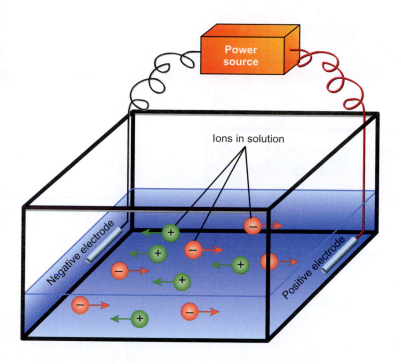

Ions in solution

Power source

Negative electrode

Positive electrode

FIGURE 4.18
Principle of Electrophoresis

Creating an electrical field in a solution of positively- and negatively-charged ions allows the isolation of the ions with different charges. Since DNA has a negative charge due to its phosphate backbone, electrophoresis will isolate the negatively-charged DNA from other components.

form increases with the level of negative supercoiling and the length of the inverted repeat. In practice, the four to eight base sequences recognized by most regulatory proteins and restriction enzymes are too short to yield stable cruciform structures. Palindromes of 15 to 20 base pairs will produce cruciform structures. Their existence can be demonstrated because they allow single strand specific nucleases to cut the double helix. (A nuclease is an enzyme that cuts nucleic acid strands.) Cutting occurs within the small single-stranded loop at the top of each hairpin.

6. Separation of DNA Fragments by Electrophoresis

Perhaps the most widely used physical method in all of molecular biology is **gel electrophoresis**. This technique separates and purifies fragments of DNA or RNA as well as proteins. The basic idea of **electrophoresis** is to separate the molecules based on their intrinsic electrical charge. Electrically-positive charges attract negative charges and repel other positive charges. Conversely, negative charges attract positive charges and repel other negative charges. Two electrodes, one positive and the other negative, are connected to a high voltage source. Positively-charged molecules move towards the negative electrode and negatively charged molecules move towards the positive electrode (Fig. 4.18).

Since DNA carries a negative charge on each of the many phosphate groups making up its backbone, it will move towards the positive electrode during electrophoresis. The bigger a molecule, the more force required to move it. However, the longer a DNA molecule, the more negative charges it has. In practice, these two factors cancel out because all fragments of DNA have the same number of charges per unit length. Consequently, DNA molecules in free solution will all move toward the positive electrode at the same speed, irrespective of their molecular weights.

In order to separate the DNA pieces by size, the fragments must travel through a region with a sieving action, rather than a free solution. For example, in addition to the chromosome, bacteria often contain plasmids, smaller rings of DNA.

> Electrophoresis separates DNA and RNA on the basis of charge.

---

electrophoresis   Movement of charged molecules due to an electric field; Used to separate sand purify nucleic acids and proteins
gel electrophoresis   Electrophoresis of charged molecules through a gel meshwork in order to sort them by size

FIGURE 4.19
Agarose Gel Electrophoresis of DNA

Agarose gel electrophoresis separates fragments of DNA by size. Negatively-charged DNA molecules are attracted to the positive electrode. As the DNA migrates, the fragments of DNA are hindered by the cross-linked agarose meshwork. The smaller the piece of DNA, the less likely it will be slowed down. Therefore, smaller fragments of DNA migrate faster.

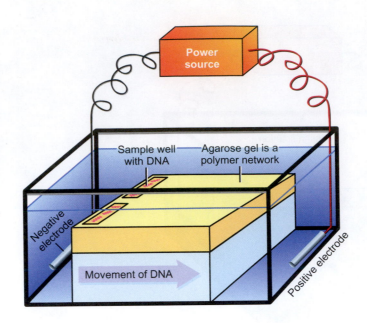

Gel electrophoresis separates nucleic acid molecules according to their molecular weight.

DNA can be detected by staining with a dye such as ethidium bromide, methylene blue, or SYBR® green.

Gel electrophoresis will separate the two different sized molecules of DNA. In free solution these would reach the positive pole at the same time, but when they travel through a matrix of cross-linked polymer chains, the larger DNA molecule moves slower than the smaller plasmid. The larger molecules find it more difficult to squeeze through the gaps in contrast to the small pieces that can wiggle through the meshwork with ease. The result is that the DNA fragments separate in order of size (Fig. 4.19). In our example, the plasmid DNA will move farther in the gel than the chromosome.

Most DNA is separated using **agarose gel electrophoresis**. **Agarose** is a polysaccharide extracted from seaweed. When agarose and water are mixed and boiled, the agarose melts into a homogeneous solution. As the solution cools, it gels to form a meshwork, which has small pores or openings filled with water. The cooled gel looks much like a very concentrated mixture of gelatin without the food coloring or flavoring. The pore size of agarose is suitable for separating nucleic acid polymers consisting of several hundred nucleotides or longer. Shorter fragments of DNA are usually separated on gels made of **polyacrylamide**. The meshwork formed by this polymer has smaller pores than agarose polymers. In either gel matrix, samples of DNA are loaded into a slot or sample well at the end of the gel closest to the negative electrode. DNA molecules move through the gel away from the negative electrode and towards the positive electrode.

Agarose gels are normally square slabs that allow multiple samples to be run side by side. Because DNA is naturally colorless, some way of visualizing the DNA after running the gel is needed. The gel can be stained with **ethidium bromide**, which binds tightly and specifically to DNA or RNA. Ethidium bromide intercalates between the base pairs of DNA and RNA, therefore, DNA binds much more molecules of ethidium bromide than RNA. The gel must then be examined under UV light, where ethidium bromide bound to DNA glows bright orange. RNA also fluoresces under UV light, but not as intensely so it is seen in gels, but not as distinctly. Since ethidium bromide is a DNA mutagen, less toxic DNA dyes are sometimes used; the most common of these are methylene blue and SYBR® green. These stains are less sensitive, but still dye the DNA molecules embedded inside the agarose.

Agarose gel electrophoresis can be used to purify DNA for use in genetic engineering or it can be used to measure the sizes of different fragments. To find the size of an unknown piece of DNA or a protein, a set of standards of known sizes is run

---

agarose   A polysaccharide from seaweed that is used to form gels for separating nucleic acids by electrophoresis
agarose gel electrophoresis   Technique for separation of nucleic acid molecules by passing an electric current through a gel made of agarose
ethidium bromide   A stain that specifically binds to DNA or RNA and appears orange if viewed under ultraviolet light
polyacrylamide   Polymer used in separation of proteins or very small nucleic acid molecules by gel electrophoresis

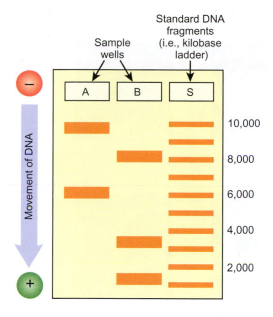

Sample wells

Standard DNA fragments (i.e., kilobase ladder)

Movement of DNA

10,000
8,000
6,000
4,000
2,000

▶ **FIGURE 4.20**
Agarose Gel Separation of DNA: Staining and Standards

To visualize DNA, the agarose gel containing the separated DNA fragments is soaked in a solution of ethidium bromide, which intercalates between the base pairs of the DNA. Excess ethidium bromide is removed by rinsing in water, and the gel is placed under a UV light source. The UV light excites the ethidium bromide and causes it to fluoresce orange. In sample A, there are two fragments of DNA each of a different size and each forming a separate band in the gel. To determine the size of fragments, a standard set of DNA fragments of known sizes is run alongside the sample to be analyzed.

alongside, on the same gel (Fig. 4.20). For DNA fragments, a set of DNA fragments that are exact multiples of 1000 bp, called the **kilobase ladder**, are often used.

As stated above, the mobility of a DNA molecule in such a gel depends on its molecular weight, but the conformation of the DNA also is a large factor in how easily the DNA passes through the agarose. Consequently, for DNA, supercoiled cccDNA moves farther than linearized DNA, which in turn moves farther than open circular DNA. For small circular molecules of DNA, it is even possible to separate topoisomers with different numbers of superhelical twists. The more superhelical twists, the more compact the molecule is and the faster it moves in an electrophoretic field (Fig. 4.21). Cruciform structures partially straighten supercoiled DNA and so the molecule is not folded up as compactly and thus travels slower during gel electrophoresis.

7. Alternative Helical Structures of DNA Occur

Several double helical structures are actually possible for DNA. Watson and Crick (Nature (1953) 171: 737) described the most stable and most common of these. It is right-handed with 10 base pairs per turn of the helix. The grooves running down the helix are different in depth and referred to as the major and minor grooves. The standard Watson and Crick double helix is referred to as the **B-form** or as **B-DNA** to distinguish it from the other helical forms, **A-DNA**, and Z-DNA (Fig. 4.22). Most of these structures apply not only to double-stranded DNA (dsDNA), but also to RNA when it is double-stranded.

The **A-form** of the double helix is shorter and fatter than the B-form and has 11 base pairs per helical turn. In the A-form, the bases tilt away from the axis, the minor groove becomes broader and shallower, and the major groove becomes narrower and deeper. Double-stranded RNA or hybrids with one RNA and one DNA strand usually form an A-helix, since the extra hydroxyl group at the 2′ position of ribose prevents double-stranded RNA from forming a B-helix. Double-stranded DNA tends to form an A-helix only at a high salt concentration or when it is dehydrated. The tendency to form an A-helix also depends on the sequence. The physiological relevance, if any, of A-DNA is obscure. However, double-stranded regions of RNA exist in this form *in vivo*.

Alternative helical structures are sometimes found for DNA in addition to the common known as B-DNA.

The A-form helix is found in dsRNA or DNA/RNA hybrids. It has 11 bp per turn—one more than B-DNA.

A-DNA   A rare alternative form of double-stranded helical DNA
A-form   An alternative form of the double helix, with 11 base pairs per turn, often found for double-stranded RNA, but rarely for DNA
B-form or B-DNA   The normal form of the DNA double helix, as originally described by Watson and Crick
kilobase ladder   Standard set of DNA fragments with known length used to compare DNA fragments of unknown size in gel electrophoresis

FIGURE 4.21
Separation of Supercoiled DNA by Electrophoresis

A) Supercoiled DNA molecules, all of identical sequence, were electrophoresed to reveal multiple bands, with each band differing in the number of supercoils, which is shown beside the band. Zero refers to relaxed or open circular DNA. B) Actual ethidium bromide stained electrophoresis gels of four differing concentrations (A: 6.1 nM; B: 9.2 nM; C: 12.2 nM; D: 18.3 nM) of negatively supercoiled *E. coli* plasmid DNA (pXXZ06) incubated with pure DNA topoisomerase I. In each panel, lanes 1-7 contain DNA samples incubated at 37°C for 0.5, 1, 2, 5, and 15 mins. Symbols: R = relaxed or open circular; S = supercoiled. *(From Xu, X. and Leng, F. (2011) A rapid procedure to purify Escherichia coli DNA topoisomerase I. Prot. Exp. Purification 77: 214-219.)*

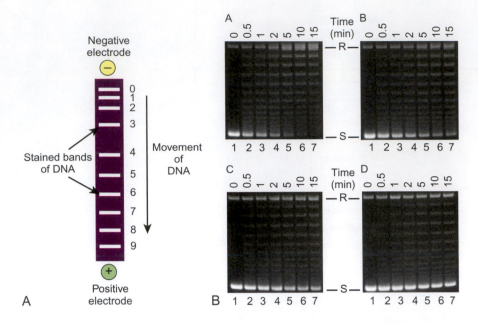

FIGURE 4.22
Comparison of B-DNA, A-DNA, and Z-DNA

Several structurally different versions of the double helix exist. Shown here is the normal Watson-Crick double helix, the B-form (middle), together with the rarer A-form (left) and Z-DNA form (right).
(Photo courtesy of Richard Wheeler; Wikicommons.)

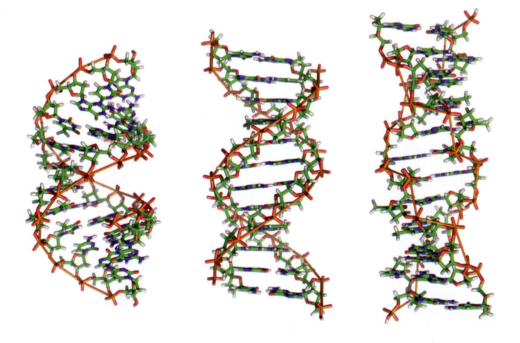

Z-DNA is a left-handed double helix with 12 base pairs per turn. It is thus longer and thinner than B-DNA and its sugar phosphate backbone forms a zigzag line rather than a smooth helical curve (Fig. 4.23). High salt favors Z-DNA as it decreases repulsion between the negatively charged phosphates of the DNA backbone. Z-DNA is formed in regions of DNA that contain large numbers of alternating GC or GT pairs, such as

```
GCGCGCGCGCGCGC   or   GTGTGTGTGTGTGT
CGCGCGCGCGCGCG        CACACACACACACA
```

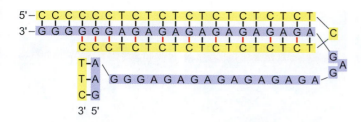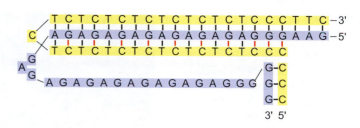

FIGURE 4.23
Structure of H-DNA

This triple helix is formed by GA- and TC-rich regions of a plasmid and is composed of triads of bases.

Such tracts may be abbreviated as (GC)n·(GC)n and (GT)n·(AC)n. Note that the sequence of each individual strand is written in the 5'- to 3'-direction. That Z-DNA depends specifically on G (not A) alternating with C or T is shown by the fact that DNA with many alternating AT pairs forms cruciform structures, not Z-DNA.

Since Z-DNA is a left-handed helix, its appearance in part of a DNA molecule helps to remove supercoiling stress. As negative supercoiling increases, the tendency for GC- or GT-rich tracts of DNA to take the **Z-form** increases. Its existence can be demonstrated in small plasmids by changes in electrophoretic mobility. Single-strand specific nucleases can cut DNA at the junction between segments of Z-DNA and normal B-DNA.

Short artificial segments of double-stranded DNA made solely of repeating GC units take the Z-form even in the absence of supercoiling, provided that the salt concentration is high. (This was how Z-DNA was originally discovered.) Antibodies to Z-DNA can be made by immunizing animals with such linear (GC)n fragments. These antibodies can tag or mark the presence of Z-form (helix) regions in natural DNA, provided it is highly supercoiled.

> Z-DNA is a left-handed double helix with 12 base pairs per turn.

It has been suggested that regions of Z-helix may be specifically recognized by certain enzymes. An example is the RNA editing enzyme ADAR1, which modifies bases in dsRNA (see Ch. 12 for the details of RNA processing). ADAR1 stands for adenosine deaminase (RNA) type I and it removes the amino group off adenosine, so converting it to **inosine**. It requires a double-stranded RNA substrate that is formed by folding an intron back onto the neighboring exon inside the eukaryotic cell nucleus. ADAR1 contains separate binding motifs for both DNA and for dsRNA. It has been postulated that the DNA-binding domain recognizes Z-DNA because base modification must occur before cutting and splicing of the introns and exons. When RNA polymerase moves along, transcribing DNA into RNA, it generates positive supercoils ahead and negative supercoils behind. Negative supercoiling induces the formation of Z-DNA, especially in GC- or GT- regions. Consequently, a zone of Z-DNA will be found just behind the RNA polymerase. Binding to Z-DNA would ensure that ADAR1 works on newly synthesized RNA.

H-DNA is even more peculiar—it is not a double but a *triple* helix. It depends on long tracts of purines in one strand and, consequently, only pyrimidines in the other strand; for example,

```
GGGGGGGGGGGGGGG  or  GAGAGAGAGAGAGAG
CCCCCCCCCCCCCCC      CTCTCTCTCTCTCTC
```

Two such segments are required and may interact forming H-DNA when the DNA is highly supercoiled (Fig. 4.23). In addition, the overall region must be a mirror-like palindrome. H-DNA contains a triple helix, consisting of one purine-rich strand and two pyrimidine-rich strands. The other purine-rich strand is displaced and left unpaired.

---

inosine A purine nucleoside, found most often in transfer RNA, that contains the unusual base hypoxanthine
Z-form An alternative form of double helix with left-handed turns and 12 base pairs per turn.; both DNA and dsRNA may be found in the Z-form

FIGURE 4.24
Chromatin Territories

Each of the different colors represents a different chromosome from chicken situated in a separate chromatin territory. Since the chicken is diploid, there are two chromatin territories for each chromosome. The model proposed by this paper suggests that the chromatin territory is separated from another by an interchromatin space. This organization could partly explain how some eukaryotic genes are expressed and others are not expressed.
(Credit: Fig. 2 in Cremer, T and Cremer C., 2001. Chromosome territories, nuclear architecture and gene regulation in mammalian cells. Nature Rev Genetics 2: 292-301.)

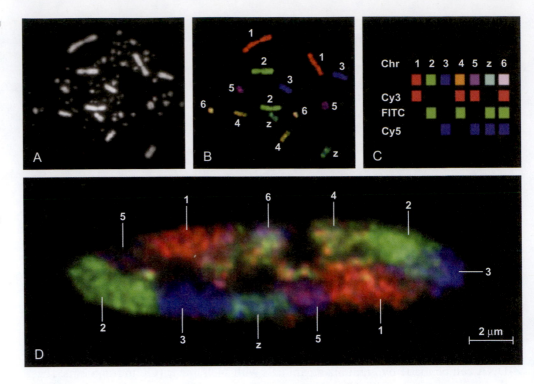

> H-DNA involves strange "sideways" base pairing to form a triple helix.

In H-DNA, adenine pairs with two thymines and guanine pairs with two cytosines. In each case, one pairing is normal, the other sideways (so-called **Hoogsteen base pairs**—hence the name H-DNA). Furthermore, to form the C=G=C triangle, an extra proton (H^+) is needed for one of the hydrogen bonds. Consequently, formation of H-DNA is promoted by acidic conditions. High acidity also tends to protonate the phosphate groups of the DNA backbone, so decreasing their negative charges. This reduces the repulsion between the three strands and helps form a triple helix.

Despite these complex sequence requirements, computer searches of natural DNA have shown that potential sequences that might form triplex H-DNA are much more frequent than expected on a random basis. These are called **PIT (potential intrastrand triplex)** elements. For example, the *E. coli* genome contains 25 copies of a 37 base PIT element. Isolated PIT element DNA does form a stable triplex even at neutral pH. Not surprisingly, the presence of artificial triplexes has been shown to block transcription. This suggests that H-DNA does have some real biological function, although what this is remains obscure.

8. Packaging DNA in Eukaryotic Nuclei

Understanding the structure of the eukaryotic nucleus has become a very hot topic in molecular biology research. The structure of the nucleus at interphase was originally thought to be an amorphous mess of chromatin. With the advent of fluorescence *in situ* hybridization (FISH) (see Ch. 5), now it is known that some organisms keep their different chromosomes in distinct territories within the interphase nucleus (Fig. 4.24).

The DNA at interphase of the cell cycle has the loosest conformation, yet the chromosomes are still quite compact. Eukaryotic chromosomes may be as much as a centimeter long and must be folded up to fit into the cell nucleus, which is five microns across. The folding requires that the DNA be compacted 2,000-fold. However, eukaryotic chromosomes are not circular, and instead of supercoiling using DNA gyrase, the mechanism of packaging involves several levels of twisting and protein interactions.

Hoogsteen base pair   A type of non-standard base pair found in triplex DNA, in which a pyrimidine is bound sideways on to a purine
potential intrastrand triplex (PIT)   Stretch of DNA that might be expected from its sequence to form H-type triplex DNA

In the first level of compaction, DNA is wound around special proteins called the **histones**. Eight histones comprise the core unit: two of H2A, two of H2B, two of H3, and two of H4. The DNA wraps around the ball of histones two times, and this structure is called a **nucleosome** (Fig. 4.25). The entire length of nucleosomes is called **chromatin** and because of its appearance, it is also referred to as "beads on a string" (Fig. 4.26). In between each nucleosome is a short span of free DNA, which is partially protected by the ninth histone, H1. The exposed DNA can be cut with nucleases specific for dsDNA and which make double-stranded cuts. In the laboratory, micrococcal nuclease is often used to perform this job. This enzyme cuts the linker DNA in three stages. First, single nucleosomes with 200 bp of DNA are released, and then the linker region is cut off, leaving about 165 bp. Finally, the ends of the DNA wound around the core are nibbled away, leaving about 146 bp that are fully protected by the core particle from further digestion.

The core histones, H2A, H2B, H3, and H4, are small, roughly spherical proteins with 102 to 135 amino acids. However, the linker histone, H1, is longer, having about 220 amino acids. H1 has two arms extending from its central spherical domain. The central part of H1 binds to its own nucleosome and the two arms bind to the nucleosomes on either side (Fig. 4.27). The core histones have a body of about 80 amino acids and a tail of 20 amino acids at the N-terminal end. This tail contains several lysine residues that may have acetyl groups added or removed (Fig. 4.28). This is thought to partly control the state of DNA packaging and hence of gene expression. Thus, in active chromatin, the core histones are highly acetylated (see Ch. 17 for further discussion). In addition to histone modification, proteins like Heterochromatin protein 1 (HP1), Polycomb group (PcG) proteins, and chromatin remodeling complexes modulate the level of nucleosome packaging. During both replication and transcription, the histones and other chromatin remodeling enzymes are temporarily displaced from short regions of the DNA. After the synthetic enzymes have passed by, the histone cores reassemble on the DNA.

In the second level of compaction, the chain of nucleosomes is wound into a giant helical structure with six nucleosomes per turn known as the **30 nanometer fiber** (Fig. 4.26). The nucleosomes may form a tubular solenoid shape, or they may zigzag back and forth. The ability to discern these two alternate forms is limited

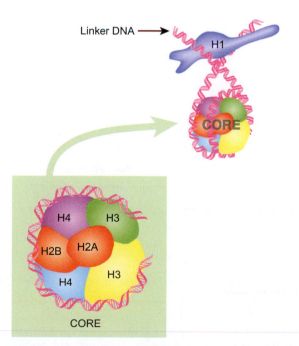

Linker DNA ⟶

H1

CORE

H4 H3

H2B H2A

H4 H3

CORE

FIGURE 4.25
Nucleosomes and Histones

The basic unit in the folding of eukaryotic DNA is the nucleosome as shown here. A nucleosome is composed of eight histones comprising a core and one separate histone (H1) at the site where the wrapped DNA diverges. The enlarged region shows the packing of histones in the core. The H3-H4 tetramer dictates the shape of the core. Only one of the H2A and H2B dimers is shown; the other is on the other side, hidden from view.

---

chromatin Complex of DNA plus protein which constitutes eukaryotic chromosomes
histone Special positively-charged protein that binds to DNA and helps to maintain the structure of chromosomes in eukaryotes
nucleosome Subunit of a eukaryotic chromosome consisting of DNA coiled around histone proteins
30 nanometer fiber Chain of nucleosomes that is arranged helically, approximately 30 nm in diameter

TABLE 4.04	Summary of Chromosome Folding	
Level of Folding	**Consists of**	**Base Pairs per Turn**
DNA double helix	nucleotides	10
Nucleosomes	200 base pairs each	100
30 nanometer fiber	6 nucleosomes per turn	1,200
Loops	50 helical turns per loop	60,000
Chromatid	2,000 loops	

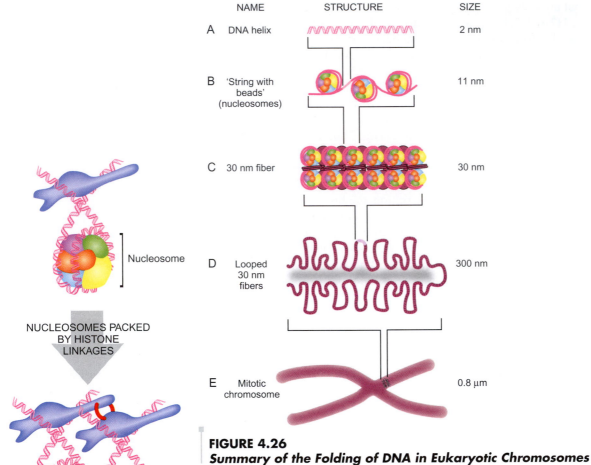

FIGURE 4.26
Summary of the Folding of DNA in Eukaryotic Chromosomes

The DNA helix (A) is wrapped around (B) eight histones (the core). The linker DNA regions unite the nucleosomes to give a "string with beads." This in turn is coiled helically (C) (not clearly indicated) to form a 30 nm fiber. The 30 nm fibers are further folded by looping and attachment to a protein scaffold. Finally, during mitosis the DNA is folded yet again to yield very thick chromosomes.

FIGURE 4.27
Histone H1 Links Nucleosomes

The positioning of H1 (blue) above the DNA wrapped around the core particles allows one H1 to bind to another along a linear chain of nucleosomes. This helps in the tighter packing of the nucleosomes.

because the ionic conditions of the DNA may alter the actual structure present inside the interphase nucleus. In the third level of compaction, the 30 nm fibers loop back and forth. The loops vary in size, averaging about 50 of the helical turns (i.e., about 300 nucleosomes) per loop. The ends of the loops are attached to a protein scaffold, or chromosome axis. **Heterochromatin** forms when these loops are highly condensed (Fig. 4.29). The rest of the chromatin, the **euchromatin**, is in the more

euchromatin   Normal chromatin, as opposed to heterochromatin
heterochromatin   A highly-condensed form of chromatin that cannot be transcribed because it cannot be accessed by RNA polymerase

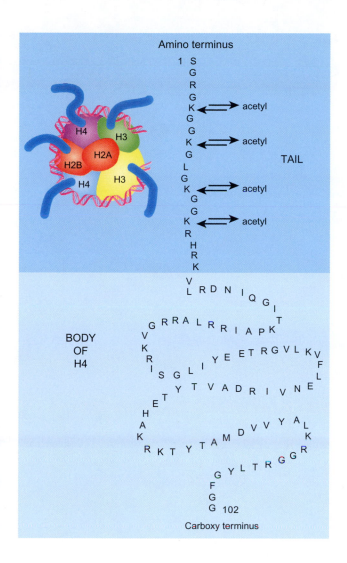

■**FIGURE 4.28**
Histone Tails May Be Acetylated

The N-terminal domains of some of the histone proteins are free for acetylation as indicated by "acetyl." The single letter system for naming amino acids is used.

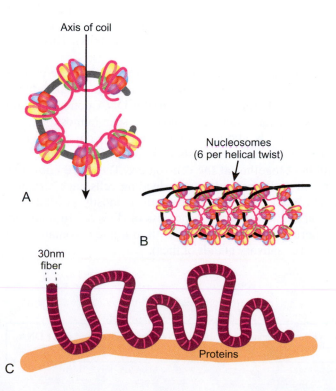

■**FIGURE 4.29**
Looping of 30 Nanometer Fiber on Chromosome Axis

A) A chain of nucleosomes is coiled further with six nucleosomes forming each turn. B) The coiled nucleosomes form a helix, known as a 30 nm fiber. C) The 30 nm fibers form loops that are periodically anchored to protein scaffolding.

FIGURE 4.30
Interphase and Metaphase Chromosomes

Between rounds of cell division, chromosomes consist of single chromatids and are referred to as interphase chromosomes. Before the next cell division, the DNA is replicated and each chromosome consists of two DNA molecules or chromatids linked at the centromere. Just prior to mitosis, condensation occurs, making the chromosomes (and chromatids) visible. The chromosomes are best viewed while spread out during the middle part (metaphase) of mitosis. Each daughter cell will acquire one of the chromatids and the process begins anew.

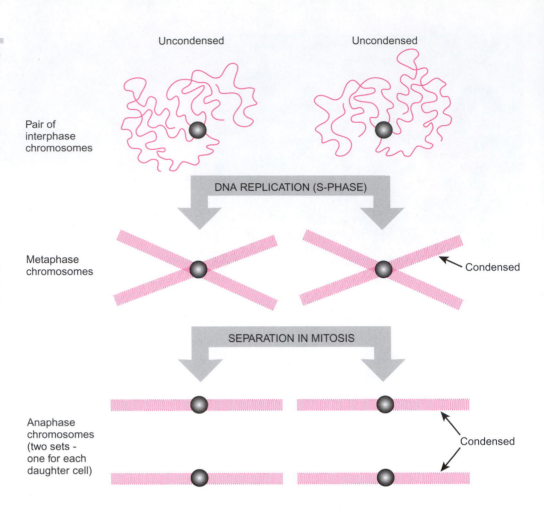

Eukaryotes package their DNA by winding it around specialized proteins, known as histones.

A nucleosome contains around 200 bps of DNA wrapped around a ball of eight histone proteins.

Eukaryotic DNA is so long that it needs several successive levels of folding to fit into the nucleus.

extended form shown as a string of beads in panels B and C of Figure 4.26. About 10% of this euchromatin is even less condensed and is either being transcribed or is accessible for transcription in the near future (see Ch. 17 for details). This is the "active chromatin."

Further folding of chromosomes occurs in preparation for cell division. The precise nature of this is uncertain, but condensed mitotic chromosomes are 50,000 times shorter than fully extended DNA. Just before cell division, the DNA condenses and folds up, as described above. The typical metaphase chromosome, seen in most pictures, has replicated its DNA previously, and is about to divide into two daughter chromosomes, as shown in Figure 4.30. It therefore consists of two identical double helical DNA molecules that are still held together at the centromere. These are known as **chromatids**. Note that between cell divisions and in non-dividing cells each chromosome consists of only a single chromatid. The term chromatid is mostly needed only to avoid ambiguity when describing chromosomes in process of division. In addition, there are a few unusual cases where giant chromosomes with multiple chromatids are found in certain organisms (e.g., in the salivary glands of flies).

chromatid Single double-helical DNA molecule making up whole or half of a chromosome; a chromatid also contains histones and other DNA-associated proteins

Khochbin S (2011) In the heart of a dynamic chromatin. Chem & Biol 18: 410-412.

This associated preview article discusses how two new probes are being used to ascertain whether or not histone H4 is acetylated on lysine 5 (K5), lysine 8 (K8), and lysine 12 (K12). The group of Minoru Yoshida created a genetic fusion of the bromodomain gene, Brdt, and the gene for histone H4 connected by a linker region for flexibility. When expressed, the histone amino terminal domain has an extension that forms four alpha-helices bundled together called a bromodomain. This structure is typically found in proteins that bind acetylated histone tails. This fusion was then connected to two fluorescent proteins (Venus and cyan-fluorescent protein (CFP)) on each side of the fusion protein. The two fluorescent proteins will undergo FRET (or fluorescence resonance energy transfer) when the fusion protein changes shape. So, for example, if the bromodomain flips to bind to the acetylated K5 and K8, then the fluorescent signal changes and can be recorded by changes in emitted light.

The authors describe two different probes with this basic structure that differ in the bromodomain protein. The first probe, Histac, used a bromodomain protein that recognizes double acetylation of K5 and K8, whereas the second probe, Histac-K12, has a bromodomain protein that recognizes acetylation at K12. These probes allow scientists to monitor the amount of histone H4 acetylation by live cell imaging under a variety of circumstances. These probes are useful tools to ascertain the effects of acetylation on chromosome compaction into chromatin and how this changes through cell division. The probes also elucidate the efficacy of new drugs

FOCUS ON RELEVANT RESEARCH

that alter DNA acetylation or deacetylation. The continued development and use of such probes will be useful to determine the landscape of histone modifications in all parts of the genome and hopefully lead to a greater understanding of how such molecular changes affect DNA compaction and gene expression.

Box 4.04: Histones are highly conserved and originated among the archaebacteria

Of all known proteins, the eukaryotic core histones, especially H3 and H4, are the most highly conserved during evolution. For example, only two out of 102 amino acids are different between the H4 sequence of cows and peas. The linker histone, H1, is more variable in composition.

Typical bacteria (i.e., the eubacteria) do not possess histones. (Large numbers of "histone-like proteins" are found bound to bacterial chromosomes. Despite the name, these are not homologous in sequence to true histones nor do they form nucleosomes for packaging DNA.) However, some members of the genetically-distinct lineage of archaebacteria (e.g., the methane bacteria) do possess histones. Archaeal histones vary significantly from each other. They are 65–70 amino acids long and are missing the tails characteristic of eukaryotic histones. Archaeal nucleosomes accommodate a little under 80 bp of DNA and contain a tetramer of the archaeal histone. They are probably homologous to the $(H3-H4)_2$ tetramers found in the core of the eukaryotic nucleosome.

Key Concepts

- Bacterial genomes are mostly circular with densely packed genes, some of which are grouped into operons (i.e., where multiple genes are controlled by one promoter).
- Viral genomes are smaller than most bacterial genomes and often are missing key genes for survival since they can rely on their host cell to provide these gene products.
- Organelle genomes are circular and only have some of the genes necessary for their function within the cell. Often, host encoded proteins are used in combination with their proteins to function effectively. Since many organellar

genes are similar to bacteria, it is thought that organelles are degenerate symbiotic organisms.

- Eukaryotic genomes are large and contain much more intervening or noncoding DNA than bacterial, viral, or organellar genomes. Intervening or intron DNA interrupts the coding region of most eukaryotic genes.
- A consensus sequence is determined by comparing sets of related sequences and determining which nucleotide or amino acid is most prevalent at that particular position.
- Eukaryotic DNA also contains a variety of repeat elements from pseudogenes, which are repeated only a few times, to moderately repetitive sequences, which are multiple copies of highly used genes or LINEs, to highly repetitive DNA, which includes SINEs. VNTRs are tandem repeats that are variable in number. This satellite DNA has a different density than the remaining genome and can induce unequal crossing over during meiosis. Human centromeres consist of 171 bp repeats called alpha DNA.
- The structure of the DNA helix is affected by the order of nucleotides. Inverted repeats of nucleotides can create stem-loop structures. When two longer inverted repeats are found in supercoiled DNA, the area can form into a cruciform structure. Repeats of adenine from 3-5 nucleotides in length induce the DNA helix to bend.
- Electrophoresis separates DNA fragments by size, but the actual structure of the DNA also affects movement though the gel. Multiple A-tracts cause DNA to move more slowly than linear DNA without multiple adenines. Supercoiling causes the DNA to move more quickly than its open circle or a linear form since it is more compact.
- Bacterial DNA is compacted by supercoiling and binding to a DNA scaffold. The compacted DNA is kept within the nucleoid region where specific genes are found in specific locations. During bacterial cell division, special proteins (topoisomerases and gyrase) unwind and rewind the supercoils so that the DNA can be replicated. Shortly after replication initiation, ParA and ParB recognize the *parS* site near the origin and then assemble filaments that move the two chromosomes into each daughter cell.
- DNA helices can exist as the standard Watson and Crick B-form, A-form, Z-form, and a triple helix called H-DNA.
- Eukaryotic DNA is compacted into "beads on a string" by wrapping the DNA around a core of proteins called histones. The beads or nucleosomes are then compacted into a 30 nm fiber by forming a solenoid or zigzag conformation. This form loops from a protein scaffold and then condenses even further during mitosis.
- The amino terminus of histones juts from the core proteins and can be modified by adding acetyl groups, methyl groups, or phosphate groups. These alterations affect the DNA structure and whether or not a particular gene is expressed.

Review Questions

1. What was the evidence for Watson and Crick's structural model of DNA?
2. How can parasitic bacteria survive with a very small genome?
3. What is non-coding DNA? Name and describe two types of non-coding DNA in eukaryotes.
4. What is the difference between intergenic DNA and introns?
5. What are exons?

6. What are repeated sequences and where are they commonly found?
7. Name the two genes that are usually found in multiple copies in a prokaryotic cell.
8. What are the two main types of repetitive sequences in mammalian DNA? How are they formed?
9. What is the difference between SINEs and tandem repeats?
10. What is the importance of VNTRs?
11. What are the two types of palindromes found in DNA?
12. What are inverted repeats in DNA and why are they important?
13. How are bent DNAs formed? How can they be separated from unbent DNAs?
14. What is the importance of supercoiling?
15. Why is negative supercoiling preferred over positive supercoiling?
16. Name the enzyme that negatively supercoils bacterial chromosomes.
17. What is a linking number (L)?
18. What are topoisomerases? What is the difference between type I and type II topoisomerases?
19. Name two antibiotics that kill bacteria by affecting DNA gyrase. Describe how each one works.
20. What are catenanes? How are they unlinked?
21. When are positive supercoils created and how are they removed?
22. Which DNA moves faster during electrophoresis, supercoiled or open circular DNA, and why?
23. What are the alternate structures formed due to strain caused by supercoiling and how do they differ?
24. What is the most stable and common form of the DNA double helix?
25. What are the two alternate and rare forms of the DNA double helix and what are the conditions for their formation?
26. How is H-DNA formed and what are Hoogsteen base pairs?
27. How are the DNA molecules packaged in eukaryotes?
28. Define chromatin and nucleosome.
29. Name the core histones and the linker histone and describe how they are arranged in a nucleosome.
30. What are the two types of chromatin and how do they differ?
31. Do histones exist in prokaryotes, and if so where are they found?
32. Describe the folding of DNA in eukaryotic chromosomes.
33. What constitutes a 30 nm fiber?
34. What are chromatids?

Conceptual Questions

1. Researchers identified a new gene they called ICOSAPETALS that doubles the number of petals on the purple coneflower, *Echinacea purpurea*. The researchers want to identify other genes from other flowers that might be similar in sequence. Using BLAST sequence alignment, the researchers identified the following region of ICOSAPETALS that is similar to only five other sequenced flower genes. Determine the consensus sequence for this region.

```
ICOSAPETALS   5'AGGCGCCCATTACTGATCCAAATTTGACTCTGG3'
MUMPTLA       5'AGGCACCCTAATGAGATCCAAATTTGACTGACC3'
DAISY556DA    5'AGGCGCAATAATGTCTTCCAAATTTGACTGTCC3'
PETUNPETL     5'AGGCGCAATATCGTGTTCCAAATTTGACTGTGG3'
VINCANOPE     5'AGGCCCTTAAACGTGTTCCAAATTTGACTGTGG3'
SHASTAPET     5'AGGCGCCCTTACGTCTTCCAAATTTGACTCTGG3'
```

2. The entire ICOSAPETALS gene was cloned and sequenced. The region of similarity to the other genes was determined to be within an exon. Using your knowledge of gene structure, propose an explanation of why this region of ICOSAPETALS was similar to other genes.

3. Upstream of the gene for ICOSAPETALS is a region of tandem repeats. When the purple coneflower has 20 or more petals, there are over 50 tandem repeats upstream of ICOSAPETALS. When the coneflower has 10 or fewer petals, there are less than 30 repeats. Explain why the number of repeats upstream of ICOSAPETALS is different from plant to plant. Propose a mechanism for how ICOSAPETALS controls the number of petals on the flower.

4.   In the laboratory, a brand new single-celled eukaryotic organism was finally able to be grown in large quantities. The researchers were interested in sequencing the genome to determine if this new life form was similar to other single-celled eukaryotes. The researchers roughly purified the genomic DNA and decided to size fractionate the DNA using density gradient centrifugation. They got the following result when assessing the amount of DNA in each fraction from cesium chloride gradient. Describe the two peaks and explain what types of DNA are found in each.

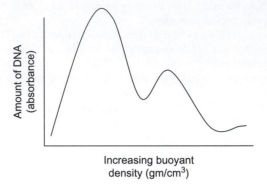

Manipulation of Nucleic Acids

Chapter 5

This chapter surveys several of the fundamental techniques that are used to manipulate DNA and RNA in molecular biology and biotechnology. One of the techniques that allowed the molecular cloning of DNA and hence initiated the whole modern realm of DNA technology is the use of restriction enzymes to cut DNA into shorter segments. Small segments can then be purified, analyzed, and rejoined in a variety of configurations. Other basic methods include measuring the amount of DNA or RNA by absorption of UV radiation and the use of radioactive or, more recently, fluorescent labels to monitor the presence of nucleic acids. Another technological breakthrough is the creation of artificial segments of DNA and RNA. This discovery is essential for techniques such as sequencing the human genome, assessing the origin of a piece of DNA via DNA fingerprinting, and diagnosing various infections fast and cheap via the polymerase chain reaction (PCR). Here, we describe the methods for the synthesis of both natural nucleic acids and some useful chemical derivatives. Finally, the use of nucleic acid hybridization in a variety of situations is introduced to facilitate the further understanding of more modern molecular biology techniques.

1. Manipulating DNA

Basic research on model organisms has provided new insights on how our own bodies function, provided insights into how we and other organisms have adapted to live in our environment, and has launched many new technologies that are essential to our economy. The first example in this chapter is the discovery of restriction enzymes and other nucleases in bacteria. The ability to isolate DNA from an organism and then cut that DNA into small pieces is essential for finding new genes, especially in the days before sequencing whole genomes. In particular, the discovery of restriction enzymes that only recognize specific nucleotide sequences has been essential for mapping the location of the various fragments in reference to the other fragments. Being able to order the fragments of DNA was an essential discovery for sequencing the human genome, and is still used for organisms in which the entire genome has not been sequenced.

1.1. Restriction and Modification of DNA

Nucleases are enzymes that degrade nucleic acids. These are categorized as **ribonucleases** (RNases) that attack RNA and **deoxyribonucleases** (DNases) that attack DNA. Most nucleases are specific, though the degree of specificity varies greatly. Some nucleases will only attack single-stranded nucleic acids, others will only attack double-stranded nucleic acids, and a few will attack either kind. **Exonucleases** attack at the end of nucleic acid molecules and usually remove just a single nucleotide, or sometimes a short piece of single-stranded DNA. Exonucleases can attack either the 3′-end or the 5′-end but not both. **Endonucleases** cleave the nucleic acid chain in the middle. Some endonucleases are non-specific; others, in particular the restriction enzymes, are extremely specific and will only cut DNA after binding to specific recognition sequences.

In nature, foreign DNA entering a bacterial cell is most likely due to virus infection, and natural defense systems have evolved to prevent the infection. When viruses attack bacteria, the virus coat is left outside and only the virus DNA enters the target cell (see Ch. 1). The viral DNA will take over the victim's cellular machinery and use it to manufacture more virus particles unless the bacteria fight back. Bacteria produce **restriction enzymes** to destroy foreign DNA. These are endonucleases that recognize a specific sequence of four to eight nucleotides in length on foreign DNA and cut both phosphate backbones to break the single piece into two. This sequence of bases on DNA is called a **recognition site**.

Restriction enzymes must recognize the bacterial DNA because the key to successful defense is to degrade the foreign DNA without endangering the bacterial cell's own DNA. Consequently, bacteria need a mechanism that distinguishes their DNA from foreign DNA. To protect their own DNA, **modification enzymes** recognize the same recognition site and transfer a methyl group onto the DNA. Modification enzymes usually add the methyl group to adenine or cytosine within the recognition site, and this group prevents the corresponding restriction enzyme from recognizing the site (Fig. 5.01). Therefore, the bacterial DNA, either chromosomal or plasmid, is immune to that cell's own restriction enzymes. In contrast, incoming, unmodified DNA will be degraded by the restriction enzyme.

Endonucleases and exonucleases cut DNA or RNA in the middle or remove single nucleotides from the ends, respectively.

Restriction enzymes are endonucleases that cut DNA at specific nucleotide sequences.

Modification enzymes protect DNA from the corresponding restriction enzymes by adding methyl groups at the recognition site.

---

deoxyribonuclease (DNase)   Enzyme that cuts or degrades DNA
endonuclease   Enzyme that cleaves nucleic acid molecule in the middle
exonuclease   Enzyme that cleaves nucleic acid molecule at the end and usually removes just a single nucleotide
modification enzyme   Enzyme that binds to the DNA at the same recognition site as the corresponding restriction enzyme but methylates the DNA
nuclease   Enzyme that cuts or degrades nucleic acids
recognition site   Sequence of bases on DNA that is recognized by a specific protein, such as a restriction enzyme
restriction enzyme   Type of endonuclease that cuts double-stranded DNA at a specific sequence of bases, the recognition site
ribonuclease (RNase)   Enzyme that cuts or degrades RNA

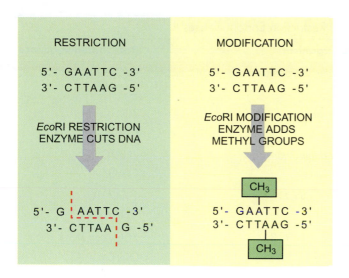

► **FIGURE 5.01**
Restriction and Modification Systems

Restriction enzymes recognize non-methylated double-stranded DNA and cut it at specific recognition sites. For example, *Eco*RI recognizes the sequence 5'-GAATTC-3', and cuts after the G. Since this sequence is an inverted repeat, the enzyme also cuts the other strand after the corresponding G, giving a staggered cut. Modification enzymes are paired with restriction enzymes and recognize the same sequence. Modification enzymes methylate the recognition sequence, which prevents the restriction enzyme from cutting it.

1.2. Recognition of DNA by Restriction Endonucleases

Due to their ability to recognize specific sites in DNA, restriction endonucleases have become one of the most widely used tools in genetic engineering. Restriction enzyme recognition sites are usually four, six, or eight bases long and the sequence forms an inverted repeat. Thus, the sequence on the top strand of the DNA is the same as the sequence of the bottom strand read in the reverse direction (as shown in Fig. 5.01 above).

Several hundred different restriction enzymes are now known and each has its own specific recognition site. Some recognition sites require a specific base at each position. Others are less specific and may require only a purine or a pyrimidine at a particular position. Some examples are shown in Table 5.01.

Since any random series of four bases will be found quite frequently, four base-recognizing enzymes cut DNA into many short fragments. Conversely, since any particular eight-base sequence is less likely, the eight-base-recognizing enzymes cut DNA only at longer intervals and generate fewer larger pieces. The six-base enzymes are the most convenient in practice, as they give an intermediate result.

> Most recognition sites for restriction enzymes are inverted repeats of 4, 6, or 8 bases.

1.3. Naming of Restriction Enzymes

Restriction enzymes have names derived from the initials of the bacteria they come from. The first letter of the genus name is capitalized and followed by the first two letters of the species name (consequently, these three letters are in italics). The strain is sometimes represented (e.g., the R in *Eco*RI refers to *E. coli* strain RY13). The roman letter indicates the number of restriction enzymes found in the same species. For example, *Moraxella bovis* has two different restriction enzymes called *Mbo*I and *Mbo*II. Some examples are shown in Table 5.01.

If two restriction enzymes from different species share the same recognition sequence they are known as **isoschizomers**. Note that isoschizomers may not always cut in the same place even though they bind the same base sequence. For example, the sequence GGCGCC is recognized by four enzymes, each of which cuts in different places: *Nar*I (GG/CGCC), *Bbe*I (GGCGC/C), *Ehe*I (GGC/GCC), and *Kas*I (G/GCGCC).

> Isoschizomers are two restriction enzymes that share the same recognition sequence although they do not necessarily cut at precisely the same place.

isoschizomers Restriction enzymes from different species that share the same recognition sequence

TABLE 5.01	Examples of Restriction Endonucleases	
Enzyme	**Source Organism**	**Recognition Sequence**
HpaII	Haemophilus parainfluenzae	C/CGG GGC/C
MboI	Moraxella bovis	/GATC CTAG/
NdeII	Neisseria denitrificans	/GATC CTAG/
EcoRI	Escherichia coli RY13	G/AATTC CTTAA/G
EcoRII	Escherichia coli RY13	/CC(A or T)GG GG(T or A)CC/
EcoRV	Escherichia coli J62/pGL74	GAT/ATC CTA/TAG
BamHI	Bacillus amyloliquefaciens	G/GATCC CCTAG/G
SauI	Staphylococcus aureus	CC/TNAGG GGANT/CC
BglI	Bacillus globigii	GCCNNNN/NGGC CGGN/NNNNCCG
NotI	Nocardia otitidis-caviarum	GC/GGCCGC CGCCGG/CG
DraII	Deinococcus radiophilus	RG/GNCCY YCCNG/GR

/ = position where enzyme cuts   N = any base, R = any purine, Y = any pyrimidine

1.4. Cutting of DNA by Restriction Enzymes

It might seem logical for the DNA to be cut at the recognition site where the restriction enzyme binds. This is often true, but not always. There are two major classes of restriction enzyme that differ in where they cut the DNA, relative to the recognition site.

Type I restriction enzymes cut the DNA a thousand or more base pairs away from the recognition site. This is done by looping the DNA around so that the enzyme binds both at the recognition site and the cutting site (Fig. 5.02). Type I restriction systems consist of a single protein with three different subunits. One subunit recognizes the DNA, another subunit methylates the recognition sequence, and the third subunit cuts the DNA at a distance from the recognition sequence. Since the exact length of the loop is not constant, and since the base sequence at the cut site is not fixed, these enzymes are of little practical use to molecular biologists. Even more bizarre is that type I restriction enzymes are suicidal. Most enzymes carry out the same reaction over and over again on a continual stream of target molecules. In contrast, each molecule of a type I restriction enzyme can cut DNA only a single time and then it is inactivated!

Type II restriction enzymes cut the DNA in the middle of the recognition site. Since the exact position of the cut is known, these are the restriction enzymes that

Type I restriction enzymes cut the DNA more than 1000 base pairs from the recognition sequence.

Type II restriction enzymes cut DNA within the recognition sequence. Some generate blunt ends, others create sticky ends.

type I restriction enzyme   Type of restriction enzyme that cuts the DNA a thousand or more base pairs away from the recognition site
type II restriction enzyme   Type of restriction enzyme that cuts the DNA in the middle of the recognition site

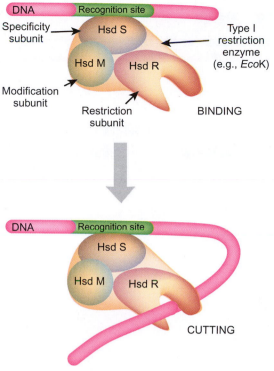

■ **FIGURE 5.02**
Type I Restriction Enzyme

Type I restriction enzymes have three different subunits. The specificity subunit recognizes a specific sequence in the DNA molecule. The modification subunit adds a methyl group to the recognition site. If the DNA is non-methylated, the restriction subunit cuts the DNA, but at a different site, usually over 1000 base pairs away. In the *Eco*K restriction enzyme, the subunits are HsdS, HsdM, and HsdR.

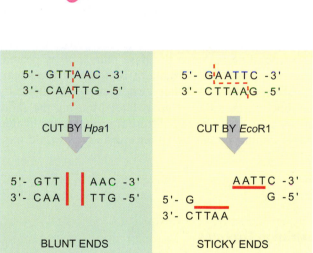

■ **FIGURE 5.03**
Type II Restriction Enzymes—Blunt Versus Sticky Ends

*Hpa*I is a blunt end restriction enzyme, that is, it cuts both strands of DNA in exactly the same position. *Eco*RI is a sticky end restriction enzyme. The enzyme cuts between the G and A on both strands, which generates a four base pair overhang on the ends of the DNA. Since these bases are free to base pair with any complementary sequence, they are considered "sticky."

are normally used in genetic engineering. There are two different ways of cutting the recognition site in half. One way is to cut both strands of the double-stranded DNA at the same point. This leaves **blunt ends** as shown in Figure 5.03. The alternative is to cut the two strands in different places, which generates overhanging ends. The ends made by such a staggered cut will base pair with each other and consequently are known as **sticky ends**. In type II restriction systems the restriction endonuclease only cuts DNA unlike Type I restriction enzymes that have multiple enzymatic functions. The corresponding modification enzymes are completely separate.

Enzymes that generate sticky ends are the most useful. If two different pieces of DNA are cut with the same restriction enzyme or even with different enzymes that

Sticky ends are more convenient than blunt ends when joining together fragments of DNA using DNA ligase.

blunt ends   Ends of a double-stranded DNA molecule that are fully base paired and have no unpaired single-stranded overhang
sticky ends   Ends of a double-stranded DNA molecule that have unpaired single-stranded overhangs, generated by a staggered cut

FIGURE 5.04
Matching of Compatible Sticky Ends

*Bam*HI and *Bgl*II generate the same overhanging or sticky ends. *Bam*HI recognizes the sequence 5′-GGATCC-3′ and cuts after the first 5′ G, which generates the 5′-GATC-3′ overhang on the bottom strand. *Bgl*II recognizes the sequence 5′-AGATCT-3′ and cuts after the first 5′ A, which generates a 5′-GATC-3′ overhang on the top strand. If these two pieces are allowed to anneal, the complementary sequences will hydrogen bond together, allowing the nicks to be sealed more easily by DNA ligase.

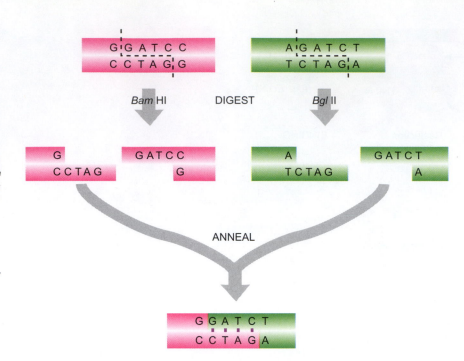

generate the same overhang, the same sticky ends are generated. This allows fragments of DNA from two different original DNA molecules to be bound together by matching the sticky ends. Such pairing is temporary since the pieces of DNA are only held together by hydrogen bonding between the base pairs, not by permanent covalent bonds. Nonetheless, this assists the permanent bonding of the sugar-phosphate backbone by DNA ligase (see below). When two sticky ends made by the same enzyme are ligated, the junction may be cut apart later by using the same enzyme again. However, if two sticky ends made by two different enzymes are ligated together, a hybrid site is formed that cannot be cut by either enzyme (as would happen with *Bam*HI and *Bgl*II in Fig. 5.04).

Box 5.01 Star Activity of Restriction Enzymes

Restriction enzymes recognize and cut at specific nucleotide sequences. Under certain reaction conditions, restriction enzymes exhibit **star activity**, which means they begin cutting DNA at nucleotide sequences other than the exact recognition sites. In addition, the restriction enzyme can also create single nucleotide substitutions in the DNA or place random cuts into the phosphate backbone of one DNA strand. Some of the conditions that can cause star activity include too much glycerol in the reaction, incorrect ion concentrations, and contaminants such as organic solvents. In addition, star activity occurs by simply having too much enzyme for the amount of DNA or simply letting the reaction incubate for too long. Although these activities do not occur in bacteria that are defending themselves against a virus, the altered activity occurs in the laboratory. Most restriction enzymes are sold with a specific buffer to include in the reaction, which provides the optimal conditions for the enzyme. As long as this buffer is used with a sufficient amount of DNA, star activity is also avoided in the laboratory.

---

star activity   Imprecise or random cleavage of DNA by restriction enzymes that only occurs under specific reaction conditions

Pingoud A and Wende W (2007) A sliding restriction enzyme pauses. Structure 15: 391–393.

The reaction mechanism for a restriction enzyme recognizing its target site and cleaving the backbone is based on the actual structure of the enzyme/DNA complex. If restriction enzymes simply diffused to the target DNA, assayed the sequence for an exact match to its recognition site, and released the DNA when the sequence did not match, it would take a long time before the target DNA was actually cut. This reaction mechanism would be too slow, and the invading viral DNA would be able to initiate transcription and translation before the enzyme could protect the bacteria. Restriction enzymes must work faster and more efficiently. One way to increase the speed of finding the recognition site is called facilitated diffusion, where the enzyme first homes in on the target DNA, and then finds the recognition sequence. Restriction enzymes could do this in one of three different ways. They could slide along the DNA strand without letting go. The enzymes could hop or jump from one DNA sequence to another, as long as the two sites are near each other. Finally, the enzyme could have two or more DNA binding domains so it can assay more than one DNA sequence at a time.

Previous work in the 1980s and 1990s has identified that restriction enzyme *Eco*RI uses the first mechanism. The enzyme slides along the DNA, and whenever it finds a sequence that closely resembles the recognition site, it pauses. If the site is an exact match, the enzyme clamps onto the DNA tightly and cuts the backbone of both strands. There are considerable changes in the protein's 3D shape and the DNA shape during the transition from loosely associated with DNA to tightly associated with

DNA. These conformational changes have been studied and include the expulsion of water molecules from in between the enzyme and DNA. In contrast, how the enzyme pauses at sites that are close to the recognition site is still unknown. The prevailing theory is that the nucleotide bases that are exactly like the recognition site contact the enzyme, but the incorrect nucleotides do not fit. This causes the pause, but since the fit isn't exact, the enzyme releases and moves on. This review summarizes the results of a recent study that supports this theory. The paper crystallized the restriction enzyme *Bst*YI with DNA that had a recognition site that differed from the exact recognition sequence by one nucleotide. *Bst*YI is a homodimer that recognizes DNA, scans along the helix, and cuts at 5'-RGATCY-3', where R is a purine and Y is any pyrimidine. This crystal structure demonstrated that the half of the recognition sequence that was a perfect match bound tightly to the homodimer of *Bst*YI, whereas the mismatch half of

FOCUS ON RELEVANT RESEARCH

the DNA was loosely associated with the restriction enzyme. The structure supports the idea that the entire recognition site needs to match the restriction enzyme perfectly in order for the DNA to attach tightly. When the DNA does not fit exactly, then the restriction enzyme does not change its 3D shape properly and the DNA is not cut.

1.5. DNA Fragments Are Joined by DNA Ligase

Although sticky ends bring two cut pieces together, the bond is temporary. In order to keep the pieces together, the enzyme **DNA ligase** is used. DNA ligase operates during DNA replication where it covalently links the fragments of the lagging strand (see Ch. 10). If DNA ligase finds two DNA fragments touching each other end to end, it will ligate them together (Fig. 5.05). In practice, segments of DNA with matching sticky ends will tend to stay attached much of the time and consequently DNA ligase will join them efficiently. Since DNA fragments with blunt ends have no way to bind each other, they drift apart most of the time. Ligating blunt ends is very slow and requires a high concentration of DNA ligase as well as a high concentration of DNA. In fact, bacterial ligase cannot join blunt ends at all. In practice, **T4 ligase** is normally used in genetic engineering as it can join blunt ends if need be. T4 ligase originally came from bacteriophage T4, although nowadays it is manufactured by expressing the gene that encodes it in *E. coli*.

1.6. Making a Restriction Map

A diagram that shows the location of restriction enzyme cut sites on a segment of DNA is known as a **restriction map**. The first step in generating such a map is to digest the DNA with a series of restriction enzymes, one at a time. The fragments of digested DNA are separated by agarose gel electrophoresis (as described in Ch. 4).

> A restriction map is a diagram showing the location of cut sites on DNA for a variety of restriction enzymes.

DNA ligase   Enzyme that joins DNA fragments covalently, end to end
restriction map   A diagram showing the location of restriction enzyme cut sites on a segment of DNA
T4 ligase   Type of DNA ligase from bacteriophage T4 and which is capable of ligating blunt ends

FIGURE 5.05
DNA Ligase Joins Fragments of DNA

T4 DNA ligase connects the sugar-phosphate backbone of two pieces of DNA. In the example, overlapping sticky ends connect a double-stranded piece of DNA, but the backbone of each strand has not been connected. T4 DNA ligase recognizes these nicks or breaks in the backbone and uses energy from the hydrolysis of ATP to drive the ligation reaction.

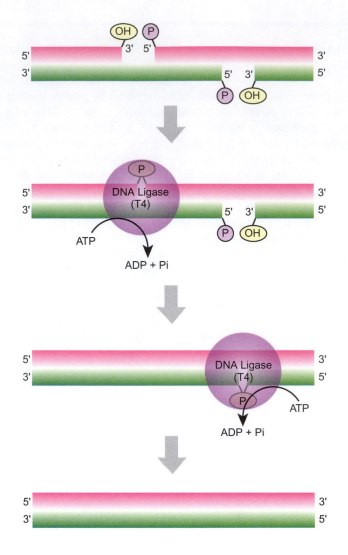

Restriction maps are deduced by cutting the target DNA with a selection of restriction enzymes, both alone and in pairs.

Comparison with appropriate standards allows the sizes of the fragments to be estimated. This reveals how many recognition sites each enzyme has in the DNA and their distances apart. What remains unknown is the relative order of the fragments.

For example, suppose we start with a 5,000 base pair (bp) piece of DNA that is cut twice by the restriction enzyme *Bam*HI giving three fragments of 3,000 bp, 1,500 bp, and 500 bp. There are three alternative arrangements for three fragments (Fig. 5.06A). You might think there should be six possible arrangements, but the other three theoretical arrangements are merely the first three, turned back to front. They are not genuinely different physical molecules. Fig. 5.06A shows the backward arrangement just for fragment number III.

To decide which of the three possible arrangements are correct, double digests using two restriction enzymes are performed. The DNA is cut with both enzymes simultaneously (Fig. 5.06B). The results of an example gel electrophoresis result are shown for two single digests and a double digest. When the DNA fragment from Fig. 5.06A is cut with *Eco*RI alone, it cuts just once to give two fragments of 4,000 bp and 1,000 bp. Thus, for *Eco*RI alone there is only one possible arrangement. In the double digest, the largest fragment seen in the *Bam*HI lane has disappeared. This means that there is an *Eco*RI cut site within this 3,000 bp *Bam*HI fragment. Since, in this example, there is only one *Eco*RI cut site, only one of the *Bam*HI fragments disappears in the double digest. This allows us to reduce the possibilities to the restriction maps shown in Figure 5.06B.

To refine the restriction enzyme map even more, the same DNA must be digested with a third enzyme. Double digests with *Bam*HI plus enzyme III and of *Eco*RI plus

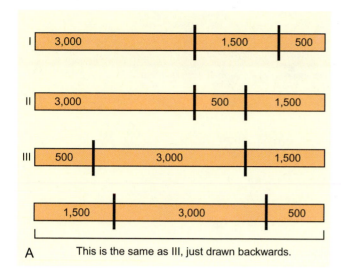

A   This is the same as III, just drawn backwards.

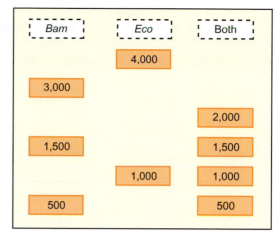

GEL ELECTROPHORESIS OF SINGLE AND
DOUBLE DIGEST

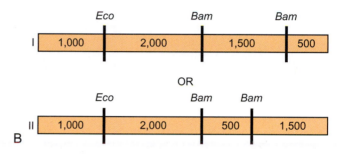

B

FIGURE 5.06
Restriction Mapping

A) To determine the location and number of restriction enzyme sites, a segment of DNA is digested with a restriction enzyme. In this example, the piece of DNA is 5,000 base pairs in length. Cutting with *Bam*HI gave three fragments: 3,000 bp, 1500 bp, and 500 bp. The figure shows the three possible arrangements of these three fragments. The fourth arrangement shown is not really different but is merely the third possible arrangement drawn in the opposite orientation. B) Double digestion is the next step in compiling a restriction map. Cutting the DNA with *Eco*RI alone would give two fragments: 4000 bp and 1000 bp. When the DNA is cut with both *Eco*RI and *Bam*HI simultaneously, four fragments are resolved by gel electrophoresis. Two of these are identical to the 1500 bp and 500 bp fragments from the *Bam*HI single digest; therefore, no *Eco*RI sites are present within these fragments. The remaining two fragments, 2000 bp and 1000 bp, combine to give the 3000 bp *Bam*HI fragment. Therefore, the single *Eco*RI site must be within the 3000 bp *Bam*HI fragment. Of the three possible arrangements shown in part A), the third arrangement is ruled out (if it was cut with *Eco*RI alone, it could not give two fragments of 4000 bp and 1000 bp).

enzyme III would be analyzed, as above. Eventually, this approach allows the construction of a complete restriction map of any segment of DNA. This may then be used as a guide for further manipulations.

Box 5.02 Gene Disruption by Engineered Insertional Mutagenesis

A variety of techniques have been used to construct mutations utilizing genetic engineering technology. These techniques are usually known as site-directed mutagenesis (see Ch. 6). In particular, mutations that serve to completely inactivate a gene are useful in genetic analysis. So, genes may be deliberately disrupted by the insertion of foreign DNA. To do this, it is first necessary to clone the gene onto some convenient vector such as a bacterial plasmid (see below). To disrupt the gene, a deliberately designed segment of DNA is used. Known as a **gene cassette**, it usually carries with it a gene for resistance to some

Continued

gene cassette   Deliberately designed segment of DNA that is flanked by convenient restriction sites and usually carries a gene for resistance to an antibiotic or some other easily observed characteristic

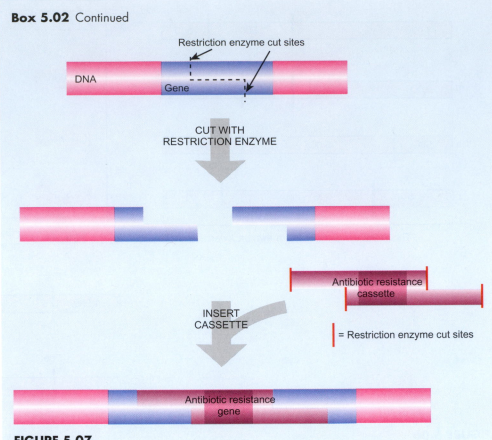

Box 5.02 Continued

Restriction enzyme cut sites

DNA

Gene

CUT WITH
RESTRICTION ENZYME

Antibiotic resistance
cassette

INSERT
CASSETTE

= Restriction enzyme cut sites

Antibiotic resistance
gene

FIGURE 5.07
Gene Disruption Using a Cassette

A gene to be disrupted is cut with a restriction enzyme. An artificially constructed cassette that confers antibiotic resistance is inserted into the cut site and ligated into the gene. The new DNA construct formed can be detected easily since it provides resistance to antibiotics.

antibiotic such as chloramphenicol or kanamycin. This way, the inserted DNA cassette can easily be detected, because cells carrying it will become resistant to the antibiotic. At each end, the cassette has several convenient restriction enzyme sites. The target gene is cut open with one of these restriction enzymes and the cassette is cut from its original location with the same enzyme. The cassette is then ligated into the middle of the target gene (Fig. 5.07). The disrupted gene is then put back into the organism from which it came.

1.7. Restriction Fragment Length Polymorphisms (RFLPs)

Related molecules of DNA, such as different versions of the same gene from two related organisms, normally have very similar sequences. Consequently, they will have similar restriction maps. However, occasional differences in base sequence will result in corresponding differences in restriction sites. Each restriction enzyme recognizes a specific sequence (usually of 4, 6, or 8 bases). If even a single base within this recognition sequence is altered, the enzyme will no longer cut the DNA (Fig. 5.08). Consequently, restriction sites that are present in one version of a sequence may be missing in its close relatives.

If two such related but different DNA molecules are cut with the same restriction enzyme, segments of different lengths are produced. Consequently, a difference

If two related DNA molecules differ in sequence at a restriction enzyme recognition sequence, fragments of different sizes will result after a restriction enzyme digest.

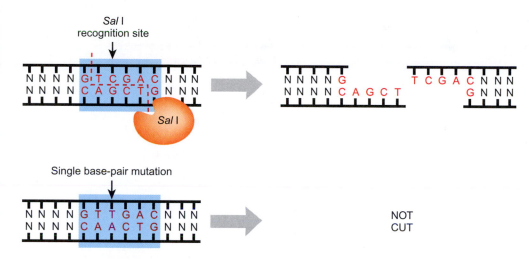

► **FIGURE 5.08**
Single Base Changes Prevent Cutting by Restriction Enzymes

The recognition sequence for a particular restriction enzyme is extremely specific. Changing a single base will prevent recognition and cutting. The example shown is for *Sal*I whose recognition sequence is GTCGAC.

between two DNA sequences that affects a restriction site is known as a **restriction fragment length polymorphism (RFLP)**. When these are separated on a gel, bands of different sizes will appear (Fig. 5.09). RFLPs may be used to identify organisms or analyze relationships even though we do not know the function of the altered gene. In fact, since we are examining the DNA directly, the alteration may be in non-coding DNA or an intervening sequence. It does not need to be in the coding region of a gene. RFLPs are widely used in forensic analysis.

2. Chemical Synthesis of DNA

An alternative to isolating DNA from natural sources is to synthesize it artificially. Molecular biologists routinely use artificially manufactured lengths of DNA for a variety of purposes. Short stretches of single-stranded DNA are used as probes for hybridization (see below), primers in PCR (see Ch. 6), and primers for DNA sequencing (see Ch. 8). Short lengths of double-stranded DNA are made by synthesizing two complementary single strands and allowing them to anneal. Such pieces may be used as linkers or adaptors in genetic engineering (see Ch. 7). It is also possible to synthesize whole genes, although this is more complicated (see below).

> Short to medium lengths of DNA are routinely made by chemical synthesis.

The first step in chemical synthesis of DNA is to anchor the first nucleotide to a solid support. **Controlled pore glass (CPG)** beads that have pores of uniform sizes are most commonly used. The beads are packed into a column and the reagents are poured down the column one after another. Nucleotides are added one by one and the growing strand of DNA remains attached to the glass beads until the synthesis is complete (Fig. 5.10). Chemical synthesis of DNA is performed by an automated machine (Fig. 5.11). After loading the machine with chemicals, the required sequence is typed into the control panel. Gene machines take a few minutes to add each nucleotide and can make pieces of DNA 100 nucleotides or more long. Modern DNA synthesizers are also usually capable of adding fluorescent dyes, biotin, or other groups used in labeling and detection of DNA (see below).

> Chemical synthesis of DNA occurs by attaching nucleosides, one by one, onto a solid support.

Since the DNA is made only with chemical reagents, the process requires some special modifications not necessary if biological enzymes were to manufacture the DNA. The first problem is that each deoxynucleotide has two hydroxyl groups, one used for bonding to the nucleotide ahead of it and the other for bonding to the nucleotide behind it in the nucleic acid chain. Chemical reagents cannot distinguish between these two hydroxyls. Therefore, each time a nucleotide is added, one of its hydroxyl groups must first be chemically blocked and the other must be activated.

controlled pore glass (CPG) Glass with pores of uniform sizes that is used as a solid support for chemical reactions such as artificial DNA synthesis

restriction fragment length polymorphism (RFLP) A difference in restriction enzyme sites between two related DNA molecules that results in production of restriction fragments of different lengths

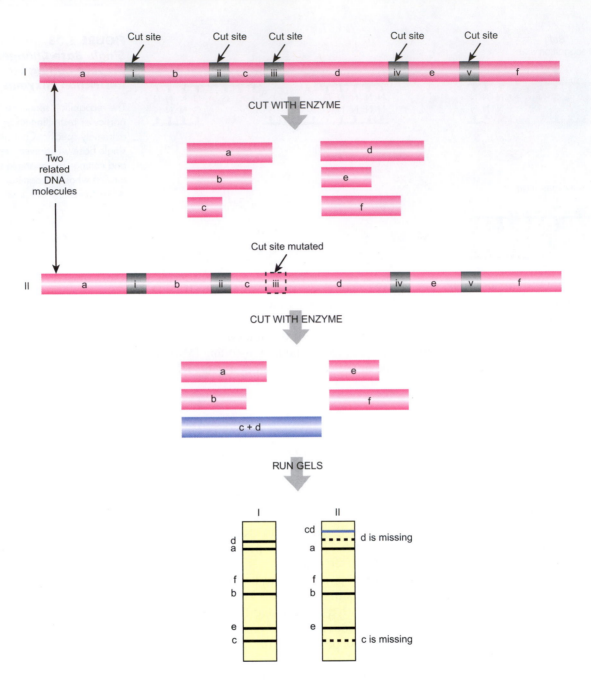

FIGURE 5.09
Restriction Fragment Length Polymorphism (RFLP)

DNA from related organisms shows small differences in sequence that result in changes in restriction map patterns. In the example shown, cutting a segment of DNA from the first organism yields six fragments of different sizes (labeled a–f on the gel). If the equivalent region of DNA from a related organism is digested with the same enzyme we would expect a similar pattern. Here, a single nucleotide difference is present, which eliminates one of the restriction sites. Consequently, digesting this DNA only produces five fragments, since site iii has been mutated and the original fragments c and d are no longer separated. Instead, a new fragment, the size of c plus d, is seen.

The standard **phosphoramidite method** for artificial chemical synthesis of DNA proceeds in the 3′ to 5′ direction. Consequently, before adding any new nucleotide to the chain, the 5′-hydroxyl of the previous nucleotide is blocked with a **dimethoxytrityl (DMT) group** and the 3′-hydroxyl is activated with a phosphoramidite. Note that the

dimethoxytrityl (DMT) group   Group used for blocking the 5′-hydroxyl of nucleotides during artificial DNA synthesis
phosphoramidite method   Method for artificial synthesis of DNA that utilizes the reactive phosphoramidite group to make linkages between nucleotides

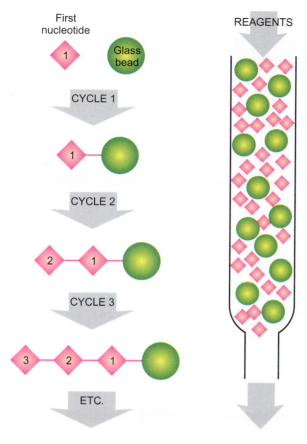

First
nucleotide

REAGENTS

Glass
bead

CYCLE 1

CYCLE 2

CYCLE 3

ETC.

►**FIGURE 5.10**
Chemical Synthesis of DNA on Glass Beads—Principle

DNA is synthesized and attached to porous glass beads in a column. Chemical reagents are trickled through the column one after the other. The first nucleoside is linked to the beads and each successive nucleoside is linked to the one before. After the entire sequence has been assembled, the DNA is chemically detached from the beads and eluted from the column.

►**FIGURE 5.11**
DNA Synthesizer

Biologist programs an automated DNA synthesizer to produce a specific oligonucleotide for her research. *(Credit: Hank Morgan, Photo Researchers, Inc.)*

FIGURE 5.12

FIGURE 5.12

Phosphoramidite Nucleosides Are Used for Chemical Synthesis of DNA

During chemical synthesis of DNA, modifications must be added to each nucleoside to ensure that the correct group reacts with the next chemical reagent. Each nucleoside has a blocking DMT group attached to the 5′-OH. The 3′-OH is activated by attaching a phosphoramidite group.

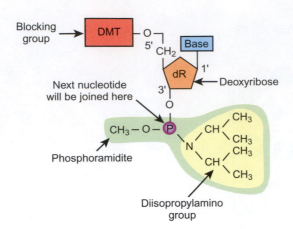

FIGURE 5.13

Addition of Spacer Molecule and First Base to Glass Bead

The first nucleoside is linked to a glass bead via a spacer molecule attached to its 3′-OH group.

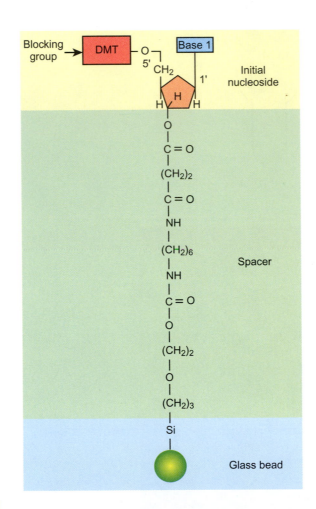

In the most common method of making DNA, phosphoramidite groups react to link nucleosides together.

Blocking agents are used throughout the procedure to prevent the wrong groups from reacting with the next nucleoside.

reagents for chemical synthesis are phosphoramidite nucleotides (with a single phosphorus), not nucleoside triphosphates as in biosynthesis (Fig. 5.12). Also, chemical synthesis occurs in the 3′ to 5′ direction, the opposite of biological DNA synthesis, which always assembles in a 5′ to 3′ direction.

The first base is anchored to a glass bead via its 3′-OH group. The first base is actually added as a nucleoside without any phosphate groups. It is bound to the glass bead via a spacer molecule (Fig. 5.13). The spacer helps prevent the bases in the growing nucleotide chain from reacting with the glass bead surface.

Next, acid (often trichloroacetic acid, TCA) is poured through the column to remove the DMT blocking group and expose the 5′-hydroxyl group on the first nucleoside. Then the second phosphoramidite nucleoside is added to the column.

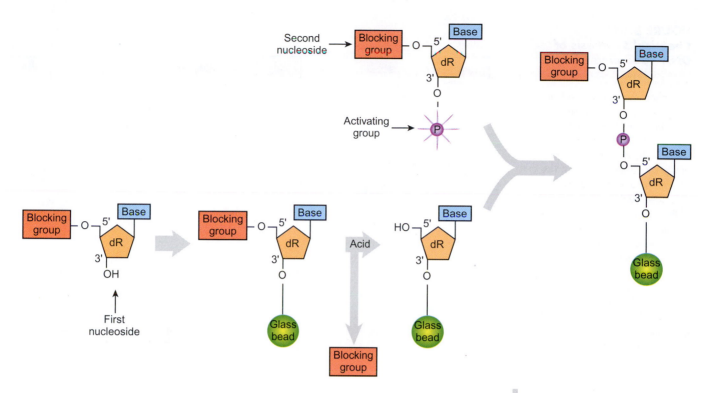

FIGURE 5.14
*Chemical Synthesis
of DNA—Nucleoside
Addition*

During chemical synthesis of
DNA, the DNA is added in a 3'
to 5' direction. In order to add
the successive bases correctly, the
3'-OH of incoming bases must
be activated by phosphoramidite
(purple), but the 5'-OH must be
blocked with a dimethoxytrityl
(DMT) group. After the first
nucleoside is anchored to the glass
bead its DMT blocking group
is removed by acid. The next
nucleoside can then link to the
exposed 5'-OH by a phosphate
linkage. Notice that the second
nucleoside still has its 5'-OH
blocked with DMT. The process
continues by using acid to remove
DMT from the second nucleoside,
adding the third nucleoside and so
on (not shown here).

This nucleoside links to the first one via the single phosphate in the phosphoramidite moiety (Fig. 5.14). After each reaction step, the column is washed by acetonitrile to remove unreacted reagents and then flushed with argon to remove any traces of acetonitrile. The cycle continues until a full-length sequence is manufactured.

The activating group on the incoming nucleotide is a phosphoramidite and is stable by itself. So, in order to attach this group to the oligonucleotide chain, tetrazole is added to activate the phosphoramidite group. The phosphoramidite then forms a bond with the exposed 5'-hydroxyl group of the previous nucleoside (Fig. 5.15). Since this coupling reaction is not 100% efficient, any remaining unreacted 5'-hydroxyl groups must be blocked by acetylation before adding the next nucleotide. This is done using acetic anhydride plus dimethylaminopyridine. Without this step, unused and unblocked 5'-hydroxyl groups would react in the next synthetic cycle and give rise to incorrect DNA sequences. After coupling, nucleotides are linked by a relatively unstable phosphite triester (Fig. 5.15). This is oxidized by iodine to a phosphate triester, or phosphotriester. The methyl group that occupies the third position of the phosphotriester is removed to give a phosphodiester linkage only after the whole DNA strand has been synthesized. After the phosphite oxidation step, the column is washed and treated with acid to expose the 5'-hydroxyl of the nucleotide that was just added. It is then ready for another nucleotide.

Since the amino groups of the bases are reactive they must also be protected throughout the whole reaction sequence. Benzoyl groups are normally added to protect the amino groups of adenine and cytosine, whereas guanine is protected by an isobutyryl group (thymine needs no protection as it has no free amino group). These groups are not shown in Figures 5.12 through 5.15 and are only removed after synthesis of the whole strand of DNA. When all nucleotides have been added, the various protective groups are removed and the 5'-end of the DNA is phosphorylated, either chemically or by ATP plus polynucleotide kinase. After detaching from the column, the final product is purified by HPLC or gel electrophoresis to separate it from the defective shorter molecules that are due to imperfect coupling.

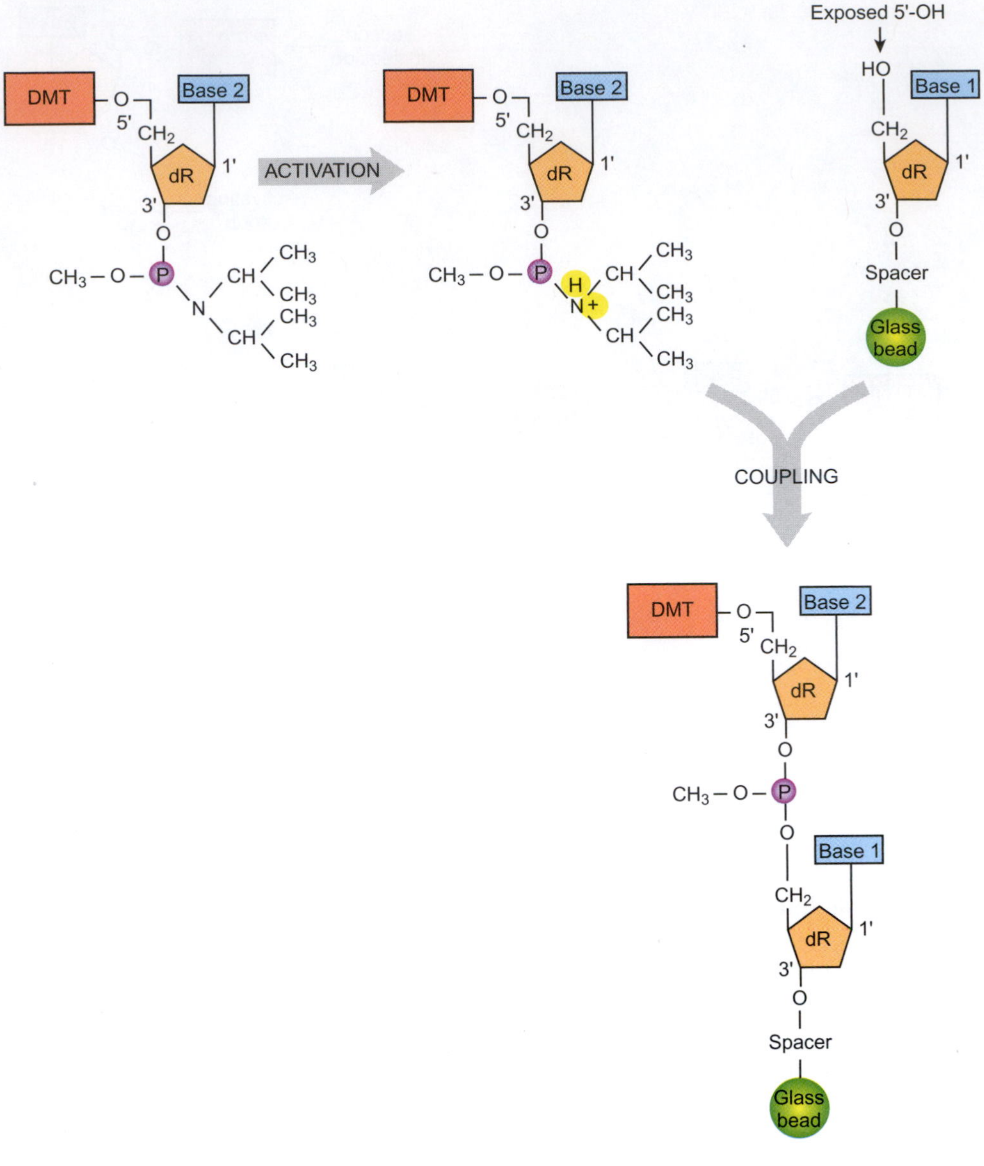

FIGURE 5.15
Chemical Synthesis of DNA—Coupling

In order to couple a phosphoramidite nucleoside to the growing chain of DNA, the phosphoramidite moiety must be activated. Tetrazole activates the N of the diisopropylamino group by adding a proton. The diisopropylamino group is then displaced by the exposed 5'-OH of the acceptor nucleoside. The coupling reaction results in two nucleosides linked by a phosphite triester. Further reaction with iodine oxidizes this to a phosphate triester, which is then hydolyzed to a phosphodiester link by removal of the methyl group in the third position.

Whole genes have been made by synthesizing smaller fragments and then assembling them.

2.1. Chemical Synthesis of Complete Genes

The coupling efficiency limits the length of DNA that can be chemically synthesized. For example, a coupling efficiency of 98% would give a yield of around 50% for a 40-mer oligonucleotide and 10% for a 100-mer. Very short complete genes with sequences of up to 80 or 100 nucleotides can therefore be synthesized as single pieces of DNA. Longer sequences must be made in segments and assembled.

To chemically synthesize a complete gene, a series of overlapping fragments, representing both strands of the gene to be assembled, are manufactured. These are purified and annealed together as shown in Figure 5.16. Two alternatives are possible. In the first case, the fragments comprise the whole gene with only nicks remaining between the assembled fragments. DNA ligase is then used to seal the nicks between the segments. In the second case, only part of each strand is made and large single-stranded gaps remain after annealing the fragments. In this case, DNA polymerase I fills in the gaps before ligase joins the segments. In either variant, regulatory sequences may be included along with the coding sequence of the gene. In addition, artificial restriction sites are often added flanking the artificial gene to allow its insertion into a cloning vector.

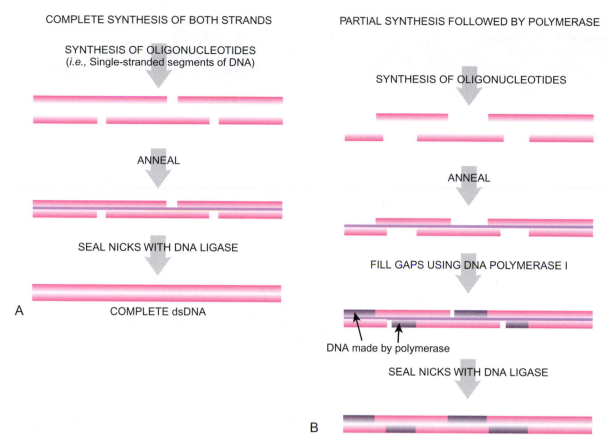

FIGURE 5.16
Synthesis and Assembly of a Gene

A) Complete synthesis of both strands. Small genes can be chemically synthesized by making overlapping oligonucleotides. The complete sequence of the gene, both coding and non-coding strands, is made from small oligonucleotides that anneal to each other forming a double-stranded piece of DNA with nicks at intervals along the backbone. The nicks are then sealed using DNA ligase. B) Partial synthesis followed by polymerase. To manufacture larger stretches of DNA, oligonucleotides are synthesized so that a small portion of each oligonucleotide overlaps with the next. The entire sequence is manufactured, but gaps exist in both the coding and non-coding strands. These gaps are filled using DNA polymerase I and the remaining nicks are sealed with DNA ligase.

2.2. Peptide Nucleic Acid

Peptide nucleic acid (PNA) is a totally artificial molecule that is used as a DNA analog in genetic engineering. PNA is just what its name indicates, consisting of a polypeptide backbone with nucleic acid bases attached as side chains. Note that the polypeptide backbone of PNA is not identical to that of natural proteins (Fig. 5.17). PNA is designed to space out the bases that it carries at the same distances as found in genuine nucleic acids. This enables a strand of PNA to base pair with a complementary strand of DNA or RNA.

> Peptide nucleic acid is useful since the cellular nucleases and proteases do not recognize its unusual structure, yet it is to bind to DNA.

PNA is deliberately designed with an uncharged backbone. The objective is that single strands of complementary PNA form a triple helix with double-stranded DNA. What actually happens is that one of the DNA strands is displaced and a triple helix is formed from two single strands of PNA plus one strand of DNA (Fig. 5.18). Not only is the repulsion due to phosphate in the backbone absent in PNA, but extra H-bonds form between the N atoms of the PNA peptide backbone and the phosphate of the single DNA strand. If two identical PNA strands are joined by a flexible linker (a "**PNA clamp**"), this binds even more avidly to its target DNA and forms a virtually irreversible triple helix.

---

peptide nucleic acid (PNA)   Artificial analog of nucleic acids with a polypeptide backbone
PNA clamp   Two identical PNA strands that are joined by a flexible linker and are intended to form a triple helix with a complementary strand of
DNA or RNA

FIGURE 5.17
Structure of Peptide Nucleic Acid Compared to DNA

A) The PNA polypeptide backbone.
B) The sugar and phosphate backbone of normal DNA.
B = nucleic acid base. The brackets with "n" indicate a polymer with multiple repeats.

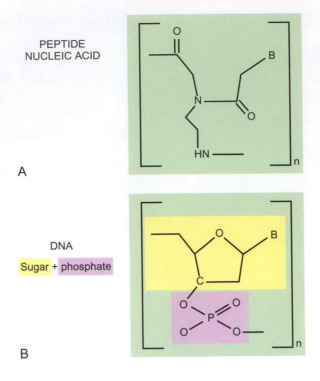

FIGURE 5.18
Triple Helix of Peptide Nucleic Acid with DNA

PNA displaces one of the strands of a DNA double helix. Two strands of PNA pair with the adenine-rich strand of DNA to form a triple helix. Regions of DNA that are bound by PNA cannot be transcribed.

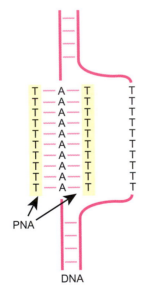

The PNA backbone is very stable and is not degraded by any naturally occurring nucleases or proteases. PNA can be used to bind to and block target sequences of DNA in purine-rich regions and prevent transcription of DNA to give mRNA. It also bind to RNA and prevent translation of mRNA into protein. PNA is useful for laboratory applications and may also be used clinically in the near future. For example, antisense PNA that inhibits the translation of *gag-pol* mRNA of HIV-1 has been shown to reduce virus production by 99% in tissue culture. Similarly, antisense PNA can stop the *in vitro* translation of *Ha-ras* and *bcl-2* mRNA (both from cancer cells). In the case of *bcl-2*, PNA was also shown to inhibit gene expression by binding to the DNA at a purine-rich tract.

The major problem with using PNA clinically is that it penetrates cells poorly—much worse than natural nucleic acids. Recent developments suggest that PNA can enter cells effectively if it is coupled with other molecules that are taken up readily or if it is carried by positively-charged liposomes.

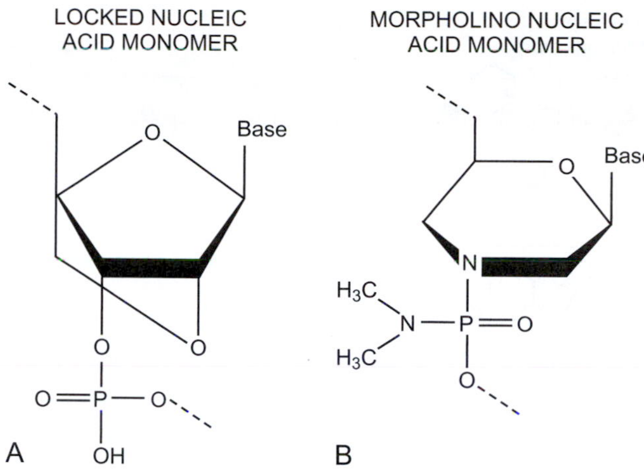

LOCKED NUCLEIC
ACID MONOMER

MORPHOLINO NUCLEIC
ACID MONOMER

A   B

FIGURE 5.19
Other Nucleic Acid Mimics

Locked nucleic acids and morpholinos mimic the structure of DNA, yet are resistant to cellular enzymes that degrade foreign DNA.

2.3. Other Nucleic Acid Mimics

In addition to peptide nucleic acid, several other nucleic acid mimics are now in use. Two of the more useful are locked nucleic acids (LNA) and morpholino nucleic acids. Like PNA they can form both duplexes and triplexes and have greater stability than natural nucleic acids. LNA has a methylene group locking together the 2'-oxygen and the 4'-carbon atom of the ribose ring. LNA is recommended for use in hybridization assays that need high specificity between the target and probe. Morpholino nucleic acids substitute a morpholino ring for deoxyribose (Fig. 5.19). They are cheaper to make than natural DNA. They are extremely resistant to nuclease degradation and highly soluble. They are especially useful in relatively long-term *in vivo* studies.

3. Measuring the Concentration of DNA and RNA with Ultraviolet Light

The aromatic rings of the bases found in DNA and RNA absorb ultraviolet light with an absorption maximum at 260 nm. If a beam of UV light is shone through a solution containing nucleic acids, the proportion of the UV absorbed depends on the amount of DNA or RNA. This approach is widely used to measure the concentration of a solution of DNA or RNA. The amount of UV light absorbed by a series of standard DNA concentrations is measured to calibrate the technique. The amount of UV light absorbed by the unknown DNA is then determined and plotted on the standard curve to deduce the concentration.

Proteins absorb UV at 280 nm, largely due to the aromatic ring of tryptophan. The relative purity of a preparation of DNA may be assessed by measuring its absorbance at both 260 and 280 nm and computing the ratio. Pure DNA has an A260/A280 ratio of 1.8. If protein is present, the ratio will be less than 1.8 and if RNA is present, the ratio will be greater than 1.8. Pure RNA has an A260/A280 ratio of approximately 2.0.

In a solution of unlinked nucleotides, the bases are more spread out and absorb more radiation. In a DNA double helix, the bases are stacked on top of each other and relatively less UV light is absorbed (Fig. 5.20). In single-stranded RNA (or in single-stranded DNA), the situation is intermediate. UV light is preferentially absorbed by the purine and pyrimidine rings, not the sugar or phosphate components of nucleic acids. The stacking of bases in a DNA double helix tends to shield them from UV radiation, as compared to free nucleotides in solution. In single-stranded nucleic acids (whether DNA or RNA), the bases are only partially shielded and they absorb an intermediate amount of light. (UV absorption by nucleic acids is due to energy transitions of the delocalized electrons of the aromatic rings of the bases. When the bases stack, these electrons interact and no longer absorb UV radiation so readily. Hence, the shielding effect.) So for a given amount of nucleotides, dsDNA absorbs less than ssRNA, which absorbs less than free nucleotides.

The concentration of DNA or RNA is usually measured by absorption of ultraviolet light.

FIGURE 5.20

Absorption of UV Radiation by Nucleic Acids

All nucleic acids absorb UV light by the aromatic rings of the bases. The phosphate backbone (pink line or black dots) is not involved in UV absorption. The structure of the nucleic acid dictates how much light the aromatic rings absorb. On the right side of the light bulb, free nucleotides are shown spread out such that each ring can absorb the UV light. Overall, the free nucleotides absorb more UV. In contrast, as shown on the left, the aromatic rings are stacked along the phosphate backbone in a nucleic acid polymer. In this configuration the rings shield each other and absorb less UV light.

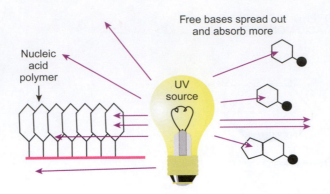

> Radioactive isotopes of sulfur or phosphorus are used to label DNA or RNA.

4. Radioactive Labeling of Nucleic Acids

Various sources of DNA can be distinguished by adding a specific label to one source. For example, when a virus attacks bacteria, the viral DNA enters the bacteria. A scientist who is studying this process will have to distinguish the viral DNA from the bacterial DNA by labeling one of the two. Originally, incorporating **radioisotopes** into the DNA of the virus or bacteria was used to distinguish the two. For example, radioactive nucleotides would be provided to the bacteria so that during the next round of DNA synthesis, some of the radioactive molecules would be incorporated into the bacterial chromosome. The bacterial DNA would be "**hot**," or radioactive, and the viral DNA would be "**cold**," or non-radioactive.

Radioisotopes are the radioactive forms of an element. In molecular biology, two are especially important: the radioactive isotopes of phosphorus, ^{32}P, and sulfur, ^{35}S. Nucleic acids consist of nucleotides linked together by phosphate groups, each containing a central phosphorus atom. If ^{32}P is inserted at this position we have radioactive DNA or RNA (Fig. 5.21). The half life of ^{32}P is 14 days, which means that half of the radioactive phosphorus atoms will have disintegrated during this time period, so experiments using ^{32}P need to be done fast!

The sulfur isotope, ^{35}S, is also widely used. Since sulfur is not a normal component of DNA or RNA, we use **phosphorothioate** derivatives of nucleotides. A normal phosphate group has four oxygen atoms around the central phosphorus. In a phosphorothioate one of these is replaced by sulfur (Fig. 5.21). To introduce ^{35}S into DNA or RNA, phosphorothioate groups containing radioactive sulfur atoms are used to link together nucleotides. Despite these complications, ^{35}S is usually preferable to ^{32}P in most molecular biology applications. There are two reasons: first, the half life of ^{35}S is 88 days, so it doesn't disappear so fast; second, the radiation emitted by ^{35}S is of lower energy than for ^{32}P. Therefore, the radiation doesn't travel so far and the radioactive bands are more precisely located and not so fuzzy. In short, ^{35}S is more accurate.

4.1. Detection of Radio-Labeled DNA

The two methods most widely used in molecular biology to measure radioactivity are **scintillation counting** and **autoradiography**. If a sample is in liquid or on a strip of filter paper, the amount of radioactivity is measured using scintillation counting. If the

---

autoradiography   Laying a piece of photographic film on top of a gel in order to identify the exact location of the radioactive DNA
"cold"   Slang for non-radioactive
"hot"   Slang for radioactive
phosphorothioate   A phosphate group in which one of the four oxygen atoms around the central phosphorus is replaced by sulfur
radioisotope   Radioactive form of an element
scintillation counting   Detection and counting of individual microscopic pulses of light

POSITIONING OF ^{32}P IN DNA

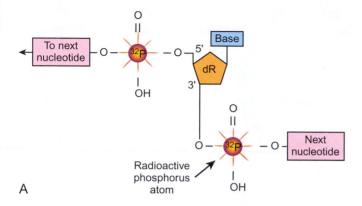

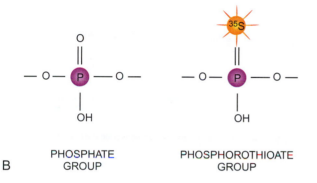

PHOSPHATE VERSUS PHOSPHOROTHIOATE

PHOSPHATE GROUP   PHOSPHOROTHIOATE GROUP

B

FIGURE 5.21

Use of ^{32}P and ^{35}S to Label Nucleic Acids

A) Positioning of ^{32}P in DNA. To make radioactive DNA, the phosphorus atom in the phosphate backbone is replaced with its radioactive isotope, ^{32}P. B) Phosphate versus phosphorothioate. Instead of replacing the phosphorus atom of the phosphate group, one of the oxygen atoms can be replaced by the radioactive isotope of sulfur, ^{35}S, giving a phosphorothioate.

sample is flat, such as an agarose gel, scientists use autoradiography to pinpoint the location of radioactive bands or spots.

Scintillation counting relies on special chemicals called **scintillants**. These emit a flash of light when high energy electrons known as beta-particles are released by the radioactive isotopes in DNA. The light pulses from the scintillant are detected by a photocell (Fig. 5.22). A **scintillation counter** is simply a very sensitive device for recording faint light pulses. To use the scintillation counter, radioactive samples to be measured are added to a vial containing scintillant fluid and loaded into the counter. The counter prints out the number of light flashes it detects within a designated time. The researcher can then compare the number of flashes with a series of standards to determine the relative amount of radioactive DNA in the sample.

Scintillation counters can also be used to measure light generated by chemical reactions. In this case, the light is emitted directly so no scintillant fluid is needed and the luminescent sample is merely inserted directly. (The detection of light emitted by luciferase and lumi-phos in genetic analysis is described in Ch. 19.)

Autoradiography is used for detecting radioactively-labeled DNA or RNA in a gel after separation by electrophoresis. Electrophoresis gels containing RNA or DNA are transferred onto membranes by blotting to allow more convenient handling during autoradiography, or the gel can be dried onto filter paper. In addition, autoradiography can detect radioactive DNA bound to filter paper during hybridization experiments. Whether the radioactive DNA is in a dried gel, on a membrane or piece of filter paper, the radioactive isotope emits beta-particles. The particles will

Scintillation counters measure radioactivity in liquid samples, whereas autoradiography is used to locate radioactive molecules on gels or membranes.

scintillant   Molecule that emits pulses of light when hit by a particle of radioactivity
scintillation counter   Machine that detects and counts pulses of light

FIGURE 5.22
Scintillation Counter is Used to Measure Radioactivity

Radioactive DNA is mixed with a liquid scintillant. The scintillant molecules absorb the β-particles emitted by the ^{32}P in the DNA, and in turn emit a flash of light. The photocell counts the number of light pulses in a specific time period.

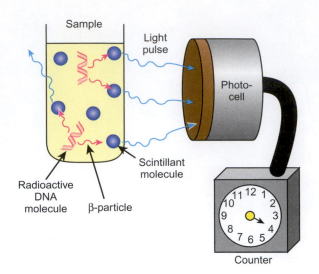

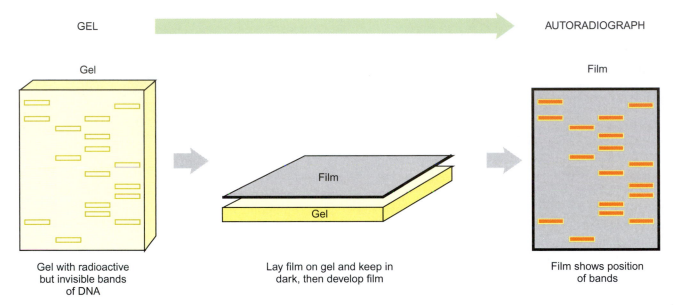

GEL   AUTORADIOGRAPH

Gel with radioactive
but invisible bands
of DNA

Lay film on gel and keep in
dark, then develop film

Film shows position
of bands

FIGURE 5.23
Autoradiography to Detect Radio-Labeled DNA or RNA

A gel containing radioactive DNA or RNA is dried and a piece of photographic film is laid over the top. The two are loaded into a cassette case that prevents light from entering. After some time (hours to days), the film is developed and dark lines appear where the radioactive DNA was present.

DNA or RNA may be labeled with fluorescent dyes.

turn regular photographic film black, exactly the same way that light turns photographic film black. To see the location of radioactively-labeled DNA, a sheet of photographic film is laid on top of the gel or filter and left for several hours or sometimes even for days. The film darkens where the radioactive DNA bands or spots are found (Fig. 5.23). Exposing the film must be carried out in a dark room to avoid visible light.

5. Fluorescence in the Detection of DNA and RNA

Most classic work in biochemistry and molecular biology was done with radioactive tracers and probes. Although the low levels of radioactivity used in laboratory analyses are little real hazard, the burden of storage and proper disposal of the waste has made it relatively cheaper and quicker to use other detection methods. Some of the newer DNA detection methods use **fluorescence**, chemical tagging, and hybridization.

fluorescence   Process in which a molecule absorbs light of one wavelength and then emits light of another, longer, lower energy wavelength

FLUORESCENT TAGGING OF DNA

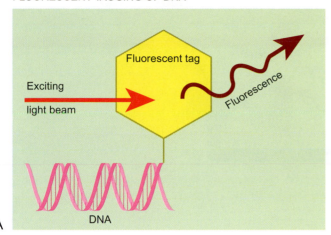

A

ENERGY LEVELS IN FLUORESCENCE

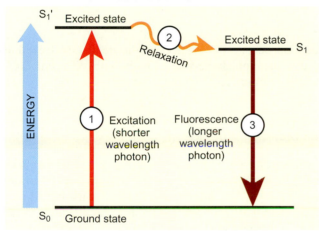

B

■─**FIGURE 5.24**

Fluorescence Detection

A) Fluorescent tagging of DNA. During DNA synthesis, a nucleotide linked to a fluorescent tag is incorporated at the 3' end of the DNA. A beam of light excites the fluorescent tag, which in turn releases light of a longer wavelength (fluorescence). B) Energy levels in fluorescence. The fluorescent molecule attached to the DNA has three different energy levels, S_0, S_1', and S_1. The S_0 or ground state is the state before exposure to light. When the fluorescent molecule is exposed to a light photon of sufficiently short wavelength, the fluorescent tag absorbs the energy and enters the first excited state, S_1'. Between S_1' and S_1, the fluorescent tag relaxes slightly, but doesn't emit any light. Eventually the high-energy state releases its excess energy by emitting a longer wavelength photon. This release of fluorescence returns the molecule back to the ground state.

Fluorescence occurs when a molecule absorbs light of one wavelength and emits light of lower energy at a longer wavelength (Fig. 5.24). Detection of fluorescence requires both a beam of light to excite the dye and a photo-detector to detect the fluorescent emission. A variety of fluorescent dyes are available to attach to DNA, and each one has slightly different absorption and emission wavelengths. The detectors can readily discern and record the different tags making it possible to have multiple fluorescent tags in one single sample.

Another instrument that uses fluorescence is the **fluorescence activated cell sorter** or **FACS**. Its job, originally, was sorting cells labeled with a fluorescent tag from those that were untagged. The new generation of more sensitive FACS machines is capable of sorting chromosomes labeled by hybridization with fluorescent tagged DNA probes (Fig. 5.25). Another novel use for a FACS machine is in sorting whole small organisms such as the nematode worm *C. elegans*, which is widely used in genetic analysis. Since these are transparent, any sort of fluorescent tag is readily identified by the detector.

Another recent and growing use for FACS technology is in fluorescent bead sorting. Many reactions in modern high-throughput screening involve anchoring one or more reagents to microscopic polystyrene beads. In some reaction schemes, fluorescently-labeled molecules may bind to colorless reagents, such as DNA or proteins, previously attached to the beads. In other cases, the beads may be color-coded

fluorescence activated cell sorter (FACS)   Instrument that sorts cells (or chromosomes) based on fluorescent labeling

FIGURE 5.25
FACS Machine Can Sort Chromosomes

FACS machines can separate fluorescently-labeled chromosomes from unlabeled ones. Liquid carrying a mixture of labeled and unlabeled chromosomes passes by a laser, which excites the fluorescent tags. Whenever the photo-detector detects fluorescence, the controller module directs that drop into the test tube on the left. When no fluorescence is emitted, the controller module directs the drop into the test tube on the right. This sorting procedure allows the separation of fluorescently-labeled particles from unlabeled ones.

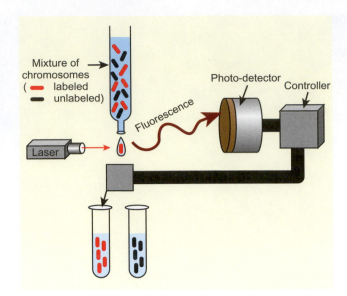

by fluorescent dyes before reaction occurs. In either case, the beads are sorted after reaction. Modern equipment can distinguish between beads with different colored fluorescent dyes, not merely separate fluorescent beads from unlabeled ones based on brightness.

5.1. Chemical Tagging with Biotin or Digoxigenin

Biotin (a vitamin) and **digoxigenin** (a steroid from foxglove plants) are two molecular tags widely used for labeling DNA. Both biotin and digoxigenin are linked to uracil, which is normally a component of RNA. In order to label DNA with biotin or digoxigenin, the uracil nucleotide is modified from uridine triphosphate (UTP) to deoxyuridine triphosphate (deoxyUTP). If deoxyUTP labeled with biotin or digoxigenin is added to a DNA synthesis reaction, DNA polymerase will incorporate the labeled uridine where thymidine is normally inserted. The biotin or digoxigenin tags stick out from the DNA without disrupting its structure (Fig. 5.26).

Biotin and digoxigenin are not colored or fluorescent molecules, but they can be detected in a two-stage process. The first step is to bind a molecule that tags the location of biotin or digoxigenin. Biotin is a vitamin required both by animals and many bacteria. Scientists learned the biology of this vitamin to find a molecule that binds to biotin. Chickens lay highly nutritious eggs that would be a paradise for invading bacteria. One of the defense mechanisms to protect the egg from bacterial attack is a protein known as **avidin**. This protein, found in egg white, binds biotin so avidly that invading bacteria become vitamin deficient. Molecular biologists use avidin to bind the biotin tag. Digoxigenin also requires a second detectable molecule. In this case, an antibody that recognizes and binds digoxigenin is used. (Antibodies are proteins made by the immune system, which specifically recognize foreign molecules—see Ch. 15)

The antibody or avidin molecules are easily visualized by a variety of different methods. The first option is to attach an enzyme that generates a colored product to the avidin or the antibody. For example, avidin can be conjugated to **alkaline phosphatase**, an enzyme that snips phosphate groups from a wide range of molecules. One substrate for alkaline phosphatase, an artificial **chromogenic substrate** known

> Biotinylated DNA is recognized by avidin conjugated to alkaline phosphatase.

> Digoxigenin labeled DNA is recognized by an antibody to digoxigenin.

---

alkaline phosphatase   An enzyme that cleaves phosphate groups from a wide range of molecules
avidin   A protein from egg white that binds biotin very tightly
biotin   One of the B family of vitamins that is also widely used for chemical labeling of DNA molecules
chromogenic substrate   Substrate that yields a colored product when processed by an enzyme
digoxigenin   A steroid from foxglove plant widely used for chemical labeling of DNA molecules

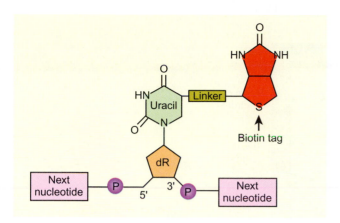

Uracil can be incorporated into a strand of DNA if the nucleotide has a deoxyribose sugar. Prior to incorporation, the uracil is tagged with a biotin molecule attached via a linker, which allows the biotin to stick out from the DNA helix without disrupting its structure.

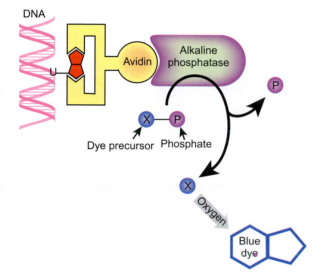

DNA that has biotin attached via uracil can be detected with a two-step process. First, avidin is bound to the biotin. The avidin is conjugated to an enzyme called alkaline phosphatase, which cleaves phosphate groups from various substrates. Second, a substrate such as X-phos (shown) or lumi-phos (not shown) is added. Alkaline phosphatase removes the phosphate group from either substrate. In the case of X-phos, cleavage releases a precursor that reacts with oxygen to form a blue dye. If the substrate is lumi-phos, cleavage allows the unstable luminescent group to emit light.

as "**X-phos**," produces a blue dye when cleaved. X-phos consists of a dye precursor bound to a phosphate group, and when alkaline phosphatase cleaves the phosphate off, the dye precursor is converted to a blue dye by oxygen in the air (Fig. 5.27).

Another option to detect biotin/avidin or digoxigenin/antibody is to use **chemiluminescence**. In this method, an enzyme that produces light by a chemical reaction is used to identify the labeled DNA. Alkaline phosphatase is still conjugated to the avidin or antibody, but a different substrate, called "**lumi-phos**," is added. Lumi-phos consists of a light-emitting group bound to the phosphate group. When alkaline phosphatase removes the phosphate, the unstable luminescent group emits light. Detecting and recording the light is accomplished by using photographic film if the DNA is on a filter or in a gel, or by an instrument capable of detecting light emissions if the DNA is in a solution.

> Colored or luminescent products are released when alkaline phosphatase removes a phosphate group from X-phos or lumi-phos, respectively.

6. The Electron Microscope

With an ordinary light microscope, objects down to approximately a micron (a millionth of a meter) in size can be seen. Typical bacteria are a micron or two long by about half a micron wide. Although bacteria are visible under a light microscope, their internal details are too small to see. The resolving power of a microscope depends on the wavelength of the light. In a light microscope, if two dots are less than about half a wavelength apart, they cannot be distinguished. Visible light has wavelengths in

> Bacteria are just visible under a light microscope.

chemiluminescence   Production of light by a chemical reaction
lumi-phos   Substrate for alkaline phosphatase that releases light upon cleavage
X-phos   Substrate for alkaline phosphatase that is cleaved to release a blue dye

FIGURE 5.28

FIGURE 5.28
Principle of the Electron Microscope

Electron microscopy can reveal the substructures of animal cells, viruses, and bacteria. A beam of electrons is emitted from a source and is focused on the sample using electromagnetic lenses. When the electrons hit the sample, components such as cell walls, membranes, etc., absorb electrons and appear dark. The image is viewed on a screen or may be transferred to film for a permanent record. Since air molecules also absorb electrons, the entire process must be done in a vacuum chamber.

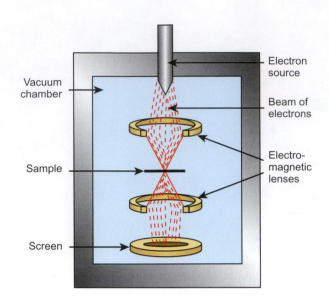

FIGURE 5.29
Metal Shadowed DNA Molecules Are Visible under an Electron Microscope

A hot metal filament releases vaporized metal atoms into the chamber containing a sample of DNA. The sample is rotated around the filament and metal ions attach to the exposed surface of the DNA. Once the DNA has a coat of metal atoms, it can be visualized by electron microscopy.

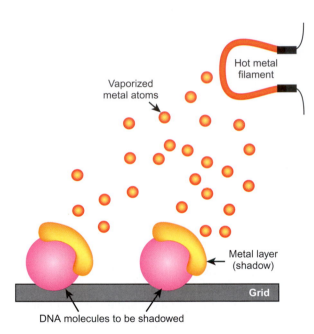

the range of 0.3 (blue) to 1.0 (red) micron, so bacteria are just at the limits of visible detection and most viruses cannot be seen.

A beam of electrons has a much smaller wavelength than visible light and so can distinguish detail far beyond the limits of resolution by light. Electron beams may be focused like visible light except that the lenses used for electron beams are not physical (glass absorbs electrons) but electromagnetic fields that alter the direction in which the electrons move. Using an electron microscope allows visualization of the layers of the bacterial cell wall and of the folded-up bacterial chromosome, which appears as a light patch against a dark background. When an electron beam is fired through a sample, materials that absorb electrons more efficiently appear darker. Because electrons are easily absorbed, even by air, an electron beam must be used inside a vacuum chamber and the sample must be sliced extremely thin (Fig. 5.28).

Electron microscopes allow viruses, subcellular components, and even single macromolecules to be visualized.

To improve contrast, cell components are usually stained with compounds of heavy metals such as uranium, osmium, or lead, all of which strongly absorb electrons. Individual, uncoiled DNA molecules can be seen if they are shadowed with metal atoms to increase electron absorption (Fig. 5.29). Shadowing is done by spreading the

DNA out on a grid and then rotating it in front of a hot metal filament. Metal atoms evaporate and cover the DNA. Gold, platinum, or tungsten is typically used for shadowing. This is the method used to demonstrate that eukaryotic genes have introns that interrupt the actual coding sequences (see Ch. 12).

Peckys DB, Mazur P, Gould KL, de Jonge, N (2011) **Fully hydrated yeast cells imaged with electron microscopy. Biophys J. 100: 2522–2529.**

Electron microscopy is one of the key techniques used to understand the structure of cellular organelles, nuclear structures, and cell walls. Each electron micrograph is very informative, yet each cell has to be processed by first preserving the intracellular architecture as well as possible, adding a substance such as metal atoms to increase the electron density of the cellular structures, and then embedding the sample into an epoxy resin for ultrathin sectioning. How these processes alter the actual structure from the living cell is unknown. In this associated article, the authors have developed a method to visualize water hydrated living fission yeast cells (*Schizosaccaromyces pombe*) for scanning transmission electron microscopy (STEM). As the name implies, STEM is simply transmission electron microscopy as described in the text, but the electron beam can scan across the sample to give a better picture of the image. To keep the cells alive, the yeast cells are added to a microfluidic chamber created by using two pieces of SiN membranes that allow electrons to pass without distortion. Next, a dye is added to the fluid which fluoresces when taken into yeast vacuoles. Only living yeast cells will have red fluorescence within the cell. After confirming which cells were living, the yeast are visualized with the electron beam.

FOCUS ON RELEVANT RESEARCH

The resulting images showed many structures within the yeast cell such as lipid droplets, vacuoles, peroxisomes, and cell wall septum. The results are at a higher resolution than light microscopy, and will be a useful method to study cellular processes in living cells.

---

7. Hybridization of DNA and RNA

As discussed in Ch. 3, the complementary nature of DNA allows the two strands to melt apart into single strands as temperature increases, and then as the temperature is slowly cooled back to normal, the two strands reanneal matching every adenine to thymine and every guanine to cytosine. Suppose on the other hand, two closely-related DNA molecules are used. Although the sequences may not match perfectly, if they are similar enough, some base pairing will occur. The result will be the formation of **hybrid DNA** molecules.

The formation of hybrid DNA molecules has a wide variety of uses. The relatedness of two DNA molecules can be assayed. In Fig. 5.30, a sample of the first DNA molecule is heated to melt it into single-stranded DNA. The single strands are then attached to a filter. Next, the filter is treated chemically to block any remaining sites that would bind DNA. Then, a second single-stranded DNA sample is poured on top of the filter. If the second DNA has any related sequences, then a hybrid will form on the membrane. (As discussed above, the second DNA must be labeled by radioactivity, fluorescence, or some other way to enable its detection.)

The more closely related the two molecules are, the more hybrid molecules will be formed and the higher the proportion of the labeled DNA will be bound to the filter. For example, if the DNA for a human gene, such as hemoglobin, was fully melted and bound to a filter, then DNA for the same gene but from different animals could be tested. We might expect gorilla DNA to bind strongly, frog DNA to bind weakly, and mouse DNA to be intermediate.

Another use for **hybridization** is in isolating genes for cloning. Suppose we already have the human hemoglobin gene and want to isolate the corresponding gorilla gene. First, the human DNA is bound to the filter as before. Then gorilla DNA is cut into short segments with a restriction enzyme. The gorilla DNA is heated to

> Heating of the DNA double helix melts it into single strands. Single strands will base pair to recreate a double helix if slowly cooled.

> Hybrid double helices may be formed by annealing single strands that are related in sequence.

---

hybridization Formation of double-stranded DNA molecule by annealing of two single strands from two different sources
hybrid DNA Artificial double-stranded DNA molecule made by two single strands from two different sources

FIGURE 5.30
Relatedness of DNA by Filter Hybridization

A) DNA 1 is denatured and attached to a filter. B) When DNA 2 is added to the filter, some of the DNA strands will hybridize, provided that the sequences are similar enough. If the sequence is identical, all of the single-stranded DNA 1 should hybridize to strands of DNA 2 (green). If the sequences are very different, little or none of DNA 2 will hybridize with DNA 1.

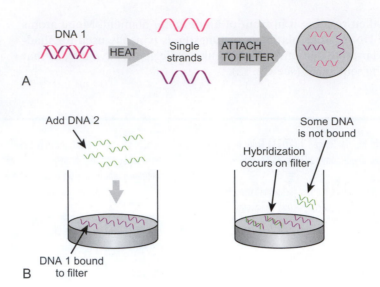

Probes are labeled molecules of DNA (or RNA) that are used to detect complementary sequences by hybridization.

melt it into single strands and poured over the filter. The DNA fragment that carries the gorilla gene for hemoglobin will bind to the human hemoglobin gene and remain stuck to the filter. Other, unrelated genes will not hybridize. This approach allows the isolation of new genes provided a related gene is available for hybridization.

A wide range of methods based on hybridization is used for analysis in molecular biology. The basic idea in each case is that a known DNA sequence acts as a "**probe.**" Generally, the **probe molecule** is labeled by radioactivity or fluorescence for ease of detection. The probe is used to search for identical or similar sequences in the experimental sample of target molecules. Both the probe and the target DNA must be treated to give single-stranded DNA molecules that can hybridize to each other by base pairing. In the previous example, the probe DNA would be the human hemoglobin DNA since the sequence is already known. The gorilla DNA would be the sample of target molecules.

7.1. Southern, Northern, and Western Blotting

Isolating new genes from related species provides a wealth of information for a scientist. Many times, scientists that are studying human genes need to find similar genes in other organisms such as yeast or *Drosophila*. Many scientists use **Southern blotting** to identify the gene in a different organism. Southern blotting is a technique in which one DNA sample is hybridized to another DNA sample. Suppose we have a large DNA molecule, such as the yeast chromosome, and we wish to locate the particular gene whose sequence is similar to the human gene of interest. First, the target or yeast DNA is cut with a restriction enzyme, and the fragments are separated by gel electrophoresis. The double-stranded fragments are melted into single-stranded fragments by soaking the gel in alkaline denaturing solution such as sodium hydroxide. Then the DNA fragments are transferred to a nylon membrane. Finally, the membrane is dipped in a solution of labeled DNA probe molecules—in this example, a radioactively-labeled piece of the human gene (Fig. 5.31). The probe binds only to those fragments with similar sequence. When the probe hybridizes to the corresponding DNA, the filter will be "hot" or radioactive in that area, and if a piece of photographic film is placed over the

---

probe molecule   Molecule that is tagged in some way (usually radioactive or fluorescent) and is used to bind to and detect another molecule
probe   Short for probe molecule
Southern blotting   A method to detect single-stranded DNA that has been transferred to nylon paper by using a probe that binds DNA

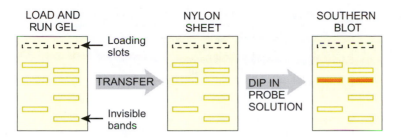

FIGURE 5.31
Southern Blotting: Hybridizing DNA to DNA

Southern blotting requires the target DNA to be cut into smaller fragments and run on an agarose gel. The fragments are denatured chemically to give single strands, and then transferred to a nylon membrane. Notice that the DNA is invisible both in the gel and on the membrane. A radioactive probe (also single-stranded) is passed over the membrane. When the probe DNA finds a related sequence, a hybrid molecule is formed. Surplus probe that has not bound is washed away. Photographic film is placed on top of the membrane. The location of radioactive hybrid molecules is revealed by black bands on the film.

TABLE 5.02	Different Types of Blotting	
Type of Blotting	**Molecule on Membrane**	**Probe Molecule**
Southern blotting	DNA	DNA
Northern blotting	RNA	DNA
Western blotting	Protein	Antibody
South-Western blotting	Protein	dsDNA

filter, a black spot corresponding to the hybrid molecule will appear. Southern blotting only refers to hybridization of DNA to DNA.

Although Southern blotting was actually named after its inventor, Edward Southern, it set a geographical trend for naming other types of hybridization techniques (Table 5.02). **Northern blotting** refers to hybridization that uses RNA as the target molecule and DNA as a probe. For example, DNA probes may be used to locate messenger RNA molecules that correspond to the same gene. The mixture of RNA is run on the gel and transferred to the filter. The filter is then probed just as above.

Western blotting does not even involve nucleic acid hybridization. Proteins are separated on a gel, transferred to a membrane, and detected by antibodies. Since Western blotting applies to proteins it is described more fully in Chapter 15, Proteomics.

7.2. Zoo Blotting

Zoo blotting is not a distinct method but a neat trick using Southern blotting. One problem encountered when cloning human genes is that most DNA in higher animals is non-coding DNA, yet most scientists want the coding regions. The question is how to identify coding regions in the large amount of non-coding DNA. During evolution, the base sequence of non-coding DNA mutates and changes rapidly, whereas coding

Using probes to detect DNA sequences by hybridization can be carried out on a membrane and is then referred to as "blotting."

Northern blotting   Hybridization technique in which a DNA probe binds to an RNA target molecule
Western blotting   Detection technique in which a probe, usually an antibody, binds to a protein target molecule
zoo blotting   Comparative Southern blotting using DNA target molecules from several different animals to test whether the probe DNA is from a coding region

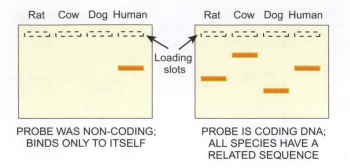

FIGURE 5.32
Zoo Blotting Reveals Coding DNA

A specialized form of Southern blotting, called zoo blotting, is used to distinguish coding DNA from non-coding regions. The target DNA includes several samples of genomic DNA from different animals, hence the term "zoo." The probe is a segment of human DNA that may or may not be from a coding region. Blotting is carried out as usual. On the left, the only hybrid seen was between the probe and the human DNA. Therefore, related sequences were not found in other species and the probe is probably non-coding DNA. In the example on the right, the probe binds to related sequences in other animals; therefore, this piece of DNA is probably from a coding region.

sequences change much more slowly and can still be recognized after millions of years of divergence between two species, especially in essential genes (see Ch. 26).

Therefore, DNA is extracted from a series of related animals, such as a human, monkey, mouse, hamster, cow, etc. Samples in this DNA "zoo" are each cut up with a suitable restriction enzyme and the fragments are run on a gel and transferred to a nylon membrane. They are probed using DNA that is suspected of being human-coding DNA. If a DNA sample really does include a coding sequence, it will probably hybridize with some fragment of DNA from most other closely-related animals (Fig. 5.32). If the DNA is non-coding DNA, it will probably hybridize only to the human DNA.

7.3. Fluorescence *in Situ* Hybridization (FISH)

All the previous techniques require the scientist to isolate the DNA, RNA, or protein from its cellular environment. In contrast, **fluorescence *in situ* hybridization**, better known as **FISH**, detects the presence of a gene, or the corresponding messenger RNA within the actual cell (Fig. 5.33). DNA sequences from the gene of interest must first be generated for use as a probe. These may be obtained by cloning the gene or, more usually nowadays, amplified by PCR (see Ch. 6 for details). In practice, it is rarely necessary to use the whole gene sequence as a probe, unless distinguishing between closely related genes is essential. As the name indicates, the DNA probe is labeled with a fluorescent dye whose localization will eventually be observed under a fluorescence microscope. The tissue or cell must also be treated to denature the chromosomal DNA, but this is done on the actual tissue section, leaving the DNA within the nuclei.

A thin section of tissue from a particular animal, a mouse for example, may be treated with a DNA probe for a known mouse gene. In this case, the mouse probe will hybridize to the mouse DNA in the nucleus of all the cells. This tells us that the genes are in the nucleus, which we knew anyway. Some more useful applications are as follows:

1. Using a virus gene as a probe reveals which cells contain virus genes, and whether the virus genes are in the cytoplasm or have penetrated the nucleus.

DNA or RNA sequences may be detected in their natural location inside the cell by hybridization to fluorescent probes.

FISH   See Fluorescence *in situ* hybridization
fluorescence *in situ* hybridization (FISH)   Using a fluorescent probe to visualize a molecule of DNA or RNA in its natural location

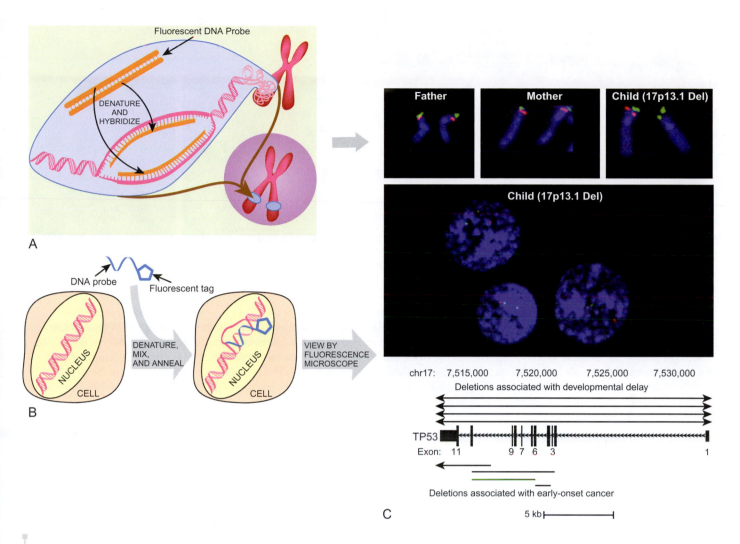

FIGURE 5.33
Fluorescence **In Situ** *Hybridization*

A) FISH can localize a gene to a specific place on a chromosome. First, metaphase chromosomes are isolated and attached to a microscope slide. The chromosomal DNA is denatured into single-stranded pieces that remain attached to the slide. The fluorescent probe hybridizes to the corresponding gene. When the slide is illuminated, the hybrid molecules fluoresce and reveal the location of the gene of interest. B) A cell with intact DNA in its nucleus is treated to denature the DNA, forming single-stranded regions. The fluorescently-labeled DNA probe is added, and the single-stranded probe can anneal with the corresponding sequence inside the nucleus. The hybrid molecule will fluoresce when the light from a fluorescence microscope excites the tag on the probe. This technique can localize the gene of interest to different areas of the nucleus. C) A copy number variation (CNV) for TP53 (red) and 17ptel (green) probes is visible in the child. There is a hemizygous deletion of TP53 in the child, which is visible in both the metaphase (top) and interphase nuclei (bottom). Notice the presence of four green and four red spots in the father and mother, whereas, the child has four green and two red spots. *(Credit: Shlien, et al. A common molecular mechanism underlies two phenotypically distinct 17p13.1 microdeletion syndromes. Am J Hum Gen 87: 631–642)*

2. Besides DNA within the nucleus, FISH can be used to identify where a particular gene is in metaphase chromosomes. First, a chromosome smear is made on a microscope slide, and then probed with a fluorescently-labeled gene of interest. The place where the probe binds reveals which chromosome carries the gene corresponding to the probe. With sufficiently sophisticated equipment, the gene may be localized to a specific region on the chromosome (Fig. 5.33).

3. A DNA probe can be used to detect mRNA within the target tissue since one of the two strands of the denatured DNA will bind to the RNA. Since mRNA is already single-stranded, the cells do not have to be treated with high heat or chemical denaturants. Cells actively transcribing the gene of interest will have high levels of the corresponding mRNA, which will bind the probe and light up

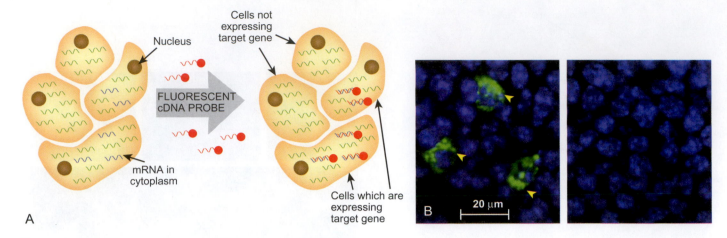

FIGURE 5.34
FISH To Detect and Measure mRNA Levels

A) A variety of different mRNA species are expressed at any one time within a tissue. Not all cells express the same genes or express them at the same level. If the target cells are probed with fluorescently-labeled DNA (red), this will bind to the corresponding mRNA (blue). Notice that the probe DNA does not bind to the nuclear DNA because in this procedure the cells were not treated to denature the chromosomal DNA. The target gene in this example was only expressed in two of the cells. B) Actual fluorescence micrograph showing *Arc* mRNA expression in the dentate gyrus of the mouse brain after sound stimulation (yellow arrow heads on left). This probe was an antisense copy of the *Arc* cDNA, and the right side shows the same procedure using a sense probe to the mRNA. *(Credit: Ivanova, et al. (2011) Arc/Arg3.1 mRNA expression reveals a subcellular trace of prior sound exposure in adult primary auditory cortex. Neuroscience 181: 117–126.)*

(Fig. 5.34). The greater the gene expression, the brighter the cell will fluoresce. Identifying the level of a particular mRNA can be very helpful when comparing tissues. For example, comparing the amount of mRNA for a particular gene in liver cells versus heart cells can help suggest the function of the gene of interest. Levels of many mRNA molecules are low and hence would give only a weak signal by FISH. In practice, such mRNA is often amplified by RT-PCR before detection (see Ch. 6). A longer probe can also be used to increase the fluorescent signal. Alternatively, techniques such as microarrays and quantitative RT-PCR can be used for mRNA detection (see Ch. 19).

Paré A, Lemons D, Kosman D, Beaver W, Freund Y, McGinnis W (2009) Visualization of individual *Scr* mRNAs during *Drosophila* embryogenesis yields evidence for transcriptional bursting. Curr. Biol. 2037–2042.

Microscopy of fluorescently-labeled cells is a common technique, and recent years have seen advances in the resolution of the images. In addition, microscopy is now able to optically slice the image, analyze each slice independently, and then reconstruct the images into a 3D picture. Besides advances in microscopy, the quality of FISH probes has increased dramatically. Taken together, it is now possible to detect single mRNA transcripts with a single cell using FISH. Previous to this article, most single cell FISH analysis was performed on single yeast cells that normally are found apart from other cells, or mammalian cells kept in a monolayer on a plate. The associated paper extends this research to identify mRNAs within cells of whole *Drosophila* embryos. These cells are surrounded by other cells, and analysis of mRNA localization within the cell in its 3D environment is a better estimate of the true state of mRNA localization without any laboratory manipulations.

This paper uses single-cell FISH to identify the location of single mRNA from a gene called *Sex combs reduced* (*Scr*), which is a transcription factor that turns on genes to make the posterior mouthparts and the first thoracic segment (T1) during *Drosophila* development. This gene is a member of the HOX gene family, which are evolutionarily conserved regulators of body pattern development. In *Drosophila*, the HOX members are expressed in segments along the embryo length, such that the HOX member that controls the head is expressed most anteriorly, and the member that controls the abdomen segments is found most posteriorly (Fig. 5.35). Expression of *Scr* mRNA is stably expressed in parasegment 2 (PS2) and then transiently in PS3 during late stage 10 through late stage 11 of embryonic development (in the head region).

The first part of this paper uses different probes to ensure that the FISH signal is truly from a single mRNA transcript. Three different methods can ensure that the FISH signal is from a single mRNA. First, two different probes from the same transcript can be used to localize the mRNA. If one probe is from the beginning of the mRNA and the other probe is from the end of the mRNA, the FISH signal for these two should be shifted, and this shift should occur for all the different dots seen in the FISH analysis. In this paper, the authors found that a probe to the coding region spatially shifted in comparison to the probe from the 3' UTR of the mRNA. Second, when using a series of oligonucleotide probes, each probe should bind to the same spot, and the binding should be reproducible. For such high resolution of FISH signals, the probes must fluoresce much brighter than any background fluorescence. Oligonucleotide probes are too short to fluoresce that bright, and therefore, this technique is not technically feasible. Instead, the authors used a third method to ensure each fluorescent spot represented one mRNA. In this method, the authors used two probes that bound to the same exact sequence on the mRNA. Each probe was labeled with a different tag and could be identified by the microscope. This experiment demonstrated that the two probes were rarely found in the same position (<10%), and that means that each fluorescent spot was only able to bind one of the probes at a time. If more than one mRNA was present in the fluorescent spot, then there would have been considerable overlap of signal, but since the two probes rarely bound to the same spot, the authors concluded that each spot represents one single mRNA.

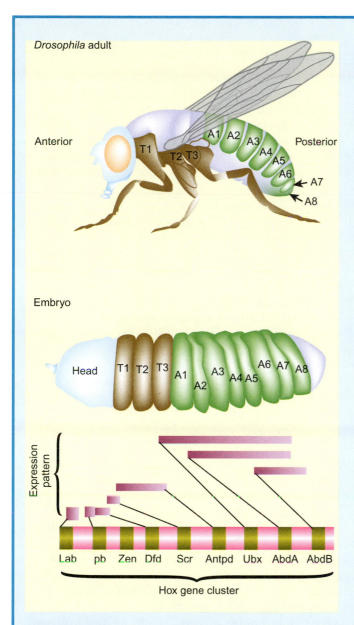

FIGURE 5.35
Cluster of Hox *Genes in Drosophilia*

Different *Hox* genes control development of each segment of the *Drosophilia* fruit fly embryo. For example, the embryo segments T1, T2, and T3 express the *antennapedia* gene, encoding a transcription factor of the Hox family. This protein controls the development of the adult legs from the T1, T2, and T3 embryonic segments. The Antennapedia transcription factor is expressed as a gradient in these three segments with the highest concentration in T1. If this protein is expressed in a difference segment of the embryo, a leg will grow in the wrong place. When this transcription factor was expressed in the head region, the fly's antenna grew an extra leg.

FOCUS ON RELEVANT RESEARCH

Since each FISH signal represented one *Scr* mRNA, the authors determined the number of transcripts in each cell. Their research suggests that *Scr* mRNA is found in variable numbers in adjacent cells, that is, one cell could have over 200 transcripts, yet its neighboring cell has less than 100. This was especially true in PS2 cells that are stably expressing *Scr* throughout the development period analyzed. By using a probe to an intron, which would only label nuclear mRNA, and comparing this to the probe to the middle of the transcript after splicing, the authors could compare the number of mRNAs in the nucleus to the cytoplasmic transcripts in PS2. These results were interesting because the amount of mRNA in the nucleus did not correlate with the transcripts within the cytoplasm. In fact, in single cells, nuclear transcripts for *Scr* were low, yet the cytoplasmic transcripts were high, or vice versa. Taken together, the authors conclude that transcription may occur in bursts, and then stop. After transcription stops, then the mRNA is processed and sent to the cytoplasm. In addition to this pattern of transcription, the authors found that the amount of nuclear and cytoplasmic *Scr* mRNA did correlate in the PS3 region of the embryo. If there was a lot of nuclear mRNA, there was a lot of cytoplasmic mRNA. This region does not appear to have bursts of transcription as seen in the PS2 region. High-resolution single-cell FISH analysis on cells within a whole embryo is a new technique, and further advances in this technique will be useful to analyze mRNA transcription at a single-cell level.

Key Concepts

- Nucleases are enzymes that degrade nucleic acids. Ribonucleases degrade RNA; deoxyribonucleases degrade DNA. Exonucleases degrade DNA starting at the ends; endonucleases degrade DNA starting in the middle.
- Restriction enzymes are endonucleases that bind to a specific DNA sequence called a recognition site. Type I restriction enzymes cut 1000 or more bases away from the recognition site and are not used very often in molecular biology. Type II restriction enzymes cut DNA within the recognition site and are used extensively in molecular biology.
- DNA ligase covalently links the phosphate backbone of DNA fragments. It is easier for ligase to link the two pieces if they have complementary single-stranded overhangs or sticky ends than two DNA fragments that have blunt ends.
- Restriction enzymes can be used for creating maps of DNA pieces and are used to assess if two pieces of DNA are from related individuals using restriction enzyme length polymorphisms (RFLPs).
- Small pieces of DNA can be chemically synthesized and used for various molecular biology techniques such as sequencing, PCR, and hybridization experiments.
- Chemical synthesis of DNA occurs in a 3' to 5' direction, which is the opposite of DNA synthesis in the cell. Since chemicals cannot discern which group is the correct one for addition of the next nucleotide, these reactive groups are protected and deprotected as needed.
- Chemical synthesis of DNA can be used to synthesize whole genes or can be modified to use DNA analogs that are not subject to cellular degradation.
- DNA and RNA can be measured and visualized by exposure to UV light at 260 nm.
- DNA and RNA can also be labeled by incorporating radioisotopes such as 32P or 35S into the DNA during chemical synthesis or during synthesis by DNA polymerase. Radioactively-labeled DNA can be visualized by scintillation counting or autoradiography.
- DNA can also be labeled with fluorescent tags or chemical tags such as biotin and digoxigenin, which are detected with different methods.
- Electron microscopy uses electron diffraction through ultrathin sections of a biological sample to visualize the intracellular structures.
- DNA hybridization is a common technique used in Southern blots, Northern blots, and fluorescence *in situ* hybridization (FISH).

Review Questions

1. Describe the differences between exonucleases and endonucleases.
2. During modification events, which bases of DNA most frequently become modified?
3. What does a typical restriction site look like?
4. What term is used to describe two restriction enzymes from different species that share the same recognition site?
5. Do restriction enzymes always cut the DNA at the recognition sequence? Describe the two major classes of restriction enzymes.
6. Which class of restriction enzymes is most useful for genetic engineering? Why?
7. What two types of "ends" do restriction enzymes produce? What is the difference between them?
8. What does DNA ligase do? What can T4 DNA ligase do that bacterial DNA ligase does not?
9. What is a restriction map and how is it produced?
10. What does RFLP stand for? What is an RFLP?

11. What are some practical uses for RFLPs? Must the function of the gene that contains the RFLP be known?
12. How are DNA molecules synthesized chemically? Name two uses for chemically synthesized DNA.
13. How can you chemically synthesize a whole gene?
14. Name two enzymes used in the chemical synthesis of complete genes.
15. What is peptide nucleic acid (PNA)? How does it work?
16. What are morpholino nucleic acids? What are they used for?
17. How does ultraviolet light measure the concentration of nucleic acid?
18. Name the two radioactive isotopes most often used to label nucleic acid. How is the radioactivity incorporated?
19. Name the isotope that is preferred to label nucleic acid. Why?
20. Describe two methods to detect radio-labeled nucleic acid. How do they differ?
21. What is fluorescence? How is fluorescence used in automated DNA sequencing?
22. What is the fluorescence activated cell sorter (FACS) used for?
23. What is chemical tagging? Name two molecular tags used to label DNA.
24. What protein acts as a defense mechanism of chicken eggs from bacterial attack? How does this work?
25. How is DNA tagged with biotin?
26. How are chemical tags detected?
27. Explain the principle of an electron microscope.
28. Define denaturation of DNA, melting temperature, and annealing.
29. What are probe molecules? How are they used in molecular biology?
30. Why does denaturation of DNA increase UV absorption?
31. How are hybrid DNA molecules formed?
32. What are the different types of hybridizations or blotting? How do they differ from one another?
33. What is the principle of FISH technology? Name two uses.

Conceptual Questions

1. Many methods for labeling molecules and then visualizing their location have been presented. Fill in the following table to ensure you understand each of these labeling methods.

Label	What molecule does it tag?	Where does it bind?	How do you visualize the label?
Ethidium Bromide			
Radioisotope			
Fluorescence			
Biotin			
Digoxigenin			

2. Interpret the following zoo blots. Does the DNA probe identified in each gel belong to a non-coding region of DNA or a coding region of DNA?

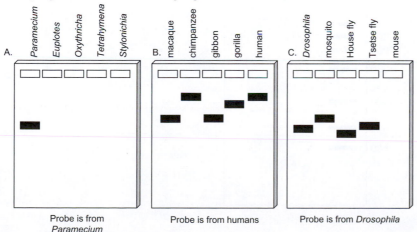

3. The following mother wanted to confirm to her husband that the two children were actually his genetic children. She decided to have a paternity test done. The test was performed by first collecting cheek cell samples from each of the children, the mother, and the father in question. The researcher isolated the DNA and performed RFLP with two different probes. The results of the RFLP are shown below. Is the father the genetic father for the children?

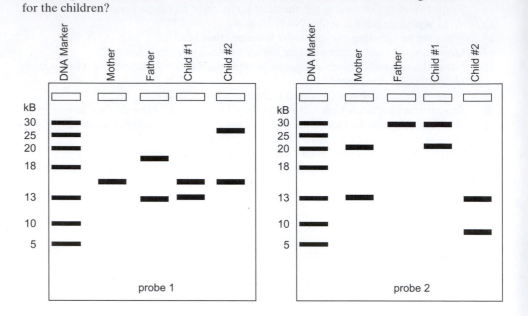

4. Predict the size of each DNA fragment of the following piece of DNA after cutting with an EcoRI and BamHI double enzyme digest. The nucleotide position of each restriction enzyme site is above, and the name of each enzyme is below.

A. Predict the size of each DNA fragment for an EcoRI and BamHI double digest if the DNA fragment is circularized at the cos site.

B. If you separated the restriction enzyme digests in part A and B on an agarose gel using electrophoresis, sketch the pattern you would expect to get on the following picture.

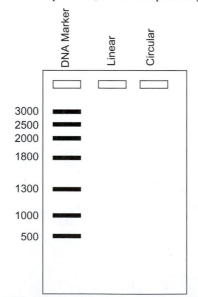

5. Interpret the following data by constructing a restriction map for this plasmid. The first
   *Pst*I site is at position 1 in the plasmid map and the second *Pst*I site is at position 1676. The
   first *Sac*I site is at position 1117.

Restriction Enzyme	Fragment sizes in base pairs
*Pst*I	1676, 1456, 1245, 664
*Sac*I	2567, 1578, 896
*Pst*I and *Sac*I	1245, 1117, 1019, 559, 459, 437, 205

Polymerase Chain Reaction

Chapter 6

Of all the technical advances in modern molecular biology, the **polymerase chain reaction (PCR)** is one of the most useful. PCR provides a means of amplifying DNA sequences, starting with incredibly tiny amounts of DNA. Then purified DNA polymerase makes copy after copy of the DNA in the test tube. This generates amounts of DNA sufficient for a variety of analyses, including both sequencing and cloning. Indeed, PCR is sensitive enough to amplify the DNA from a single cell to yield amounts sufficient for cloning or sequencing. Although PCR requires prior knowledge of sequence information at the two ends of the target sequence, a variety of strategies have been invented that get around this limitation. Not surprisingly, PCR has found applications in every field of molecular biology and biotechnology, both basic and applied. These include DNA cloning and manipulation, constructing transgenic plants and animals, forensic science, medical diagnosis, gene therapy, and environmental analysis. Recent technical advances use fluorescent probes and allow PCR analyses to be run in real time (rather than requiring running gels to analyze the PCR products).

polymerase chain reaction (PCR) Amplification of a DNA sequence by repeated cycles of strand separation and DNA replication

1. Fundamentals of the Polymerase Chain Reaction

PCR allows trace amounts of a DNA sequence to be amplified, giving enough DNA for cloning, sequencing, or other analyses.

PCR has revolutionized the whole area of recombinant DNA technology. Previously, the only way to amplify DNA was to use bacteria to make more copies. This time-consuming process involved cloning the fragment into a plasmid, transforming it into bacteria, growing the bacteria, and isolating the DNA once again. PCR allows the rapid generation of large amounts of specific DNA sequences that are easier to purify without the need for cloning and transformation. As the name indicates, DNA polymerase is used to manufacture DNA using a pre-existing DNA molecule as template. Each new DNA molecule synthesized becomes a template for generating more, thus creating a chain reaction. PCR actually amplifies only a chosen segment (the **target sequence**) within the original DNA template, not the whole template DNA molecule (Fig. 6.01).

The components used in PCR are as follows:

1. The original DNA molecule that is to be copied is called the template and the segment of it that will actually be amplified is known as the target sequence. A trace amount of the DNA template is sufficient.

2. Two **PCR primers** are needed to initiate DNA synthesis. These are short pieces of single-stranded DNA that match the sequences at either end of the target DNA segment. PCR primers are made by chemical synthesis of DNA as described in Chapter 5.

3. The enzyme DNA polymerase is needed to manufacture the DNA copies. The PCR procedure involves several high temperature steps so a heat resistant DNA polymerase is required. Thermostable polymerases are isolated from

FIGURE 6.01
Polymerase Chain Reaction (PCR)

During PCR, two primers anneal to complementary sequences at either end of a target sequence on a piece of denatured template DNA. DNA polymerase synthesizes a complementary strand of DNA from the primers, resulting in two new strands of DNA. In further cycles, the newly made DNA molecules are denatured in turn and duplicated by the same sequence of events, resulting in multiple copies of the original target sequence.

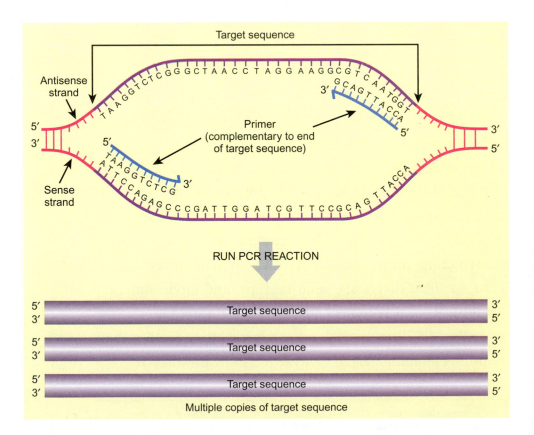

PCR primers   Short pieces of single-stranded DNA that match the sequences at the ends of the target DNA segment, which are needed to initiate DNA synthesis in PCR

target sequence   Sequence within the original DNA template that is amplified in a PCR reaction

bacteria living in hot springs at temperatures up to 90°C. ***Taq* polymerase** from ***Thermus aquaticus*** is most widely used.

4. A supply of nucleotides is needed by the polymerase to make the new DNA. These are supplied as the nucleotide triphosphates.

5. Finally, a **PCR machine** is needed to keep changing the temperature (Fig. 6.02). The PCR process requires cycling through several different temperatures. Because of this, PCR machines are sometimes called **thermocyclers**.

PCR requires template DNA, primers, *Taq* polymerase, nucleotides, and a thermocycler to alternate the temperature of the reaction.

The requirement for primers means that some knowledge of the sequence of the DNA template is needed. Now that the human genome is fully sequenced, as well as many different bacterial and model organism genomes, this information is easily obtained. Unknown sequences are dealt with in a variety of ways (for some specialized approaches see below). However, since binding of a primer need not be perfect, related sequences can often be used successfully, especially if longer primers are used.

> PCR needs primers to start DNA synthesis, which means that we must know some DNA sequence in or close to the region of interest.

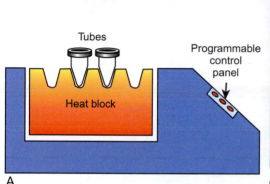

A   B

FIGURE 6.02
PCR Machine or Thermocycler

A) The thermocycler or PCR machine can be programmed to change temperature rapidly. The heat block typically changes from a high temperature such as 90°C (for denaturation) to 50°C (for primer annealing), then back to 70°C (for DNA elongation) in a matter of seconds. This may be repeated for many cycles. B) Rows of GeneAmp PCR machines copying human DNA at the Joint Genome Institute, in Walnut Creek, California, which is a collaboration between three of the U.S. Department of Energy's National Laboratories. (*Credit: David Parker, Science Photo Library.*)

Box 6.01: Kary Mullis Invents PCR After a Vision

Science, like nothing else among the institutions of mankind, grows like a weed every year. Art is subject to arbitrary fashion, religion is inwardly focused and driven only to sustain itself, law shuttles between freeing us and enslaving us."—Kary Mullis

Kary Mullis won the Nobel Prize in Chemistry in 1993 for developing the polymerase chain reaction. PCR is one of modern biology's most useful techniques and has been used in virtually every area of molecular biology and biotechnology. Kary Mullis is one of science's true eccentrics. In addition to molecular biology he has also contributed to other areas of science. While a doctoral candidate working on bacterial iron transport, he published an article entitled "The Cosmological Significance of Time Reversal" (*Nature 218*:663 (1968)), which deals with his notion that about half of the mass in the universe is going backward in time.

PCR machine   See thermocycler
***Taq* polymerase**   Heat-resistant DNA polymerase from *Thermus aquaticus* that is used for PCR
thermocycler   Machine used to rapidly shift samples between several temperatures in a pre-set order (for PCR)
Thermus aquaticus   Thermophilic bacterium found in hot springs and used as a source of thermostable DNA polymerase

Kary Mullis invented PCR while working as a scientist for the Cetus Corporation. He conceived the idea while cruising in a Honda Civic on Highway 128 from San Francisco to Mendocino in April 1983. Mullis recalls seeing the polymerase chain reaction as clear as if it were up on a blackboard in his head, in lurid pink and blue (Fig. 6.03). He pulled over and started scribbling. One basic ingredient of the PCR is that it amplifies DNA by constant repetition—rather like the computer programs Mullis was then involved in writing. Kary Mullis was given a $10,000 bonus by Cetus, who at first failed to realize the significance of the discovery. Later, they sold the technology to Roche for $300,000,000.

In 1999, Kary Mullis mentioned the computer DNA connection again, "It is interesting that biochemistry developed alongside computers. If computers had not come along at about the same time as the structure of DNA was discovered, there would be no biochemistry. You always needed the computer to process the information. Without it we would have rooms and rooms full of monks writing out the sequences."

FIGURE 6.03
Kary Mullis Sees PCR in a Vision

1.1. Cycling Through PCR

Each PCR cycle has three steps: template denaturation, primer annealing, and elongation of new strands of DNA.

There are three basic steps in each cycle of PCR. First, the template DNA is denatured by heating to 90°C or so for a minute or two. Although the primers are present from the beginning, they cannot bind to the template DNA at 90°C. So, in the second step, the temperature is dropped to around 50°C to 60°C, allowing the primers to anneal to the complementary sequences on the template strands (Fig. 6.04). Although the illustration shows 10 base primers, in real life they would be longer, say 15 to 20 bases. A longer primer is more specific for binding to the exact target sequence. In the third step, the temperature is increased to 70°C for a minute or two to allow the thermostable polymerase to elongate new complementary DNA strands starting from the primers (Fig. 6.05). Note that DNA synthesis goes from 5' to 3' for both new

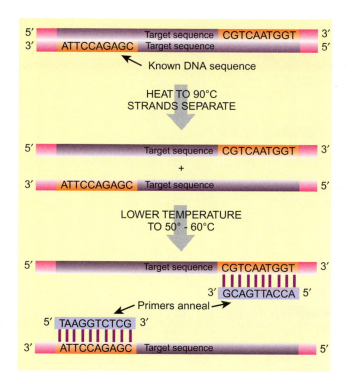

▪▬ **FIGURE 6.04**

Denaturing the Template and Binding the Primers

In the steps of PCR, a very small amount of template DNA is heated to 90°C, which separates the two strands of the double helix. When the temperature is lowered to 50–60°C, the primers can anneal to the ends of the target sequence. Since the primer is present in large excess over the template DNA, essentially all template strands will bind to primers rather than re-annealing to each other.

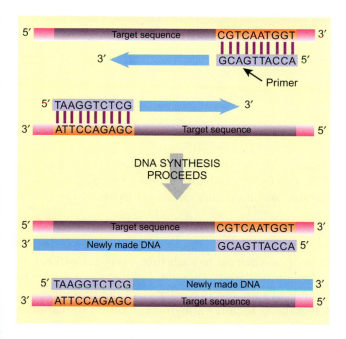

▪▬ **FIGURE 6.05**

Elongation of New Strand by Taq Polymerase

Once the primers have annealed to the template, the temperature is increased to 70°C. This is the optimum temperature for the thermostable *Taq* polymerase to elongate DNA. The polymerase synthesizes new strands of DNA using the 3′ end of the primer as a starting point. A pool of nucleotide precursors is also necessary for this step of the reaction.

strands. This gives two partly double-stranded pieces of DNA. Notice that the two new strands are not as long as the original templates. They are each missing a piece at the end where synthesis started. However, they are double-stranded over the region that matters, the target sequence. The three steps are repeated in each cycle of PCR.

After repeating the same three steps, the second cycle produces four partly double-stranded pieces of DNA (Fig. 6.06). Note again that although they vary in length, they all include double-stranded DNA from the target region. As the cycles continue, the single-strand overhangs are rapidly outnumbered by segments of DNA containing only the target sequence. During the third cycle (Fig. 6.07), the first two pieces of double-stranded DNA that correspond exactly to the target sequence are

FIGURE 6.06
The Second Cycle of the PCR

The entire cycle is repeated starting with the two DNA pieces produced in the first cycle. The two double-stranded pieces of DNA are denatured into four single-stranded pieces at 90°C. The temperature is dropped to 50°C in order for the primers to anneal. Finally, the temperature is heated to 70°C, the optimal temperature for thermostable polymerase to copy the templates. Notice how one DNA molecule has given rise to four molecules.

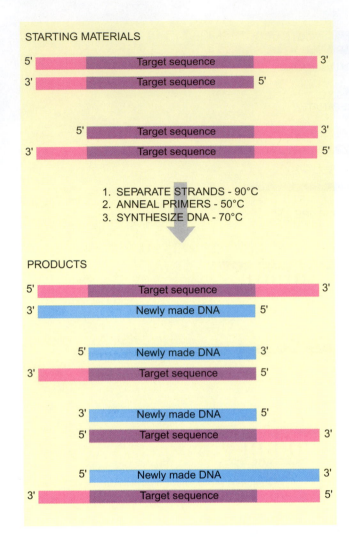

made. These do not have any dangling single-stranded ends. Once past the first two or three cycles, the vast majority of the product is double-stranded target sequence with flush ends. Finally, the DNA generated is run on an agarose gel to assess the size of the PCR fragment.

When Kary Mullis invented PCR in 1987, he used normal DNA polymerase. Since the temperature needed to separate DNA into single strands destroys this enzyme, he had to add a fresh dose of polymerase to each tube every cycle! Luckily, heat resistant DNA polymerase was purified from *Thermus aquaticus* just a year or two later. *Taq* polymerase can be added to the reaction mixture at the beginning and survives all of the heating steps. It actually requires a high temperature to manufacture new DNA.

1.2. PCR Primers

PCR is a wonderful technique, but it may fail if the primers are not correctly designed. PCR primers must be long enough to bind to the specific target DNA, and they must preferentially bind the target DNA rather than themselves. In other words, primers must not have any complementary sequences that would form hairpin or cloverleaf structures. Computer programs have been developed to aid in PCR primer design and are available through the Internet.

Another major problem with PCR is obvious. In order to make the PCR primers, some sequence information is required, at least at the ends of the target sequence. When only a partial sequence is available, one strategy to amplify the target sequence

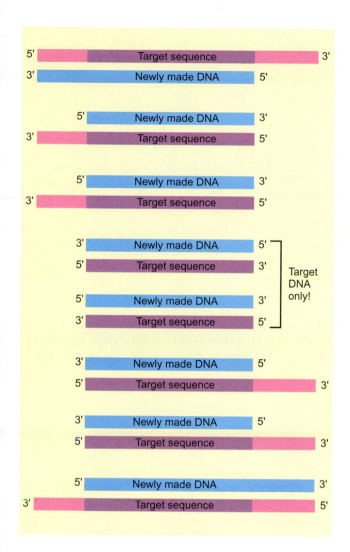

► **FIGURE 6.07**
The Third Cycle of the PCR

The products from the second cycle go through the same process as before. The four double-stranded pieces are denatured into eight single-stranded pieces. The primers anneal and DNA polymerase makes the complementary strands. After this cycle, the number of sequences containing only the target DNA grows exponentially, far exceeding any other product shown in this figure.

is to create **degenerate primers**. For example, let's say the sequence for a gene from one organism is known, and the corresponding gene from another organism is unknown. If two organisms are related, their DNA sequences for a particular gene will be close, although rarely identical. Therefore, degenerate or redundant DNA primers are made that have a mixture of all possible bases in specific positions. Although a mixture of nucleotides is added, each individual primer has only one of the four possible bases, and therefore, degenerate primers represent a mixture of different individual primers. Presumably one of the primers in this mixture will recognize the DNA of the gene of interest. In addition, a perfect match is not really necessary. If, say, 18 of 20 bases match the template, a primer will work quite well. Many segments of DNA have been amplified successfully by PCR using sequence data from close relatives.

Degenerate DNA primers can also be used if only a protein sequence is available. In this case, the protein sequence is translated backwards to give the corresponding DNA sequence (Fig. 6.08). Due to the degeneracy of the genetic code, several possibilities will exist for the sequence of DNA that corresponds to any particular polypeptide sequence. Most of the ambiguity is in the third codon position since most amino acids can be encoded by more than one triplet codon. This ambiguous sequence may be used to make degenerate primers, as before. Although proteins are rarely sequenced in their entirety, short stretches of N-terminal sequence are often obtained.

degenerate primer   Primer with several alternative bases at certain positions

FIGURE 6.08
Designing Degenerate DNA Primers

Degenerate primers are used if only partial DNA sequence information is available. Often, a short amino acid sequence from a protein is known. Because many amino acids are encoded by several alternative codons, the deduced DNA coding sequence is ambiguous. For example, the amino acid tyrosine is encoded by TAC or TAT. Hence, the third base is ambiguous, and when the primer is synthesized a 50:50 mixture of C and T will be inserted at this position. This ambiguity occurs for all the bases shown in red, resulting in a pool of primers with different, but related sequences. Hopefully, one of these primers will have enough complementary bases to anneal to the target sequence that is to be amplified.

Artificial restriction sites are often added to the ends of PCR primers so that the final PCR product can be used directly for cloning into a vector.

Partial sequence of polypeptide:

```
Met---Tyr---Cys---Asn---Thr---Arg---Pro---Gly
```

Possible codons in DNA:

```
ATG   TAC   TGT   AAT   ACT   AGA   GCT   GGT
      TAT   TGC   AAC   ACC   AGG   GCC   GGC
                        ACA         GCA   GGA
                        ACG         GCG   GGG
```

Corresponding redundant primer:

```
ATG   TAC   TGT   AAT   ACT   AGA   GCT   GGT
      T     C     C     C     G     C     C
                        A           A     A
                        G           G     G
```

Bases in the third codon position are shown in red. The redundant primer consists of a mixture of primers with these bases varied as shown.

After separating proteins, automated N-terminal sequencing may yield a sequence of the first dozen or more amino acids. This is often sufficient to allow design of a degenerate probe for hybridization screening of a gene library (see Ch. 7) or degenerate primers for PCR.

1.3. Adding Artificial Restriction Sites

Once a segment of DNA has been amplified by PCR it may be sequenced or cloned (Ch. 7 and 8). For cloning it is often convenient to use restriction enzymes to generate sticky ends on the PCR product. However, it is unlikely that such sites will be located just at the ends of any particular target sequence. One way to create convenient restriction cut sites at the end of PCR fragments is to incorporate them into the primers. When designing the primers, artificial restriction enzyme recognition sites are added at the far ends of the primers (Fig. 6.09). As long as the primer has enough bases to match its target site, adding a few extra bases at the end will not affect the PCR reaction. The bases making up the restriction site get copied and appear on the ends of all newly manufactured segments of DNA. After the PCR reaction has been run, the PCR fragment is cut with the chosen restriction enzyme to generate sticky ends. The fragment is then cloned into a convenient plasmid for cloning or sequencing.

1.4. Alternative Polymerases and PCR Modifications

Since the introduction of *Taq* polymerase, a variety of other thermostable polymerases from other thermophilic organisms have also been used in PCR. *Pfu* polymerase from *Pyrococcus furiosus* is used when increased accuracy is desired because it possesses proofreading ability—unlike *Taq* polymerase. *Tli* polymerase isolated from *Thermococcus litoralis* (Vent™ DNA polymerase) also has 3′ to 5′ proofreading exonuclease activity to create accurate copies.

PCR has been adapted to a variety of different situations. **Long PCR** (or **long range PCR**) modifies PCR so that longer pieces of DNA can be amplified. Normal PCR is limited in the length of target sequence that can be successfully amplified and when the target sequence is longer than 5 kB no PCR product is usually made. A couple of simple modifications are used to amplify long target sequences. First, the polymerase elongation time, or third step in each cycle, is increased from a few

long PCR   PCR reaction used specifically to amplify longer target sequences than standard PCR
long range PCR   See long PCR

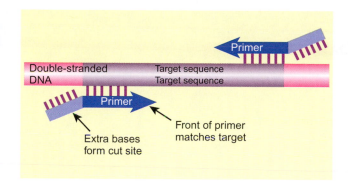

Double-stranded DNA

Target sequence
Target sequence

Primer

Primer

Extra bases
form cut site

Front of primer
matches target

▶**FIGURE 6.09**

Incorporation of Artificial Restriction Sites

Primers for PCR can be designed to have non-homologous regions at the 5′ end that contain the recognition sequence for a particular restriction enzyme. After PCR, the amplified product has the restriction enzyme site at both ends. If the PCR product is digested with the restriction enzyme, this generates sticky ends that are compatible with a chosen vector.

minutes to 10–20 minutes. This gives the polymerase more time to make DNA. Second, a mixture of two different thermostable polymerases is used. Standard *Taq* polymerase cannot proofread whether or not the base it inserted is correct. For short target sequences, this is not an issue, but longer sequences have a higher chance of accumulating mistakes. However, other thermostable polymerases (e.g., *Pfu* or *Tli* polymerase) are known that do possess proofreading ability. If a small amount of *Pfu* or *Tli* polymerase is added, these proofread any mistakes made by *Taq* and replace the incorrect nucleotide with a correct one. Third, long target sequences are often damaged by heat and the denaturation step can cause depurination. Therefore, in long PCR the denaturation times are decreased from one minute to 2–10 seconds to protect the purine bases. Fourth, the buffer conditions are modified to provide better pH and ionic concentrations for the alternative polymerases.

Hot start PCR is a variant of regular PCR that reduces non-specific amplification at the lower temperature steps of the PCR cycle, especially during the original setup. The key idea is to prevent *Taq* polymerase activity at low temperatures during setup and initial annealing stage. Several different approaches have been used that improve specificity. The physical approach simply keeps the reagents apart until activated by heat. Other approaches inhibit DNA polymerase by antibodies or blocking proteins that are released at high temperatures. Finally, thermolabile inactive derivatives of either primers or dNTPs that break down at high temperatures may be used to prevent premature DNA polymerization.

> Using alternate thermostable DNA polymerases is useful to increase the fidelity of PCR amplification.

2. Inverse PCR

Another approach that uses incomplete sequence information to amplify a target gene is **inverse PCR**. In this case a sequence of part of a long DNA molecule, say a chromosome, is known. The objective is to extend the analysis along the DNA molecule into the unknown regions. To synthesize the primers for PCR, the unknown target sequence must be flanked by two regions of known sequence. The present situation is exactly the opposite of that. To circumvent this problem, the target molecule of DNA is converted into a circle. Going around a circle brings you back to the beginning. In effect, even though only one small stretch of sequence is known, the circular form allows you to have that one region on both sides of the target sequence.

A restriction enzyme, usually one that recognizes a six-base sequence, is used to make the circle. This enzyme must not cut into the known sequence, but it will cut upstream and downstream from the known region. The resulting fragment will have unknown sequence first, the known sequence in the middle, followed by more unknown sequence. The two ends of the fragment will have compatible sticky ends that are easily ligated together to make a circle of DNA (Fig. 6.10). Two primers corresponding to the known region and facing outwards around the circle are used

> Performing PCR on a circularized DNA template amplifies neighboring regions of unknown sequence.

hot start PCR PCR in which *Taq* polymerase is sequestered by antibodies or blocking proteins from the remaining ingredients until the template DNA is fully denatured

inverse PCR Method for using PCR to amplify unknown sequences by circularizing the template molecule

FIGURE 6.10
Inverse PCR

Inverse PCR allows unknown sequences to be amplified by PCR provided that they are located next to DNA in which the sequence is already known. The DNA is cut with a restriction enzyme that does not cut within the region of known sequence, as shown in Step 1. This generates a fragment of DNA containing the known sequence flanked by two regions of unknown sequence. Since the fragment has two matching sticky ends, it may be easily circularized by DNA ligase. Finally, PCR is performed on the circular fragments of DNA (Step 2). Two primers are used that face outwards from the known DNA sequence. PCR amplification gives multiple copies of one linear product that includes unknown DNA from both left and right sides.

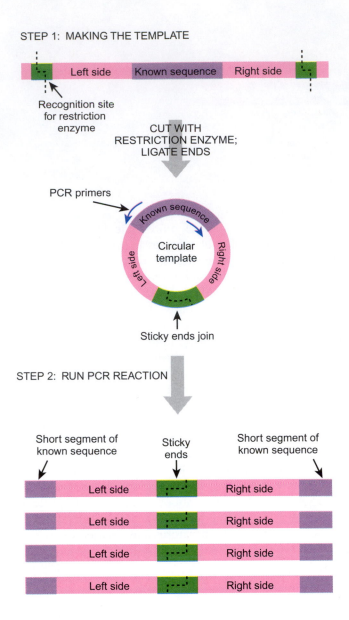

STEP 1: MAKING THE TEMPLATE

Recognition site for restriction enzyme

CUT WITH RESTRICTION ENZYME; LIGATE ENDS

PCR primers

Circular template

Sticky ends join

STEP 2: RUN PCR REACTION

Short segment of known sequence

Sticky ends

Short segment of known sequence

for PCR. Synthesis of new DNA will proceed around the circle clockwise from one primer and counter-clockwise from the other. Overall, inverse PCR gives multiple copies of a segment of DNA containing some DNA to the right and some DNA to the left of the original known region.

3. Randomly Amplified Polymorphic DNA (RAPD)

Randomly Amplified Polymorphic DNA, or **RAPD**, is usually found in the plural as RAPDs and is pronounced "rapids," partly because it is a quick way to get a lot of information about the genes of an organism under investigation. The purpose of RAPDs is to test the relatedness of two organisms. In practice, DNA samples from unknown organisms are compared with DNA from a previously characterized organism. For example, traces of blood from a crime scene may be compared to possible suspects, or disease-causing microorganisms may be related to known pathogens to help trace an epidemic.

randomly amplified polymorphic DNA (RAPD)   Method for testing genetic relatedness using PCR to amplify arbitrarily chosen sequences

The principle of RAPDs is statistically based. Given any particular five-base sequence, such as ACCGA, how often will this exact sequence appear in any random length of DNA? Since there are four different bases to choose from, one in every 4^5 (or $4 \times 4 \times 4 \times 4 \times 4 = 1{,}024$) stretches of five bases will—on average—be the chosen sequence. Any arbitrarily chosen 11-base sequence will be found once in approximately every 4 million bases. This is approximately the amount of DNA in a bacterial cell. In other words, any chosen 11-base sequence is expected to occur by chance once only in the entire bacterial genome. For higher organisms, with much more DNA per cell, a longer sequence would be needed for uniqueness.

For RAPDs, the arbitrarily-chosen sequence should be rare but not unique. PCR primers are made using the chosen sequence and a PCR reaction is run using the total DNA of the organism as a template. Every now and then a primer will find a correct match, purely by chance, on the template DNA (Fig. 6.11). For PCR amplification to occur there must be two such sites facing each other on opposite strands of the DNA. The sites must be no more than a few thousand bases apart for the reaction to work well. The likelihood of two correct matches in this arrangement is quite low.

In practice, the length of the primers is chosen to give 5 to 10 PCR products. For higher organisms, primers of around 10 bases are typical. The bands from PCR are separated by gel electrophoresis (see Ch. 4) to measure their sizes. The procedure is repeated several times with primers of different sequences. The result is a diagnostic pattern of bands that will vary in different organisms, depending on how closely they are related. Although we do not know in which particular genes the PCR bands originate, this does not matter in measuring relatedness. Diagnosis therefore relies on having a primer (or set of primers) that reliably give a band of a particular size with the target organism and give different bands with other organisms, even those closely related. RAPD results using such a primer are shown in Figure 6.12. Grey mold, due

> PCR may be performed with arbitrary primers. Comparing results from two samples of DNA reveals their relatedness.

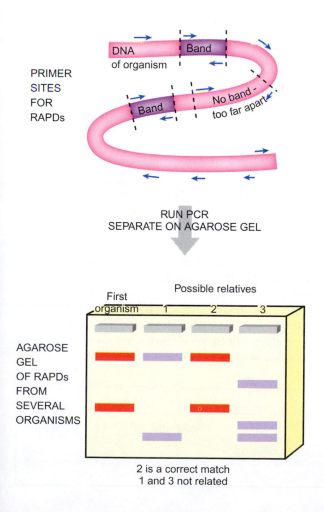

PRIMER SITES FOR RAPDs

RUN PCR
SEPARATE ON AGAROSE GEL

AGAROSE GEL OF RAPDs FROM SEVERAL ORGANISMS

2 is a correct match
1 and 3 not related

■ **FIGURE 6.11**
Randomly Amplified Polymorphic DNA

The first step of RAPD analysis is to design primers that will bind to genomic DNA at random sites that are neither too rare nor too common. In this example, the primers were sufficiently long to bind the genomic DNA at a dozen places. For PCR to be successful, two primers must anneal at sites facing each other but on opposite strands. In addition, these paired primer sites must be close enough to allow synthesis of a PCR fragment. In our example, there are three pairs but only two of these pairs were close enough to actually make the PCR product. Consequently, this primer design will result in two PCR products as seen in the first lane of the gel (marked "First organism"). The same primers are then used to amplify genomic DNA from other organisms that are suspected of being related. In this example, suspect #2 shows the same banding pattern as the first organism and is presumably related. The other two suspects do not match the first organism and are therefore not related.

FIGURE 6.12
Identification of Fungal Pathogens by RAPD

RAPD banding patterns generated using the 10-base primer, AACGCGCAAC, on genomic DNA of five closely related strains of the pathogen *Botrytis cinerea* (lanes 1–5), three other strains from the genus *Botrytis* (lanes 6, 7, & 8) and several less related harmless fungi *Alternaria* (9), *Aspergillus* (10), *Cladosporium* (11), *Epicoccum* (12), *Fusarium* (13), *Hainesia* (14), *Penicillium* (15), *Rhizoctonia* (16 & 17), and the host plant, strawberry (18). Lane 0: negative control (no DNA). Lane M: molecular mass marker. (*Credit: Rigotti et al., FEMS Microbiology Letters (2002) 209: 169–174.*)

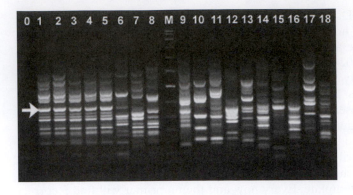

to *Botrytis cinerea*, is one of the most destructive infections of strawberries and also attacks other plants. Classical diagnosis involves culturing the fungus on nutrient agar. It is slow and difficult due to the presence on the plants of other harmless fungi, which often grow faster in culture. As can be seen, RAPD analysis clearly identifies the pathogens from other related fungi, including other species from the genus *Botrytis*.

4. Reverse Transcriptase PCR

The coding sequence of most eukaryotic genes is interrupted by intervening sequences, or introns (see Ch. 12 for introns and RNA processing). Consequently, the original version of a eukaryotic gene is very large, difficult to manipulate, and virtually impossible to express in any other type of organism. Since mRNA has had the introns removed naturally, it may be used as the source of an uninterrupted coding sequence that is much more convenient for engineering and expression. This involves converting the RNA back into a DNA copy, known as **complementary DNA (cDNA)** by **reverse transcriptase**. Thus, when amplifying eukaryotic genes by PCR the cDNA version is often used (rather than the true chromosomal gene sequence) since this lacks the introns.

Reverse transcriptase is an enzyme found in retroviruses that converts the RNA genome carried in the retrovirus particle into double-stranded DNA. Reverse transcriptase first transcribes a complementary strand of DNA to make an RNA:DNA hybrid. Next, reverse transcriptase or RNase H degrades the RNA strand of the hybrid. The single-stranded DNA is then used as a template for synthesizing double-stranded DNA (cDNA). Once the cDNA has been made, PCR can be used to amplify the cDNA and generate multiple copies (Fig. 6.13). This combined procedure is referred to as **reverse transcriptase PCR (RT-PCR)** and allows genes to be amplified and cloned as intron-free DNA copies starting from mRNA.

Performing RT-PCR on an organism under different growth conditions reveals when a gene of interest is expressed (i.e., when the corresponding mRNA is present) and what environment induces gene expression. To compare two different conditions, mRNA is extracted from cells growing in both conditions. RT-PCR is performed on the two samples of mRNA using PCR primers that match the particular gene of interest. If the gene is expressed, a PCR product will be produced, whereas if the gene is switched off, its mRNA will be missing (Fig. 6.14).

> Reverse transcription followed by PCR allows cloning of genes starting from the messenger RNA, and thus, identifying the expressed exons of the eukaryotic gene.

complementary DNA (cDNA) Version of a gene that lacks the introns and is made from the corresponding mRNA by using reverse transcriptase

reverse transcriptase Enzyme that starts with RNA and makes a DNA copy of the genetic information

reverse transcriptase PCR (RT-PCR) Variant of PCR that allows genes to be amplified and cloned as intron-free DNA copies by starting with mRNA and using reverse transcriptase

Original gene

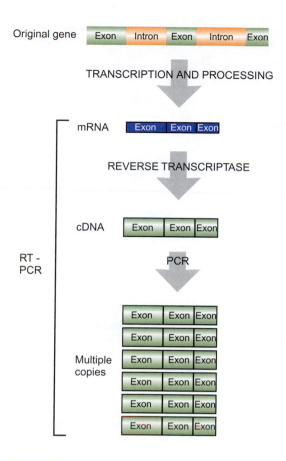

■ **FIGURE 6.13**
Reverse Transcriptase PCR

RT-PCR is a two-step procedure that involves making a cDNA copy of the mRNA, then using PCR to amplify the cDNA. First, a sample of mRNA (which lacks introns) is isolated. Reverse transcriptase is used to make a cDNA copy of the mRNA. The cDNA sample is then amplified by PCR. This yields multiple copies of cDNA without introns.

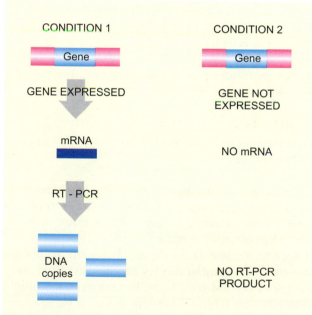

■ **FIGURE 6.14**
RT-PCR for Gene Expression

RT-PCR can determine the amount of mRNA for a particular gene in two different growth conditions. In this example, the gene of interest is expressed in condition 1 but not in condition 2. Therefore, in condition 1 mRNA from the gene of interest is present, reverse transcriptase generates a cDNA, and PCR amplifies this cDNA into many copies. In condition 2 the mRNA is absent and so the RT-PCR procedure does not generate the corresponding DNA.

5. Differential Display PCR

The mixture of mRNA made by a cell can be surveyed by a PCR-based approach.

Differential display PCR is used to specifically amplify mRNA from eukaryotic cells. The technique is valuable because it allows the researcher to assess the expression of many different mRNA molecules simultaneously. This technique is a combination of

differential display PCR   Variant of RT-PCR that specifically amplifies messenger RNA from eukaryotic cells using oligo(dT) primers

FIGURE 6.15
Differential Display PCR

Differential display PCR allows simultaneous measurement of the expression of many different mRNA molecules. The example shows that under the conditions used, three of the genes are turned on and three are turned off. Those mRNA molecules that are expressed are converted to cDNA by reverse transcriptase using an oligo(dT) primer, then amplified by PCR. The reverse PCR primer is oligo(dT) and so binds to the poly(A) sequences. The forward PCR primer is a mixture of random sequences, calculated to anneal approximately once per cDNA. These primers ensure that many different cDNA molecules are amplified rather than just one. In this example, three PCR products are produced, corresponding to the original genes that are expressed.

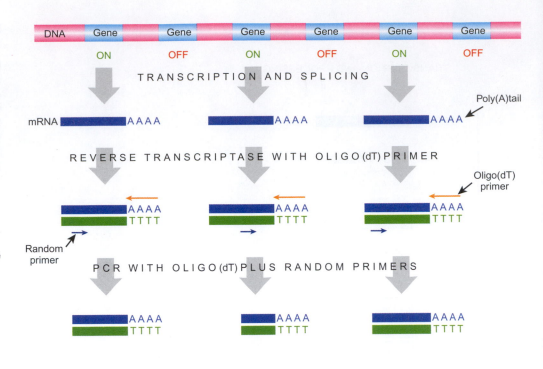

RAPD (see above) with RT-PCR and has one clever modification of its own, the use of oligo(dT) primers. Since almost all eukaryotic mRNA molecules have a 3'-tail of poly(A), an artificial primer made only of dT will base pair to this tail. This method of PCR allows the researcher to compare two different growth conditions on many different genes, rather than just one gene as in RT-PCR.

As in RT-PCR the RNA is extracted from the cells and the corresponding cDNA is made by the use of reverse transcriptase using a primer that recognizes the poly(A) tail. Then a PCR reaction is run with two primers (Fig. 6.15):

1. An oligo(dT) primer that binds to the 3' end of each cDNA
2. A mixture of random primers (similar to those used in RAPDs) with an optimal length that anneals to many cDNAs

These two primers ensure that there are not too many or too few amplified fragments. Then, as usual, the PCR products are separated by size using gel electrophoresis. This gives a series of DNA bands corresponding to each of the mRNAs being made in the cells that were analyzed. The pattern is specific for those mRNAs present at that moment in time, and under the specific growth conditions. This pattern changes depending upon the conditions, therefore, differential display can be used to identify new genes that are only expressed under certain conditions. In many cases multiple bands will appear or disappear, thus allowing multiple random genes to be analyzed rather than just a single gene of interest as in RT-PCR. This type of analysis may be used to compare expression patterns in development, the effect of stresses such as heat shock, normal versus cancer cells, etc.

Rescuing the "lost" ends of cloned genes may be done by a rapid amplification of cDNA ends (RACE). The 5' end of the primers used is modified by adding anchor sequences for subsequent reactions.

6. Rapid Amplification of cDNA Ends (RACE)

Using only reverse transcriptase, full-length cDNA copies are hard to get, especially from mRNA that is present only in very low amounts or unusually long. Reverse transcriptase often fails to reach the end of a long RNA template due to hindrance by RNA secondary structure, thus the 5'-end is often incomplete. Consequently, some means of recovering the complete cDNA is needed. The **RACE** technique generates the complete cDNA in two halves; hence, the name **rapid amplification of cDNA ends**. It is necessary to know part of the internal sequence of the mRNA/cDNA in order to design the internal primers; therefore, the technique is generally used when an incomplete cDNA was isolated by other techniques such as library screening (see Ch. 7) or incomplete PCR amplification. The RACE procedure is essentially a modification of RT-PCR, but unique so-called **anchor sequences** are added to each end of the cDNA to facilitate the PCR portion of the reaction (Fig. 6.16).

RACE can amplify either the 3′ end or 5′ end of a gene depending upon the primers used. The 3′-reaction of RACE-PCR primes reverse transcriptase to synthesize a DNA copy from the poly(A) tail of the mRNA by using an oligo(dT) primer that has a unique anchor sequence at the 5′ end. Since the internal sequence is known, an internal primer is designed so that PCR will amplify from the middle of the gene to the anchor sequence. In the 5′-reaction, the internal primer is used to initiate DNA synthesis using reverse transcriptase. Next, an artificial poly(A) tail is added to the 3′ end of the DNA by terminal transferase and dATP. An alternative to terminal transferase is to ligate a different anchor primer onto the 3′ end and use a PCR primer complementary to this sequence in the subsequent PCR reaction. The same oligo(dT)/anchor primer as used to initiate the 3′-reaction is then used again during the PCR amplification cycles for the 5′-reaction. The anchor sequence primer and internal primers are generally designed to include convenient restriction sites to allow further cloning and sequencing.

Collins RWJ, Littink KW, Klevering BJ, van den Born LI, Koenekoop RK, Zonneveld MN, Blokland EAW, Strom TM, Hoyng CB, den Hollander AI, Cremers FPM (2008) Identification of a 2 Mb human ortholog of *Drosophila eyes shut/spacemaker* that is mutated in patients with retinitis pigmentosa. Am J Hum Genetics, 83: 594–603.

The focus for this chapter of the textbook is to familiarize the reader with a widely used technique, PCR. This paper uses a variety of PCR-based methods to isolate the entire mRNA transcript of a gene implicated in a hereditary human disease called retinitis pigmentosa (RP). This is an autosomal recessive genetic disease that is caused by accumulation of phytanic acid in the eye, which damages the retina. Early stages of the disease are characterized by reduced night vision and loss of peripheral vision. As the disease progresses, central vision is also lost. In the human genome, locus RP25 has been implicated in some forms of RP, although other mutations can also cause the disease. The RP25 region contains several genes, but the exact genetic mutation that induces RP is still unknown. To identify a potential mutation that may cause RP, the authors of this paper used homozygosity mapping to find regions within RP25 that are homozygous in a large number of different individuals, especially families affected by RP. Since RP is autosomal recessive, both alleles of the causative gene must have the same mutation. In addition, comparing the unaffected sibling should show that they either do not contain the same markers, or more likely that these individuals will be heterozygous for the alleles. Comparing the sequence of RP25 among a family afflicted with RP identified a region (6q12-q11.1) that was homozygous in two siblings with RP. Therefore, this was a good candidate region to examine for defective genes. This DNA region has five predicted genes of which only one is expressed in the eye according to previous analyses. Using RT-PCR confirmed that this gene, *EGFL11*, was expressed in the human retina.

The authors then identified all the exons in this gene. Mature mRNA was isolated from retinal cells and subjected to RT-PCR. Because the cDNA was very long, sequences on the 5′ and 3′ ends were missing from the RT-PCR product, so the authors used 5′ and 3′ RACE to obtain a full length transcript. They found the mRNA contained 44 different exons that spanned 2 Mb of genomic DNA. Some exons were not even predicted by computer analysis, confirming that experiments are essential to confirm—or revise—computer predictions. The entire cDNA/mRNA was determined to be 10,475 nucleotides long!

FOCUS ON RELEVANT RESEARCH

The authors then compared this sequence to genes from other organisms. The most closely related gene is *eyes shut* or *spacemaker* from *Drosophila*. This gene is essential for photoreceptor development in the insect eye, and the authors decided to name the human gene *eyes shut homolog* (*EYS*). Next, the tissue distribution of *EYS* mRNA was confirmed by RT-PCR analysis of various human tissues. The *EYS* gene is expressed primarily in the retina, confirming that this gene could be a cause of RP. In some patients with RP, the *EYS* gene had a mutation that caused premature termination of protein synthesis, resulting in a protein 10 amino acids shorter than the wild-type version. The same mutation was found in a completely unrelated family. However, in another patient, the same gene was mutated, but in a different location. Taken together, this evidence suggests that defects in the *EYS* gene of humans could be one cause of retinitis pigmentosa.

anchor sequence   Sequence added to primers or probes that may be used for binding to a support or may incorporate convenient restriction sites, primer binding sites for future manipulations, or primer binding sites for subsequent PCR reactions
RACE   See rapid amplification of cDNA ends
rapid amplification of cDNA ends (RACE)   RT-PCR-based technique that generates the complete 5′ or 3′ end of a cDNA sequence starting from a partial sequence

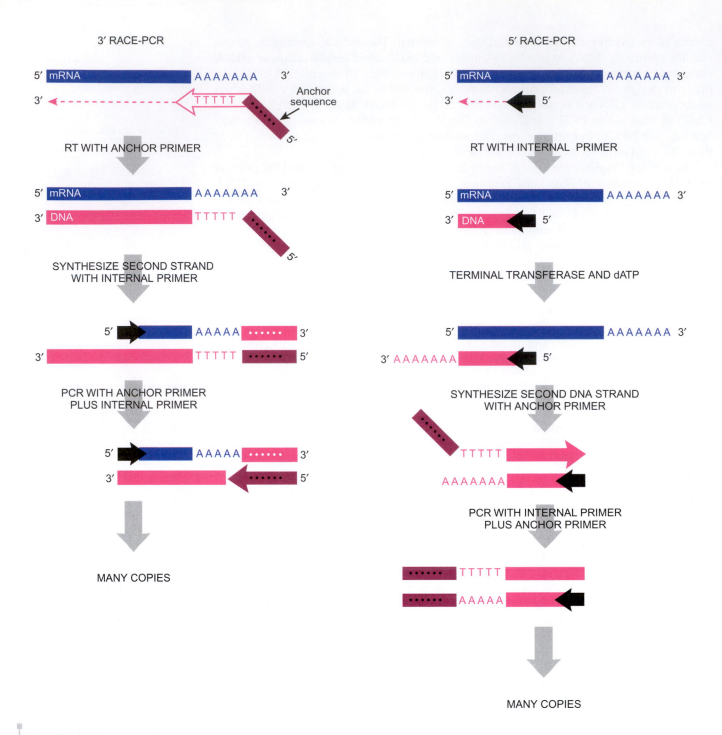

FIGURE 6.16
Rapid Amplification of cDNA Ends (RACE)

RACE is used to isolate the 5′ and/or 3′ ends of an incomplete cDNA. The method to amplify the 3′ end of the cDNA is shown on the left. This requires an oligo(dT) primer that has an anchor sequence at the 5′ end. This primer is used with reverse transcriptase to make the mRNA:DNA hybrid molecule. The mRNA portion of this is removed and a second strand of DNA is synthesized using an internal primer. The same internal primer and anchor sequence primer are then used in a standard PCR reaction to amplify the 3′ end. The right side shows the 5′ RACE. An internal primer is designed to prime reverse transcriptase and make a hybrid molecule of mRNA:DNA. In order to add a primer binding site upstream of the end of the hybrid molecule, the enzyme terminal transferase is added together with dATP. This enzyme adds a run of adenines to the 3′ end of the DNA half of the hybrid. The mRNA half of the hybrid is then removed and replaced with DNA by using an oligo(dT) primer carrying the anchor sequence. The oligo(dT) binds the newly synthesized poly(A) stretch on the DNA and primes the polymerase to make a cDNA. Subsequent PCR using the internal primer and the anchor primer amplify only the 5′ end of the cDNA.

7. PCR in Genetic Engineering

A whole plethora of modifications has been made to the basic PCR scheme. Some of these are used to alter genes rather than to analyze them. We can divide these modifications into two broad categories: a) rearranging large stretches of DNA and b) changing one or two bases of a DNA sequence. The latter is discussed below as "directed mutagenesis."

As already illustrated, PCR can amplify any segment of DNA provided there are primers that match its ends. Such segments of DNA may be joined or rearranged in a variety of ways. In order to make a hybrid gene, segments of two different genes can be amplified by PCR and then joined together. There are several protocols that vary in their details. But the crucial point is to use an **overlap primer** that matches part of both gene segments (Fig. 6.17).

Instead of just two primers, overlap PCR reactions use a primer for the front end of the first gene, a primer for the rear end of the second gene, and an overlap primer. When the amplification is complete, the front end of the first gene is connected to the second gene. Some variants of this **"molecular sewing"** make the two halves separately and mix and join them later; other versions of this technique mix all three primers plus both templates in a single large reaction.

By making hybrid genes using components from various sources it is sometimes possible to work out in detail which regions of a gene or protein are responsible for precisely which properties. The approach can also be used in biotechnology to construct artificial genes made up of genetic modules from different sources.

> Overlap primers connect two segments of DNA that are not normally next to each other.

8. Directed Mutagenesis

The term **directed mutagenesis** refers to a wide variety of *in vitro* techniques that are used to deliberately change the sequence of a gene. Several of these techniques use a PCR approach. The most obvious way to change one or two bases in a segment of DNA is to synthesize a PCR primer that carries the required alterations. Consider the sequence AAG CCG **GAG** GCG CCA. Suppose we wish to alter the A in the middle of this sequence to a T. Then a PCR primer with the required base alteration is made; that is, AAG CCG **GTG** GCG CCA. This mutant primer is used as one of a pair of PCR primers to amplify the appropriate segment of DNA using wild type DNA as the template. The PCR product will contain the desired mutation, close to one end.

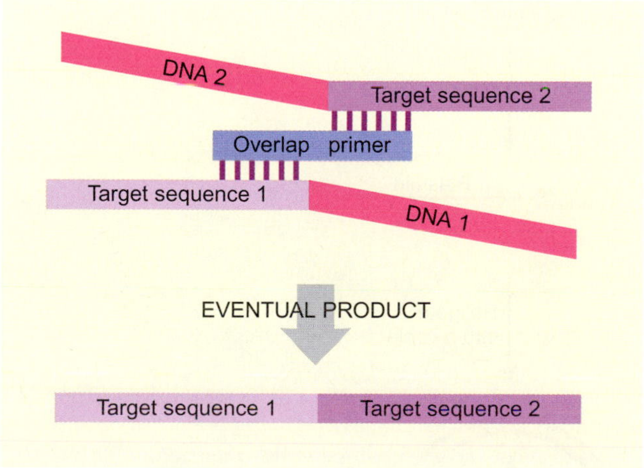

► FIGURE 6.17

Synthesis of Hybrid Gene by Using Overlap Primers

Overlapping primers can be used to link two different gene segments. In this scheme, the overlapping primer has one end with sequences complementary to target sequence 1 and the other half similar to target sequence 2. The PCR reaction will create a product with these two regions linked together.

---

directed mutagenesis   Deliberate alteration of the DNA sequence of a gene by any of a variety of artificial techniques
molecular sewing   Creation of a hybrid gene by joining segments from multiple sources using PCR
overlap primer   PCR primer that matches small regions of two different gene segments and is used in joining segments of DNA from different sources

As long as this primer is long enough to bind to the correct location on either side of the mutation, the DNA product will incorporate the change made in the primer. The mutant PCR product must then be reinserted into the original gene at the correct place.

Alternatively, if the target gene is on a plasmid vector, the mutational primer can be used with a primer that anneals to the opposite strand, just upstream of the mutation. The PCR product will have the mutation at one end and encompass the entire gene and plasmid (Fig. 6.18). The mutant gene is reassembled by ligating the ends of this PCR product.

Both alternatives yield PCR products with the majority containing the mutation, but a few copies of the original DNA without the mutation remain. One strategy to get rid of these is to digest them with methylation-sensitive restriction enzymes. The original non-mutated DNA is normally methylated because it was isolated from a living organism. In contrast, PCR products made *in vitro* are not methylated. If the PCR reaction products are digested with a restriction enzyme that only cuts methylated recognition sequences, then any original non-mutated DNA molecules are cut but PCR products with the mutation are unaltered.

Another less controlled way to introduce mutations into PCR products is to use manganese. *Taq* polymerase requires magnesium ions for proper function. Replacement of magnesium by manganese allows the enzyme to continue to synthesize DNA, but the accuracy is greatly reduced. This approach introduces random base changes and yields a mixture of different mutations from a single PCR reaction. The error rate depends on the manganese concentration, so it is possible to get single or multiple mutations as desired.

> Mutations may be inserted artificially into DNA by altering a few bases in a PCR primer.

FIGURE 6.18
Directed Mutagenesis

On the left, a mutagenic PCR primer is paired with a reverse primer at the end of the gene. The PCR product has a portion of the gene with the mutation, but this piece must be rejoined to the remaining gene fragment. On the right, the mutagenic primer and reverse primer recognize sequences next to each other in the gene. After PCR, the entire plasmid is amplified from the mutagenic site to the other end of the gene. The gene is recovered by ligating the PCR product.

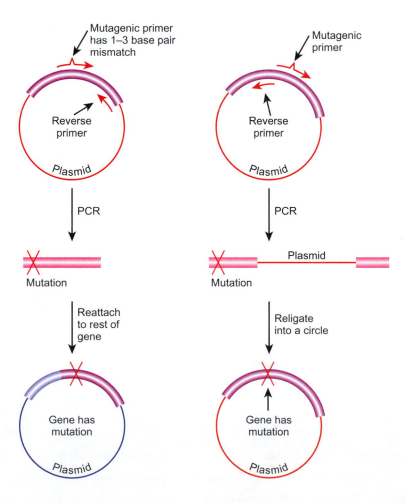

9. Engineering Deletions and Insertions by PCR

PCR is also used to generate DNA cassettes that can be inserted into chromosomes by homologous recombination (see Ch. 24). Such cassettes contain a convenient marker gene, usually an antibiotic resistance gene, flanked by DNA identical to the sequence of the chosen insertion site. Cassettes are created by using PCR primers that overlap the resistance gene and also contain about 40–50 bp of DNA homologous in sequence to the target location (Fig. 6.19). The cassette is transformed into the host organism and is inserted into the chromosome by homologous recombination. Antibiotic resistance is then used to select those organisms that have gained the cassette. If the cassette DNA flanking the antibiotic resistance gene recognizes adjacent sequences in the host genome, then the cassette is inserted into the chromosome. If the cassette DNA is homologous to sequences that are far apart in the chromosome, or perhaps flank a gene of interest, then the cassette replaces that DNA, and the original chromosomal fragments are lost.

A collection of yeast strains deleted for all known genes has been generated by this procedure. Each strain has had a single coding sequence replaced by a cassette

PCR primers may be used to precisely insert segments of foreign DNA into a chromosome or other DNA molecule.

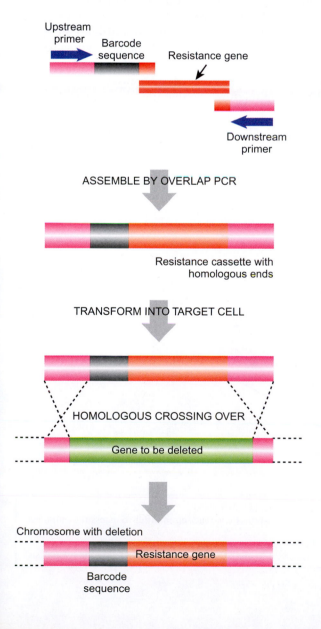

ASSEMBLE BY OVERLAP PCR

Resistance cassette with homologous ends

TRANSFORM INTO TARGET CELL

HOMOLOGOUS CROSSING OVER

Gene to be deleted

Chromosome with deletion

FIGURE 6.19

Generation of Insertion or Deletion by PCR

In the first step, a specifically-targeted cassette is constructed by PCR. This contains both a suitable marker gene and upstream and downstream sequences homologous to the chromosomal gene to be replaced. The engineered cassette is transformed into the host cell and homologous crossing over occurs. Recombinants are selected by the antibiotic resistance carried on the cassette.

comprising the *npt* gene plus a **barcode** or **zipcode sequence**. The *npt* gene encodes neomycin phosphotransferase, which confers resistance to neomycin and kanamycin on bacteria and resistance to the related antibiotic geneticin on eukaryotic cells, such as yeast. A barcode sequence is a unique sequence of around 20 base pairs that is included as a molecular identity tag. Each insertion has a unique barcode sequence allowing it to be tracked and identified. Such barcode or zipcode sequences are increasingly being used in high volume DNA screening projects where it is necessary to keep track of many similar constructs (Fig. 6.20).

Clearly, if the cassette carried another gene in addition to the antibiotic resistance marker, then the above procedure can also generate gene insertions. Thus, any foreign gene may be inserted by this approach. There is no need to delete a resident gene if the objective is the insertion of an extra gene. All that is necessary is that the incoming gene must be flanked with appropriate lengths of DNA homologous to some location on the host chromosome.

10. Real-Time Fluorescent PCR

Recently methods have been developed that allow PCR reactions to be followed in real time by monitoring the increased emission from fluorescent probes (see Ch. 5). Instruments for combined PCR and fluorescence detection carry out the PCR reaction in glass capillary tubes. Glass allows light to pass and activate the fluorophore and monitor the fluorescence emissions as the reaction occurs. In **multiplex PCR**, the thermocycler monitors several different fluorescent dyes simultaneously, allowing several reactions to be run in the same tube.

DNA-binding fluorescent probes whose fluorescence increases upon binding to DNA are included in the PCR reaction mixture. As the amount of newly synthesized target DNA increases, the probe binds to the target DNA and the fluorescence emission increases. The simplest DNA-binding fluorescent probes are not sequence specific. An example is the dye **SYBR Green I** (Molecular Probes, Eugene, OR),

> PCR has been modified for rapid diagnosis by using fluorescent dyes to follow double-stranded DNA accumulation.

FIGURE 6.20
Barcode or Zipcode Sequences

Barcodes are small regions of DNA with sequences that are unique and not found anyplace else in the host genome.

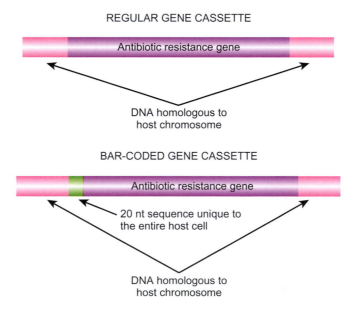

barcode sequence   A unique 20 base pair sequence added to a cassette to mark the gene with a tag for subsequent analysis
multiplex PCR   Using different fluorescent dyes on different probes in a quantitative or real-time PCR reaction in order to assess the amplification of more than one target sequence
SYBR Green I   A DNA-binding fluorescent dye that binds only to double-stranded DNA and becomes fluorescent only when bound
zipcode sequence   See barcode sequence; A unique 20 base pair sequence added to a cassette to mark the gene with a tag for subsequent analysis

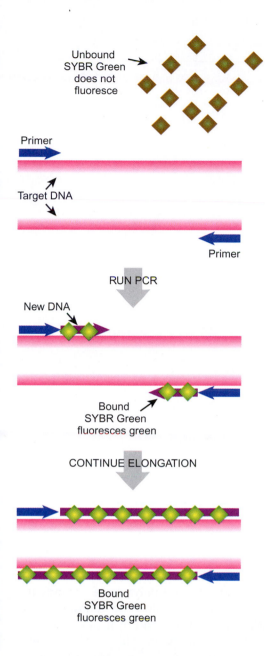

Unbound
SYBR Green
does not
fluoresce

Primer

Target DNA

Primer

RUN PCR

New DNA

Bound
SYBR Green
fluoresces green

CONTINUE ELONGATION

Bound
SYBR Green
fluoresces green

■ **FIGURE 6.21**

***Real-Time Fluorescent PCR
with SYBR Green***

When the fluorescent probe SYBR
green is present during a PCR
reaction, it binds to the double-
stranded PCR product and emits light
at 520 nm. The SYBR Green dye
only fluoresces when bound to DNA,
therefore, the amount of fluorescence
correlates with the amount of PCR
product produced. The accumulation
of PCR product is followed through
many cycles by measuring the
amount of fluorescence.

with fluorescence emission at 520 nm. This binds only to double-stranded DNA and becomes fluorescent only when bound (Fig. 6.21).

SYBR Green monitors the total amount of double-stranded DNA but cannot distinguish between different sequences. To be sure that the correct target sequence is being amplified a sequence-specific fluorescent probe is needed. An example is the **TaqMan® probe** (Applied Biosystems, Foster City, CA). The TaqMan probe consists of two fluorophores linked by a DNA sequence that will hybridize to the middle of the target DNA. **Fluorescence resonance energy transfer (FRET)** transfers the energy from the short-wavelength fluorophore on one end to the long-wavelength fluorophore at the other end. This quenches the short wave emission (Fig. 6.22).

---

fluorescence resonance energy transfer (FRET)   Transfer of energy from short-wavelength fluorophore to long-wavelength fluorophore so quenching the short wave emission

TaqMan probe   Fluorescent probe consisting of two fluorophores linked by a DNA probe sequence. Fluorescence increases only after the fluorophores are separated by degradation of the linking DNA

FIGURE 6.22
Real-Time Fluorescent PCR with TaqMan Probe

The TaqMan probe has three elements: a short-wavelength fluorophore on one end (diamond), a sequence that is specific for the target DNA (blue), and a long-wavelength fluorophore at the other end (circle). The two fluorophores are so close that fluorescence is quenched and no green light is emitted. This probe is designed to anneal to the center of the target DNA. When *Taq* polymerase elongates the second strand during PCR, its nuclease activity cuts the probe into single nucleotides. This releases the two fluorophores from contact and abolishes quenching. The short-wavelength fluorophore can now fluoresce and a signal will be detected that is proportional to the number of new strands synthesized. (hν = Light)

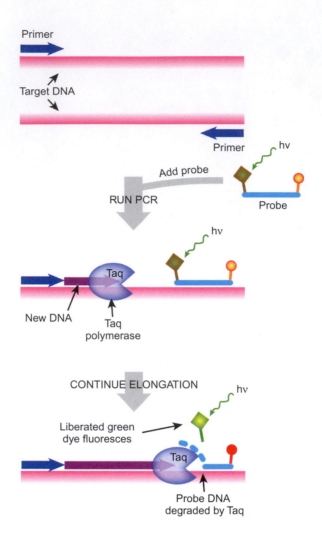

Specific probes can be included in fluorescent PCR procedures to ensure that only specific target DNA sequences give rise to fluorescence.

During PCR the TaqMan probe binds to the target sequence after the denaturation step that separates the two DNA strands. As *Taq* polymerase extends the primer during the next PCR cycle it will eventually bump into the TaqMan probe. *Taq* polymerase is not only capable of displacing strands ahead of it but also has a 5′-nuclease activity that degrades the DNA strand of the probe. This breaks the linkage between the two fluorophores and disrupts the FRET. The short-wavelength fluorophore is free from quenching and its fluorescence increases. In this case, the increase in fluorescence is directly related to the amount of the specific target sequence that has been amplified.

11. Molecular Beacons and Scorpion Primers

A **molecular beacon** is a fluorescent probe molecule that is designed to fluoresce only when it binds to a specific DNA target sequence. The probe contains both a fluorescent group, or **fluorophore**, and a **quenching group** at opposite ends

fluorophore   fluorescent group
molecular beacon   A fluorescent probe molecule that contains both a fluorophore and a quenching group and that fluoresces only when it binds to a specific DNA target sequence
quenching group   Molecule that prevents fluorescence by binding to the fluorophore and absorbing its activation energy

of a DNA sequence of 20–30 bases. The central region of the probe is complementary to the target sequence. The terminal half dozen bases at each end of the probe are complementary and form a short double-stranded region as shown in Fig. 6.23. In the stem and loop conformation the quenching group is next to the fluorophore and so prevents fluorescence. When the molecular beacon binds to the target sequence it is linearized. This separates the quenching group from the fluorophore, which is now free to fluoresce. Care is needed to avoid disrupting the short stem structure. For example, high temperatures will cause unpairing and give a false positive response.

Molecular beacons may be used in conjunction with PCR primers to give a highly specific amplification plus detection system. **Scorpion primers** consist of a molecular beacon joined to a single-stranded DNA primer by an inert linker molecule (e.g., hexethylene glycol). When the beacon is in its hairpin structure, the quencher (e.g., methyl red) binds to the fluorophore (e.g., fluorescein) and prevents fluorescence. The loop portion of the stem and loop structure has sequences complementary to the target DNA and constitutes the probe segment. When the probe sequence binds to target DNA, the hairpin is disrupted, the quencher and fluorophore are separated, and fluorescence occurs.

During PCR, the Scorpion primer binds to the target DNA and is elongated by the *Taq* polymerase. The two strands are separated in the next denaturation cycle. The Scorpion probe sequence then hybridizes to the single-stranded DNA in the middle of the target sequence. This releases the fluorophore from the quencher and promotes fluorescence (Fig. 6.24).

> Fluorescent probes may be designed that only fluoresce after binding to the target DNA sequence.

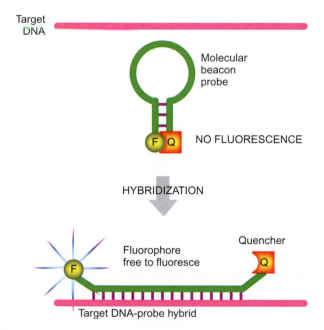

■ **FIGURE 6.23**
Molecular Beacon

A molecular beacon is a probe that has two engineered regions at the ends of the probe sequence. On the 5′ side, a fluorescent tag is added (F), and on the 3′ side, a quenching group is added (Q). Just inside the two tags are six base pairs that can form a stem-loop structure. In this state the probe cannot fluoresce. When the probe binds to the target sequence, the stem-loop structure is lost. Since the quenching group is no longer next to the fluorescent tag, the probe can now fluoresce.

Scorpion primer   DNA primer joined to a molecular beacon by an inert linker. When the probe sequence binds target DNA, the quencher and fluorophore are separated allowing fluorescence

FIGURE 6.24
Scorpion Primer with Combined Fluorescent Probe

Scorpion primers provide another method to detect the PCR product by fluorescence. The Scorpion probe has a stem loop structure that keeps the fluorophore molecule (diamond) in close proximity to the quencher (circle). The loop has a sequence complementary to the target DNA. The stem/loop is linked to a regular PCR primer designed to amplify the target DNA. During the extension step of PCR, the primer portion of the probe anneals to the template and *Taq* polymerase makes new DNA. During the next denaturation step, the whole probe plus new DNA strand become single-stranded. The loop (blue) can now anneal to the single-stranded target DNA, releasing the fluorophore from the quencher. The fluorescence emitted is a direct measure of the amount of PCR product produced.

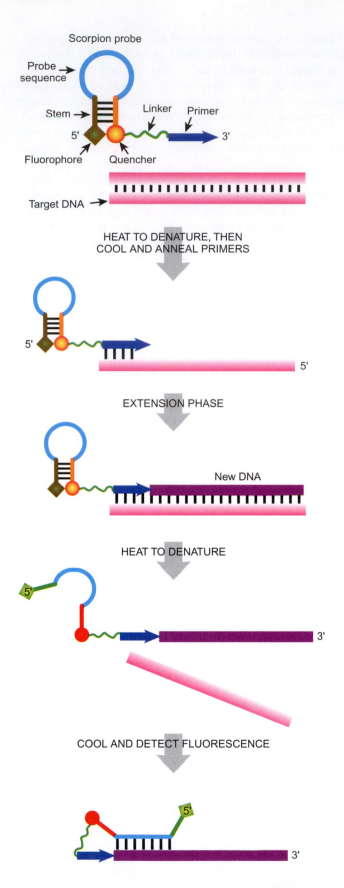

Box 6.02: On-Site Diagnosis of Plant Disease by Real-Time PCR

The development of real-time PCR has resulted in a great decrease in the time needed for detection of DNA from infectious microorganisms. Classical methods often required 3 or 4 days to isolate the microorganism and another week to confirm its identity—always assuming the pathogen can be cultured. Standard PCR methods not only cut the time needed to 2–3 days but also work directly on tissue samples without requiring that the microorganisms should be cultured. Real-time PCR with fluorescent detection has cut the time required for diagnosis even further, although the technique was originally lab-based and relatively expensive. However, portable real-time PCR machines have been developed recently that allow DNA identification in a couple of hours. An example is the Smart Cycler® TD made by Cepheid Corporation of Sunnyvale, California (Fig. 6.25).

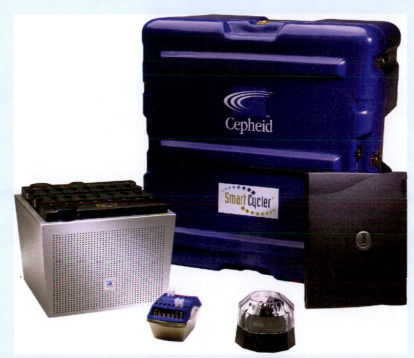

FIGURE 6.25
Portable Real-Time PCR

The Cepheid Smart Cycler® system has been used for rapid on-site detection of plant diseases. (Image courtesy of Cepheid Corporation.)

The Smart Cycler has been used to diagnose plant diseases, on-site in the fields where the crops are growing. For example, watermelon fruit blotch is a bacterial disease that causes major economic losses of watermelon crops worldwide. The causative agent, *Acidovorax avenae* subsp. *citrulli,* requires 10–14 days for diagnosis by classical procedures. Portable, real-time PCR allows on-site identification within an hour of taking samples from plants with suspected infections. This rapid diagnosis is of great value both in managing crop diseases and also in deciding whether quarantine is required when facing a possible outbreak of a transmissible plant disease that could threaten crops in other locations.

12. Use of PCR in Medical Diagnosis

Traces of DNA can be amplified by PCR and used for diagnosis of an infectious disease or forensic analysis of crime scene evidence.

PCR can be used to identify an unknown sample of blood, tissue, or hair. Two primers are designed to amplify a region of the genome that is different from one person to another or one organism to another. One of the PCR primers is designed to recognize nucleotide polymorphisms found in the person or organism. If the unknown DNA is from that person or organism, the PCR primers will anneal to that polymorphism and work (e.g., Fig. 6.26; unknown sample No. 2). If the test DNA is not from the same person or organism, the primer will not match the DNA, and no fragment will be generated (e.g., Fig. 6.26; unknown sample No. 1). The key to this experiment is the primers and how well they anneal to the target sequence. The primers may bind to closely-related sequences, but the DNA made by PCR can be sequenced to determine whether or not this occurred. Increasing the temperature in which the primers anneal to the template can also increase the stringency for this reaction.

Clearly, PCR can be used in a variety of diagnostic tests. For example, visible symptoms of AIDS only appear several years after infection. However, using PCR primers specific for sequences found only in the HIV genome, scientists can test for HIV DNA in blood samples, even when no symptoms are apparent. Another example is tuberculosis. Unlike many bacteria, *Mycobacterium*, which causes this disease, grows very slowly. Originally, to test for tuberculosis, the bacteria were cultured on nutrient plates, but this test took nearly a month. In contrast, PCR identification of mycobacterial DNA can be done in a day. Faster medical diagnoses are critical to help prevent the spread and progression of these diseases.

PCR is a powerful tool for amplifying small amounts of DNA. The DNA from 1/100th of a milliliter of human blood contains about 100,000 copies of each chromosome. If the target sequence for PCR is 500 base pairs, then there is about one-tenth of a picogram (10^{-12} gram) by weight of a target sequence. A good PCR run will amplify the target sequence and yield a microgram (10^{-6} gram) or more of DNA. A microgram may not seem much but is plenty for complete sequencing or cloning. Obviously, it is possible to identify an organism from an extremely small trace of DNA-containing material. In fact, the DNA from a single cell can be used to amplify a specific target sequence. This technique has revolutionized the criminal justice system by allowing highly accurate identification of individuals from very small samples.

FIGURE 6.26
PCR is Used to Diagnose Genetic Relatedness

PCR can determine the identity of an unknown DNA sample. In this figure, two unknown DNA samples are isolated and amplified using PCR. The primers used to test these DNA samples are specific for a known sequence (pink). In sample 1, the primers did not anneal and no PCR product was made. Therefore, the sample must not contain any DNA with the known sequence (pink). Sample 2 showed a PCR product of the predicted size for the known sequence, therefore, the primers must have annealed and amplified this sequence.

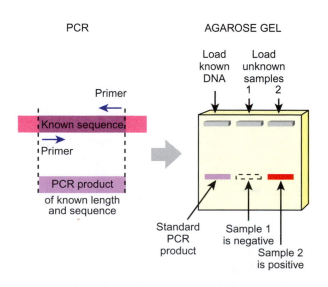

Niemz A, Ferguson TM, Boyle DS (2011) Point-of-care nucleic acid testing for infectious diseases. Trends Biotech. 29(5): 240–250.

This article presents the state of current technology for the detection of infectious disease using nucleic acid testing (NAT) such as PCR. In most cases, diagnosing the presence or absence of an infectious disease involves taking a specimen or sample, sending that to a central laboratory, analysis of the material for the presence of a disease by culturing any microorganisms, and finally relaying the results back to the patient. The turnaround time for this procedure can take days or even longer for slow growing organisms like *Mycobacterium tuberculosis*. Using techniques such as PCR has revolutionized the manner in which the diagnosis of an infectious disease is accomplished, and future hope is to create on-site analyses for diseases that are fast, accurate, and of reasonable cost by simply amplifying part of the genomic sequence of the infectious organism. As described in the textbook, the Cepheid portable real-time PCR machine has been used for plant analyses, and Cepheid has now marketed a new version called GeneXpert. As of 2010, the World Health Organization has endorsed the use of the Cepheid GeneXpert to diagnose drug-resistant tuberculosis infections, but the cost is still prohibitive to small rural areas. In the United States, only the Cepheid system has been approved for use as a point-of-care test.

This article describes a variety of ways in which samples can be tested for nucleic acid from an infectious organism. The use of PCR-based technology still leads the way, but in addition, isothermic nucleic acid amplification technologies are described. These technologies are based upon rolling circle replication as seen in plasmids and circular viral genomes (see Chs. 20 and 21), which means that the polymerase displaces the new strand of DNA as it laps itself around the single-stranded template. The temperature is kept constant, and therefore, less sophisticated equipment is necessary, thus reducing the cost.

In addition, the article discusses the method of detecting the results of the nucleic acid amplification step. Traditionally, methods of detecting PCR products include agarose gel electrophoresis and real-time fluorescence detection. A newer method of detection is called nucleic acid lateral flow (NALF). In this method, the DNA from an infectious agent is recognized by an antibody connected to a colored particle. This complex is then detected by DNA hybridization (see Ch. 5) to a complementary strand of DNA attached to a specific spot along a lateral flow strip. In a more complex method, the complementary DNA is not attached directly to the strip. Instead, an antibody that recognizes the end of the complementary DNA is attached to the strip. In either scenario, the hybridization of the two DNAs results in an aggregation of colored particles in an area that denotes a positive result. Lack of color in this area indicates a negative result.

FOCUS ON RELEVANT RESEARCH

Finally, the paper summarizes some applications of the new technologies. Tuberculosis is a difficult disease to diagnose since the bacteria require strong methods to break open the cells. The Cepheid GeneXpert uses a sonication step that breaks apart the bacterial cell wall using sound waves. In other areas, PCR-based technology has been very useful to the diagnosis of HIV infections. The viral load, that is, the amount of replicating virus within a patient can be easily determined using PCR. In addition, HIV infections in infants can be identified much quicker using NALF, and therefore, there is a surge to create new technology to quickly and easily test infants for the virus. These technologies are rapidly progressing, and soon there will be even quicker and easier techniques to diagnose individuals for such diseases.

13. Environmental Analysis by PCR

It is possible to extract DNA directly from environmental samples, such as soil or water, without bothering to isolate and culture the living organisms that contain it first. Such environmental samples of DNA may be amplified by PCR. Because PCR is so sensitive, DNA can be amplified and sequenced from microorganisms that are present in such very low numbers that they cannot be detected by other means. Furthermore, it is not necessary for the microbial cells to be culturable or even viable. If specific PCR primers are used, it is possible to amplify genes from a single bacterium out of the billions that might be present in an environmental sample. As explained in Chapter 26, molecular-based classification of organisms is based primarily on the sequence of the ribosomal RNA. Consequently, for identification of microorganisms from environmental samples by PCR, primers to the gene for 16S rRNA are usually used. One fascinating result of environmental PCR analysis has been the discovery of many novel microorganisms that have never been cultured or identified by any other means. Such "microorganisms" are known only as novel rRNA sequences, and it is presumed that the corresponding organisms do actually live and grow in the environment even though they do not grow in culture in the laboratory.

By using primer sets specific for any chosen gene, PCR also allows us to check whether or not that particular gene is present in the environment being sampled. For example, suppose that we wish to know whether or not there are microorganisms capable of photosynthesis in a lake. A sample of lake water would be analyzed by PCR using primers specific for a gene that encodes an essential component of

DNA sequences may be amplified by PCR directly from environmental samples.

the light-harvesting mechanism. (Primer sets that are based on genes involved in a specific metabolic pathway are called "metabolic primers.") If a positive result is obtained we may conclude that there are organisms capable of photosynthesis in the lake. However, this approach does not tell us whether these organisms are actually growing by this mechanism or even still alive. A further step in analysis is to extract RNA from the environment and subject it to RT-PCR. This converts any messenger RNA present in the sample into the corresponding cDNA, which is then amplified. This reveals whether the corresponding genes are being actively transcribed, although, strictly speaking, we still do not know if the corresponding enzyme or protein is present.

Community profiling by PCR involves assessing the abundance and diversity of bacteria in an environment. In the simplest approach, total genomic DNA is first isolated from an environmental sample. All of the bacterial 16S rRNA genes in that sample are then amplified by PCR, cloned, and sequenced. The sequences are analyzed to identify the bacteria present in that environment. Metabolic profiling using primers specific for genes in particular metabolic pathways may also be performed. The relative abundance of the various organisms in the environment may then be estimated by the number of rRNA sequences for each organism.

It is also possible to isolate useful genes directly by environmental PCR—an approach sometimes referred to as **eco-trawling**. DNA isolated directly from the environment is amplified by PCR using primers that correspond to a known gene. The PCR fragments are then cloned into a suitable plasmid that will allow expression of any successfully captured genes. The plasmids are transformed into a suitable bacterial host cell and the captured genes are expressed. The PCR products are real genes used in that environment and may contain certain mutations or alterations that make its host organism a better fit for that niche. For example, primers corresponding to the ends of DNA polymerase could be used to amplify DNA extracted from a sample of water from a hot spring. The desired result would be genes encoding novel DNA polymerases able to function at high temperatures. This approach is obviously well suited to finding variants of known enzymes that function under novel or extreme conditions.

14. Rescuing DNA from Extinct Life Forms by PCR

Since any small trace of DNA can be amplified by PCR and then cloned or sequenced, some scientists have looked for DNA in fossils. Stretches of DNA long enough to yield valuable information have been extracted from museum specimens such as Egyptian mummies and fossils of various ages. In addition, DNA has been extracted from mammoth and plant remains frozen in the Siberian permafrost. This data has helped in studying molecular evolution and is discussed more fully in Chapter 26.

In the sci-fi best seller "Jurassic Park," the DNA was not obtained directly from fossilized dinosaur bones. Instead, it was extracted from prehistoric insects trapped in amber (Fig. 6.27). The stomachs of bloodsucking insects would contain blood cells complete with DNA from their last victim, and if preserved in amber, this could be extracted and used for PCR. DNA has indeed been extracted from insect fossils preserved in amber. However, the older the fossil the more decomposition that has occurred. Normal rates of decay should break the DNA double helix into fragments less than 1,000 bp long in 5,000 years or so. So, though we will no doubt obtain gene fragments from an increasing array of extinct creatures, it is unlikely that any extinct animal will be resurrected intact, although certain ancient genes could be added to extant creatures to determine their function.

RT-PCR allows detection of RNA from environmental samples and so reveals whether the gene is being transcribed by a living organism.

Genes encoding useful proteins may be cloned from environmental samples without knowing their host organism's identity.

DNA may be amplified from fossil material and used in identification.

community profiling Assessing the abundance and diversity of bacteria in an environment using PCR
eco-trawling Isolating useful genes from the environment by PCR

FIGURE 6.27
Mosquito Preserved in Amber

(Photo by Karen Fiorino.)

Key Concepts

- PCR uses a small amount of template DNA, two primers that flank the target sequence, nucleotides, and thermostable DNA polymerase to amplify a specific region of DNA, thus creating a large amount of DNA from a very small sample.

- PCR cycles include three different steps: denaturation of the template DNA into single strands at about 90°C, a moderate temperature (around 50–60°C) to anneal the primers to the target sequence, and an extension step in which DNA polymerase elongates a second strand of DNA complementary to each strand of the target.

- During the third cycle of PCR, the products include pieces of DNA containing only the target sequence.

- PCR primers are the key element for amplifying a specific target sequence from an entire DNA sample. To create PCR primers, a certain amount of sequence information needs to be known.

- Degenerate PCR primers can be created such that the primers contain a mixture of bases at specific positions. The degenerate positions for the primers tend to fall in the nucleotide that is in the third position of the codon sequence.

- PCR primers do not have to match the target sequence at the 5′ end, and in some scenarios, the 5′ end can have the recognition site for a restriction enzyme or anchor sequences for subsequent PCR reactions.

- Longer target sequences can be amplified by varying the time of elongation and the type of DNA polymerase.

- Inverse PCR amplifies unknown sequences by creating a circular template DNA.

- Randomly amplified polymorphic DNA or RAPD uses random primers to amplify genomic DNA from different individuals or organisms. PCR products are only formed when the two primers anneal reasonably close and in opposite orientations on different strands. If there are different sequences between two organisms, the pattern of PCR products will be altered.

- Reverse transcriptase PCR (RT-PCR) amplifies processed mRNA by first converting the mRNA into complementary DNA with reverse transcriptase and then amplifying the cDNA with regular PCR. The technique identifies what regions of a gene are exons and what regions are introns.

- Differential display PCR compares the mRNA expression patterns of two different organisms or the same organism in two different conditions. The mRNA is amplified by using an oligo(dT) primer and a mixture of random primers.

- Rapid amplification of cDNA ends uses a modified PCR reaction to amplify the 5' and 3' ends of an mRNA molecule.
- PCR is also used to manipulate genes. For example, PCR can connect two different target sequences with an overlap primer; PCR can create mutations within a gene with mismatched primers; and PCR can create deletions and insertions into target genes.
- Barcodes or zipcodes are unique sequences that mark a specific PCR product to make it unique from other sequences within a cell or DNA sample.
- Real-time or quantitative PCR (qPCR) measures the amount of PCR product as it is being produced. One method to ascertain the amount of product is to include a non-specific dye (i.e., SYBR green I) that binds to double-stranded DNA. The more PCR product, the more the fluorescence. Another method uses fluorescence resonance energy transfer (FRET) between two fluorophores attached to each end of a probe that binds to the middle of the target sequence. When DNA polymerase displaces each nucleotide of the probe during DNA synthesis, the two fluorophores are released and fluoresce.
- Molecular beacons and scorpion primers use FRET technology to ascertain the amount of PCR product produced in real time by measuring the amount of fluorescence produced over time.
- Real-time PCR or qPCR are used in agriculture, medicine, and environmental studies to determine if there are any infectious agents in a sample of leaf tissue, root tissue, blood, water, air, or many other types of samples.
- Environmental PCR can be used to identify the types of microorganisms that inhabit an environment even if the actual organism cannot be grown in the lab.

Review Questions

1. What does PCR stand for? Who invented it?
2. What is the underlying principle of PCR?
3. Name the different components required for PCR.
4. What is the enzyme used in PCR? Explain why this enzyme is used for the reaction.
5. What are the different steps involved in one cycle of a PCR reaction?
6. What are degenerate primers? Name two uses for degenerate primers.
7. What modifications to the standard PCR are made for long-range PCR?
8. Describe hot start PCR.
9. How does an inverse PCR work?
10. What is the purpose of RAPD? How does it work?
11. What is reverse transcriptase?
12. What are cDNAs? How are they made?
13. What are the different uses of RT-PCR?
14. How is differential display PCR different from other types of PCR?
15. What is molecular sewing?
16. Why is it not usually possible to get full length cDNA using reverse transcriptase? How are cDNA ends rescued?
17. What are two ways to make base alterations in DNA by PCR?
18. What are the steps involved in insertion or deletion of DNA by PCR?
19. What is multiplex PCR? How is this procedure accomplished?
20. Name two uses of PCR in medical diagnosis.
21. What are the advantages of environmental PCR analysis?
22. What is community profiling?
23. What is the principle behind real-time PCR?
24. What are the two parts of scorpion primers? How are they used?
25. What is eco-trawling?

Conceptual Questions

1. Discuss the similarities and differences between RAPD analysis and RFLP analysis.
2. Your graduate advisor gave you a sample of genomic DNA from *C. elegans*. He wants you to amplify the gene for transposase. The following is a diagram of the gene. Using the sequence information given, design two 18 nucleotide PCR primers to amplify the gene. Both sequences are shown from the non-coding strand of DNA.

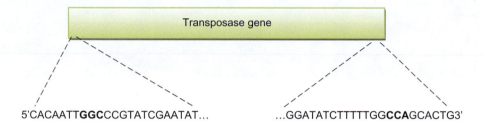

5'CACAATT**GGC**CCGTATCGAATAT...   ...GGATATCTTTTTGG**CCA**GCACTG3'

3. Using your knowledge of PCR methods to induce mutations, design PCR primers that will inactivate the above transposon by mutating the key nucleotides (as shown in bold) in the inverted repeats.
4. Design degenerate primers that are 9 nucleotides in length that will amplify the gene corresponding to the following protein sequence listed in single-letter amino acid abbreviations:

    ```
    mikdtsvepe ganfiaeffg fvfeldpdtd asprplaphl eirvnvdtli
    dlalrespra algpsgpvat ftdkvearml rfwpktrrrr sttpggqrgl fda
    ```

5. You have just started as a genetic counselor, and there is a couple that would like to have children, but the man has a brother with Friedreich's Ataxia. The woman has a distant relative that had the same disease. This is an autosomal recessive genetic disorder that is cause by trinucleotide (GAA) expansion within the first intron of the *FXN* gene, which encodes the protein frataxin. This intron is 500 base pairs long without the GAA repeats. The expansion of this repeat interrupts the splicing of the intron, and therefore, inactivates frataxin. A normal person has between 5 and 35 GAA repeats, a person with premutation has between 34 and 65 repeats, and an affected individual has between 66 and 1700 uninterrupted GAA repeats. It is possible to be a carrier for the disease without any symptoms because the normal allele is dominant. Explain how you would design a PCR experiment to determine if either couple is a carrier. What is the chance of having a child with Friedreich's Ataxia if both parents are carriers?

Chapter 7

Cloning Genes for Analysis

Molecular cloning refers to the isolation and cloning of individual genes or other segments of DNA. This has two general stages: firstly, the DNA region of interest must be located, cut out, and purified; secondly, this segment of DNA must be mounted on a convenient carrier molecule or **cloning vector** (**vector**, for short) that allows it to be moved around to whatever final location is desired. The term **chimera** refers to any hybrid molecule of DNA, such as a vector plus a cloned gene, which has been engineered from two different sources of DNA. A variety of DNA molecules may be used as vectors. However, by far the most popular are plasmids and small virus genomes. Most modern vectors consist of plasmids or viruses that have been modified to make them more convenient to use. In particular, the most popular modifications allow the easy insertion of cloned DNA and mechanisms for detecting if the cloned DNA has been successfully inserted. Specialized vectors are used to move genes between multiple organisms or to

---

chimera Hybrid molecule of DNA that has DNA from more than one source or organism

cloning vector Any molecule of DNA that can replicate itself inside a cell and is used for carrying cloned genes or segments of DNA. Usually a small multicopy plasmid or a modified virus

vector (a) In molecular biology, a vector is a molecule of DNA which can replicate and is used to carry cloned genes or DNA fragments; (b) in general biology, a vector is an organism (such as a mosquito) that carries and distributes a disease-causing microorganism (such as yellow fever or malaria)

Molecular Biology.

allow efficient high-level expression of cloned genes. Other specialized approaches are needed to cope with the enormous amounts of DNA found in higher organisms and to deal with the inserted regions, or introns, found in many eukaryotic genes.

1. Properties of Cloning Vectors

In principle, any molecule of DNA that can replicate itself inside a cell could work as a cloning vector. For convenience in manipulation, the following factors must be considered:

1. The vector should be a reasonably small and manageable DNA molecule.
2. Moving the vector from cell to cell should be relatively easy.
3. Generating and purifying large amounts of vector DNA should be straightforward.

In addition to these basic requirements, most vectors have been designed to provide some convenient means to perform the following:

1. Detect the presence of the vector.
2. Directly select for cells that contain the vector.
3. Insert genes into the vector.
4. Detect the presence of an inserted gene on the vector.

In practice, bacterial plasmids come closest to these requirements and are the most widely used vectors. Many viruses are also used as vectors, especially when attempting the engineering of higher organisms. For some special purposes, where very large fragments of DNA are to be cloned, whole chromosomes are sometimes used as vectors.

> The term "vector" refers to self-replicating DNA molecules used to carry cloned genes (or any other pieces of cloned DNA).

1.1. Multicopy Plasmid Vectors

Vectors derived from the small multicopy plasmids of bacteria (see Ch. 20) were the first to be used and are still the most widespread. The **ColE1 plasmid** of *E. coli* is a small circular DNA molecule that forms the basis of many vectors widely used in molecular biology. It exists in up to 40 copies per cell so obtaining plenty of plasmid DNA is relatively easy and it can be moved from cell to cell by transformation as described in Chapter 25.

> Multicopy plasmids are convenient as vectors since they make plenty of plasmid DNA and express cloned genes at high levels.

Although the original ColE1 plasmid was once used directly as a vector, most modern ColE1-based vectors contain a range of artificial improvements. First, the genes for colicin E1, a toxic protein for killing bacteria (see Ch. 20), were removed, since these are obviously not needed. Next, a gene for resistance to an antibiotic was added. The most popular antibiotic for this purpose is **ampicillin**, a widely used penicillin derivative (Fig. 7.01). The ampicillin resistance gene is known as ***amp*** or ***bla***, which refers to **beta-lactamase**. This enzyme degrades penicillins and related antibiotics.

> Vectors containing genes for antibiotic resistance are easily selected when expressed in bacteria.

When such a vector is transformed into bacterial cells, the plasmid ampicillin resistance gene allows the bacteria to grow in a medium containing ampicillin. Those bacteria that did not get a plasmid are killed by the antibiotics.

---

ampicillin   A widely used antibiotic of the penicillin family
***amp* gene**   Gene conveying resistance to ampicillin and related antibiotics and encoding beta-lactamase. Same as *bla* gene
beta-lactamase (β-lactamase)   Enzyme that degrades beta-lactam antibiotics, including penicillins and cephalosporins
***bla* gene**   Gene conveying resistance to ampicillin and related antibiotics and encoding beta-lactamase. Same as *amp* gene
ColEI plasmid   Small multicopy plasmid of *Escherichia coli* that forms the basis of many cloning vectors widely used in molecular biology

FIGURE 7.01
Antibiotic Resistant Cloning Plasmid

The ColE1 plasmids of *E. coli* have been modified for use as cloning vectors. The original colicin genes have been deleted so that the bacteria carrying these plasmids no longer produce these antibacterial toxins. In addition, a gene for antibiotic resistance has been added. This provides an easily identifiable phenotype to the bacteria that carry the altered plasmid.

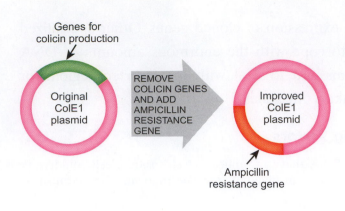

FIGURE 7.02
Insertion of DNA into Cloning Vector

To insert the gene of interest into a plasmid vector, both the vector and the gene of interest should have compatible sticky ends. To achieve this both the gene and the vector must be digested with the same restriction enzyme. The two pieces are mixed together with DNA ligase, which joins the ends yielding a closed double-stranded circular plasmid carrying the gene of interest.

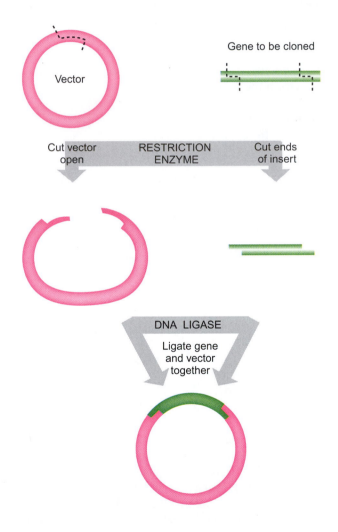

Vectors are often engineered to contain large numbers of convenient restriction enzyme cut sites, called a polylinker or multiple cloning site.

1.2. Inserting Genes into Vectors

When the vector and the target DNA are cut with the same restriction enzyme, the ends are compatible for cloning. If a restriction enzyme that generates sticky ends is used, the vector and the insert will have matching overhangs. A mixture of the two is treated with DNA ligase, which links together DNA strands. The result is the ligation of the target DNA fragment into the vector as shown in Figure 7.02. If a restriction enzyme that generates blunt ends is used, ligation is more difficult and T4 ligase is used because it has the ability to join blunt ends (unlike bacterial ligase).

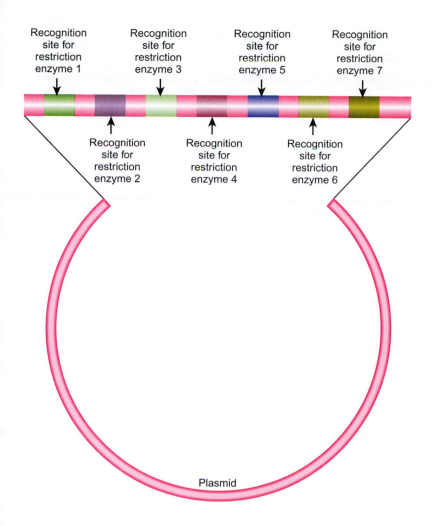

Recognition site for restriction enzyme 1

Recognition site for restriction enzyme 3

Recognition site for restriction enzyme 5

Recognition site for restriction enzyme 7

Recognition site for restriction enzyme 2

Recognition site for restriction enzyme 4

Recognition site for restriction enzyme 6

Plasmid

■ **FIGURE 7.03**
Polylinker or Multiple Cloning Site

Many plasmid vectors contain an artificial region of DNA that has many different restriction enzyme sites. Such polylinkers or multiple cloning sites are designed so that all the restriction enzyme sites in the polylinker are unique, and the corresponding enzymes only cut the plasmid once.

This procedure relies on the vector possessing only one site for the chosen restriction enzyme. If there were more than one cut site in the vector the restriction enzyme would cut it into multiple fragments. Furthermore, we must avoid inserting the cloned gene into any of the genes needed by the plasmid for its own replication and survival within the cell. Moreover, since there are many different restriction enzymes, it would be convenient to have a wide range of restriction recognition sites in the vector. These issues are all resolved by inserting a **polylinker**, or **multiple cloning site (MCS)**, into the cloning vector. This is a stretch of artificially-synthesized DNA, about 50 base pairs long, that contains cut sites for seven or eight commonly used restriction enzymes. This not only allows a wide choice of restriction enzymes, but ensures that the insert does not damage the plasmid and goes into more or less the same location each time (Fig. 7.03).

The remainder of the plasmid should not contain any cut sites for any of the enzymes represented in the polylinker. One way to ensure this is to choose only enzymes with zero cut sites in the original plasmid. Alternatively, we can get rid of unwanted cut sites by the approach shown in Figure 7.04. Due to spontaneous mutation (see Ch. 23 for mutations), occasional plasmids will suffer a base change within the cut site that needs to be eliminated. This will abolish recognition of the site by

multiple cloning site (MCS) A stretch of artificially-synthesized DNA that contains cut sites for seven or eight widely used restriction enzymes. Same as polylinker
polylinker A stretch of artificially-synthesized DNA that contains cut sites for seven or eight widely used restriction enzymes. Same as multiple cloning site (MCS)

FIGURE 7.04
Eliminating Unwanted Restriction Sites

The restriction site shown in blue is unwanted. During normal DNA replication, occasional mutations occur. Consequently, a very small percentage of the plasmids will carry a random mutation (red) that alters this particular restriction recognition sequence. A sample of the plasmid DNA is isolated from a bacterial culture. The plasmids are treated with the appropriate restriction enzyme. All will be cut, except those with mutant restriction sites. The mixture is then transformed back into bacterial cells. Bacteria receiving cut, linearized plasmids will degrade them. Only mutant plasmids that remain circular will survive.

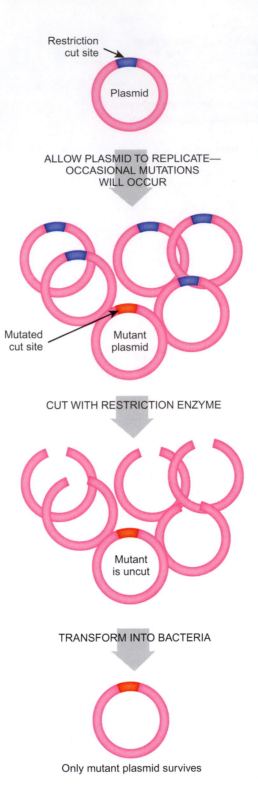

the restriction enzyme. The question is how to find this one rare mutant plasmid. First plasmid DNA is prepared and treated with the restriction enzyme in question. The plasmid DNA is then transformed into fresh bacterial cells without re-ligating the break. Wild-type bacteria rapidly degrade incoming linear DNA; therefore, the majority of plasmids will be destroyed by this procedure. Those few that have lost the cut site by mutation will remain circular and survive.

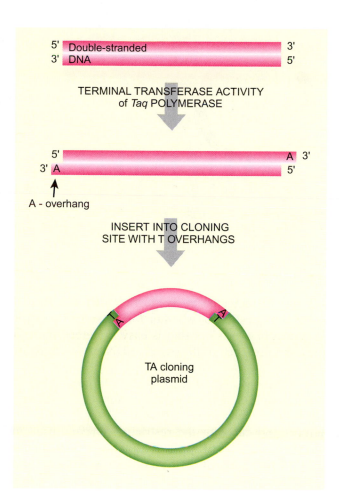

■ **FIGURE 7.05**
TA Cloning

When *Taq* polymerase amplifies a piece of DNA during PCR, the terminal transferase activity of *Taq* adds an extra adenine at the 3′ end of the PCR product. The TA cloning vector was designed so that when linearized it has single 5′ thymidine overhangs at each end. The PCR product can be ligated into this vector without the need for special restriction enzyme sites.

1.3. TA Cloning of PCR Products

A related approach to cloning applies to PCR products generated by *Taq* polymerase (see Ch. 6). This enzyme has terminal transferase activity, which means it adds a single adenosine to the 3′-ends of double-stranded DNA. This reaction does not depend on the sequence of the template or primers. The **TA cloning** procedure exploits this terminal transferase activity, which is shared by *Taq* polymerase and several other thermophilic DNA polymerases. Thus, most of the DNA molecules amplified by *Taq* polymerase possess single 3′-A overhangs (Fig. 7.05). Consequently, they can be directly cloned into a vector that has matching single 3′-T overhangs on both ends. The same **TA cloning vector** can be used to clone any segment of amplified DNA. For that matter, DNA from other sources can have single 3′-A overhangs added to its ends by using *Taq* polymerase and can then be cloned by the same mechanism. This procedure is especially useful when convenient restriction sites are not available.

Taq polymerase generates single 3′-A overhangs with its terminal transferase activity. These can be used for cloning PCR products into TA cloning vectors.

2. Detecting Insertions in Vectors

Once a gene or other fragment of DNA has been cloned into a plasmid vector and transformed into a bacterial cell, an assay to detect whether it is present or not must

TA cloning   Procedure that uses *Taq* polymerase to generate single 3′-A overhangs on the ends of DNA segments that are used to clone DNA into a vector with matching 3′-T overhangs
TA cloning vector   Vector with single 3′-T overhangs (in its linearized form) that is used to clone DNA segments with single 3′-A overhangs generated by *Taq* polymerase

Antibiotic
resistance gene #1

Vector

LIGATE
INSERT INTO
ANTIBIOTIC
RESISTANCE
GENE #2

Vector
plus
insert

Cut site

Antibiotic
resistance gene #2

Insert

CELLS CARRYING VECTOR
ARE RESISTANT TO
BOTH ANTIBIOTICS

CELLS CARRYING VECTOR
WITH INSERT ARE
RESISTANT TO FIRST
ANTIBIOTIC ONLY

be performed. The plasmid itself may be detected by conferring antibiotic resistance on the host cell, but this leaves the question of whether the presumed insert is present. If the cloned gene itself codes for a product that is easy to detect, there is no problem. In most cases, however, the presence of the inserted DNA itself must be directly monitored.

The least sophisticated and most tedious method is to screen a large number of suspects for the inserted DNA. In this method, many separate bacterial colonies that received the plasmid vector are grown in separate vials. Plasmid DNA is extracted from each of these bacterial cultures and cut with the restriction enzyme used in the original cloning experiment. If there is no insert in the plasmid, this merely converts the plasmid from a circular to a linear molecule of DNA. If the vector contains inserted DNA, two pieces of DNA are produced, one being the original plasmid and the other the inserted DNA fragment. To see how many fragments of DNA are present, the cut DNA is separated by agarose gel electrophoresis. If enough transformed colonies are tested, sooner or later one carrying a plasmid with the inserted DNA fragment will be found. This approach was of necessity used in the early days of genetic engineering. Today, modified vectors are available that facilitate screening by a variety of approaches.

> Inserts in a vector can be
> checked by isolating DNA,
> cutting with a restriction
> enzyme, and seeing how many
> fragments are generated.

> Inserts are sometimes screened
> by the change in growth
> properties due to disrupting a
> gene on the vector.

Rather less laborious is to use a plasmid with two antibiotic resistance genes. One antibiotic resistance gene is used to select for cells that have received the plasmid vector itself. The second is used for the insertion and detection of cloned DNA (Fig. 7.06). The cut site for the restriction enzyme used must lie within this second antibiotic resistance gene. When the cloned fragment of DNA is inserted this antibiotic resistance gene will be disrupted. This is referred to as **insertional inactivation**. Consequently, cells that receive a plasmid without an insert will be resistant to both antibiotics. Those receiving a plasmid with an insert will be resistant to only the first antibiotic.

2.1. Reporter Genes

Genes that are used in genetic analysis because their products are easy to detect are known as **reporter genes**. They are often used to report on gene expression (see Ch. 19), although they may also be used for other purposes, such as detecting the location of a protein or, as here, the presence of a particular segment of DNA in a cloning vector.

> Genes whose products are
> convenient to assay are used as
> "reporters."

---

insertional inactivation   Inactivation of a gene by inserting a foreign segment of DNA into the middle of the coding sequence
reporter gene   Gene that is used in genetic analysis because its product is convenient to assay or easy to detect

One common reporter is *lacZ* **gene** encoding β-**galactosidase**. This enzyme normally splits lactose, a compound sugar found in milk, into the simpler sugars glucose and galactose. However, β-galactosidase will also split a wide range of galactose compounds (i.e., **galactosides**) both natural and artificial (Fig. 7.07). The two most commonly used artificial galactosides are ONPG and X-gal. **ONPG (***o*-**nitrophenyl galactoside**) is split into *o*-nitrophenol and galactose. The *o*-nitrophenol is yellow and soluble, so it is easy to measure quantitatively in solution. **X-gal (5-bromo-4-chloro-3-indolyl β-D-galactoside)** is split into galactose plus the precursor to an indigo type dye. Oxygen in the air converts the precursor to an insoluble blue dye that precipitates out at the location of the *lacZ* gene. Consequently, X-gal is used to monitor β-galactosidase expression in bacterial colonies on agar.

> Beta-galactosidase cleavage of X-gal generates a blue dye, whereas, cleavage of ONPG generates a yellow product. These colored products correlate with the amount of β-galactosidase produced in the bacterial cell.

2.2. Blue/White Color Screening

The most convenient and widely used method to check for inserts in cloning vectors uses color screening. The most common procedure uses β-galactosidase and X-gal to produce bacterial colonies that change color when an insert is present within the vector. The process, called **blue/white screening**, uses a vector that carries the 5′-end of the *lacZ* gene. This truncated gene encodes the **alpha fragment** of β-galactosidase, which consists of the N-terminal region or first 146 amino acids. A specialized bacterial host strain is required whose chromosome carries a *lacZ* gene missing the front portion but encoding the rest of the β-galactosidase protein. If the plasmid and chromosomal gene segments are active they produce two protein fragments that associate to give an active enzyme. This is referred to as **alpha complementation** (Fig. 7.08). Note that assembling an active protein from fragments made separately is normally not possible. Fortunately, β-galactosidase is exceptional in this respect. The reason for splitting *lacZ* between plasmid and host is that the *lacZ* gene is unusually large (approximately 3000 bp—almost as large as many small plasmids) and it greatly helps if cloning plasmids are small.

In order to utilize this unique protein for cloning, a polylinker is inserted into the *lacZα* coding sequence on the plasmid, very close to the front of the gene. Luckily, the beginning part of the β-galactosidase protein is inessential for enzyme activity. As long as the polylinker is inserted without disrupting the reading frame, the small addition does not affect the enzyme. However, if a foreign segment of DNA is inserted into the polylinker, the alpha fragment of β-galactosidase is disrupted and no active enzyme can form (Fig. 7.09). The active form of β-galactosidase splits X-gal, which produces a blue color. Plasmids without a DNA insert will produce β-galactosidase and the bacterial cell that carries them will turn blue. Plasmids with an insert will be unable to make β-galactosidase and the cells will stay white.

> Inserts are often detected by blue and white screening. When an insert is within the plasmid polylinker, it disrupts production of the beta-galactosidase alpha-fragment. Without active beta-galactosidase, the bacteria form white (rather than blue) colonies when grown with the reporter substance X-gal.

3. Moving Genes Between Organisms: Shuttle Vectors

The plasmid vectors we have discussed so far are designed to work in bacteria. Even when investigating genes from animals or plants, they are normally cloned first onto bacterial plasmids. Eventually, however, cloned genes are often moved from one organism to another. This may be done using a **shuttle vector**. As its name implies, this

---

alpha fragment   N-terminal fragment of β-galactosidase
alpha complementation   Assembly of functional β-galactosidase from N-terminal alpha fragment plus rest of protein
beta-galactosidase (β-galactosidase)   Enzyme that cleaves lactose and other β-galactosides so releasing galactose
blue/white screening   Screening procedure based on insertional inactivation of the gene for β-galactosidase
galactoside   Compound of galactose, such as lactose, ONPG, or X-gal
lacZ **gene**   Gene encoding β-galactosidase; widely used as a reporter gene
o-**nitrophenyl galactoside (ONPG)**   Artificial substrate that is split by β-galactosidase, releasing yellow *o*-nitrophenol
shuttle vector   A vector that can survive in and be moved between more than one type of host cell
X-gal (5-bromo-4-chloro-3-indolyl β-D-galactoside)   Artificial substrate that is split by β-galactosidase, releasing a blue dye

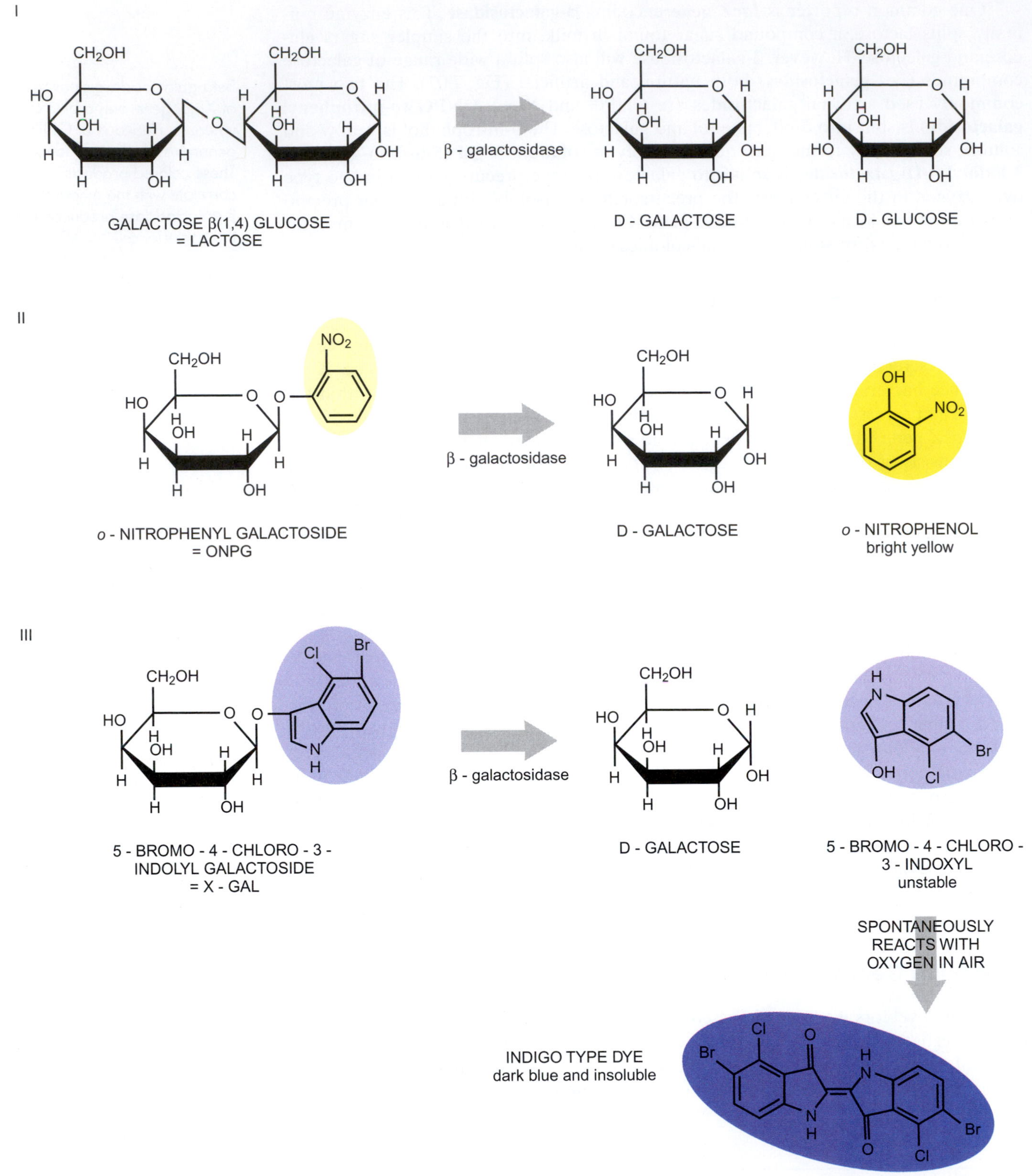

I

GALACTOSE β(1,4) GLUCOSE
= LACTOSE

β - galactosidase

D - GALACTOSE

D - GLUCOSE

II

o - NITROPHENYL GALACTOSIDE
= ONPG

β - galactosidase

D - GALACTOSE

o - NITROPHENOL
bright yellow

III

5 - BROMO - 4 - CHLORO - 3 -
INDOLYL GALACTOSIDE
= X - GAL

β - galactosidase

D - GALACTOSE

5 - BROMO - 4 - CHLORO -
3 - INDOXYL
unstable

SPONTANEOUSLY
REACTS WITH
OXYGEN IN AIR

INDIGO TYPE DYE
dark blue and insoluble

FIGURE 7.07
Substrates Used by β-Galactosidase

The enzyme β-galactosidase normally cleaves lactose into two monosaccharides, glucose and galactose. β-Galactosidase also cleaves two artificial substrates, ONPG and X-Gal, releasing a group that forms a visible dye. ONPG releases a bright yellow substance called *o*-nitrophenol, whereas X-gal releases an unstable group that reacts with oxygen to form a blue indigo dye.

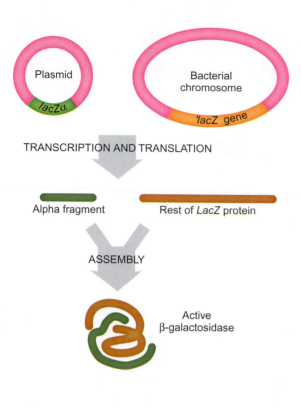

■ **FIGURE 7.08**
Alpha Complementation

The β-galactosidase protein is unique since it can be expressed as two pieces that come together to form a functional protein. The two protein fragments can be encoded on two different molecules of DNA within the bacterial cell. The alpha fragment can be expressed from a plasmid and the remainder of the β-galactosidase can be expressed from the chromosome.

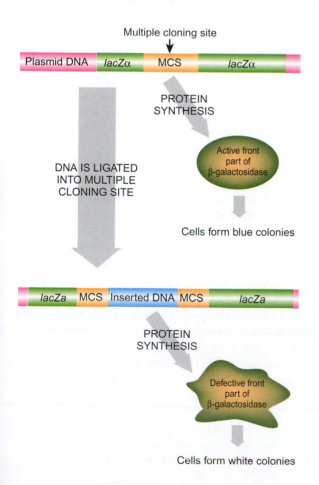

■ **FIGURE 7.09**
Blue/White Screening for β-Galactosidase

In order to screen for inserts in a plasmid with the *lacZα* gene, a small polylinker or multiple cloning site is inserted into the extreme N-terminal portion of *lacZα*. The small insertion is cloned in-frame; therefore, the modified alpha fragment is still active when complexed with the remainder of β-galactosidase expressed from the chromosomal gene. Bacteria containing this construct will turn blue in the presence of X-gal. However, if a large segment of DNA, such as a cloned gene, is inserted into the multiple cloning site, the alpha fragment is disrupted and β-galactosidase is no longer active. Bacteria harboring this plasmid cannot cleave X-gal, and therefore remain white.

FIGURE 7.10
Shuttle Vector for Yeast

In order for a shuttle vector to grow in both yeast and *E. coli*, it must have several essential elements: two origins of replication, one for *E. coli* and one for yeast; a yeast centromere sequence so that it is partitioned into the daughter cells during yeast replication; selectable markers for both yeast and *E. coli*; and a multiple cloning site for inserting the gene of interest.

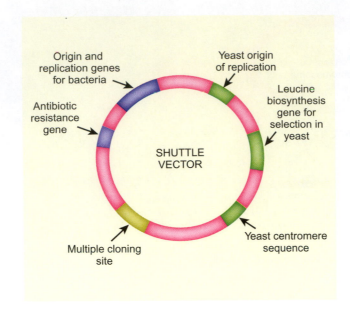

Shuttle vectors can replicate in more than one organism. This allows the same gene to be expressed in different hosts.

is a vector that can survive in more than one type of host cell. Obviously, the detailed requirements for vectors will vary depending on the host organism, but the general ideas are the same.

The earliest shuttle vectors were designed to shuttle between bacteria, such as *E. coli*, and yeast, a simple eukaryote (Fig. 7.10). Starting with a bacterial plasmid vector, several extra components are needed to create such a shuttle vector:

1. An **origin of replication** that works in yeast. The origin of replication is the site where replication of DNA is initiated. It contains recognition sequences for several proteins that are specific for different groups of organisms (for more details see Ch. 10). Prokaryotic replication origins do not work in eukaryotes or vice versa, since the required DNA sequences differ substantially. However, the sequences of replication origins are rather similar in different eukaryotes, and so the yeast origin will work in many other higher organisms, at least to some extent.

2. A **centromere (Cen) sequence** to allow correct partition of the plasmid in yeast. When a yeast cell divides, the duplicated chromosomes are pulled apart by microtubules attached to their centromeres, so that each daughter cell gets a full set. Shuttle vectors must also be segregated correctly at cell division. To achieve this, the shuttle vector must contain a segment of DNA from the centromere of the yeast chromosomes, the Cen sequence. This is recognized by the microtubules that drag new chromosomes apart.

Shuttle vectors must have separate origins of replication and separate selection mechanisms for each host organism.

3. A gene to select for the plasmid in yeast. The problem here is that yeast is not affected by most of the antibiotics that kill bacteria. In practice, a less satisfactory selection technique is used. A yeast host strain that has a defect in a gene for making an amino acid, say leucine, is used. The corresponding biosynthetic gene is present on the vector. In the absence of leucine the yeast will starve. Only if it obtains the plasmid carrying the *leu*$^+$ gene will it survive.

---

centromere (Cen) sequence   Sequence at centromere of eukaryotic chromosome that is needed for correct partition of chromosomes during cell division

origin of replication   Site on chromosome where DNA replication starts

Cen sequence   See centromere sequence

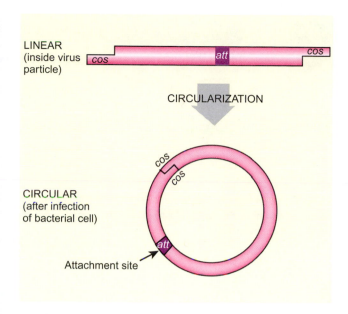

FIGURE 7.11
Lambda—Linear and Circular Genomes

In the lambda phage particle, the genome is a linear DNA molecule with two *cos* sequences at each end. After the phage injects its DNA into the bacterial host, the DNA circularizes. The two cohesive ends base pair and are ligated together by bacterial enzymes so forming a circle.

4. Bacteriophage Lambda Vectors

Bacteriophage lambda, a bacterial virus that infects *E. coli*, has been widely used as a cloning vector. As described in Ch. 21, lambda is a well-characterized virus with both lytic and lysogenic alternatives to its life cycle. Although lambda DNA circularizes for replication and insertion into the *E. coli* chromosome, the DNA inside the phage particle is linear (Fig. 7.11). At each end are complementary 12 bp long overhangs known as *cos* **sequences (cohesive ends)**. Once inside the *E. coli* host cell, these pair up and the cohesive ends are ligated together by host enzymes forming the circular version of the lambda genome.

Only DNA molecules of between 37 and 52 kb can be stably packaged into the head of the lambda particle. Small fragments of extra DNA may be inserted into the lambda genome without preventing packaging. However, to accommodate longer inserts it is necessary to remove some of the lambda genome. The left-hand region has essential genes for the structural proteins and the right-hand region has genes for replication and lysis. The middle region (~15 kb) of the lambda genome is non-essential and may be replaced with approximately 23 kb of foreign DNA (Fig. 7.12). Since the middle region of lambda has the genes for integration and recombination, such lambda replacement vectors cannot integrate into the host chromosome and form lysogens by themselves. To generate lysogens it is necessary to use a helper phage to provide the integration and recombination functions.

If foreign DNA is inserted into the middle of lambda, the result is a linear DNA molecule with two cohesive ends. To get such constructs into an *E. coli* host cell efficiently requires *in vitro* **packaging** (Fig. 7.13). In this technique, a mixture of lambda proteins is mixed with the recombinant lambda DNA *in vitro* to form phage particles. Infecting two separate *E. coli* cultures with two different defective lambda mutants generates the necessary lambda proteins. Each of the two mutants lacks an essential head protein and cannot form particles containing its own DNA. A mixture of the two lysates gives a full set of lambda proteins and when mixed with lambda DNA can generate infectious phage particles.

Viruses may be used as vectors, especially if they can adopt a lysogenic or latent state where they replicate in step with the host cell.

If essential genes are removed from a virus vector, a helper virus is needed for viral replication.

bacteriophage lambda Virus of *E. coli* with both lytic and lysogenic alternatives to its life cycle, which is widely used as a cloning vector
***cos* sequences (lambda cohesive ends)** Complementary 12 bp long overhangs found at each end of the linear form of the lambda genome
***in vitro* packaging** Procedure in which virus proteins are mixed with DNA *in vitro* to assemble infectious virus particles. Often used for packaging recombinant DNA into bacteriophage lambda

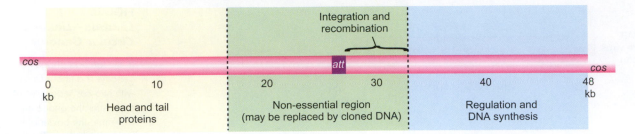

FIGURE 7.12
Lambda Replacement Vector

Since lambda phage is easy to grow and manipulate, the genome has been modified to accept foreign DNA inserts. The green region of the genome has genes that are non-essential for lambda growth and packaging. This region can be replaced with large inserts of foreign DNA (up to about 23 kb). When used with a helper phage, such modified lambdas provide useful cloning vectors.

FIGURE 7.13
In Vitro Packaging of Lambda Replacement Vector

A lambda cloning vector containing cloned DNA must be packaged in a phage head before it can infect *E. coli*. Before the DNA can be packaged, the phage head proteins must be isolated. To do this, a culture of *E. coli* is infected with a mutant lambda that lacks the gene for one of the head proteins called E. A different culture of *E. coli* is infected with a different lambda mutant that lacks the phage head protein D. Both *E. coli* cultures are grown with the mutant lambdas and the viruses are induced to enter the lytic cycle. Although the *E. coli* are lysed by the phage, they cannot form complete heads. Instead, a soluble mixture of phage proteins is isolated. Each lysate contains phage tails, assembly proteins, and components of the heads, except either D or E. These two lysates are mixed along with the lambda vector containing the cloned DNA. Although mixing is done *in vitro*, the components can self-assemble into a functional phage that can infect *E. coli*.

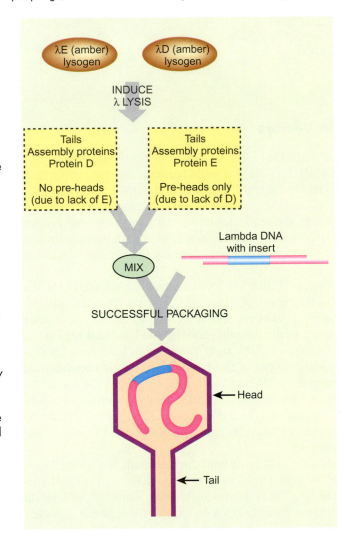

5. Cosmid Vectors

In vitro packaging using lambda lysates is a powerful technique. Packaging of DNA into a lambda head does not require the lambda genes; in fact, it is possible to fill almost the whole of a lambda particle with cloned DNA by using **cosmid** vectors. Cosmids themselves are small multicopy plasmids that carry *cos* sites (Fig. 7.14).

cosmid   Small multicopy plasmid that carries lambda *cos* sites and can carry around 45 kb of cloned DNA

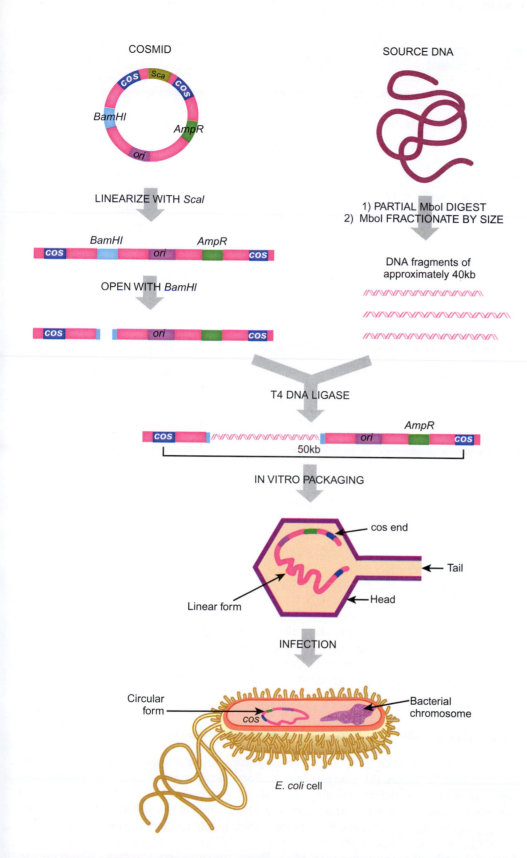

COSMID

SOURCE DNA

LINEARIZE WITH *Sca*I

1) PARTIAL MboI DIGEST
2) MboI FRACTIONATE BY SIZE

OPEN WITH *Bam*HI

DNA fragments of
approximately 40kb

T4 DNA LIGASE

50kb

IN VITRO PACKAGING

cos end

Tail

Linear form

Head

INFECTION

Circular form

cos

Bacterial chromosome

E. coli cell

FIGURE 7.14
Cosmid Vector

To clone large pieces of DNA into cosmid vectors, both must have compatible sticky ends. The cosmid vector is first linearized so that each end has a *cos* site. Then the linear cosmid is cut with *Bam*HI, which generates sticky ends with the overhang sequence GATC. The genomic DNA from the source of interest is also digested. Instead of *Bam*HI, this DNA is partially digested with *Mbo*I, which also generates a GATC overhang. Partial digestion leaves some sites uncut and allows large segments of a genome to be isolated. These segments are mixed with the two halves of the cosmid and joined using ligase. The final constructs are packaged into lambda particles *in vitro* and are used to infect *E. coli*.

The cosmid is first linearized so that each end has a *cos* sequence. In order to clone a gene of interest into the cosmid, both the gene and cosmid are cut either with the same restriction enzyme or with two enzymes that give identical sticky ends (e.g., *Bam*HI and *Mbo*I, as in Fig. 7.14). The target DNA is often only partially digested (i.e., some sites remain uncut). First, this allows large segments of a genome to be isolated. Second, if a cut site lies within a gene of interest, some fragments will

TABLE 7.01	Insert Sizes and Cloning Vectors	
Vector		**Maximum Insert Size**
Multicopy plasmid		10 kb
Lambda replacement vector		20 kb
Cosmid		45 kb
P1 plasmid vector		100 kb
PAC (P1 artificial chromosome)		150 kb
BAC (bacterial artificial chromosome)		300 kb
YAC (yeast artificial chromosome)		2,000 kb

still carry the intact gene. Ligation of the two cosmid pieces to either side of the target DNA results in a length of DNA with a *cos* site at each end. This construct can be packaged into lambda particles *in vitro* and then used to infect *E. coli*. Using a small cosmid, of say 4 kb, allows inserts of up to about 45 kb to be cloned.

6. Yeast Artificial Chromosomes

Artificial chromosomes are vectors designed to carry large amounts of DNA. Yeast artificial chromosomes carry the most DNA, around 2000 kb.

Analysis of the genomes of higher organisms requires the cloning of much larger fragments than for bacteria. Because eukaryotic genes contain introns they may be hundreds of kilobases in length. Such large DNA fragments require special vectors. The largest capacity vectors derived from bacteriophage can handle at most 100 kb (Table 7.01). Consequently, "artificial chromosomes" have been developed to carry huge lengths of eukaryotic DNA.

Huge segments of DNA, up to 2,000 kb or 2 million base pairs, may be carried on **yeast artificial chromosomes or YACs** (Fig. 7.15). For any replicon, whether plasmid or chromosome, to survive in yeast, the vector must have a yeast specific origin of replication and a centromere recognition sequence (Cen sequence). The YAC has both of these elements. In addition, as required by all eukaryotic chromosomes, telomere sequences are present on both ends. A yeast cell will treat this structure, although artificial, as a chromosome. Of course, for practical use a selectable marker and a suitable multiple cloning site are also included.

Colossal amounts of cloned DNA can be inserted into a YAC and may thus be replicated inside yeast cells. Because the recognition sequences for replication origins, centromeres, and telomeres are so similar among higher organisms, an added bonus is that YACs will survive in mice and are even passed on from parent to offspring. Admittedly, not every baby mouse inherits the YAC, but this opens the way for cloning the huge DNA sequences needed for engineering higher animals and for sequencing their genomes.

7. Bacterial and P1 Artificial Chromosomes

Multicopy vectors, such as ColE1-derived plasmids, are valuable because they give higher yields of DNA than single copy vectors. However, they also have disadvantages. In particular, the inserts may be unstable especially if they are very long and contain repeated sequences. Many times, unstable inserts are deleted from the plasmid by recombination events. Eukaryotic DNA is particularly unstable in plasmids. Therefore, cloning large segments of eukaryotic DNA in bacteria is now done using **bacterial artificial chromosomes (BACs)**. These are single copy vectors based on the

bacterial artificial chromosome (BAC) Single copy vector based on the F-plasmid of *E. coli* that can carry very long inserts of DNA. Widely used in the human genome project
yeast artificial chromosome (YAC) Single copy vector based on yeast chromosome that can carry very long inserts of DNA. Widely used in the human genome project

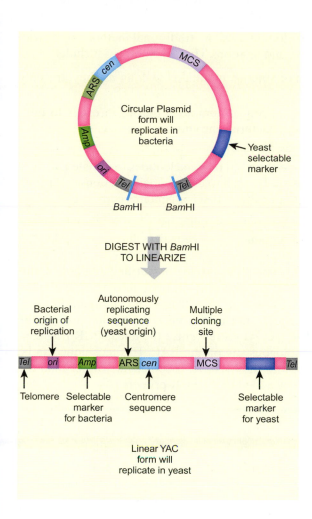

FIGURE 7.15
***Yeast Artificial
Chromosome (YAC)***

The YAC has two forms, a circular form for growing in bacteria, and a linear form for growing in yeast. The circular form can be manipulated and grown like any other plasmid in bacteria since it has a bacterial origin of replication and an antibiotic resistance gene. In order to use this in yeast, the circular form is isolated and linearized such that the yeast telomere sequences are on each end. This form can accommodate up to 2,000 kb of cloned DNA inserted into its multiple cloning site (MCS).

F-plasmid of *E. coli*. They can accept inserts of 300 kb or more. Electroporation is necessary to transform these large constructs into *E. coli* host cells, and the yields of DNA are low. Nonetheless, bacterial artificial chromosomes have been widely used in the human genome project and other eukaryotic genome sequencing projects.

Another cloning vector used for larger eukaryotic DNA segments is the **P1 artificial chromosome (PAC)**. This cloning vector is derived from bacteriophage P1 and has been used to carry inserts of up to 150 kb. Just like the lambda derived vectors (see above), these PACs require *in vitro* packaging.

8. Recombineering Increases the Speed of Gene Cloning

During homologous recombination, two DNA molecules exchange segments in a precise manner, using regions of homology (for more details see Ch. 24). DNA engineering via homologous recombination is known as **recombineering**. The recombination system most used is actually that of Lambda (λ) phage. Lambda is a bacterial virus that can enter a lysogenic (latent) state. It uses recombination via the RED system to integrate its genome into the genomes of host bacteria, such as *E. coli*. The proteins that integrate λ can be transiently expressed and will integrate any linear piece of DNA. They are now used to create new vector combinations. The λ phage RED

P1 artificial chromosome (PAC)   Single copy vector based on the P1-phage/plasmid of *E. coli* that can carry very long inserts of DNA
recombineering   Technique that uses homologous recombination to insert a piece of DNA into a vector

system has many advantages for gene cloning over the traditional methods of restriction enzyme digests, fragment isolation, and ligations. These advantages include:

1. RED proteins do not cause spontaneous mutations as long as they are only expressed for a short time.
2. The proteins only need a small region of homology (45 bp) in order to integrate a foreign piece of DNA into bacterial chromosomes or BACs.
3. The laboratory procedure is quick and easy.
4. The system recognizes short single-stranded oligonucleotides to initiate a recombination event. Therefore, these can be used to create small deletions, insertions, or even single nucleotide changes in either a BAC or genomic copy of a gene.

To perform recombineering, the first step is to amplify the gene or DNA fragment of interest by PCR. In addition to the sequence that recognizes the target DNA, the PCR primers also include around 50 base pairs of sequence that is complementary to a homologous site on a BAC (Fig. 7.16). This PCR product is purified and added into bacteria that have the λ phage integration enzyme genes in their chromosome and the BAC in the cytoplasm. The integration enzymes are controlled by a temperature sensitive promoter, so at 32°C the enzymes are not made, but after a shift to 42°C, the integration enzymes are produced. The λ enzymes recognize the linear PCR fragment ends and integrate them into the homologous site on the BAC. This technique can be adapted to integrate the PCR product into the bacterial chromosome rather than the BAC by changing the region of homology added to the PCR primers.

Song H, Chung S-K, and Xu Y (2010) Modeling disease in human ESCs using an efficient BAC-based homologous recombination system. Cell Stem Cell 6: 80–89.

In this associated paper, the authors use recombineering to disrupt two different genes, ATM and p53, so they can determine what effect these mutations have on cellular function. Rather than working with a mouse model, the authors chose to work on human embryonic stem cells (ESCs), which are undifferentiated and have the potential to form many different cell types. The authors hypothesized that these cells would be a good model system to study ATM and p53 and would have a similar phenotype as is seen in people. For people, mutations in ATM result in

FOCUS ON RELEVANT RESEARCH

Ataxia-telangiectasia, a disease characterized by retarded growth and neural degeneration that causes uncoordinated movements, small dilated blood vessels (telangiectasia), a poor immune system, and a predisposition to cancer. The ATM gene helps repair DNA damage, so when it is defective, people progressively accumulate DNA mutations in their somatic cells, hence the propensity for cancer and neural defects. For the p53 gene, mutations are commonly found in human cancers. p53 encodes a protein that controls progression through the cell cycle, so when the gene is defective, the cells fail to stop dividing, resulting in uncontrolled growth and eventually, a tumor.

To disrupt the ATM gene in the ESCs, the researchers constructed a BAC-based vector by recombineering in E. coli. The ATM gene is very large with over

40 exons. In order to disable this gene, the researchers created two different targeting vectors, one for each ATM allele. They first isolated a 12.5 kb NcoI restriction enzyme fragment from the ESCs that has exons 39 through 45 of ATM. This fragment was put into a BAC and then recombineering was used to create the targeting vectors, one for each ATM allele. The first targeting vector has a neomycin resistance gene with a splice site integrated into the neomycin promoter. After inserting this recombineered fragment into the ATM gene of the ESCs, the splice site causes part of the intron to be included in the ATM mRNA transcript. The second targeting vector induces the same mistake, except it has a gene for puromycin resistance instead of neomycin. The altered mRNA will no longer express the correct protein.

To create ATM knockout ESCs, the first targeting vector and ESCs were incubated together and then an electrical charge was applied, a process called **electroporation**. This opens the cytoplasmic and nuclear membranes of the ESCs and allows the targeting vector to enter the nucleus. Once in the nucleus, the regions that are homologous to the ATM gene recombine, and the targeting vector replaces one of the copies of ATM (see Ch. 24 for more information about recombination). These cells are selected by resistance to neomycin. Next, the heterozygous cells were electroporated with the second targeting vector and selected by resistance to neomycin and puromycin. These cells have both copies of ATM disrupted. The absence of this gene and its encoded protein was confirmed by PCR, Southern blot analysis, and Western blots. Similar methods were used to disrupt the p53 gene.

The rest of the paper characterizes the ATM knockout and p53 knockout cell phenotypes in respect to the disease traits. The authors found that ATM knockout cells showed neural defects when they were grown as tumors in mice. The p53 knockout ESCs did not stop dividing when there was damage to the genome, just as is seen in human cancers. This method to create disease cell lines is much more efficient than other approaches and will help in trying to find a way to treat the associated diseases.

---

electroporation   Inducing small pores or openings in a cellular membrane with an electrical current; used for bacteria, mammalian cells, yeast, and other small organisms

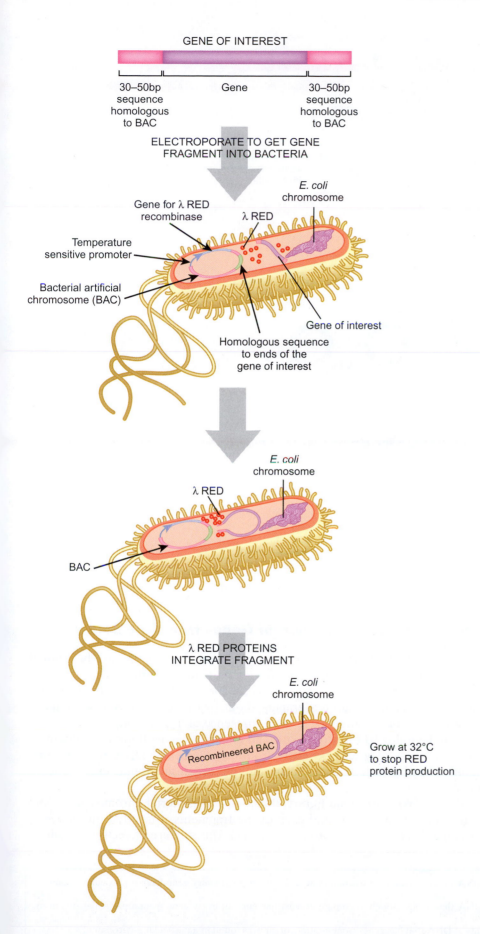

GENE OF INTEREST

30–50bp sequence homologous to BAC | Gene | 30–50bp sequence homologous to BAC

ELECTROPORATE TO GET GENE FRAGMENT INTO BACTERIA

Gene for λ RED recombinase

Temperature sensitive promoter

Bacterial artificial chromosome (BAC)

E. coli chromosome

λ RED

Gene of interest

Homologous sequence to ends of the gene of interest

E. coli chromosome

λ RED

BAC

λ RED PROTEINS INTEGRATE FRAGMENT

E. coli chromosome

Recombineered BAC

Grow at 32°C to stop RED protein production

FIGURE 7.16
Recombineering

PCR is used to create linear fragments of DNA that have ends that are homologous to regions of a bacterial artificial chromosome (BAC) lambda integration site. Bacteria containing the gene for the RED protein are grown at a high temperature for a few hours to induce production of the RED protein. The large fragment is inserted into a bacteria. The RED proteins recognize the ends of the linear DNA and initiate homologous recombination of the fragment and BAC at the region of homology. The final bacteria are grown at 32°C to inhibit any further RED gene expression.

FIGURE 7.17
Creating a DNA Library

A DNA library contains as many genes from the organism of interest as possible. The genomic DNA from the organism of interest is isolated and digested with a restriction enzyme. Usually, the restriction enzyme used has a recognition sequence of four base pairs; therefore, the DNA would be cut into fragments much smaller than the average gene. Therefore, the digestion is carried out for a brief period that leaves many of the restriction sites uncut. A suitable vector for the required insert size is chosen and is cut with a restriction enzyme that produces compatible sticky ends. The digested genomic DNA and the vector are ligated together and transformed into bacterial host cells. A large number of transformed bacterial colonies must be isolated and kept to ensure that all possible genes from the genome of interest are represented on at least one vector.

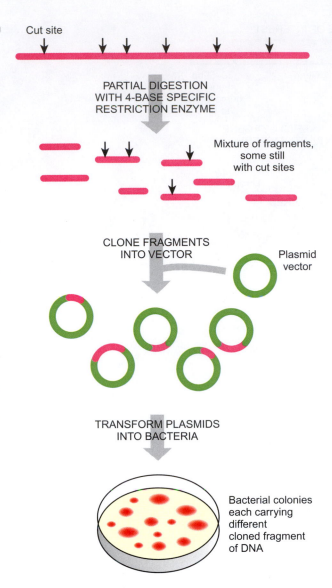

Cut site

PARTIAL DIGESTION WITH 4-BASE SPECIFIC RESTRICTION ENZYME

Mixture of fragments, some still with cut sites

CLONE FRAGMENTS INTO VECTOR

Plasmid vector

TRANSFORM PLASMIDS INTO BACTERIA

Bacterial colonies each carrying different cloned fragment of DNA

9. A DNA Library is a Collection of Genes from One Source

Collections of cloned genes carried in vectors are called libraries. DNA libraries have all the genes from one organism, whereas metagenomic libraries have genes from multiple organisms that inhabit a particular environment.

Gene libraries or **DNA libraries** are collections of cloned genes that are big enough to contain at least one copy of every gene from a particular organism. The size of the genes and the organism dictate which vector is used for holding the inserts. The genes of prokaryotes are relatively short, averaging about 1000 bp each. In contrast, eukaryotic genes are much longer, largely due to the presence of introns. Different strategies must therefore be followed for prokaryotic and eukaryotic gene libraries as discussed below. Gene libraries may also be made from environmental DNA samples. Such **metagenomic libraries** include genes from multiple organisms found in a particular environment.

To make a prokaryotic gene library, the complete bacterial chromosomal DNA is cut with a restriction enzyme and each of the fragments is inserted into a vector, usually a simple ColE1-derived plasmid (Fig. 7.17). This mixture of vectors containing

DNA library   Collection of cloned segments of DNA that is big enough to contain at least one copy of every gene from a particular organism. Same as gene library

Gene library   Collection of cloned segments of DNA that is big enough to contain at least one copy of every gene from a particular organism. Same as DNA library

metagenomic library   Collection of cloned segments of DNA that has genes from multiple organisms found in a particular environment

a different piece of the bacterial chromosome is transformed into a suitable bacterial host strain and a large number of colonies, each containing a single vector plus insert, are kept. These must then be screened for the gene of interest. If the gene has an observable phenotype, this may be used. Otherwise, more general methods such as hybridization or immunological screening are necessary.

Gene libraries are often made using a 4-base specific restriction enzyme to cut the genomic DNA. This cuts DNA every 256 bases on average. Since this is shorter than an average gene, the DNA is only partially digested by only allowing a short amount of time for the restriction enzyme to cut the DNA. This generates a mixture of fragments of various lengths, many of which still have restriction sites for the enzyme used. The hope is that an intact copy of every gene, even those cut by the enzyme used, will be present on at least some fragments of DNA (Fig. 7.17). However, because restriction sites are not truly distributed at random, some fragments will be too large to be cloned and some genes will contain clustered multiple restriction sites and will be destroyed even in a partial digest. For total coverage, another library should be made with another restriction enzyme.

9.1. Screening a Library by Hybridization

After cloning all the possible genes from an organism into a library, the next step is to identify the genes of interest. Libraries are often screened by DNA/DNA hybridization using DNA probes. The probes themselves are generally derived from two sources. Cloned DNA from a related organism is often used to screen a library. The stringency of the hybridization conditions must be adjusted to allow for a greater or lesser percentage of mismatches, depending on the relatedness of the two organisms. Another possibility is to synthesize an artificial probe, using the base sequence deduced from the amino acid sequence of the corresponding protein. This assumes that the protein has been purified and that a partial amino acid sequence from the N-terminal region is available. The DNA probe is labeled for detection by autoradiography, fluorescence, or chemical tagging as described in Chapter 5. Probes generally range from 100 to 1000 bases long, although shorter probes may sometimes be used. At least 80% matching over a 50-base stretch is needed for acceptable hybridization and identification.

> DNA probes for a specific gene are used to identify which bacteria contain the DNA insert complementary to the probe.

The **target DNA** (i.e., the DNA from the library to be probed) is denatured and bound to a nitrocellulose or nylon membrane. The membrane is then incubated with the labeled probe. After washing away excess probe, the membrane is screened by the chosen detection system (e.g., autoradiography as illustrated in Fig. 7.18).

9.2. Screening a Library by Immunological Procedures

Instead of looking for DNA/DNA hybrids to identify the gene of interest from a library, the protein itself can be identified by **immunological screening**. This method relies on the production of the protein encoded by the gene of interest and therefore assumes that the cloned gene is efficiently expressed under the experimental conditions. That is, each of the library inserts must have transcriptional and translational start sequences as well as stop sequences. Usually, the library vector supplies these sequences, since the promoters from the genomic DNA will not usually be cloned still attached to the genes they control (see below). When the protein is expressed, it may be detected by binding to an **antibody**. This means the antibody to the encoded protein (or a closely related protein from another organism) must be available.

> Gene libraries can be expressed into proteins, and these can be screened using antibodies and secondary antibodies that are linked to a detection system.

In order to screen an expression library, the bacteria expressing the library inserts are grown on master plates and samples of each bacterial colony are transferred to

---

antibody   Protein made by the immune system to recognize and bind to foreign proteins or other macromolecules
immunological screening   Screening procedure that relies on the specific binding of antibodies to the target protein
target DNA   DNA that is the target for binding by a probe during hybridization or the target for amplification by PCR

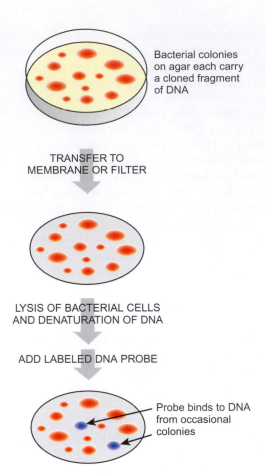

Bacterial colonies on agar each carry a cloned fragment of DNA

TRANSFER TO MEMBRANE OR FILTER

LYSIS OF BACTERIAL CELLS AND DENATURATION OF DNA

ADD LABELED DNA PROBE

Probe binds to DNA from occasional colonies

FIGURE 7.18
Screening a DNA Library by Probing

The first step in screening a DNA library is to grow colonies of bacteria containing the library inserts on agar. A large number of different transformed bacteria are grown, so that all genes in the library have a reasonable chance of being present. Next, the bacterial colonies are transferred to a membrane or filter. The filter is applied to the top of the bacterial colonies and carefully lifted off. A portion of each bacterial colony will stick to the filter while the rest of the colony stays on the agar plate. Once on the filter, the bacteria are lysed open and the DNA is denatured. The single-stranded DNA stays bound to the filter, and the majority of the bacterial components are washed away. The library filters are covered with a solution of a radioactively labeled single-stranded DNA probe, allowed to hybridize, and then the excess probe is washed away. Placing a piece of photographic film over the filter identifies the hybrid molecules. When the probe hybridizes to a library insert, a black spot appears on the photographic film. By lining up the original bacterial colonies with the photographic film, the corresponding library insert can be isolated from the bacteria.

a suitable membrane. The cells are lysed and the released proteins are attached to the membrane. The membrane is then treated with a solution of the appropriate antibody. After excess primary antibody is washed away, a second antibody that is specific for the primary antibody is added. This will bind any primary antibody it encounters (Fig. 7.19). This secondary antibody carries the detection system, such as alkaline phosphatase, which converts a colorless substrate, such as X-phos, to a colored product (see Ch. 19). If X-phos is used, the region on the membrane where the secondary antibody is bound turns blue. The blue spots must be aligned with the original bacterial colonies. The DNA from the bacteria containing the insert encoding the protein of interest can then be isolated.

The reason for using two different antibodies is to allow flexibility and amplify the signal. Antibodies to the protein of interest are made by injecting a rabbit with the

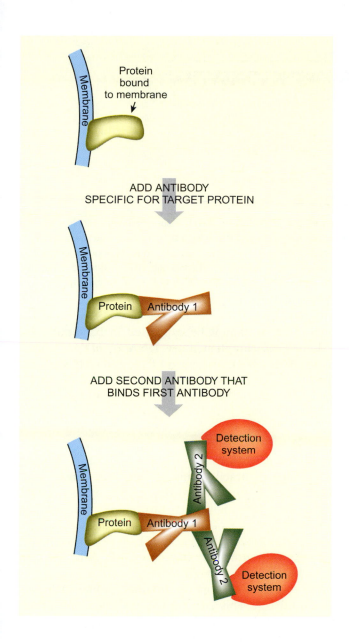

■ **FIGURE 7.19**
Immunological Screening of a DNA Library

Bacteria carrying a library are grown on agar, transferred to a membrane, and lysed. Released proteins are bound to the membrane. This figure shows only one attached protein, but in reality, a large number of different proteins will be present. These proteins include both those from the library as well as many bacterial proteins. The membrane is incubated with a primary antibody that only binds the protein of interest. Any non-specifically bound antibody is washed away. Finally, a second antibody that binds the primary antibody and that also carries a detection system is added.

protein and isolating all the antibodies from a sample of the rabbit's blood. Producing an antibody is costly and a long process, so instead of directly conjugating this antibody to the detection system, a second antibody is produced in another animal, such as a goat. The secondary antibody recognizes all rabbit antibodies; therefore, it can be used for any primary antibody made in a rabbit. The secondary antibodies are available commercially and are relatively inexpensive. Secondary antibodies also amplify the signal, since usually two secondary antibody molecules bind to each primary antibody. Double color intensity will be generated using the two antibody system.

10. Cloning Complementary DNA Avoids Introns

Most eukaryotic genes have intervening sequences of non-coding DNA (introns) between the segments of coding sequence (exons). In higher eukaryotes, the introns are often longer than the exons and the overall length of the gene is therefore much larger than the coding sequence. This creates two problems. First, cloning large segments of DNA is technically difficult; plasmids with large inserts are often unstable and transform poorly. Secondly, bacteria cannot process RNA to remove the introns

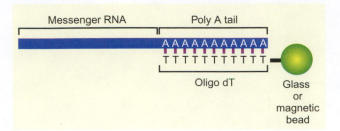

FIGURE 7.20
Purifying mRNA by Oligo(dT)

In order to isolate only messenger RNA from a sample of eukaryotic tissue, the unique features of the mRNA molecule are used. Only mRNA has a poly(A) tail, a long stretch of adenines following the coding sequence. The poly(A) tail of the mRNA will bind by base pairing to an oligonucleotide consisting of a long stretch of deoxythymidine residues—oligo(dT). The oligo(dT) is attached to glass or magnetic beads, which consequently bind mRNA specifically. Other RNAs will not bind to the beads and can be washed from the column.

> Reverse transcriptase makes cDNA copies of mRNA. These are valuable for making libraries from eukaryotic organisms since they do not contain any intron sequences.

and so eukaryotic genes containing introns cannot be expressed in bacterial cells. Using a DNA copy of mRNA, known as **complementary DNA or cDNA**, solves both problems since the mRNA has already been processed by the cell so that all the introns are removed.

To make a cDNA library, the messenger RNA must be isolated and used as a template. The library generated therefore reflects only those genes expressed in the particular tissue under the chosen conditions. First, total RNA is extracted from a particular cell culture, tissue, or specific embryonic stage. The mRNA from eukaryotic cells is normally isolated from the total RNA by taking advantage of its poly(A) tail. Since adenine base pairs to thymine or uracil, columns containing **oligo(U)** or **oligo(dT)** tracts bound to a solid matrix are used to bind the mRNA by its poly(A) tail. The mRNA is retained and the other RNA is washed through the column. The mRNA is then released by eluting with a buffer of high ionic strength that disrupts the H-bonding of the poly(A) tail to the oligo(dT) (Fig. 7.20). A variation of this approach is the use of magnetic beads with attached oligo(dT) tracts. After binding the mRNA the beads are separated magnetically.

To generate cDNA the enzyme reverse transcriptase, originally found in retroviruses (see Ch. 21), is added to the mRNA. This enzyme will make a complementary DNA (cDNA) strand using the mRNA as template. Several further steps are required to generate a double-stranded cDNA copy of the original mRNA (Fig. 7.21). Ribonuclease H, which only degrades the RNA strand in an RNA/DNA hybrid, is used to remove the mRNA strand of the mRNA/cDNA hybrid molecule leaving a single-stranded cDNA. DNA polymerase I is then used to synthesize the second DNA strand. Alternatively, some reverse transcriptases are multifunctional and are able to remove the mRNA and synthesize the complementary strand of DNA. Any remaining single-stranded ends are trimmed off by S1 nuclease, which is an exonuclease specific for single-stranded regions of DNA. (Such single-stranded ends mostly result from oligo(dT) primers binding in the middle of the mRNA poly(A) tail.) The resulting double-stranded cDNA molecules can be isolated and cloned into an appropriate vector, resulting in a **cDNA library**. Since each mRNA has a different sequence, convenient restriction sites are generally added at each end. This is done by attaching linkers—short pieces of DNA that have restriction sites compatible with those in the

complementary DNA (cDNA)   DNA copy of a gene that lacks introns and therefore consists solely of the coding sequence. Made by reverse transcription of mRNA

cDNA library   Collection of genes in their cDNA form, lacking introns

oligo(dT)   Stretch of single-stranded DNA consisting solely of dT or deoxythymidine residues

oligo(U)   Stretch of single-stranded RNA consisting solely of U or uridine residues

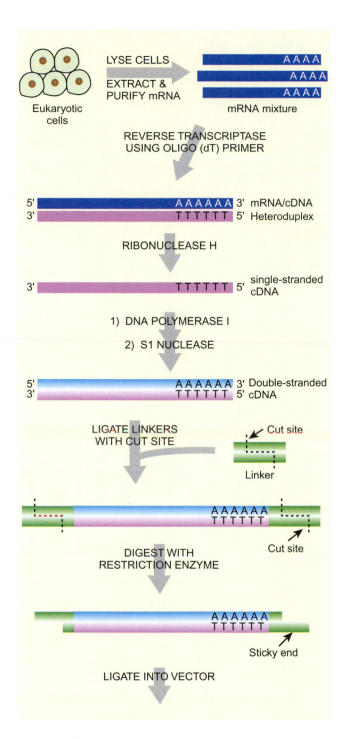

LYSE CELLS

EXTRACT &
PURIFY mRNA

Eukaryotic
cells

mRNA mixture

REVERSE TRANSCRIPTASE
USING OLIGO (dT) PRIMER

5' AAAAAA 3' mRNA/cDNA
3' TTTTT 5' Heteroduplex

RIBONUCLEASE H

3' TTTTT 5' single-stranded
cDNA

1) DNA POLYMERASE I

2) S1 NUCLEASE

5' AAAAAA 3' Double-stranded
3' TTTTT 5' cDNA

LIGATE LINKERS
WITH CUT SITE

Cut site

Linker

AAAAAA
TTTTT

Cut site

DIGEST WITH
RESTRICTION ENZYME

AAAAAA
TTTTT

Sticky end

LIGATE INTO VECTOR

FIGURE 7.21
*Making a cDNA Library
from Messenger RNA*

First, eukaryotic cells are lysed
and the mRNA is purified. Next,
reverse transcriptase plus primers
containing oligo(dT) stretches are
added. The oligo(dT) hybridizes to
the adenine in the mRNA poly(A)
tail and acts as a primer for reverse
transcriptase. This enzyme makes
the complementary DNA strand so
forming an mRNA/cDNA hybrid
molecule. The mRNA strand is
digested with ribonuclease H and
DNA polymerase I is added to
synthesize the opposite DNA strand,
thus creating double-stranded
cDNA. S1 nuclease trims off single-
stranded ends. Since each mRNA
has a different sequence, linkers
must be ligated to the ends of the
cDNA to allow convenient insertion
into the cloning vector. After
addition, the linkers are digested
with the appropriate restriction
enzyme and the cDNA is ligated into
the vector. The resulting hybrid DNA
molecules are then transformed into
bacteria, so giving the final cDNA
library.

multiple cloning site of the vector. Not only is cDNA easier to handle, because the cloned fragments are much shorter than the original eukaryotic genes, but the cDNA versions of eukaryotic genes can often be successfully expressed in bacteria.

11. Chromosome Walking

As a result of genetic investigation, we may know the approximate chromosomal location of a particular gene. Using this information to clone the gene is referred to as positional cloning. One of the simplest versions of this is **chromosome walking**,

chromosome walking Method for cloning neighboring regions of a chromosome by successive cycles of hybridization using overlapping probes

FIGURE 7.22
Chromosome Walking

Chromosome walking utilizes overlapping fragments of a particular chromosome to isolate genes upstream and downstream from the original DNA fragment. The first step is to identify the region of the chromosome to which the probe hybridizes. In this example, probe #1 hybridizes to the purple region of the chromosome. When the chromosome is cut with restriction enzyme #1, fragment 1A will hybridize to probe #1 at one end. This allows fragment 1A to be isolated and sequenced and its downstream sequence is used to generate probe #2. To find the next segment of the chromosome, a different restriction enzyme is used to cut the DNA (step 2). This time probe #2 will hybridize to fragment 2B. Once again, the probe recognizes only the first half of this fragment. The downstream sequence of fragment 2B can then be determined, and this information can be used to make probe #3. In step 3, the chromosome is cut with restriction enzyme #1 again. Now probe #3 will hybridize with fragment 1B, whose downstream sequence can therefore be determined, and another probe, called probe 4 can be made. This procedure can be continued as far as desired, working in either direction.

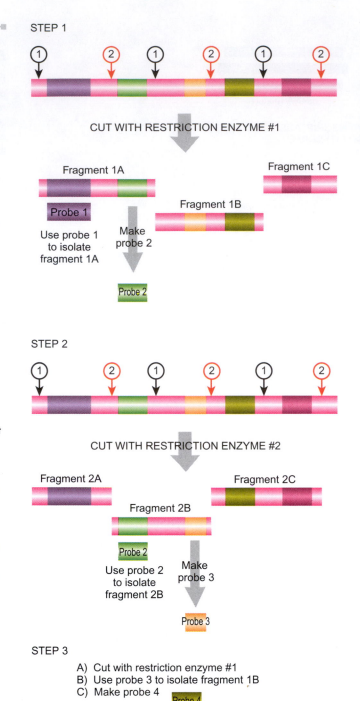

a method based on hybridization. This approach is used when one segment of DNA has already been cloned and the neighboring genes are of interest.

The chromosome that has the target gene can be identified and isolated by fluorescence *in situ* hybridization (FISH) followed by fluorescence activated sorting as described in Chapter 5. The chromosome of interest is then cut into manageable fragments with a suitable restriction enzyme. The original cloned fragment of DNA is used as the first probe. Each fragment is tested for hybridization to an initial probe made from the segment that was already cloned. Obviously, this will overlap at least one or two neighboring fragments. These overlap fragments are then cloned and used as probes in a second cycle of hybridization, and so on. For simplicity, we have shown a walk in one direction only in Figure 7.22, though obviously in real life we would walk along the chromosome in both directions from the original starting point.

Each cycle of hybridization identifies segments of DNA that overlap the fragments isolated earlier and used as probes. By moving outwards from the starting position step by step, the whole chromosome can be mapped and cloned (slowly!).

12. Cloning by Subtractive Hybridization

Subtractive hybridization is a technique used to isolate a DNA segment that is missing from one particular sample of DNA. Obviously, a second DNA sample that contains the fragment of interest is necessary. Suppose that a hereditary defect is due to the deletion of the DNA for a particular gene. A sample of DNA from the appropriate chromosome of the afflicted individual will lack this particular segment of DNA. To find the missing DNA, the corresponding chromosome from a healthy (wild-type) individual is isolated. For example, the *dmd* gene, for **Duchenne muscular dystrophy**, is located in the Xp21 band, close to the middle of the short or p-arm of the X-chromosome. Using light microscopy to analyze chromosomal banding patterns, a patient was found who had a deletion large enough that the Xp21 band was missing. Subtractive hybridization of the mutant chromosome with a normal chromosome allowed the *dmd* gene to be cloned.

To do subtractive hybridization, both the mutant and wild-type DNA samples are cut into fragments of convenient size using a restriction enzyme. Then the two sets of fragments are hybridized together. This will give hybrid molecules for all regions of the DNA except the region of the deletion, which is present only in the wild-type chromosome. If a large surplus of mutant DNA is used, all fragments of the wild-type chromosome will be hybridized to mutant fragments except the region corresponding to the deletion, which will be left over. The single strands of this lone fragment will have to pair with each other. Thus, we have subtracted out all the segments of DNA that are not wanted.

In practice, some means to obtain the left over "deletion fragment" is required. One approach is to cut the two batches of DNA with different restriction enzymes. If the normal DNA is cut with restriction enzyme 1 and the mutant DNA is cut with restriction enzyme 2, any hybrid molecule will have non-matching ends. If the mutant DNA hybridized with itself, the ends will match, and will be cut with restriction enzyme 2. If the gene of interest from the normal DNA self-hybridizes, then this will have ends compatible with restriction enzyme 1. This fragment can then be cloned into a vector using restriction enzyme 1. To make the procedure even easier, restriction enzyme 2 could leave a blunt end after cutting. Since blunt ends are so much harder to ligate, only a self-hybrid molecule flanked by sticky ends 1 would be cloned into the vector. Only the self-paired fragment of wild-type DNA would have sticky ends due to cutting by restriction enzyme 1 (Fig. 7.23).

Subtractive hybridization can also be used to isolate a set of genes that are expressed under particular conditions. Two batches of cells are grown, one under standard conditions and the other under the conditions being investigated. For example, one batch of mouse cells can be grown with all the necessary nutrients and another set of mouse cells can be grown with only limited nutrients. The total RNA is isolated from both samples, then the mRNA is purified by hybridization to oligo(dT) as described above. The standard sample will contain mRNA from genes expressed under normal nutrient conditions. The experimental sample will contain mRNA from genes only expressed when nutrients are limited. Limited nutrients may stimulate cells to manufacture their own nutrients, thus some mRNAs would be produced in a higher abundance than the other sample.

> Cloned genes can sometimes by found by a negative approach. Hybridization is used to remove genes shared by two organisms, leaving behind only those that are unique.

> Subtractive hybridization can also be performed with two samples of mRNA from the same organism grown under different conditions.

Dmd gene   Gene responsible for Duchenne muscular dystrophy
Duchenne muscular dystrophy   One of several inherited diseases affecting muscle function
subtractive hybridization   Technique used to remove unwanted DNA or RNA by hybridization so leaving behind the DNA or RNA molecule of interest

FIGURE 7.23
Cloning by Subtractive Hybridization

The key to subtractive hybridization is to hybridize all the wild-type or "healthy" DNA fragments (pink) with an excess of mutant DNA (purple). In this example, the mutant DNA is digested with restriction enzyme 2 and the wild-type DNA is digested with restriction enzyme 1. Both samples are heated to separate the strands, forming a pool of single-stranded fragments. In order to ensure all the wild-type DNA is hybridized to mutant DNA and not to itself, a large surplus of mutant DNA is mixed with a small amount of wild-type DNA. The DNA is allowed to anneal yielding double-stranded DNA consisting of a mixture of mutant:mutant, mutant:wild-type, and rare wild-type:wild-type molecules. Since the ratio of wild-type DNA to mutant DNA is so low, theoretically the only molecule where two wild-type strands anneal should be the missing fragment in the mutant. Because two different restriction enzymes were originally used to digest the different samples of DNA, these desired DNA molecules are the only ones that can be digested with restriction enzyme 1. This allows them to be cloned into a vector cut with restriction enzyme 1 and captured.

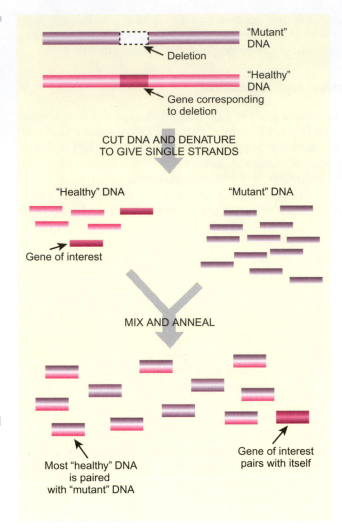

The basic idea is that the standard mRNA is used to subtract out the corresponding mRNA molecules from the experimental sample. However, two mRNA molecules of the same sequence obviously cannot hybridize together directly. Therefore, the standard mRNA is first converted to the corresponding double-stranded cDNA by reverse transcriptase. The cDNA is bound to a filter and the experimental sample of mRNA is incubated with the filter (Fig. 7.24). Messenger RNA corresponding to genes in the cDNA is retained by hybridization. Only the mRNA from genes that are expressed under the specific conditions of interest remains unbound. This, in turn, can now be converted to cDNA so giving a sample of those genes expressed under the particular conditions chosen.

13. Expression Vectors

Once a gene has been cloned into a vector it may or may not be expressed. If both the structural gene and promoter were cloned on the same segment of DNA the gene may well be expressed. On the other hand, if only the structural gene was cloned then expression will depend on whether a promoter is provided by the plasmid. Vectors that use blue and white screening (see above) place the cloned gene under control of the *lac* promoter, which lies upstream of the multiple cloning site.

Often, the objective of cloning a gene is to isolate high levels of the encoded protein. Purification of proteins has long been complicated because each protein folds up in an individualized manner and consequently behaves differently. To get around this problem the target protein is often tagged with another peptide that is easy to

Vectors may carry promoters and ribosome binding sites to mediate the expression of cloned genes.

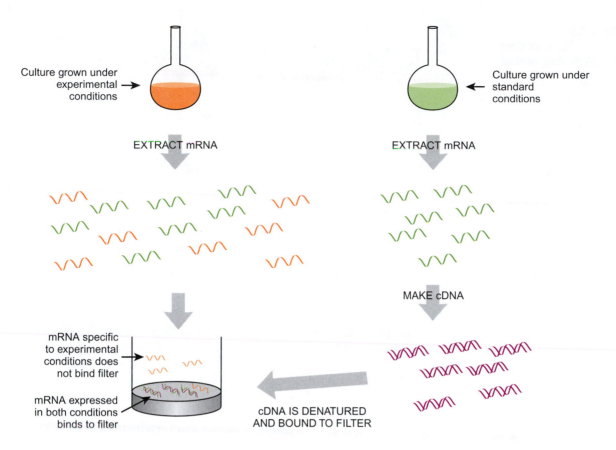

FIGURE 7.24
Subtractive Hybridization Captures mRNA Expressed Under Specific Conditions

Two different cultures are grown; one under standard conditions (green) and one under experimental conditions (orange). The mRNA is isolated from each culture. To allow hybridization, one set of mRNA must be converted into double-stranded DNA by reverse transcriptase. The double-stranded cDNA is then denatured and bound to a filter. In this example, cDNA corresponding to the standard mRNA (green) is bound to the filter. The experimental mRNA (orange) is incubated with the filter, where it binds to complementary single-stranded DNA from the standard conditions. If a gene is expressed to a greater extent under experimental conditions but not expressed (or present only in low amounts) during standard conditions, its mRNA will not be bound as no corresponding cDNA will be present on the filter. In practice, the mRNA that does not hybridize is pooled and then rehybridized to cDNA from the standard conditions. Repeating this step ensures that all the isolated mRNA is truly unique to the experimental condition.

detect and/or purify. This allows purification and manipulation of many different proteins by the same procedures. Tagging is generally done at the genetic level—that is, an extra segment of DNA that codes for the tag is inserted next to the DNA coding for the target protein. (This topic is discussed in detail in Ch. 15, Proteomics.) For now we should remember that a "cloned gene" often includes extra sequences that specify tags or alter regulation to facilitate later analysis.

It is often helpful to deliberately control or enhance expression of a cloned gene, especially if high levels of the encoded protein are needed. **Expression vectors** are specifically designed to place the cloned gene under control of a plasmid-borne promoter. In practice, the gene under investigation is normally first cloned in a general cloning vector and then transferred to the expression vector.

A variety of expression vectors exist with different promoters. The two basic alternatives are very strong promoters and tightly regulated promoters. Strong promoters are used when high levels of the gene product are required. Tightly regulated promoters are useful in physiological experiments where the effects of gene expression are to be tested under a variety of conditions.

Strong promoters are used to express high levels of proteins from cloned genes.

expression vector	Vector specifically designed to place a cloned gene under control of a plasmid-borne promoter

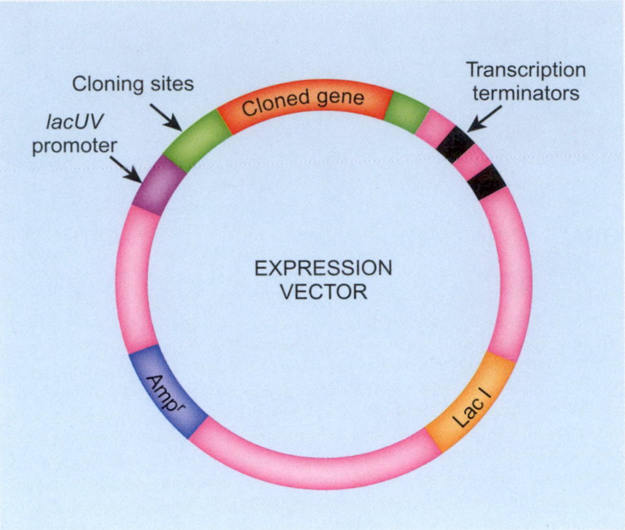

FIGURE 7.25
Expression Vectors Can Have Tightly Regulated Promoters

An expression vector contains sequences upstream of the cloned gene that control transcription and translation of the cloned gene. The expression vector shown uses the *lacUV* promoter, which is very strong, but inducible. To stimulate transcription, an artificial inducer molecule called IPTG is added. IPTG binds to the LacI repressor protein, which then detaches from the DNA. This allows RNA polymerase to transcribe the gene. Before IPTG is added to the culture, the LacI repressor prevents the cloned gene from being expressed.

Cloning sites
lacUV promoter
Cloned gene
Transcription terminators
EXPRESSION VECTOR
Amp^r
Lac I

Strong virus promoters, such as lambda or T7 promoters, are useful for controlling the expression of cloned genes.

Some promoters are both strong and strictly regulated. These are useful when expressing large amounts of a foreign protein in a bacterial cell. Even if the foreign proteins are not actually toxic, the large amounts produced interfere with bacterial growth. Consequently, the bacteria are allowed to grow for a while before the foreign gene is turned on by addition of inducer. The bacteria then devote themselves to manufacture of the foreign protein.

The *lac* promoter of *E. coli* is inducible and certain mutant versions exist that are extremely strong promoters, such as the *lacUV* promoter. IPTG is an artificial inducer that turns on the *lac* promoter (see Ch. 19). However, repression by LacI, the *lac* repressor, is leaky (i.e., incomplete). Including the *lacI* gene on a multicopy cloning vector results in high levels of repressor, which turn off the cloned gene more effectively (Fig. 7.25).

The ***tet* operon** confers resistance to the antibiotic tetracycline. The *tet* system is regulated in a similar manner to the *lac* system. The TetR repressor protein binds to an operator site and so prevents expression of the tetracycline resistance genes. When tetracycline is present, it binds to the TetR protein, which is therefore released from the DNA. Consequently, the *tet* operon is induced by tetracycline. The *tet* system may be used in higher organisms like the *lac* system. If the cloned gene has a *tetO* operator site inserted into its promoter it may then be induced by tetracycline.

Another strictly regulated promoter is the **lambda left promoter, p_L**. The **lambda repressor** or **cI protein** represses this promoter. If host cells contain a temperature sensitive version of the **cI gene**, such as *cI857*, then raising the temperature can alleviate repression. At 30°C the repressor is functional but at 42°C the repressor is inactivated.

A third popular method is to place the gene under control by a strong promoter from **bacteriophage T7**. Such promoters are not recognized at all by bacterial RNA polymerase but only by T7 RNA polymerase. Transcription will only occur in specialized host cells that contain the gene for T7 RNA polymerase. Another regulated promoter, such as the *lac* promoter, in turn controls the expression of T7 RNA polymerase. Induction of the *lac* promoter induces synthesis of T7 RNA polymerase, which in turn, transcribes the cloned gene (Fig. 7.26). This provides both strict regulation and high-level expression.

bacteriophage T7   A bacteriophage that infects *E. coli* whose promoters are only recognized by its own RNA polymerase
***cI* gene**   Gene encoding the lambda repressor or cI protein
cI protein   Lambda repressor protein responsible for maintaining bacteriophage lambda in the lysogenic state
lambda left promoter (p_L)   One of the promoters repressed by binding of the lambda repressor or cI protein
lambda repressor (cI protein)   Repressor protein responsible for maintaining bacteriophage lambda in the lysogenic state
***tet* operon**   Bacterial genes that produce proteins that confer resistance to the antibiotic tetracycline

SPECIALIZED BACTERIAL HOST CELL

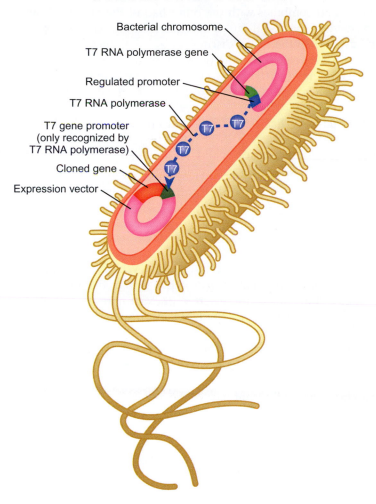

Bacterial chromosome

T7 RNA polymerase gene

Regulated promoter

T7 RNA polymerase

T7 gene promoter
(only recognized by
T7 RNA polymerase)

Cloned gene

Expression vector

■ **FIGURE 7.26**
T7 RNA Polymerase System

Specialized promoters can be used
to control the expression of cloned
genes. In the T7 RNA polymerase
system, the cloned gene cannot be
expressed unless the bacterial cell
makes T7 RNA polymerase. The
polymerase is produced by certain
genetically engineered bacteria,
which have the gene encoding
it inserted into the chromosome.
Expression of the T7 RNA
polymerase gene is under control of
the *lac* promoter, as described in the
previous figure.

Key Concepts

• Taking a gene from one organism and expressing it in a different one requires the use of cloning vectors that are relatively small rings of DNA.
• ColE1 plasmids of *E. coli* are the most common and widely used vectors. The original plasmids have had their genes for colicin production removed and replaced with a gene for antibiotic resistance so that a bacterium harboring this plasmid will become resistant to that antibiotic. This phenotype distinguishes bacteria containing the plasmid versus those without the plasmid.
• DNA can be inserted into a vector by digesting the piece of DNA and vector with the same restriction enzyme and ligating the two. Polylinkers or multiple cloning sites in a vector have a series of unique restriction enzyme sites to use. Alternatively, PCR amplified DNA inserts have a single adenine extension onto the 3' end of each strand that can be cloned into a TA vector that has a single thymine overhang.
• Insertional inactivation is a method to detect the presence of an insert in a vector, whereby the DNA insert is cloned so that it disrupts a gene for antibiotic resistance. The bacterium harboring the vector with insert is no longer resistant to that antibiotic and can be discerned from those bacteria harboring the vector without an insert.
• Beta-galactosidase is a common reporter gene used to detect the presence of an insert in a vector. When the vector has no insert, the alpha fragment of

β-galactosidase is made and combines with the other half of the enzyme. The active enzyme then converts the X-gal into a precursor that reacts with oxygen to create a blue dye. When the insert disrupts *lacZ*, no alpha fragment is made, and the bacterial colony remains white on X-gal plates.

- Shuttle vectors can survive in two different organisms and include two origins of replication (one for each organism), and two genes for selection (one for each organism).

- The genome from lambda virus has been converted into a vector for large DNA inserts (about 23 kb) by removing the central region of the genome, which contains the genes for integration and recombination into the *E. coli* chromosome. The large DNA insert can replace this region and can be inserted into *E. coli* bacteria using *in vitro* packaging. Cosmid vectors are about 45 kb of DNA flanked by the *cos* ends of the lambda genome. These are also inserted into *E. coli* using *in vitro* packaging.

- Artificial chromosomes from yeast, bacteria, or P1 bacteriophage are used for even larger DNA inserts (up to 150 kb).

- Recombineering inserts specific pieces of DNA into a vector or artificial chromosome by homologous recombination. The RED system from bacteriophage lambda recognizes the ends of the insert with exact homology to the insertion site on the vector and recombines the DNA insert with the vector to make the two pieces one.

- DNA libraries are constructed by partially cutting the genome of interest with a restriction enzyme to generate large fragments, inserting each of the fragments into a vector and then putting each vector into a bacterial cell. Each bacterium in a library has a different part of the genome.

- DNA libraries can be screened by hybridizing a labeled probe to the library DNA. Both the probe and the library DNA must be single-stranded for hybridization to occur. Rather than screening for DNA sequences, antibodies can be used to screen the library by expression of the library DNA into protein.

- Genomic DNA from eukaryotes cannot be made directly into an expression library since the genes contain introns. Using cDNA circumvents this problem.

- Hybridization of one library clone to another can find overlapping regions of DNA, and hence can be used for identifying upstream and downstream sequences from the region of interest. This is called chromosome walking.

- In subtractive hybridization, two different mRNA or DNA samples are isolated and hybridized. Any mRNA or DNA found in both samples will hybridize and any unique sequences will not. The unhybridized mRNA or DNA can then be purified and cloned into a vector for analysis.

- Expression vectors have promoters for the DNA insert that are inducible; that is, cloned genes are only expressed under certain conditions or with certain polymerases.

Review Questions

1. Give three properties that make vectors useful for cloning. Describe each.
2. Why are multicopy plasmids more desirable for use as vectors? Give an example of multicopy plasmids.
3. What is the easiest way to select for the presence of a vector in a cell population? How does this selection mechanism work?
4. What is a polylinker or "MCS"? Why is it useful?
5. What would happen if a cloning vector contained more than one recognition site for a particular restriction enzyme? How could this be overcome?

6. How is insertional inactivation used for detecting vectors that carry the cloned DNA?
7. Describe blue/white color screening in cloning. How does it work?
8. Describe alpha-complementation. What are the necessary components and where are the genes located?
9. What are the necessary features of shuttle vectors that allow transfer of cloned DNA from one organism to another on the vector?
10. When is a centromere sequence needed in a shuttle vector? Would this sequence be needed to transfer a gene from bacteria species "A" to bacterial species "B"? Why or why not?
11. What modifications are made to bacteriophage lambda DNA before it can be used as a vector? What consequences do these modifications have on the properties of the lambda phage itself?
12. What is the purpose of the cohesive ends or *cos* sequences?
13. What is *in vitro* packaging and why is it necessary?
14. Why is a helper phage needed for using bacteriophage lambda as a vector?
15. What is the advantage of using cosmids over lambda phage?
16. What are YACs and BACs? What is the difference between them?
17. List in increasing order, the approximate amount of DNA that each vector can accommodate.
18. What is the lambda RED protein? How is it useful for cloning DNA into vectors?
19. How are gene libraries constructed in prokaryotes? Describe two methods to screen these libraries.
20. What is most useful about gene libraries?
21. What is a major problem associated with cloning eukaryotic genes that can be overcome by using cDNA?
22. How is cDNA generated? Why is it not necessary to generate cDNA in prokaryotes?
23. Why can mRNA not be isolated from prokaryotes using oligo(U) or oligo(dT) columns?
24. Why is chromosome walking useful?
25. Describe subtractive hybridization and its uses.
26. What is the difference between a cloning vector and an expression vector?
27. What are some of the promoters used to control expression of genes cloned onto expression vectors? How do these work?
28. Why is it sometimes necessary to control expression of cloned genes?
29. Why is the T7 promoter system host-specific?

Conceptual Questions

1. You are working as a technician in a lab, and your supervisor has asked you to isolate the gene for super sweet corn (*ssc1*) from the sample of corn genomic DNA. She gives you the following map, and she also gives you a radioactively labeled probe that is specific for the *ssc1* gene. Devise an experimental protocol to isolate the gene from the genomic DNA. What enzymes would you use to digest the DNA? How would you isolate the specific fragment from all the other genomic fragments?

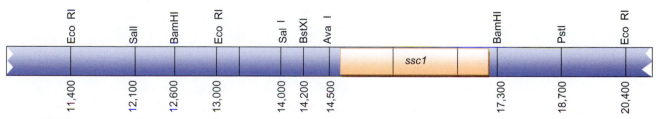

Genomic DNA surrounding the *ssc1* gene (the bars represent restriction enzyme sites and numbers represent position in the chromosome as determined by restriction enzyme site mapping).

2. Luckily, before you were able to complete the experiments for question 1, the entire corn genome sequence became available, so now you know both the restriction enzyme map of the *ssc1* gene and the entire sequence. Your advisor still wants you to clone the *ssc1* into a vector. Devise a different experiment to clone this gene using PCR.

3.   A colleague down the hall from your laboratory has just informed you about the technique called recombineering. How would you clone the gene presented in question 1 into a vector using recombineering?

4.   The plant breeder identified two different corn varieties that had different amounts of sweetness. Extensive genomic analysis of both varieties has only found one difference in the gene *ssc1* for super sweet corn. The difference is found in the RNA polymerase binding site, and your supervisor's hypothesis is that the corn variety with more sweetness produces more *ssc1* mRNA. Devise an experiment that will support or refute her hypothesis. Be specific in how you will determine if the sweet variety has excess *ssc1* mRNA and how you are going to specifically identify this mRNA in the mixture of mRNAs.

5.   Use the following reagents to create a recombinant plasmid with *araH* in the *Eco*RI restriction enzyme site. You have the maps below, a small sample of pure plasmid DNA, and the gene for the arabinose transporter, *araH* from *E. coli*. Design this experiment and describe each of the steps necessary to perform this task.

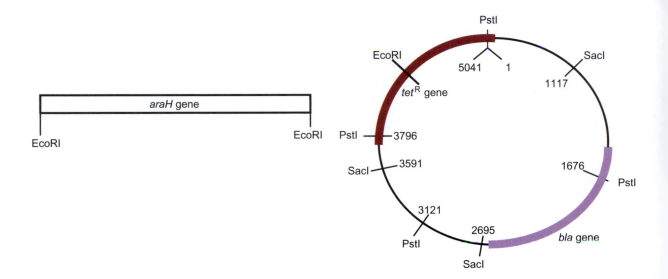

DNA Sequencing

Those investigatory approaches that deal with the whole genome, rather than single genes, define the emerging field of **genomics**. Obviously, genomics is based on having the complete DNA sequence of the organism under study available. We are presently in the middle of an information explosion in DNA sequences. Hundreds of whole bacterial genomes and many eukaryote genomes have been fully sequenced. A large and growing number of human genomes are now available including those of extinct branches of the human family such as the Neanderthals. In this chapter we will survey the methods used to sequence DNA, and in the following chapter we will consider the assembly of whole genome sequences. DNA sequencing technology is advancing so rapidly that it is impossible for a textbook to keep up. New methods emerge roughly every six months, although few of these actually find widespread use or survive for any length of time. Perhaps the best illustration of this is decreasing cost. Between July 2001 and January 2011 the cost of sequencing a million base pairs of DNA (1 megabase) dropped from $5,000 to 50 cents!

---

genomics   Study of genomes as a whole rather than one gene at a time

FIGURE 8.01
Sequencing—Fragments of All Possible Lengths

Original	These are grouped as follows:			
8 fragments	Ending in A	Ending in G	Ending in T	Ending in C
ACGATTAG		ACGATTAG		
ACGATTA	ACGATTA			
ACGATT			ACGATT	
ACGAT				ACGAT
ACGA	ACGA			
ACG		ACG		
AC				AC
A	A			

Chain termination DNA sequencing actually involves synthesizing DNA subfragments of all possible lengths and separating them on a gel.

1. DNA Sequencing—General Principles for Chain Termination Sequencing

Before searching for genes or comparing different DNA sequences, DNA must first be sequenced. The overall approach first involves generating a reasonably-sized template DNA by cloning or PCR. The actual sequencing involves generating subfragments of all possible lengths from this template. The subfragments are generated by DNA polymerase to differ in length by only one base pair so a subset of fragments end at each of the base pairs in the original fragment. Therefore, if the template were 200 base pairs in length, there would be 200 different subsets of fragments ranging from one base pair to 200 base pairs. These are then grouped according to which base they end in, and are separated by gel electrophoresis. Let's illustrate this using the eight-base sequence ACGATTAG as an example (Fig. 8.01). DNA polymerase generates eight subsets of this example, ranging from the entire eight base pair piece to only a single nucleotide found at the beginning.

The four groups of fragments are separated by gel electrophoresis by running them on the same gel in four parallel lanes. Those fragments ending in A are run in the first lane, those ending in G in the second lane, and so on. The fragments are then separated according to their lengths to create a separate band for each subset of fragments (Fig. 8.02). Starting at the bottom of the gel and reading upwards, we can read off the sequence directly.

1.1. The Chain Termination Method for Sequencing DNA

How are these fragments actually made, in particular, how are they separated into four groups depending on the last base? The method routinely used is known as **chain termination sequencing** or **dideoxy sequencing**. Both names refer to the fact that dideoxy analogs of normal DNA precursors cause premature termination of a DNA strand being made by DNA polymerase. Four separate reactions are established, one for each of the **dideoxynucleotides** (ddATP, ddTTP, ddCTP, and ddGTP).

Sequencing reactions need a variety of different components. Sequencing reactions use DNA polymerase to make DNA (see Ch. 10 for details), and therefore, the reaction also requires a template strand of DNA to copy. This template must be single-stranded because DNA polymerase cannot unwind a double-stranded helix. Thirdly, the reaction needs a primer to which nucleotides are added since DNA polymerase cannot initiate DNA synthesis without a pre-existing 3′-OH group. DNA

---

chain termination sequencing   Method of sequencing DNA by using dideoxynucleotides to terminate synthesis of DNA chains. Same as dideoxy sequencing
dideoxy sequencing   Method of sequencing DNA by using dideoxynucleotides to terminate synthesis of DNA chains. Same as chain termination sequencing
dideoxynucleotide   Nucleotide whose sugar is dideoxyribose instead of ribose or deoxyribose

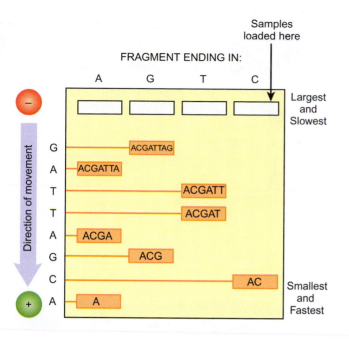

■ **FIGURE 8.02**
Principle of DNA Sequencing

The sequence of a DNA fragment can be determined by generating subfragments from the original sequence as shown in Fig. 8.01. The subfragments are generated in four separate reactions, one for each of the four bases. Each reaction mixture is then separated by size using gel electrophoresis. The sequence can be read off, starting at the bottom of the gel and reading upwards.

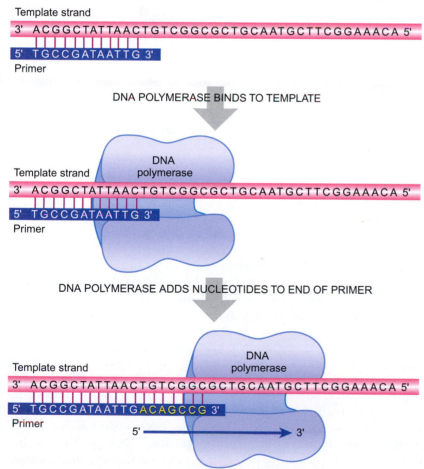

■ **FIGURE 8.03**
Synthesis of DNA—Priming and Elongation

During normal DNA synthesis, DNA polymerase reads the antisense strand and makes a new complementary strand of DNA. To get DNA synthesis started, a short oligonucleotide primer must anneal to the 3′ end of the template. DNA polymerase recognizes the 3′ end of the primer and adds incoming nucleotides to the 3′ end; hence, synthesis occurs in a 5′ to 3′ direction.

polymerase elongates the primer and makes a new DNA strand complementary to the template strand (Fig. 8.03). In addition to the dideoxy nucleotides, each normal deoxynucleotide (dATP, dTTP, dCTP, and dGTP) is also needed. These are supplied as nucleoside triphosphates (NTP; a base linked to a sugar and three phosphate groups). The outermost two phosphate groups are lost when each nucleotide is added to the tip of the growing DNA chain (Fig. 8.04). When nucleotides are joined, the

FIGURE 8.04
Synthesis of DNA— Phosphodiester Bonding

DNA polymerase links nucleotides via phosphodiester bonds. When adding another nucleotide, DNA polymerase breaks the bond between the first and second phosphates of the deoxynucleoside triphosphate. The incoming nucleotide is then joined to the free 3′ hydroxyl of the growing DNA chain.

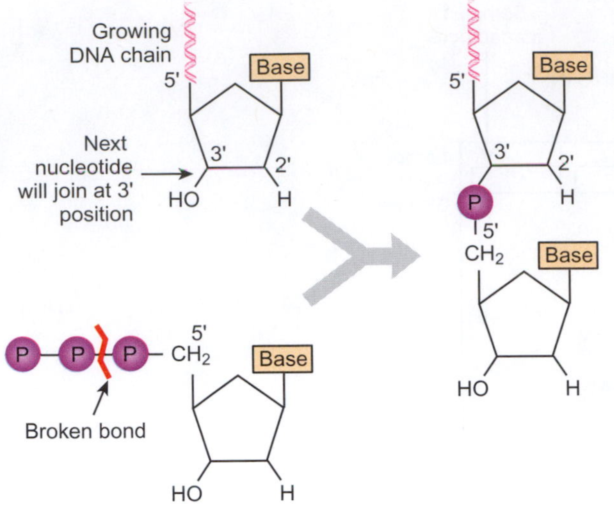

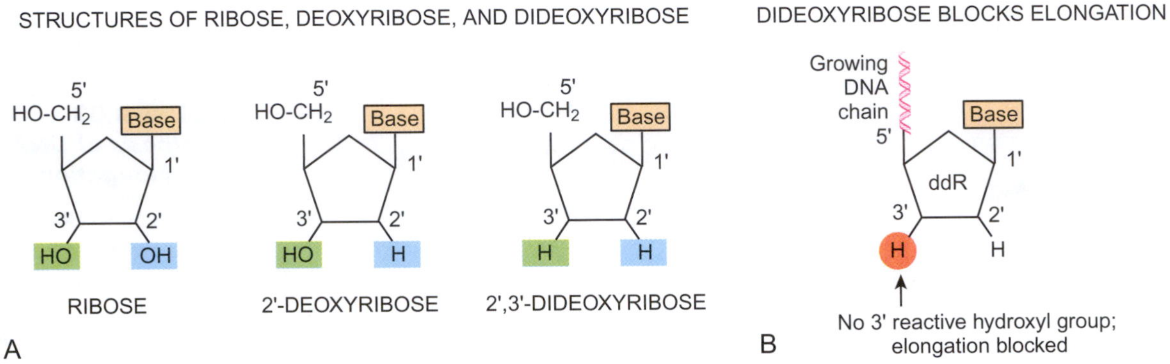

STRUCTURES OF RIBOSE, DEOXYRIBOSE, AND DIDEOXYRIBOSE

DIDEOXYRIBOSE BLOCKS ELONGATION

A

B

FIGURE 8.05
Dideoxyribose, Deoxyribose, and Ribose

A) The structures of ribose, deoxyribose, and dideoxyribose differ in the number and location of hydroxyl groups on the 2′ and 3′ carbons.
B) DNA polymerase cannot add another nucleotide to a chain ending in dideoxyribose because its 3′ carbon does not have a hydroxyl group.

Dideoxy base analogs are used to terminate growing DNA chains.

phosphate group attached to the 5′-carbon atom of the sugar of the incoming nucleotide is linked to the 3′-hydroxyl group of the sugar belonging to the previous nucleotide. Or, in brief, DNA is polymerized in the 5′ to 3′ direction.

The structural differences between dideoxy and deoxynucleotides are key to understanding chain termination sequencing. The sugars of DNA and RNA differ by a hydroxyl group at the 2′ carbon. Nonetheless, both DNA and RNA have 3′-hydroxyl groups on deoxyribose and ribose, respectively. However, **dideoxyribose** is a related sugar that is missing <u>both</u> the 2′- and 3′-hydroxyl groups (Fig. 8.05). If a nucleotide with this sugar is added to a growing chain during DNA synthesis, no 3′-hydroxyl group is available for further elongation. DNA polymerase cannot add any more nucleotides. In essence, this sugar terminates DNA synthesis because of its missing hydroxyl group. During a sequencing reaction, DNA polymerase has a choice of ddGTP and dGTP when the polymerase senses a C along the template. If it uses ddGTP instead of dGTP, no other nucleotides can be added even though there is more template sequence (Fig. 8.06). The amount of ddGTP relative to dGTP is adjusted to yield a mixture of chains that are terminated by ddGTP at all positions across from a C in the template DNA.

Dideoxy analogs are missing the hydroxyl group on both the 2′ and 3′ carbon of the sugar.

dideoxyribose   Derivative of ribose that lacks the oxygen of both the 2′ and the 3′ hydroxyl groups

RANDOM TERMINATION AT "G" POSITIONS

Original sequence:
T C G G A C C G C T G G T A G C A

Mixture of chains terminated at G
using mixtures of dGTP and ddGTP:

1. T C G
2. T C G G
3. T C G G A C C G
4. T C G G A C C G C T G
5. T C G G A C C G C T G G
A   6. T C G G A C C G C T G G T A G

RUN ON SEQUENCING GEL

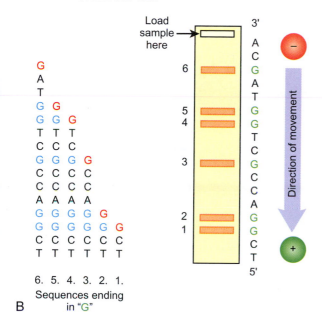

B

FIGURE 8.06

Chain Termination by Dideoxynucleotides

A) During the sequencing reaction, DNA polymerase makes multiple copies of the original sequence. Sequencing reaction mixtures contain artificial dideoxynucleotides that terminate growing DNA chains. The example here shows the G reaction, which includes triphosphates of both deoxyguanosine (dG) and dideoxyguanosine (ddG). Whenever ddG is incorporated (shown in red), it causes termination of the growing chain. If dG (blue) is incorporated, the chain can continue growing. B) When the sequencing reaction containing the ddG is run on a polyacrylamide gel, the fragments are separated by size. Each band directly represents a guanine in the original sequence.

The other three sequencing reactions are set up similarly. In practice, the template, primers, radioactive dNTPs for all four bases, and DNA polymerase are all mixed together. This mixture is then distributed among four tubes, each with a different dideoxynucleotide.

Each of the four sequencing reactions is then separated according to size by gel electrophoresis (Fig. 8.07A). Since the largest DNA subfragments generated by sequencing reactions are usually only 200–300 base pairs in length and each fragment differs in size by as few as one nucleotide, the large pores of agarose are too big, and therefore, polyacrylamide gels are used instead. The shortest pieces are closer to the bottom since they move fastest during electrophoresis. These short pieces are generated when the ddNTP is incorporated shortly after DNA polymerase begins synthesis at the primer. Each band corresponds to a piece of DNA of a particular length. The lengths reveal the positions of each base in the original DNA. All four samples are loaded side by side onto a gel to give four ladders representing each of the bases (Fig. 8.07A).

Some way is needed to detect the DNA bands. In one method, radioactive nucleotide precursors or radioactively labeled primers are incorporated into the subfragments generated by the polymerase. In practice, the sequencing reaction contains one radioactive deoxynucleotide (usually ^{32}P-dATP), all four regular deoxynucleotides, and one of the four dideoxynucleotides. After separating the subfragments by electrophoresis, the polyacrylamide gel is attached to filter paper and dried. Then, a sheet of

DNA fragments are separated according to size on a polyacrylamide gel so that a separation occurs in fragments that only differ by one nucleotide in length.

DNA fragments are detected by incorporating radioactive nucleotides and then determining the position of each band by autoradiography.

FIGURE 8.07
**Separation and Detection
of Fragments on Gel During
DNA Sequencing**

A) The products of the four separate
sequencing reactions are run side
by side on a polyacrylamide gel
and the fragments of different sizes
are separated by electrophoresis.
B) To detect the fragments, the gel
is transferred to a piece of paper to
give it strength, and dried so that
the polyacrylamide does not stick to
the film. After the gel is completely
dry, a piece of photographic film is
placed over the gel. The positions
of the radioactive DNA fragments
are revealed by the dark bands they
produce on the film.

SEPARATION OF BANDS ON GEL

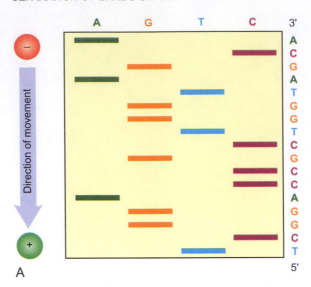

DETECTION OF BANDS BY AUTORADIOGRAPHY

Film on gel

Film

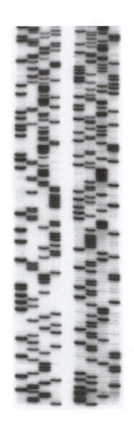

Lay film on gel and keep in
dark, then develop film

Film shows position
of bands

FIGURE 8.08
**Autoradiograph of Real
Sequencing Gel**

photographic film is laid on top of the gel. The radioactive bands leave a black mark
on the film, thus allowing the researcher to visualize the position of the original bands
(Fig. 8.07B). As before, the position of each band corresponds to a chain of DNA of a
particular length and reveals the position of one base. The sequence is read from the
bottom since these are the smallest fragments, and hence are closest to the primer
binding site. The complete sequence is determined by combining results from all four
bases. Several hundred bases of sequence can usually be obtained from one gel. Part
of a real sequencing gel is shown in Figure 8.08. Instead of radioactivity, newer DNA
sequencing techniques use fluorescently labeled nucleotide precursors for detection
(see below).

The sequencing of two different DNA templates is shown. The two sequences
each consist of four lanes that represent the four different bases. The sequence is
read from the bottom of the gel toward the top. This gel was run by Kiswar Alam in
Dr. David Clark's laboratory at Southern Illinois University.

1.2. DNA Polymerases for Sequencing DNA

All DNA polymerases can elongate a primer that is annealed to a single-stranded
DNA template. However, the characteristics needed for use in sequencing are more

E. COLI

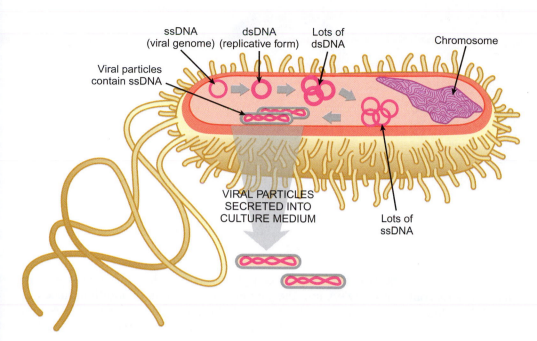

Viral particles contain ssDNA

ssDNA (viral genome)

dsDNA (replicative form)

Lots of dsDNA

Chromosome

VIRAL PARTICLES SECRETED INTO CULTURE MEDIUM

Lots of ssDNA

FIGURE 8.09

Single-Stranded DNA from Bacteriophage M13

When M13 infects *E. coli*, the single-stranded viral DNA is converted to a double-stranded replicative form (RF). This RF then replicates so making many double-stranded copies. After the double-stranded form becomes abundant, large numbers of single-stranded copies are made. These are eventually packaged into viral particles that are secreted into the culture medium.

rigorous. First, the polymerase must have high processivity; that is, it must move a long way along the DNA before dissociating. Premature dissociation would give strands that ended at random before the dideoxynucleotide was incorporated. In addition, many DNA polymerases possess exonuclease activities that interfere with accurate sequencing. 5′ to 3′ exonuclease activity is used to remove a strand of DNA ahead of the replication point. In contrast, 3′ to 5′ exonuclease activity is used to remove incorrect bases during proofreading. Such activities shorten the length of strands already synthesized.

In practice, no natural DNA polymerase is entirely suitable for sequencing. The first DNA polymerase used was **Klenow polymerase**, which is DNA polymerase I from *E. coli* that lacks the 5′ to 3′ exonuclease domain. Klenow polymerase was originally obtained by protease digestion of purified DNA polymerase I but was later made by expression of a modified gene. Because Klenow polymerase has relatively low processivity, it can only be used to sequence around 250 bases per reaction. Another commonly used enzyme is a genetically modified DNA polymerase from bacteriophage T7. This is marketed as "**Sequenase**" and has high processivity, a rapid reaction rate, negligible exonuclease activity, and the ability to use many modified nucleotides as substrates, thus making it perfect for sequencing reactions.

Genetically-engineered DNA polymerase with higher processivity and less exonuclease activity is now used for DNA sequencing.

1.3. Producing Template DNA for Sequencing

High-quality DNA sequencing requires purified single-stranded DNA to which the sequencing primer can bind. Originally, template DNA was obtained from the bacterial virus **M13** that was engineered to contain the template sequences (Fig. 8.09). The M13 virus is rod-shaped and contains a circle of single-stranded DNA (ssDNA). Upon infecting an *E. coli* cell, the single-stranded viral DNA is converted to a double-stranded form, the **replicative form (RF)**. After replicating itself for a while, the

Klenow polymerase   DNA polymerase I from *E. coli* that lacks the 5′ to 3′ exonuclease domain
M13   Rod-shaped bacteriophage that infects *E. coli*, contains a circle of single-stranded DNA, and is used to manufacture DNA for sequencing
replicative form (RF)   Double-stranded form of the genome of a single-stranded DNA (or RNA) virus. The RF first replicates itself and is then used to generate the ssDNA (or ssRNA) to pack into the virus particles
Sequenase®   Genetically-modified DNA polymerase from bacteriophage T7 used for sequencing DNA

FIGURE 8.10
Sequencing Using M13-Based Vectors

The use of M13 vectors allows the easy production of single-stranded template DNA. The DNA to be sequenced is inserted into the multiple cloning site (MCS) within the M13 vector. The MCS is located within the alpha fragment of the *lacZ* gene. When no insert is present functional β-galactosidase is made, which turns the *E. coli* host blue in the presence of X-gal. When an insert disrupts *lacZ*, no functional β-galactosidase is made and the cells stay white. This allows simple identification of cells carrying M13 vectors that have received DNA inserts. Sequencing is carried out using primers corresponding to M13 sequences just outside the cloned DNA.

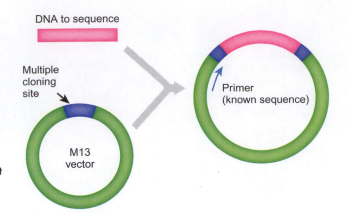

> The M13 virus uses a double-stranded replicative form of DNA to create single-stranded genomes. These are packaged into viral capsids and secreted from the bacteria without killing the host.

> Adding template DNA for sequencing into an M13 vector provides a convenient way to make lots of template DNA in a single-stranded form.

> M13 origins of replication turn an ordinary plasmid into one that can produce large quantities of single-stranded DNA.

> PCR products can be used for sequencing without the need for cloning.

RF then turns its efforts to manufacturing large numbers of single-stranded circles of DNA to pack into newly made virus particles.

Not only does M13 generate single-stranded DNA, it also purifies it. Unlike most viruses, M13 doesn't destroy the bacterial cells. Instead, the cells continuously secrete virus particles containing ssDNA into the surrounding medium. In addition, since the viral DNA does not integrate into the bacterial chromosome, only viral DNA gets packaged into the particles. Since the viral particles are secreted, they are easily isolated from the bacterial cells, and the DNA they contain can be extracted.

To create single-stranded template DNA with M13, the DNA to be sequenced is first cloned into the double-stranded replicative form of M13. Normally, an M13 vector that has already been engineered to contain a convenient multiple cloning site is used (Fig. 8.10). This multiple cloning site is contained within the N-terminal fragment of the *lacZ* gene of *E. coli*, which allows the use of blue/white screening to monitor insertion of the template DNA into the M13 vector (see Ch. 7 for details). Furthermore, the sequence to the side of the inserted DNA is already known and provides a starting point. This is essential, as the primer for sequencing must be complementary to a known sequence on the template strand in order to hybridize in the correct position. This engineered virus is used to infect *E. coli*, and virus particles containing single strands are manufactured in large quantities. Nowadays, bacterial plasmids containing the M13 origin of replication are used to manufacture single-stranded DNA. The use of intact virus is avoided and improved yields of DNA can be obtained more conveniently.

A variety of technical improvements have made DNA sequencing a little less tedious. Using double-stranded DNA (dsDNA) directly for sequencing is more convenient than generating single strands. In reality, "double-stranded" DNA sequencing involves a preliminary step, either heat or alkali treatment, to denature the dsDNA into single strands. Therefore, the actual sequencing reactions use single-stranded DNA just as described above.

In fact, it is now possible to completely avoid cloning DNA into either M13 or a plasmid vector by using PCR to generate segments of DNA (see Ch. 6). PCR products are linear double-stranded lengths of DNA, and they can be directly sequenced after separation into single strands.

2. Primer Walking Along a Strand of DNA

Sequencing moderately long pieces of DNA was originally done by cutting the DNA into smaller segments with restriction enzymes and then subcloning each fragment separately into M13 or another vector. **Primer walking** is a quicker and easier

---

primer walking Approach to sequencing a long cloned DNA molecule by using successive primers located at stages along the molecule

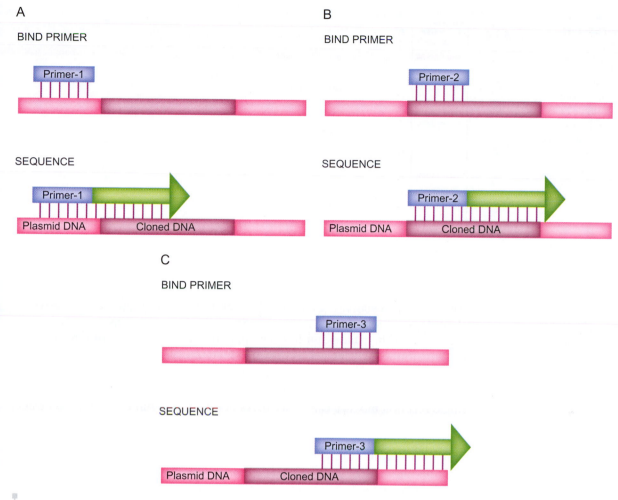

FIGURE 8.11
Primer Walking Along a DNA Molecule

When the DNA to be sequenced is too long to be sequenced by a single reaction, primer walking is used. First (A), the cloned DNA is sequenced as far as possible starting from a primer-binding site within the vector. The sequence information obtained allows a second primer to be made that lies close to the far end of the known sequence. A second sequencing reaction with this primer provides a further stretch of sequence (B). This process is continued, using as many primers as necessary to cross the inserted DNA. Eventually, the sequence obtained corresponds to the vector (C). This tells the experimenter that the unknown DNA has been completely crossed and is now fully sequenced.

method of sequencing long stretches of DNA (Fig. 8.11). This involves first sequencing the cloned DNA as far as possible using the primer belonging to the M13 or plasmid vector. Next, the newly obtained sequence information is used to design another sequencing primer. This primer is then used to sequence more DNA. Then another primer is made to the end of this sequence information and used to sequence the template, and so on until the end of the cloned DNA is reached.

3. Automated Sequencing

Today, the majority of sequencing is done using automated techniques. The main difference between automated and manual sequencing is to use fluorescent labeling instead of radioactivity. The reactions are carried out as described above, with modifications for fluorescent labeling. Each base (G, A, T, and C) is represented by a different color. There are two ways to perform the labeling:

1. Dye primer sequencing in which the fluorescent dyes are attached to the 5′ end of the primer. In this case four separate reactions are run, one for each base. A different colored dye is attached to the primer for each of the four

FIGURE 8.12
Automated Fluorescent DNA Sequencing

Automated sequencing uses four different fluorescent dyes, one for each of the four bases. All four reactions are run in a single lane of the gel since the four bases are easily distinguished by their colors.

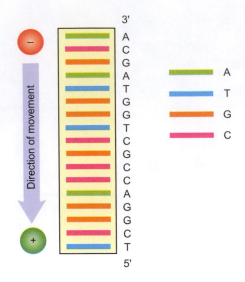

> Automated sequencing relies on using four fluorescent dyes of different colors, one for each base. This allows all fragments to be run in a single gel lane, where they are scanned by a laser.

base-specific reactions. (Since the primer is labeled, fluorescent dideoxynucleoside triphosphates are not used.) Since the four reactions are color-coded according to their bases, all four completed reactions can be run in the same track on the sequencing gel, as shown in Figure 8.12.

2. Dye terminator sequencing in which four different fluorescent dyes are attached to the four different dideoxynucleoside triphosphates. In this case, a single reaction mixture can be used. As before, this is run on a single gel track. Because it is simpler, this version has tended to predominate.

Rather than running the gel for a fixed period and then examining the bands afterwards, the automated sequencing reaction is monitored while running. As the bands travel down the gel they pass a laser and detector assembly. The laser beam scans the band and determines which of the four bases is at the end of that subfragment by the fluorescent color. A computer records the color of each band and compiles the data into actual sequence. The first bands to be recorded run right through the gel and off the end while later bands are still passing the laser. Consequently, more bases can be read from a single sequencing reaction by the continuous flow approach. Automated sequencers have been improved by using capillary separation. This improves speed but more importantly has allowed the assembly of machines that may have as many as 96 sequencing reactions running simultaneously.

4. Cycle Sequencing

Most sequencing reactions that are used for automated sequencing are a combination of PCR and regular sequencing. As in automated sequencing, a mixture of double-stranded DNA template, primer, and deoxynucleotides (dNTPs) are added. In addition, fluorescently labeled ddNTPs are mixed in a reaction tube. Instead of using Sequenase or Klenow polymerase, a heat resistant DNA polymerase, such as a modified *Taq* polymerase, is used. The reactions are cycled through three different temperatures to generate the labeled fragments. Just as in PCR, the first step is to denature the template DNA at a high temperature (90°C). The next step is to anneal the primer at a lower temperature (50–60°C) and then finally increase the temperature to the optimum for *Taq* polymerase (70°C). These three steps are repeated over and over in order to generate a large number of labeled fragments for automated sequencing. As before, each of these fragments varies by size in single base increments due to the random incorporation of the dideoxynucleotides. The final sequencing products are separated by size and recorded for the fluorescent dye as described in automated sequencing (above).

5. The Emergence of DNA Chip Technology

Earlier DNA technology was largely based on gel electrophoresis, an approach that is both difficult to automate and labor intensive. **DNA chips** were developed to allow automated side-by-side analysis of multiple DNA sequences. In practice, the simultaneous analysis of thousands of DNA sequences is possible. The first chip was introduced by a company called Affymetrix in California in the early 1990s. Since then, DNA chips have been used for a variety of purposes including sequencing, detection of mutations, and gene expression analysis. DNA chips all rely on hybridization between single-stranded DNA permanently attached to the chip and DNA (or RNA) in solution. Many different DNA molecules are attached to a single chip forming an array of spots on a solid support (the chip). The DNA or RNA to be analyzed must be labeled, usually with fluorescent dyes. Hybridization at each spot is scanned and the signals are analyzed by appropriate software to generate colorful data arrays. Two major variants of the DNA chip exist. Earlier chips mostly used short oligonucleotides. However, it is also possible to attach full-length cDNA molecules. Prefabricated cDNA or oligonucleotides may be attached to the chip. Alternatively, oligonucleotides may be synthesized directly onto the surface of the chip by a modification of the phosphoramidite method described in Chapter 5. Modern arrays may have 100,000 or more oligonucleotides mounted on a single chip.

> DNA arrays are used for a variety of purposes, including sequencing. Large numbers of probes are bound to the chip in a grid-like pattern, and then, hybridization with labeled target DNA occurs on the chip surface.

5.1. The Oligonucleotide Array Detector

The **oligonucleotide array detector** simultaneously detects and identifies lots of short DNA fragments (i.e., oligonucleotides). It can be used both for diagnostic purposes and for large-scale DNA sequencing. The key principle involved is DNA-DNA hybridization (see Ch. 3).

> DNA arrays can detect the presence of multiple small fragments of DNA sequence. A computer then compiles the overall sequence.

Consider a piece of DNA of unknown sequence. This is denatured to give single strands and one of these is tested for hybridization to a known probe sequence of say, eight bases (an octanucleotide; e.g., CGCGCCCG). If the unknown DNA binds to the probe, then the probe sequence occurs somewhere in the complementary strand of the unknown DNA. The unknown DNA is then tested for hybridization to all other possible stretches of eight bases, one at a time, to see which are found.

In practice, the hybridizations are all carried out at once. There are actually 65,536 possible eight-base sequences. Samples of each of the eight-base sequences to be used as probes are arranged in a square array and anchored to the surface of a glass chip. The glass chip can then be dipped in a solution of the target DNA, which will hybridize simultaneously to all those eight-base sequences with which it has complementary sequences. Instead of nucleotide array detector, the array is simply called a **DNA chip**. The technology is so precise that an array about 1 cm square can carry up to a million nucleotide probe sequences.

For example, if the unknown sequence of DNA is TCCAACGATTAGTCG, then its complementary strand will be AGGTTGCTAATCAGC. Consequently, of all 65,536 possible eight-base sequences, only the following can hybridize with the original sequence:

```
AGGTTGCT  TAATCAGC  TGCTAATC  GCTAATCA
GTTGCTAA  GGTTGCTA  TTGCTAAT  CTAATCAG
```

Given this information, a computer program can test all possible overlaps for these eight-base sequences and generate the solution as shown in Figure 8.13.

DNA chip   Chip used to simultaneously detect and identify many short DNA fragments by DNA-DNA hybridization. Also known as DNA array or oligonucleotide array detector

oligonucleotide array detector   Chip used to simultaneously detect and identify many short DNA fragments by DNA-DNA hybridization. Also known as DNA array or DNA chip

FIGURE 8.13
Deducing Sequence From Oligonucleotide Overlaps

By aligning all of the eight base pair probe sequences that hybridize to the unknown DNA, the computer can determine the sequence of the DNA.

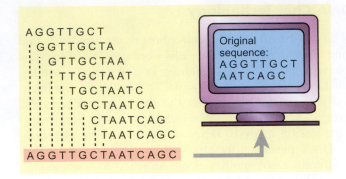

FIGURE 8.14
Sequencing by Oligonucleotide Array

This example shows an oligonucleotide array that has every four base pair combination possible. The unknown DNA fragment is fluorescently tagged and allowed to hybridize to all the possible oligonucleotides on the chip. The first spot that hybridized to the unknown DNA has the sequence, ACTG. The second has the sequence, CTGG, and the third, TGGC. A computer assembles these into the correct overlapping order. The sequence is then determined based on this information.

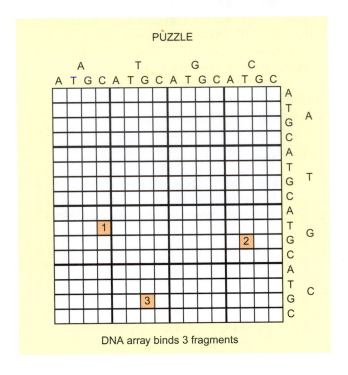

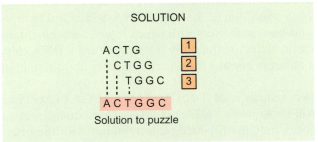

To sequence a large piece of DNA, it is first broken into relatively small pieces and each is tagged with a fluorescent dye. The unknown DNA will bind to just a few of the many eight-base probes on the oligonucleotide array. The chip is then scanned by a laser, which locates the fluorescently-tagged DNA. The positions to which it has bound are recorded. The computer then calculates the complete sequence of the unknown DNA. For simplicity, the oligonucleotide array illustrated in Figure 8.14 is designed to detect all possible four-base sequences. The "unknown" sequence, ACTGGC, contains three overlapping four-base sequences: ACTG (No. 1), CTGG (No. 2), and TGGC (No. 3). Their positions are shown on the array.

The oligonucleotide array runs into difficulties if the target DNA contains repeated sequences. Therefore, conventional sequencing is performed on totally new DNA sequences. However, for diagnostic tests to check for hereditary defects (i.e., mutations in known genes) and for forensic analysis, the oligonucleotide array is faster and simpler. The first **GeneChip® array**, made by Affymetrix Corporation, was designed to detect mutations in the reverse transcriptase gene of the AIDS virus. A variety of others are now available both for genome analysis and for diagnostic purposes, such as checking for mutations in the *p53* or *BRCA1* cancer genes. DNA arrays are also used for the whole genome analysis of gene expression as described in Chapter 19.

6. Pyrosequencing

Pyrosequencing is a "mini-sequencing" method that can be automated. In practice it only sequences short pieces of DNA. In this method, DNA polymerase adds nucleotides to a primer as usual, but the different bases are identified as they are added onto the growing chain. The detection scheme relies on the generation of a light pulse by a coupled reaction each time a base is incorporated (Fig. 8.15). This means that each of the four different nucleotide triphosphates (dNTPs) must be added in turn, one at a time, to the reaction mix. If a light pulse is seen, the added base was incorporated (and therefore the sequence being analyzed contains the complementary base at this point). The coupled reaction generates light each time a nucleotide is incorporated and pyrophosphate is released. The pyrophosphate and added adenosine phosphosulfate (APS) are then converted to ATP by ATP sulfurylase. ATP provides the energy for firefly **luciferase** to oxidize luciferin and this reaction releases a pulse of light. Unused dNTP and ATP are degraded by the enzyme apyrase before the next dNTP is added. Since both ATP and dATP are used by luciferase, dATP cannot be used for nucleotide incorporation. Instead, an analog that is used by DNA polymerase but not by luciferase is added. This is typically α-thio-dATP (in which the first phosphate is replaced by sulfate).

Pyrosequencing is especially useful if the DNA sequence is already known and variants that differ in one or a few bases are being compared. In this case, bases that correspond to the known sequence are added until one of them gives no light signal—indicating that a sequence alteration is present at that point. Then the other three bases are tried until one gives a response and reveals the base present in the particular sample or individual being analyzed.

> Pyrosequencing is used to analyze short regions of DNA by monitoring light release in a two-step reaction that results in the conversion of luciferase into light.

7. Second-Generation Sequencing

Recently, a slew of new methods have been developed to increase the speed and lower the cost of DNA sequencing. These have been divided (so far) into second- and third-generation sequencing. It has been estimated that new sequencing methods appear every six months on average. In view of this rapid change we will refrain from too much detail on methods many of which will probably be obsolete by the time you read this book!

The key characteristic of second-generation sequencing is the use of massively parallel methods. Simply put, this means that large numbers of samples are sequenced side-by-side on the same apparatus. In practice this requires substantial miniaturization.

In second-generation sequencing, the template DNA is produced from genomic DNA. Instead of using fragments of larger sizes or DNA templates cloned in vectors, these methods use pure genomic DNA samples directly. The chromosomes are simply sheared into smaller pieces before amplification by PCR. Two main methods for shearing chromosomal DNA include endonuclease treatment with a non-specific enzyme so that small random fragments are created. The second method is to sonicate the DNA, a technique that uses sound waves to break the DNA into pieces. Since the DNA fragments have different sequences, a known sequence must be added to the end of the sheared DNA in order for PCR primers to anneal. The most common method is to use some sort of linker DNA that has a defined sequence.

> Second-generation sequencing uses massively parallel methods where thousands or even millions of reactions occur simultaneously.

> Template DNA for second-generation sequencing is prepared by shearing genomic DNA into small pieces with endonucleases or sonication.

GeneChip array   The first brand of DNA chip, made by Affymetrix Corporation
luciferase   Enzyme that emits light when provided with a substrate known as luciferin
pyrosequencing   Sequencing method based on the generation of light pulses when a base is added onto a growing nucleotide chain by DNA polymerase

FIGURE 8.15

Principle of Pyrosequencing

A) During each sequencing reaction, DNA is elongated by one nucleotide and pyrophosphate is released. B) The pyrophosphate is used together with adenosine phosphosulfate (APS) by ATP sulfurylase to generate ATP. Luciferase uses ATP plus luciferin and emits light. C) Apyrase removes unused triphosphates. D) An example of a short sequence generated by pyrosequencing. *(Modified after material kindly provided by Pyrosequencing AB, Uppsala, Sweden.)*

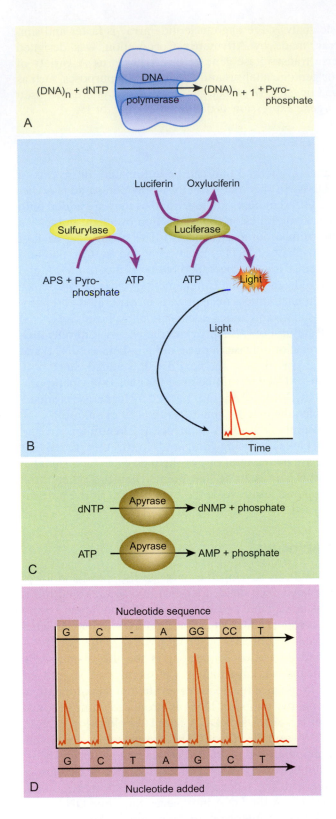

Initially, the most widespread second-generation method was **454 sequencing**, which uses pyrosequencing (see above) in order to determine the next base added during DNA chain extension. The name refers to 454 Life Sciences who invented the technique and have since merged with Roche Diagnostics. Like all second-generation

454 sequencing   Second-generation method of sequencing that uses pyrosequencing to determine what nucleotide is added by DNA polymerase

methods it uses tiny reaction volumes (only picoliters). DNA is amplified by PCR inside microscopic water droplets suspended in oil; each droplet contains a single DNA template. The massively parallel apparatus has many sequencing wells, in each of which one sequencing reaction is carried out. The small pieces of sequence data are assembled into one long sequence with computer by comparing these to the referenced genome for that organism. In reality, the small tidbits of sequence data alone cannot be assembled together unless they are compared to a previously sequenced genome. The *454* method's two most notable achievements are the sequencing of the first million base pairs of the Neanderthal genome and the complete genome sequence of James Watson—the first human genome costing less than $1 million, both of which were referenced to the human genome sequence data.

> *454* sequencing uses pyrosequencing technology to identify which nucleotide is added onto the growing DNA chain.

Two other second-generation methods are the **Illumina/Solexa sequencing** method and the SOLiD/Applied Biosystems method. Both give shorter reads than *454* pyrosequencing and therefore rely more heavily on computerized fragment assembly. The Illumina/Solexa method depends on DNA synthesis and reversible dye terminators. Each of the four nucleotides carries a different fluorescent tag that also acts as a blocking group for the critical 3′–OH. This ensures only one nucleotide is added in each reaction cycle. The computer analyzes the fluorescent signal and records the identity of the added nucleotide for each DNA chain being extended. The blocking group/fluorescent tag is then removed, and another round of DNA extension proceeds. As before massively parallel technology enables sequencing of many DNA chains simultaneously.

> Illumina/Solexa sequencing uses reversible dye terminators to stop DNA polymerase from adding nucleotides and record the nucleotide as a G, A, T, or C. The dye termination is reversible, and after removal, DNA polymerase can add another nucleotide.

The SOLiD method of Applied Biosystems differs from the other second-generation methods by using ligation rather than chain extension. An eight base oligonucleotide is ligated at each sequencing step, instead of adding a nucleotide with a single base, as in other methods. A set of eight nucleotide long oligonucleotides (8mer) is used but only one will complement and thus hybridize to the exposed primer. The 8mer that binds depends on the sequence. Ligase then links the two together. A fluorescent tag is present on the 8mer and identifies the critical base. The 8mer is then cleaved between bases 5 and 6

Johnston JJ, Teer JK, Cherukuri PF, Hansen NF, Loftus SK. NIH Intramural Sequencing Center, Chong K, Mullikin JC, Biesecker LG. (2010) Massively parallel sequencing of exons on the X chromosome identifies *RBM10* as the gene that causes a syndromic form of cleft palate. Am J Hum Gen 86: 743–748.

Second-generation sequencing has revolutionized research identifying the molecular basis for human genetic disease. This paper used Illumina sequencing to identify the gene that causes TARP syndrome. This is a pleiotropic disease that is characterized by clubfoot (talipes equinovarus), atrial septal defects, Robin sequence (lower ears, cleft palate, small jaw), and defects in the vena cava. Standard haplotype analysis on one family narrowed the probable location of the mutation to the X chromosome between Xp11.23–Xq13.3. Unfortunately, there are over 200 genes in this region, and the disease is rare, so locating the exact gene was challenging.

Rather than looking at each of the 200 candidate genes independently, the authors used Illumina sequencing of the exons in this region of the X chromosome to identify all differences between individuals with the disease and maternal carriers. Those regions suspected of carrying the mutation were sequenced 110–115 times over. All predicted exons in this region of the DNA from heterozygous female carriers were compared to the human reference genome. Computer analysis then predicted whether each change would cause silent, missense, nonsense, splice site, or frameshift mutations. In addition, the sequences were screened for insertions and deletions.

Silent mutations were excluded since these would not cause disease, as were known variants in the SNP database for normal (healthy) humans (see Ch. 9). The sequences were also compared to those from related but different diseases to rule out more general mutations. Finally, only heterozygous positions were retained since the sequences were from female carriers (whose two alleles for the candidate gene must be different). This left just one candidate mutation. Two female carriers from different families both had a mutation in the *RBM10* gene, which encodes a protein predicted to contain an RNA binding motif.

FOCUS ON
RELEVANT RESEARCH

Using mouse embryos, the authors then determined that the mouse homolog of *RBM10* was expressed in the same locations affected by the human disease, such as the areas that forms the jaw and lower limbs, and the region around the heart. This suggests that the *RBM10* gene is essential for proper human—and mouse—development.

Illumina/Solexa sequencing   Second-generation method of sequencing DNA that uses reversible dye terminators to identify the nucleotide that is added by DNA polymerase

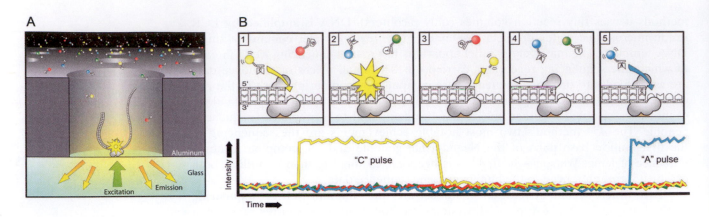

FIGURE 8.16
Principle of Zero-Mode Waveguide Sequencing

A) Single molecules real-time (SMRT) DNA sequencing occurs within small wells called zero-mode waveguides. These chambers allow detection of fluorescent light flashes only far enough to see a single molecule of DNA polymerase adding nucleotides onto a single DNA chain. The rest of the well is invisible to the fluorescent detector. B) DNA polymerase adds fluorescently labeled nucleotides that diffuse into the zero-mode waveguide. Each different nucleotide has a different fluorescent tag. As the nucleotide is added onto the growing DNA chain, the flash of light is detected through the bottom of the zero-mode waveguide. C) The fluorescence detector measures the wavelength of the four different fluorescent tags simultaneously. When DNA polymerase adds a particular nucleotide, the fluorescence intensity for that channel increases. For example, this data shows a peak for C and A in the time shown. A computer compiles the information into completed sequence data. *(Credit: Eid, J., et al. (2009). Real-Time DNA Sequencing from Single Polymerase Molecules. Science, 323, 133.)*

to remove the fluorescent tag and allow another cycle. Reads are relatively short, but vast numbers of reads can be performed cheaply and in parallel.

8. Third-Generation Sequencing

The key characteristic of third-generation sequencing is the sequencing of single molecules of DNA. At present (early 2011) there are three main contenders:

A. Nanopore sequencing has been under development for some time, but has yet to be commercialized. It takes advantage of advances in nanotechnology—see below.

B. Helicos Biosciences has developed the Heliscope Single Molecule Sequencer. Single-stranded fragments of DNA around 32 bases long are attached to a glass slide. As the complementary strand is synthesized, fluorescent tags on the incoming nucleotides are monitored on a microscope. The machine can simultaneously monitor a billion DNA fragments. A computer then assembles the fragments into a complete sequence.

C. Pacific Biosciences **SMRT (single-molecule real-time) sequencing** (Fig. 8.16) uses **zero-mode waveguides** and is probably the leader at present. Apart from its other advantages it has impressive terminology! As in several other sequencing methods, DNA polymerase extends a growing chain by adding nucleotides tagged with four alternative fluorescent dyes. Incoming nucleotides emit a flash of light as they are linked in place. The fluorescent tag is then washed away to allow the next cycle. The sequence of colors reveals the order of the bases.

SMRT sequencing (for single-molecule real-time) Third-generation sequencing method that identifies the nucleotide added by DNA polymerase onto a growing strand of DNA using fluorescently labeled pyrophosphate
zero-mode waveguides Small nanosized metal cylindrical wells that reduce background light so that only a very small portion of the cylinder can be visualized for fluorescent light flashes

Two novel features are critical. The reactions are carried out inside nanocontainers—that is, within hollow metal cylindrical wells 20 nanometers across. These are the zero-mode waveguides. Their small size reduces background light enough for individual flashes from single reacting nucleotides to be detected. Several thousand zero-mode waveguides are assembled onto a single chip. Future plans are for a million per chip. If successful this would allow human genomes to be sequenced in 30 minutes for less than $1000 each.

The second breakthrough is in attaching the fluorescent tag. Instead of linking it to the part of the incoming nucleotide that will be incorporated into the growing chain, it is attached to the pyrophosphate group that is discarded. Thus, the DNA does not accumulate tags. Instead each extension reaction gives a brief burst of color.

> SMRT sequencing has nanocontainers that permit the visualization of a single DNA polymerase adding a nucleotide onto the growing DNA. The phosphate molecules are labeled with a fluorescent tag that emits light upon release from the DNA.

Box 8.01 Hydrogen Ion Sequencing

Although new, the Ion Torrent DNA sequencing method does not use single-molecule sequencing. Hence, whether it counts as second- or third-generation sequencing is debatable. Ion Torrent takes a new approach to DNA sequencing chemistry. Instead of using labeled nucleotides, it relies on the fact that hydrogen ions are released whenever a new nucleotide is added to a growing strand of DNA. A silicon chip detects the hydrogen ions. Sequencing is fast but relatively expensive since (at present—early 2011) a new chip is needed for each run. If this problem is solved it may become a serious contender.

9. Nanopore Detectors for DNA

Nanotechnology is based on microscopic machinery that operates at the level of single molecules. **Nanopore detectors** for DNA contain extremely narrow pores that permit single strands of DNA to pass through one at a time. As the DNA molecule transits the pore, a detector records its chemical characteristics (hopefully in enough detail to give the base sequence). The advantages of nanopore technology are its high speed and its ability to handle long DNA molecules. In addition, many nanopores can be assembled into a very small region, and many long fragments of DNA sequence can be determined simultaneously.

A practical nanopore detector consists of a channel in a membrane that separates two aqueous compartments. When a voltage is applied across the membrane, ions flow through the open channel. Since DNA is negatively charged, the DNA is pulled through the nanopore to the positive side. The DNA molecules enter the pore and are pulled through in extended conformation, one at a time (Fig. 8.17). During the time the channel is occupied by the DNA, the normal ionic current is reduced. The amount of reduction depends on the base sequence (G > C > T > A); therefore, a computer can measure the current and decipher the sequence based on the differences.

> Nanopore detectors permit a single DNA strand to pass through a tiny pore and sequence it as it passes.

Initial nanopore detectors have used alpha-hemolysin from *Staphylococcus* as the channel and a lipid bilayer as the membrane. The mouth of the alpha-hemolysin channel is about 2.5 nm wide—roughly 10 atomic diameters. Double-stranded DNA can enter the pore mouth, but toward the middle, the channel narrows to less than 2 nm, which prevents dsDNA from going any further. The dsDNA remains stuck until the strands separate, allowing single-stranded DNA to pass through the length of the pore.

At present, nanopore detectors can tell apart two 20 nucleotide DNA strands that differ in a single base. It takes approximately a microsecond per base for DNA to transit the pore. Although single-stranded DNA molecules 1000 bases long have been

---

nanopore detector   Detector that allows a single strand of DNA through a molecular pore and records its characteristics as it passes through

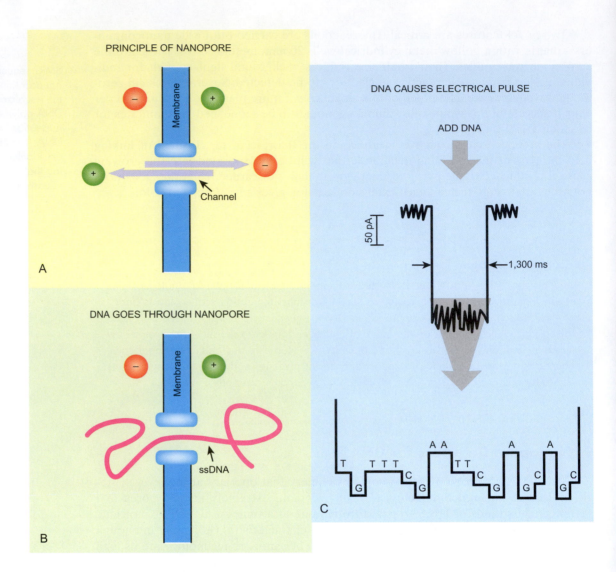

PRINCIPLE OF NANOPORE

Membrane

Channel

A

DNA GOES THROUGH NANOPORE

Membrane

ssDNA

B

DNA CAUSES ELECTRICAL PULSE

ADD DNA

50 pA

1,300 ms

A A   A   A

T   T T T   C   T T   C   C   C

G   G   G   G G   G

C

FIGURE 8.17
Principle of the Nanopore Detector

A) Nanopores are small openings in a membrane that only allow one molecule through at a time. The nanopore membrane separates two compartments of different charge. B) Since there is a charge separation between compartments, negatively-charged molecules like DNA can pass through the pore in an extended conformation. C) While the DNA is passing through the pore, a detector measures how much the current, due to normal ion flow, is reduced. Since each base alters the current by different amounts, the detector can determine the sequence as the DNA passes through the pore.

successfully pulled through nanopores, they move so fast through the pores that it is difficult to detect individual bases. Estimates based on future technical improvements suggest the possibility of chips with 500 pores each reading 1,000 bases/second. This could in theory read a bacterial genome in around a minute and read the entire human genome (3×10^9 bases) in less than two hours. This technology will reduce the cost of sequencing even more, and may make genomic sequencing available to the general public.

Key Concepts

- Chain termination sequencing requires that the original template DNA be copied such that the copies vary in size by one nucleotide.
- Dideoxynucleotides are missing the 3'-OH group, and therefore, DNA polymerase is unable to add another nucleotide onto the strand of DNA, even if there is a template sequence.
- Chain termination sequencing involves mixing a single-stranded template DNA, labeled deoxynucleotides, dideoxynucleotides, and DNA polymerase. The resulting fragments are then separated by polyacrylamide gel electrophoresis.
- One way to label the fragments in a sequencing reaction is to add one radioactively labeled deoxynucleotide that is incorporated during DNA synthesis. Since all the fragments have the same label, the sequencing reaction must be

done so that each dideoxynucleotide is added to a separate tube. The resulting fragments are visualized using autoradiography.

- DNA polymerases have been genetically modified to have higher processivity and less exonuclease activity so as to provide better, longer, and more accurate fragment subsets of the template.
- Making single-stranded template DNA can be done by cloning the template into M13 bacteriophage double-stranded replicative form (RF) and expressing this in bacteria. The viral genome is packaged as a single-stranded DNA surrounded by a phage coat that is released without lysing the host bacterium.
- Primer walking is sequencing an entire cloned DNA insert by sequencing the beginning, making a new primer based upon that sequence, and then sequencing with that primer. Each new sequence overlaps the prior sequence data until the entire insert is sequenced.
- Automatic sequencing still uses the chain termination method of sequencing, but the dideoxynucleotides are each labeled with a different dye. Since each has a different label, all four dideoxynucleotides can be mixed in one tube and separated by size on a single lane of an electrophoresis gel. Each dye is automatically recorded by a sequencing machine containing a fluorescent detection system and assembled into complete sequence using a computer.
- Cycle sequencing is the same as automatic sequencing, except the subsets of fragments are generated with a thermostable polymerase and amplified with PCR.
- DNA chips have an array of oligonucleotide sequences attached to a glass slide. Sequence information is determined by hybridizing a labeled template DNA to the chip. Wherever the labeled DNA binds to a specific oligonucleotide, that particular oligonucleotide sequence is present within the DNA sample. A computer then determines the overlap and assembles all the oligonucleotide sequences into one long sequence.
- Pyrosequencing is a technique for short pieces of DNA or mutation detection. The reaction emits a pulse of light when the correct nucleotide is added onto the elongating DNA by DNA polymerase. The light pulse is generated in a two-step reaction where the released pyrophosphate reacts with ATP sulfurylase and phosphosulfate to form ATP and then ATP and luciferase oxidize luciferin to release the light.
- Second-generation sequencing assembles the sequence from a large number of samples using miniaturized chips and picoliter amounts of reagents. The template DNA in second-generation sequencing is usually entire genomic DNA that has been sheared into small fragments using either nucleases or sonication. 454 sequencing technology uses pyrosequencing methodology in small water droplets suspended in an oil solution. Illumina sequencing uses reversible dye terminators to first record the identity of the nucleotide, which are then removed to add the next nucleotide. The SOLiD method uses oligonucleotide hybridization to identify the sequence followed by removal of the fluorescent tag and ligation of the next oligonucleotide sequence.
- Third-generation sequencing focuses on single DNA molecule analysis. The leading technology as of now uses zero-mode waveguides to minimize background light. As the next nucleotide is added by DNA polymerase, a fluorescently-labeled pyrophosphate flashes a wave of light that identifies the nucleotide as a G, A, T, or C. A computer records these flashes and assembles the data into sequence.
- Another third-generation sequencing method uses synthetic or biological nanopores, which are just big enough for a single strand of DNA to pass. Since each individual nucleotide reduces the voltage across the membrane to a different extent, the actual sequence is determined by measuring the current across the membrane.

Review Questions

1. What is the basic principle of traditional DNA sequencing?
2. What is the Sanger method for sequencing DNA and how does it work?
3. How does a dideoxynucleotide terminate a growing DNA chain?
4. How are DNA fragments separated and detected after traditional chain termination sequencing?
5. What is the advantage of Sequenase over DNA polymerase and Klenow polymerase?
6. What are the different methods of producing single-stranded DNA templates for sequencing?
7. How are long strands of DNA sequenced?
8. How does automatic sequencing work?
9. What is the advantage of capillary separation in automated sequencers?
10. What is the principle of DNA chip technology?
11. What are the major uses of DNA chip technology?
12. How is template DNA generated for second-generation sequencing?
13. How does *454* sequencing detect whether the sequence has a G, A, T, or C?
14. How does automated sequencing and Illumina sequencing differ? How are they the same?
15. What two properties of single-molecule real-time sequencing (SMRT) are novel to this third-generation sequencing technique?
16. What is the principle of pyrosequencing?
17. When is pyrosequencing used?
18. What is a picoliter?
19. What is meant by third-generation sequencing?
20. What is a zero-mode waveguide?
21. What is the basis for nanopore technology?
22. How can DNA be sequenced by nanopore detectors? What is the advantage of nanopore technology?

Conceptual Questions

1. Read the following sequencing gels from the autoradiograms of chain termination sequencing. Be sure to write your sequence in a 5′ to 3′ direction. Based on your sequence, what special DNA feature or pattern is found in the sequence of part B?

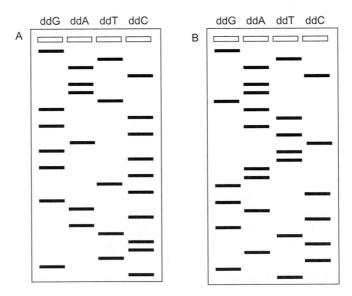

2. Discuss the advantages of using M13 bacteriophage to make the DNA used for sequencing.
3. A graduate student wanted to design her own sequencing chip to determine the sequence of the viral genome she has been studying. Based on your knowledge of oligonucleotide arrays, help her decide the best method to design her own 9 nucleotide

long oligonucleotide array. How many different oligonucleotides need to be created? How does she use this new array to sequence the viral genomic DNA?

4. Many different techniques require the use of oligonucleotide primers, including sequencing and PCR. Discuss why primers are used in these experiments. What other component or reagent is needed for each of these techniques?

5. The following chromatogram was obtained by research using cycle sequencing. Determine the sequence by reading the peaks, where green is A, blue is C, black is G, and red is T.

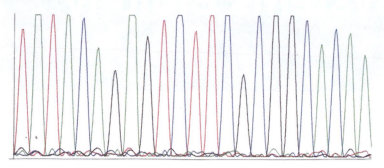

Genomics & Systems Biology

A vast amount of DNA sequence data is now available. In this chapter we will consider what useful information can be extracted from these sequences, especially in those cases where whole genome sequences are known. **Genomics** refers to the analysis of whole genomes, as opposed to individual genes, whether by experimentation or data analysis. Our own human genomes contain between 20,000 and 25,000 genes. Several genomes from plants and even a few protozoa have more than 25,000 genes, so humans (and other mammals that have approximately the same number of genes as we do) do not have the largest genomes. Although typical bacteria have just a few thousand genes (4,000 for the model bacterium *Escherichia coli*, for example) some bacteria have as many as 10,000 genes—almost half as many as humans. After discussing how complete genome sequences are assembled from experimental sequence data, we survey the human genome. Unambiguously identifying protein coding genes is more difficult than might be imagined, for a variety of reasons, including the presence of genes whose final products are RNA molecules rather than proteins.

> **genomics** Study of genomes as a whole rather than one gene at a time

Molecular Biology.

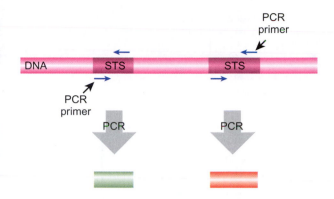

■ **FIGURE 9.01**
PCR Detection of Sequence Tagged Site

Mapping large genomes requires many genetic markers. Those based solely on their unique sequences are known as sequence tagged sites (STS). These unique sequences may be amplified by PCR and mapped relative to each other. In practice, each STS is defined by a pair of specific PCR primers at its ends that generate a PCR fragment of defined length.

1. Large-Scale Mapping with Sequence Tags

Ultimately, all the DNA sequence fragments of a genome must be correlated with a genetic map showing the location of the genes. However, in higher organisms, genes only comprise a small proportion of the DNA. Thus, it is necessary to have a series of genetic markers other than genes themselves. We have already discussed sequence motifs such as RFLPs (Ch. 5) and VNTRs and microsatellites (see Ch. 4). However, even the combination of these motifs is not sufficiently specific to map a large genome. Therefore, the human genome was mapped largely by the use of **sequence tagged sites (STSs)** including **expressed sequence tags (ESTs)**.

A sequence tagged site consists of 100–500 base pairs of unique sequence that can be amplified by polymerase chain reaction (PCR) using sequence specific primers (see Chapter 6 for information on PCR). STSs are located in non-repetitive regions of the genome and have specific lengths (Fig. 9.01).

An expressed sequence tag (EST) is a special type of STS derived from a region of the genome that is expressed (i.e., transcribed into mRNA). As before, ESTs are amplified by PCR using specific primers, but instead of amplifying genomic DNA, the EST is amplified from cDNA. An EST represents only a small section of a gene, thus it is possible to have several ESTs from the same gene. To avoid such duplication, ESTs are usually derived from the 3'-untranslated regions of the mRNA since an oligo(dT) primer is used in the reverse transcriptase step.

> Sequence tags are merely short regions of known sequence in known locations. They are needed when genomes contain vast amounts of non-coding DNA.

> Expressed sequence tags come from transcribed regions of the DNA.

1.1. Mapping of Sequence Tagged Sites

To sequence the human genome, the relative location of each STS and EST was originally determined by mapping. As with mapping any type of gene, STS and EST mapping is performed by determining how often two different STSs are found on the same DNA fragment (Fig. 9.02). The fragments may be derived from a single chromosome or the whole genome. The chance that any two STSs will be on the same fragment depends how close they are on the original chromosome. Neighboring STSs will tend to be found together on many fragments whereas those further apart will only rarely be found on the same fragment. This type of data can be used to construct a linkage map for the STS sites examined.

The chromosome fragments to be examined were originally derived by cloning large segments of DNA into high capacity vectors such as yeast artificial chromosomes (YACs). Unfortunately, YAC clones often contain two or more segments of DNA from different original locations. In practice, locating STS and EST sites

> Sequence tags are mapped relative to each other by analyzing how frequently tags are found together on the same chromosome fragments.

---

expressed sequence tag (EST) A special type of STS derived from a region of DNA that is expressed by transcription into mRNA
sequence tagged site (STS) A short sequence (usually 100–500 bp) that is unique within the genome and can be easily detected, usually by PCR

FIGURE 9.02
Mapping of Sequence Tagged Sites

STS mapping is shown for four STS sites on a single chromosome. A variety of restriction enzyme digests are performed to cut the chromosome into fragments of many different sizes. The number of times two STS sequences are found on the same fragment reveals how close the two markers are to each other. In this example, the two purple STSs are found on the same fragment six times and must be close to each other on the chromosome. The two green STSs are only found on the same fragment two times and are therefore further apart. The purple STSs are never found on the same fragment as the green STSs and therefore they must be far apart on the chromosome.

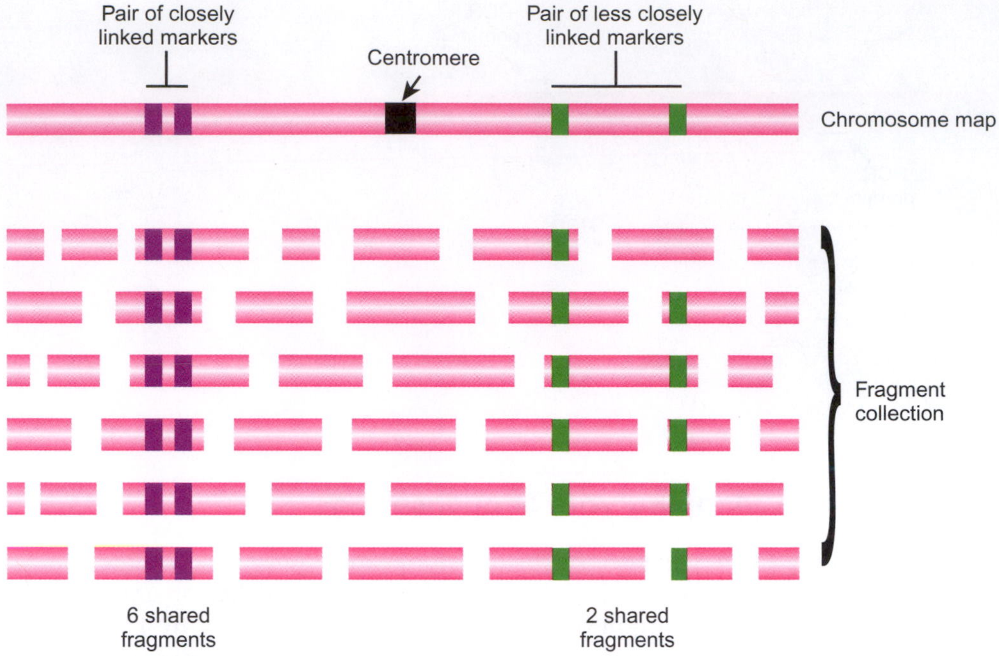

Pair of closely linked markers

Pair of less closely linked markers

Centromere

Chromosome map

Fragment collection

6 shared fragments

2 shared fragments

Radiation hybrids are cell lines that contain fragments of chromosomes from other eukaryotic cells.

relative to each other has mostly been done by radiation hybrid mapping (Fig. 9.03). A **radiation hybrid** is a cell (usually from a rodent) that contains fragments of chromosomes from another species.

To make a radiation hybrid, cultured human cells are irradiated with a lethal dose of X-rays or γ-rays, which breaks the chromosomes into fragments. The dying human cells are then fused with hamster cells. Cell fusion is promoted with polyethylene glycol or by using Sendai virus. The resultant hybrid cells contain random selections of the human chromosome fragments. Typical fragments are 5–10 Mbp in size and each hamster cell contains about 15–35% of the human genome. The hybrid cells are screened to see which STSs or ESTs are found together—that is, are on the same human chromosome fragments. The more often two STSs are found in the same hybrid cell, the closer they are linked on the original human chromosome before fragmentation. By the late 1990s, STS-based maps of over 30,000 sites had been constructed for the human genome. This gives a density of approximately one marker per 100 kbp of DNA.

2. Assembling Small Genomes by Shotgun Sequencing

As described in Chapter 8, individual dideoxy sequencing reactions give lengths of sequence that are several hundred base pairs long. A whole genome must be assembled from vast numbers of such short sequences. There are three approaches to whole genome assembly: shotgun sequencing, cloned contig sequencing, and the directed shotgun approach, which is really a mixture of the first two.

In **shotgun sequencing** the genome is broken randomly into short fragments (1 to 2 kbp long) suitable for sequencing. The fragments are ligated into a suitable vector and then partially sequenced. Around 400–500 bp of sequence can be generated from each fragment in a single sequencing run. In some cases, both ends of a fragment are sequenced. Computerized searching for overlaps between individual sequences then assembles the complete sequence. Overlapping sequences are assembled to generate

radiation hybrid   A cell (usually from a rodent) that contains fragments of chromosomes (generated by irradiation) from another species
shotgun sequencing   Approach in which the genome is broken into many random short fragments for sequencing. The complete genome sequence is then assembled by computerized searching for overlaps between individual sequences

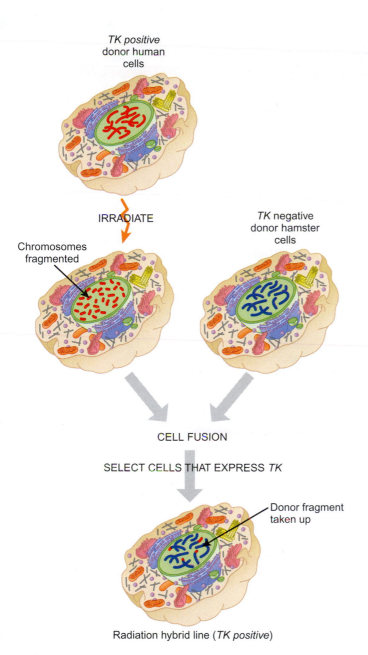

TK positive
donor human
cells

IRRADIATE

Chromosomes
fragmented

TK negative
donor hamster
cells

CELL FUSION

SELECT CELLS THAT EXPRESS *TK*

Donor fragment
taken up

Radiation hybrid line (*TK positive*)

■ **FIGURE 9.03**
Radiation Hybrid Mapping

To determine how close STSs and ESTs are to each other, many large chromosome fragments must be analyzed. First, the human chromosomes, which have the gene for thymidine kinase (TK⁺), are fragmented by irradiation. The human cells are then fused with hamster cells, which are TK⁻. If a human cell and hamster cell fuse successfully, the hybrid should express thymidine kinase and can be selected by plating on selective medium. Random loss of human chromosome fragments occurs during this process. Consequently, each radiation hybrid cell line will contain a different set of human chromosome fragments, which can be screened for the presence of STSs and ESTs.

contigs (Fig. 9.04). The term contig refers to a known DNA sequence that is contiguous and lacks gaps.

Since fragments are cloned at random, duplicates will quite often be sequenced. To get full coverage the total amount of sequence obtained must therefore be several times that of the genome to allow for duplications. For example, 99.8% coverage requires a total amount of sequence that is 6- to 8-fold the genome size. In principle, all that is required to assemble a genome, however large, from small sequences is a sufficiently powerful computer. No genetic map or prior information is needed about the organism whose genome is to be sequenced. The original limitation to shotgun sequencing was the massive data handling that is required. The development of faster computers overcame this problem.

The first bacterial genome to be sequenced was *Haemophilus influenza*. The sequence was deduced from just under 25,000 sequences averaging 480 bp each.

Sequencing very large numbers of small fragments provides enough information to assemble a complete genome sequence—if your computer is powerful enough.

---

contig   A stretch of known DNA sequence that is contiguous and lacks gaps

FIGURE 9.04
Shotgun Sequencing

The first step in shotgun sequencing an entire genome is to digest the genome into a large number of small fragments suitable for sequencing. All the small fragments are then cloned and sequenced. Computers analyze the sequence data for overlapping regions and assemble the sequences into several large contigs. Since some regions of the genome are unstable when cloned, some gaps may remain even after this procedure is repeated several times.

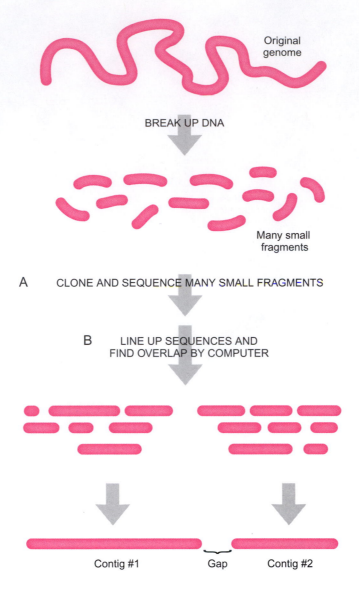

Original genome

BREAK UP DNA

Many small fragments

A   CLONE AND SEQUENCE MANY SMALL FRAGMENTS

B   LINE UP SEQUENCES AND
FIND OVERLAP BY COMPUTER

Contig #1   Gap   Contig #2

The bacterium *Haemophilus* had the honor of being the first organism to be totally sequenced.

This gave a total of almost 12 million bp of sequence—six times the genome size. Computerized assembly using overlaps resulted in 140 regions of contiguous sequence—that is, 140 contigs.

The gaps between the contigs may be closed by more individualistic procedures. The easiest method is to re-screen the original set of clones with pairs of probes corresponding to sequences on the two sides of each gap. Clones that hybridize to both members of such a pair of probes presumably carry DNA that bridges the gap between two contigs. Such clones are then sequenced in full to close the gaps between contigs. However, many of the gaps between contigs are due to regions of DNA that are unstable when cloned, especially in a multicopy vector. Therefore, a second library in a different vector, often a single copy vector such as a lambda phage, is often used during the later stages of shotgun cloning. Pairs of end-of-contig probes are used to screen the new library for clones that hybridize to both probes and carry DNA that bridges the gap between the two contigs (Fig. 9.05A). A third approach, which avoids cloning altogether, is to run PCR reactions on whole genomic DNA using random pairs of PCR primers corresponding to contig ends. A PCR product will result only if the two contig ends are within a few kb of each other (Fig. 9.05B).

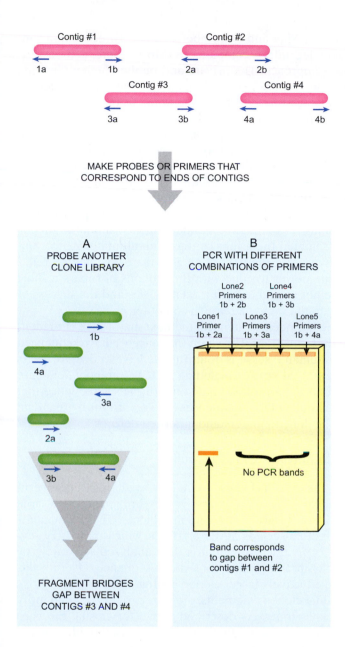

■▪ **FIGURE 9.05**
Closing Gaps between Contigs

To identify gaps between contigs, probes or primers are made that correspond to the ends of the contigs (pink). In (A) a new library of clones (green) is screened with end-of-contig probes. Clones that hybridize to probes from two sides of a gap are isolated. In this example, a probe for the end of contig #3 (3b) and the beginning of contig #4 (4a) hybridize to the fragment shown. Therefore, the sequence of this clone should close the gap between contig #3 and #4. (B) The second approach uses PCR primers that correspond to the ends of contigs to amplify genomic DNA. If the primer pair is within a few kilobases of each other, a PCR product is made and can be sequenced.

3. Race for the Human Genome

As described above, the first completely-sequenced genomes of living cells were from bacteria with genomes consisting of a few million base pairs. The genomes of more complex animals and plants contain a thousand-fold more DNA and their sequencing therefore presents greater problems. The original aim of the **Human Genome Project** was to sequence the DNA making up the entire human genome. When it first began in 1990, the official Human Genome Project was to be completed by 2005. In 1990, it cost $10 to determine each base of sequence, but technological advances reduced the cost of sequencing to 50 cents per base by the late 1990s. Therefore, the Human Genome Project was to cost roughly 1.5 billion dollars.

However, in 1998 an upstart private venture, Celera Genomics, led by Craig Venter, claimed that it would finish the job by the end of 2001 at a cost of a mere $200,000,000. To prove it was serious, Celera Genomics sequenced the entire genome

Human Genome Project Program to sequence the entire human genome

of the fruit fly, *Drosophila*, between May and December of 1999. Three million random sequence reads were assembled into a complete 180 Mb genome, demonstrating the feasibility of the shotgun approach. The private and public Human Genome Projects jointly announced that the first draft of the human genome was complete in June 2000. In fact, the Celera sequence was 99% complete, whereas the public project was only 85% done. A working draft was published in February of 2001.

The official Human Genome Project started by cloning large fragments of human DNA, mostly in yeast artificial chromosomes (YACs) and bacterial artificial chromosomes (BACs), and mapping them to their chromosomal locations. Only then were the large mapped fragments broken up for shotgun sequencing. Although this makes assembly easier, cloning consumes time and money. Instead, Celera used the directed shotgun sequencing approach described below. Automated sequencers sequenced vast numbers of small fragments. Finally, a computer examined all the random segments of sequence for overlaps and assembled them into contigs and, ultimately, into complete chromosomal sequences. Even with the help of an STS map, this method relies on colossal amounts of computing power to cope with the millions of separate short sequences. Such computers were not available until the late 1990s and the success of the Celera method is largely due to the rapid increase in computer power.

Now that the human genome has been sequenced, the next major task is to identify all the genes and elucidate their functions. The claim that once we have sequenced ourselves we will understand all human disease is doubtful. We have known for years the complete gene sequences of several viruses, including HIV, yet no complete cure has emerged. Deducing the function of a protein given only the DNA sequence that encodes it is hazardous at best. Although DNA sequences are very useful, a great deal of experimental work must also be done to understand inherited defects.

3.1. Assembling a Genome from Large Cloned Contigs

Using large cloned contigs was the approach that was taken by the official government sponsored human genome project. The genome is first broken up into large fragments that are cloned to give a library of overlapping pieces. These fragments are inserted into high-capacity vectors such as YACs or BACs, which may carry several hundred kb of DNA (see Ch. 7). Each fragment is then analyzed separately by shotgun sequencing, which results in a complete contiguous sequence. Overall, this approach yields a set of what are effectively large "cloned contigs."

After sequencing the individual cloned fragments, the next problem is to identify the overlapping regions among the clones. As described above for closing the gaps left after shotgun sequencing, hybridization and PCR methods are used to identify the overlapping fragments. Other methods include screening the cloned contigs for similarities in restriction profiles and repetitive elements.

However, comparing large numbers of clones by such methods is slow and tedious when tackling very large genomes. The human genome of 3×10^9 bp would give 10,000 cloned fragments of 300,000 bp (the maximum size for BAC/YAC inserts)—even without the 6- to 8-fold redundancy necessary to ensure complete coverage. In order to sequence each of these large clones, approximately 600 reactions would have to be performed, assuming about 500 base pairs per reaction. If 80,000 clones were constructed, about 48 million different sequences would have to be assembled into the complete genome.

3.2. Assembling a Genome by Directed Shotgun Sequencing

For random shotgun sequencing of the human genome it would be necessary to sequence about 70 million small stretches of about 500 bp. This would give the necessary redundancy for 99.8% coverage of 3×10^9 bp. With 100 automatic sequencers generating 1000 sequences per day, this could be done in 700 days—that is, roughly two years.

The critical issue is the assembly of these sequences into contigs and ultimately into complete chromosomes. The vast amount of computer time plus the uncertainties

---

> The first draft of the human genome appeared in early 2001.

> Very large genomes may be broken up into large fragments that are cloned using YACs or BACs and then sequenced by the shotgun approach.

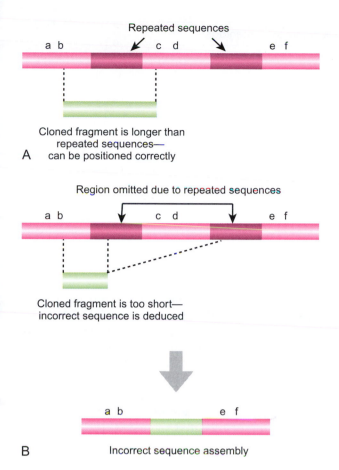

Repeated sequences

Cloned fragment is longer than
repeated sequences—
A   can be positioned correctly

Region omitted due to repeated sequences

Cloned fragment is too short—
incorrect sequence is deduced

B   Incorrect sequence assembly

■**FIGURE 9.06**

*Avoiding Incorrect Overlaps
between Repetitive
Sequences*

(A) If a clone is large enough
to completely contain repetitive
sequences, then the unique flanking
sequences are able to position
that clone within the genome.
(B) If a cloned fragment is short
relative to repetitive sequences,
errors may result. Here, the end of
the green cloned fragment could
align with either the first or second
repetitive sequence; therefore,
the sequence labeled c-d may be
omitted. Determining the spacing
and number of repetitive elements is
virtually impossible with short clones.

due to repetitive sequences make this approach prohibitive as it stands. However, if an STS map is used as a framework, then assembly becomes possible. In fact, this is the successful approach taken by Craig Venter of Celera genomics to complete the human genome ahead of schedule.

Sixty million sequences were generated from a library of fragments averaging 2 kb inserted into a multicopy plasmid vector. Another ten million sequences were from another library of larger pieces (10 kb) in a different vector. The 10 kb library is especially important in dealing with repeated sequences, since most of these are around 5 kb in size (or smaller) and can be entirely contained within a 10 kb fragment. The 2 kb library would not contain the entire repetitive region, and correctly aligning the numerous clones for this region would be impossible. Using the end sequences of the 10 kb fragments allows the assembly process to avoid making incorrect overlaps between two identical repetitive sequences that are actually in different locations (Fig. 9.06).

Using a map of sequence
tagged sites greatly helps
in the computations needed
to assemble a genome from
shotgun sequencing.

4. Survey of the Human Genome

The sequence of the human genome still has a few gaps. These are mostly in the highly-repetitive and highly-condensed heterochromatin, which contains few coding sequences (see Ch. 4). The total estimated size of the human genome is 3,200 million (3.2×10^9) base pairs of DNA or 3.2 **gigabase pairs** (Gbp; 1 Gbp = 10^9 base pairs) of which 2.95 Gb is euchromatin. A typical page of text contains about 3,000 letters. So the human genome would fill about a million pages. Most DNA from

gigabase pair (Gbp) 10^9 base pairs

TABLE 9.01	The Human Genome by Percentages	
Type of DNA		**Total Percentage**
Protein Coding Genes		1.5
Introns		25.9
LINEs		20.4
SINEs		13.1
DNA transposons		2.9
Long Terminal Repeat Retrotransposons		8.3
Segmental duplication		5
Simple sequence repeats		3
Unique sequences		11.6
Heterochromatin		8

Mammalian genomes have about three 3 gigabase pairs, but only 1.5% codes for proteins.

Repetitive sequences account for half the human genome.

Identifying genes unambiguously is especially difficult in genomes with large amounts of non-coding DNA.

higher organisms is non-coding DNA, including intergenic regions, introns, repetitive sequences, and so forth. About 28% of human DNA is transcribed into RNA but, since primary transcripts include introns, only a mere 1.5% is sequence that actually codes for proteins (Table 9.01). On average, the introns are longer in human DNA than in other organisms sequenced so far. There are both A/T-rich regions and G/C-rich regions in the human genome. Curiously, the zones of G/C-rich sequence have a higher density of genes and the introns are shorter. The significance of this is unknown.

Over half of the human genome consists of repeated sequences (these are discussed in Ch. 4). Some 45% is DNA that makes up SINEs—13%; LINEs—20%; defunct retroviruses—8%; and DNA-based transposons—3%. Repeats of just a few bases (microsatellites, VNTRs, etc.) account for 3% and duplications of large genome segments for 5%. Much of the genome resembles a retro-element graveyard, with only scattered outcrops of human information. This type of DNA tends to accumulate near the ends and close to the centromeres of the chromosomes.

How many genes the human genome contains may seem to be a simple question; however, computer algorithms to find genes are far from perfect, especially when surveying DNA that is mostly non-coding. The consensus at the time of writing is that the genome contains 20,000 to 25,000 genes. However, some predicted genes may be inactive pseudogenes and conversely, genes may be overlooked especially if they consist of small exons interrupted by many long introns. Furthermore, different computer analyses may assign a particular exon sequence to different genes. Although all protein encoding genes must be transcribed, unfortunately the converse is not true. Non-coding sequences are transcribed relatively frequently and thus the presence of a transcript does not confirm the existence of a genuine gene. Determining the exact number of human genes will require very detailed analysis using a combination of laboratory and computer methods.

Are humans really more complex than other organisms? The revelation that humans only have around 25,000 genes rather than the previously estimated 100,000 upset many people. (Actually, the estimate of 100,000 human genes was little more than a guess based on inflated self-importance.) The lowly nematode worm *Caenorhabditis elegans,* with approximately 18,000 genes, therefore has two-thirds as much genetic information as humans. The mouse and presumably most of our fellow mammals have essentially the same number of genes as humans. Those who apparently feel that human pride depends on having more genetic information than other organisms have retreated behind the claim that humans have more gene products than other organisms. This claim is based on the observation that alternative splicing generates multiple proteins from single genes and is more frequent in higher animals (see Ch. 12). Even so, most genes do not undergo alternative splicing, and there is no particular reason to believe that humans indulge in more alternative splicing

TABLE 9.02	Homology of Predicted Human Proteins	
		%
No homology		1
Prokaryotes only		<1
Eukaryotes plus prokaryotes		21
Eukaryotes (including animals)		32
Animals (including vertebrates)		24
Vertebrates only		22

than other mammals—especially the chimpanzee with which we share some 98.5% of our DNA sequence. Quibbling about which animal has most genes is in any case now moot. Sequencing has revealed that the genome of the rice plant contains around 40,000—some 15,000 more than humans. So it seems the flowering plants are more highly evolved than us! Even worse is that several single-celled protozoans have very high numbers of genes. At present the record holder is *Trichomonas vaginalis*, a flagellated protozoan that causes human infections, with approximately 60,000 genes.

Comparing the sequences of the 20,000–25,000 predicted human genes with other organisms yields some surprising results (Table 9.02). Few of our protein families are specific to vertebrates. More than 90% of the identifiable domains that make up human proteins are related to those of worms and flies. Most novel genes include previously evolved domains and thus appear to result from the re-shuffling of ancient modules.

Comparing the sequence of the human genome with other genomes originally suggested that just over 200 human genes were apparently borrowed from bacteria during relatively recent evolution. Homologs of the genes were absent from the genomes of flies, worms, and yeast, but were often found in bacteria as well as other vertebrates. Several independent horizontal transfer events were proposed as responsible. However, further phylogenetic analysis suggests that most of these are actually present in more ancient eukaryotes. In particular, homologs of many of these genes have been found in EST databases from lower eukaryotes whose whole genomes had not been fully sequenced at the time, such as the slime mold *Dictyostelium*.

Comparison of gene families is beginning to reveal insights into behavior. Apes and monkeys have a poor sense of smell compared to most other mammals, and humans are the most defective. Most mammals have over a thousand closely-related genes for olfactory receptors. These are the detector proteins that bind and detect molecules responsible for odors in the nasal lining. In the mouse, essentially 100% of olfactory receptor genes are intact and functional, in chimps and gorillas 50%, and in humans only 30%.

Not all genes code for proteins. Several thousand human genes produce non-coding RNA (rRNA, tRNA, snRNA, snoRNA, etc.—see Ch. 12). Such non-coding RNA genes lack open reading frames and are often short. Such genes are therefore difficult to identify by computer searches (unless of course the sequence of the non-coding RNA or a homolog from another organism is already known). In addition, some non-coding RNAs lack the poly(A) tails characteristic of mRNA so are absent from cDNA or EST libraries. There are about 500 human tRNA genes—fewer than in the worm *C. elegans*! The human rRNA genes for 18S, 28S, and 5.8S rRNA are found as cotranscribed units. Tandem repeats of these are found on the short arms of chromosomes 13, 14, 15, 21, and 22, giving a total of about 200 copies of these rRNA genes. The 5S rRNA gene is found separately, but also in small tandem repeats, the longest cluster being on chromosome 1, near the telomere of the long, q-arm. There are several hundred 5S rRNA genes but only a few are functional.

Both locating their genes and telling apart related sequences and pseudogenes from true functional copies are even more difficult for the other non-coding RNAs.

The highest numbers of genes are found in protozoa and flowering plants, not animals.

It has been long known that humans have a poor sense of smell. The genome sequence has revealed why—many of our smell sensors are defective.

Genes for non-translated RNA are hard to find just by computer searching.

At present several genes for expected non-coding RNAs are still missing, yet at the same time, many related sequences of uncertain function have been located.

4.1. Sequence Polymorphisms: SSLPs and SNPs

Although many genomes have been sequenced, analysis of different genomes from individuals within the species can lead to a variety of discoveries. A **polymorphism** is simply a difference in DNA sequences between two related organisms (e.g., two individual humans). Polymorphisms may be divided into those consisting of base changes and those where there is a difference in the length of the corresponding region of DNA. Polymorphisms could explain why people have different appearances, different susceptibility to diseases, and even different personality traits.

An **SNP** ("snip") or **single nucleotide polymorphism** is genomics terminology for a single base change between two individuals. In the human genome, there are several hundred thousand SNPs within genes and vastly more in non-coding DNA. SNPs are generally detected by use of hybridization using DNA chips (see Ch. 8). SNPs can also be identified by RFLP (restriction fragment length polymorphism; see Ch. 5), since a single base pair difference prevents restriction enzymes from cutting its recognition sequence. However, most SNPs do not give rise to RFLPs because they do not lie within a restriction cut site.

SSLP stands for **simple sequence length polymorphism**. This is a general term that applies to any DNA region consisting of tandem repeats that vary in number from individual to individual. It includes VNTRs, microsatellites, and other tandem repeats that have already been discussed (Ch. 4).

Different human individuals differ by approximately one base change every 1,000–2,000 bases. This amounts to around 2.5 million SNPs over the whole genome. About 60,000 known SNPs are found within the exons of genes. Therefore, the genetic diversity in the human population is much smaller than would be expected. Despite their smaller population, chimpanzees show much more genetic diversity than humans. The most likely explanation is that after splitting off from chimps about 5 million years ago, humans went through a genetic bottleneck. Modern humans probably emerged about 100,000 years ago from a small initial population (see African Eve, Section 8.2, Ch. 26); therefore, the genetic diversity in the beginning was very low.

SNP analysis is used increasingly to screen for possible hereditary defects and also to test for individual variation in genes that affect the response to pharmaceuticals. A variety of methods are used to identify SNPs. One of these techniques uses PCR to amplify the region of DNA that contains the polymorphism from a variety of individuals. The sequence must then be determined—but only for a single base at one precise location. Thus, it is not necessary to sequence the whole PCR fragment. Instead, a single base extension reaction is performed using a primer that binds just in front of the polymorphism site, plus specifically labeled dideoxynucleotides. Thus, each of A, T, G, and C is labeled with fluorescent dyes of different colors. Using dideoxynucleotides ensures that the primer is only elongated by a single base—the one that is complementary to the base at the polymorphism site (Fig. 9.07). The elongated primer is fluorescently labeled, which reveals which base was present in that individual.

In practice, large numbers of SNP analyses are often run in parallel. One way to sort these out is to use so-called Zipcode sequences attached to the primers (Ch. 6). Each SNP is allocated a different Zipcode sequence that can be specifically bound by using the complementary sequence or cZipcode. The cZipcode sequence is attached to a solid support or a polystyrene bead. Different cZipcode sequences may be

> Many single base differences are found between the genomes of individuals of the same species.

> SNP analysis looks for single base changes in different regions of the genome.

---

polymorphism   A difference in DNA sequence between two related individual organisms
single nucleotide polymorphism (SNP)   A difference in DNA sequence of a single base change between two individuals
simple sequence length polymorphism (SSLP)   Any DNA region consisting of tandem repeats that vary in number from individual to individual, including VNTRs, microsatellites, and other tandem repeats

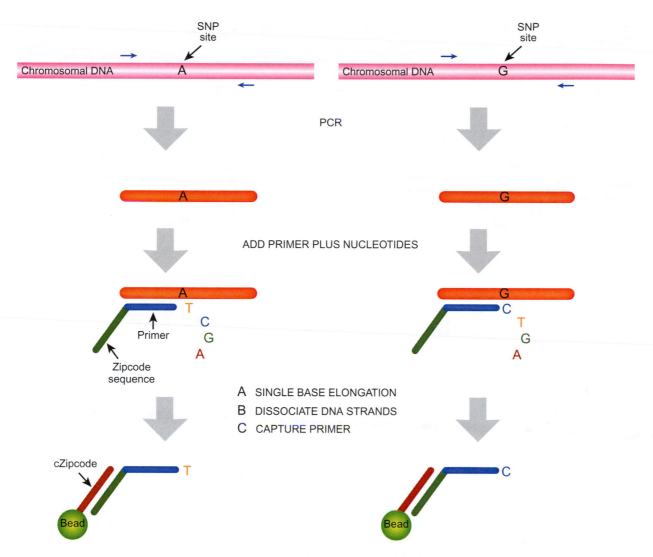

FIGURE 9.07
Zipcoded SNP Analysis by Single Base Extension

A segment of DNA that includes a SNP is amplified by PCR (only a single strand of the DNA is shown here, for simplicity). Next, single base extension is performed with a primer that binds one base in front of the SNP. Person I has an A at the SNP site and therefore T is added; in person II, a C is added. Each nucleotide is labeled with a different fluorescent dye. The elongated primer is then trapped by binding of its Zipcode sequence to the complementary cZipcode, which is attached to a bead or other solid support. The color of the fluorescent label is then determined to figure out what nucleotide is present in each individual.

attached to color-coded beads that are later separated by a FACS (fluorescence activated cell sorter—see Ch. 5) or attached to a solid surface forming an array.

> Artificial Zipcode sequences are used to keep track of the vast number of different SNPs.

4.2. Gene Identification by Exon Trapping

In eukaryotes, the actual coding sequences only account for a minority of the DNA. Given a large stretch of DNA sequence, how are the genes identified? Although computer algorithms exist to analyze sequences, the method known as **exon trapping** allows the experimental isolation of coding sequences. This method relies on the fact that exons are flanked by splice recognition sites that are used during RNA

exon trapping   Experimental procedure for isolating exons by using their flanking splice recognition sites

FIGURE 9.08
Exon-Trapping Vector

The pSPL vector is used to identify exons within regions of suspected coding DNA. The vector has both bacterial and eukaryotic origins of replication so that it can be grown in both *E. coli* and animal cells. The multiple cloning site is within an intron sequence that is flanked by two exons. This region of the vector can be transcribed into RNA because it contains eukaryotic promoter and eukaryotic poly(A) tail sequences.

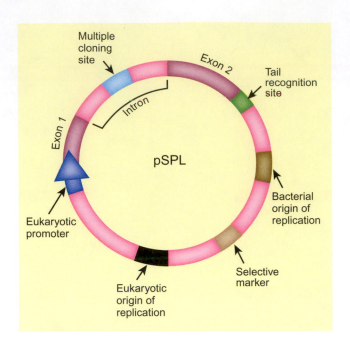

FIGURE 9.08
Exon-Trapping Vector

The pSPL vector is used to identify exons within regions of suspected coding DNA. The vector has both bacterial and eukaryotic origins of replication so that it can be grown in both *E. coli* and animal cells. The multiple cloning site is within an intron sequence that is flanked by two exons. This region of the vector can be transcribed into RNA because it contains eukaryotic promoter and eukaryotic poly(A) tail sequences.

Exons can be experimentally isolated and identified by using their flanking splice sites.

processing to splice out the introns (see Ch. 12 for details of splicing). Introns can be spliced using an *in vitro* system; therefore, a length of DNA containing the splice recognition site can be identified. Consequently, exon trapping can be used even if the sequence of the DNA is unknown, although in this case we will not know the relative order in the original DNA of the exons that are isolated.

During exon trapping, the DNA to be analyzed must first be cloned into a special vector that can replicate both in *E. coli* and in suitable animal cells. The vector carries an artificial mini-gene consisting of just two exons and an intervening intron, together with a promoter and poly(A) tail recognition site (Fig. 9.08). The intron contains a multiple cloning site for cloning lengths of unknown DNA. The pSPL vectors, as they are called, use a simian virus 40 (SV40) origin of replication as well as an SV40 promoter and tail site for the mini-gene. These vectors can replicate in modified monkey cells (COS cells) that contain a defective SV40 genome integrated into a host chromosome.

DNA containing the exons to be trapped is cut into segments using an appropriate restriction enzyme. These segments are inserted into the multiple cloning site within the intron on the pSPL vector (Fig. 9.09). The plasmid is then transformed into the COS monkey cells where the mini-gene is transcribed into a primary transcript and spliced. If an extra exon is in the middle of the mini-gene, it will be present in the spliced mRNA, which will therefore be longer. To isolate the trapped exon, the mRNA is converted to cDNA and then PCR is used to amplify the region containing the trapped exon. This technique will have to be used in conjunction with sequence analysis to identify all the different exons within the human genome.

4.3. The Evolution of Junk DNA

Sequencing the human genome multiple times has led to a surprising discovery about junk DNA. This part of the genome was originally identified as "junk" since it has no protein coding sequences and consists primarily of repeats. These regions of the genome also interfered with hybridization experiments and made sequencing the genome difficult, hence the term "junk." With the advent of second-generation sequencing, DNA microarray analysis, and the sequencing of multiple human genomes, some of the so-called "junk" has been transformed from "useless" sequence into a key element in genome-to-genome variation. In each human genome, there

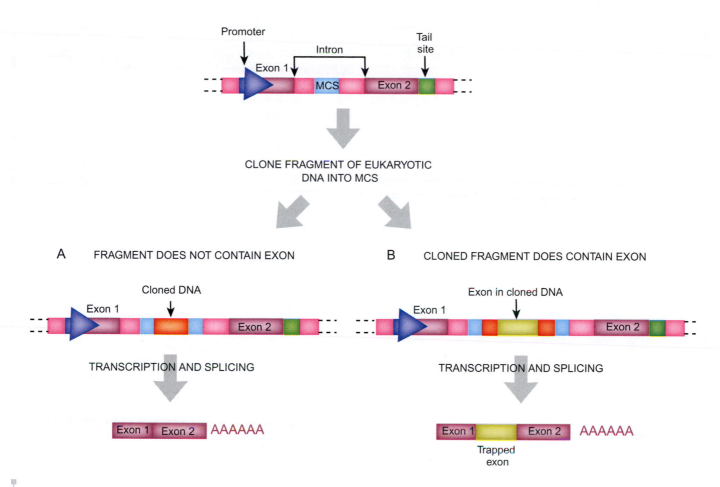

FIGURE 9.09
Exon-Trapping Procedure

In order to determine if a length of DNA includes an exon, the unknown DNA is cloned into the pSPL vector. The multiple cloning site is within a mini-gene that is transcribed and spliced when the vector is in animal cells. If the unknown DNA does not contain an exon, the mRNA transcript will be the same size as in the original vector. If the unknown DNA does contain an exon, the mRNA transcript will be longer. Any exons discovered by this method can be amplified using PCR for cloning and/or sequencing.

are approximately $3.0–3.5 \times 10^6$ single nucleotide variations from the reference human genome. In addition, each genome has about 1000 differences in the number of repeats, called **copy number variations** (**CNVs**), which primarily refer to tandemly repeated sequences. There are also a number of highly repetitive sequences such as LINE-1 and *Alu* elements (see Ch. 4). In the same category of LINE-1 and *Alu* elements, there is variation among other insertion sequences most of which are random defunct retroelements. Although these sequences appear to have no relevance to phenotypes, some recent research has suggested that the elements move around the genome much more often than originally believed. In addition, these retroelements are often found in introns and may affect the mRNA splicing and expression. Thus, "junk" DNA is actually a large source of individual variation among humans and may play a role in genetic diseases and perhaps cancer, especially if retroelements move into an important gene during mitosis.

> Genomes have many structural variants, including SNPs and CNVs. In addition, the genome has a large number of scattered repeated elements, which include defunct retroelements and transposons.

copy number variation (CNVs) A form of structural variation in which one genome will have either an insertion or deletion relative to another genome from a different individual

Mobile interspersed repeats are major structural variants in the human genome

Huang CRL, Schneider AM, Lu U, Niranjan T, Shen P, Robinson MA, Steranka JP, Valle D, Civin CI, Wang T, Wheelan SJ, Ji H, Boeke JD, and Burns KH. (2010). Cell 141: 1171–1182.

Now that multiple human genomes have been sequenced, the view of our genomic organization has been changed from primarily "protein-coding DNA" and "junk DNA" to a greater understanding that the "junk DNA" plays an essential role by providing genetic diversity. This part of our genome has many structural variations such as copy number variations. Another less studied variation in the human genome is the high copy interspersed repeats or insertion sequence variations, including the number and distribution of LINE-1 and *Alu* elements. In this associated paper, the authors examine the repetitive LINE-1 elements and determine their location in multiple human genomes in order to identify if these provide genetic diversity in the human population. The authors hypothesize that this repetitive element is not stagnant and may contribute to our genetic diversity and possibly even cause some genetic diseases.

The authors identified the location of L1(Ta), a transcribed LINE-1 element that belongs to "subset a." To identify the exact location for each L1(Ta), the authors used a novel method called **transposon insertion profiling (TIP)-chip**, which combines PCR amplification and sequence identification by gene array (Fig. 9.10). The method begins by digesting genomic DNA with restriction

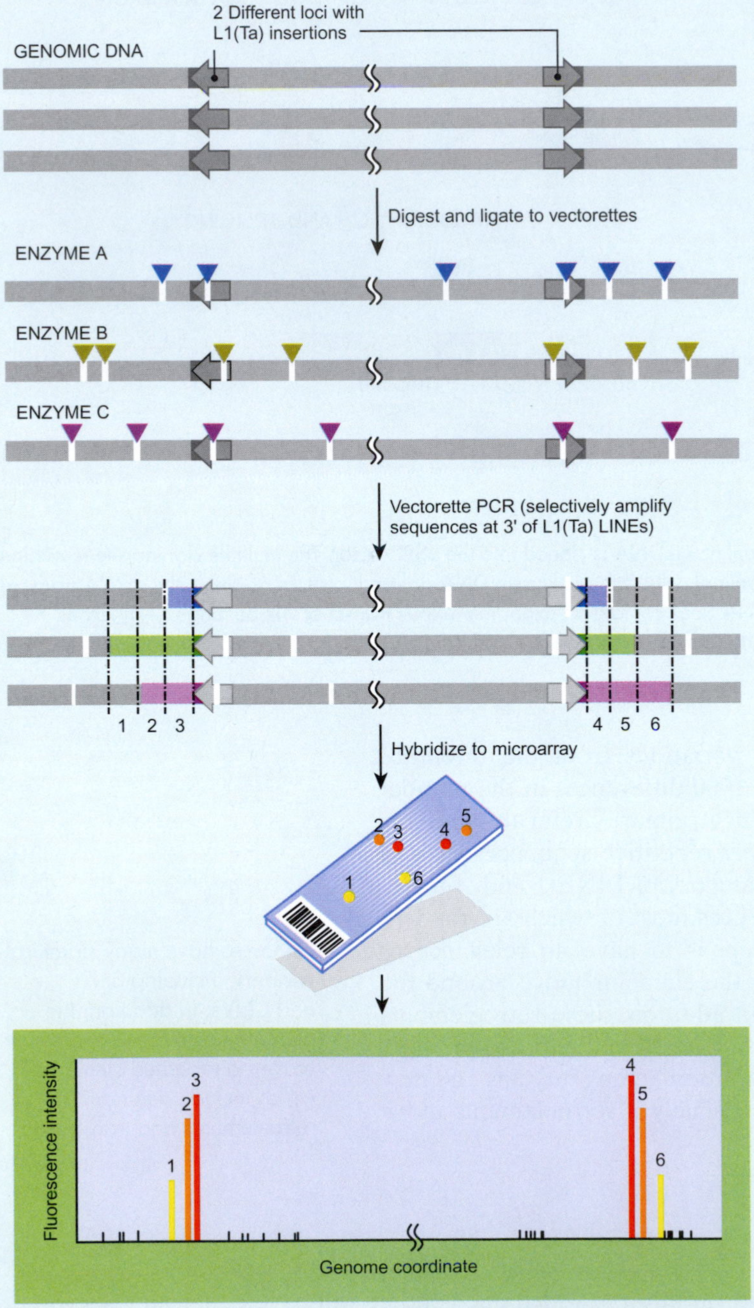

FIGURE 9.10
Transposon Insertion Profiling by Microarray

Transposon insertion profiling array (TIP-chip) begins by isolating genomic DNA from a person and digesting with different restriction enzymes in parallel. Each restriction enzyme (A, B, or C) recognizes one site within the L1(Ta) repetitive element. (The arrow represents L1(Ta) insertions from a 5' to 3' orientation.) Next, the fragments are ligated to vectorette primers, and PCR amplified with primers that recognize the L1(Ta) element and the vectorette sequence 3' to the L1(Ta)LINEs. These amplified DNAs are labeled with a fluorescent tag and then hybridized to a genomic microarray to identify the sequence 3' to the L1(Ta) insertion. The spots that fluoresce after hybridization are then analyzed by computer analysis to identify the exact sequence 3' to the L1(Ta).

transposon insertion profiling (TIP)-chip   A method to determine the location of scattered repeated elements within a genome such as LINEs or SINEs

enzymes that have one recognition sequence within the L1(Ta). By chance, recognition sequences for the same restriction enzymes will be located either upstream or downstream of L1(Ta). The restriction enzyme site within the L1(Ta) element has a known flanking sequence, but the sequence around the next restriction enzyme site is unknown so special linkers, called "vectorette" linkers, are added onto the ends of the restriction enzyme fragment. Next, PCR primers are used to amplify the fragment. The forward primer recognizes L1(Ta) so only genomic fragments with L1(Ta) elements are amplified. The reverse primer is complementary to the "vectorette" linker. This combination of primers amplifies the genomic DNA 3' or downstream from the L1(Ta). The sequence and genomic location of the DNA 3' to L1(Ta) is then determined by hybridization to a genomic array (see Ch. 8).

The associated paper identifies L1(Ta) elements in the X-chromosome and then determines the location of each of these elements in the entire genome for a number of different individuals. The researchers estimated that LINE-1 elements move in 1 out of 108 live births, as opposed to the previous estimate of 1 in 225 births. The distribution of elements is much more widespread than the reference human genome sequence suggests. The researchers located 34 elements on the X-chromosomes, of which 13 were polymorphic. These elements were inherited in an X-linked manner. They also found great variation in the number and location of L1(Ta) from one person to another. Several other recent studies substantiate the results of this paper.

FOCUS ON RELEVANT RESEARCH

LINE-1 elements may also cause genetic diseases. Some males with X-linked intellectual disability have an insertion of LINE-1 inside an intron of the *DACH2* gene. This gene is an ortholog to *dachshund*, a *Drosophila* gene that regulates neuronal differentiation. More evidence is needed to definitively prove that this LINE-1 insertion causes X-linked intellectual disability, but this finding suggests the possibility.

TABLE 9.03   Allele Frequencies for Cytochrome P450 CYP2D6 (Percentages)

		European	East Asian	Sub-Saharan Africa
*CYP2D6*UM*	Duplications (2-13 alleles)	1	12	28
*CYP2D6*5*	Deletion	3	6	6
*CYP2D6*10*	Unstable enzyme	3	4	4
*CYP2D6*17*	Lower affinity for drugs	0	0	12

5. Pharmacogenomics—Genetically-Individualized Drug Treatment

The genetic differences between individuals may cause significant differences in their reactions to certain drugs or clinical procedures. The new and rapidly expanding field that relates individual genotypes to pharmaceutical treatment is known as **pharmacogenomics**. Related to this term is **pharmacogenetics**, which refers to specific genes that affect drug reactions. A major goal of this research is to correlate a specific SNP pattern or gene with a particular drug reaction. For example, cytochrome P450 plays a major role in the oxidative degradation of many foreign molecules, including a wide range of pharmaceuticals. Cytochrome P450 is actually a family of several closely-related enzymes whose substrate range varies, providing protection against a wide range of different foreign molecules. One member of the family, the CYP2D6 isoenzyme, oxidizes drugs of the tricyclic antidepressant class and may be responsible for metabolizing about 25% of the drugs commonly used by doctors. Any given cytochrome P450 may have multiple allelic variants, some of which may show altered activity. Thus, the *CYP2D6* gene has several alleles with lowered activity and also duplications with increased activity (Table 9.03).

pharmacogenomics   A field of research that correlates individual genotypes relative to the person's reaction to a pharmaceutical agent

pharmacogenetics   Studying the particular genes that affect how a person reacts to drugs

Such alleles may be present at different frequencies in different human populations. Patients who possess low activity alleles metabolize the corresponding drugs much more slowly and are consequently more likely to show toxic side effects because the drug accumulates in their system. On the other hand, multiple copies of this gene causes rapid metabolism of the drug, and therefore, the drug has little time to work. In such cases, individual SNP analysis of patients can reveal which allele is present before administering the drug. The drug dosage can then be adjusted to the individual patient's genetic constitution.

> Pharmacogenomics correlates the genotype of an individual with the reaction that individual has to certain drugs.

FOCUS ON RELEVANT RESEARCH

Copy number variants in pharmacogenetics genes

He Y, Hoskins JM, McLeod HL (2011) Trends Mol. Med. 17(5): 244–251.

This associated article reviews the current state of pharmacogenomics for copy number variants. The use of copy number variations (CNVs) works the same way as using SNPs. Researchers are trying to correlate the number of repeats in a particular gene associated with drug metabolism with the phenotype of drug metabolism. In the article, the correlation between the numbers of *CYP2D6* genes a person has in their genome can differentiate between people who are ultrarapid metabolizers (UMs), extensive metabolizers (EMs), or poor metabolizers (PMs). Patients with more than two copies of this gene are considered UMs. Since codeine is a commonly used analgesic and this drug is metabolized by CYP2D6, the UMs metabolize the same dose of codeine ultra fast, leading to a very high concentration of the metabolite, morphine. In pediatric patients, high levels of morphine can cause respiratory depression and potentially death. In addition to *CYP2D6*, the review article discusses CNVs associated with glutathione S-transferase which detoxifies carcinogens, therapeutic chemicals, and environmental toxins. Deletions of this gene leave the patient with an elevated risk of various cancers due to environmental exposure. The article asserts the need to screen patients for these types of genetic differences so as to provide more accurate care and better information to a patient.

6. Personal Genomics and Comparative Genomics

To take full advantage of pharmacogenomics requires knowledge of individual DNA sequence differences, at least for those genes directly relevant to the clinical treatment proposed. At present, specific genes can be sequenced on a need-to-know basis. However, it has been suggested that everyone should get their complete genome sequenced individually.

The three factors involved are time, technology, and cost. The Human Genome Project took about 10 years, cost $3,000,000, and has given a consensus sequence based on 10 different people. Today, second-generation sequencing technology has lowered the cost of sequencing the human genome to about $50,000, and as third-generation sequencing technology emerges, the cost could drop to less than that of a flat screen TV.

So what will your personal DNA sequence reveal? We know a reasonable amount about hereditary defects due to single genes (such as cystic fibrosis or sickle cell anemia). However, the genetic factors involved in conditions such as heart disease, obesity, cancer, life expectancy, and mental disorders are more complex and due to multiple interacting genes, many of which have yet to be identified. Even though interpretation will be a problem, it will doubtless be more economical to get your whole genome sequenced than pay for lots of individual tests for each gene whose effect is understood.

Once specific genetic markers are known for different inherited disorders, parents can have their children's genomes sequenced to check if the defective gene was passed on to the next generation. This can be very useful to prevent the disease from appearing when there are environmental factors that can be controlled. Another benefit from comparing many whole genomes will be to reveal other genetic changes,

TABLE 9.04	Some Bioinformatics Websites

GenBank and linked databases

National Center for Biotechnology Information (all databases)	http://www.ncbi.nlm.nih.gov/Entrez/
Human Genome Resources	http://www.ncbi.nlm.nih.gov/genome/guide/human/
Institute for Genomics Research (TIGR)	http://www.tigr.org/tdb
Genome Database (GDB) (human genome)	http://gdbwww.gdb.org
European Bioinformatics Institute (Including EMBL & Swissprot)	http://www.ebi.ac.uk/
Flybase (*Drosophila* genome)	http://flybase.bio.indiana.edu:82
Wormbase (*C. elegans* genome)	http://wormbase.org
RCSB Protein Data Bank	http://www.rcsb.org/pdb/
PIR Protein Information Resource (PIR)	http://www-nbrf.georgetown.edu/pir/searchdb.html

not just those that occur in protein-encoding exons or known SNPs. In diseases like autism, there is emerging evidence of changes that occur in the non-coding regions of the genome. Finally, sequencing whole genomes from cancerous tumors will show how many somatic mutations have occurred and in what genes. Comparing different cancer genomes could tease out the genetic mutations that cause cancer rather than those mutations that occur because the cells are cancerous, a subtle but important difference.

Whole individual genome sequencing will have other benefits besides personalized medicine. The study of multiple human genomes can identify the rate at which our genomes change each generation. Similar comparative studies are used to determine the evolutionary relatedness of different species (see Ch. 26).

7. Bioinformatics and Computer Analysis

The field of **bioinformatics** deals with the computerized analysis of large amounts of genetic sequence data. A variety of websites are now available for online searching and manipulation of sequences (Table 9.04).

Both in molecular biology and other areas, vast amounts of information are accumulating in computer data banks. **Data mining** is the use of computer programs to find useful information by filtering or sifting through the data. Hence, intelligent software designed for data mining is sometimes known as "siftware." **Genome mining** is the application of this approach to genomic data banks. There are several stages to genome mining:

1. Selection of the data of interest.
2. Pre-processing or "data cleansing." Unnecessary information is removed to avoid slowing or clogging the analysis.
3. Transformation of the data into a format convenient for analysis.
4. Extraction of patterns and relationships from the data.
5. Interpretation and evaluation.

Now that the cost of sequencing an entire genome has dropped dramatically, the idea of individualized medicine is more conceivable.

Vast amounts of genetic data are now available. Computer analysis of this data has essentially created a new field of enquiry called bioinformatics.

bioinformatics The computerized analysis of large amounts of biological sequence data
data mining The use of computer analysis to find useful information by filtering or sifting through large amounts of data
genome mining The use of computer analysis to find useful information by filtering or sifting through large amounts of biological sequence data

TABLE 9.05	Databases of Information Available from Genomes and Maps Section at NCBI
Database of Genomic Structural Variation (dbVar)	archives information about large insertions, deletions, translocations, and inversions, and correlates these to the phenotype
Genome	contains sequence and map data for whole genomes of over 1000 organisms from all three domains of life
Genome Project	collection of all in-progress and complete sequencing, assembly, and mapping projects for different organisms
Nucleotide Database	collection of nucleotide sequences from other databases like GenBank and RefSeq
Sequence Read Archive (SRA)	stores sequencing data from second-generation sequencing platforms
UniSTS	database of sequence tagged sites (STSs)

A variety of analyses may be performed on DNA sequences. Some simple examples are as follows:

A. Searching for related sequences. Any DNA sequence may be compared with other sequences available in the data banks. Searches can also be run on protein sequences after translation of coding DNA. If another protein is found with a related sequence this may give some idea of the function of the protein under investigation. Of course, this assumes that the function of the other protein has already been deciphered! Another major use of sequence comparisons is to trace the evolution both of individual genes and of the organisms that carry them (see Ch. 26).

B. Codon bias analysis can locate coding regions. Due to third-base redundancy and the preferential use of some codons over others (in coding regions but not in random, intergenic DNA), there are differences in codon frequency between coding and non-coding DNA. A codon bias index can be computed that gives a reasonable first estimate of whether a stretch of DNA is likely to be coding or non-coding.

C. Searching for known consensus sequences. A variety of short consensus sequences or sequence motifs are known. Analysis of DNA sequences may reveal promoters, ribosome-binding sites (in prokaryotes only), terminators, and other regulatory regions. Inverted repeats in DNA imply possible stem and loop structures, which are often sites for the binding of regulatory proteins. Analysis of protein sequences may indicate binding sites for metal ions cofactors, nucleotides, DNA, etc.

Despite the vast amount of information available from the analysis of DNA sequences, we still need to investigate how genes are regulated at the genome level and how the encoded proteins function. Just as the totality of genetic information is known as the genome, so the sum of the transcribed sequences is the transcriptome and the total protein complement of an organism is the proteome. These are discussed in Chapters 19 and 15, respectively.

A vast amount of bioinformatics information is available free on the website for the National Center for Biotechnology Information (NCBI), http://www.ncbi.nlm.nih.gov/. See Table 9.05 for the list of different databases available under genomes and maps. There are many more databases than listed, but these are the most useful in genomics research.

8. Systems Biology

Systems biology is a term used extensively in recent years and refers to the integration of many different fields of research to give an overview of an organism. Until the

systems biology   A term that refers to the integration of many different types of research on an organism with the goal of defining the biological state of an organism within a certain environment

advent of genomics, the only way to study genetics was to focus on a single gene or group of genes. This approach has garnered massive amounts of information about the biology of organisms, yet the picture is not complete. Such studies provide single snapshots of the organism. Systems biology tries to study the complete organism. Using genomics and individual gene studies in combination with bioinformatics, the systems biologist tries to understand how one particular environment or condition affects the expression of all the different genes in a particular organism. The information is then integrated with other types of research, such as proteomics and metabolomics (Ch. 15). The overall goal is to provide a complete picture rather than a simple snapshot. The ability to store and analyze large amounts of information by computer is essential to systems biology, and just like the race for the human genome, understanding an entire system is going to evolve hand-in-hand with computer power and storage capabilities.

One of the first reports of systems biology research was by Thorsson, et al. in 2001. They used this approach to study galactose-utilization in yeast. The entire system included 997 mRNAs that responded to galactose. In addition, the studies identified 15 proteins that were regulated post-transcriptionally together with several physical interactions among different proteins. The results provided a larger view of how galactose utilization is integrated into other metabolic pathways and yeast metabolism as a whole.

> No concise definition of systems biology exists, but the overall goal is to understand how an organism responds to certain environments or conditions.

Integrative genomic profiling of human prostate cancer

Taylor BS, Schultz N, Hieronymus H, Gopalan A, Xiao Y, Carver BS, Arora VK, Kaushik P, Cerami E, Reva B, Antipin Y, Mitsiades N, Landers T, Dolgalev I, Major JE, Wilson M, Socci ND, Lash AE, Heguy A, Eastham JA, Scher HI, Reuter VE, Scardino PT, Sander C, Sawyers CL, Gerald WL. (2010) Cancer Cell 18: 11–22.

Systems biology integrates a lot of different information about how a particular organism responds to a specific environment. The application of systems biology to cancer has received a lot of attention by researchers since this is essentially a cell in a different environment or condition. Ever since U.S. President Richard Nixon declared war on cancer in 1971, researchers have been trying to find answers to what causes cancer, what makes cancer spread, and how we can prevent or stop the growth of cancerous cells. Gene-by-gene analyses have provided a lot of information about individual genes that can lead to cancer, but studying cancer with a systemic approach offers a new perspective.

Previous studies have looked for common genomic mutations in a variety of different cancers including breast, lung, colon, thyroid, and ovarian cancer. These studies have identified specific genetic changes common to many different tumors. One of the goals of systemic or genomics approaches may distinguish which mutations are "drivers"; that is, mutations that initiate the cancer, and which are "passenger" mutations that occur secondarily to the driver mutations. This distinction can be made by first looking at a number of individual tumors for the changes that occur due to becoming cancerous, and then comparing this information to multiple tumors. This associated paper examines 218 different samples of prostate cancer, from a combination of excised tumors, cell lines, and metastatic tumors to identify mutations that are found most often in prostate cancer. Then, the researchers asked whether or not the different mutations can predict if the cancer is aggressive or slow growing.

These cancers were studied by a variety of methods. Each cancer genome was studied by array comparative genomic hybridization (aCGH) using a microarray of human exons. aCGH identifies genomic deletions and duplications, also known as copy number variations. These results confirmed previous work that chromosome 8 was commonly altered in prostate cancer. The authors then sequenced all the exons in chromosome 8 for eighty of the tumors to look for somatic mutations that might cause missense mutations in specific proteins. The authors also used iPLEX Sequenome assays to find any SNPs common among the cancers. iPLEX assays use standard SNP analysis with single-nucleotide extension. However, to identify which nucleotide was added, the different extension products are separated by charge and mass using MALDI TOF mass spectroscopy (see Ch. 15), which analyzes multiple SNPs simultaneously.

The combined results revealed that about 40% of the examined prostate cancers had alterations in phosphoinositol-3-kinase (PI3K) and about 56%

FOCUS ON RELEVANT RESEARCH

of the primary tumors and 100% of the metastatic tumors had defects in androgen receptor signaling. Most importantly, the tumors with many copy number variations were more likely to be aggressive, whereas those with few copy number variations were less aggressive, and had better survival rates. This finding will help doctors determine the proper course of treatment for each individual patient. In addition, this study will hopefully direct the development of drugs that counteract the errors in PI3K or androgen receptor signaling.

9. Metagenomics and Community Sampling

Metagenomics is the study of the genomes of whole biological communities from a particular habitat. Normally it is applied to microorganisms. Metagenomic data sometimes allows us to identify microorganisms, viruses, or free DNA that exist in the natural environment by identifying genes or DNA sequences from the organisms.

Metagenomics applies the knowledge that all creatures contain nucleic acids; therefore, organisms do not have to be cultured, but can be identified by a particular gene sequence, or its derived protein, or even metabolite. The term "meta" is derived from meta-analysis, which is the process of statistically combining separate analyses. Metagenomics uses the same approach as genomics. It differs in the nature of the sample. Genomics focuses on a single organism, whereas metagenomics deals with multiple organisms, "gene creatures" (i.e., viruses, viroids, plasmids, etc.) and/or free DNA. Though rarely mentioned, many habitats contain relatively large fractions of DNA that is free rather than inside organisms. Metagenomic researchers isolate DNA and RNA from a sample of the habitat, such as soil or sea-water, without isolating or identifying individual organisms. The DNA or RNA is then analyzed by various genomic procedures, including shotgun DNA sequencing, PCR, RT-PCR, and so forth.

Most microorganisms have never been cultured or previously identified. Using metagenomics, researchers can analyze microbial diversity and also identify new proteins, enzymes, and biochemical pathways. Metagenomics has been used to identify new beneficial genes from the environment, together with novel antibiotics, enzymes that biodegrade pollutants, and enzymes that make novel products. Enzymes that can reduce the toxic effects of oil- and petroleum-based contaminants are found in bacteria that utilize the pollutants as an energy source. Even bacteria that thrive in environments contaminated with radioactivity have been identified. The knowledge gained from metagenomics has the potential to affect how we use the environment to our benefit or harm.

> Metagenomics is the sequence analysis of samples from the environment without culturing or even identifying the organisms.

10. Epigenetics and Epigenomics

Epigenetics refers to the inheritance of altered characteristics in the absence of alterations in the DNA sequence. Changes in gene expression are indeed involved, and these must be passed on from one cell to its descendents to qualify as epigenetic. In the early days of genetics such events were regarded as exceptions to the laws of Mendelian genetics and often ignored as awkward. Today as we begin to understand the molecular basis for epigenetics it is now best viewed as an "extra" level of inheritance, superimposed on top of the DNA sequence.

The most straightforward examples of epigenetics depend on methylation of the DNA. Although this is indeed a chemical alteration to DNA it does *not change the base sequence* of the DNA. Note that DNA methylation may affect gene expression during the lifetime of a cell (see Ch. 19).

Another mechanism for epigenetics relies on modifications to the histone code. The histone proteins around which eukaryotic DNA is wound are subject to a variety of chemical modifications, especially methylation and acetylation (see Ch. 17 for details). Alterations in these modifications may strongly influence gene expression.

These alterations do not give true epigenetic effects unless the altered expression state is inherited by another generation of cells. For single-celled organisms, this is unambiguous, but in multicellular organisms epigenetic inheritance may occur at two levels: between cells within the same organism or across generations via the gametes and sexual reproduction.

> Altered gene expression can sometimes be inherited even when the base sequence DNA is unchanged.

---

epigenetics Inheritance of phenotypic differences that occur without alteration of the nucleotide DNA sequences; often refers to the methylation pattern of DNA or the post-translational modification pattern on histones

metagenomics The genome level study of whole biological communities

Two examples of epigenetic inheritance that depend largely on DNA methylation will be discussed in more detail in Chapter 17. In genetic imprinting, inheritance occurs across generations, whereas X-chromosome inactivation occurs within a single multicellular animal.

Another example similar to X-chromosome inactivation is nucleolar dominance. Eukaryotic cells possess several hundred or thousand rRNA genes, depending on the organism. These are clustered into nucleolar organizer regions that, in humans for example, lie on five separate chromosomes. However, only around half of the rRNA genes are expressed. In any individual cell, only one parental copy of each nucleolar organizer region is expressed. This is referred to as nucleolar dominance and resembles X-chromosome inactivation, except that it applies to the clusters of genes for ribosomal RNA. Early on during development, the DNA of one copy is methylated in the promoter region. This methylation pattern is passed on through subsequent cell divisions. In addition, there are accompanying alterations in histone methylation and acetylation. In particular, multiple acetylations of histones H3 and H4 are found for the active copies of the rRNA genes.

Another complex example that involves both DNA methylation and both methylation and acetylation of histones is the effect of maternal nutrition during pregnancy on the future health of the offspring. When rat fetuses are poorly nourished due to the mother being poorly fed, they become epigenetically adapted to a future life of poor nutrition. Such rats grow to smaller sizes than unstressed rats. They are also more likely to suffer from a variety of diseases including diabetes, obesity, and cardiovascular problems. These effects are largely mediated by epigenetic modification of the gene for insulin growth factor-1 (IGF-1).

A fascinating role of epigenetics is how it may generate differences between identical twins. Comparison of identical twins has shown that the levels of DNA methylation may vary considerably. How these differences arise is presently unknown. However, such differences can have severe health consequences. DNA methylation has been surveyed in identical twins of which one suffered from an auto-immune disease. In the case of systemic lupus erythematosus, the sick twin had lower levels of DNA methylation on around 50 out of 800 genes surveyed. This implies that over-expression of certain genes is involved in lupus since DNA methylation usually lowers gene expression. In contrast, no significant differences in methylation have so far been found for the auto-immune diseases multiple sclerosis and rheumatoid arthritis.

Methylation of DNA in animals occurs largely on cytosine that is followed by guanine; that is, on CpG sequences. Methylation must be sequence specific to affect gene regulation correctly. There are two categories of DNA methyltransferase (DNMT). Maintenance methyltransferase (DNMT1 in mammals) methylates newly-synthesized strands of DNA at previously methylated CpG sites. In contrast, de novo methyltransferases add methyl groups to new sites. The DNMT enzymes are probably recruited to the DNA sites by other DNA-binding proteins that are sequence specific. However, the precise sequences that identify methylation target sites are still largely unknown.

Key Concepts

- Sequence tagged sites (STSs) and expressed sequence tagged sites (ESTs) are unique regions of DNA sequence that are used in mapping the human genome and other large genomes.
- STSs and ESTs are mapped relative to each other using linkage analysis on yeast artificial chromosomes (YAC) clones or in radiation hybrid cells.
- Shotgun sequencing is when a genomic library is constructed by ligating random genomic DNA fragments into a vector and then randomly sequencing

these clones. The sequence data is then assembled into contigs using computers that determine regions of overlap.

- Closing the gap between contigs can be done by screening for library clones that hybridize to probes from the ends of previously-identified contigs or by PCR amplifying with primers that anneal to the ends of the two known contigs.

- The official Human Genome Project cloned human DNA into YACs and bacterial artificial chromosomes (BACs) and then mapped these relative to each other on the chromosomes. Next, the YACs and BACs were subjected to sequencing.

- Celera Genomics used directed shotgun sequencing to generate 70 million sequences that were ordered into what is now called the reference human genome sequence. Although the bulk of the information came from pure shotgun sequencing, the alignment was done with the aid of the STS and EST map.

- The human genome has enough sequence data to fill a million pages of text with 3,000 letters per page. About half of this is non-coding repetitive sequences such as tandem repeats, SINEs, LINEs, and defunct retroviruses and transposons.

- The actual number of human genes is predicted to fall between 20,000 and 25,000, some of which produce non-coding RNA such as rRNA, tRNA, snRNA, and snoRNA.

- Between the genome of one individual and the next are variations or polymorphisms such as different bases, insertions, and/or deletions. A single nucleotide polymorphism (SNP) is a single base change. A simple sequence length polymorphism (SSLP) is any insertion and/or deletion different between two individuals. Copy number variations (CNVs) are variations in the number of repeats.

- SNP analysis is used to screen for hereditary defects and test individuals for phenotypic variations in response to pharmaceuticals. Many SNPs are identified by single nucleotide extension from a Zipcoded primer.

- Exon trapping identifies a protein coding exon *in vitro* by cloning random human genome fragments into a multiple cloning site of a eukaryotic expression vector. If the random genome fragment has an exon, it will be included in the expressed mRNA from the vector and increase its size accordingly. This exon can then be identified by sequencing the genome fragment.

- The non-coding parts of the human genome were originally called "junk," but further analysis has indicated that these are a major source of human genomic variation that perhaps can cause some diseases.

- SNP analysis is useful to identify different alleles for a specific gene, particularly, the cytochrome P450 gene, which is used to degrade different pharmaceutical drugs.

- Bioinformatics uses computers to analyze the large amounts of genetic sequence data. Data mining filters or sifts the data, whereas genome mining specifically applies to identifying the genomic sequence data of interest, cleaning the data of unnecessary information, reformatting the data into convenient forms for analysis, and then interpreting the genomic data for different patterns or relationships.

- Systems biology studies how a particular organism adapts its entire genetic expression patterns in response to a different condition.

- Metagenomics is the study of genomes from multiple organisms that inhabit a particular environment.

- Epigenetics refers to modifications in DNA that do not involve changes in sequence. These types of changes include methylation patterns, histone modification patterns, and inactivation of certain genes or chromosomes by converting the area into heterochromatin.

Review Questions

1. What is genomics?
2. What were the most important things revealed by the study of the human genome?
3. What are sequence tags? How are they used?
4. How are STS mapped?
5. What is a radiation hybrid?
6. How are radiation hybrids used in mapping?
7. What were the first prokaryotic and eukaryotic genomes to be sequenced?
8. What is shotgun sequencing?
9. What is a contig?
10. What is the procedure for sequencing the human genome? What technique has changed how genomes are sequenced?
11. What are the different repetitive sequences that make up half of the human genome?
12. What is polymorphism? What is an SNP? What is an SSLP?
13. What are the uses of SNP analysis?
14. How are genes identified by exon trapping?
15. What is the original definition of "junk DNA"? How did it originate?
16. How has the view of "junk DNA" changed since many human genomes have been sequenced?
17. What is a copy number variant?
18. What is pharmacogenomics?
19. How may personal genomics help in future medical treatment?
20. What is bioinformatics? What type of information can be learned at the NCBI website?
21. What is systems biology?
22. What is metagenomics? What practical use does it have?
23. What is epigenetics?
24. Give two mechanisms by which epigenetics can operate.

Conceptual Questions

1. The following set of three contigs is identified from sequencing data for a newly-discovered organism from the Amazon River. You were instructed to find the sequence information between the contigs using PCR. Describe how you would do this experiment. First, draw the location of the primers you will have synthesized using arrows to describe their orientation and then describe which primers pairs you would use in PCR.

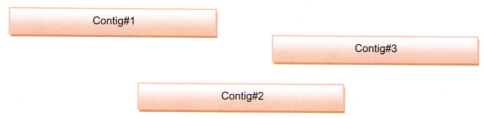

2. The following set of genomic clones were identified by chromosome walking (not primer walking) and contain a candidate gene for a rare genetic disease. Explain a method you would use to identify the exons for this gene. The corresponding genomic segment is diagrammed below.

clone 1.1		clone 15.0

clone 2.7

clone 8.3

human chromosome 17p34.1

3. Describe why repetitive sequences are so troublesome for assembling shotgun sequences by computer. What is one method to overcome the misinterpretation of repetitive sequences?

4. PubMed's Entrez website can search for protein or nucleotide sequences that are similar in sequence to your sequence of interest. Go to the NCBI website

(http://www.ncbi.nlm.nih.gov) and select the dropdown option "HomoloGene." In the search box enter myoD to find the entry for this transcription factor. Click on the entry 18404, and scroll down the page to find Protein Alignments. Generate an alignment of *Homo sapiens* and *Pan troglodytes* (chimpanzee). Repeat the analysis by comparing the sequence identity between *H. sapiens* and *Mus musculus* (mouse) and between *H. sapiens* and *Danio rerio* (zebrafish).

a. What is the percent identity between each set of two sequences? What is the subject length?

b. Which of these three species has the greatest similarity to humans?

5. Use the following clone map of a chromosome segment to answer the following questions:

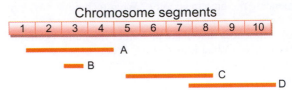

a. If the gene for *ssc1* hybridizes to clone C and D, what gene segment must contain this gene?

b. If an STS hybridizes to clone B and A, what segment of the chromosome has this genetic marker?

c. Will the same STS hybridize to both clone A and C?

The Central Dogma of Molecular Biology

Cell Division and DNA Replication

When living cells divide, each daughter cell must receive a copy of the genome. For this to occur, the DNA must be replicated either before or during cell division. DNA synthesis and cell division must therefore be carefully coordinated. The way in which DNA is replicated underlies the mechanism of heredity. The two strands of the DNA helix must first be separated. Then a complementary strand is made using each of the two original strands as template. This yields two identical copies of the original DNA. Finally, the two copies must be physically distributed between the two progeny cells.

A variety of technical issues must be resolved during DNA replication. For example, DNA molecules are compacted to fit into the cell. Consequently, they must be unfolded for replication to proceed. Other problems include the initiation of new strands of DNA (priming) and how to coordinate synthesis of two new strands of DNA that run in opposite directions (discontinuous synthesis and the lagging strand). In higher organisms, the presence of multiple chromosomes wrapped around histones and surrounded by a nuclear membrane complicates DNA replication and cell division. But the basic principles are much the same as in bacteria.

1. Cell Division and Reproduction Are Not Always Identical

Since each cell needs a complete set of genes, a parental cell must duplicate its genome before dividing. Each of the two new cells then receives one copy of the genome. Because the genes are made of DNA and are located on the chromosomes, this means that each chromosome must be accurately copied. When a bacterial cell with a single chromosome divides, each daughter cell receives a copy of the parental chromosome. Division of eukaryotic cells is more complex, as each cell has multiple chromosomes. Not only must all of the chromosomes be duplicated, but a mechanism is needed to ensure that both daughter cells receive identical sets of chromosomes at cell division or mitosis (see Ch. 2).

When a single-celled organism divides, the result is two new organisms, each consisting of one cell. However, in multicellular organisms cell division does not automatically result in the creation of new organisms. When the cells composing a multicellular organism divide, they increase the size and/or complexity of the original organism. A distinct process called meiosis is needed to form gametes that develop into new organisms after fertilization (see Ch. 2 for details). The term reproduction is used to signify the production of a new individual organism. Thus, in unicellular organisms, cell division and reproduction occur simultaneously, whereas in multicellular organisms cell division and reproduction are two different processes.

In many plants and fungi, clumps of cells may break off or single-celled spores may be released from the parental organism and give rise to new individual multicellular organisms. This is known as **asexual** or **vegetative reproduction** as these new individuals will be genetically identical to their parents. This contrasts with sexual reproduction, where each new individual receives roughly equal amounts of genetic information from two separate parents and is therefore a novel genetic assortment. Sexual reproduction is especially characteristic of animals and also occurs in more complicated plants and many fungi. Some organisms, particularly plants and fungi, possess the ability to reproduce both sexually or asexually. Although they are inextricably entwined in humans and many other animals, it is important to realize that sex and reproduction are two distinct processes from a biological viewpoint.

Strictly speaking, bacteria do not reproduce sexually since new bacteria always result from the division of a single parental cell. But all is not lost; mixing of genes from two individuals does occur in bacteria. However, this occurs in the absence of cell division and involves transfer of a relatively small segment of DNA from one cell (the donor) to another cell (the recipient) (see Ch. 25 for details). Such sideways transfer of DNA, without reproduction, is sometimes referred to as **horizontal gene transmission**. In contrast, **vertical gene transmission** is when genes are transmitted from the previous generation to the new generation. Vertical transmission thus includes all forms of cell division and reproduction that create a new copy of the genome, whether sexual or not.

> When cells divide, the genome must be replicated so each new cell gets a complete set of genes.

> Reproduction creates new organisms. Cell division creates new cells. These two processes are only the same for organisms that are single-celled.

2. DNA Replication Occurs at the Replication Fork

Replication is the process by which the DNA of the ancestral cell is duplicated, prior to cell division. Upon cell division, each of the descendants receives one complete copy of the DNA that is identical to its predecessor. The first stage in replication is to separate the two DNA strands of the parental DNA molecule. The second stage is to build two new strands. Each of the separated parental strands of DNA serves as a **template strand** for the synthesis of a new complementary strand. The incoming

> DNA is replicated semi-conservatively; that is, each parental strand of DNA is separated and copied. The two daughter strands have one from the parent and one newly synthesized.

---

asexual or vegetative reproduction   Form of reproduction in which there is no re-shuffling of the genes between two individuals
horizontal gene transfer   Movement of genes sideways from one organism to another without one being parent of the other
replication   Duplication of DNA prior to cell division
template strand   Strand of DNA used as a guide for synthesizing a new strand by complementary base pairing
vertical gene transmission   Transfer of genetic information from an organism to its descendents

FIGURE 10.01
Template Strand and Base Pairing in DNA Replication

Incoming nucleotides line up on the antisense or template strand and are then linked together to form the new strand of DNA. The arriving nucleotides are positioned by base pairing in which A pairs with T and G binds to C. These base pairs are held together with hydrogen bonds.

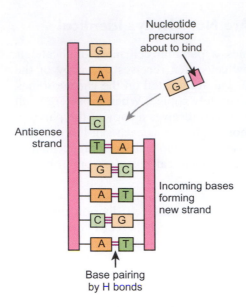

Nucleotide precursor about to bind

Antisense strand

Incoming bases forming new strand

Base pairing by H bonds

nucleotides for the new strand recognize their partners by base pairing and so are lined up on the template strand (Fig. 10.01). Since A pairs only with T, and since G pairs only with C, the sequence of each original strand dictates the sequence of the new complementary strand.

Synthesis of both new strands of DNA occurs at the **replication fork** that moves along the parental molecule. The replication fork consists of the zone of DNA where the strands are separated, plus an assemblage of proteins that are responsible for synthesis, sometimes referred to as the **replisome**. The result of replication is two double-stranded DNA molecules, both with sequences identical to the original one. One of these daughter molecules has the original left strand and the other daughter has the original right strand. This pattern of replication is **semi-conservative**, since each of the progeny conserves half of the original DNA molecule (Fig. 10.02). Amazingly, in *E. coli*, DNA is made at nearly 1,000 nucleotides per second.

Replication is similar, but not exactly the same, in prokaryotes and eukaryotes. DNA replication in bacteria will be covered initially, as this process is less complicated than the process in eukaryotes.

> The two strands of the double helix are separated by a complex of enzymes called the replisome. The position of the replisome marks a replication fork.

2.1. Supercoiling Causes Problems for Replication

Several major problems must be solved to accomplish bacterial DNA replication. First, there are the topological problems due to DNA being not only a double helix but also supercoiled. Because the two strands forming a DNA molecule are held together by hydrogen bonding and are twisted around each other to form a double helix, they cannot simply be pulled apart. The higher level supercoiling further complicates the problem of separating the strands (see Ch. 4 for a description of supercoiling). Consequently, before new DNA can be made, first the supercoils must be unwound, and then the double helix must be untwisted (see below). In addition, since the vast majority of bacterial chromosomes are circular, it is important to untangle the two new circles of DNA.

The supercoiled bacterial chromosome is circular, and two replication forks proceed in opposite directions around the circle (Fig. 10.03). This process is known as

> The strands of the parent DNA molecule cannot be separated until the supercoils and helical twisting have been removed.

---

replication fork   Region where the enzymes replicating a DNA molecule are bound to untwisted, single-stranded DNA
replisome   Assemblage of proteins (including primase, DNA polymerase, helicase, SSB protein) that replicates DNA
semi-conservative replication   Mode of DNA replication in which each daughter molecule gets one of the two original strands and one new complementary strand

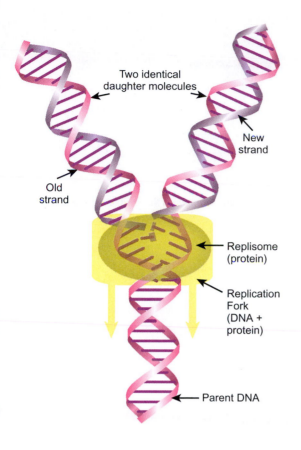

Two identical daughter molecules

New strand

Old strand

Replisome (protein)

Replication Fork (DNA + protein)

Parent DNA

■ **FIGURE 10.02**
Semi-conservative Replication and the Replisome

The replication fork is the site of DNA replication and, by definition, includes both the DNA and associated proteins. The assembled proteins, known as the replisome, facilitate the unwinding of the helix and the addition of new nucleotides. The arrows indicate the direction of movement of the replication fork. The synthesis of two DNA helices results from adding a new complementary strand to each one of the separated old strands.

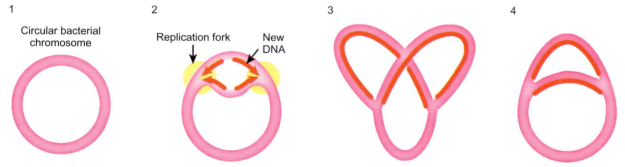

1

Circular bacterial chromosome

2

Replication fork   New DNA

3

4

FIGURE 10.03
Theta-Replication

Successive steps in DNA replication are shown for a circular bacterial chromosome. The chromosome (1) begins to replicate using two replication forks (2). Continued replication results in division of the chromosome (3) and its apparent resemblance to θ, the Greek letter, theta (4).

bi-directional replication. A circle observed half-way through division looks like the Greek letter theta (θ) and so this mode of replication is also called **theta-replication**.

In *E. coli*, the two type II topoisomerases, **DNA gyrase** and **topoisomerase IV**, solve the supercoiling problem. As the replication fork proceeds along the DNA, it over-winds the DNA and generates positive supercoils ahead of the replisome. Since the bacterial chromosome is negatively supercoiled, the over-winding introduced by replication is at first cancelled out. However, after replication of about 5% of the chromosome, the pre-existing negative supercoils are all unwound, and the continued unwinding of the helix starts creating positive supercoils. For DNA replication to proceed, the over-winding must be removed. DNA gyrase binds to the DNA ahead of the replication fork and introduces negative supercoils that cancel out the positive supercoiling. The net result is that DNA gyrase "removes" supercoiling ahead of

bi-directional replication   Replication that proceeds in two directions from a common origin
DNA gyrase   An enzyme that introduces negative supercoils into DNA, a member of the type II topoisomerase family
theta-replication   Mode of replication in which two replication forks go in opposite directions around a circular molecule of DNA
topoisomerase IV   A particular topoisomerase involved in DNA replication in bacteria

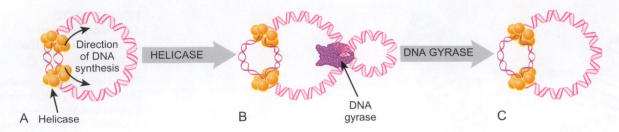

FIGURE 10.04
Unwinding of Double Helix and of Supercoils

For the replication fork to proceed, both the double helix and the supercoils must be unwound. Helicase unwinds the double helix and DNA gyrase removes the supercoiling.

FIGURE 10.05
Helicase Unwinds the Double Helix

To unwind DNA, helicase first binds to DNA and then cleaves the hydrogen bonds connecting base pairs to separate the strands of the helix. Then single-stranded binding protein attaches to the single-stranded DNA.

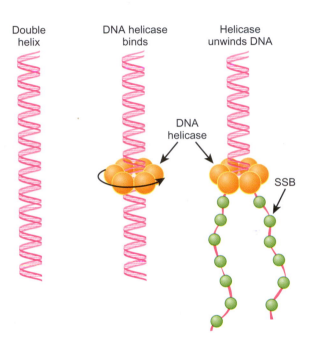

the replication fork (Fig. 10.04). Topoisomerase IV may help in this process to some extent, but its main function is disentangling daughter molecules after replication has finished, as described below.

These topoisomerases are absolutely essential to bacterial survival. The quinolone antibiotics (e.g., nalidixic acid and ciprofloxacin) inhibit type II topoisomerases, in particular DNA gyrase. Inhibited DNA gyrase remains bound to the DNA at one location and blocks movement of the replication fork. The net effect is cell death, which is good for treating bacterial infections in people, but not good from the bacterium's perspective.

2.2. Strand Separation Precedes DNA Synthesis

The second problem, the hydrogen bonds holding the double helix together, is solved by another enzyme, **DNA helicase** (Fig. 10.05). The major helicase of *E. coli* is DnaB protein, which forms hexamers. Helicase does not break the DNA backbone; it simply disrupts the hydrogen bonds holding the base pairs together. Energy is needed for this process and helicase cleaves ATP to supply this energy.

> Quinolone antibiotics kill bacteria by inhibiting DNA gyrase. This blocks replication of the DNA.

> Helicase unwinds the DNA helix and SSB protein keeps the strands apart.

DNA helicase   Enzyme that unwinds double-helical DNA

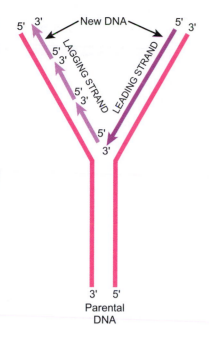

FIGURE 10.06

Continuous and Discontinuous Synthesis of DNA at the Replication Fork

The protein at the replication fork responsible for DNA synthesis, DNA polymerase, always synthesizes DNA in the 5′ to 3′ direction. Therefore, one new strand (leading strand) can be made continuously, while the other (lagging strand) must be made discontinuously (i.e., in short segments).

The two separated strands of the parental DNA molecule are complementary and so prefer base pairing to each other. In order to manufacture the new strands, the two original strands must be kept apart. This is done by **single strand binding protein**, or **SSB protein**, which binds to the unpaired single-stranded DNA and prevents the two parental strands from reannealing. In reality, the single-stranded region between helicase and the lagging strand is longer than that between helicase and the leading strand, due to the 3D arrangement of the replication fork.

One new DNA strand is made continuously. The other is made as a series of fragments, called Okazaki fragments, which must be linked together after synthesis has occurred.

3. Properties of DNA Polymerase

Now that the two strands are open, replication enzymes can enter the area. The key component is a **polymerase**, which is a general name for an enzyme that joins nucleotides together. Bacterial cells contain several DNA polymerases that have different roles both in DNA replication and in DNA repair (see Ch. 23). In *E. coli*, **DNA polymerase** III replicates most of the DNA. But this enzyme has some peculiarities that pose problems for DNA replication. Firstly, DNA polymerase will only synthesize DNA in a 5′ to 3′ direction. Since the strands in a double helix are anti-parallel, and since a single replication fork is responsible for duplicating the double helix, this means that one of the new strands can be made continuously, but that the other cannot (Fig. 10.06). The strand that is made in one piece is called the **leading strand** and the strand that is made discontinuously is the **lagging strand**.

Secondly, DNA polymerases lack the ability to initiate a new strand and can only elongate a pre-existing strand; that is, it needs a pre-existing 3′-OH on a sugar in order to attach the new nucleotide. Consequently, a special mechanism for strand initiation is needed. This involves synthesis of a short **RNA primer** whenever a new DNA strand is started. Unlike DNA polymerases, **RNA polymerases** can start

DNA polymerase can only make new DNA in one direction. Even stranger, it cannot start new strands of DNA.

New strands of DNA must be started with short segments of RNA known as primers.

DNA polymerase   Enzyme that synthesizes DNA
lagging strand   The new strand of DNA that is synthesized in short pieces during replication and then joined later
leading strand   The new strand of DNA that is synthesized continuously during replication
polymerase   Enzyme that synthesizes nucleic acids
RNA primer   Short segment of RNA used to initiate synthesis of a new strand of DNA during replication
RNA polymerase   Enzyme that synthesizes RNA
single strand binding protein (SSB protein)   A protein that keeps separated strands of DNA apart

FIGURE 10.07
Strand Initiation Requires an RNA Primer

DNA polymerase cannot begin a new strand but can only elongate. Therefore, DNA replication requires an RNA primer to initiate strand elongation. One RNA primer is needed to start the 5' to 3' leading strand. In contrast, multiple RNA primers are needed for the 3' to 5' lagging strand because this is made in short stretches each running 5' to 3'. DNA polymerase will then add nucleotides to the end of each RNA primer. Later, the short RNA primers will be removed and replaced by DNA.

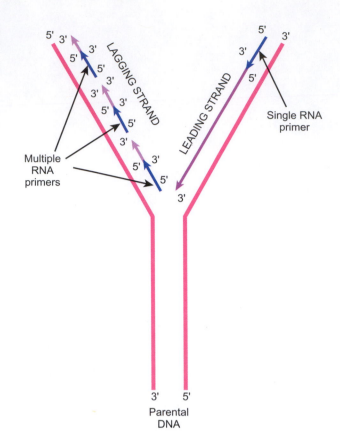

new strands without any pre-existing 3'-OH. A special RNA polymerase, known as **primase** (DnaG protein), makes the RNA primers that are responsible for strand initiation during DNA synthesis in bacteria. Although the leading strand only needs to be started once, the lagging strand is made in short sections and a new RNA primer must be created each time a new portion is made. DNA polymerase then builds new strands of DNA starting from each RNA primer (Fig. 10.07).

4. Nucleotides Are the Precursors for DNA Synthesis

All DNA polymerases and RNA polymerases synthesize nucleic acids in a 5' to 3' direction. Incoming nucleotides are added to the hydroxyl group at the 3'-end of the growing chain. The precursors for DNA synthesis are the **deoxyribonucleoside 5'-triphosphates** (**deoxy-NTPs**), dATP, dGTP, dCTP, and dTTP. Proceeding from the deoxyribose outwards, the three phosphate groups are designated the α-, β-, and γ-phosphates. Upon polymerization, the high-energy bond between the α and β phosphates is cleaved, releasing energy to drive the polymerization. The outermost two phosphates (the β- and γ- phosphates) are released as a molecule of pyrophosphate. A new bond is made between the innermost phosphate (the α-phosphate) of the incoming nucleotide and the 3'-OH of the previous nucleotide at the end of the growing chain (Fig. 10.08).

The DNA precursors, which contain deoxyribose, are made from the corresponding ribose-containing nucleotides (Fig. 10.09). Reduction of ribose to deoxyribose is catalyzed by the enzyme **ribonucleotide reductase**. This acts on the diphosphate

deoxyribonucleoside 5'-triphosphate (deoxyNTP)   Precursor for DNA synthesis consisting of a base, deoxyribose, and three phosphate groups
primase   Enzyme that starts a new strand of DNA by making an RNA primer
ribonucleotide reductase   Enzyme that reduces ribonucleotides to deoxyribonucleotides

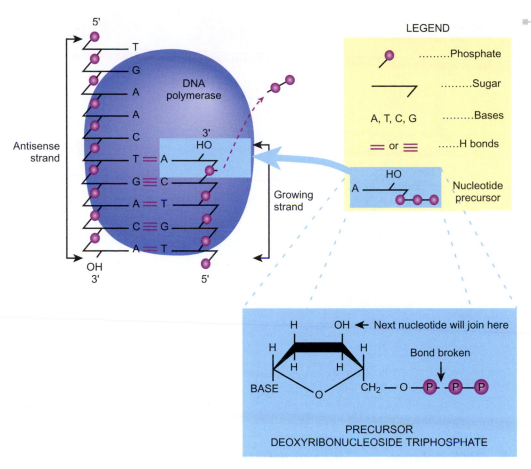

LEGEND

........Phosphate

.........Sugar

A, T, C, G ....Bases

= or ≡ ......H bonds

Nucleotide precursor

PRECURSOR DEOXYRIBONUCLEOSIDE TRIPHOSPHATE

H   OH ← Next nucleotide will join here

H   H

H   H   Bond broken

BASE   O   CH₂ — O —P—P—P

FIGURE 10.08
Polymerization of Nucleotides

Details of the growth of the new DNA strand are shown. At each step, a nucleoside triphosphate is added. During chain elongation, the two outermost phosphate groups of the precursor are cleaved off, releasing pyrophosphate. This provides energy for the reaction. The 3'-hydroxyl group of the precursor triphosphate remains available for the next addition to the growing strand.

derivatives (ADP, GDP, CDP, and UDP), removing the 2'-OH group from ribose. Next, a **kinase** adds the third phosphate onto dADP, dGDP, and dCDP, thus creating three of the four precursor nucleotides, dATP, dGTP, and dCTP.

dUDP follows a different route, since DNA does not contain uracil but instead has thymine, the methyl derivative of uracil. Before methylation, dUDP is converted to dUMP by removal of a phosphate. Then **thymidylate synthetase** adds the methyl group, so converting dUMP to dTMP. Finally, two phosphates are added to give dTTP. The methyl group of thymine is donated by the **tetrahydrofolate (THF)** cofactor, which is oxidized to **dihydrofolate (DHF)** during the reaction. The DHF must be reduced back to THF by **dihydrofolate reductase** for DNA synthesis to proceed.

Some well-known pharmaceuticals act by inhibiting nucleotide synthesis. **Methotrexate (amethopterin)** inhibits eukaryotic dihydrofolate reductase. Since growth of tumors involves rapid cell division and DNA replication by cancer cells, methotrexate is used as an anti-tumor agent. An antibiotic called **trimethoprim** also prevents dihydrofolate reductase from reducing DHF to THF and therefore preventing the methylation of dUMP to dTMP. Without any THF, bacteria are also unable to synthesize adenine and guanine from their precursors, so in essence, trimethoprim

Precursors for DNA are made from ribonucleotides by oxidizing the ribose to deoxyribose.

The fact that all new strands of nucleic acid start with a piece of RNA plus the fact that ribonucleotides are made first supports the idea that RNA came first in evolution. This theory, the RNA world, is discussed further in Chapter 26.

dihydrofolate (DHF)   Co-factor with a variety of roles including making precursors for DNA and RNA synthesis
dihydrofolate reductase   Enzyme that converts dihydrofolate back to tetrahydrofolate
kinase   Enzyme that attaches a phosphate group to another molecule
methotrexate (or amethopterin)   Anti-cancer drug that inhibits dihydrofolate reductase of animals
tetrahydrofolate (THF)   Reduced form of dihydrofolate cofactor that is needed for making precursors for DNA and RNA synthesis
thymidylate synthetase   Enzyme that adds a methyl group, so converting the uracil of dUMP to thymine
trimethoprim   Antibiotic that inhibits dihydrofolate reductase of bacteria

FIGURE 10.09
Synthesis of Precursors

First, ribonucleotide reductase converts the ribonucleoside diphosphates to the deoxy form. Second, a kinase adds a phosphate to form the deoxyribonucleoside triphosphates. The precursor for the thymidine nucleotides is made from a uridine derivative by adding a methyl group that is transferred by the carrier tetrahydrofolate (THF).

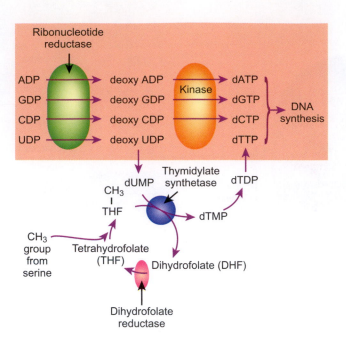

blocks bacterial replication in two pathways. The **sulfonamide** class of antibiotics inhibits synthesis of the **folate** co-factor itself. Animals do not make folate, but require it in their diet, so sulfonamides are harmless to human patients in reasonable doses, but are fatal to bacteria that need to synthesize folate for survival.

5. DNA Polymerase Elongates DNA Strands

DNA **polymerase III (Pol III)** is the major enzyme involved in elongating DNA during chromosome replication. DNA polymerase III has several subunits that combine into a **holoenzyme**. First, the **core enzyme (Pol III)** synthesizes DNA and consists of three subunits, DnaE (α-subunit; polymerization), DnaQ (ϵ-subunit; proofreading), and HolE (θ-subunit; function uncertain, but probably helps stabilize the ϵ-subunit) (not shown in Fig. 10.10). Next, the "**sliding clamp**" is shaped like a doughnut and is made from two semi-circular subunits of DnaN protein. This is also called the β-subunit. It slides up and down like a curtain ring on each template strand of DNA (Fig. 10.10). Several accessory proteins (δ, χ, ψ and τ plus the γ subunit) known as the **clamp-loading complex** are required to load the sliding clamp onto the DNA—a process requiring energy (see Focus on Relevant Research). This is also called the γ/τ complex for short. Therefore, the holoenzyme includes two DNA polymerase core enzymes, a pair of tau subunits in the single sliding clamp-loader complex, the other subunits of the clamp-loader complex, and two sliding clamps (Fig. 10.10).

Although hydrogen bonding alone would match bases correctly, the great majority of the time this is not accurate enough for replication of the genome. Consequently, many DNA polymerases possess **kinetic proofreading** ability (Fig. 10.11). This refers to the ability to make corrections on the fly. **Mismatches** are sensed

> DNA polymerase is kept attached to the DNA by the clamp-loader complex and sliding clamp protein assemblies.

> DNA polymerase not only synthesizes new DNA, but also checks for mismatched base pairs and corrects any errors it has made.

---

clamp-loading complex Group of proteins that loads the sliding clamp of DNA polymerase onto the DNA
core enzyme The part of DNA or RNA polymerase that synthesizes new DNA or RNA (i.e., lacking the recognition and/or attachment subunits)
DNA polymerase III (Pol III) Enzyme that makes most of the DNA when bacterial chromosomes are replicated
folate Co-factor involved in carrying one carbon group in DNA synthesis
holoenzyme An active enzyme complex consisting of multiple functional subunits that are made of multiple proteins
kinetic proofreading Proofreading of DNA that occurs during the process of DNA synthesis
mismatch Wrong pairing of two bases in a double helix of DNA
sliding clamp Subunit of DNA polymerase that encircles the DNA, thereby holding the core enzyme onto the DNA
sulfonamide Antibiotic that inhibits the synthesis of the folate cofactor

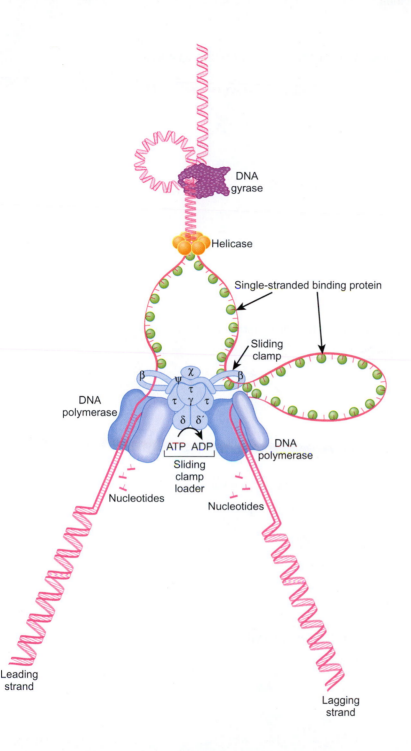

■ **FIGURE 10.10**
*DNA Polymerase III
Replication Assembly*

During replication, the sliding clamp loader makes contacts with single-stranded binding proteins and the sliding clamps in order to stabilize the template DNA as it enters into DNA polymerase III, which polymerizes new DNA as the template passes.

because they cause minor distortions in the shape of the double helix. The polymerase halts upon sensing a mismatch and removes the last nucleotide added. As this was added to the 3' end of the growing DNA strand, the enzyme activity that removes the offending nucleotide is a **3'-exonuclease**. In the case of DNA polymerase III, proofreading is due to a separate subunit, the DnaQ protein (ε-subunit). (In some other DNA polymerases, proofreading ability resides on the same protein as polymerase activity.) In addition, immediately after replication, the new DNA is checked and if necessary repaired by the **mismatch repair** system (see Ch. 23).

3'-exonuclease An enzyme that degrades nucleic acids from the 3' end
mismatch repair DNA repair system that recognizes and corrects wrongly-paired bases

ε SUBUNIT DETECTS MISMATCH DUE TO BULGE

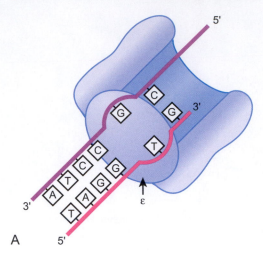

A

ε SUBUNIT CUTS OUT MISMATCHED BASE (G)

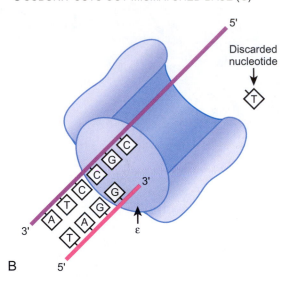

Discarded nucleotide

B

POLYMERASE GOES BACK,
α SUBUNIT REPAIRS MISMATCH

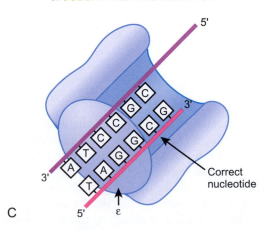

Correct nucleotide

C

FIGURE 10.11
DNA Polymerase III—Proofreading

The portion of DNA polymerase responsible for checking the accuracy of new nucleotides is the DnaQ protein (ε-subunits). A mismatch is detected by a bulge in the new chain. The incorrect nucleotide is discarded and the appropriate nucleotide added.

Park AH, Jergic S, Politis A, Ruotolo BT, Hirshberg D, Jessop LL, Beck JL, Barsky D, O'Donnell M, Dixon NE, Robinson CV (2010) A single subunit directs the assembly of the *Escherichia coli* DNA sliding clamp loader. Structure 18: 285–292.

The assembly of the sliding clamp loader is essential for replication and therefore an understanding of how the clamp loader comes together is fundamental to our understanding of replication. In addition, since many antibiotics target various components of replication, understanding what parts of the complex are essential can help researchers identify potential targets for new antibiotics. The clamp-loader complex of proteins stabilizes the position of each core enzyme and is responsible for loading and unloading the sliding clamp, which holds the two strands of unwrapped DNA apart. The sliding clamp loader has seven different protein subunits that unite to form a functional complex. A fully-functional sliding clamp-loader has two τ, one γ, one δ, one δ', one ψ, and one χ subunit. The loader can also form with three τ rather than two τ and a γ-subunit. The τ-, γ-, and δ-subunits form a pentameric ring that uses ATP to load the sliding clamp. The ψ- and χ-subunits form a dimer that attaches to the ring with a connection between the ψ-subunit and the τ- or γ-subunits. The χ-subunit then connects this complex to the single-stranded binding proteins.

In this paper, the authors used electrospray ionization (ESI) mass spectroscopy (see Ch. 15 for an explanation) to determine the order in which the seven subunits of the clamp-loader complex assemble. ESI identifies non-covalent protein interactions as seen in multiprotein complexes. The technique determines the stoichiometry, or specific ratio of subunits. By determining the ratio of subunits and what subunit attaches to another, the authors elucidated the possible steps for clamp-loader

FOCUS ON RELEVANT RESEARCH

complex assembly. These experiments demonstrate that the clamp-loader complex initiates by τ self-assembly, since the entire complex would not assemble without these first coming together. After self-assembly, the δ'-subunit joins the complex and induces a conformational change that allows δ to close the ring and χ/ψ to connect to single-stranded binding protein. The assembled sliding clamp loader can attach the sliding clamp and associate the two DNA polymerase core enzymes at the replication fork.

6. The Complete Replication Fork Is Complex

The replication fork is defined as all the structural components in the region where the DNA molecule is being duplicated. It includes the zone where the DNA is being untwisted by gyrase and helicase, together with the stretches of single-stranded DNA held apart by single strand binding protein (SSB). It also includes DNA polymerase III holoenzyme, which is making two new strands of DNA (Fig. 10.10). Table 10.01 summarizes the name and function of these enzymes. Although the leading strand is made continuously, the lagging strand is made in short segments of 1,000–2,000 bases in length, known as **Okazaki fragments** after their discoverer.

As noted above, during DNA replication, the two new strands must be synthesized in opposite directions. A linear representation of this would imply that the two polymerase assemblies might move apart. In fact, the two molecules of Pol III that are making these two strands are held together by the tau subunits of the clamp loader complex. In order for them both to make new DNA simultaneously, the strands of DNA must be looped, as shown in Figure 10.10.

> One of the new strands of DNA, the "lagging strand," is made in short fragments that are joined up later.

Box 10.01 Molecular Biology Rarity

Both the tau (τ) and gamma (γ) subunits of DNA polymerase III clamp-loader complex are encoded by the same gene—*dnaX*. The tau subunit is a normal, full-length product of translation. However, the gamma subunit is made by frame-shift (see Ch. 23) followed by premature termination during translation of the same mRNA. Such differential synthesis of two proteins from a single gene is extremely rare in living cells, but this particular case appears to be widespread among bacteria, including *E. coli*. However, frame-shifting to generate two alternative proteins is not so rare among RNA viruses with small genomes. For example, both retroviruses (e.g., HIV) and filoviruses (e.g., Ebola virus) use this approach (see Ch. 21).

Okazaki fragments   The short pieces of DNA that make up the lagging strand

TABLE 10.01	Proteins Involved in DNA Replication in *E. coli*	
Protein	**Gene**	**Function**
DnaA	*dnaA*	Initiates chromosome division; binds to replication origin
Helicase	*dnaB*	Unwinds the double helix
DnaC	*dnaC*	Loading of DNA helicase
SSB	*ssb*	Single strand binding protein
Primase	*dnaG*	Synthesis of RNA primers
RNase H	*rnhA*	Partial removal of RNA primers
Pol I	*polA*	Polymerase I; fills gaps between Okazaki fragments
Polymerase III		DNA polymerase III holoenzyme
α	*dnaE*	Strand elongation
ε	*dnaQ*	Kinetic proofreading
θ	*holE*	Unknown; part of core enzyme
β	*dnaN*	Sliding clamp
τ	*dnaX*	Sliding clamp loader; dimerization of Pol III core enzyme
γ	*dnaX*	Sliding clamp loader
δ	*holA*	Sliding clamp loader
δ'	*holB*	Sliding clamp loader; ensures only three τ or two τ and one γ
χ	*holC*	Sliding clamp loader; connects to SSB proteins
ψ	*holD*	Sliding clamp loader
DNA Ligase	*lig*	Seals nicks in phosphate backbone, especially in the lagging strand
DNA Gyrase		Introduces negative supercoils
α	*gyrA*	Makes and seals double-stranded breaks in DNA
β	*gyrB*	ATP-using subunit
Topoisomerase IV		Decatenation
A	*parC*	Makes and seals double-stranded breaks in DNA
B	*parE*	ATP-using subunit

7. Discontinuous Synthesis of the Lagging Strand

Although the leading strand is synthesized continuously, the lagging strand is composed of multiple pieces, the Okazaki fragments. When synthesis of each new Okazaki fragment is begun, it needs a fresh RNA primer. Priming occurs in three steps. First, **PriA** protein displaces SSB proteins from a short stretch of DNA. Then primase (DnaG) binds to PriA. Finally, primase synthesizes a short RNA primer of 11–12 bases. This priming complex is sometimes known as the **primosome** (Fig. 10.12).

Each time a new Okazaki fragment is begun, the Pol III assembly that is making the lagging strand releases its grip on the DNA and relocates to start making a new strand of DNA starting from the 3′ end of the RNA primer. This involves disassembly and relocation of the sliding clamp, which is performed by the clamp-loading complex. Note that the replisome contains two Polymerase III core enzymes, each with its own sliding clamp, but only one sliding clamp is released and re-attached at a new location. This is because only the lagging strand needs constant clamp removal and reloading. It is worth repeating that in *E. coli*, all of this happens about 1000 times per second.

PriA   Protein of the primosome that helps primase bind
primosome   Cluster of proteins (including PriA and primase) that synthesizes a new RNA primer during DNA replication

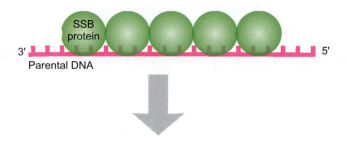

A   PriA DISPLACES SSB PROTEIN

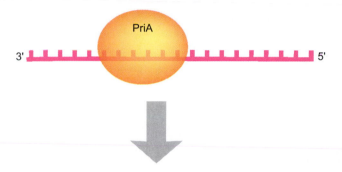

■**FIGURE 10.12**

Three Steps in Starting a Primer for a New Okazaki Fragment

Prior to primer formation, the bases of the parental DNA strand are covered with SSB proteins.
A) First, the PriA protein displaces the SSB proteins. B) Second, a primase associates with the PriA protein. C) Lastly, the primase makes the short RNA primer needed to initiate the Okazaki fragment.

B   PRIMASE BINDS

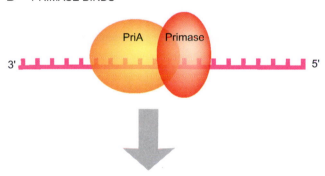

C   PRIMASE MAKES SHORT RNA PRIMER

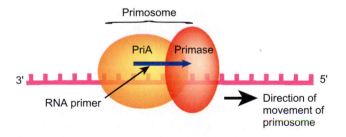

7.1. Completing the Lagging Strand

After the replication fork has passed by, the lagging strand is left as a series of Okazaki fragments with **gaps** (i.e., spaces from which one or more nucleotides are missing) between them. In addition, there are short pieces of RNA primer alternating with newly-synthesized DNA. Joining the Okazaki fragments to give a complete strand of DNA is accomplished by two—or perhaps three—enzymes working in

The fragments of the discontinuous lagging strand must be finished by removing the RNA primers (RNase H or Pol I), filling the gaps with DNA (Pol I) and, finally, joining the ends (DNA ligase).

gap   A break in a strand of DNA or RNA where bases are missing

FIGURE 10.13
Three Steps in Joining the Okazaki Fragments

When first made, the lagging strand is composed of alternating Okazaki fragments and RNA primers. The first step in replacing the RNA primer with DNA is the binding of DNA polymerase I to the primer region. As Pol I moves forward it degrades the RNA and replaces it with DNA. Lastly, DNA ligase seals the nick that remains.

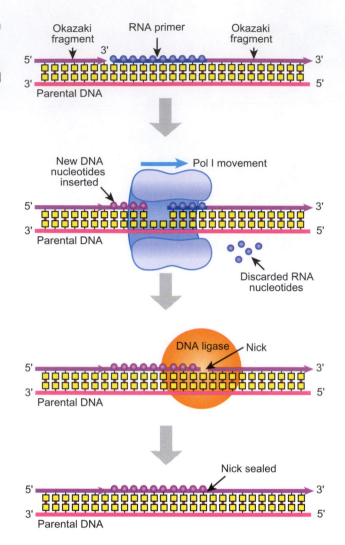

succession: **Ribonuclease H (RNase H), DNA polymerase I (Pol I)**, and **DNA ligase**. Only the last two enzymes are involved in the classical model. DNA polymerase I degrades the RNA primers and fills the resulting gaps. Finally, DNA ligase joins the sugar phosphate backbone (Fig. 10.13). In an alternate model, RNase H, which typically degrades the RNA strand of DNA:RNA double helixes, removes most of each RNA primer and DNA polymerase I only removes the last few bases of the RNA primers.

Despite being a single polypeptide chain, DNA polymerase I possesses kinetic proofreading ability like Pol III. Pol I also has the unique ability to start replication at a nick in the DNA. The term "**nick**" refers to a break in the nucleic acid backbone with no missing nucleotides. When Pol I finds a nick, it cuts out a small stretch of DNA—or RNA—approximately 10 bases long. It then fills in the gap with new DNA. Pol I completes the lagging strand and also functions during DNA repair (see Ch. 23). This property of Pol I is used in the laboratory for labeling small fragments of

---

DNA ligase   Enzyme that joins up DNA fragments end to end
DNA polymerase I (Pol I)   Bacterial enzyme that makes small stretches of DNA to fill in gaps between Okazaki fragments or during repair of damaged DNA
nick   A break in the backbone of a DNA or RNA molecule (but where no bases are missing)
ribonuclease H (RNase H)   Enzyme that degrades the RNA strand of DNA:RNA hybrid double helixes. In bacteria it removes the major portion of RNA primers used to initiate DNA synthesis

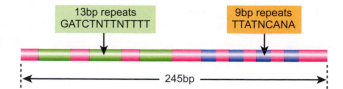

Sequences at the Origin of DNA Replication

Sequence repeats at the origin of replication are of two varieties, both being AT-rich.

DNA with radioactive nucleotides by a procedure known as "**nick translation**." Nicks are introduced into one strand of the DNA by DNase I (deoxyribonuclease I). DNA polymerase I then starts at the nick and moves along the DNA removing the nucleotides ahead of it and replacing them with the radioactive nucleotides provided.

DNA polymerases I and II were discovered before DNA polymerase III, hence the numbering. In retrospect, it is easy to understand why Pol III, with its complex requirements and multiple subunits, took longer to discover than the relatively simple enzymes Pol I and Pol II, which are involved in DNA repair.

8. Chromosome Replication Initiates at *oriC*

So far we have discussed the processes involved in the duplication of the DNA double helix. In addition, DNA replication must be synchronized with cell division. This involves starting replication at a specific location on the chromosome and stopping when the chromosome has been successfully copied. Prokaryotic DNA replication starts at a unique site on the chromosome, known as the **origin of replication**, and proceeds in both directions around the circle. The **initiation complex** contains five proteins: DnaA, DnaB, DnaC, DNA gyrase, and SSB. Of these, only DnaA is unique to chromosome initiation; the others are also involved in starting new Okazaki fragments. The origin, *oriC*, has three 13-base repeats, each consisting of GATCTNTTNTTTT, followed by four 9-base repeats, each consisting of TTATNCANA. Note that both sequences are AT-rich, a feature that aids strand separation since it takes less energy to break two hydrogen bonds in an AT pair than the three bonds in a GC pair. These repeats are scattered over a 245 base-pair region that is required for chromosome initiation (Fig. 10.14).

The first event of initiation is binding of **DnaA protein** to the four 9-base sequences. A cluster of 20–30 or so DnaA proteins binds and the whole *oriC* region wraps around them. Next, the DnaA proteins open the DNA at all three of the 13-base repeats. Next, six DnaB helicase subunits join the partially open DNA region with the help of six DnaC proteins, which help helicase load correctly. Helicase displaces DnaA from the single-stranded 13-base repeats and begins to unwind the DNA and so create a replication fork (Fig. 10.15). A second helicase hexamer creates a second replication fork moving in the opposite direction. Helicase also activates primase, which then makes an RNA primer at each of the two points where the leading and lagging strands are initiated. DNA gyrase then promotes further unwinding and SSB proteins attach in order to keep the DNA single-stranded.

Plasmids have been constructed that carry the chromosomal origin sequences of *E. coli*. Using purified plasmid and purified initiation proteins has allowed analysis of initiation in a cell-free system. In fact, the binding of the DnaA/DnaB/DnaC complex to *oriC* has been seen under the electron microscope (Fig. 10.16).

Replication of a bacterial chromosome starts at a specific point, the origin of replication, which encompasses 245 base pairs of DNA that contain three 13 bp repeats and four 9 bp repeats.

DnaA protein   Protein that binds to the origin of bacterial chromosomes and helps initiate replication
initiation complex (for replication)   Assemblage of proteins that binds to the origin and initiates replication of DNA
nick translation   The removal of a short stretch of DNA or RNA, starting from a nick, and its replacement by newly-made DNA
origin of replication   Site on a DNA molecule where replication begins

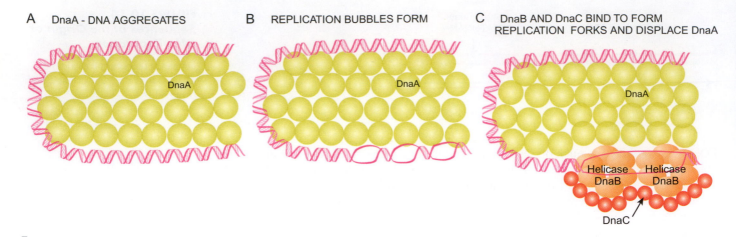

A   DnaA - DNA AGGREGATES

B   REPLICATION BUBBLES FORM

C   DnaB AND DnaC BIND TO FORM REPLICATION FORKS AND DISPLACE DnaA

FIGURE 10.15
Three Steps in the Initiation of Replication by DnaA

A) DnaA protein binds first to the four 9-base repeats and then to the three 13-base repeats. B) As more DnaA binds, the DNA folds and the three 13-base repeats are unwound. C) Two complexes of DnaB and DnaC bind to the three 13-base repeats. This pushes DnaA away and causes the DNA strand to open all along the AT-rich region. The two DnaB complexes now start two replication forks, each headed in opposite directions around the circular DNA.

FIGURE 10.16
Initiation Complex Bound to oriC

Supercoiled DNA of plasmid pCM959 was mixed with the proteins DnaA, DnaB, DnaC, plus HU. The complexes were cross-linked and the plasmid DNA was cut with *Ban1* to produce six fragments. The origin, *oriC*, was asymmetrically situated on the 703-bp fragment shown here. *(Credit: Funnell, Baker and Kornberg, In vitro assembly of a pre-priming complex at the origin of the* Escherichia coli *chromosome. Journal of Biological Chemistry, 262 (1987) 10327–10334.)*

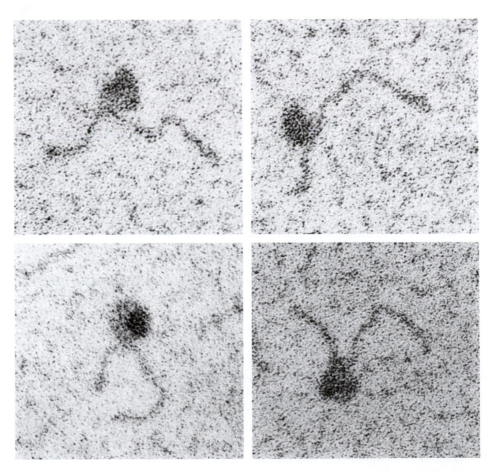

8.1. DNA Methylation and Membrane Attachment Control Initiation of Replication

The control of initiation, especially the correct timing of new rounds of replication, is a highly-regulated process. In prokaryotes, two factors appear to be involved: methylation of DNA at the origin and attachment to the cell membrane. The *oriC* region

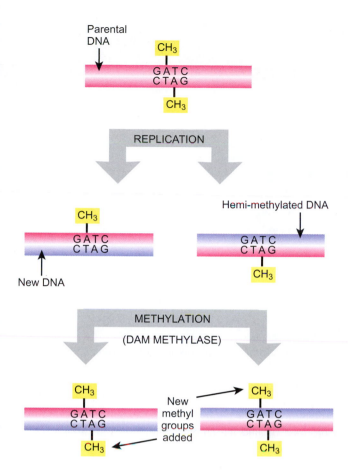

Parental DNA

REPLICATION

Hemi-methylated DNA

New DNA

METHYLATION

(DAM METHYLASE)

New methyl groups added

◼–**FIGURE 10.17**
Methylation after DNA Duplication

Dam methylase recognizes GATC palindromic sites and methylates them prior to DNA strand separation. The complementary sequences are synthesized during DNA replication but are not immediately methylated. Later, Dam methylase adds methyl groups to the newly-synthesized sequences.

contains a total of 11 GATC sequences, which are recognized by the Dam methylase (see Ch. 17). Dam methylase recognizes the GATC palindrome and transfers a methyl group onto the adenine of both strands. Before replication, each GATC on the *E. coli* chromosome, including those in the origin of replication, are fully methylated. Immediately after replication, the old strand is methylated, but the new strand has not yet been methylated. The DNA is thus hemi-methylated.

Over most of the chromosome, full methylation is restored within a minute or two of replication (Fig. 10.17). This brief period allows for use of hemi-methylation as a guide for the mismatch repair system (see Ch. 23). However, the origin of replication is slow to regain full methylation; it takes 10–15 minutes. There is a corresponding lag in re-methylating the promoter region of the *dnaA* gene. Transcription of this gene is repressed while hemi-methylated; consequently, there is also a drop in the amount of DnaA protein available for initiation. Hemi-methylated sites cannot be used to initiate chromosome replication. Hemi-methylated DNA binds to the cell membrane, helped by **SeqA (sequestration protein)**, whereas fully methylated DNA does not.

The preceding implies that methylation and membrane binding are necessary controlling factors for initiation of replication. This is by no means the whole story, as *dam* mutants of *E. coli*, lacking the ability to add methyl groups, are viable and grow quite well. Thus, an origin with no methylation at all must be functional. Mutants lacking SeqA protein initiate replication more frequently than normal but are also viable. The factor(s) that control membrane binding are still obscure, as is the timing mechanism that oversees how long the origin is bound to the membrane and hidden from Dam methylase.

The methylation state of the DNA is involved in controlling new rounds of DNA synthesis, but this is not the whole story.

sequestration protein (SeqA)   Protein that binds the origin of replication, thereby delaying its methylation

FIGURE 10.18
Termination of replication by Tus and Ter Sites

The circular bacterial chromosome has a termination region, or terminus, with several sites that stop replication forks moving clockwise (*TerF, TerB,* and *TerC*) and counterclockwise (*TerE, TerD,* and *TerA*).

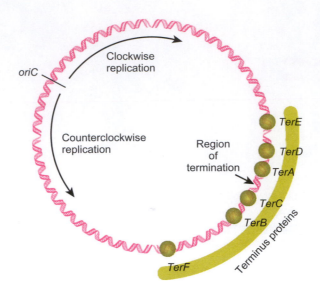

Box 10.02 When Forks Collide

Both replication and transcription rely on protein complexes that travel along the DNA. Since genes may be transcribed in either direction, there will be occasions when a transcription fork collides head on with one of the two replication forks. When they collide, the replication fork stalls but does not disassemble. RNA polymerase is displaced from the DNA with the help of the Mfd protein or TRCF (transcription repair coupling factor). The replication fork then resumes its progress around the chromosome. The Mfd protein also releases RNA polymerase, which is stalled due to DNA damage or other reasons.

9. Chromosome Replication Terminates at *terC*

Replication finishes when the two replication forks meet at the **terminus** of the chromosome. This region is surrounded by several *Ter* **sites** that prevent further movement of replication forks (Fig. 10.18). Since replication proceeds in two directions in prokaryotes, the *Ter* sites act asymmetrically. *TerC, TerB,* and *TerF* prevent clockwise movement of forks and *TerA, TerD,* and *TerE* prevent counterclockwise movement. The two innermost sites (*TerA* and *TerC*) are most frequently used, and the outer sites presumably serve as back-ups in case a fork manages to slip past *TerA* or *TerC*.

The *Ter* sites have a 23bp consensus sequence that binds **Tus protein**. This blocks the movement of the DnaB helicase and brings movement of the replication fork to a halt. Tus protein binds asymmetrically and stops movement from one direction only. It can be displaced from the DNA by a fork coming from the other direction. The meeting of the two replication forks is thus controlled by Tus proteins bound to multiple *Ter* sites within the termination region. However, the whole terminus region of *E. coli* (including the gene for the Tus protein, which is located next to *TerB*) can be deleted without apparent ill effects. This implies that the replication forks do not absolutely need to meet at a *Ter* site and can terminate successfully wherever they collide.

9.1. Disentangling the Daughter Chromosomes

When a circular chromosome finishes replicating, the two new circles may be physically interlocked or *catenated* (see Ch. 4). Such catenanes must be separated so that

DNA replication finishes at special sites in the terminus region of the chromosome. *TerC, TerB,* and *TerF* stop the clockwise replication and *TerA, TerD,* and *TerE* stop the counterclockwise replication.

***Ter* site**   Site in the terminus region that blocks movement of a replication fork
terminus   Region on a chromosome where replication finishes
Tus protein   Bacterial protein that binds to *Ter* sites and blocks movement of replication forks

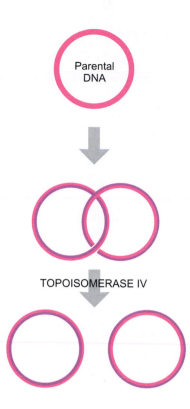

FIGURE 10.19

Decatenation by Topoisomerase IV

Topoisomerase IV aids in unlinking two newly replicated circles of DNA. Replication of the circular bacterial chromosome may give two catenated circles of DNA. Topo IV unlinks these to give decatenated circles.

each daughter cell receives a single chromosome upon cell division (Fig. 10.19). Decatenation of interlocked circles is carried out by topoisomerase IV. Although the terminology is confusing, Topo IV is actually a type II topoisomerase whose mode of action is similar to DNA gyrase. Topo IV is found behind the replication fork, where it untangles the newly formed DNA as replication proceeds. It can also decatenate finished DNA circles, both of chromosomes and plasmids.

A related problem sometimes results from recombination, which as in eukaryotes, can occur between the two daughter chromosomes. The two growing circular chromosomes may recombine even during the process of replication. Each pair of crossovers or exchanges of genetic material may cause the growing circles to become interlocked. If there is an even number of crossovers, Topo IV can disentangle the circles and no harm is done. However, an odd number of crossovers will covalently join the two circles of DNA (Fig. 10.20). This is estimated to occur about one in every six replication cycles. The covalent dimer must be separated by the **crossover resolvase**, XerCD, which uses the *dif* **sites** on the two chromosomes to introduce a final crossover. This gives, in effect, an even number of crossovers. The dif site is approximately opposite the origin of replication, between *TerA* and *TerC*.

If circular molecules of DNA become interlocked, specific enzymes are needed to untangle them. Topoisomerase IV decatenates interlocked circles; crossover resolvase detaches covalently-linked dimers.

10. Cell Division in Bacteria Occurs after Replication of Chromosomes

Bacteria divide by **binary fission**, or splitting. Bacteria have only one chromosome, which lies within the cytoplasm. Bacterial cell division is thus relatively simple and may be divided for convenience into four stages, although some of the processes overlap slightly:

1. Replication of the chromosome
2. Partition of the daughter chromosomes

---

crossover resolvase   Bacterial enzyme that separates covalently-fused chromosomes
dif **site**   Site on bacterial chromosome used by crossover resolvase to separate covalently-fused chromosomes
binary fission   Simple form of cell division, by splitting down the middle, found among bacteria

FIGURE 10.20
Recombination Causes Problems

The number of recombination sites determines how chromosomes are unwound. Topo IV can readily untangle circles with even numbers of crossovers, whereas Resolvase XerCD is additionally needed to separate circles with odd numbers of crossovers.

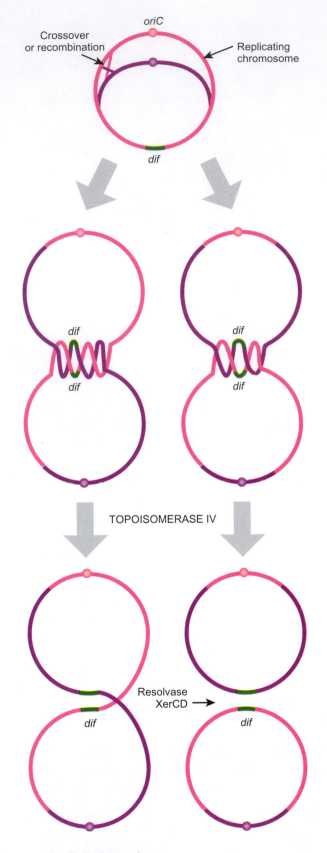

3. Cell elongation

4. Separation of the two daughter cells by formation of a cross-wall

Replication proceeds in both directions at once around the circular bacterial chromosome. Eventually, the two replication forks meet and merge, yielding two new

Bacterial cells grow longer and replicate their DNA simultaneously, then they divide.

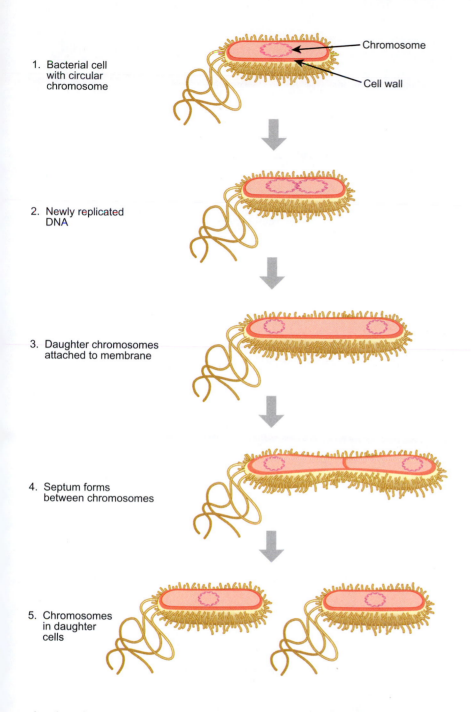

1. Bacterial cell with circular chromosome

Chromosome

Cell wall

2. Newly replicated DNA

3. Daughter chromosomes attached to membrane

4. Septum forms between chromosomes

5. Chromosomes in daughter cells

■► **FIGURE 10.21**

Elongation of the Cell Separates Chromosomes

Segregation of chromosomes is caused by elongation of the cell. Subsequently, a partition, or septum, is formed that completes cell division.

circular chromosomes. These are attached to the cell membrane at their origins. As the cell elongates, the chromosomes are pulled apart (Fig. 10.21). The final step of cell division is the formation of a cross-wall, or **septum**.

10.1. How Long Does It Take for Bacteria to Replicate?

The time required for an *E. coli* cell to divide, the **generation time**, ranges from 20 minutes to several hours, depending on the conditions. Despite this, duplication of the chromosome always takes 40 minutes and the time from termination of replication to completion of cell division takes 20 minutes. If the generation time is less than

generation time   The time from the start of one cell division to the start of the next
septum   Cross-wall that separates two new bacterial cells after division

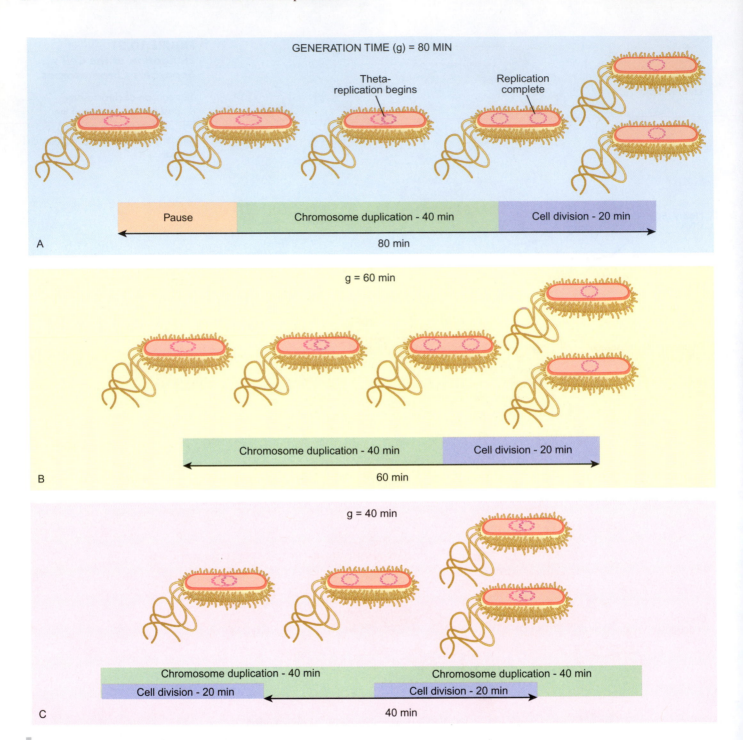

FIGURE 10.22
Cell Division and Chromosome Replication

A) Cell division in more than 60 minutes allows for a time gap after cell division and the start of the next DNA replication. B) Cell division in 60 minutes requires 40 minutes for replication and 20 minutes for completion of cell division without pausing. C) Cell division in less than 60 minutes requires that a new cycle of replication be initiated before the last cell division is completed.

60 minutes, one or more rounds of chromosome replication must overlap. This means that a new cycle of replication may start before the previous one has finished. Cells in rapidly-dividing cultures of bacteria therefore contain multiple but incomplete copies of the chromosome. If the generation time is longer than 60 minutes, there is a gap between division of the cell and initiation of the next round of chromosome replication (Fig. 10.22).

Jonas K, Chen YE, Laub MT (2011) Modularity of the bacterial cell cycle enables independent spatial and temporal control of DNA replication. Curr Biol 21: 1092-1101.

One of the essential questions asked by researchers that study bacterial replication is how a bacterium controls initiation of DNA replication. Obviously, if you start replicating too often relative to cell division, then how do you divide the four or more daughter DNAs into only two cells? If you divide the cell before replicating DNA, then one side will not have any genetic information. One of the most interesting systems used to garner answers to these questions is *Caulobacter* bacteria, where every cell division is asymmetric; that is, one of the daughter cells is stalked just like the parent and the other daughter cell is a swarmer. Stalked cells are immobile and immediate enter another round of replication and cell division. Swarmer cells are motile and move to a new location before differentiating into a stalked cell and then entering replication and cell division. Thus, the initiation of replication is delayed in the swarmer cell. Unlike *E. coli*, *Caulobacter* do not initiate another round of replication until cell division is complete.

Interestingly, much is known about the control of initiation of replication, but what is still not completely understood is how this is repressed in the daughter cells until cell division has completed, especially in the swarmer cells which first have to move to a new location and then differentiate into a stalked cell. Like *E. coli*, *Caulobacter* DNA replication is controlled by DnaA binding to the origin of replication (see Fig. 10.15). In order to block another round of DNA replication, the swarmer cells make CtrA, a transcription factor, which physically blocks DnaA from binding to the origin. CtrA also induces transcription of about 100 genes that ensure swarmer cells attach to a new location before initiating replication. This associated paper discusses the balance between CtrA and DnaA in the initiation of replication.

The paper provides evidence that DnaA is the only true modulator of initiation of replication. The presence of CtrA only inhibits initiation and has no effect on the cycling of DnaA. In other words, DnaA amounts increase every 65 min-

FOCUS ON RELEVANT RESEARCH

utes whether or not CtrA is present or absent. This result contrasts to previous studies that suggested that the two molecules were transcriptionally connected such that CtrA controls the amount of DnaA produced. In addition, the finding that these two systems are regulated independent of each other suggests that the control of replication initiation by DnaA evolved first in prokaryotes. Then as *Caulobacter* evolved to form a differentiated swarmer cell, a different control system evolved (CtrA) to control replication initiation. How these two systems function in *Caulobacter* provides a greater understanding of cellular regulatory circuits.

11. The Concept of the Replicon

A **replicon** is any DNA (or RNA) molecule that is capable of surviving and replicating itself inside a cell. A replicon must possess an origin of replication. A replicon must also be an intact, "complete" molecule of DNA (or RNA) that is either circular or has ends that are protected from attack by a cell's defense systems. Although chromosomes are clearly replicons, they are by no means the only ones. Virus genomes replicate when inside their host cell. Consequently, their nucleic acid qualifies as a replicon. Since some virus genomes consist of RNA, this means that the definition of replicon must include both DNA and RNA.

> Nucleic acid molecules that survive and divide must have both an origin of replication and be circular (or have ends that are protected).

In prokaryotes, replicons are usually closed circles of DNA that have no ends. In most bacteria, linear molecules of DNA are degraded by **exonucleases**. These are enzymes that degrade nucleic acids one nucleotide at a time, starting from one end or the other. Consequently, linear segments of DNA that enter a bacterial cell during conjugation or transformation (see Ch. 25) will eventually be degraded. This DNA can survive, but it must circularize or integrate itself into the pre-existing circular DNA.

Despite this, a few bacteria contain linear chromosomes. These have a variety of individual adaptations to protect the ends from endonucleases. *Borrelia burgdorferi*, which causes **Lyme disease**, has hairpin sequences at the ends of its linear chromosome. *Streptomyces lividans*, a soil organism, has proteins covalently attached to the ends of its DNA.

> Extra DNA molecules, known as plasmids, are found in many bacteria. They are usually circular and much smaller than chromosomes.

Plasmids are another group of replicons. They are extra self-replicating molecules of DNA that are not necessary for survival of the host cell (see Ch. 20). Plasmids

exonuclease   Enzyme that cleaves nucleic acid molecules at the end and usually removes just a single nucleotide
Lyme disease   Infection caused by *Borrelia burgdorferii* and transmitted by ticks
plasmid   Accessory molecule of nucleic acid capable of self-replication. Does not normally carry genes needed for existence of host cell. Usually consists of double-stranded circular DNA but occasional plasmids that are linear or made of RNA exist
replicon   Molecule of DNA or RNA that contains an origin of replication and can self-replicate

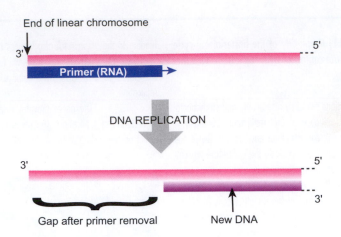

are usually circular, although linear plasmids occur in *Borrelia* and *Streptomyces*, the same bacteria that contain linear chromosomes. Circular replicons are occasionally found in eukaryotic cells, including plasmids such as the 2μ circle of yeast. Mitochondria and chloroplasts also contain their own genomes, or replicons, which are circular molecules of self-replicating DNA.

12. Replicating Linear DNA in Eukaryotes

Replicating a linear molecule of DNA requires certain adaptations. Since DNA polymerases can only elongate, not initiate, new strands of DNA must be initiated with an RNA primer. Since synthesis always proceeds from 5' to 3', one of these RNA primers must be located right at the 5' end of each new strand when replicating linear DNA (Fig. 10.23). When this terminal RNA primer is removed, it cannot be replaced with DNA, as there is no pre-existing 3'-OH for DNA polymerase to elongate. If nothing was done to overcome this problem, the molecule of DNA would grow shorter, by the length of an average RNA primer, with each round of replication. In fact, some theorize that the continued shortening of telomeres acts as a clock, controlling how long a cell can divide before dying. Circular prokaryotic DNA molecules do not have ends and so do not have this problem.

Eukaryotes have solved the problem of replicating linear DNA by using structures known as **telomeres**, which are located at the ends of their chromosomes. Telomeres consist of multiple tandem repeats (from 20 to several hundred) of a short sequence, usually of six bases (TTAGGG, in vertebrates including humans). During each replication cycle, the chromosomes are indeed shortened and several of the telomere repeats are lost. However, no unique coding information is lost. Furthermore, in cells where the enzyme **telomerase** is present, the lost DNA is later replaced by adding several of the six-base-pair units to the 3' end after each replication cycle (Fig. 10.24). Telomerase carries a small segment of RNA, complementary to the six-base pair telomere repeat. This allows it to recognize the telomeres and provides a template for elongating the telomere.

After telomerase has elongated the 3' ends, the complementary strand can be filled in by normal RNA priming followed by elongation by DNA polymerase and joining by ligase. The telomere repeats also protect the ends of chromosomes against degradation by exonucleases.

Telomere repeat sequences have been remarkably conserved throughout evolution, although some variation is seen. The characteristic TTAGGG repeat of vertebrates is also found in the protozoan *Trypanosoma*, while the sequence in the

> **Eukaryotic DNA is linear and needs special structures, the telomeres, to protect its ends.**

---

telomere   Specific repetitive sequence of DNA found at the end of linear eukaryotic chromosomes
telomerase   Enzyme that adds DNA to the telomere of a eukaryotic chromosome

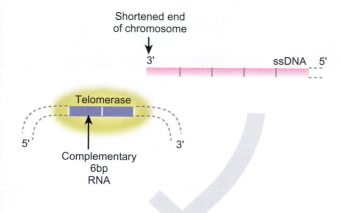

Shortened end
of chromosome

Telomerase

Complementary
6bp
RNA

TELOMERASE RNA RECOGNIZES
6bp REPEAT

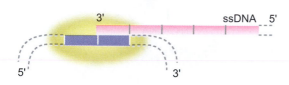

TELOMERASE MAKES A NEW DNA
6bp REPEAT

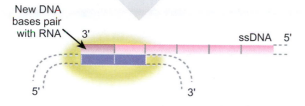

New DNA
bases pair
with RNA

■ **FIGURE 10.24**

***Telomerase Replaces
Repeats at the Ends of
Chromosomes***

Telomerase RNA recognizes the
tandem repeats at the end of linear
DNA. The RNA of telomerase
sticks out beyond the chromosome
ends and serves as a template for
addition of new DNA repeats that
will repair the segment lost during
the last DNA replication.

protozoans *Paramecium* and *Tetrahymena*, TTGGGG, differs by only one base. Many insects have the five-base repeat, TTAGG, whereas the flowering plant *Arabidopsis* has a seven-base repeat, TTTAGGG. However, recent data indicate that this is not typical of all plants; indeed, several monocotyledonous plants have the same TTAGGG repeat as vertebrates. Among the fungi, *Aspergillus nidulans* has TTAGGG, whereas its close relative *Aspergillus oryzae* has a double-length repeat— TTAGGGTCAACA. One strange exception to this general pattern is the fruit fly, *Drosophila*, which has telomeres consisting of tandem sequences generated by successive transposition of two retrotransposons (HeT-A and TART) instead of being synthesized by telomerase.

Box 10.03 Protein Primers for Replicating DNA Ends

One solution to the problem of initiation of replication in linear DNA is to use a **protein primer** at the ends. We normally think of DNA polymerase as only being able to extend nucleic acid chains. However, strictly speaking, DNA polymerase can add nucleotides only to a free hydroxyl group. Although this free hydroxyl group is normally furnished by DNA itself, or by an RNA primer, some DNA polymerases *can* add nucleotides to a free hydroxyl group on specific proteins. This solution is used by several viruses and for the linear plasmids and chromosomes of *Streptomyces* (Fig. 10.25).

Continued

protein primer   Protein used instead of RNA as a primer for DNA synthesis in some bacteria and viruses

Box 10.03 Continued

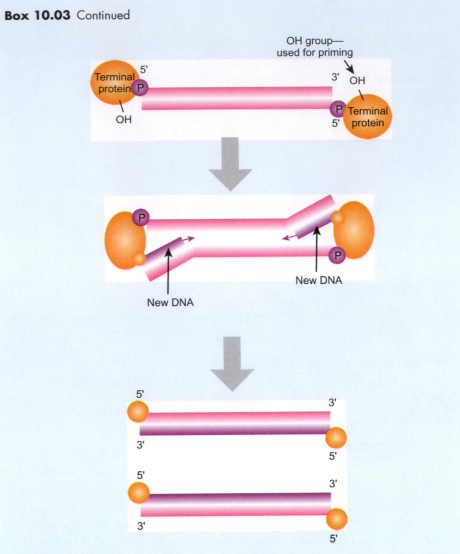

FIGURE 10.25
Protein Primers for the Ends of Linear DNA

The terminal proteins of some viruses and plasmids bind to the 5′ end of linear DNA. These proteins have special hydroxyl groups (–OH) that allow priming of DNA synthesis. The result is complete replication of the linear DNA, without "end-shortening."

12.1. Eukaryotic Chromosomes Have Multiple Origins

Eukaryotic chromosomes are often very long and have numerous replication origins scattered along each chromosome. Replication is bi-directional as in bacteria. A pair of replication forks starts at each origin of replication, and the two forks then move in opposite directions (Fig. 10.26). The bulges where the DNA is in the process of replication are often called **replication bubbles**.

A vast number of replication origins function simultaneously during eukaryotic DNA replication. For example, there are estimated to be between 10,000 and 100,000 replication origins in a dividing human somatic cell. This creates major problems in synchronization. Synthesis at each origin must be coordinated to make sure that each chromosome is completely replicated. Conversely, each origin must initiate once and

> Eukaryotic chromosomes are much longer than bacterial ones and have multiple replication origins.

replication bubble (replication eye) Bulge where DNA is in the process of replication

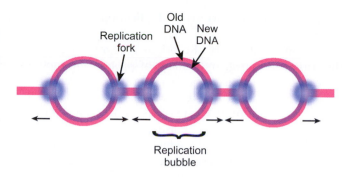

Old
DNA
Replication New
fork DNA

Replication
bubble

■ **FIGURE 10.26**
Eukaryotic Chromosome
Replication Bubbles

Numerous openings in the DNA, or replication bubbles, occur at the sites of replication in eukaryotic chromosomes. The longer replication continues, the larger the bubbles. The bubbles eventually merge together, which separates the newly replicated DNA molecules (not shown).

Kumagai A, Shevchenko A, Shevchenko A, Dunphy WG (2010) Treslin collaborates with TopBP1 in triggering the initiation of DNA replication. Cell 140: 349–359.

Just like prokaryotes, eukaryotes have a specific order of proteins that load onto the origin of replication to control replication initiation. The eukaryotes have large numbers of proteins that are coordinately regulated to drive the cell through synthesis and then the completion of cell division by mitosis. The cell cycle in eukaryotes consists of a rest period called interphase, alternating with mitosis. Mitosis consists of prophase where the chromosomes condense and attach to the spindle apparatus, metaphase where the chromosomes align at the center of the cell, and finally anaphase and telophase where the chromosomes migrate to the two sides of the parental cell and form two nuclei (see below). The synthesis of DNA occurs during the interphase stage, which actually has a rest period, G1, followed by DNA synthesis, followed by a rest period, G2. Synthesis cannot occur at any other point of the eukaryotic cell cycle.

Regulation of DNA synthesis is due to the accumulation and degradation of proteins called **cyclins**. In brief, the entry into synthesis occurs after G1 and is due to G1-CDK (cyclin-dependent kinase) activation. Activated G1-CDK then activates S-phase specific CDK (S-CDK), which starts the assembly of proteins at the origins of replication. (Note: Protein activations occur by transferring phosphate groups from one protein to the next. Phosphorylated proteins change their shape to open new binding sites for substrates, or in other cases release bound inhibitors.) In yeast, S-CDK transfers a phosphate to Sld2 and Sld3. These two phosphorylated proteins bind to Dbp11, which acts as a scaffolding protein that holds the replication origin proteins in position. The key stage of DNA synthesis initiation occurs next, where cdc45 associates with the origin of replication to form the pre-loading complex

FOCUS ON
RELEVANT RESEARCH

G1 ends

Ⓟ
G1-CDK ——————→ G1-CDK ~ Ⓟ

Sld - 2
Sld - 3

Sld3 ~ Ⓟ   Sld2 ~ Ⓟ

Dbp11

cdc45   cdc45   cdc45

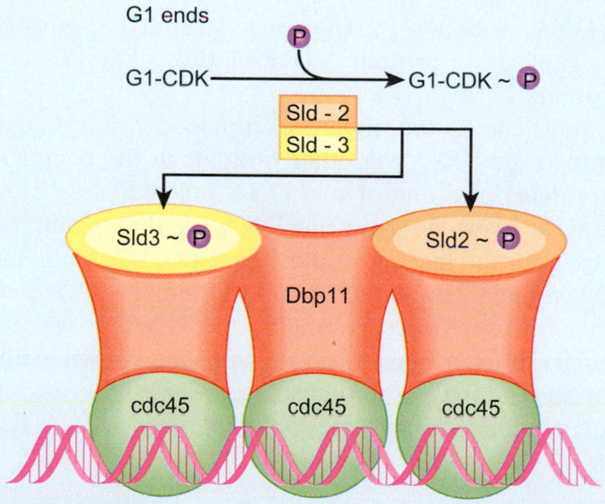

FIGURE 10.27
Yeast Cell Initiation of Replication

At the origin of replication in yeast cells, Dbp11 acts as a scaffolding protein that is activated by binding of phosphorylated Sld2 and Sld3. This complex can start assembly of the replication enzymes, which begins by binding of cdc45.

(pre-LC), and along with a large number of different proteins, initiates unwinding of the DNA helix (Fig. 10.27).

Although the identity and function of all these proteins is known in yeast, in vertebrates the process is still not understood clearly. The vertebrate homolog of the scaffolding protein, Dbp11, is called TopBP1, and it performs the same function; that is, pulling cdc45 into the origin of replication. What proteins activate TopBP1 is still unknown. In this paper, the authors have isolated a potential activator of TopBP1, called Treslin, from *Xenopus* egg extracts. This protein was found attached to TopBP1 in the frog egg nuclei, and when the eggs are depleted of Treslin, DNA replication was reduced to only 20% of the control amount. In addition, the egg nuclei that were depleted of Treslin no longer loaded cdc45 into the pre-LC. Further experiments in the article show that Treslin is phosphorylated before it binds to TopBP1, and without the phosphate group, Treslin cannot bind TopBP1. Taken together, these experiments suggest that Treslin is phosphorylated, and this form attaches to TopBP1, which then can recruit cdc45 into the origin of replication and start the unwinding of eukaryotic DNA for replication.

Cyclins   Proteins that regulate the stages of the eukaryotic cell cycle by controlling the addition of phosphate to multiple other proteins once only during each replication cycle in order to avoid duplication of DNA segments that have already been replicated.

This process is best understood in yeast where the origin recognition complex (ORC) binds to each replication origin and triggers a chain of protein interactions. First, Cdc6, Cdt1 (also known as replication licensing factor), and ORC recruit MCM complex to form the **pre-replicative complex (pre-RC)**, which only forms in the beginning of G1. This ensures that replication only occurs one time in each cell cycle. Pre-RC is then activated by S-phase cyclin-dependent kinase (S-CDK), which in turn activates Sld2 and Sld3. These two associate with Dpb11, which in turn brings in cdc45 and DNA polymerase ε. This is called the **pre-loading complex** (pre-LC). The MCM is the helicase that initiates unwinding of the helix at the origin and triggers the beginning of DNA elongation.

12.2. Synthesis of Eukaryotic DNA

The same general principles of bacterial replication apply to eukaryotic replication, although there are differences in detail from the bacterial scheme. In eukaryotes, semi-conservative replication occurs, and as was seen in bacteria, one new strand is made continuously and the other in fragments. Both strands are made simultaneously by a replisome consisting of a helicase plus DNA polymerase holoenzyme. Within the holoenzyme, the sliding clamp loader holds each core enzyme and two sliding clamps. In addition, an RNA primer is required to initiate each new DNA strand.

In animal cells, the double helix first needs to be separated into two single strands at the origin of replication. The pre-loading complex (see Fig.10.27) recruits the helicase, **minichromosome maintenance (MCM)**, which then moves along the helix in a 3′ to 5′ direction, separating the two strands of DNA (Fig. 10.28). MCM is similar to the bacterial helicase DnaB, since both molecules consist of multiple subunits. Three DNA polymerases (α, δ, and ε) are involved in eukaryotic chromosome replication. **DNA polymerase α (Polα-primase)** and two associated small proteins are responsible for initiation of new strands. First, the complex makes an RNA primer. Then, polymerase α elongates the RNA primer by making a short piece of DNA about 20 nucleotides long (the **initiator DNA**, or "iDNA"). The single-stranded regions of the replication fork are covered by replication protein A (RPA), the eukaryotic equivalent of single-strand binding protein.

The bacterial functional equivalent to the sliding clamp loader is called **replication factor C (RFC)** and binds to the iDNA via other proteins in the complex and loads a sliding clamp (**PCNA protein**) plus one of two DNA polymerases onto each strand of DNA. **DNA polymerase ε** is loaded onto the DNA for the leading strand, whereas **DNA polymerase δ** is used for the lagging strand. These two DNA polymerase assemblies elongate the two new strands. The sliding clamp of animal cells is a trimer (not a dimer as in bacteria) that forms a ring surrounding the DNA. It was named PCNA, for proliferating cell nuclear antigen, before its role was fully known. Although two different polymerases are usually found in the replisome, at least in yeast, replication can take place without polymerase ε—apparently polymerase δ can substitute if

DNA polymerase α   Enzyme that makes short segment of initiator DNA during replication of animal chromosomes
DNA polymerase ε   Enzyme that makes most of the leading strand DNA when animal chromosomes are replicated
DNA polymerase δ   Enzyme that makes most of the lagging strand DNA when animal chromosomes are replicated
initiator DNA (iDNA)   Short segment of DNA made just after the RNA primer during replication of animal chromosomes
minichromosome maintenance (MCM)   Helicase that separates the two strands of DNA during replication of eukaryotic DNA
replication factor C (RFC)   Eukaryotic protein that binds to initiator DNA and loads DNA polymerase δ plus its sliding clamp onto the DNA
PCNA protein   The sliding clamp for the DNA polymerase of eukaryotic cells (PCNA = proliferating cell nuclear antigen)
pre-replicative complex (pre-RC)   A complex of enzymes (ORC, Cdc6, Cdt1, and MCM) that assemble at the origin of replication during eukaryotic DNA replication
pre-loading complex (pre-LC)   Complex of proteins that forms prior to binding to the origin of replication, but is essential to promoting the correct association of the pre-RC

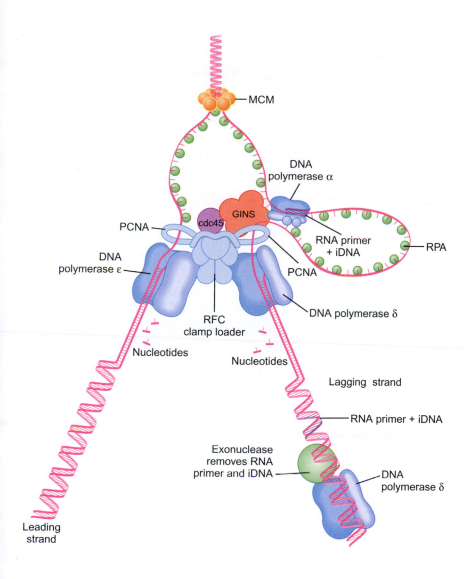

■**FIGURE 10.28**
The Eukaryotic Replisome

The eukaryotic replication fork is created by MCM unwinding the two strands of DNA, and RPA coating these to prevent reannealing. Next, DNA polymerase α, also called primase, including its two smaller subunits makes an RNA primer followed by iDNA. Next, RFC clamp-loader complex assembles the sliding clamp (PCNA) around the single-stranded DNA and recruits DNA polymerase. DNA polymerase ε loads onto the leading strand, and DNA polymerase δ loads onto the lagging strand to synthesize new DNA from the 3′ end of iDNA. On the lagging strand, sections of RNA primer followed by iDNA are removed by an exonuclease and then refilled with DNA by DNA polymerase δ.

necessary. Eukaryotic replisomes also contain other regulatory proteins such as Cdc45 and a complex of four proteins called GINS for the four protein names, Go, Ichi, Nii, and San. The function of these within the replisome is still under investigation.

Linking of the Okazaki fragments differs significantly between animal and bacterial cells. In animals, there is no equivalent of the dual function DNA polymerase I of bacteria. The RNA primers are removed by the exonucleases Fen1 and/or Dna2, and the gaps are filled by the DNA polymerase δ that is already working on the lagging strand. As in bacteria, the nicks are sealed by DNA ligase.

Eukaryotic replisomes are more complex than bacterial replisomes. For example, eukaryotic replisomes have more regulatory proteins (cdc45 and GINS), and the replisome has different forms of DNA polymerase (α, ε, and δ) to synthesize the new DNA.

12.3. Histones Are Remodeled and Replaced During Replication

Histone remodeling during replication is one of the areas of intense research on replication in eukaryotes. Eukaryotic DNA is tightly wound around histones to condense the genome and maintain the chromosomal integrity. In addition, histones are critical to gene expression and therefore many modifications of these proteins are essential to proper cellular function. Tightly-condensed histone and DNA complexes form heterochromatin and certain post-translational modifications mark the promoter region of expressed genes (see Ch. 17). Typical animations of replication depict the DNA as a free-floating double helix, but in reality, the strands are wrapped around the histones tightly which inhibits the binding of other proteins. Histones, therefore, must be moved or removed from the DNA in order for the replicative machines to open up the helix and make the new strands. After the region is

Box 10.04 DNA Polymerase Families

DNA polymerases all perform the same basic function, catalyzing the connection of a 5'-phosphate group from the incoming nucleotide to the 3'-OH group on the ribose or deoxyribose of the upstream nucleotide. Yet, there are so many different variations of structure and other functions that DNA polymerases actually fall into seven different families:

TABLE 10.02	DNA Polymerases Are Grouped into Different Families	
Family	Notable Family Member	Function
A	Bacterial DNA Polymerase I	removes RNA primers from lagging strand
B	Eukaryotic DNA polymerase α, δ, and ϵ	major polymerases used during replication
C	Bacterial DNA Polymerase III	major polymerase used in replication
D	PolD polymerase	replication in Archaea
X	Eukaryotic DNA polymerase β	removes and repairs altered or defective nucleotides
Y	Bacterial DNA polymerase V	replicates damaged DNA in the SOS repair system (see Ch. 23)
RT	Reverse transcriptase and telomerase	make DNA from RNA templates

Histones are removed and replaced during replication. In addition, the exact methylation and acetylation patterns are maintained during each replication, even though there are new histones created to complement the old histones.

replicated, then the histones must be replaced, but since the amount of DNA is double, new histones are also made and incorporated into the two genomes. The process is dynamic, but the proper replacement of the histones is essential to gene expression and nuclear structure.

FOCUS ON RELEVANT RESEARCH

Falbo K, Shen X. (2010) The tango of histone marks and chaperones at replication fork. Molecular Cell 37: 595–596.

One of the most important events during eukaryotic replication is the removal and replacement of histones. These proteins are essential to condensing the genome and also contain various post-translational modifications that prevent or allow access of transcription factors to genes (see Ch. 17 for a complete discussion). Histone PTMs (post-translational modifiers) and histone chaperones modulate the removal and replacement of histones after replication. Histone PTMs faithfully replicate the chromatin structure from parental cell to daughter cell by adding acetyl groups onto key amino acids of the histone tails. Histone chaperones physically interact with the MCM helicase at the replication fork and use either the newly-synthesized or the displaced parental histone proteins to reassemble the chromatin. The picture of what occurs during replication of eukaryotic DNA is slowly becoming clearer and further research into the specific PTMs and chaperones will provide a better idea of the complexity of this system. This short preview is a quick summary of two other primary research articles that address the issue of histone removal and replacement during replication. Jasencaova et al. (2010) analyze the post-translation modifications of histone H3 and H4 during replication and then determine what happens when replication is halted. Their results suggest that replication stress can result in epigenetic changes. In the other paper, Burgess et al. (2010) find that cells without two specific lysine acetyltransferase (Gcn5 and Rtt109) are more sensitive to DNA-damaging agents and have altered nucleosome reassembly after replication. Interestingly, Gcn5 is also important for transcription. The continued research in this field of molecular biology is bound to uncover a well-orchestrated and tightly-regulated series of molecular interactions that control the process of replication and chromatin reassembly.

13. Cell Division in Higher Organisms

The eukaryotic cells of higher organisms face further problems during cell division. Not only do they have multiple chromosomes, but these are inside the nucleus, separated from the rest of the cell by the nuclear envelope. Consequently, an elaborate process is needed to disassemble the nucleus, replicate the chromosomes, partition them among the daughter cells, and finally, reassemble the nuclear envelope. This process is mitosis and involves several operations:

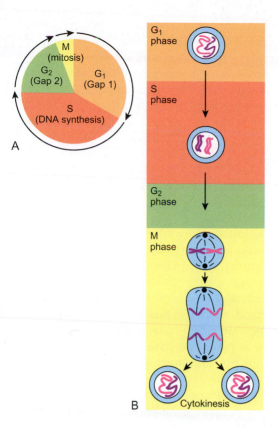

■**FIGURE 10.29**
The Eukaryotic Cell Cycle

DNA replication occurs during the S-phase of the cell cycle, but the chromosomes are actually separated later, during mitosis or M-phase. The S- and M-phases are separated by G1 and G2.

1. Disassembly of the nuclear membrane of the mother cell
2. Alignment of the chromosomes at the center axis
3. Partition of the chromosomes
4. Reassembly of nuclear membranes around each of the two sets of chromosomes
5. Final division of the mother cell, or **cytokinesis**

The presence of a nucleus complicates cell division in eukaryotes. The result is a complex cell cycle that includes dissolution and reassembly of the nucleus as well as duplication of the chromosomes.

Mitosis itself is only one of several phases of the eukaryotic **cell cycle** (Fig. 10.29). The process of DNA replication described above takes place in the synthetic, or **S-phase**, of the cell cycle. The S-phase is separated from the actual physical process of cell division (**mitosis** or M-phase) by two gap phases, or **G-phases**, in which nothing much appears to happen (except the normal processes of cellular activity and metabolism). Together, G1, S, and G2 constitute the **interphase**.

Key Concepts

- DNA replication and cell division in multicellular organisms simply creates more cells, whereas the same process in single-celled organisms creates new organisms. When DNA replication and cell division occur during mitosis, the two daughter cells are genetically identical to the parent. In meiosis, however, genetic recombination and independent assortment of chromosomes occur after DNA replication, which creates new genetic combinations in the gametes. Then two gametes from two different individuals merge to form a new organism.

cell cycle   Series of stages that a cell goes through from one cell division to the next
cytokinesis   Cell division
G1 phase   Stage of the eukaryotic cell cycle following cell division; cell growth occurs here
G2 phase   Stage of the eukaryotic cell cycle between DNA synthesis and mitosis: preparation for division
interphase   Part of the eukaryotic cell cycle between two cell divisions and consisting of G1, S, and G2 phases
mitosis   Division of eukaryotic cell into two daughter cells with identical sets of chromosomes
S-phase   Stage in the eukaryotic cell cycle in which chromosomes are duplicated

- In DNA replication, two strands of DNA are pulled apart by a complex of enzymes called a replisome to create a replication fork or zone of single-stranded template DNA used for synthesis of a second strand. Replication forks can travel in either direction along a double-stranded template.

- To create a replication fork, DNA supercoiling must be relaxed by DNA gyrase and DNA topoisomerase and then the two strands are pulled apart by DNA helicase. The strands are prevented from re-attaching by single strand binding protein (SSB).

- DNA polymerase synthesizes new DNA, but has two limitations: it cannot start synthesis without the presence of a pre-existing 3′-OH and it only synthesizes DNA in a 5′ to 3′ direction, meaning it cannot synthesize the lagging strand in one piece.

- Primase resolves the problem of strand initiation by synthesizing a short RNA primer at the origin in the leading strand and multiple primers along the lagging strand for DNA polymerase. Fortunately, primase does not need a pre-existing 3′-OH to create the RNA primers.

- DNA polymerase adds the incoming nucleotide by linking its 5′-phosphate group to the 3′-OH group of the preexisting strand, which is called the 5′ to 3′ direction.

- Nucleotide precursors have three phosphate groups attached to the 5′-carbon. The two outer phosphates are released as pyrophosphate by DNA polymerase and the innermost phosphate links to the previous nucleotide.

- DNA polymerase III is a holoenzyme, which has two core enzymes (Pol III), each consisting of three subunits (α, ϵ, and θ), a sliding clamp that has two beta subunits, and a clamp-loading complex that has multiple subunits (δ, τ, γ, ψ, and χ). The ϵ-subunit (DnaQ) of the core enzyme has proofreading ability and 3′-exonuclease activity to ensure that replication is accurate from generation to generation.

- The lagging strand synthesis is done discontinuously. Okazaki fragments are initiated by creation of a new RNA primer by the primosome. To restart DNA synthesis, the DNA clamp loader releases the lagging strand from the sliding clamp, and then reattaches the clamp at the new RNA primer. Then DNA polymerase III can synthesize the segment of DNA.

- DNA polymerase I recognizes a "nick" or break in the phosphate backbone and then removes each RNA primer and fills the gaps with DNA. DNA ligase then covalently links the phosphate backbone.

- In bacteria, there is one origin of replication that has three 13-base pair repeats and four 9-base pair repeats. A cluster of 30 DnaA proteins open the helix at the 13-base pair repeats by first binding to the 9-base pair repeats and bending the DNA. DnaC proteins then help DNA helicase (DnaB) load onto the origin correctly.

- *E. coli* mark the parental strand of DNA by adding methyl groups onto GATC by Dam methylase. The newly-synthesized strand of DNA is not methylated immediately, so that the mismatch repair enzymes can double-check for mistakes. This complex of enzymes removes the new incorrect nucleotide and replaces it with the correct one.

- Hemi-methylation of the origin of replication helps control how often bacterial chromosomes are replicated.

- Replication finishes at the terminus, which has several termination sites and the Tus protein. These block the movement of DNA helicase.

- Circular chromosomes of bacteria can tangle during replication or even become covalently joined due to odd numbers of crossovers. Topoisomerase IV decatenates tangled circles and resolvase separates circles with odd numbers of crossovers.

- Bacterial cell division occurs separately from DNA replication and involves the movement of the two daughter chromosomes to opposite sides of the parent cell, the building of a septum, and finally, separation into two cells. In *E. coli*, if growth conditions are good, a new cycle of DNA replication can be initiated before the cell has even divided.

- Replicons are pieces of DNA that have the components necessary for DNA replication (origin of replication) and are circular or have their ends protected from exonuclease degradation.
- Eukaryotic chromosomes have special structures such as telomeres, centromeres, and histones that create unusual problems for DNA polymerase. In addition, the very long pieces require multiple origins of replication.
- Telomeres are shortened in each round of replication because there is no 3'-OH to prime DNA polymerase. To overcome this problem, an enzyme called telomerase uses an RNA template to lengthen telomeres.
- Some of the similarities between bacterial and eukaryotic DNA replication are: both are semi-conservative; both have leading and lagging strand synthesis; both use RNA primers to initiate synthesis; and both use sliding clamps and sliding clamp-loader complexes.
- Eukaryotic DNA replication is different than bacterial replication in the following ways: primase consisting of DNA polymerase α and two smaller proteins creates an RNA primer and initiator DNA; two different DNA polymerases synthesize the leading and lagging strands, which are DNA polymerase ϵ and DNA polymerase δ, respectively; and RNA primers are removed by an exonuclease and then filled by DNA polymerase δ, not a dual function protein like DNA polymerase I.
- Histones must be removed in order to replicate eukaryotic DNA and then when the two new strands are complete, the DNA must be rewound around a combination of recycled and new histones. The post-translational modifications to histones are also faithfully reconstructed by methyltransferases, acetyltransferases, and other histone modification enzymes.
- Eukaryotes control the timing of replication and cell division with a cell cycle. DNA replication occurs in the S-phase, then there is a gap (G1), followed by cell division or mitosis (M), and finally there is another gap (G2) before DNA replication occurs again. Interphase constitutes G1, S, and G2.

Review Questions

1. What is asexual or vegetative reproduction?
2. What is the difference between horizontal gene transfer and vertical gene transfer?
3. What is a replication fork?
4. Replication is semi-conservative. Explain.
5. What is theta-replication?
6. Name the two type II topoisomerases involved in DNA replication and explain their functions.
7. How do quinolone antibiotics work?
8. What are the functions of DNA helicase and SSB protein during replication?
9. What is the major enzyme that replicates prokaryotic DNA and how does it work?
10. What is the function of primase in DNA replication?
11. How does polymerization of nucleotides take place during DNA replication?
12. How are the precursors for DNA synthesis made?
13. What are the two functions of the THF cofactor during DNA replication?
14. How does the antibiotic trimethoprim inhibit DNA replication?
15. How does methotrexate function as an anti-tumor agent?
16. Why is the sulfonamide class of antibiotic harmless to animals?
17. What components make up the core enzyme of DNA Pol III and what are their functions?
18. Why is the lagging strand made as short fragments?
19. How are Okazaki fragments joined to form a continuous strand?
20. What are the functions of DNA Pol I?
21. What is the origin of replication? What are the three major steps involved in the initiation of replication?
22. What is the function of Dam methylase?
23. What is hemi-methylated DNA?
24. How is replication terminated in a bacterial chromosome?
25. How are catenated chromosomes unlinked?

26.   What are the four major steps of bacterial cell division?
27.   What is a replicon? Give three examples.
28.   What is the fate of linear chromosomes in most bacteria and how are they protected in some bacteria?
29.   What are telomeres and why are they important?
30.   What prevents the shortening of linear DNA after each round of replication in eukaryotes?
31.   What is telomerase? How does it work?
32.   What is the role of pre-replicative complex (pre-RC) in eukaryotic DNA replication?
33.   What are the main similarities between bacterial and eukaryotic DNA replication?
34.   List the major differences between bacterial and eukaryotic DNA replication.
35.   What are the four phases of the eukaryotic cell cycle?

Conceptual Questions

1.   One of the key experiments that elucidated the mechanism of DNA replication was by Matthew Meselson and Franklin Stahl in 1958. They used *E. coli* to demonstrate that replication occurred semi-conservatively. They grew *E. coli* in nutrient broth that contained ^{15}N, which is a heavy isotope of nitrogen, for many generations (cell divisions) and then transferred the cells into broth containing normal nitrogen (^{14}N) for one generation. They then isolated the DNA from these bacteria and found that ^{15}N had incorporated into the DNA (remember each of the bases have nitrogens in their structure). They purified the DNA using CsCl equilibrium density gradient centrifugation, which can separate different densities of DNA. The results are shown below:

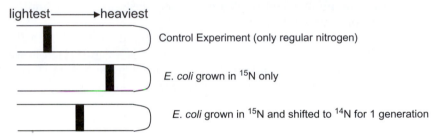

lightest ⟶ heaviest

Control Experiment (only regular nitrogen)

E. coli grown in ^{15}N only

E. coli grown in ^{15}N and shifted to ^{14}N for 1 generation

   a.   Explain these results using your knowledge of replication. Predict what the bands would look like if the *E. coli* were left in ^{14}N for two generations. What would the bands look like in the third generation in ^{14}N?

2.   *Drosophila melanogaster*, the common fruit fly, has four chromosomes. The entire genome has been sequenced and chromosome 4 is a small chromosome that contains 1,351,857 nucleotides of DNA. If each replication fork travels at 2000 nucleotides per minute, how many replication forks would be needed to copy the entire chromosome 4 in 20 minutes?

3.   Label each of the numbered ends of the diagram with either 5' or 3', and indicate which strand is a leading strand and which strand is a lagging strand for the replication fork.

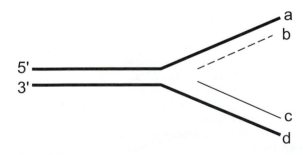

4.   A researcher has synchronized two different *E. coli* cultures so that all of the cells enter cell division simultaneously. One culture is wild-type *E. coli* cells, and the other culture is a mutant *E. coli* that does not have the gene for Dam methylase. He isolates the DNA from both wild-type and mutant *E. coli* 20 minutes after replication is complete. What is the methylation state for each strain of *E. coli*? What would be the methylation state for each bacterial strain if you isolated the DNA immediately after replication termination?

Transcription of Genes

In order to use the genetic information encoded in DNA, genes must be expressed. The first step is to make an RNA copy of the DNA sequence. The term **transcription** refers to the synthesis of an RNA copy of information encoded on DNA. In some cases, such as ribosomal RNA (rRNA) and transfer RNA (tRNA), the RNA is the final product of gene expression. More often, the RNA serves as an intermediate carrier of genetic information—messenger RNA (mRNA). This is then used to make proteins, the ultimate gene products, by the process of translation, as described in Chapter 13.

Individual genes, or small clusters of genes, are transcribed separately. Thus only short regions of the DNA are transcribed at a time. This allows different genes to be expressed under different conditions and allows the organism to adapt to its surroundings. Consequently, it is necessary for a cell to know where on the DNA to start transcription, where to stop transcription, and when to turn genes on and off. Regulating gene expression may be extremely complex, especially in higher organisms, and several later chapters (Ch. 16–19) consider gene regulation in more detail. Here, we limit ourselves to discussing the basic role of regulatory proteins in turning genes on or off.

---

transcription Process by which information from DNA is converted into its RNA equivalent

---

FIGURE 11.01
Transcription in Its Simplest Form

The two strands of the DNA to be transcribed are separated locally. The antisense strand (at top) serves as a template for building a new RNA molecule.

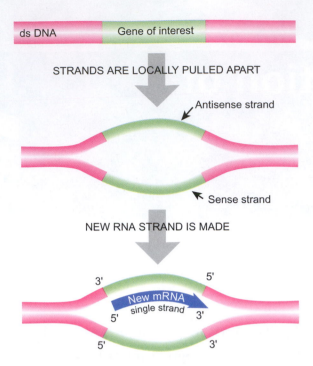

1. Genes Are Expressed by Making RNA

DNA "merely" stores genetic information. Putting the information to use requires RNA and (usually) protein.

For a cell to operate, its genes must be expressed. The word "expressed" means that the gene products, whether proteins or RNA molecules, must be made. The DNA molecule that carries the original copy of the genetic information is used to store genetic information but is not used as a direct source of instructions to run the cell. Instead, working copies of the genes, made of RNA, are used. The transfer of information from DNA to RNA is known as transcription, and RNA molecules are therefore sometimes referred to as transcripts. Genes may be subdivided into two major groups: those whose final product is an RNA molecule (e.g., tRNA, rRNA, assorted regulatory RNAs—see below) and those whose final product is protein. In the latter case, the RNA transcript acts as an intermediary and a further step converts the information carried by the RNA to protein as discussed in Chapter 13. The type of RNA molecule that carries genetic information encoding a protein is known as **messenger RNA**, or mRNA. Since the great majority of genes encode proteins, we will deal with these first.

Messenger RNA (mRNA) carries the information for making proteins from the genes to the cytoplasm.

For a gene to be transcribed, the DNA, which is double-stranded, must first be pulled apart temporarily, as shown in Figure 11.01. Then, RNA is made by **RNA polymerase**. This enzyme binds to the DNA at the start of a gene and opens the double helix. Finally, it manufactures an RNA molecule.

The DNA double helix must be opened up for RNA polymerase to read the template strand and make RNA.

The sequence of the RNA message is complementary to the **antisense strand** of the DNA from which it is synthesized. Apart from the replacement of thymine in DNA with uracil in RNA, this means that the sequence of the new RNA molecule is identical to the sequence of the **sense strand** of DNA; that is, the strand not actually used as a template during transcription. Note that RNA, like DNA, is synthesized in the 5′ to 3′ direction (Fig. 11.02). Other names for the antisense strand are the *non-coding* or **template strand**; other names for the sense strand are *non-template* or **coding strand**. Only one of the strands of DNA is copied in any given transcribed

---

antisense strand   Strand of DNA used as a guide for synthesizing a new strand by complementary base pairing
coding strand   The strand of DNA equivalent in sequence to the mRNA (same as plus strand)
messenger RNA (mRNA)   The molecule that carries genetic information from the genes to the rest of the cell
RNA polymerase I   Eukaryotic RNA polymerase that transcribes the genes for the large ribosomal RNAs
sense strand   The strand of DNA equivalent in sequence to the mRNA (same as plus strand)
template strand   Strand of DNA used as a guide for synthesizing a new strand by complementary base pairing

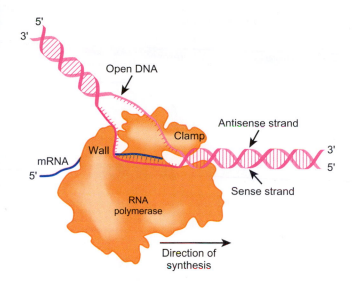

► **FIGURE 11.02**
Naming the Basic Components Involved in Transcription

The DNA is shown in its double helical form. After local separation of the strands, the new RNA is synthesized so that it base pairs with one of the DNA strands—the antisense or template strand. The other DNA strand is inactive and is called the sense or coding strand. The enzyme RNA polymerase synthesizes single-stranded RNA in the 5' to 3' direction. The sequence of bases in the RNA is the same as in the sense strand of DNA and complementary to the antisense strand of DNA (except that uracil substitutes for thymine).

region. (But note that the two different strands of the DNA may each be used as templates in different regions of the chromosome.)

1.1. Short Segments of the Chromosome Are Turned into Messages

Although a chromosome carries hundreds or thousands of genes, only a fraction of these are in use at any given time. In a typical bacterial cell, about 1000 genes, or about 25% of the total, are expressed under any particular set of growth conditions. Humans have around 22,000 protein coding genes whose expression varies under different conditions and in different tissues. Some genes are required for the fundamental operations of the cell and are therefore expressed under most conditions. These are known as **constitutive** or **housekeeping genes**. Other genes vary in expression in response to changes in the environment. During cell growth and metabolism, each gene or small group of related genes is used to generate a separate RNA copy when, and if, it is needed. Consequently, each cell contains many different RNA molecules, each carrying the information from a short stretch of DNA.

In the cells of more complex organisms, which have many more genes than do bacteria, the proportion of genes in use in a particular cell at a particular time is much smaller. Different cells of multicellular organisms express different selections of genes depending on their specialized roles in the organism. In addition, gene expression varies with the stage of development. To create a functional adult, embryonic genes are often expressed only at certain times and in a highly-orchestrated order. Thus, the control of gene expression is much more complex in more complex organisms, although the basic principles are the same.

Each mRNA carries information from only a short stretch of the DNA.

1.2. Terminology: Cistrons, Coding Sequences, and Open Reading Frames

In early bacterial genetics a **cistron** was defined as a **structural gene**; in other words, a coding sequence or segment of DNA encoding a polypeptide. It was defined originally as a genetic unit by complementation using the *cis/trans* test. Nowadays, the terms cistron and structural gene also include DNA sequences with a non-coding RNA as an end product (e.g., rRNA, tRNA, snRNA, etc.). An **open reading frame** (**ORF**) is

cistron   Segment of DNA (or RNA) that encodes a single polypeptide chain
constitutive gene   Gene that is expressed all the time
housekeeping genes   Genes that are switched on all the time because they are needed for essential life functions
open reading frame (ORF)   Sequence of bases (either in DNA or RNA) that can be translated (at least in theory) to give a protein
structural gene   Sequence of DNA (or RNA) that codes for a protein or for an untranslated RNA molecule

FIGURE 11.03 ─────
Monocistronic Versus Polycistronic mRNA

The typical situation in eukaryotes is to have one structural gene produce monocistronic RNA and this, in turn, is translated into a single protein. In bacteria, it is common to see several structural genes transcribed under the control of a single promoter. The RNA produced is polycistronic and yields several separate proteins after translation.

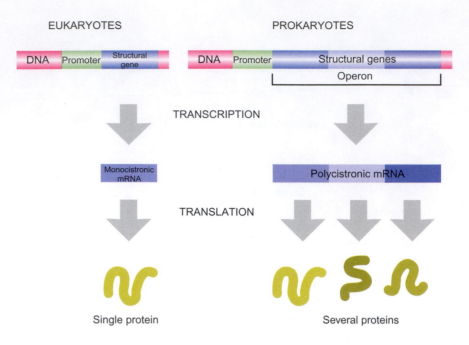

In eukaryotes, each mRNA carries only a single gene. In prokaryotes, several genes may be carried on the same mRNA, which is therefore polycistronic.

any sequence of bases (in DNA or RNA) that could, in theory, encode a protein. The ORF is "open" in the sense that it does not contain any stop codons that would interrupt its translation into a polypeptide chain (although, of course, every ORF ends in a stop codon). (A cistron that encodes a protein must also be an ORF, whereas a cistron that encodes a nontranslated RNA is not an ORF.)

In eukaryotes, the majority of genes are transcribed to give a separate mRNA, and each mRNA molecule therefore encodes the information for only a single protein and is known as **monocistronic mRNA** (Fig. 11.03). In bacteria, clusters of related genes, known as **operons**, are often found next to each other on the chromosome and are transcribed together to give a single mRNA, which is therefore called **polycistronic mRNA**. Thus, a single bacterial mRNA molecule may encode several proteins, usually with related functions, such as the enzymes that oversee the successive steps in a metabolic pathway.

2. How Is the Beginning of a Gene Recognized?

Transcription will first be described in bacteria because it is simpler than in eukaryotes. The principles of transcription are similar in higher organisms, but the details are more complicated, as will be shown below. The major differences between prokaryotes and eukaryotes occur in the initiation and regulation of transcription, rather than in the actual synthesis of RNA. In front of each gene is a region of regulatory DNA that is not transcribed. This contains the **promoter**, the sequence to which RNA polymerase binds (Fig. 11.04), together with other sequences involved in the control of gene expression. This stretch of DNA in front of a gene (i.e., at the 5′ end) is often referred to as the **upstream region**. Note that the first base of the mRNA for a protein-encoding gene is not the first base of the protein coding sequence. Between these two points there is a short stretch known as the **5′-untranslated region**, or 5′-UTR, meaning it will

---

monocistronic mRNA   mRNA carrying the information of a single cistron, which is a coding sequence for only a single protein
operon   A cluster of prokaryotic genes that are transcribed together to give a single mRNA (i.e., polycistronic mRNA)
polycistronic mRNA   mRNA carrying the information of multiple cistrons, which is coding sequences for several proteins
promoter   Region of DNA in front of a gene that binds RNA polymerase and so promotes gene expression
upstream region   Region of DNA in front (i.e., beyond the 5′ end) of a structural gene; its bases are numbered negatively counting backwards from the start of transcription
5′-untranslated region (5′-UTR)   Region of an mRNA between the 5′ end and the translation start site

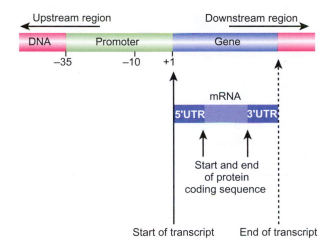

► **FIGURE 11.04**
Upstream and Downstream Regions

Upstream and downstream are two general terms used by molecular biologists. Upstream regions are before the point of reference and downstream are after that point in the linear DNA. The start point of mRNA transcription is a common point of reference. Transcription begins at the end of the promoter, designated +1, and continues until the end of the gene. The resulting mRNA has more information than needed for making a protein. The 5'-UTR and 3'-UTR are regions in mRNA that are not used to make the final protein, but often contain important regulatory elements.

not be translated to form protein. At the far end of the mRNA there is another short region, beyond the end of the protein-coding sequence that is not translated. This is the **3′-untranslated region**, or 3′-UTR.

Bacterial RNA polymerase consists of two major components, the **core enzyme** (itself made of five subunits—α, α, β, β', ω) and the **sigma subunit**. Together, these make up the holoenzyme. The core enzyme is responsible for RNA synthesis, whereas the sigma subunit is largely responsible for recognizing the promoter. The sigma subunit recognizes two special sequences of bases in the promoter region of the coding (non-template) strand of the DNA (Fig. 11.05). These are known as the **−10 sequence** and the **−35 sequence** because they are found by counting backward approximately 10 and 35 bases, respectively, from the first base that is transcribed into mRNA. (Previously, the −10 sequence was known as the **Pribnow box**, after its discoverer. This name is rarely used nowadays.)

The consensus sequence for the −10 sequence is TATAAT and the consensus sequence at −35 is TTGACA. (Consensus sequences are found by comparing many sequences and taking the average—see Ch. 4.) Although a few highly-expressed genes do have the exact consensus sequences in their promoters, the −10 and −35 region sequences are rarely perfect. However, as long as they are wrong by only up to three or four bases, the sigma subunit will still recognize them. *The strength of a promoter depends partly on how closely it matches the ideal consensus sequence.* Strong promoters are highly expressed and are often close to consensus. Promoters further away from the consensus sequence will be expressed only weakly (in the absence of other factors—but see below).

In practice, consensus sequences for regulatory sites on DNA such as promoters will vary from one group of organisms to another. Thus, the −10 and −35 consensus sequences given above are for *E. coli* and related bacteria. Both the consensus sequences and the proteins that recognize them will diverge in more distantly related organisms. This is of practical importance when genes from one organism are expressed in another as a result of biotechnological manipulations. Consequently, it is often helpful to supply consensus promoters (or other regulatory sequences) that work well in the host organism to achieve high expression of a cloned gene from a

Before starting transcription, RNA polymerase binds to the promoter, a recognition sequence in front of the gene.

The sigma subunit of bacterial RNA polymerase recognizes the promoter. The core enzyme makes RNA.

Strong promoters usually have sequences close to consensus.

Promoter sequences vary in different organisms.

---

core enzyme   Bacterial RNA polymerase without the sigma (recognition) subunit
3′-untranslated region (3′-UTR)   Sequence at the 3′ end of mRNA, downstream of the final stop codon, which is not translated into protein
−10 region   Region of bacterial promoter 10 bases back from the start of transcription that is recognized by RNA polymerase
Pribnow box   Another name for the −10 region of the bacterial promoter
sigma subunit   Subunit of bacterial RNA polymerase that recognizes and binds to the promoter sequence
−35 region   Region of bacterial promoter 35 bases back from the start of transcription that is recognized by RNA polymerase

FIGURE 11.05
Sigma Recognizes the −10 and −35 Sequences

A) The sigma protein binds to both the −10 and −35 sequences of the promoter, thereby establishing a constant position with respect to the start of transcription. B) Actual RNA polymerase structure showing the core enzyme subunits, sigma factors, and DNA.

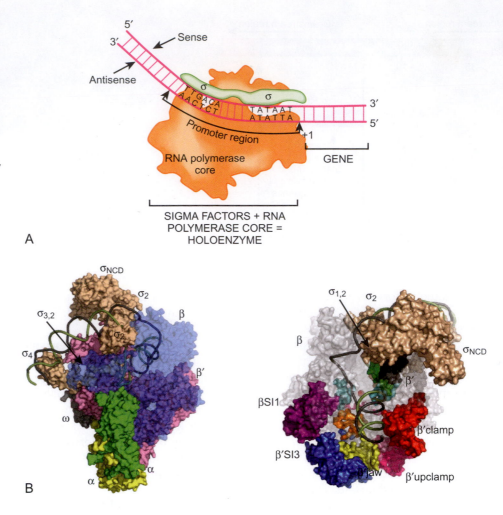

foreign source. The use of "expression vectors" to optimize gene expression during cloning was discussed in more detail in Chapter 7.

3. Manufacturing the Message

Once the sigma subunit has bound to a promoter, the RNA polymerase core enzyme opens up the DNA double helix locally to form the **transcription bubble**. Note that the −10 sequence, TATAAT, consists of AT base pairs. Because these are weaker than GC base pairs, this assists in the melting of the DNA into single strands. After the DNA helix has been opened, a single strand of RNA is generated using one of the DNA strands as a template for matching up the bases. Once the RNA polymerase has bound to the DNA and initiated a new strand of RNA, the sigma subunit is no longer needed and often (though not always) detaches from the DNA, leaving behind the core enzyme. The RNA polymerase actually remains at the promoter until the new strand is eight or nine bases long. At this point, the sigma subunit leaves, and the core enzyme travels along the DNA, elongating the mRNA (Fig. 11.06).

The first transcribed base of the mRNA is normally an A. This special A is usually flanked by two pyrimidines, most often giving the sequence CAT. Sometimes the first transcribed base is a G, but almost never a pyrimidine. Synthesis of mRNA is from 5′

RNA polymerase opens up the DNA to form the transcription bubble.

transcription bubble   Region where DNA double helix is temporarily opened up so allowing transcription to occur

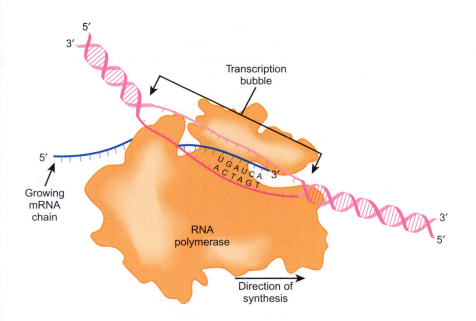

FIGURE 11.06
Elongation of the mRNA

The beginning of RNA synthesis is shown. The DNA strands have separated at the transcription bubble. Synthesis of six bases of RNA complementary to those of the DNA template (antisense) strand occurs while RNA polymerase remains at the promoter site.

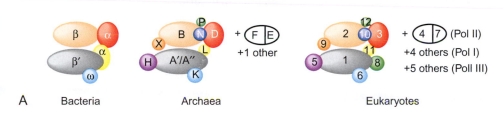

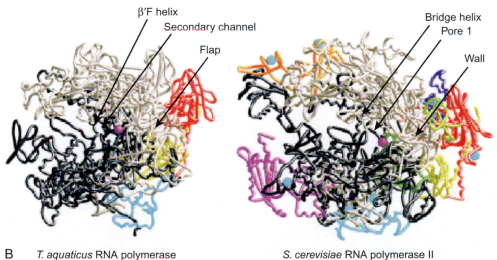

FIGURE 11.07
Structure of RNA Polymerase

A) Bacterial RNA polymerase has five types of subunits with four functional specificities. The subunits from bacteria, archaea, and eukaryotes are compared schematically here. Each color specifies functionally similar units from one species to another. B) Three-dimensional structure of *T. aquaticus* RNA polymerase and *S. cerevisiae* RNA polymerase II shown in the same orientation as the top and also colored according to the top. The active site contains a metal ion (shown in a pink sphere). Zinc ions are shown as blue spheres. *(Credit: Cramer, P. 2002 Curr Op Struct Biol 12: 89–97.)*

to 3′ and proceeds at about 40 nucleotides per second. This is much slower than DNA replication (~1,000 bp/sec), but roughly equivalent to the rate of polypeptide synthesis (15 amino acids per second).

The core enzyme of RNA polymerase consists of five subunits, two α plus β and β′ and ω (Fig. 11.07). The β- and β′-subunits comprise the catalytic site of the enzyme. The α-subunit is required partly for assembly and partly for recognizing promoters. The ω- (omega) subunit binds to the β′-subunit, stabilizes it and aids in its assembly into the core enzyme complex. RNA polymerase has a deep groove through the middle that can accommodate about 16 bp of DNA in the case of bacteria and about 25 bp in the case of eukaryotes such as yeast whose RNA polymerase is larger. A thinner groove, roughly at right angles to the first, may hold the newly constructed strand of RNA.

> The core enzyme moves ahead, manufacturing RNA and leaving sigma behind at the promoter.

The negative supercoiling of the chromosome promotes opening of DNA during transcription. As RNA polymerase moves along the DNA, it winds the DNA more tightly ahead of itself, creating positive supercoils. It also leaves partly-unwound DNA behind, which generates negative supercoils. To restore normal levels of supercoiling, DNA gyrase inserts negative supercoils ahead of RNA polymerase and topoisomerase I removes negative supercoils behind RNA polymerase (see Ch. 4).

4. RNA Polymerase Knows Where to Stop

Just as there is a recognition site at the front of each gene, so there is a special **terminator** sequence at the end. The terminator is in the template strand of DNA and consists of two inverted repeats separated by half a dozen bases and followed by a run of adenines (A's). The sequence of the mRNA will be the same as the non-template strand of DNA except for the substitution of U for T. Thus, the string of As in the DNA template (antisense) strand gives rise to a run of uracils (U's) at the 3' end of the mRNA (Fig. 11.08). Note that in the DNA, the two inverted repeat sequences are on opposite strands. Although researchers often talk as if the mRNA has inverted "repeats," its second "repeat" is actually the complement of the first. Because of this, such inverted repeats on the same strand of an RNA molecule can pair up to generate a stem and loop or "hairpin" structure.

> The end of a gene is marked by a terminator sequence that forms a hairpin structure in the RNA.

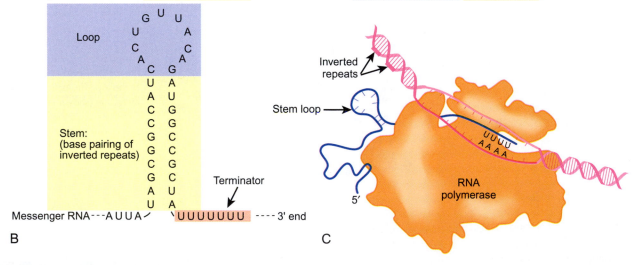

FIGURE 11.08
The Terminator Sequence Is Transcribed into RNA

A) The signal for RNA polymerase to stop is shown in both the DNA and the RNA transcribed from it. The terminator consists of two complementary sequences separated by approximately 10 bases from a run of U's. B) The complementary bases form the stem of the hairpin, with the intervening bases forming the loop. C) During transcription, the stem loop of the mRNA destabilizes the interaction of the mRNA with RNA polymerase. In addition, the heteroduplex of mRNA:DNA within RNA polymerase is not as tightly linked since A:U base pairs have only two hydrogen bonds. The combination of these two creates instability in the complex, ultimately resulting in mRNA and DNA release.

---

terminator   DNA sequence at end of a gene that tells RNA polymerase to stop transcribing

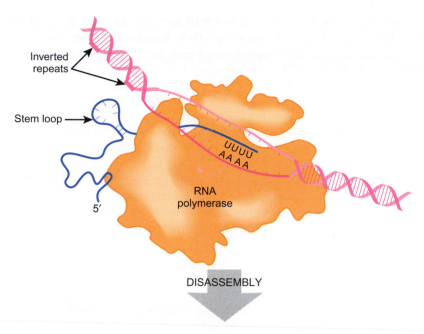

■ **FIGURE 11.09**
Termination of mRNA

When the mRNA reaches the hairpin of the terminator it pauses; when it reaches the AAAAA sequence it falls off the template strand along with the newly synthesized RNA due to instability of the complex.

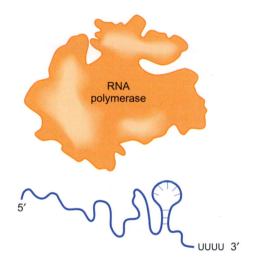

    Once the RNA polymerase reaches the stem and loop, it pauses. Long RNA molecules contain many possible hairpin structures that cause RNA polymerase to slow down or stop briefly, depending on the size of the hairpin. This provides an opportunity for termination, but if there is no string of U's, the RNA polymerase will start off again. However, a string of U's paired with a string of A's in the template strand of DNA is a very weak structure, and the RNA and DNA fall apart while the RNA polymerase is idling (Fig. 11.09). Pausing varies in length, but is around 60 seconds for a typical terminator.

    Termination may actually occur at several possible positions in the middle or end of the run of U's. In other words, the RNA polymerase "stutters" and the precise location of termination may vary slightly between different molecules of the same mRNA. Once the DNA and RNA have separated at the terminator structure, the RNA polymerase falls off and departs to find another gene (Fig. 11.09).

Two classes of terminators exist. **Rho-dependent terminators** need **Rho (ρ) protein** to separate the RNA polymerase from the DNA. **Rho-independent**, or "intrinsic," terminators do not need Rho or any other factor to cause termination. Most terminators in *E. coli* do not need Rho. In contrast, Rho-dependent terminators are relatively frequent in bacteriophages.

Rho protein is a specialized helicase that uses energy from ATP to unwind a DNA/RNA hybrid double helix. It consists of a hexamer of six identical subunits that recognizes and binds to a sequence of 50–90 bases located upstream of the terminator in the mRNA. The Rho hexamer does not form a closed ring, but instead is split open and resembles a lock washer in structure. The RNA sequence for Rho binding is poorly defined but is high in cytosines and low in guanines. Rho can only bind to the growing mRNA chain once the RNA polymerase has synthesized the C-rich/G-poor recognition region and moved on. It was originally thought that Rho moved along the RNA transcript and caught up with the RNA polymerase at the terminator stem and loop structure where the RNA polymerase pauses (Fig. 11.10). However, recent data suggest that Rho accompanies RNA polymerase throughout the transcription cycle. When Rho binds to its recognition site it causes allosteric changes in the catalytic subunits of the RNA polymerase that result in termination. Rho then unwinds the DNA/RNA helix in the transcription bubble and separates the two strands, leading to disassembly.

5. How Does the Cell Know Which Genes to Turn On?

Some genes, known as housekeeping or constitutive genes, are switched on all the time; that is, they are expressed constantly. In bacteria, these often have both their −10 and −35 region promoter sequences very close or identical to consensus. Consequently, they are always recognized by the sigma subunit of RNA polymerase and are expressed under all conditions. Other constitutive promoters are further from consensus and expressed less strongly. Nonetheless, if only relatively low amounts of the gene product are needed, this is acceptable.

Genes that are only needed under certain conditions sometimes have poor recognition sequences in the −10 and −35 regions of their promoters. In such cases, the promoter is not recognized by sigma unless another accessory protein is there to help (Fig. 11.11). These accessory proteins are known as gene **activator proteins** and are different for different genes. Each activator protein may stimulate the transcription of one or more genes. A group of genes that are all recognized by the same activator protein will be expressed together under similar conditions, even if the genes are at different places on the DNA. Higher organisms have many genes that are expressed differently in different tissues. As a result, eukaryotic genes are often controlled by multiple activator proteins, more specifically known as **transcription factors** (see below).

5.1. What Activates the Activator?

Long ago, the Greek philosopher Plato pondered the political version of this question: "Who will guard the guardians?" In living cells, especially in more complex higher organisms, there may indeed be a series of regulators, each regulating the next. What is the initial event? The cell must respond to some outside influence or must be influenced by other internal processes. The regulation of gene expression will be considered in more detail in Chapters 16 and 17. This chapter will be limited to a discussion of the basic mechanisms needed for a promoter to be functional.

---

activator protein   Protein that switches a gene on
Rho-dependent terminator   Transcriptional terminator that depends on Rho protein
Rho-independent terminator   Transcriptional terminator that does not need Rho protein
Rho (ρ) protein   Protein factor needed for successful termination at certain transcriptional terminators
transcription factor   Protein that regulates gene expression by binding to DNA in the control region of the gene

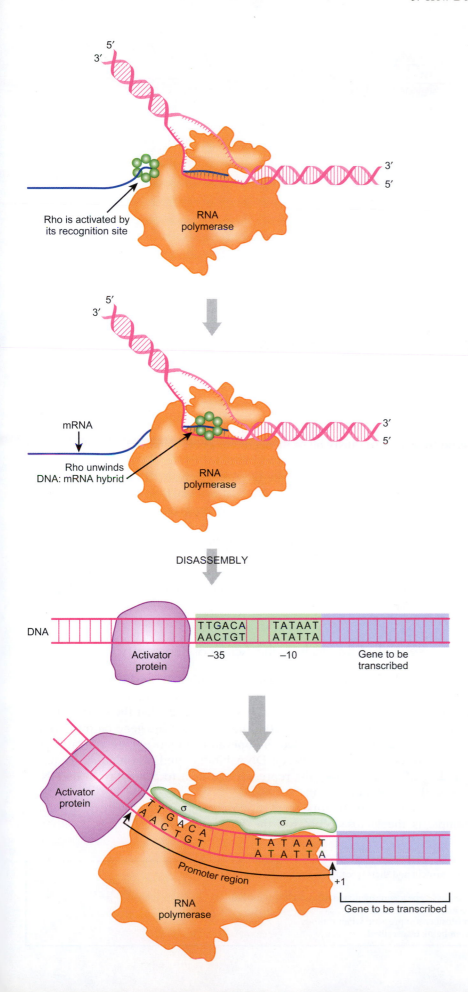

■ **FIGURE 11.10**
Termination by Rho

Rho accompanies the RNA polymerase as it synthesizes the growing mRNA. Once the RNA polymerase has made the Rho recognition site, Rho binds to this and causes a change in conformation of the RNA polymerase, which then pauses at the termination site. Rho then untwists the newly-formed mRNA strand from the DNA. Subsequently, the mRNA and RNA polymerase fall off the DNA and Rho detaches from the mRNA.

■ **FIGURE 11.11**
Gene Activator Proteins

The activator protein first binds to the promoter region of the gene. Once bound, the activator protein facilitates the binding of the sigma subunit of RNA polymerase. Gene transcription then commences.

FIGURE 11.12
MalT Changes Shape upon Binding Maltose

The MalT protein has a binding site complementary in shape to the sugar maltose. In step 1, MalT binds to maltose, which causes MalT to change shape. In step 2, the new conformation of MalT protein allows it to bind to DNA at a specific sequence found only in certain promoters. The activated gene is involved in the metabolism of maltose.

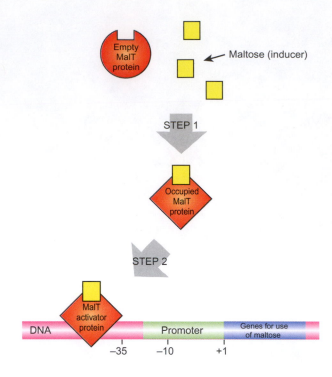

As a simple example of an activator, consider the use of maltose by *E. coli*. Maltose is a sugar made originally from the starch in malt and many other sources. It can be used by *E. coli* to satisfy all of its needs for energy and organic material. An activator protein, MalT, detects maltose and binds to it (Fig. 11.12). This causes the MalT protein to change shape, exposing its DNA-binding site. The original "empty" form of MalT cannot bind to DNA. The active form (MalT + maltose) binds to a specific sequence of DNA found only in the promoter region of genes needed for growth on maltose. The presence of MalT helps RNA polymerase bind to the promoter and transcribe the genes. The small molecule, in this case maltose, which causes gene expression, is known as the **inducer**. The result of this is that the genes intended for using maltose are only induced when this particular sugar is available. The same general principle applies to most nutrients, although the details of the regulation may vary from case to case.

5.2. Negative Regulation Results from the Action of Repressors

Genes may be controlled by positive or negative regulation. In **positive regulation**, an activator protein binds to the DNA only when the gene is to be turned on. In **negative regulation**, a **repressor** protein binds to the DNA and insures that the gene is turned off. Only when the repressor is removed from the DNA can the gene be transcribed. The site where a repressor binds is called the **operator** sequence. Like activator proteins, repressor proteins alternate between DNA-binding and non-binding forms. In this case, binding of the inducer to the repressor causes it to change from its DNA-binding form to the non-binding form.

Historically, negative regulators were discovered before activators. The best known example is the lactose repressor, the **LacI protein** (Fig. 11.13). Lactose is

Repressors are proteins that switch genes off.

---

inducer   Small signal molecule that binds to a regulatory protein and thereby causes a gene to be switched on
LacI protein   Repressor that controls the *lac* operon
negative regulation   Regulatory mode in which a repressor keeps a gene switched off until it is removed
operator   Site on DNA to which a repressor protein binds
positive regulation   Control by an activator that promotes gene expression when it binds
repressor   Regulatory protein that prevents a gene from being transcribed

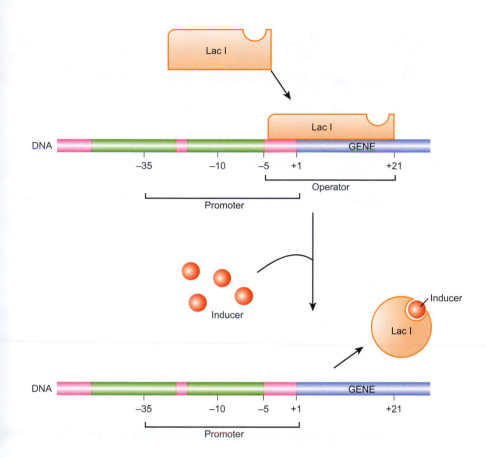

FIGURE 11.13
Principle of Negative Regulation by a Repressor

The LacI protein is bound to the operator site, within the promoter region of a gene that affects lactose metabolism. The inducer binds to LacI, changing its conformation and causing its release from the DNA. The RNA polymerase is then free to transcribe the gene.

another sugar, found in milk, on which bacteria such as *E. coli* can grow. When no lactose is available, the LacI protein binds to its operator sequence, which overlaps part of the promoter and the front part of the coding region for the genes for using lactose. When lactose is present, the LacI protein changes shape and is released from the DNA and the lactose genes are induced. Overall, the result is the same as for maltose: when lactose is available, the genes for using it are switched on and when there is no lactose, the genes are turned off.

The detailed mechanism by which repressors prevent transcription varies considerably and is often unknown. The repressor sometimes blocks the binding of RNA polymerase to the promoter, simply by getting in the way (steric hindrance), as is the case of the well-studied CI repressor of bacteriophage lambda. Another way to inhibit expression is for the repressor to bind further downstream, inside the structural gene. In this case, RNA polymerase can still bind to the promoter but is prevented from moving forward and transcribing the gene. Sometimes, even though their binding sites in the DNA sequence overlap, the RNA polymerase and the repressor both bind the DNA simultaneously. (Remember that the DNA double helix is 3D and that two proteins may therefore bind to the same linear segment if they occupy separate locations around its surface.) Indeed, an example of this is the LacI repressor (see Focus on Relevant Research). In this case the RNA polymerase actually binds more tightly in the presence of repressor, but is locked in place and cannot open the DNA to initiate transcription.

5.3. Many Regulator Proteins Bind Small Molecules and Change Shape

Whether a regulator protein is an activator or a repressor, it needs a signal of some sort. One of the most common ways to do this is by using some small molecule that fits into a binding site on the regulatory protein (Fig. 11.14). This is called the

FIGURE 11.14
*Allosteric Protein Binds
a Signal Molecule and
Changes Shape*

The two subunits shown have a
signal-binding site and a DNA-
binding site. When the signal
molecule binds to the subunits, they
pair and change conformation. They
are then able to bind to DNA.

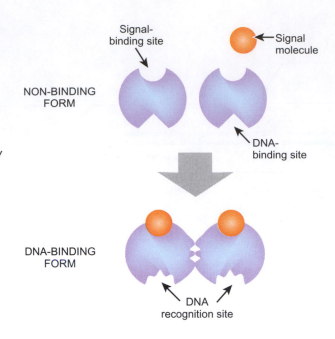

FOCUS ON
RELEVANT RESEARCH

La Penna G, Perico A (2010) Wrapped-around models for the *lac* operon complex. Biophys J. 98:2964-2973.

The *lac* operon is the classic system used to investigate gene regulation in bacterial genetics. The *lac* repressor protein (LacI) forms a tetramer when binding DNA. The size and charge of the LacI tetramer are important in bending the DNA around this protein. Several slightly different structural models have been proposed for wrapping DNA around the LacI tetramer. Many DNA-binding regulatory proteins bind two separate sequences located a few base pairs apart. This is true for LacI whose tetramer is arranged as a dimer of dimers and forms a V-shape. It makes contact twice with the DNA, at each of the tips of the V.

The authors perform a variety of calculations and structural simulations to investigate the possible structural variations for the LacI-DNA complex. These support a model where the DNA is wrapped closely around the LacI protein, rather than looping out of the complex. The DNA maintains its classical double helical structure throughout most of the complex. Some distortion of the DNA occurs downstream of the bound tetramer.

Small molecules may control
gene expression by binding to
regulatory proteins.

signal molecule. In the case of using a nutrient for growth, an obvious and common choice is the nutrient molecule itself. (In prokaryotes, the DNA-binding protein often binds the signal molecule directly. In eukaryotes, where the DNA is inside the nucleus, things are often more complex, and multiple proteins are involved. The signal molecule is often bound by proteins in the cell membrane or cytoplasm and the signal is then transmitted to the nucleus. The DNA-binding protein itself normally stays in the nucleus, and upon receiving the signal, is converted to its DNA-binding form, often by phosphorylation.)

When a regulator protein binds its signal molecule, it changes shape (Fig. 11.14). Regulator proteins have two alternative forms, the DNA-binding form and the non-binding form. Binding or loss of the signal molecule causes the larger protein to flip-flop between its two alternative shapes. Proteins that change in activity by changing shape in this manner are called **allosteric proteins**. Examples include some enzymes,

allosteric protein   Protein that changes shape when it binds a small molecule
signal molecule   Small molecule that exerts a regulatory effect by binding to a regulatory protein

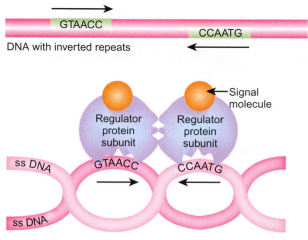

GTAACC

CCAATG

DNA with inverted repeats

Signal molecule

Regulator protein subunit | Regulator protein subunit

ss DNA GTAACC CCAATG

ss DNA

DNA-binding regulator proteins

FIGURE 11.15
Regulator Binds at an Inverted Repeat—Principle

At sites where regulator proteins bind there is often an inverted repeat with both DNA strands participating as shown. If the subunits of regulator protein are identical, they each recognize one of the inverted repeats and pair so that the same regions of each subunit face each other.

LAC OPERATOR

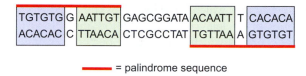

| TGTGTG | G | AATTGT | GAGCGGATA | ACAATT | T | CACACA |
| ACACAC | C | TTAACA | CTCGCCTAT | TGTTAA | A | GTGTGT |

───── = palindrome sequence

FIGURE 11.16
Lac Operator

The *lac* operator site has two imperfect palindrome sequences that are on opposite strands. These are also inverted relative to each other; therefore, these are referred to as inverted repeats.

> Recognition sites on DNA are often inverted repeats. Separate subunits of the regulator protein each bind one of the repeat sequences.

transport proteins, and regulators. Allosteric proteins have multiple subunits that change shape in concert. Usually there is an even number of subunits, most often two or four. All bind the signal molecule and then they all change shape together (see Ch. 14 for more discussion).

Since there is an even number of protein subunits for regulator proteins, the recognition site on the DNA is often duplicated. In this case, the recognition site is usually an inverted repeat, often referred to as a palindrome. This is because the subunits of the regulator protein bind to each other head to head rather than head to tail (Fig. 11.15). Consequently, the two protein molecules are pointing in opposite directions. Because they have identical binding sites for DNA, they recognize the same sequence of bases but in opposite directions on the two strands of the DNA. The two half-sites are usually separated by a spacer region of several bases whose identity is free to vary. The two half-sequences of such recognition sites are not always exact matches.

One regulator protein subunit binds to the recognition sequence on the template strand of the DNA double helix and its partner binds to the same sequence but on the non-template strand of the DNA pointing in the opposite direction. This is simpler in practice than it sounds, precisely because the DNA molecule is helical. Although the two recognition sequences are on different strands of DNA, they end up on the same face of the DNA molecule due to its helical twisting (Fig. 11.15).

An example of a palindromic recognition site is the *lac* operator sequence (Fig. 11.16), which is bound by the LacI repressor (a tetramer). This sequence runs from −6 to +28 relative to the start of transcription. It is not exactly symmetrical. The two half-sites are TGTGTGgAATTGTgA and, running in the opposite direction on the other strand, TGTGTGaAATTGTtA (capital letters indicate matching bases). The two half-sites are separated by five-base pairs. The left-hand half of this site binds the LacI protein more strongly than the right-hand side. A stronger operator sequence can be generated by artificially changing the right-hand half-site to exactly match the left.

FIGURE 11.17
Clusters of rRNA Genes

The genes for rRNA are located at multiple sites along the DNA. A single transcribed unit of DNA yields an initial RNA molecule of 45S. The 45S RNA is processed to yield the final 18S and 28S subunits of rRNA.

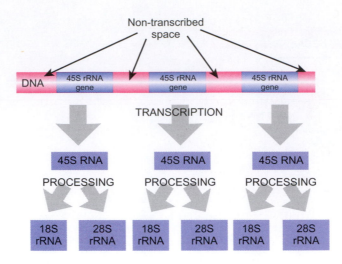

6. Transcription in Eukaryotes Is More Complex

Since typical eukaryotic cells have 10 times as many genes as do bacteria, the whole process of transcription and its regulation is more complex. To start, eukaryotes have three different RNA polymerases, unlike bacteria which have just one. The three RNA polymerases transcribe different categories of nuclear genes. In addition, mitochondria and chloroplasts have their own RNA polymerases, which resemble the bacterial enzyme.

RNA polymerase I transcribes the genes for the two large rRNA molecules and **RNA polymerase III** transcribes the genes for tRNA, 5S rRNA, and several other small RNA molecules. **RNA polymerase II** transcribes most eukaryotic genes that encode proteins and as a result is subject to the most complex regulation. Since rRNA and tRNA are needed all the time by all types of cells, RNA polymerases I and III operate constitutively in most cell types.

A variety of proteins, known as **transcription factors**, are also needed for the correct functioning of the eukaryotic RNA polymerases. Transcription factors may be divided into general transcription factors and specific transcription factors. General transcription factors are needed for the transcription of all genes transcribed by a particular RNA polymerase, and are typically designated TFI, TFII, TFIII followed by individual letters. The I, II, and III refer to the corresponding RNA polymerase (see below). Specific transcription factors are needed for transcription of particular gene(s) under specific circumstances. (Proteins such as the sigma subunit of bacterial RNA polymerase may also be regarded as transcription factors; however, this terminology is usually only used for eukaryotes.)

6.1. Transcription of rRNA and tRNA in Eukaryotes

The genes for the two large rRNAs are present in multiple copies, from seven in *E. coli* to several hundred in higher eukaryotes (Fig. 11.17). In bacteria, the copies are dispersed, but in eukaryotes they form clusters of tandem repeats. In humans, there are clusters of rRNA genes on five separate chromosomes. The 18S and the 28S rRNA are transcribed together as a single large RNA (45S RNA) by RNA polymerase I. The longer transcript is then cleaved to release the two separate rRNA molecules. Between the transcription units for 45S RNA are non-transcribed spacer regions.

Synthesis of rRNA by RNA polymerase I is localized to a special zone of the nucleus known as the nucleolus. Here, the rRNA precursor is both transcribed and

Eukaryotes have three RNA polymerases that specialize in which type of genes they transcribe.

Many transcription factors are involved in controlling gene expression in eukaryotes.

Eukaryotes contain many copies of the genes for ribosomal RNA. These are found in clusters and are transcribed by RNA polymerase I.

RNA polymerase   Enzyme that synthesizes RNA using a DNA template
RNA polymerase II   Eukaryotic RNA polymerase that transcribes the genes encoding proteins
RNA polymerase III   Eukaryotic RNA polymerase that transcribes the genes for 5S ribosomal RNA and transfer RNA
transcription factor   Protein that regulates gene expression by binding to DNA and/or to RNA polymerase

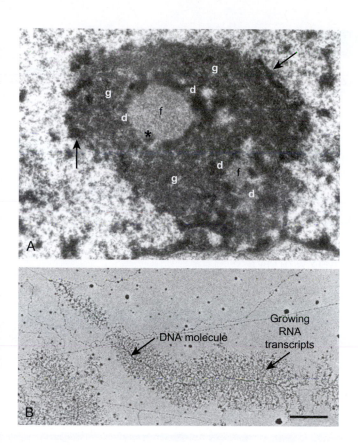

FIGURE 11.18
Ribosomal RNA is Made in the Nucleolus

A) Electron micrograph of a thin-sectioned nucleolus from a mouse cell fixed *in situ*. Black arrows indicate peri-nucleolar condensed chromatin and the asterisk shows dense fibrillar components (d) clumping around fibrillar centers (f). Granular regions (g) of newly-made ribonucleoproteins are also marked. (Credit: Ulrich Scheer, University of Würzburg.) B) Repetitive structures consisting of elongating strands of rRNA are known as Christmas trees. Those shown (4 microns long) are from a mouse cell and are at the same magnification as (A). The bar represents 0.5 micron. *(Credit: Raska I., Oldies but goldies: searching for Christmas trees within the nucleolar architecture. Trends in Cell Biology 13 (2003) 517–525.)*

processed into 18S and 28S rRNA. These rRNA molecules then bind proteins, giving ribonucleo-protein particles. This yields a dense granular region when seen under the microscope (Fig. 11.18). The segments of chromosomes associated with the nucleolus were previously named "**nucleolar organizers**." It is now known that these correspond to the clusters of rRNA genes.

Although most promoters are A/T-rich, presumably because the weaker base pairs help in opening up the DNA, the promoter for RNA polymerase I contains many G/C pairs. There are two G/C-rich regions, the core promoter and the upstream control element, that are 80–90% identical in sequence (Fig. 11.19). Both are recognized by protein UBF1 (Upstream Binding Factor 1), a single polypeptide that binds as a homodimer. UBF binds to distinct regions in the coding region. After UBF1 has bound, another protein, selectivity factor SL1, binds. SL1 consists of four polypeptides, one of which, TBP (TATA Binding Protein), is also required for RNA polymerases II and III (see below). Once UBF1 and SL1 are in place, RNA polymerase I can bind with the help of TIF1A (also called Rrn3). Several other transcription factors are needed for RNA polymerase I that are related to those for polymerase II (see Focus on Relevant Research). It is uncertain how the binding of UBF1 and SL1 at the upstream control element helps initiation in the case of RNA polymerase I. However, in similar cases, the DNA is known to bend around, bringing the upstream element into direct contact with the promoter region.

RNA polymerase III is responsible for making 5S rRNA and tRNA. It also makes some **small nuclear RNAs (snRNAs)**, while other snRNAs are transcribed by RNA polymerase II (see below). The promoters for 5S rRNA and tRNA are unique and somewhat bizarre in being internal to the genes. Transcription of these genes requires the binding of either of two proteins known as TFIIIA and TFIIIC to a

RNA polymerase III transcribes genes for small non-coding RNAs, in particular tRNA and 5S rRNA.

---

nucleolar organizer   Chromosomal region associated with the nucleolus; actually, a cluster of rRNA genes
small nuclear RNA (snRNA)   Small RNA molecules that are found only in the nucleus of eukaryotes where they oversee the splicing of mRNA

FIGURE 11.19 ─────
RNA Polymerase I
Transcribes rRNA Genes

The promoter for RNA polymerase I has an upstream control element and a core promoter, the latter rich in G/C sequences. The UBF1 protein recognizes and binds to both the upstream control element and the core promoter. Subsequently, SL1 binds to the DNA in association with UBF1. Finally, RNA polymerase I binds and transcription commences.

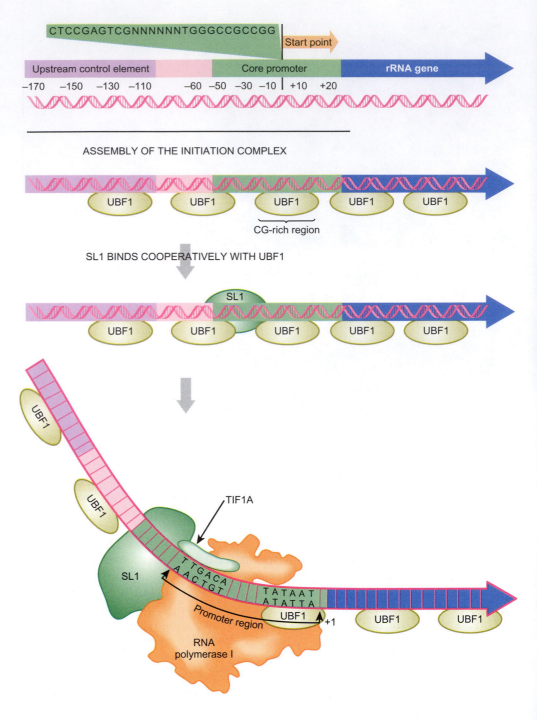

Geiger SR et al., (2010) RNA polymerase I contains a TFIIF-related DNA-binding subcomplex. Mol. Cell 39: 583–594.

FOCUS ON
RE LEVANT RESEARCH

The difference in promoter usage between eukaryotic RNA polymerases I, II, and III depends on different initiation factors. For RNA Pol II, these include TFIIF and TFIIE. The authors demonstrate here that the Pol I-specific initiation factors known as subunits A49 and A34.5 are in fact related to TFIIF and TFIIE. The authors determined the structure of A49 and A34.5 by X-ray diffraction. They also carried out DNA-binding studies. They concluded that these factors form a subcomplex that binds DNA via its "tandem winged helix" domain. Similar domains are seen in the corresponding complex used by RNA Pol II.

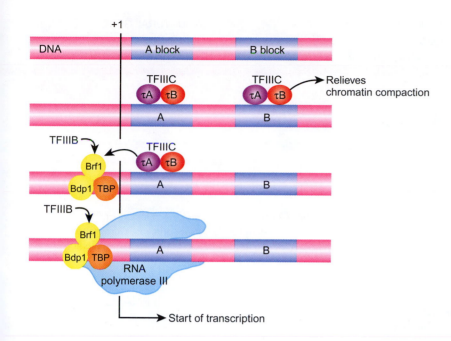

■- **FIGURE 11.20** ——————
Internal Promoter for RNA Polymerase III

The gene for 5S rRNA is transcribed using a promoter located within the gene itself. The recognition sites are downstream of the start site. TFIIIC consists of two subunits τA and τB, and binds to the A and B blocks. This induces TFIIIB, which consists of three subunits (Brf1, Bdp1, and TBP), to bind to the promoter near the start site. Only after TFIIIB binds can RNA polymerase III bind.

region over 50 bp downstream from the start site (Fig. 11.20). Once these have bound, they enable TFIIIB to bind to the region around the start of transcription. TFIIIB consists of three polypeptides, including TBP, and positions RNA polymerase III correctly at the start site.

As the promoters for RNA polymerase I and RNA polymerase III illustrate, recognition factor sites may be upstream or downstream from the start of transcription. However, in both cases, a positioning factor (SL1 or TFIIIB, respectively) is required to make sure that the polymerase starts transcribing at the correct place. These positioning factors thus play a similar role to that of the sigma factor in bacteria.

6.2. Transcription of Protein-Encoding Genes in Eukaryotes

RNA polymerase II transcribes most eukaryotic genes that encode proteins. Recognition of the promoter and initiation of transcription by RNA polymerase II requires a number of **general transcription factors**. In addition, since many protein-encoding genes vary markedly in expression, a variety of **specific transcription factors** are needed for expression of certain genes under particular circumstances. For example, in a multicellular organism, different cell types produce different types of proteins. Thus, red blood cells produce hemoglobin, whereas white blood cells make antibodies. Further, protein production often varies during development. Fetal hemoglobin is different from the adult version because two different genes are expressed at the two different stages of development.

The assorted transcription factors bind to and recognize specific sequences on the DNA. These DNA sequences are of two major classes, those comprising the promoter itself and a variety of **enhancer** sequences (Fig. 11.21). The general transcription factors for RNA polymerase II (TFII factors) bind to the promoter region. However, although some of the specific transcription factors also bind to the promoter region, others bind to the enhancer. A summary of the general transcription factors is presented in Table 11.01.

> RNA polymerase II transcribes eukaryotic genes that code for proteins.

> Some transcription factors bind to the promoter region, others to distant enhancer sequences.

---

enhancer   Regulatory sequence outside, and often far away from, the promoter region that binds transcription factors
general transcription factor   Transcription factor required for expression of most eukaryotic genes
specific transcription factor   Transcription factor needed for expression of certain specific genes under specific conditions

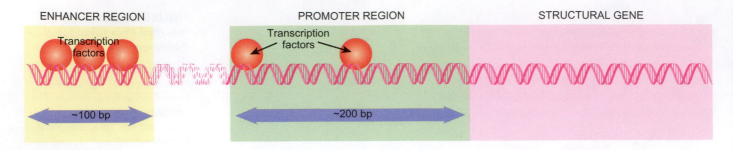

FIGURE 11.21
Promoter and Enhancer

Although one RNA polymerase is used to transcribe most protein encoding genes, specificity is controlled by transcription factors and their recognition sequences. The promoter region is close to the start site and usually binds several transcription factors. In addition, extra transcription factors bind to regions known as enhancers. These may be far upstream of the promoter, as shown, or may be located downstream. Binding of the transcription factors to their recognition sequences influences polymerase activity and gene expression.

TABLE 11.01	General Transcription Factors Associated with RNA Polymerase II
TBP	binds to TATA box, part of TFIID
TFIID	includes TBP, recognizes Pol II specific promoter
TFIIA	binds upstream of TATA box; required for binding of RNA Pol II to promoter
TFIIB	binds downstream of TATA box; required for binding of RNA Pol II to promoter
TFIIF	accompanies RNA Pol II as it binds to promoter
TFIIE	required for promoter clearance and elongation
TFIIH	phosphorylates the tail of RNA Pol II, retained by polymerase during elongation
TFIIJ	required for promoter clearance and elongation

> The TATA box is the critical sequence that allows RNA polymerase II to recognize the promoter.

In eukaryotes, many protein-encoding genes are interrupted by introns. These are removed at the RNA stage. Consequently, transcription of DNA to give RNA does not yield mRNA directly. The RNA that results from transcription is known as the **primary transcript** and must be processed as described in Chapter 12 to give mRNA. The present discussion will therefore be limited to the transcription of genes by RNA polymerase II to give the primary transcript.

Promoters for RNA polymerase II consist of three regions, the **initiator box**, the **TATA box,** and a variety of **upstream elements** (Fig. 11.22). The initiator box is a sequence found at the site where transcription starts. The first transcribed base of the mRNA is usually A with a pyrimidine on each side, as in bacteria. The consensus is weak: YYCAYYYYY (where Y is any pyrimidine). About 25 base pairs upstream from this is the TATA box, an A/T-rich sequence, which is recognized by the same factor TBP (**TATA binding protein** or **TATA box factor**) that is needed for binding of RNA polymerases I and III. TBP is unusual in binding in the minor groove of DNA. (Almost all DNA-binding proteins bind in the major groove.) On both sides of the TATA box are G/C-rich regions (Fig. 11.22).

TBP is found in three different protein complexes, depending on whether RNA polymerase I, II, or III is involved. In the present case, TBP forms part of a transcription factor complex known as TFIID that is needed to recognize promoters specific for RNA polymerase II. The binding of TFIID to the TATA box via TBP

---

initiator box   Sequence at the start of transcription of a eukaryotic gene
primary transcript   RNA molecule produced by transcription before it has been processed in any way
TATA box   Binding site for a transcription factor that guides RNA polymerase II to the promoter in eukaryotes
TATA binding protein (TBP)   Transcription factor that recognizes the TATA box
TATA box factor   Another name for TATA-binding protein
upstream element   DNA sequence upstream of the TATA box in eukaryotic promoters that is recognized by specific proteins

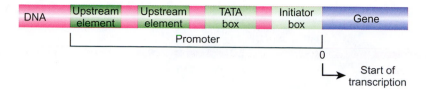

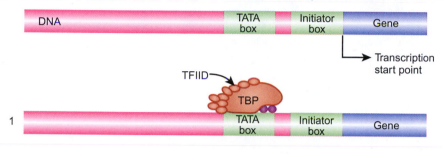

FIGURE 11.22
Eukaryotic Promoter Components—Initiator and TATA Boxes

The promoter for RNA polymerase II has an initiator box at the start site and a TATA box slightly upstream of this. Further upstream there are normally several upstream elements (two are shown here).

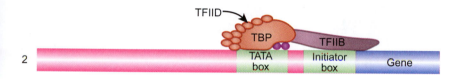

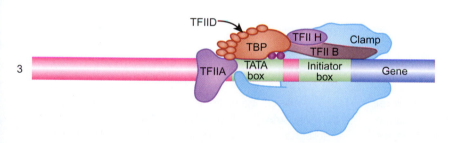

■ **FIGURE 11.23**
Binding of RNA Polymerase II to Promoter

Starting with TFIID, which contains TATA-binding protein, the components of the TFII complex bind one after another. Finally, TFIIF helps RNA polymerase II to bind to the DNA.

is the first step of transcription initiation. Several other TFII complexes are also needed for RNA polymerase II function. TFIIA and TFIIB bind next. Then at last RNA polymerase II itself arrives, accompanied by TFIIF, which probably helps RNA polymerase bind (Fig. 11.23). At this point RNA polymerase II can initiate synthesis of RNA. However, it is not yet free to move away from the promoter. This situation is called RNA pausing, and recent evidence suggests that many genes are kept in this state, even when they are not needed immediately for protein production.

Release of RNA polymerase II from the promoter and elongation of the RNA requires three more TFII complexes: TFIIF, TFIIH, and TFIIJ. In particular, TFIIH must phosphorylate the tail of RNA polymerase before it can move (Fig. 11.24). The tail, or **CTD (carboxy-terminal domain)**, consists of a seven-amino acid sequence (Tyr Ser Pro Thr Ser Pro Ser) repeated approximately 50 times. This may be phosphorylated on the serine or threonine residues. All of the TFII complexes except for TFIIH are left behind as RNA polymerase moves forward (see Focus on Relevant Research).

carboxy-terminal domain (CTD)   Repetitive region at the C-terminus of RNA polymerase II that may be phosphorylated

FIGURE 11.24
RNA Polymerase II Moves Forward from the Promoter

Before RNA polymerase II can move forward, the binding of other factors must occur. One of these, TFIIH, phosphorylates the tail of RNA polymerase II. The tail changes position with respect to the body of RNA polymerase II. The other factors leave and RNA polymerase moves along the DNA and begins the process of transcription.

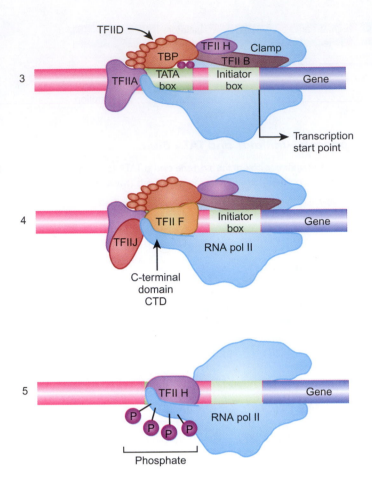

FOCUS ON RELEVANT RESEARCH

Buratowski S (2009) Progression through the RNA polymerase II CTD cycle. Mol. Cell 36: 541–546.

This article reviews the multiple roles of the carboxy-terminal domain (CTD) of RNA polymerase II (Pol II). Evidence shows that the CTD controls transcription, mRNA processing (see Ch. 12 for review), and even histone modifications. There are multiple repeats of the same sequence within the CTD, and in the repeated sequences, the two serines at position 2 and 5 of the repeat (Ser2 and Ser5) control many of these functions. In addition, Ser7 has been implicated in Pol II function.

The CTD of Pol II is unphosphorylated during transcription initiation. In this form, Pol II binds to a multiprotein complex called the mediator (see Ch. 17). This complex receives signals from the upstream activator proteins and then signals Pol II to bind to the promoter. After binding, TFIIH transfers a phosphate group to Ser5 of the repeated domains in the CTD tail. This ejects the mediator complex from Pol II and readies Pol II to make mRNA.

During elongation, the CTD is positioned so it lies close to the exit point of the new mRNA so it controls mRNA modification. Because of its location, Ser5 phosphorylation activates the capping enzyme that adds a 7meG-cap onto the 5' end of the mRNA. The cap stabilizes the mRNA and prevents its degradation by nucleases.

The CTD also controls histone structure of the nucleosomes behind RNA polymerase. Ser5-P CTD attracts histone methyltransferases, which add methyl groups to histones. During transcription, the DNA is unwrapped from histone cores ahead of Pol II and re-wound around the histones after Pol II finishes transcribing the region. The replaced histones have a specific methylation and acetylation pattern (see Ch. 17 for discussion) that marks the gene regions. The methyltransferases and demethylases that establish the pattern of histone modification are attracted to Ser5-P CTD, meaning that this region of RNA polymerase is also essential for genomic integrity.

As elongation continues, the number of Ser5 phosphorylations decrease and the number of Ser2 phosphorylations increase. In conjunction with the increase of Ser2 phosphorylation, the reattached histones receive new methyl groups on lysine 36 of histone H3 in addition to the lysine4 methylations. Now, the histones at the beginning of a gene are physically distinct from the histones toward the center or end of a gene. Ser2-P keeps increasing as Pol II moves away from the promoter region, and this modification may move Pol II down the DNA faster than during initiation by promoting interactions of elongation factors with Pol II.

At the end of the gene, Ser2 phosphorylations are also involved in the polyadenylation of mRNA after the termination of transcription. The exact mechanism of these two processes is not well understood, but CTD phosphorylations are critical to all stages of transcription in eukaryotes, and also help mark the genome with the correct histone modifications.

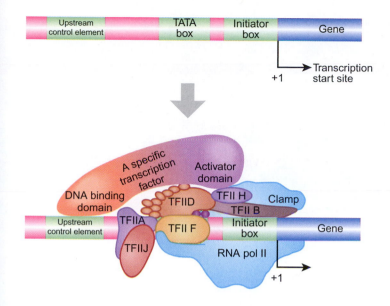

► **FIGURE 11.25**
*Upstream Elements
Facilitate Transcription*

The upstream elements make
contact with one domain of an
activator protein. The activator also
binds to the transcription apparatus
near the start site.

Like bacterial RNA polymerase, the eukaryotic RNA polymerases all have multiple subunits. RNA polymerase II has more than 10 subunits and shares three of these with RNA polymerases I and III. The largest subunit of RNA polymerase II is related to the β'-subunit of bacterial RNA polymerase and possesses the CTD tail. In addition, the assorted TFII complexes each consist of several polypeptide chains. Thus, the initiation complex for RNA polymerase II includes over 20 polypeptides.

6.3. Upstream Elements Increase the Efficiency of RNA Polymerase II Binding

RNA polymerase II can bind and initiate transcription at a minimal promoter consisting of just initiator and TATA boxes. However, this is extremely inefficient unless upstream elements are also present. There are many different upstream elements. They are typically 5 to 10 base pairs long and located from 50 to 200 bases upstream of the start site. There may be more than one upstream element in a given promoter and the same upstream element may be found at different places in different promoters.

The TFII proteins are *general* transcription factors because they are always required. In contrast, *specific* transcription factors affect only certain genes and are involved in regulating gene expression in response to a variety of signals (Fig. 11.25). The upstream elements are the recognition sites for many of the specific transcription factors. These usually make contact with the transcription apparatus via TFIID, TFIIB, or TFIIA, not by directly touching RNA polymerase II itself. Most commonly, binding is to TFIID. Binding of the specific transcription factors helps assembly of the transcription apparatus and therefore increases the frequency of initiation.

Common upstream elements include the GC box, CAAT box, AP1 element, and Octamer element. The GC box (GGGCGG) is often present in multiple copies. Despite being non-symmetrical, the GC box works in either orientation and is recognized by the SP1 transcription factor. Some upstream elements are recognized by more than one protein. In these cases, different transcription factors are often present in different tissues. For example, the Oct-1 and Oct-2 proteins both recognize the Octamer element. Oct-1 is found in all tissues, but Oct-2 only appears in immune cells where it helps activate genes encoding antibodies.

During mRNA elongation, RNA polymerase is also subject to negative regulation. Two proteins, DSIF (DRB-sensitivity inducing factor) and **negative elongation factor** (**NELF**), bind to Pol II and pause transcription of the gene after the transcript

Upstream elements close to
the promoter bind a range of
specific transcription factors.

In eukaryotes, genes are
controlled by positive and
negative regulation.

negative elongation factor (NELF)   A complex of proteins that inhibits RNA polymerase elongation in eukaryotes

FIGURE 11.26
Negative Regulation of RNA Polymerase II

Some genes initiate transcription, but due to the binding of NELF and DSIF, RNA polymerase pauses. When the conditions are favorable for the particular gene expression, then p-TEFb phosphorylates the C-terminal domain (CTD) of RNA polymerase, DSIF, and NELF, causing NELF to release and the remaining complex to re-initiate transcription.

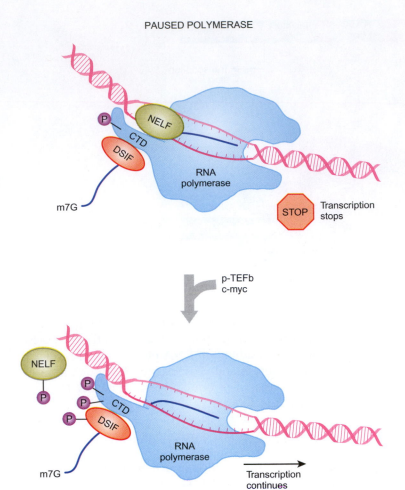

has been initiated. When these two proteins are phosphorylated, along with the carboxy-terminal domain of Pol II, transcription resumes (Fig. 11.26). Recent evidence suggests that this mechanism of regulation is more common than previously thought. Indeed, the majority of genes appear to have paused Pol II due to unphosphorylated DSIF and NELF. This may provide another level of gene control. Initiation may occur more frequently, but productive elongation of the transcript occurs only when the gene product is definitely needed.

6.4. Enhancers Control Transcription at a Distance

Enhancers are sequences that are involved in gene regulation, especially during development or in different cell types. Enhancers do exactly what their name indicates—they enhance the initiation of transcription as a result of binding specific transcription factors. Enhancers often consist of a cluster of recognition sites and therefore bind several proteins. Some recognition sites (e.g., Octamer and AP1) are found in both enhancers and as upstream elements in promoters.

Although enhancers are sometimes close to the genes they control, more often they are found at a considerable distance, perhaps thousands of base pairs away. Enhancers may be located either upstream or downstream from the promoter and the position varies considerably from case to case. In addition, enhancers function equally well in either orientation. Experiments in which enhancers have been moved have shown that an enhancer will increase transcription from any promoter within

> Enhancer sequences are located far away from the genes they control.

Enhancer   Regulatory sequence outside, and often far away from, the promoter region that binds transcription factors

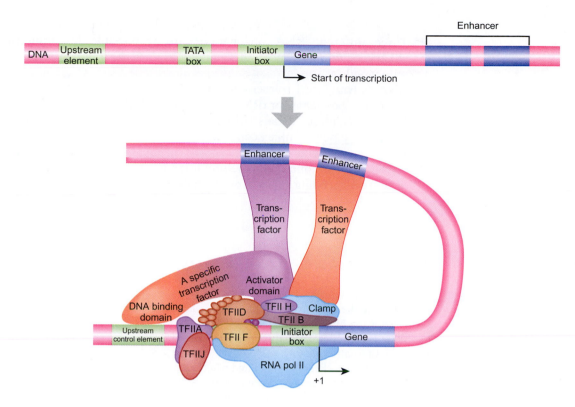

FIGURE 11.27
Looping Model for Enhancer

The enhancer shown here is located downstream of the start site. To enhance transcription, the enhancer first binds several transcription factors. Subsequently, the DNA forms a loop allowing the enhancer to make contact with the transcription apparatus via the bound transcription factors.

its neighborhood. These properties imply that the enhancer must make contact with the transcription apparatus. When an enhancer switches a gene on, the DNA between it and the promoter loops out as shown in Figure 11.27. One question that arises from this model is whether the transcription factors cause the loops to form, or do the loops form due to the chromatin structure at interphase; that is, do the proteins that induce the condensed structures during interphase establish the pattern of gene expression (see Ch. 4 for details on the chromatin structure)?

Key Concepts

- Transcription is the process by which genetic information carried by DNA is converted to an RNA copy.
- Individual messages are made by transcribing short sections of the genome including one or a few genes.
- Sequences in the DNA that are recognized by specific proteins allow the transcription apparatus to locate the correct start point.
- Synthesis of RNA is carried out by a protein complex known as RNA polymerase.
- RNA polymerase knows where to stop transcription due to terminator sequences that are located downstream from units of transcription.
- Regulatory proteins bind to specific sequences in the DNA and control which genes to turn on under any particular conditions.
- Regulatory proteins themselves often receive information by binding small signal molecules, whereupon they change shape, which alters their ability to bind DNA.

- Three different RNA polymerases are responsible for transcription of eukaryotic nuclear genes and separate enzymes are found in chloroplasts and mitochondria.
- RNA polymerase I transcribes genes for large rRNA and RNA polymerase III transcribes genes for tRNA and many small RNA molecules.
- RNA polymerase II transcribes genes that encode proteins and as a result is under extremely complex control.
- Upstream elements and enhancers are specific DNA sequences that may be a considerable distance from the promoter, but nonetheless regulate expression of protein encoding genes in eukaryotes.

Review Questions

1. What is transcription? What is the enzyme that makes the transcripts?
2. What are the two major groups of RNA that are transcribed?
3. What is the function of mRNA?
4. What is the difference between antisense strand and sense strand of DNA? What are their alternative names?
5. What are housekeeping genes and what are constitutive genes?
6. What is the difference between a cistron and an ORF?
7. What are monocistronic and polycistronic mRNAs?
8. What are the 5′ and 3′ UTRs?
9. What is a promoter? What is a strong promoter?
10. Name the two special sequences of a bacterial promoter.
11. What are the two major components of bacterial RNA polymerase?
12. How does bacterial RNA polymerase synthesize RNA?
13. What are the four subunits of the core enzyme of bacterial RNA polymerase?
14. What is the terminator for transcription? What are the two classes of terminators (in bacteria)?
15. How does Rho-protein help in the termination of transcription?
16. Describe one specific activator protein and one repressor protein. How do each of these affect gene expression?
17. What are signal molecules and allosteric proteins?
18. What are the three different RNA polymerases in eukaryotes? What are their functions?
19. What are transcription factors?
20. What are the structural components of a eukaryotic ribosome and how and where are they transcribed?
21. How is transcription of tRNA different from rRNA?
22. How is the specificity controlled in eukaryotic mRNA transcription?
23. Name the two major classes of DNA sequence that transcription factors recognize (in eukaryotes).
24. List the three major regions of an RNA Pol II promoter.
25. What is the importance of the TATA box?
26. What are the events that help RNA Pol II to bind to the promoter?
27. How does RNA Pol II synthesize RNA?
28. What is the difference between a general transcription factor and a specific transcription factor?
29. What are upstream elements? What are the common upstream elements found in eukaryotes?
30. What is the difference between bacterial and eukaryotic repressors?
31. What are enhancers? How do they work?

Conceptual Questions

1. *E. coli* bacteria can become infected with viruses called bacteriophages. When a bacteriophage injects its DNA into the *E. coli* cell, it uses the bacterial proteins to transcribe its own genome into new virus particles. Some bacteriophage prevent the bacterial genome from being expressed into proteins, whereas others co-exist with the bacteria and the mRNA transcripts for both the virus and bacteria are made simultaneously. Researchers have identified a brand new type of bacteriophage they have named γ8788 and want to determine if it allows bacterial transcription or not. You have been taught laboratory techniques to grow bacteria with radioactively labeled ^{3}H-uridine, which incorporates into RNA; to isolate ^{3}H-uridine RNA from bacteria; to isolate DNA from γ8788; to isolate DNA from *E. coli*; and to create labeled RNA/DNA hybrids by attaching RNA or DNA to nylon membrane. Design a hybridization experiment that will determine if *E. coli* cells are able to synthesize RNA after infection with γ8788. Include a chart with the results for both possible conditions.

2. RNA polymerase I, II, and III have varying levels of sensitivity to the poison called α-amanitin, which is from the mushroom *Amanita phalloides*. RNA polymerase II is completely sensitive to the poison; RNA polymerase III has intermediate sensitivity; and RNA polymerase I is insensitive to the poison. What would happen to transcription of the rRNA genes, tRNA genes, and the gene for the glucose transporter if a eukaryote was poisoned with α-amanitin?

3. Describe each of the following features of promoters. Be sure to characterize whether the feature is in eukaryotic promoters or prokaryotic promoters.
   TATA box
   -10 region
   -35 region
   initiator box
   upstream elements

4. A researcher has isolated a new gene and wants to transcribe the sequence into RNA *in vitro* (in a test tube). She has combined bacterial RNA polymerase and other nuclear proteins, the bacterial gene, and a free pool of ribonucleotides. After incubating the components, she is unable to isolate any RNA. After studying the bacteria, she realizes that this gene is only expressed when the bacteria have lactose sugar in the medium. After adding lactose to her *in vitro* system, she is finally able to purify RNA. Propose a mechanism for how lactose stimulated transcription.

5. The following sequence was found upstream of the -10 and -35 region for the arabinose transporter gene. When this region is removed from the promoter, the arabinose transporter gene is transcribed constitutively. When this region is present, the arabinose transporter gene is only expressed when arabinose is present. Why does this region regulate transcription of this gene? Propose a mechanism for this control.

5' GATTCGTTC   3'
3'   CTTGCTTAG   5'

Most genes have final gene products that are proteins. They are first transcribed to give messenger RNA (mRNA) and this is then translated to give protein. In a smaller but significant number of cases, the RNA itself is the final gene product. In either case, the RNA molecule that is the initial result of transcription, the primary transcript, may be chemically altered before fulfilling its role. This is referred to as RNA processing and may sometimes be extremely complex. The chemical nature of the modifications varies greatly.

Almost all RNA molecules are processed in some way. The major exception is bacterial mRNA, although even in this case a few of these mRNA molecules are processed. Not surprisingly, RNA processing is more complicated in eukaryotic cells, where it mostly occurs inside the nucleus before the RNA molecule is released into the cytoplasm. In some cases, such as mRNA, modification of RNA is largely regulatory in effect. In other cases, especially ribosomal RNA (rRNA) and transfer RNA (tRNA), the modifications improve the performance of the RNA in its final role.

1. RNA Is Processed in Several Ways

RNA is made by RNA polymerase, using a DNA template, in the process known as transcription (see Ch. 11). Sometimes the RNA molecule is ready to function immediately after it has been transcribed (e.g., most bacterial mRNAs). However, in many cases, the RNA needs further processing before it is functional. In these cases, the original

Molecular Biology.

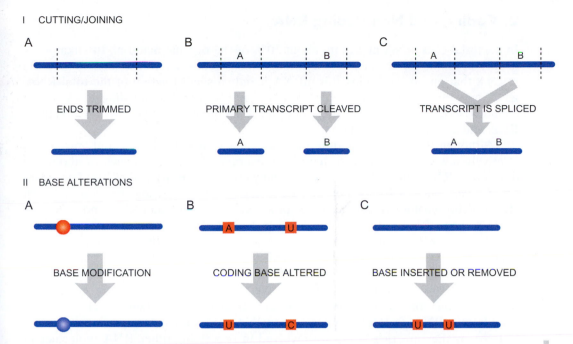

I   CUTTING/JOINING

A	B	C
ENDS TRIMMED	PRIMARY TRANSCRIPT CLEAVED	TRANSCRIPT IS SPLICED

II   BASE ALTERATIONS

A	B	C
BASE MODIFICATION	CODING BASE ALTERED	BASE INSERTED OR REMOVED

FIGURE 12.01
Types of RNA Processing

RNA processing can be divided into cutting/joining RNA segments or base alteration of the ribonucleotides.

RNA molecule, before any further processing occurs, is known as the **primary transcript**. For specific classes of RNA, the precursor (i.e., primary transcript) may be referred to as pre-mRNA, pre-tRNA, etc. The term hnRNA (heterogeneous nuclear RNA) was also used previously, before the relationship of precursor RNA to the final processed RNA product was understood.

All classes of RNA are subject to processing by base modification and cleavage. In addition, eukaryotic mRNA undergoes capping and tailing as well as **splicing** (Fig. 12.01). Base modifications occur primarily in tRNA and rRNA, and occur after the RNA is transcribed. These modifications are essential for their proper function in protein translation (see Ch. 13). Certain RNA molecules such as prokaryotic and eukaryotic rRNA are modified by cleavage; that is, the RNA is made as a longer precursor that is trimmed to the correct length. In other, related cases, several RNA molecules are included in the same primary transcript, which is then cleaved into several parts. In eukaryotes, the primary transcript for mRNA contains segments called introns or intervening sequences that are not used to encode the final protein product (see Ch. 4). Splicing involves the removal of these introns and rejoining of the ends to create a streamlined mRNA with an uninterrupted coding sequence that is translated into a protein.

Most mRNA processing relies on typical enzymes consisting of proteins to catalyze the reaction. However, as shown below, more complex RNA processing involves other RNA molecules. These RNAs are involved both in sequence recognition and in the actual chemical reactions of cutting and splicing. In fact, certain introns are self-splicing; that is, they cut themselves out in a reaction that does not require any protein components (see below). Such RNA enzymes are known as **ribozymes**. As will be discussed in Chapter 13, the formation of peptide bonds during protein synthesis is actually catalyzed by ribosomal RNA, not a protein. The involvement of RNA in such fundamental processes as protein synthesis and RNA processing has led to the idea that ribozymes were more common in early life. Indeed the "**RNA world**" hypothesis suggests that the original enzymes were all RNA and that protein only assumed this role later in evolution. The RNA world scenario is discussed in more detail in Chapter 26, "Molecular Evolution."

Many RNA molecules are modified in a variety of ways after being synthesized.

RNA processing sometimes requires other RNA molecules, either as guides or as actual enzymes—ribozymes.

primary transcript   The original RNA molecule obtained by transcription from a DNA template, before any processing or modification has occurred

ribozyme   An RNA molecule that shows enzymatic activity

RNA world   Theory that early life depended largely or entirely on RNA for both enzyme activity and for carrying genetic information and that DNA and protein emerged later in evolution

splicing   Removal of intervening sequences and re-joining the ends of a molecule; usually refers to removal of introns from RNA

2. Coding and Non-Coding RNA

In bacterial cells, RNA makes up about 20% of the organic material. In eukaryotes, it only accounts for about 3–4%. Although most genes are transcribed to give the mRNA that encodes proteins, this mRNA is only a small fraction of the total RNA. RNA may be divided into coding RNA (i.e., mRNA) and **non-coding RNA**, which includes tRNA, rRNA, and a variety of other RNA molecules that function directly as RNA and are not translated into protein.

Although there are many different molecules of mRNA, each is only present in relatively few copies. In *E. coli* there is an average of 3–4 copies each of about 400 different mRNAs. In contrast, there are many copies of rRNA and tRNA. For example, *E. coli* contains 10–20 thousand ribosomes, each possessing one copy of each rRNA. Ribosomal RNA thus accounts for about 80% of the total RNA and tRNA for 14–15%. The mRNA only makes up 4–5% by weight of the RNA.

Ribosomal RNA and tRNA are found in all living cells. The other types of non-coding RNA vary from organism to organism. Bacteria contain several small regulatory RNAs (see Ch. 18) as well as tmRNA (a hybrid transfer and messenger RNA) that rescues ribosomes trapped by defective messages (see Ch. 13). In eukaryotes, there are **small nuclear RNA (snRNA)**, **small nucleolar RNA (snoRNA),** and **small cytoplasmic RNA (scRNA)** molecules. snRNA and snoRNA (occasionally called U-RNA as they are rich in U) are involved in processing other RNA molecules in the eukaryotic nucleus (see below). The small cytoplasmic RNA is a miscellaneous group that comprises molecules with various functions. An increasing number of small regulatory RNA molecules are being found in eukaryotes and, to a lesser extent, in prokaryotes. In eukaryotes the two major classes are siRNA (short interfering RNA), involved in RNA interference, and miRNA (microRNA), short RNA molecules involved in regulating gene expression (see Ch. 18). Although most regulatory RNA is short (less than 200 nucleotides), an increasing number of "long non-coding RNA" or lncRNA are coming to light. These have a variety of roles in the regulation of eukaryotic genes.

3. Processing of Ribosomal and Transfer RNA

The three rRNA molecules of bacteria are transcribed together to give a single pre-rRNA. This contains 16S rRNA, 23S rRNA, and 5S rRNA joined by linker regions (Fig. 12.02). In bacteria, this pre-rRNA transcript also includes some tRNAs. Most bacteria have several copies of the rRNA genes; *E. coli* has seven, for example.

The mature rRNAs are made by cleavage of the precursor by ribonucleases. This occurs in two stages (Fig. 12.02). First, internal cuts are made, separating the three rRNAs. Ribonucleases III, P, and F recognize sites where the pre-rRNA is folded into double-stranded regions held together by base pairing. After this cleavage, the ends are trimmed by several exonucleases.

In eukaryotes, there are four ribosomal RNAs. The 5S rRNA is made separately and does not need processing. The other three (18S rRNA, 28S rRNA, and 5.8S rRNA) are made as a single pre-rRNA and processed much as in bacteria. Transfer RNAs are transcribed as longer precursors that also need processing (Fig. 12.03). Some tRNAs are made singly; others are transcribed together; and in bacteria, some

Coding RNA (i.e., mRNA) is used only to carry information, whereas non-coding RNA is not translated but performs a variety of active roles as RNA.

Many non-coding RNA molecules are found inside the eukaryotic nucleus.

Ribosomal RNAs are transcribed together as one long transcript that must be cut apart and trimmed at the ends.

---

non-coding RNA   RNA molecule that functions without being translated into protein; includes tRNA, rRNA, snRNA, snoRNA, scRNA, tmRNA, and some regulatory RNA molecules
small cytoplasmic RNA (scRNA)   Small RNA molecules of varied function found in the cytoplasm of eukaryotic cells
small nuclear RNA (snRNA)   Small RNA molecules that are involved in RNA splicing in the nucleus of eukaryotic cells
small nucleolar RNA (snoRNA)   Small RNA molecules that are involved in ribosomal RNA base modification in the nucleolus of eukaryotic cells

PRE-rRNA

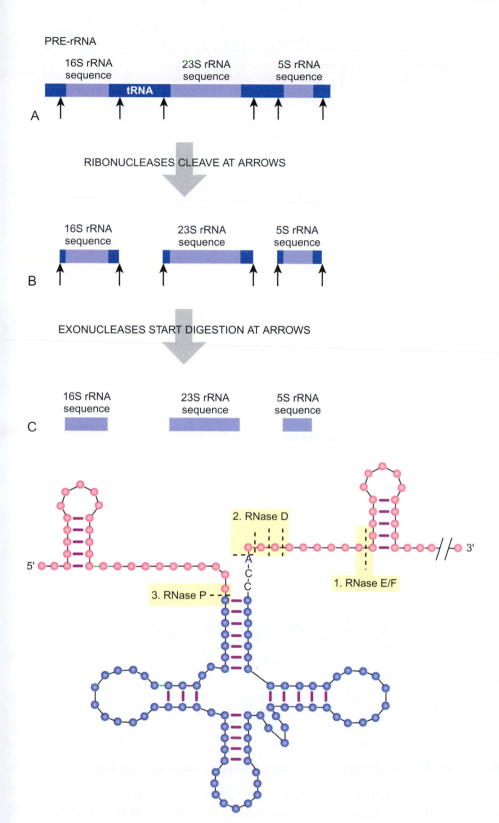

A

RIBONUCLEASES CLEAVE AT ARROWS

B

EXONUCLEASES START DIGESTION AT ARROWS

C

2. RNase D

3. RNase P

1. RNase E/F

5'

3'

A
C
C

■ **FIGURE 12.02**
Cleavage of rRNAs from Their Precursor in Prokaryotes

The pre-ribosomal RNA contains sequences for all three rRNA molecules as well as one or two tRNA molecules. Initial processing involves ribonucleases that cut the primary transcript at the sites shown by arrows. The ends must then be further trimmed. (Only the processing of the rRNA molecules is shown in full here; the tRNA is also trimmed after release.)

■ **FIGURE 12.03**
Processing of tRNA

Nucleotides shown in red are removed. First (1), ribonuclease E or F cleaves the precursor RNA near the 3′ end. Second (2), ribonuclease D chews off bases from the new 3′ end leaving the CCA at the end of the acceptor stem. Third (3), ribonuclease P cleaves the 5′ end precisely.

Transfer RNA precursors must be processed to give functional tRNA molecules.

are included in the pre-rRNA transcript. The 5′ end of bacterial tRNA is trimmed by **ribonuclease P**. This enzyme is of note because it is a ribozyme. Ribonuclease P consists of both an RNA molecule and a protein, but the catalytic activity is due to the RNA. The protein component merely modulates the activity of the RNA.

ribonuclease P   A ribonuclease involved in processing tRNA in bacteria that consists of an RNA ribozyme plus an accessory protein

FIGURE 12.04 ────────
Splicing Out of the Introns and Maturation of Eukaryotic mRNA—An Overview

The DNA, containing exons and introns, is transcribed to the RNA primary transcript, which contains the RNA version of introns and exons plus a tail signal. Processing the RNA involves removing the introns, adding a cap and a poly(A) tail. Translation of the fused exons forms the protein. B) The cap consists of guanosine triphosphate (attached in the reverse direction) preceding a small 5' untranslated region; the tail consists of a poly-A tract and is preceded by a brief 3' untranslated segment. When translated, only the information from the exons is used to build the protein.

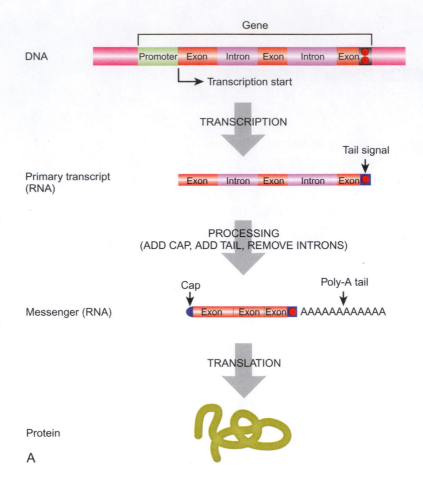

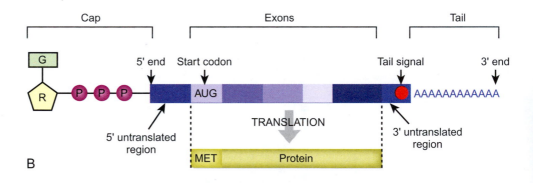

In eukaryotes, the primary transcript is converted to mRNA in three steps:
1) adding a cap at the front;
2) adding a tail to the end;
3) removing the introns.

4. Eukaryotic Messenger RNA Contains a Cap and a Tail

In eukaryotic cells, transcription of genes to give mRNA is much more complex than in prokaryotes. First, eukaryotic genes are inside the nucleus, not free in the cytoplasm. Second, most eukaryotic genes are interrupted by segments of non-coding DNA, the introns.

The RNA molecule resulting from transcription of a eukaryotic gene is known as the primary transcript. This transcript is not mRNA because it still needs to be processed. If the primary transcript were translated as is, it would result in a huge, dysfunctional protein containing many extra stretches of random amino acids due to the intron regions. The primary transcript is trapped inside the nucleus until it is completely processed. First, there is the addition of a cap structure to the front and a tail

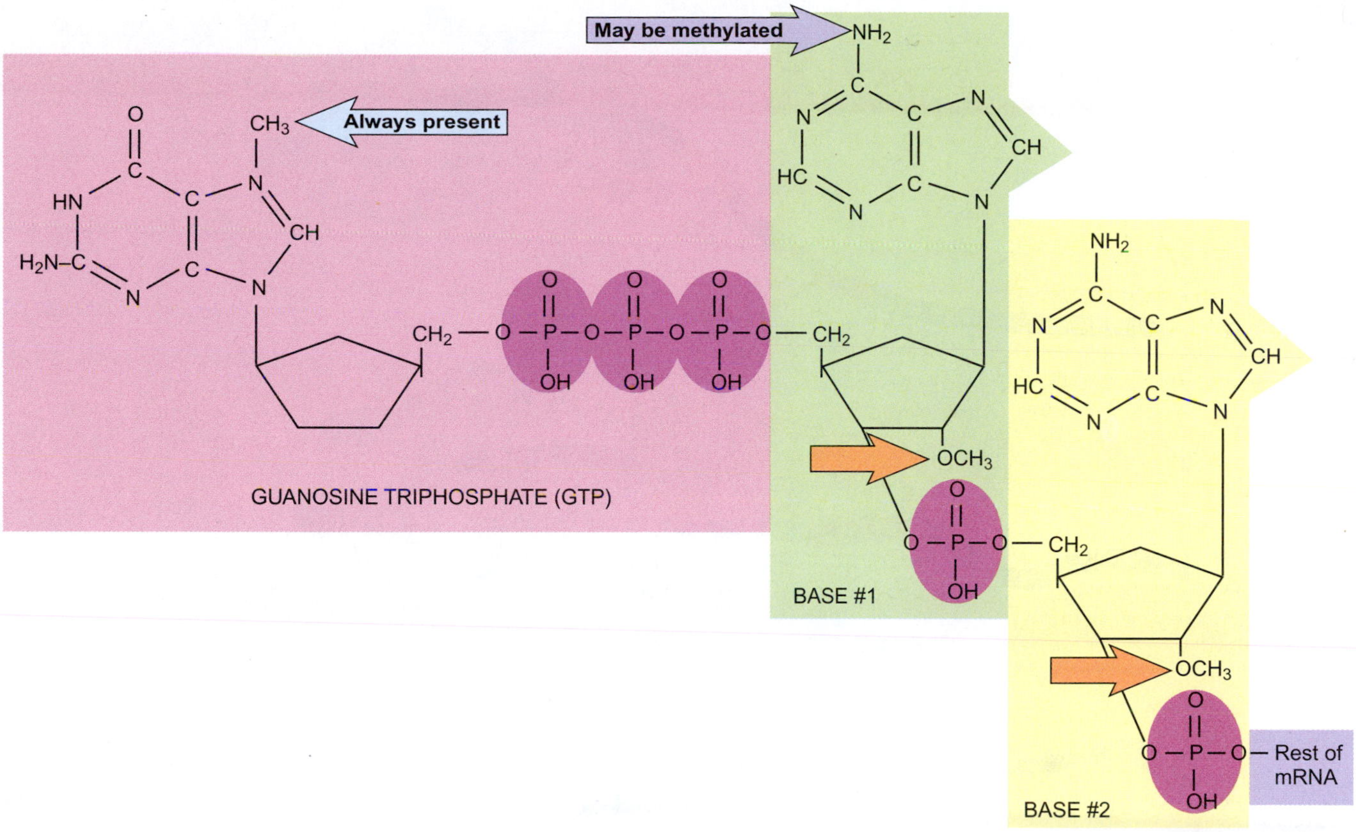

FIGURE 12.05
Capping of Eukaryotic mRNA

A capping enzyme is responsible for adding a guanosine triphosphate to the 5′ end of mRNA. The guanine of the cap is methylated at the 7 position. Possible additions of other methyl groups to the ribose of nucleotides #1 and #2 are indicated.

of many adenines to the rear of the RNA molecule. Next, the introns are removed, which is a process known as **splicing**. This involves cutting out the introns and joining the ends of the exons to generate an RNA molecule that has only the exons; that is, it contains an uninterrupted coding sequence (Fig. 12.04). After the processing is complete, the mRNA exits the nucleus to be translated by ribosomes.

4.1. Capping Is the First Step in Maturation of Eukaryotic mRNA

Before leaving the nucleus, RNA molecules destined to become messenger RNA have a **cap** added to their 5′ ends and a tail added to their 3′ ends. This occurs inside the nucleus and before splicing. Shortly after transcription starts, the 5′ end of the growing RNA molecule is capped by the addition of a guanosine triphosphate (GTP) residue (Fig. 12.05). This is added in a backward orientation relative to the rest of the nucleotides in the RNA. After addition of the GTP, the guanine base has a methyl group attached at the 7 position. This structure is known as a "cap0" structure. Lower eukaryotes and plants only proceed this far.

Further methyl groups may be added to the ribose sugars of the first one or two nucleosides of the original mRNA in some higher eukaryotes (Fig. 12.05). This gives respectively the "cap1" and "cap2" structures. The cap2 structure is especially typical of mammalian mRNA and appears to increase the efficiency of translation. If the first base of the original mRNA is adenine this is sometimes methylated at the N^6 position. Whether or not a particular mRNA always has exactly the same cap structure is uncertain.

The cap on a eukaryotic mRNA consists of GTP in reverse orientation.

---

cap Structure at the 5′ end of eukaryotic mRNA consisting of a methylated guanosine attached in reverse orientation
Splicing Removal of intervening sequences and re-joining the ends of a molecule; usually refers to removal of introns from RNA

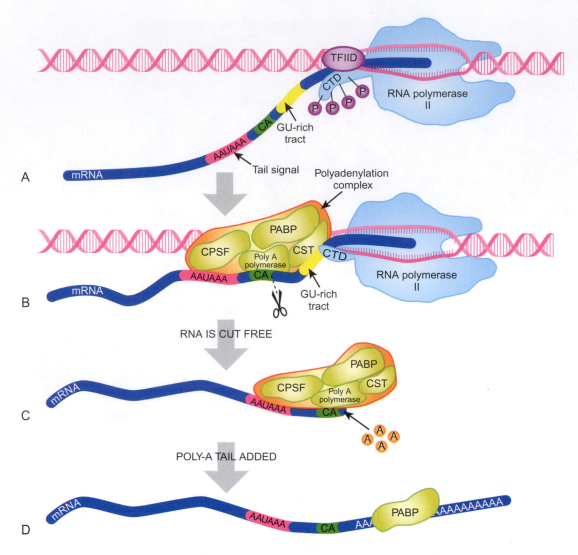

FIGURE 12.06
Addition of Poly(A) Tail to Eukaryotic mRNA

A) During transcription RNA polymerase continues on beyond the end of the coding sequence until the RNA is cut free. Three important sequences beyond the end of the coding sequence are involved in cutting and tail addition: the tail signal (AAUAAA), a CA dinucleotide a few bases downstream, and a GU-rich tract. B) The polyadenylation complex consists of several proteins that bind to these sequences, including poly(A) polymerase itself, cleavage factor, and poly(A)-binding protein (PABP). C) The growing RNA is cut just beyond the CA site by cleavage factor. D) Poly(A) polymerase adds the poly(A) tail to the free 3′ end of the mRNA. The completed poly(A) tail is bound by PABP.

4.2. A Poly(A) Tail Is Added to Eukaryotic mRNA

After being capped, the growing RNA has a **poly(A) tail** added (Fig. 12.06). Transcripts destined to become mRNA have a tail recognition sequence—AAUAAA—close to the 3′ end. The RNA polymerase that is making the RNA molecule continues past this point. However, a specific endonuclease recognizes this sequence and cuts the growing RNA molecule 10–30 bases downstream, just after a 5′-CA-3′ dinucleotide. Beyond the cutting site is a GU-rich tract that is also involved in recognition, although it is lost after cutting.

The AAUAAA and GU-rich sites both bind proteins. Cleavage and polyadenylation specificity factor (CPSF) binds to the AAUAAA sequence and cleavage stimulation factor (CST) binds to the GU-rich tract. These two proteins provide a platform

> The tail of eukaryotic mRNA is a long string of adenines at the 3′ end.

poly(A) tail   A stretch of multiple adenosine residues found at the 3′ end of mRNA

for assembly of cleavage factor and **poly(A) polymerase** as well as the **poly(A)-binding protein (PABP)**. Once this **polyadenylation complex** is assembled, the RNA is cut by the cleavage factor (an endonuclease) and a poly(A) tail is added by the poly(A) polymerase. The tail consists of 100–200 adenine residues (Fig. 12.06).

PABP stays associated with the mRNA and binds to the poly(A) tail. PABP binds both ends of the mRNA as it also appears to protect the cap from being cut off (see below). Whether the presence of the poly(A) tail and PABP make the mRNA more stable is doubtful; some stable mRNAs have very short tails. The poly(A) tail is required for translation, though. Certain mRNAs in early embryos are stored without a poly(A) tail and cannot be translated. When they are needed for translation, the poly(A) tail is added. Exceptions do occur; for example, the genes and mRNA that encode histones are regarded as "special" and do not get a poly(A) tail but are processed differently (see Focus on Relevant Research).

Box 12.1 Trapping mRNA by its Tail

Eukaryotic mRNA may be isolated by taking advantage of its poly(A) tail (Fig. 12.07). Artificial strands of **oligo(U)** or **oligo(dT)** will base pair with poly(A) tracts. Generally, the oligo(U) or oligo(dT) is immobilized on a column and the mixture containing mRNA is poured through the column. The mRNA is trapped by binding of its poly(A) tail (Fig. 12.07). Other molecules, in particular non-coding RNA, pass through.

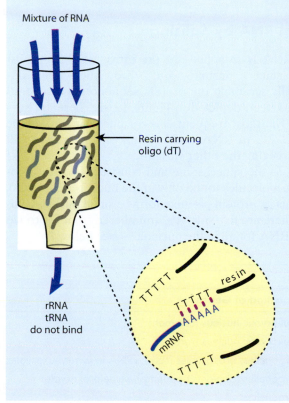

Mixture of RNA

Resin carrying oligo (dT)

rRNA
tRNA
do not bind

FIGURE 12.07
Binding of Poly(A) Tail Allows Isolation of Eukaryotic mRNA

Oligo (dT) attached to a resin binds the mRNA molecules by base pairing with their poly(A) tails. Ribosomal RNA and transfer RNA molecules are not bound and exit the column. The mRNA is then eluted in purified form.

---

oligo(dT)   DNA strand consisting only of thymidine
oligo(U)   RNA strand consisting only of uridine
poly(A) polymerase   Enzyme that adds the poly(A) tail to the end of mRNA
poly(A)-binding protein (PABP)   Protein that binds to mRNA via its poly(A) tail
polyadenylation complex   Protein complex that adds the poly(A) tail to eukaryotic mRNA

Some mRNAs of bacteria are also polyadenylated. However, the role of the poly(A) tail is quite different in prokaryotes. In fact, the poly(A) tail triggers degradation of prokaryotic mRNA. The bacterial poly(A) polymerase is associated with the ribosomes and the tails are relatively short (10–40 bases). Addition of a poly(A) tail to mRNA in chloroplasts also promotes its degradation. This is another indication of the prokaryotic ancestry of these organelles (see Ch. 26).

FOCUS ON RELEVANT RESEARCH

Yang XC, Burch BD, Yan Y, Marzluff WF, and Dominski Z. FLASH, a proapoptotic protein involved in activation of caspase-8, is essential for 3' end processing of histone pre-mRNAs. Mol Cell. (2009) 23:267–78.

The histones are some of the most highly conserved proteins throughout evolution. This is due to their critical function in binding DNA in eukaryotic organisms. Perhaps it is not too surprising that the processing of histone mRNA is different from that of other genes. In particular histone mRNA lacks the poly(A) tail characteristic of most eukaryotic mRNA. Instead, the pre-mRNA ends in a highly conserved stem-loop structure that is then removed before the mRNA can be translated. The endonuclease CPSF73 cleaves the pre-mRNA for histones a few bases downstream of the stem-loop. The small nuclear RNP (snRNP) U7 plus several protein factors are required for recognition of the cleavage site and binding of the nuclease.

In this paper the authors demonstrate that one of the necessary protein factors, in both mammals and insects, is FLASH. The authors used the two hybrid protein screening system (see Ch. 15 for details) to find novel proteins that bound to those proteins already known to be involved in processing histone pre-mRNA. FLASH was bound and further work demonstrated that it was necessary for processing. If insufficient FLASH is present, histone mRNA ends up getting polyadenylated like other eukaryotic mRNA.

FLASH protein is known from its role in promoting apoptosis, suggesting some as yet uncertain link between cell death and histone synthesis.

5. Introns Are Removed from RNA by Splicing

The third step in processing pre-mRNA is to splice out the introns. Splicing must be accurate to within a single base since a mistake would throw the whole coding sequence out of register and totally scramble the protein sequences. The overall result of this cutting and rejoining has been depicted in Figure 12.01. There are several classes of introns (Table 12.01). The most frequent class of intron in eukaryotic nuclear genes is the GT-AG (or GU-AG in RNA code) group of introns. We will therefore discuss these first, before surveying the other variants.

Introns are removed and the exons forming the coding sequence are joined together by the spliceosome.

snRNA molecules recognize the ends of the intron as well as the future branch site.

The splicing machinery is known as the **spliceosome** and consists of several proteins and some specialized, small RNA molecules found only in the nucleus (Fig. 12.08). Each small nuclear RNA (snRNA) plus its protein partners forms a **small nuclear ribonucleoprotein (snRNP)** or **"snurp."** There are five snRNPs—numbered from U1 to U6 with U3 missing! (U3 is actually a snoRNA found in the nucleolus—see below.)

TABLE 12.01	Classes of Intron
Class of Intron	**Location of Genes**
GT-AG (or GU-AG) introns	eukaryotic nucleus (common)
AT-AC (or AU-AC) introns	eukaryotic nucleus (rare)
Group I introns	organelles, prokaryotes (rare), rRNA in lower eukaryotes
Group II introns	organelles (of plants and fungi), some prokaryotes
Group III introns	organelles
Twintrons	organelles
Pre-tRNA introns	tRNA of eukaryotic nucleus
Archeal introns	archaebacterial tRNA and rRNA

small nuclear ribonucleoprotein (snRNP)   Complex of snRNA plus protein
snurp   snRNP or small nuclear ribonucleoprotein
spliceosome   Complex of proteins and small nuclear RNA molecules that removes introns during the processing of messenger RNA

The snRNAs of the snurps recognize three sites on the pre-mRNA. These are the **5′ and 3′ splice sites** and the **branch site**. The vast majority of introns initiate with GU and terminate with AG. Recognition is due to base pairing between the snRNA and the primary transcript. The protein part of the snurp supervises the cutting and joining reactions. This is shown in detail in Figure 12.09 for the **U1** snurp which recognizes the 5′ splice site. In the middle of the intron is a special adenine residue used as a branch site during splicing. The consensus recognition sequences for the 5′ splice site, 3′ splice site, and branch site are as follows (residues in bold are most highly conserved as they are involved in the splicing mechanism):

5′ splice site: 5′–AG ↓ **GU**AAGU-3′

3′ splice site: 5′–YYYYYYNC**AG** ↓ −3′ (Y = any pyrimidine; N = any nucleotide)

branch site: 5′–UAC**U**AAC-3′

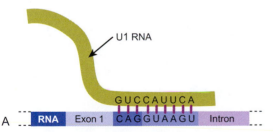

BASE PAIRING OF U1 TO SPLICE SITE

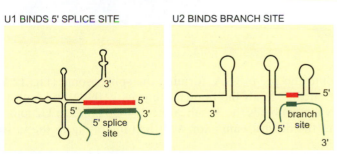

A

FIGURE 12.08

Spliceosome Recognizes Intron/Exon Boundaries

The spliceosome consists of several ribonucleoproteins (U1 to U6), also known as "snurps," which are involved in splicing. These assemble at the splice sites at the intron/exon boundaries.

FIGURE 12.09

Recognition of 5′ Splice Site by U1 Snurp

A) The 5′ splice site at the beginning of the intron is detected by base pairing with the RNA of the U1 snurp. B) The overall binding of U1 at the 5′ splice site and of U2 at the branch site are shown.

3′ splice site	Recognition site for splicing at the downstream or 3′ end of the intron
5′ splice site	Recognition site for splicing at the upstream or 5′ end of the intron
branch site	Site in the middle of an intron where branching occurs during splicing
U1	Snurp (snRNP) that recognizes the upstream splice site

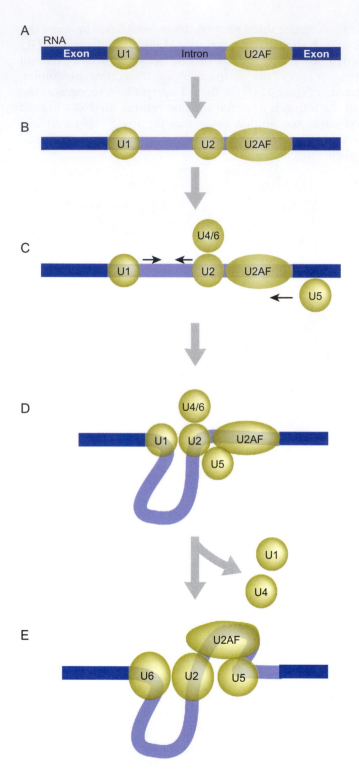

FIGURE 12.10
Stages in Spliceosome Assembly

A) U1 binds the 5′ splice site and U2AF binds the 3′ splice site. B) U2 binds the branch site. C) U4 and U6 bind to U2 and U5 binds to the downstream exon. D) A loop forms by association of U1 and U2. E) U6 displaces U1 from the spliceosome and U4 departs as well.

The spliceosome is made of several snRNA molecules plus accompanying proteins.

The snurps assemble onto the pre-mRNA, forming the spliceosome (Fig. 12.10). U1 recognizes the 5′ splice site, **U2** binds the branch site (Fig. 12.09B), and a protein called **U2AF (U2 accessory factor)** binds the 3′ splice site. U4/U6 binds to U2 and U5 then arrives, binding first to the downstream exon and then migrating to the intron/

U2   Snurp (snRNP) that recognizes the branch site
U2AF (U2 accessory factor)   Protein involved in splicing of introns that recognizes the downstream splice site

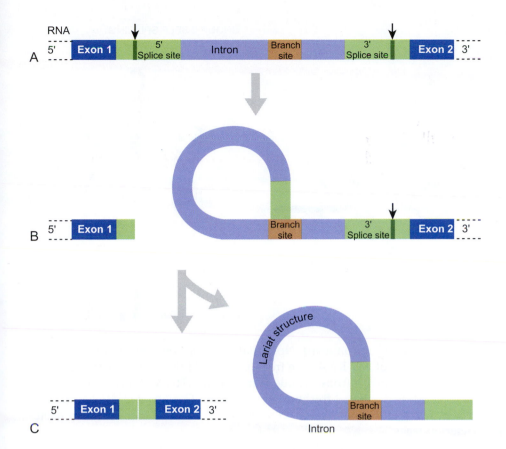

■ **FIGURE 12.11**
The Splicing Reactions

A) Two exons and the intervening intron with their 5′ and 3′ splice sites and branch site are shown prior to splicing. B) The 5′ splice site is cut first and the free end of the intron loops back to bind the branch site. C) The 3′ splice site is severed next and the ends of the two exons are joined. The intron is released as a lariat structure.

exon boundary. This makes the intron RNA loop appear as shown in Figure 12.10. The complex then rearranges itself. In particular, U6 displaces U1 from the 5′ splice site and U1 and U4 are lost from the complex.

Finally, splicing proceeds in two steps (Fig. 12.11). First, the intron and exon are cut apart at the 5′ splice site and the free 5′ end of the intron loops around and is joined to the adenine at the branch site. Second, the free 3′ end of the upstream exon displaces the intron from the 3′ splice site and the two exons are joined together. The intron is released as a branched **lariat structure** that is later degraded.

> The intron is removed and forms a loop with a branch. The exons are joined together.

5.1. Different Classes of Intron Show Different Splicing Mechanisms

There are several classes of introns (Table 12.01). The GT-AG (or GU-AG) introns described above are by far the most frequent in eukaryotic nuclear genes. The AT-AC (or AU-AC) introns are extremely similar to the GT-AG introns except for their different intron boundary sequences. They are processed in an almost identical manner, by a different, but closely related, set of splicing factors.

> Certain introns act as ribozymes and splice themselves out.

Group I introns are **self-splicing**. The RNA itself provides the catalytic activity and thus acts as an RNA enzyme or ribozyme. No proteins are required for splicing. Folding of the RNA into a series of base-paired stem and loop structures is needed for ribozyme activity. The 3D structure is folded so as to bring the two splice sites together and to strain the bonds that will be broken. The reaction pathway starts with the guanosine of any of GMP, GDP, or GTP attacking the 5′ splice site (Fig. 12.12)

lariat structure Branched, lariat-shaped segment of RNA generated by splicing out an intron
self-splicing Splicing out of an intron by the ribozyme activity of the RNA molecule itself without the requirement for a separate protein enzyme

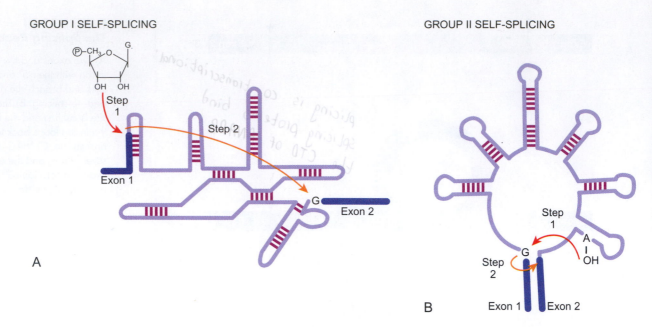

GROUP I SELF-SPLICING   GROUP II SELF-SPLICING

FIGURE 12.12
Group I and Group II Intron Self-Splicing Reactions

In both Group I and Group II self-splicing introns, the intron folds up so as to bring the ends of the two exons together (not shown). For clarity, here we have indicated only the first level of intron folding by base pairing. A) In Group I introns, the 5′ splice site is attacked by a soluble guanosine nucleotide that cuts the exon and intron apart. Next, the free 3′ OH group of the exon reacts with the 3′ splice site and promotes the joining of the two exons (which are actually held close together). B) The self-splicing reaction in Group II introns is similar to Group I, except that an internal adenosine initiates splicing.

> Introns are found in Archaea but are removed by simple ribonucleases without needing a splicesome.

and cutting the exon and intron apart. Note that the guanosine nucleotide is free in solution and is not part of any RNA. The free exon-3′-OH then reacts with the downstream splice site. Group I introns include those in the rRNA of lower eukaryotes, such as the single-celled, ciliated freshwater protozoan, *Tetrahymena*. However, most are found in genes of mitochondria and chloroplasts. Occasional cases occur in bacteria and bacteriophages.

Group II introns are found in the organelles of fungi and plants and occasional examples occur in prokaryotes. Group III introns are found in organelles. Both classes are also self-splicing. However, the reaction is started by attack of an internal adenosine (not a free nucleotide as in Group I introns) (Fig. 12.12). This results in a lariat structure being formed, as in the typical nuclear pre-mRNA introns described above. These three types of intron may thus have a common evolutionary origin. Group III introns are similar to Group II introns, but are smaller and have a somewhat different 3D structure.

Twintrons are complex arrangements in which one intron is embedded within another. They consist of two or more Group I, Group II, or Group III introns. Since introns are embedded within other introns, they must be spliced out in the correct order, innermost first, rather like dealing with parentheses in algebra.

Archeal introns are found in tRNA and rRNA and are similar in some respects to eukaryotic pre-tRNA introns. No complex splicing occurs; no snurps are needed and no ribozymes are involved. The tRNA and rRNA precursors fold up into their normal 3D structures with the intron forming a loop. This loop is cut out by a ribonuclease and the ends joined by an RNA ligase. Their stable 3D structures hold the two halves of the tRNA and rRNA molecules together during cleavage and ligation, and there is no need for extra factors such as snRNPs for recognition or processing.

5.2. R-Loop Analysis Determines Intron and Exon Boundaries

Electron microscopy has been used for direct visualization of eukaryotic introns. Both DNA and mRNA contain the exons that comprise the coding sequence, but the final mRNA lacks the introns (non-coding regions). If mRNA is hybridized to single-stranded DNA from the corresponding gene, the results are regions of base pairing (the exons) interrupted by loops due to the extra intron sequences in the DNA. This is called **R-loop analysis** (Fig. 12.13).

> **R-loop analysis**   Hybridization of the DNA copy of a gene to the corresponding mRNA that results in the appearance of loops that represent the intervening sequences or introns in the DNA

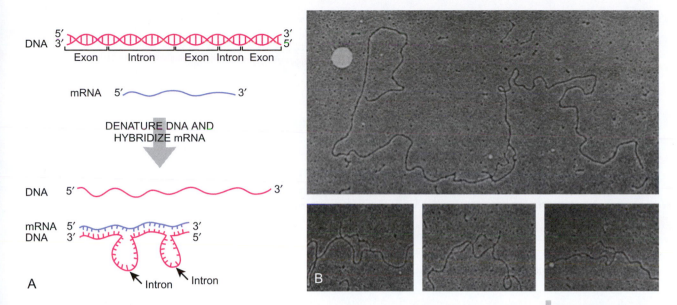

A

B

FIGURE 12.13
R-Loop Analysis to Visualize Introns

A) R-loop analysis can be used to visualize the introns of eukaryotic genes. First, a double-stranded DNA molecule is denatured into two single strands. One strand is annealed to the corresponding mRNA. Because RNA lacks introns the DNA and RNA anneal only in the coding regions. The introns, which are only found in the DNA, remain single-stranded and loop out from the heteroduplex. B) The entire complex can be visualized by electron microscopy after shadowing with metal ions. (Credit: Thomas, M et al., (1976) Hybridization of RNA to double-stranded DNA: Formation of R-loops. Proc. Natl. Acad. Sci. USA 73: 2294–2298.)

6. Alternative Splicing Produces Multiple Forms of RNA

Although any particular splice junction must be made with total precision, eukaryotic cells can sometimes choose to use different splice sites within the same gene. Generally, **alternative splicing** is used by different cell types within the same animal. This allows a single original DNA sequence to be used to make several different proteins that have distinct but overlapping functions. At first glance it might seem that alternative splicing provides, at least in theory, a way for each gene to encode multiple proteins, hence increasing the total number of different proteins available to an organism. However, selection of which alternative splice site to use in a particular cell or tissue must itself be controlled, and this often requires several additional proteins.

6.1. Alternative Promoter Selection

Alternative promoter selection occurs when two alternative promoters are available. The choice of which promoter to use depends on cell-type specific transcription factors. As Figure 12.14 shows, two alternative transcripts result in two different mRNAs.

6.2. Alternative Tail Site Selection

Alternative tail site selection may occur when alternative sites for adding the poly(A) tail are possible. The choice between them again depends on cell type. In this case, cleavage at the earlier poly(A) site results in loss of the distal exon (Fig. 12.15). If the later poly(A) site is chosen, then the earlier poly(A) site and the exon just in front of it are spliced out. This mechanism is used to produce antibodies that recognize the same invading, foreign molecule but that have different rear ends. One type of antibody is secreted into the blood, whereas the other type remains attached to the cell surface.

6.3. Alternative Splicing by Exon Cassette Selection

Alternative splicing by **exon cassette selection** involves a genuine choice between actual splicing sites. Depending on the choice made, a particular exon may or may not

Different proteins can be made from the same gene by alternative splicing during formation of the mRNA.

---

alternative splicing   Alternative ways to make two or more different final mRNA molecules by using different segments from the same original gene

exon cassette selection   Type of alternative splicing that makes different mRNA molecules by choosing different selections of exons from the primary transcript

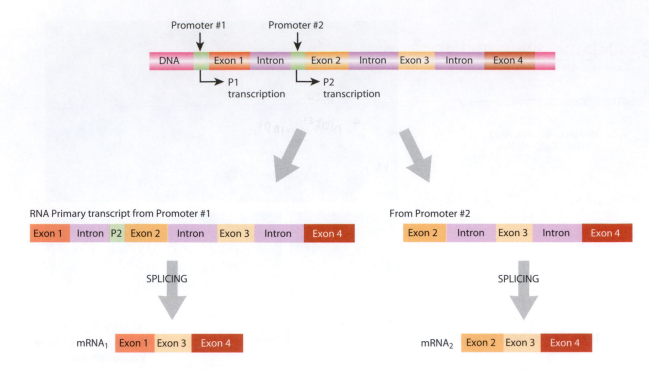

FIGURE 12.14
Alternative Promoter Selection

The DNA contains two promoters that can produce two primary transcripts. If promoter #1 is used, then the segment containing promoter #2 and exon #2 is spliced out. If promoter #2 is used, then exon #1 is not even part of the primary transcript and is therefore not in the mRNA.

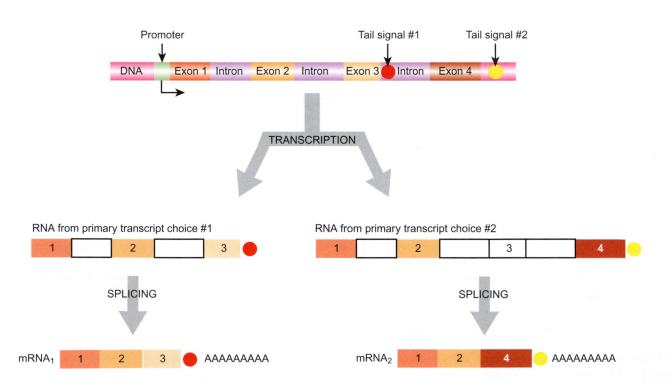

FIGURE 12.15
Alternative Tail Site Selection

In the DNA shown there are two tail signals. Use of the first tail signal results in an mRNA including exons #1, #2, and #3. Use of the downstream tail signal results in an mRNA containing exons #1, #2, and #4.

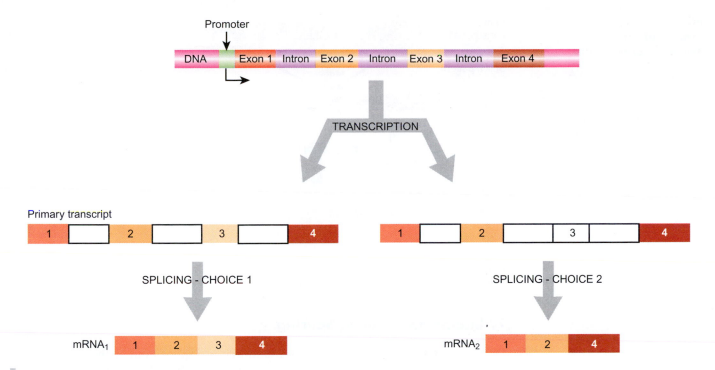

FIGURE 12.16
Alternative Splicing by Exon Cassette Selection

In this example, the DNA contains four exons with intervening introns. Transcription yields a single primary transcript that is spliced in two alternative ways. The same primary transcript is drawn twice in different ways to illustrate the two splicing plans. In one case all four exons are found in the final mRNA, whereas in the other case exon #3 plus the two surrounding introns is removed as a unit.

appear in the final product as shown in Figure 12.16. Here, the primary transcripts are actually the same. They are drawn differently in Figure 12.16 to illustrate the splicing plans. Some cell-type specific factor that recognizes the different possible splice sites must come into play here, but the details are still obscure.

Exon cassette selection occurs in the gene for the skeletal muscle protein troponin T. In the rat this gene has 18 exons. Of these, 11 are always used. Five (exons 4 through 8) may be used in any combination (including none used) and the final two (exons 17 and 18) are mutually exclusive, and one or the other must be chosen. This gives a theoretical mind-boggling 64 possible final mRNAs. The result is that muscle tissue contains multiple forms of this structural protein. The details of troponin splicing vary substantially among different vertebrates. Other muscle proteins quite often show similar multiple forms.

6.4 Trans-Splicing

Trans-splicing, although rare, splices together segments from two different primary transcripts. **Trypanosomes** are parasitic single-celled eukaryotes that cause sleeping sickness and other tropical diseases. They evade immune surveillance by constantly changing the proteins on their cell surfaces by the genetic trick of shuffling gene parts. In addition they indulge in the trans-splicing of many genes (Fig. 12.17). On the other hand, trypanosomes do not appear to have introns and so do not have normal splicing! Although it has only occasionally been reported in vertebrates, trans-splicing of

trans-splicing   Splicing of a segment from one RNA molecule into another distinct RNA molecule
trypanosome   Type of single-celled eukaryotic microorganism that lives as a parasite in higher animals and causes diseases such as sleeping sickness

FIGURE 12.17
Trans-Splicing of Trypanosome mRNA

Two primary transcript RNA molecules are depicted. The splicing shown combines two of the exons from the first RNA with one from the second RNA to give the final messenger RNA.

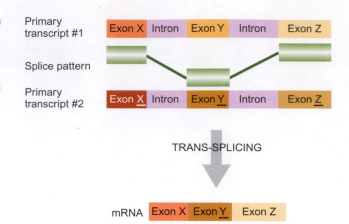

segments from one RNA molecule into another also occurs reasonably often in nematodes and in the chloroplasts of plant cells.

7. Inteins and Protein Splicing

Occasional intervening sequences are found that are spliced out at the protein level. Such protein splicing is rare, which is why it was only noticed relatively recently. **Inteins** and **exteins** are the protein analogs of the introns and exons found in the DNA and RNA. In other words, inteins are intervening sequences in proteins that are present when the protein is first made, but are later spliced out. The final protein is made of the exteins that are now joined together (Fig. 12.18). Inteins have been found in yeasts, algae, bacteria, and archaea (archaebacteria).

Intein splicing involves no accessory enzymes. The intein segment catalyzes its own release as a free polypeptide. Certain specific amino acids must be present at the extein/intein boundaries for the splicing reaction to work. The splicing occurs in two steps, with a branched intermediate. Serine (or cysteine) must be the first amino acid of the downstream extein, as its hydroxyl group (or sulfhydryl if cysteine is used) is needed to carry the upstream extein during the branched stage (Fig. 12.19).

Usually there is just a single intein per protein, but examples are known where multiple inteins are inserted into the same host protein. More bizarre is the case of the *dnaE* gene of *Synechocystis* (a blue-green bacterium). This gene is split in two, and each half has part of the DNA sequence for an intein attached. These two half-genes are transcribed and translated separately to give two proteins (Fig. 12.20). These fold up together and even though the intein is separated into two segments it still manages to cut itself out. The two halves of the DnaE protein are joined together as the intein splices itself out. The intein is released as two fragments.

If the DNA segment that codes for the intein were deleted, the protein would be made in one step, with no intein and no need for protein splicing, and the intein itself would be extinct. However, some spliced-out intein polypeptides are not just a waste product; they are site-specific deoxyribonucleases (DNases). Their role is to protect the existence of the intein. If a mutation occurs that deletes the intein DNA sequence from the middle of the host gene, the previously formed intein DNase cuts the cell's DNA at this point—a potentially lethal move. Thus, any cell with a single copy of DNA that deletes the useless intein DNA will be killed by the intein protein. Only cells that keep the intein survive. Inteins appear unnecessary for cell survival and intein-encoding DNA may be therefore regarded as a form of selfish DNA. The origin of inteins is obscure.

extein   A segment of a protein that remains after the splicing out of any inteins
intein   An intervening sequence in a protein—a segment of a protein that can splice itself out

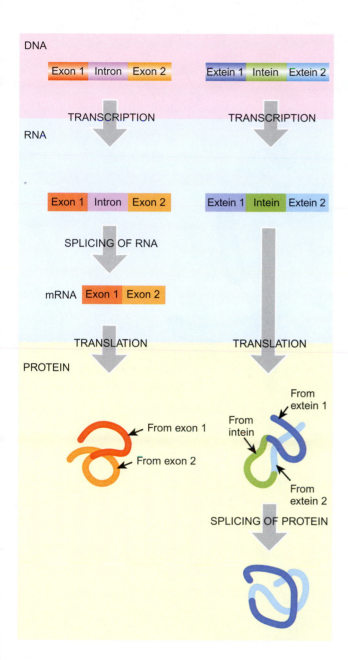

■- **FIGURE 12.18**
Inteins and Exteins in Proteins

On the left is the standard scheme by which introns are eliminated during RNA splicing. The intron is eliminated at the level of RNA and is never translated into protein. On the right is the scheme for removal of intervening sequences at the protein level. Regions remaining in the final protein are called exteins and those destined to be lost are called inteins. The major difference from RNA splicing is that in protein splicing the inteins are cut out after the protein is made.

In eukaryotes there are two copies of each gene. If one copy loses the intein DNA sequence, it will be cut into two fragments by the intein DNase. Yeast and many other eukaryotes can mend double-stranded breaks in their DNA by a special form of recombination. The second (undamaged) copy of the gene where the break occurred lends one strand of its DNA to mend the break. After the single-stranded regions are filled in, the result is a repaired copy of the gene that is identical to the undamaged copy. Now both copies again have the intein DNA sequence inserted. This type of repair process is known as **gene conversion** (Fig. 12.21).

Although most inteins have DNase activity, some shorter inteins exist that do not. Possibly they are defective and have lost the original sequences that encoded the nuclease. Inteins are not alone in their attempts to kill cells that delete their encoding DNA sequence. Certain introns use the same tactic. In this case, the DNA of the

gene conversion   Recombination and repair of DNA during meiosis that leads to replacement of one allele by another. This may result in a non-Mendelian ratio among the progeny of a genetic cross

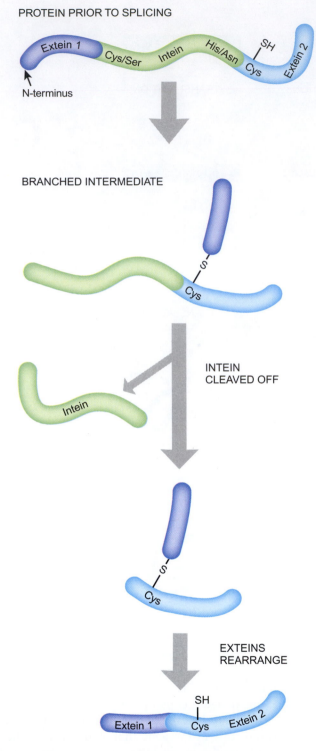

PROTEIN PRIOR TO SPLICING

BRANCHED INTERMEDIATE

INTEIN
CLEAVED OFF

EXTEINS
REARRANGE

FINAL PROTEIN FORMED

FIGURE 12.19
Branched Intermediate during Intein Splicing

The intervening intein segment splices itself out in two stages. The intein has a Cys or Ser at the boundary with extein 1 and a basic amino acid at its boundary with extein 2. The downstream extein (#2) has a Cys residue at the splice junction. Extein 1 is cut loose and attached to the sulfur side chain of the cysteine at the splice junction. This forms a temporary branched intermediate. Next, the intein is cut off and discarded and the two exteins are joined to form the final protein.

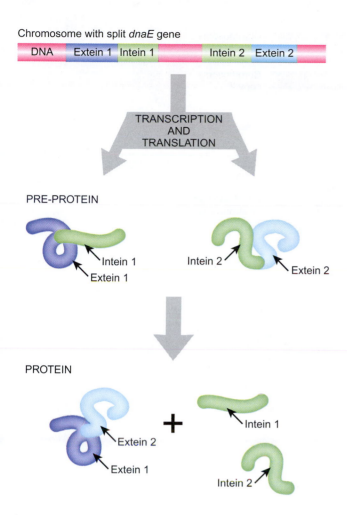

Chromosome with split *dnaE* gene

PRE-PROTEIN

Intein 1
Extein 1

Intein 2
Extein 2

TRANSCRIPTION
AND
TRANSLATION

PROTEIN

Extein 2
Extein 1

+

Intein 1

Intein 2

■ **FIGURE 12.20**
*Intein Splicing
Reconstitutes the DnaE
Protein*

The DNA coding for the DnaE
protein of *Synechocystis* is
transcribed and translated into two
separate proteins, each containing
an intein and an extein. The exteins
of the two proteins are spliced
together by the inteins. During
splicing both inteins are lost.

intron is not just meaningless, but encodes a DNase that cuts in two any copies of the host gene that have lost the intron. Certain plasmids also have mechanisms to kill any cell that loses the plasmid. They use a different and more complicated approach, as plasmids are not inserts in host DNA and so there is no insertion site to recognize. In all such cases, the idea is that only host cells that carry the selfish DNA will survive.

8. Base Modification of rRNA Requires Guide RNA

RNA molecules often contain modified bases. These are made by chemical modification of pre-existing bases. This is especially true of tRNA, which contains a relatively high percentage of many different modified bases (Ch. 13). However, rRNA has several modified bases and even occasional mRNA molecules may have a modified base or two.

In the case of tRNA, individual enzymes are sufficient for the various base modifications. These recognize particular bases in specific regions of various tRNA molecules and modify them. This scenario also applies to bacterial rRNA, which is modified in only a handful of locations. In the case of eukaryotic rRNA, modification occurs at multiple sites and requires small RNA guide molecules called **guide RNA (gRNA)** in addition to the modification enzymes. The gRNA molecules locate the correct sites for modification by base pairing over a short region with the rRNA. Synthesis and processing of rRNA in eukaryotes occurs in the **nucleolus** and so the gRNAs are known as small nucleolar RNAs (snoRNAs).

Modified bases are frequently found in tRNA and rRNA.

Short RNA guide molecules are used to locate the sites for base modification in eukaryotic rRNA.

guide RNA (gRNA) Small RNA used to locate sequences on a longer mRNA during RNA editing
nucleolus Region of the nucleus where ribosomal RNA is made and processed

FIGURE 12.21
Gene Conversion Repairs Broken Chromosomes in Eukaryotes

A double-stranded break in one member of a pair of chromosomes can be repaired in eukaryotes. Repair uses the intact chromosome and involves base pairing of the fragmented DNA with the intact DNA. After recognition, the base pairs in the gap are filled in and, when complete, the chromosomes separate.

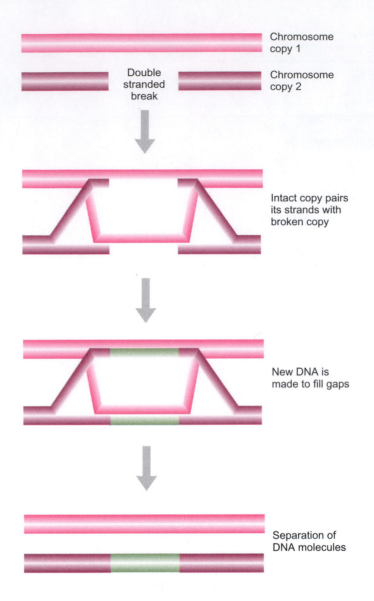

Chromosome copy 1

Double stranded break

Chromosome copy 2

Intact copy pairs its strands with broken copy

New DNA is made to fill gaps

Separation of DNA molecules

Methylated bases and pseudo-uridine are the most common modifications in eukaryotic rRNA.

Many gRNA molecules (snoRNA) are encoded inside introns.

Nucleotides in eukaryotic rRNA are modified by methylation of the 2′-OH group of the ribose or by converting uridine to **pseudouridine**. Although the number of different types of modification is limited, the number of sites is very large. Thus human pre-rRNA is methylated at 106 positions and pseudo-uridylated at 95. The base sequences around the modification sites are rarely related and so there is no consensus sequence for a modifying enzyme to use. Instead, there is a different snoRNA for each modification site. Each snoRNA is 70–100 nucleotides long and has a unique sequence that recognizes the modification site on the rRNA. In addition, all snoRNAs that recognize potential methylation sites share a sequence motif that is recognized by the methylase that modifies the bases (Fig. 12.22). Similarly, a family of snoRNAs that recognize pseudo-uridylation sites have a motif that binds the pseudo-uridylation enzyme.

Interactions between rRNA and snoRNA often involve G-U base pairing. This non-standard base pair is stable in double-stranded RNA and also occurs in the base pairing between the gRNA and mRNA in RNA editing (see below).

Since there are many base modifications there are several hundred different snoRNAs per cell in eukaryotes. Only a few snoRNAs are transcribed from standard genes. Most snoRNAs are encoded in the introns of other genes. The snoRNA is made by cutting up the intron after it has been spliced out of the mRNA (Fig. 12.23).

pseudouridine   An isomer of uridine that is introduced into some RNA molecules by post-transcriptional modification

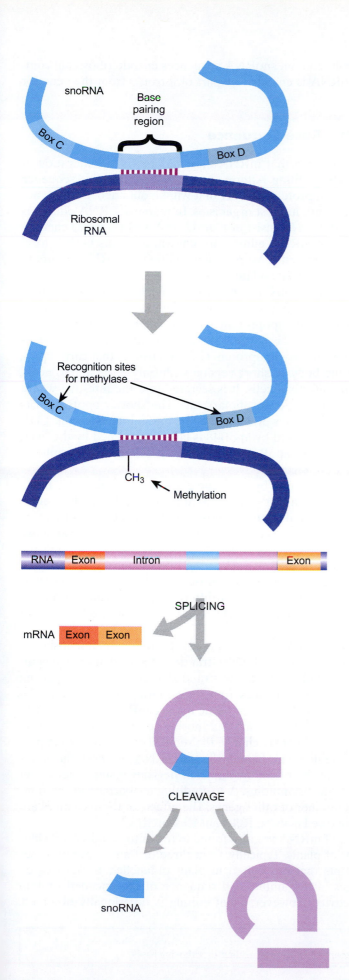

◾ **FIGURE 12.22**
Recognition of Modification Sites on rRNA by snoRNA

The site on the rRNA to be modified is identified by base pairing to specific sequence on the snoRNA. After snoRNA/rRNA pairing, the methylase binds to the *BoxC* and *BoxD* sequences and then methylates one of the bases of the rRNA.

◾ **FIGURE 12.23**
Generation of snoRNA from Intron

After splicing the mRNA, the intron assumes a lariat structure. The snoRNA is spliced from the intron, leaving smaller fragments of RNA.

Many of the genes whose introns contain snoRNA sequences encode ribosomal components; for example, U16 snoRNA is encoded by part of intron 3 from the gene for L1 ribosomal protein.

9. RNA Editing Alters the Base Sequence

Perhaps the most bizarre modification that mRNA may undergo is the alteration of its base sequence—known as **RNA editing**. Altering bases within the coding sequence of mRNA usually changes the final protein product that will be obtained. Perhaps not surprisingly RNA editing is very rare in most organisms. In mammals, RNA editing is restricted to base substitution and may consist of C-to-U or A-to-I (inosine) changes and relatively few specific cases of mRNA editing are known. In plants both C-to-U and U-to-C editing occur quite frequently. More radical editing of mRNA occurs in certain protozoa where bases are inserted and deleted.

An example of C-to-U editing occurs in the human gene for apolipoprotein B, which encodes a protein of 4,563 amino acids, one of the longest polypeptide chains on record. The full-length protein, apolipoprotein B100, is made in liver cells and secreted into the bloodstream. ApoB100 is required for the assembly of very low density lipoprotein (VLDL) and low density lipoprotein (LDL) particles that carry lipids, including cholesterol, around the body. A short version with only 2,153 amino acids, apolipoprotein B48, is made by intestinal cells. It is secreted into the intestine where it plays a role in the intestinal absorption of dietary fats. The shorter apoB48 lacks the region of apoB100 (between residues 3,129 to 3,532) that is bound by the LDL receptor. Consequently, dietary fat carried by apoB48 is largely absorbed by the liver, whereas VLDL or LDL particles with apoB100 can deliver cholesterol to peripheral tissues whose cells possess the LDL receptor.

The short version, apoB48, is encoded by the same gene as apoB100 and is made by editing the mRNA. The CAA codon (for glutamine) at position 2,154 is changed to a UAA stop codon by an enzyme that deaminates this specific cytosine, so converting it to uracil. Several accessory proteins are needed to make sure that the **deaminase** binds only at the correct site (Fig. 12.24).

Although only a single gene is needed to encode the two versions of apolipoprotein B, editing needs several extra proteins to recognize the editing site and convert the C to U. Having two apolipoprotein B genes of different lengths would surely be more economical, and there is no known reason why the more complicated procedure of mRNA editing is used.

A-to-I editing of mRNA also occurs in mammals. In this case adenosine is converted by deamination into inosine by double-stranded RNA adenosine deaminase. Recognition is due to formation of double-stranded regions by base pairing between the modification sites and sequences from neighboring introns. Thus, the intron sequence affects the final coding sequence of the mature mRNA. Consequently, this editing must occur before removal of the introns. Inosine acts as guanosine during translation and therefore A-to-I editing may change the sequence of the resulting protein if it occurs within a coding region. This happens in the mRNAs for both the glutamate and serotonin receptor proteins of the mammalian nervous system. Defects in editing result in severe neurological symptoms. A-to-I editing also occurs in the noncoding regions of a significant number of other genes. The effects of these changes are for the most part still uncertain (see Focus on Relevant Research).

C-to-U and U-to-C editing of mRNA are also found in the mitochondria and chloroplasts of most major groups of plants. Typically, from three or four to twenty bases are changed in those transcripts that are edited in plant organelles. In most cases, such editing results in changes in the amino acid sequence of the encoded protein that are necessary for full activity. However, silent editing is occasionally observed.

RNA editing involves changing the actual base sequence of messenger RNA.

Several mRNAs encoding proteins of the mammalian nervous system undergo A-to-I editing in the coding sequences.

deaminase   An enzyme that removes an amino group
RNA editing   Changing the coding sequence of an RNA molecule after transcription by altering, adding, or removing bases

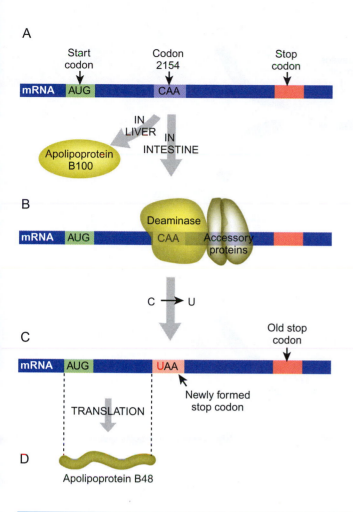

A

Start codon   Codon 2154   Stop codon

mRNA AUG   CAA

IN LIVER   IN INTESTINE

Apolipoprotein B100

B

mRNA AUG   CAA

Deaminase   Accessory proteins

C → U

C

Old stop codon

mRNA AUG   UAA

Newly formed stop codon

TRANSLATION

D

Apolipoprotein B48

■ **FIGURE 12.24**
Editing of Apolipoprotein B mRNA

A) The apolipoprotein B gene normally makes apolipoprotein B100 in the liver. B) In the intestine the mRNA is altered by base editing to make apolipoprotein B48. A deaminase binds to a CAA codon in association with accessory proteins. C) The cytosine in the CAA codon is converted to uracil giving UAA. D) The UAA serves as a new stop codon that halts translation in the middle of the coding region. This results in a truncated protein, apolipoprotein B48.

Farajollahi S and Maas S (2010). Molecular diversity through RNA editing: a balancing act. Trends in Genetics 26: 221–231.

RNA editing by deamination of adenosine to inosine affects multiple processes. Since the translation machinery regards inosine as equivalent to guanosine, some edited protein coding sequences insert different amino acids from those expected from the DNA sequence of the gene. Alternative splicing may also be affected if editing changes the splice recognition sites. Micro RNA sequences are sometimes altered by RNA editing that may result in changes in the level of expression of genes regulated by those miRNAs (see Ch. 18). In addition, there is often widespread editing of repetitive sequences derived originally from retrotransposons (see Ch. 22).

In many cases the precise effects of adenosine to inosine editing are still uncertain. Nonetheless, cases are known where defects in editing cause human disease. Nerve cells, especially in the brain, show high levels of A-to-I editing. As a result the behavior of mammals is more affected by A-to-I RNA editing than is basic physiology. An example is the 5HT2C serotonin receptor that affects depression and schizophrenia. Humans with depression show changes in the editing pattern of this receptor.

FOCUS ON RELEVANT RESEARCH

The ultimate in pointlessness appears to be the case of the tobacco chloroplast *atpA* gene where a CUC codon is edited to CUU in the mRNA. Both codons encode serine. Conceivably, such silent editing might be an adjustment to the differential availability of tRNAs with different anticodons, however there is no evidence for this.

Trypanosomes frequently practice RNA editing. Moreover, they do not merely modify bases chemically but actually insert or remove them. Some of the primary transcripts of trypanosomes, especially those from the mitochondrial genes, are altered by insertion or removal of multiple uridine nucleotides, one at a time, before the final mRNA is generated (Fig. 12.25). In these cases, the coding sequences found on the DNA

Trypanosomes frequently edit their mRNA by inserting or removing bases.

FIGURE 12.25
Editing of Trypanosome mRNA

The gRNA base pairs along a specific region of the trypanosome mRNA. The extra adenine (A) of the AGA sequence in the guide RNA is used as a template to insert uracil (U) into the mRNA. (Note: some slightly distorted duplex RNA structures allow guanine, as in the AGA sequence shown, to pair with uracil.)

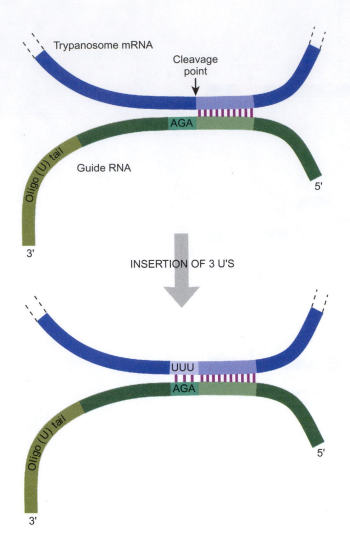

have incorrect reading frames. If the trypanosome did not edit its mRNA, the result would be defective, frame-shifted proteins made from out of phase coding sequences.

Multiple insertions of U in trypanosome mRNAs occur at positions specified by short guide RNAs. These are complementary to short stretches of the mRNA but have an extra A. The U residues are inserted into the mRNA opposite the extra A on the guide RNA.

10. Transport of RNA out of the Nucleus

The nucleus is surrounded by a double membrane. Each nucleus has many pores that allow macromolecules in or out in a carefully controlled manner (Fig. 12.26). Each **nuclear pore** is surrounded by a cluster of proteins known as the nuclear pore complex that controls entry and exit. The nuclear pore complex is the largest protein complex in the cell, with a mass of about 65 MDa in yeast and about double that in higher eukaryotes. It consists of about 30 different proteins (known as nucleoporins or nups). The pore complex has 8-fold symmetry (Fig. 12.27).

The details of precisely which molecules are allowed in or out are still murky. We do know that once messenger RNA has received its cap and tail and had its introns spliced out, it is free to exit the nucleus. Binding of the spliceosome to the mRNA prevents it from leaving until splicing is finished. However, the final exit of mRNA requires a variety of other protein factors. Exportins and importins are protein factors known to control the exit and entry of specific classes of molecules through the

nuclear pore   Pore in nuclear membrane that allows proteins and RNA into and out of the nucleus

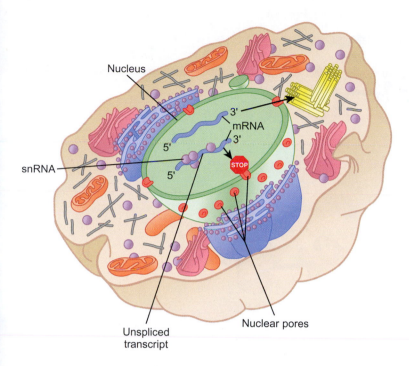

Nucleus

mRNA

snRNA

Unspliced

Nuclear pores

FIGURE 12.26
***Transport of Eukaryotic
mRNA out of the Nucleus***

Processed mRNA is free to leave
the nucleus, whereas primary
transcript still attached to snRNA is
prevented from leaving.

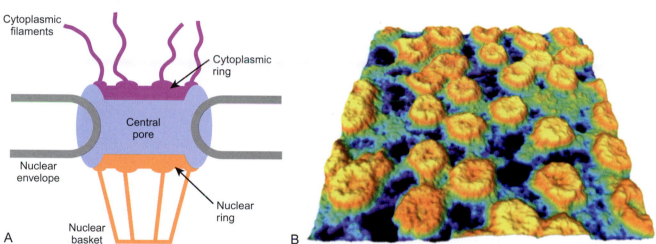

Cytoplasmic
filaments

Cytoplasmic
ring

Central
pore

Nuclear
envelope

Nuclear
ring

Nuclear
basket

A

B

FIGURE 12.27
Nuclear Pore Complex

A) Structure of nuclear pore complex showing 8-fold symmetry. B) Nuclear envelope of *Xenopus laevis* oocyte, a "biological nanoarray," imaged
with atomic force microscopy. The area is about 600 × 600 nm. The image height (z-axis) is 15 nm. Nuclear pore complexes are the yellow
structures. The lipid bilayer membrane visible between the pores is more or less dark-blue.
(Credit: Shahin, Schillers and Oberleithner, Institute of Physiology II, University of Muenster, Germany.)

nuclear pores. For example, exportin-t is specific for export of tRNA and exportin 5
for microRNA precursors. Transport out of the nucleus of large molecules like RNA
and proteins require energy. This is obtained by the hydrolysis of GTP.

11. Degradation of mRNA

Messenger RNA molecules are relatively short-lived and in bacteria the half-life is
generally only a few minutes. Molecules of mRNA that are not bound to ribosomes
are especially vulnerable to degradation. Bacteria contain multiple **ribonucleases** that

ribonuclease   A nuclease that cuts RNA

FIGURE 12.28
Degradation of Prokaryotic mRNA

A) The mRNA being translated has a 5' end unprotected by ribosomes. An endonuclease (RNase E) cuts near the 5' end. B) The released fragment is then cut by exonuclease, starting at its 3' end. C) The mRNA is shortened and the severed segment is degraded.

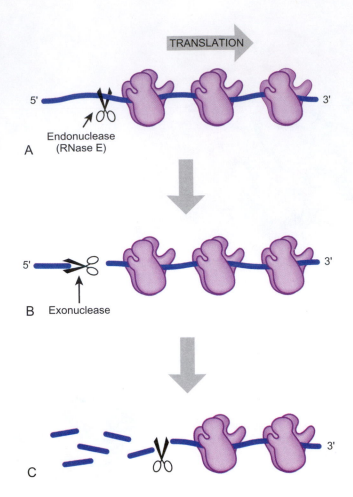

Messenger RNA is degraded after a relatively short lifetime.

Eukaryotic mRNA must have the tail and cap removed before degradation can proceed.

are involved both in processing the precursors to tRNA and rRNA and in the degradation of mRNA. These ribonucleases can often substitute for one another at least to some extent and so mutants that have lost only one ribonuclease are usually still viable. Bacterial mRNA is degraded in two stages (Fig. 12.28). First, an **endonuclease**, usually ribonuclease E, cleaves regions that are unprotected by ribosomes. Next, **exonucleases** that move in a 3'-to-5' direction degrade the fragments. Note that *overall* degradation moves in a 5'-to-3' direction due to the initial endonuclease following the ribosomes.

Degradation of mRNA follows a different route in eukaryotes such as yeast. First, the poly(A) tail and then the cap must be removed before actual nuclease digestion. Once the poly(A) tail is shortened to 10–20 bases the poly(A)-binding protein (PABP) is released. Only after the PABP has gone can the cap be removed. Once the cap has also gone, an exonuclease, Xrn1, degrades the mRNA in the 5'-to-3' direction (Fig. 12.29).

The stability of eukaryotic mRNA depends on the presence or absence of destabilizing sequences. Short-lived mRNA often contains an AU-rich sequence of about 50 bases (known as an ARE) in its 3'-UTR. The consensus for the ARE is multiple repeats of the 5-base sequence, AUUUA (hence, ARE = AUUUA repeat element). An ARE-binding protein recognizes the ARE and promotes deadenylation and degradation (Fig. 12.30).

11.1. Nonsense-Mediated Decay of mRNA

Eukaryotic cells possess a special RNA surveillance mechanism that destroys mRNA molecules that contain premature stop codons. Such defective mRNA molecules may result from expression of genes with nonsense mutations, in which a codon that codes

endonuclease A nuclease that cuts a nucleic acid in the middle
exonuclease A nuclease that cuts a nucleic acid at the end

A

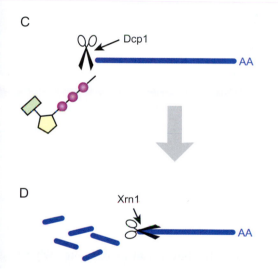

■ **FIGURE 12.29**
Degradation of Eukaryotic mRNA

A) The mRNA is shown with the poly(A) tail bound to poly(A)-binding protein (PABP). B) An exonuclease sequentially removes the poly(A) tail. C) A decapping protein (Dcp1) removes the cap. D) Degradation of the mRNA proceeds in the 5'-to-3' direction due to exonuclease Xrn1.

for an amino acid is mutated into a stop codon, also known as a nonsense codon. Consequently, the mechanism for destroying these mRNAs is known as **nonsense-mediated decay** or **NMD**.

NMD plays a protective role in eukaryotic organisms that are heterozygotes with one functional allele and one allele carrying a nonsense mutation. Full expression of the nonsense allele would result in production of a truncated protein. Sometimes this is merely a waste of resources. However, many polypeptides form multi-subunit complexes (either with themselves or with other polypeptides). In this case, aberrant forms of the protein may still bind to the complex and interfere with normal function. Hence, some truncated proteins are actively dangerous. Degradation of mRNA carrying the nonsense allele prevents the synthesis of the aberrant truncated proteins and so protects heterozygotes from possible deleterious effects.

Despite its name, nonsense-mediated decay probably evolved to deal with defective mRNAs that result from errors in the expression of normal genes rather than from inherited mutations. In particular, errors during the complex RNA splicing process that removes introns can lead to defective mRNA molecules. It is notable that

Special mechanisms detect and destroy defective mRNA molecules.

nonsense-mediated decay (NMD)   Mechanism used to destroy mRNAs that contain a premature stop codon that is found in eukaryotes

FIGURE 12.30
**_ARE Sequence Facilitates
Digestion of Eukaryotic
mRNA_**

A) The structure of undegraded
eukaryotic mRNA shows the AUUUA
repeat sequence, the cap, and the
poly(A) tail. B) ARE-binding protein
recognizes a repeated AUUUA
sequence. C) Poly(A) ribonuclease
degrades the poly(A) tail.
D) Endonucleases sever the mRNA
at multiple sites.

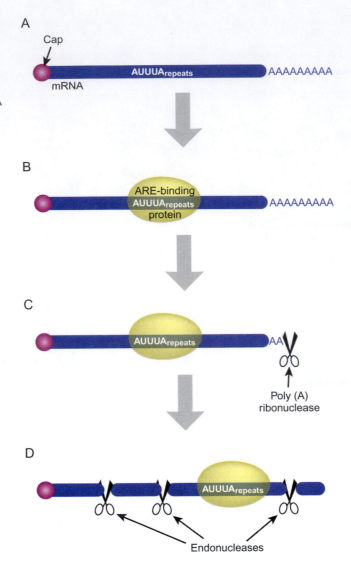

nonsense-mediated decay is only found in eukaryotes but not in prokaryotes where
splicing is largely absent.

Defective mRNA with premature stop codons may result from several events:

1. Expression of a mutant gene with an internal nonsense mutation.

2. Errors during expression of normal genes that create nonsense mutations.

3. Errors during transcription that insert incorrect bases resulting in a premature
   stop codon.

4. Errors during splicing that alter the reading frame and so create in frame stop
   codons.

5. Errors during splicing that result in retaining all or part of an intron whose
   sequence includes in frame stop codons.

6. Errors during RNA editing (presumably, although this has not been directly
   demonstrated).

Nonsense-mediated decay is triggered whenever there is a stop codon more than
50–55 nucleotides upstream of the final exon-exon junction created during splicing.
This requires that the location of the exon-exon splice junctions should somehow be
marked on the mature mRNA. In animal cells exon-exon junctions are labeled when
the primary transcript is converted to the mature mRNA during the splicing process.
A complex of proteins, known as the exon junction complex (EJC), is bound to the
mRNA about 20–24 nucleotides upstream of each exon-exon junction (Fig. 12.31).

Nonsense-mediated decay
of mRNA is triggered by the
presence of premature stop
codons.

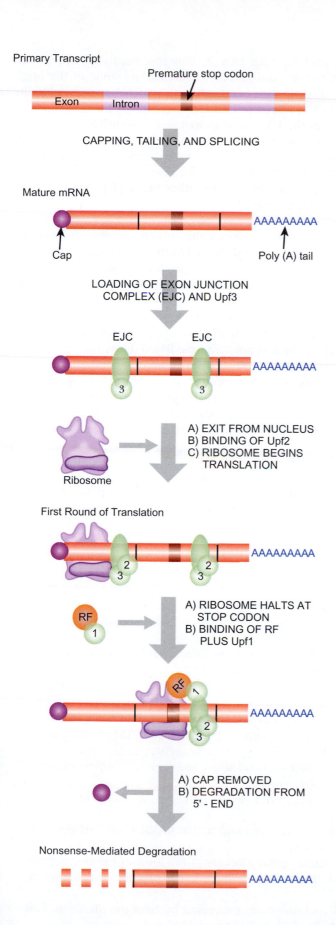

Primary Transcript

Premature stop codon

Exon   Intron

CAPPING, TAILING, AND SPLICING

Mature mRNA

Cap

Poly (A) tail

LOADING OF EXON JUNCTION
COMPLEX (EJC) AND Upf3

EJC   EJC

3   3

Ribosome

A) EXIT FROM NUCLEUS
B) BINDING OF Upf2
C) RIBOSOME BEGINS
   TRANSLATION

First Round of Translation

2   2
3   3

RF
1

A) RIBOSOME HALTS AT
   STOP CODON
B) BINDING OF RF
   PLUS Upf1

RF
1

2
3

A) CAP REMOVED
B) DEGRADATION FROM
   5' - END

Nonsense-Mediated Degradation

AAAAAAAAA

FIGURE 12.31
***Nonsense-Mediated Decay
of Eukaryotic mRNA***

The primary transcript is capped
and tailed and the introns are
spliced out to give mature mRNA.
In this case the mRNA contains
a premature termination codon.
During splicing the mRNA
is loaded with exon junction
complexes (EJC) upstream of each
exon-exon junction. Upf3 protein
binds to the EJC. The mRNA leaves
the nucleus and Upf2 protein
binds to the EJC. The ribosome
loads onto the mRNA and the
first round of translation occurs. If
translation does not remove all of
the EJCs, then nonsense-mediated
decay is triggered by binding of
release factor (RF) plus Upf1 to the
remaining EJC. Numbered circles
represent Upf1, Upf2, and Upf3.

The UP-frameshift (Upf) proteins take part in nonsense-mediated decay. Two of the three Upf proteins bind to the exon junction complex and some of the original members of the EJC are lost. Upf3 binds first, while the mRNA is still inside the nucleus. Upf2 binds after the mRNA has exited the nucleus. During the first round of translation, the ribosome displaces the EJCs as it moves along the mRNA. If there is a premature stop codon, the ribosome finishes translating before all of the EJC complexes have been bumped off the mRNA. In this case, the termination complex, which includes the release factor plus Upf1, interacts with the remaining EJC (Fig. 12.31). This apparently involves the binding of Upf1 by the other two Upf proteins. This in turn triggers destruction of the mRNA molecule.

In nonsense-mediated decay, the first step is removal of the cap from the mRNA. (This is opposite of normal mRNA degradation where the poly (A) tail must be removed before the cap—see above). Next, the mRNA is degraded from the exposed 5′ end.

In yeast, less than 5% of genes contain introns. Therefore, most yeast genes do not require splicing of the primary transcript to generate the mRNA. Consequently, there are no exon-exon junctions to serve as markers for nonsense-mediated decay. Instead, most yeast mRNA contains downstream sequence elements (DSE). The DSE sequences are rather ill-defined but are AU rich. The DSE sequences serve as binding sites for the proteins involved as markers in nonsense-mediated decay.

Nonsense-mediated decay in yeast also differs from animals in another respect. In both mammals and in the roundworm, *Caenorhabditis elegans*, NMD is regulated by phosphorylation of protein Upf1. This does not occur in yeast. Phosphorylation probably occurs during the translation termination process and is necessary for NMD to proceed. Addition and removal of phosphate from Upf1 requires several other proteins. These are absent in yeast.

Yeast mutants with knockouts in the *Upf* genes grow nearly normally on many media, but show impairment of mitochondrial function. *C. elegans* with impaired NMD is viable. However, the reproductive system develops abnormally and fertility is greatly reduced. In mammals, defects in the *Upf* genes appear to be lethal.

Key Concepts

- The original RNA molecule that results from transcription is known as the primary transcript. Before performing its biological role, the primary transcript is often processed.
- RNA processing mechanisms include cutting, splicing, base modification, and addition of extra nucleotides.
- Ribosomal and transfer RNA are derived from their primary transcripts by cutting and trimming and also by base modification.
- Eukaryotic messenger RNA is processed in three major stages. Firstly, a cap is added at the 5′ end. Secondly, a poly(A) tail is added at the 3′ end. Finally, the introns are spliced out.
- The splicing out of introns takes several steps and requires a ribonucleoprotein complex known as the spliceosome.
- Several different classes of intron exist that follow different splicing mechanisms.
- Some primary transcripts may undergo alternative splicing patterns that generate multiple forms of mRNA.
- A few cases are known where splicing occurs at the level of the polypeptide chain, rather than RNA. Intervening sequences in proteins are known as inteins.
- Many non-coding RNA molecules are processed by base modification. This sometimes requires the help of guide RNAs (gRNA).
- Altering the base sequence is known as RNA editing and may change the sequence of the encoded protein when performed on mRNA.

- The transport of RNA into the eukaryotic cytoplasm from the nucleus is controlled by a protein complex at the nuclear pore.
- Several systems exist to degrade RNA molecules once they are no longer needed or if they were made defectively in the first place.

Review Questions

1. What are coding and non-coding RNAs?
2. What are snRNA, snoRNA, and scRNA?
3. What are ribozymes?
4. List the ribosomal RNAs found in both prokaryotes and eukaryotes.
5. Since ribosomal RNAs are transcribed with linker regions the pre-rRNA must be cut to yield the mature rRNAs; outline this process.
6. What is the main difference between the enzyme that cuts pre-rRNA and the enzyme that processes tRNA?
7. What three steps are needed to produce mRNA in eukaryotes?
8. What is needed to make a cap?
9. What is meant by cap0, cap1, and cap2?
10. What enzymes and proteins are needed to add the poly (A) tail?
11. How is the poly (A) tail added?
12. What is the difference between the role of the poly (A) tail in eukaryotes and in prokaryotes?
13. What is the spliceosome? What are snurps? Name all five of them.
14. What sites on the pre-mRNA are recognized by the snRNA of the snurps?
15. How is the spliceosome assembled?
16. What are the two major steps in the final splicing of an intron?
17. How are group I introns different than the GT-AG introns?
18. What is meant by self-splicing? How does this happen?
19. How do group II and group III introns differ from group I introns?
20. What are some benefits of alternative splicing?
21. What are the different types of alternative splicing? How do they differ?
22. What is trans-splicing and how is it useful?
23. What are inteins and exteins?
24. How are inteins like group I introns?
25. What are the steps of intein splicing?
26. Why is the DNA that encodes the intein called selfish DNA?
27. What is gene conversion?
28. How are base modifications made to rRNA?
29. How are snoRNA molecules made?
30. What is RNA editing? Give an example.
31. How does RNA degradation in eukaryotes differ from RNA degradation in prokaryotes?
32. Why must defective mRNA be removed?
33. What is the purpose of nonsense-mediated decay?
34. What is the trigger for nonsense-mediated decay?

Conceptual Questions

1. A piece of eukaryotic DNA is hybridized to its mRNA and the hybrid structure is visualized by electron microscopy. The structure has multiple single-stranded loops like the diagram below. Are these loops made of RNA or DNA? Why are the sequences of the mRNA and DNA different in eukaryotes? Would this look different if the hybridization occurred between the primary RNA transcript and its DNA?

2. What are the specific nucleotide sequences and proteins necessary for correct splicing of introns from the primary transcript?

3. List the different types of RNA molecules found in eukaryotes and their functions.

4. Discuss why degradation of prokaryotic mRNAs moves in a 5′-to-3′ direction overall, but the endonuclease that starts the degradation works in a 3′ to 5′ direction.

5. In a recent experiment, a researcher was using Northern blot analysis to determine the size of mRNA for the gene she is studying called RHG3 from soybeans. She isolated total cellular RNA (found in the nucleus and cytoplasm) and she isolated mRNA from the cytoplasm from soybean seeds. She ran the isolated RNA on an agarose gel to separate the fragments by size, and then used capillary action to move the RNA fragments onto a membrane. The RNA fragments stay in the same position and location during the capillary transfer although the membrane is a mirror image of the gel. She then took a section of the gene and labeled the DNA with radioactivity. This is called a probe, and when she denatures the double-stranded DNA and incubates it with the membrane, it can form a hybrid to any complementary RNA. This procedure is called a Northern blot. Her results are shown below. Interpret the meaning of the bands in both lanes. Lane A = Total cellular RNA; lane (B) = mRNA.

Protein Synthesis

Approximately two-thirds of the organic matter of a typical cell consists of protein. Proteins are the final products of gene expression in most cases, and the term **proteome** refers to the complete set of proteins encoded by a genome, or the total number of proteins found in an organism at one point in time. They are made by translating **messenger RNA (mRNA)** using a molecular decoding machine known as a ribosome. **Translation** refers to the conversion of genetic information in nucleic acid "language" to protein "language." Proteins are polymers made of amino acids, and each of the three bases of DNA or RNA corresponds to a single amino acid in the protein chain. Not only is the genetic information carried to the ribosome by mRNA, but other RNA molecules take part in the process of translation itself. **Ribosomal RNA (rRNA)** catalyses the formation of the peptide bonds between the individual amino acids and **transfer RNA (tRNA)** acts as an adaptor between the nucleic acid and protein "languages." Thus, RNA lies at the very heart of gene expression to give proteins.

> **messenger RNA (mRNA)** The type of RNA molecule that carries genetic information from the genes to the rest of the cell
> **proteome** The total set of proteins encoded by a genome or the total protein complement of an organism
> **ribosomal RNA (rRNA)** Class of RNA molecule that makes up part of the structure of a ribosome
> **translation** Making a protein using the information provided by mRNA
> **transfer RNA (tRNA)** RNA molecules that carry amino acids to the ribosome

FIGURE 13.01
How Many Proteins Per Gene?

A) Normally each gene is transcribed giving one mRNA and this is translated into a single protein. Variations in the normal theme are B) alternative splicing, C) polyproteins, and D) multiple proteins due to the use of different reading frames.

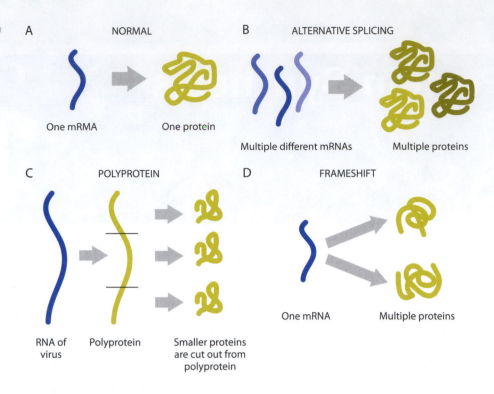

A   NORMAL

One mRMA   One protein

B   ALTERNATIVE SPLICING

Multiple different mRNAs   Multiple proteins

C   POLYPROTEIN

RNA of virus   Polyprotein   Smaller proteins are cut out from polyprotein

D   FRAMESHIFT

One mRNA   Multiple proteins

1. Overview of Protein Synthesis

Each protein in every organism is made using the genetic information stored in the genome. The genetic information is transmitted in two stages. First, the information in the DNA is transcribed into mRNA. Next, the information carried by the mRNA is read to give the sequence of amino acids making up a polypeptide chain. This overall flow of information in biological cells from DNA to RNA to protein is known as the central dogma of molecular biology (see Ch. 3, Fig. 3.20) and was first formulated by Sir Francis Crick.

The decoding of mRNA is carried out by the **ribosome**, which binds the mRNA and translates it with tRNA. The ribosome moves along the mRNA reading the message, recruiting the proper tRNA at each step, and synthesizing a new polypeptide chain.

The correspondence between genes and mRNA dictates the way in which proteins are made. An early rule of molecular biology was Beadle and Tatum's dictum: "one gene—one enzyme" (see Ch. 2). This rule was later broadened to include other proteins in addition to enzymes. Proteins are therefore often referred to as "**gene products**." However, it must be remembered that some RNA molecules (such as tRNA, rRNA, small nuclear RNA) are never translated into protein and are therefore also gene products.

Furthermore, instances are now known where one gene may encode multiple proteins (Fig. 13.01). Two relatively widespread cases of this are known—alternative splicing and polyproteins. In eukaryotic cells, the coding sequences of genes are often interrupted by non-coding regions, the introns. These introns are removed by splicing at the level of mRNA. Alternative splicing schemes may generate multiple mRNA molecules and therefore multiple proteins from the same gene. This is especially frequent in higher eukaryotes, in particular vertebrates (see Ch. 12). A set of proteins generated in this manner shares much of their sequence and structure.

Ribosomes use the information carried by mRNA to make proteins.

Gene products include proteins as well as non-coding RNA.

gene product   End product of gene expression; usually a protein but includes various untranslated RNAs such as rRNA, tRNA, and snRNA
ribosome   The cell's machinery for making proteins

In eukaryotic cells, most mRNAs only carry information from a single gene and therefore can only be translated into a single protein. This causes problems for certain viruses that infect eukaryotic cells and that have RNA genomes (see Ch. 21). To circumvent the problem, these viruses make a huge "polyprotein" from an extremely long coding sequence in their RNA. This polyprotein is then cut up into several smaller proteins.

Finally, there are occasional oddities, such as the generation of two proteins from the same gene due to frame-shifting (see below). Despite these exceptions, it is still generally true that most genes give rise to a single protein.

To help understand the very complex process of translation, this chapter is divided into different sections. First, the key components (the genetic code, amino acids, tRNA, and ribosomes) are described in detail. Then the process of translation is broken into four steps: initiation, elongation, termination, and ribosome recycling. The simpler bacterial process of translation is first explained, and then the more complex eukaryotic version is presented. Finally, protein synthesis in mitochondria and chloroplasts is discussed. The final part of the chapter deals with getting proteins to the correct location in the cell and some of the post-translational modifications and amino acid substitutions that affect a protein's ultimate function.

> Although there are exceptions, most genes give rise to a single protein.

2. Proteins Are Chains of Amino Acids

Proteins consist of linear chains of monomers, known as amino acids, and are folded into a variety of complex 3D shapes. A chain of amino acids is called a polypeptide chain. What is the difference between a polypeptide chain and a protein? Firstly, some proteins consist of more than one polypeptide chain and secondly, many proteins contain additional components such as metal ions or small organic molecules known as co-factors in addition to their polypeptide portion (see Ch. 14).

Twenty different **amino acids** are incorporated when polypeptide chains are synthesized. (Strictly speaking, there are 22, but the other two are inserted only under special conditions—see below.) They all have a central carbon atom, the **alpha carbon**, surrounded by an amino group, a carboxyl group, a hydrogen atom, and a side chain or **R-group**, as shown in Figure 13.02A (proline is an exception—see below). The simplest amino acid is **glycine** (Fig 13.02B) in which the R-group is just a single hydrogen atom. In solution, under physiological conditions, the amino group and the carboxyl group of amino acids are both ionized to give a **zwitterion** or **dipolar ion** with one positive and one negative charge (Fig. 13.02C).

Amino acids are joined together by **peptide bonds** (Fig. 13.03) to give a **polypeptide chain**. The first amino acid in the chain retains its free amino ($-NH_2$) group, and this end is therefore called the **amino-** or **N-terminus** of the polypeptide chain. The last amino acid to be added is left with a free carboxyl ($-COOH$) group, so this end is the **carboxy-** or **C-terminus**. When synthesized, the polypeptide is elongated from the amino terminus toward the carboxy terminus.

Apart from glycine, amino acids have four different chemical groups surrounding the alpha carbon atom. This is called a **chiral** or **asymmetric center** (Fig. 13.04).

> Proteins are linear polymers made from amino acids. Most proteins fold into complex 3D structures.

---

alpha (α) carbon   Central carbon atom of an amino acid that carries both the amino group and the carboxyl group
amino acid   Monomer from which polypeptide chains are built
amino- or N-terminus   The end of a poly peptide chain that is made last and has a free carboxyl group
asymmetric center   Carbon atom with four different groups attached. This results in optical isomerism.
carboxy- or C-terminus   The end of a poly peptide chain that is made last and has a free carboxy-group
chiral center   Same as asymmetric center
dipolar ion   Same as zwitterion; a molecule with both a positive and a negative charge
glycine   The simplest amino acid
peptide bond   Type of chemical linkage holding amino acids together in a polypeptide chain
polypeptide chain   A polymer that consists of amino acids
R-group   Any unspecified chemical group; in particular, the side chain of an amino acid
zwitterion   Same as dipolar ion; a molecule with both a positive and a negative charge

FIGURE 13.02
General Structure of Amino Acids

A) A generalized amino acid contains an alpha carbon atom, an alpha carbon atom, an R-group, an —NH₂ group, and a —COOH group. B) Glycine is the simplest amino acid with an H atom as the R-group. C) When glycine is placed in solution at neutral pH it ionizes to form a zwitterion.

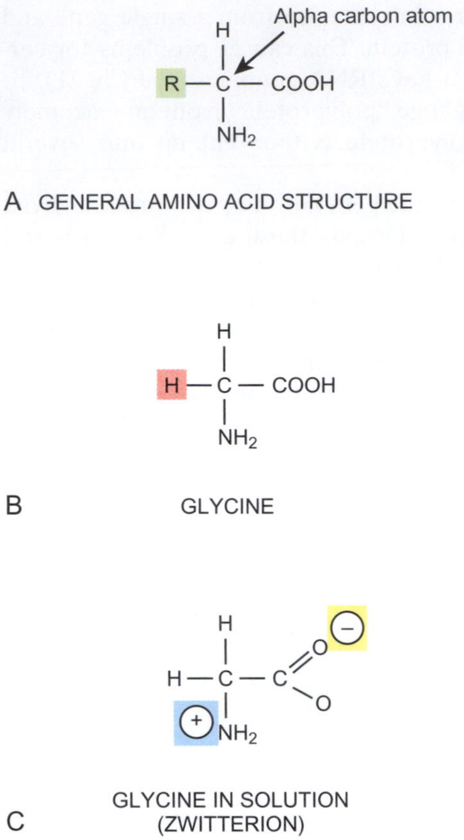

A   GENERAL AMINO ACID STRUCTURE

B   GLYCINE

C   GLYCINE IN SOLUTION
    (ZWITTERION)

Amino acids occur as pairs of optical isomers. Natural proteins are made from the L-isomers.

The D-isomers of amino acids are found in bacterial cell walls and some antibiotics.

Consequently, such amino acids exist as two alternative mirror-image isomers with different "chirality" or "handedness." A pair of mirror-image isomers is known as **enantiomers** or **optical isomers**. They are referred to as the **L-** and **D-forms** because solutions of chiral molecules rotate the plane of polarization of light in either a left-handed (L- = levorotatory) or right-handed (D- = dextrorotatory) direction. The amino acids found in proteins are all of the L-form. Although "L-amino acid" are sometimes referred to as the "natural" isomers, D-amino acids do exist in nature. The **peptidoglycan** that is found in bacterial cell walls contains several different D-amino acids and several peptide antibiotics, also made by prokaryotes (e.g., bacitracin, polymixin B, actinomycin D), contain D-amino acids.

2.1. Twenty Amino Acids Form Biological Polypeptides

The 20 amino acids found in proteins possess a variety of different chemical groups (Fig. 13.05). This wide choice of possible monomers makes proteins very versatile, with a wide range of properties and capabilities, including a great variety of possible 3D structures. The amino acids may be classified into groups depending on their physical and chemical characteristics. The major division is between those with **hydrophilic** (water-loving) and those with **hydrophobic** (water-hating) R-groups. Glycine has only a single hydrogen atom as its side chain, so it does not really fit into either group.

The 20 amino acids that comprise proteins vary greatly in their chemical and physical properties.

enantiomers   A pair of mirror-image optical isomers (i.e., D- and L-isomers)
hydrophilic   Water-loving; readily dissolves in water
hydrophobic   Water-hating; repelled by water and dissolves in water only with great difficulty
L- and D-forms   The two isomeric forms of an optically active substance; also called L- and D-isomers
optical isomers   Isomers where the molecules differ only in their 3D arrangement and consequently affect the rotation of polarized light
peptidoglycan   Polymer that makes up eubacterial cell walls; consists of long chains of sugar derivatives, cross-linked at intervals with short chains of amino acids

TWO AMINO ACIDS

AMINO ACID   AMINO ACID

```
       H   O                    H   O
       |   ||                   |   ||
  H — N — C — C — OH      H — N — C — C — OH
       |   |                    |   |
       H   R₁                   H   R₂
```

A

HOH
water

PEPTIDE BOND LINKAGE

```
       H   O       H   O
       |   ||      |   ||
  H — N — C — C — N — C — C — OH
       |   |       |   |
       H   R₁      H   R₂
```

B

N - terminus   C - terminus

FORMATION OF MANY PEPTIDE BONDS

POLYPEPTIDE

```
  H₂N - - AA₁ - - AA₂ - - AA₃ - - AA₄ - - AAₙ₋₂ - - AAₙ₋₁ - - AAₙ - - COOH
```

C   N - terminus   C - terminus

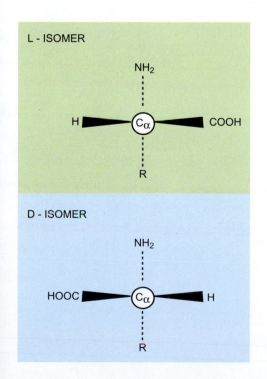

L - ISOMER

NH₂

H ◄ Cα ► COOH

R

D - ISOMER

NH₂

HOOC ◄ Cα ► H

R

FIGURE 13.03

Polypeptide Chain Is Made of Amino Acids

A) Two generic amino acids are shown. The R-groups, R_1 and R_2, represent the side chains of any of the 20 different amino acids that make up proteins. Each amino acid has an amino ($-NH_2$) and carboxyl ($-COOH$) group.
B) A peptide bond is formed between one $-NH_2$ and one $-COOH$ group, with water eliminated in the process.
C) Successive amino acids are joined in a similar manner by peptide bonds to form a polypeptide. The polypeptide contains an amino (or N-) terminus and a carboxy (or C-) terminus.

FIGURE 13.04

The L- and D-Forms of an Amino Acid

The four groups are arranged around the alpha carbon differently in the L-form and the D-form of an amino acid. Although they share the same molecular formula, one is the mirror image of the other.

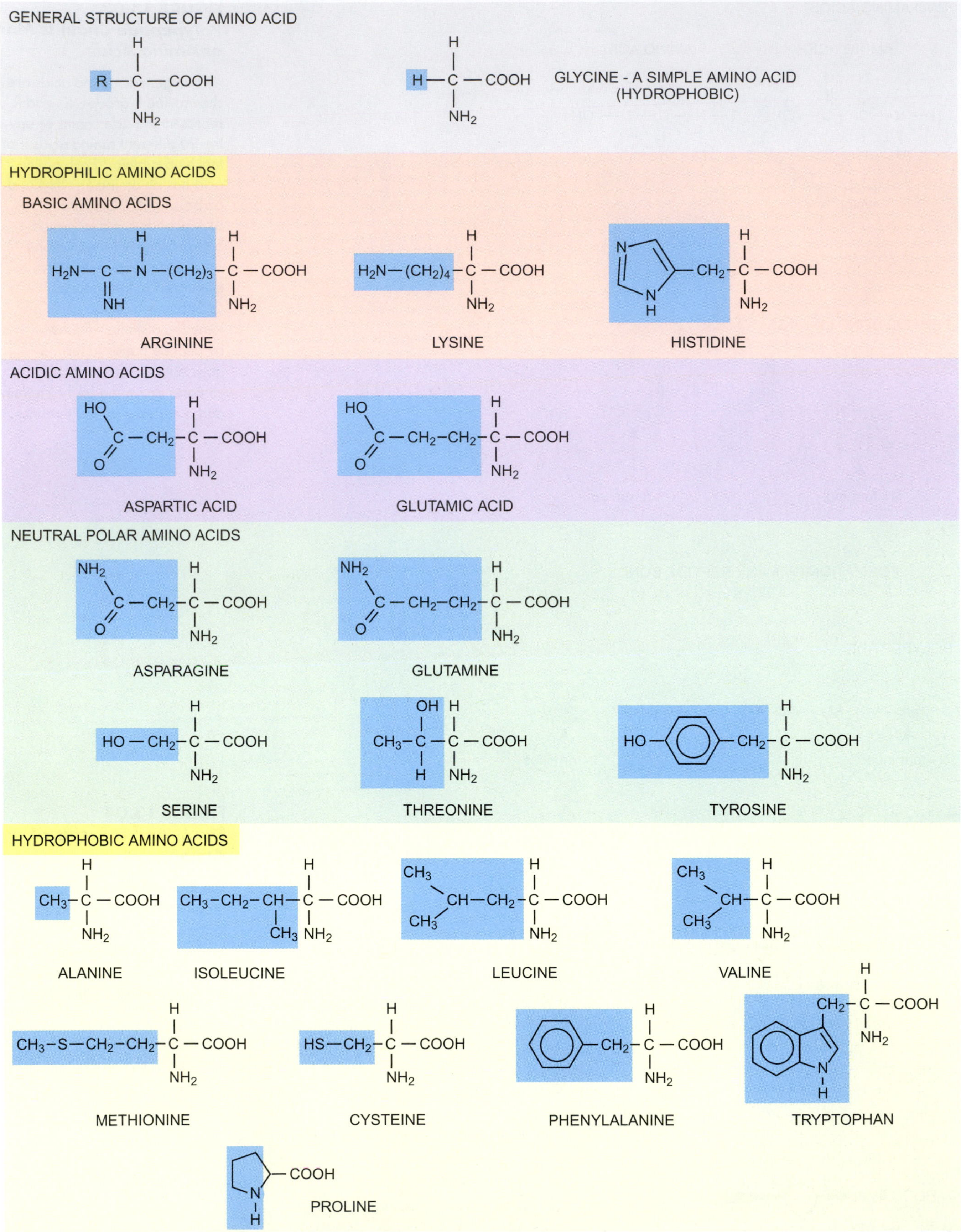

FIGURE 13.05
The 20 Amino Acids Found in Proteins

Amino acids can be grouped by their physical and chemical properties. The R-group for each amino acid is highlighted.

Amino Acid	3-Letter Code	1-Letter Code	Physical Properties
Alanine	Ala	A	hydrophobic
Arginine	Arg	R	basic
Asparagine	Asn	N	neutral polar
Aspartic acid	Asp	D	acidic
Cysteine	Cys	C	hydrophobic
Glutamic acid	Glu	E	acidic
Glutamine	Gln	Q	neutral polar
Glycine	Gly	G	—
Histidine	His	H	basic
Isoleucine	Ile	I	hydrophobic
Leucine	Leu	L	hydrophobic
Lysine	Lys	K	basic
Methionine	Met	M	hydrophobic
Phenylalanine	Phe	F	hydrophobic
Proline	Pro	P	hydrophobic
Serine	Ser	S	neutral polar
Threonine	Thr	T	neutral polar
Tryptophan	Trp	W	hydrophobic
Tyrosine	Tyr	Y	neutral polar/hydrophobic
Valine	Val	V	hydrophobic
Aspartic acid or Asparagine	Asx	B	
Glutamic acid or Glutamine	Glx	Z	
Unspecified amino acid		X	

TABLE 13.01 Amino Acids and Their Properties

The hydrophilic amino acids may be subdivided into basic, acidic, and neutral. Basic amino acids contribute a positive charge to the protein, whereas acidic residues provide a negative charge. Strictly, this refers to the situation in solution within the physiological pH range. Neutral polar residues have side chains that are capable of forming hydrogen bonds. The side chains of the hydrophilic amino acids carry chemical groups that can take part in reactions. The active sites of enzymes (see below) often contain serine, histidine, and basic or acidic amino acids.

The hydrophobic amino acids may be subdivided into those that are aliphatic (Ala, Leu, Ile, Val, Met) and those containing aromatic rings (Phe, Trp, and Tyr). These amino acid residues are largely structural in function, except for tyrosine, which has a hydroxyl group attached to its aromatic ring and can therefore take part in a variety of reactions. The classification of tyrosine is ambiguous as its hydroxyl group is polar in nature. Proline has a non-aromatic ring and contains a secondary amino group (—NH—) rather than a primary amino group (NH$_2$) like the other imino acids. Such structures are sometimes referred to as imino acids (rather than amino acids).

Two of the hydrophobic amino acids, Met and Cys, contain sulfur. When a polypeptide chain is first synthesized, methionine is always the first amino acid, although it may be trimmed off later. Cysteine is important for 3D structure as it forms disulfide bonds. The free **sulfhydryl group** of cysteine is highly reactive and is often used in enzyme active sites or to attach various chemical groups to proteins (see Ch. 14).

The 20 different amino acids normally found in proteins may be represented by both three-letter and one-letter abbreviations (Table 13.01). These mostly correspond

sulfhydryl group   -SH; Chemical group of sulfur and hydrogen

FIGURE 13.06
The Genetic Code

The 64 codons as found in mRNA are shown with their corresponding amino acids. As usual, bases are read from 5′ to 3′ so that the first base is at the 5′ end of the codon. Three codons (UAA, UAG, UGA) have no cognate amino acid but signal stop. AUG (encoding methionine) and, less often, GUG (encoding valine) act as start codons. To locate a codon, find the first base in the vertical column on the left, the second base in the horizontal row at the top, and the third base in the vertical column on the right.

1st base	2nd (middle) base				3rd base
	U	**C**	**A**	**G**	
U	UUU Phe UUC Phe UUA Leu UUG Leu	UCU Ser UCC Ser UCA Ser UCG Ser	UAU Tyr UAC Tyr UAA stop UAG stop	UGU Cys UGC Cys UGA stop UGG Trp	U C A G
C	CUU Leu CUC Leu CUA Leu CUG Leu	CCU Pro CCC Pro CCA Pro CCG Pro	CAU His CAC His CAA Gln CAG Gln	CGU Arg CGC Arg CGA Arg CGG Arg	U C A G
A	AUU Ile AUC Ile AUA Ile AUG Met	ACU Thr ACC Thr ACA Thr ACG Thr	AAU Asn AAC Asn AAA Lys AAG Lys	AGU Ser AGC Ser AGA Arg AGG Arg	U C A G
G	GUU Val GUC Val GUA Val GUG Val	GCU Ala GCC Ala GCA Ala GCG Ala	GAU Asp GAC Asp GAA Glu GAG Glu	GGU Gly GGC Gly GGA Gly GGG Gly	U C A G

to the first letter(s) of the name, but since several amino acids start with the same letter of the alphabet, the others need a little imagination. They are used especially when writing out protein sequences. The amides, asparagine and glutamine, are relatively unstable and break down easily into the corresponding acids, aspartate and glutamate. Consequently, some analyses do not distinguish the acids from their amides. The abbreviations Asx and Glx were invented for these ambiguous pairs.

3. Decoding the Genetic Information

There are 20 amino acids in proteins but only four different bases in the mRNA, so one base of a nucleic acid cannot code for a single amino acid when making a protein. During translation, the bases of mRNA are read off in groups of three, which are known as **codons**. Each codon represents a particular amino acid. Four different bases gives 64 possible groups of three bases; that is, there are 64 different codons in the **genetic code**. Because there are only 20 different amino acids, some are encoded by more than one codon. In addition, three of the codons are used for punctuation. Those are the **stop codons** that signal the end of a polypeptide chain. Figure 13.06 shows nature's genetic code.

To read the codons, tRNAs recognize the codon on the mRNA. At one end, each tRNA has an **anticodon** consisting of three bases that are complementary to the three bases of the corresponding codon on the mRNA. The codon and anticodon recognize each other by base pairing and are held together by hydrogen bonds (Fig. 13.07). At its other end, each tRNA carries the amino acid encoded by the codon it recognizes. This amino acid is sometimes known as the tRNA's "cognate" amino acid.

The genetic code is not quite universal. Despite this, the term "**universal genetic code**" is used to refer to the codon table shown above (Fig. 13.06), since it applies to almost all organisms. Rarely, exceptions to the code are found in some protozoans and mycoplasmas and in the mitochondrial genome of animals and fungi (Table 13.02). Mycoplasmas are parasitic bacteria with unusually small genomes. *Paramecium* and *Euplotes* are ciliated protozoans and *Candida* is a type of yeast.

Each amino acid in a protein is encoded by three bases in the DNA or RNA sequence.

The anticodon of tRNA recognizes the codon on mRNA by base pairing.

Each tRNA carries one particular amino acid.

anticodon Group of three complementary bases on tRNA that recognize and bind to a codon on the mRNA
codon Group of three RNA or DNA bases that encodes a single amino acid
genetic code System for encoding amino acids as groups of three bases (codons) of DNA or RNA
stop codon Codon that signals the end of a protein
universal genetic code Version of the genetic code used by almost all organisms

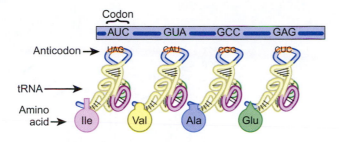

FIGURE 13.07
Transfer RNA Recognizes Codons

Several tRNAs are seen bound to mRNA codons by their anticodons. Each tRNA carries a different amino acid at the end of the adaptor stem. This diagram is intended to show the principle of mRNA decoding. It does NOT illustrate the actual mechanism of protein synthesis. In real life, the codons are contiguous and there are no spacers in between and only two tRNAs are bound at any given time.

TABLE 13.02	Exceptions to the Universal Genetic Code

Exceptions in the Chromosomal Genome

Codon	Universal	Mycoplasma	Paramecium	Euplotes	Candida
UGA	Stop	**Trp**	Stop	**Cys**	Stop
UAA/UAG	Stop	Stop	**Gln**	Stop	Stop
CUG	Leu	Leu	Leu	Leu	Ser

Exceptions in the Mitochondrial Genome

Codon	Universal	Fungi	Protozoa	Mammals	Flatworm
UGA	Stop	**Trp**	**Trp**	**Trp**	**Trp**
UAA	Stop	Stop	Stop	Stop	Tyr
AUA	Ile	**Met**	**Met**	**Met**	Ile
AGA/AGG	Arg	Arg	Arg	**Stop**	**Ser**
AAA	Lys	Lys	Lys	Lys	Asn
CUA	Leu	**Thr**	Leu	Leu	Leu

Note that there is no general mitochondrial genetic code. Although fungal and animal mitochondria share similarities (e.g., UGA = Trp), there are also differences (e.g., CUA = Thr in fungi but Leu in animals). Mammalian mitochondria also have a reduced set of tRNAs that results in an increase in the number of tRNAs that read multiple codons (see Focus on Relevant Research). However, plant mitochondria and chloroplasts use the universal genetic code.

> Minor variations in the genetic code are found in mitochondria and certain microorganisms.

3.1. Transfer RNA Forms a Folded "L" Shape with Modified Bases

Transfer RNA molecules are about 80 nucleotides in length. About half the bases are paired to form double helical segments. A typical tRNA has four short base-paired stems and three loops (Fig. 13.08). This is shown best in the **cloverleaf structure**, intended to reveal details of base pairing, which shows the tRNA spread out flat in only two dimensions. (Such a diagram is sometimes called a secondary structure map). The tRNA is in reality folded up further to give an L-shaped 3D structure, in which

---

cloverleaf structure   2D structure showing base pairing in a tRNA molecule

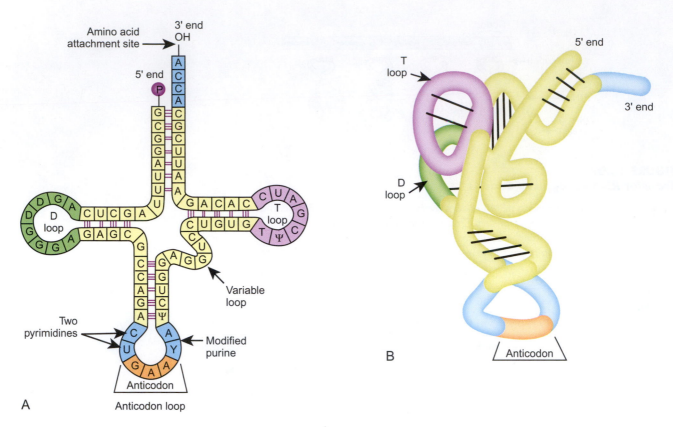

FIGURE 13.08
Structure of tRNA

A) A planar view (secondary structure) of a tRNA shows its cloverleaf structure comprised of the 3′ and 5′ acceptor stem, the T- (or TψC) and D-loops, and the anticodon loop. A variable loop, which varies in length in different tRNA molecules, is also found. B) The folded (tertiary structure) configuration resembles an "L."

the TψC-loop (or T-loop) and the D-loop are pushed together. The anticodon and attached amino acid are located at the two ends of the L-structure. Different tRNA molecules vary considerably in sequence, but they all conform to this same overall structure. Variations in length (from 73–93 nucleotides) occur, due mostly to the variable loop.

The **acceptor stem** is made by pairing of the 5′ end, which almost always ends in G and is phosphorylated, and the 3′ end, which ends in CCA-OH. The amino acid is bound to the 3′-hydroxyl group of the adenosine at the free 3′ end of the acceptor stem. The anticodon is about halfway round the sequence, in the **anticodon loop**. This consists of seven bases with the three anticodon bases in the middle. The anticodon is always preceded, on the 5′ side, by two pyrimidines and followed by a modified purine (Fig. 13.08).

The other two loops of tRNA are named after **modified bases**. These bases are modified after the RNA has been transcribed and are especially common in tRNA, although they are also found in other classes of RNA. The TψC-loop contains "ψ" (spelled "psi" but pronounced "sigh"), which stands for pseudouracil; and the D-loop or DHU-loop has "D" for dihydrouracil (Fig. 13.09). In pseudouridine, the uracil itself is not altered, but is attached to ribose by carbon-5 instead of nitrogen-1, as in normal uridine. Note that thymine (= 5-methyl uracil), which is normally only found in DNA, is also found in the TψC-loop of tRNA, where it is attached to ribose and is made by

acceptor stem   Base-paired stem of tRNA to which the amino acid is attached
anticodon loop   Loop of tRNA molecule that contains the anticodon
modified base   Nucleic acid base that is chemically altered after the nucleic acid has been synthesized

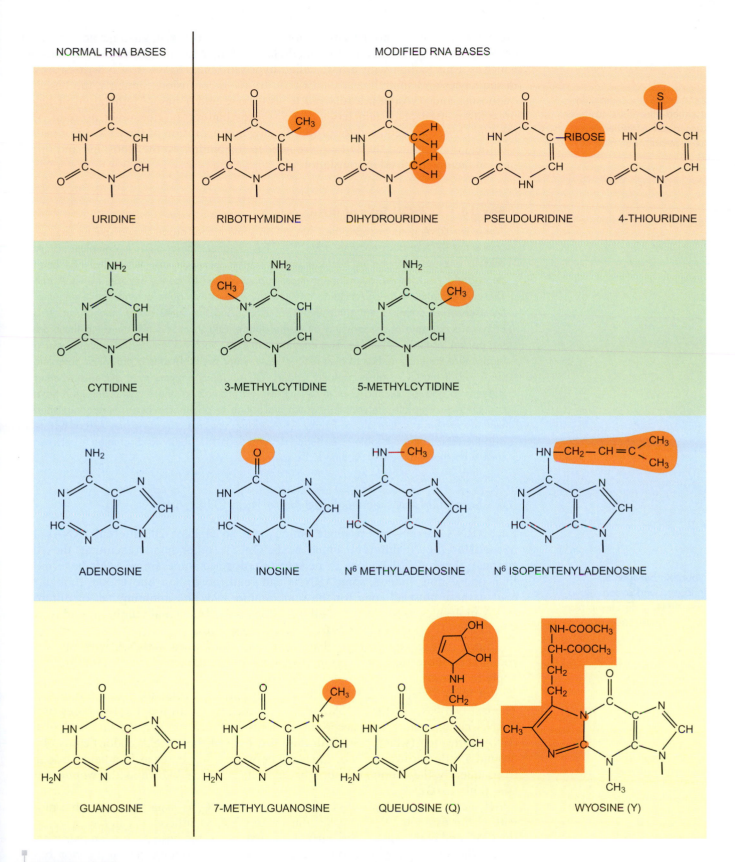

FIGURE 13.09
Modified Bases in tRNA

All four bases normally found in RNA have modified derivatives that may be found in tRNA. The names given are those of the corresponding nucleosides (i.e., base plus ribose).

methylation of uracil after transcription. In addition to uracil, guanine, adenine, and cytosine may also be modified by methylation. Other modifications include **inosine**, which is a nucleotide with a base called hypoxanthine. However, it is written as "I" in sequences and sometimes called I-base to avoid confusion. Similarly, the bases of queuosine and wyosine are referred to as Q-base and Y-base. Methylation of bases prevents pairing of certain bases and also aids binding of ribosomal proteins. The modified bases are required for proper folding and operation of the tRNA. For example, the TψC-loop and the D-loop are needed for binding to the ribosome and other protein factors involved in translation.

Box 13.1 CCA-OH Tails of tRNA

Not every tRNA gene encodes the CCA tail. In some higher organisms, a special enzyme called the CCA-adding enzyme adds 5′ –CCA 3′ after transcription is complete. CCA-adding enzyme is found in bacteria and functions to repair any tRNA that has been degraded by nucleases, even though most eubacterial tRNA genes encode the CCA tail. The crystal structure of CCA-adding enzyme shows that it is quite a remarkable enzyme. Most enzymes catalyze the same exact reaction using the same exact mechanism. This enzyme is capable of recognizing all of the different tRNA molecules, meaning the binding pocket for tRNA is flexible. This enzyme also adds each nucleotide individually, which means that the enzyme has to also recognize whether the tRNA has a partial or complete CCA tail. If the CCA tail is partially added, CCA-adding enzyme then discerns whether it needs to add the second C or third A. Even more amazing, the enzyme does not have a nucleotide template, determining which nucleotide to add and when it is accomplished with the protein structure alone. Finally, CCA-adding enzyme can use either ATP or CTP to add the tRNA tail but is able to exclude UTP and GTP. The ability to discriminate between such subtleties is a mystery waiting to be solved.

3.2. Some tRNA Molecules Read More Than One Codon

Each tRNA carries only a single amino acid, so at least 20 different tRNAs are needed for the 20 different amino acids. On the other hand (excluding the stop codons), there are 61 codons to be recognized, since some amino acids have more than one codon. Indeed, some tRNAs can read more than one codon, though, of course, these must all code for the same amino acid. The minimum set of different tRNA molecules needed to read all 61 codons is 31. The actual number found is usually somewhat higher and varies from species to species.

Since only complementary bases can pair, how does a tRNA with one anticodon read more than one codon? Remember that the standard base pairing rules apply to bases that form part of a DNA double helix. Since the codon and anticodon do not form a standard double helix, slightly different rules for base pairing apply. The last two bases of the tRNA anticodon, which pair with the first two bases of the mRNA codon, pair strictly according to normal rules. However, the first base of the tRNA anticodon (which pairs with the third base of the mRNA codon) can wobble around a little because it is not squeezed between other bases as in a helix structure. Consequently, the codon/anticodon basepairing rules are known as the **wobble rules** (see Table 13.03).

If the first anticodon base is G it can pair with C, as usual, or, in wobble mode, with U. For example, tRNA for histidine, with GUG as anticodon, can recognize both the CAC and CAU codons. Similarly, if the first anticodon base is U, it can pair with A or G. Whenever an amino acid is encoded by a pair of codons, the third codon bases are U and C (e.g., histidine, tyrosine) or A and G (e.g., lysine, glutamic acid), but never

inosine An unusual modified nucleoside derived from guanosine
wobble rules Rules allowing less rigid base pairing but only for codon/anticodon pairing

TABLE 13.03	Wobble Rules for Codon/Anticodon Pairing	
	Pairs with Third Codon Base	
First Anticodon Base	**normal**	**by wobble**
G	C	U
U	A	G
I	—	C or U or A
C	G	no wobble
A	U	no wobble

other combinations. Similarly, those privileged amino acids with four or six codons may be regarded as having two or three such pairs. Due to wobble pairing, only a single tRNA is needed to read each such pair of codons. It is possible for a single tRNA to read three codons by making use of inosine. The I-base is occasionally used as the first anticodon base because it can pair with any of U, C, or A.

3.3. Charging the tRNA with the Amino Acid

For each tRNA there is a specific enzyme that recognizes both the tRNA and the corresponding amino acid. These enzymes, known as **aminoacyl tRNA synthetases**, attach the cognate amino acid to the tRNA. This is called "charging the tRNA." Empty tRNA is known as **uncharged tRNA** while tRNA with its amino acid is **charged tRNA**.

Charging occurs in two steps (Fig. 13.10). First the amino acid reacts with ATP to form aminoacyl-AMP (also known as aminoacyl-adenylate). Next, the aminoacyl-group is transferred to the 3′ end of the tRNA.

a. amino acid + ATP → aminoacyl-AMP + PPi

b. aminoacyl-AMP + tRNA → aminoacyl-tRNA + AMP

The amino-acyl tRNA synthetases are highly specific for both the correct amino acid and the correct tRNA. In some cases they recognize the correct tRNA by its anticodon and in others by the sequence of the acceptor stem. Some amino-acyl tRNA synthetases recognize both regions of the tRNA. Figure 13.11 shows an amino-acyl tRNA synthetase bound to its cognate tRNA.

> A specific enzyme, called aminoacyl tRNA synthetase, attaches the correct amino acid to the correct tRNA.

4. The Ribosome: The Cell's Decoding Machine

The decoding process is carried out by the ribosome, a submicroscopic machine that binds mRNA and charged tRNA molecules. The mRNA is translated into protein starting at the 5′ end. After binding to the mRNA, the ribosome moves along it, adding a new amino acid to the growing polypeptide chain each time it reads a codon. Each codon on the mRNA is actually read by an anticodon on the corresponding tRNA and so the information in the mRNA is used to synthesize a polypeptide chain from the amino acids carried by the tRNAs.

The ribosome and its components were originally analyzed by ultracentrifugation. Consequently, sizes are referred to in Svedberg units (S-value), which measure sedimentation velocity. Although higher S-values indicate larger particles, the S-value is

> Ribosomes consist of RNA plus proteins and their role is to synthesize new proteins.

aminoacyl tRNA synthetase   Enzyme that attaches an amino acid to tRNA
charged tRNA   tRNA with an amino acid attached
uncharged tRNA   tRNA without an amino acid attached

FIGURE 13.10
Charging tRNA with the Amino Acid

This two-step procedure begins (A) by attachment of the amino acid to adenosine monophosphate (AMP) to give aminoacyl-AMP or aminoacyl-adenylate. This involves splitting ATP and the release of inorganic pyrophosphate. Then, in the second step (B), the amino acid is transferred to the hydroxyl group of the ribose at the 3' end of the tRNA, yielding AMP as a byproduct.

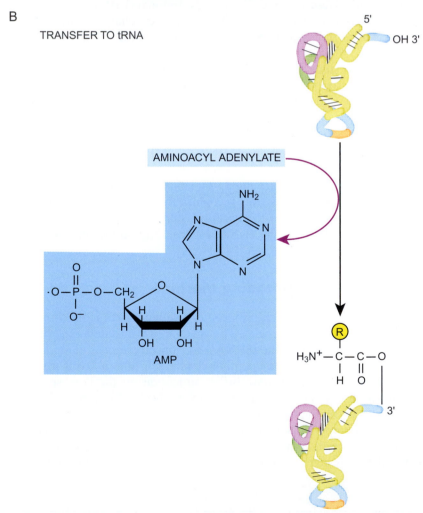

4. The Ribosome: The Cell's Decoding Machine **383**

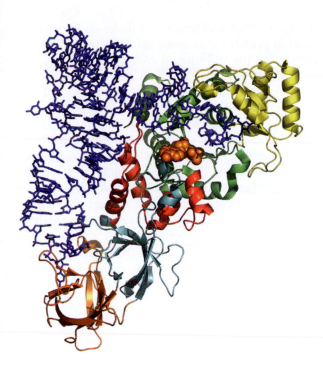

■ **FIGURE 13.11**

Glutamine tRNA Bound to its Aminoacyl tRNA Synthetase

Structure of glutaminyl-tRNA synthetase bound to tRNA(Gln) and a glutaminyl adenylate analog. The analog is in orange and is shown in a space-filling representation. The tRNA is depicted in dark blue. Domains of the enzyme are color-coded as follows: active-site Rossman fold, green; acceptor-end binding domain, yellow; connecting helical subdomain, red; proximal beta-barrel, light blue; distal beta-barrel, orange. *(The image was made in PyMol by John Perona, Department of Chemistry and Biochemistry, University of California at Santa Barbara.)*

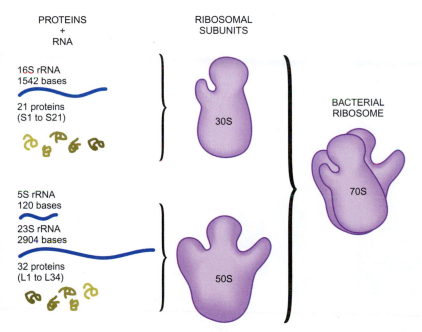

■ **FIGURE 13.12**

Components of a Bacterial Ribosome

The ribosome is composed of 30S and 50S subunits. The 30S subunit consists of one 16S rRNA together with 21 proteins, and the 50S subunit has a 5S and 23S rRNA plus 34 proteins.

not directly proportional to molecular weight. The **bacterial (70S) ribosome** consists of two subunits, the **50S** or **large subunit** and the **30S** or **small subunit** (Fig. 13.12). **Eukaryotic (80S) ribosomes** are somewhat larger, consisting of **60S** and **40S subunits** (see below).

---

bacterial (70S) ribosome   Type of ribosome found in bacterial cells
eukaryotic (80S) ribosome   Type of ribosome found in cytoplasm of eukaryotic cell and encoded by genes in the nucleus
large subunit   The larger of the two ribosomal subunits, 50S in bacteria, 60S in eukaryotes
30S subunit   Small subunit of a 70S ribosome
40S subunit   Small subunit of an 80S ribosome
50S subunit   Large subunit of a 70S ribosome
60S subunit   Large subunit of an 80S ribosome
small subunit   The smaller of the two ribosomal subunits, 30S in bacteria, 40S in eukaryotes

The bacterial ribosome contains three rRNA molecules that make up about two-thirds of its weight and about 50 smallish proteins that make up the remaining third. The 30S subunit contains the 16S rRNA and the 50S subunit contains the 5S and 23S rRNA (Fig. 13.12). The 3D structure of a 70S ribosome is shown in Figures 13.13 and 13.14. Recently, the structure of the eukaryotic ribosome was also solved (Fig. 13.15).

The rRNA molecules have highly defined secondary structures with many stems and loops (Fig. 13.16). Although it was originally believed to have a largely structural role, recent work indicates that the rRNA is responsible for most of the critical reactions of protein synthesis. In particular, the 23S rRNA of the large subunit is a **ribozyme** that catalyzes the synthesis of the peptide bonds between the amino acids; that is, it is the **peptidyl transferase**. Indeed, X-ray crystallography of the 50S subunit has shown that no ribosomal proteins are close enough to the catalytic center to take part in the reaction. Alteration by mutation of the catalytic residues in typical ribozymes either abolishes activity completely or reduces it by many-fold. However, the peptidyl-transferase center of 23S rRNA behaves in an atypical manner. Alteration of A2451 or G2447 (*E. coli* numbering) did not greatly reduce catalytic activity, although these residues are present in the catalytic center.

> The peptide bond linking amino acids in the growing protein is made by the largest rRNA, which acts as a ribozyme.

FIGURE 13.13
3D Structure of a Ribosome by EM

This structure was deduced from negatively-stained electron microscope images of a bacterial 70S ribosome.

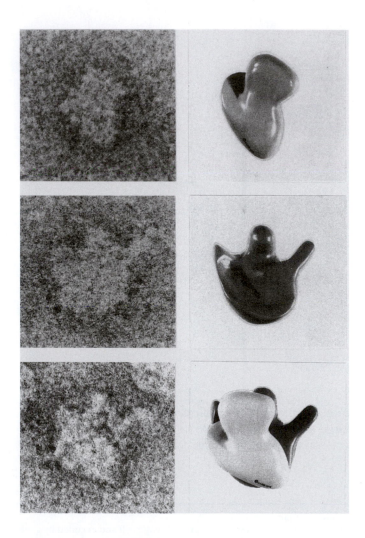

---

peptidyl transferase   Enzyme activity on the ribosome that makes peptide bonds; actually 23S rRNA (bacterial) or 28S rRNA (eukaryotic)
ribozyme   RNA molecule that acts as an enzyme

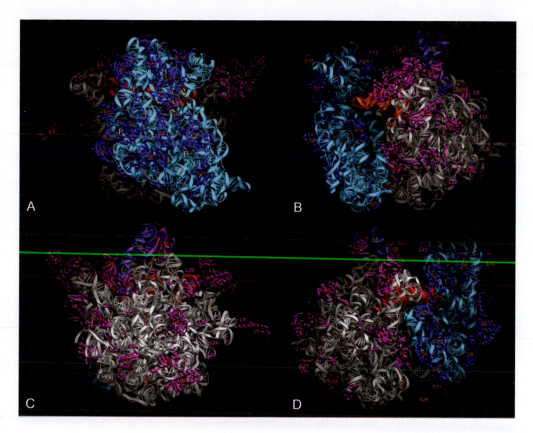

FIGURE 13.14
3D Structure of a Ribosome by X-Ray

Views of the structure of the *Thermus thermophilus* 70S ribosome. A, B, C, and D are successive 90° rotations about the vertical axis. A) view from the back of the 30S subunit. H, head; P, platform; N, neck; B, body. B) view from the right-hand side, showing the subunit interface cavity, with the 30S subunit on the left and the 50S on the right. The anticodon arm of the A-tRNA (gold) is visible in the interface cavity. C) View from the back of the 50S subunit. EC, the end of the polypeptide exit channel. D) View from the left-hand side, with the 50S subunit on the left and the 30S on the right. The anticodon arm of the E-tRNA (red) is partly visible. The different molecular components are colored for identification: cyan, 16S rRNA; grey, 23S rRNA; light blue, 5S rRNA; dark blue, 30S proteins; magenta, 50S proteins. *(Credit: Yusupov et al., Crystal Structure of the Ribosome at 5.5 Å Resolution. Science 292 (2001) 883–96.)*

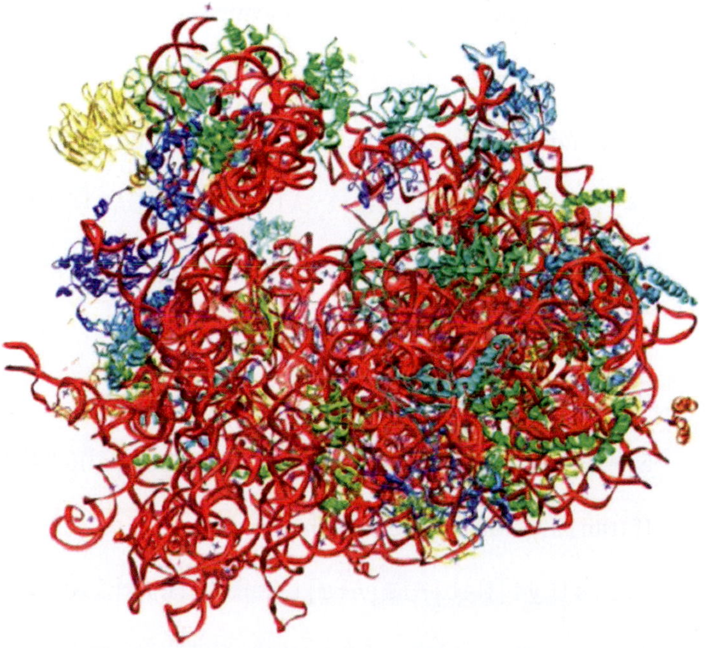

FIGURE 13.15
3D Structure of a Eukaryotic Ribosome

Crystal structure of the eukaryotic ribosome. Image from the RCSB PDB (www.pdb.org) of PDB ID 3O30 *(Ben-Shem, A, et al. (2010) Science 330: 1203–1209).*

These results suggest that the ribosome does not operate via direct chemical catalysis. Rather, the ribosome acts by correctly positioning the two substrates. The activated aminoacyl-tRNA then reacts spontaneously with the end of the growing polypeptide chain.

FIGURE 13.16
Secondary Structure of an rRNA

The 16S rRNA from the small ribosomal subunit of *E. coli* is complex with extensive secondary structure, forming loops and stems. Red indicates regions of base pairing.

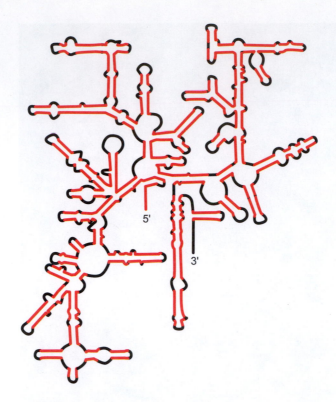

5. Three Possible Reading Frames Exist

Before mRNA is translated into protein, the issue of **reading frame** must be dealt with. The bases of mRNA are read off in groups of three, with each codon corresponding to one amino acid. But how is the sequence divided into codons? For any given nucleotide sequence there are three alternatives, depending on where the start is considered to be. Consider the following sequence:

GAAAUGUAUGCAUGCCAAAGGAGGCAUCUAAGG

If we start at base #1 we get the following codons:

GAA|AUG|UAU|GCA|UGC|CAA|AGG|AGG|CAU|CUA|AGG

If translated this would give the following amino acid sequence:

Glu|Met|Tyr|Ala|Cys|Gln|Arg|Arg|His|Leu|Arg

If we start at base #2 we get the following codons:

G|AAA|UGU|AUG|CAU|GCC|AAA|GGA|GGC|AUC|UAA|GG

If translated this would give the following amino acid sequence:

—|Lys|Cys|Met|His|Ala|Lys|Gly|Gly|Ile|Stop|—

And if we start at base #3 we get the following codons:

GA|AAU|GUA|UGC|AUG|CCA|AAG|GAG|GCA|UCU|AAG|G

---

reading frame   One of three alternative ways of dividing up a sequence of bases in DNA or RNA into codons

If translated this would give the following amino acid sequence:

```
-|Asn|Val|Cys|Met|Pro|Lys|Glu|Ala|Ser|Lys|-
```

Each set of codons gives a translation completely out of step with each of the others. These three possibilities are known as reading frames. As there are three bases in a codon, there are only three possible reading frames. Changing the reading frame by three (or a multiple of three) provides the same sequence as the first example above.

Any sequence of DNA or RNA, beginning with a start codon, and which can, at least theoretically, be translated into a protein, is known as an **open reading frame**, often abbreviated to (and pronounced!) **ORF**. Since ORFs are derived by examining nucleic acid sequences, deciding whether an ORF is a genuine protein coding sequence requires further information. Any mRNA will possess several possible ORFs. The correct one is what matters. Note that the message on an mRNA molecule does not start exactly at the 5′ end. Between the 5′ end and the coding sequence is a short region that is not translated—the **5′-untranslated region** or **5′-UTR** (sometimes 5′-nontranslated region or 5′-NTR). Hence, the reading frame cannot be defined simply by starting at the front end of the mRNA.

One way to define the reading frame is by choosing the **start codon**. The first codon is almost always AUG, encoding methionine. This will define both the start of translation and the reading frame. In the example considered above, there are three possible start codons (underlined), each of which starts at a slightly different point and gives a different reading frame:

```
GAAAUGUAUGCAUGCCAAAGGAGGCAUCUAAGGA
```

> Since the genetic code is read in groups of three bases, any nucleic acid sequence contains three possible reading frames.

> Between the very front of the mRNA and the coding sequence is a short untranslated region called the 5'-UTR.

> The start codon begins the coding sequence and is read by a special tRNA that carries methionine.

5.1. The Start Codon Is Chosen

A special tRNA, the **initiator tRNA**, is charged with methionine and binds to the AUG start codon (Fig. 13.17). In prokaryotes, chemically-tagged methionine, **N-formyl-methionine (fMet)**, is attached to the initiator tRNA, whereas in eukaryotes unmodified methionine is used. Consequently, all polypeptide chains begin with methionine, at least when first made. Sometimes the initial methionine (in eukaryotes), or N-formyl-methionine (in prokaryotes), is snipped off later, so mature proteins do not always begin with methionine. In bacteria, even when the fMet is not removed as a whole, the N-terminal formyl group is often removed leaving unmodified methionine at the N-terminus of the polypeptide chain.

AUG codons also occur in the middle of messages and result in the incorporation of methionines in the middle of polypeptide chains. So how does the ribosome know which AUG codon to start with? Near the front (the 5′ end) of the mRNA of prokaryotes is a special sequence, the **ribosome-binding site (RBS)**, often called the **Shine-Dalgarno** or **S-D sequence**, after its two discoverers (Fig. 13.18). The sequence complementary to this, the **anti-Shine-Dalgarno sequence**, is found close to the 3′ end of the 16S rRNA. Consequently, the mRNA and the 16S rRNA of the ribosome

> To choose the correct start codon, the mRNA binds to 16S rRNA at a specific sequence.

5′-untranslated region (5′-UTR)   Short sequence at the 5′ end of mRNA that is not translated into protein
anti-Shine-Dalgarno sequence   Sequence on 16S rRNA that is complementary to the Shine-Dalgarno sequence of mRNA
initiator tRNA   The tRNA that brings the first amino acid to the ribosome when starting a new polypeptide chain
N-formyl-methionine or fMet   Modified methionine used as the first amino acid during protein synthesis in bacteria
open reading frame (ORF)   Sequence of mRNA or corresponding region of DNA that can be translated to give a protein
ribosome binding site (RBS)   Same as Shine-Dalgarno sequence; sequence close to the front of mRNA that is recognized by the ribosome; only found in prokaryotic cells
Shine-Dalgarno (S-D) sequence   Same as RBS; sequence close to the front of mRNA that is recognized by the ribosome; only found in prokaryotic cells
start codon   The special AUG codon that signals the start of a protein

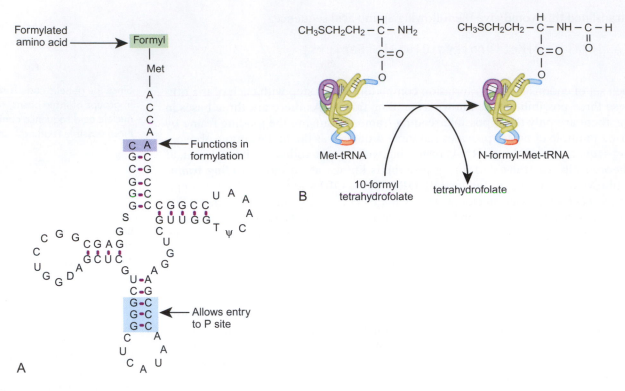

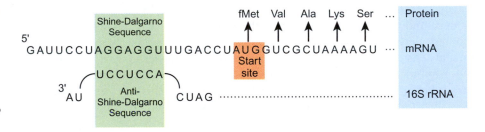

FIGURE 13.17
Initiator tRNA Carries N-Formyl-Methionine

A) The structure of the initiator tRNA, fMet-tRNA, is unique. A CA base pair at the top of the acceptor stem is needed to allow formylation (violet). The initiator tRNA must enter the P-site directly (see below), which requires the three GC base pairs in the anticodon stem (blue).
B) The initiator tRNA is first charged with unmodified methionine. Then a formyl group carried by the tetrahydrofolate cofactor is added to the methionine.

FIGURE 13.18
Shine-Dalgarno Sequence of mRNA Binds to 16S rRNA

The Shine-Dalgarno sequence on the mRNA is recognized by base pairing with the anti-Shine-Dalgarno sequence on the16S rRNA. The first AUG downstream of the S-D/anti-S-D site serves as the start codon.

bind together by base pairing between these two sequences. The start codon is the next AUG codon after the ribosome-binding site. Typically, there are about seven bases between the S-D sequence and the start codon. In some cases, the S-D sequence exactly matches the anti-S-D sequence and the mRNAs are translated efficiently. In other cases, the match is poorer and translation is less efficient. (Note that eukaryotes do not use an S-D sequence to locate the start of translation; instead, they scan the mRNA starting from the 5'-cap—see below.)

Occasionally, coding sequences even start with GUG (normally encoding valine) instead of AUG. This leads to inefficient initiation and is mostly found for proteins required only in very low amounts, such as regulatory proteins, for example, LacI, the repressor of the *lac* operon (see Ch. 16). Note that when GUG acts as the start codon, the same initiator fMet-tRNA is used as when AUG is the start codon. Consequently, formyl-Met is the first amino acid, even for proteins that start with a GUG codon. This is apparently due to the involvement of the initiation factors, especially IF3 (see below).

5.2. The Initiation Complex Assembles

Before protein synthesis starts, the two subunits of the ribosome are floating around separately. Because the 16S rRNA, with the anti-Shine-Dalgarno sequence, is in the small subunit of the ribosome, the mRNA binds to a free small subunit. Next the initiator tRNA, carrying fMet, recognizes the AUG start codon. Assembly of this **30S initiation complex** needs three proteins (IF1, IF2, and IF3), known as **initiation factors**, which help arrange all the components correctly. IF2 physically contacts the acceptor stem of the fMet-tRNA, and this interaction is essential to stabilize the initiation complex.

IF3 recognizes the start codon and the matching anticodon end of the initiator tRNA. IF3 prevents the 50S subunit from binding prematurely to the small subunit before the correct initiator tRNA is present. Once the 30S initiation complex has been assembled, IF3 departs and the 50S subunit binds. IF1 and IF2 are now released, resulting in the **70S initiation complex**. This process consumes energy in the form of GTP, which is split by IF2 (Fig. 13.19).

> Proteins known as initiation factors help the ribosomal subunits, mRNA and tRNA assemble correctly.

6. The tRNA Occupies Three Sites During Elongation of the Polypeptide

After the large subunit of the ribosome has arrived, the polypeptide can be made. Amino acids are linked together by the peptidyl transferase reaction, which is catalyzed by the 23S rRNA of the large subunit. The amino acids are carried to the ribosome attached to tRNA. The ribosome has three sites for tRNA: the **A (acceptor) site**, the **P (peptide) site,** and the **E (exit) site**. However, only two charged tRNA molecules can be accommodated on the ribosome at any given instant (Fig. 13.20).

> Only two tRNA molecules can occupy the ribosome at any instant.

The fMet initiator tRNA starts out in the P-site. Another tRNA, carrying the next amino acid, arrives and enters the A-site. Next, fMet is cut loose from its tRNA and bonded to amino acid #2. So tRNA #2 now carries two linked amino acids, the beginnings of a growing polypeptide chain. The enzyme activity that joins two amino acids together is referred to as the peptidyl transferase activity, as the growing peptide chain is transferred from the tRNA carrying it at each step. The ribosome is necessary for this enzyme activity because it provides an environment that shields the active site from water and provides the proper charged residues to orient the incoming substrates.

> After peptide bond formation the tRNA carrying the growing polypeptide chain moves sideways between sites on the ribosome.

After peptide bond formation, the two tRNAs are tilted relative to the A- and P-sites, a process called ratcheting (Fig. 13.21). The tRNA carrying the growing polypeptide chain now occupies part of the A-site on the 30S subunit but part of the P-site on the 50S subunit. This movement is facilitated by the **elongation factor** EF-G, which uses energy from GTP hydrolysis to maintain the ratcheted state and prevent the ribosome from returning to the original conformation.

The next step is **translocation**, in which the mRNA moves one codon sideways relative to the ribosome (Fig. 13.21). This moves the two tRNAs into the P- and E-sites, leaving the A-site empty. Once the two tRNAs are in the E- and P-sites, EF-G exits the complex. This dislodges the tRNA in the E-site and relocks the ribosome with tRNA #2 now in the P-site. The A- and E-sites cannot be simultaneously

---

A- (acceptor) site   Binding site on the ribosome for the tRNA that brings in the next amino acid
E- (exit) site   Site on the ribosome that a tRNA occupies just before leaving the ribosome
elongation factors   Proteins that are required for the elongation of a growing polypeptide chain
initiation factors   Proteins that are required for the initiation of a new polypeptide chain
30S initiation complex   Initiation complex for translation that contains only the small subunit of the bacterial ribosome
70S initiation complex   Initiation complex for translation that contains both subunits of the bacterial ribosome
P- (peptide) site   Binding site on the ribosome for the tRNA that is holding the growing polypeptide chain
translocation   a) Transport of a newly-made protein across a membrane by means of a translocase; b) sideways movement of the ribosome on mRNA during translation; and c) removal of a segment of DNA from a chromosome and its reinsertion in a different place

FIGURE 13.19
Formation of 30S and 70S Initiation Complexes

A) The small subunit and the mRNA bind to each other at the Shine-Dalgarno sequence. The start codon, AUG, is just downstream of this site. B) The initiator tRNA becomes tagged with fMet and binds to the AUG codon on the mRNA. C) The large ribosomal subunit joins the small subunit and accommodates the tRNA at the P-site.

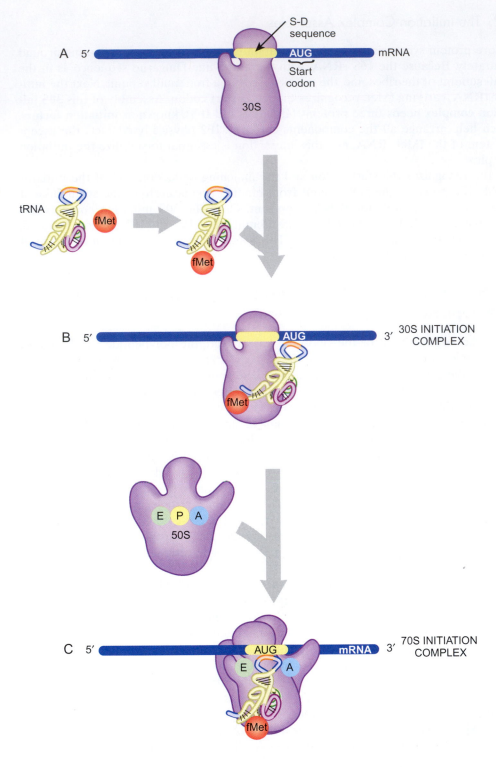

The empty tRNA leaves the ribosome at the E-site and a new, charged tRNA enters at the A-site.

occupied. Once the previous tRNA has exited, another charged tRNA can enter the A-site. As the peptide chain continues to grow, it is constantly cut off from the tRNA holding it and joined instead to the newest amino acid brought by its tRNA into the A-site; hence the name "acceptor" site. This process is repeated for each codon until the stop codon.

Bacterial elongation requires another elongation factor, EF-T, which uses energy in the form of GTP. EF-T actually consists of a pair of proteins, EF-Tu and EF-Ts.

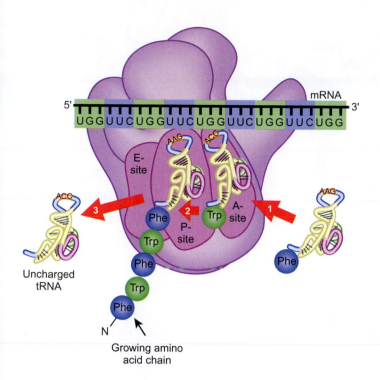

▬ **FIGURE 13.20**
Overview of the Elongation Cycle on the Ribosome

1) The incoming charged tRNA first occupies the A-site. 2) The peptide bond is formed between the amino acid at the A-site and the growing polypeptide chain in the P-site. 3) The uncharged tRNA exits the ribosome.

ACCEPTANCE OF A NEW tRNA

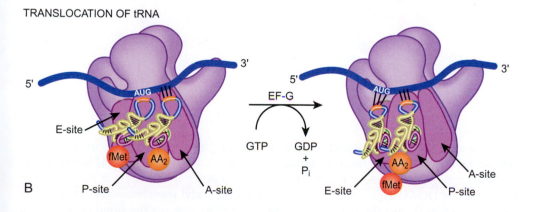

▬ **FIGURE 13.21**
Elongation Factors and Site Occupancy

A) The EF-T factor helps charged tRNA to occupy the A-site. B) The EF-G factor helps translocate the tRNAs from the A- and P-sites to the P- and E-sites, respectively. Note that during translocation the tRNA temporarily binds "diagonally" across two sites.

TRANSLOCATION OF tRNA

Incoming charged tRNA is delivered to the ribosome and installed into the A-site by elongation factor EF-Tu. This requires energy from the hydrolysis of GTP. EF-Ts is responsible for exchanging the GDP left bound to EF-Tu for a fresh GTP (Fig. 13.21).

6.1. Termination of Translation and Ribosome Recycling

Eventually the ribosome reaches the end of the message. This is marked by one of three possible stop codons, UGA, UAG, and UAA. As no tRNA exists to read these

FIGURE 13.22 ───────
Termination and Release of Finished Polypeptide

In prokaryotes, after the ribosome has added the final amino acid, release factors (RF1 and RF2) recognize the stop codon and cause the ribosome complex to dissociate.

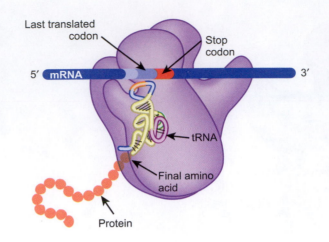

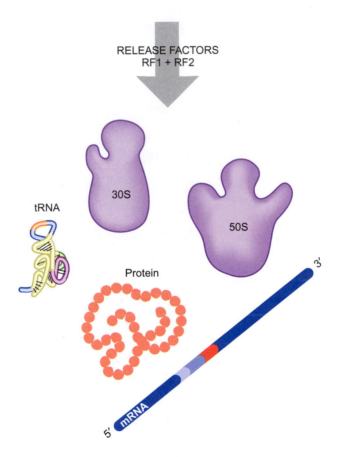

three codons, the polypeptide chain can no longer grow. Instead, proteins known as **release factors (RF)** read the stop signals (Fig. 13.22). RF1 recognizes UAA or UAG and RF2 recognizes UAA or UGA. The completed polypeptide chain is now released from the last tRNA. This is actually done by the peptidyl transferase. Binding of the release factor activates the peptidyl transferase that hydrolyzes the bond between the finished polypeptide chain and the tRNA in the P-site. Then, RF3 releases RF1 and RF2 from the ribosome using GTP as an energy source.

After polypeptide release, the ribosome complex is dissociated and recycled to translate a new mRNA. Two factors aid in dissociation: **ribosome recycling factor**

> The stop codon is read by a protein, the release factor, not by a tRNA.

───────

release factor   Protein that recognizes a stop codon and brings about the release of a finished polypeptide chain from the ribosome
ribosome recycling factor (RRF)   Protein that dissociates the ribosomal subunits after a polypeptide chain has been finished and released

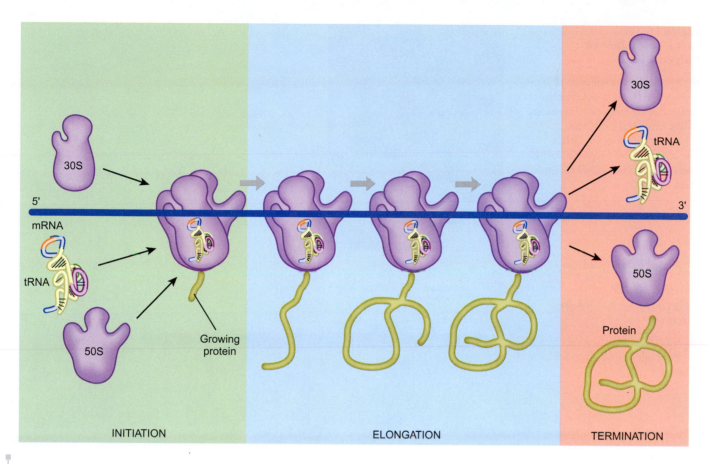

FIGURE 13.23
Polysome

A single mRNA molecule is associated with several ribosomes. At initiation the two subunits assemble; during chain elongation several ribosomes are at different stages in reading the same mRNA message; at termination the ribosome complex disassembles.

(RRF) and EF-G, which remove the large 50S subunit. Next, IF3 dissociates the last tRNA and mRNA from the small subunit. All the components are now free to be used again.

6.2. Several Ribosomes Usually Read the Same Message at Once

Once the first ribosome has begun to move, another can associate with the same mRNA and travel along behind. In practice, several ribosomes will move along the same mRNA about a hundred bases apart (Fig. 13.23). An mRNA with several attached ribosomes is called a **polysome** (short for polyribosome).

Electron microscope observations have suggested that the polysomes of eukaryotic cells are circular (Fig. 13.24). Apparently, the 3′ end of the mRNA is attached to the 5′ end by protein-protein contact between the poly(A) binding protein (attached to the 3′-poly(A) tail) and the eukaryotic initiation factor, eIF4 (attached to the cap at the 5′ end). In prokaryotic cells, such circularization cannot occur as the 3′ end of the mRNA is still being elongated by RNA polymerase while ribosomes have begun translating from the 5′ end.

> Messenger RNA is long enough for several ribosomes to translate it simultaneously.

---

polysome   Group of ribosomes bound to and translating the same mRNA

FIGURE 13.24
False Color TEM of Polysome

This false-color transmission electron micrograph (TEM) shows a polysome from a human brain cell. Polysomes consist of several individual ribosomes connected by slender strands of mRNA. Magnification: ×240,000. (Credit: CNRI/Science Photo Library.)

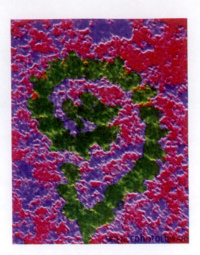

FIGURE 13.25
Polycistronic mRNA in Bacteria

The mRNA contains several cistrons or ORFs, each of which codes for a protein.

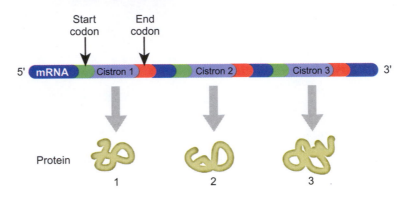

7. Bacterial mRNA Can Code for Several Proteins

In bacteria, several genes may be transcribed to give a single mRNA. The term **operon** refers to clusters of genes that are co-transcribed. The result is that several proteins may be encoded by the same mRNA. As long as each ORF has its own Shine-Dalgarno sequence in front of it, the ribosome will bind and start translating. ORFs that are translated into proteins are sometimes known as cistrons; consequently, mRNA which carries several of these is called **polycistronic mRNA** (Fig. 13.25).

In higher organisms operons are rare and neighboring genes are usually not co-transcribed. Each individual gene is transcribed separately to give an individual molecule of RNA. Apart from a few exceptional cases, each molecule of eukaryotic mRNA only carries a single protein coding sequence. Furthermore, eukaryotic mRNA does not make use of the Shine-Dalgarno sequence. Instead, the front (5′ end) of the mRNA molecule is recognized by its cap structure. Consequently, in eukaryotes only the first ORF would normally be translated, even if multiple ORFs were present (see below for further discussion).

> Messenger RNA in bacteria often carries several coding sequences.

> Eukaryotic mRNA molecules each code for a single protein.

7.1. Transcription and Translation Are Coupled in Bacteria

When mRNA is transcribed from the original DNA template, its synthesis starts at the 5′ end. The mRNA is also read by the ribosome starting at the 5′ end. In prokaryotic cells, the chromosome and ribosomes are all in the same single cellular

---

operon   A cluster of prokaryotic genes that are transcribed together to give a single mRNA (i.e., polycistronic mRNA)
polycistronic mRNA   mRNA carrying multiple coding sequences that may be translated to give several different protein molecules; only found in prokaryotic (bacterial) cells

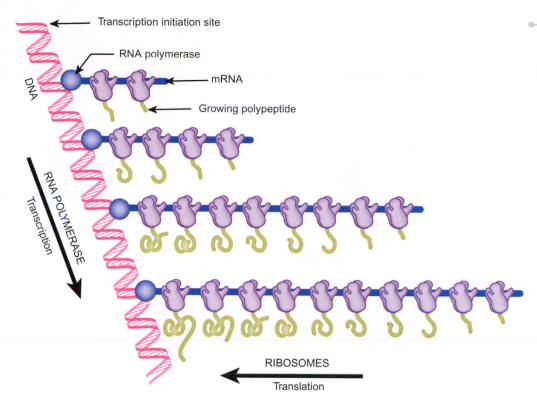

Transcription initiation site

RNA polymerase

mRNA

Growing polypeptide

DNA

RNA POLYMERASE
Transcription

RIBOSOMES
Translation

FIGURE 13.26
Coupled Transcription-Translation in Bacteria
Even as the DNA is being transcribed to give mRNA, ribosomes sequentially attach to the growing mRNA and initiate protein synthesis.

compartment. Therefore, ribosomes can start translating the message before synthesis of the mRNA molecule has actually been finished. The result is that partly finished mRNA, still attached to the bacterial chromosome via RNA polymerase, may have several ribosomes already moving along it making polypeptide chains. This is known as **coupled transcription-translation** (Fig. 13.26). This is impossible in higher eukaryotic cells because the DNA is inside the nucleus and the ribosomes are outside, in the cytoplasm.

How are the ribosomes and the RNA polymerase kept in sync? It now appears that a complex of two proteins, NusE and NusG, binds the small subunit of the ribosome directly to the RNA polymerase (Fig. 13.27). When a ribosome binds mRNA and begins to make a polypeptide chain, this stimulates the RNA polymerase to increase the speed of RNA synthesis in order to keep up. Conversely, if bacteria are treated with an antibiotic that halts protein synthesis, RNA polymerase will slow down.

> In prokaryotes, the ribosomes can begin to translate a message before the RNA polymerase has finished transcribing it.

8. Some Ribosomes Become Stalled and Are Rescued

Cellular metabolism is not perfect and cells must allow for errors. One problem ribosomes sometimes run across is defective mRNA that lacks a stop codon. Whether synthesis of the mRNA was never completely finished or whether it was mistakenly snipped short by a ribonuclease, problems ensue. In the normal course of events, a ribosome that is translating a message into protein will, sooner or later, come across a stop codon. Even if an mRNA molecule comes to an abrupt end, ribosomes may be released only by release factor and this in turn needs a stop codon. If the mRNA is defective and there is no stop codon, a ribosome that reaches the end could just sit there forever and the ribosomes behind it will all be stalled, too.

Bacterial cells contain a small RNA molecule that rescues stalled ribosomes. This is named **tmRNA** because it acts partly like tRNA and partly like mRNA. Like a

> Ribosomes that have stalled due to defective mRNA can be rescued by a special RNA—tmRNA.

coupled transcription-translation   When ribosomes of bacteria start translating an mRNA molecule that is still being transcribed from the DNA
tmRNA   Specialized RNA used to terminate protein synthesis when a ribosome is stalled by damaged mRNA

FIGURE 13.27
NusEG Couples Transcription with Translation

Ribosomes are directly attached to RNA polymerase via the NusEG complex.

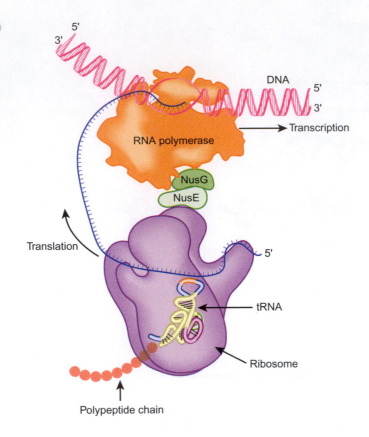

FIGURE 13.28
Stalled Ribosome Liberated by tmRNA

Binding of a tmRNA carrying alanine allows the translation of a damaged message to continue. First, alanine is added, then a short sequence of about 10 amino acids encoded by the tmRNA. Finally, the stop codon of the tmRNA allows proper termination of the polypeptide chain.

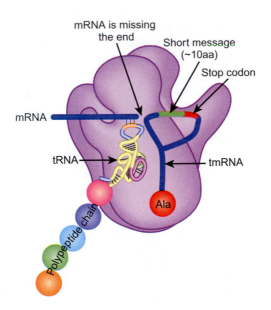

tRNA, the tmRNA carries alanine, an amino acid. When it finds a stalled ribosome, it binds beside the defective mRNA (Fig. 13.28). Protein synthesis now continues, first using the alanine carried by tmRNA, and then continuing on to translate the short stretch of message that is also part of the tmRNA. Finally, the tmRNA provides a proper stop codon so that release factor can disassemble the ribosome and free it for continued protein synthesis. The tRNA domain of tmRNA lacks an anticodon loop and a D-loop. A protein known as SmpB (not shown in Fig. 13.28 for clarity) binds to the tRNA domain and makes contacts to the ribosome that would normally be made by the missing D-loop.

TABLE 13.04 Comparison of Protein Synthesis	
Prokaryotes	**Eukaryotes (cytoplasm)**
Polycistronic mRNA	Monocistronic mRNA
Coupled transcription and translation	No coupled transcription and translation for nuclear genes
Linear polyribosomes	Circular polyribosomes
No cap on mRNA	5′ end of mRNA is recognized by cap
Start codon is next AUG after ribosome binding site	No ribosome binding site so first AUG in mRNA is used
First amino acid is formyl-Met	First Met is unmodified
70S ribosomes made of: 30S and 50S subunits	80S ribosomes made of: 40S and 60S subunits
Small 30S subunit: 16S rRNA and 21 proteins	Small 40S subunit: 18S rRNA and 33 proteins
Large 50S subunit: 23S and 5S rRNA plus 31 proteins	Large 60S subunit: 28S, 5.8S and 5S rRNA plus 49 proteins
Elongation factors: EF-T (2 subunits) and EF-G	Elongation factors: eEF1 (3 subunits) and eEF2
Three initiation factors: IF1, IF2 and IF3	Multiple initiation factors: eIF2 (3 subunits), eIF3, eIF4 (4 subunits), eIF5
Shut-off by dimerization of ribosomes in non-growing cells	Control via eIF sequestration

Clearly, the protein that has just been made is defective and should be degraded. As might be supposed, the tmRNA has signaled that the protein that was made is defective. The short stretch of 11 amino acids specified by the message part of tmRNA and added to the end of the defective protein acts as a signal, known as the ssrA tag. (SsrA stands for small stable RNA A, a name used for tmRNA before its function was elucidated.) The ssrA tag is recognized by several proteases (originally referred to as "**tail specific proteases**"), which degrade all proteins carrying this signal. These include the Clp proteases and the HflB protease involved in the heat shock response (see Ch. 16). Eukaryotic cells lack tmRNA, but do have a process called nonsense-mediated decay that degrades defective mRNA (see Ch. 12).

9. Differences between Eukaryotic and Prokaryotic Protein Synthesis

The overall scheme of protein synthesis is similar in all living cells. However, there are significant differences between bacteria and eukaryotes. These are summarized in Table 13.04 and discussed in the following sections. Note that eukaryotic cells contain mitochondria and chloroplasts, which have their own DNA and their own ribosomes. The ribosomes of these organelles operate similarly to those of bacteria and will be considered separately below. In eukaryotic protein synthesis, it is usually the cytoplasmic ribosomes that translate nuclear genes. Several aspects of eukaryotic protein synthesis are more complex. The ribosomes of eukaryotic cells are larger and contain more rRNA and protein molecules than those of prokaryotes. In addition, eukaryotes have more initiation factors and a more complex initiation procedure.

A few aspects of protein synthesis are actually less complex in eukaryotes. In prokaryotes, mRNA is polycistronic and may carry several genes that are translated to give several proteins. In eukaryotes, each mRNA is monocistronic and carries only a single gene, which is translated into a single protein. In prokaryotes, the genome and the ribosomes are both in the cytoplasm, whereas in eukaryotes the genome is in the nucleus. Consequently, coupled transcription and translation is not possible for eukaryotes (except for their organelles; see below).

Both prokaryotes and eukaryotes have a special initiator tRNA that recognizes the start codon and inserts methionine as the first amino acid. In prokaryotes, this first

Eukaryotic ribosomes are larger and more complex than those of prokaryotes.

tail specific protease   Enzyme that destroys defective proteins by degrading them tail-first; that is, from the carboxy terminal end

TABLE 13.05	Translation Factors: Prokaryotes vs. Eukaryotes	
	Prokaryotes	**Eukaryotes**
Initiation	IF1	eIF1A
	IF2	eIF5B (GTPase)
	IF3	eIF1
		eIF2 (α, β, γ) (GTPase)
		eIF2B (α, β, γ, δ, ϵ)
		eIF3 (13 subunits)
		eIF4A (RNA helicase)
		eIF4B (activates eIF4A)
		eIF4E (cap binding protein)
		eIF4G (eIF4 complex scaffold)
		eIF4H
		eIF5
		eIF6
		PABP (Poly(A)-binding protein)
Elongation	EF-Tu	eEF1A
	EF-Ts	eEF1B (2–3 subunits)
		SBP2
	EF-G	eEF2
Termination	RF1	eRF1
	RF2	
	RF3	eRF3
Recycling	RRF	
	EF-G	
		eIF3
		eIF3j
		eIF1A
		eIF1

Functionally homologous factors are in the same row. Adapted from Table 1 of Rodnina MV and Wintermeyer W. (2009) Recent mechanistic insights into eukaryotic ribosomes. Curr. Op. Cell Biol. 21: 435–443.

methionine has a formyl group on its amino group (i.e., it is N-formyl-methionine), but in eukaryotes unmodified methionine is used.

9.1. Initiation, Elongation, and Termination of Protein Synthesis in Eukaryotes

Initiation of protein synthesis differs significantly between prokaryotes and eukaryotes. Eukaryotic mRNA has no ribosome binding site (RBS). Instead recognition and binding to the ribosome rely on a component that is lacking in prokaryotes: the cap structure at the 5′ end, which is added to eukaryotic mRNA before it leaves the nucleus (see Ch. 12). Cap-binding protein (one of the subunits of eIF4) binds to the cap of the mRNA.

Eukaryotes also have more initiation factors than prokaryotes and the order of assembly of the initiation complex is different (see Table 13.05). Two different complexes assemble before binding to mRNA. The first is the 43S pre-initiation complex. This is an assembly of the small 40S subunit of the ribosome attached to several eukaryotic initiation factors (eIFs). These include eIF1, eIF1A, eIF3, and eIF5. This binds the charged initiator tRNA, Met-tRNA$_i^{Met}$, plus eIF2. The second complex, the cap-binding complex, contains cap-binding protein (eIF4E), eIF4G, eIF4A, eIF4B, and poly(A)-binding protein (PABP).

Eukaryotic mRNA is recognized by its cap structure (not by base pairing to rRNA).

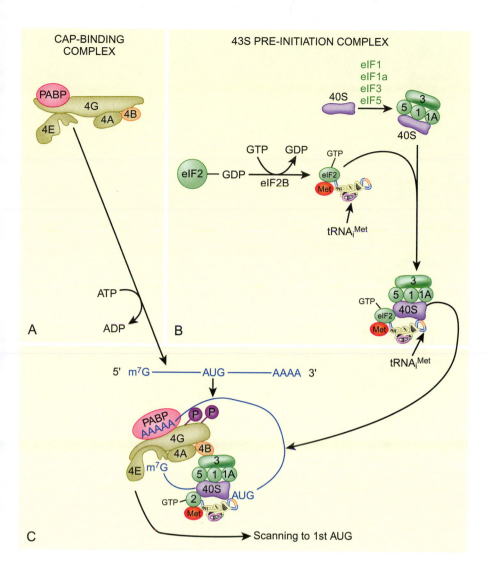

■- **FIGURE 13.29**

Assembly of the Eukaryotic Initiation Complex

A) The cap-binding complex includes poly(A)-binding protein (PABP), eIF4A, eIF4B, eIF4E, and eIF4G, which is in an unphosphorylated state when unbound to mRNA. ATP transfers phosphates to the complex to make it competent for binding the mRNA. B) The 43S initiation complex forms bringing the small ribosomal subunit together with the tRNA$_i$^{met}. This complex uses GTP to attach the tRNA to the 40S subunit via eIF2. In addition, initiation factors eIF1, eIF1A, eIF3, eIF5, and eIF2B guide and make the complex competent to bind to the 5'-UTR of mRNA. C) The mRNA is recognized by the cap-binding complex via the connections between eIF4E and PABP which bind the 5' and 3' ends of the mRNA, respectively. These two connections cause the rest of the mRNA to loop out. When this is established, then the 43S pre-initiation complex can attach and start scanning for the first AUG. After pausing at the first AUG, then the 50S subunit of the ribosome can bind and initiate translation.

During eukaryotic initiation, cap-binding complex first attaches to the mRNA via its cap. Next, the poly(A) tail is bound by PABP so that the mRNA forms a ring. This structure can now bind the 43S assembly. In order to align the Met-tRNA$_i$^{Met} with the correct AUG codon, the two structures work together to scan each codon from the 5' end. This scanning process uses energy from ATP (Fig. 13.29). Normally, the first AUG is used as the start codon, although the sequence surrounding the AUG is important. The consensus is GCCRCCAUGG (R = A or G). If its surrounding sequence is too far from consensus an AUG may be skipped. Once a suitable AUG has been located, eIF5 joins the complex, which in turn allows the 60S subunit to join and the cap binding protein, eIF2, eIF1, eIF3, and maybe eIF5 to depart. eIF5 uses energy from GTP to accomplish this remodeling of the ribosome.

The next stage is elongation (Fig. 13.30). Of all the stages of translation, elongation in bacteria and eukaryotes is the most similar. As in bacteria, elongation factors work to decode the mRNA and bind the tRNA into the A-site of the ribosome. Rather than EF-Tu and EF-Ts, eukaryotes use eEF1A to deliver the tRNA using GTP hydrolysis for energy and eEF1B to replace the depleted GDP with fresh GTP. The only difference is that eukaryotic elongation factors include more subunits. The remaining steps are the same. The peptidyl transferase activity of the 28S rRNA of the large subunit links the incoming amino acid to the polypeptide chain. Then elongation factor eEF2 (direct counterpart to bacterial EF-G) uses GTP to drive the conformational changes in the ribosome and ratchet the tRNAs from the P- and A-sites into the E- and P-sites. Elongation continues until a stop codon enters the A-site.

FIGURE 13.30
Beginning Eukaryotic Translation Elongation

Once the eukaryotic 40S subunit complex finds the first AUG, then the remaining 60S subunit and associated factors combine to form the final 80S ribosome.

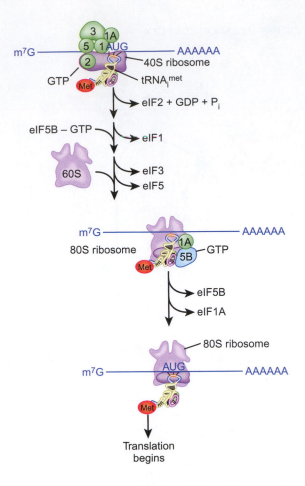

Eukaryotic termination differs from prokaryotic termination in two ways. First, rather than having two different release factors (RF1 and RF2) to recognize different stop codons, eukaryotes have a single release factor (eRF1) that recognizes all three stop codons. eRF1 binds the stop codon, but this does not affect peptide bond formation. Instead, eRF3 carrying a GTP molecule binds to eRF1. GTP hydrolysis then rearranges the factors and the final amino acid attaches to the polypeptide. Therefore, eukaryotes require GTP for polypeptide completion, whereas in bacteria, RF1 or RF2 is sufficient.

Finally, as in bacteria, eukaryotic ribosomes are recycled. eIF3 triggers the release of the 60S subunit, and then eIF1 releases the final tRNA. An additional factor, eIF3j, then removes the mRNA. The components are then recycled.

Box 13.2 Internal Ribosome Entry Sites

Although most eukaryotic mRNA is scanned by the 40S subunit to find the first AUG, exceptions do occur. Sequences known as internal ribosome entry sites (IRES) are found in a few mRNA molecules. As the name indicates, these allow ribosomes to initiate translation internally, rather than at the 5′ end of the mRNA. IRES sequences were first found in certain viruses that have polycistronic mRNA despite infecting eukaryotic cells. In this case, the presence of IRES sequences in front of each coding sequence allows a single mRNA to be translated to give multiple proteins. The best known examples are members of the Picornavirus family, which includes poliovirus (causative agent of polio) and rhinovirus (one of the agents of common cold).

More recently, it has been found that a few special mRNA molecules encoded by eukaryotic cells themselves also possess IRES sequences. During major stress situations, such as heat shock or energy deficit, synthesis of the majority of proteins is greatly

Continued

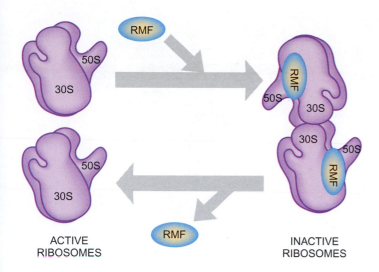

►■FIGURE 13.31
***Bacterial Ribosomes on
Standby During Bad
Conditions***

Active bacterial ribosomes
can become inactive when the
RMF protein binds to them. The
ribosomes form dimers with the
30S subunits attached to one
another. When conditions are
favorable, dissociation occurs.

ACTIVE
RIBOSOMES

INACTIVE
RIBOSOMES

Box 13.2 Continued

decreased. Much of this regulation occurs at the initiation stage of translation (see below).
However, a few proteins are exempted from this down-regulation as they are needed
under stress conditions. The mRNAs encoding these proteins often contain an IRES
sequence. In these cases the mRNA carries only a single coding sequence and the IRES
is located in the 5'-UTR, between the 5' end of the mRNA and the start of the coding
sequence. This allows translation to be initiated at the IRES even in the absence of the
standard initiation/scanning procedure.

10. Protein Synthesis Is Halted When Resources Are Scarce

Proteins make up about two-thirds of the organic matter in a cell and their synthesis
consumes a major part of the cell's energy and raw materials. Clearly, when cells run
low on nutrients or energy they cannot continue to synthesize proteins at the normal rate. In bacteria, ribosomes are taken out of service during stationary phase or
periods of slow growth. A small basic protein, **ribosome modulation factor (RMF)**,
binds to ribosomes and inactivates them (Fig. 13.31). The inactive ribosomes exist as
dimers. When favorable conditions return, the inactive dimers are disassembled and
the ribosomes are reactivated. In *E. coli*, starvation induces the **stringent response**,
where the cell only transcribes genes that are essential for survival and virulence.
Most other genes are turned off at the transcriptional level. This response is triggered,
in part, when an uncharged tRNA loads into the A-site of the ribosome. This will only
occur if charged tRNA is in short supply—due to lack of energy or amino acids. When
this happens, RelA, a protein associated with the ribosome during translation, starts
making pppGpp, which binds to RNA polymerase to modulate what genes are transcribed. This response is also important for bacteria such as *Mycobacterium tuberculosis*, which causes tuberculosis. These bacteria are highly resistant to antibiotic
treatment and the human immune system and can live for years inside immune cells.
They are able to persist because of their stringent response.

Higher organisms also stop protein synthesis when nutrients or energy run low.
However, they do so by inactivating the initiation factors rather than the ribosomes
(Fig. 13.32). Initiation factor eIF2 uses energy by hydrolyzing GTP to GDP. After

---

ribosome modulation factor (RMF)   Protein that inactivates surplus ribosomes during slow growth or stationary phase in bacteria
stringent response   Decreasing transcription of non-essential genes when nutrients are in limited supply

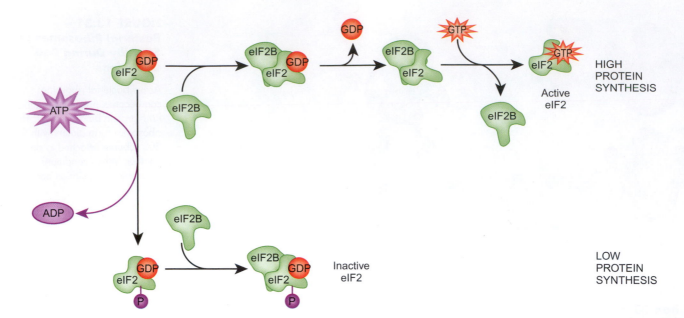

FIGURE 13.32
Recycling of Initiation Factor eIF2 is Controlled

When eukaryotes down-regulate the level of protein synthesis, a protein kinase phosphorylates eIF2/GDP. This prevents eIF2B from removing the GDP and eIF2/GDP stays locked in an inactive complex with eIF2B. Absence of active eIF2 decreases the rate of initiation.

initiation is over, it is released from the ribosome with the GDP still bound. It then binds to eIF2B, which exchanges GDP for GTP, so recycling the eIF2. In times of stress, a kinase phosphorylates eIF2 and prevents the removal of GDP. The GDP bound form of eIF2 cannot initiate translation and protein synthesis is halted. Some viruses have taken advantage of this mechanism and use phosphorylation of eIF2 as a way to shut down host protein synthesis (see Focus on Relevant Research).

As usual, eukaryotes have multiple pathways of control. Another way the cell controls protein synthesis is to regulate the cap-binding complex. When the cell is stressed or starved, mTOR (target of the antibiotic rapamycin) is prevented from phosphorylating eIFs 4E, 4G, and 4B. Without these phosphates, the eIF complex cannot bind to mRNA, and translation is halted.

FOCUS ON RELEVANT RESEARCH

Estaban M. (2009) Hepatitis C and evasion of the interferon system: a PKR paradigm. Cell Host Microbe 6:495–497.

Hepatitis C virus (HCV) is a major health concern since it causes hepatitis, liver cirrhosis, and cancer, and has infected 170 million people in the world.

This virus is resistant to our immune system and infections can become chronic. Treatment of HCV infections requires antiviral therapy with two different drugs: type I interferon (IFN), which stimulates the immune system, and ribavirin, which is a nucleoside analog. Ribavirin is incorporated into DNA during replication, but it causes chain termination. Thus ribavirin prevents virus replication. Both types of drug are essential for the proper treatment of HCV because neither alone works well. IFN alone will initially block HCV infection, but after some exposure, the virus becomes resistant. The nucleoside analog blocks DNA replication, but is not effective without the immune support.

Recent work has shown how HCV resists the immune system and IFN treatment. HCV does not block transcription of immune genes as demonstrated by the observation that the mRNA levels were not changed. Instead, the virus blocks translation. In uninfected cells IFN binds to a cell surface receptor, which in turn, activates a variety of cellular proteins. These transmit the signal from the cell membrane into the nucleus where a transcription factor turns on a variety of genes that stimulate the immune system to block and destroy viruses. This pathway is stimulated when a patient takes IFN. However, when HCV is present the resulting mRNAs are not translated. This paper shows that the virus is able to modulate translation by phosphorylating the dsRNA-dependent protein kinase R (PKR), which then phosphorylates eIF2α. This form of eIF2α is unable to activate the 43S pre-initiation complex, and translation is blocked. The virus also has another trick up its sleeve. The virus has IRES elements in its mRNA rather than the normal capped mRNA structure. IRES elements do not require eIF2α for translation. Thus the virus has successfully evolved a way to block the cell's viral defenses and simultaneously subvert the cell into translating the viral mRNAs.

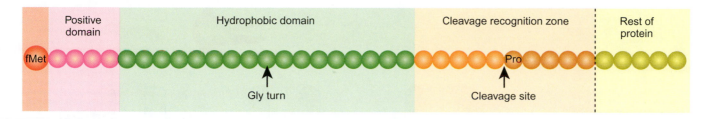

FIGURE 13.33
Standard Signal Sequence for Exported Proteins

The signal sequence contains a positively-charged domain (containing lysine and/or arginine), a α-helical hydrophobic domain (rich in alanine, leucine and valine), and a cleavage site preceded by a glycine or serine and followed by a proline. A reverse turn due to glycine is found approximately half-way through the hydrophobic domain.

11. A Signal Sequence Marks a Protein for Export from the Cell

Once a protein has been made, it must find its correct location within the cell. Although cytoplasmic proteins are made in the cell compartment where they belong, other proteins, which do not reside in the cytoplasm, must be transported. Proteins destined to be exported to the exterior of the cell must be exported through the cell membrane. Similar systems exist in bacterial and eukaryotic cells. Proteins destined for export are tagged at the N-terminus with a **signal sequence**. This is cut off after export by proteases attached to the outside of the membrane and is therefore not present in the mature protein. The signal sequence consists of approximately 20 amino acids that form an α-helix. There is little specific sequence homology between signal sequences from different exported proteins. A positively-charged, basic N-terminus of two to eight amino acids is followed by a long stretch of hydrophobic amino acids. The amino acid just before the cleavage site has a short side chain (Fig. 13.33).

> Exported proteins have a signal sequence at the front.

A polypeptide destined for export is recognized by its signal sequence. In bacteria, the signal recognition protein (SecA) binds the signal sequence and guides it to the **translocase** complex in the cell membrane. The rest of the protein being exported is synthesized and follows the signal sequence into and through the membrane via the translocase. This is known as **cotranslational export**, since the protein is exported as it is made. The signal sequence is cut off by the **leader peptidase** (or signal peptidase) after translocation (Fig. 13.34).

> After export of a protein across the cell membrane via the translocase, the signal sequence is cut off.

There are approximately 500 translocases per *E. coli* cell. Each cell exports about 1×10^6 proteins from the cytoplasm prior to dividing. In a cell that doubles in 20 minutes, 100 proteins are exported per minute per translocase. Protein export is 10-fold faster than protein synthesis. So the demand for a growing protein chain will allow the translocase to be ready for a new chain as fast as the ribosome can make it. Note that in gram-negative bacteria such as *E. coli*, most of these exported proteins are structural components of the outer membrane that are being made constantly, rather than enzymes being excreted outside the cell for digestive purposes.

In eukaryotes, cotranslational export occurs across the membranes of the endoplasmic reticulum. In multi-cellular eukaryotes proteins involved in digestion, such as amylases and proteases, must be exported. So must proteins located in blood and other body fluids, such as antibodies, albumins, and circulating peptide hormones. When the animal genes for preproinsulin or ovalbumin are put into *E. coli*, correct export across the cell membrane occurs and cleavage of the signal sequence by the *E. coli* leader peptidase happens at the correct position. Conversely, yeast cells correctly

cotranslational export Export of a protein across a membrane while it is still being synthesized by a ribosome
leader peptidase Enzyme that removes the leader sequence after protein export
signal sequence Short, largely hydrophobic sequence of amino acids at the front of a protein that label it for export
translocase Enzyme complex that transports proteins across membranes

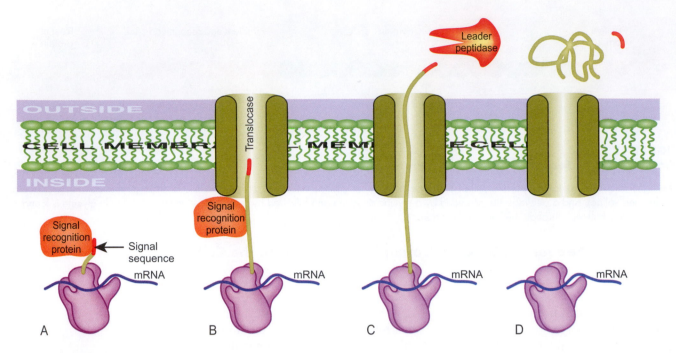

FIGURE 13.34
Cotranslational Export of Proteins

A) The ribosome making the polypeptide chain approaches the cell membrane. The polypeptide with its signal sequence binds to the signal recognition protein. B) The signal recognition protein recognizes the translocase and binds to it, allowing the polypeptide chain to begin its journey through the membrane. C) After the signal sequence exits the translocase, leader peptidase cuts the polypeptide chain, liberating the signal peptide. D) Final folding of the protein occurs outside the cell.

process and excrete bacterial β-lactamase. Thus, the export machinery is highly conserved between diverse organisms.

11.1. Molecular Chaperones Oversee Protein Folding

Molecular **chaperones**, or **chaperonins**, are proteins that oversee the correct folding of other proteins. Many chaperonins belong to the family of **heat shock proteins (HSPs)**, as their levels increase at high temperature (see Ch. 16). Chaperonins may be divided into two main classes: "holders" and "folders"—respectively, those that prevent premature folding and those that attempt to rectify misfolding. Obviously, chaperonins cannot "know" the correct 3D structure for several other proteins. Mechanistically, they act to prevent incorrect folding, rather than actively creating a correct structure.

During bacterial protein export, the secretory chaperonin SecB keeps the polypeptide chain from folding up prematurely. Secreted proteins must travel through a narrow translocase channel and so must remain unfolded until they reach the other side of the membrane. The Hsp70 set of chaperonins tends to bind to newly synthesized or highly uncoiled proteins (Fig. 13.35).

The more complex GroE (= Hsp60/Hsp10) chaperonin machine attempts to refold damaged or misfolded proteins. When polypeptide chains unfold, they expose hydrophobic regions that are normally clustered in the center of the folded protein. Left to themselves, many proteins could refold. However, inside a cell, there is a high concentration of protein. Consequently, exposed hydrophobic regions from multiple

> Chaperonins are proteins that promote the correct folding of other proteins.

---

chaperone Sometimes "molecular chaperone"; same as chaperonin
chaperonin Protein that oversees the correct folding of other proteins
heat shock protein (HSP) Protein induced in response to high temperature. Many heat shock proteins are chaperonins

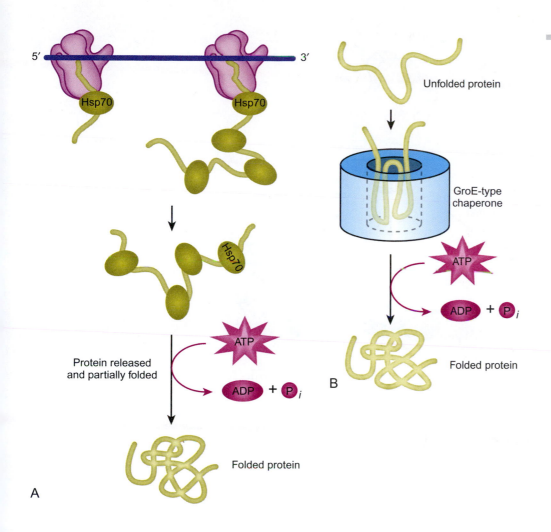

5′

3′

Hsp70

Hsp70

Hsp70

Unfolded protein

GroE-type chaperone

ATP

ADP + P $_i$

Folded protein

B

Protein released and partially folded

ATP

ADP + P $_i$

Folded protein

A

FIGURE 13.35
Chaperonins Act by Two General Mechanisms

A) Chaperonins of the Hsp70 type act during protein formation by binding to hydrophobic patches of the protein. Once chaperonins are released, the protein automatically folds. B) Large chaperonins, such as GroE, act after translation by sequestering misfolded protein in a central cavity. Freed from the influences of other molecules in the cytoplasm, the protein will fold correctly.

proteins bind to each other and the proteins aggregate together. The GroE chaperonin machine forms a cavity in which a single polypeptide can refold on its own, protected from interactions with other polypeptide chains.

Newly-made proteins face a folding problem. The N-terminus of the growing polypeptide has already left the ribosome, while the C-terminal region is still being made. Consequently, the N-terminal region does not yet have access to any folding information that resides in later regions of the protein. To prevent misfolding at this stage, the emerging protein is sheltered by a chaperonin known as trigger factor (Fig. 13.36). This binds to the large subunit of the ribosome close to the polypeptide exit tunnel.

12. Protein Synthesis Occurs in Mitochondria and Chloroplasts

Mitochondria and chloroplasts are thought to be of prokaryotic origin. The symbiotic hypothesis of organelle origins argues that symbiotic prokaryotes evolved into organelles by specializing in energy production and progressively losing their genetic independence (see Ch. 4 for further details). Both mitochondria and chloroplasts contain circular DNA that encodes some of their own genes and they divide by binary fission. They contain their own ribosomes and make some of their own proteins. Organelle ribosomes resemble the ribosomes of bacteria rather than the ribosomes of the eukaryotic cytoplasm. The initiation and elongation factors of organelles are also bacterial in nature. Nonetheless, there are differences in composition between organelle and bacterial ribosomes, as shown in Table 13.06.

Protein synthesis in mitochondria and chloroplasts resembles that of bacteria in many respects.

FIGURE 13.36
*Trigger Factor and the
Ribosome*

Trigger factor shelters newly made
polypeptides as they emerge from
the ribosome. Trigger factor protects
about 70% of newly made proteins.
The other 30% need the DnaJK
or GroEL chaperonins. *(Credit:
Hoffmann A, Bukau B, Kramer G.
(2010) Structure and function of
the molecular chaperone Trigger
Factor. Biochim Biophys Acta
1803:650–61.)*

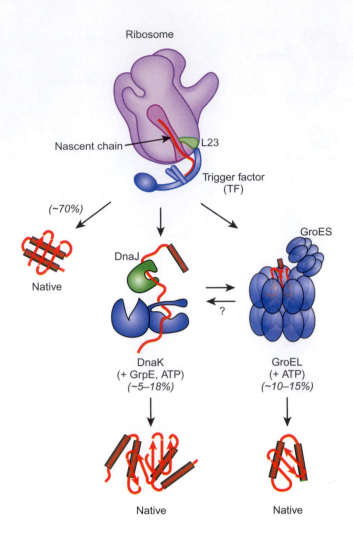

TABLE 13.06	Components of Cytoplasmic, Organelle, and Bacterial Ribosomes		
Location	**Subunits**	**Ribosomal RNA**	**Proteins**
Animal Cytoplasm	40S	18S	33
	60S	28S, 5.8S, 5S	49
Animal Mitochondria	28S	12S	31
	39S	16S	48
Plant Cytoplasm	40S	18S	~35
	60S	28S, 5.8S, 5S	~50
Plant Chloroplast	30S	16S	22–31
	50S	23S, 5S, 4.5S	32–36
Plant Mitochondria	30S	18S	>25
	50S	26S, 5S	>30
Bacterial	30S	16S	21
	50S	23S, 5S	31
Archaeal	30S	16S	26–27
	50S	23S, 5S	30–31

Box 13.3 Hybrid Ribosomes

Although they are larger, the cytoplasmic/nucleus-encoded ribosomes of eukaryotes resemble those from Archaea in the way they operate. A similar relation holds for the associated initiation and elongation factors. In fact, it is possible to make hybrid ribosomes containing one subunit from yeast and one from *Sulfolobus* (an archaeon). These still make protein, albeit less efficiently than native ribosomes. In contrast, hybrid ribosomes made by mixing subunits from yeast and *E. coli* are totally functionless.

12.1. Proteins Are Imported into Mitochondria and Chloroplasts by Translocases

The size of organelle genomes varies considerably from organism to organism. Generally, the more advanced eukaryotes have smaller organelle genomes. The mitochondria of mammals make only around 10 proteins and in higher plants the chloroplasts make approximately 50 proteins. The other organelle proteins are encoded by nuclear genes and made on the cytoplasmic ribosomes. They are then transported into the organelles.

Proteins for import into mitochondria have a leader sequence at the N-terminus. This consists of 20 or more amino acids with a positively-charged lysine or arginine every three or four residues and no negatively-charged residues. The leader forms an α-helix with a positively-charged face and a hydrophobic face. This is recognized by a receptor on the mitochondrial surface. The protein is imported successively through two translocase complexes known as TOM (translocase, outer mitochondrial) and TIM (translocase, inner mitochondrial) that lie in the outer and inner membranes of the mitochondria, respectively. After importing the protein, its leader sequence is trimmed off.

Plant cells are more complex than animal cells as they possess not only mitochondria but also chloroplasts. The principle of protein import is similar. The chloroplast contains two translocases equivalent to TIM and TOM, which are known as TIC and TOC (C for chloroplast). The leader sequences for chloroplast proteins resemble those for mitochondria, and in fact only plant cells can tell them apart. Thus, the mitochondria of fungi will import chloroplast proteins if genes encoding these are artificially introduced into the fungal cell. It is still unclear how plants decide between chloroplast and mitochondrial leader sequences; however, it seems that the leaders for the two kinds of organelle form different secondary structures.

Protein import by organelles also needs chaperonins on both sides of the membrane. An imported protein must travel through the narrow translocase channel in an uncoiled conformation. To avoid premature folding, newly synthesized organelle proteins are kept in a loosely folded conformation by chaperonins. Later, when the imported protein emerges from the translocase into the inside of the organelle, it is bound by another set of chaperonins. In particular, an Hsp70-type chaperonin is responsible for hauling in the incoming protein. The Hsp70 acts as a ratchet, binding to successive segments of unfolded polypeptide chain. Each binding and release of Hsp70 consumes energy in the form of ATP.

> Many mitochondrial and chloroplast proteins are made in the eukaryotic cytoplasm and enter the organelle after synthesis.

13. Mistranslation Usually Results in Mistakes in Protein Synthesis

Ribosomes are not perfect and make occasional mistakes. Perhaps 1 in 10,000 codons is misread and results in the wrong amino acid being incorporated. Two amino acids whose codons differ by only one base are most likely to be confused. Other possible errors are shifts in the reading frame ("**frameshift**") or reading through stop codons. These assorted errors are collectively known as **mistranslation**.

frameshift   Alteration in the reading frame during polypeptide synthesis
mistranslation   Errors made during translation

Although such events are rare, a few weird genes actually require such errors for proper expression. For example, the *pol* gene of retroviruses (see Ch. 21) is only translated if the ribosome frameshifts or reads through a stop codon while translating the preceding *gag* gene. In Chapter 10, it was noted that the *dnaX* gene of *E. coli* gives rise to two proteins, tau and gamma, both subunits of DNA polymerase. The gamma protein is made only as a result of frameshifting. Release factor 2 (RF2) of *E. coli* also requires a frameshift for its synthesis.

14. Many Antibiotics Work by Inhibiting Protein Synthesis

Many well-known antibiotics work by inhibiting protein synthesis. Most of these are specific for prokaryotic ribosomes. However, very high concentrations of these agents will inhibit the ribosomes of mitochondria and chloroplasts, which are of prokaryotic ancestry.

Aminoglycoside antibiotics bind to the 30S subunit. **Streptomycin** binds to the 16S rRNA near where the two ribosomal subunits touch. The presence of streptomycin distorts the A-site and hinders binding of incoming charged tRNA. In particular, binding of initiator tRNA-Met is inhibited and so initiation of translation is prevented. Streptomycin-resistant mutants have alterations in nucleotide 523 of 16S rRNA or in ribosomal protein S12 (RpsL), which assists antibiotic binding. Many of the other aminoglycosides, such as gentamycin and kanamycin, bind to multiple sites on the 30S subunit and mainly inhibit the translocation step of protein synthesis. Streptomycin and other aminoglycosides also cause misreading of the mRNA.

Tetracyclines inhibit both bacterial and eukaryotic ribosomes. They bind to the 16S (or 18S) rRNA of the small subunit and block the attachment of charged tRNA. Despite inhibiting both types of ribosome, tetracyclines inhibit bacteria preferentially due to the fact that bacteria actively take them up, whereas eukaryotic cells actively export them.

Chloramphenicol binds to the 50S subunit to the loop of 23S rRNA that interacts with the acceptor stem of the tRNA and inhibits the peptidyl transferase. **Cycloheximide** binds to the 60S subunit of eukaryotic ribosomes and inhibits the peptidyl transferase. **Erythromycin** and related macrolide antibiotics bind to the 23S rRNA of bacterial ribosomes and inhibit the translocation step.

Fusidic acid is a steroid derivative that binds to prokaryotic elongation factor EF-G. In the presence of fusidic acid, EF-G, with its bound GDP, is frozen in place on the ribosome. Fusidic acid also inhibits the corresponding eukaryotic elongation factor EF-2; however, in practice, animal cells are unaffected as they do not take up the antibiotic.

15. Post-Translational Modifications of Proteins

Although the genetic code has codons for only 20 amino acids, many other amino acids are occasionally found in proteins. Apart from selenocysteine and pyrrolysine (see below), these extra amino acids are made by modifying genetically encoded amino acids after the polypeptide chain has been assembled. This is known as **post-translational modification**. See Fig. 13.37 for a summary of some common post-translational modifications.

Streptomycin and related antibiotics bind to rRNA in the small subunit of the bacterial ribosome.

Tetracycline binds rRNA in the small subunit of both prokaryotic and eukaryotic ribosomes.

Chloramphenicol binds to 23S rRNA and prevents peptide bond formation.

---

aminoglycosides   Class of antibiotics that inhibits protein synthesis; includes streptomycin, neomycin, kanamycin, amikacin, and gentamycin
chloramphenicol   An antibiotic that inhibits bacterial protein synthesis
cycloheximide   An antibiotic that inhibits eukaryotic protein synthesis
erythromycin   An antibiotic that inhibits bacterial protein synthesis
fusidic acid   An antibiotic that inhibits protein synthesis
post-translational modification   Modification of a protein or its constituent amino acids after translation is finished
streptomycin   An antibiotic of the aminoglycoside family that inhibits protein synthesis
tetracyclines   Family of antibiotics that inhibit protein synthesis

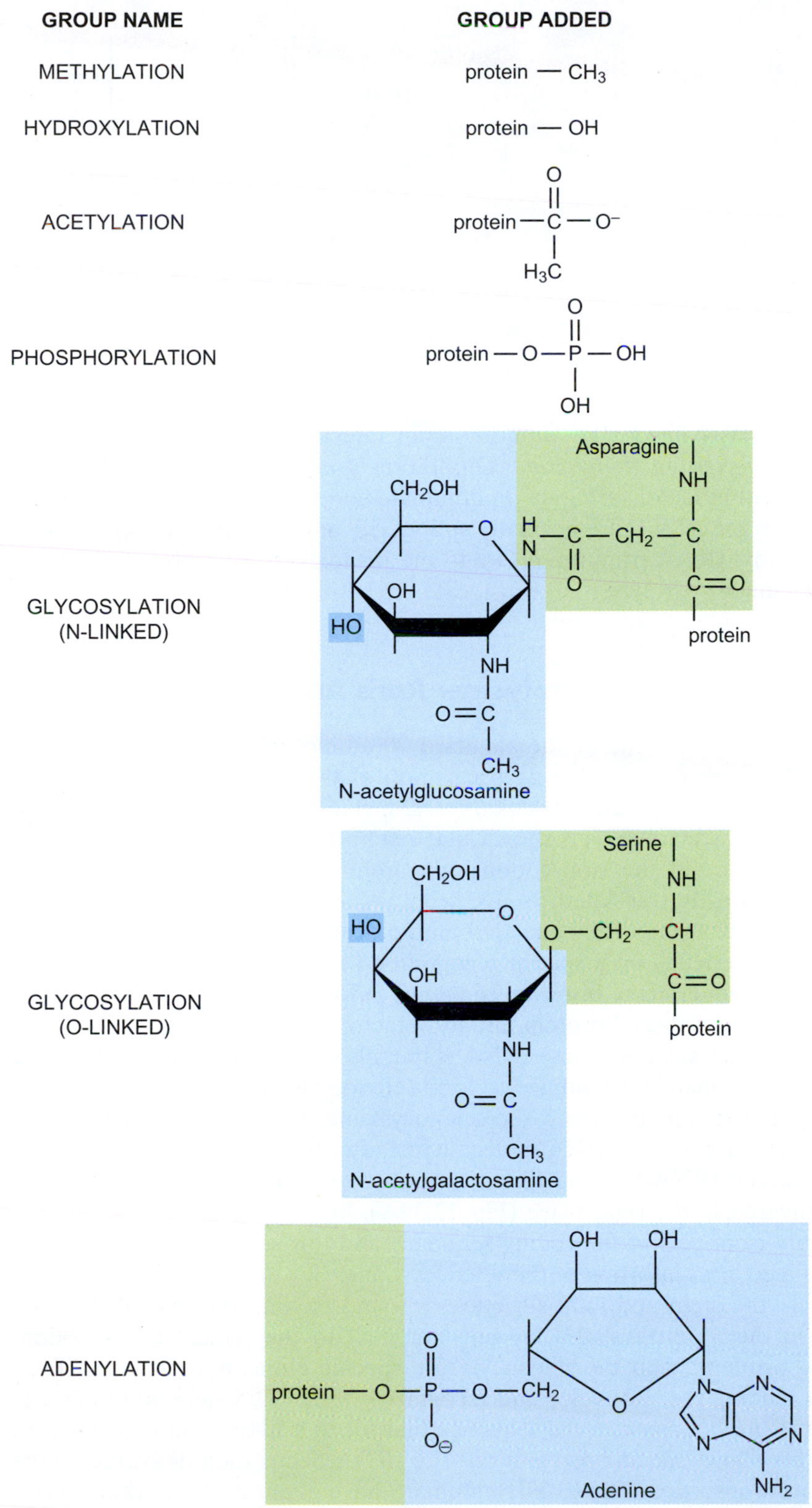

GROUP NAME	GROUP ADDED

METHYLATION

protein — CH$_3$

HYDROXYLATION

protein — OH

ACETYLATION

$$\text{protein} - \overset{\overset{\displaystyle O}{\|}}{\underset{\underset{\displaystyle H_3C}{|}}{C}} - O^-$$

PHOSPHORYLATION

$$\text{protein} - O - \overset{\overset{\displaystyle O}{\|}}{\underset{\underset{\displaystyle OH}{|}}{P}} - OH$$

GLYCOSYLATION (N-LINKED)

N-acetylglucosamine

Asparagine

protein

GLYCOSYLATION (O-LINKED)

N-acetylgalactosamine

Serine

protein

ADENYLATION

Adenine

■ **FIGURE 13.37**

Common Post-Translational Modifications

The structure of some of the more common protein modification groups are shown.

An example of medical importance is **diphthamide**, which is derived from histidine by post-translational modification (Fig. 13.38). It is found only in elongation factor eEF2 of eukaryotes and Archaea, in a region of the amino acid sequence that is highly conserved. The corresponding bacterial factor, EF-G, does not contain diphthamide.

---

diphthamide   Modified amino acid found only in eukaryotic elongation factor eEF2 that is the target for diphtheria toxin

HISTIDINE   DIPHTHAMIDE

Diphthamide was named after diphtheria, an infectious disease caused by the bacterium *Corynebacterium diphtheriae*. Diphtheria toxin attaches an ADP-ribose fragment to elongation factor eEF2 via diphthamide and this inhibits protein synthesis and kills the target cells. eEF2 normally splits GTP and uses the energy released to move the peptidyl-tRNA from the A-site to the P-site. ADP-ribosylated eEF2 still binds GTP but cannot hydrolyze it or translocate the peptidyl-tRNA.

16. Selenocysteine and Pyrrolysine: Rare Amino Acids

Selenocysteine (Sec) is not one of the standard 20 amino acids and yet it is incorporated into a few rare proteins during translation of the mRNA by the ribosome. This occurs both in bacteria and in eukaryotes, including humans. Sequencing of the genes and proteins involved has shown that selenocysteine is encoded by UGA. However, UGA is one of the stop codons. Apparently, UGA is normally read as "stop" but is occasionally translated to give selenocysteine, which therefore has the honor of being the 21st genetically encoded amino acid. The choice between "stop" and selenocysteine depends on a special recognition sequence in the following part of the gene—the **selenocysteine insertion sequence (SECIS element)**. Selenocysteine has its own tRNA and a special protein initiation factor to escort charged tRNA-Sec to the ribosome. In fact, selenocysteine-tRNA is initially charged with serine. Then the attached serine is enzymatically modified to form selenocysteine.

When bacteria use selenocysteine, the selenocysteine insertion sequence forms a stem and loop structure in the mRNA molecule just after the UGA. SelB protein recognizes both charged tRNA-Sec and this stem and loop. Thus, selenocysteine bound to tRNA is delivered to the right place (Fig. 13.39A). In bacteria, the stem and loop form temporarily from part of the coding sequence and this section of the mRNA is therefore translated after insertion of the selenocysteine.

In mammals, the stem and loop structure is found beyond the end of the coding sequence, in the 3′-untranslated region—not next to the critical UGA codon! SECIS binding protein 2 and the selenocysteine specific elongation factor, eEFsec, are needed for binding the tRNA-Sec and recognizing the SECIS stem and loop. The stem of this structure has a conserved sequence that forms a K-turn via non-standard base pairing. This allows specific recognition. The tRNA-Sec is then delivered to the correct position for insertion (Fig. 13.39B). Humans have about 25 genes that encode selenoproteins. These are divided into two groups for regulation. Some are expressed under all conditions and others only when needed to combat oxidative stress (see Focus on Relevant Research).

Very rarely, the stop codon UGA is read as the unusual amino acid selenocysteine.

---

selenocysteine (Sec)   Amino acid resembling cysteine but containing selenium instead of sulfur
selenocysteine insertion sequence (SECIS element)   Recognition sequence that signals for insertion of selenocysteine at a UGA stop codon

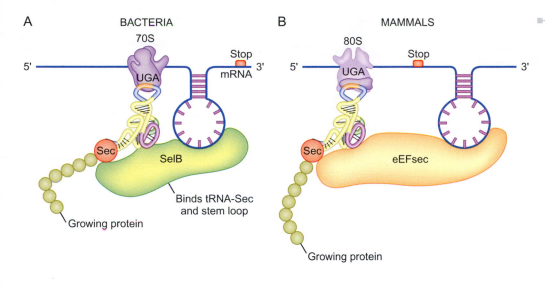

A   BACTERIA
70S

B   MAMMALS
80S

FIGURE 13.39
Delivery of tRNA with Selenocysteine to an Internal UGA Stop Codon

A) In bacteria, the tRNA carrying selenocysteine (Sec) first binds to SelB and the complex then binds to a stem and loop in the mRNA. This aligns the tRNASec with a UGA codon within the coding sequence on the mRNA. Selenocysteine is then inserted as part of the growing polypeptide. Only the fully bound complex is shown. B) In mammals, the protein that binds the stem and loop and the tRNASec is called eEFsec. In addition, the stem and loop are more distant, being found after the stop codon.

Morley SJ and Willett M. (2009) Kinky binding and *SEC*sy insertions. Mol. Cell 35: 396–398.

Although selenocysteine is a rare amino acid, it is incorporated into many different proteins. Making selenocysteine requires selenium, which must therefore be supplied in trace amounts by our diet. Many selenocysteine containing proteins are essential for mammals since they are associated with oxidation-reduction reactions and cellular antioxidants. Selenocysteine containing proteins fall into two different categories: essential housekeeping enzymes that are produced all the time, or stress induced selenoproteins. In people suffering from selenium deficiency, the body discriminates between the two kinds, and only translates the essential housekeeping enzymes. Since both housekeeping and stress-related selenoprotein genes have a SECIS sequence and use the same Sec insertion system, their separate regulation was puzzling.

This paper summarizes recent research that reveals the differences between the expression of housekeeping and stress-related selenoproteins. Expression of the gene for phospholipid hydroperoxide glutathione peroxidase (PHGPx), an essential housekeeping enzyme, was compared to expression of glutathione peroxidase 1 (GPx1), a stress-related enzyme that is not made when selenium is deficient. The key is eukaryotic initiation factor eIF4a3, which has no previously known role in translation. However, it was recently found that eIF4a3 interacts with the SECIS element. The gene for this transcription factor is controlled by the level of selenium, and when selenium is low, more eIF4a3 is made. The eIF4a3 protein then binds to certain SECIS elements and prevents binding of the Sec insertion apparatus (SBP2, eEFSec, and tRNASec) to the element. However, eIF4a3

**FOCUS ON
RELEVANT RESEARCH**

does not bind to the SECIS of housekeeping genes because their SECIS elements have a slightly different structure from those of stress-related genes. The secondary structure motif, AAR, is essential to SECIS function. In housekeeping mRNAs, the AAR motif lies at the top of the stem loop, whereas, in stress-related mRNAs, the AAR is near the K-turn of the stem. AAR in the stem preferentially binds eIF4a3, which in turn prevents incorporation of selenocysteine at the UGA. This stalls translation and the mRNA is then targeted for degradation. Therefore, a slightly different structure within a stem loop is sufficient to decide whether the mRNA is essential or dispensable under conditions of low selenium.

Selenocysteine is an analog of cysteine, but has selenium instead of sulfur (Fig. 13.40). Selenium is more susceptible to oxidation than sulfur and so proteins that contain it must be protected from oxygen. Examples are the formate dehydrogenases found in many bacteria. These contain selenocysteine in their active sites and function in anaerobic metabolism. They are inactivated by oxygen and are normally made only in the absence of air. It has been suggested that the occurrence of selenoproteins in different groups of organisms is related to their oxygen sensitivity. Higher plants, which make oxygen, completely lack proteins that contain selenocysteine. Fungal genomes also totally lack selenoproteins. Conversely, fish, which live in the sea where oxygen levels are lower than on land, have more selenoproteins than typical mammals. Indeed, zebrafish selenoprotein P contains 17 Sec residues, the largest number in any known protein.

FIGURE 13.40
Selenocysteine and Cysteine

Selenocysteine is identical to cysteine except for the replacement of sulfur by selenium.

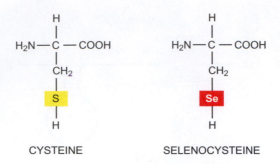

FIGURE 13.41
Pyrrolysine and Lysine

Pyrrolysine is (5R,5R)-4-methyl-pyrroline carboxylate.

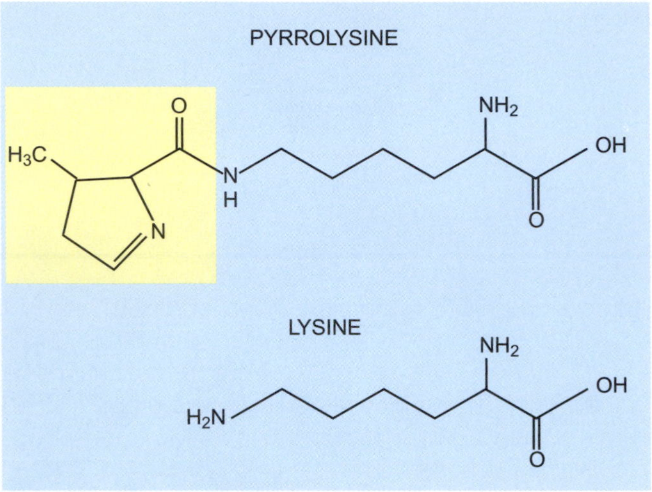

The stop codon UAG is very rarely translated as the rare amino acid, pyrrolysine.

In 2002, a 22nd genetically encoded amino acid was discovered—**pyrrolysine (Pyl)**, a derivative of lysine with an attached pyrroline ring (Fig. 13.41). This is found in a few Archaea where it is encoded by the stop codon UAG in occasional proteins. Pyrrolysine was first discovered in the active site of methylamine methyl-transferases found in methane-producing Archaea of the genus *Methanosarcina*. An unusual aminoacyl-tRNA synthase, a special tRNA, and genes for three accessory proteins are also found in organisms with pyrrolysine.

By analogy with selenocysteine, it was thought that pyrrolysine-tRNA was first charged with lysine, which was then modified to form pyrrolysine. However, this proved to be wrong. Pyyrolysine is made first as a free amino acid and then attached to tRNA-Pyl. There is no pyrrolysine specific elongation factor. Moreover, the sequence determinants that specify which UAG codons should be used for pyrrolysine insertion are unclear. Genome sequencing has found genes homologous to those for the pyrrolysine system in occasional Eubacteria suggesting that pyrrolysine may be present. However, pyrrolysine itself has not yet been identified directly in these organisms. The fact that the same five genes in the same order are found in both Archaea and Eubacteria suggests that horizontal gene transfer has occurred.

17. Degradation of Proteins

Living cells not only synthesize proteins, they also degrade them. Although protein degradation is nowhere near as complex as synthesis, it is nonetheless carefully regulated and often highly specific. **Proteases** (or proteinases) are enzymes that degrade

---

protease   Same as proteinase; an enzyme that degrades proteins
pyrrolysine (Pyl)   22nd genetically encoded amino acid, derived from lysine

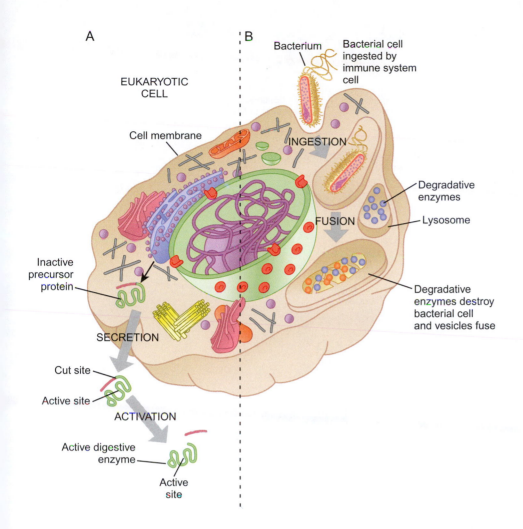

A

EUKARYOTIC
CELL

Cell membrane

Inactive
precursor
protein

SECRETION

Cut site

Active site

ACTIVATION

Active digestive
enzyme

Active
site

B

Bacterium   Bacterial cell
ingested by
immune system
cell

INGESTION

FUSION

Degradative
enzymes

Lysosome

Degradative
enzymes destroy
bacterial cell
and vesicles fuse

■ **FIGURE 13.42**
*Digestive Enzymes Are
Activated on Location*

A) Proteases destined for export
are made as precursors and
are cleaved to form the active
protease once safely outside the
cell. B) Proteases in the membrane-
bound lysosome degrade ingested
material.

proteins. They are potentially dangerous to the organism that makes them and must be carefully controlled. Proteases are often located in separate compartments where they can act without endangering other components of the organism. Alternatively, proteases may be designed so that they only accept specifically tagged proteins for degradation.

Proteases are found in three main locations: extracellular, inside special compartments, and free in the cytoplasm. Animals secrete proteases into their digestive tracts. These enzymes are usually synthesized as inactive precursors and only activated once they are safely outside the cells of the animal that made them (Fig. 13.42A). Examples are trypsin (and its precursor trypsinogen) and pepsin (and its precursor pepsinogen). Plants that catch insects, fungi that trap nematodes, and bacteria that live in rotting animal or plant tissue also secrete proteases. As with animals, these proteases are generally secreted as inactive precursors and only activated once outside the cells of the producer organism.

Lysosomes are membrane-bound organelles found in eukaryotic cells. They contain a variety of digestive enzymes, including proteases, and function in self-defense. When cells of the immune system have engulfed bacteria or virus particles, the vesicles containing the invader are merged with lysosomes and the infectious agent is, hopefully, digested (Fig. 13.42B). Bacteria do not always cooperate—for example, many pathogenic strains of *Salmonella* can survive the toxins and digestive enzymes inside lysosomes.

Enzymes that degrade proteins are dangerous. They are frequently kept in separate compartments and often made as inactive precursors.

lysosome Membrane-bound organelle of eukaryotic cells that contains degradative enzymes

FIGURE 13.43
Operation of Proteasome

Ubiquitin tags damaged proteins and is recognized by the cylindrical proteasome. After degradation, the polypeptide fragments and ubiquitin are extruded.

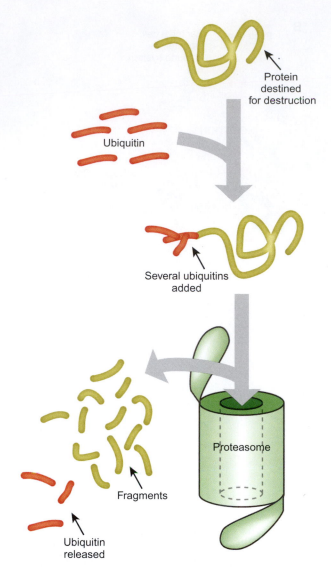

Protein destined for destruction

Ubiquitin

Several ubiquitins added

Proteasome

Fragments

Ubiquitin released

Proteases located in the cytoplasm itself must be very carefully controlled. Nonetheless, the cell needs some internal proteases to degrade damaged or misfolded proteins. Many proteases found inside bacterial cells form hollow cylinders, with the dangerous active site on the inside of the cylinder. Proteins slated for destruction are ferried to the protease cylinder and pushed into its center by accessory proteins. The number of misfolded proteins and consequently the level of protein degradation increases greatly under certain conditions, in particular when cells are exposed to uncomfortably high temperatures that tend to disrupt protein structure. This induces the heat shock response described in more detail in Chapter 16.

Eukaryotes have more sophisticated structures known as **proteasomes**. These are cylindrical, with the protease active sites inside. The top and bottom of the cylinder are covered by protein complexes that recognize and bind damaged or unwanted proteins. Proteins destined for degradation are identified by tagging with **ubiquitin**. This is a small protein that is fixed to damaged or misfolded proteins and also to certain proteins that are needed only for a brief period (Fig. 13.43). Ubiquitin tagged proteins are unfolded and then fed into the barrel of the proteasome where they are degraded into short peptides. The ubiquitin tags themselves are cleaved off and recycled.

proteasome   Protein assembly found in eukaryotic cells that degrades proteins
ubiquitin   Small protein attached to other proteins as a signal that they should be degraded; used by eukaryotic cells, not bacteria

Key Concepts

- Polypeptide chains are linear polymers of amino acids. There are 20 common genetically encoded amino acids plus two rare ones.
- Each amino acid is encoded by a codon consisting of three bases. Each codon is recognized by the corresponding anticodon on a transfer RNA (tRNA) to which the amino acid is attached.
- Translation is the synthesis of proteins using information from RNA and is carried out by the ribosome.
- Codons can be read in three possible reading frames, thus it is important to locate the correct start codon by using recognition sequences.
- During elongation of the polypeptide chain, the tRNA moves between three separate sites on the ribosome.
- Translation is terminated when the stop codon is read by a protein known as release factor. The ribosome subunits are then recycled.
- Several ribosomes usually read the same mRNA simultaneously.
- In bacteria a single mRNA may encode several proteins. In addition, since there is no nucleus, transcription and translation are coupled.
- There are differences between protein synthesis in eukaryotic and prokaryotic cells.
- Proteins to be exported from the cell have signal sequences that direct them to the export machinery.
- Molecular chaperones are specialized proteins that oversee the correct folding of other proteins.
- Some proteins are made inside mitochondria and chloroplasts, but most organelle proteins are made on cytoplasmic ribosomes and transported into the organelles.
- Many antibiotics inhibit protein synthesis. These include tetracycline, aminoglycosides (such as streptomycin), and chloramphenicol.
- The rare amino acids selenocysteine and pyrrolysine are inserted at special stop codons.
- Proteins are broken down by proteases. In eukaryotes proteins tagged with ubiquitin are degraded in cylindrical assemblies called proteasomes.

Review Questions

1. What is a gene product? Are all gene products translated?
2. Describe three gene products that are not translated.
3. How common is the generalization "one gene—one protein" true? Briefly give three examples of how one gene can yield multiple proteins.
4. Would you expect to find the same number of genes in the genome as there are proteins in the proteome? Why or why not?
5. Translate the following mRNA sequence into protein:

   `5'-AUGCUAGCUCCUGAUUUCUA-3'`

6. Name two modified bases commonly found in the TψC-loop and D loops of tRNA molecules and explain why they are necessary.
7. If there are 61 codons (excluding the stop codons), then why are only 31 different tRNA molecules needed to carry out translation?
8. How is a tRNA molecule "charged" with an amino acid?
9. Compare and contrast ribosomes and their subunits in prokaryotes and eukaryotes.
10. What role does 23S rRNA play in synthesis of protein by a ribosome?
11. Why is the correct reading frame important in translating an mRNA into protein?
12. Which mutation is more detrimental for an ORF: insertion of <u>one</u> extra nucleotide or insertion of <u>three</u> extra nucleotides? Why?

13. Summarize the initiation of translation in prokaryotes. Include the following terms: initiator tRNA, N-formyl-methionine, Shine-Dalgarno sequence, 16S rRNA, AUG start codon, and initiation factors.

14. What other codon is sometimes used as a start codon instead of AUG? What types of proteins tend to use this codon?

15. Summarize the elongation step of translation in prokaryotes. Include the following terms: E-site, P-site, A-site, peptidyl transferase, translocation, elongation factors, and GTP.

16. What are the three stop codons? What proteins recognize them?

17. What is a polysome?

18. What is polycistronic mRNA? Is it found in prokaryotes, eukaryotes, or both?

19. How does a prokaryote recognize an mRNA that needs to be translated? How about a eukaryote?

20. What is meant by "coupled transcription and translation"? Does this occur in prokaryotes, eukaryotes, or both? Explain.

21. What is tmRNA? What role does it play during translation?

22. How does tmRNA tag proteins for degradation?

23. In what ways is translation in eukaryotes simpler than translation in prokaryotes?

24. Summarize initiation of translation in eukaryotes. In what ways is it different from prokaryotes?

25. How does the halting of protein synthesis due to nutrient deprivation differ between prokaryotes and eukaryotes?

26. What is a signal sequence and what purpose does it serve?

27. Summarize the process of cotranslational export.

28. What purposes do chaperone proteins serve?

29. Compare and contrast bacterial protein export with mitochondrial and chloroplast protein import.

30. Provide an example of how some organisms utilize "mistranslation" to create a functional protein.

31. Why is the genetic code not completely "universal"?

32. What is post-translational modification?

33. Why are selenocysteine and pyrrolysine considered the 21st and 22nd amino acids, whereas diphthamide is not considered the 23rd amino acid?

34. How does a ribosome know to incorporate selenocysteine when it encounters a UGA stop codon?

35. Which stop codon occasionally encodes pyrrolysine?

36. Describe the mechanism of action of the aminoglycoside antibiotics.

37. Do you think spontaneous streptomycin resistance is likely to occur in bacteria? Why or why not?

38. How does a cell protect itself from proteases that it normally exports after synthesis?

39. What are lysosomes?

40. How do eukaryotes dispose of unwanted proteins?

Conceptual Questions

1. Since the genetic code is degenerate, more than one triplet codon adds the same amino acid during translation. List all the possible mRNA sequences that could code for these amino acids: (a) Met-Ser-Asn; (b) Val-His-Phe; (c) Trp-Glu-Tyr.

2. Reading frame is important for proper protein translation from mRNA. When the tRNA shifts one or two nucleotides from the prior triplet codon, the subsequent amino acids will be completely different. Translate the following mRNA using all three reading frames:

   5' UGC-CUU-AAU-CAC-CGU-CUA-AUG-GGC-CGC-AUU-AUC-CGG 3'

3. A researcher identified a mutant *E. coli* that was unable to take in the sugar lactose and metabolize it for growth. The researcher isolated the polycistronic mRNA that has the gene for lactose uptake and metabolism. He also isolated the same mRNA from a wild-type strain that was able to use lactose as a sugar source. He found that the mRNA from the mutant mRNA was different from the corresponding sequence found in the wild-type strain. What is missing in the mutant mRNA?

Mutant mRNA	5' GAAUCTTAAGUUCUACAAU... 3'
Wild-type mRNA	5' GAAUCTTAUGAGUUCUACAAU... 3'

4. Many antibiotics inhibit translation by binding to ribosomes. Why do antibiotics inhibit bacterial pathogens but do not harm our cellular ribosomes?

Protein Structure and Function

Proteins are linear polymers of amino acids. There are 20 common genetically encoded amino acids, each with chemically distinct side chains. The wide choice of possible monomers makes proteins extremely versatile, with a wide range of properties and capabilities. Proteins may be catalytically active (i.e., enzymes) or may act as transport carriers across membranes or in aqueous solution. In addition there are structural proteins with no catalytic or binding activities and a variety of regulatory proteins that control both gene expression and other cellular activities. In order to fulfill their diverse roles, proteins fold up into a variety of 3D shapes and structures. Several levels of folding are required to generate the final structure of proteins and these depend on a variety of chemical forces to hold them together. After folding of the protein itself, other chemical groups may need to be attached for the protein to function correctly. These range from metal ions to complex organic co-factors.

1. The Structure of Proteins Reflects Four Levels of Organization

During the process of translation, amino acids are linked together into a linear polypeptide chain. Each amino acid has a specific chemistry that dictates how that chain interacts and folds into a 3D structure to function properly. Furthermore, many proteins are assembled from more than one polypeptide chain and many also have **co-factors** or **prosthetic groups**—associated molecules that are not made of amino acids. The final shape of a protein is determined by its amino acid sequence, so proteins with similar sequences have similar 3D conformations. As more and more protein 3D structures are deciphered, it has emerged that some proteins have similar 3D structures even when their amino acid sequences are quite different.

> Linear polypeptide chains are folded up to give the final 3D structures.

Typical polypeptides are 300–400 amino acids long. Polypeptides much smaller or much larger are less common. However, many hormones and growth factors, such as insulin, do consist of relatively short polypeptide chains. Individual polypeptides with more than a thousand amino acids are very rare and very large proteins tend to consist of several separate polypeptide chains rather than a single long chain.

The structures of biological polymers, both proteins and nucleic acids, are often divided into four levels of organization:

1. **Primary structure** is the order of the monomers; that is, the sequence of the amino acids for a protein, or of the nucleotides in the case of DNA or RNA.

2. **Secondary structure** is the folding or coiling of the original polymer chains by means of hydrogen bonding. In the case of proteins, the hydrogen bonds are between the atoms of the polypeptide backbone.

3. **Tertiary structure** is the further folding that gives the final 3D structure of a single polymer chain. In the case of proteins, this involves interactions between the R groups of the amino acids.

4. **Quaternary structure** is the assembly of several separate polymer chains.

1.1. The Secondary Structure of Proteins Relies on Hydrogen Bonds

By definition, the secondary structure is folding that depends solely on hydrogen bonding. In DNA, hydrogen bonding occurs between base pairs and is the basis of the double helix. In proteins, hydrogen bonding occurs between the peptide groups that form the backbone of the polypeptide (Fig. 14.01). The polypeptide chain must be folded around to bring two peptide groups alongside each other. The hydrogen on the nitrogen of one peptide group is then bound to the oxygen of the other. (Note that hydrogen bonds may also contribute to tertiary structure, but here they are not the only or even the major forces involved.)

> Hydrogen bonding is responsible for the formation of alpha-helix and beta-sheet structures in proteins.

Most of the secondary structure found in proteins is due to one of two common secondary structures, known as the α- (alpha) helix and the β- (beta) sheet. Both structures allow formation of the maximum possible number of hydrogen bonds and are therefore highly stable.

In the α-helix (Fig. 14.02), a single polypeptide chain is coiled into a right-handed helix and the hydrogen bonds run vertically up and down, parallel to the helix axis. In fact, the hydrogen bonds in an α-helix are not quite parallel to the axis. They are slightly tilted relative to the helix axis because there are 3.6 amino acids per turn rather than a whole number. The pitch (repeat length) is 0.54 nm and the rise per residue is about 0.15 nm.

The hydrogen bonds hold successive turns of the helix together and run from the C=O group of one amino acid to the NH group of the fourth amino acid residue

---

α- **(alpha) helix**   A helical secondary structure found in proteins
β- **(beta) sheet**   A flat sheet-like secondary structure found in proteins
co-factor   Extra chemical group bound (often temporarily) to a protein but which is not part of the polypeptide chain
prosthetic group   Extra chemical group bound (often covalently) to a protein but which is not part of the polypeptide chain
primary structure   The linear order in which the subunits of a polymer are arranged
quaternary structure   Aggregation of more than one polymer chain to form a final structure
secondary structure   Initial folding up of a polymer into a regular, repeating structure, due to hydrogen bonding
tertiary structure   Final 3D folding of a polymer chain

Folded protein

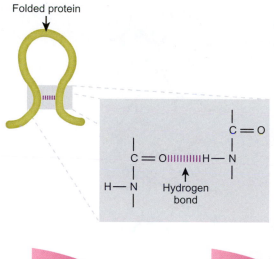

■ **FIGURE 14.01**
***Hydrogen Bonding
between Peptide Groups***

Two peptide bonds of a
polypeptide chain may be aligned
to form a hydrogen bond by
looping the polypeptide chain
around.

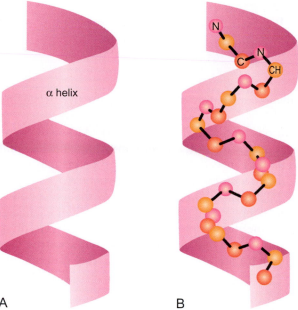

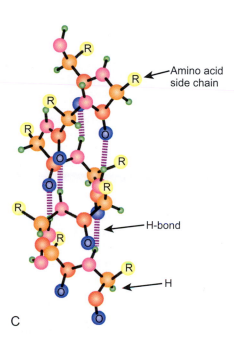

A   B   C

FIGURE 14.02
The Alpha Helix

A) The general shape of an
α-helix. B) The carbon backbone
of the polypeptide chain. C) The
hydrogen bonds between peptide
groups.

along the polypeptide chain. The α-helix is very stable because all of the peptide groups ($-CO-NH-$) take part in two hydrogen bonds, one up and one down the helix axis. A right-handed helix is most stable for L-amino acids. (A stable helix cannot be formed with a mixture of D- and L-amino acids, although a stable left-handed helix could theoretically be formed from D-amino acids).

The R-groups extend outwards from the tightly packed helical polypeptide backbone. Of the 20 amino acids, Ala, Glu, Leu, and Met are good α-helix formers but Tyr, Ser, Gly, and Pro are not. Proline is totally incompatible with the α-helix, due to its rigid ring structure. Furthermore, when proline residues are incorporated, no hydrogen atoms remain on the nitrogen atom that takes part in peptide bonding. Consequently, proline residues interrupt hydrogen-bonding patterns. In addition, two bulky residues or two residues with the same charge that lie next to each other in the polypeptide chain will not fit properly into an α-helix. Overall, the α-helix forms a solid cylindrical rod.

The β-sheet is also held together by hydrogen bonding between peptide groups but in this case the polypeptide chain is folded back on itself to give a flattish zigzag structure (Fig. 14.03). Like the α-helix, the β-sheet is very stable because all of the peptide groups (except for those on the edge of the sheet) take part in two hydrogen bonds. In the β-sheet, these go sideways from each peptide group, one to each side. The sections of the polypeptide chain which lie side by side are usually, although not

FIGURE 14.03
The β-sheet—Hydrogen Bonding

A) A polypeptide chain is shown folded back and forth three times to form a flattened sheet. B) The polypeptide backbone of the sheet is depicted demonstrating how the β-sheet forms a zigzag in three dimensions. C) A ball and stick model shows the hydrogen bonding between the strands of the β-sheet.

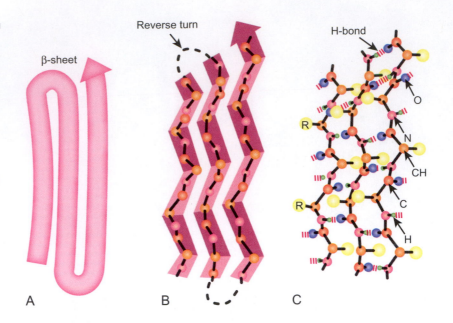

always, antiparallel and the R-groups lie alternately above and below the zigzagging plane of the β-sheet.

A variety of β-sheet conformations are known (Fig. 14.04). Although some β-sheets are flat, most known β-sheets are twisted (in a right-handed manner) so that the sheet structure as a whole is not flat but curved. In some cases, β-sheets may curve around so that the last strand bonds to the first, forming a barrel structure.

A **reverse turn** (also known as a β-turn or β-bend) is where the polypeptide chain turns back upon itself. Beta sheets have reverse turns at the ends of each segment, but such turns are also found frequently in other places. Pro and Gly are often found in reverse turns. Those regions of protein that do not form secondary structures are referred to as "**random coil**" although they are, of course, not truly random, but merely irregular.

1.2. The Tertiary Structure of Proteins

Further folding of the polymer chain constitutes the tertiary structure. In a nucleic acid this would be the supercoiling. In a protein, the polypeptide chain, with its pre-formed α-helix and β-sheet regions, is folded to give the final 3D structure. In general, polypeptide chains with similar amino acid sequences fold to give similar 3D structures. Tertiary folding depends on interactions between the side chains of the individual amino acids. Since there are 20 different amino acids, a large variety of final 3D conformations is possible, although most polypeptides form roughly spherical tertiary structures.

The α-helix and β-sheet modules form the basic structural units of the protein (Fig. 14.05). They are linked by loops of random coil of various lengths. Many of these loops are at the surface of the protein, exposed to the solvent and contain predominantly charged or polar amino acid residues. The rigid ring structure of proline causes an approximately 90° change in direction of the polypeptide backbone. Consequently, proline disrupts secondary structures and contributes to overall folding by forming bends. Examination of their 3D structures has shown that the thousands of known proteins are in fact built from relatively few structural motifs. Such motifs generally

random coil	Region of polypeptide chain lacking secondary structure
reverse turn	Region of polypeptide chain that turns around and goes back in the same direction

FLAT RECTANGULAR SHEET   TWISTED SHEET, SADDLE SHAPE   β - BARREL

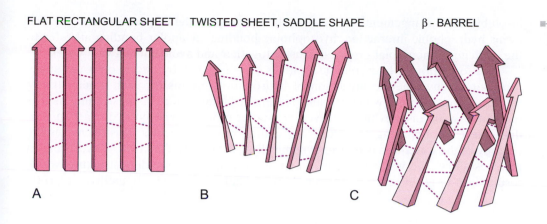

A   B   C

■ FIGURE 14.04
The β-sheet Conformations

A) Flat sheet. B) Twisted sheet.
C) Barrel.

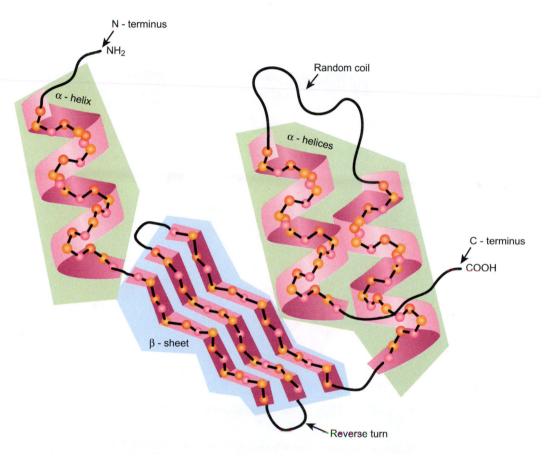

FIGURE 14.05
Modular Arrangement of α-Helices and β-Sheets

A stylized polypeptide chain. Both α-helical regions and a β-sheet segment are shown in this folded polypeptide chain. These modular segments are linked by reverse turns and regions of random coil.

consist of several α-helices and/or β-sheets joined to form a useful and recognizable structure (Fig. 14.06).

This 3D folding is largely driven by two factors acting in concert. Many of the amino acids have R-groups that are very water-soluble (hydrophilic). These side chains prefer to be on the surface of the protein so they can dissolve in the water surrounding the protein and make hydrogen bonds to water molecules. In contrast, R-groups, which are water repellent (hydrophobic), huddle together inside the protein away from the water (Fig. 14.07). Since hydrophobic molecules are greasy and

Hydrophobic amino acids tend to cluster together inside proteins so promoting 3D folding.

Hydrophilic amino acids often end up on the surface of folded proteins where they make contact with water.

insoluble, this arrangement is known as the **oil drop model** of protein structure. The terms hydrophobic interaction, hydrophobic bonding, or apolar bonding all refer to the tendency of non-polar groups to cluster together and avoid contact with water.

The formation of hydrophobic bonds is driven mostly by effects on water structure, not by any inherent attraction of hydrophobic groups for each other. Strictly, the term hydrophobic (which means "fearing/disliking water") is misleading as it is the water which dislikes the dissolved non-polar groups. Exposed hydrocarbon residues exert an organizing effect on surrounding water molecules. This decreases the entropy of the water and is thermodynamically unfavorable. Removal of hydrocarbon residues allows the water to return to its less organized H-bonding structure, which results in a large increase in entropy (approximately 0.7 Kcal per methylene group

FIGURE 14.06
Arrangement of an α/β-Barrel Domain

Schematic diagram of the α/β-barrel domain of the enzyme methylmalonyl CoA mutase. Alpha-helices are red and beta-strands are blue. The inside of the barrel is lined by small hydrophilic side chains (Ser and Thr), which allows space for the substrate coenzyme A (green) to bind along the axis of the barrel. *(Credit: Introduction to Protein Structure by Brandon & Tooze, 2nd ed., 1999. Garland Publishing, Inc., New York and London.)*

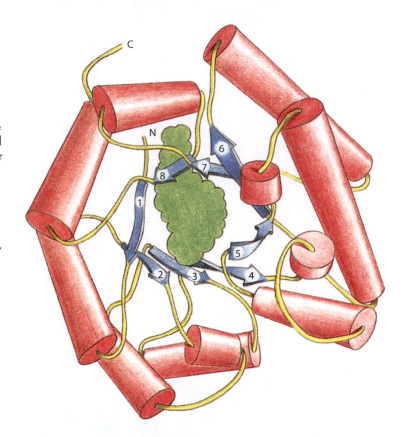

FIGURE 14.07
The Oil Drop Model of Protein Structure

A simplified model to illustrate that hydrophilic groups are exposed to the water environment and that most hydrophobic groups are centrally positioned. Note that in real life many of the hydrophilic groups will form hydrogen bonds to the surrounding water molecules. In addition some hydrophilic groups will ionize, forming charged ions.

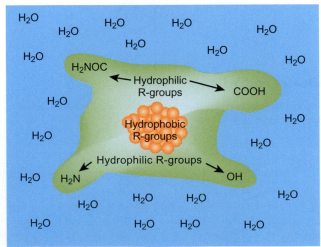

oil drop model   Model of protein structure in which the hydrophobic groups cluster together on the inside away from the water

removed). Therefore, when a leucine side chain is removed from contact with water 3.5 Kcal/mole of energy is released. Because the hydrophobic interaction depends on entropy, the strength of hydrophobic bonding increases with temperature, unlike most other forms of bonding which become less stable at higher temperatures.

Membrane proteins have an inverted structure in comparison to the standard oil drop scheme. They are hydrophobic on the surfaces where they contact the membrane lipid. Their hydrophilic residues are mostly clustered internally, but some are found at the surface in those regions where the protein emerges from the membrane (Fig. 14.08).

1.3. A Variety of Forces Maintain the 3D Structure of Proteins

In addition to the major influence of hydrophobic interactions in the core of the protein and the hydrogen bonding of hydrophilic side chains to water, a variety of other effects are important (Fig. 14.09). These include hydrogen bonds, ionic bonds, van der Waals forces, and disulfide bonds.

Hydrogen bonds may form between the R-groups of two nearby amino acids. Those amino acids with hydroxyl, amino, or amide groups in their side chains can take part in such hydrogen bonding. Similarly, ionic bonds ($-NH_3^{+}$ $^{-}OOC-$) may form between the R-groups of basic and acidic amino acid residues (Fig. 14.09). Relatively few of the possible ionic interactions occur in practice. This is because most polar groups are on the surface of the protein and form hydrogen bonds to water.

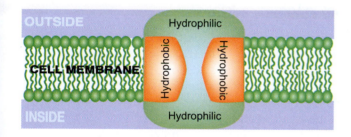

FIGURE 14.08

Inverse Conformation of a Membrane Protein

The hydrophilic regions of membrane proteins interact with the aqueous cell interior and exterior. Their hydrophobic regions face the hydrophobic phospholipids of the membrane.

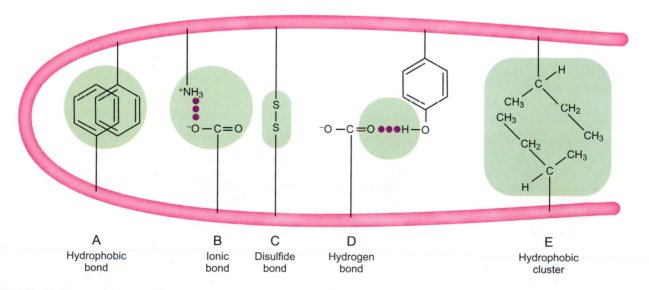

A	B	C	D	E
Hydrophobic bond	Ionic bond	Disulfide bond	Hydrogen bond	Hydrophobic cluster

FIGURE 14.09

Some Forces that Maintain 3D Structure of Proteins

Other forces that maintain the 3D structure include: A) hydrophobic ring stacking, B) ionic bonds, C) disulfide bonds, D) hydrogen bonds, and E) hydrophobic clustering.

Van der Waals forces hold molecules or portions of molecules together if they fit well enough to approach very closely. Van der Waals forces are weak and decrease very rapidly with distance. Consequently, they are only significant for fairly large regions that are complementary in shape. Disulfide bonds between cysteines are also sometimes important for 3D structure (see below).

1.4. Cysteine Forms Disulfide Bonds

Under oxidizing conditions, the sulfhydryl groups of two cysteines can form a **disulfide bond**. The dimer consisting of two cysteines (pronounced "cystEEn") joined by a disulfide bond is known as cystine (pronounced "cystYne") (Fig. 14.10). Disulfide bonds between cysteine residues are important in maintaining 3D structure in certain cases (see Fig. 14.09, above). Disulfide bonds may hold together two regions of the same polypeptide chain (intra-chain disulfide bond for tertiary structure) or may be used to hold together two separate polypeptides (inter-chain disulfide bond for quaternary structure).

Since disulfides are easily reduced to sulfhydryl groups inside cells, they are of little use in stabilizing intracellular proteins. Disulfide bonds are mostly used to stabilize extracellular proteins that are exposed to more oxidizing conditions. The classic examples are the antibodies that circulate in the blood of vertebrates; secreted enzymes, such as **lysozyme**; or hormones, such as insulin. Single-celled organisms make relatively few extracellular proteins compared to higher multicellular organisms and consequently employ disulfide bonding much less.

> Disulfide bonds between two cysteine residues can stabilize protein structures.

1.5. Multiple Folding Domains in Larger Proteins

Long polypeptide chains may contain several regions that fold up more or less independently and are joined by linker regions with little 3D structure. Such regions are known as **domains** and may be from 50–350 amino acids long (Fig. 14.11). Short proteins may have a single domain and extremely long proteins may occasionally have up to a dozen. Computerized databases of protein structure now list over 20,000 different domain structures divided into a thousand or more families.

> In long proteins folding may occur separately in different regions of the polypeptide chain.

Note that proteins begin to fold before they are completely made. As soon as a sufficient length of polypeptide chain to form a 3D structure has emerged from

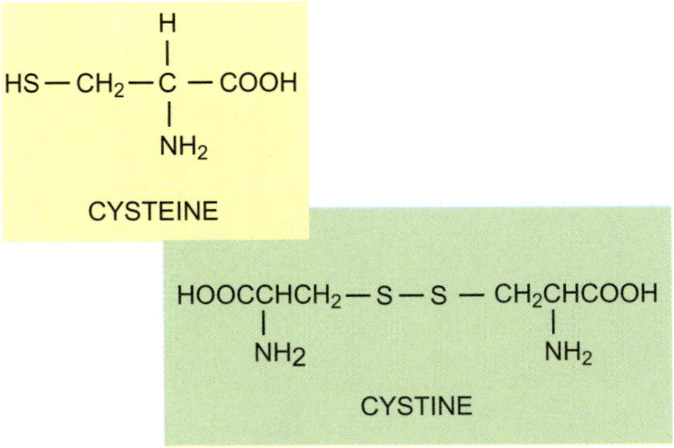

FIGURE 14.10
Cysteine and Cystine

Two cysteines can join by a disulfide bond to form cystine.

$$HS-CH_2-\overset{\overset{\displaystyle H}{|}}{\underset{\underset{\displaystyle NH_2}{|}}{C}}-COOH$$

CYSTEINE

$$HOOCCHCH_2-S-S-CH_2CHCOOH$$
$$\underset{NH_2}{|} \qquad\qquad \underset{NH_2}{|}$$

CYSTINE

---

disulfide bond   A sulfur to sulfur bond formed between two sulfhydryl groups, in particular between those of cysteine, and which binds together two protein chains

domain   (of protein) A region of a polypeptide chain that folds up more or less independently to give a local 3D structure

lysozyme   Enzyme that degrades peptidoglycan, the cell wall polymer of bacteria

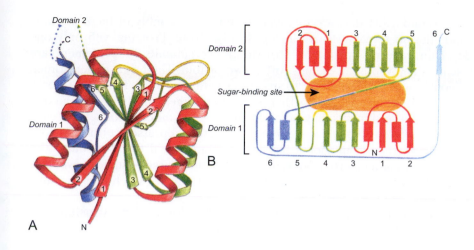

A

B

■ **FIGURE 14.11**
Two Domains in Arabinose Binding Protein

The arabinose binding protein of *E. coli* contains two open twisted α/β domains of similar structure. A) Schematic diagram of a single domain. B) Topology diagram showing the orientation of the two domains and the crevice between them in which the arabinose molecule binds. *(Credit: Introduction to Protein Structure by Brandon & Tooze, 2nd ed., 1999. Garland Publishing, Inc., New York and London.)*

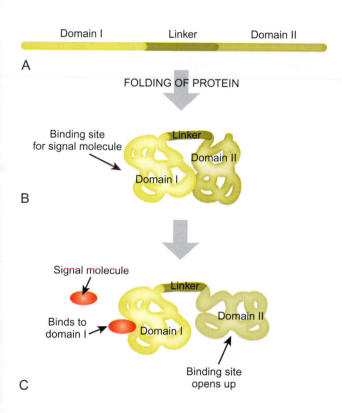

A

B

C

■ **FIGURE 14.12**
Interactions between Protein Domains to Activate a DNA-Binding Site

A) The unfolded protein shows two domains (I and II) and a linker region. B) When folded, both domain I and domain II form binding sites. C) A signal molecule binds to domain I and changes its conformation. The interaction between the two domains triggers domain II to change shape so opening up its binding site.

the ribosome, it folds. Hence, domains fold up independently, one after the other. (Because the sequence of folding is different, this also means that the refolding of a denatured polypeptide often differs significantly from the original folding.)

Many transcription factors consist of two domains—one that binds DNA and another that binds the signal molecule. When the signal molecule is bound, it changes the shape of its own domain (Fig. 14.12). The change in conformation is then transmitted to the DNA-binding domain, which also changes shape. Thus, although they fold separately, domains do interact physically.

1.6. Quaternary Structure of Proteins

Many proteins consist of several individual polypeptide chains. This is especially true of proteins whose total molecular weight is much greater than 50,000 daltons (i.e., around 400 amino acids). (Although occasional polypeptide chains are found with

1,000 or more amino acids, they are relatively rare.) The assembly of these multiple subunits yields the quaternary structure (Fig. 14.13). (Note: Proteins with only one polypeptide chain have no quaternary structure.) The subunits, or **protomers**, are usually present as an even number, most often two or four. The terms dimer, trimer, tetramer, oligomer, and multimer refer to structures with two, three, four, few/several, and multiple subunits, respectively. Less than 10% of **multimeric** proteins have an odd number of subunits. The subunits may be all identical or all different or several each of two (or more) different types. For example, the lactose repressor consists of four identical subunits, whereas hemoglobin has two α-subunits and two β-subunits. The prefixes homo- (same) and hetero- (different) are sometimes used to indicate whether the subunits are the same or different. Thus, the lactose repressor is a homo-tetramer, whereas hemoglobin is a hetero-tetramer.

> Many proteins consist of subunits, usually an even number.

The same hydrophobic forces largely responsible for tertiary structure are involved in the assembly of multiple subunits. Soluble proteins, with only a single polypeptide chain, fold so that almost all of their hydrophobic residues are hidden in the interior. In the case of subunit proteins, the polypeptide chains are folded, leaving a cluster of hydrophobic residues exposed to the water at the protein surface (Fig. 14.14). This is an unfavorable arrangement and when two polypeptide chains with exposed hydrophobic patches come into contact with each other, they tend to stick together, rather like hook-and-loop fasteners. As noted above, the hydrophobic force is weaker at lower temperatures, and many subunit proteins tend to come apart at low temperatures.

Subunits that are designed with more than one bonding region can be used to build rings or chains. Long chains of identical protein subunits are often twisted helically, as in bacterial flagella or pili. If a helix of protein subunits has wide coils that are packed close together, it will form a hollow cylinder (Fig. 14.15). The protein shell of certain viruses (e.g., tobacco mosaic virus) and the microtubules that are found in eukaryotic cells are both constructed in this manner. In some cases, merely mixing the subunits allows the final structure to form. This is known as **self-assembly** and is true of the coat of tobacco mosaic virus, for example. In other cases, these higher-level structures require other proteins and co-factors to help assembly.

FIGURE 14.13
Hemoglobin—An Example of a Heterotetramer

Two α-subunits and two β-subunits form the hemoglobin tetramer. The two α-subunits are identical as are the two β-subunits. However, the α-subunits differ from the β-subunits, although both types of chain are related in amino acid sequence and have a similar overall shape.

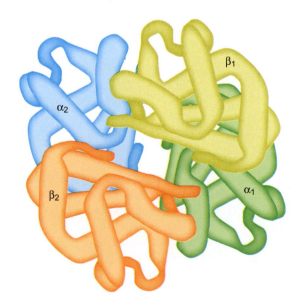

---

multimeric   Formed of multiple subunits
protomer   A single polymer chain that is itself a subunit for a higher level of assembly
self-assembly   Automatic assembly of protein subunits without need of any outside assistance

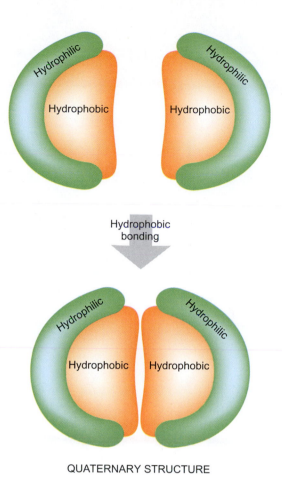

Hydrophobic bonding

QUATERNARY STRUCTURE

FIGURE 14.14
Hydrophobic Force Drives Assembly of Subunits

Two proteins with hydrophobic regions will often bind together so that their hydrophobic regions become internalized, away from the surrounding water, in the dimer.

1.7. Co-factors and Metal Ions Are Often Associated with Proteins

To function properly, many proteins need extra components, called co-factors or prosthetic groups, which are not themselves made from amino acids. Many proteins use single metal atoms as co-factors; others need more complex organic molecules. Strictly speaking, prosthetic groups are fixed to a protein, whereas co-factors are free to wander around from protein to protein. However, this classification breaks down because the same organic co-factors may be covalently attached to one enzyme but non-covalently associated with another. Consequently, the terms are often used loosely. A protein with its prosthetic group is referred to as a **holoprotein** and without its prosthetic group as an **apoprotein**.

For example, oxygen carrier proteins such as hemoglobin have a cross-shaped organic co-factor called heme that contains a central iron atom. The heme is bound in the active site of the apoprotein, in this case globin, to give hemoglobin. Oxygen binds to the iron atom at the center of the heme and the hemoglobin carries it around the body. Prosthetic groups are often shared by more than one protein; for example, heme is shared by hemoglobin and by myoglobin, which receives oxygen and distributes it inside muscle cells.

Most bacteria and plants are able to synthesize their own co-factors. However, many organic enzyme co-factors cannot be made by animals and consequently they

---

apoprotein   That portion of a protein consisting only of the polypeptide chains without any extra co-factors or prosthetic groups
holoprotein   Complete protein consisting of the polypeptide chains plus any extra metal ions, co-factors, or prosthetic groups

FIGURE 14.15
Protein Assemblies: Rings, Chains, and Cylinders

A) Joining protein subunits in a circle forms a ring. B) A helical chain, like that of actin, allows assembly of a very long thin structure from globular subunits. C) Winding protein subunits into a helix forms a cylinder, like that found in microtubules.

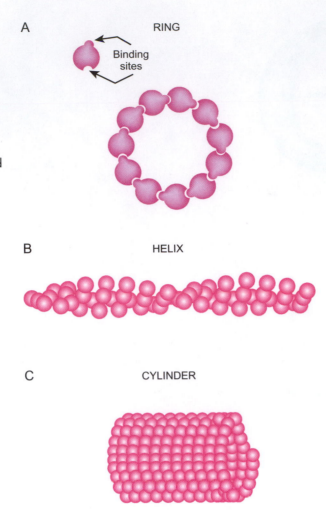

or their immediate precursors must be provided in the diet. Such co-factors and/or their precursors are then referred to as vitamins (Table 14.01). Some co-factors, such as heme, can be synthesized by animals and are therefore not vitamins. Conversely, not all vitamins are co-factors or their precursors. For example, vitamin D gives rise to a hormone. Vitamin A confuses the classification scheme as it is partly converted to retinaldehyde—a protein co-factor—and partly to retinoic acid—a hormone. Vitamin C does not act directly as a co-factor but is needed to keep metal ions (such as Cu and Fe) that do act as co-factors in their reduced states. A further complication is that certain co-factors can be made by some animals but not others. Thus, vitamin C is needed by humans but is not required by most other mammals.

2. Determining Protein Structures

X-ray crystallography, also known as X-ray diffraction, is used to solve the 3D structure of molecules, in particular proteins and nucleic acids. It was X-ray diffraction that first revealed that DNA was twisted into a double helix. Knowing 3D shapes helps us to understand how biological molecules fit together and interact.

When a beam of X-rays travels through a substance, the X-rays are scattered by the atoms they encounter. If the target is a crystal with a regular structure, the scattering of the X-rays will give rise to a regular, though complex, diffraction pattern (Fig. 14.16). In practice, the crystal is rotated into a variety of positions on a computer-controlled stage. The diffraction patterns are recorded and, after computer analysis, are used to generate a 3D atomic map of the protein molecule.

TABLE 14.01	Organic Co-factors and Vitamins	
Vitamin/Parent Compound	**Active form/Co-factor**	**Function in Animals**
Vitamins that are Enzyme Co-factors or Their Precursors		
Vitamin A = Retinol	Retinaldehyde	Vision
Vitamin B1 = Thiamine	Thiamine pyrophosphate	Decarboxylations
Vitamin B2 = Riboflavin	Flavin adenine dinucleotide; Flavin mononucleotide	Many redox reactions
Vitamin B3 = Niacin (refers to nicotinamide and/or nicotinic acid)	Nicotinamide adenine dinucleotide; Nicotinamide adenine dinucleotide phosphate	Redox reactions (degradative); redox reactions (biosynthetic)
Vitamin B5 = Pantothenic acid	Coenzyme A 4'-Phosphopantetheine	Acylation reactions and fatty acid synthesis
Vitamin B6 = Pyridoxine, pyridoxal, or pyridoxamine	Pyridoxal phosphate	Amino acid metabolism
Vitamin B12 = Cobalamin	Methyl cobalamin; Deoxyadenosyl cobalamin	Methyl group carrier Rearrangements
Biotin (a B vitamin)	Biotin	Carboxylations
Folic acid (a B vitamin)	Tetrahydrofolate	Redox reactions and one carbon carrier
Vitamin K1 = Phylloquinone	Phylloquinone	Post-translational
Vitamin K2 = Menaquinone	Menaquinone	carboxylation of
Vitamin K3 = Menadione	Menaquinone	glutamate
Vitamins that are not Enzyme Co-Factors or Their Precursors		
Vitamin A = Retinol	Retinoic acid	Hormone/Regulator
Vitamin C = Ascorbic acid	Ascorbic acid	Antioxidant
Vitamin D = Ergocalciferol (D2) or Cholecalciferol (D3)	Calcitriol	Hormone controlling Ca and P metabolism
Vitamin E = Tocopherol	Tocopherol	Antioxidant
Co-Factors that are not Vitamins (I.E., Can be Made by Animals)		
Heme	Heme	Oxygen carrier
Lipoic acid	Lipoamide	Redox reactions; 2-carbon carrier
Biopterin	Tetrahydrobiopterin	Phe metabolism

X-ray crystallography needs large well-formed crystals of highly purified protein. The proteins must form an ordered array called a lattice. Nowadays, molecular cloning and over-expression of any gene of interest makes it possible to produce enough protein to create the lattice. However, getting nice crystals is often difficult for such massive molecules as proteins, especially those whose 3D shapes are irregular. The diffraction pattern is essential to determining the central part of a protein's structure, as well as any associated ligands, inhibitors, ions, or other small parts. Some X-ray crystals actually capture enzymes in their transition states!

Another technique used to identify 3D protein structure is **NMR spectroscopy**. In this method, a solution of pure protein is bombarded with radio waves as it is held in a strong magnetic field. The radio waves cause the atomic nuclei of each molecule to spin and eventually align. The movement emits energy that is detected by a computer

NMR spectroscopy Technique to determine protein structure that uses alternating magnetic fields to change the spin of electrons within the sample

FIGURE 14.16
X-ray Crystallography

A) As light passes through an object, the light waves are distorted from their original pathway. Using a lens allows the distorted light waves to be refocused into an image of the object. B) X-rays are also distorted when they pass through an object. The structure of the proteins within a single crystal dictates the pattern in which the X-rays are diffracted. Moving the crystal around alters the pattern in which the X-rays are diffracted, and all the different patterns can be combined into one to form a model of the protein's actual structure.

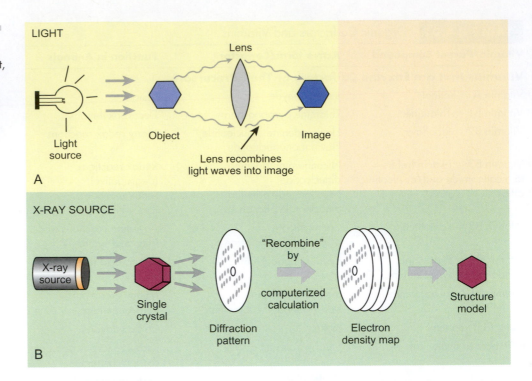

and analyzed with Fourier transformation, which allows the researcher to determine the identity of each molecule in the sample. The advantage of NMR spectroscopy is that proteins are maintained in solution, but the technique is limited to small and medium sized molecules (Fig. 14.17).

A third technique to study protein structure is using electron microscopy to determine the 3D surface contours of a protein (see Ch. 5, Section 6, Figs. 5.28 and 5.29 for details of this technique). This method works well with proteins that form a regular pattern in a small crystal or even a membrane. A major problem with electron microscopy is sample preparation. Negative staining with metal ions is widely used (as for DNA in Fig. 5.29) but may distort the sample. These limitations have largely been overcome by freezing the samples in liquid nitrogen–cooled ethane. The sample is thus embedded in a layer of ice that preserves the native structure. The advent of such cryo-electron microscopy has allowed increased resolution in imaging complex biological structures (Fig. 14.18).

Protein structure determination is sophisticated and time consuming and each protein must be purified and examined individually. Nonetheless, structures are available for a significant number of proteins. As of September 2011, the Protein Data Bank (http://www.rcsb.org/pdb/) lists approximately 75,000 structures. Since many proteins are members of related families, once a 3D structure is available for one, it provides insight into the conformation of a whole series of related molecules.

3. Nucleoproteins, Lipoproteins, and Glycoproteins Are Conjugated Proteins

Conjugated proteins are proteins that are linked to molecules of other types. For example, **nucleoproteins** are complexes of protein and nucleic acid, **lipoproteins** are proteins with lipid attached, and **glycoproteins** have carbohydrate components.

conjugated protein   Complex of protein plus another molecule
glycoprotein   Complex of protein plus carbohydrate
lipoprotein   Complex of protein plus lipid
nucleoprotein   Complex of protein plus nucleic acid

ELECTRON SPIN IN A MAGNETIC FIELD

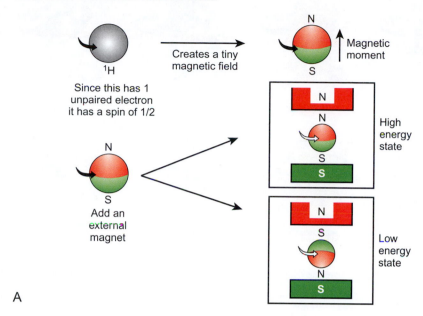

A

ENERGY STATES IN A CHANGING MAGNETIC FIELD ARE
MEASURED BY A COMPUTER

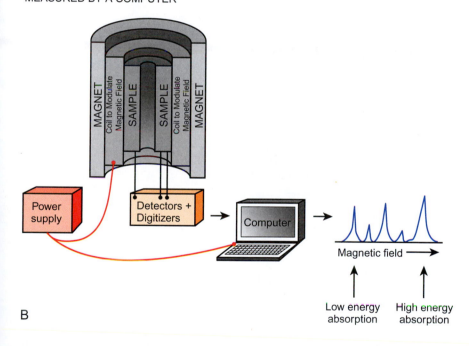

B

■▶ **FIGURE 14.17**

NMR spectroscopy

A) NMR spectroscopy is based on the idea that electrons can be induced to spin via external radio waves. Then the spin can be modulated by the presence of an external magnetic field. When a hydrogen atom is spinning its unpaired electron creates a tiny magnetic field (top). When an external magnetic field is added, the atom adopts either a high energy state or low energy state. B) An NMR spectrometer contains a very large magnet (sometimes as large as a room) with coils that modulate the magnetic field. The sample is placed in a tube in the center of the magnet and connected to a variety of detectors and a computer that measure the different energy states in the atoms as the magnetic field is changed. The computer plots the energy absorption for each change in magnetic field.

Many glycoproteins are found at the surface of cells. They carry short carbohydrate chains consisting of several sugar molecules, which usually project outward from the cell (Fig. 14.19). The sugar chain is usually linked via the hydroxyl group of serine or threonine or the amide group of asparagine. Glycoproteins often function in *cell-to-cell adhesion*, especially in organisms (animals) that lack rigid cell walls. In addition, the carbohydrate portion of glycoproteins is often the key factor in *cellular recognition*. For example, sperm recognize egg cells by binding to the carbohydrate part of a surface glycoprotein. Recognition by the immune system often depends on the precise structure of the carbohydrate chains of glycoproteins.

Certain proteins are linked to other large molecules, such as lipids, carbohydrates, or nucleic acids.

FIGURE 14.18
Cryo-EM Images of 40S Ribosome

The structure of the yeast 40S pre-initiation complex using cryo-EM shows the 40S subunit alone (A), with eIF1 and eIF1A (B), with eIF1 alone (C), and with eIF1A alone (D). The top of the molecule marks the solvent accessible side, whereas the bottom side interacts with the 60S ribosome. Abbreviations: b, beak; n, neck; sh, shoulder; pt, platform; lf, left foot; rf, right foot. Notice how a connection between the shoulder and head forms after binding of eIF1 and eIF1A in (B). *(Credit: Passmore, et al.(2007) Mol. Cell 26: 41-50.)*

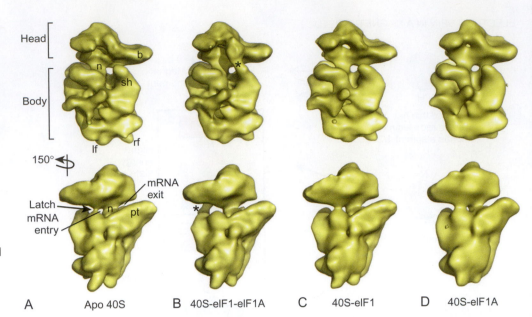

FIGURE 14.19
A Membrane Glycoprotein

A glycoprotein in the cytoplasmic membrane of an animal cell is shown protruding from both sides of the membrane. At the exterior surface, several sugar residues project from the protein into the extra-cellular space.

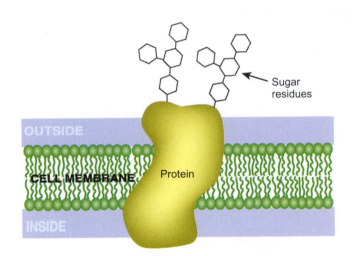

Many lipoproteins are attached to membranes by their lipid tails (Fig. 14.20). An example is the β-**lactamase** found in many gram-positive bacteria, such as *Bacillus*. The β-lactamase protects the cell by destroying antibiotics of the β-lactam family, which includes penicillin. Since the target for penicillin action is the cell wall, the protective enzyme needs to be outside the cell. The lipid tail makes sure it does not drift away into the surrounding medium.

Proteolipids are a specialized subclass of lipoproteins that are extremely hydrophobic and are insoluble in water. They are soluble in organic solvents and are found in the hydrophobic interior of membranes. These properties are not solely due to attached lipid groups. In part, their hydrophobicity is due to a high percentage of hydrophobic amino acid residues. Instead of being hidden inside the protein, many of these are exposed on the surface.

β-**lactamase**   Enzyme that destroys antibiotics of the β-lactam class that includes penicillins and cephalosporins
proteolipid   A type of lipoprotein that is extremely hydrophobic and found in the interior of membranes

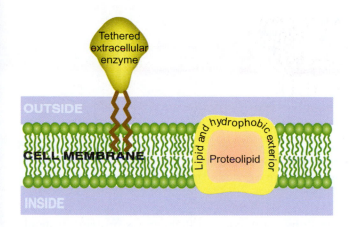

■- **FIGURE 14.20**
Extracellular Enzyme Tethered by Lipid Tail

A lipoprotein is held to the membrane surface by one or more lipid tails that penetrate into the lipid bilayer of the membrane. The lipid tails are related to those of the phospholipids that comprise the membrane itself.

4. Proteins Serve Numerous Cellular Functions

There is colossal variety in the functional role of proteins. Nonetheless, proteins may be subdivided into several major categories as noted in Chapter 3. **Enzymes** are proteins that catalyze chemical reactions. These are discussed in more detail below. Many of the characteristics of enzymes, such as the presence of binding pockets for small molecules and the ability to change shape, are shared by other proteins.

Certain proteins emit or absorb light. **Luciferases** are enzymes that send signals from one organism to another by emitting light. Luminous bacteria glow to attract deep-sea fish. After being swallowed, the bacteria take up residence in the fish intestines. Fireflies flash as part of their mating strategy. Both bacterial and insect luciferases have been used as **reporter proteins** in genetic analysis (see Ch. 19). Another widely used reporter is **green fluorescent protein** (**GFP**), which is made by fluorescent jellyfish. However, it is non-enzymatic and emits fluorescence after absorbing light of higher energy. Thus, the green fluorescence can be used to monitor gene expression and/or to localize areas where a particular gene is expressed (see Ch. 19). Rhodopsins are proteins that absorb light. They are used in the eyes of both invertebrate and vertebrate animals to detect light and by certain bacteria to harvest light energy.

Many subcellular structures consist largely or partly of **structural proteins**. Specialized proteins are known that control the structure of water. Fish that live in polar regions have **antifreeze proteins** to keep their blood from freezing. These proteins bind to ice surfaces and prevent the growth of ice crystals. Conversely, surface proteins of certain bacteria, known as **ice nucleation factors**, promote the formation of ice crystals and are important in causing frost damage to plants. Damaging plant cells releases nutrients on which the bacteria can grow and may also allow colonization of plant tissue by the bacteria.

Mechanical proteins may be classified as specialized structural proteins or as enzymes. They perform physical work at the expense of biological energy (usually by hydrolyzing ATP or GTP). Their energy consumption results in a reversible change of conformation. Proteins of the myosin family that contract upon energization are found in muscle fibers and the filaments of eukaryotic flagella. Actins are found in muscle with myosin where the two proteins participate in muscle contraction (Fig. 14.21), but they are also involved in cellular movements such as endocytosis and amoeboid motion.

> Proteins play a wide variety of functional roles in the cell.

> Most metabolic reactions are catalyzed by proteins known as enzymes.

> Movement of both muscles and eukaryotic flagella is due to proteins that contract. This uses energy from ATP.

---

antifreeze protein   Protein that prevents freezing of blood, tissue fluids, or cells of organisms living at subzero temperatures
enzyme   A protein that catalyzes a chemical reaction
green fluorescent protein (GFP)   A jellyfish protein that emits green fluorescence and is widely used in genetic analysis
ice nucleation factor   Protein found on surface of certain bacteria that promotes the formation of ice crystals
luciferase   Enzyme that consumes energy and generates light
mechanical protein   Protein that uses chemical energy to perform physical work
reporter protein   A protein that is easy to detect and gives a signal that can be used to reveal its location and/or indicate levels of gene expression
structural protein   A protein that forms part of a cellular structure

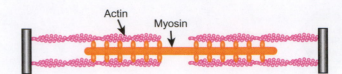

FIGURE 14.21
Mechanical Proteins

Myosin and actin interact to cause contraction. The participating proteins move relative to one another but do not shorten individually. The movements bring the end attachments of the actin filaments closer together. Numerous units such as those shown here are attached end-to-end, causing skeletal muscle to shorten.

Bacterial flagella do not operate by contraction of filamentous proteins. Instead, the base of the flagellum consists of a protein ring that rotates as it consumes energy. The filament is attached to the rotating ring and makes a helical lashing motion that drives the bacterium along.

Chaperone proteins, or **chaperonins**, assist other proteins in folding correctly. Some chaperonins are involved in protein export and prevent premature folding of proteins that are to be secreted through membranes (see Ch. 13). Other chaperonins attempt to re-fold proteins that have become denatured due to high temperature or other environmental stresses that damage proteins (see the heat shock response, Ch. 16).

Binding proteins bind small molecules but unlike enzymes they do not carry out a chemical reaction. Nonetheless, they also need "**active sites**" to accommodate the small molecules. They may be subdivided into:

a) **Permeases** are located in membranes and transport their target molecules across the membrane. Nutrients must be transported into the cells of all organisms, whereas waste products are deported. Many permeases consist of a bundle of α-helical segments (often 7 or 11 or multiples thereof) that cross the membrane and are joined by regions of random coil (Fig. 14.22). Most permeases require energy to operate.

b) **Carrier proteins** are soluble, that is, not membrane bound, and move their substrates around without crossing membranes. Most transport nutrients within the bodies of multicellular organisms and many are consequently extra-cellular and found in the bloodstream or other body fluids. Other carrier proteins are found inside specialized cells that travel around the body—such as red blood cells. Classic examples are the oxygen carriers hemoglobin and myoglobin, located in red blood cells and muscle tissue, respectively.

c) Storage proteins sequester small molecules, but store them instead of transporting them. For example, **ferritin** in animal cells, and the corresponding bacterioferritin in bacteria, store iron. **Metallothionein** binds assorted heavy metal ions, and its role is largely protective (Fig. 14.23). The metallothionein gene is induced by traces of heavy metals and has a very strong promoter that is widely used in genetic engineering. Protective metal-binding proteins are also found in certain bacteria. They are sometimes exploited in biotechnology for extraction of metals such as gold or uranium from ore or industrial waste streams.

> Some proteins carry nutrients or other molecules across membranes or around the organism.

---

active site   Special site or pocket on a protein where other molecules are bound and the chemical reaction occurs
binding protein   Protein whose role is to bind another molecule
carrier protein   Protein that carries other molecules around the body or within the cell
chaperonin   Protein that helps other proteins to fold correctly
ferritin   An iron storage protein
metallothionein   Protein that protects animal cells by binding toxic metals
permease   A protein that transports nutrients or other molecules across a membrane

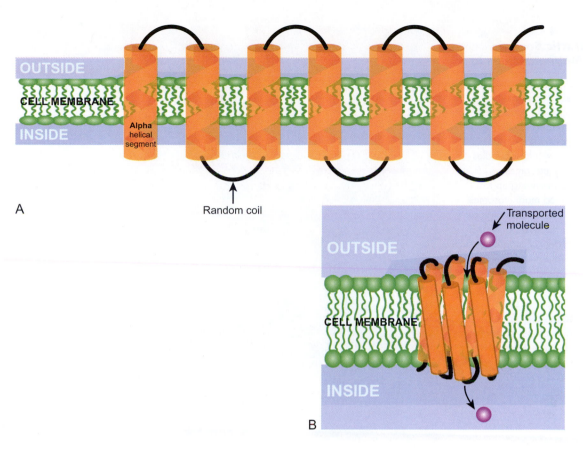

FIGURE 14.22
Membrane Permease Made of Helical Segments

A) Seven α-helical segments traverse the membrane and are joined by random coil regions. B) The seven helical segments are actually bundled together to form a transport channel, with the transported molecule passing through the center of the bundle.

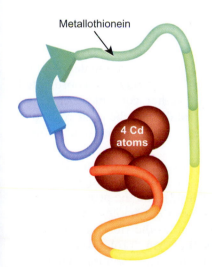

FIGURE 14.23
Metallothionein Is a Metal-Binding Protein

This short protein is capable of sequestering four cadmium atoms.

Binding and recognition/regulation are functions that overlap. Thus, the immune system of higher animals has many highly-specialized binding proteins that protect against infection by recognizing invading bacteria and viruses. These include antibodies and T-cell receptors. Regulatory proteins include cell surface receptors, signal transmission proteins, and DNA-binding proteins that control the expression of genes at various levels. The functions of these proteins are discussed in detail in the appropriate chapters (especially Ch. 16 through Ch. 18). In this chapter, the 3D structural motifs of proteins that bind DNA are discussed further below.

FIGURE 14.24

Magnetotactic Bacteria Contain Magnetosomes to Bind Iron

Transmission electron micrograph of *Magnetobacterium bavaricum*, a rod-shaped magnetotactic bacterium from Lake Chiemsee (Upper Bavaria) with four bundles of chains of magnetosomes. Individual cells containing up to 1000 hook-shaped magnetosomes yield magnetic moments as high as $(10–60) \times 10^{12}$ Gauss ccm, which is one to two orders of magnitude more than the values characteristic of other magnetotactic bacteria. The large electron-opaque bodies inside the cell consist of sulfur. The magnetosomes are made of magnetite and measure, on average, 100 nm. *(Credit: Prof. Michael Winklhofer, Institut für Geophysik, Theresienstrasse 41, D-80333 München, Germany.)*

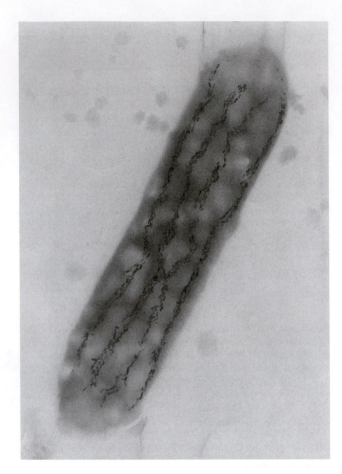

Subcellular nano-machines are assemblies of protein and, often, RNA, that carry out complex tasks in the cell.

5. Protein (Nano)-Machines

Many cellular processes such as DNA synthesis, RNA splicing, or protein degradation are performed by groups of a dozen or more proteins that operate in a coordinated manner. When such groups of proteins stay associated together, consume energy, and carry out carefully controlled movements as well as catalyzing chemical reactions, they are sometimes referred to as protein machines or, more fashionably, as protein nano-machines (see Focus on Relevant Research).

There is a trend to name these assemblies to rhyme with ribosome. Thus, the replisome moves along the chromosome while synthesizing new DNA (see Ch. 10), the spliceosome is responsible for RNA splicing (see Ch. 12), and the proteasome carries out protein degradation (see below). However, as seen in Chapter 13, ribosomal RNA plays a more critical part in protein synthesis than do the ribosomal proteins. RNA is also involved in spliceosome function. Thus, the term "protein machine" is too narrow. And how should we classify the magnetosome? This assembly binds iron oxide and orients bacteria in magnetic fields (Fig. 14.24).

Referring to biological components as machines may be due to influence from the field of nanotechnology. This refers to the ability to manipulate matter by precisely placing individual atoms and molecules. Nanotechnology involves molecular-sized nano-machines. These machines would be programmed to reproduce themselves in the millions and then place atoms precisely to build other molecules. These molecules could then be assembled together into whatever components are needed. Perhaps to take into account the role of RNA as well as protein, the term bionanomachine should be introduced.

Maillard RA, et al. (2011) ClpX(P) Generates Mechanical Force to Unfold and Translocate Its Protein Substrates. Cell 145:459–469.

FOCUS ON RELEVANT RESEARCH

The ClpXP complex from *Escherichia coli* consists of ClpX unfoldase and ClpP peptidase and is an example of a simple protein nanomachine. The complex consists of six subunits of ClpX arranged as a ring attached to two rings, each of seven ClpP monomers. As its name indicates, ClpX unfolds the proteins destined for degradation. This requires energy and ClpX hydrolyzes ATP to generate the needed mechanical force. ClpX threads unfolded polypeptide substrates through its own central pore and into the cavity formed by ClpP. The ClpP protein is the actual protease and degrades the incoming unfolded polypeptide chains.

To monitor the mechanical force exerted by ClpX, the authors of this paper used optical tweezers in combination with a single molecule assay system. Both the ClpX enzyme and the substrate (green fluorescent protein, GFP) were attached, via linkers, to polystyrene beads. ClpX proceeds in discrete steps (power strokes) when unfolding a polypeptide substrate. Sometimes it fails in unfolding and slips. The presence of the ClpP module improves unfolding and reduces slipping. ClpX typically tugs on GFP several times before it succeeds in unfolding a domain consisting of several beta-sheet strands. The rest of GFP then spontaneously unravels. The unfolded GFP is then moved into the cavity formed by ClpP and degraded.

6. Enzymes Catalyze Metabolic Reactions

Enzymes are proteins that catalyze chemical reactions but are not consumed in the process. Virtually all metabolic reactions depend on enzymes. An enzyme first binds the reacting molecule, known as its **substrate**, and then performs a chemical operation upon it. Some enzymes bind only a single substrate molecule; others may bind two or more, and combine them together to give the final product. Many enzyme-catalyzed reactions are reversible; that is, the enzyme speeds up reaction in either direction.

The most famous enzyme in molecular biology is β-**galactosidase**, encoded by the *lacZ* gene of the bacterium *E. coli*. This enzyme is so easy to assay that it is widely used in genetic analysis (see Ch. 19 for details). One of the natural substrates of β-galactosidase is the sugar lactose, made by linking together the two simple sugars glucose and galactose. β-galactosidase hydrolyses lactose into the two simpler sugars (Fig. 14.25).

Substrates bind to the enzyme at the active site, a pocket or cleft in the protein, where the reaction occurs. The active site is the result of precise folding of the polypeptide chain so that amino acid residues that may have been far apart in the linear sequence can come together to cooperate in the enzyme reaction (Figs. 14.26 and 14.27).

Many enzymes rely on co-factors or prosthetic groups to assist in catalysis. These may be organic molecules, such as NAD (nicotinamide adenine dinucleotide), which carries the hydrogen atoms added to (or removed from) the substrate by many enzymes. Metal ions are also common. Zinc ions are found at the active site of most enzymes that synthesize or degrade nucleic acids. The Zn^{2+} binds to the negatively-charged phosphate groups and weakens the critical bonds.

In order to make enzyme names consistent, all newly discovered enzymes end in "ase." Some enzymes, such as trypsin or lysozyme, were named before this convention was introduced and so have irregular names. Some major categories of enzymes of interest in molecular biology are listed in Table 14.02.

> Beta-galactosidase splits a range of molecules that consist of galactose linked to another component.

> Both the substrate and co-factors (if needed) bind to the active site of the enzyme.

6.1. Enzymes Have Varying Specificities

Some enzymes are extremely specific and will use only a single substrate. Aspartase catalyzes the inter-conversion of aspartic acid and fumaric acid (Fig. 14.28). Aspartase

---

β-galactosidase   Enzyme that splits lactose and related molecules to release galactose
substrate   Molecule that binds to an enzyme and is the target of enzyme action

FIGURE 14.25
β-*Galactosidase Splits Lactose*

Lactose is split by the enzymatic action of β-galactosidase into glucose and galactose.

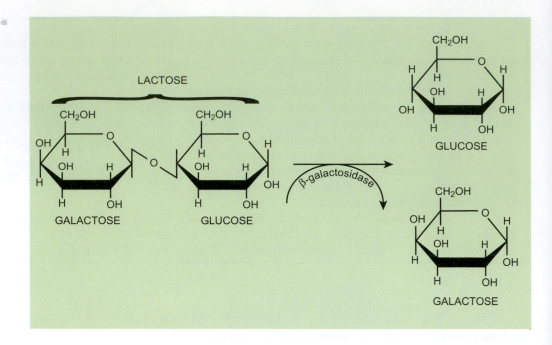

FIGURE 14.26
Active Site Consists of Residues Far Apart in the Sequence

Before a polypeptide chain is folded, the amino acid residues that cooperate to carry out the enzymatic reaction are often far apart in the sequence. Upon correct folding, the critical amino acids are brought together and form the active site.

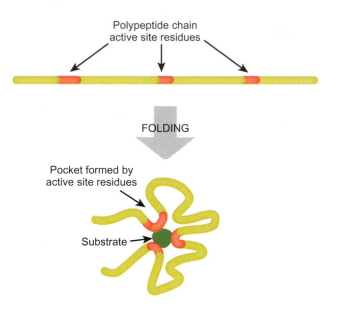

Polypeptide chain active site residues

FOLDING

Pocket formed by active site residues

Substrate

Biosynthetic enzymes are often highly specific, unlike degradative enzymes, which often show wide specificity.

will use only aspartate (not similar amino acids such as glutamate) and it uses only the L-**isomer** of aspartate. Aspartase can also catalyze the reverse reaction, but it will add NH_3 only to fumarate and will not use maleate, the *cis*-isomer of fumarate.

Other enzymes have much broader substrate specificity. Broad specificity is shown by many degradative enzymes; for example, alkaline phosphatase, which removes phosphates from a wide range of molecules, or carboxypeptidase, which snips the C-terminal amino acid off many polypeptide chains. Most enzymes have intermediate specificity. For example, alcohol dehydrogenase from *E. coli* will act on C3 and C4 alcohols as well as its natural substrate, ethanol. Again, β-galactosidase uses several compounds in which galactose is linked to another chemical group, such as lactose, ONPG, and X-gal (see below).

L-isomer   That one of a pair of optical isomers that rotates light in an anticlockwise direction

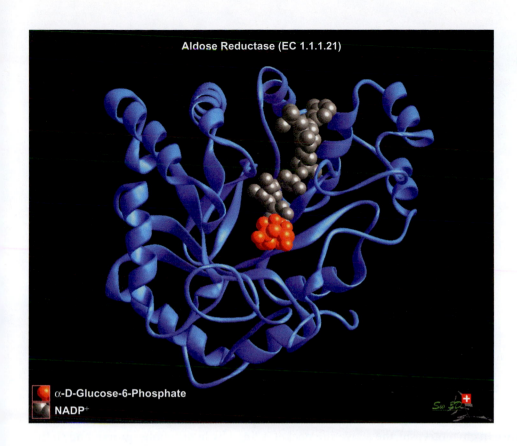

Aldose Reductase (EC 1.1.1.21)

α-D-Glucose-6-Phosphate
NADP⁺

FIGURE 14.27
Enzyme Substrate Complex of Aldose Reductase

A 3D computer model of aldose reductase (EC 1.1.1.21) shown binding its substrates, glucose-6-phosphate (orange) and NADP (gray). *(Credit: Dr. Manuel C. Peitsch at the Glaxo Institute for Molecular Biology in Geneva, Switzerland, using coordinates from the Brookhaven Protein Data Bank.)*

TABLE 14.02 Types of Enzymes and Their Roles

Enzyme Type	Type of Reaction Catalyzed
Hydrolase	Splits substrate by hydrolysis. Includes many subcategories, including nucleases, proteases, glycosidases, and phosphatases.
Nuclease	Cleaves nucleic acids by hydrolyzing phosphodiester bonds between nucleotides. Includes restriction enzymes that cut DNA at specific sequences.
Protease	Cleaves polypeptide chains by hydrolyzing peptide bonds between amino acids.
Glycosidase	Cleaves polysaccharides by hydrolyzing glycoside bonds between sugars. Includes β-galactosidase, lysozyme, and cellulase.
Phosphatase	Removes phosphate groups by hydrolysis.
β-Lactamase	Hydrolase that inactivates antibiotics of the β-lactam group (penicillins and cephalosporins) by opening the lactam ring.
ATPase	General name for enzymes that hydrolyse ATP so releasing energy that is used to drive a reaction.
Synthase	General name for a biosynthetic enzyme. Includes ligases, polymerases, transferases, and other classes.
Ligase	Forms new bonds by joining fragments together. The term "ligase" alone is usually understood to refer to DNA ligase.
Polymerase	Type of ligase that synthesizes polymers by linking the subunits. Includes DNA polymerase and RNA polymerase.
Transferase	Transfers chemical groups from one molecule to another. Includes methylases, acetylases, kinases, and others.
Methylase	Transferase that adds methyl groups to a molecule. Includes modification enzymes that methylate DNA.
Acetylase	Transferase that adds acetyl groups to a molecule. Includes histone acetyl transferase (HAT), which acetylates histones.
Kinase	Transferase that adds phosphate groups to a molecule. Includes protein kinases, which attach phosphate groups to proteins.
Isomerase	Interconverts the isomers of a molecule.
Racemase	Specialized isomerase that interconverts the D- and L-isomers of an optically-active molecule such as an amino acid.
Oxidoreductase	Catalyzes oxidation and reduction reactions. Includes dehydrogenases, which remove hydrogen atoms from molecules.

FIGURE 14.28
Some Enzymes, Such as Aspartase, Can Distinguish Isomers

L-Aspartate is transformed to fumarate plus NH_3 by the enzyme aspartase. This enzyme will not transform "look-alike" substrates such as glutamic acid nor structural isomers like maleic acid. Furthermore, aspartase can distinguish optical isomers and only uses the L-form of aspartate.

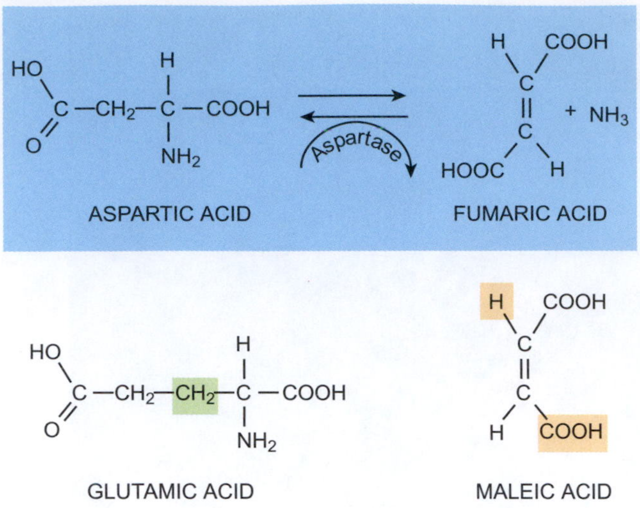

Differences in specificity between similar enzymes are largely determined by the nature of the active site. The active site pocket can vary in size and shape and in the chemical nature of the amino acids comprising it. For example, the three digestive enzymes trypsin, chymotrypsin, and elastase all split polypeptide chains by the same catalytic mechanism (Fig. 14.29). However, the active site of trypsin has a negative charge at the bottom, so trypsin cuts after positively-charged residues (e.g., Lys or Arg). In chymotrypsin, the active site pocket is lined by hydrophobic groups and so this enzyme cuts after hydrophobic residues (e.g., Phe or Val). In elastase, the active site pocket is very small and so elastase cuts after residues with small side chains (e.g., Ala).

Two different models have been proposed to explain the binding of substrate in the active site of enzymes. In the **lock and key model**, the active site of the enzyme fits the substrate precisely. In contrast, the **induced fit** model proposes that the binding of the substrate induces a change in enzyme conformation so that the two fit together better (Fig. 14.30).

Enzymes may be found that operate by either mechanism. For example, the enzyme chymotrypsin shows almost no detectable structural change when it binds its substrate. In contrast, when carboxypeptidase binds substrate, this causes a tyrosine (at position 248) to move 12 Angstroms to a new position where it physically contacts the substrate and can now play a direct catalytic role.

6.2. Enzymes Act by Lowering the Energy of Activation

In any chemical reaction, the reactants are converted to the products via the reaction intermediate or **transition state**. In an enzyme catalyzed reaction, the substrate is bound by the enzyme and the reaction occurs within the active site. Energy is often needed to prime the reaction. This **transition state energy**, $\Delta G^{\ddagger}$, is the difference in free energy between the starting materials and the transition state (Fig. 14.31). When $\Delta G^{\ddagger}$ is positive, as shown, energy must be supplied to reach the transition state. Even if this energy is released once the overall reaction is over, the higher the transition state energy, the slower the reaction. Enzymes cannot change the overall free energy of a reaction; that is, the energy difference between reactants and products. The role of an enzyme is to *lower the transition state energy*, either by stabilizing the

---

induced fit When the binding of the substrate induces a change in enzyme conformation so that the two fit together better
lock and key model Model of enzyme action in which the active site of an enzyme fits the substrate precisely
transition state Another term for the activated intermediate in a chemical reaction
transition state energy Energy difference between the reactants and the activated reaction intermediate or transition state

GENERAL MECHANISM

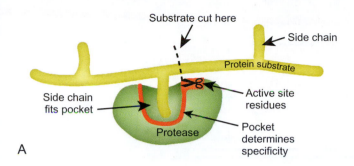

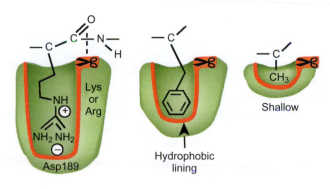

SPECIFIC ACTIVE SITES

B   TRYPSIN   CHYMOTRYPSIN   ELASTASE

LOCK AND KEY MODEL   INDUCED FIT MODEL

Substrate   Substrate

+   +

Active site   Active site

Enzyme   Enzyme

+   +

A   Enzyme-substrate   B   Enzyme-substrate
    complex   complex

▶ **FIGURE 14.29**
Active Site Specificity of Proteases

A) In general, a side chain of an amino acid fits into the pocket of a protease and a clipping mechanism nearby breaks the amino acid chain. B) The pockets have different properties in different protease enzymes. A negatively-charged pocket as in trypsin binds positively-charged residues such as arginine. The hydrophobic pocket of chymotrypsin attracts hydrophobic residues and the shallow pocket of elastase only fits small residues such as glycine or alanine.

▶ **FIGURE 14.30**
Lock and Key Versus Induced Fit

A) The lock and key model says that the active site and the substrate must fit perfectly. B) The induced fit model proposes that the conformation of the active site will change upon binding to the substrate.

·FIGURE 14.31

Energy Is Required to Reach the Transition State

Initiation of a chemical reaction requires an input of energy to reach the transition state. The required energy is called the energy of activation or transition state energy, $\Delta G^{\ddagger}$, and is needed even if the reaction is exothermic (as shown) and will eventually release more energy overall than originally put in. The net energy released is ΔG.

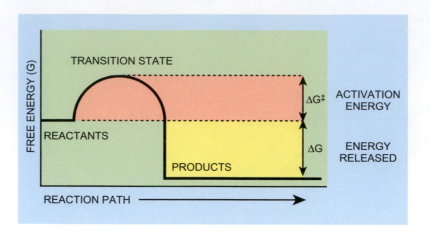

intermediates in the active site, or by providing an alternative reaction mechanism that proceeds by a pathway of lower energy.

The rate of a reaction may be slow without an enzyme. Rate increases by enzymes range from 10^8 to 10^{20} relative to the uncatalyzed, spontaneous reaction (e.g., 10^9 for alcohol dehydrogenase; 10^{16} for alkaline phosphatase). A high enzymatic rate occurs when enzymes position the substrate correctly relative to the catalytic groups in the active site. Several factors are involved in enzyme rate increases:

1. _Proximity_—(up to 10^6-fold increase). The enzyme binds the substrate so that the susceptible bond is very close to the catalytic group in the active site. The local concentration of substrate in the active site may be as much as 50 Molar, whereas its concentration in the cytoplasm may be less than 1 mM. Chemical reaction rates are proportional to the concentrations of the reactants.

2. _Orientation_—(up to 10^2-fold). When the substrate is bound, the reacting groups must be properly oriented. The orbital steering hypothesis suggests that the binding of the substrate(s) to the enzyme aligns the reactive groups so that the relevant molecular orbitals involved in bond formation overlap. This increases the probability of forming the transition state.

3. _Covalent intermediates_—(around 10^{10}-fold). The strategy is to lower the transition state energy "hump" by taking an alternative reaction pathway. Consider the transfer of a chemical group "X" from molecule A to molecule B. Here, "X" is a molecular fragment such as a phosphate group, acyl group, or glycosyl group.

$$AX + B \rightarrow A + BX$$

The enzyme may react in two steps, picking up group "X" from molecule A and then transferring it to molecule B in a second reaction:

$$AX + Enz \rightarrow Enz\text{-}X + A$$

$$Enz\text{-}X + B \rightarrow Enz + BX$$

4. _Acid-base catalysis_—(around 10^{10}-fold). Acid-base catalysis is due to proton donors (acids) or proton acceptors (bases) that donate protons to or remove protons from the reaction intermediate. Enzymes use the side chains of acidic or basic amino acids to attack the substrate. An acidic group in one part of the active site may donate a proton and a basic group in another part of the active site may remove another proton from the reaction intermediate. This is referred to as concerted acid-base catalysis. Such simultaneous treatment with acid plus base obviously cannot occur in free solution as the acid and base would neutralize each other. Most hydrolytic enzymes use acid-base catalysis.

Enzymes provide alternative reaction routes of lower energy for organic reactions.

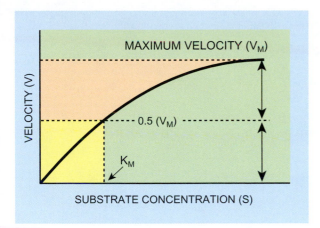

■ **FIGURE 14.32**
Saturation Kinetics

As the substrate concentration increases, the velocity of the reaction reaches a maximum (V_m). The K_m is the substrate concentration that yields half the maximum velocity.

5. *Metal ion catalysis*—(around 10^{10}-fold). Many enzymes have metal ions at the active site. Sometimes these are used as redox centers, as in cytochromes. Often, however, the metal ion is not reduced or oxidized, but acts to stabilize negative charges on the reaction intermediate. Thus, Zn^{2+} in the active site of carboxypeptidase polarizes the C=O of the peptide bond that is about to be broken.

6. *Distortion of the substrate*—(up to 10^8-fold). Binding to the enzyme may distort the substrate. The active site of the enzyme may fit the transition state intermediate better than the substrate. Once the substrate is bound it will be forced into the shape of the reaction intermediate. Distortion is difficult to demonstrate since the strain is imposed only after the substrate has bound to the enzyme.

6.3. The Rate of Enzyme Reactions

The principles of chemical kinetics apply to enzyme reactions, with certain modifications. Typically, reaction rates are proportional to the concentrations of the reactants. For enzymes, the rate is proportional to substrate concentration, [S], only at low substrate concentrations. At higher substrate concentrations all the active sites of the enzyme will be filled by substrate and the enzyme can work no faster however much more substrate is available. The enzyme is said to be **saturated** and has reached its **maximum velocity**, or V_{max} or V_m. The resulting relation between [S] and rate is a hyperbolic curve (Fig. 14.32).

The maximum velocity (V_m) and the Michaelis constant (K_m) summarize an enzyme's kinetic properties.

The V_m depends on the nature of the chemical reaction and hence depends on the chemical properties of the substrate and the enzyme active site. The V_m may vary with pH and temperature. The more enzyme, the more substrate it can handle, so V_m is also proportional to the amount of enzyme. The K_m, or **Michaelis constant**, is the substrate concentration that gives half maximal velocity. It largely depends on the affinity of the substrate for the active site and is independent of the enzyme concentration. The lower the K_m, the higher the affinity of the substrate for the enzyme and the faster the reaction (for low substrate; at high substrate concentrations K_m is less important). These factors are summarized in the **Michaelis-Menten equation**:

$$\text{Rate } (V) = (V_m \times [S])/(K_m + [S])$$

K_m See Michaelis constant
saturated   (Referring to enzymes) When all the active sites are filled with substrate and the enzyme cannot work any faster
maximum velocity (V_m or V_{max})   Velocity reached when all the active sites of an enzyme are filled with substrate
Michaelis constant (K_m)   The substrate concentration that gives half maximal velocity in an enzyme reaction. It is an inverse measure of the affinity of the substrate for the active site
Michaelis-Menten equation   Equation describing relationship between substrate concentration and the rate of an enzyme reaction

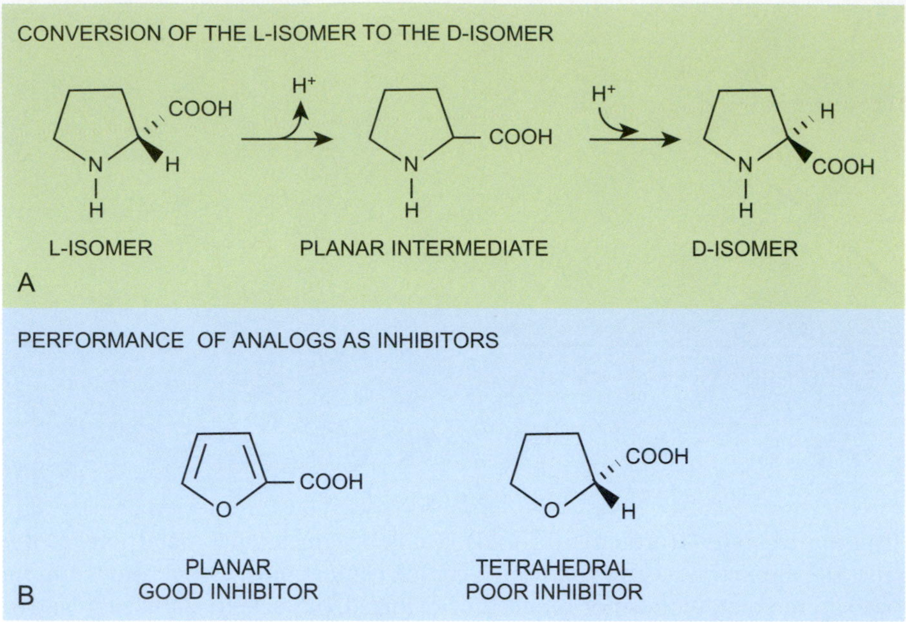

FIGURE 14.33
Proline Racemase Produces a Planar Intermediate

A) In the conversion of the L-isomer to the D-isomer of proline, the intermediate product is flat. This reaction is reversible. B) The best analogs for inhibiting the reaction are planar since these most closely resemble the flat transition state intermediate.

CONVERSION OF THE L-ISOMER TO THE D-ISOMER

L-ISOMER   PLANAR INTERMEDIATE   D-ISOMER

A

PERFORMANCE OF ANALOGS AS INHIBITORS

PLANAR   TETRAHEDRAL
GOOD INHIBITOR   POOR INHIBITOR

B

6.4. Substrate Analogs and Enzyme Inhibitors Act at the Active Site

> Substrate analogs bind to the enzyme active site instead of the substrate. Some inhibit the enzyme, others react in a way similar to the natural substrate.

Analogs are molecules that resemble natural substrates sufficiently well to bind to the enzyme active site. Some analogs act as alternative substrates for the enzyme. Other analogs bind to the active site, but instead of reacting, they block the active site and inhibit the enzyme. Such analogs are known as **competitive inhibitors**, as they compete with the true substrate for binding to the enzyme. The extent of inhibition depends both on the relative concentrations and the relative affinities of the substrate and the inhibitor.

The active site of many, perhaps most, enzymes is designed to fit the reaction intermediate or transition state better than the substrate itself. This tends to distort the substrate in the required direction and promote the reaction. Consequently, some of the best competitive inhibitors are molecules that mimic the reaction intermediate more than the substrate itself. These are sometimes known as "**transition state analogs**." Proline racemase interconverts the L- and D-isomers of proline by removing an H-atom and replacing it in a different configuration. The substrate and product are both tetrahedral about the α-carbon but the transition state is planar. The best competitive inhibitors are flat ring compounds rather than ones that look most like the substrate (Fig. 14.33).

> Beta-galactosidase splits ONPG giving a yellow color and X-gal to release a blue dye.

The enzyme β-galactosidase splits many molecules in which galactose is linked to another molecule (see Fig. 14.25, above). Researchers take advantage of this by using a substance called **ONPG (*ortho*-nitrophenyl galactoside)**, which consists of *ortho*-nitrophenol linked to galactose. When ONPG is split, it yields galactose, which is colorless, and *ortho*-nitrophenol, which is bright yellow (Fig. 14.34). Using ONPG allows researchers to monitor the level of β-galactosidase by measuring the appearance of the yellow color. Similarly, **X-gal** is split by β-galactosidase into a blue dye plus galactose (see Ch. 19 for applications). Compounds that are themselves

---

analog   A chemical substance that mimics another well enough to be mistaken for it by biological macromolecules, in particular enzymes, receptor proteins, or regulatory proteins

competitive inhibitor   Chemical substance that inhibits an enzyme by mimicking the true substrate well enough to be mistaken for it

***ortho*-nitrophenyl galactoside (ONPG)**   Artificial substrate for β-galactosidase that yields a yellow color upon cleavage

transition state analog   Enzyme inhibitor that mimics the reaction intermediate or transition state, rather than the substrate

X-gal   Artificial substrate for β-galactosidase that yields a blue color upon cleavage

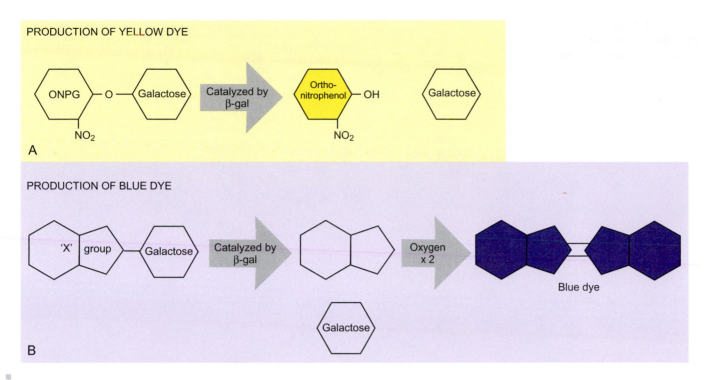

FIGURE 14.34
β-Galactosidase Splits ONPG and X-Gal

β-galactosidase is an enzyme that splits off galactose from other molecules to which it is attached. A) The substrate, ONPG, is used in molecular biology to measure the level of β-galactosidase activity since the reaction product, ortho-nitrophenol, is yellow and can be easily measured. B) X-Gal is another chromogenic substrate for β-galactosidase that is split releasing galactose and a fragment that reacts with oxygen in air yielding a blue dye.

colorless (or only pale) but react to release strongly colored products are known as **chromogenic substrates**.

Irreversible inhibition occurs when an inhibitor covalently modifies an enzyme, often at the active site. Covalent inhibitors are not always analogs of the substrate. Some irreversible inhibitors react with components of the active site; other inhibitors react with other regions of the protein and may affect protein conformation, solubility, or other properties. For example, the nerve agent sarin, used in chemical warfare, covalently inhibits enzymes that have serine in the active site (Fig. 14.35). Sarin reacts with the serine and blocks the active site permanently, inactivating the enzyme. Many hydrolytic enzymes, including proteases, rely on active serine residues. So does acetylcholine esterase, the enzyme that splits the neurotransmitter acetylcholine. Inhibition of this causes paralysis of neuromuscular junctions and ultimately death from respiratory failure.

> Some inhibitors react with the enzyme and inactivate it permanently.

6.5. Enzymes May Be Directly Regulated

The rate at which most enzymes work depends on the level of substrate and how much enzyme protein is present. The latter, of course, depends on the expression level of the gene encoding the enzyme. Thus, the activity of an enzyme or other protein may be controlled at the genetic level by deciding how much protein to synthesize. Such genetic regulation is discussed in later chapters.

More rapid cellular responses are possible if the activity of an existing protein is controlled. Consequently, a significant proportion of enzymes are also *directly*

---

chromogenic substrate   Colorless or pale substrate that is converted to a strongly colored product by an enzyme
irreversible inhibition   Type of inhibition in which an enzyme is permanently inactivated by a chemical change

FIGURE 14.35
Covalent Inhibition of Serine Enzyme by Sarin

The "nerve gas" sarin interacts with serine residues in the active site of proteins. It blocks the active site and prevents entry of the true substrate into the active site.

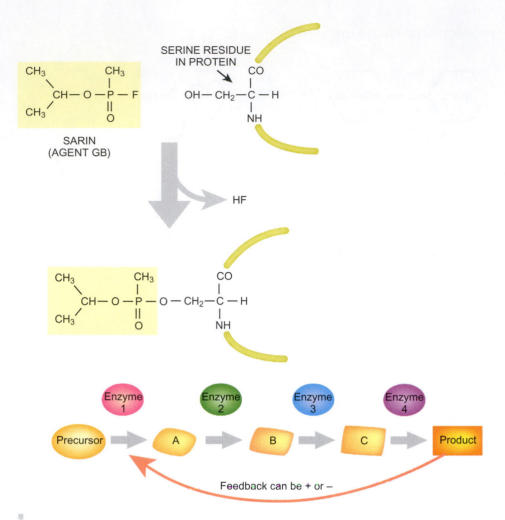

FIGURE 14.36
Feedback Inhibition of Metabolic Pathways

Feedback occurs when the product of an enzyme-catalyzed pathway interacts with one of the enzymes in the pathway, usually the first, to inhibit (negative feedback) or promote (positive feedback) the activity of that enzyme. The activities of the other enzymes in the pathway (#2, #3, and #4) are usually unaffected.

Enzymes located at critical points in metabolic pathways are often regulated by altering their activity.

End products of metabolic pathways often inhibit the enzyme that performs the first step of the pathway.

regulated in a variety of ways. This has the advantage that it is more or less instantaneous. Such regulation is usually reversible, so that an enzyme that is temporarily inactive is not destroyed, but can be used again later if needed. Although most of the following examples apply to enzymes, all types of proteins, including regulatory proteins, **transport proteins**, and mechanical proteins, can be modified to regulate their activity.

Regulated enzymes are usually found at the beginnings, ends, or branch points of metabolic pathways. Control is exerted at these critical points and the enzymes in between merely operate automatically. Many biosynthetic pathways are regulated by **negative feedback** (Fig. 14.36). The final product of the pathway inhibits the first enzyme in the pathway. This way, when sufficient product has accumulated, the cell does not waste materials making any more.

Many biosynthetic pathways are branched and lead to several final products. In this case, the enzymes at each branch point may be controlled by the products of

---

negative feedback   Form of negative regulation where the final product of a pathway inhibits the first enzyme in the pathway
transport protein   Protein that transports another molecule across membranes or from one cell to another

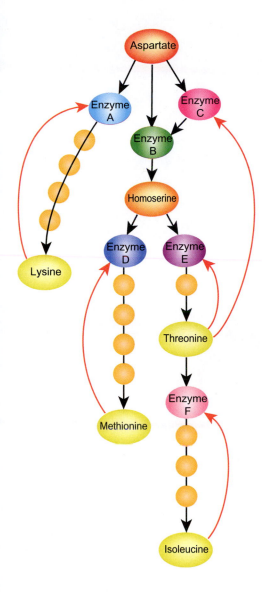

■- **FIGURE 14.37**
*Feedback Inhibition of
Aspartate Family Pathway*

Aspartate can be converted into
four other amino acids. Negative
feedback, as shown by arrows, is
exerted on the production of the
four amino acids shown in yellow.

the separate branches. An example is the synthesis of the aspartate family of amino acids. Aspartate is the precursor to four other amino acids (Lys, Met, Thr, and Ile), all of which inhibit, by feedback, both their own branch and the pathway as a whole (Fig. 14.37).

6.6. Allosteric Enzymes Are Affected by Signal Molecules

In Chapter 11 we discussed regulatory proteins (**allosteric proteins**) that change shape upon binding small signal molecules. Not surprisingly, some enzymes are regulated by the same mechanism. Such **allosteric enzymes** bind small molecules at a site distinct from the active site (the allosteric site) and alternate between two different 3D conformations. Interconversion between active and inactive forms involves both a shape change and usually also assembly or disassembly of the protein subunits that make up the enzyme. In the active form, the active site is available for substrate, whereas in the inactive form, the active site is blocked or altered in shape so that substrate cannot bind (Fig. 14.38).

Allosteric enzymes may be inhibited or activated upon binding small signal molecules.

| allosteric enzyme | Enzyme that changes shape when it binds a small molecule |
| allosteric protein | Protein that changes shape when it binds a small molecule |

FIGURE 14.38
Allosteric Enzymes Change Shape

An enzyme that has both an active site and an allosteric site is altered in shape when a signal molecule binds. The altered shape also affects the configuration of the active site.

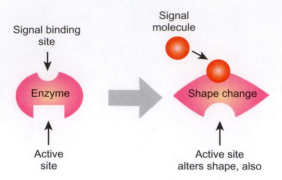

FIGURE 14.39
Allosteric Enzymes Have Multiple Subunits

A) Four subunits of an allosteric enzyme are induced to bind together forming the active enzyme when the signal molecule binds to the allosteric site. B) An allosteric enzyme with four subunits already associated binds the signal molecule and undergoes a concerted shape change.

SEPARATE SUBUNITS ASSEMBLE

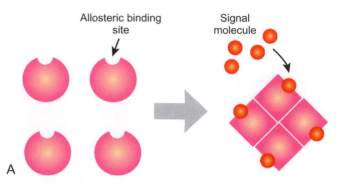

CONCERTED SHAPE CHANGE

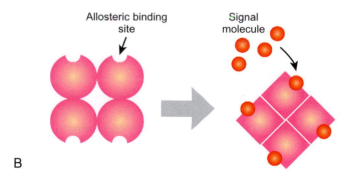

Allosteric enzymes do not follow standard Michaelis-Menten enzyme kinetics. Plots of rate versus substrate concentration are S-shaped or sigmoid, not hyperbolic. Allosteric changes may alter the K_m or the V_m or both. Allosteric effectors are usually unrelated chemically to the substrate of the enzyme they control. As noted above, they are often the end products of metabolic pathways in which the enzyme is involved.

Allosteric enzymes may be regulated negatively or positively or both. For example, phosphofructokinase (PFK) is a major control point for glycolysis. When there is plenty of ATP, this inhibits PFK, whereas the build-up of AMP and/or ADP signals that the cell is running low on energy and activates PFK. Thus, some allosteric enzymes have two different allosteric sites, one for an activator and the other for an inhibitor. In the case of PFK, ATP is the negative allosteric effector and AMP is the positive allosteric effector.

All allosteric enzymes consist of multiple subunits. When the allosteric effector binds, it changes the shape of the subunit to which it binds. This often affects subunit assembly (Fig. 14.39). One form of the allosteric protein exists as monomers and the

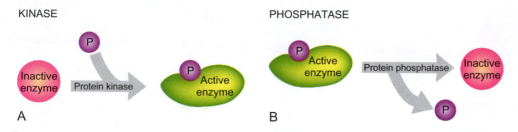

FIGURE 14.40
Addition and Removal of Phosphate to and from Proteins

A) An inactive enzyme may be made active by the addition of a phosphate group by a protein kinase. A shape change occurs that makes the phosphorylated enzyme active. B) The active enzyme is altered back to the inactive conformation when the phosphate group is removed by a phosphatase.

other as multimers. In other cases, the subunits stay together. In this case, the shape change may be transmitted from one subunit to the next (which, as a bonus, can now bind the allosteric effector more easily). The subunits thus undergo a *concerted shape change*.

6.7. Enzymes May Be Controlled by Chemical Modification

Allosteric proteins change conformation after binding a small molecule non-covalently. In other cases, a shape change is caused by modifying the protein chemically. The most common mechanism is to add a chemical group—one that can be removed again later. Phosphate groups are most common, but other groups used include acetyl, methyl, and adenyl (AMP). Phosphate groups are attached to proteins by enzymes known as **protein kinases** and are removed by protein **phosphatases** (Fig. 14.40). Control of enzyme activity by covalent modification is relatively rare in bacteria but extremely common in animal cells. About a third of all the 10,000 or so different proteins in an animal cell are phosphorylated at any given instant. When animal cells receive signals from outside, they often respond by phosphorylating a particular set of proteins that includes both enzymes and transcription factors.

> The activity of many proteins is altered by adding or removing phosphate groups.

The classic example of control by phosphorylation is the synthesis and breakdown of **glycogen** by animal cells. Glycogen is a storage carbohydrate that is split to give glucose when cells need energy. It is made by glycogen synthase and broken down by glycogen phosphorylase (Fig. 14.41). Both enzymes are controlled by phosphorylation. Glycogen phosphorylase is active when phosphorylated, whereas glycogen synthase is inactive. As is often the case, there is a cascade of reactions involving two protein kinases. When the cells need energy, protein kinase A phosphorylates both glycogen synthase and phosphorylase kinase, which in turn phosphorylates glycogen phosphorylase.

Some enzymes are activated by cleavage of a precursor protein to yield active enzyme. The digestive enzymes trypsin, chymotrypsin, and pepsin are synthesized as longer precursors known as trypsinogen, chymotrypsinogen, and pepsinogen. Only in the intestine, and safely outside the cells that made them, are they activated by cleavage of the polypeptide chain. Unlike control by binding small molecules or phosphorylation, this sort of activation is not reversible.

> Potentially dangerous enzymes are often activated by cleavage of inactive precursors.

---

glycogen Storage carbohydrate found both in bacteria and in the livers of animals
phosphatase An enzyme that removes phosphate groups
protein kinase An enzyme that adds phosphate groups to another protein

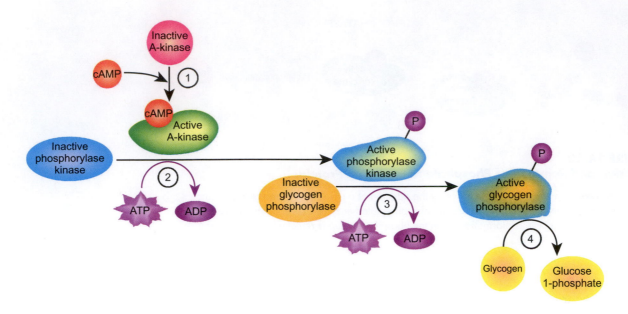

FIGURE 14.41
Control of Glycogen Synthesis and Breakdown

A four-step process is necessary to break down glycogen to release glucose as glucose 1-phosphate. 1) Inactive protein kinase A is activated upon binding to cyclic AMP (cAMP). 2) Activated protein kinase A uses ATP to convert the inactive phosphorylase kinase to the active phosphate-bound form. 3) Activated phosphorylase kinase converts inactive glycogen phosphorylase to the active phosphorylated form. 4) Ultimately, active glycogen phosphorylase converts glycogen to glucose 1-phosphate, which is the first substrate for glycolysis.

7. Binding of Proteins to DNA Occurs in Several Different Ways

> Many DNA-binding proteins recognize specific base sequences. Such recognition sequences are often (but not always) inverted repeats.

A wide range of proteins bind to DNA. These proteins are involved in DNA replication, in gene expression and its control, in protection and repair of DNA, and various other processes. Understanding the properties of DNA-binding proteins is of major practical importance, since they are used to control the expression of cloned genes. DNA-binding proteins are also important in molecular medicine, especially in such areas as cancer and aging. Despite the great variety of DNA-binding proteins, there are some common themes in how these proteins interact with DNA.

Although some DNA-binding proteins are relatively non-specific, many recognize and bind to specific base sequences in the DNA. Almost all DNA-binding proteins fit into the major groove of DNA as this allows them to recognize and make contact with the bases. When several sites recognized by a particular DNA-binding protein are compared, they are found to have very similar, though rarely identical, sequences. Many DNA-binding proteins, including both regulatory proteins and enzymes that cut or modify DNA, recognize palindromes or inverted repeats in the DNA. In this case, proteins that consist of single subunits often bind to inverted repeats that are 4–8 bp long overall. Proteins that consist of paired subunits usually bind to inverted repeats that have two 5- or 6-base repeats separated by half a dozen bases whose sequence is relatively unimportant (Fig. 14.42). These relatively short palindromes do not form hairpins or stem and loop structures.

> Just a few structural motifs are responsible for binding DNA in a large number of different DNA-binding proteins.

A vast number of different transcription factors and other regulatory proteins bind DNA by means of a relatively small number of DNA-binding domains. The best known of these motifs are the helix-turn-helix, helix-loop-helix, leucine zipper, and zinc finger.

Both the **helix-turn-helix (HTH)** and the similar but distinct **helix-loop-helix (HLH)** motifs consist of two α-helices joined by a loop (Fig. 14.43). The turn or loop

helix-loop-helix (HLH)   One type of DNA-binding motif common in proteins
helix-turn-helix (HTH)   One type of DNA-binding motif common in proteins

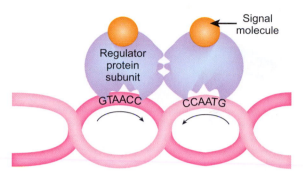

A   DNA with inverted repeats

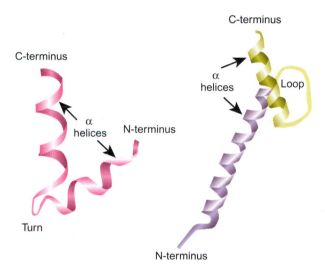

B   DNA-binding regulator proteins

■ **FIGURE 14.42**
Binding of Proteins to Inverted Repeats on DNA

A) Double-stranded DNA with a 5-base inverted repeat. B) A protein dimer has bound to the inverted repeat sequences on the two different strands of DNA. Note how the helical twisting of the DNA brings the two recognition sequences together and so allows the two protein subunits to bind side-by-side.

A   HELIX-TURN-HELIX   B   HELIX-LOOP-HELIX

■ **FIGURE 14.43**
Helix-Turn-Helix (HTH) and Helix-Loop-Helix (HLH) Motifs

A simple bend versus a loop in the protein is the structural feature distinguishing between these DNA-binding proteins.

is shorter for the HTH motif and longer for the HLH domain. In the case of the HTH domain, one of the α-helices fits into the major groove of the DNA double helix and makes contact with the bases. In the HTH motif, it is the second of the two helices (counting from the N-terminal end) that is responsible for DNA binding. In the HLH motif, the two α-helices are responsible for dimerization of the protein, not DNA binding per se. Sometimes identical subunits form dimers; in other cases, the HLH motif is used to bind another protein of a DNA-binding complex, such as the Mre11/Rad50 complex involved in DNA repair (see Focus on Relevant Research). DNA binding is due to a basic region just in front of the first α-helix. Proteins with both the HTH and HLH motifs usually bind as dimers to inverted repeats in the DNA (Fig. 14.44).

The HTH domain is widely used by both prokaryotes and eukaryotes whereas the HLH motif is found mostly in eukaryotes. For example, the HTH motif is found in the Crp global activator of *E. coli* and in both the CI and Cro regulatory proteins of bacteriophage lambda. Eukaryotic transcription factors that recognize homeobox sequences and control development in multicellular animals use an HTH motif. (Homeobox sequences are found in the regulatory regions of genes involved in overseeing spatial and temporal development in animals; see Ch. 19.) Although the rest of the DNA-binding domain is different, the HTH motif that actually binds to the DNA is almost identical to that in the lambda CI repressor.

FIGURE 14.44
Binding of Helix-Turn-Helix (HTH) Motif to DNA

A typical HTH protein is a dimer with two sets of α-helices, labeled α2 and α3, which actually bind to the DNA. B) The pairs of α-helices fit into two adjacent major groves in the DNA. In panel B only the α2 and α3 helices are indicated to show how they interact with the DNA. The HTH shown is the phage lambda CI repressor.

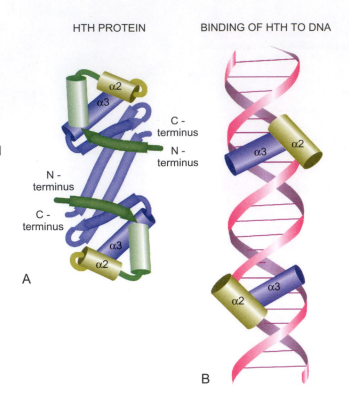

HTH PROTEIN

BINDING OF HTH TO DNA

The **leucine zipper** is found in many eukaryotic transcription factors, including the Fos, Jun, and Myc proteins that are involved in control of cell division and carcinogenesis. A leucine zipper motif consists of an α-helix with leucine residues every seventh amino acid. In addition, the amino acids halfway between the leucines are usually hydrophobic. Because there are 3.6 amino acids per turn, these hydrophobic residues form a strip down the side of the α-helix (Fig. 14.45). Two such α-helices can bind together by their hydrophobic strips forming a zipper structure. Thus, the leucine zipper is a dimerization domain like the HLH motif. The actual binding of DNA is due to basic residues in front of the zipper region.

A **zinc finger** consists of a central zinc atom with a segment of 25–30 amino acid residues arranged around it (Fig. 14.46). In the classic version of the Zn finger, the Zn is bound to two cysteines, which lie in a very short piece of β-sheet—a β-hairpin—and

Zinc fingers are widely distributed in DNA-binding proteins. Each zinc finger recognizes three bases in the DNA.

Lammens K, et al. (2011) The Mre11:Rad50 Structure Shows an ATP-Dependent Molecular Clamp in DNA Double-Strand Break Repair. Cell 145:54–66.

FOCUS ON RELEVANT RESEARCH

The complex between Mre11 nuclease and Rad50 ATPase is highly conserved through evolution. It acts as a sensor for double-stranded breaks in DNA. This complex can recognize and process such broken ends of DNA even if they are chemically blocked or mis-folded.

The authors determined the crystal structure of the complex from the microorganism, *Thermotoga maritima*. They found that the C-terminal helix-loop-helix domain of Mre11 binds Rad50 and attaches flexibly to the Mre11 nuclease domain. This allows large conformational changes of the complex. When ATP is bound to the two Rad50 subunits, the Mre11 helix-loop-helix domains rotate. The resulting structure shows enhanced ability in binding DNA.

leucine zipper   One type of DNA-binding motif common in proteins
zinc finger   One type of DNA-binding motif common in proteins

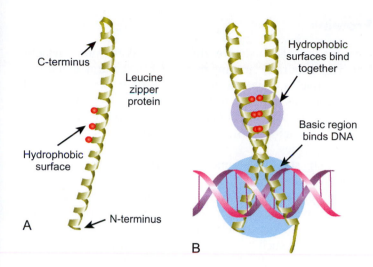

■ **FIGURE 14.45**
Leucine Zipper Protein Binding DNA

A) The leucine zipper consists of two α-helixes that have hydrophobic zones and basic ends. B) The helices of the leucine zipper bind to each other by their hydrophobic regions and to DNA by their basic regions. The basic end region fits into the major groove of the DNA. Because the basic regions are roughly parallel and open up around the DNA, the two helical segments resemble a zipper.

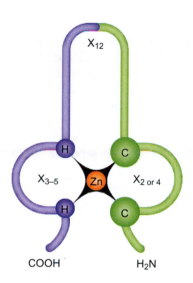

■ **FIGURE 14.46**
Zinc Finger DNA-Binding Protein

A central zinc atom is bound to the sulfurs of cysteine (C) and the nitrogens of histidine (H). Chains of amino acids of varying lengths (x = chain length) extend from these binding regions. The zinc finger is a component of a much larger protein and binds the protein to DNA.

two histidines, which lie in a short α-helix. The far end of the α-helix protrudes into the major groove of the DNA. Over a thousand zinc finger proteins are known, and many of them have multiple fingers. The first zinc finger protein discovered was the general eukaryotic transcription factor TFIIIA, from the toad *Xenopus*, which has nine zinc fingers.

Each zinc finger unit usually recognizes three bases in the DNA. Less often, four or five bases are recognized by a single zinc finger. The sequence specificity of each zinc finger depends on the amino acid sequence of the polypeptide chain between the His and Cys residues that bind the zinc. Amino acids in this region make hydrogen bonds with bases in the DNA.

Several modified versions of the zinc finger motif have been found. For example, the **steroid receptor** family of transcription factors has a DNA-binding domain that contains two Zn atoms, each surrounded by four cysteines. A short α-helix that lies between the two zincs binds to the DNA. Several fungal transcription factors, including GAL4 of yeast, contain fingers built around a cluster of two Zn atoms bound to six cysteines.

steroid receptor Protein that binds steroid hormones

8. Denaturation of Proteins

Denaturation is the loss of correct 3D structure due to breaking non-covalent bonds. Loss of biological activity generally accompanies such structural denaturation. Enzymes are especially sensitive to denaturation. When proteins are denatured they often precipitate out of solution, as happens to the proteins of an egg when it is boiled. Heat, extremes of pH, and a variety of chemical agents destroy non-covalent structure and denature proteins. Even agitation may denature some proteins as when egg whites are whipped to give meringues.

Proteins vary greatly in their stability. Some are very sensitive and even slight changes in pH or temperature may inactivate them. This often causes problems in both the purification of proteins and their biotechnological applications. Consequently, researchers have often searched for natural proteins that are unusually resistant, or have modified proteins to increase their stability. Bacteria that live under extreme conditions are a rich source of such resistant proteins. For example, the Taq polymerase used in PCR (see Ch. 6) is a highly heat-resistant enzyme made by bacteria that live naturally at temperatures so high they would kill most organisms.

The hydrophobic forces responsible for maintaining much of the 3D structure of proteins are disrupted by **detergents** and **chaotropic agents**. Detergents consist of a hydrophobic tail joined to a highly water-soluble group. They act by directly binding to the hydrophobic regions of other molecules and solubilizing them. In the case of proteins, detergents can bind to the hydrophobic groups that are normally buried deep inside and also to the relatively hydrophobic polypeptide backbone. This destabilizes the 3D folding of the polypeptide chain.

The detergent **sodium dodecyl sulfate (SDS)** is widely used to solubilize and denature proteins before running them on polyacrylamide gels to separate them by molecular weight (see Ch. 15). SDS has a long hydrocarbon tail that binds to the polypeptide and a negatively charged sulfate group that sticks out into the water and solubilizes the protein/SDS complex (Fig. 14.47). SDS binds to polypeptides along

> The correct 3D structure of proteins can be destroyed by heating, detergents, acids, bases, and certain chemicals.

> Detergents and chaotropic agents help solubilize hydrophobic groups in water.

> Sodium dodecyl sulfate is widely used to solubilize proteins before running them on gels.

FIGURE 14.47
Structure and Function of Sodium Dodecyl Sulfate (SDS)

A) SDS contains a strongly hydrophobic carbon chain and a strongly hydrophilic sodium sulfate region. B) When a folded protein is boiled in the presence of SDS, the hydrophobic region winds around the polypeptide backbone and the negatively-charged sulfate protrudes. The negative charges repel each other which helps in straightening out the protein.

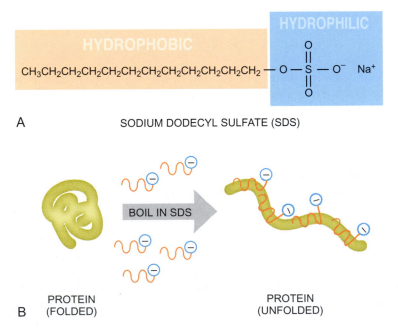

chaotropic agent   Chemical compound that disrupts water structure and so helps hydrophobic groups to dissolve
detergent   Molecule with both hydrophobic and hydrophilic regions that can solubilize hydrophobic molecules including fats, grease, and lipids
denaturation   Loss of correct 3D structure of proteins or nucleic acids
sodium dodecyl sulfate (SDS)   A detergent widely used to denature and solubilize proteins before separation by electrophoresis

their long axes and converts them to an extended rod-shaped conformation. The precise nature of SDS-polypeptide binding is disputed. One theory is that it binds to the non-polar R-groups of hydrophobic amino acids. However, SDS binds in a ratio of one SDS to every two amino acid residues for the vast majority of proteins and this suggests that it binds mostly to the polypeptide backbone. From a practical viewpoint, it is important that the number of negative charges contributed by bound SDS is proportional to the length of the polypeptide; that is, to its molecular weight.

Chaotropic agents (e.g., thiocyanate, perchlorate) also destabilize proteins by promoting the exposure of hydrophobic groups, but by an indirect mechanism. When exposed to water, hydrophobic groups induce the formation of regular cages of water molecules around themselves. This decreases the disorder of the water; that is, it causes an increase in entropy, which is thermodynamically unfavorable. Hydrophobic groups tend to cluster together to avoid contact with water, rather than because of any positive attraction. Chaotropes disrupt water structure and so allow hydrophobic groups to dissolve more readily.

> Chaotropes alter the structure of water, allowing hydrophobic groups to dissolve more easily.

Protein **denaturants** are molecules that disrupt the hydrogen bonds that maintain secondary structure. Examples are **urea, guanidine**, and **guanidinium chloride** (Fig. 14.48). They act by forming hydrogen bonds between their own CO or NH$_2$ groups and all the groups on the protein that can take part in hydrogen bonding. High temperatures and extremes of pH also destroy hydrogen bonding.

> Urea and guanidine disrupt hydrogen bonds.

Denaturation is also aided by breaking the disulfide bonds of those proteins that rely on them to stabilize their structures. In the laboratory, β-**mercaptoethanol** or **BME** (HOCH$_2$CH$_2$SH) is often used to reduce disulfide bonds to two -SH groups, so breaking the linkage.

 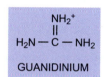

FIGURE 14.48
Protein Denaturants Disrupt Hydrogen Bonds

Urea, guanidine, and guanidinium form hydrogen bonds with the peptide groups of proteins. This disrupts the hydrogen bonds within the protein that help maintain secondary structure of the protein. The result is unfolding (denaturation) of the polypeptide chain.

Key Concepts

- The primary structure of proteins is the amino acid sequence of the polypeptide chains.
- The secondary structure of proteins results from hydrogen bonds.
- The tertiary structure of proteins is due to interactions between amino acid side chains. Larger proteins may have multiple folding domains.
- The quaternary structure of proteins refers to the assembly of multiple polypeptide chains to form proteins with subunits.

β-**mercaptoethanol (BME)** A small molecule with free sulfhydryl groups often used to break disulfide bonds in proteins
denaturant Chemical compound that destroys the 3D structure of proteins, especially by breaking hydrogen bonds
guanidine Non-ionized form of guanidinium
guanidinium chloride A widely-used denaturant of proteins
urea A nitrogen waste product of animals; also widely used as a denaturant of proteins

- Protein structures are investigated using X-ray crystallography, NMR spectroscopy, and the electron microscope.
- Proteins often contain other chemical components in addition to the polypeptide chains. These may include metal ions, organic co-factors, and attached sugar and lipid residues.
- Proteins comprise about 60% of the organic matter of a typical cell and carry out many functions. Complex protein assemblies may be regarded as nanomachines.
- Enzymes are proteins that catalyze chemical reactions. They act by lowering the energy of activation.
- Different enzymes bind their substrates with different specificities.
- The rate of enzyme reactions depends on the substrate concentration, the affinity of the enzyme for the substrate, and the inherent properties of the active site (reflected by the Vmax).
- Some enzymes are directly regulated by binding small signal molecules, others are controlled by chemical modification.
- Some proteins bind to DNA at specific sequences. Several different structural motifs are common in DNA-binding proteins.
- Proteins may be denatured by heat, detergents, and assorted chemicals.

Review Questions

1. What are proteins made from?
2. What are the N- and C-terminus of a protein?
3. What are zwitterions? Give an example.
4. What is the general structure of an amino acid? Which is the simplest amino acid?
5. Name the different groups of amino acids based on their physical and chemical properties.
6. Name the two amino acids that contain sulfur.
7. What is the chiral or asymmetric center of an amino acid?
8. What are enantiomers or optical isomers? What are the two types of optical isomers for amino acids?
9. How is the secondary structure of a protein formed? What are the two common secondary structures?
10. Describe the differences between an alpha-helix and a beta-sheet.
11. Define reverse turn and random coil.
12. How are the tertiary structures of a protein formed?
13. What is the oil drop model of protein structure? What are the two major factors affecting this model?
14. What are the different bonds that affect the 3D structure of proteins?
15. How are disulfide bonds formed? Why are they important for protein structure?
16. What is the difference between cysteine and cystine?
17. How are domains formed? How do they function?
18. Define the quaternary structure of proteins. What are protomers?
19. What is the difference between a homotetramer and a heterotetramer? Give an example of each.
20. How do hydrophobic forces influence tertiary and quaternary structures?
21. What is self-assembly? List three major types of self-assembly.
22. What is an apoprotein?
23. What are vitamins? What are their biological roles?
24. What is X-ray crystallography? How does it work?
25. What are conjugated proteins? List three examples of conjugated proteins.
26. Give five major categories of cellular functions performed by proteins.
27. What is the difference between antifreeze proteins and ice nucleation factors?

28. What are binding proteins? List three major types of binding proteins and give examples for each.
29. What are structural proteins? Give two examples.
30. How do mechanical proteins work? What supplies the energy?
31. What are the functions of chaperonins?
32. What are luciferases, GFP, and rhodopsins?
33. What are enzymes? How do they work? Give one example.
34. Why must polypeptide chains be folded to make the protein active?
35. What are co-factors or prosthetic groups?
36. How is the specificity of an enzyme determined?
37. Describe the two alternative mechanisms for substrate binding. Give examples.
38. Define free energy of a reaction, transition state, and transition state energy.
39. What are the factors involved in the enzyme rate increase?
40. What is the Michaelis-Menten equation? Explain the equation.
41. What are substrate analogs? How do they behave as competitive inhibitors?
42. What are chromogenic substrates? Give two examples.
43. What is irreversible inhibition?
44. How are enzymes regulated directly?
45. What is negative feedback? How does it apply to enzymes?
46. How does an allosteric enzyme differ kinetically from the standard Michaelis-Menten enzyme kinetics?
47. Name the enzymes that add or remove phosphates.
48. Give an example where control of an enzyme is due to phosphorylation.
49. How are enzymes like trypsin, chymotrypsinogen, and pepsin activated?
50. What are the most common structural motifs for DNA-binding proteins?
51. What is the difference between the HTH and HLH motifs?
52. What is a leucine zipper motif? Give two examples of proteins with this motif.
53. How are zinc finger motifs formed? Give one example.
54. What is denaturation of proteins? What happens during denaturation? What are some different ways of denaturing a protein?
55. How do detergents disrupt the 3D structure of protein? Give one example.
56. What is the mechanism by which chaotropic agents denature protein?
57. What are the different ways of disrupting the secondary structure of proteins?
58. What is the difference between the denaturation mechanism of urea and beta-mercaptoethanol?

Conceptual Questions

1. The active site of the phosphatase, Fcp1, has several essential amino acids. The most important is an aspartic acid at position 170 because it forms an acylphosphate intermediate that removes the phosphate from its substrate. Nearby amino acids in the 3D structure are important, but some can tolerate certain amino acid substitutions. In particular, the valine found four amino acids away from asp170 can be replaced with isoleucine, leucine, and alanine, but any other substitution abolishes Fcp1 phosphatase activity. Why does this enzyme tolerate these three amino acids at this location, but no other amino acid can replace the valine?
2. Antibodies are multimeric proteins that have two small light chains and two larger heavy chains or protein subunits. The four subunits are connected together by disulfide bonds. Draw a disulfide bond and then propose a way to disrupt the antibody structure so the light and heavy chains are no longer attached.
3. A researcher has been given a pure sample of enzyme JJ99 isolated from the *E. coli* cytoplasm which controls the conversion of substrate B into two subunits called Ba and Bb. When the pure enzyme is mixed with substrate B, very little product is made. Suggest reasons why the pure enzyme and pure substrate do not work in her assay?
4. The following mutations in the primary structure of the gene for enzyme JJ99 were found to inhibit the activity of JJ99 in *E. coli*. Using your knowledge of protein structure, propose a hypothesis of how these mutations affect the activity of the enzyme.

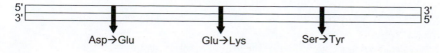

5. The following biochemical pathway was discovered in *E. coli*, and the final product C inhibits the activity of enzyme X. What is this type of inhibition called? Why is this method of control better than transcriptional regulation for some pathways?

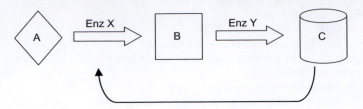

Proteomics: The Global Analysis of Proteins

Chapter 15

Today we know the complete genome sequences for many microorganisms and quite a few higher organisms including ourselves. However, we still have no idea what the function of most genes, or their encoded proteins, might be. The availability of this novel ocean of ignorance to explore has led some scientists to designate the twenty-first century as the post-genomic or proteomic era. Just as genomics refers to the global analysis of genes, so proteomics refers to the same approach for proteins. Although they are "mere" gene products, proteins are in many ways more complicated than the nucleic acids that encode them. Proteins are made of 22 different genetically encoded amino acids and vary greatly in their 3D structure. They also vary in both function and stability, as described in Chapter 14. Consequently, elucidating the role of proteins, especially on a large scale, is in many ways more difficult than for nucleic acids. Nonetheless, methods have been developed or modified to analyze many different proteins simultaneously and we even have protein arrays. A further complication is that many proteins act as complexes that contain multiple different proteins. Thus it is necessary to survey protein interactions—hence, the "interactome" which represents all the protein interactions within an organism.

1. The Proteome

The term **proteome** was originally defined as the total set of proteins encoded by a genome. Alternatively, it may be viewed as the total protein complement of an organism. The term **translatome** is sometimes used to refer to all the proteins that are present in a cell under any particular set of conditions, that is, those that have actually been translated. This is distinct from the proteome, which consists of all those proteins that are potentially available. The relationship between genome and proteome/translatome is not simply linear. Many proteins are processed and modified by other proteins; therefore, the final protein complement depends on complex interactions between proteins that may vary with the growth conditions.

Although gene expression is often assessed by monitoring transcription, mRNA levels do not always correspond to the levels or activity of the final gene products, the encoded proteins. The disparity between transcriptome and proteome may be summarized as follows:

a. Some RNA molecules are non-coding and do not give rise to any protein products.

b. Some primary RNA transcripts undergo alternative splicing; therefore, the same gene may give rise to multiple protein products.

c. Levels of mRNA may not correlate with protein levels due to differential rates of mRNA translation or degradation.

d. The activity of many proteins is regulated after translation by addition or removal of acetyl, phosphate, AMP, ADP-ribose, or other groups.

e. The activity of many proteins is altered after translation by chemical modification of amino acid residues.

f. Many proteins are processed after translation; for example, by proteolytic cleavage or addition of sugar or lipid residues to give glycoproteins or lipoproteins.

g. Proteins themselves may be degraded and vary greatly in stability.

The various modifications listed above may all vary depending on the growth conditions and the activities of other genes and/or proteins. Thus, in practice it may be necessary to monitor not only the level but also the activity of proteins.

The fundamental problem in proteomics is the individuality of different proteins. Since proteins differ in structure, stability, solubility, charge, activity, etc., traditionally each individual protein had to be purified and assayed by a different procedure. Proteomics requires the parallel analysis of multiple proteins and relies on methods that are applicable to many different proteins and are not affected by variations in 3D structure or enzyme activity. In practice, separation of proteins for proteomics usually relies on denaturation followed by 2D gel electrophoresis. Identification is often based on recently developed improvements in mass spectrometry, especially the MALDI/TOF approach. Generic methods for purification of proteins rely on attaching the same tag molecule (e.g., His-tag, FLAG, or GST) to many different proteins and then binding the tag.

1.1. Isolating and Quantifying Proteins

The best method to isolate protein depends upon the source, but generally, the first step is to break the cells open. As described in Chapter 1, the nature of the organism dictates which method of cell lysis is used, since the cell envelope may be very strong as in bacteria or plant cells or simple lipid bilayers as in mammalian cells. Cells may be broken open by lysing the membranes with detergents and/or destroying tough walls

---

Proteomics involves surveying the global protein composition of a cell or organism.

Proteins differ from one another much more in both structure and function than nucleic acid molecules.

---

proteome   The total set of proteins encoded by a genome or the total protein complement of an organism
translatome   The total set of proteins that have actually been translated and are present in a cell under any particular set of conditions

by mechanical methods such as glass beads or blenders. Other methods to break open cells include forcing the cell suspension or tissue though a very small slit, or sonication, which uses high frequency sound waves. Perhaps the easiest method is to freeze and thaw the cells. As ice crystals form, they break the cells open and then when the cells thaw, the cellular contents flow out. The DNA and RNA are then removed from the samples by adding non-specific DNase and RNase to digest the nucleic acids.

Isolating specific cellular compartments from eukaryotic cells is also possible. Traditionally, the required components, such as mitochondria, chloroplasts, endoplasmic reticulum, or nuclei, are separated from the other cell parts using centrifugation through density gradients. Nowadays, there are enough specific detergents and chemicals for cell lysis that separation can be accomplished chemically. These organelles and subcellular compartments are then able to analyzed away from the rest of the cell.

Determining the amount of total protein in a sample whether it is a whole cell extract or a subcellular extract is accomplished by a variety of colorimetric assays. Two widely-used colorimetric assays rely on copper chemistry and a protein-binding dye, respectively. In the first type called **biuret reagent**, the proteins react with cupric ions (Cu^{+2}) and sodium potassium tartrate in an alkaline environment. The reaction produces a blue to violet complex that is quantified by measuring absorbance at 540 nm. The commonly used BCA assay combines the biuret reagent with a second reaction. Bicinchoninic acid (BCA) is added to reduce the cupric ion to cuprous ion (Cu^{+1}), which creates a stronger purple color and increases the assay sensitivity. The Lowry assay instead adds Folin-phenol reagent to enhance the color and sensitivity, which is measured at a higher absorbance (between 650 and 750 nm). The second type of protein assay uses a non-specific protein dye called **Coomassie Blue** and is sometimes called the Bradford assay. The protein sample is mixed with the reagent and color develops as Coomassie Blue reacts with positively charged amino acids in those proteins that are at least 3 kD in size. In all these assays, the amount of color produced is proportional to the amount of protein in the sample. The actual amount of protein is determined by comparing the assay value to a set of known protein concentrations.

1.2. Gel Electrophoresis of Proteins

Because nucleic acids are all negatively charged they all move towards the positive electrode during electrophoresis (Ch. 4). However, proteins are not so convenient. Some of the amino acids of proteins have a positive charge, some have a negative charge, and most are neutral. So, depending on its overall amino acid composition, a protein may be positive, negative, or neutral.

If a mixture of native (i.e., non-denatured) proteins is run on a gel, some move towards the positive electrode, others towards the negative electrode whereas neutral proteins scarcely move at all. Consequently, the samples are usually started in the middle of the gel, rather than at the end. This is referred to as native protein electrophoresis and is sometimes used to purify proteins without inactivating them. After running, the gel is stained with reagents specific to the protein of interest. For example, an enzyme that gives a colored product may be located in this manner.

To separate proteins on the basis of molecular weight, they are first boiled in a solution of the detergent **sodium dodecyl sulfate (SDS)**. Boiling in detergent destroys the folded 3D structure of the protein; that is, the protein is denatured. The SDS molecule has a hydrophobic tail with a negative charge at the end. The tail wraps around the backbone of the protein and the negative charge dangles in the water. The protein becomes unrolled and covered from head to toe with SDS molecules, which gives it an overall negative charge (Fig. 15.01). Furthermore, the number of negative charges

When proteins are separated by size using gel electrophoresis, they are first denatured and coated with negatively-charged detergent molecules.

biuret reagent A solution containing potassium hydroxide, hydrated copper (II) sulfate, and sodium potassium tartrate used to assess the total amount of protein in a sample
Coomassie Blue A blue dye used to stain proteins
sodium dodecyl sulfate (SDS) Detergent used to unfold proteins and cover them with negative charges for electrophoresis

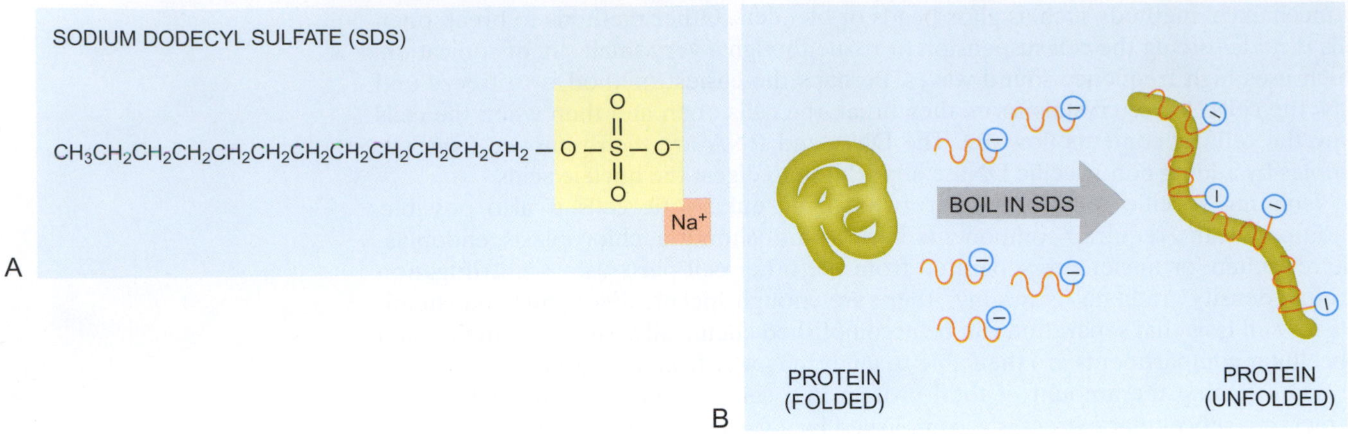

FIGURE 15.01
Denaturation of Protein by SDS

A) The structure of SDS is amphipathic; that is, the long hydrocarbon tail is hydrophobic and the sulfate is hydrophilic. B) The first step to separate a mixture of proteins by size is to boil the sample. Boiling proteins in a solution of SDS destroys the tertiary structure of the protein. The hydrophobic portion of SDS coats the polypeptide backbone and prevents the protein from refolding. The hydrophilic group of SDS keeps the protein soluble in water, and the negative charges repel each other, which also helps to keep the protein from refolding. The result is an unfolded protein with a net negative charge that is proportional to its molecular weight.

bound is proportional to the length of the protein. In addition, the disulfide bonds that help maintain the tertiary structure of some proteins and/or hold protein subunits together must be disrupted for proper denaturation. This is done by adding small sulfhydryl reagents, usually β-mercaptoethanol (HS—CH$_2$CH$_2$OH):

$$\text{Protein-S—S-Protein} + 2\,\text{SH—CH}_2\text{CH}_2\text{OH} \rightarrow$$
$$2\,\text{Protein-SH} + \text{HO—CH}_2\text{CH}_2\text{—S—S—CH}_2\text{CH}_2\text{OH}$$

After denaturation, proteins can be separated by size by electrophoresis through a gel (Fig. 15.02). Because proteins are a lot smaller on average than DNA or RNA, the gel is made of the polymer polyacrylamide, which gives smaller gaps in its meshwork than agarose. The technique is thus known as **PAGE** or **polyacrylamide gel electrophoresis**. After running, the gel is stained to visualize the protein bands. The two favorite choices are Coomassie Blue, or silver compounds. Silver atoms bind very tightly to proteins and yield black or purple complexes. Silver staining is more sensitive and, of course, more expensive.

1.3. 2D PAGE of Proteins

Separation of large numbers of proteins is normally done by 2D polyacrylamide gel electrophoresis (2D PAGE). The proteins are separated by charge in the first dimension and then by size in the second dimension. **Isoelectric focusing** is used in the first dimension and separates native proteins based on their original charge. A pH gradient is set up along a cylindrical gel and proteins migrate until they find a position where their native charges are neutralized—the isoelectric point. Standard SDS-PAGE as described above is used in the second dimension and separates denatured proteins based on their molecular weight (Fig. 15.03).

Early 2D gels were able to resolve 1,000 or so protein spots and were used to characterize the protein complement of bacteria such as *Escherichia coli* where about

isoelectric focusing   Technique for separating proteins according to their charge by means of electrophoresis through a pH gradient
polyacrylamide gel electrophoresis (PAGE)   Technique for separating proteins by electrophoresis on a gel made from polyacrylamide

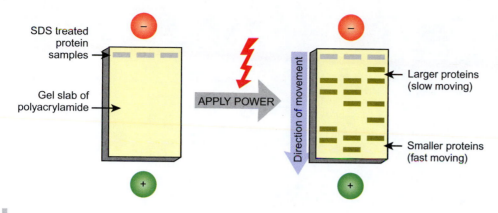

FIGURE 15.02
SDS Polyacrylamide Gel Electrophoresis

Proteins treated with SDS can be separated by size using gel electrophoresis. Since all the proteins have a net negative charge, the proteins are repelled by the negative cathode and attracted to the positive anode. As the proteins move toward the anode, the polyacrylamide meshwork obstructs and slows the larger proteins but allows the smaller proteins to move faster. In consequence, the distance traveled in a given time is proportional to the log of the molecular weight. After separation, the proteins are visualized with a dye such as Coomassie blue or a silver compound.

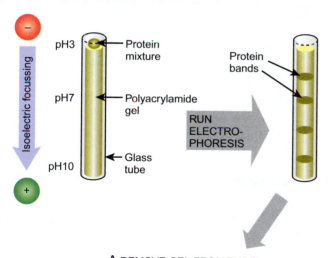

A REMOVE GEL FROM TUBE
B TREAT WITH SDS
C PLACE TUBE GEL ONTO SLAB GEL

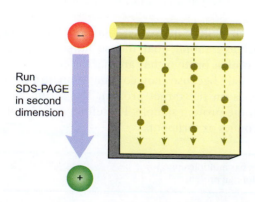

FIGURE 15.03
2D Polyacrylamide Gel Electrophoresis

The first step in separating large numbers of proteins in two dimensions is to separate them according to their inherent charge. The mixture of proteins is loaded onto a gel that has a gradient of increasing pH. An electric field is applied and the proteins move along the pH gradient until they reach the point at which their charges are neutralized. At this point, each band in the gel contains several different proteins with the same (or very similar) isoelectric point. The tube gel is removed from its tube and exposed to SDS to denature the proteins. It is then placed on a slab of polyacrylamide gel and traditional SDS-PAGE is run in the second dimension to separate the proteins by size. After staining, the result of 2D-PAGE is a square with small scattered dots representing individual proteins.

FIGURE 15.04
2D Protein Gel of Mouse Brain Tissue

Soluble proteins were extracted from mouse brain and separated by 2D-PAGE. The first dimension used isoelectric focusing and the second dimension used standard SDS-PAGE. The proteins were visualized by silver staining. Each spot on the gel is due to a separate protein. However, because proteins are frequently modified after synthesis, multiple spots sometimes arise from variants of the same original protein. (Credit: Prof. Dr. Joachim Klose, Institut für Humangenetik, Humboldt-Universität, Berlin.)

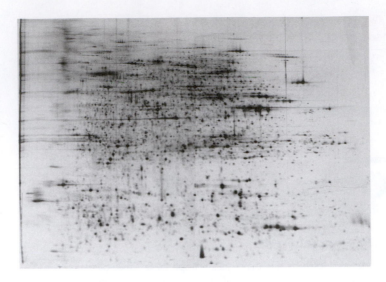

A large number of proteins can be separated by gel electrophoresis in two directions.

Proteins are visualized with a variety of dyes.

1,000 of the 4,000 genes are expressed at any given time (see Ch. 16, Fig. 16.01). More recently, large 2D gels with higher resolution have been developed that allow separation of over 10,000 spots and can be used to analyze the proteome of higher organisms (Fig. 15.04). After separation, the protein spots are cut out from the gel, digested by protease treatment, and the resulting peptides are analyzed by mass spectrometry (see below). This allows unambiguous identification of each protein spot.

To detect the presence or absence of a particular protein, a sensitive silver stain is used. For quantization of protein spots either Coomassie blue or fluorescent dyes are used to stain the gel. The gel is then scanned with a laser. A variety of fluorescent dyes are used, especially in large-scale proteomics work. SYPRO Orange and Red dyes bind proteins non-covalently. These are non-fluorescent in aqueous solution but fluoresce when in a non-polar environment, including when bound to protein-SDS complexes. It is also possible to covalently label two protein samples with differently colored dyes, say methyl-Cy5 (red) and propyl-Cy3 (green), and run them on the same gel. Differences between the samples may then be directly visualized (pure red and green spots indicate that a protein was found in one or the other sample and a yellow spot indicates its presence in both).

2. Antibodies Are Essential Proteomics Tools

One vital tool to study proteins is the use of antibodies, which are naturally produced by the immune system. Antibodies are Y-shaped proteins that are made of four separate polypeptide chains linked by disulfide bonds (Fig. 15.05). Each antibody has a highly variable region that specifically recognizes foreign proteins, often those of viruses or bacteria. When a foreign substance enters the body, some circulating antibodies will recognize and bind the invader. This triggers the B cells to make more of the same antibody. Each B cell makes antibodies against only one **epitope**, or specific 3D portion of one protein. For proteomics research, antibodies are used to recognize a specific protein of interest.

To make **polyclonal antibodies** to a protein of interest, a small amount of the protein is used to immunize an animal such as a rabbit, goat, mouse, horse, or sheep. The animal makes antibodies against the foreign protein that are harvested by collecting blood serum. These are called "polyclonal" because they contain a mixture of different antibodies against different epitopes on the target protein. The serum will also contain antibodies to any impurities in the protein sample. Therefore, polyclonal antibodies are not suitable for many proteomics experiments.

epitope   Specific 3D portion of a protein that is recognized by an antibody. Every protein has multiple epitopes
polyclonal antibodies   Antibodies from an animal that recognize multiple epitopes of a foreign protein

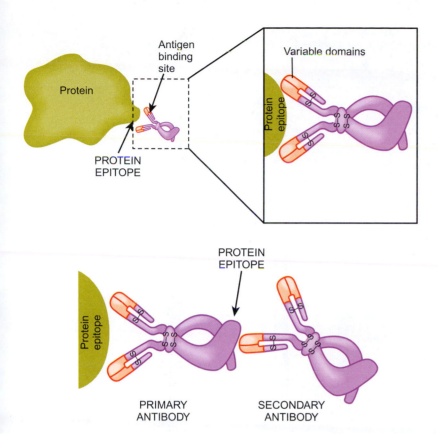

►**FIGURE 15.05**
***Antibodies Recognize
Specific Epitopes***

Antibodies have variable regions
that can bind to a particular part
of a protein. Secondary antibodies
are antibodies that recognize
another antibody. The specific point
at which an antibody recognizes
its target is called an epitope, even
if the recognition site is another
antibody.

In contrast, **monoclonal antibodies** recognize only one specific epitope of one
specific protein. In practice, an animal, such as a mouse, is injected with the protein
against which antibodies are needed. When antibody production has peaked, a sample
of antibody-secreting B cells is removed from the animal. These are fused to immortal
myeloma cells to give a mixture of many different so-called "hybridoma" cells that
can live indefinitely *in vitro*. The tedious part comes next. Many individual hybridoma
cell lines must be screened to find one that recognizes the target protein. Once found,
the hybridoma is grown in culture to give large amounts of the monoclonal antibody.

The antibody that binds the protein of interest directly is referred to as the **primary antibody**. In some applications, a **secondary antibody** is also used that recognizes the primary antibody (Fig. 15.05). Secondary antibodies may be monoclonal
or polyclonal, and are created by immunizing an animal with antibodies from a different species. The animal used for the secondary antibody must obviously be different from that used for the primary antibody. Secondary antibodies are named after
the two species, for example, goat anti-rabbit is created by immunizing a goat with
rabbit antibodies. So if the primary antibody was created in a mouse, the secondary
antibody would be goat anti-mouse or rabbit anti-mouse. Secondary antibodies are
often tagged to make them easy to detect. Common tags include enzymes that make
colored products or fluorescent dye molecules.

3. Western Blotting of Proteins

Much as a Southern or Northern blot identifies one specific DNA or RNA (see Ch. 5), a
Western blot allows detection of a single protein within a sample of many proteins. First,

---

monoclonal antibody   Antibody produced by a single clonal B cell from a mouse such that it only recognizes one specific epitope on the protein
    of interest
primary antibody   The antibody that binds directly to the protein of interest
secondary antibody   The antibody that binds to the primary antibody that often contains the detection system
Western blot   Detection method in which an antibody is used to identify a specific protein

FIGURE 15.06

Electrophoretic Transfer of Proteins from Gel onto Nitrocellulose

A "sandwich" is assembled to keep the gel in close contact with a nitrocellulose membrane while in a buffer solution. The sandwich consists of the gel (gray) and nitrocellulose (green) between layers of thick paper and a sponge (yellow). The entire stack is squeezed between two solid supports so that none of the layers can move. The "sandwich" is transferred to a large tank filled with buffer to conduct the current. As in SDS-PAGE, the proteins are repelled by the negatively-charged cathode and attracted to the positively-charged anode. As the proteins move out of the gel, they travel into the nitrocellulose where they adhere.

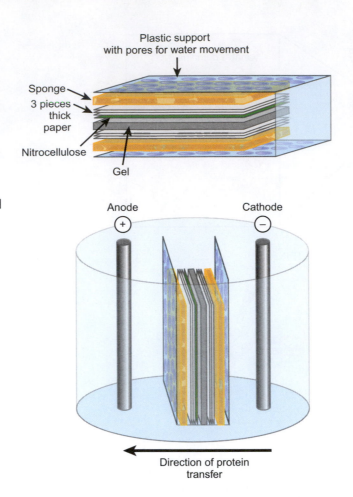

Proteins may be detected by binding to a specific antibody.

the sample of proteins is denatured with heat, SDS, and a reducing agent, then separated by size using SDS-PAGE or 2D SDS-PAGE. The proteins are then electrophoretically transferred to a solid membrane such as nitrocellulose. Electrophoresis moves the proteins from the gel onto the nitrocellulose where the proteins adhere (Fig. 15.06).

To detect a specific protein, an antibody to that protein must be available. An antibody can either be produced for the protein of interest or sometimes purchased commercially. The nitrocellulose membrane itself has many non-specific sites that can bind antibodies, so before adding the primary antibody, these sites must be blocked with a non-specific protein solution such as re-hydrated powdered milk or bovine serum albumin (BSA). The primary antibody is then added to the membrane where it only binds the protein of interest. The antibody protein complex is finally detected using a secondary antibody that has a detection system attached (Fig. 15.07).

4. Isolating Proteins with Chromatography

Chromatography is a general term for techniques that separate mixtures of components by using a mobile phase (liquid or gas) to carry the mixture over a stationary phase (solid or liquid). In **liquid chromatography** of proteins, mixtures of dissolved proteins (mobile phase) are separated into fractions using columns packed with various solid materials (resins).

In **ion exchange chromatography**, the resin is either positively charged (anion exchange) or negatively charged (cation exchange). The ions on the resin bind to oppositely charged proteins in the sample mixture and hold them on the column. Any proteins with the same or neutral charge pass through the column without binding. The

ion exchange chromatography   Technique that separates a mixture of protein based upon their native charge
liquid chromatography   Technique that separates a mixture of proteins through a column containing different solid materials

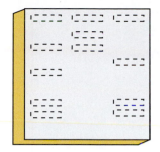

Nitrocellulose
membrane
with attached
proteins

ADD NON-SPECIFIC MILK PROTEINS
AND PRIMARY ANTIBODY

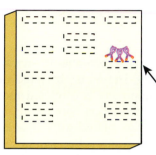

Primary antibody
binds to specific
protein

ADD SECONDARY ANTIBODY
CONJUGATED TO ALKALINE PHOSPHATASE

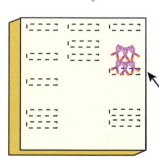

Secondary antibody
binds to
primary antibody

ADD X-PHOS TO DETECT LOCATION
OF SECONDARY ANTIBODY

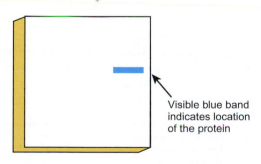

Visible blue band
indicates location
of the protein

■ **FIGURE 15.07**
Western Blot

After a mixture of proteins have
adhered to the nitrocellulose
membrane, one specific protein can
be detected using an antibody. The
antibody is added with a solution
of milk proteins or BSA and
incubated with the nitrocellulose
membrane. The milk proteins
or BSA bind to and block those
regions of the nitrocellulose that do
not have any protein attached. The
primary antibody attaches only to
the protein of interest. A secondary
antibody with alkaline phosphatase
attached binds specifically to the
primary antibody and allows
the single protein band to be
visualized.

FIGURE 15.08
HPLC Chromatogram

Reverse phase HPLC was performed on sesame seed extract. Six peptides (marked with vertical red lines) were identified. Each peptide eluted from the column at different timepoints, and the amount of peptide was determined by recording the absorption at 220 nm. *(Credit: Das, R., Dutta, A., & Bhattacharjee, C. (2012). Preparation of sesame peptide and evaluation of antibacterial activity on typical pathogens. Food Chemistry, 131(4), 1504–1509.)*

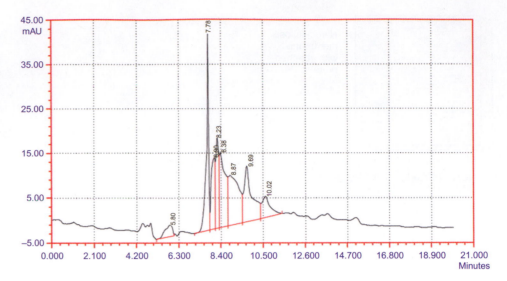

attached proteins can be later removed by changes in pH. **Hydrophobic interaction columns** (HIC) have resins that bind hydrophobic proteins. These may be removed by adjusting the salt concentration. Reverse phase chromatography is a related method since the resin also binds hydrophobic proteins, but the interaction between the resin and hydrophobic proteins is much stronger. The proteins are eluted with organic solvents rather than salts. Finally, affinity chromatography uses resins with a ligand specific for the protein of interest. Consequently, only the protein of interest binds to the resin, while the remaining proteins pass through. Antibodies are used as affinity ligands in order to isolate the protein of interests. The protein of interest is then isolated with treatments that disrupt the ligand:protein binding, such as pH changes or salt. Another example of affinity chromatography is the purification of His-tagged proteins (see below).

One widely-used modification of liquid chromatography is to use high pressure to propel the sample through the column. This technique is known as **high-pressure liquid chromatography (HPLC)**, and since the sample does not rely on gravity to move it, the columns for HPLC can be much longer and more densely packed. This results in much better separation of proteins.

For all chromatography methods, the bound proteins are eluted into multiple fractions. Because proteins absorb light at 280 nm the amount of protein in each sample can be monitored with UV light. As fractions emerge an attached computer graphs the UV absorption values to create a chromatogram (Fig. 15.08). The peaks represent fractions with pure protein.

5. Mass Spectrometry for Protein Identification

Analysis of proteins and peptides by **mass spectrometry** relies on several recently developed techniques that are both extremely accurate and may be automated. The two most important are **matrix-assisted laser desorption-ionization (MALDI)** and **electrospray ionization (ESI).** Mass spectrometry (MS) measures the mass to charge ratio (*m/z*) of ions and allows derivation of the molecular weight. Before MALDI and ESI large heat-labile molecules such as proteins could not be analyzed by mass spectrometry.

In MALDI, gas-phase ions are generated from a solid sample by a pulsed laser. First, the sample protein or peptide is crystallized along with a matrix that absorbs at

Automated mass spectrometry may be used to identify proteins in large numbers of samples.

---

electrospray ionization (ESI)   Type of mass spectrometry in which gas-phase ions are generated from ions in solution
high pressure liquid chromatography (HPLC)   Chromatography technique that separates a mixture of protein through a column containing different solid materials. The mixture of proteins is forced through the column with a pump rather than by gravity
hydrophobic interaction columns (HIC)   Type of chromatography column that contains a resin that binds to hydrophobic proteins
matrix-assisted laser desorption-ionization (MALDI)   Type of mass spectrometry in which gas-phase ions are generated from a solid sample by a pulsed laser
mass spectrometry   Technique for measuring the mass of molecular ions derived from volatilized molecules

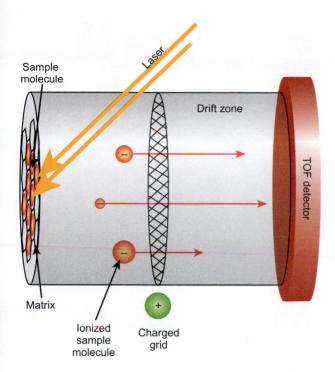

Sample molecule

Laser

Drift zone

TOF detector

Matrix

Ionized sample molecule

Charged grid

■ **FIGURE 15.09**
MALDI/TOF Mass Spectrometer

Mass spectrometry can be used to determine the molecular weight of proteins. The proteins are crystallized in a solid matrix and exposed to a laser, which releases ions from the proteins. These travel along a vacuum tube, passing through a charged grid, which helps separate the ions by size and charge. The time it takes for ions to reach the detector is proportional to the square root of their mass to charge ratio (m/z). The molecular weight of the protein can be determined from this data.

the wavelength of the laser. Matrix materials are usually aromatic acids such as 4-methoxy cinnamic acid. The laser excites the matrix material, which transfers the energy to the crystallized protein. The energy then releases ions, the size and charge of which are unique to each protein. The ions are accelerated by a high voltage electric field and travel in a vacuum through a tube to a detector. The **time-of-flight (TOF)** detector measures the time for an ion to fly from the ion source to the detector (Fig. 15.09). The time-of-flight is proportional to the square root of m/z. Typically, molecular ions up to 100,000 daltons may be measured by MALDI/TOF. However, advances in instrumentation will probably increase this limit substantially in the near future.

The amount of sample needed for MALDI/TOF is now less than a picomole (10^{-12} mole). Mass resolution is 1 in 10,000 or better. Thus, proteins separated by 2D-PAGE may be routinely identified by MALDI/TOF often after preliminary digestion to give a characteristic set of peptides. Post-translational modification of proteins may also be detected by shifts in molecular weight. Phosphate or sugar residues yield characteristic ion fragments and analysis after protease digestion may reveal the location of such groups within the protein.

Electrospray ionization (ESI) refers to the generation of gas-phase ions from ions in solution. A narrow capillary tube allows droplets of liquid to emerge into a strong electrostatic field (Fig. 15.10). The solvent evaporates and the droplet breaks up. Repeated evaporation and splitting of droplets eventually releases separate ions (either with single or multiple charges) that are accelerated towards a mass analyzer by an electric field. Mass analyzers such as quadrupole or ion-trap detectors are normally used with ESI mass spectrometers. The typical range for a singly-charged ion is up to 5,000 daltons, but multiple charges allow heavier ions to be analyzed.

An advantage of ESI is that it can be directly coupled to liquid separation techniques such as capillary electrophoresis or HPLC (high performance liquid chromatography). Also, a parent ion can be isolated and fragmented into daughter ions, so allowing more detailed analysis of molecules. This is known as **tandem mass spectrometry (MS/MS)**. It allows two parent ions with the same mass (e.g., two peptides with the same amino acid composition but different sequence) to be distinguished.

tandem mass spectrometry (MS/MS)   Two successive rounds of mass spectrometry in which a parent ion is first isolated and then fragmented into daughter ions for more detailed analysis
time-of-flight (TOF)   Type of mass spectrometry detector that measures the time for an ion to fly from the ion source to the detector

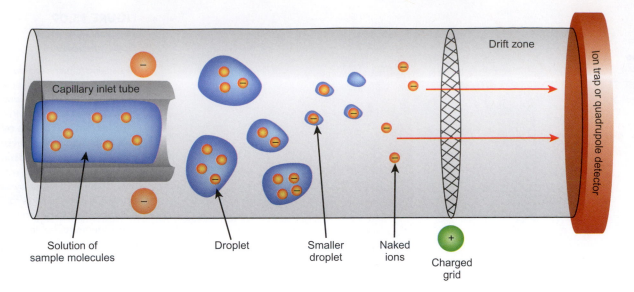

FIGURE 15.10
Electrospray Ionization Mass Spectrometer

ESI mass spectrometry uses a liquid sample of the protein held in a capillary tube. After exposure to a strong electrostatic field, small droplets are released from the end of the capillary tube. A flow of heated gas within the drift zone helps evaporate the solvent and release small charged ions. The charged ions vary in size and charge and the pattern of ions produced is unique to each protein. The ions are further separated by size using a charged grid to either impede or help the flow toward the detector.

6. Protein-Tagging Systems

Tagging of proteins with another easily recognized molecule allows a researcher to purify or identify the protein of interest without creating a specific antibody for it. Protein tagging is usually done genetically, that is, the DNA encoding the protein is engineered to add an extra segment that codes for the tag. The gene must therefore first be cloned and carried on a suitable vector. The hybrid gene is then expressed in a suitable host organism, such as *E. coli* or cultured mammalian cells, and the protein is purified by a method that binds to and isolates the tag sequence. Since the protein of interest is synthesized attached to the tag, it is purified along with the tag molecule. In many instances, the tag portion may then be removed. Alternatively, a protein array can by constructed by attaching the tagged protein to a chip via the tag (see below).

The first protein tag to be widely used was the **polyhistidine tag (His tag)**, consisting of six tandem histidine residues. This may be added to the target protein at either the amino or carboxy terminus. The **His tag** binds very tightly to nickel ions; therefore, His-tagged proteins are purified on a column to which Ni^{2+} ions are attached by a metal chelator (metal-chelate column chromatography) (Fig. 15.11).

Other widely-used short tags are the **FLAG®** and "Strep" tags. The FLAG tag is a short peptide (AspTyrLysAspAspAspAspLys) that is bound by a specific antibody. Anti-FLAG® antibody is available commercially and may be attached to a suitable resin for use in column purification. The "Strep" tag is a 10 amino acid peptide that mimics the 3D structure of biotin. It is bound by the biotin-binding proteins **avidin** or **streptavidin** and was originally selected from a random oligopeptide library for its ability to bind to streptavidin. Improved and modified tags based on FLAG have recently been used to investigate protein interactions (see Focus on Relevant Research).

> A convenient way to identify and purify proteins is to tag them using a genetic approach.

> The His tag allows proteins to be purified by binding to a resin containing nickel ions.

avidin   Protein from egg-white that binds biotin very tightly
FLAG tag   A short peptide tag (AspTyrLysAspAspAspAspLys) that is bound by a specific anti-FLAG antibody that may be attached to a resin for use in column purification of proteins
His tag   Six tandem histidine residues that are fused to proteins so allowing purification by binding to nickel ions that are attached to a solid support. Also known as polyhistidine tag
polyhistidine tag (His tag)   Six tandem histidine residues that are fused to proteins so allowing purification by binding to nickel ions that are attached to a solid support
streptavidin   Protein from *Streptococcus* that binds biotin very tightly

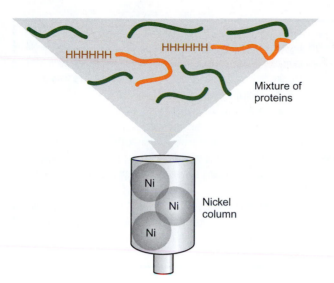

Mixture of proteins

Nickel column

PROTEINS POURED THROUGH COLUMN

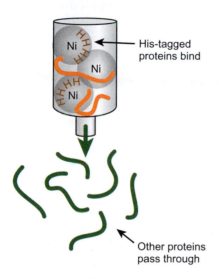

His-tagged proteins bind

Other proteins pass through

ELUTE WITH HISTIDINE OR IMIDAZOLE

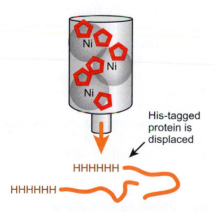

His-tagged protein is displaced

HHHHHH

HHHHHH

■ **FIGURE 15.11**
Nickel Purification of His6-Tagged Protein

To isolate a pure sample of one specific protein, the gene for the protein is linked to a tag sequence. In this example, the tag sequence encodes six histidines in a row. This engineered gene is then expressed in either bacteria or mammalian cells. The cells are harvested and lysed to release all the proteins. To isolate the tagged protein, the mixture of proteins is added to a nickel column. The nickel-coated beads bind the histidine tag, allowing all the other proteins to pass through the column. Next, a solution containing histidine or imidazole is added to the column. The histidine or imidazole binds to the nickel-coated beads thus releasing the histidine-tagged protein.

FOCUS ON RELEVANT RESEARCH

The Oct4 protein (also called Pou5f1) is a key transcription factor in the development and reprogramming of embryonic stem cells into differentiated adult cells, such as liver cells, kidney cells, red blood cells, etc. A variety of other proteins are associated with Oct4 and take part in stem cell reprogramming. About half of these associated proteins are transcriptionally regulated by Oct4 itself or other stem cell transcription factors. About a third of these associated proteins show major changes in expression during differentiation of the stem cells.

These authors wished to define the set of proteins associated with Oct4 more accurately. They therefore used affinity purification, combined with identification by mass spectrometry. An improved protein tag (FTAP) was assembled that contains three FLAG motifs plus a calmodulin-binding peptide. This was inserted at the C-terminus of the Oct4 coding sequence. The tagged Oct4 was expressed under control of its natural promoter. The resulting Oct4 interactome was much cleaner and better defined than in previous work. Most Oct4 associated proteins are essential to early development since mutations in these are lethal. In addition, mutations in some of the human equivalents are associated with cancer or developmental disorders.

6.1. Full-Length Proteins Used as Fusion Tags

> Proteins may be fused to a second protein chosen for its convenient properties.

Longer tags, consisting of entire proteins, are sometimes used to tag a specific target protein. The tag protein coding sequence is normally placed 5' to the gene of interest in order to ensure good translation initiation. Three of the most popular are **protein A** from *Staphylococcus*, **glutathione-S-transferase (GST)** from *Schistosoma japonicum*, and **maltose-binding protein (MBP)** from *E. coli*. These three proteins generally give fusion protein products that are stable, soluble, and behave well during purification.

The fusion proteins are purified on columns containing materials that specifically bind the tag protein. For protein A fusion proteins, a specific commercially-available antibody binds protein A. After purification on an antibody column, the fusion protein is eluted at pH 3. Glutathione-S-transferase binds to glutathione, a tripeptide. GST fusions are bound by glutathione-agarose columns and eluted with free glutathione. Maltose-binding protein binds maltose and the maltose polymer, amylose. An amylose column purifies proteins with an MBP tag. The bound fusion protein is then eluted with maltose, which displaces the protein from the amylose column, and releases purified fusion protein of interest (Fig. 15.12).

The pMAL vectors (New England Biolabs Inc., MA) are used to create MBP fusion proteins (Fig. 15.13). These vectors carry the *malE* gene of *E. coli*, encoding MBP. A spacer sequence encoding 10 Asn residues lies between *malE* gene and the **polylinker**, which is used to add the gene encoding the protein of interest. The spacer insulates the maltose-binding protein from the protein of interest. Adjoining the spacer is the recognition site, (Ile Glu (or Asp) Gly Arg), for a highly-specific protease of the blood clotting system, called Factor Xa. After purification, the fusion protein is treated with Factor Xa and the two proteins are separated. Factor Xa itself must then be removed by binding to benzamidine-agarose. The pMAL vector contains a strong *tac* promoter to transcribe the protein fusion. Variants exist with or without a signal sequence thus allowing for either secretion or cytoplasmic expression.

Sometimes, fusion proteins (or for that matter unmodified proteins) are degraded by host-cell proteases during purification. Protease inhibitors may be included, or in the case of a bacterial host such as *E. coli*, protease-deficient mutants may be used. Bacteria defective in several protease genes are useful, even though not all *E. coli* protease genes can be inactivated since some are essential for survival.

---

glutathione-S-transferase (GST)   Enzyme that binds to the tripeptide glutathione. GST is often used in making fusion proteins
maltose-binding protein (MBP)   Protein of *E. coli* that binds maltose during transport. MBP is often used in making fusion proteins
polylinker   A stretch of artificially synthesized DNA that contains cut sites for seven or eight widely used restriction enzymes
protein A   Antibody-binding protein from *Staphylococcus* that is often used in making fusion proteins

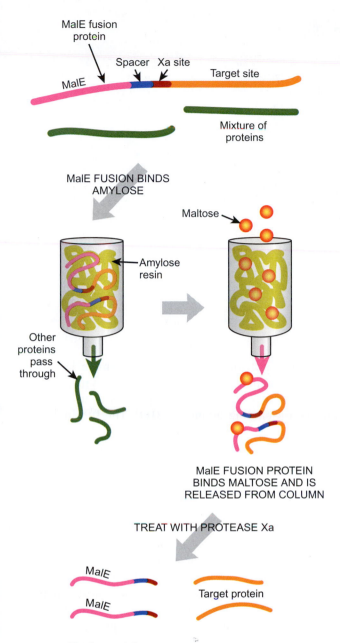

MalE FUSION BINDS
AMYLOSE

Amylose
resin

Maltose

Other
proteins
pass
through

MalE FUSION PROTEIN
BINDS MALTOSE AND IS
RELEASED FROM COLUMN

TREAT WITH PROTEASE Xa

MalE

MalE

Target protein

■- **FIGURE 15.12**

***Protein Isolation Using
Fusions to Maltose-Binding
Protein***

In order to isolate one specific
protein from a mixture, the gene
for maltose-binding protein can be
genetically fused to the protein of
interest. The fusion gene can then
be expressed in an organism such
as *E. coli*. The bacteria are lysed,
and the mixture of proteins (top
of figure) is isolated. The protein
of interest has three new regions
attached to it: the entire maltose-
binding protein, a cleavage site
for Factor Xa, and a small spacer
region. The protein mixture is
added to an amylose column,
where only the fusion protein binds
via the maltose-binding protein.
Adding free maltose elutes the
fusion protein. The eluted protein
is treated with Factor Xa, which
cleaves the maltose-binding protein
section from the protein of interest.

6.2. Self-Cleavable Intein Tags

A recent improvement in tagging systems eliminates the protease cleavage and purification steps. Instead of using a protease plus recognition site, the fusion protein
cleaves itself after purification of the target protein. This approach is based on the
properties of **inteins**, self-splicing intervening sequences that are found in proteins
(see Ch. 12). The advantages are a reduction in the number of steps required for purification and the avoidance of expensive proteases that may sometimes cleave the protein of interest at other sites.

The Intein Mediated Purification with Affinity Chitin-binding Tag (IMPACT™)
system from New England Biolabs depends on the self-splicing of an intein originally
from the *VMA1* gene of *Saccharomyces cerevisiae*. This intein has been modified to
undergo self-cleavage only at its N-terminus. This is triggered at low temperatures by
thiol reagents such as dithiothreitol (DTT). At the C-terminal end of the intein is a
small chitin-binding domain (CBD) from the carboxy terminus of the chitinase A1 gene

Tags based on inteins are
convenient because they can be
designed to cleave themselves
when no longer required.

intein   Self-splicing intervening sequence that is found in a protein

FIGURE 15.13
Maltose-Binding Protein Fusion Vector

In order to manufacture an MBP-tagged protein, the target gene is cloned into the pMAL vector. The vector has a polylinker downstream from the *malE* gene for inserting the gene of interest. For correct translation, the target gene must be cloned in frame with the *malE* gene. The entire region between the strong *tac* promoter and the strong terminator produces a fusion protein consisting of MBP (MalE protein) linked to the target protein via a short spacer and a protease cleavage site. The *lacZα* sequence allows for blue/white color screening to detect DNA insertion (see Ch. 7).

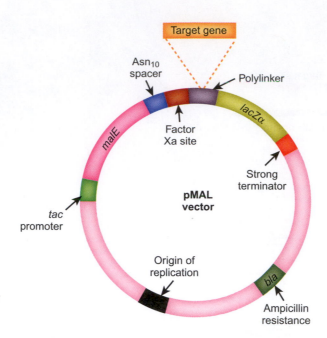

of *Bacillus circulans*. (Chitin is a structural polymer of N-acetyl glucosamine. It is found in most fungi and is the major structural component of the exoskeletons of insects.)

The gene encoding the target protein is inserted into a polylinker upstream of the gene encoding the intein plus chitin-binding domain (CBD). The fusion protein is purified by binding to a chitin column. While the fusion protein is still attached to the column, intein self-cleavage is induced by incubation at 4°C with DTT. The target protein is released while the intein plus the chitin-binding domain remain bound to the column (Fig. 15.14).

7. Selection by Phage Display

Most procedures in molecular biology deal with either genetic information (i.e., DNA or RNA) or gene products (usually proteins). Display protocols are designed to identify both the gene and its encoded protein at the same time. The most common of these is the **phage display** technique. Here, a full-length protein or a shorter peptide is fused to a coat protein of a bacteriophage so as to be displayed on the outer surface of the virus particle. Meanwhile, the DNA encoding the fusion protein is carried inside the bacteriophage (Fig. 15.15).

Filamentous phage M13 is a popular choice for phage display but other bacteriophage, such as Lambda, T7, and T4 are also used. M13 is preferred since it is non-lytic and does not destroy the host bacteria during phage production. Instead, phage particles are secreted through the bacterial cell envelope. The absence of cell debris simplifies purification of the phage. M13 structural proteins such as gene III and gene VIII have been used as expression platforms for the proteins of interest with gene III protein most popular. There are about 2,500 copies of the major coat protein (gene VIII protein) on the phage surface but only five copies of the minor coat protein (gene III protein). Having fewer copies of the displayed peptide on the surface of the phage avoids artifacts due to simultaneous binding of multiple polypeptides. Since the N-terminus of the M13 coat proteins is external and the C-terminal region interacts with the DNA inside the phage particle, the displayed peptide must be attached to the N-terminus of gene III.

The DNA encoding the protein or peptide is fused to the gene for the phage coat protein by a PCR-based technique. The linear PCR product is amplified and

> Phage display allows us to find a gene by identifying the protein it encodes.

> The proteins to be screened are fused to virus proteins so that they appear on the outside surface of the virus particle.

phage display   Fusion of a protein or peptide to the coat protein of a bacteriophage whose genome also carries the cloned gene encoding the protein. The protein is displayed on the outside of the virus particle and the corresponding gene is carried on the inside

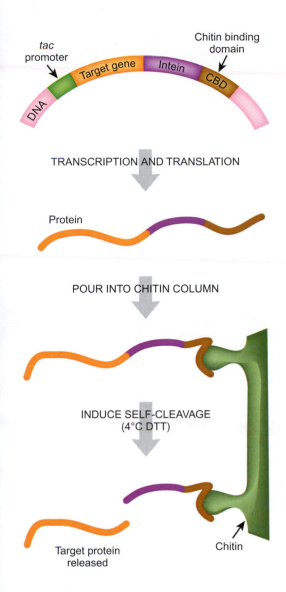

tac promoter

Target gene

Intein

Chitin binding domain

CBD

DNA

TRANSCRIPTION AND TRANSLATION

Protein

POUR INTO CHITIN COLUMN

INDUCE SELF-CLEAVAGE
(4°C DTT)

Target protein released

Chitin

■ **FIGURE 15.14**
Intein-Mediated Purification System

Inteins that can self-cleave at the N-terminus allow specific proteins to be purified and cleaved from a fusion protein in one step. First, the target gene is cloned upstream of the intein sequence and a chitin-binding domain. The fusion gene is transcribed and translated in bacteria. The bacteria are lysed and release a mixture of proteins that are passed through a chitin column. The proteins with the chitin-binding domain bind to the chitin and the remaining proteins pass through. Adding DTT and incubating at 4°C activates the intein to cleave itself from the target protein, which is therefore released from the column.

circularized to give a viral genome that is transformed into *E. coli* cells. The phages that are produced are slightly longer than the original wild-type M13; nonetheless, they are packaged by extending the filamentous protein coat. The fusion proteins are expressed and the intruding peptides displayed on the surface of the M13 virus particles.

Many proteins have regions that bind to other proteins. These regions can be large or small, but usually a few amino acids are critical for the two proteins to interact properly. In order to identify the peptide sequence these binding sites recognize, **phage display libraries** are constructed. These consist of a large number of modified phages displaying a library of different peptide sequences. The first such libraries consisted of large collections of short random peptides. They were screened to find peptides that bind to specific molecules (the "target"), such as a particular antibody, enzyme, or cell-surface receptor. The peptide of interest is found by a selection procedure referred to as **biopanning**. The phage display library is incubated with target molecules that are attached to a solid support (beads or membranes, etc.). Unbound phage is washed away. Bound phage is eluted and amplified by re-infecting cells of *E. coli*. Several cycles of binding and amplification will enrich for the phage that carries the peptide that binds most tightly to the target. Finally, individual clones are characterized by DNA sequencing (Fig. 15.16).

Biopanning is a procedure in which a mixture of bacteriophage particles with different proteins displayed is screened for binding to an antibody or other specific binding protein.

biopanning Method of screening a phage display library for a desired peptide of interest by binding to a bait molecule attached to a solid support
phage display library Collection of a large number of modified phages displaying different peptide or protein sequences

FIGURE 15.15
Principle of Phage Display

In order to display a peptide on the surface of a bacteriophage, the DNA sequence encoding the peptide must be fused to the gene for a bacteriophage coat protein. In this example, the chosen coat protein is encoded by gene III of phage M13. Here, the N-terminal portion will be on the outside of the phage particle, whereas the C-terminus will be on the inside. Therefore, the peptide must be fused in frame at the N-terminus to be displayed on the outside of the phage.

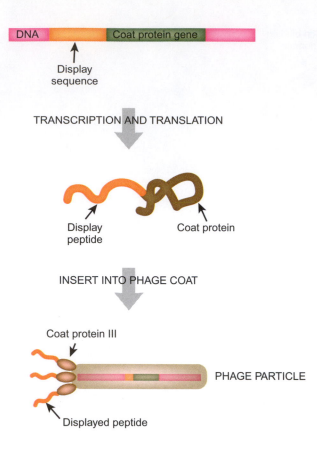

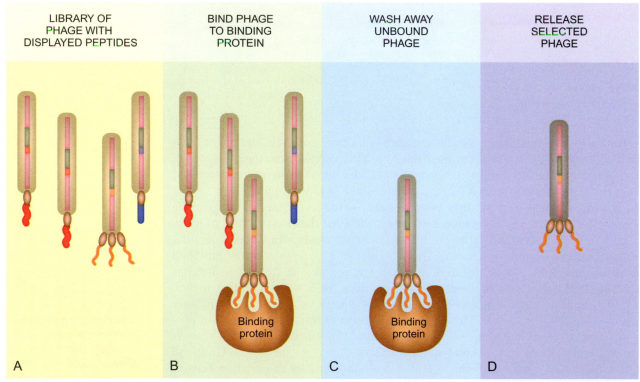

FIGURE 15.16
Biopanning to Screen a Phage Display Library

Biopanning is used to isolate peptides that bind to a specific target protein, which is usually attached to a solid support such as a membrane or column. The phage display library (A) is added to the binding protein (B). Those phage that display peptides that bind to the target protein will be retained (C), but the others are washed away. The phage that does recognize the binding protein can then be released, isolated, and purified.

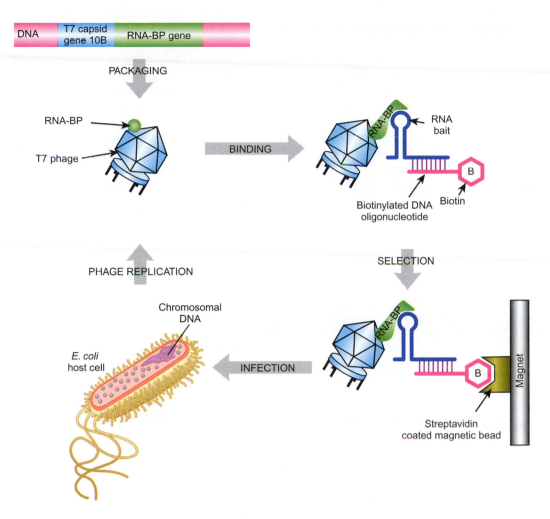

PACKAGING

PHAGE REPLICATION

BINDING

RNA-BP

T7 phage

RNA-BP

RNA bait

Biotinylated DNA oligonucleotide

Biotin

B

SELECTION

Chromosomal DNA

E. coli host cell

INFECTION

RNA-BP

B

Magnet

Streptavidin coated magnetic bead

Several commercially-available peptide libraries are marketed by New England Biolabs under the trade name Ph.D. (for *Phage Display*!). The Ph.D.-7 library consists of 2.0×10^9 random heptapeptide clones. It probably contains most of the theoretically possible amino acid heptamers (of which there are $20^7 =$ approximately 1.3×10^9). In contrast, the Ph.D.-12 library, also with 2.0×10^9 independent clones, only represents a small fraction of the $20^{12} = 4.1 \times 10^{15}$ 12-mers.

Full-length proteins can also be fused to phage coat proteins to produce a full-length phage display library. In principle, a gene library from any organism could be converted into a phage display library by insertion into a suitable phage coat protein gene. M13 is not practical for this type of library since the insert must be positioned between the signal sequence, which is needed for secretion and phage assembly, and the N-terminus of the coat protein. Therefore, both ends of the insert must thus be in frame, and in addition, no stop codons must be present in the insert. In contrast, in T7 the C-terminal region of the coat protein is exposed on the outside. Using C-terminal coat protein fusions avoids the above problems and allows complete coding sequences to be inserted. Using the T7Select™ system from Novagen, several cDNA libraries have now been biopanned to find proteins that bind to some chosen target molecule. For example, phage that display RNA-binding proteins have been isolated using RNA anchored to a solid support as bait (Fig. 15.17).

Other display systems use whole bacterial cells to carry the protein of interest. DNA sequences encoding polypeptides to be screened can be fused to the flagellin or pilin genes of *E. coli*. The polypeptide library is then exposed on the cell surface attached to either the flagella or the pili. The phage display libraries are generally more convenient but one advantage of using bacteria is that they are large enough for a fluorescence-activated cell sorter (FACS) to sort the cells, provided that the peptide target is labeled with a fluorescent dye.

FIGURE 15.17
Biopanning for RNA-Binding Proteins

To identify RNA-binding proteins (RNA-BP), a bait RNA (blue) is used that is linked to a biotinylated oligonucleotide. The bait RNA is incubated with a full-length phage display library. In this example, T7 gene 10B DNA is fused to a gene library that includes RNA-binding protein genes (shown here in green). Those phage that express the full-length RNA-binding protein on the outside will bind to the RNA bait. This in turn is bound via the biotinylated oligonucleotide to magnetic beads coated with streptavidin. The captured phage is eluted from the magnetic beads by free biotin and is used to infect *E. coli*. Isolating the T7 DNA and sequencing the insert identifies the gene for the RNA-binding protein.

DNA | T7 capsid gene 10B | RNA-BP gene

8. Protein Interactions: The Yeast Two-Hybrid System

Many proteins recognize and bind to other proteins. The total of all protein-protein interactions is sometimes referred to as the **protein interactome** by those enthusiastic about "omics" terminology. Mass screening of such interactions has proven possible by means of the **"two-hybrid"** system. Interactome analysis is based on the idea of guilt by association. It is assumed that the binding of a novel protein to one that is well characterized may provide some hint as to function of the unknown partner.

Two-hybrid analysis depends on the modular structure of transcriptional activator proteins. Many of these proteins consist of two domains, a DNA-binding domain and an activator domain. The DNA-binding domain (DBD) recognizes a specific sequence in the DNA upstream of a promoter and the activator domain (AD) stimulates transcription by binding to RNA polymerase (Fig. 15.18). Provided that the two

FIGURE 15.18
Principle of Two-Hybrid Analysis

A) Transcription of a yeast gene involves activation of RNA polymerase by a transcription factor with two different domains. The DBD (purple) recognizes upstream regulatory sites, and the AD (red) activates RNA polymerase to start transcription of the reporter gene. For two-hybrid analysis, two proteins (bait and prey) are fused separately to the DBD and AD domains of the transcription factor. The bait protein is joined to the DBD and the prey protein to the AD. B) Here, the bait protein and prey protein do not interact and the reporter gene is not turned on. C) Here, the bait binds the prey, thus bringing the transcription factor halves together. The complex activates the RNA polymerase and the reporter gene is expressed.

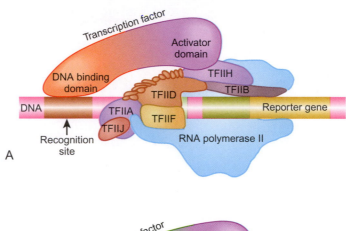

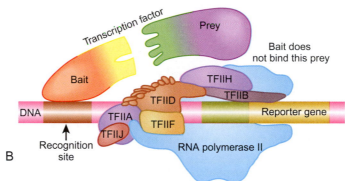

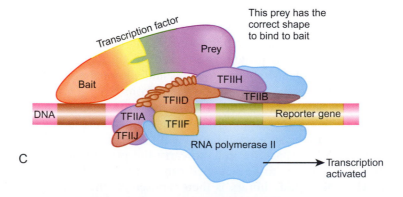

protein interactome   The total of all the protein-protein interactions in a particular cell or organism
two-hybrid system   Method of screening for protein-protein interactions that uses fusions of the proteins being investigated to the two separate domains of a transcriptional activator protein

domains interact, they will activate transcription. It is not usually necessary for the two domains to be covalently joined to form a single protein.

In the two-hybrid system, both the DBD domain and the AD domain are fused to two other proteins (X and Y). These two hybrid proteins are referred to as the "**bait**" (DBD-X) and the "**prey**" (AD-Y). If the bait captures the prey—that is, if proteins X and Y interact—a complex will form and the gene will be activated. A convenient reporter gene is used to monitor for a successful interaction (Fig. 15.18).

Two-hybrid analysis was developed in yeast and is being used to generate a complete list of interactions between all 6,000 or so yeast proteins. It is thus necessary to examine 6,000 × 6,000 combinations. To examine these potential interactions, each open reading frame in the yeast genome was amplified by PCR and cloned into two separate vectors, one carrying the DBD domain and one with the AD domain. Thus, each yeast protein is tested as both bait and prey. The vectors are designed to give in-frame gene fusions of each ORF with the DBD domain and AD domains of a suitable transcriptional activator, such as GAL4 (Fig. 15.19). One vector has a multiple cloning site downstream of the GAL4-DBD and thus gives a 3'-fusion of GAL4-DBD to protein X (GAL4-DBD-X). The other vector has its MCS upstream of the GAL4-AD and gives a 5'-fusion of GAL4-AD and protein Y (Y-GAL4-AD).

The bait and prey fusion plasmids are transformed into yeast cells of different mating types. This results in two sets of approximately 6000 transformants. All possible matings are carried out between the two sets using a laboratory robot to manipulate the colonies. When the two yeasts mate, the diploid cell will have a bait plasmid and a prey plasmid. If the two fusion proteins X and Y interact, the reporter gene is switched on. In yeast, the *HIS3* or *URA3* genes are usually used. If the reporter gene is not activated, the yeast strain cannot grow unless provided with histidine or uracil, respectively. If the reporter gene is turned on the cells can grow on medium without histidine or uracil. Thus, the diploid cells from the 6,000 × 6,000 matings are selected on medium lacking the chosen nutrient (Fig. 15.20). Only those combinations where proteins X and Y interact yield viable colonies.

The original two-hybrid system has several limitations. For example, it relies on proteins interacting within the nucleus. Membrane proteins often misfold when localized in the nucleus. Conversely, other proteins are only correctly modified when present in the cytoplasm. Toxic effects and steric problems with very large proteins may also cause some interactions to be missed. Furthermore, many proteins bind RNA and/or rely on small molecules to alter their conformation so promoting protein-protein interactions.

A variety of modified two-hybrid systems have been developed to deal with these issues. One of the most interesting is the RNA three-hybrid system (Fig. 15.21). In this case, the bait and prey proteins (DBD-X and Y-AD) are brought together by an intervening RNA molecule that is bound by both X and Y. This can be used to screen for genes encoding RNA-binding proteins.

Libraries of genes from other organisms can also be used for two-hybrid screening provided they are expressed in yeast. In addition, a two-hybrid screening system (BacterioMatch™ from Stratagene Corporation) has recently been devised for use in the bacterium *E. coli*. This uses two tandem reporter genes, *bla* and *lacZ*, that encode β-lactamase (ampicillin resistance) and β-galactosidase respectively.

> The test proteins (bait and prey) are fused separately to the two halves of a transcription factor. If the bait and prey bind each other they will reassemble the transcription factor and activate the genes it controls.

prey The fusion between the activator domain of a transcriptional activator protein and another protein as used in two-hybrid screening
bait The fusion between the DNA-binding domain of a transcriptional activator protein and another protein as used in two-hybrid screening

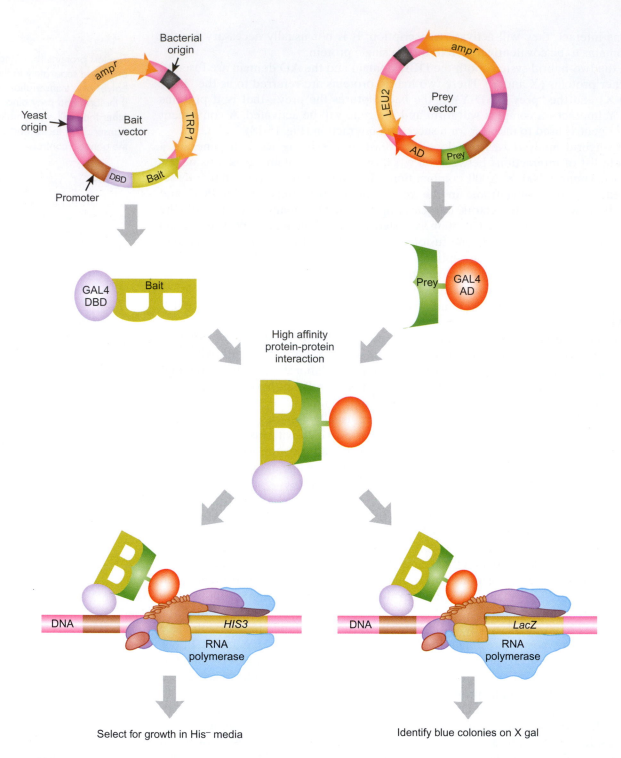

FIGURE 15.19
Vectors for Two-Hybrid Analysis

Two different vectors are necessary for two-hybrid analysis. The bait vector has the coding regions for the DBD and for the bait protein. The prey vector has the coding regions for the AD and for the prey protein. These two different constructs are expressed in the same yeast cell. If the bait and prey interact, the reporter gene is expressed. Two possible reporter systems are shown here. If the yeast *His3* gene is used, yeast expressing the reporter gene will be able to make histidine and hence to grow in media without histidine provided. If the *lacZ* gene from *E. coli* is used, the yeast cells will turn blue on plates containing X-gal.

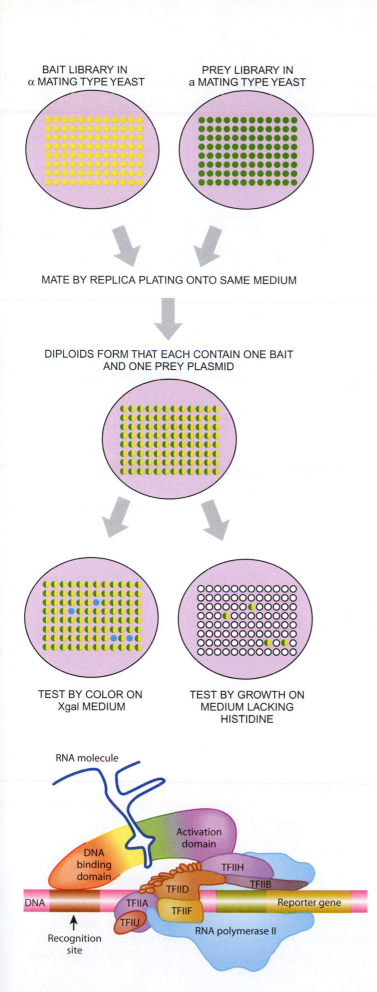

BAIT LIBRARY IN
α MATING TYPE YEAST

PREY LIBRARY IN
a MATING TYPE YEAST

MATE BY REPLICA PLATING ONTO SAME MEDIUM

DIPLOIDS FORM THAT EACH CONTAIN ONE BAIT
AND ONE PREY PLASMID

TEST BY COLOR ON
Xgal MEDIUM

TEST BY GROWTH ON
MEDIUM LACKING
HISTIDINE

RNA molecule

Activation
domain

DNA
binding
domain

TFIIH

TFIID

TFIIB

DNA

TFIIA

TFIIF

Reporter gene

TFIIJ

RNA polymerase II

Recognition
site

■ FIGURE 15.20
Two-Hybrid Analysis: Mass Screening by Mating

To identify all possible protein interactions using the two-hybrid system, haploid α yeasts are transformed with the bait library, and haploid a yeasts are transformed with the prey library. When the two yeast types are mated with each other, the diploid cells will each contain a single bait fusion protein and a single prey fusion protein. If the two proteins interact, they activate the reporter gene, which allows the yeast to grow on media lacking histidine (yeast *His3* gene) or turns the yeast cells blue when grown on X-gal media (*lacZ* gene from *E. coli*). This process can be done for all 6,000 yeast proteins using automated techniques.

■ FIGURE 15.21
RNA Three-Hybrid System

The RNA three-hybrid system identifies proteins that interact through an intermediary RNA molecule. Two fusion proteins are used, one (yellow/red) includes the DBD and the other (green/purple) includes the AD of the transcription factor. When these two fusion proteins interact via an RNA molecule, they activate the reporter gene.

Box 15.1 Tethering Technology Allows Small Molecule Screening

Another problem for proteomics researchers is to isolate small molecules that bind to a given protein. In particular, a protein may have been chosen as a potential drug or antibiotic target. Consequently small molecules that bind to and inhibit the protein of interest are wanted. A library of small chemical molecules will be synthesized and screened for those that bind to the protein. This may be achieved by a variety of tedious procedures. However, a new approach known as tethering technology, developed by Sunesis Corporation, allows greatly improved screening. This approach involves modifying the small molecule library by adding a chemical group that will act as a tether. The target protein is immobilized and also modified with a tethering group. The two tethering groups are chosen so that they will form a cross link if they come into close contact. In the diagram shown (Fig. 15.22), a disulfide linkage is formed between the two tethering groups. The small molecules carry a short side chain that ends in a masked sulfhydryl group, and the protein has a cysteine residue engineered into it whose sulfhydryl group is exposed on the surface close to the binding pocket. When a small molecule fits the binding site on the protein, a cross link can form between the tethering groups and the small molecule is trapped. The small molecule is later released for identification.

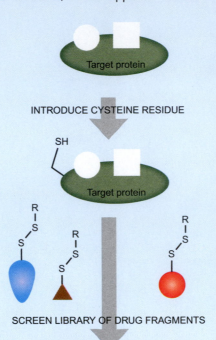

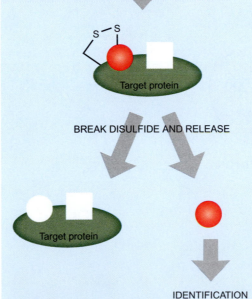

FIGURE 15.22
Tethering Technology

A library of small molecules is modified by addition of a short side chain carrying a blocked sulfhydryl group. The target protein has a cysteine situated so as to provide another sulfhydryl group close to the binding site. The protein is immobilized and treated with the small molecule library. After binding of a small molecule to the protein, the tethering groups react to form a disulfide linkage. The other small molecules are rinsed away. The trapped small molecules are then released for identification.

9. Protein Interaction by Co-Immunoprecipitation

In mammalian cells, protein interactions can be identified by **co-immunoprecipitation** (Fig. 15.23). The gene for the protein of interest is first expressed in mammalian cells and then isolated from the cytoplasm using antibodies. If no antibody is available for the protein of interest, the gene can be tagged with the FLAG peptide (or some other convenient tag). In this scenario, the antibody to the FLAG tag is used to isolate the

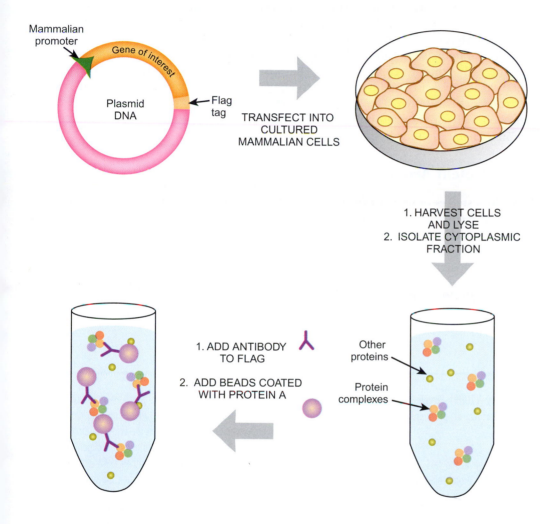

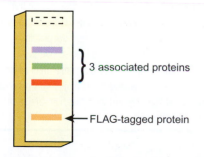

■**FIGURE 15.23**
*Principle of
Co-Immunoprecipitation*

To determine how many other proteins bind to a target protein, they are isolated by precipitating them together using an antibody. An antibody specific to the target protein is needed. If no specific antibody is available, a tag (such as FLAG) is added to the coding sequence. The target protein (orange) is cloned behind a mammalian promoter and expressed in mammalian cells where it will bind some other proteins (red, green, purple). The cytoplasmic fraction is isolated. Antibody to the target protein is added and the complexes are isolated by binding the antibody to beads coated with protein A. The beads are spun down. The components are then separated by SDS-PAGE to identify the number and size of the other proteins that bond to the protein of interest.

| **co-immunoprecipitation** | Method of identifying protein-protein interaction by using antibodies to one of the proteins |

protein from the remaining cellular contents. The protein is isolated under conditions in which it stays associated with its intracellular binding partners. Protein A from *Staphylococcus* binds tightly to antibodies and immobilized protein A is therefore used to isolate the antibodies plus any attached proteins. The protein complexes are then separated by electrophoresis on an SDS-PAGE gel to see how many individual proteins are associated in the complex. The identity of the associated proteins can be determined using such techniques as mass spectroscopy or protein sequencing.

Co-immunoprecipitation can confirm protein interactions detected by the two-hybrid system (Fig. 15.24). First, the two proteins of interest are genetically linked to two different tags, such as the FLAG and His6 tags. The two constructs are both expressed in the same cultured mammalian cells. Antibody to one of the tags is added to the cell extract to capture one of the proteins with its binding partners. The antibody complex is isolated by binding to beads coated with protein A, and the fraction is run on SDS-PAGE. The gel is transferred to nitrocellulose and the membrane is probed with separate antibodies to each tag. If the two proteins of interest interact in mammalian cells, then both proteins/tags will be present on the Western blot.

> If two proteins are associated in the cell and one is precipitated by an antibody, the other should accompany it.

10. Protein Arrays

Previous studies of proteins generally examined a single protein at a time. With the recent sequencing of whole genomes, proteome analysis has turned to methods that allow simultaneous monitoring of multiple proteins. Microarrays have been used for DNA for some time, but the variable structures and properties of proteins made such an array approach more difficult. Nonetheless, new technologies have been developed that allow high-throughput analysis of proteins. As a result, **protein microarrays** have recently become available for proteome analysis. Genuine protein arrays should not be confused with the use of DNA microarrays to survey DNA-binding proteins (see Focus on Relevant Research).

> Protein arrays can be constructed if all the proteins carry tags that allow binding to the array support.

FOCUS ON RELEVANT RESEARCH

Kerschgens J, Egener-Kuhn T, and Mermod N. (2009) Protein-binding microarrays: probing disease markers at the interface of proteomics and genomics. Trends in Mol. Med. 15: 352–358.

In contrast to the protein arrays described in the text, this associated review article describes the latest technology for protein-binding arrays. These are combination arrays that have DNA sequences bound to the surface in an ordered manner, which are then probed with various proteins. This research tool identifies DNA-binding proteins, such as transcription factors, which this paper defines as the genome-proteome interface. These arrays are relatively simple, but recent advances have created new approaches to using these arrays. First, a method called MITOMI, for mechanically-induced trapping of molecular interactions, uses microfluidic reaction chambers to reduce the reaction volume. The flow of fluid through the chambers is controlled by pumps. Small volumes ensure that there is less diffusion, and reactants are not diluted, which can stabilize weaker interactions between the protein and its binding site on the array. Another method of assessing DNA-protein binding is to use aptamers rather than oligonucleotides or double-stranded DNA on the grid. An aptamer is a synthetic RNA with sufficient complementary nucleotides so that it forms a 3D shape. When the array is probed with a set of proteins, only the proteins with a complementary 3D shape bind specifically. This provides a very useful method to diagnose infections. For example, an aptamer that specifically recognizes HIV or the flu virus could be used in a clinic to screen patients.

Other potential uses for protein-binding microarrays would be for screening potential drugs. The array in this example consists of a specific protein/RNA binding pair attached to the membrane. Each spot is then incubated with a different drug, and then assessed to see if the drug disrupted the RNA/protein binding. For example, hepatitis C virus has an RNA genome that attaches to the endoplasmic reticulum (ER) of an infected cell in order to be replicated. This RNA-protein interaction was mimicked *in vitro* by attaching the ER protein to a solid support in an array and then adding the hepatitis C RNA genome. Different possible drugs were added to each of the wells, and the binding of the RNA genome to the protein was assessed. Several new drug leads were discovered that disrupted the interaction. One of these was found to inhibit hepatitis C replication *in vivo*.

protein microarray Microarray of immobilized proteins used for proteome analysis and normally screened by fluorescent or radioactive labeling

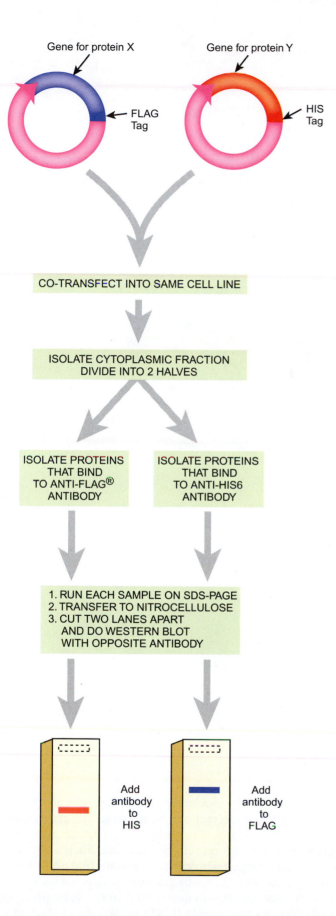

■ **FIGURE 15.24**
Confirmation of Protein Interaction using Co-Immunoprecipitation

To determine if protein X and Y interact in cultured mammalian cells, they must be fused to two different tag sequences (His6 tag and FLAG tag in this example). The two expression plasmids are transfected into the same mammalian cells. The cytoplasmic protein fraction is isolated and divided into two samples. Antibody to the FLAG tag is added to one sample and antibody to the His6 tag is added to the other. The antibody/protein complexes are isolated using protein A beads. Each sample is then tested for the presence of the other protein by antibody directed against its tag. That is, the proteins precipitated along with protein X-FLAG are tested for the presence of protein Y. And conversely the proteins precipitated along with protein Y-His6 are tested for protein X.

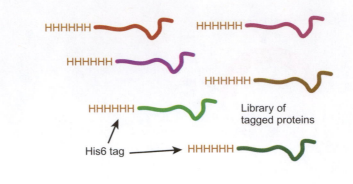

FIGURE 15.25
Protein Microarray Principle

To assemble a protein microarray, a library of His-tagged proteins is incubated with a nickel-coated glass slide. The proteins adhere to the slide wherever a nickel ion is attached.

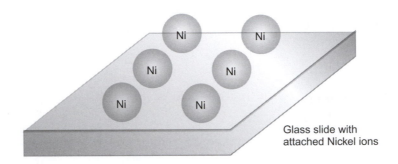

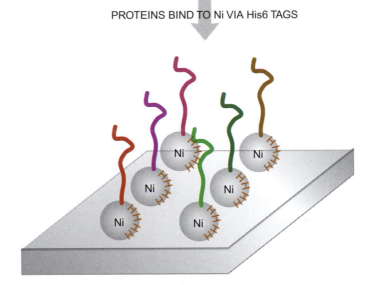

Protein microarrays have been used for the biochemical and enzymatic analysis of proteins as well as to survey protein–protein interactions. Some proteome arrays have used the yeast *Saccharomyces cerevisiae* as model organism. A complete proteome analysis needs an array of approximately 6,000 proteins in this case. Such arrays are assembled using proteins that have been tagged with chemical groups that allow binding to solid supports such as 96-well microtiter dishes or glass microscope slides.

Yeast protein arrays have been constructed with nearly 90% of the yeast proteins fused to a suitable tag. The glutathione-S-transferase (GST) tag allows binding to a solid support via glutathione and the His tag allows binding via nickel (Fig. 15.25). To construct the array, each protein must be genetically fused to the tag-coding sequence.

These constructs have been expressed under control of the *GAL1* (galactose inducible) or *CUP1* (copper inducible) promoters. Such protein libraries may be pooled or distributed individually into the wells of microtiter dishes. Simpler and less expensive screening is usually done individually, whereas complex or expensive assays are more often run first on pooled protein samples that are subdivided for further analysis if positive results are found.

The functional assays must be designed so that the arrays can be screened conveniently, usually for fluorescence, less often for radioactivity. For example, the yeast proteome has been screened for those proteins that bind **calmodulin** (a small calcium-binding protein) or phospholipids in the laboratory of Michael Snyder at Yale University. The His-tagged proteins were attached to nickel-coated glass slides. Both calmodulin and phospholipid were tagged with biotin. After binding of calmodulin or phospholipids to the proteome array, the biotin was detected by streptavidin carrying a Cy3 fluorescent label (Fig. 15.26). This revealed 39 calmodulin-binding proteins of which six were previously known. Some 150 phospholipid-binding proteins were also found.

11. Metabolomics

By analogy with genome and proteome, the **metabolome** is the totality of small molecules and metabolic intermediates. NMR of extracts from cells labeled with ^{13}C-glucose has allowed simultaneous measurement of multiple metabolic intermediates. An alternative is the separation of metabolites from cells labeled with ^{14}C-glucose by thin layer chromatography. However, these methods are limited in both sensitivity and in the number and chemical types of compounds that can be readily separated.

> Global surveys of low molecular weight metabolites can be done with NMR or high-resolution mass spectrometry.

Nearly complete metabolome analysis may be achieved by mass spectrometry. This approach is not limited to particular classes of molecule and is extremely sensitive. Carbon-12 is defined as having a mass of exactly 12 daltons. However, the masses of other atoms, such as ^{14}N or ^{16}O, are not exact integers. Consequently, mass spectrometry using extremely high mass resolution (EHMR; that is, to 1 ppm or less) allows the unambiguous determination of the molecular formula of any metabolite. Isomers have the same molecular formula, but may be distinguished by the different fragmentation patterns of their molecular ions.

Metabolome analysis is especially useful for analysis of plants, which typically make many secondary metabolites including pigments, scents, flavors, alkaloids, and other commercially-important products. For example, using both aqueous and organic extracts from strawberries has allowed the measurement of nearly 7,000 different metabolites. If printed out, the complete EHMR mass spectra for such mixtures would be a couple of miles long. Comparison of white mutants with wild-type red strawberries directly revealed differences in the levels of several intermediates in the pathway for pigment synthesis as well as of the red pigment itself (Fig. 15.27).

Metabolome analysis may be approached by a combination of methods, such as liquid chromatography followed by mass spectrometry (LC-MS). This has been used to compare the metabolome of mutant and wild-type bacteria, thus allowing the metabolic effects of specific genes or mutations to be analyzed (see Focus on Relevant Research).

---

calmodulin   A small calcium-binding protein of animal cells
metabolome   The total complement of small molecules and metabolic intermediates of a cell or organism

FIGURE 15.26
Screening Protein Microarray using Biotin/Streptavidin

Protein microarrays can be screened to find proteins that bind to phospholipids, for example. The protein microarray is incubated with phospholipid bound to biotin. Then the bound phospholipid is visualized by adding avidin conjugated to a fluorescent dye. Spots that fluoresce represent specific proteins that bind phospholipids.

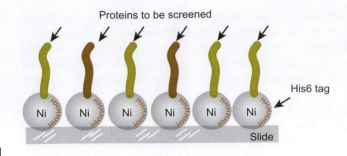

Proteins to be screened

His6 tag

PHOSPHOLIPID TAGGED WITH BIOTIN

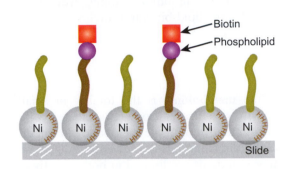

Biotin

Phospholipid

AVIDIN WITH Cy3 FLUORESCENT LABEL

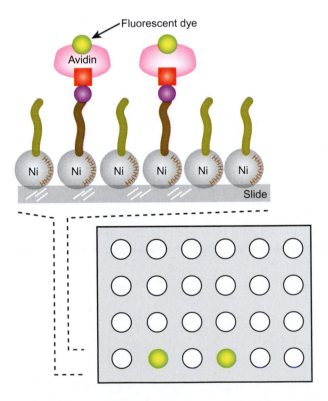

Fluorescent dye

Avidin

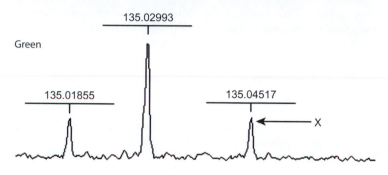

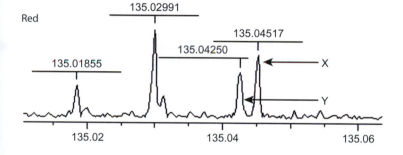

135.02
135.04
135.06

►■ **FIGURE 15.27**
**_Metabolome Analysis of
Strawberries_**

Non-targeted metabolic analysis
in strawberry. A) Four consecutive
stages of strawberry fruit
development (G, green; W, white;
T, turning; R, red) were subjected
to metabolic analysis using FTMS.
Similar fruit samples were used
earlier to perform gene expression
analysis using cDNA microarrays.
B) An example of high-resolution
(>100,000) separation of very
close mass peaks in data obtained
from the analysis of green and
red stages of fruit development.
Peaks marked with an X have
the same mass, while peak Y is
different by a mere 3 ppm. (*Credit:
Phenomenome Discoveries Inc.,
Saskatoon, Canada.*)

Duckworth BP, and Aldrich CC. (2010) Assigning enzyme function from the metabolic milieu. Chem. & Biol. 17: 313–314.

This review summarizes recent research on metabolomics. The research uses liquid chromatography/mass spectrometry (LC-MS) to assign a definitive function to a gene from the *Mycobacterium tuberculosis* genome. The gene was previously shown to encode an α-ketoglutarate decarboxylase based on sequence homology and an *in vitro* enzymatic assay. Rather than presenting the enzyme with just its suspected substrate, the authors of this study used a physiologically-relevant substrate pool. The results were analyzed using LC-MS and computer analysis to measure the production of any small molecules. This demonstrated that the decarboxylase activity was actually a side reaction. Instead, the enzyme required two co-factors, Mg^{+2} and thiamine diphosphate, and a second substrate, glyoxylate. The physiological reaction formed 5-hydroxylevulinic acid (a precursor to heme). Such analyses demonstrate that metabolic profiling is useful in determining the real physiological function of many of the genes in an organism. These results are especially useful to study enzymes that, at least in the test tube, share reactions with other enzymes.

FOCUS ON RELEVANT RESEARCH

Key Concepts

- Proteomics is the global study of proteins. The proteome is the total set of proteins possessed by an organism.

- Proteins are often purified by gel electrophoresis. More sophisticated separations are done using 2D electrophoresis.
- Proteins may be identified and purified by binding to specific antibodies.
- Several types of chromatography, especially high-pressure liquid chromatography, are used to isolate proteins.
- Mass spectrometry is now commonly used in proteomics for the identification of purified proteins.
- Tagging with another easily recognized molecule allows proteins to be purified much more easily. Short peptides, whole proteins, and inteins are all used as tags.
- Phage display allows genes to be isolated by detecting the proteins they encode. The proteins are displayed on the surface of phage particles.
- Proteins are screened by two-hybrid analysis to see which proteins bind to each other in the cell.
- Another method to screen protein interactions is co-immunoprecipitation.
- Protein arrays are built using tagged proteins and are screened by a variety of approaches.
- Metabolomics is the global study of metabolic intermediates. High-resolution mass spectrometry is the major approach used.

Review Questions

1. What is the distinction between the proteome and the translatome?
2. Why is denaturing proteins important when running proteins on a gel?
3. What detergent is used to denature proteins? How does it aid in separating proteins based on molecular weight?
4. What is isoelectric focusing?
5. How does 2D PAGE differ from standard SDS-PAGE?
6. What are the two dimensions of a 2D PAGE?
7. What is Western blotting? How is it performed?
8. What is mass spectrometry?
9. What are the two most important techniques used in the mass spectrometry of proteins? Describe them.
10. What is time-of-flight (TOF)?
11. What are the differences between MALDI/TOF mass spectrometry and electrospray ionization mass spectrometry?
12. What is tandem mass spectrometry?
13. How are proteins tagged? What are the most widely-used short tags?
14. What are the three most common full-length tag proteins, and how are they eluted from columns?
15. How can proteins be isolated using a His tag or maltose-binding protein?
16. What makes intein tags convenient?
17. What is phage display? What is the mechanism of phage display?
18. Why is M13 the preferred phage for phage display?
19. What is a phage library?
20. What is biopanning? How are peptides screened in biopanning?
21. What is a protein interactome? How are protein interactomes screened?
22. Upon what principle is interactome analysis based?
23. In the two-hybrid system for analysis what are the "bait" and the "prey"?
24. What is two-hybrid analysis used for?
25. What is tethering technology and how is it used in protein analysis?
26. What are the drawbacks of the original two-hybrid system and what has been done to make the process easier?
27. What is co-immunoprecipitation? How is it performed?
28. What is one method to monitor many proteins at one time?

29. How are protein microarrays made? What is the best way to detect desired proteins?
30. What is the metabolome?
31. What are three ways by which the metabolome may be analyzed? Which method is the best and why?
32. What is metabolome analysis used for?

Conceptual Questions

1. Discuss why a protein array might provide more information for proteomics research than a traditional yeast two-hybrid screen.
2. Gene expression can be monitored transcriptionally, but there is often a difference between mRNA levels and gene product levels; what factors may cause this difference?
3. The following experiment was done by a research team, but they need help interpreting the results. The researchers identified 25 putative transcription factors by looking for genes with the basic helix-loop-helix motifs in the human genome. They cloned each of these putative transcription factor genes in an expression vector that has a polyhistidine tag, and then purified each of the 25 proteins on a nickel column. The purified and tagged proteins were then attached to a membrane in small spots to form a grid. Each protein was spotted at three different concentrations. The membrane was then incubated with an oligonucleotide probe called bHLHwt-Cy3 that contained a consensus basic helix-loop-helix sequence with a Cy3 fluorescent tag that appears green. In addition, to ascertain any non-specific interactions, the protein array was incubated with an oligonucleotide probe called bHLHmut-Cy5 that has nucleotide substitutions in the important bases that mediate the binding of the bHLH transcription factors to DNA. This probe was labeled with Cy5, a red fluorophore.

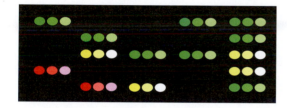

   How many different putative bHLH transcription factors bind to the bHLHwt-Cy3 oligonucleotide probe? How many bind both probes non-specifically? How many bind just the mutated probe? Discuss why there may be putative transcription factors that bind to bHLHmut-Cy5.
4. The following sequence data were obtained for the 5′ end of gene X cloned into a vector that adds a polyhistidine tag onto the N-terminus of gene X. Gene X encodes a protein called X that is known to be 74 amino acids long. Do you think this vector will produce a His-tagged protein X? Why or why not?

5. Protein X was finally properly cloned into the polyhistidine tag vector as shown in question 4. The recombinant vector is transformed into *E. coli*. Devise an experiment to isolate any proteins that interact with protein X inside the cytoplasm by using nickel column chromatography.

Regulation of Transcription in Prokaryotes

Chapter 6

The functioning of living creatures depends on the expression of genetic information. Although some genes are expressed all the time, most are expressed only under conditions where they are needed. Consequently, gene expression is heavily regulated. Most genes encode proteins although a minority of genes encode non-translated RNA molecules that function without being translated into proteins. In this chapter, we will focus on the expression of genes that encode proteins. Transcription is the first step in gene expression and, not surprisingly, is the site of a great deal of regulation, which is the topic of the present chapter. Regulation at some later stages of gene expression is covered in Chapter 18. Much of cell growth and metabolism depends on the functioning of proteins. However, the mere presence of a protein is not sufficient to ensure a correct physiological response and the activity of many proteins is regulated after they are made, as discussed in Chapter 14.

Chapter 16

1. Gene Regulation Ensures a Physiological Response

In bacteria such as *Escherichia coli*, about 1,000 of the 4,000 genes are expressed at any given time. If conditions change, some genes are turned off and others are switched on (Fig. 16.01). A major change of growth conditions, such as a shift in temperature, may result in altered expression of 50–100 genes.

FIGURE 16.01

2D Protein Gels of E. coli under Different Conditions

E. coli was grown in 50 mM acetate or 20 mM formate. Three gels were run for each condition and the figure shows a layered view of two three-gel composites. The pink and green spots are proteins induced in acetate or formate, respectively. The circled spots are those that were statistically validated, based on a pair-wise comparison of all the individual gels. (*Credit: Joan L. Slonczewski and Christopher Kirkpatrick, Kenyon College, Gambier, Ohio.*)

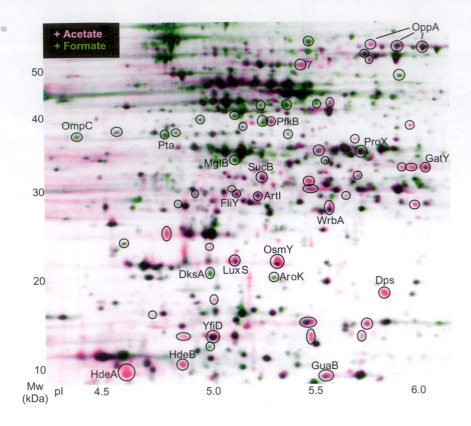

Single-celled organisms regulate their genes in response both to changes in the environment (such as temperature, osmotic pressure, or availability of nutrients) and to the internal state of the cell (such as readiness for cell division). In multicellular organisms, one must also consider communication between cells and the developmental progression of the organism as a whole.

Gene expression, to yield a functional protein, may be regulated at a number of different stages:

- *Transcription of the gene* to give the primary transcript
- *Processing of the primary transcript* to give mRNA
- *Stability of mRNA* to degradation
- *Translation of mRNA* to give polypeptide chains
- *Processing and assembly of polypeptide chains* and any necessary co-factors to give a functional protein
- *Control of activity* of an enzyme or other protein
- *Degradation of protein*

Each of the steps listed above may be regulated, but not with equal frequency. For example, transcription is more frequently regulated than is translation. Each of the major steps listed above may be subdivided further. Thus, transcription involves the following steps: access of transcription apparatus to DNA, recognition of promoter sequences, initiation of RNA synthesis, elongation of RNA, and termination. Any of these individual steps may be regulated, but in practice, regulation of certain steps is much more common than the regulation of others. For example, the initiation of transcription is more often controlled than its elongation phase.

Nature has evolved strategies to optimize regulation. Clearly, it is less wasteful to control the initial synthesis of mRNA rather than wait until the mRNA has been made before deciding whether or not to translate it into protein. However, factors

> Cells respond actively to their environments by switching genes on or off.

> Efficient regulation: control mRNA synthesis. Rapid regulation: control protein activity.

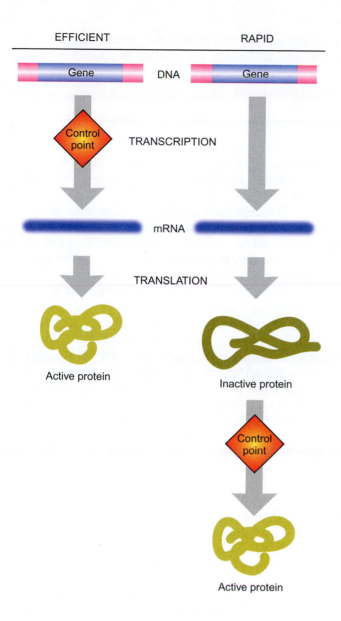

EFFICIENT   RAPID

Gene — DNA — Gene

Control point — TRANSCRIPTION

mRNA

TRANSLATION

Active protein   Inactive protein

Control point

Active protein

■ **FIGURE 16.02**
Efficiency Versus Rapid Response

Two common regulatory mechanisms serve two separate needs. For an efficient response—in other words, one requiring less energy—the gene is often regulated at the level of transcription. When a rapid response is needed, a precursor protein is produced in advance and then rapidly converted to an active protein when the conditions warrant.

other than efficiency may intrude. If a rapid response is critical to the survival and propagation of a species, making an inactive enzyme and holding it on standby is a good strategy. Bacteria must respond not only to the "natural" environment, but those that inhabit animals or plants must also be prepared to counter the attempts of their host organisms to eliminate them (see Focus on Relevant Research). In general, regulation of transcription is most efficient for the organism, but control of enzyme activity allows for a fast response (Fig. 16.02). Not surprisingly, these are the most common targets of regulation. Nonetheless, examples of regulation may be found for virtually every step between DNA and the final active gene product.

Both positive and negative regulation control mechanisms exist. In **positive regulation**, a gene is incapable of expression unless it receives a positive signal of some sort. In **negative regulation**, a gene is inherently active but is prevented from

negative regulation   Control by a repressor that prevents expression of a gene unless it is somehow removed
positive regulation   Control by an activator that promotes gene expression when it binds

Genes may be regulated positively or negatively.

expressing itself unless certain inhibitory factors are removed. Some genes are regulated positively, others negatively, and still others by multiple regulators, including both types.

FOCUS ON RELEVANT RESEARCH

Wade WF and O'Toole GA. (2010) Antibodies and immune effectors: shaping Gram-negative bacterial phenotypes. Trends Microbiol. 18: 234–239.

Understanding how bacteria regulate gene expression may seem like a simple academic endeavor, but medically speaking, bacteria are very important. There are many different diseases that are caused by bacteria, and although antibiotics have changed the way most infections are treated, many bacteria are resistant to antibiotics. Thus, some relatively benign bacterial infections can cause serious illness. Although some bacteria cause disease, a large number are actually beneficial to humans. The intestine is filled with a large number of different bacteria that aid digestion of different foods. These bacteria are essential to nutrient absorption. Because of the long-standing relationship between humans and

bacteria, humans have developed elaborate immune systems to fight off harmful bacteria and nourish good bacteria. To balance these two, human cells have extensive proteins, metabolites, and small peptides that interact in order to activate the appropriate genes.

So much attention is devoted to studying how humans react and respond to bacteria, yet there is a corollary. Bacteria must also respond to different environments and defend themselves. Bacteria must have genes that help them withstand the human immune system. The bacterial cells must have regulatory proteins that mediate their gene expression in response to human immune effectors (IE) and antibodies. As this review points out, these are only relevant if they fit five different criteria. First, the bacteria must have a receptor that binds the IE or antibody. Second, the receptor must communicate with transcription factors that activate and repress gene expression. Third, the gene expression must respond to the antibody or IE. Fourth, the change in gene expression must make the bacteria more resistant to the immune response. And finally, there must be some sort of target in the pathway that can be used to block bacterial response. In other words, there must be some way to use the information to develop a way to modulate bacterial growth. If the bacterial gene regulation pathway fits all these criteria, then understanding the way the bacteria respond may provide a future strategy to kill the pathogenic bacteria, or even enhance the beneficial ones.

---

2. Regulation at the Level of Transcription Involves Several Steps

In Chapter 11, the basic requirements for transcribing a gene were described, including the need for activator proteins to help RNA polymerase bind to the promoter and the possibility of repressor proteins blocking access of RNA polymerase. Here, the regulatory aspects of transcription will be examined in more detail. It is useful to subdivide regulation into subprocesses such as *access*, *recognition*, *initiation*, *elongation*, and *termination* as follows:

- *Access to coding DNA*: Coding DNA may be inaccessible under some circumstances. This is especially true of eukaryotes (see next chapter) where DNA is often condensed into **heterochromatin** while not being transcribed. In prokaryotes, access has not been discovered to be a major issue.

- *Recognition*: Obviously many regulatory proteins recognize binding sites on the DNA. Here, we are referring to the recognition of the promoter by RNA polymerase itself. In eukaryotes, there are three different RNA polymerases that recognize and transcribe different categories of genes. In bacteria, there is a single RNA polymerase, but there are multiple different sigma factors.

- *Initiation*: Even if recognition is successful, RNA polymerase may be unable to initiate RNA synthesis. In some cases, activator proteins that bind upstream of RNA polymerase may be needed. In other cases, a repressor that blocks movement of RNA polymerase prevents transcription.

---

heterochromatin   A highly condensed form of chromatin that cannot be transcribed because it cannot be accessed by RNA polymerase

- *Elongation*: Once transcription has been initiated, it usually continues without interruption. Regulatory effects at the stage of elongation are uncommon. They may be subdivided into *slowing down of the elongation rate* and *premature termination*.
- *Termination*: Normally, RNA polymerase stops at terminator sites. However, in a few rare cases, termination may be over-ridden by **anti-terminator proteins**. This allows for the expression of those genes downstream of the terminator and their regulation.

3. Alternative Sigma Factors in Prokaryotes Recognize Different Sets of Genes

In bacteria, the sigma subunit of RNA polymerase recognizes the promoter. However, several major groups of genes have promoters lacking the −10 and −35 recognition sequences to which the standard sigma factor binds (see Fig. 11.05 in Ch. 11). These genes have distinct promoter recognition sequences that are recognized by **alternative sigma factors**.

> Different sigma factors recognize different groups of genes.

The various sigma factors are named either as σ followed by the molecular weight in kd or as "RpoX," where Rpo refers to RNA polymerase and X signifies the function. The default or "housekeeping" sigma factor of *E. coli* is σ70 or RpoD protein. Each alternative sigma factor is needed for the expression of a large suite of genes (typically 50 or more) that is required under particular conditions such as stationary phase, nitrogen starvation, etc. Some examples of alternative sigma factors found in *E. coli* are shown in Table 16.01.

3.1. Heat Shock Sigma Factors in Prokaryotes Are Regulated by Temperature

Almost all organisms that have been studied respond to **heat shock**, so this provides a good example of an environmental stimulus. Many of the genes expressed at high temperatures are highly conserved throughout evolution. However, the regulatory circuits that control the heat shock genes vary greatly from one group of organisms to another. In bacteria such as *E. coli*, two alternative sigma factors, RpoH and RpoE, control the heat shock response.

At increasingly high temperatures, the 3D structure of proteins begins to unravel. Unfolded or mis-folded proteins not only lose their own activity, but they may also bind to other functional proteins and create insoluble aggregates. Thus, the **heat shock response** is primarily concerned with protecting the cell from damaged and/or mis-folded proteins.

> Overheating provokes a protective response from almost all living cells.

E. coli is optimized for growth at body temperature (37°C). It grows happily up to about 43°C but almost stops growing at 46°C. At 46°C, about 30% of all the proteins made by *E. coli* are **heat shock proteins**. Most of these fall into two categories: some are **chaperonins** that help other proteins fold correctly and prevent aggregation, while others are **proteases** that degrade heat-damaged proteins that are past rescue (Fig. 16.03).

To adapt the level of heat shock proteins to the environment, RpoH protein is regulated. At 30°C, RpoH is not needed, and therefore it is rapidly degraded by a membrane bound protease, FtsH. When the temperature increases to 42°C, called the induction phase, RpoH is stabilized and no longer degraded as rapidly. This results in elevated expression of those heat shock genes that depend on RpoH for

---

alternative sigma factor   A non-standard sigma factor needed to recognize a specialized subset of genes
anti-terminator protein   Protein that allows transcription to continue through a transcription terminator
chaperonin   A protein that helps other proteins fold correctly
heat shock proteins   A set of proteins that protect the cell against damage caused by high temperatures
heat shock response   Response to high temperature by expressing a set of genes that encode heat shock proteins
protease   Enzyme that degrades proteins

TABLE 16.01	Alternative Sigma Factors of *E. coli*				
	Sigma Factor	Name	Consensus Sequence		
			−35	Spacing	−10
Housekeeping	σ70	RpoD	TTGACA	16–18	TATAAT
Stationary phase	σ38	RpoS	CCGGCG	16–18	CTATACT
Nitrogen control	σ54	RpoN	TTGGNA	6	TTGCA
Flagellar motion	σ28	FliA	CTAAA	15	GCCGATAA
Heat shock	σ32	RpoH	CTTGAA	13–15	CCCCATNT
Extracytoplasmic heat shock	σ24	RpoE	GAACTT	16	TCTGAT

(N = any base)

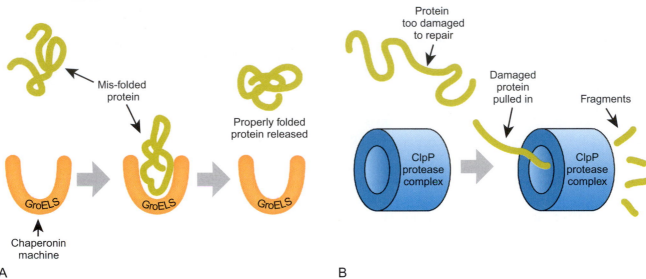

REPAIR, IF POSSIBLE

Mis-folded protein

Properly folded protein released

GroELS

GroELS

GroELS

Chaperonin machine

DESTROY, IF CANNOT REPAIR

Protein too damaged to repair

Damaged protein pulled in

Fragments

ClpP protease complex

ClpP protease complex

A

B

FIGURE 16.03
Heat Shock Response in E. coli

The cell responds to heat shock either by repairing mis-folded proteins or by degrading them.

transcription. Finally, after induction, RpoH undergoes rapid inactivation and the level of heat shock protein gene expression reaches a steady state. Control of the level of RpoH itself is very complex and modulated by a variety of minor factors; however, the main signal is the level of mis-folded proteins present in the cell. This is monitored by two heat shock proteins, DnaK (a chaperonin) and DnaJ (a co-chaperonin). When the level of mis-folded proteins is low, DnaK/DnaJ bind to RpoH, and transfer the protein to FtsH, which is free to degrade it. In addition, they bind to partly synthesized RpoH protein, even before it is finished by the ribosome, and block further translation. When the level of mis-folded proteins rises, DnaK and FtsH bind to these instead and are unable to affect RpoH levels (Fig. 16.04).

Transcription of the *rpoH* gene from its main promoter requires the standard sigma factor σ70 or RpoD. At temperatures above 50°C, σ70 is inactivated and synthesis of RpoH would come to a halt, thus undermining the heat shock response. This is prevented by the presence of a second promoter for the *rpoH* gene that is recognized by the RpoE sigma factor. Transcription can continue until 57°C, when the core enzyme of RNA polymerase is inactivated.

The level of RpoE (E for extra-cytoplasmic) is controlled in response to the level of mis-folded proteins in the outer membrane and periplasmic space, rather than in the cytoplasm, as in the case of RpoH. In addition to *rpoH*, another group of a dozen or so heat shock genes requires RpoE for their transcription.

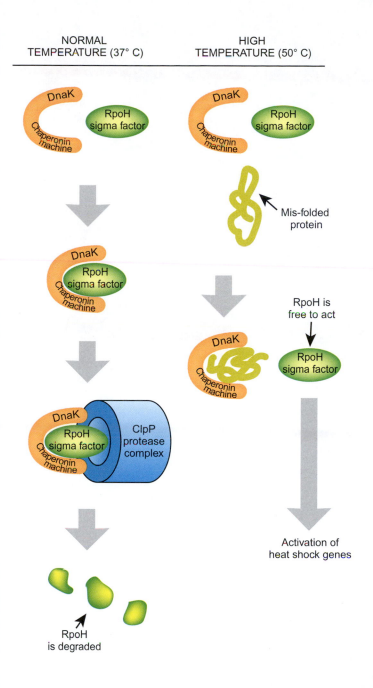

NORMAL TEMPERATURE (37° C)

HIGH TEMPERATURE (50° C)

Mis-folded protein

RpoH is free to act

Activation of heat shock genes

RpoH is degraded

■-**FIGURE 16.04**

Regulation of the Heat Shock Response

At normal temperatures, the RpoH sigma factor binds the chaperonin and is presented to the protease for digestion. At high or extreme temperatures, the chaperonin binds instead to mis-folded proteins, leaving RpoH free to activate the heat shock genes.

3.2. Cascades of Alternative Sigma Factors Occur in *Bacillus* Spore Formation

The requirement of RpoE for transcription of the *rpoH* gene illustrates, at a simple level, that the expression of one sigma factor may depend upon another. Indeed, in some complex processes, a series of alternative sigma factors may depend on each other. The classic case of a cascade of alternative sigma factors is the regulation of **spore** formation in the gram-positive bacterium, *Bacillus*. When nutrients are scarce, *Bacillus* forms spores designed to survive bad times. Spores are formed by an asymmetric division that gives a full-sized mother cell and a much smaller spore. The spore is surrounded by the mother cell until it is fully developed. The mother cell then bursts, releasing the spore. This represents cellular differentiation at its most primitive level (Figs. 16.05 and 16.06).

Some bacteria survive hard times by making spores.

spore A cell specialized for survival under adverse conditions and/or designed for distribution

FIGURE 16.05
Spore Formation in **Bacillus**

Bacillus (A) first duplicates its DNA (B), then walls off the new DNA into a spore that lies within the cell (C and D). The spore is released from the original mother cell as it bursts and dies (E).

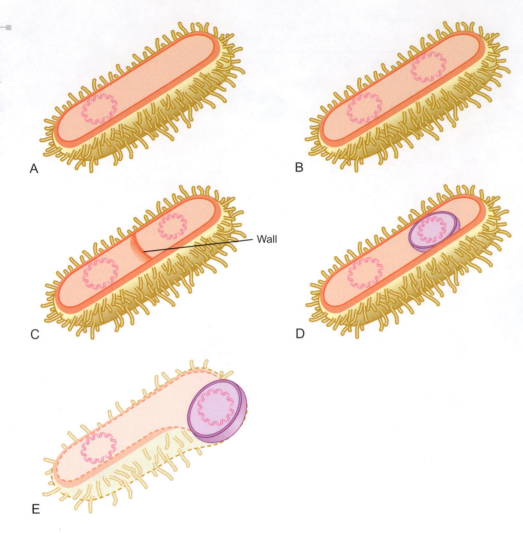

A

B

C

Wall

D

E

FIGURE 16.06
Spore Formation in **Bacillus anthracis**

Transmission electron micrographic image of *Bacillus anthracis* from an anthrax culture showing cell division (A) and spores (B). (*Credit: Public Health Image Library (CDC) by Dr. Sherif Zaki and Elizabeth White.*)

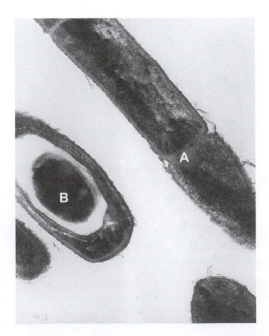

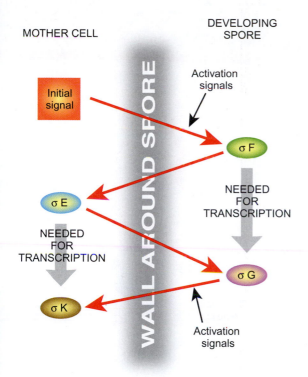

MOTHER CELL

DEVELOPING SPORE

Initial signal

Activation signals

σ F

σ E

NEEDED FOR TRANSCRIPTION

NEEDED FOR TRANSCRIPTION

σ G

σ K

WALL AROUND SPORE

Activation signals

FIGURE 16.07

Outline of Spore Formation Regulatory Cascade

A cascade of four sigma factors is involved in the stepwise development of the spore in *Bacillus*. An external signal activates synthesis of σF in the spore. This is required for transcription of the gene for σG (inside spore) and for factors that cross into the mother cell (red arrow) and convert pre-σE into active σE (in mother cell). Active σE is required for synthesis of the precursor, pre-σK. Finally, σG allows synthesis of factors that cross into the mother cell (red arrow) and convert pre-σK into active σK (in mother cell).

Spore formation is controlled by four alternative sigma factors, σE and σK in the mother cell, and, once sporulation has started, σF and σG in the developing spore. Two of these, σE and σK, are first synthesized as inactive precursor proteins—pre-σE and pre-σK—that must be activated by specific proteases at the correct point in the pathway. First, an environmental signal activates the synthesis of pre-σE in the mother cell and σF in the spore. The presence of σF allows transcription of early sporulation genes in the spore. These include the gene for the sigma factor, σG, as well as protein sporulation factors that move into the mother cell and split pre-σE protein to give active σE. The activated σE switches on several genes in the mother cell, including the gene that encodes pre-σK. As a result, pre-σK accumulates in the mother cell. The presence of σG in the spore allows transcription of late sporulation genes in the spore, including factors that move into the mother cell and activate pre-σK to active σK (Fig. 16.07). Each of these sigma factors is responsible for the transcription of a group of genes needed for successive stages in the development and release of the spore.

The cascade of regulators seen in spore formation ensures that steps in a complex series of events follow each other in the correct sequence. Each regulator controls one stage in the process and also controls the regulator for the next stage. In addition, regulatory signals are exchanged between more than one cell. Thus, the developing spore and mother cell cross-regulate each other. The development and differentiation of higher organisms use much the same principles as spore formation, but the regulatory schemes are vastly more complex.

3.3. Anti-sigma Factors Inactivate Sigma; Anti-anti-sigma Factors Reactivate Sigma

Sigma factors may be inhibited by proteins known as **anti-sigma factors**. These bind to specific sigma factors and prevent them from associating with RNA polymerase (Fig. 16.08). When σF is first made in the developing spore, it is inactive. Unlike σE and σK, which need to be activated by the proteolysis of an inactive precursor

anti-sigma factor   Protein that binds to a sigma factor and blocks its role in the initiation of transcription

FIGURE 16.08
Anti-Sigma Factor

The anti-sigma factor SpoIIAB binds to σF and inactivates it. When the cell receives an external signal, the phosphorylated form of SpoIIAA, an anti-anti-sigma factor, loses its phosphate and engages SpoIIAB. This releases σF, which is then free to activate the sporulation cascade shown above in Figure 16.07.

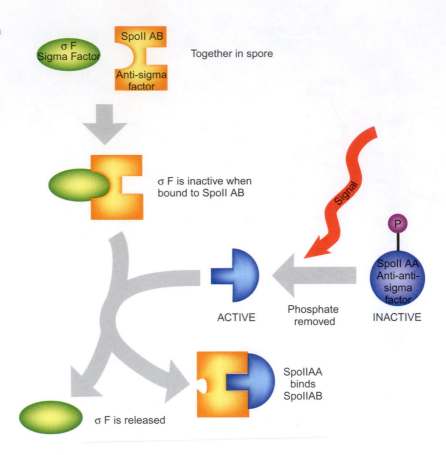

Positive regulatory factors are often opposed by negative factors.

protein, σF is kept inactive by an anti-sigma factor (SpoIIAB). This anti-sigma factor is, in turn, displaced from σF by an **anti-anti-sigma factor** (SpoIIAA). This event triggers the cascade of gene activation described above.

Regulation of alternative sigma factors by binding to anti-sigma factors and their release by anti-anti-sigma factors is not especially common, but several examples are known. Assembly of flagella in *E. coli* and *Salmonella* is under control of the FliA sigma factor and the FlgM anti-sigma factor. A clinically-important example is the production of mucus by *Pseudomonas aeruginosa*. This bacterium often infects the lungs of cystic fibrosis patients, where it switches on the genes for mucus production. These are under control of the alternative sigma factor AlgU and the anti-sigma factor MucA. The bacterial mucus clogs the patient's airways and is a major contributor to the symptoms of the disease. Strictly speaking, the material made by *Pseudomonas* is alginate, an acidic polysaccharide that is a repeating polymer of mannuronic and glucuronic acids. This is chemically distinct from the true mucus made by animal cells, which consists of glycoproteins with many short side chains of galactose, N-acetyl-galactosamine, and N-acetyl-neuraminic acid.

4. Activators and Repressors Participate in Positive and Negative Regulation

Activator proteins turn genes on. Repressor proteins turn genes off.

Some promoters, both in lower and higher organisms (see Ch. 11), function poorly or not at all in the absence of extra proteins known as gene activator proteins, or transcription factors. In addition, there is a class of gene regulator proteins known as *repressors* that act to turn genes off.

anti-anti-sigma factor   Protein that binds to an anti-sigma factor and so prevents the anti-sigma factor from binding to and inhibiting a sigma factor

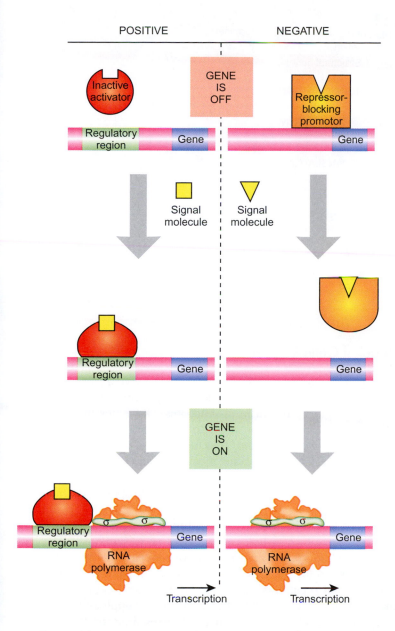

POSITIVE   NEGATIVE

Principle of Positive and Negative Regulation

In positive regulation, a signal changes the conformation of an inactive regulator, which then becomes active and binds to the regulatory region of a gene. Its presence aids the binding of the RNA polymerase and helps switch on the gene. In negative regulation, a repressor molecule blocks the promoter of the gene. A signal changes the conformation of the repressor, releasing it from the gene and allowing the RNA polymerase to bind.

In **positive control**, an activator is required to turn a gene on, in response to a signal of some kind. In **negative control**, a gene is switched off by a repressor and is only expressed in the presence of a signal that removes the repressor from the gene. Positive and negative control may be exerted at the level of transcription or at later stages in gene expression. Furthermore, although most activators and repressors are proteins, cases are known in which regulation is due to regulatory RNA or even small molecules.

In both positive and negative control, a small **signal molecule**, the **inducer**, typically binds to the regulatory protein and induces gene expression. In the standard model of positive regulation, an inactive activator protein binds the signal molecule and is converted to its DNA-binding form, which then turns on the gene (Fig. 16.09). Similarly, in typical negative regulation, the DNA-binding form of a repressor protein is converted to its inactive form by binding the signal molecule.

---

inducer   A signal molecule that turns on a gene by binding to a regulatory protein
negative control or regulation   Regulatory mode in which a repressor keeps a gene switched off until it is removed
positive control or regulation   Control by an activator that promotes gene expression when it binds
signal molecule   A small molecule that triggers a regulatory response by binding to a regulatory protein

FIGURE 16.10
Transcription of an Operon to Give Polycistronic mRNA

Several structural genes are located side-by-side and are transcribed into a single length of mRNA. During translation, several proteins, usually functionally related, are made using the single mRNA.

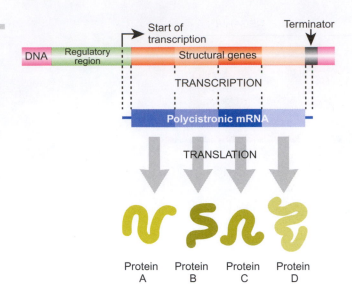

4.1. The Operon Model of Gene Regulation

An **operon** is a cluster of genes that are transcribed together to give a single messenger RNA (mRNA) molecule, which therefore encodes multiple proteins (Fig. 16.10). Such **polycistronic mRNA** is typically found in prokaryotes. The genes in an operon are often related functionally, so it makes good sense to regulate them as a group. For example, an operon may encode several enzymes that take part in the same biochemical pathway. Some operons have only a single gene; most have two to half a dozen and a few have more.

The operon model for regulating bacterial genes was first proposed by François Jacob and Jaques Monod using the negatively-regulated lactose genes of *E. coli* as an example. Since then a vast number of bacterial genes, including those with activators as well as those with repressors, have been fitted to this model or variants of it. The lactose operon, like many bacterial operons, is controlled at two levels. **Specific regulation** refers to regulation in response to factors specific for a particular operon, in this case the availability of lactose. **Global regulation**, discussed later, is regulation in response to more general conditions, such as the overall carbon and energy supply of the cell.

The lactose or *lac* operon consists of three structural genes, *lacZ*, *lacY*, and *lacA*, together with an upstream regulatory region (Fig. 16.11). The *lacZ* structural gene encodes β-**galactosidase**, the enzyme that degrades lactose. The *lacY* gene encodes **lactose permease** (a transport protein) and *lacA* encodes lactose acetylase, whose role is not known (it is not needed for growth on lactose by *E. coli)*. The *lac* operon is regulated by the **LacI** repressor protein, which is encoded by the *lacI* gene. This lies upstream of *lacZYA* and is transcribed in the opposite direction.

The upstream region of the *lac* operon contains a recognition sequence for the repressor protein, known as the **operator** (*lacO* in Fig. 16.11). If no inducer is present, LacI protein binds to the operator. This blocks the movement of RNA polymerase at the promoter. When lactose is present, it induces the *lac* operon. The actual inducer

Operons are clusters of genes that are controlled as a unit.

The lac operon includes genes for lactose uptake and metabolism.

---

beta-galactosidase or β-galactosidase (LacZ)   Enzyme that splits lactose and related molecules to release galactose
global regulation   Regulation of a large group of genes in response to the same stimulus
lactose permease (LacY)   The transport protein for lactose
LacI   The lactose repressor protein
operon   A cluster of prokaryotic genes that are transcribed together to give a single mRNA (i.e., polycistronic mRNA)
operator   The site on DNA where a repressor binds
polycistronic mRNA   mRNA that carries several structural genes or cistrons
specific regulation   Regulation that applies to a single gene or operon or to a very small number of related genes

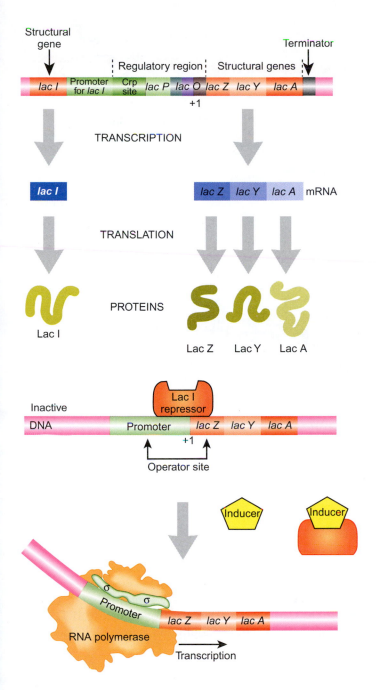

■ **FIGURE 16.11**
Components of the lac Operon

The *lac* operon consists of three structural genes, *lacZYA*, all transcribed from a single promoter, designated *lacP*. The promoter is regulated by binding of the repressor at the operator, *lacO*, and of Crp protein at the Crp site. Note that in reality the operator partly overlaps both the promoter and the structural gene. The single *lac* mRNA is translated to produce the LacZ, LacY, and LacA proteins. The *lacI* gene that encodes the LacI repressor has its own promoter and is transcribed in the direction opposite to the *lacZYA* operon.

■ **FIGURE 16.12**
Specific Regulation of the lac Operon

When LacI binds to the operator site, no transcription takes place. The presence of the inducer can remove LacI from the operator, allowing RNA polymerase to move forward and transcribe the operon. The inducer may be *allo*-lactose, derived from lactose or an artificial compound, such as IPTG.

is not lactose itself, but ***allo*-lactose**, an isomer of lactose that is made from lactose by β-galactosidase. *Allo*-lactose binds to LacI repressor, which changes its shape so it cannot bind DNA. RNA polymerase is now able to move forward from the promoter and transcribe the *lac* operon (Fig. 16.12). LacI protein exists as a tetramer that can, in fact, bind three different DNA recognition sites in the promoter region. This will result in looping of the DNA—see below. (Unlike many allosteric proteins, the LacI protein does not alternate between monomer and tetramer forms, but exists as a tetramer of unusual stability under all physiological conditions.)

In the laboratory, the *lac* operon is often induced by the compound **IPTG** (***iso*-propyl-thiogalactoside**) (Fig. 16.13). This artificial compound is known as a **gratuitous inducer** because it is not metabolized by the products of the genes it induces.

IPTG is an artificial inducer of the lactose operon.

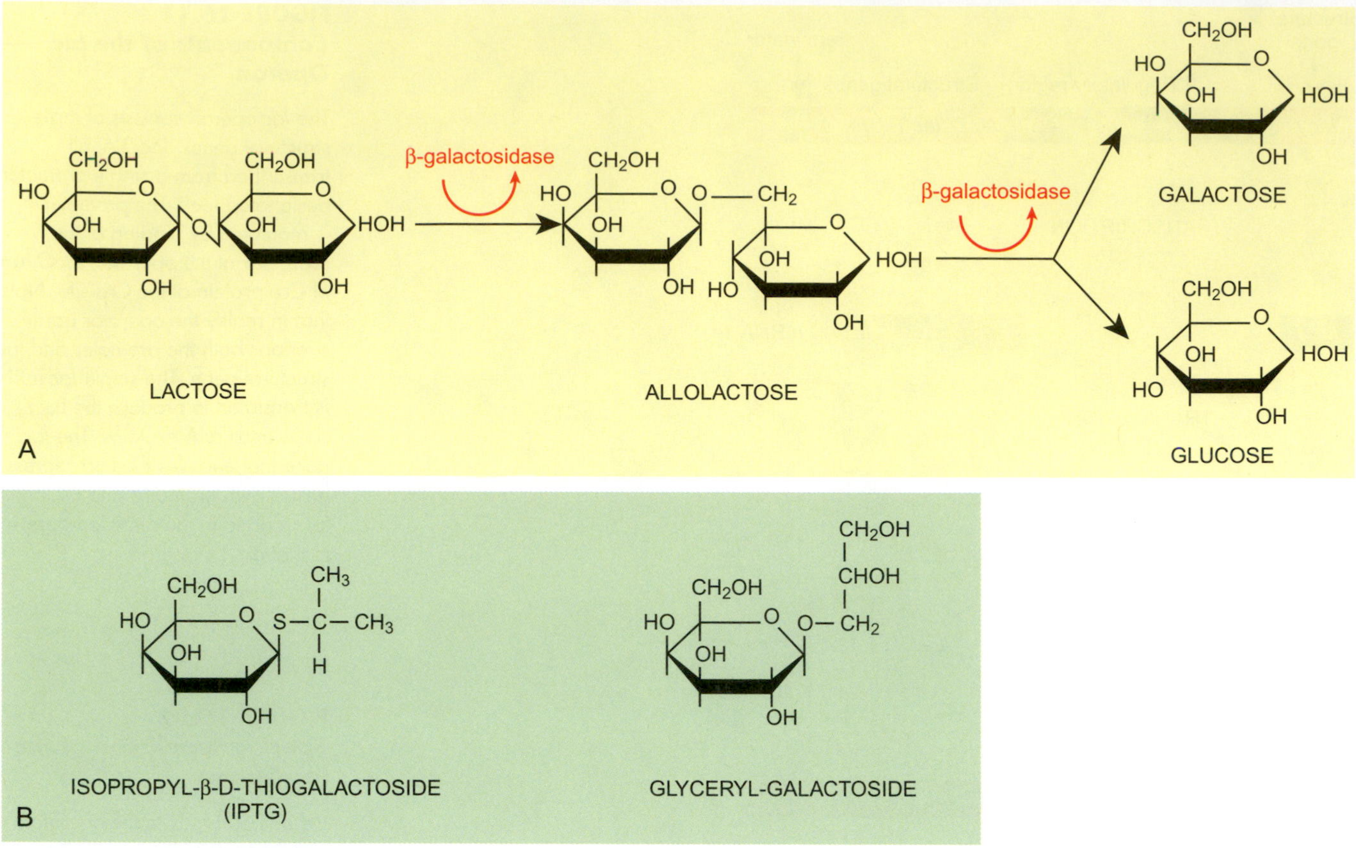

FIGURE 16.13
Lactose and Related Galactoside Derivatives

A) The enzyme β-galactosidase splits lactose into galactose plus glucose. β-galactosidase can also interconvert lactose with its isomer, *allo*-lactose, which is the true inducer of the *lac* operon. B) The structures of the gratuitous inducer, IPTG, and of the most likely natural substrate for β-galactosidase, glyceryl-galactoside, are also shown.

In this particular case, although IPTG induces the *lacZYA* genes, it is not broken down by β-galactosidase, the enzyme that degrades lactose. Consequently, IPTG continues to induce the *lac* operon long-term, whereas, natural inducers only induce for a short period of time before they are broken down.

Box 16.01 The Lactose Operon Is Not Really Typical

Although the lactose operon genes were the first whose regulation was characterized in detail, and although they are often cited as a typical example, they are aberrant in several ways. Curiously, lactose itself is not the inducer. Lactose, which consists of glucose linked to galactose, is converted to *allo*-lactose, an isomer in which the same two sugars are linked differently. This transformation is carried out by β-galactosidase, which normally splits lactose, but makes a small amount of *allo*-lactose as a side reaction. It is *allo*-lactose that actually binds to the LacI protein and acts as an inducer.

Lactose is present in the milk consumed by babies and children, but adult diets often contain very little. Moreover, lactose is almost all absorbed in the small intestine, so in nature *E. coli*, which lives in the large intestine, will never get any lactose. In reality, the lactose genes are probably intended to digest glyceryl-galactoside, a compound derived from breakdown of the lipids of animal cells. This is released into the large intestine as the cells lining it are

Continued

Box 16.01 Continued

sloughed off and disintegrate. Glyceryl-galactoside is both a genuine inducer, which binds to LacI protein, and a substrate for β-galactosidase, which splits it into glycerol plus galactose.

Furthermore, the segment of DNA containing the lactose genes is missing from *Salmonella* and several other bacteria of the enteric family that are close relatives of *E. coli*. It seems likely that this segment of DNA is a relative newcomer to the *E. coli* genome and came originally from some source outside the enteric bacteria.

In retrospect, it was both fortuitous and fortunate that Jacob and Monod chose a rather anomalous gene rather than a typical one. The regulation of the *lac* operon is simpler than that of many genes that are more fully integrated into the central metabolism of *E. coli*.

4.2. Some Proteins May Act as Both Repressors and Activators

Activators generally bind upstream of the promoter and help RNA polymerase to bind. Conversely, repressors bind downstream of the promoter and either block the binding of RNA polymerase or prevent it from moving forward and transcribing the gene. Not surprisingly, the same DNA-binding protein can act as an activator for one gene and a repressor for another if it binds at different locations in the two genes (Fig. 16.14).

Certain regulatory proteins may alternate between two different forms, both of which bind DNA. This is rather different from the binding proteins already discussed, which alternate between an active, DNA-binding form, and an inactive, non-binding form. Here, the two forms of the protein act as an activator and a repressor and both bind DNA, but at different recognition sites. The AraC regulatory protein controls the transport and metabolism of the 5-carbon sugar **arabinose**. When arabinose binds to AraC, it converts it from a repressor to an activator. The ***araBAD* operon** for arabinose metabolism and the *araFG* operon for arabinose uptake are repressed by AraC in the absence of arabinose and activated by AraC plus arabinose (Fig. 16.15).

Quite often regulatory proteins control their own production. This is known as **autogenous regulation**, or auto-regulation. For example, the AraC protein represses the *araC* gene and the Mlc protein (see below) represses transcription of the *mlc* gene.

> The same regulatory protein can sometimes turn genes on or off depending on where it binds on the DNA.

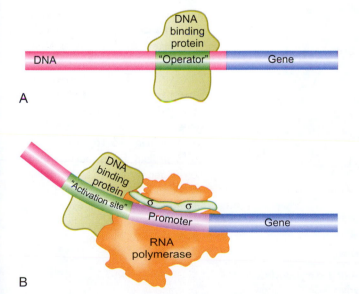

A

B

■ **FIGURE 16.14**
Binding Site Determines Action: Repressor or Activator

A) The DNA-binding protein shown in orange acts as a repressor when it binds to the operator region of the promoter, thus preventing binding of the RNA polymerase. B) The same protein may also bind to an activation site on the DNA of another operon, thus facilitating the binding of the RNA polymerase and promoting gene transcription. The recognition sequences for the DNA-binding protein are identical; only their position relative to RNA polymerase has changed.

---

arabinose   A 5-carbon sugar often found in plant cell wall material that can be used as a carbon source by many bacteria
***araBAD* operon**   Operon that encodes proteins involved in metabolism of the sugar arabinose
autogenous regulation   Self regulation; that is, when a DNA-binding protein regulates the expression of its own gene

FIGURE 16.15
AraC Repressor and Activator

A) AraC dimers are either activators or repressors depending on whether arabinose is bound or not. B) When AraC binds arabinose, the dimer changes configuration and binds to DNA at sites 1 and 2. Here, it acts as an activator, allowing the RNA polymerase to bind. C) When the repressor form of AraC binds DNA, it occupies sites 2 and 3, forming a loop in the DNA and causing gene inactivation.

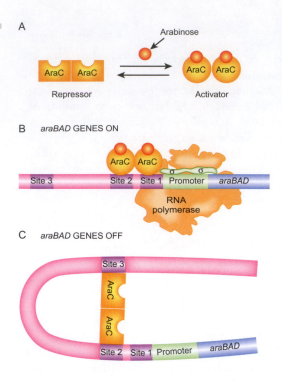

Box 16.02 FadR—An Example of a Repressor and an Activator

The FadR protein of *E. coli* provides a good example of differential binding by a DNA-binding protein. FadR represses the genes for fatty acid breakdown, but also activates certain genes involved in fatty acid synthesis. FadR responds to the availability of long-chain fatty acids in the growth medium. The cell can incorporate pre-made fatty acids into its lipids and can also break them down for energy. Fatty acids are taken up as coenzyme A derivatives, not free fatty acids; hence, the signal molecule recognized by FadR is a long-chain acyl-CoA. In the absence of acyl-CoA, FadR represses the operons for fatty acid degradation and also activates *fabA*, a gene involved in fatty acid biosynthesis (Fig. 16.16). When the FadR protein binds acyl-CoA, it no longer binds to DNA. Fatty acid degradation is induced and in addition the level of expression of *fabA* decreases, so fewer fatty acids are manufactured.

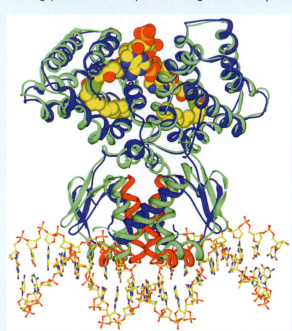

Current opinion in structural biology

FIGURE 16.16
FadR Structure and Binding

Overlay of FadR bound to DNA and to myristoyl-CoA (in yellow, at top). Atoms of myristoyl-CoA are shown as spheres. The HTH motif is colored red in the DNA-bound structures. (*Credit: Huffman & Brennan, Current Opinion in Structural Biology 12 (2002) 98–106.*)

4.3. Nature of the Signal Molecule

The substrate specificity and the inducer specificity of an operon need not be identical. One is determined by which molecules fit the active site(s) of the enzyme(s) of the pathway and the other by which molecules fit the binding site on the regulatory protein. In the case of the lactose operon *allo*-lactose, glyceryl-galactoside and IPTG are true inducers (i.e., they bind to the LacI protein). Lactose itself is only an apparent inducer, as it does not bind directly to LacI and must first be converted to *allo*-lactose. However, lactose, *allo*-lactose, and glyceryl-galactoside are all substrates of β-galactosidase, whereas IPTG is not (Fig. 16.13, above).

For induction of the *lac* operon by lactose, low levels of both LacY (transport protein) and LacZ (β-galactosidase) proteins are necessary. A small amount of lactose must be transported into the cell and be converted by β-galactosidase to *allo*-lactose before it can bind to LacI and induce. In practice, when a gene is "switched off," it is not utterly inactive. Even when the *lac* operon is not induced, occasional mRNA molecules are made and a few LacY and LacZ proteins are present.

The maltose system allows transport and metabolism of maltose and longer oligosaccharides also made of glucose subunits. Again, maltose itself is not the true inducer. The maltose system is under positive control and the MalT activator protein actually binds maltotriose, a trisaccharide consisting of three glucose residues.

Some repressors are only active when they bind a small signal molecule called a **co-repressor**. This situation is often found when regulating biosynthetic pathways. If an amino acid, such as arginine, is present in the culture medium, then the cell does not need to make it. On the other hand, if the amino acid is not present in sufficient amounts, the pathway for synthesis needs to be turned on. In general, the cell should turn biosynthetic pathways off when their products are present in the medium or have been synthesized in sufficient amounts. Thus, biosynthetic pathways respond to the corresponding nutrient. An example is the ArgR repressor of *E. coli*, which binds the amino acid arginine as its co-repressor (Fig. 16.17).

The signal molecule itself is not always small. Occasionally, repressors or activators bind other proteins, rather than small metabolites. For example, Mlc is a repressor that regulates glucose transport and a variety of other genes involved in the uptake and metabolism of monosaccharides. The Mlc protein does not bind glucose, yet responds to its presence indirectly. When glucose is absent, phosphate groups accumulate on the glucose transporter or PtsG protein. When glucose enters the cell, phosphate transfers from PtsG to glucose, thus converting it to glucose-6-phosphate. Most PtsG protein is therefore non-phosphorylated when there is plenty of glucose. Unphosphorylated PtsG binds to Mlc, which sequesters the transcription factor at the cell membrane (Fig. 16.18). This form of Mlc cannot repress genes; therefore, the presence of glucose indirectly induces expression of genes involved in glucose uptake and metabolism.

> Biological signals are often carried by small molecules. These signals are detected by proteins that bind to them.

4.4. Activators and Repressors May Be Covalently Modified

Some regulatory proteins do not bind a separate independent signal molecule. Instead, some activators and repressors are chemically modified. Most often this is done by the attachment of a chemical group, usually phosphate (see below). Less commonly, the regulatory protein is altered chemically in some other way, for example, by oxidation or reduction.

Examples of bacterial regulatory proteins that are altered by oxidation or reduction are the activators OxyR and Fnr. OxyR is converted to its active form by hydrogen peroxide or related oxidizing agents that oxidize sulfhydryl groups to disulfides (Fig. 16.19A). It then activates a set of genes involved in protecting bacterial cells against oxidative damage.

---

co-repressor In prokaryotes—a small signal molecule needed for some repressor proteins to bind to DNA; in eukaryotes—an accessory protein, often a histone deacetylase, involved in gene repression

FIGURE 16.17
ArgR Repressor Uses Arginine as a Co-repressor

In the absence of high levels of arginine the ArgR repressor cannot bind to DNA. Therefore, RNA polymerase transcribes the genes for the synthesis of arginine. When sufficient arginine is present, the arginine acts as a co-repressor by binding to ArgR. The complex then binds to the double operator sites and represses the operon.

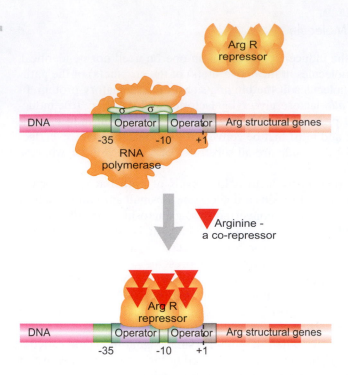

FIGURE 16.18
Sequestration of Mlc by PtsG Protein

A) The membrane transporter PtsG has an open channel for glucose when it is phosphorylated (P). B) Glucose enters the cell and is converted into glucose-6-phosphate so removing the phosphate. C) The dephosphorylated PtsG protein binds the Mlc repressor. This inactivates the Mlc repressor as it can no longer bind DNA when trapped by PtsG. If the supply of glucose runs out, PtsG will be able to retain its phosphate and Mlc is released.

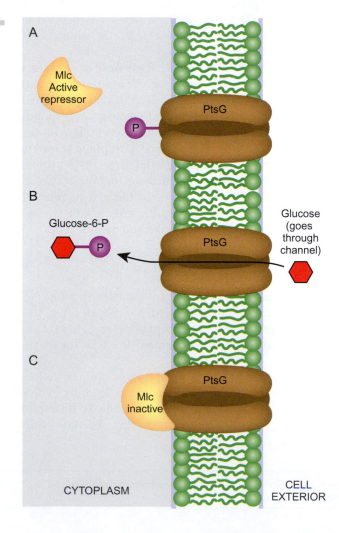

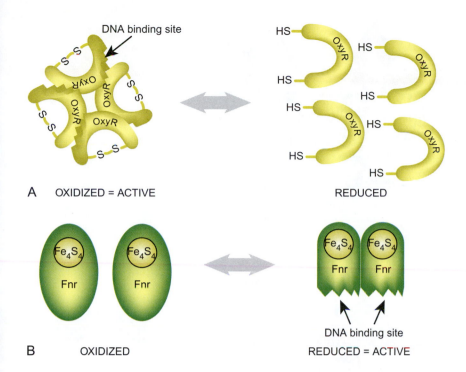

■ **FIGURE 16.19**
Regulatory Proteins That Respond to Oxidation or Reduction

A) OxyR changes from the DNA-binding disulfide form to the inactive sulfhydryl form. B) The reduced state of Fnr actively binds DNA, whereas the oxidized state is inactive. Note that the disassembled form of the protein is inactive in both instances.

In contrast, Fnr is inactive when oxidized and becomes an activator when reduced. In this case, an Fe_4S_4 **iron sulfur cluster** in the N-terminal domain of Fnr is reduced under anaerobic conditions. This results in the formation of dimers and a change in shape of the C-terminal DNA-binding domain (Fig. 16.19B). The Fnr activator then activates genes involved in **anaerobic respiration**, such as those for nitrate reductase, fumarate reductase, and formate dehydrogenase.

5. Two-Component Regulatory Systems

Covalent addition of a chemical group (as opposed to binding an entire signal molecule) is widely used to control the activity of both enzymes and DNA-binding proteins. Phosphate is the most common group used, but methyl, acetyl, AMP, and ADP-ribose moieties may also be used.

One large class of regulatory systems that use a phosphate group is the **two-component regulatory systems**. Although often regarded as characteristic of bacteria, they have also been found in lower eukaryotes, including yeast and slime molds. As the name implies, two-component regulatory systems consist of two proteins that cooperate to regulate gene expression. The first component is a DNA-binding regulator protein that only binds DNA when phosphorylated. The second is a **sensor kinase** that senses a change in the environment and changes shape. This causes the sensor kinase to phosphorylate itself using ATP and then transfer the phosphate group to the DNA-binding regulator (Fig. 16.20).

There are many different two-component regulatory systems in *E. coli* (see Table 16.02 for examples). Usually the sensor kinase is a trans-membrane protein that senses either physical conditions of some sort (e.g., aeration, osmotic pressure) or a nutrient (e.g., phosphate, nitrate). The DNA-binding form of the regulator may

> Some signals consist of chemical alterations to protein molecules.

> A sensor protein plus a regulator protein often act together as a two-component regulatory system.

---

anaerobic respiration   Respiration using other oxidizing agents (e.g., nitrate) instead of oxygen
iron sulfur cluster   Group of iron and sulfur atoms found in proteins and involved in oxidation/reduction reactions
sensor kinase   A protein that phosphorylates itself when it senses a specific signal (often an environmental stimulus, but sometimes an internal signal)
two-component regulatory system   A regulatory system consisting of two proteins, a sensor kinase and a DNA-binding regulator

FIGURE 16.20
Model of Two-Component Regulatory System

The two-component regulatory system includes a membrane component and a cytoplasmic component. Outside the cell, the sensor domain of the kinase detects an environmental change, which leads to phosphorylation of the transmitter domain. The response regulator protein receives the phosphate group, and as a consequence, changes configuration so as to bind the DNA.

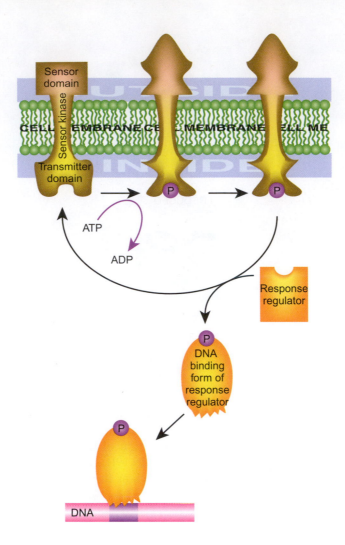

TABLE 16.02	Two-Component Regulatory Systems in *E. coli*	
Stimulus/Function	**Sensor**	**Regulator**
Lack of oxygen	ArcB	ArcA
Osmolarity, envelope proteins	EnvZ	OmpR
Osmolarity, potassium transport	KdpD	KdpE
Phosphate deprivation	PhoR	PhoB
Nitrogen metabolism	NtrB	NtrC
Nitrate respiration	NarX	NarL
Nitrate and nitrite respiration	NarQ	NarP

act as an activator or a repressor. The ArcAB system senses aerobic versus anaerobic conditions. Under anaerobic conditions, ArcB phosphorylates itself and then phosphorylates ArcA. The ArcA~P regulator then represses about 20 genes that are only required for aerobic metabolism and activates half a dozen genes needed when oxygen is absent or very low.

Two component systems are present in many other bacteria. In *Mycobacterium tuberculosis*, the causative agent of tuberculosis, some virulence factors are controlled by two-component regulators. PhoP is a DNA-binding regulator that induces the

Gora KG, Tsokos CG, Chen YE, Srinavasan BS, Perchuk BS, Laub MT (2010) A cell-type-specific protein-protein interaction modulates transcriptional activity of a master regulator in *Caulobacter crescentus*. Mol. Cell 39:455–467

C. crescentus is a good model prokaryotic organism to study differentiation because the crescent-shaped bacteria divide into two different types of cells. The mother cell keeps a stalk that sticks to the soil or solid surfaces in fresh or sea water. After cell division, her daughter cell, called a swarmer cell, forms a flagellum and swims away to a new location where the cell then reattaches to the surface and converts into a stalked cell. The steps from swarmer cell to stalked cell are highly regulated in step with the cell cycle. In G1-phase, a swarmer cell differentiates into a stalked cell. After attachment, the stalked cell is competent to progress in the cell cycle. First, the cell replicates and partitions its DNA during S-phase. Next, the stalked cell divides into two during M-phase, forming another swarming daughter cell that stays in G1 until it attaches to a surface.

These events are controlled by a two-component regulatory system. In *C. crescentus*, the response regulator, CtrA, is a master regulator of cell cycle progression. Phosphorylated CtrA (CtrA~P) is abundant in G1 swarmer cells, and silences the origin of replication, perhaps by precluding the binding of DnaA protein. As the cell attaches and differentiates into a stalked cell during G1, CtrA is dephosphorylated and degraded, exposing the origin of replication to initiate DNA synthesis. During S-phase, the gene for CtrA (*ctrA*) is transcribed and translated into more protein that is phosphorylated. At this point of the cell cycle, CtrA~P acts as an activator protein that induces transcription of genes used in G2 and M phase. Interestingly, CtrA~P does not activate these genes during G1; therefore, some other factor controls whether or not CtrA~P acts as a transcription activator.

In this paper, the authors identified a possible candidate that determines whether or not CtrA~P can promote transcription. The gene, called *sciP* (small CtrA inhibitory protein), encodes a protein that contains a helix-turn-helix motif, suggesting that it binds to DNA. The paper provides evidence that this protein prevents CtrA~P from activating gene transcription during G1-phase. SciP is only found during G1-phase, and if SciP is depleted during G1, then CtrA activated genes that are normally repressed during G1 become activated. The authors confirmed that SciP physically associates with CtrA~P, and that the complex prevents RNA polymerase from binding to the promoters of CtrA activated genes.

FOCUS ON RELEVANT RESEARCH

These results suggest a model of cell cycle dependent gene expression. In G1, SciP is abundant, and binds to CtrA~P at the promoter of CtrA-regulated genes. This complex precludes RNA polymerase from binding to the promoter, and none of these genes are expressed. In this state, any newly synthesized CtrA~P is free to bind to the origin of replication, thus preventing replication. As cells enter S-phase, CtrA is dephosphorylated and degraded so that the repressor complex of SciP and CtrA~P are removed from the promoters. In addition, DnaA can bind to the origin of replication and DNA replication can begin without CtrA~P present. After this time, more CtrA protein is made and phosphorylated, but there is no SciP present during these stages of the cell cycle. Without SciP, CtrA~P binds to CtrA-controlled genes, recruits RNA polymerase, and transcription is activated for any gene that has the CtrA~P binding site upstream from the RNA polymerase binding site. In other genes, CtrA~P binds to the RNA polymerase binding site and inhibits transcription. This moves the cell from S into G2 and converts the newly made daughter cell into a swarmer (Fig. 16.21).

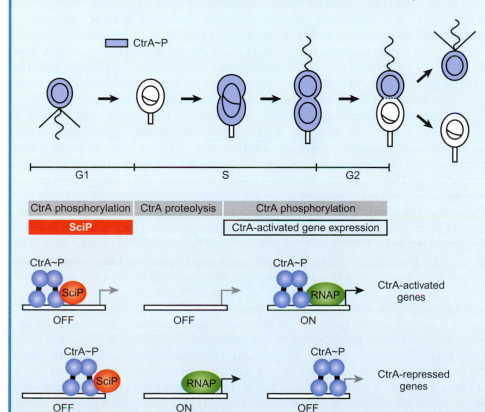

FIGURE 16.21
Control of* Caulobacter *Cell Division

The *Caulobacter* cell cycle summary is diagrammed at the top. In the cell at G1, the CtrA protein is phosphorylated and SciP is expressed. These two regulatory events combine to prevent transcription of CtrA-activated genes and CtrA-repressed genes. At the onset of S-phase (synthesis), the CtrA protein is degraded by proteolysis and the absence of CtrA allows RNA polymerase to transcribe CtrA-repressed genes. At the midpoint of S-phase, CtrA is remade and phosphorylated as earlier, but the absence of SciP allows CtrA to act as an activator protein, and CtrA-activated genes are turned on.

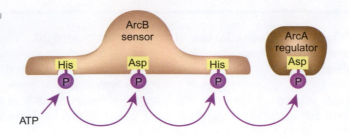

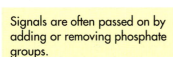

Signals are often passed on by adding or removing phosphate groups.

transcription of a virulence factor gene in response to its partner sensor kinase, PhoR. This sensor kinase autophosphorylates when it senses that it is inside the host organism. Mutations in PhoP cause decreased virulence and help treatment of tuberculosis. Two-component systems are also involved in those rare cases where bacteria show differentiation between different types of cell (see Focus on Relevant Research on the previous page).

5.1. Phosphorelay Systems

The pathway of phosphate transfer in two-component regulatory systems actually involves four protein domains. These domains are highly conserved among different regulatory proteins and are of two types, those where the phosphate is attached to a histidine residue and those where it is carried by an aspartate. In a typical phosphorelay, the phosphate passes from His to Asp to His to Asp. In the case of the ArcAB system, the first three sites are on the ArcB protein and the fourth is on ArcA (Fig. 16.22).

In addition to the two-component regulatory systems, other control systems use phosphorelays. The number of proteins, the total number of phosphate-binding domains, and their arrangement varies in different regulatory systems. For example, the widespread Rcs system has five main proteins (see Focus on Relevant Research on the following page). The regulator at the end of the line may bind DNA upon being phosphorylated or it may activate/deactivate one or more enzymes. In eukaryotic cells, especially in multicellular organisms, there are many highly complex signal transmission pathways, which often include one or more phosphorelays.

Two-component regulatory systems and phosphorelays are also found in cyanobacteria, a category of algae that use photosynthesis to generate energy. Cyanobacteria are found in every niche on the Earth and have been around for eons. In fact, plant chloroplasts are believed to be degenerate cyanobacteria that became symbiotic within the eukaryotic host cells. During evolution, the photosynthetic genes were maintained and the remaining genes for independent growth were lost.

Some cyanobacteria can switch color from brick red to green in a process called complementary chromatic adaptation (CCA). The bacteria turn red in green light and green in red light. The bacteria use antennae called phycobilisomes to absorb the available light from the environment. These are made of proteins called phycobiliproteins, and depending upon the available light, these change from a reddish protein (phycoerythrin, PE) to a green protein (phycocyanin, PC). Red light activates expression of PC, which absorbs red light (and appears green because this light is reflected). In contrast, in green light, the gene for PE is activated. The genes are controlled by a sensor histidine kinase, RcaE, which has a light sensing pigment on its extracellular surface. Light activation of RcaE triggers autophosphorylation, and then the phosphate group is transferred to RcaF. In red light, the phosphate is then transferred to RcaC, which activates transcription of PC by binding to its promoter. The discovery of two-component systems in cyanobacteria suggests that this type of gene regulation has been used throughout evolution.

Schmöe K, Rogov VV, Rogova NY, Löhr F, Güntert P, Bernhard F and Dötsch V (2011) Structural insights into Rcs phosphotransfer: the newly identified RcsD-ABL domain enhances interaction with the response regulator RcsB. Structure 19:577–587.

FOCUS ON RELEVANT RESEARCH

The Rcs system was originally discovered as a regulator of capsule formation. However, it is involved in a wide range of other phenomena including cell division, motility, biofilm formation, and virulence. The Rcs system is widely distributed among the gram-negative bacteria including enteric bacteria such as *E. coli* and *Salmonella*.

The Rcs phosphorelay system has five components: RcsF, RcsC, RcsD, RcsB, and RcsA. The sensor, RcsF, is an outer membrane lipoprotein that triggers the cascade by autophosphorylation. The phosphate group then travels via RcsC and RcsD to the DNA-binding protein, RcsB. The RcsB protein binds to the DNA either alone or in combination with protein RcsA, depending on level of phosphorylation.

The authors have identified a new domain on the RcsD protein that interacts with RcsB. They determined its structure and investigated the details of its interaction with RcsB. Their main approach was the use of sophisticated NMR spectroscopy accompanied by computer modeling. Finally, they generated a structural model of the RcsB/RcsD complex.

6. Specific versus Global Control

Many bacteria can grow on a wide range of sugars, such as fructose (fruit sugar), lactose (milk sugar), and maltose (from starch breakdown), as well as glucose. When a preferred sugar, such as glucose, is present, less favored sugars, such as fructose, lactose, or maltose, are not used. Only when glucose runs out will the other sugars be consumed. In molecular terms, this means the genes for using these other sugars are switched off when glucose is available.

A **regulon** is a group of several genes or operons that are turned on or off in response to the same signal by the same regulatory protein. The members of a regulon have separate promoters and are widely separated on the chromosome. Two examples of regulons in *E. coli* are the genes for using maltose, which are divided into several operons, and the genes for the synthesis of arginine. The arginine regulon consists of a dozen genes for biosynthesis and transport scattered over nine locations on the chromosome. They are controlled by a repressor, ArgR, which binds arginine as co-repressor, and is unlinked to any of the genes it controls.

As noted above, specific regulation refers to control by a signal specific for a small group of genes. Thus, *allo*-lactose induces the *lac* operon, maltotriose induces the *mal* regulon, etc. In contrast, **global regulators** control large numbers of genes in response to a more general signal or stimulus. Most genes respond to both specific and global signals. Thus, in addition to specific control by the *lac* repressor, the *lac* operon is also regulated by the global activator protein **Crp (Cyclic AMP Receptor Protein)**. Similarly, the maltose genes are regulated by the specific activator protein MalT and also by Crp. Thus, Crp responds to overall carbon source availability and mediates the choice among different sugars.

Global regulators control large families of genes.

6.1. Crp Protein Is an Example of a Global Control Protein

Crp is a global activator protein that is required for switching on the genes for using maltose, lactose, and other nutrients less favored than glucose. The Crp protein is allosteric. In order to bind DNA and activate genes, it must first bind its signal molecule, cyclic AMP. When Crp binds cyclic AMP, it forms dimers and these can bind to

CRP (cyclic AMP receptor protein)   Bacterial protein that binds cyclic AMP and then binds to DNA
global regulator   A regulator that controls a large group of genes, generally in response to some stimulus or developmental stage
regulon   A set of genes or operons that are regulated by the same regulatory protein even though they are at different locations on the chromosome

FIGURE 16.23
Cyclic AMP and the Crp Global Regulator

Individual Crp units bind cyclic AMP to form a dimer that has the ability to bind DNA.

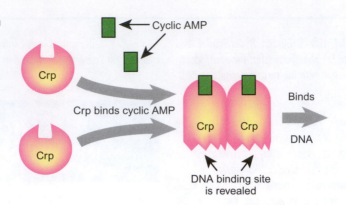

The global regulator Crp binds the signal molecule, cyclic AMP.

a recognition site in the DNA upstream of the promoter. The presence of Crp helps RNA polymerase bind to the promoter (Fig. 16.23).

Cyclic AMP is a global signal that the bacterial cell has run out of glucose, its favorite energy source. Only when this has occurred can the genes for using less favored nutrients be switched on. Consequently, in order to switch on genes for using any individual sugar, say, lactose, both an individual signal (the availability of lactose) and a global signal that indicates the need for nutrition (cyclic AMP) are required. Because of this role, Crp has also been called CAP, for catabolite activator protein.

Whether or not the *lac* operon is switched on or off thus depends on the two regulator proteins, LacI and Crp. The various possibilities are illustrated in Figure 16.24. Only when the repressor, LacI, is absent and the Crp protein is present to help it bind can RNA polymerase bind to the promoter and make mRNA.

6.2. Regulatory Nucleotides

Cyclic AMP is a key signal molecule in many control systems, both in bacteria and higher organisms. Because it is derived from a nucleic acid precursor, it is known as a **regulatory nucleotide**. Other regulatory nucleotides include cyclic guanosine monophosphate (cyclic GMP; important mostly in animals), cyclic diguanosine monophosphate (c-di-GMP; important in biofilm formation in bacteria), and guanosine tetraphosphate (ppGpp; controls stringent/starvation response in bacteria).

Cyclic AMP is synthesized from ATP by the enzyme adenylate cyclase. In bacteria, glucose inhibits the synthesis of cyclic AMP and also stimulates cyclic AMP transport out of the cell. When glucose enters the cell, the cyclic AMP level drops, Crp protein cannot bind DNA, and RNA polymerase fails to bind to the promoters of operons subject to catabolite repression. Thus, catabolite repression is the indirect result of the presence of a better energy source (glucose). The direct mediator of catabolite repression is a low level of cyclic AMP (Fig. 16.25).

Cyclic di-GMP regulates the transition between the free-swimming and biofilm lifestyles in many bacteria. These control networks are unusually complex for bacteria and combine inputs from several sources. Biofilms allow bacteria to maintain a foothold in hazardous situations. For example, forming films that tightly bind to surfaces allows bacteria to live in rapidly running streams by adhering to rocks. Bacteria, as biofilms, are also capable of clogging fuel lines in airplanes and colonizing clinical apparatus such as catheters, ventilators, and stents.

Quite apart from contaminating health equipment, biofilms are also directly important in bacterial virulence. One of the most fascinating cases is that of bubonic plague. The bacterium that causes plague, *Yersinia pestis*, is carried by fleas and transmitted when the flea bites a new animal host. This differs greatly from other, less deadly, species of *Yersinia* that are transmitted by contaminated food and water and

regulatory nucleotide   A nucleic acid base that is modified and used as a signaling molecule

A

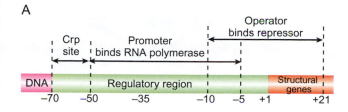

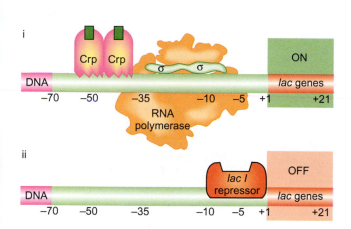

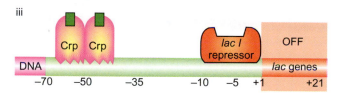

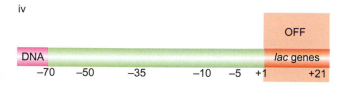

■ **FIGURE 16.24**
Overall Regulation of the lac Operon

A) The important regulatory elements are shown in relationship to the structural genes in the *lac* operon. B) Only the first of the four scenarios induces expression of the *lac* operon. In scenario (i) glucose is absent, hence CRP (plus cyclic AMP) binds, AND lactose is present, hence the LacI repressor is removed from the DNA by binding the inducer. In (ii) glucose is present, hence CRP is absent and lactose is also absent, hence LacI is still bound. In (iii) both glucose and lactose are absent and so although CRP is present, the LacI repressor still blocks transcription. In (iv) lactose is present, hence the LacI repressor is removed from the DNA. However, glucose is present, hence CRP is absent and RNA polymerase still cannot transcribe the genes.

cause intestinal ailments. *Yersinia pestis* blocks the digestive tracts of fleas by forming biofilms. This provokes the fleas to bite repeatedly and thus infect new hosts. The biofilm is held together by a polymer of N-acetyl-D-glucosamine, as is the case for many bacterial biofilms. Cyclic-di-GMP controls biofilm formation in *Yersinia pestis* as for many other bacteria; yet only *Y. pestis* can form biofilms inside fleas.

7. Accessory Factors and Nucleoid-Binding Proteins

In bacteria such as *E. coli*, there are several rather non-specific DNA-binding proteins that are sometimes referred to as accessory factors or **histone-like proteins**. They function partly by affecting chromosome structure and partly in gene regulation. Some tend to have negative effects on gene expression (e.g., H-NS, StpA), whereas others usually act in a positive manner (e.g., HU, IHF). However, these regulatory effects tend to be indirect and rather non-specific.

---

histone-like protein   Bacterial protein that binds non-specifically to DNA and participates in maintaining the structure of the nucleoid; they do not actually have much in common with true histones

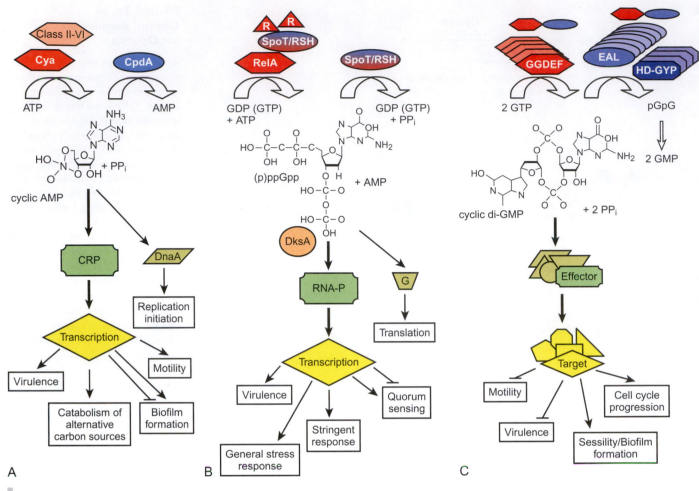

FIGURE 16.25
Bacteria Use Second Messengers to Control Gene Expression

Three basic second messenger control systems are shown. A) Cyclic AMP (cAMP) is synthesized by adenylate cyclase (Cya) when the phosphotransferase system signals an absence of glucose, and degraded by a phosphodiesterase (CpdA) when glucose is present again. Cyclic AMP activates transcription of many different genes through its interaction with the cyclic AMP receptor protein (CRP), and initiates replication via DnaA protein to adapt to the change in sugar availability. B) (p)ppGpp is created by RelA conversion of GDP and ATP in response to amino acid deprivation, and degraded by a RelA/SpoT homolog when amino acids are abundant. (p)ppGpp and DskA bind to RNA polymerase directly which changes the transcription of many different genes to adapt to the lack of amino acids by slowing growth. C) Cyclic-di-GMP (c-di-GMP) is created from two GTP molecules by enzymes that contain a protein domain called "GGDEF" and degraded by proteins that contain the domains called "EAL" and "HD-GYP." Multiple copies of enzymes with these specific domains exist. When expressed, cyclic-di GMP alters gene expression via a variety of targets that are still being investigated. This second messenger specifically activates genes associated with biofilm formation and cell cycle regulation. (*Credit: Pesavento C and Hengge R (2009) Bacterial nucleotide-based second messengers Curr Op Microbiol 12:170–176*).

The primary role of the **H-NS (histone-like nucleoid structuring)** protein of *E. coli* and related bacteria is to maintain the structure of the bacterial DNA within the nucleoid. The term nucleoid refers to the compact structure formed by the bacterial chromosome together with its accessory proteins. H-NS protein binds in a relatively non-specific manner, although it prefers regions of bent DNA as illustrated in Figure 16.26. H-NS consists of two domains, one for DNA-binding and one for protein-protein interaction. These domains are joined by a linker region. H-NS binds to DNA and then the H-NS proteins bind to each other, forming aggregates of four or more H-NS units. As they aggregate, the DNA loops out from the core of proteins.

> Bacterial DNA is covered with non-specific binding proteins.

H-NS protein (histone-like nucleoid structuring protein) A bacterial protein that binds non-specifically to DNA and helps maintain the higher level structure of the nucleoid

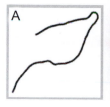
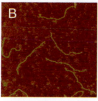
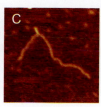
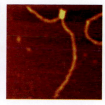
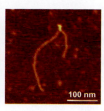

FIGURE 16.26
H-NS Binds Preferentially to Curved DNA

A) Structure of the linear DNA fragment used as predicted by the CURVATURE program. The fragment is shown in the plane of its intrinsic curvature. B) Atomic force microscope (AFM) image of naked linear DNA molecules with the curved region at one-third of their length. The image shows an area 900 × 900 nm. C) AFM image of DNA after incubation with H-NS (1 monomer per 20 bp). H-NS-DNA complexes are specifically formed at the position of the curved region only. The image shows a 300 × 300 nm surface area. The color scale ranges from 0.0 to 3.0 nm (from dark to bright). (*Credit: Structural basis for preferential binding of H-NS to curved DNA Dame RT, Wyman C, Goosen N, Biochimie 83 (2001) 231–234.*)

In addition, H-NS binds with higher affinity to the regulatory regions of a wide range of genes scattered throughout the bacterial chromosome. H-NS prefers to bind to about 350 A/T rich regions in the *E. coli* chromosome (about 10–15% of the genome), where it represses gene expression. Most of these genes respond to some sort of environmental conditions, but apart from this they are unrelated. In contrast to genuine global regulators, H-NS does not control a specific response nor does it respond to any particular signal; therefore, this effect is referred to as **silencing**. Induction of these genes requires specific transcriptional activators to overcome the silencing and remodel the H-NS binding.

The StpA protein is very similar to H-NS but is present in smaller amounts and appears predominantly under certain stressful conditions (e.g., high temperature or high osmotic pressure). The protein-binding domains of StpA and H-NS bind to each other so that mixed aggregates are formed. StpA silences fewer genes than H-NS. In particular, it allows expression of genes induced by stress. For example, the *proU* gene, expressed at high osmotic pressure, is silenced by H-NS but not by StpA.

7.1. Action at a Distance and DNA Looping

The **HU** (**heat-unstable nucleoid protein**) and **IHF** (**integration host factor**) proteins are often required as positive factors in gene expression. Both are **heterodimers**, consisting of two different subunits. The four subunits (two from each protein) are all similar in sequence and 3D structure and HU and IHF can, to some extent, substitute for each other. HU is relatively non-specific, whereas IHF is more specific. Both HU and IHF are examples of proteins that take part in bending DNA (as opposed to H-NS, which binds to DNA already bent as a result of its sequence). They help in integration, inversion, and recombination events by bending DNA into the appropriate conformation (see Ch. 24). They also affect the expression of certain genes that need the DNA in their upstream regions to be bent, in order to be transcribed. A variety of other accessory proteins, activator proteins, and repressor proteins are also involved in the looping of DNA.

> DNA may be bent into a loop by some regulatory proteins.

Many genes for nitrogen metabolism require the alternative sigma factor RpoN (=NtrA = σ54) for their transcription. In addition, they are regulated by activator proteins that bind far upstream. Most activator proteins, at least in bacteria, bind just upstream of the promoter and make direct contact with RNA polymerase, so helping it bind to the promoter. For the activators of RpoN-dependent promoters to

---

HU protein (heat-unstable nucleoid protein)   A bacterial protein that binds to DNA with low specificity and is involved in bending of DNA
heterodimer   Dimer composed of two different subunits
IHF (integration host factor)   A bacterial protein that bends DNA so helping the initiation of transcription of certain genes; named after its role in helping the integration of bacteriophage lambda into the chromosome of *E. coli*
silencing   In genetic terminology, refers to switching off genes in a relatively non-specific manner

FIGURE 16.27
Looping of DNA in RpoN-dependent Promoter

The sites for binding of transcription factors and the alternative sigma factor, RpoN, are shown in A). The IHF protein induces a bend in the DNA, which brings the NtrC sites close to the binding site for the alternative sigma factor RpoN. Because the DNA loops around, the RNA polymerase can be bound by two sets of NtrC dimers as well as by the RpoN protein.

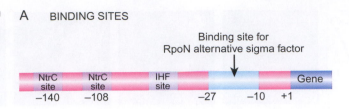

A   BINDING SITES

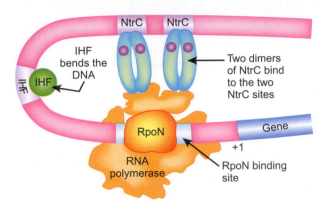

B   ACTIVE STATE

touch the RNA polymerase, the DNA must be bent around, forming a loop. The bend results from IHF binding between the promoter and activator sites (Fig. 16.27).

Genes for using many alternative nitrogen sources are regulated by the NtrBC two-component regulatory system; although NtrB is not a membrane protein as is typically the case (see above). In the absence of ammonia, the NtrB protein phosphorylates NtrC protein. NtrC~P then binds to the upstream region of nitrogen-source genes and activates transcription. Similarly, the genes for nitrogen fixation, in *Klebsiella* and related bacteria, require the RpoN sigma factor and the activator NifA. In both cases, IHF must bend the DNA into a loop for activation to work.

The enhancers that activate eukaryotic promoters also act at a distance and depend on looping of DNA (see Ch. 17). Because of this similarity, the activator sites of RpoN-dependent promoters have sometimes been called "bacterial enhancers." However, the bacterial activator sites are only about 100 bp upstream and occupy a fixed position. In contrast, eukaryotic enhancers may lay several kilobases upstream or downstream of their target genes and can work in either orientation.

8. Anti-Termination as a Control Mechanism

Anti-termination factors are proteins that prevent termination of transcription at specific sites. The RNA polymerase therefore continues on its way and transcribes the region of DNA beyond the terminator (Fig. 16.28). This mechanism for controlling gene expression is common in bacterial viruses and is also found for a few bacterial genes. Anti-termination factors attach themselves to the RNA polymerase before it reaches the terminator. The recognition sequences for anti-termination are found in the DNA well upstream of the terminator. As the RNA polymerase passes by, the anti-termination factors are loaded on. They remain attached and allow the RNA polymerase to travel past the stem and loop region of the terminator without pausing. Consequently, termination is suppressed.

In genetics there are examples of anti-everything—including anti-termination.

anti-termination factor   Protein that allows transcription to continue through a transcription terminator

A   START OF TRANSCRIPTION

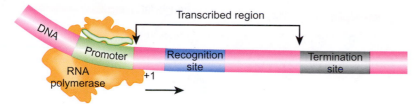

B   ABSENCE OF ANTI-TERMINATION FACTOR

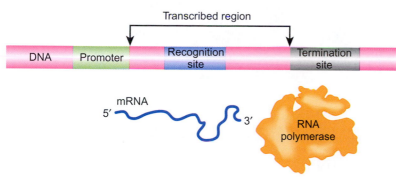

C   PRESENCE OF ANTI-TERMINATION FACTOR

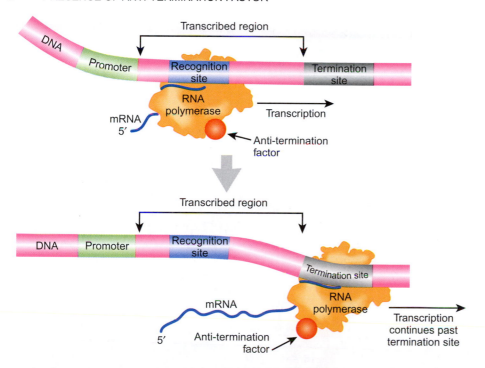

■▶ **FIGURE 16.28**
Operation of Anti-Termination Factor

A) Transcription of the DNA begins with the RNA polymerase bound to the promoter. B) In the absence of an anti-terminator factor, the RNA polymerase reaches the terminator region and drops off, having transcribed a short RNA. C) In the presence of an anti-termination factor, the RNA polymerase binds the factor when it reaches the recognition site. The factor allows the RNA polymerase to transcribe through the termination site.

Anti-termination in *E. coli* involves several **Nus proteins**. **NusA protein** is probably attached to the core RNA polymerase shortly after the sigma factor is lost just after initiation. NusA itself actually promotes termination, apparently by increasing the duration of pauses at hairpin structures. NusA and sigma cannot both bind to the core enzyme simultaneously. As long as RNA polymerase is attached to DNA, the Nus proteins cannot be dislodged. However, addition of sigma displaces NusA from free RNA polymerase (Fig. 16.29). Thus, RNA polymerase cycles between initiation mode (with sigma bound) and termination mode (with NusA bound).

Nus proteins   A family of bacterial proteins involved in termination of transcription and/or in anti-termination
NusA protein   A bacterial protein involved in termination of transcription

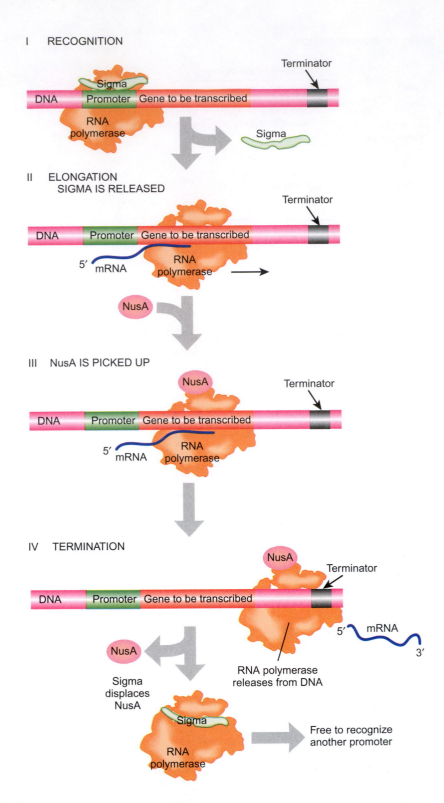

FIGURE 16.29
Sigma NusA Cycle of RNA Polymerase

RNA polymerase needs the sigma subunit to recognize and bind to the promoter. Once RNA polymerase moves forward and starts transcribing, it releases sigma and picks up NusA protein. After termination, NusA protein is displaced by sigma.

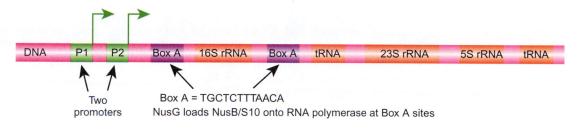

Box A = TGCTCTTTAACA
NusG loads NusB/S10 onto RNA polymerase at Box A sites

Two promoters

FIGURE 16.30
Operation of Anti-Termination in E. coli rrn Genes

As RNA polymerase passes the *boxA* sequences, NusG protein loads NusB plus NusE (=S10 = RpsJ) onto the polymerase. These Nus proteins prevent premature termination.

The other Nus proteins are involved in anti-termination. Two of these proteins, NusB plus RpsJ (=NusE), are attached to the RNA polymerase as it passes a "*boxA*" anti-termination sequence (Fig. 16.30). RpsJ (S10) is also found in the small subunit of the ribosome. The connection between its two roles is still obscure. NusG protein probably helps in loading. The presence of NusA is required for NusB plus RpsJ to bind to RNA polymerase. The best known genes in *E. coli* that show anti-termination are the *rrn* genes for synthesis of ribosomal RNA.

Key Concepts

- Genes are regulated to ensure appropriate physiological responses.
- Regulation of transcription is widespread. Several possible stages of transcription may be regulated.
- In prokaryotes, alternative sigma factors recognize different sets of genes intended for different environmental conditions.
- In prokaryotes, response to heat shock depends on sigma factors that are regulated by temperature.
- Spore formation in *Bacillus* is regulated by cascades of alternative sigma factors.
- Anti-sigma factors are proteins that inactivate sigma factors and anti-anti-sigma factors reactivate the sigma factors.
- In prokaryotes, many genes are clustered into operons and transcribed and regulated as a group.
- Activator and repressor proteins take part in positive and negative transcriptional regulation
- Some regulatory proteins may function as either a repressor or an activator depending on the situation.
- Regulatory proteins are often controlled in turn by binding small signal molecules.
- Regulatory proteins may also be controlled by covalent modification, most often by addition of a phosphate group.
- Two-component regulatory systems consist of a sensor protein that monitors conditions and a regulator protein that binds DNA and regulates transcription of genes.
- Some more complex regulatory systems operate by phosphorelays, in which phosphate groups are transferred between multiple regulatory proteins.
- Global control is the regulation of a large number of genes in response to the same signal.

- Regulatory nucleotides, such as cyclic AMP, often act as signal molecules.
- A variety of nucleoid-binding proteins cover the DNA in bacterial cells.
- When regulatory proteins bind DNA far upstream of a gene, the control mechanism often involves DNA looping.
- Anti-termination of transcription is a rare form of gene regulation.

Review Questions

1. List the stages at which gene regulation can take place.
2. What is positive regulation? What is negative regulation?
3. What steps are involved in regulation at the level of transcription?
4. What recognizes bacterial promoters that lack normal -10 and -35 recognition sequences but have distinct promoter recognition sequences?
5. During sporulation the spore and the mother cell cross regulate each other. How is this accomplished, which sigma factors are involved, and where do they originate (the mother cell or the spore)?
6. What are anti-sigma factors?
7. What is an inducer?
8. What is an operon?
9. What is polycistronic mRNA?
10. What is a benefit of having genes organized into operons?
11. What do the structural genes *lac*Z, *lac*Y, and *lac*A encode?
12. What is an operator?
13. What is a gratuitous inducer? Give an example.
14. What is the repressor for the lactose operon, how is it removed?
15. How can a DNA-binding protein act as both an activator and a repressor?
16. What is autogenous regulation?
17. What is a co-repressor?
18. Name ways in which activators and repressors can be chemically modified (naturally, in living cells). Give examples.
19. What is a two-component regulatory system?
20. What is a sensor kinase?
21. Which two amino acids are phosphorylated in a two-component regulatory system?
22. What is specific regulation? What is global regulation?
23. What is the global activator protein for the *lac* operon?
24. What is a regulon?
25. What signal molecule must Crp bind before it can bind DNA?
26. What two conditions must be met for RNA polymerase to bind to the *lac* operon?
27. What are histone-like proteins?
28. What is H-NS protein, and what is its function?
29. What are the two domains of H-NS protein and how does H-NS protein aid in the condensation of DNA?
30. Define silencing as it applies to genes.
31. How can HU and IHF aid in activation?
32. What are anti-terminator factors? How do they function?

Conceptual Questions

1. The lactose operon was the first operon studied, and many mutations have been created to understand how the operon functions. In genetic nomenclature, a small superscript " + " denotes that the gene is normal, whereas a superscript "-" denotes the gene is defective or mutated. When a gene is deleted, the "Δ" symbol precedes the gene designation. In the following chart, fill in the expected phenotype for each of the genetic mutants using your knowledge of *lac* operon control. The + + + + designates enzyme activity.

Genotype	LacZ (β-galactosidase) activity		LacY (lactose permease) activity	
	with lactose	without lactose	with lactose	without lactose
I⁺P⁺O⁺Z⁺Y⁺ (wild-type)	++++	None	++++	none
IP⁺O⁺Z⁺Y⁺				
I⁺P⁻O⁺Z⁺Y⁺				
I⁺P⁺O⁺ΔZY⁺				
I⁺P⁺O⁺Z⁺ΔY				

2. In *E. coli*, partial diploids can be created by adding an extrachromosomal ring of DNA called a plasmid that has an origin and terminator regions so it is replicated and maintained throughout each cell division. The plasmid can be modified in the lab to contain different genes such as all the genes in the *lac* operon. When the plasmid is introduced into the bacteria, there are two copies of each of the genes in the *lac* operon, and this is denoted as $I^+P^+O^+Z^+Y^+/I^+P^+O^+Y^+Z^+$ where the genes before the "/" are chromosomal and those genes after the "/" are on the plasmid. The following partial diploid, $I^-O^+Z^+Y^-/I^-O^+Z^+Y^+$, has β-galactosidase activity and lactose permease activity with or without an inducer. Why?

3. Two different types of *E. coli* bacteria were grown in rich broth. The first type of *E. coli* is wild-type for H-NS and the second type has a deletion that removes the gene for H-NS. The mRNA from each culture was isolated and analyzed for the total amount of mRNA for each of the following genes using quantitative PCR. Each mRNA is compared to an internal control mRNA, and the fold induction is presented. Based upon this data, what genes are regulated by H-NS? What type of DNA structure is most likely present in the promoter for these genes?

Gene	Wild-type *E. coli*	H-NS Deletion
LacY	2.0	2.0
YgeH	18.5	3.1
TetR	28.8	25.7
GadA	16.0	1.0
AppY	33.9	1.8
YdeO	80.0	9.0

4. The *trp* operon is under control of an aporepressor called TrpR that binds to the operator site upstream from the biosynthetic genes of tryptophan biosynthesis. How would you expect the presence or absence of tryptophan to affect transcription of the operon? Unlike the inducible LacI repressor, aporepressors do not bind the operator site unless bound by a co-repressor. If tryptophan is the co-repressor, describe the binding of TrpR to the operator with and without tryptophan.

5. Look at the following promoter sequence. Underline the sequences necessary for σ^{54} factor binding based on the information in Table 16.01.

   −40 GATCGCAGCCGGATTGGCAATATCCTTGCAATACTTAAATC **+1**

The same principles of transcriptional regulation apply to both prokaryotes and eukaryotes. It is therefore assumed that the basic concepts from the previous chapter have been understood before continuing into eukaryotic transcriptional regulation. Transcriptional regulation in eukaryotes, especially multicellular organisms, is more complex than in prokaryotes. Protein-encoding genes in higher organisms are regulated by a wide selection of transcription factors. Signals from multiple transcription factors are transmitted to the RNA polymerase by a protein complex known as the "mediator." The enormous amounts of DNA and lengthy intergenic regions found in eukaryotic genomes introduce extra problems for gene regulation. In addition, DNA packaging by histones must be taken into account when expressing eukaryotic genes.

1. Transcriptional Regulation in Eukaryotes Is More Complex Than in Prokaryotes

Higher eukaryotes have many more genes than bacteria and regulate their expression differently in different tissues of the body and at different stages of development. In general, expression of a eukaryotic gene requires the presence of several activators. These may bind to the upstream region of the promoter or to enhancer sequences that may be several kilobases away from the promoter, as described briefly in Chapter 11. Furthermore, eukaryotic enhancers may lie downstream of their target genes and work in either orientation.

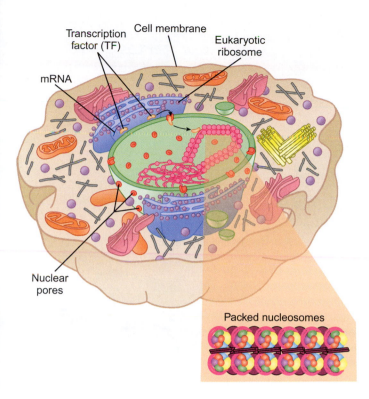

Transcription
factor (TF)

Cell membrane

Eukaryotic
ribosome

mRNA

Nuclear
pores

Packed nucleosomes

■ FIGURE 17.01

***Eukaryotic Genes Are
Difficult to Access***

Compartmentalization of the
eukaryotic cell into nucleus and
cytoplasm means that transcription
factors are made in the cytoplasm
and must then be transported into
the nucleus. The DNA in the nucleus
is often highly condensed and
difficult to access.

Controlling gene expression
in eukaryotes is complicated
by the high number of genes
and the segregation of the
chromosomes in the nucleus.

Eukaryotic genes are sequestered in the nucleus. Since transcription factors are proteins, they are made by ribosomes in the cytoplasm, but to act they must enter the nucleus. Although both bacterial and eukaryotic DNA are condensed and covered with protein this is much more pronounced in eukaryotic cells. Here, the DNA is highly condensed into nucleosomes and protected by histones. Nucleosomes condense into a 30 nm fiber, and then the fiber folds into looped domains that bring DNA sequences thousands of base pairs apart in the linear sequence of DNA into close contact. As discussed in Chapter 11, transcription factors control the looping of far away enhancers in close proximity to gene promoters. In parts of the genome, long sections of DNA are frequently condensed tightly into heterochromatin and are therefore inaccessible to RNA polymerase (Fig. 17.01). This makes access to the DNA difficult, both for RNA polymerase and transcription factors. For transcription to occur, the DNA must first be exposed. Nuclear pores also affect gene expression in yeast and are found associated with highly expressed genes.

2. Specific Transcription Factors Regulate Protein-Encoding Genes

This section deals with the regulation of genes that encode proteins and that are transcribed by eukaryotic RNA polymerase II. As already discussed in Chapter 11, several general transcription factors are required for expression of these genes. Expression also requires specific transcription factors that only affect certain genes in response to specific stimuli or signals. Transcription factors may bind to upstream elements in the promoter region or to enhancer elements that lie far away from the promoter.

Typical specific transcription factors share four general properties:

1. They respond to a stimulus which signals that one or more genes should be turned on.

2. Unlike most proteins, transcription factors are capable of entering the nucleus where the genes reside.

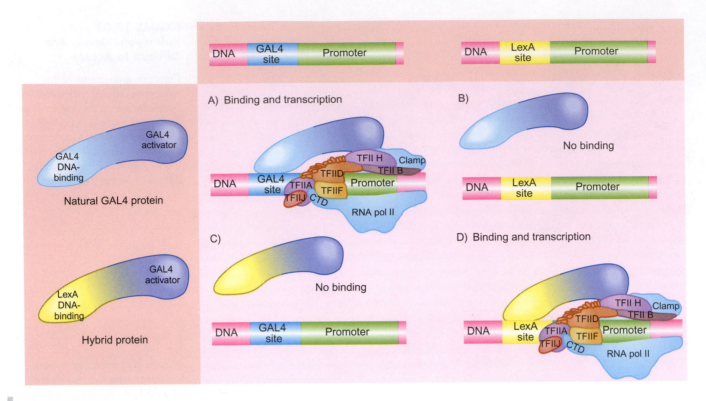

FIGURE 17.02
Transcription Factors Have Two Independent Domains

A) One domain of the GAL4 transcription factor normally binds to the GAL4 DNA recognition sequence and the other binds to the transcription apparatus. B) If the LexA recognition sequence is substituted for the GAL4 site, GAL4 does not recognize the DNA and no binding occurs. C) An artificial protein made by combining a LexA-binding domain with a GAL4 activator domain will not recognize the GAL4-binding site, but D) will bind to the LexA recognition sequence and activate transcription. Thus, the GAL4 activator domain acts independently of any particular recognition sequence.

3. They recognize and bind to a specific sequence on the DNA.

4. They also make contact with the transcription apparatus, either directly or indirectly.

> Protein-encoding genes of eukaryotes are regulated by transcription factors that respond to specific signals.

Certain DNA-binding proteins may respond directly to a stimulus, a situation that is common in prokaryotes. In contrast, the transcription factors of higher organisms are more often separated from the original signal by several intervening steps. Here, the DNA-binding proteins and their effects on transcription will be considered first.

Transcription factors usually have at least two domains, one that binds to DNA (**DNA binding domain**) and another that interacts with the transcription apparatus (**activation or activator domain**). This may be illustrated by using artificial hybrid proteins consisting of the DNA-binding domain from one protein plus the activation domain of another (Fig. 17.02). A hybrid with the DNA-binding domain from a bacterial protein plus the activation domain of yeast GAL4 activator will no longer activate transcription from the original yeast promoter. However, it will activate transcription from an artificial promoter into which the recognition sequence for the bacterial protein has been inserted. The bacterial DNA-binding protein used in these experiments was LexA, which is actually a repressor. These hybrid proteins are especially useful for controlling expression of genes introduced into an organism by genetic engineering, by transformation in bacteria, or by transfection of cultured eukaryotic cells.

activation or activator domain   The part of a transcription factor that interacts with the transcription apparatus
DNA binding domain   The part of a transcription factor that binds to DNA

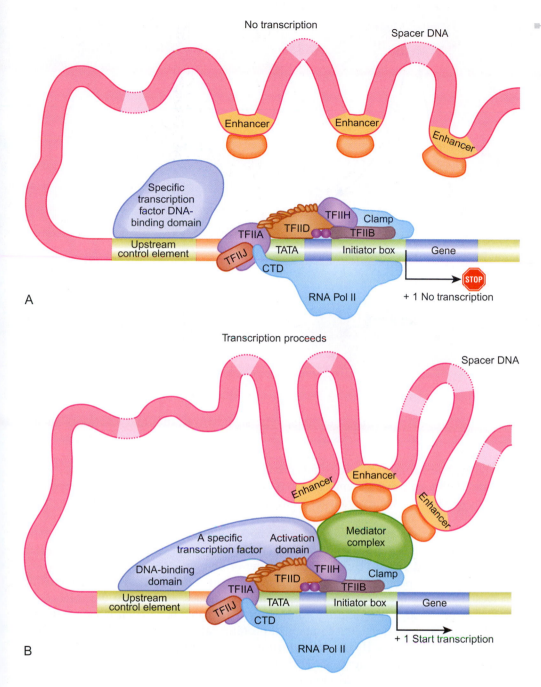

■ **FIGURE 17.03**
Activator Proteins and the Mediator

A) The folding of DNA allows numerous activators that are bound to enhancer sequences to approach the transcription apparatus. B) The mediator complex allows contact of the activators and/or repressors with the RNA polymerase.

2.1. The Mediator Complex Transmits Information to RNA Polymerase

In prokaryotes, sigma factors recognize the promoter and activator proteins help RNA polymerase to bind to the promoter. In eukaryotes, recognition and binding to the promoter are both functions of the general transcription factors. Activators in eukaryotes may be viewed as granting RNA polymerase permission to proceed forward from the promoter. Some eukaryotic activators make contact with the general transcription factors TFIIB, TFIID, and TFIIH. However, this is not sufficient to initiate transcription.

The **mediator** is a protein complex that sits on top of RNA polymerase II (Pol II) and provides a site of contact for activators, especially those that are bound at enhancer sequences (Fig. 17.03). In humans, the mediator is comprised of 26 subunits and has a

The mediator complex combines multiple signals to regulate transcription of genes by RNA polymerase II.

mediator   A protein complex that transmits the signal from transcription factors to the RNA polymerase in eukaryotic cells

FIGURE 17.04
***Insulator Sequences
Restrict the Range of
Enhancer Action***

A large loop of DNA is shown
with an enhancer that may interact
with gene X or gene Y. Insulator
sequences at the base of the loop
are recognized by an insulator-
binding protein (IBP) that prevents
the enhancer from acting outside of
the loop.

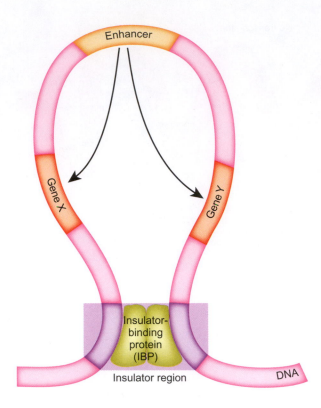

mass of 1.2 MDa. Apparently, it combines the signals from multiple activators and/or
repressors and sends the final result to the RNA polymerase II enzyme. Some subunits
of the mediator act in a positive manner while others act in a negative manner.

The mediator consists of a constant core that is similar in yeast and higher eukaryotes.
Attached to this are other subunits that vary between organisms and also between different tissues within the same organism. Many individual mediator proteins were identified
as "co-activator" proteins before it was realized that they belong together in a complex.

One role for the mediator is to recruit and stabilize Pol II at the promoter. The
large complex acts as a scaffold, holding the general transcription factors (TFIIB,
TFIID, TFIIE, and TFIIH) close to Pol II at the promoter. Many genes have this complex poised at the promoter, yet the genes are not yet transcribed. After the mediator
receives signals from the cytoplasm, then Pol II continues to elongate the mRNA. In
support of this model, two different forms of the mediator have been isolated. In one
form, the mediator is bound to a complex of proteins called CDK8, and these are found
in genes with paused Pol II. The second form of the mediator is free of CDK8 and is
found in actively transcribing genes. Perhaps removal and replacement of CDK8 alternates the mediator from transcriptional scaffold complex into a transcriptional activator.

2.2. Enhancers and Insulator Sequences Segregate DNA Functionally

Enhancers may be found up to several kilobases away from the promoter and either
upstream or downstream from the promoters they control. They work by looping the
DNA around so that the activator proteins bound at the enhancer can make contact with
the transcription apparatus via the mediator complex as discussed above (Fig. 17.03).

This looping mechanism allows a single enhancer to control several genes in its
vicinity. But how is an enhancer prevented from activating genes further along the
chromosome, which are supposed to be under control of another, closer enhancer?
It appears that chromosomes are divided into regulatory neighborhoods by special
sequences known as "boundary elements," or **insulators** (Fig. 17.04). An enhancer is

Enhancer sequences loop
around to contact the
transcription apparatus.

insulator A DNA sequence that shields promoters from the action of enhancers and also prevents the spread of heterochromatin

prevented from controlling a gene if an insulator sequence lies between them on the chromosome.

Insulators are regions of DNA consisting of clusters of sequences that bind multiple copies of special zinc-finger proteins known as **insulator-binding proteins** (**IBPs**). In vertebrates, the best known of these is CTCF (CCCTC-binding Factor). These must bind to the insulator sequences to block enhancer action. (In some organisms, such as the fruit fly *Drosophila*, there are not only fixed insulator sequences, but also mobile ones. An example is the so-called gypsy element, which is a retro-transposon and can move from place to place within the genome (see Ch. 22 for information on transposons.)

CTCF has many other potential roles in the cell beside creating physical boundaries, including stimulating and repressing gene expression, and controlling imprinting and X-chromosome inactivation, but more research is needed to fully understand the role CTCF plays in these other roles. Studies in which CTCF genes are deleted or mutated in the genome have dire consequences. In fact, mice with no gene for CTCF die shortly after fertilization, even before the zygote implants. Studies that deplete all mRNA transcripts for CTCF in cultured cells change cell growth, differentiation, and apoptosis (programmed cell death). Bioinformatics studies have found that there are about 15,000 CTCF-binding sites in the human genome. Taken together, the role of CTCF in the cell is clearly important.

The role CTCF plays in the *H19/Igf2* mouse gene locus has been studied in detail. The insulator sequence at this locus is converted between operational or nonoperational forms. Insulators are GC-rich and, as described below, CG sequences may be methylated. When the insulator element is methylated, it no longer binds the CTCF protein and no longer functions (Fig. 17.05). The *Igf2* gene (encoding insulin-like growth factor-II) and the *H19* gene are close together and face in the same direction. The maternal copy of the *Igf2* gene is normally silenced but the paternal copy is active. Conversely, the maternal *H19* gene is normally active, whereas the paternal copy is silenced. This is due to differing methylation patterns on the maternal and paternal chromosomes (i.e., an imprinting mechanism—see below).

The reason insulators were also called "boundary elements" is because they form boundaries to regions of heterochromatin. In addition to blocking the action of enhancers, insulators also prevent the spread of heterochromatin and the resultant silencing of genes (see below).

> Insulator sequences prevent enhancers from interfering with the wrong genes.

> Insulators can be inactivated by methylating their CG sequences.

2.3. Matrix Attachment Regions Allow DNA Looping

Both in bacteria and in eukaryotes, the DNA is arranged in giant loops attached at intervals to the chromosomal scaffold (see Ch. 4). In bacteria the loops consist of about 40 kbp of DNA, whereas the eukaryotic loops are somewhat longer, about 60–100 kbp.

During interphase, a filamentous web of proteins, the **nuclear matrix**, appears just on the inside of the nuclear membrane. DNA is attached to the proteins of the matrix by sites known as **matrix attachment regions** (**MARs**). Because the same DNA sites are used for attachment to the chromosomal scaffold during replication as for attachment to the nuclear matrix during interphase, they are sometimes also called **scaffold attachment regions** (**SARs**).

> DNA forms giant loops that are attached to scaffold proteins by special AT-rich sequences.

---

insulator-binding protein (IBP)   Protein that binds to insulator sequence and is necessary for the insulator to function
matrix attachment region (MAR)   Site on eukaryotic DNA that binds to proteins of the nuclear matrix or of the chromosomal scaffold—same as SAR sites
nuclear matrix   A mesh of filamentous proteins found on the inside of the nuclear membrane and used in anchoring DNA
scaffold attachment region (SAR)   Site on eukaryotic DNA that binds to proteins of the chromosomal scaffold or of the nuclear matrix—same as MAR sites

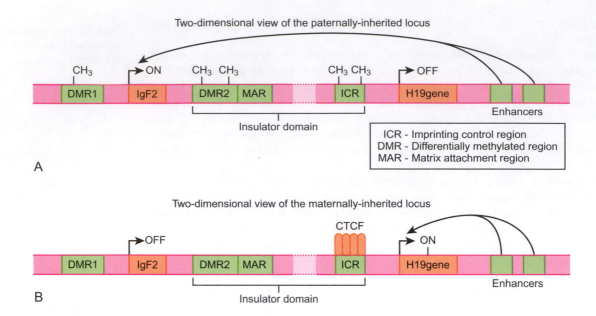

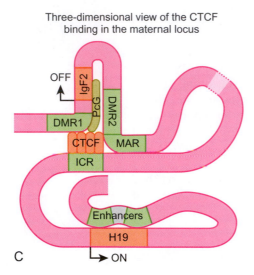

FIGURE 17.05
Methylation of Insulator Sequences and Binding

A) In the paternally-inherited locus, the methylation of DMR1, DMR2, and the ICR elements prevent CTCF from binding to the insulator domain. Without CTCF, the upstream enhancer elements can reach the *Igf-2* promoter and turn on gene expression. B) In the maternally-inherited locus, DMR1, DMR2, and ICR are NOT methylated; therefore, four CTCF proteins can bind to the ICR. This prevents the enhancer elements from reaching the *Igf-2* gene, and so they loop around to interact with the *H19* gene. C) The CTCF proteins bind to the unmethylated DMR1, MAR, and ICR region at the same time. This buries the *Igf-2* gene in a three-dimensional knot of chromatin. The structure is actually more complex than shown since MAR domains interact with the nuclear matrix proteins, and a member of the polycomb repressor complex (PcG) is also present. These proteins are implicated in histone methylation and hence chromatin silencing.

These MAR/SAR sites are 200–1000 bp long and AT-rich (70% AT), but otherwise share no obvious consensus. DNA with multiple runs of adenines (As) is inherently bent (see Ch. 4) and the nuclear proteins that bind the MAR sites recognize the bent DNA rather than a specific sequence. Topoisomerase II recognition sites are often found next to MAR sites, implying that the supercoiling of each giant loop is adjusted independently. Enhancers and other regulatory elements are often associated with MAR sites, and, at least in some cases, chromatin remodeling (see below) starts from a MAR site and then affects the whole chromatin loop (Fig. 17.06).

For example, a MAR site lies between *Igf2* and *H19*. The MAR site can attach to CTCF, thus linking the nuclear matrix scaffold to this gene locus. Studies using 3C

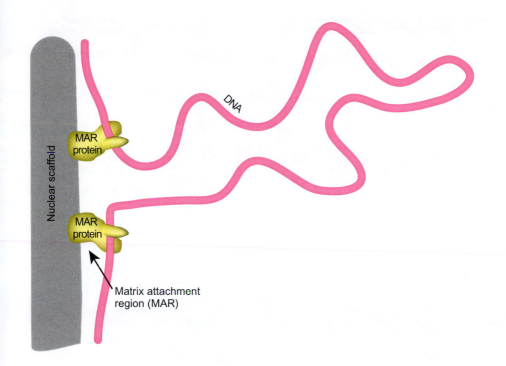

FIGURE 17.06
Looped Domains between MAR Sites

Although many loops are present, only a single loop of histone-free DNA is drawn coming from a region of the nuclear scaffold. The matrix attachment regions contain matrix attachment proteins (MAR protein) that anchor the DNA to the scaffold.

(chromatin conformation), a technique that crosslinks DNA to a particular protein, support the idea that the insulator sequence is near the MAR. The interaction of MAR, insulator sequence, and the nuclear scaffold can explain how very distant enhancer elements can remain associated with a particular promoter, but not others. The interesting caveat to remember in this gene system is that the condensed looping and associations of the DNA locus occurs in only the maternally-derived allele, and therefore in only one of the homologous chromosomes found in the nucleus. The other allele is methylated and found in a looser less condensed chromatin state (Fig. 17.05).

3. Negative Regulation of Transcription Occurs in Eukaryotes

Simple repressors are common in prokaryotes, but rarely found in eukaryotes. Those examples of repressors that do occur are usually found in simpler, single-celled eukaryotes, such as yeast. Although repressors are rare in eukaryotes, this does not mean that **negative regulation** itself is uncommon. On the contrary, some form of **negative control** is vital to most of the complex regulatory circuits found in higher organisms. Generally, negative signals act by hindering activator proteins or RNA polymerase itself in some manner.

> Negative regulators in eukaryotes often act by interfering with activators (rather than by obstructing the movement of RNA polymerase).

One obvious way to obstruct an activator is to occupy its recognition site on the DNA and so prevent the activator from binding. An example of this concerns the **CAAT box**, often found in eukaryotic promoters. Activation of the sea urchin *H2B* gene occurs in the testis only and requires, among other things, the binding of the activator protein CTF to the CAAT sequence. However, the CAAT-displacement protein (CDP) may also occupy the CAAT box and prevent binding of the activator. This occurs in embryonic tissue and prevents premature expression of testis-specific genes. The presence of CDP prevents assembly of the transcriptional apparatus. Note,

---

CAAT box   A sequence often found in the upstream region of eukaryotic promoters that binds transcription factors
negative control   See negative regulation
negative regulation   Control by a repressor that prevents expression of a gene unless it is somehow removed

FIGURE 17.07
Blocking the CAAT Box in Sea Urchins

Several recognition sequences must bind their appropriate transcription factors before transcription occurs. Transcription is prevented if CAAT displacement protein (CDP) binds to the site that CAAT-binding factor (CTF) normally fills. This prevents the mediator complex from activating the transcriptional apparatus, and therefore the gene is not expressed.

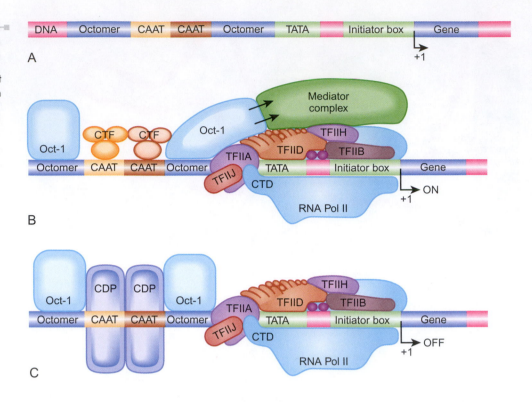

> The activity of some transcription factors is controlled by forming mixed dimers with different partners.

however, that CDP does not block the binding site for RNA polymerase as a classical bacterial repressor would do (Fig. 17.07).

Another example of negative regulation involves the **MyoD** transcription factor, which induces a set of genes specifically needed for formation of muscle cells. MyoD is produced only in cells destined to differentiate into muscle tissue. It is a member of the large class of basic helix-loop-helix (bHLH) proteins. As discussed in Chapter 14, the helix-loop-helix is a widespread motif found in DNA-binding proteins. The basic HLH proteins share a stretch of basic amino acids, located next to the first helix, which helps in binding DNA.

HLH proteins bind to DNA as dimers. If both partners have a basic region, the dimer can bind to DNA. Basic HLH proteins usually function as **heterodimers** consisting of a tissue-specific bHLH protein plus one of the widely expressed bHLH proteins known as E-proteins. By itself, MyoD dimerizes poorly. In order to bind DNA, MyoD must form mixed dimers with E12 or E47. These are also basic HLH proteins that are alternative splicing products from the same gene, *E2A*. They are similar in shape and structure to MyoD, but unlike MyoD, they are expressed in all tissues. The MyoD/E12 or MyoD/E47 heterodimer binds to the DNA sequence CAAATG and activates muscle-specific genes.

Other HLH proteins lack the basic region and cannot bind DNA. An example is the Id protein. This binds to MyoD and E12 or E47. The heterodimers formed by Id cannot bind DNA (Fig. 17.08). Thus, the presence of Id protein *inhibits differentiation*. Id therefore plays a negative role, but without binding to DNA like a true repressor. Id protein is present in large amounts in precursor myoblasts, where it plays a role in restraining MyoD activity. During myoblast differentiation, the level of Id falls and this allows the activation of MyoD.

---

heterodimer   Dimer composed of two different subunits
MyoD   A eukaryotic transcription factor that takes part in muscle cell differentiation

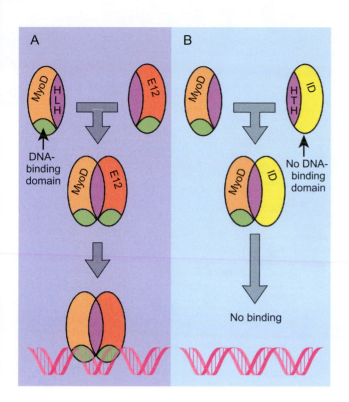

■ **FIGURE 17.08**
MyoD and Its Alternative Partners

A) MyoD and E12 both possess basic DNA-binding domains. When MyoD dimerizes with E12 the dimer therefore binds to DNA. B) In contrast, ID protein lacks a basic region. When MyoD dimerizes with Id this dimer cannot bind DNA.

Box 17.01 DNA Twist Can Affect Eukaryotic Gene Expression

The *GATA2* gene is conserved in many different species and GATA2 protein is a key regulator of red blood cell development in mice and humans. The homologous transcription factor is found in *Xenopus* and is a key mediator of zygote development. In the oocyte, the *GATA2* gene is transcribed and the mRNA is stored until after fertilization when it mediates the formation of ventral mesoderm. This tissue ultimately develops into blood cells.

GATA2 protein can only be expressed at a defined moment in development, and therefore, is subject to regulation. The control of *GATA2* expression in *Xenopus* relies on two key promoter elements, a CCAAT box and a stretch of A/T-rich DNA. A multisubunit transcription factor called CBTF binds to the CCAAT box to activate transcription of *GATA2*. When CBTF is not bound to the promoter, *GATA2* is not expressed. One of the subunits of CBTF is the Xilf3 protein, which contains two double-stranded RNA-binding domains (dsRBDs). Before *GATA2* is expressed, Xilf3 is bound to cytoplasmic RNA. Just before *GATA2* needs to be expressed, Xilf3 leaves the cytoplasm, enters the nucleus, and attaches to the *GATA2* promoter as a subunit of CBTF. The dsRBD of Xilf3 is essential to DNA binding, and analysis of the DNA structure in the A/T-rich region reveals that this helical region is twisted into an A-form helix, which is very similar in structure to a double-stranded RNA helix (see Ch. 4). Mutational studies have confirmed that both the A-form helical region of the promoter and the CCAAT element are essential for *GATA2* expression and provide evidence that even the DNA shape can affect gene expression (Fig. 17.09).

Continued

Box 5.01 Continued

GATA2 gene expression is OFF

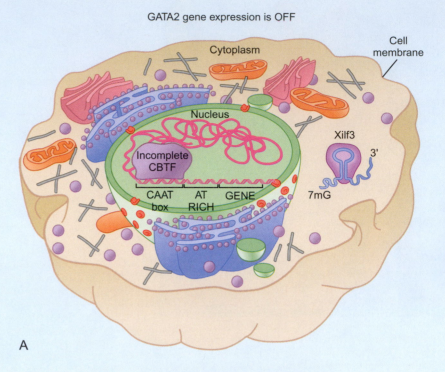

A

GATA2 gene expression is ON

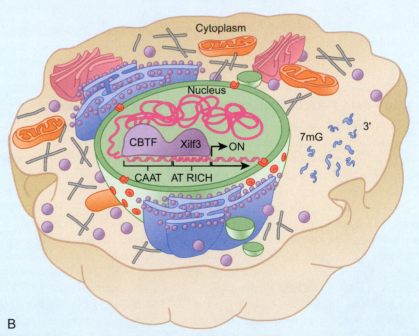

B

FIGURE 17.09
GATA2 Expression Is Controlled by DNA Structure

A) The *GATA2* gene is not expressed when only part of the CBTF protein is bound to its promoter. The other part of CBTF (Xilf3) is sequestered in the cytoplasm. B) At the appropriate time in development, the mRNA is degraded which releases Xilf3. Xilf3 moves into the nucleus, binds to CBTF, and binds to the A/T-rich region of the promoter by inducing the DNA into an A-form. These events activate transcription by RNA polymerase.

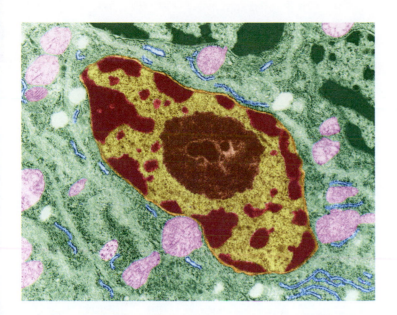

The nucleus shown contains regions of densely packed heterochromatin and of less densely packed euchromatin. This electron micrograph is of a neuroglial cell nucleus from an insect nervous system, magnified by 3980. Note the nucleolus (brown) and condensed DNA (dark red) in the nucleus. Mitochondria (pink) can be seen in the surrounding cytoplasm. (Credit: Dennis Kunkel.)

4. Heterochromatin Blocks Access to DNA in Eukaryotes

In bacteria, the DNA is freely accessible to RNA polymerase and regulatory proteins. However, in eukaryotes, the DNA is coiled around the histones, forming nucleosomes, as discussed in Chapter 4. The nucleosomes are wound into a helix and held close together largely by interactions due to the histone proteins. Densely-packaged DNA is referred to as **heterochromatin** and cannot be transcribed, because RNA polymerase cannot gain access to the promoters. Generally, DNA that is visible in electron micrographs is heterochromatin. DNA that is not dense is called euchromatin (Fig. 17.10). Recent work has shown that each of these two types of chromatin can be further subdivided (see Focus on Relevant Research).

Filion GJ, van Bemmel JG, Braunschweig U, Talhout W, Kind J, Lucas D. Ward LD, Wim Brugman, de Castro IJ, Kerkhoven RM, Bussemaker HJ, and van Steensel B (2011). Systematic protein location mapping reveals five principal chromatin types in *Drosophila* cells. Cell 143: 212–224.

The DNA wrapped around nucleosomes is much less accessible to many proteins, including transcription factors. Consequently, the positioning of nucleosomes along the DNA can greatly affect gene accessibility and hence transcription. The original classification of DNA into heterochromatin, with closely-packed nucleosomes that effectively silence gene expression versus euchromatin with more access appears to be oversimplified.

In this paper, the authors divide chromatin into five major categories. These are defined by their protein compositions and may extend for considerable

distances (over 100 kb). To achieve this the authors surveyed the distribution of around 50 components of chromatin. Euchromatin was divided into two classes. Both

FOCUS ON RELEVANT RESEARCH

were transcriptionally active, but they differed in the methylation pattern of their histones and the type of genes they comprised.

The linker histone, H1, has two arms extending from its central spherical domain. The central part of H1 binds to its own nucleosome and the two arms are thought to bind to the nucleosomes on either side; however, the exact arrangement is uncertain. The histones of the nucleosome core (H2A, H2B, H3, and H4) have a body of about

---

heterochromatin   A highly-condensed form of chromatin that cannot be transcribed because it cannot be accessed by RNA polymerase

A AGGREGATED

B DIS AGGREGATED

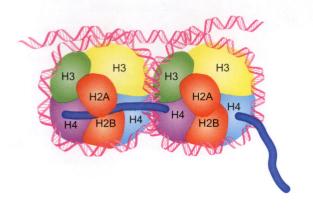
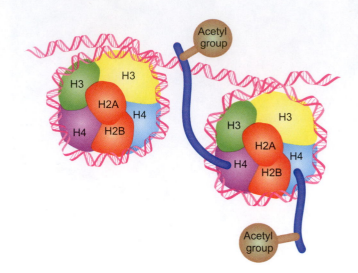

FIGURE 17.11
Acetylation of Histone Tails Disaggregates Nucleosomes

A) Closely-packed nucleosomes are stabilized by binding of histone tails to histones in the next nucleosome. B) When the tail of H4 is acetylated, it no longer binds to histones in an adjacent nucleosome. This promotes disaggregation of neighboring nucleosomes. (Histone H1 binds to the linker DNA between the nucleosomes but is not shown in this figure for the sake of clarity.)

Acetylation of histones controls access of regulatory proteins to the DNA.

80 amino acids and a tail of 20 amino acids at the N-terminal end that faces outwards from the core. Interactions due to these tails are believed to be important in nucleosome aggregation and higher level folding of the chromatin.

The histone tails contain several lysine residues that may have acetyl groups added or removed. All four of the core histones may be acetylated, although H3 and H4 are most often modified. The degree of **acetylation** affects the state of nucleosome aggregation and therefore of gene expression. Non-acetylated histones form highly condensed heterochromatin, whereas acetylated histones form less condensed chromatin. Note that the nucleosomes themselves are not disassembled by acetylation, but their clustering is loosened (Fig. 17.11).

Enzymes known as **histone acetyl transferases (HATs)** add acetyl groups and **histone deacetylases (HDACs)** remove them. Several proteins previously known as co-activators are actually HATs. Examples include the human CBP and p300 proteins involved in cell cycle control and differentiation. Similarly, several so-called **co-repressor** proteins are histone deacetylases. Co-activators and co-repressors do not bind to the DNA itself, but bind to transcription factors that have already bound to the DNA (Fig. 17.12).

Access to eukaryotic DNA involves moving or restructuring the nucleosomes.

In addition to disaggregating the nucleosomes by acetylation, a further step is needed to provide access to the DNA itself. This is performed by **chromatin remodeling complexes**. These carry out two main types of remodeling. Firstly, they can slide nucleosomes along a DNA molecule, so exposing sequences for transcription. Secondly, they are able to rearrange the histones, so remodeling nucleosomes into looser structures that allow access to the DNA. ATP is used to provide energy for this remodeling.

---

acetylation   Addition of an acetyl (CH_3CO) group
co-repressor   In prokaryotes—a small signal molecule needed for some repressor proteins to bind to DNA; in eukaryotes—an accessory protein, often a histone deacetylase, involved in gene repression
chromatin remodeling complex   A protein assembly that rearranges the histones of chromatin in order to allow transcription
histone acetyl transferase (HAT)   Enzyme that adds acetyl groups to histones
histone deacetylase (HDAC)   Enzyme that removes acetyl groups from histones

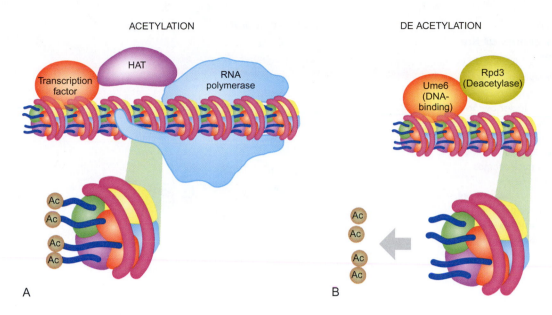

FIGURE 17.12
Acetylation and Deacetylation of Histones

A) Histone tails are acetylated by co-activators known as histone acetyl transferases (HATs). B) Deacetylation of histone tails is due to a repressor complex containing both a DNA-binding subunit and a deacetylase.

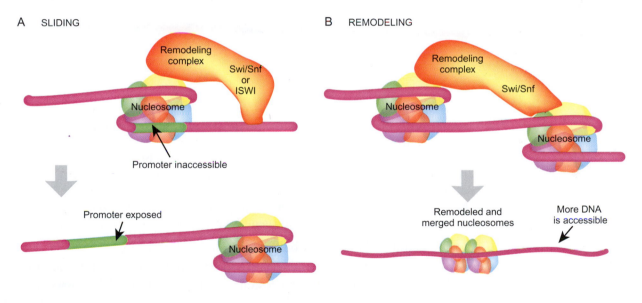

FIGURE 17.13
Sliding and Remodeling of Nucleosomes

A) Sliding of the nucleosome relative to the DNA exposes a previously inaccessible promoter. B) A remodeling complex such as Swi/Snf can merge two nucleosomes and loosen the winding of DNA making more of the DNA accessible.

There are two families of chromatin remodeling complexes. The larger **Swi/Snf** ("switch sniff") complexes consist of 8 to 12 proteins and bind to DNA strongly (Fig. 17.13). Swi/Snf can both slide and remodel nucleosomes. Apparently, Swi/Snf merges two nucleosomes into a new, looser structure. The smaller **ISWI** ("imitation switch") complexes contain 2 to 6 polypeptides and can slide nucleosomes but cannot

ISWI ("imitation switch") complex   Smaller type of chromatin remodeling complex
Swi/Snf ("switch sniff") complex   Larger type of chromatin remodeling complex

FIGURE 17.14
Sequence of Events at the HO Promoter

A) The HO endonuclease gene of yeast is covered by nucleosomes. B) The transcription factor Swi5p binds to the DNA. C) This is followed by binding of the Swi/Snf complex to Swi5p. D) The remodeled nucleosomes allow binding of an acetyl transferase, SAGA, to the Swi/Snf complex and to a nucleosome. E) As the acetylated histones become less compact the SBF transcription factor binds. F) The transcription apparatus binds.

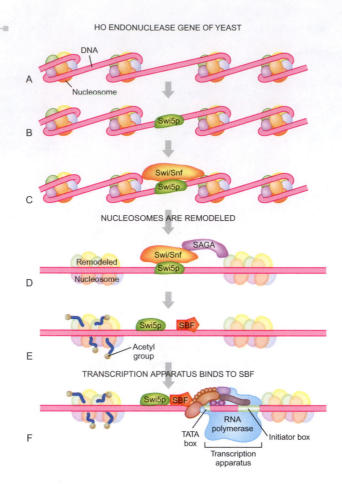

HO ENDONUCLEASE GENE OF YEAST

rearrange them. They bind to histones rather than to DNA. (The Swi factors were named after the switching of mating type; Snf factors refer to sucrose non-fermenting mutants. Both were found first in yeast.)

Binding of the chromatin-remodeling complexes by transcription factors targets them to the stretch of DNA that needs opening up. The precise order in which transcription factors, histone acetyl transferases, and chromatin-remodeling complexes bind appears to vary from promoter to promoter. A generalized sequence of events for activating a typical eukaryotic gene is as follows:

1. A transcription factor binds to the DNA.
2. A histone acetyl transferase (HAT) binds to the transcription factor.
3. The HAT acetylates the histones in the vicinity and the association of the nucleosomes is loosened.
4. The chromatin-remodeling complex slides or rearranges the nucleosomes, allowing access to the DNA.
5. Further transcription factors bind.
6. RNA polymerase binds to the DNA.
7. Initiation requires a positive signal to be transmitted via the mediator complex from one or more specific transcription factors.

Consider, for example, the yeast *HO* gene, which encodes an endonuclease required for the switching of mating type in yeast. First, the Swi5p transcription factor binds. Then the Swi/Snf complex binds to Swi5p. Apparently, the Swi/Snf complex moves along the DNA using one nucleosome to push others aside (see Focus on Relevant Research). Next to arrive is SAGA, an HAT whose binding depends on the presence of Swi/Snf. The histones in the promoter region are then acetylated. This allows another transcription factor, SBF, to bind. This then allows the general transcription factors to bind, followed by the RNA polymerase (Fig. 17.14).

Liu N and Hayes JJ. (2010) When push comes to shove: SWI/SNF uses a nucleosome to get rid of a nucleosome. Mol. Cell 38: 484–486.

This review article summarizes the current state of knowledge on how SWI/SNF complexes actually remove histones from DNA. The review summarizes recent research that examined the mechanism of nucleosome remodeling of DNA wrapped around two histones, that is, a dinucleosome structure. The paper uses new techniques to determine the movement for one single molecule of SWI/SNF. The results suggest that SWI/SNF wraps around one nucleosome and uses ATP to loosen the DNA from the histone core. As the DNA loosens, then SWI/SNF shifts the loosened complex to hit the next nucleosome. As the nucleosomes collide, H2A and

H2B dimers are ejected from the second nucleosome. Further sliding of SWI/SNF knocks the remaining core histones, H3 and H4, out of the DNA. The original nucleosome remains with SWI/SNF (Fig. 17.15).

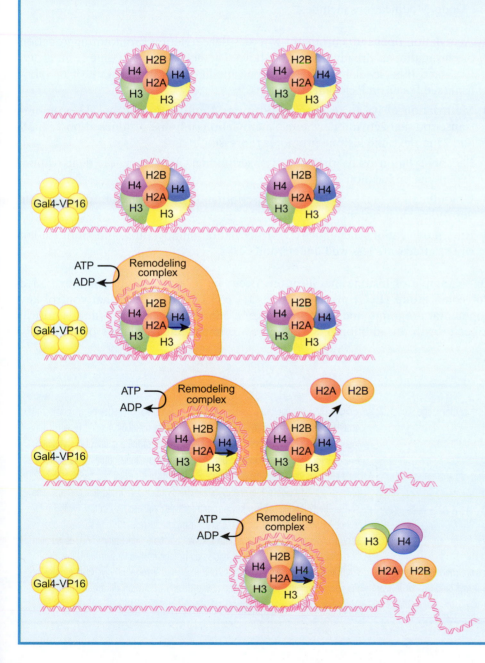

FIGURE 17.15
***Proposed Mechanism of SWI/
SNF Nucleosome Eviction***

Swi/Snf complexes can evict nucleosomes. In one proposed mechanism, Swi/Snf wraps around one nucleosome and loosens the DNA. This structure travels down the DNA, and collides into another nucleosome, which releases histone H2A and H2B. A second collision displaces the remaining histones, and releases the DNA.

4.1. The Histone Code

As noted above, several enzymes that modify histones were originally discovered as "co-activators" or "co-repressors" of transcription. Many chemical modifications of the histone proteins are now known. These modify chromatin structure and affect the binding of a variety of transcription factors and other proteins to DNA. Although the histones are folded to form relatively compact structures, their N-terminal regions form tails that project outwards and are exposed on the nucleosome surface. These tails, in particular the lysine residues, are the sites of the histone modifications (Fig. 17.16). For example, regions of chromatin that lack transcriptional activity may be condensed into heterochromatin. Such regions show very low acetylation but have a high incidence of methylation of lysine residues at position 9 in histone 3 ("H3-K9").

The wide range of possible histone modifications is sometimes referred to as the Histone Code. Modifications include:

a) Acetylation opens up chromatin whereas deacetylation promotes aggregation of chromatin, as already discussed above. The acetyl groups come from acetyl CoA and their availability links histone modification to the state of metabolism (see Focus on Relevant Research).

b) Methylation of lysine residues at positions 4 and 36 in histone H3 promotes transcriptional activation, whereas methylation of lysine residues at positions 9 and 27 in the same histone promotes repression.

c) Phosphorylation may promote transcriptional activation or deactivation, depending on location.

d) Ubiquitination may promote transcriptional activation or deactivation, depending on location.

e) Rare modifications including ADP-ribosylation, sumoylation, and biotinylation whose effects are less well understood.

One example of histone modifications occur by a large protein complex called the **Polycomb group (PcG) proteins**. These are conserved proteins that control key developmental programs, and their role in gene expression and repression is critical to proper growth. In addition, PcG proteins are important regulators of stem cells,

---

Ladurner AG (2009). Chromatin places metabolism center stage. Cell 138: 18–20.

FOCUS ON RELEVANT RESEARCH

The acetyl groups used to modify histones are critical to proper gene expression. In most genetic analyses, the source of the acetyl group is not investigated. This review article describes recent research describing the source of acetyl groups. These groups are found in the nucleus attached to coenzyme A (CoA). The acetyl-CoA molecules are actually products of citrate metabolism, and research has found that the cell's metabolic state is a key regulator of the amount of acetyl groups available for histone modification. During quiescence, available nutrients are sufficient for a cell, and there is ample acetyl-CoA. But during stages of rapid growth, metabolic processes must balance the energy requirements for the cell and the nucleus. One part of the balancing act is a key enzyme for lipid synthesis, ATP-citrate lyase (ACL). Normally thought to only be in the cytoplasm, recent research has shown that it is very abundant in the nucleus where it converts citrate into acetyl-CoA. The amount of available acetyl-CoA fluctuates, and so do histone acetylations. In turn, the histone modifications alter gene expression, and the genes involved in glycolysis are one of the primary targets. So, the results show a clear link between glucose availability and histone acetylation.

Polycomb group (PcG) proteins   A large protein complex that controls developmental expression of genes important for growth by methylating histones

Methylation

Acetylation

Phosphorylation

Ubiquitinization

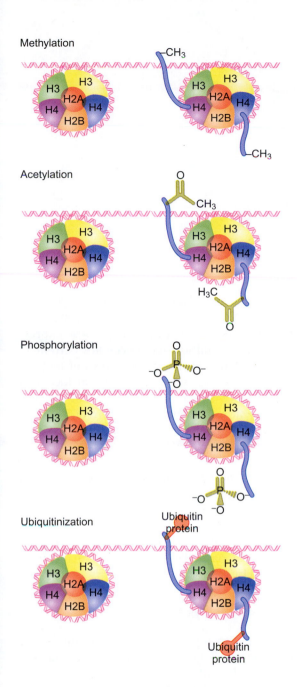

►**FIGURE 17.16**
Histone Modifications

Histone tails can be modified by addition of methyl, acetyl, phosphate, and ubiquitin groups.

and their deregulation are also implicated in cancer. PcG complexes bind PRE, or Polycomb response elements, in DNA regulatory regions and silence the associated genes. When they bind, one subunit of the complex adds methyl groups on histone H3, lysine 27. Another subunit ubiquinates histone H2A, which pauses RNA polymerase II. These modifications ensure that the genes around the PRE are repressed.

5. Methylation of Eukaryotic DNA Controls Gene Expression

Methylation of bases in DNA occurs in both prokaryotes and eukaryotes, although the purpose is generally quite different. Prokaryotes use methylation to distinguish newly synthesized DNA as discussed in Chapter 10. In eukaryotes, newly synthesized DNA is recognized by other means that are still unclear. Nonetheless, many eukaryotes do methylate their DNA as a marker for regulating gene expression.

Methylation of DNA is often used to control gene expression during development of higher organisms.

FIGURE 17.17
Control of Methylation of DNA in Eukaryotes

Three enzymes control methylation of DNA. *De novo* methylase adds methyl groups to non-methylated CG-islands. Maintenance methylase adds a second methyl group on the opposite strand of hemi-methylated sites. Demethylase removes methyl groups.

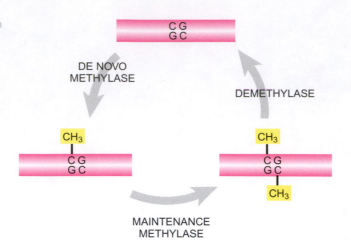

Methylation of DNA is rare in lower eukaryotes. Higher animals methylate up to 10% of their cytosines and higher plants methylate up to 30%. In these multicellular organisms DNA methylation is often used as a marker for genes whose expression is involved in tissue differentiation. The recognition sequences are extremely short; typically CG for animals and CNG for plants. There are two types of methylases. **Maintenance methylases** add methyl groups to newly-made DNA at locations opposite methyl groups on the old, parental DNA strand. This ensures that the pattern of methylation is inherited during cell division. Changing the pattern of methylation involves ***de novo*** **methylases** that add new methyl groups and **demethylases** that remove methyl groups (Fig. 17.17).

Methylation in eukaryotes silences gene expression, as discussed below. In animals, about half the genes are located close to **CG-islands** (i.e., clusters of CG sequences). **Housekeeping genes**, which are expressed in all tissues, possess non-methylated CG-islands. In contrast, the CG-islands of tissue specific genes are only non-methylated in those particular tissues where the genes are expressed. The maintenance of the pattern of methylation therefore makes sure that the pattern of gene expression stays constant among the cells of a particular tissue. In plants, certain transposable elements may also be inactivated by methylation.

5.1. Silencing of Genes Is Caused by DNA Methylation

Silencing is a somewhat vague term that refers to repression of a large number of genes in a relatively nonspecific manner. Limited silencing is known in bacteria (see Ch. 16). However, silencing is widespread in eukaryotes, where it involves the covalent modification of both DNA and of the histones. Silencing may affect a single gene, a cluster of genes, substantial regions of a chromosome or even a whole chromosome. The most spectacular example is the almost total silencing of a whole X-chromosome in female mammalian cells (see below).

Silencing results from methylation of the cytosine in 5'-CG-3' sequences of eukaryotic DNA. Occasionally, 5'-CNG-3' sequences are also methylated. The methyl groups project into the major groove of the DNA and thus hinder the binding of most transcription factors. In addition, methylated CG sequences are recognized by **methylcytosine-binding proteins (MeCPs)**. Bound MeCPs are, in turn, recognized by other proteins that

Genes are silenced by methylation of the DNA followed by removal of acetyl groups from the histones.

CG-islands   Region of DNA in eukaryotes that contains many clustered CG sequences that are used as targets for cytosine methylation
***de novo* methylase**   An enzyme that adds methyl groups to wholly non-methylated sites
demethylase   An enzyme that removes methyl groups
housekeeping genes   Genes that are switched on all the time because they are needed for essential life functions
maintenance methylase   Enzyme that adds a second methyl group to the other DNA strand of half-methylated sites
methylcytosine-binding protein (MeCP)   Proteins in eukaryotes that recognize methylated CG islands

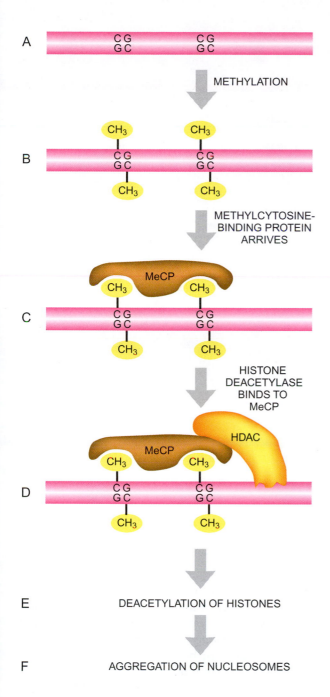

■ **FIGURE 17.18**
Silencing Starts with Methylation

A) CG regions on the DNA are
B) methylated. C) A methylcytosine-binding protein (MeCP) is attracted to the methylated sites and
D) histone deacetylase (HDAC) is bound to both MeCP and DNA.
E) and F) The consequence is deacetylation of the histone tails and aggregation of nucleosomes to form heterochromatin.

remove acetyl groups from the histones (especially H4). This results in the condensation of the DNA to form heterochromatin that is no longer accessible for transcription (Fig. 17.18). The genes in such regions are said to be "**silenced**."

The pattern of methylation in adult, differentiated cells is duplicated after each round of cell division. This is done by the maintenance methylases that add methyl groups to newly-made DNA at locations opposite methyl groups on the old, parental DNA strand. *De novo* methylases and demethylases change the pattern of methylation when necessary, especially during development. Just after fertilization of an egg to form a zygote, the methylation patterns of most of the DNA are erased. New methylation patterns are then laid down in a tissue-specific manner by processes still poorly understood. Those genes whose promoter regions are methylated are silenced.

DNA methylation patterns are reprogrammed when a new zygote is formed.

silencing   In genetic terminology, refers to switching off genes in a relatively non-specific manner

FIGURE 17.19
Imprinting and Development

The two copies of the same gene inherited from each parent are not always methylated in an identical manner. A) Here the *Igf2* gene from the mother has methyl groups and is inactive, whereas the non-methylated gene from the father is active. B) After fertilization, the embryo has one active and one inactive *Igf2* gene. C) Most methylation patterns are reprogrammed in the early embryo but a few survive, resulting in imprinting.

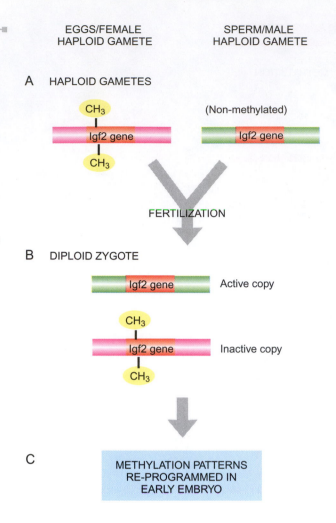

A few special genes retain their methylation patterns through fertilization.

5.2. Genetic Imprinting in Eukaryotes Is Due to DNA Methylation Patterns

Imprinting occurs when methylation patterns that are present in the gametes survive the formation of the zygote and affect gene expression in the new organism. This only applies to very few genes, although examples are known from mammals, fungi, and plants. Imprinting is a mechanism to ensure that only one of a pair of genes in a diploid cell is expressed. The second copy is silenced by methylation. The choice as to which allele of a gene to express depends on its parental origin.

Methylation patterns are set up during the formation of the gametes. Most genes in eggs and sperm cells are inactive and therefore silenced by methylation. However, some genes remain active, and there are a few differences in methylation patterns between male and female gametes. This causes an initial difference in expression of alleles inherited from the father and mother (Fig. 17.19). Let's imagine that this diploid zygote develops into a female. This woman will use her father's copy of *Igf-2* in all her cells, except when she creates new gametes or eggs. During meiosis, each gamete will have a 50% chance of getting her active paternal allele or the methylated copy. When she is a mother, the zygote receiving the unmethylated allele is remethylated. If for some reason, there is a mistake and her *Igf2* gene is not methylated, the baby will have both alleles active. People with two active *Igf2* genes develop Beckwith-Wiedemann Syndrome which is characterized by growth defects and tumor formation.

imprinting   When the expression of a particular allele depends on whether it originally came from the father or the mother (imprinting is a rare exception to the normal rules of genetic dominance)

About 70 imprinted genes are known in the mouse. For example, IGF-II (insulin-like growth factor II) from the father is expressed, whereas the maternal allele is not. Usually, it is the paternal allele that is expressed in the zygote, but not always. An example of the expression of the maternal allele is IGF-IIR, the receptor for IGF-II.

> A catalog of imprinted genes is kept online at www.otago.ac.nz/IGC.

6. X-Chromosome Inactivation Occurs in Female XX Animals

X-inactivation is a special form of imprinting found in animals. Females possess two X-chromosomes, whereas males have one X-chromosome plus a much shorter Y-chromosome. Consequently, females have two copies of genes carried on the X-chromosomes, whereas males only have one copy of most. Expression of both alleles in females would double the amount of proteins or RNA produced in comparison to males. As is seen above in over-expression of *Igf2*, too much of one protein can have dire consequences.

> Only one of the two X-chromosomes in a female mammalian cell is active and expresses its genes.

Evolution has developed a variety of mechanisms for gene dosage compensation in order to avoid different levels of gene expression in male and female. In nematodes, such as *C. elegans*, the expression level of genes on both X-chromosomes is halved. Conversely, in insects, such as *Drosophila*, the expression of genes on the single X-chromosome in the male is doubled. In mammals, one of the pair of X-chromosomes in each female cell is silenced (except for a few loci that are exempted and are referred to as "pseudo-autosomal" regions). In worms and insects, protein complexes that bind specifically to the X-chromosomes are responsible for decreasing (worms) or increasing (insects) transcription from genes located on the X-chromosomes. In female mammals, a mechanism involving non-coding RNA inactivates just one of the X-chromosomes (see below).

In *C. elegans*, there is no Y-chromosome and males have a single unpaired X-chromosome (this situation is designated XO). Furthermore, XX animals are actually hermaphrodites and possess both male and female sex organs. Dosage compensation relies on a protein, Sdc2, which is only expressed in XX animals. The Sdc2 protein binds to specific sites on the X-chromosomes and the dosage compensation complex, consisting of half a dozen proteins, assembles on Sdc2 and decreases gene expression. The mechanism used by *Drosophila* is essentially a mirror image of that in *C. elegans*. In flies, a protein, Msl2, that is only expressed in XY animals binds to specific sites on the X-chromosome. The dosage compensation complex assembles around Msl2 and increases gene expression. The dosage compensation complex of *Drosophila* includes two non-coding RNAs as well as several proteins.

> Worms and flies use different mechanisms from mammals to regulate X-chromosome expression.

In mammals, X-inactivation is controlled by methylation of the **Xist** gene, which is located on the X-chromosome. The *Xist* gene of the active X-chromosome is inactivated by methylation, and the *Xist* gene on the inactivated X-chromosome is transcribed. Once established during gamete development, this methylation pattern is inherited at cell division; thus, the same X-chromosome homologue will remain active in the daughter cells. The expression of the *Xist* gene on the active X-chromosome is regulated by the antisense RNA, Tsix, which is transcribed from the *Xist* locus, but in the reverse direction. Consequently, Tsix is involved in choosing which X-chromosome to inactivate. However, the timing of expression of Tsix RNA varies among different mammals and the regulation of Tsix itself is presently unclear.

> The DNA of inactivated X-chromosomes is highly condensed.

Expression of the *Xist* gene causes the inactivation of the X-chromosome that carries it. A long non-translated RNA is transcribed from the *Xist* gene. This *Xist* RNA coats the inactive X-chromosome. Starting from the *Xist* gene and proceeding along the X-chromosome in both directions, the DNA is converted into heterochromatin, a condensed form of DNA that cannot be transcribed (Fig. 17.20). Highly-condensed X-chromosomes are visible in the cells of female mammals and are

---

Xist **gene**   A gene that causes the inactivation of the X-chromosome that carries it
X-inactivation   The condensation and complete shutting down of gene expression of one of the two X-chromosomes in cells of female mammals

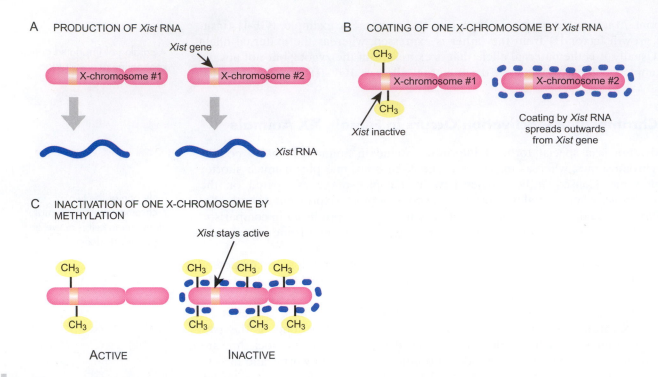

A) PRODUCTION OF *Xist* RNA

B) COATING OF ONE X-CHROMOSOME BY *Xist* RNA

C) INACTIVATION OF ONE X-CHROMOSOME BY METHYLATION

FIGURE 17.20
X-inactivation Involves the Xist Gene and Xist RNA

A) Originally, both X-chromosomes transcribe *Xist* RNA from the *Xist* gene. B) The X-chromosome that will remain active is methylated in the *Xist* gene region, which inactivates the *Xist* gene. The *Xist* RNA coats the other X-chromosomes and inactivates it. C) The inactive X-chromosome is almost entirely methylated, except for the *Xist* gene (together with a few aberrant loci that are exempted from X-inactivation and are not shown here). This causes the X-chromosome to transform largely into heterochromatin. Only the *Xist* gene itself remains active.

> *Xist* RNA is involved in silencing X-chromosomes.

called **Barr bodies**, after Murray Barr, who discovered them in 1948. The presence or absence of Barr bodies has sometimes been used to check whether female Olympic athletes are indeed genetic females.

If an active *Xist* gene is inserted into another chromosome, this is only partly inactivated. So another factor(s) is needed to explain X-inactivation. The X-chromosome has twice as many LINE-1 elements (see Ch. 4) per unit length as other chromosomes, and it has been suggested that these may somehow promote the binding of *Xist* RNA. The converse theory argues that more LINE-1 elements have accumulated on the X-chromosomes precisely because they are often inactivated!)

The mechanism of Xist-induced silencing is only partly understood. After *Xist* RNA binds, it recruits proteins that are responsible for the actual transcriptional silencing and heterochromatin formation. Changes occur in the histones of the inactive X-chromosome. First, histone 3 becomes methylated on Lys7 instead of on Lys4 as in active chromatin. Next, histone H4 loses most of its acetyl groups. Somewhat later, an unusual histone, macroH2A, an H2A variant with an extra C-terminal domain, is found solely on the inactive X-chromosome. Finally, methylation of CpG islands occurs along the chromosome. Once silencing has been established, the *Xist* RNA is no longer required for its maintenance.

In those rare cases where three or more X-chromosomes are present in female mammals, only one remains active. Moreover, mice with a single X-chromosome (and no Y-chromosome) are healthy and fertile, implying that the second X-chromosome is not even necessary. In marsupials, the X-chromosome from the father is always inactivated.

In other mammals, the choice is random. Furthermore, which X-chromosome is active varies in different cell lines. Consequently, female mammals consist of a genetic mosaic, in which different alleles of genes borne on the X-chromosome are expressed

---

Barr body   Inactive and highly condensed X-chromosome as seen in the light microscope

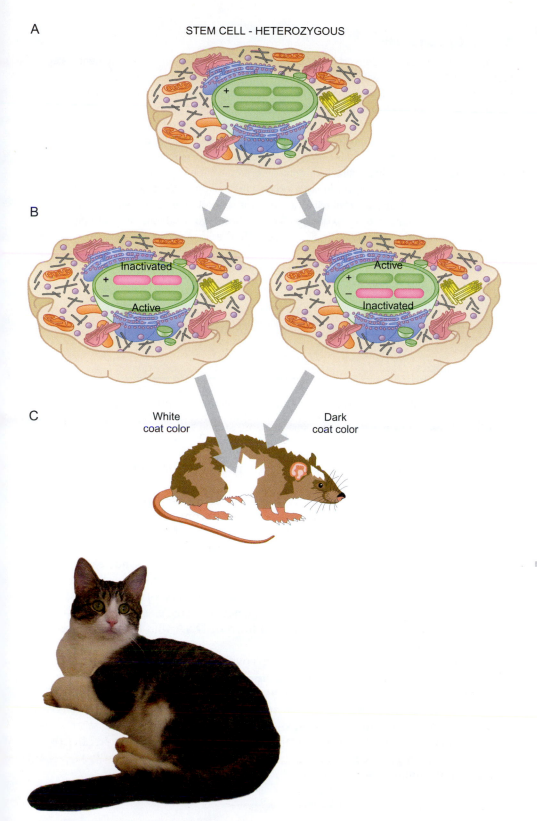

A
STEM CELL - HETEROZYGOUS

B
Inactivated
+
−
Active

Active
+
−
Inactivated

C
White
coat color

Dark
coat color

■-**FIGURE 17.21**

***X-Inactivation Causes Skin
Patterning in Mice***

A female mouse is shown that is
heterozygous for an X-linked gene
involved in hair pigmentation. A) All
cells contains two X-chromosomes,
one with a functional copy (+)
of the hair color gene and the
other with a defective copy (−).
B) During development, different
X-chromosomes are inactivated at
random in different ancestral cells.
Each ancestral cell divides and gives
rise to a patch of cells on the body
surface. C) The result is a mixture of
zones of different skin colors. The
white (mutant) zones appear when
the active X-chromosome carries
the defective hair color gene. The
dark zones appear when the active
X-chromosome carries a wild-type
hair color gene.

■-**FIGURE 17.22**

A Calico Cat

Calico coloration is seen only in
female cats. It occurs because genes
for fur color are carried on the
X-chromosomes. If a female cat is
heterozygous for mutant and wild-
type alleles, random inactivation of
the two X-chromosomes in different
regions creates the pattern. White
patches are due to cells where the
activated X-chromosome contains
the mutant allele.

in different regions of tissues. This is illustrated by the variegated coat color seen in female
mice that are heterozygous for a coat color mutation in an X-linked gene (Fig. 17.21). In
any given cell, only one X-chromosome is active and so only one of the two alleles will be
expressed. The descendents of a particular ancestral cell stay together and form regions
of skin with the same color. Hence, some regions of the coat are wild-type and others
show the mutant color. A similar effect is seen in calico cats (Fig. 17.22).

Box 17.02 Epigenetics

The term epigenetics was derived by merging epigenesis (the intricate developmental steps that create an organism) with genetics in order to describe inherited phenotypic changes that occur in a species outside of genomic changes. The term implies environmental effects on gene expression, but only in a very restricted way. Today, the term, epigenetics refers to those inherited changes that are NOT due to any change in the base sequence of DNA. Epigenetic inheritance is often due to modification of DNA, usually by methylation or by modifications to the histones, although other obscure mechanisms exist. Examples include the formation of heterochromatin, X-chromosome inactivation, and imprinting. In each of these cases, the environment originally triggers the change in chromatin structure. Such changes can be transient, only affecting the cell for a short time, or stable, and affecting the cell for the remainder of its existence. Strictly, these two situations are not epigenetic although they are often discussed in this context. True epigenetic inheritance is when the changes are passed onto the daughter cells after division. The mechanisms and role of these kinds of alterations are still being investigated, and future experiments will hopefully delineate the effect of the environment more accurately.

Key Concepts

- Regulation of transcription in eukaryotes is much more complex than in bacteria.
- Specific transcription factors regulate protein-encoding genes in response to a variety of signals.
- The mediator complex compiles information from multiple sources and transmits the result to RNA polymerase.
- Enhancer sequences are found far from the genes they control. The DNA loops around to bring them close to the mediator complex.
- Insulator sequences shield genes from the effects of regulatory sequences intended for other genes in the general neighborhood.
- Matrix attachment regions are sites where the DNA binds nuclear matrix proteins.
- In eukaryotes, negative regulation occurs by a variety of mechanisms, but rarely by the direct binding of repressor proteins to DNA.
- Heterochromatin is a highly-condensed form of DNA that blocks the access of transcription factors in eukaryotes.
- The histones that bind DNA may be covalently modified in a variety of ways. Some modifications activate gene expression, whereas others repress expression.
- A variety of enzymes are known that add or remove methyl groups from DNA.
- DNA methylation causes silencing of genes in eukaryotes.
- In eukaryotes, the difference in DNA methylation patterns between the two parents causes genetic imprinting.
- Female mammals inactivate one of their pair of X-chromosomes.

Review Questions

1. What factors make regulation of genes more complicated in eukaryotes than prokaryotes?
2. What are the four general properties of transcription factors?
3. What is a mediator complex? What is it made of?
4. How does the mediator complex communicate with RNA polymerase II?
5. What are enhancers and how do they aid in transcription?

6. What is an insulator?
7. How does the insulator block enhancer function?
8. How does negative regulation differ in prokaryotes and eukaryotes?
9. What is the CAAT box?
10. What is a hetero-dimer? Why are they needed in some transcription factors?
11. What are the nuclear matrix, MARs, and SARs?
12. What do MAR sites bind?
13. What is heterochromatin and how does it interfere with transcription?
14. How does acetylation of histones affect gene expression in eukaryotes?
15. What are HATs and HDACs? What role do they play in gene expression?
16. What role do chromatin remodeling complexes play in gene expression?
17. What are the two major families of chromatin remodeling complexes and how do they differ?
18. Describe the sequence of events needed to activate a eukaryotic gene.
19. What are the three types of DNA methylases and what are their roles?
20. How do prokaryotes and eukaryotes differ in their use of DNA methylation?
21. How does methylation differ between house keeping genes and tissue-specific genes?
22. What is gene silencing?
23. What role does methylation play in silencing?
24. What is imprinting?
25. How does imprinting occur? What are its benefits?
26. What is X-inactivation?
27. What is the *Xist* gene? What does it encode?
28. How does the same X-chromosome of the pair remain active after cell division?
29. What are Barr bodies and how are they formed?
30. What series of events lead to the formation of heterochromatin?

Conceptual Questions

1. Many eukaryotic genes are only expressed in certain tissues, such as *APETALA1* (*AP1*) which directs the flower meristem to create sepals and petals in the flowers of *Arabidopsis*. Many scientists study promoters, trying to identify the important elements necessary for tissue-specific expression. For the *AP1* gene, the researcher took 500 nucleotides of DNA upstream or before the transcription start site and fused it to a reporter gene called GUS for β-glucuronidase. The GUS gene encodes an enzyme that converts X-gluc into a blue dye, so when the *AP1* gene promoter turns on the expression of GUS that tissue in the plant will turn blue. Different regions of the promoter can be fused to the reporter gene to determine the elements important for transcription initiation. Interpret the following table of information by identifying what regions of the *AP1* promoter are important for transcription initiation:

Promoter Construct	Gus expression in flower meristem
−500 to +1 AP1::GUS	+
−407 to +1 AP1::GUS	+
−336 to +1 AP1::GUS	+
−288 to +1 AP1::GUS	+
−198 to +1 AP1::GUS	+
−100 to +1 AP1::GUS	+
−39 to +1 AP1::GUS	−

2. A new gene that has a DNA-binding domain (leucine zipper) has been discovered and named UPT7. The gene is thought to control the transcription of a variety of genes that control flower development in pansies. What would happen to this plant if UPT7 had a frameshift mutation in the ORF region?

3. Researchers use Gal4 promoters to drive the transcription and translation of different genes. In the laboratory, the gene of interest is removed from its natural promoter and fused to the Gal4-binding site and promoter. Describe what happens when the following construct is placed into one of the chromosomes of *Saccharomyces cerevisiae* and grown on plates containing galactose and plates lacking galactose. How can you determine if the GUS gene is expressed?

Gal4-binding site	RNA polymerase-binding site	GUS gene

4. In doing research on a new *Arabidopsis* gene, the researcher notices that a region 1,000 base pairs downstream from the transcriptional start site contains what appears to be a new gene that has the ability to increase transcription 5-fold. However, the effect is only seen in an *Arabidopsis* mutant missing the gene for *de novo* methylase. Upstream from the new gene is an insulator sequence that has various CG islands. Based on your knowledge of *H19* expression in humans, propose a mechanism for the transcriptional control of the new *Arabidopsis* gene.

5. Transcription factors usually have one or more motifs that affect their function. What is the function of the following: helix-turn-helix, zinc finger, and leucine zipper?

Regulation at the RNA Level

Regulating gene expression at the level of transcription is most efficient in conserving materials and energy, whereas regulation of enzyme activity provides the most rapid response. Since regulation at the level of translation is neither the most efficient nor the most rapid, it is consequently less frequent than these other forms of regulation. Although not as common as either alternative, regulation at the level of translation is not as rare as once thought. Early work on the regulation of the expression of messenger RNA (mRNA) attributes this mostly to regulatory proteins. However, more recently a number of regulatory mechanisms have been discovered that involve regulation by RNA molecules. When antisense RNA binds to functional RNA it blocks activity. This mechanism underlies some individual cases of regulation of translation by specific antisense RNA molecules. More important by far are the RNA interference response and regulation by micro RNA. In both cases, double-stranded RNA gives rise to small derivatives that operate by an antisense mechanism.

1. Regulation at the Level of mRNA

Generally, once mRNA has been made, it quickly moves to the ribosome where it is translated. This is especially true in prokaryotes where there is no nuclear membrane restricting access between the transcription machinery and the ribosomes. It was once thought that translational regulation was very rare. This was partly due to measuring translation and mRNA stability being more difficult than assaying transcription or protein levels. However, more recent work has revealed a growing number of cases of regulation at the level of translation, especially in eukaryotes. In almost every multicellular eukaryote, translational regulation via the use of small regulatory RNA molecules is a common mechanism for both cellular defense and developmental control. Recent experiments have also identified small regulatory RNAs that are transcribed from intergenic regions of bacterial genomes.

In addition to controlling the translation of mRNA after it has been made, there is also the possibility of aborting the synthesis of mRNA after transcription has been initiated and only a short stretch of RNA has been made. This somewhat ambiguous mechanism is referred to as transcriptional attenuation. It is sometimes classified as a form of transcriptional regulation. However, it has been included in this chapter as it is closely related to true translational regulation in the sense that alternative structures of mRNA are involved in both cases.

Many of the known cases of translational regulation occur as extra steps in highly-complex regulatory cascades that also include regulation at the level of transcription and of protein activity. Examples include the heat shock response in both bacteria and animals and the control of cell growth and differentiation in higher animals. In this chapter, we have attempted to illustrate regulation at the RNA level using examples where other regulatory mechanisms do not overly complicate the issue.

Given a pre-made mRNA molecule, there are several ways in which the translation of the message may be regulated:

1. Control over the rate of degradation of the mRNA
2. Modification of untranslatable mRNA to a form that can be translated
3. Control of mRNA translation by regulatory proteins
4. Binding of antisense RNA to mRNA to prevent its translation
5. Activating RNases to degrade specific viral or cytoplasmic mRNAs by creating small interfering RNAs
6. Control of mRNA translation by riboswitches
7. Preferential translation of certain classes of mRNA due to alteration of the ribosome

1.1. Binding of Proteins Controls the Rate of mRNA Degradation

Unlike the DNA that constitutes the cell's genome, mRNA is relatively short-lived. All cells contain a series of **ribonucleases** whose role is to remove mRNA once it has served its function. The half-life of a typical mRNA in bacteria such as *E. coli* is 2–3 minutes. The susceptibility of mRNA to degradation depends on its secondary structure and thus some mRNA molecules are inherently more stable than others. In this chapter, only cases where the susceptibility of mRNA to degradation is altered in response to regulatory signals will be discussed. For example, the degradation of mRNA may be hindered or hastened by the binding of RNA-specific regulatory proteins. Interpretation of the precise mechanism is often difficult because binding to ribosomes protects mRNA from degradation. Thus, mRNA that is not being translated will be degraded faster. There are two main ways in which the binding of a protein might affect mRNA stability. First, it could directly alter susceptibility to ribonuclease attack. Second, the protein might help or hinder binding to the ribosome, which would alter the rate of translation and affect mRNA stability indirectly.

---

ribonuclease   An enzyme that degrades RNA

Components of Csr translational control system

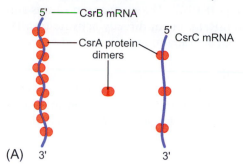

(A)

CsrA prevents glycogen synthesis

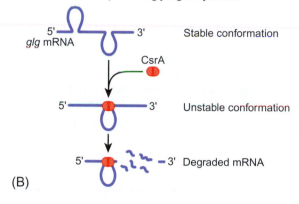

(B)

CsrA stabilizes *flhDC* mRNA for flagellar synthesis

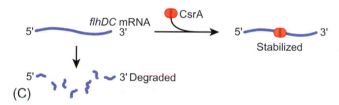

(C)

▬ **FIGURE 18.01**

Control of mRNA Degradation by CsrA

A) The Csr translational control system of *E. coli* consists of two small non-coding RNAs (ncRNA) called CsrB and CsrC, and an RNA-binding protein dimer called CsrA. The two ncRNAs sequester CsrA away from binding to mRNAs. B) When free CsrA binds to *glg* mRNA, the conformation of *glg* is destabilized and targeted for degradation. C) Binding of CsrA to *flhDC* mRNA has the opposite effect. It stabilizes the mRNA, resulting in increased translation and more flagellar protein synthesis.

The CsrABC regulatory system of *E. coli* consists of an RNA-binding protein, CsrA, and two **non-coding regulatory RNAs (ncRNA)**, also known as **small RNA (sRNA)** called CsrB and CsrC (Fig. 18.01A). CsrB sequences are found in many different eubacteria and are thought to function in regulating carbon metabolism, making extracellular components, cell movement, biofilm formation, and sensing of bacterial populations. Because this family of genes affect cell movement and biofilm formation, they also play a role in bacterial pathogenesis. Each CsrB sRNA has about 22 CsrA protein-binding sites that sequester about nine CsrA dimers from affecting expression of other mRNAs. CsrC sRNA works the same way but has fewer binding sites.

The CsrAB regulatory system controls the balance between sugar storage as glycogen, which accumulates in bacteria, and the breakdown of sugars by glycolysis (Csr = carbohydrate storage regulator). Overall, CsrA activates glycolysis and represses glucose synthesis and glycogen synthesis. CsrA protein binds to mRNA carrying genes involved in glycogen synthesis (Fig. 18.01B). The binding of CsrA hastens the decay of *glgC* mRNA so preventing its translation. *GlgC* is involved in glycogen synthesis. The CsrA-binding sites on *glgC* overlap with the Shine-Dalgarno sequences, so that ribosomes are unable to bind. In response, the cell rapidly degrades *glgC* transcripts, which prevents glycogen synthesis.

RNA-binding proteins may change the conformation and stability of mRNA.

non-coding regulatory RNA An RNA molecule that sequesters regulatory proteins from functioning as translational repressors (e.g., CsrB and CsrC)

small RNA (sRNA) An RNA molecule that sequesters regulatory proteins from functioning as translational repressors (e.g., CsrB and CsrC)

In contrast, the CsrAB system activates the *flhDC* operon, which is involved in flagella synthesis, by stabilizing the mRNA (Fig. 18.01C). The mechanism of this stabilization is unknown. Binding of CsrA to the mRNA might directly activate translation. Alternatively protection of the mRNA from degradation by ribonucleases would indirectly increase translation.

Other bacteria have similar proteins with ncRNA partners to control gene expression. In *Pseudomonas fluorescens*, there are two CsrA homologs (RsmA and RsmE) and three ncRNAs (RsmX, RsmY, and RsmZ) that function in a similar mechanism. Other bacteria such as *P. aeruginosa*, *Salmonella enterica*, and *Vibrio cholerae* also have homologous systems.

1.2. Some mRNA Molecules Must Be Cleaved Before Translation

Eukaryotic mRNA requires processing that removes the introns and adds a cap and tail to the **primary transcript** (see Ch. 12 for details). Generally, prokaryotic mRNA is not processed in this manner. The primary transcript, as made by RNA polymerase, is generally the mRNA in prokaryotes. However, in a few rare cases, further processing of a prokaryotic mRNA is needed before it can be translated.

In *Escherichia coli*, **ribonuclease III** is one of several ribonucleases that take part in processing the precursors to tRNA and rRNA. In addition, a few mRNA molecules must be processed by RNase III before they can be translated. The mRNA from the *speF* gene that encodes ornithine decarboxylase is translated about four-fold better if cut by RNase III. However, mRNA from the *adhE* gene, encoding alcohol dehydrogenase, has an absolute requirement for processing by RNase III. In such cases, the original mRNA molecule is folded so that the ribosome-binding site and start codon are inaccessible. Cleavage of the mRNA upstream of the ribosome-binding site by RNase III frees these sequences for recognition by the ribosome (Fig. 18.02). In *rnc* mutants, which lack RNase III, the *adhE* mRNA cannot be translated and the cells are unable to grow anaerobically by fermentation.

The degradation of *adhE* message is also controlled by ribonuclease activity. The mature *adhE* mRNA is specifically degraded by ribonuclease G, another ribonuclease normally involved in processing rRNA. In *rng* mutants that lack RNase G, the half-life of *adhE* mRNA increases from 4 minutes to 10 minutes, levels of the mRNA rise, and AdhE protein is over-produced.

1.3. Some Regulatory Proteins Cause Translational Repression

Just as regulatory proteins may bind to DNA and either promote or hinder transcription, proteins may also bind to specific sequences on mRNA and regulate its translation. Examples of both positive- and negative-translational regulation are known. As discussed before, CsrA has been shown to bind the Shine-Dalgarno sequence to block ribosome-binding. The response of mRNA to iron is another example of translational repression.

Iron is an essential nutrient and is the co-factor for many proteins, such as cytochromes and hemoglobin. However, free iron generates toxic free radicals and is therefore dangerous. Consequently, surplus iron atoms are stored by **ferritin** and its prokaryotic counterpart, **bacterioferritin**. Ferritin is a hollow spherical protein consisting of 24 subunits. Up to 5,000 iron atoms may be stored as a hydroxyphosphate complex inside this sphere.

---

bacterioferritin   The bacterial analog of ferritin, an iron-storage protein
ferritin   An iron-storage protein
primary transcript   The original RNA molecule obtained by transcription from a DNA template, before any processing or modification has occurred
ribonuclease III   A ribonuclease of bacteria whose main function is processing rRNA and tRNA precursors

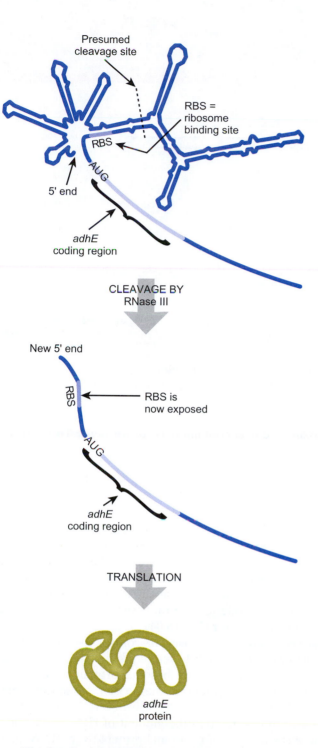

■**FIGURE 18.02**
*Cleavage of adhE mRNA
by RNase III*

The ribosome-binding site of *adhE*
mRNA cannot bind to the ribosome
due to folding of the pre-mRNA.
RNAase III cleaves the *adhE* pre-
mRNA so exposing the ribosome-
binding site.

The level of ferritin is regulated in response to the iron supply. In plants, ferritin
levels are regulated at the level of transcription. However, in both animals and bacteria, ferritin levels depend on translational regulation. When iron is scarce, translation
of ferritin mRNA is reduced. When iron is plentiful, more ferritin is made to sequester it from creating toxic free radicals. In animals, an RNA-binding protein is responsible, whereas in bacteria control is by antisense RNA (see below).

The **5′-untranslated region (5′-UTR)** of ferritin mRNA of animals contains a special recognition sequence known as an **iron-responsive element (IRE)**, which forms

In animals, translation of genes
involved in iron uptake and
storage is controlled by an
iron-regulatory protein that
binds to the mRNA.

iron-responsive element (IRE)   Site on mRNA where the IRP binds
5′-untranslated region (5′-UTR)   The untranslated sequence between the 5′ end of an mRNA and the start codon

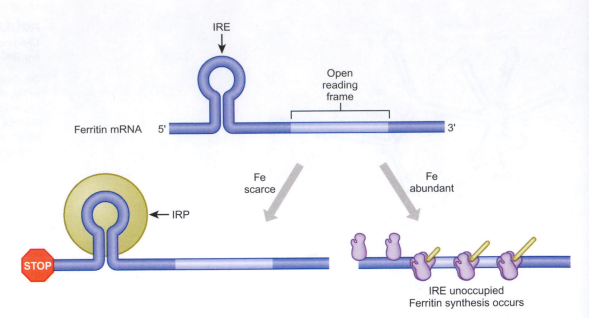

FIGURE 18.03
Regulation of Ferritin mRNA Translation by IRP

The ferritin mRNA has a stem-loop structure (the iron-responsive element or IRE) in the 5' untranslated region. When iron is scarce, the iron-regulatory protein (IRP) binds, preventing the small subunit of the ribosome from binding to the cap structure of the mRNA to initiate translation. When iron is abundant, no IRP is bound and translation occurs.

a stem-loop structure. When iron is scarce, an **iron-regulatory protein (IRP)** binds to the IRE stem and loop and prevents translation. Surplus iron results in the detachment of IRP from the ferritin mRNA, which can then be translated (Fig. 18.03).

In animals, the enzyme δ-aminolevulinic acid synthase catalyses the rate limiting step in the pathway for synthesizing the iron-containing co-factor heme. Not surprisingly, its mRNA contains an iron-responsive element and is also under IRP translational control. The receptor for the iron transport protein transferrin, which is found in blood, is also regulated by IRP binding to an IRE on the mRNA, but in this case mRNA stability is affected.

The free iron concentration in the cytoplasm is directly monitored by the IRP. Although there are several IRPs, the major one, IRP1, is identical to the cytoplasmic enzyme, **aconitase**. This enzyme has an **Fe4S4 cluster** that is needed for enzyme activity. When iron is scarce, one of the four iron atoms is lost from the Fe4S4 cluster. The aconitase/IRP1 then loses enzyme activity and changes conformation, exposing its RNA-binding site. Thus, when iron is plentiful, aconitase/IRP1 acts as aconitase (an enzyme in the Krebs cycle that converts citrate to isocitrate) and when iron is scarce, it acts as an RNA-binding **translational repressor** (Fig. 18.04).

Regulation by **translational repression** is also used to control the synthesis of ribosomal proteins in bacteria such as *E. coli*. The ribosomal proteins are grouped in several operons. For each operon, one of the encoded ribosomal proteins binds to the mRNA and so auto-regulates synthesis of all proteins in the operon. These ribosomal proteins bind to rRNA preferentially and only bind to their own mRNA when there is no rRNA available. This mechanism ensures that the amount of rRNA and ribosomal proteins is balanced. If there is an excess of ribosomal protein over rRNA, then translation will be decreased (Fig. 18.05).

1.4. Some Regulatory Proteins Activate Translation

Positive regulation of translation is used to control protein synthesis in chloroplasts after light stimulation. Synthesis of many chloroplast proteins is induced as much as a

---

aconitase   An enzyme of the Krebs cycle that, in animals, also acts as an iron-regulatory protein
Fe4S4 cluster   A group of inorganic iron and sulfur atoms found as a co-factor in several proteins
iron-regulatory protein (IRP)   Translational regulator that controls expression of mRNA in animals in response to the level of iron
translational repressor   A protein that binds to mRNA and prevents its translation
translational repression   Form of control in which the translation of a messenger RNA is prevented

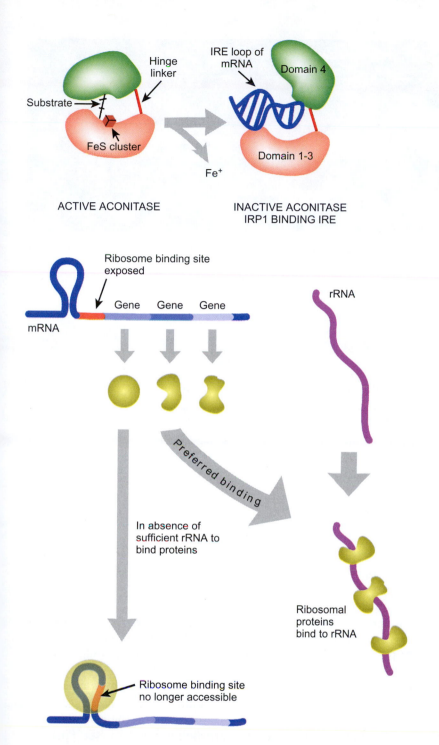

■- **FIGURE 18.04**

Aconitase Activity versus RNA-binding of IRP1

IRP1 acts as an enzyme (aconitase) when iron is plentiful. Upon losing an iron atom from the central Fe4S4 cluster during situations of iron scarcity, the two major domains open, which exposes an RNA-binding site. When aconitase/IRP1 binds to mRNA, transferrin synthesis is blocked.

■- **FIGURE 18.05**

Regulation of Synthesis of Ribosomal Proteins in Bacteria

The mRNA shown at the top of the figure contains a loop, a ribosome-binding site (RBS), and structural genes coding for several ribosomal proteins. These proteins prefer to bind rRNA. In situations where there is insufficient rRNA to bind the ribosomal proteins, one of them binds to its own mRNA and alters its configuration, thus blocking the ribosomal-binding site. This prevents synthesis of excess ribosomal proteins.

hundred-fold by light. The levels of some of these proteins are controlled by transcription, others by translation, and others by protein degradation.

For example, the mRNA for the large subunit of **Rubisco** accumulates in developing chloroplasts even in the dark. (Rubisco is ribulose bisphosphate carboxylase, a critical enzyme in the fixation of carbon dioxide during photosynthesis. It is the most abundant protein on Earth.) Another example is PsbA (=D1 protein), a component of photosystem II. Translation of these mRNAs is controlled by proteins encoded by the nucleus that act as **translational activators**. These proteins bind to an adenine-rich

Rubisco (ribulose bisphosphate carboxylase) A critical enzyme in the fixation of carbon dioxide during photosynthesis
translational activator A protein that binds to mRNA and promotes its translation

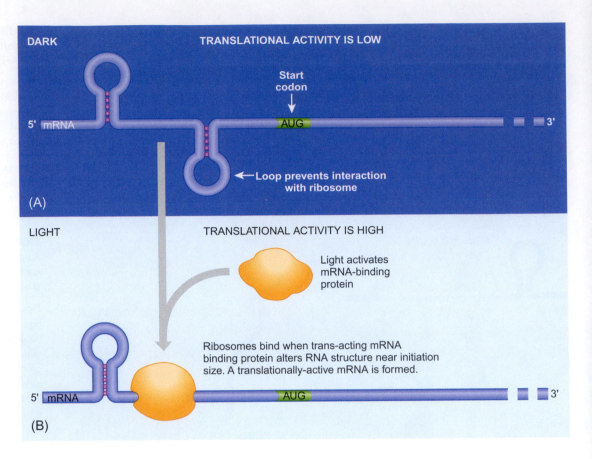

FIGURE 18.06
Translational Activation of Chloroplast mRNA

In the dark, the central loop in the mRNA prevents the mRNA from binding to the ribosome. When there is sufficient light, an mRNA-binding protein straightens out the RNA to allow ribosome-binding.

region in the 5′-UTR of their target mRNA. The activators bind to the mRNA in the light and allow translation. In the dark, they do not bind to the mRNA which therefore cannot be translated due to its unfavorable secondary structure (Fig. 18.06).

The translational activator **cPABP (chloroplast polyadenylate-binding protein)** exists in two conformations, only one of which can bind RNA. The interconversion of the two forms of cPABP is controlled by light. Energized electrons from photosystem I are passed down a short electron transport chain to cPABP. The electrons reduce the disulfide form of cPABP to the sulfhydryl form. The reduced sulfhydryl form can bind to RNA and activate translation, whereas the disulfide form cannot (Fig. 18.07). Another example where translation is controlled by association of the ribosome with a protein complex is the synthesis of proteins in growing nerve cells (see Focus on Relevant Research).

1.5. Translation May Be Regulated by Antisense RNA

Messenger RNA is transcribed using only one DNA strand as the template. This is referred to variously as the template strand, non-coding strand, or antisense strand. The mRNA produced is consequently **sense RNA**. The other strand of DNA (the coding strand or sense strand) is not normally used as a template for transcription.

cPABP (chloroplast polyadenylate-binding protein)   A translational activator protein that controls expression of chloroplast mRNA
sense RNA   Normal RNA that has been produced from the non-coding strand of DNA

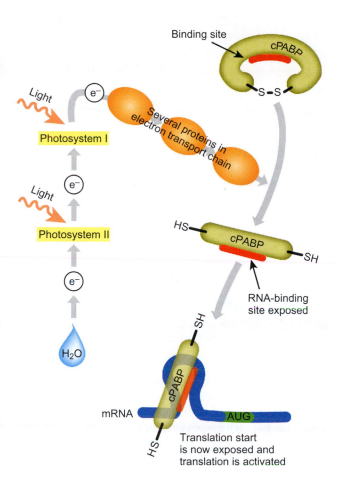

Binding site

cPABP

S–S

Light

e⁻

Photosystem I

Several proteins in electron transport chain

Light

e⁻

Photosystem II

e⁻

H₂O

HS

cPABP

SH

RNA-binding site exposed

SH

cPABP

mRNA

AUG

HS

Translation start is now exposed and translation is activated

■ **FIGURE 18.07**

The Conformation of Translational Activator cPABP Responds to Light Intensity

Light initiates a chain reaction in photosystem II (PS II) and I (PS I) of plant chloroplasts whereby an electron transfer takes place through the electron transport chain. The electron reduces the disulfide bond of cPABP so changing its conformation and allowing it to bind mRNA and activate translation.

Tcherkezian J, Brittis PA, Thomas F, Roux PP, Flanagan JG (2010) **Transmembrane receptor DCC associates with protein synthesis machinery and regulates translation.** Cell 141:632–644.

This article describes the specific association of a transmembrane protein, DCC, with ribosomal subunits and initiation factors. DCC is found in the cytoplasmic membrane of nerve cells where it receives signals from outside. The nerve cell body, containing the nucleus, is separated from the axon terminus by an axon. Many neurons have long thin axons that can be meters in length. How the nucleus maintains control at such distances is explained by axonal transport proteins that move up and down axon filaments with protein cargo. But this does not fully explain rapid axon growth during development, which requires massive protein synthesis. Movement of all these components so far seems too difficult for a quick and accurate assembly.

This paper first explores the location of the translation machinery and DCC. Ribosome subunit S6, translation initiation factor, eIF4E, and DCC were shown to overlap when antibodies to these proteins were visualized by fluorescence microscopy suggesting that ribosomes are found associated with DCC at the cytoplasmic membrane. Electron microscopy and immunoprecipitation experiments support this idea. The authors confirmed these associations by density gradient centrifugation of ribosomal components followed by identification of proteins in the various subunit fractions. Interestingly, DCC was found with individual ribosomes but not with polysomes. Further, when the extracellular ligand netrin binds to DCC the ribosomes are released from DCC. Netrin is known to stimulate protein synthesis, and its release of ribosomes from DCC may be one mechanism for this.

FOCUS ON RELEVANT RESEARCH

Finally, the paper addresses the functional significance of DCC associating with ribosomes. Their results suggest that this interaction is essential for axonal outgrowth from the neural cell body. Furthermore, the location of DCC in dendrites corresponds to active protein translation as determined by incorporation of labeled amino acid analogs.

FIGURE 18.08
Antisense RNA can Base Pair with mRNA

mRNA is normally made using the non-coding strand of DNA as a template. Such mRNA is also known as sense RNA. If RNA is made using the coding strand as a template, it will be complementary in sequence to mRNA and is known as antisense RNA. The sense and antisense strands of RNA can base pair.

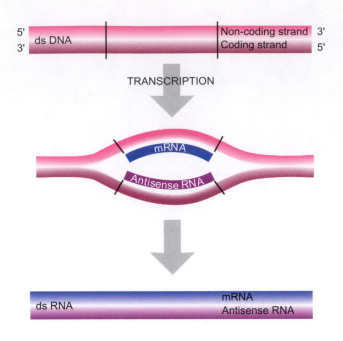

Antisense RNA binds to mRNA and prevents its translation.

If RNA were transcribed using the coding strand as a template we would produce an RNA molecule complementary in sequence to the original mRNA. This is known as **antisense RNA** and can base pair with its complementary mRNA, just as the two strands of DNA in the original gene base pair with each other (Fig. 18.08). (Note that uracil pairs with adenine in duplex RNA.)

Antisense RNA is occasionally used in gene regulation both by bacteria and eukaryotes. If antisense RNA is made, it will base pair with the mRNA and prevent it from being translated. In practice, antisense RNA can be made by transcribing the wrong strand of the gene, or by transcribing an entirely different gene with complementary sequences to the one it regulates. When the wrong strand of the gene is transcribed, the antisense RNA is a perfect match, whereas when a separate "antisense gene" is transcribed, the sense/antisense pair may only be partially complementary.

Antisense RNA works by a variety of mechanisms to control gene expression. Some antisense transcripts induce conformational changes in their target RNA. One example is the control of replication of a staphylococcal plasmid. Here, the antisense RNA (RNAIII) binds to *repR* RNA and creates a terminator stem loop that causes RNA polymerase to fall off. This is unusual for antisense control because it blocks transcription rather than translation. The more common mechanism of antisense control is to block translation. For example, *tisB* RNA (encoding a toxic peptide in *E. coli*) is transcribed until its antisense partner, IstR-1, binds to its 5′-UTR far upstream from the ribosome-binding site. This creates a double-stranded region of RNA that is cut by ribonuclease III, thus destroying the mRNA. By far the most common antisense mechanism is preventing ribosomes from initiating translation by physically blocking the ribosome-binding site. Some less common antisense RNA activities are the activation of translation by inducing the cutting of a polycistronic message into two different transcripts. This cleavage exposes a second ribosome-binding site on the downstream gene and enhances its translation.

Bacterioferritin is the protein used by bacteria to store surplus iron atoms. The *bfr* gene encodes bacterioferritin itself and the anti-*bfr* gene encodes the antisense RNA (Fig. 18.09). Since only a relatively short piece of antisense RNA is needed to block the mRNA, the anti-gene is similar in sequence but shorter than the original

antisense RNA   An RNA molecule that is complementary to mRNA

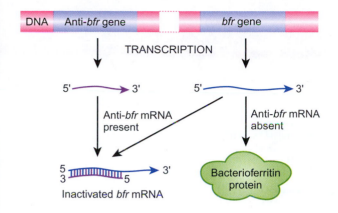

◄ **FIGURE 18.09**

Antisense RNA Regulates Bacterioferritin Synthesis

The bacterial chromosome contains genes for both *bfr* mRNA and anti-*bfr* RNA. If both RNA molecules are transcribed the anti-*bfr* RNA pairs with the *bfr* mRNA and prevents it from being translated. When iron is plentiful the anti-*bfr* gene is not expressed and only the *bfr* mRNA is produced. Under these conditions translation of the *bfr* mRNA to give bacterioferritin can take place.

gene. When the iron concentration in the culture medium is low, bacterioferritin is not needed, but it is made if the iron level goes up. The *bfr* gene itself is transcribed to give mRNA whether there is iron or not.

The anti-*bfr* gene is controlled by a regulatory protein known as **Fur (ferric uptake regulator)**, which senses iron levels. When plenty of iron is present, Fur acts as a repressor and turns off the transcription of a dozen or more operons needed for adapting the cell to iron scarcity. These include genes for several iron uptake systems designed to capture trace levels of this essential nutrient. In addition, Fur plus iron turns off the anti-*bfr* gene, which turns on the production of bacterioferritin. In low iron, the anti-*bfr* gene is transcribed to give antisense RNA. This prevents synthesis of the bacterioferritin protein when iron is scarce. The anti-*bfr* gene is now known to control several genes and has been renamed *ryhB*. Thus, by using antisense RNA, several genes can be regulated the opposite way to a group of others, although all respond to the same stimulus.

Artificially-synthesized antisense RNA will interfere with gene expression or any other cell process involving RNA. For example, antisense RNA is being tested experimentally to suppress cancer growth by stopping chromosome division. Antisense therapy is also used to treat retinitis due to cytomegalovirus (CMV). CMV is normally present in almost everyone, but when the person's immune system is compromised due to organ transplantation or AIDS, then the virus is activated. The virus attacks the retina and eventually will cause blindness if left untreated. The antisense therapy prevents the virus from replicating and thus damaging the retina.

1.6 Regulation of Translation by Alterations to the Ribosome

The same ribosomes have to translate many different messages, expressed under many different conditions. Not surprisingly, the ribosomes themselves are rarely modified, except in such general cases as putting whole ribosomes on standby to reduce the overall rate of protein synthesis in non-growing cells, as discussed in Chapter 13.

One rare exception is the phosphorylation of protein S6 of the small ribosomal subunit in the cells of mammals. The S6 protein may be phosphorylated up to five times on a cluster of serine residues close to its C-terminus. (Lower eukaryotes such as yeast lack these phosphorylation sites.) The phosphorylation occurs after cells in a growing tissue receive a signal to proliferate. The modified ribosomes preferentially translate mRNA molecules that possess **5′-terminal oligopyrimidine (5′-TOP)** tracts (Fig. 18.10). These are long pyrimidine-rich sequences located at the 5′ end of the mRNA, just upstream of the start codon. It turns out that most of the favored mRNAs encode ribosomal proteins and elongation factors. In other words,

---

Fur (ferric uptake regulator)   Global regulatory protein that senses iron levels in bacteria
5′-terminal oligopyrimidine tract (5′-TOP)   Long pyrimidine-rich tracts located between the 5′ end of mRNA and the start codon

FIGURE 18.10
Phosphorylation of S6 Favors mRNA with 5'-TOP

The S6 protein of the small ribosome subunit can be phosphorylated by an activated protein kinase. Those mRNA molecules possessing a 5'-TOP tract then bind better to the ribosome and commence translation.

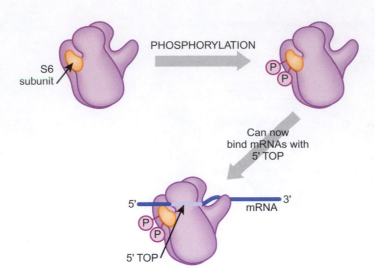

phosphorylation of S6 protein stimulates the ribosome to make parts for more new ribosomes—an evolved response to an increased rate of cell growth and division.

2. Basic Principles of RNA Interference (RNAi)

RNA interference (RNAi) is a mechanism for gene silencing that is induced by double-stranded RNA (dsRNA). It is sequence specific and involves the degradation both of dsRNA and of single-stranded RNA (ssRNA) molecules—usually mRNA—that are homologous in sequence to the dsRNA that triggered the response.

RNA interference is triggered by dsRNA that is fully base-paired and is at least 21–23 base pairs in length. Longer molecules of dsRNA are cleaved into fragments of 21–23bp by a nuclease known as "**Dicer**" (Fig. 18.11). These RNA fragments are referred to as **siRNA (short interfering RNA)** and are recognized by proteins of the **RNA-induced silencing complex (RISC)**. The RISC complex then separates the two strands of the siRNA and scours the cytoplasm for complementary RNA sequences. The nuclease activity of the RISC complex, sometimes referred to as "**Slicer**," but more commonly known as an **Argonaut (AGO) family** member, then degrades the complementary RNAs.

RNA interference operates in a wide range of eukaryotes, including protozoa, invertebrates, mammals, and plants. It is not found in prokaryotes. However, bacteria do possess ribonuclease III, an enzyme homologous to Dicer that rapidly degrades dsRNA molecules as short as 12bp. Moreover, bacteria use the CRISPR system to destroy both RNA and DNA of incoming viruses (see below). Recent investigations have complicated the simple idea of RNAi operating solely through RNA degradation. RNAi-related mechanisms can also silence transcription of targeted genes by altering chromatin structure and promoting DNA methylation.

> Double-stranded RNA is regarded as alien and normally destroyed by living cells of all organisms.

> RNA interference destroys mRNA that has the same sequence as double-stranded RNA detected in the cell.

Argonaut (AGO) family   Enzymes found within the RISC complex that degrade any cellular RNAs that are complementary to the guide strand of an siRNA
Dicer   Ribonuclease that cleaves double-stranded RNA into segments of 21–23bp
RNA interference (RNAi)   Response that is triggered by the presence of double-stranded RNA and results in the degradation of mRNA or other RNA transcripts homologous to the inducing dsRNA
RNA-induced silencing complex (RISC)   Protein complex induced by siRNA that degrades single-stranded RNA corresponding in sequence to the siRNA
short interfering RNA (siRNA)   Double-stranded RNA molecules of 21–23 nucleotides involved in triggering RNA interference in eukaryotes
Slicer   Ribonuclease activity of the RISC complex

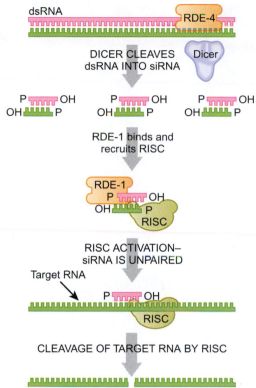

FIGURE 18.11
Mechanism of RNA Interference

Intruding double-stranded RNA (dsRNA) is recognized as foreign by RDE-4 and other proteins. Dicer cleaves the dsRNA into segments of 21 or 22 nucleotides with one or two base overhangs—pieces called short interfering RNA (siRNA). This is recognized by RDE-1 which recruits the RNA-induced silencing complex (RISC). The strands of the siRNA are separated during RISC activation. Finally, RISC cleaves target RNA that corresponds to the siRNA.

2.1. Sources of Double-Stranded RNA that Trigger RNAi

The prevailing theory is that RNAi developed as a defense mechanism against viruses, transposons, and transgenes. During infection by most RNA viruses, the virus genome passes through a dsRNA intermediate (the replicative intermediate). This is true both for viruses that carry their genomes as ssRNA and those that use dsRNA (see Ch. 21). Consequently, dsRNA is regarded as a sign of infection and triggers an anti-viral response. To protect the cell from viral damage, the viral genome is cut into siRNA by Dicer enzymes, and then all the mRNA homologous to the viral genome is efficiently removed by RISC complexes containing the Argonaut enzymes, subsequently keeping the cell virus free.

In a similar way, RNAi defends the genome against transposon movement (see Ch. 22) and transgenes. In fact, one of the first reports of gene silencing occurred when researchers tried to create a deep purple petunia. The researchers reasoned that by adding more copies of the gene for purple pigment, the flower would become a deeper purple. But adding an extra copy of the purple color gene into a purple petunia instead created a white flower! The transgene was not expressed, and neither was the endogenous purple gene. RNAi had effectively destroyed all the mRNA for purple color because this transgene was transcribed with a hairpin structure that created dsRNA.

2.2. MicroRNA is Short Regulatory RNA

Another source of small RNAs is from the genome itself. **Micro RNA (miRNA)** molecules are short RNA molecules that share several properties in common with siRNA. Micro RNA regulates the gene expression of the eukaryotic cell itself rather

micro RNA (miRNA)   Small regulatory RNA molecules of eukaryotic cells

FIGURE 18.12
Micro RNA

Micro RNA (miRNA) is made by processing a longer precursor that folds into a stem loop. Processing occurs in two steps using the nucleases Drosha and Dicer. After binding to miRISC and strand separation, one strand of the miRNA binds to the target mRNA and prevents translation.

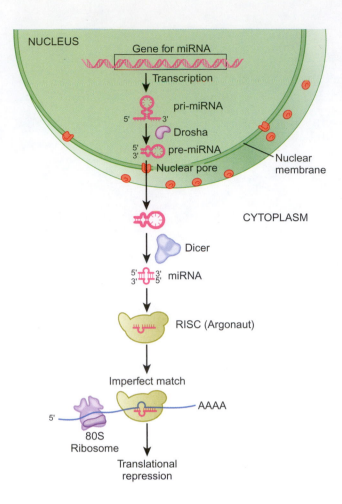

Micro RNA is a kind of short regulatory RNA that blocks translation of developmentally-regulated mRNA.

than acting as a defense mechanism against intruders. The miRNA acts primarily by blocking translation of mRNA. Translation is blocked by the binding of miRNA to mRNA, usually in the 3′ untranslated region (3′-UTR) and less commonly in the coding region.

Micro RNA molecules are produced from longer RNA precursors of approximately 70 nucleotides that are transcribed from chromosomal genes. These precursor RNA molecules ("primary miRNA" or pri-miRNA) fold into stem-loop structures (Fig. 18.12). The double-stranded stem region is then cut by Drosha, an enzyme that is related to Dicer, to generate pre-miRNA. The pre-miRNA exits the nucleus and is finally trimmed by Dicer (the same nuclease that generates siRNA) to form miRNA. In general, the miRNA molecules have 1–3 unpaired bases in the middle and thus differ from siRNA, which is completely base-paired. The miRNA is then bound by a protein complex, miRISC, which is analogous to RISC. Here, the strands are separated in a manner similar to what happens to siRNA in RISC. One strand of the miRNA then binds to the target mRNA. However, base-pairing is not usually perfect. The result is that translation of the target mRNA is blocked, but the mRNA is not degraded. A single miRNA may repress translation of hundreds of targets, but the effect is usually mild. Thus, miRNA tends to modulate the level of protein synthesis rather than switch it completely off. Although this is a general mechanism of miRNAs, there are examples of miRNAs that match the target mRNA completely and work much like siRNAs.

Micro RNA is found in worms, insects, mammals, and plants—that is, the same organisms that display RNA interference. The first miRNAs were found in *C. elegans* where they were called small temporal RNA (stRNA) because they regulate the timing of worm development during the conversion of the larva into the adult. Many miRNAs appear to be involved in the regulation of development, and many of the targets for miRNA are mRNAs that encode transcription factors, which in turn regulate the expression of other genes.

An assortment of other small RNA molecules that are involved in regulation have been discovered recently. Repetitive sequences have been shown to create a large number of small RNAs. Repetitive sequences (many of which are defective transposons) are clustered into tandem repeats around the centromere and telomere. In addition, scattered repeats, also including transposons, are found throughout the eukaryotic genome. Some of these repeats include genes that are transcribed (see Ch. 4). When small RNAs were isolated from *Drosophila,* many of the sequences were similar to transposons and repeat elements, and these were slightly longer than the previously-characterized siRNA and miRNA (about 23–26 nucleotides long). These are now known as **Piwi-interacting RNA (piRNA)** for the Argonaut-like protein that uses them to suppress gene expression (see below). These are encoded by the genome in mammals, but are not generated by Dicer/Drosha-style cleavage of dsRNA. In *Drosophila,* piRNA are derived from clusters of tandem repeats around the centromere, which is tightly coiled into heterochromatin. These repeats are filled with transposon sequences, so many *Drosophila* piRNA suppress transposon expression and subsequent movement around the genome. In eukaryotes, piRNA protect against the spread of repeated DNA sequences and are especially active in the reproductive cells of higher animals. Their full biological roles and mechanisms of action are still being discovered.

Box 18.01 Hybrid Dysgenesis in *Drosophila*

When a male fruit fly that has P-element transposons is mated to a female without this transposable element, the resulting progeny are sterile due to chromosomal breaks and rearrangements. These rearrangements are due to the movement of the P-element transposon during development. Interestingly, this effect is not seen when the female has P-elements. These offspring are fine. This phenomenon is called hybrid dysgenesis. Although the cause of P-element movement is known, what is not understood is the mechanism by which P-element movement is suppressed when the female is P-element positive.

Recent evidence suggests that RNAi may suppress this transposon movement. *Drosophila* oocytes have thousands of mRNA messages (maternal mRNAs) that are present before fertilization. These mRNAs are inactive until fertilization. Included in this population of mRNA are piRNA. When the female has P-elements, piRNAs are created during oogenesis to suppress P-element transcription and movement and then deposited in the egg. When the egg develops, the P-elements are unable to make transposase because piRNA suppresses its expression. When the female does not have the P-element, her eggs are devoid of piRNA complementary to the transposase. When she mates with a P-element positive male, the transposase is not repressed, and the transposon is free to move around the genome of the progeny causing chromosomal breaks and rearrangements, and ultimately sterile offspring.

2.3. Dicer Cuts Double-Stranded RNA into Small RNAs

Dicer is a general name for a family of enzymes that generate short pieces of RNA that are about 21–23 nucleotides in length. This protein family includes the single Dicer found in mammals and *C. elegans* and the multiple Dicers found in plant species and *Drosophila.* The enzyme is subdivided into different protein domains that all work to find, bind, and dice double-stranded RNA. There are two RNase III domains

Piwi-interacting RNA (piRNA)   Small RNA molecule slightly longer than siRNA that are derived from genomic tandem repeats clustered at the centromere and prevent the spread of transposons through an Argonaut–like protein

FIGURE 18.13
Structure of Dicer

Dicer has multiple domains including a dsRNA-binding domain to hold the target mRNA, a PAZ domain that binds to the 3'-nucleotide overhang on the target, and two RNase III domains that cut a 21–23 nucleotide siRNA.

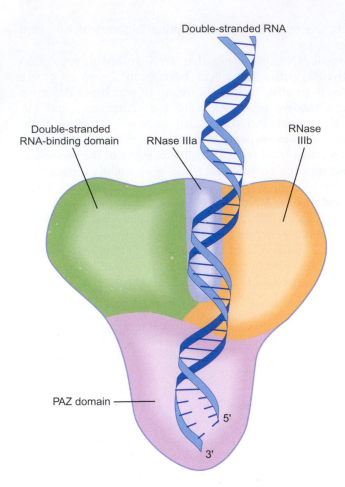

Double-stranded RNA

Double-stranded RNA-binding domain

RNase IIIa

RNase IIIb

PAZ domain

5'

3'

that work on each backbone of the double helix to create the specific cuts. The PAZ domain binds to RNA ends, preferring a 2-nucleotide overhang on the 3' end of the RNA. The distance between the PAZ domains and the enzyme-cutting site within the RNase III domains dictates the final length of the siRNA. On the opposite side of the PAZ domain is a double-stranded RNA-binding domain that holds the remaining RNA strands in place until RNase III makes the cuts (Fig. 18.13). When modified, Dicer may also take part in degrading DNA (rather than RNA) during programmed cell death (see Focus on Relevant Research).

Okamura K, Lai EC (2010) A deathly DNase activity for Dicer. Cell 18: 692–694.

FOCUS ON RELEVANT RESEARCH

Dicer generates siRNA for RNA interference, but recent evidence suggests that this enzyme may also function as a mediator of programmed cell death or apoptosis. A hallmark feature of apoptosis is the fragmentation of DNA into approximately 200bp long pieces. Once activated, DNases cut in between histones to create the 200bp fragments. In mammals, the DNase, called DFF45 (or ICAD), is activated by caspase-3 by removal of a small piece of the protein. Once active, it cuts the genomic DNA. In *C. elegans*, there is no similar protein, and the DNase responsible for cutting the DNA was unknown until recently. By screening genetic mutations in *C. elegans* that were unable to fragment DNA during apoptosis, it was discovered that loss of Dicer also prevented DNA fragmentation. Research confirmed that Dicer was activated by Ced-3 (*C. elegans* homolog to caspase-3) by cleaving a small piece of Dicer from the full-length protein. The PAZ domain and part of the first RNase III domain are removed in the process. The shortened form of Dicer no longer cuts RNA to make siRNA, but instead cuts DNA into pieces. Thus, simple changes to protein structure can establish an entirely new function for an enzyme.

2.4. The Argonaut Family of Proteins Destroys the Target mRNA

There are three different families of Argonaut-like proteins. The two most widely distributed are the Piwi group, which binds to piRNAs, and the Ago group, which binds to miRNAs and siRNAs. The third group is found in nematodes and has only been identified and not studied. The Ago protein is the central molecule in the RISC complex, and it unwinds the small RNAs created by Dicer to make a single-stranded template called the guide strand. The other strand of the siRNA, miRNA, and piRNA is discarded. The guide strand is loaded into Ago protein, and then the enzyme searches the cytoplasm for complementary sequences. Ago proteins have a PAZ domain that binds to the 3′ ends of RNA, a Mid domain that binds the 5′ end of RNA, and a tract of positively-charged amino acids in between that attract the negatively-charged phosphates of the guide RNA. When this complex finds a complementary mRNA sequence in the cytoplasm, it cuts the middle of the target mRNA with its RNase III-like domain. Once the target mRNA is cut by RISC, the message is further degraded by RNases. When the guide strand only partially binds to the target mRNA, then Ago proteins do not cut the message, and instead block the mRNA from translation. These activities actually occur in a small part of the cytoplasm called a P body, a place filled with mRNA degradation enzymes. Other proteins in RISC help unwind the siRNA, miRNA, and piRNA, and maintain the interaction with the target mRNAs. Most species contain multiple types of Argonaut proteins and variations in the accessory RISC proteins, suggesting that these different types define the different functions for small RNAs.

2.5. Amplification and Spread of RNAi

RNAi is remarkably potent. Less than 50 molecules of siRNA can silence target RNA that is present in thousands of copies per cell. This results from creation of more siRNA copies via an **RNA-dependent RNA polymerase (RdRP)** (Fig. 18.14). Cutting of the target mRNA by Ago gives two aberrant and unstable RNA molecules, one capped but without the poly(A) tail and the other with a tail but no cap. One or both of these aberrant RNA molecules are apparently used as template by RdRP to generate dsRNA. The dsRNA then acts as a substrate for Dicer, which generates more siRNA, called secondary siRNAs, so amplifying the RNA interference effect.

In plants and lower animals, RNA interference can be amplified and can spread throughout the organism, after being triggered in a localized zone.

In addition to amplification, the RNAi effect is capable of spreading from cell to cell and may travel considerable distances through an organism. This effect is especially noticeable in plants. Spreading of the siRNA signal throughout the body is also seen in certain animals. Remarkably, in the worm *C. elegans* the RNAi effect is passed on for several generations (without alterations in the genomic DNA sequence of the target gene occurring). Mammals do not possess the RdRP responsible for RNAi amplification, hence RNAi remains relatively localized. It is thought that the development of the specific immune system has made RNAi less important in mammals.

2.6. Formation of Heterochromatin by RNAi Proteins

RNAi has also been implicated in silencing gene expression by creating heterochromatin. This step occurs in conjunction with degradation of the mRNA target. A RISC-like complex containing a guide RNA sequence enters the nucleus and identifies any DNA sequences complementary to the guide RNA. When found, the complex attracts chromatin remodeling and histone-modifying enzymes that condense the DNA into heterochromatin. In the purple petunia example, RNAi not only destroyed the mRNA to purple pigment, but also completely silenced the transgene and the endogenous gene. This alternate function of RNAi is best understood in the fission yeast, *Schizosaccharomyes pombe*. A RISC-like complex, called RITS

RNA-dependent RNA polymerase (RdRP) RNA polymerase that uses RNA as a template and is involved in the amplification of the RNAi response

FIGURE 18.14
Amplification of RNA Interference by RdRP

Anomalous RNA generated by RISC-mediated cleavage is used as a template by RNA-dependent RNA polymerase (RdRP). This generates more dsRNA, which in turn is converted into more siRNA by Dicer.

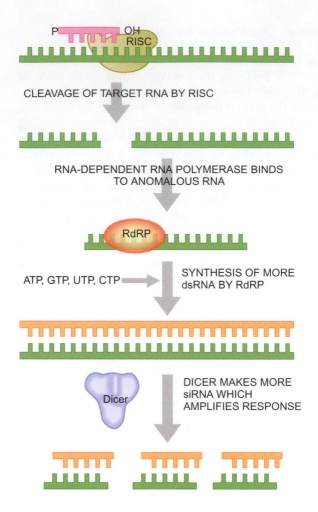

CLEAVAGE OF TARGET RNA BY RISC

RNA-DEPENDENT RNA POLYMERASE BINDS TO ANOMALOUS RNA

RdRP

ATP, GTP, UTP, CTP → SYNTHESIS OF MORE dsRNA BY RdRP

Dicer — DICER MAKES MORE siRNA WHICH AMPLIFIES RESPONSE

(RNA-induced transcriptional silencing complex), captures an siRNA created by its Dicer homolog. The complex containing a guide RNA enters the nucleus and finds the DNA sequence complementary to the guide. When RITS attaches to the DNA, it promotes methylation of histone H3 and association of Swi6, which remodel the area into heterochromatin and abolish transcription of the gene. This complex also establishes the heterochromatin around the yeast's centromere after a cell divides, and without the RITS complex, heterochromatin does not reform. Since the centromeric repeats contain a lot of transposon sequences, these become active in RITS mutants.

> In fission yeast, centromeric heterochromatin is created by the RITS complex, which is related to RISC, and contains a member of the Ago family of RNases.

2.7. Experimental Administration of siRNA

RNAi is widely used in laboratory research to prevent expression of cellular genes (as opposed to virus genetic information). RNAi in animals was first seen with the roundworm *C. elegans* where injection of dsRNA into the worm destroys any mRNA with complementary sequences. This is a useful tool for assaying gene function. Small interfering RNAs can be created artificially by looking at the sequence of a particular gene. When the dsRNA is injected, the gene of interest is no longer expressed, and the resulting phenotype can be assayed.

> RNA interference is now being widely used to investigate gene function in animals and plants.

Using RNAi allows investigation of gene function without the need to make mutants with altered or inactivated versions of a particular gene. This is especially useful for eukaryotes, most of which are diploid and where classic genetic analysis therefore requires introducing mutations into both copies. However, RNAi by its very nature prevents expression of all copies of the target gene. Indeed, RNAi may be used in organisms where multiple gene copies are present. For example, protozoa such as *Paramecium* contain two types of nuclei, a germline micronucleus and a highly-polyploid somatic nucleus. This situation makes it difficult to use standard genetic analysis. RNAi does not depend on where the mRNA is produced, it simply destroys it.

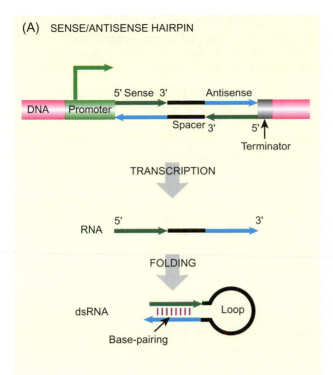

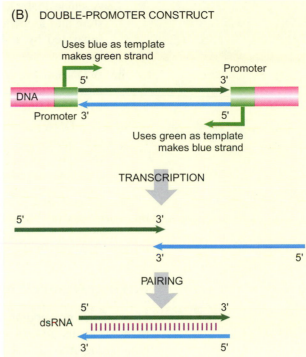

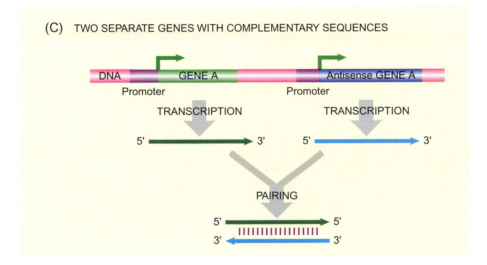

FIGURE 18.15
Experimental Induction of RNA Interference

RNA interference occurs when both the sense and antisense RNA of a gene are present and form dsRNA. Three constructs are shown that direct the synthesis of a dsRNA molecule. The first construct (A) has a sense region and an antisense region that base pair. A spacer separates the sense and antisense regions and forms a loop at the end of the hairpin. The double promoter construct (B) has a promoter that directs the transcription of the sense strand and another promoter for the antisense strand. The two resulting RNA molecules are complementary and form a dsRNA molecule. In (C), dsRNA is created from two separate genes and two separate promoters.

Experimentally, RNAi may be induced by providing long molecules of dsRNA that are cut into siRNA by the Dicer enzyme (Fig. 18.11). Single-stranded antisense RNA against cellular genes may also trigger RNAi by base-pairing with the corresponding plus RNA strand. This generates dsRNA inside the cell. Alternatively, short dsRNA molecules of 21–23 nucleotides in length may be administered directly and will act as siRNA.

Although dsRNA may be used directly, it is often more convenient to provide a DNA construct that generates dsRNA *in vivo*. The dsRNA generated from the DNA constructs may be long (and rely on Dicer to generate siRNA) or short, giving siRNA directly. This may be done by three main variations (Fig. 18.15):

i. A single DNA segment transcribed from a single promoter that generates a stem and loop structure. Here, the plus and minus strands are in tandem but separated by a short stretch of DNA that remains unpaired and forms the loop.

ii. A DNA segment flanked by two opposing promoters. Consequently, one promoter transcribes the template and the other transcribes the sense strand from the same dsDNA segment.

iii. Two DNA segments, one being the inverse of the other and both having separate promoters. Consequently, one promoter transcribes the plus strand from the sense version of the DNA and the other transcribes the minus strand from the other, inverted, antisense, DNA segment.

In the case of *C. elegans*, dsRNA may be injected into the worm. Alternatively, the worm may be fed bacteria, such as *E. coli*, carrying plasmids with DNA constructs as described above that generate dsRNA *in vivo*. (Conveniently enough, bacteria are the natural diet of *C. elegans*.)

3. Long Non-coding Regulatory RNA

Most of the non-coding regulatory RNAs so far discussed are relatively short (less than 200 nucleotides). However, recent work has revealed that a large number of **long non-coding RNA (lncRNA)** molecules are produced by eukaryotic cells. In eukaryotes with large amounts of non-coding DNA, it now appears that up to 20 times as much of the DNA sequence is transcribed into **non-coding RNA** as into coding mRNA! Some of this is probably "background noise," that is, occasional RNA transcripts made by mistake since RNA polymerase is not perfectly regulated. However, it is clear that a significant number of lncRNA molecules play a genuine physiological role.

The precise mechanism of action of most lncRNAs is still mysterious, even for many lncRNAs that have known biological effects. However, a possible unifying factor is that, due to their length, most lncRNA molecules seem to act by bringing multiple DNA sites and/or regulatory proteins together into a functional complex (see Focus on Relevant Research). An example is the class of lncRNA molecules that promote the action of enhancers (see Ch. 17, Section 2.2 for information on enhancers). The lncRNA probably acts by helping the enhancer and its associated proteins bind to the transcription factors or mediator complex at the promoter.

Long non-coding RNA includes a variety of longer regulatory RNA molecules whose precise roles are still uncertain in many cases.

Nagano T and Fraser P (2011) No-Nonsense Functions for Long Non-coding RNAs. Cell 145: 178–181.

FOCUS ON RELEVANT RESEARCH

This review surveys the possible roles that long non-coding RNA (lncRNA) plays in regulating gene expression in eukaryotic cells. LncRNA may bind one or more chromatin-modifying proteins and help localize them correctly to specific chromosomal sites. LncRNA may bind simultaneously to sites on more than one chromosome and thus facilitate regulatory interactions between loci on different chromosomes. As noted above, lncRNA probably helps in the operation of enhancers by DNA looping. In addition, formation of large chromatin loops probably often involves lncRNA. Similarly, the formation of large nuclear structures and complexes may depend on lncRNA acting as scaffolds.

long non-coding RNA (lncRNA) Longer regulatory RNA molecules (>200 bases) of eukaryotic cells
non-coding RNA Any RNA molecule that is not translated to give protein

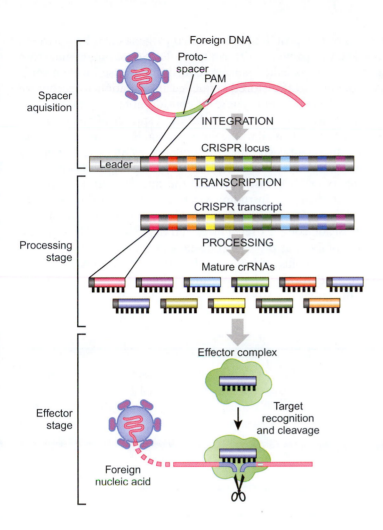

■ **FIGURE 18.16**
The CRISPR System

An overall model of CRISP/Cas
activity. During spacer acquisition,
sequence elements from invading
nucleic acids become incorporated
at the leader-proximal end of the
CRISPR locus. Chosen invader
sequences are found close to
protospacer adjacent motifs
(PAMs). In the processing stage, the
locus is transcribed and processed
into mature crRNAs containing an
8nt repeat tag (gray) and a single
spacer unit (multiple colors). During
the effector stage, the mature
crRNAs associate with Cas proteins
to promote the degradation of
complementary nucleic acids
(two separate effector complexes
are used for DNA and RNA
targets). (Credit: Terns MP and
Terns RM (2011) CRISPR-based
adaptive immune systems. Curr Op
Microbiol 14: 321–327)

4. CRISPR: Anti-Viral Defense in Bacteria

Although bacteria do not have RNAi, they do have another defense program to
destroy invading virus genomes. CRISPR stands for Clustered Regularly Interspaced
Short Palindromic Repeats. The CRISPR system consists of a memory bank of hostile virus sequences plus a mechanism for identification and destruction of incoming
virus DNA or RNA. Note that, unlike the RNAi system of eukaryotes, the CRISPR
system protects against DNA viruses as well as RNA viruses. The CRISPR system is
widespread in bacteria and Archaea but not in eukaryotes. Approximately 90% of
Archaea and 70% of bacteria possess the CRISPR system and some have multiple
CRISPR regions on their chromosomes. However, in many cases the CRISPR system appears to be incomplete or defective. This has important clinical consequences.
Several pathogenic bacteria, for example *Streptococcus pyogenes*, possess virulence
factors that are actually carried by bacterial viruses that have integrated into their
chromosomes. Strains with intact CRISPR defense systems are rarely infected with
these viruses, whereas strains of *Streptococcus* with defective CRISPR systems are
much more likely to pick up viruses carrying virulence factors.

The CRISPR memory bank is found on the bacterial chromosome and consists of many different segments of virus sequence alternating with identical
repeated sequences (Fig. 18.16). It provides resistance to any viruses that contain
the same or very closely related sequences. The proteins of the **CRISPR system**

CRISPR system A bacterial defense system that uses an enzyme complex associated with a short single-stranded RNA derived from the
CRISPR locus to find and destroy any RNA with complementary sequences

(CRISPR-associated proteins or Cas proteins) are encoded by genes that lie upstream of the CRISPR sequences. The Cas proteins perform two roles. Some obtain and store segments of virus sequence, a process called spacer acquisition, whose mechanism is still obscure. Other Cas proteins use the stored sequence information to recognize intruding virus genomes and destroy them. There is considerable variation in the nature and arrangement of the Cas proteins between different species of bacteria.

The **CRISPR locus** is transcribed as a whole into a long RNA molecule (pre-crRNA) that is cleaved in the middle of each repeated sequence by an assembly of Cas proteins. This converts the pre-crRNA into individual virus-specific segments known as crRNA. When these segments of crRNA base pair with the nucleic acid of an invading virus, other Cas proteins destroy the intruding viral genome. Two different Cas complexes degrade viral nucleic acids, one specific for RNA and the other for DNA. For example, in *E. coli*, the Cascade complex recognizes and degrades double-stranded DNA targets. Cascade consists of five Cas proteins (one each of CasA, CasD, and CasE, two copies of CasB, and six of CasC) plus a 61-nucleotide crRNA. The RNA-degrading protein, Cas6, from the Archeon *Pyrococcus* has recently had its structure investigated and appears to wrap around the target RNA (see Focus on Relevant Research).

FOCUS ON RELEVANT RESEARCH

Wang R, Preamplume G, Terns MP, Terns RM and Li H (2011) Interaction of the Cas6 riboendonuclease with CRISPR RNAs: recognition and cleavage. Structure 139:19:257–264.

The CRISPR system protects prokaryotes from viruses by destroying viral nucleic acid. This paper describes the structure and RNA-binding mechanism of the Cas6 protein from *Pyrococcus furiosus*, a member of the Archaea. X-ray crystallography at 3.2 Å resolution was used to give a crystal structure.

As mentioned above there is great variation in the Cas proteins in different bacteria and Archaea. Unlike other families of Cas endonucleases, the Cas6 protein binds nucleotides 2–9 of the RNA target using two ferredoxin-like domains. This brings the cleavage site of the target RNA close to the active site of this ribonuclease.

---

5. Premature Termination Causes Attenuation of RNA Transcription

Transcriptional **attenuation** is a regulatory mechanism that involves premature termination of mRNA synthesis. The basic principle of attenuation is that the first part of the mRNA to be made, the **leader region**, can fold up into two alternative secondary structures. One of these allows continued transcription, but the other secondary structure causes premature termination. The status of attenuation is somewhat ambiguous. It is often viewed as regulation at the level of transcription. However, it does involve mRNA that has already been partly transcribed. Consequently, it is sometimes regarded as "post-transcriptional." Since attenuation is closely related to other mechanisms based on alternative RNA stem and loop structures, we have chosen to include it along with other forms of RNA-based regulation.

Typically, the leader region contains four subregions (sequences 1 through 4) that may base pair in two different ways. When no other factors intervene, sequence 1 pairs with 2 and sequence 3 pairs with 4, so forming two stem and loop structures (Fig. 18.17). The second of these stem and loop structures, containing paired sequences 3 and 4,

Regulation by attenuation involves alternative stem and loop structures in the mRNA.

---

attenuation   Type of transcriptional regulation that works by premature termination and depends on alternative stem and loop structures in the leader region of the mRNA
CRISPR locus   Region of the bacterial genome that has a series of small sequence elements acquired from invading nucleic acids of viruses
leader region   The region of an mRNA molecule in front of the structural genes, especially when involved in regulation by the attenuation mechanism

A. PREMATURE TERMINATION   B. READ THROUGH

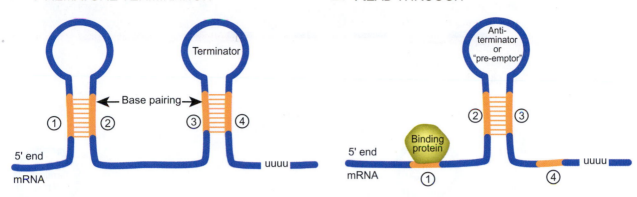

FIGURE 18.17
Alternative Secondary Structures of mRNA Leader Region

The sequences in the leader region of the mRNA designated by 1, 2, 3, and 4 can base pair in two alternative ways. A) The structure for premature termination of mRNA is due to base-pairing that forms two stem and loop structures in the mRNA. The second loop, 3 plus 4, causes termination. B) Termination can be prevented if a protein binds to the mRNA at site 1, allowing sites 2 and 3 to pair off. This creates the "pre-emptor" and prevents the terminator from forming.

acts as a terminator that stops RNA polymerase transcription shortly after starting. However, sequence 2 may instead pair with 3. For this to happen, a protein must bind to sequence 1 and sequester it from its complementary sequence 2. The net result is that the terminator loop, normally consisting of sequences 3 and 4, no longer forms and transcription of the mRNA can continue.

Attenuation is used to regulate the genes for biosynthesis of amino acids in both gram-negative bacteria, such as *E. coli*, and gram-positive bacteria, such as *Bacillus*. If the supply of amino acid is plentiful, then the genes for its biosynthesis should be turned off. Conversely, if the level of the amino acid is low, the biosynthetic genes should be transcribed.

In *E. coli*, attenuation is complicated and usually involves the binding of ribosomes to the mRNA leader region where they translate a **leader peptide**. These ribosomes act as the protein that blocks region 1 from pairing with region 2. The leader peptide is encoded by a short open reading frame and only consists of 14 or 15 amino acids. It lies close to the 5′ end of the mRNA, upstream of the structural genes for the enzymes of the biosynthetic pathway (Fig. 18.18A).

The leader peptide contains several tandem codons for the amino acid in question. For example, in the leader peptide of the *his* operon of *E. coli* there are seven codons for histidine in a row. In the leader peptide of the *thr* operon, there are 11 clustered codons for threonine and isoleucine. Since threonine is the precursor for isoleucine, codons for both of these amino acids are included in attenuation control. When an amino acid is in short supply, the ribosome has difficulty finding a charged tRNA carrying that particular amino acid and it slows down. When several codons for a scarce amino acid follow each other, the ribosome grinds to a halt. The stalled ribosome covers sequence 1, loop 2/3 forms, and the pre-emptor is made (Fig. 18.18B). Without the region 3/4 stem loop, RNA polymerase carries on, transcribing the rest of the operon.

In *Bacillus*, ribosomes are not involved and there is no leader peptide. Nonetheless, the leader region of the mRNA possesses four sequences that can pair up to give two alternative structures. An **attenuation protein** binds the amino acid in question. In the presence of the amino acid, the attenuation protein binds to the mRNA leader region and promotes termination. In *Bacillus*, the 5′ region of the leader of *trp* mRNA contains

attenuation protein   Regulatory protein involved in attenuation and that binds to the leader region of mRNA
leader peptide   The short protein produced in attenuated genes that has codons that correspond to the amino acid in which the operon synthesizes

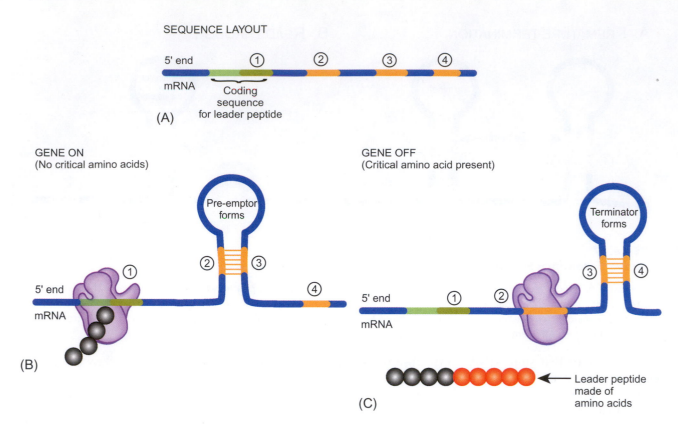

FIGURE 18.18
Stalled Ribosome Prevents Formation of Terminator Loop

Attenuation controls whether synthesis of mRNA is completed or aborted. (The RNA polymerase is not shown in this figure, just the attenuation mechanism.) A) The leader region of the mRNA contains the coding sequence for the leader peptide and four specific sequences (1, 2, 3, and 4) that can base pair to form stem and loop structures. B) When there is a shortage of the corresponding amino acid, the ribosome slows down at region 1, allowing the pre-emptor structure to form and transcription of the mRNA by RNA polymerase to continue. Note that the leader peptide is not completed. C) When there is an abundance of the corresponding amino acid, the leader peptide is made and the ribosome quickly moves to region 2, allowing regions 3 and 4 to form the terminator loop. This prevents further elongation of the mRNA.

a run of 11 UAG or GAG triplets separated from each other by two or three other bases. Eleven subunits of the tryptophan attenuation protein (TRAP) bind to these, forming an eleven-membered ring (Fig. 18.19) that terminates transcription.

6. Riboswitches—RNA Acting Directly As a Control Mechanism

One of the most fascinating recent stories in molecular biology is that RNA can carry out many of the functions that were previously believed to need proteins. Ribozymes, that is to say catalytically-active RNA, are involved in RNA splicing, protein synthesis, and viroid replication. Antisense RNA and a variety of small regulatory RNA molecules, such as siRNA, are involved in gene regulation. Most recently, it has been found that RNA domains at the front of mRNA, referred to as **riboswitches**, can directly interact with small molecules and can control gene expression. The vast majority of riboswitches have been found in bacteria and so far it is only in bacteria that translation control by riboswitch operation exists. However, sequence analysis has revealed that the genomes of certain fungi and plants contain thiamine riboswitches. However, these work by regulating the alternative splicing of mRNA.

Biosynthetic pathways that make metabolites such as amino acids and vitamins are generally induced when the metabolite is in short supply but are shut down when there is a plentiful supply of the metabolite. The genes for such pathways are often controlled by repressors or by attenuation and are repressed in response to high concentrations of

> A few mRNA molecules can control their own translation via riboswitch domains at their 5′ ends that bind small molecules.

riboswitch   Domain of messenger RNA that directly senses a signal and controls translation by alternating between two structures

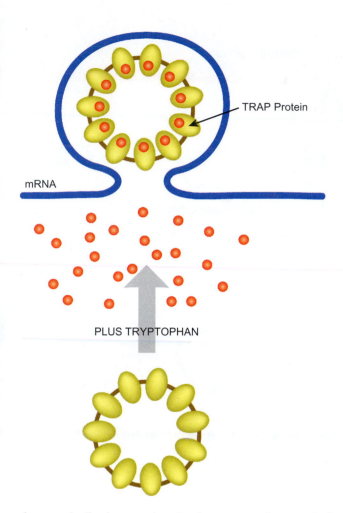

FIGURE 18.19
Attenuation by RNA-Binding Protein

When tryptophan is present, it binds avidly to the tryptophan attenuation protein. This binds to the RNA and alters its structure. The alternative RNA structure possesses a stem and loop that causes premature termination.

the metabolite in question. In these cases, the metabolite is bound by a regulatory protein as already described, such as the ArgR repressor of *E. coli* that binds arginine or the tryptophan attenuation protein (TRAP) of *Bacillus* that binds tryptophan.

In riboswitches, the metabolite is directly bound by an RNA sequence at the 5′ end of the mRNA. For example, the thiamine riboswitch of *E. coli* contains a sequence that binds the vitamin/co-factor thiamine pyrophosphate with great specificity and is known as the THI box. Similarly, the RFN box of the riboflavin riboswitch in *Bacillus subtilis* binds flavin mononucleotide. When these vitamins are in short supply the biosynthetic genes are turned on without the intervention of any regulatory protein. Conversely, when the vitamin is present at high levels the genes are turned off. Riboswitches are presently known for several vitamins, amino acids, glucosamine, magnesium, and the purine bases adenine and guanine.

Binding of the metabolite to its RNA box changes the conformation of the whole riboswitch domain. Riboswitches exist in two alternative conformations that have different stem and loop structures. This in turn controls gene expression by one of two related mechanisms, premature termination of transcription (i.e., attenuation) or translational inhibition.

In the attenuation mechanism (Fig. 18.20A), the riboswitch controls whether or not the mRNA for the biosynthetic genes will be prematurely terminated. Here, the riboswitch sequesters the terminator sequences in the absence of the signal metabolite and transcription continues. When the metabolite binds, the riboswitch switches to a conformation that allows the formation of a terminator stem and loop. This causes premature termination of the mRNA. Consequently, the genes are not expressed.

In the translational inhibition mechanism (Fig. 18.20B), the riboswitch controls whether or not the mRNA will be translated. When the signal metabolite is absent the Shine-Dalgarno sequence is free to bind to ribosomal RNA and translation proceeds. When the signal metabolite binds, the riboswitch sequesters the Shine-Dalgarno sequence and translation is prevented.

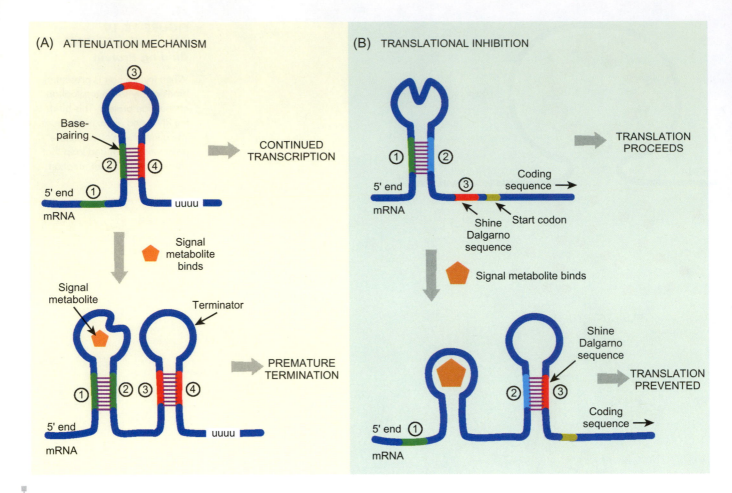

FIGURE 18.20
Riboswitch Mechanisms

Riboswitches alternate between two alternative stem and loop structures depending on the presence or absence of the signal metabolite. A) In the attenuation mechanism, the presence of the signal metabolite results in formation of the terminator structure and transcription is aborted. B) In the translational inhibition mechanism, the presence of the metabolite results in sequestration of the Shine-Dalgarno sequence, which prevents translation of the mRNA.

Riboswitches can respond to temperature as well as to metabolite concentrations.

In addition to metabolites, riboswitches are known that respond to the regulatory nucleotide cyclic-di-GMP. This nucleotide regulates lifestyle changes, such as conversion between free-swimming and biofilm formation, in many bacteria. The effects of cyclic-di-GMP on gene expression are exerted partly by regulatory proteins that bind to DNA and partly by riboswitches.

Riboswitches can also respond to physical conditions as opposed to small molecules. The **RNA thermosensor** is a specialized kind of riboswitch that responds to temperature and controls mRNA translation by sequestration of the Shine-Dalgarno sequence. The principle is the same as above, but formation of the alternative stem and loop structures depends directly on temperature. At high temperature one of the stems is unstable and the riboswitch flips to its high temperature form. The *rpoH* gene of *E. coli* is involved in the heat shock response, as described in Chapter 16. In addition to the regulation described there, translation of *rpoH* mRNA is prevented by an RNA thermosensor at low temperature but allowed as the temperature increases. Other genes whose translation is controlled by RNA thermosensors include certain activator genes involved in the virulence of pathogenic bacteria such as *Yersinia* and *Listeria*. Outside the host, these genes are switched off by translational inhibition. Inside the hot-blooded mammalian host the temperature is warmer and the activator

RNA thermosensor A specialized riboswitch that responds to temperature to control mRNA translation

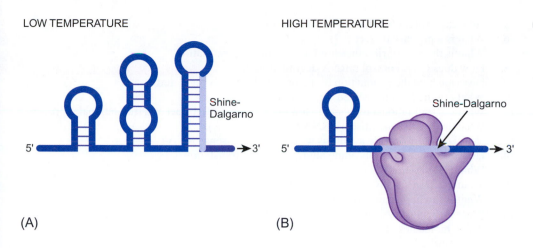

LOW TEMPERATURE

Shine-Dalgarno

5' ———→ 3'

(A)

HIGH TEMPERATURE

Shine-Dalgarno

5' ———→ 3'

(B)

FIGURE 18.21
RNA Thermosensors

Unstable stem-loop structures sequester the Shine-Dalgarno from access to ribosomes in lower temperatures. In higher temperatures, the stem-loop pairing melts, and ribosomes gain access to the Shine-Dalgarno.

proteins are expressed. The activator proteins then proceed to activate several other genes involved in bacterial virulence (Fig. 18.21).

Key Concepts

- Genes may be regulated at the level of RNA as well as DNA.
- The rate of mRNA degradation may be controlled by the binding of certain proteins.
- Some mRNA molecules cannot be translated into protein unless they are first cleaved.
- Translation of mRNA is sometimes activated or repressed by RNA-binding proteins.
- Translation of mRNA may also be controlled by antisense RNA.
- Occasionally, translation of mRNA is controlled by alterations to ribosomal proteins.
- Double-stranded RNA is regarded as foreign and destroyed by all living cells.
- RNA interference (RNAi) destroys mRNA that matches the sequence of previously-detected double-stranded RNA.
- The nuclease Dicer cuts double-stranded RNA into short segments (siRNA) that trigger the RNAi response.
- The target mRNA is degraded by proteins of the Argonaut family.
- In some organisms, the RNAi response is amplified by RNA-dependent RNA polymerase and may spread throughout the organism.
- In certain cases RNAi also silences genes by the formation of heterochromatin.
- RNAi is now widely used to study the function of genes in eukaryotes. This avoids the need to make labor-intensive gene knockouts.
- The CRISPR system of bacteria destroys hostile RNA or DNA from intruding viruses.
- Transcriptional attenuation is a control mechanism that relies on premature termination of mRNA synthesis.
- Riboswitches are sequences in the mRNA itself that bind small regulatory molecules and regulate translation.

Review Questions

1. Why is regulation at the transcriptional level usually more beneficial than regulation at the translational level?
2. Name the ways in which translation can be regulated.
3. What is a ribonuclease?
4. What factor(s) determine(s) the susceptibility of mRNA to degradation?

5. How can protein binding affect mRNA stability?
6. What is a primary transcript?
7. What is the role of ribonuclease III in regulating translation?
8. How does CsrA control mRNA degradation?
9. Describe how the production of the protein ferritin is regulated.
10. What are the 5′ untranslated region, IRP, and IRE?
11. What is a translational repressor? How does aconitase/IRP1 act as a translational repressor?
12. What are sense and antisense RNA?
13. How is bacterioferritin synthesis regulated by antisense RNA?
14. How and when does ribosome alteration affect translation?
15. What is RNA interference?
16. What are Dicer, RISC, siRNA, and Argonaut?
17. What is the importance of RNAi?
18. What is the mechanism of RNA interference?
19. A few molecules of siRNA can silence thousands of copies of target RNA; how is this accomplished?
20. How can RNAi be used experimentally to prevent gene expression?
21. What are three ways to generate dsRNA from a DNA construct *in vivo*?
22. How can dsRNA be introduced into a cell without using a DNA construct?
23. What is micro RNA?
24. What are the differences and similarities between micro RNA and siRNA?
25. What are long non-coding RNAs? How can they promote the action of enhancers?
26. How does the CRISPR system protect bacteria from viruses?
27. What is attenuation? What is an attenuation regulatory protein?
28. The leader region of an RNA transcript may base pair in two ways; what effect may these two secondary structures have on transcription?
29. What are riboswitches?
30. What are the two main mechanisms by which riboswitches operate?
31. How do riboswitches respond to metabolite concentrations?
32. How do riboswitches respond to physical conditions like temperature?

Conceptual Questions

1. What would happen if there was a mutation that prevented the ribosomes from translating the leader peptide of an attenuated operon in *E. coli*?
2. The pivotal experiment that identified double-stranded RNA as a potent inhibitor of gene expression was done by Andrew Fire, Craig Mello, and their colleagues in 1998. The researchers were studying the *mex-3* gene of *C. elegans* and were trying to see what would happen if they injected antisense *mex-3* mRNA into the developing *C. elegans* embryo. First, they created two versions of *mex-3* in a plasmid vector, one version had the promoter and gene arranged as usual, and the RNA transcript from this vector would make normal sense RNA. In the other version, the *mex-3* gene was inverted so that the promoter initiated the transcription of the complementary strand, and this would make antisense RNA. These two versions were used to make RNA in a test tube by adding nucleotides and RNA polymerase. Next, the RNA was injected directly into a *C. elegans* egg using an ultrafine needle. The researchers injected no RNA into some eggs (control experiment), antisense *mex-3* into other eggs, and a combination of sense and antisense *mex-3* into a third group. The eggs were incubated with radioactively-labeled *mex-3* mRNA to find the location of any remaining *mex-3* transcripts. The results showed a reduction of *mex-3* mRNA in the eggs treated with antisense in comparison to the control. The eggs injected with the antisense/sense combination had no *mex-3* mRNA. How does this data suggest that double-stranded RNA is responsible for *mex-3* gene silencing?
3. What types of molecules can trigger RNAi?
4. A new gene has been identified in the bacterium *Streptococcus pyogenes* that has been linked to a particular ribosomal protein subunit. This new gene is named rProU for ribosomal protein U and has been shown to produce a primary transcript of 2300 nucleotides. During experiments on this gene, the researchers noticed another RNA of 1800 nucleotides that is identical to the end of the larger RNA. Even more interesting results are seen when this gene is examined in *S. pyogenes* mutants lacking the enzyme ribonuclease III. In this bacterial strain, there is only the larger RNA. Based on your knowledge of *adhE* regulation in *E. coli*, suggest a reason why there are two different transcripts in wild-type *S. pyogenes* and why there is only one transcript for the mutant *S. pyogenes* strain. How would you test your theory?
5. Compare and contrast transcriptional attenuation in the *E. coli his* operon and the *Bacillus trp* operon.

Analysis of Gene Expression

Gene expression may be examined in a variety of ways, both at the level of individual genes and, increasingly in recent years, at the level of the whole genome. By analogy with genomics, the sum total of an organism's RNA transcripts are sometimes referred to as the **transcriptome**. Here, we will first consider individual genes and then cover approaches to screening the expression of large numbers of genes simultaneously. This is known as transcriptome analysis and, with proteomics and metabolomics (see Ch. 15) makes up the area of **functional genomics**. Of the plethora of newly-coined terms ending in -ome, perhaps the nicest is the "unknome" proposed by Mark Gerstein of Yale University. This consists of the large proportion of genes with no currently known function!

1. Monitoring Gene Expression

Expression of most genes results in their transcription to give RNA followed by translation of the RNA to give the final gene product, protein. In addition there are a few genes that produce non-coding RNA (such as transfer RNA (tRNA), ribosomal RNA (rRNA), and microRNA (miRNA)) and so have RNA as the final gene product. Although housekeeping genes are needed all the time, most genes

---

functional genomics   The study of the whole genome and its expression
transcriptome   The total sum of the RNA transcripts found in a cell, under any particular set of conditions

are expressed only under certain environmental conditions or in particular tissues or at certain stages of the developmental cycle, as discussed in Chapters 16 and 17. Measurement of gene expression means estimating the level of gene product synthesized. Since most genes vary in expression under different conditions it is necessary to measure the level of gene expression under a variety of conditions.

It is possible to monitor gene expression directly by measuring the levels of protein or RNA. Proteins may be detected by running cell extracts on polyacrylamide gels or by antibody-based assays. If the protein is an enzyme the enzyme activity may be assayed. Direct detection and assay of proteins is gene specific.

Here, we will consider the monitoring of gene expression at the transcriptional level. The transcriptional expression of a gene may be estimated by measuring the level of messenger RNA (mRNA) directly. This may be done by hybridization (Northern blotting) using a DNA probe specific for the sequence of the gene under investigation. Hybridization in Northern blots has already been discussed in Chapter 5. The use of fluorescent probes has greatly increased the sensitivity of Northern hybridization; nonetheless, for accurate measurement of the expression of individual genes under many different conditions, using gene fusions with reporter genes is preferable.

> Gene expression is measured by monitoring the RNA or protein products that are made.

2. Reporter Genes for Monitoring Gene Expression

Genes that are used in genetic analysis because their products are easy to detect are known as **reporter genes**. They are often used to report on gene expression, although they may also be used for other purposes, such as detecting the location of a protein or the presence of a particular segment of DNA.

> Genes whose products are convenient to assay are used as "reporters."

Suppose that a DNA molecule, such as a cloning plasmid, has been inserted into a new bacterial host cell or a transgene has been inserted into the chromosome of a new animal host. Antibiotic resistance genes are often used to monitor whether the DNA is indeed in the intended location. Thus, antibiotic resistance genes may be regarded as reporter genes (Fig. 19.01). As already discussed in Chapter 7, after transformation of the plasmid into the target bacteria, they are treated with the antibiotic. Those that receive the plasmid become antibiotic resistant; those not getting the antibiotic resistance gene are killed. An antibiotic resistance gene can be used to keep track of transgenes in other organisms, such as yeast, cultured mammalian cells, or viruses.

2.1. Easily Assayable Enzymes as Reporters

One of the first reporter genes for monitoring gene expression was the *lacZ* **gene** encoding β-**galactosidase**. This enzyme normally splits lactose, a compound sugar found in milk, into the simpler sugars glucose and galactose. However, β-galactosidase will also split a wide range of galactose compounds (i.e., **galactosides**) both natural and artificial (Fig. 19.02). The two most commonly-used artificial galactosides are ONPG and X-gal. **ONPG (*o*-nitrophenyl galactoside)** is split into *o*-nitrophenol and galactose. The *o*-nitrophenol is yellow and soluble, so it is easy to measure quantitatively. **X-gal (5-bromo-4-chloro-3-indolyl β-D-galactoside)** is split into galactose plus the precursor to an indigo type dye. Oxygen in the air converts the precursor to an insoluble blue dye that precipitates out at the location where the *lacZ* gene is expressed.

> β-galactosidase and alkaline phosphatase can be assayed using substrates that generate colored or fluorescent products.

β-galactosidase   Enzyme that splits lactose and other compounds of galactose
galactoside   Compound of galactose, such as lactose, ONPG, or X-gal
***lacZ* gene**   Gene-encoding β-galactosidase; widely used as a reporter gene
ONPG (*o*-nitrophenyl galactoside)   Artificial substrate that is split by β-galactosidase, releasing yellow *o*-nitrophenol
reporter gene   Gene that is used in genetic analysis because its product is convenient to assay or easy to detect
X-gal (5-bromo-4-chloro-3-indolyl β-D-galactoside)   Artificial substrate that is split by β-galactosidase, releasing a blue dye

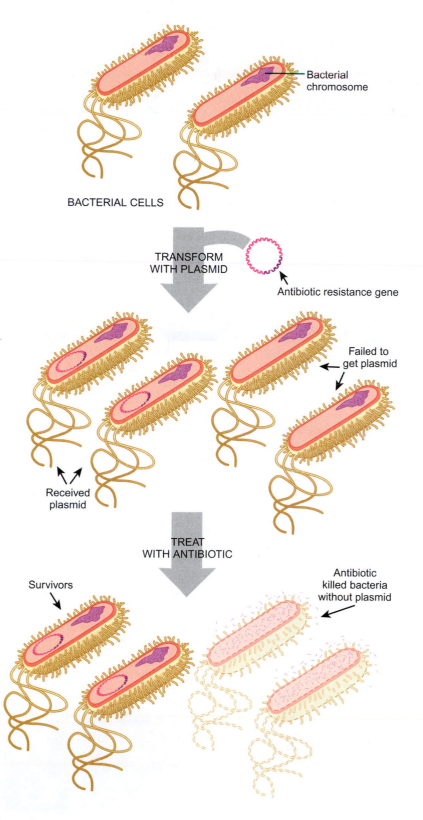

FIGURE 19.01
Antibiotic Resistance as a Reporter Gene

Antibiotic resistance genes are included on plasmids in order to determine whether the plasmids are present in a cell. When bacteria are transformed with plasmid DNA those that get a plasmid that carries an antibiotic resistance gene will survive when treated with the antibiotic, whereas those cells that fail to get a plasmid will be killed.

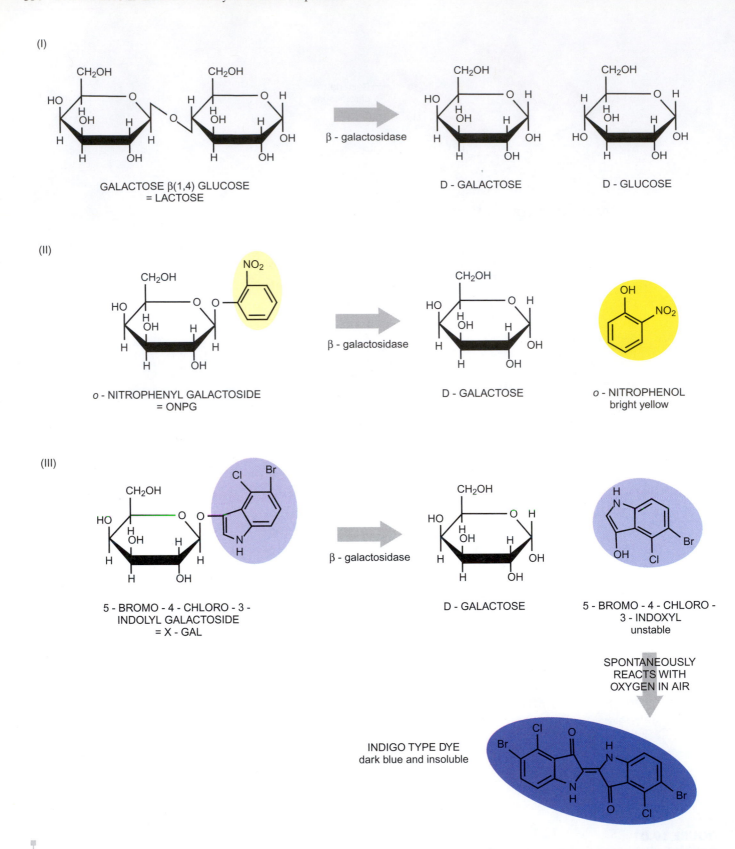

FIGURE 19.02
Substrates Used by β-Galactosidase

The enzyme β-galactosidase normally cleaves lactose into two monosaccharides, glucose and galactose. β-galactosidase also cleaves two artificial substrates, ONPG and X-gal, releasing a group that forms a visible dye. ONPG releases a bright yellow substance called o-nitrophenol, whereas X-gal releases an unstable group that reacts with oxygen to form a blue indigo dye.

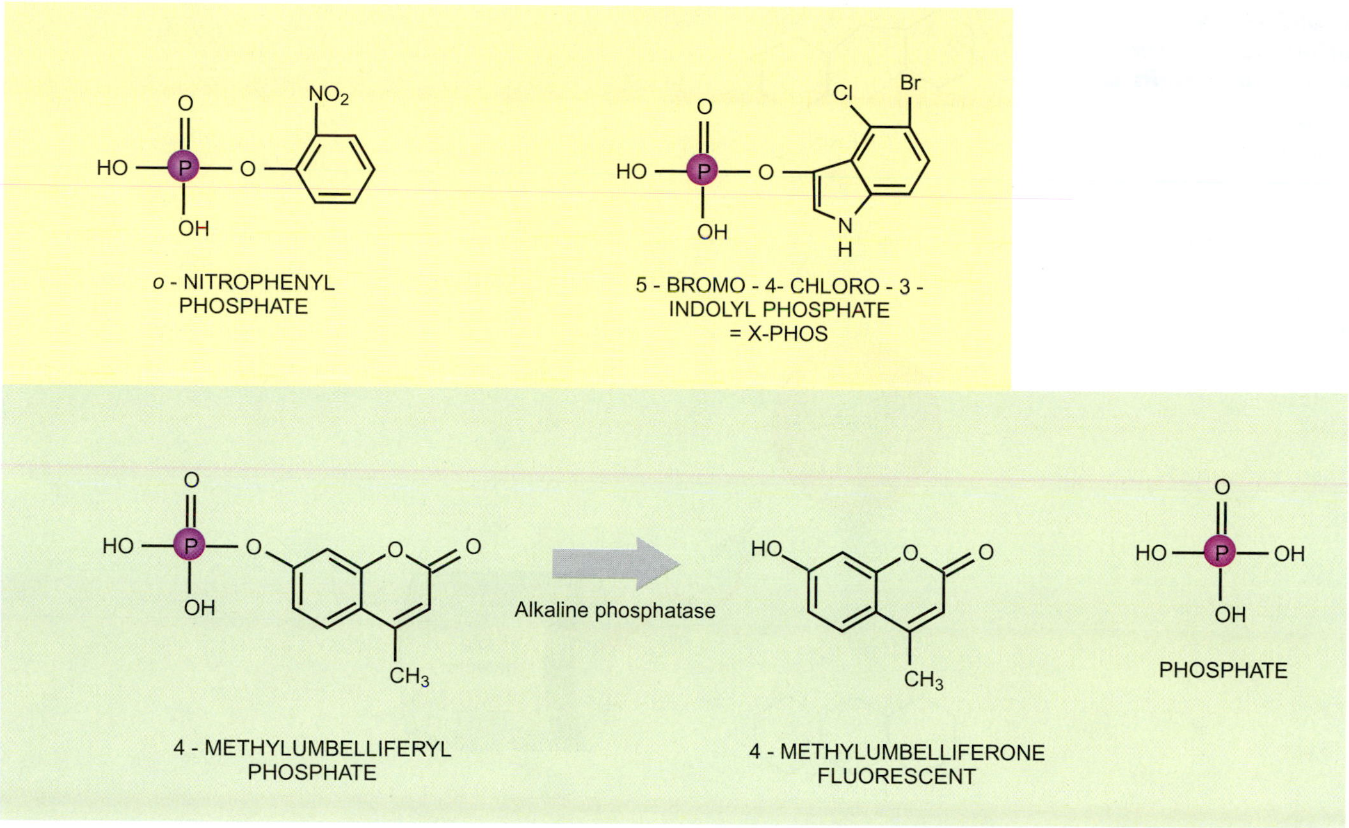

FIGURE 19.03
Substrates Used by Alkaline Phosphatase

Alkaline phosphatase removes phosphate groups from various substrates. When the phosphate group is removed from *o*-nitrophenyl phosphate, a yellow dye is released. When the phosphate is removed from X-phos, further reaction with oxygen produces an insoluble blue dye as for X-gal. Additionally, alkaline phosphatase releases a fluorescent molecule when the phosphate is removed from 4-methylumbelliferyl phosphate.

Another reporter gene is the ***phoA* gene** that encodes **alkaline phosphatase**. This enzyme cleaves phosphate groups from a broad range of substrates (Fig. 19.03). Like β-galactosidase, alkaline phosphatase will use a variety of artificial substrates:

1. ***o*-Nitrophenyl phosphate** is split, releasing yellow *o*-nitrophenol.

2. **X-phos** (5-bromo-4-chloro-3-indolyl phosphate) consists of an indigo dye precursor joined to phosphate. After the enzyme splits this, exposure to air converts the dye precursor to a blue dye, as in the case of X-gal.

3. **4-Methylumbelliferyl phosphate** releases a fluorescent compound when the phosphate is removed.

2.2. Light Emission by Luciferase as a Reporter System

A more sophisticated reporter gene encodes **luciferase** (Fig. 19.04). This enzyme emits light when provided with a substrate known as **luciferin**. Luciferase is found naturally

---

alkaline phosphatase   Enzyme that cleaves phosphate groups from a broad range of substrates
luciferase   Enzyme that emits light when provided with a substrate known as luciferin
luciferin   Chemical substrate used by luciferase to emit light
***o*-nitrophenyl phosphate**   Artificial substrate that is split by alkaline phosphatase, releasing yellow *o*-nitrophenol
***phoA* gene**   Gene encoding alkaline phosphatase; widely used as a reporter gene
4-methylumbelliferyl phosphate   An artificial substrate that is cleaved by alkaline phosphatase, releasing a fluorescent molecule
X-phos   5-bromo-4-chloro-3-indolyl phosphate, an artificial substrate that is split by alkaline phosphatase, releasing a blue dye

FIGURE 19.04
Luciferase Degrades Luciferin and Emits Light

Luciferase is an enzyme that alters the structure of luciferin. When the structure is altered, a pulse of light is emitted, which is detected by a photodetector. The luciferin shown in this figure is FMN (flavin mononucleotide), which is used by bacterial luciferases.

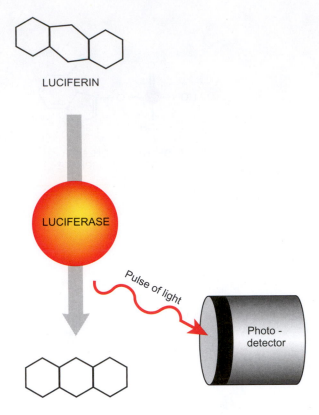

in assorted luminous creatures from bacteria to deep-sea squid. The **lux genes** from bacteria and the **luc genes** from fireflies produce different brands of luciferase, but both work well as reporter genes. The luciferins used by the different types of luciferase are chemically different. Bacterial luciferase uses the reduced form of the cofactor FMN (flavin mononucleotide) as its luciferin. Oxygen and a long chain aldehyde (R-CHO) are also needed. Both the reduced FMN and the aldehyde are oxidized:

$$R\text{-}CHO + FMNH_2 + O_2 \rightarrow R\text{-}COOH + FMN + H_2O + h\upsilon$$

Different groups of eukaryotes make several chemically distinct luciferins that are used solely for light emission. Firefly luciferase requires ATP as well as oxygen and firefly luciferin:

$$luciferin + O_2 + ATP \rightarrow oxidized\ luciferin + CO_2 + H_2O + AMP + diphosphate + h\upsilon$$

The enzyme luciferase emits light when it reacts with its substrate, luciferin.

If DNA carrying a gene for luciferase is incorporated into a target cell, it will emit light only when the appropriate luciferin is added. Although high-level expression of luciferase can be seen with the naked eye, usually the amount of light is small and must be detected with a sensitive electronic apparatus such as a luminometer or a scintillation counter.

2.3. Green Fluorescent Protein as Reporter

The products of most reporter genes are enzymes that must be assayed in some manner. Unlike the products of most reporter genes, **green fluorescent protein (GFP)** is

---

green fluorescent protein (GFP)   A protein, originally from a jellyfish, whose green fluorescence makes it useful as a reporter molecule
luc gene   Gene-encoding luciferase from eukaryotes
lux gene   Gene-encoding luciferase from bacteria

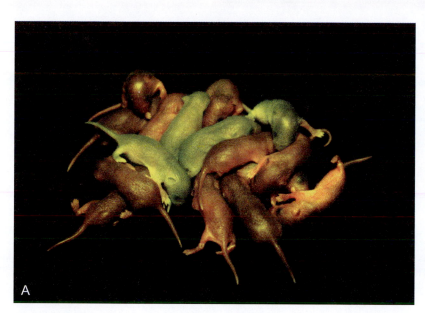
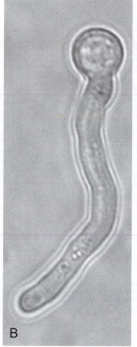
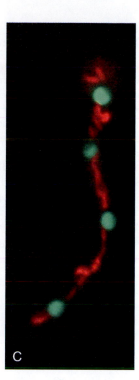

FIGURE 19.05
Transgenic Organisms with Green Fluorescent Protein

The gene for GFP has been integrated into the genome of animals, plants, and fungi. After exposure to long wavelength UV the organisms emit green light. A) Transgenic mice with GFP among normal mice from the same litter. The *gfp* gene was injected into fertilized egg cells to create these mice. GFP is produced in all cells and tissues except the hair. *(Credit: Eye of Science, Photo Researchers, Inc.)* B) Phase contrast and C) Fluorescence emission of germlings of the fungus *Aspergillus nidulans*. Original GFP was used to label the nucleus and a red GFP variant (DsRed) for the mitochondria. *(Credit: Toews et al., Current Genetics 45 (2004): 383–389.)*

not an enzyme, and it does not need a non-protein co-factor for it to fluoresce. GFP is a stable and non-toxic protein from jellyfish that can be visualized by its inherent green fluorescence. Consequently, GFP can be directly observed in living tissue without the need for adding any reagents. Nearly 2,000 years ago, the Roman author Pliny noted that the slime from certain jellyfish would generate enough light when rubbed on his walking stick to help guide his steps in the dark.

The cloned gene for GFP originally came from the jellyfish *Aequorea victoria*. This form of GFP is excited by long wavelength UV light (excitation maximum 395 nm) and emits at 510 nm in the green. A variety of genetically engineered variants of GFP are also in use. Many of these are available as the Living Colors™ series from Clontech Corporation. Some of these were chosen for showing higher fluorescence and/or emitting at a different wavelength. Thus, enhanced GFP (EGFP), enhanced yellow fluorescent protein (EYFP), and enhanced cyan fluorescent protein (ECFP) can be detected simultaneously using an argon-ion laser plus a detector with appropriate filters. A recent addition is a "Red GFP" (DsRed2—really a shade of orange) and blue FP (BFP). Other modifications include adapting GFP for high-level expression in different organisms by altering the codon usage (i.e., by changing bases in the redundant third codon position). Humanized versions of GFP exist that are adapted for use in cultured human cell lines.

Fusions of regulatory regions and promoters to the *gfp* gene have been used to monitor the expression of many genes, especially in living animals. The nematode, *Caenorhabditis elegans*, and the zebrafish are both transparent and so GFP can be used to follow differential gene expression in different internal tissues of living animals. Transgenic mice, rabbits, monkeys, and several plants have been engineered that have the *gfp* gene inserted into the host genome (Fig. 19.05).

Green fluorescent protein does not need a substrate or a co-factor. It emits green light after illumination with long-wave UV.

GFP can be used to follow gene expression or to localize proteins inside the cell.

FIGURE 19.06
GFP for Protein Localization

GFP can be used to reveal where a protein is localized within the cell. The first step is to fuse the GFP gene in frame with all or part of the structural gene that encodes the protein of interest. The fused construct is then expressed in a host cell. The cells are excited with long wavelength UV light and visualized under the microscope. If the protein is normally located in the membrane, as in this example, the cell membrane will fluoresce green in the microscope.

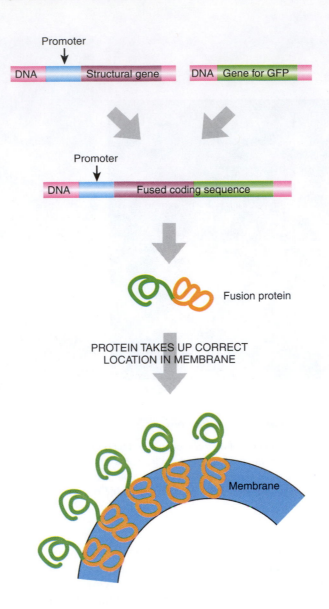

In addition to monitoring gene expression, GFP is widely used to localize proteins within the cell (Fig. 19.06). Gene fusions are constructed that yield a hybrid protein. These are normally designed so that the GFP protein is attached either to the C-terminal end or the N-terminal end of the protein under investigation. The fluorescence due to GFP will reveal the subcellular location of the target. For example, a fusion of Red GFP (DsRed) to the targeting sequence from subunit VIII of cytochrome c oxidase is located in the mitochondrial inner membrane. Fusions between actin or tubulin and GFP are used to study cell architecture. An important point to remember about these types of experiments is the intrinsic function of GFP itself. Some GFP variants work in dimers and tetramers and can cause the fused protein to aggregate, even though the protein of interest does not naturally do this. The addition of GFP may also mask important regulatory and targeting sequences of the protein of interest.

2.4. Gene Fusions

Reporter genes can be used to track the physical location of a segment of DNA or to monitor gene expression. In particular, reporter genes are often incorporated into gene fusions where they are used to follow the level of expression of the target gene. Many genes have products that are complicated or tedious to assay by direct

Dube A, Gupta R, Singh N (2009) Reporter genes facilitate discovery of drugs targeting protozoan parasites. Trends in Parasitology 25(9):432–439.

One area of molecular biology research is devising new methods to eradicate disease. In tropical regions of the world, parasitic diseases are prevalent and kill many people. Some of the more common protozoan parasites are *Plasmodium, Leishmania, Trypanosoma,* and *Toxoplasma.* Each of these parasites is carried by a vector organism, usually an insect, and when these bite humans, the disease is passed along. Malaria is caused by *Plasmodium,* a single-celled eukaryote that lives in mosquitoes. When the mosquito bites a human the parasite enters the bloodstream and infects the red blood cells. Sleeping sickness is caused by *Trypanosoma,* a small single-celled eukaryote that lives in insects such as flies. The parasites are secreted into the insect's saliva. When the insect bites a human, the eukaryote enters the bloodstream and invades the dendritic cells of the immune system. A different species of trypanosome causes Chaga's disease in South America. Trypanosomes can evade the immune system by changing their surface proteins and the immune system has to constantly create new antibodies to kill the invader.

One major obstacle in parasitic disease research is finding ways to kill the parasite without harming the infected person. Current treatments are limited, do not work well, and often are toxic to humans, so new drugs to kill these parasites are sorely needed. In order to determine whether a new drug kills the parasite, the researcher needs a way to monitor the parasite in the laboratory setting. Most of the parasites cannot be cultured alone and require host cells to survive. Assaying the number of parasites is difficult because of the presence of host cells. This review article summarizes the use of transgenic parasites that have a reporter gene in their DNA. Previously, the parasites were detected by counting under a microscope, staining with various dyes, radioactive nucleotide uptake or ELISA assays. Reporter genes provide an easier way to identify the parasite versus the host cell.

FOCUS ON RELEVANT RESEARCH

Each of the reporter genes used to make transgenic parasites is discussed. These include β-galactosidase, β-lactamase, luciferase, and GFP. The reporter assays are easily done with microscopy, ELISA, or flow cytometry. The true value of these transgenic parasites is the ability to see them easily in the host organisms. Therefore, the numbers of parasites in the host organism can be determined. The ability to visualize the parasite infecting its host enhances our understanding of these parasites and provides an improved way to screen new drugs for efficacy of killing the parasite without harming the host.

---

measurement or may even be unknown. To avoid this, the original gene product is replaced by fusing its regulatory region to the structural region of a reporter gene. To create this fusion, the target gene is cut between its regulatory region and coding region. The same is done with the reporter gene. Then the regulatory region of the gene under investigation is joined to the coding region of the reporter gene (Fig. 19.07). This hybrid structure is a **gene fusion**.

The regulatory sequences control the expression of the reporter gene in the same manner that the original gene is controlled. Once the fusion gene is present in the organism, then the researcher can alter the environment, treat the organism with different substances, or even simply determine reporter gene expression at different stages of development. This approach, especially using *lacZ* and β-galactosidase, is widely used in determining what regulatory DNA sequences are important for gene expression (Fig. 19.08). Many bacteria, such as *E. coli,* already possess a wild-type copy of reporter genes such as the *lacZ* or *phoA* genes. In these cases the wild-type version of the gene must be deleted from the chromosome before the gene fusions are used. Strains of *E. coli* deleted for the *lac* operon or for *phoA* are readily available. In eukaryotes, gene fusions use different reporter genes. For example, yeast-reporter genes include *CUP1,* a gene that enables yeast to grow on copper-containing media, *URA3,* a gene that kills yeast when growing on 5-fluorouracil, and *ADE1* and *ADE2,* two genes that synthesize adenine. *Ade1* and *Ade2* mutants produce a red pigment when grown on regular media and are easily visualized.

Much information can be learned about gene regulation using gene fusions. Gene fusions may be used to test for possible effects of regulatory genes on gene expression. A mutation that inactivates a regulatory gene can be introduced into the cell carrying the gene fusion. Expression is then measured under appropriate conditions. A series of regulatory genes suspected of controlling the target gene may be rapidly surveyed by this approach.

> Gene fusions are used to monitor genes whose products are difficult to assay. Reporter genes are fused to the regulatory region of the target gene.

---

gene fusion   Structure in which parts of two genes are joined together, in particular when the regulatory region of one gene is joined to the coding region of a reporter gene

FIGURE 19.07
Construction of a Gene Fusion

Creating gene fusions help in investigating how genes are regulated. The regulatory region of the target gene is joined to the reporter gene coding region. The reporter enzyme will now be made under conditions where the target gene would normally be expressed.

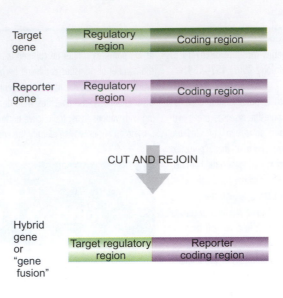

FIGURE 19.08
Using Gene Fusions to Survey Regulation

The regulatory region of the target gene (green) controls the expression of the reporter structural gene (purple). Therefore, assaying the level of reporter enzyme reveals how the target gene would be expressed under the chosen conditions. In this example, the reporter gene is *lacZ* and the level of expression is monitored by the breakdown of ONPG to release yellow nitrophenol.

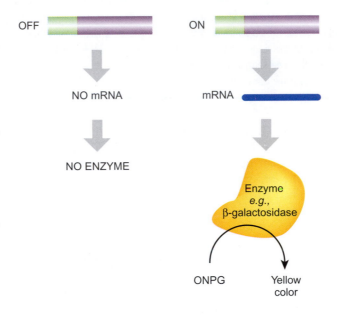

Regulatory regions may be analyzed by deleting segments of the DNA and checking for effects on gene expression.

3. Deletion Analysis of the Upstream Region

The upstream regulatory region of a gene often contains several sites where regulatory proteins such as transcription factors bind as well as the promoter region where RNA polymerase binds. These regulatory sites often enhance or suppress the expression of the gene under a variety of conditions. To determine the function of the regulatory elements, it is often helpful to construct a series of altered upstream regions in which presumed binding sites have been eliminated. The simplest way to do this is to remove successive segments from the 5' end of the upstream region. Originally, restriction enzymes were used to create the deletions. However, finding convenient restriction sites was always a problem. PCR offers a much better alternative because of its specificity. A variety of PCR primers can be designed to amplify different areas within the upstream region of the gene of interest.

These engineered upstream regions are then tested for possible alterations in gene expression and regulation by creating a gene fusion with a reporter gene. They are then examined by assaying the expression of the reporter gene. For example, suppose we have an upstream region whose sequence reveals a binding motif for Crp, the *E. coli* cAMP receptor protein (Fig. 19.09). If this region is removed in the deletion

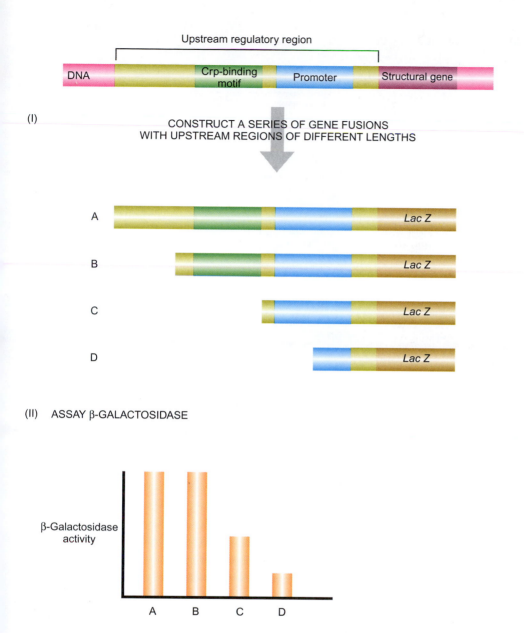

FIGURE 19.09
Deletion Analysis of Upstream Region

In order to determine which parts of the upstream regulatory region affect the expression of a gene, a series of deletions of the upstream region are made. The upstream region is usually also surveyed for possible binding sites or regulatory motifs. This example shows two potential regulatory regions, a Crp-binding site and a putative promoter region. A set of deletions are constructed such that smaller and smaller segments of the upstream region are present. These are fused with the reporter gene, *lacZ*. Next, the constructs are expressed in cells, and the activity of β-galactosidase is assayed. In this example, the whole upstream region (construct A) has high activity. Removal of the far end (B) has negligible effect, suggesting that no important sequences lie in this region. When the Crp-binding site is removed, the activity decreases by half, suggesting that Crp regulates gene expression (C). When half of the putative promoter is deleted (D), the β-galactosidase activity is almost zero. These results confirm that the two presumed sites do control the activity of the reporter gene, and therefore also control the original structural gene.

analysis, the effect of losing the regulatory motif is assayed by monitoring the expression of *lacZ*. Without the Crp-binding site in this example, the reporter gene expression is about half of normal, indicating that Crp must enhance gene expression.

3.1. Locating Protein-Binding Sites in the Upstream Region

The upstream region of a gene usually contains sequences in the DNA that control gene expression. Most of these regulatory sequences are sites where regulatory proteins bind, although others may be involved in bending DNA or forming stem and loop structures in either the DNA itself or the RNA. After sequencing a gene and its regulatory region, computer analysis identifies any potential binding sites based on homology with previously identified sites. However, computer-predicted binding sites often do not function *in vivo*. Deletion analysis of these upstream sites determines how they affect gene expression, but do not show if a protein actually binds to the site. Consequently, even after a presumed binding site has been found, the binding of the regulatory protein must be confirmed experimentally.

If a regulatory protein binds to a segment of DNA it will slow the migration of the DNA through a gel.

FIGURE 19.10
Gel Retardation to Assess Protein Binding to DNA

Gel retardation assays determine whether or not a specific regulatory protein actually binds to the DNA in the regulatory region of a gene. A) The regulatory and coding regions of a gene are cut into various fragments by a restriction enzyme. The fragments are divided into two samples. In one (I), no regulatory protein is added, whereas in the other sample (II), purified transcription factor is mixed with the DNA.
B) The DNA fragments are separated by size using non-denaturing agarose gel electrophoresis. If the regulatory protein binds to the DNA, that fragment is heavier and travels slower. It is therefore retarded relative to its position in the absence of protein. In this example, fragment c has a binding site for the regulatory protein and its band is retarded.

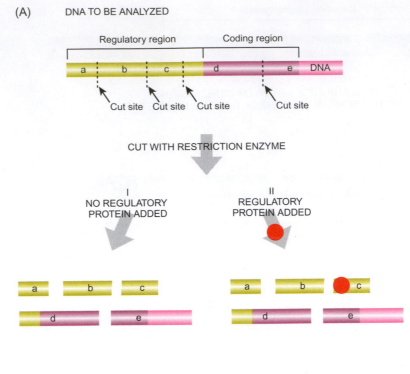

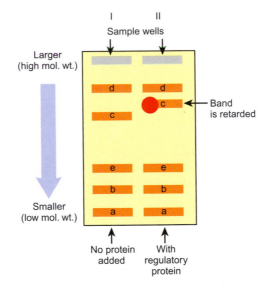

The electrophoretic **mobility shift**, **bandshift**, or **gel retardation** assay tests if a suspected protein binds to DNA from the upstream region (Fig. 19.10). First, the DNA carrying the gene and its upstream region is labeled with digoxygenin or radioactivity (see Ch. 5) and then cut with a convenient restriction enzyme to get a series of fragments. After cutting, the DNA is separated into two tubes. To the experimental sample, the suspected DNA-binding protein is added. The other sample, or control sample, is

bandshift assay   Method for testing binding of a protein to DNA by measuring the change in mobility of DNA during gel electrophoresis. Same as gel retardation or mobility shift assay

gel retardation   Method for testing binding of a protein to DNA by measuring the change in mobility of DNA during gel electrophoresis. Same as bandshift assay or mobility shift assay

mobility shift assay   Method for testing binding of a protein to DNA by measuring the change in mobility of DNA during gel electrophoresis. Same as bandshift assay or gel retardation

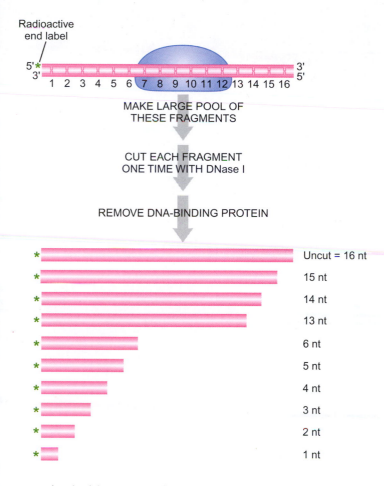

MAKE LARGE POOL OF
THESE FRAGMENTS

CUT EACH FRAGMENT
ONE TIME WITH DNase I

REMOVE DNA-BINDING PROTEIN

■- **FIGURE 19.11**
*Footprint Analysis—
Procedure*

To identify the exact location of a protein-binding site, the fragment that contains the binding site is first isolated. In this example, a small piece of DNA 16 nucleotides in length has a protein-binding site that protects nucleotides 7-12. After DNase I treatment, the pool of DNA has a piece representing every length of labeled DNA except those protected by the DNA-binding protein. The only fragments that are visible have a labeled 5' end, and although the other fragments are present, they are not visible after autoradiography.

not mixed with any proteins. Both samples are then run side by side on a non-denaturing agarose gel. If the protein binds to one of the DNA fragments, the complex will be larger and run slower than the original DNA (i.e., that fragment will be retarded).

An average protein has a molecular weight of about 40,000. A segment of DNA of 1,000 base pairs has a molecular weight of about 700,000. If a typical protein is bound to a length of DNA much bigger than this, the relative change in size, and therefore in mobility, would be 5% or less. Such a small change would be impossible to observe using electrophoresis. Consequently, for gel retardation analysis, restriction enzymes are chosen to give segments of DNA in the range of 250–1,000 base pairs. An alternative to using restriction enzymes is to use PCR to generate fragments from the upstream region of a gene. Primers can be chosen to generate any segment suspected of harboring a binding site. The PCR fragments can then be examined one at a time for binding to the suspected DNA-binding protein.

Gel retardation reveals which segment of DNA binds a protein. To locate the binding site more precisely, a **footprint** analysis is performed. In footprinting, the fragment of DNA that binds the protein is labeled at one end with radioactivity or fluorescence. As before, the sample of DNA is split into two portions and the protein is mixed with one batch. Both portions of the DNA are then treated with a small amount of a reagent that breaks DNA strands. **Deoxyribonuclease I (DNase I)** is often used because it is relatively non-specific and cuts DNA between any two nucleotides. Other chemical reagents that attack DNA may also be used. In either case, the DNA is attacked and degraded except in the region covered, and thus protected, by the protein (Fig. 19.11). Although there are many other fragments of DNA in the sample after DNase I treatment, the labeled fragment is the only one that is visible in the gel.

When proteins are bound to DNA they protect the region on the DNA that they cover from chemical attack.

---

deoxyribonuclease I (DNase I)   Non-specific nuclease that cuts DNA between any two nucleotides. Often used in footprint analysis
footprint   Method for testing binding of a protein to DNA by its protection of DNA from chemical degradation

FIGURE 19.12
Footprint Analysis—Results

The DNase I treated samples are run on a sequencing gel to separate the fragments, which differ by as little as one base pair. When Lane B (no protein) and Lane C (plus protein) are compared, it can be seen that Lane C shows no bands in the boxed region. Alignment with a sequencing ladder (lane A) allows the precise region of binding to be deduced.

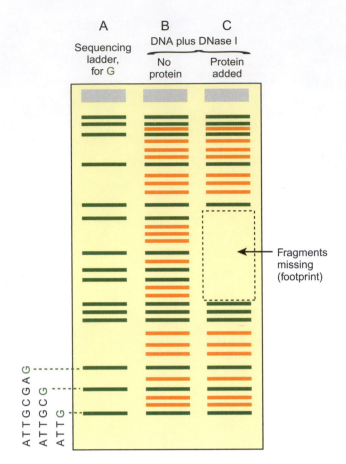

Only a small amount of DNase is used, just enough to cut each molecule of DNA once on average, in a random position. Consequently, the sample of protected DNA will have certain fragments missing. In contrast, cutting a sample of unprotected DNA will give rise to a series of fragments of all possible lengths, varying by a single base pair. When the two samples are run on a gel side by side, a region without any fragments appears as a "footprint" (Fig. 19.12). In practice, the footprint is run side by side with a sequencing reaction (see Ch. 8), which allows matching the footprint with the DNA sequence.

4. DNA-Protein Complexes Can Be Isolated by Chromatin Immunoprecipitation

Chromatin immunoprecipitation (ChIP) assays are used to identify the DNA sequence to which a particular protein attaches (Fig. 19.13). To perform chromatin immunoprecipitation, cells are first treated with a protein crosslinking reagent that freezes all the proteins in their original location by creating covalent bonds between different amino acid side chains. These crosslinking reagents also fuse proteins to DNA, so the proteins (including transcription factors, histones, etc.) will remain attached to their DNA-binding sites. Next, the long strands of DNA are broken into smaller pieces. One approach uses sonication, a method that uses sound waves to break apart the cell into small pieces. The frequency and length of exposure to the sound waves determines the size of chromatin fragments. The longer the cells are

Chromatin immunoprecipitation (ChIP) Technique that identifies the DNA binding site for a particular transcription factor by crosslinking the DNA to the transcription factor, and then immunoprecipitating the transcription factor

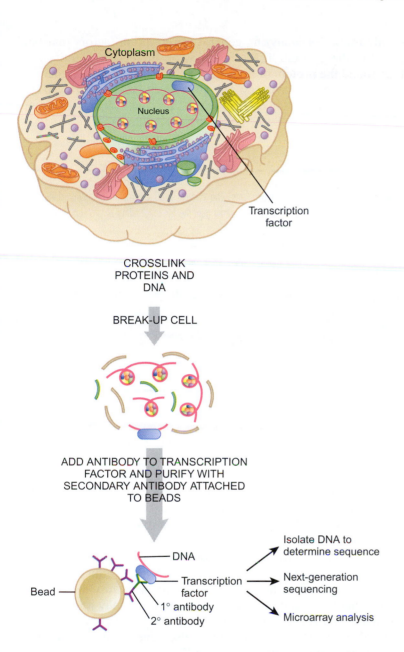

► **FIGURE 19.13**
Chromatin Immunoprecipitation (ChIP)–The Principle

Covalent crosslinks join any proteins that are attached to each other, and also any proteins attached to DNA. These crosslinked segments are then sheared or cut into smaller fragments, and then the transcription factor of interest is isolated from the remaining cellular components with immunoprecipitation. The last panel of this picture is expanded to show the details of the immunoprecipitation complex, and is not drawn to scale.

sonicated, the smaller the DNA fragments. Another method commonly used is nuclease digestion. DNases cut the exposed DNA into smaller and smaller fragments, depending upon the specificity and concentration of the DNase.

In the next stage of ChIP, the protein of interest must be isolated from the rest of the cellular components. This procedure relies on antibodies (see Ch. 15). When an antibody to the protein of interest is mixed with the fragmented cellular contents, it specifically binds to the protein of interest. In order to isolate the antibody:protein complex, agarose or magnetic beads that are coated with a secondary antibody (which binds the first antibody) are added. The beads are heavier than the rest of the proteins and cell contents, so the entire complex (secondary antibody-primary antibody-protein of interest-DNA fragment) can be isolated using centrifugation.

The isolated complexes are studied in a variety of ways. First, the DNA sequence can be determined with traditional sequencing methods. This information can identify the exact genomic location that the protein of interest binds. To sequence the DNA by traditional methods, the DNA fragment must first be cloned into a vector so a sequencing primer complementary to the vector can be used to sequence the insert. To avoid the cloning step, the DNA can be subjected to ChIP-Seq, where one of the second-generation DNA sequencing methods such as Illumina or *454* (see Ch. 8)

Chromatin immunoprecipitation (ChIP) isolates the DNA sequence that proteins bind to within the living nucleus.

is used. This is especially useful for analyzing general transcription factors that bind multiple DNA locations in the genome. These methods of sequencing identify each site in the genome that bound the protein of interest. Additionally, the DNA segments can be used as a target for a DNA microarray. This procedure, called ChIP-CHIP or ChIP-on-chip, reveals if the DNA-binding domain for the protein of interest is found in any other regions in the genome (see Focus on Relevant Research).

Lenstra TL, et al. (2011) The Specificity and Topology of Chromatin Interaction Pathways in Yeast. Mol. Cell 42:536–549.

FOCUS ON RELEVANT RESEARCH

To understand how chromatin alterations influence transcription in the yeast *Saccharomyces cerevisiae*, the authors analyzed over 100 mutations affecting chromatin components. The authors used chromatin immunoprecipitation and ChIP-on-chip to survey the yeast genome for the binding of proteins to specific sites on the DNA.

Almost all DNA-binding components were associated with others in a series of overlapping complexes. Elimination of individual DNA-binding proteins was much more specific in effect on gene expression than predicted from the number of locations where they were bound. Clearly, there are additional gene-specific mechanisms that determine these effects.

There are many related techniques that start with ChIP. In ChIA-PET (chromatin interaction analysis-paired end-tag sequencing), the protein of interest is crosslinked to DNA and isolated from the rest of the cellular components with immunoprecipitation (Fig. 19.14). This procedure is especially useful to analyze protein complexes that create DNA loops as is seen with enhancers found thousands of base pairs away from the promoter. Rather than a single piece of DNA, multiple DNA segments are attached to this type of complex. After sonication and immunoprecipitation, the isolated complexes are mixed with two different oligonucleotide half linkers (A and B). These anneal to the end of the DNA segments and have complementary sequences to pair with each other. The linkers are ligated to each other with very dilute ligase, ensuring that the DNA segments within the complex are preferentially joined before unrelated DNA segments. When two DNA segments from different complexes bind, the linker is a chimera and easily identified. The linkers also have a restriction enzyme site for MmeI, which recognizes a sequence within the linker and cuts 20 nucleotides into the DNA piece. These small tags are then analyzed by paired end-tag sequencing, a next-generation sequencing method. Sequencing results elucidate what region of the genome was associated with the protein complex, uncovering potential enhancer elements that were not previously identified.

5. Location of the Start of Transcription by Primer Extension

Identifying the transcription start site is a key piece of information about gene expression. **Primer extension** allows precise location of the start of transcription to the exact nucleotide. This approach involves binding an oligonucleotide primer to mRNA. The primer is then extended by reverse transcriptase to synthesize DNA that is complementary to the mRNA (Fig. 19.15).

primer extension   Method to locate the 5' start site of transcription by using reverse transcriptase to extend a primer bound to mRNA so locating the 5' end of the transcript

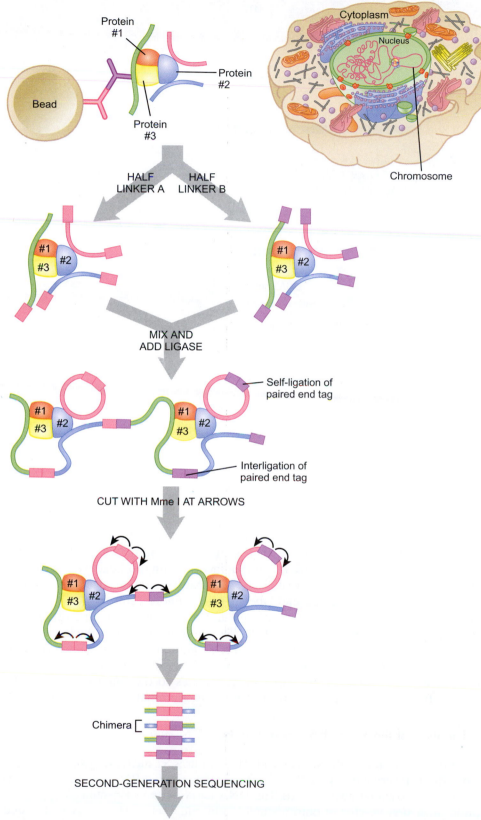

► **FIGURE 19.14**
ChIA-PET Procedure

Inside the nucleus, DNA protein interactions are 3D and involve DNA loops. After crosslinking, different regions of a chromosome are often associated with a single protein complex. After ChIP, each of these DNA sequences can be determined with paired end-tag sequencing. First, the immunoprecipitated DNA:protein complex is divided into two samples and each DNA end is connected to a different linker DNA. Then the two samples are recombined and mixed with very dilute ligase. The linkers anneal preferentially within the same complex, but occasionally there are interligations of paired tags. The ligated tags have a restriction enzyme site for *Mme*I, which recognizes its sequence in the tag, but cuts 20 nucleotides away in the DNA sequence. These small pieces of DNA are then sequenced using paired end-sequencing technology.

First, the cells are grown in conditions where the gene of interest is expressed, and then mRNA is isolated from the cells. An artificial DNA primer is synthesized that is complementary to a sequence close to the suspected start of transcription. The primer is specific for the gene of interest and therefore, hybridizes only with mRNA

FIGURE 19.15
Primer Extension Reveals Start of Transcription

First, mRNA is isolated from cells that are expressing the gene of interest. A primer specific to the gene of interest is added and anneals to the mRNA. Reverse transcriptase makes a complementary DNA strand from the primer to the 5' end of the mRNA (i.e., the start of transcription). The exact transcription start site is determined by comparing the size of the primer extension DNA strand to a sequencing ladder of the same region of DNA.

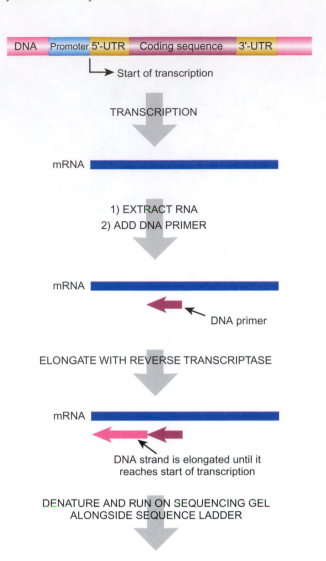

from this gene. Reverse transcriptase then synthesizes a strand of DNA from the primer to the beginning of the mRNA strand, incorporating radioactive nucleotides or fluorescent tags, so it can be visualized. Alternatively, rather than labeling the entire strand, the primer may be end-labeled with a fluorescent or radioactive tag. The resulting DNA/RNA hybrid is denatured and run on the same type of denaturing gel used in DNA sequencing (see Ch. 8). To determine the exact nucleotide with which the mRNA starts, a sample of DNA is also sequenced using the same primer as used for extension. This sequencing ladder is run side by side with the primer extension fragment. The primer extension product migrates to the same location as the sequencing fragment that represents the precise transcriptional start site.

5.1. Location of the Start of Transcription by S1 Nuclease

Another method to locate the start of transcription uses **S1 nuclease**. This is an endonuclease from *Aspergillus oryzae* that cleaves single-stranded RNA or DNA but does not cut double-stranded nucleic acids. The DNA carrying the suspected start of transcription must first be cloned onto a suitable plasmid vector. The DNA is digested with a restriction enzyme that yields a fragment containing the presumed start site. A more recent approach is to generate the fragment using PCR (see Ch. 6), thus

S1 nuclease   Endonuclease from *Aspergillus oryzae* that cleaves single-stranded RNA or DNA but does not cut double-stranded nucleic acids

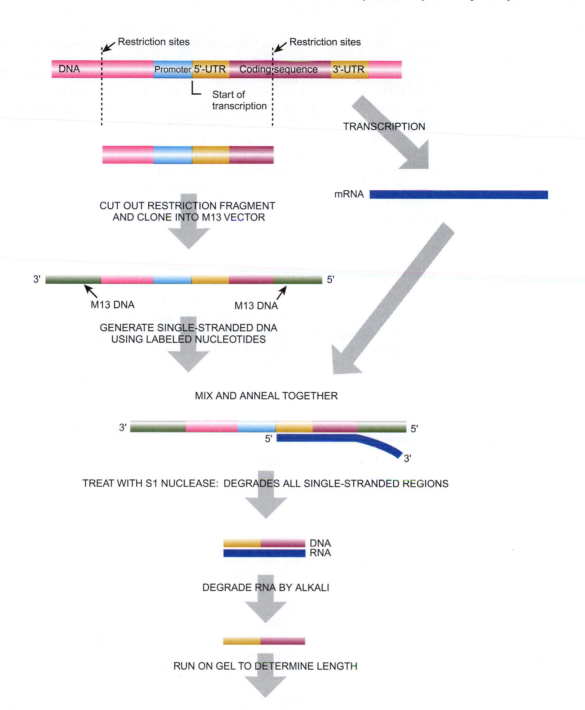

FIGURE 19.16
Locating Start of Transcription by S1 Nuclease

The first step in mapping the transcriptional start site by S1 nuclease treatment is to clone the upstream region of the gene into an M13 vector. Next, single-stranded M13 DNA is prepared using labeled nucleotide precursors for use as a probe. The labeled single-stranded DNA is mixed with the total cellular mRNA. The mRNA with sequence complementary to the DNA will hybridize with the DNA. S1 nuclease is added to the mixture to digest all the single-stranded RNA and DNA. All that is left is the DNA:RNA hybrid, which is isolated from the degraded nucleotides by precipitation. The DNA portion of the hybrid is isolated by alkali treatment and the length determined by comparing the fragment size to the entire gene.

avoiding the need for cloning. In either case the DNA must be denatured to give single strands before S1 analysis. An alternative is to clone the DNA fragment into an M13 vector, which gives single-stranded DNA directly (see Ch. 8 for the use of M13 to make single-stranded DNA for sequencing).

The single-stranded DNA fragment is labeled on its 5' end and then denatured and hybridized to the corresponding mRNA (Fig. 19.16). The 5' end of the mRNA

Another way to find the start of transcription is by hybridizing the mRNA to the corresponding DNA and cutting away the single-stranded overhangs with S1 nuclease.

corresponds to the site where transcription started. DNA beyond this point remains single-stranded. The sample is next split into two portions. S1 nuclease is mixed with one half and degrades the single-stranded overhangs (DNA at one end, RNA at the other). The two samples of DNA, with and without nuclease treatment, are compared by running side-by-side on a denaturing gel. This allows the difference in length, and hence the location of the start site, to be estimated.

Modification of this technique allows **S1 nuclease mapping** to be used for locating the 3' end of a transcript. In this case, the DNA probe includes the fragment with the suspected transcriptional stop site. S1 nuclease may sometimes degrade the ends of the RNA/DNA hybrid slightly or may not fully digest the single-stranded regions. Consequently, the S1 nuclease method is not as accurate as primer extension. However, primer extension cannot locate the 3' end of a transcript, and S1 nuclease mapping is the best way to achieve this.

6. Transcriptome Analysis

The transcriptome refers to all of the various RNA molecules that result from transcription in a particular cell. Often, interest focuses on the mRNA, but the transcriptome also includes non-coding RNA. Unlike the genome, the transcriptome varies as different genes are expressed under different conditions. Transcriptome analysis attempts to measure the levels of all transcribed RNAs simultaneously. The number and types of mRNA are captured at one moment in time and provide a snapshot of what is being expressed in the cell. Several techniques are available for monitoring multiple RNA molecules.

> The transcriptome is the mixture of RNA that results from transcribing the genome.

Differential display PCR has already been described (Ch. 6). This method isolates mRNA in two different conditions and then converts the sample into cDNA. In theory, if the correct primer combinations are used, differential display PCR can identify all the transcripts that are present in each sample. Although widely used, the results are often not quantitative and since PCR fragments of the same size may be generated from more than one gene this approach may provide ambiguous results.

The advent of newer sequencing technologies (such as SOLiD, *454,* or Illumina—see Ch. 8) is transforming the field of transcriptomics, since an entire cDNA library can be sequenced quickly and cheaply. **RNA-Seq**, or whole transcriptome shotgun sequencing (WTSS), creates a cDNA library from fragmented mRNA, and then every cDNA is sequenced (Fig. 19.17). These sequences are then aligned with the genome for the organism. The relative copy number of each cDNA sequence is an indication of gene expression levels. Besides simply assessing the amount of mRNA in a sample, RNA-Seq has clinical applications. In cancer research, gene fusions occur due to chromosomal rearrangements. RNA-Seq can reveal if the fused genes are expressed into mRNA and estimate the relative abundance of the fusion product. In addition, the technique can identify expressed single nucleotide polymorphisms (SNPs). This type of information can identify the genes responsible for a particular disease by comparing the expression of SNPs from affected individuals and their healthy family members. Using expressed sequences eliminates any irrelevant SNPs found in the DNA, since many SNPs are not in expressed areas of the genome. RNA-Seq can also identify post-translational editing of mRNA that is not evident from looking at just the DNA sequence. This can suggest a new function for a gene.

6.1. Assessing the Purity of RNA

RNA purity is essential for all the different transcriptomic procedures. Most RNA samples are obtained from the cells of interest by either isolating total RNA or

---

RNA-seq   The use of high-throughput cDNA sequencing to characterize an RNA sample
S1 nuclease mapping   Method using S1 nuclease to locate the 5' end or 3' end of a transcript

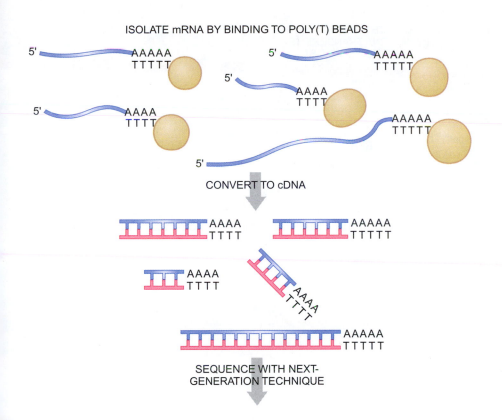

ISOLATE mRNA BY BINDING TO POLY(T) BEADS

CONVERT TO cDNA

SEQUENCE WITH NEXT-GENERATION TECHNIQUE

■ **FIGURE 19.17**
RNA-Seq

The entire transcriptome can be identified by sequencing a cDNA library in its entirety. Next-generation sequencing makes this process possible, resulting in the identification of each and every mRNA that was expressed.

mRNA with a poly(A) tail. In either case, rRNA accounts for the majority of all the RNA in a cell. Its abundance can mask the other types of RNA, and therefore, rRNA must be removed. One efficient method to remove the ribosomal RNA is to hybridize an rRNA probe labeled with a **biotin** tag to the sample of total RNA. The hybrids are then removed from the remaining RNA by binding to **streptavidin**-coated magnetic beads. The remaining RNA is enriched for mRNA and provides a better sample for transcriptome analysis (Fig. 19.18).

To ensure that the RNA sample is free of degradation and contamination, a small amount of the sample may be analyzed using lab-on-a-chip methods (see Ch. 8). For example, the RNA 6000 Pico LabChip from Agilent Technologies can analyze nanogram quantities of RNA by electrophoresis through a gel/dye matrix in a small chip (Fig. 19.19). Just as in traditional electrophoresis, the RNA fragments move based on size. The bands are visualized with a fluorescence detector and graphed via an attached computer. The graph can determine if the sample is contaminated with rRNA, and based on the sizes of the peaks, whether or not the sample is degraded.

7. DNA Microarrays for Gene Expression

We have outlined the principles of the **DNA microarray** in Chapter 8 while discussing its use in sequencing DNA and in the diagnostic detection of particular DNA sequences. In the previous experiments, DNA immobilized on the chip hybridizes to the target DNA fragments in the sample to be analyzed. Since DNA microarrays work by hybridization, they can also be used to monitor RNA. Microarrays are fairly

---

biotin   Vitamin that is widely used to label or tag nucleic acids in molecular biology because it may be bound very tightly by avidin or streptavidin
DNA microarray or DNA array   Chip-carrying array of DNA segments used to simultaneously detect and identify many short RNA or DNA fragments by hybridization. Also known as DNA chip or oligonucleotide array
streptavidin   Protein from *Streptomyces* that binds biotin extremely tightly and specifically. Used in detection procedures for molecules labeled with biotin

FIGURE 19.18 ───────■
Removing Unwanted rRNA from an RNA Sample

Although most rRNAs are not polyadenylated, a fraction of the transcripts do have poly(A) tails. These can contaminate RNA for transcriptome analysis, and therefore, need to be removed. One method uses biotinylated single-stranded probes that have complementary sequences to rRNA. These hybridize to the rRNA in the sample and are removed by binding to avidin-coated beads followed by centrifugation.

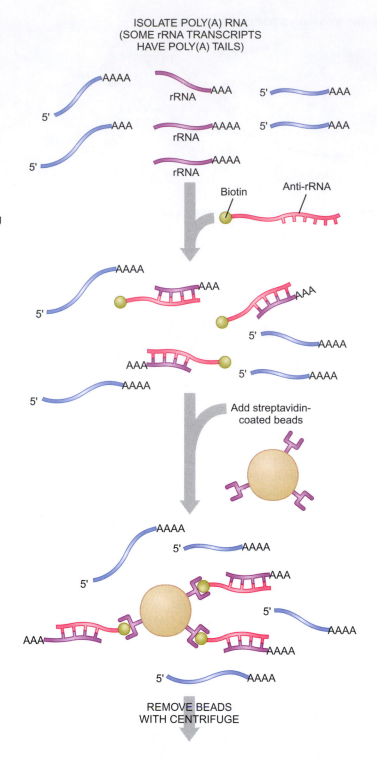

ISOLATE POLY(A) RNA
(SOME rRNA TRANSCRIPTS
HAVE POLY(A) TAILS)

Add streptavidin-coated beads

REMOVE BEADS
WITH CENTRIFUGE

DNA microarrays can be used to detect gene expression by hybridizing the array to messenger RNA.

expensive and analysis of the data is highly labor intensive, despite computerized analysis. If only one or a few genes are the objects of interest, other methods such as Northern hybridization to detect mRNA or using a reporter gene to measure the level of transcription are more appropriate.

For total transcriptome analysis, the solid support (i.e., the "chip") has DNA sequences complementary to all possible mRNA molecules that a cell might express. The DNA is robotically printed onto a nylon membrane or a glass slide. Current technology can print about 100,000 spots of DNA per cm², with glass slides capable of carrying higher densities than nylon membranes. Next, mRNA is extracted from cells

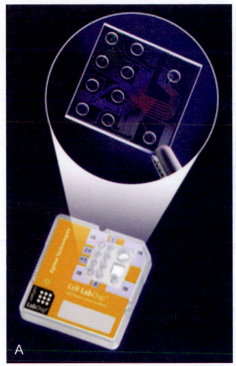

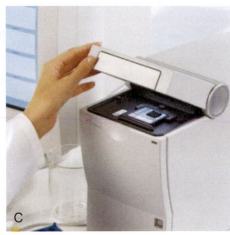

FIGURE 19.19
Pico LabChips Separate Nanogram Quantities of DNA or RNA

A) The lab on a chip has small holes in a glass piece that hold the DNA or RNA sample, the size comparison ladder, and the gel materials. The holes are connected via microfluidic channels. B) The scientist adds the experimental sample of RNA or DNA into the specific hole. C) The analyzer then performs the analysis and records the data on an attached computer (not shown). *(Credit: Reproduced with Permission of Agilent Technologies.)*

and labeled, either with a radioactive isotope or more often with a fluorescent dye. Next, the labeled mRNA is placed on the DNA array in conditions that favor binding of complementary sequences. After binding to the chip, the intensity of the label in each spot correlates to the amount of that particular mRNA. Most gene expression studies compare two different conditions, one "control" set or untreated cells, and one "experimental" set where the cells are exposed to a different environment. Both mRNA samples can be hybridized to the chip at the same time if two different fluorescent dyes (e.g., Cy3, which is green and Cy5, which is red) are used for each mRNA set. Red spots will show genes expressed under "control" conditions and green spots will show genes expressed under "experimental" conditions. When the same mRNA is expressed in both conditions that spot will fluoresce yellow (Figs. 19.20 and 19.21). Determining the intensity of green, red, and yellow for each spot is accomplished by computer analysis, which determines the mean of the pixels or median value for the pixels, and normalizes these to a set of internal controls. Rather than simply presenting a table of numbers, the computerized analysis is often presented as a "heat map" grid. The gene sequences for the control set of data are listed on the x-axis and the experimental genes are listed along the y-axis. Each square of the grid is colored, where red indicates an increase in expression and blue represents a decrease in expression over the control. Shading of either red or blue from light to dark indicates relative increases or decreases of gene expression (Fig. 19.22).

In practice, two types of DNA microarray are used for binding mRNA, arrays of cDNA or arrays of oligonucleotides. For a cDNA array, cDNA is generated by PCR amplification of each gene in the organism. One problem is the existence of gene families (e.g., the globin family) whose individual members are highly homologous and may cross-hybridize. To avoid this, it is normal to use sequences from the 3' end of the cDNA, which often include part of the 3'-UTR of the mRNA transcript. Non-coding sequences

FIGURE 19.20
DNA Chip Showing Detection of mRNA by Fluorescent Dyes

DNA chips can monitor many different mRNAs at one time. Each spot on the grid has a different DNA sequence attached. To determine which genes are expressed under which conditions, mRNA is isolated. In this case, mRNA isolated from cells grown under two different conditions is labeled with two different fluorescent dyes. Under condition one (red dye), eight different mRNAs hybridized to DNA spots on the chip. Under condition two, 19 different mRNAs were seen (green dye). Since two different color dyes were used, both samples can be analyzed on the same chip. In this case, the mRNAs that are expressed under both conditions give yellow spots.

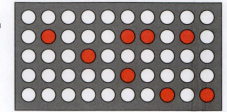

Array treated with RNA from cells grown under condition 1 and labeled with red flourescent dye

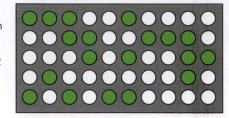

Array treated with RNA from cells grown under condition 2 and labeled with green dye

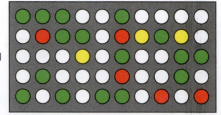

Array treated with both samples of RNA; yellow spots reveal genes expressed under both conditions

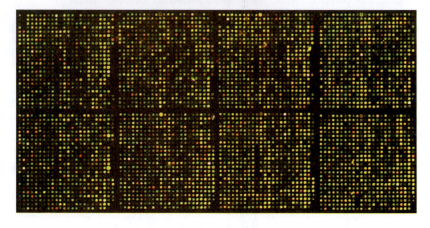

FIGURE 19.21
Hybridization of mRNA to a 19,200 Element Array

RNA from related human colon carcinoma and reference cell lines was reverse transcribed and the cDNA was labeled with Cy-5 (red) and Cy-3 (green), respectively. The cDNA was then hybridized to a microarray containing 19,200 distinct human cDNA clones. Genes expressed by the cancer cells are shown in red and those from the normal cells are green. Yellow spots indicate expression in both cell lines. *(Credit: Hegde et al. (2000) A concise guide to cDNA microarray analysis. Biotechniques 29: 548–562. The Institute for Genomic Research, Rockville, MD.)*

> cDNA arrays use the cDNA versions of whole genes.

diverge much more than coding sequences, and so are much less likely to cross-hybridize. Each amplified cDNA is attached to the nylon sheet or glass slide for use.

The immobilized DNA molecules in **oligonucleotide arrays** are synthetic segments of single-stranded DNA usually 20–25 nucleotides long. Oligonucleotides are synthesized for each gene in a genome (see Ch. 5). Determining the sequence for

oligonucleotide array DNA array used to simultaneously detect and identify many short RNA or DNA fragments by hybridization. Also known as DNA array or DNA chip

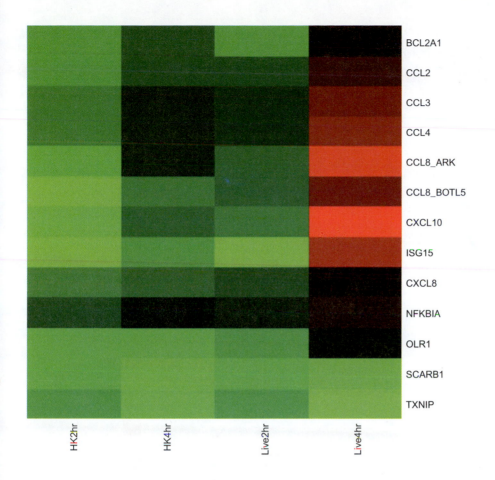

BCL2A1
CCL2
CCL3
CCL4
CCL8_ARK
CCL8_BOTL5
CXCL10
ISG15
CXCL8
NFKBIA
OLR1
SCARB1
TXNIP

HK2hr   HK4hr   Live2hr   Live4hr

■ FIGURE 19.22
Heat Map Representation of Microarray Data

The amounts of various mRNA transcripts (listed at right) in cow alveolar macrophages were measured using microarray analysis. The amount of each transcript was determined in four different conditions: HK2hr compares the mRNA amount in uninfected macrophages vs. infected with heat killed *Mycobacterium bovis*, the causative agent of bovine tuberculosis, for 2 hours; HK4hr compares the same after 4 hours of exposure; Live2hr used live *M. bovis* for 2 hours; and Live4hr used live *M. bovis* for 4 hours. The green represents down-regulation and red represents up-regulation. Each value was normalized. *(Credit: Widdison, S., Watson, M., & Coffey, T. (2011). Early response of bovine alveolar macrophages to infection with live and heat-killed* Mycobacterium bovis. *Developmental & Comparative Immunology, 35(5), 580–591.)*

each oligonucleotide requires some investigation. A particular sequence of n bases will occur simply by chance every 4^n bases. For a mammalian genome with 3×10^9 bases, n must be at least 16 for a sequence to be unique. It is safer to make oligonucleotides longer than this minimum and, for example, the GeneChip® arrays made by Affymetrix Corporation use 25-mers. Determining which 25 nucleotide sequence to use for each gene is also a consideration. The sequence of the oligonucleotide must not create any stem-loop or cloverleaf structures. In addition, the oligonucleotide must not hybridize with stable mRNA secondary structures. To overcome this obstacle, multiple different oligonucleotides for one gene are included at different locations on the chip. These also serve as controls.

Box 19.01 Tiling Arrays Survey the Whole Genome Including Non-Coding Regions

Standard microarrays contain probes against sequences known or suspected to be in coding regions or other functional sequences. However, advances in technology now allow a high enough probe density in arrays that overlapping probes can be made to completely cover both strands of the whole genome (for prokaryotes) or whole chromosomes (in the case of larger eukaryotic genomes).

Such **tiling arrays** therefore reveal binding to any region of the genome. They may reveal unidentified genes that were missed in constructing standard arrays. However, they are more often used for assessing genome-wide transcription of non-coding RNA. They are also used in ChIP-CHIP studies.

Tiling array	Type of microarray that consists of probes covering the whole genome, not just coding sequences

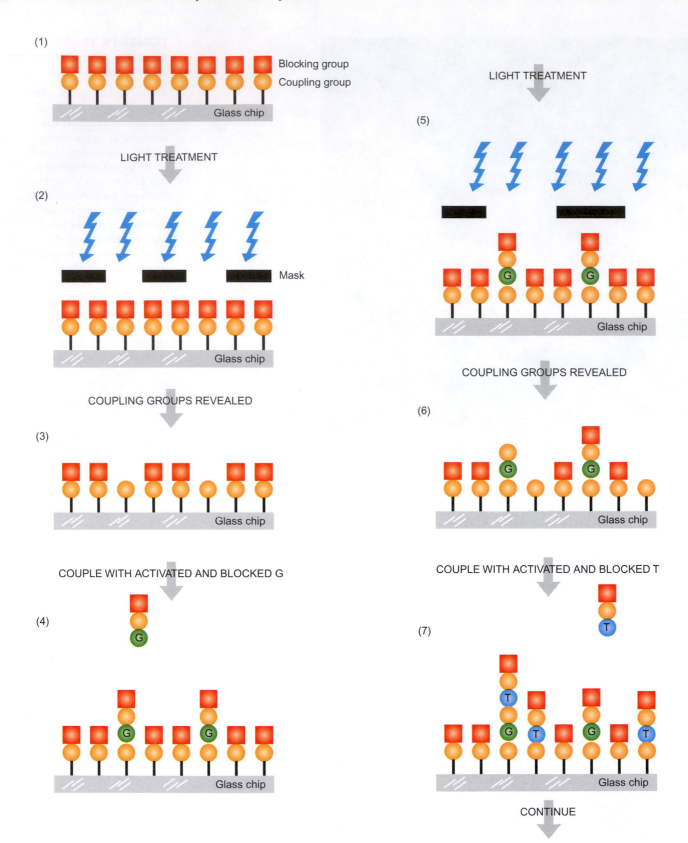

FIGURE 19.23
On-Chip Synthesis of Oligonucleotide

Arrays may be created by chemically synthesizing oligonucleotides directly on the chip. First, reactive groups are linked to the glass chip and blocked. Then each of the four nucleotides is added in turn (in this example, G is added first, then T). A mask covers the areas that should not be activated during any particular reaction. Light activates all the groups not covered with the mask, and a nucleotide is added. The cycle is repeated with the next nucleotide.

The oligonucleotides of the GeneChip array are synthesized directly on the chip (Fig. 19.23). This is done using the techniques of photolithography developed for use in building computer chips. A glass slide is first covered with a reactive group. This is then covered with a photosensitive blocking group that can be removed by light. In each synthetic cycle, those sites where a nucleotide will not be attached are covered with a mask. Those sites where a particular nucleotide (say, adenine, A) is to be attached are illuminated to remove the blocking group. The nucleotide is then added and is chemically coupled to the exposed sites. Only one kind of nucleotide can be added at a time, as it will couple to all exposed sites. Next, the other end of the newly-added nucleotide must be blocked before addition and coupling. The cycle is repeated with another nucleotide (say, thymine, T). This cycling process is repeated with different masking patterns and different nucleotides until the required oligonucleotides are finished.

Oligonucleotide arrays use short synthetic segments of single-stranded DNA.

Box 19.02 Improved On-Chip Synthesis of Oligonucleotides with Virtual Masks

The original on-chip procedure for making microarrays uses physical chrome and glass masks. A chip that uses oligonucleotides of length N needs 4^n such masks. This results in both a high cost and lengthy construction time for the array. Avoiding physical masks greatly reduces fabrication cost and allows greatly-increased flexibility in designing custom arrays. The NimbleGen Corporation has recently introduced a proprietary maskless technology into microarray synthesis. The physical mask is replaced by a computer generated "virtual mask" which controls a digital micromirror array. This is an array of tiny, individually-addressable polished mirrors that can be positioned either to direct the UV light source onto a known position in the array or direct light away from the array. By coordinating the addition of protected phosphoramidite precursors and the sequence of illumination it is possible to make custom arrays with more than 200,000 separate oligonucleotide probes (Fig. 19.24).

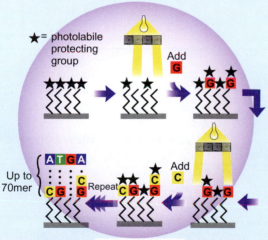

FIGURE 19.24
Virtual Mask Process for On-Chip Synthesis of Oligonucleotides

NimbleGen builds its arrays using photo deprotection chemistry with its Maskless Array Synthesizer (MAS) system. At the heart of the system is a Digital Micromirror Device (DMD; Texas Instruments, Inc.), employing a solid state array of miniature aluminum mirrors to pattern up to 786,000 individual pixels of light. The DMD creates "virtual masks" that replace the inflexible physical chromium masks used in traditional arrays.

8. TaqMan Quantitative PCR to Assay Gene Expression

The most accurate measure of gene expression is by quantitative PCR. As presented in Chapter 6, TaqMan PCR allows the researcher to monitor the amount of PCR product in real-time. These two experimental techniques can be combined to monitor mRNA abundance accurately and quickly. To review, TaqMan PCR has an additional probe in addition to the two normal PCR primers. This probe has a fluorescent molecule on the 5′ end and a quencher at the 3′ end. When these two are near each other, the quencher prevents the fluorescent molecule from emitting light. This probe recognizes the internal sequence of the amplicon and is annealed to the target DNA at the same time as the forward and reverse PCR primers. During the polymerization step, Taq polymerase displaces the probe from the DNA, and the probe is degraded. When the 5′ end separates from the quencher, it fluoresces, and the amount of fluorescence directly correlates to the amount of PCR product produced. Consequently, TaqMan quantitative PCR reactions can also be used to assay the amount of a specific mRNA in a sample.

To use TaqMan PCR to estimate mRNA abundance the only modification to the PCR reaction is in preparing the DNA sample. First, mRNA is isolated from the sample of cells under the conditions in which the gene of interest is expressed. As a control, mRNA from uninduced cells is also isolated. The mRNA samples are then incubated with reverse transcriptase and short oligonucleotide primers of random sequences. These are short enough that the random sequence is found on each mRNA, especially the target mRNA. Reverse transcriptase synthesizes a strand of DNA complementary to the mRNA. Next, the cDNA is annealed to the TaqMan probe and the specific PCR primers and amplified with Taq polymerase. If the target mRNA is abundant in the sample, then fluorescence will be initially high and grow exponentially until the reagents are exhausted. If the target mRNA is rare or not expressed, the fluorescent TaqMan signal will be initially low and take longer to be

---

Fillingham J, Kainth P, Lambert J-P, van Bakel H, Tsui K, Pe a-Castillo L, Nislow C, Figeys D, Hughes TR, Greenblatt J, Andrews BJ (2009) Two-color cell array screen reveals interdependent roles for histone chaperones and a chromatin boundary regulator in histone gene repression. Mol. Cell 35: 340–351.

FOCUS ON RELEVANT RESEARCH

Gene expression must be highly regulated for proper growth and development. *S. cerevisiae* is a unicellular eukaryote that alternates a diploid asexual stage with a sexual phase where two haploid spores combine into diploid yeast. During both types of growth, a variety of mechanisms control the progression through the cell cycle. The expression of histone proteins is regulated so that they are only expressed just before and during DNA replication in S phase. Thus, the histones are available for the newly-made DNA. Histone expression in the wrong part of the cell cycle is toxic.

In yeast, histone genes for each dimer partner (H3/H4 and H2A/H2B) are next to each other, but transcribed in opposite orientation. The promoter is sandwiched between the two genes and coordinately controls the dimer partner expression in both directions. There are two copies of the set for each dimer, so *HHT1-HHF1* and *HHT2-HHF2* encode H3/H4 dimers and *HTA1-HTB1* and *HTA2-HTB2* encode H2A/H2B. A variety of transcription factors and control elements must turn these histone genes on and off at just the right time.

This paper uses a novel reporter-synthetic genetic array (R-SGA) for yeast. In this assay, one strain of yeast is engineered to contain two reporter genes. Green fluorescent protein (GFP) is controlled by a promoter from the gene of interest. The other reporter gene for red fluorescent protein (tdTomato) is controlled constitutively and acts as an internal control. The promoter from *HTA1* was fused to GFP. Next, the authors obtained a yeast deletion library of about 5000 different yeast strains each having a different deletion. Each deletion mutant was mated to the dual-color yeast with the *HTA1* fusion. If the deletion disrupts a gene that activates *HTA1*, then GFP expression decreases, whereas if the deletion disrupts a gene that represses *HTA1*, GFP expression increases. A novel *HTA1* repressor called Rtt106, an activator called Rtt109, and a boundary element called Yta7 were found by this approach.

To confirm that the proteins are transcription factors, the authors performed quantitative PCR and ChIP analysis. The results confirmed that Rtt106 did bind the *HTA1*, *HHT1*, and *HHT2* promoters. When Yta7 was deleted Rtt106 was attached to the *HTA1* promoter as usual, but was also bound to DNA within the transcribed regions. Thus, Yta7 prevents Rtt106 from binding inside the gene, and therefore, acts as a boundary element. The combination of reporter genes, quantitative PCR, and ChIP analysis confirmed the identity of three new regulators that affected histone gene expression.

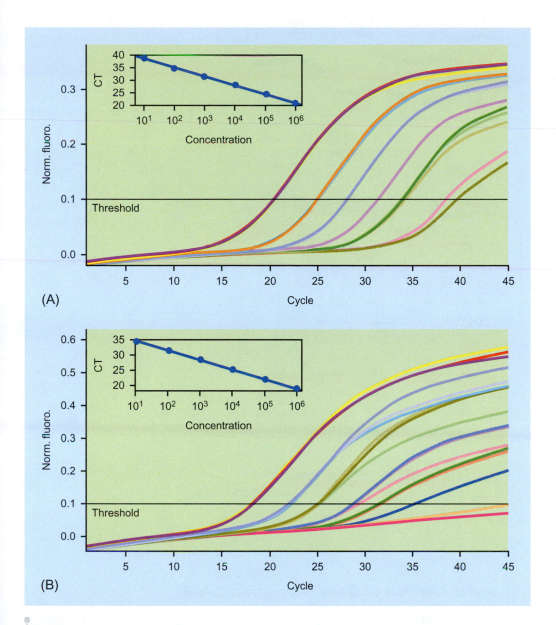

FIGURE 19.25
TaqMan Quantitative PCR Results

The amount of fluorescence from the TaqMan probe is plotted for each cycle of PCR. In the top example, plasmids containing different numbers of copies of the alcohol dehydrogenase (ADH) gene from corn were expressed in the plant and assayed for expression levels of *adh* mRNA. In the bottom panel, the same analysis was done on the 5′ flanking region of the LY038 marker of corn. This marker is used to assess the presence or absence of genetically-modified corn in a sample. *(Credit: Zhang, N., Xu, W., Bai, W., et al. (2011). Event-specific qualitative and quantitative PCR detection of LY038 maize in mixed samples.* Food Control, *22(8), 1287–1295.)*

detected in the PCR machine. The rate at which the fluorescence appears and intensifies directly determines the amount of starting mRNA in the cellular extract (Fig. 19.25). These experiments are always performed with an internal control of a housekeeping gene to ensure that there are no technical problems. The combination of quantitative PCR with other techniques, such as reporter genes and ChIP analysis, is especially effective (see Focus on Relevant Research).

TaqMan can also determine expression of multiple genes at the same time. Entire sets of probe and primer sets are available in microwell formats where each small well contains the probe and primers to a different gene. The researcher adds the cDNA to each of the wells along with Taq polymerase and buffers and then places the microwell plate into a special PCR machine that can read the fluorescence produced in each well. These plates can have up to 385 different wells, and therefore, 385 genes can be assayed at one time.

FIGURE 19.26
SAGE—The Principle

To analyze the total mRNA expressed in a cell, small sequences from each mRNA are converted to complementary DNA and linked together into one long concatemer, which is sequenced. Each of the segments represents a single mRNA; therefore, the number of repeats of each segment correlates with the level of expression of the corresponding gene in the cell.

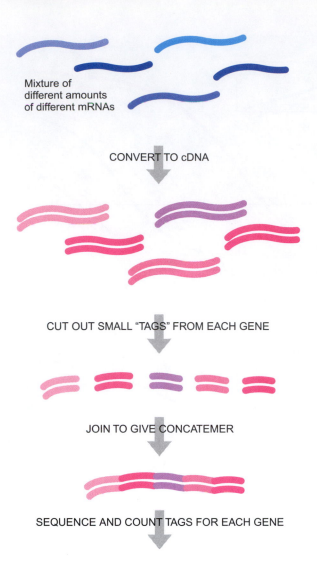

Mixture of different amounts of different mRNAs

CONVERT TO cDNA

CUT OUT SMALL "TAGS" FROM EACH GENE

JOIN TO GIVE CONCATEMER

SEQUENCE AND COUNT TAGS FOR EACH GENE

If all the mRNA molecules in a cell were joined end-to-end and then sequenced, this would reveal how many copies of each mRNA were present— hence the level of gene expression.

9. Serial Analysis of Gene Expression (SAGE)

A DNA-sequencing approach can also measure the expression level of multiple mRNA molecules simultaneously. The basic idea is to sequence all of the mRNA in a cell and then examine the accumulated sequence to see how many copies of each mRNA are represented. To actually do this the mRNA molecules must be joined end-to-end to give a single giant concatenated molecule, which is converted to DNA for sequencing. The term **serial analysis of gene expression (SAGE)** refers to this large concatemer, which contains every expressed gene. The number of copies of each repeat in the concatemer indicates the level of gene expression. To make the approach feasible, only a short sequence from each mRNA is sequenced. The DNA concatemer thus contains many linked sequence tags of approximately 10 bases each (Fig. 19.26).

The first step in SAGE is to extract all the mRNA from a eukaryotic cell and convert it into cDNA using an oligo(dT) primer that hybridizes to the poly(A) tail of the mRNA (Fig. 19.27). The oligo(dT) primer also carries a biotin tag that can be bound by the protein streptavidin. The cDNA is cleaved by a restriction enzyme (the "anchoring enzyme") with a 4 bp recognition site and which generates sticky ends.

serial analysis of gene expression (SAGE) Method to monitor level of multiple mRNA molecules by sequencing a DNA concatemer that contains many serially-linked sequence tags derived from the mRNAs

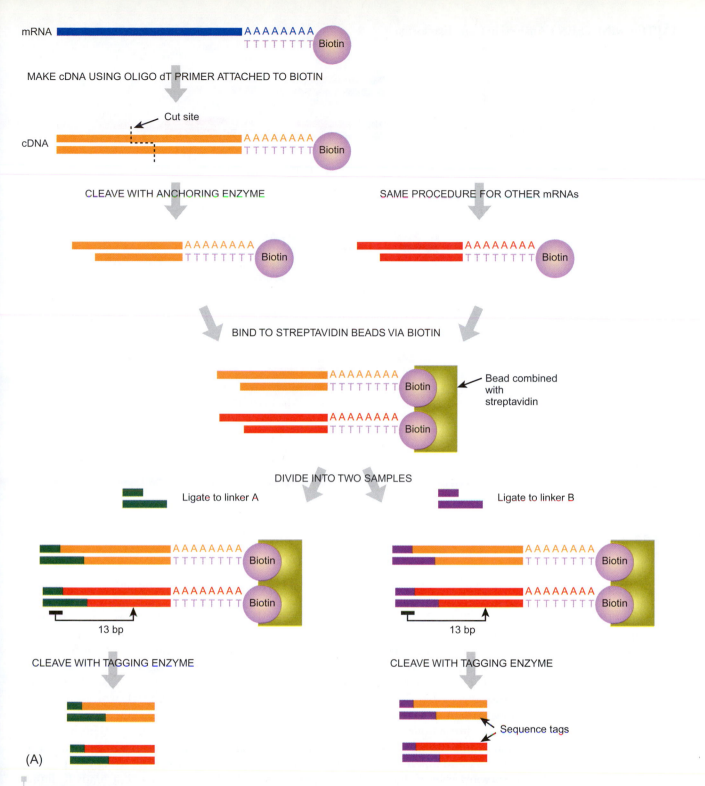

mRNA

MAKE cDNA USING OLIGO dT PRIMER ATTACHED TO BIOTIN

Cut site

cDNA

CLEAVE WITH ANCHORING ENZYME   SAME PROCEDURE FOR OTHER mRNAs

BIND TO STREPTAVIDIN BEADS VIA BIOTIN

Bead combined with streptavidin

DIVIDE INTO TWO SAMPLES

Ligate to linker A   Ligate to linker B

13 bp   13 bp

CLEAVE WITH TAGGING ENZYME   CLEAVE WITH TAGGING ENZYME

Sequence tags

(A)

FIGURE 19.27
SAGE—The Procedure

The first step in making long concatemers of expressed sequences involves isolating the total cellular mRNA and making the corresponding cDNA. The total mRNA is bound via its poly(A) tail to an oligo(dT) primer linked to biotin. It is then converted to cDNA using reverse transcriptase. The cDNAs are then truncated to short, tagged sequences. First, the cDNAs are cleaved with a restriction enzyme known as the anchoring enzyme. This generates a pool of shortened cDNA averaging 256 bp long, with some longer and others shorter. These are isolated using streptavidin, which binds to the biotin tag on the poly(A) tail end of the cDNA. This mixture is divided into two samples and each is ligated to a different linker. This linker has two features: (a) its overhang matches the overhang generated previously by the anchoring enzyme, and (b) it has a recognition site for a type II restriction enzyme (known as the tagging enzyme). Each sample is cut with the tagging enzyme. This enzyme recognizes the sequence in the linker, but actually makes a blunt end cut downstream in the cDNA sequence. This generates two pools of small cDNA sequence tags with different linkers. Finally, the sequence tags are joined into one long sequence. First, fragments are linked by blunt-end ligation. Then PCR primers complementary to the linkers are used to amplify only those ligated molecules that have linker A and linker B flanking two different sequence tags. The PCR products are digested with the anchoring enzyme to remove the linkers and generate sticky ends. These are ligated and the resulting fragment is cloned and sequenced.

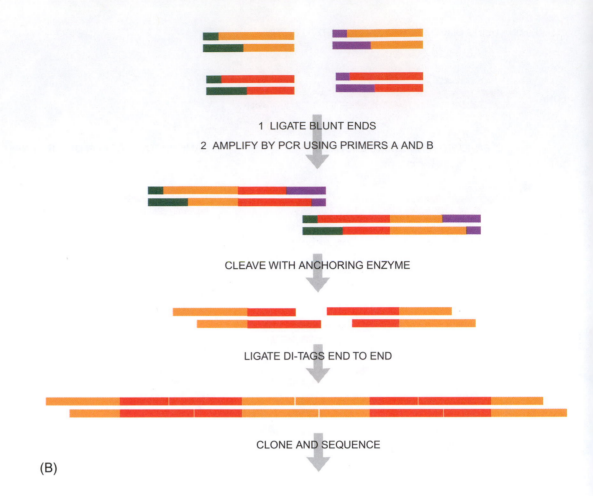

1 LIGATE BLUNT ENDS

2 AMPLIFY BY PCR USING PRIMERS A AND B

CLEAVE WITH ANCHORING ENZYME

LIGATE DI-TAGS END TO END

CLONE AND SEQUENCE

(B)

FIGURE 19.27
(Continued)

> When this approach is put into practice, the RNA is converted to DNA for sequencing and only a short segment from each RNA is sequenced.

This gives fragments of average length 256 bp, which are then bound to streptavidin-coated magnetic beads. This method generates a library of 3′ ends, many just containing the 3′-UTR. The 3′-UTR sequence is more divergent; therefore, mRNAs from highly homologous families can be more readily distinguished using this procedure.

The collected fragments are divided into two samples. The two sets of fragments are ligated to two different artificial linkers that contain recognition sites for a type II restriction enzyme (the "tagging enzyme"). Type II enzymes cut a fixed number of bases away from their recognition sites. A favorite choice is FokI, which cuts 13 bp downstream and so leaves 9 bp of the original mRNA sequence (the "tag") plus the 4 bp anchoring enzyme site attached to the linker. The fragments are then blunt-end ligated head to tail to give structures containing two mRNA-derived tags flanked by linker A and linker B. This structure is used as the target for PCR using two primers, a forward primer that binds to linker A and a backward primer that binds to linker B. (Other structures are formed during the ligation, but only those with the desired structure will be amplified by using this pair of PCR primers.) The PCR products are cleaved with the anchoring restriction enzyme to give tag-dimers with overhanging **sticky ends**, and these are ligated to give the DNA concatemer. Finally, the DNA concatemer is cloned and sequenced. The tags are identified and counted to indicate the relative levels of the original mRNA molecules.

A variation of SAGE sequencing is cap analysis gene expression (CAGE), which focuses on identifying the 5′ end of the mRNA transcriptome. As in all transcriptome analysis, total mRNA is first isolated from samples, except that in the CAGE

---

sticky ends   Ends of a double-stranded DNA molecule that have unpaired single-stranded overhangs

procedure, the mRNA is captured by the cap structure on the 5′ end. These ends are attached to linkers and cut with restriction enzymes that recognize the linker and then cut a specified number of bases from the linker. After analyzing the sequence of all the tags, the 5′ ends of genes are easily identified.

Another variation on SAGE is called massively-parallel signature sequencing (MPSS). In this variation, the small pieces generated from the 3′ ends of the mRNA are not concatenated into a large piece of DNA. Instead, these small fragments are used directly in the second-generation sequencing techniques such as SOLiD, *454*, and Illumina.

Key Concepts

- The transcriptome is the totality of RNA transcripts expressed by a cell or an organism.
- Transcriptomics refers to the global monitoring of gene expression.
- Reporter genes are widely used because their gene products are easy to assay.
- β-galactosidase is the most commonly used reporter enzyme.
- Another convenient reporter system is the emission of light by luciferase.
- Green fluorescent protein emits green light after illumination with long-wave UV and needs no co-factors or substrate.
- Gene fusions can be made by fusing reporter genes to the regulatory sequences to be investigated.
- Protein-binding sites in regulatory regions may be located by deletion analysis of the DNA.
- Gel mobility shift assays and DNA footprinting are used to localize protein-binding sites in DNA more precisely.
- Complexes of DNA with proteins can be isolated by precipitation of chromatin using antibodies.
- The start of transcription may be located by primer extension or by using S1 nuclease.
- DNA microarrays are used for global monitoring of gene expression.
- Quantitative PCR using a variety of fluorescent probes is another approach to assaying gene expression.
- Serial analysis of gene expression (SAGE) monitors global gene expression by bulk sequencing of the total mRNA (after making DNA copies).

Review Questions

1. What are reporter genes? What do reporter genes usually code for?
2. How are reporter genes used to monitor gene expression?
3. Name three enzymes widely used as reporters. How do they work?
4. The *pho* A gene can cleave a wide range of substrates. What are they and what are their products?
5. How can β-galactosidase be used as a reporter enzyme?
6. What reporter enzymes involve the use of light?
7. What is an advantage of using light in reporter systems?
8. What are the advantages of using green fluorescent protein as a reporter? How does it differ from the other reporter methods?
9. What is a gene fusion? What are two functions of gene fusions?
10. When using reporter genes why must the reporter gene be put into an organism lacking the wild-type reporter gene?
11. What are the benefits of using PCR instead of restriction enzymes in deletion analysis?
12. What methods may be used to determine if a regulatory protein binds DNA?
13. How does gel retardation work?
14. After determining to which segment of DNA a protein binds, how is the binding site located more precisely?

15. How is DNA footprinting performed? What is the principle behind it?
16. What is primer extension? What is needed to complete primer extension?
17. What is S1 nuclease? What is S1 nuclease mapping?
18. How can S1 nuclease be used to determine transcription start sites?
19. What is the transcriptome?
20. What methods are used to analyze the transcriptome?
21. What is RNA-seq and what is it used for?
22. What is the principle behind DNA microarrays?
23. If very few genes are of interest why would you not want to use a DNA microarray? What other technique could be used instead?
24. What types of arrays are used to bind mRNA?
25. In some oligonucleotide arrays the oligonucleotides are synthesized directly on the chip, how is this accomplished?
26. What is a tiling array?
27. What is SAGE?
28. In SAGE why is only a small segment of each mRNA used?
29. How is the mRNA extracted for SAGE?
30. What is the "tagging enzyme" for SAGE?

Conceptual Questions

1. You are creating a grid of oligonucleotides using On-Chip synthesis. The first three nucleotides have been added for each of the six oligonucleotides that need to be manufactured. You need to add the next two nucleotides, but first you have to create the masks. Using the sequence information and the diagram of the grid as it stands, design each mask needed for the synthesis of the next two nucleotides as labeled in bold typeface. How many masks are needed?

Oligonucleotide number	Sequence
1	ATT**CG**AGG
2	GAT**GT**ATC
3	TAG**GA**TCC
4	AAT**CG**ACA
5	GTC**CC**TCC
6	TCC**CT**ATT

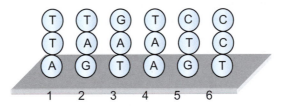

2. Devise an experiment to determine what region of the following promoter for the *AAA* gene has a binding site for the Crp protein. Determine the sequence of the binding site. You have a pure sample of Crp to use in this experiment, a sequencing primer, and pure DNA from the promoter and gene region.

Region 1	Region 2	Region 3	Region 4	AAA gene

← sequencing primer

3. All the known genes that are expressed in intestinal epithelial cells were converted into distinct overlapping oligonucleotides and attached in a grid-like pattern on a glass slide. An intestinal epithelial cell line was grown in a laboratory dish and exposed to a pathogenic strain of *E. coli*. As a control, the other dish was exposed to a non-pathogenic strain of *E. coli*. Using your knowledge of microarray analysis explain how to assay the gene expression changes that occur during exposure to pathogenic *E. coli* including all the relevant controls.
4. Compare primer extension and S1 nuclease methods for determining the transcriptional start site. Describe the advantages of each method and the disadvantages.

Chapter 20

Plasmids

In addition to chromosomes many cells, especially bacteria, contain extra genetic elements known as plasmids. These molecules of DNA are generally much smaller than chromosomes and the genetic information that they carry encodes what may perhaps be thought of as "optional extras." For example, the first plasmids to be discovered carried genes for antibiotic resistance. This ability is obviously of great use to bacteria that are treated with antibiotics. Conversely, making proteins that confer antibiotic resistance is a waste of resources for bacteria that are not exposed to antibiotics. An alternative outlook regards plasmids not merely as accessories to cells, but as genetic entities in their own right. Plasmids (along with other subchromosomal entities to be described in later chapters) may be viewed as a type of "gene creature," with their own evolutionary agenda. From this perspective the cell may be regarded as the environment in which plasmids live, multiply, and evolve.

1. Plasmids as Replicons

Plasmids are autonomous self-replicating molecules of DNA (or very rarely RNA) (Fig. 20.01). They are not chromosomes, although they do

> **plasmid** Self-replicating genetic elements that are sometimes found in both prokaryotic and eukaryotic cells. They are not chromosomes nor part of the host cell's permanent genome. Most plasmids are circular molecules of double-stranded DNA although rare linear plasmids and RNA plasmids are known

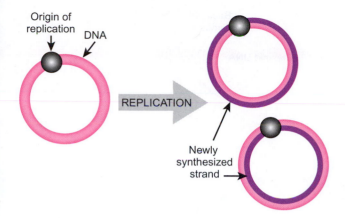

Origin of replication

DNA

REPLICATION

Newly synthesized strand

FIGURE 20.01
Plasmids are Self-Replicating Molecules of DNA

Plasmids are most often rings of double-stranded DNA found inside cells but not attached to or associated with the chromosomal DNA. The plasmid carries its own origin of replication, thus it is considered a true replicon.

reside inside living cells and carry genetic information. They are not regarded as part of the cell's genome for two reasons. First, a particular plasmid may be found in cells of different species and may move from one host species to another. Second, a plasmid may sometimes be present and sometimes absent from the cells of a particular host species. Thus, although plasmids carry genetic information that may be expressed, they are not a constant part of the cell's genetic make-up nor are they needed for cell growth and division under normal conditions.

As discussed previously, **replicons** are self-replicating molecules of nucleic acid. Chromosomes, plasmids, virus genomes (both DNA and RNA), and viroids are all replicons. Strictly speaking, a replicon is defined by the possession of its own origin of replication where DNA (or RNA) synthesis is initiated. Thus, a replicon need not carry genes that encode the enzymes needed for its own replication, nor is it necessarily responsible for generating its own nucleotide precursors or energy. This means that plasmids and viruses are replicons, even though they rely on the host cell to provide energy, raw materials, and many enzyme activities.

Plasmids may be regarded as living creatures in their own right. Just as worms wriggle through the soil and fish float in the sea, so plasmids proliferate inside their host cells. To a plasmid, the cell is its environment. So, although the plasmid is not alive in the same sense as a cell, neither is it merely part of the cell. In some ways plasmids are like domesticated viruses that have lost the ability to move from cell to cell killing as they go. Plasmids maintain some viral characteristics since the plasmid requires the host cell for replication enzymes, energy, and raw materials. Unlike viruses, though, plasmids do not possess protein coats and since they cannot leave the cell they live in, they avoid damaging it. Viruses usually destroy the cell in which they replicate and are then released as virus particles to go in search of fresh victims.

Plasmids replicate in step with their host cell (Fig. 20.02). When the cell divides, the plasmid divides and each daughter cell gets a copy of the plasmid. The plasmids are physically partitioned between the two daughter cells by the assembly of a filamentous protein polymer that pushes the two plasmids apart. The filamentous, actin-like protein is linked to the plasmid by a DNA-binding protein that binds to a partition site (*parS*) on each plasmid. As the filament increases in length, it orients the two plasmids relative to the long axis of the cell so that each daughter cell gets a copy of the plasmid.

It is easy to see how a virus that has lost the genes for its protein coat and/or for killing the host cell might evolve into a plasmid. Furthermore, certain genetic elements, such as P1 (see below), can switch between the two lifestyles and may live either as a plasmid or as a virus. It is also possible to imagine how a plasmid might

Plasmids are "extra" self-replicating molecules of DNA that are found in many cells.

Plasmids and viruses both rely on the host cell to provide energy and raw materials, but plasmids do not damage the host cell.

replicon   Molecule of DNA or RNA that contains an origin of replication and can self-replicate

FIGURE 20.02
Plasmids Replicate in Step with the Host Cell

When a bacterial cell is ready to divide, the replication machinery also duplicates the plasmid DNA. The two copies of the chromosome and two copies of the plasmid are then divided equally between the daughter cells. The replication of the plasmid does not harm the cell.

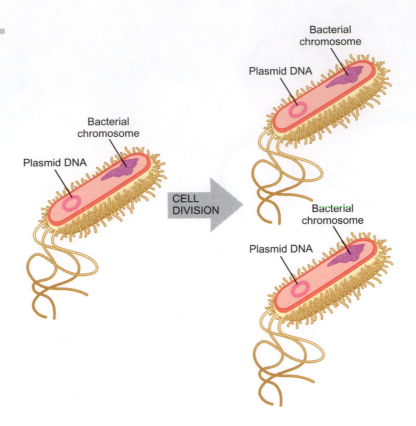

pick up coat protein genes and/or killing functions and deregulate its DNA replication so evolving into a virus. Indeed, many plasmids possess host-killing functions that they use to ensure that they are not lost by the host cell (see Section 3.2, in this chapter). So which came first, plasmids or viruses? Here, we have a true chicken and egg situation. Almost certainly some present-day plasmids are derived from viruses and equally certainly some present-day viruses are derived from plasmids. However, the ultimate origins of the majority of both kinds of element remain obscure.

Box 20.1 The Idea of a Gene Creature

When first considered, entities such as plasmids, viruses, and transposons seem well-defined and distinct. Indeed, many of these elements are easy to identify and classify. Nonetheless, as will be seen, a significant number of genetic entities appear to have "multiple personalities." Thus, P1 is a bacterial virus that can live as a plasmid inside its host cell. Mu is another bacterial virus that acts as a transposon. Moreover, many viruses are closely related in sequence to genetic elements that are undoubtedly plasmids. The concept of a **gene creature** unites those assorted genetic entities that exist at a lower level than a living cell. These genetic elements all replicate and move around, yet they do so within the confines of a host cell. We may thus regard the host cell as the environment in which gene creatures live and multiply. Although these subcellular entities can be viewed merely as parasites of cells, they can often be better understood by viewing their biology from the gene creatures' own perspective.

gene creature Genetic entity that consists primarily of genetic information, sometimes with a protective covering, but without its own machinery to generate energy or replicate macromolecules

2. General Properties of Plasmids

Plasmids are usually circular molecules of DNA, although occasionally, plasmids that are linear or made of RNA exist. They may be found as single or multiple copies and may carry from half a dozen to several hundred genes. Plasmids can only multiply inside a host cell. Most plasmids inhabit bacteria, and indeed around 50% of bacteria found in the wild contain one or more plasmids. Plasmids are also found in higher organisms such as yeast and fungi. The 2μ circle of yeast (see below) is a well-known example that has been modified for use as a cloning vector.

> Most plasmids are circular, made of DNA, and much smaller than chromosomes.

The **copy number** is the number of copies of the plasmid in each bacterial cell. For most plasmids it is one or two copies per chromosome, but it may be as many as 50 or more for certain small plasmids such as the ColE plasmids. The number of copies influences the strength of plasmid-borne characteristics, especially antibiotic resistance. The more copies of the plasmid per cell, the more copies there will be of the antibiotic resistance genes, and therefore, the higher the resulting level of antibiotic resistance.

The size of plasmids varies enormously. The F-plasmid of *Escherichia coli* is fairly average in this respect and is about 1% the size of the *E. coli* chromosome. Most multicopy plasmids are much smaller (ColE plasmids are about 10% the size of the F-plasmid). Very large plasmids, up to 10% of the size of a chromosome, are sometimes found, but they are difficult to work with and few have been properly characterized.

> Some plasmids are present in one or two copies per cell, whereas others occur in multiple copies.

Plasmids carry genes for managing their own lifecycles and some plasmids carry genes that affect the properties of the host cell. These properties vary greatly from plasmid to plasmid, the best known being resistance to various antibiotics. **Cryptic plasmids** are those that confer no identifiable phenotype on the host cell. Cryptic plasmids presumably carry genes whose characteristics are still unknown. Plasmids that are modified for different purposes are used in molecular biology research and are often used to carry genes during genetic engineering.

The host range of plasmids varies widely. Some plasmids are restricted to a few closely-related bacteria; for example, the F-plasmid only inhabits *E. coli* and related enteric bacteria like *Shigella* and *Salmonella*. Others have a wide host range; for example, plasmids of the P-family can live in hundreds of different species of bacteria. Although "P" is now usually regarded as standing for "promiscuous," due to their unusually wide host range, these plasmids were originally named after *Pseudomonas*, the bacterium in which they were discovered. They are often responsible for resistance to multiple antibiotics, including penicillins.

Certain plasmids can move themselves from one bacterial cell to another, a property known as **transferability**. Many medium-sized plasmids, such as the F-type and P-type plasmids, can do this and are referred to as Tra$^+$ (transfer-positive). Since plasmid transfer requires over 30 genes, only medium or large plasmids possess this ability. Very small plasmids, such as the ColE plasmids, simply do not have enough DNA to accommodate the genes needed. Nonetheless, many small plasmids, including the ColE plasmids have **mobilizability**, meaning they can be mobilized by self-transferable plasmids (i.e., they are Mob$^+$ [mobilization-positive]). However, not all transfer-negative plasmids can be mobilized. Some transferable plasmids (e.g., the F-plasmid) can also mobilize chromosomal genes. It was this observation that allowed the original development of bacterial genetics using *E. coli*. The mechanism of plasmid transfer and the conditions necessary for transfer of chromosomal genes are therefore discussed in Chapter 25, Bacterial Genetics.

> Some plasmids can transfer themselves between bacterial cells and a few can also transfer chromosomal genes.

---

copy number   The number of copies of a plasmid found within a single host cell
cryptic plasmid   A plasmid that confers no identified characteristics or phenotypic properties
mobilizability   Ability of a non-transferable plasmid to be moved from one host cell to another by a transferable plasmid
transferability   Ability of a plasmid to move itself from one host cell to another

Box 20.2 Plasmid or Chromosome?

When the genome of the Gram negative bacterium *Vibrio cholerae*, the causative agent of cholera, was sequenced it was found to consist of two circular chromosomes of 2,961,146 and 1,072,314 base pairs. Together, this totals approximately 4 million base pairs and encodes about 3,900 proteins—about the same amount of genetic information as *E. coli*. Many genes that appear to have origins outside the enteric bacteria, as deduced from their different base composition, were found on the smaller chromosome. Many of these lack homology to characterized genes and are of unknown function. The smaller chromosome also carries an integron gene capture system (see Ch. 22) and hosts "addiction" genes that are typically found on plasmids (see below). Furthermore, the smaller chromosome replicates by a different mechanism from the large chromosome. In fact, the smaller chromosome shares a replication system with a family of widely distributed plasmids. It seems likely that the smaller chromosome originated as a plasmid that has grown to its present size by accumulating genes from assorted external sources. Perhaps it is better to regard the smaller chromosome as a "megaplasmid" (see Focus on Relevant Research). Genome-sequence data suggest that some 10% of bacteria carry such megaplasmids, although the size varies considerably. In most of these cases, the larger chromosome carries almost all of the genes needed for vital cell functions such as protein, RNA, and DNA synthesis.

Harrison PW, Lower RPJ, Kim NKD, Young JPW (2010) Introducing the bacterial "chromid": not a chromosome, not a plasmid. Trends Microbiol. 18: 141–148.

FOCUS ON RELEVANT RESEARCH

This opinion article argues for another term to describe the "secondary chromosomes" of bacteria. This article suggests that the secondary chromosomes actually are different from the larger primary chromosome. They are true replicons with nucleotide composition (%GC) similar to that of the primary chromosome, suggesting that they have been residents in their host for a long time. Yet, they are maintained by plasmid-like replication and partitioning systems. They may contain genes from other bacterial species suggesting that they (or a smaller ancestral plasmid) inhabited other species prior to their residence inside their current host. The secondary chromosome also has unique genes that are only found within a specific phylum. Since these characteristics resemble both a chromosome and plasmid, the authors suggest the term "**chromid**," a combination of both names.

The reason for establishing a specific term is to standardize the submission of genomic sequences to the NCBI databases. This article identified 82 different bacterial species that have a chromid as of mid-2009, and these are mainly found in proteobacteria. These are identified as chromosomes for some species and plasmids for others. Having a set of distinct criteria to describe chromids will facilitate their proper labeling in the databases. In addition, a better defined identity will help understand the way in which they evolved, and the advantage they provide for their host cells.

Author's Note: While the viewpoint that a new term is needed for elements that are neither plasmids nor chromosomes has merit, the choice of the term "chromid" was unfortunate and confusing. The name "**chromID**" is already in use and is the brand name of chromogenic culture media sold by bioMérieux.

---

chromid   A term that combines chromosome and plasmid used to describe bacterial genetic material with characteristics of a chromosome and a plasmid
chromID   The brand name of a chromogenic culture media sold by bioMérieux

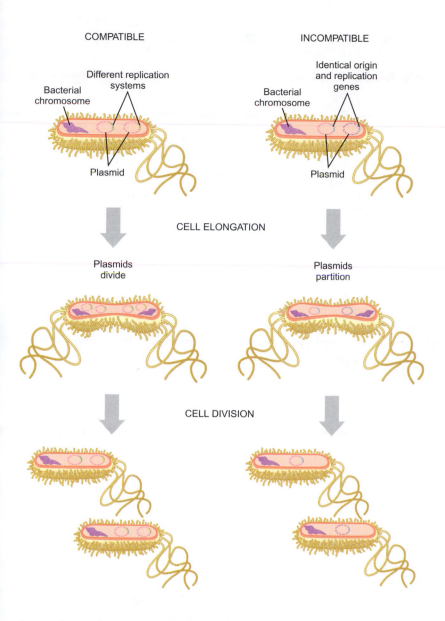

COMPATIBLE

INCOMPATIBLE

Bacterial chromosome

Different replication systems

Bacterial chromosome

Identical origin and replication genes

Plasmid

Plasmid

CELL ELONGATION

Plasmids divide

Plasmids partition

CELL DIVISION

■ **FIGURE 20.03**
Plasmid Incompatibility

Plasmids with different origins of replication and different replication genes are able to inhabit the same bacterial cell and are considered compatible (left). During cell division, both types of plasmid replicate; therefore, each daughter cell will inherit both plasmids, just like the mother cell. On the other hand, if two plasmids have identical origins and replication genes they are incompatible and will not be replicated during cell division (right). Instead, the two plasmids are partitioned into different daughter cells.

2.1. Plasmid Families and Incompatibility

Two different plasmids that belong to the same family cannot co-exist in the same cell. This is known as **incompatibility**. Plasmids were originally classified by incompatibility and so plasmid families are often known as incompatibility groups and are designated by letters of the alphabet (F, P, I, X, etc.). Plasmids of the same incompatibility group have very similar DNA sequences in their replication genes, although the other genes they carry may be very different. It is quite possible to have two or more plasmids in the same cell as long as they belong to different families. So a P-type plasmid will happily share the same cell with a plasmid of the F-family (Fig. 20.03).

Plasmids are classified into families whose members share very similar replication genes.

2.2. Occasional Plasmids are Linear or Made of RNA

Although most plasmids are circular molecules of DNA there are occasional exceptions. Linear plasmids of double-stranded DNA have been found in a variety of bacteria and in fungi and higher plants. The best-characterized linear plasmids are found in those few bacteria such as *Borrelia* and *Streptomyces* that also contain linear

incompatibility   The inability of two plasmids of the same family to co-exist in the same host cell

FIGURE 20.04
End Structures of Linear Plasmids

A) Linear plasmids of *Borrelia* form hairpin loops at the ends. B) Linear plasmids of *Streptomyces* are coated with proteins that protect the DNA ends. If linear plasmids had exposed double-stranded ends, this would trigger recombination, repair, or degradation systems.

BORRELIA HAIRPIN/LOOP ENDS

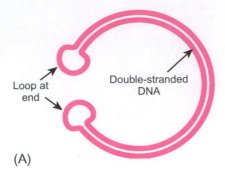

Loop at end

Double-stranded DNA

(A)

STREPTOMYCES TENNIS RACQUET ENDS

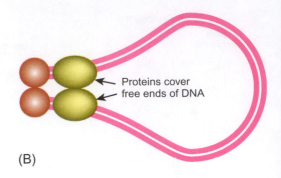

Proteins cover free ends of DNA

(B)

> Linear plasmids have special structures to protect the ends of the DNA.

chromosomes. Linear DNA replicons in bacteria are not protected by telomeres like the linear chromosomes of eukaryotes. Instead, a variety of individual adaptations protect the ends from endonucleases.

In *Borrelia* there are not actually any free DNA ends. Instead, hairpin sequences of single-stranded DNA form loops at the ends of both linear plasmids and chromosomes (Fig. 20.04A). Some animal viruses, such as the iridovirus that causes African swine fever, have similar structures. Different species of *Borrelia* cause Lyme's disease and relapsing fever. Their linear plasmids appear to encode both hemolysins that damage blood cells and surface proteins that protect the bacteria from the host immune system. Thus, as is true of many other infectious bacteria, the virulence factors of *Borrelia* are also largely plasmid-borne.

The linear plasmids of *Streptomyces* are indeed genuine linear DNA molecules with free ends. They have inverted repeats at the ends of the DNA that are held together by proteins. In addition, special protective proteins are covalently attached to the 5' ends of the DNA. The net result is a tennis racket structure (Fig. 20.04B). The DNA of adenovirus, most linear eukaryotic plasmids, and some bacterial viruses show similar structures.

Linear plasmids are also found among eukaryotes. The fungus *Flammulina velutipes*, commonly known as the enoki mushroom, has two very small linear plasmids within its mitochondria. The dairy yeast, *Kluyveromyces lactis*, has a linear plasmid that normally replicates in the cytoplasm. However, on occasion the plasmid relocates to the nucleus where it replicates as a circle. Circularization is due to site specific recombination involving the inverted repeats at the ends of the linear form of the plasmid. The physiological role of these plasmids is obscure.

RNA plasmids are rare and most are poorly characterized. Examples are known from plants, fungi, and even animals. Some strains of the yeast *Saccharomyces cerevisiae* contain linear RNA plasmids. Similar RNA plasmids are found in the mitochondria of some varieties of maize plants. RNA plasmids are found as both single-stranded and double-stranded forms and replicate in a manner similar to certain RNA viruses. The RNA plasmid encodes RNA-dependent RNA polymerase that directs its own synthesis. Unlike RNA viruses, RNA plasmids do not contain genes for coat proteins. Sequence comparisons suggest that these RNA plasmids may have evolved from RNA viruses that have taken up permanent residence after losing the ability to move from cell to cell as virus particles.

3. Plasmid DNA Replicates by Two Alternative Methods

> Most plasmids undergo bidirectional replication like bacterial chromosomes.

> Some plasmids and many viruses use the rolling circle mechanism for replication.

Typical plasmids made of circular double-stranded DNA use two alternative mechanisms for replicating their DNA. Most plasmids replicate like miniature bacterial chromosomes (see Ch. 10 for details of chromosome replication). They have an origin of replication where the DNA opens and replication begins. Then two replication forks move around the circular plasmid DNA in opposite directions until they meet (Fig. 20.05). A few very tiny plasmids have only one replication fork that moves around the circle until it gets back to the origin.

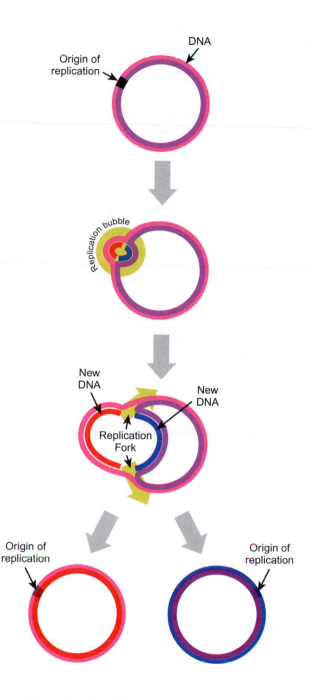

■ **FIGURE 20.05**
***Bidirectional Plasmid
Replication***

For some circular plasmids,
replication enzymes recognize
the origin of replication, unwind
the DNA, and start synthesis of
two new strands of DNA, one in
each direction. The net result is
a replication bubble. As the new
strands are synthesized, two distinct
replication forks keep moving
around the circle until they meet on
the opposite side. When both DNA
circles are complete, two distinct
plasmids are produced.

The other replication mechanism is **rolling circle replication**, which is used by some plasmids and quite a few viruses. At the origin of replication, one strand of the double-stranded DNA molecule is nicked (Fig. 20.06). The other, still circular, strand starts to roll away from the broken strand. This results in two single-stranded regions of DNA, one belonging to the broken strand and one that is part of the circular strand. DNA is then synthesized starting at the end of the broken strand. The circular strand is used as a template and the gap left where the two original strands rolled apart is filled in. This process of rolling and filling in continues, until eventually the original broken strand is completely unrolled and the circular strand is fully paired with a newly-made strand of DNA. This leaves a single strand of DNA, equal in length to the original DNA circle, dangling loose.

rolling circle replication   Mechanism of replicating double-stranded circular DNA that starts by nicking and unrolling one strand and using the other, still circular, strand as a template for DNA synthesis. Used by some plasmids and viruses

FIGURE 20.06
Rolling Circle Replication

During rolling circle replication, one strand of the plasmid DNA is nicked, and the broken strand (pink) separates from the circular strand (purple). The gap left by the separation is filled in with new DNA starting at the origin of replication (green strand). The newly-synthesized DNA keeps displacing the linear strand until the circular strand is completely replicated. The linear single-stranded piece is fully "unrolled" in the process.

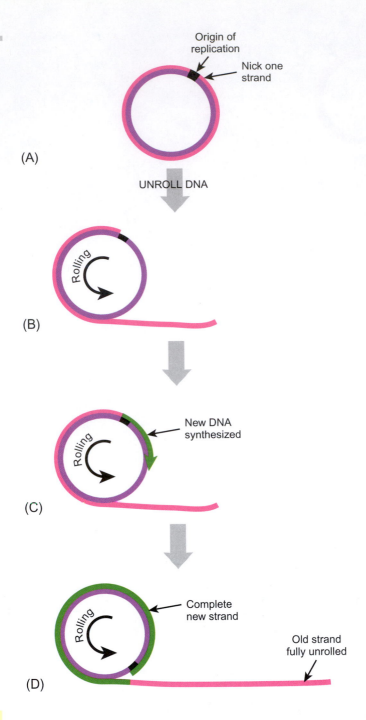

(A)

UNROLL DNA

(B)

(C)

(D)

Transferable plasmids use the rolling circle mechanism during transfer but bidirectional replication when dividing in step with the host cell.

What happens next varies, depending on the circumstances. For simple plasmid replication, the unrolled old strand is used as a template to make a complementary strand. This double-stranded region is cut free and circularized to give a second copy of the plasmid.

Some plasmids, such as the F-plasmid of *E. coli*, can transfer themselves from one bacterium to another. Such plasmids have two separate origins of replication. They divide by bidirectional replication (also known as vegetative replication) when their host cell divides, but use the rolling circle mechanism when they move from one cell to another during conjugation (see Ch. 25). Bidirectional replication starts at *oriV*, the origin of vegetative replication, which is at a different site on the plasmid from *oriT*, the origin used during transfer. All plasmids must have a vegetative origin since they must all divide to survive. But only those plasmids that can transfer themselves have a special transfer origin.

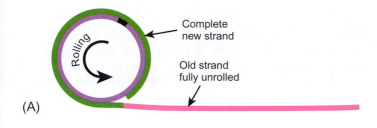

(A)

Complete
new strand

Old strand
fully unrolled

CIRCLE CONTINUES TO ROLL

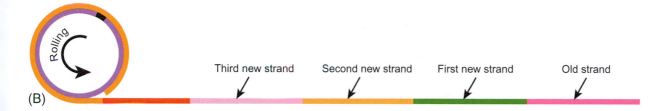

(B)

Third new strand

Second new strand

First new strand

Old strand

SYNTHESIS OF A SECOND STRAND OF DNA

(C)

New complementary strand of DNA

FIGURE 20.07
***Viruses May Use Rolling
Circle Replication***

Rolling circle replication occurs as
described in the previous figure (A),
but the replication continues around
the circular DNA (purple) for many
rounds (B). In some viruses, the long,
single-stranded piece of DNA is cut
and packaged into virus particles
as single-stranded DNA. In other
viruses, a complementary strand is
synthesized, giving double-stranded
DNA (C). The double-stranded
segments are then cut and packaged
as single genome units.

The relationship between certain plasmids and viruses is illustrated by their DNA replication mechanisms. Rolling circle replication is not only used by transferable plasmids, but also by many viruses (Fig. 20.07). Some manufacture many double-stranded molecules of virus DNA. These viruses use the dangling strand as a template to synthesize a new strand of DNA. They just keep rolling and synthesizing and end up with a long, linear double-stranded DNA many times the length of the original DNA circle. This is chopped into unit genome lengths and packaged into virus particles. (Some of these viruses convert the DNA into circles before packaging, whereas others package linear DNA and only circularize their DNA after infecting a new cell when it is time to replicate again.)

Other viruses contain single-stranded DNA. These viruses leave the dangling strand unpaired. They continue rolling and end up with a long linear single-stranded DNA (Fig. 20.07B). This is cut into unit genome lengths and packaged as before. When these viruses infect a new cell, they synthesize the opposite strand, so converting their single strand to a double-stranded DNA molecule.

Whatever the mechanism of plasmid replication, it is also necessary to make sure that each daughter cell gets a copy of the plasmid. Partition of plasmids and chromosomes in bacteria depends on a partition system of protein filaments that are part of the cytoskeleton (see Ch. 4). In addition to the ParABS system that partitions the bacterial chromosome, there are other systems that partition plasmids (see Focus on Relevant Research).

Gerdes K, Howard M, Szardenings F (2010) Pushing and pulling in prokaryotic DNA segregation. Cell 141: 927–942.

FOCUS ON RELEVANT RESEARCH

This review discusses the alternative systems for partitioning DNA (both chromosomes and plasmids) into the two daughter cells after replication. The most common Par system is the ParABS system that partitions most bacterial chromosomes and many plasmids. This includes a P loop ATPase (ParA) that forms long polymeric filaments and so distributes plasmids into the two halves of the dividing cell.

The ParM system encodes a filament-forming protein that is homologous to eukaryotic actin. This pushes plasmids apart in a process vaguely resembling eukaryotic mitosis. A third class of partition system uses a filament-forming protein (TubZ) that is homologous to eukaryotic tubulin.

3.1. Control of Copy Number by Antisense RNA

Both single-copy and multicopy plasmids regulate their copy number carefully. However, the regulatory mechanisms differ significantly for the two groups. High copy number plasmids limit the initiation of plasmid replication once the number of plasmids in the cell reaches a certain level. These plasmids are sometimes said to have "relaxed" copy number control. In contrast, single-copy plasmids have "stringent" copy number control as their division is more tightly regulated and they replicate only once during the cell cycle. The regulation of replication is much better understood for multicopy plasmids than for single-copy plasmids.

> Antisense RNA is involved in regulating the copy number of many plasmids.

The most interesting aspect of copy number regulation is the involvement of **antisense RNA** to control the initiation of plasmid replication. The details are best investigated for the multicopy plasmid ColE1, but the principle of using antisense RNA also applies to single-copy plasmids. Initiation of ColE1 replication starts with the transcription of an RNA molecule of 555 bases that can act as a primer for DNA synthesis. This pre-primer RNA (sometimes called RNAII) is cleaved by **ribonuclease H** to generate a primer with a free 3'-OH group, which can be used by DNA polymerase I (Fig. 20.08).

If ribonuclease H fails to cleave the pre-primer RNAII, no free 3' end is made and replication cannot proceed. Ribonuclease H is specific for RNA-DNA hybrids. Consequently, when an antisense RNA, known as RNAI, binds to pre-primer RNAII, this prevents cleavage (Fig. 20.09). Both RNAII and RNAI are transcribed from the same region of DNA but in opposite directions. RNAI is 108 bases long and is encoded by the opposite strand from RNAII. RNAI is complementary to the 5' end of RNAII.

The copy number is determined by the relative strengths of binding of RNAII to the DNA at the origin of replication and of RNAI to RNAII. Mutations affecting either of these interactions will change the copy number. The Rom protein (encoded by a gene on ColE1) holds RNAI and RNAII together. If the gene for Rom protein is inactivated, the copy number also rises, but is still controlled suggesting other mechanisms may also control the copy number.

3.2. Plasmid Addiction and Host-Killing Functions

Many larger plasmids possess genes whose function is to ensure that the host cell does not lose the plasmid. These "plasmid addiction" systems kill the bacterial cell if the plasmid is lost, so only cells that retain the plasmid survive. The details vary, but the scheme involves two components that are made by the plasmid. One is lethal to the host cell and the other is the antidote. The toxin is long-lived and the antidote is

---

antisense RNA   An RNA molecule that is complementary to messenger RNA or another functional RNA molecule
ribonuclease H   A ribonuclease of bacterial cells that is specific for RNA-DNA hybrids

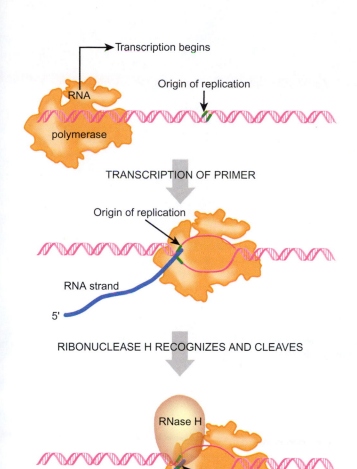

Transcription begins

RNA polymerase

Origin of replication

TRANSCRIPTION OF PRIMER

Origin of replication

RNA strand

5'

RIBONUCLEASE H RECOGNIZES AND CLEAVES

RNase H

RNA strand

5'

RNA is cleaved
by ribonuclease H
at origin

RNA PRIMER REMAINS BOUND

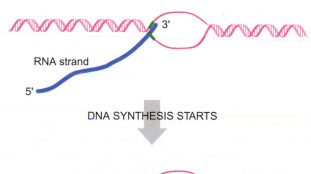

3'

RNA strand

5'

DNA SYNTHESIS STARTS

3' 5'

RNA strand

New DNA starts
from RNA primer

5'

■ **FIGURE 20.08**
Priming of ColE1 plasmid
Replication

RNA polymerase synthesizes a
strand of RNA (RNAII) near the
origin of replication. RNAII (blue
strand) is recognized and cleaved
by ribonuclease H. The free 3'-OH
created by the cleavage primes the
synthesis of DNA at the origin. The
ColE1 plasmid is then replicated.

short-lived. As long as the plasmid is present it continues to synthesize the antidote. If
the plasmid is lost, the antidote decays, but the stable toxin survives longer and kills
the cell. Many plasmids actually have two or more different systems to decrease the
chances of the host cell surviving after losing the plasmid.

Large plasmids often make
toxins that kill the host cell if,
and only if, it loses the plasmid
DNA.

FIGURE 20.09
Antisense RNA Prevents Primer Formation

A second transcript, RNAI, is also made from the same region of ColE1 as RNAII. The RNAI (green) is transcribed from the opposite DNA strand and is therefore complementary to RNAII. The complementary regions of RNAII and RNAI start to base pair. Eventually, the entire sequence aligns and a double-stranded RNA molecule is formed. Since ribonuclease H only recognizes DNA-RNA hybrid molecules, no cut is made and RNAII transcription continues. DNA synthesis fails to start and the ColE1 plasmid is not replicated.

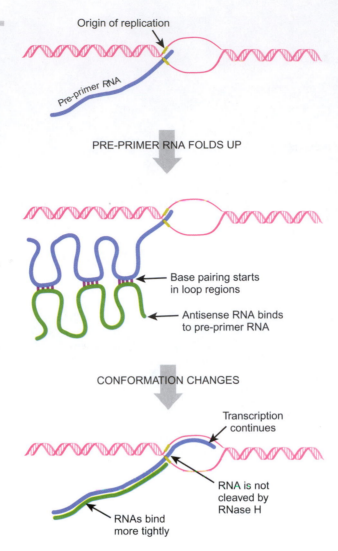

The protein-based host-killing operon of the F-plasmid consists of two genes, *ccdAB*, which are expressed to give two proteins, CcdA and CcdB. CcdA is the antidote and is readily degraded by host-cell proteases. CcdB is the toxin. As long as the plasmid is present it constantly makes a fresh supply of CcdA, which binds to CcdB and blocks its action. Consequently, the cell survives. If for some reason, the plasmid is lost, and no more CcdA is made, then CcdB kills the cell by inhibiting DNA gyrase and generating double-stranded breaks in the bacterial chromosome. Other plasmids have similar systems.

Both the F-plasmid and some R-plasmids (see below) also have a system where the antidote is an antisense RNA that prevents translation of the host-killing protein. This type of system was first found in plasmid R1. Here, the Hok (<u>H</u>ost-<u>K</u>illing) protein is not translated as long as the Sok antisense RNA is present. Sok RNA binds to *hok* messenger RNA (mRNA) and prevents ribosome binding, which in turn promotes degradation of the mRNA by ribonuclease III. Sok RNA decays relatively rapidly. If the plasmid is lost, Sok RNA decays and the *hok* mRNA is free to be translated. Hok protein then damages the cell membrane and kills the cell.

4. Many Plasmids Help Their Host Cells

If plasmids are not an essential part of the cell's genome, why do cells allow them to persist? Some plasmids are indeed useless and, as discussed above, some plasmids possess special mechanisms to protect their own survival at the expense of the host cell. Nonetheless, most plasmids do in fact provide useful properties to their host cells.

TABLE 20.01	Properties Conferred by Naturally-Occurring Plasmids

Resistance and Defense

Antibiotic resistance against aminoglycosides, β-lactams, chloramphenicol, sulfonamides, trimethoprim, fusidic acid, tetracyclines, macrolides, fosfomycin

Resistance to many heavy metal ions including Ni, Co, Pb, Cd, Cr, Bi, Sb, Zn, Cu, and Ag

Resistance to mercury and organomercury compounds

Resistance to toxic anions such as arsenate, arsenite, borate, chromate, selenate, tellurite, etc.

Resistance to intercalating agents such as acridines and ethidium bromide

Protection against radiation damage by UV and X-rays

Resistance systems that degrade bacteriophage DNA

Resistance to certain bacteriophages

Aggression and Virulence

Synthesis of bacteriocins

Synthesis of antibiotics

Crown gall tumors and hairy root disease in plants caused by *Agrobacterium*

Nodule formation by *Rhizobium* on roots of legumes

R-body synthesis due to plasmid of *Caedibacter* symbionts in killer *Paramecium*

Virulence factors of many pathogenic bacteria, including toxin synthesis, protection against immune system and attachment proteins

Metabolic Pathways

Degradation of sugars (e.g., lactose (in *Salmonella*), raffinose, sucrose)

Degradation of aliphatic and aromatic hydrocarbons and their derivatives such as octane, toluene, benzoic acid, camphor

Degradation of halogenated hydrocarbons such as polychlorinated biphenyls

Degradation of proteins

Synthesis of hydrogen sulfide

Denitrification in *Alcaligenes*

Pigment synthesis in *Erwinia*

Miscellaneous

Transport of citrate in *E. coli*

Transport of iron

Gas vacuole production in *Halobacterium*

In principle any gene can be plasmid-borne and plasmids are indeed widely used in genetic engineering to move genes between organisms. In practice, certain properties are widespread among naturally-occurring plasmids. A selection of these is given in Table 20.01.

Plasmids often carry genes for resistance to antibiotics. This protects bacteria both from human medicine and from antibiotics produced naturally in the soil. Plasmids with genes for resistance to toxic heavy metals such as mercury, lead, or cadmium protect bacteria from industrial pollution and from natural deposits of toxic minerals. Other plasmids provide genes that allow bacteria to grow by breaking down various industrial chemicals, including herbicides, or the components of petroleum. From the human perspective, such bacteria may be a nuisance or may be useful in cleaning up oil spills or other chemical pollution. Finally, some plasmids provide virulence or colonization factors needed by infectious bacteria to invade their victims and survive the countermeasures taken by the host immune system.

Many plasmids carry genes that are beneficial to their host cells, but only under certain environmental conditions.

FIGURE 20.10
Antibiotic Resistance Plasmids

Plasmids carry genes for replicating their DNA, transferring themselves from one host cell to another, and for a variety of phenotypes. Many plasmids carry genes that confer antibiotic resistance on the host cell when the genes are expressed.

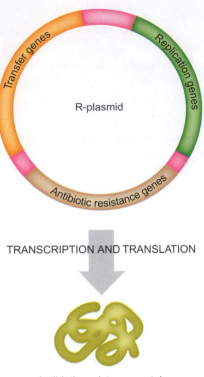

TRANSCRIPTION AND TRANSLATION

Antibiotic resistance protein

R-plasmids make bacteria resistant to antibiotics.

4.1. Antibiotic Resistance Plasmids

Plasmids were first discovered in Japan just after World War II, inhabiting the bacterium *Shigella*, which causes dysentery. The type of dysentery due to bacteria was originally treated with sulfonamides, the earliest type of antibiotic. Suddenly, strains of *Shigella* appeared that were resistant to sulfonamide treatment. The genes for resistance to sulfonamide proved to reside on plasmids, rather than the bacterial chromosome. Plasmids that confer antibiotic resistance are called **R-plasmids** (an older term is R-factor) (Fig. 20.10).

Not only did the identified *Shigella* strains carry antibiotic resistance genes, the plasmids carrying the sulfonamide resistance genes were able to transfer copies of themselves from one bacterial cell to another. Consequently, the sulfonamide resistance spread rapidly from *Shigella* to *Shigella*. Although the resistance plasmid allowed the *Shigella* to survive, transferable antibiotic resistance is highly dangerous from the human medical viewpoint. By 1953, the year Watson and Crick discovered the double helix, 80% of the dysentery-causing *Shigella* in Japan had become resistant to sulfonamides. By 1960, 10% of the *Shigella* in Japan was resistant to four antibiotics, sulfonamides, chloramphenicol, tetracycline, and streptomycin, and by 1970 this had risen to over 30%. These strains often carry resistance genes for different antibiotics on one single plasmid. Today, the transfer of multiple antibiotic resistance plasmids between bacteria has become a major clinical problem. Patients with infections after surgery, with severe burns, or with immuno-compromised systems are at highest risk to antibiotic-resistant infections.

Soil bacteria (e.g., *Streptomyces*) or fungi (e.g., *Penicillium*) produce antibiotics as a natural part of their physiology. Consequently, R-plasmids were in existence before the clinical use of antibiotics by humans, but they have spread far and wide since wide scale use of antibiotics started. A major factor in R-plasmid spread is the practice of

R-plasmid   Plasmid that carries genes for antibiotic resistance

feeding animals (e.g., pigs and chickens) antibiotics to prevent illnesses that reduce yield. Recently, some countries have banned the use of human antibiotics in animal feed and there has been a major decline in the frequency of R$^+$ bacteria carried by farm animals.

R-plasmids have similar characteristics, although they belong to a wide range of compatibility groups. Most are of moderate to large size and present in 1–2 copies per host cell. Most are self-transmissible at a low frequency, although de-repressed mutants showing high transfer frequency are sometimes found. The original F-plasmid is such a "naturally-occurring" mutant plasmid that developed an increased ability to move from bacterial cell to bacterial cell. In addition, many carry resistances to one or more antibiotics and/or toxic heavy metals and may also carry genes for colicins, virulence factors, etc.

Antibiotic-resistant mutants of bacteria may be easily isolated in the laboratory. However, the mechanism of resistance in such chromosomal mutants is usually quite distinct from that of plasmid-borne resistance. The chromosomal mutations usually alter the cell component that is the target of antibiotic action, which often causes detrimental side effects. Plasmid-borne resistance generally avoids altering vital cell components, and instead, inactivates the antibiotic or actively pumps it out of the cell. Occasionally plasmids do provide an altered (but still functional) target component. Several of the resistance genes originally found on plasmids have been used in genetic engineering. Antibiotic resistance allows scientists to screen for cells that contain a plasmid and kill all the cells that do not (see Ch. 7). Chloramphenicol, kanamycin/neomycin, tetracycline, and ampicillin resistance genes are the most widely used in laboratories.

> Plasmid-borne resistance mechanisms usually inactivate or expel the antibiotic, rather than alter vital cell components.

4.2. Resistance to Beta-Lactam Antibiotics

The β-**lactam** family includes the **penicillins** and **cephalosporins** and is the best-known and most widely-used group of antibiotics. All contain the β-lactam structure, a four-membered ring containing an amide group, which reacts with the active site of enzymes involved in building the bacterial cell wall. Crosslinking of the peptidoglycan is prevented, so causing disintegration of the cell wall and death of the bacteria. Since peptidoglycan is unique to bacteria, penicillins and cephalosporins have almost no side effects in humans, apart from occasional allergies.

> Penicillin and its relatives are the most widely-used family of antibiotics.

Resistance plasmids carry a gene encoding the enzyme β-**lactamase**, which destroys the antibiotic by opening the β-lactam ring (Fig. 20.11). Most β-lactamases prefer either penicillins or cephalosporins, though a few attack both antibiotics equally well. Resistance to **ampicillin**, a popular type of penicillin, is widely used in molecular biology, especially for selecting plasmids during cloning procedures (see Ch. 7). The same gene is referred to as either *amp* (for ampicillin) or *bla* for β-lactamase. Certain strains of *Pseudomonas* carrying the R-plasmid RP1 that encodes a high-activity, broad-spectrum β-lactamase can actually grow on ampicillin as a sole carbon and energy source!

> Penicillin and related antibiotics are destroyed by the enzyme beta-lactamase.

A vast number of penicillin and cephalosporin derivatives have been made by the pharmaceutical industry. Some of these are much less susceptible to breakdown by β-lactamase. Their development has in turn led to the emergence of altered and improved β-lactamases among bacteria-carrying R-plasmids. Another approach is to administer a mixture of a β-lactam antibiotic plus a β-lactam analog that

---

ampicillin   A widely-used antibiotic of the penicillin group
beta-lactams or β-lactams   Family of antibiotics that inhibit crosslinking of the peptidoglycan of the bacterial cell wall; includes penicillins and cephalosporins
beta-lactamase or β-lactamase   Enzyme that inactivates β-lactam antibiotics such as ampicillin by cleaving the lactam ring
bla **gene**   Gene encoding β-lactamase thereby providing resistance to ampicillin. Same as *amp* gene
cephalosporins   Group of antibiotics of the β-lactam type that inhibit crosslinking of the peptidoglycan of the bacterial cell wall
penicillins   Group of antibiotics of the β-lactam type that inhibit crosslinking of the peptidoglycan of the bacterial cell wall

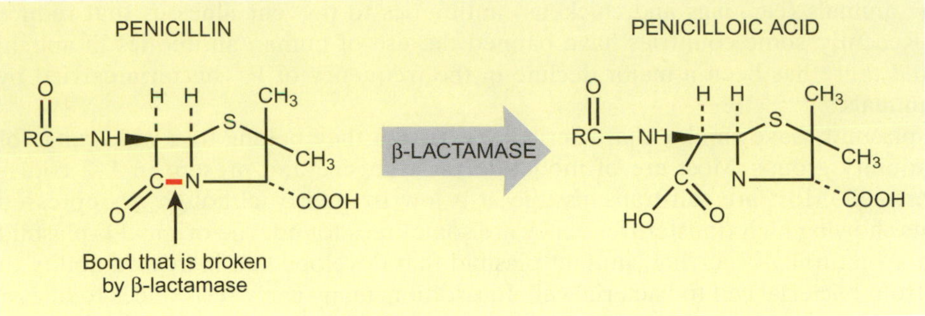

FIGURE 20.11
Inactivation of Penicillin by β-Lactamase

Penicillin is an antibiotic that attacks the cell wall of bacteria, preventing the cells from growing or dividing. The antibiotic has a four-membered β-lactam ring that binds to the active site of the enzymes that assemble the cell wall. The enzyme β-lactamase cleaves the β-lactam ring of penicillin (red bond). The penicillin is inactivated.

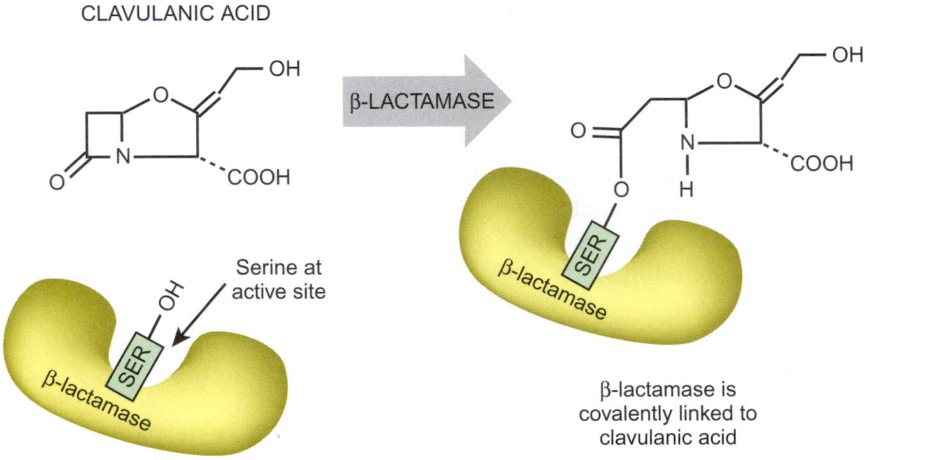

PENICILLIN → β-LACTAMASE → PENICILLOIC ACID

Bond that is broken by β-lactamase

FIGURE 20.12
Inactivation of β-Lactamase by Clavulanic Acid

In order to inactivate β-lactamase, analogs of penicillin such as clavulanic acid are added along with the antibiotic. Clavulanic acid has a four-membered ring similar to penicillin. Consequently, β-lactamase will bind and cleave this ring. When this happens, clavulanic acid is covalently bound to β-lactamase rendering it useless against penicillin. Added penicillin can now kill the bacteria, even though they contain the resistance gene.

CLAVULANIC ACID → β-LACTAMASE

Serine at active site
β-lactamase

β-lactamase is covalently linked to clavulanic acid

Chloramphenicol is inactivated by addition of acetyl groups.

inhibits β-lactamase. **Clavulanic acid** and its derivatives bind to β-lactamases and react forming a covalent bond to the amino acids in the active site that kills the enzyme (Fig. 20.12).

4.3. Resistance to Chloramphenicol

Chloramphenicol, streptomycin, and kanamycin are all antibiotics that inhibit protein synthesis by binding to the bacterial ribosomes. The difference in mechanism between resistances due to chromosomal mutations as opposed to plasmid-borne genes is especially notable for these antibiotics. Chromosomal mutants usually have altered ribosomes that no longer bind the antibiotic. Not surprisingly such mutations often cause slower or less accurate protein synthesis and the cells grow poorly. In contrast, plasmid-borne resistance to these antibiotics usually involves chemical attack on the antibiotic itself by specific enzymes encoded by the plasmid.

Chloramphenicol binds to the 23S rRNA of the large subunit of the bacterial ribosome and inhibits the peptidyl transferase reaction (see Ch. 13). R-plasmids protect the bacteria by producing the enzyme **chloramphenicol acetyl transferase (CAT)**. CAT transfers two acetyl groups from acetyl CoA to the side chain of chloramphenicol. This prevents binding of the antibiotic to the 23S rRNA (Fig. 20.13). Replacement of the terminal OH of chloramphenicol with fluorine results in non-modifiable yet still antibacterially-active derivatives. There are two major groups of chloramphenicol acetyl

CAT   Chloramphenicol acetyl transferase
chloramphenicol   Antibiotic that binds to 23S rRNA and inhibits protein synthesis
chloramphenicol acetyl transferase (CAT)   Enzyme that inactivates chloramphenicol by adding acetyl groups
clavulanic acid   And its derivatives bind to β-lactamases and react forming a covalent bond to the protein that kills the enzyme

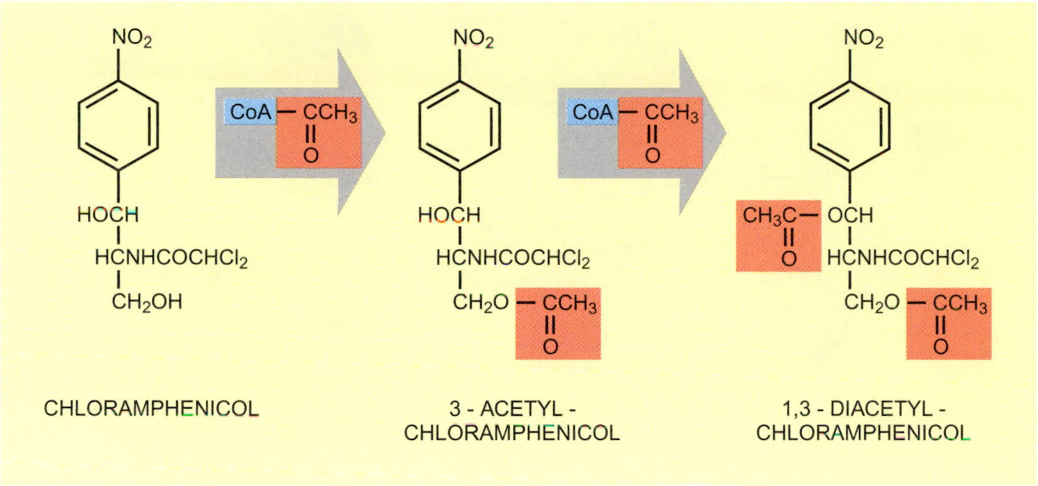

transferase: one from gram-positive and one from gram-negative bacteria. The two groups differ greatly from each other except for the chloramphenicol-binding region.

4.4. Resistance to Aminoglycosides

The **aminoglycoside** family of antibiotics includes **streptomycin, kanamycin, neomycin,** tobramycin, gentamycin, and a host of others. Aminoglycosides consist of three (sometimes more) sugar rings, at least one of which (and usually two or three) has amino groups attached. They inhibit protein synthesis by binding to the small subunit of the ribosome (see Ch. 13). Plasmid-borne resistance is due to inactivation of the antibiotics. Several alternatives exist, including modification by phosphorylation of OH groups, adenylation (i.e., addition of adenosine monophosphate (AMP)) of OH groups or acetylation of NH_2 groups. Adenosine triphosphate (ATP) is used as a source of phosphate and AMP groups, whereas acetyl-CoA is the acetyl donor (Fig. 20.14).

Modified aminoglycosides no longer inhibit their ribosomal target sites. There are many different aminoglycosides and a correspondingly wide range of modifying enzymes. The *npt* gene (**neomycin phosphotransferase**) is the most widely used and provides resistance to both kanamycin and the closely related neomycin. Aminoglycosides are made by bacteria of the *Streptomyces* group, which are mostly found in soil. These organisms need to protect themselves against the antibiotics they produce. Probably, therefore, the aminoglycoside-modifying enzymes came originally from the same *Streptomyces* strains that make these antibiotics.

Amikacin is a more recent derivative of kanamycin A in which the amino group on the middle ring that gets acetylated is blocked with a hydroxybutyrate group. This made amikacin resistant to all modifying enzymes except one obscure N-acetyl transferase. However, evolution moves on and an enzyme that phosphorylates amikacin has already appeared in some bacterial strains!

4.5. Resistance to Tetracycline

Tetracycline binds to the 16S rRNA of the small subunit and also inhibits protein synthesis. However, the mechanism of resistance is quite different from

FIGURE 20.13
Inactivation of Chloramphenicol

The side chain of chloramphenicol has two OH groups that are important for binding to the bacterial ribosomes. Chloramphenicol acetyl transferase, produced by R-plasmids, catalyzes the addition of two acetyl groups to chloramphenicol. The enzyme uses acetyl-CoA as a source for the acetyl groups. The resulting 1,3-diacetyl-chloramphenicol can no longer bind to the ribosomes.

Aminoglycoside antibiotics are inactivated by addition of phosphate, AMP, or acetyl groups.

---

aminoglycosides Family of antibiotics that inhibit protein synthesis by binding to the small subunit of the ribosome; includes streptomycin, kanamycin, neomycin, tobramycin, gentamycin, and many others
kanamycin Antibiotic of the aminoglycoside family that inhibits protein synthesis
neomycin Antibiotic of the aminoglycoside family that inhibits protein synthesis
***npt* gene** Gene for neomycin phosphotransferase. Provides resistance against the antibiotics kanamycin and neomycin
neomycin phosphotransferase Enzyme that inactivates the antibiotics kanamycin and neomycin by adding a phosphate group
streptomycin Antibiotic of the aminoglycoside family that inhibits protein synthesis
tetracycline Antibiotic that binds to 16S ribosomal RNA and inhibits protein synthesis

FIGURE 20.14
Inactivation of
Aminoglycoside Antibiotics

Much like chloramphenicol, members of the aminoglycoside family are inactivated by modification. One member, kanamycin B, can be modified by a variety of covalent modifications, such as phosphorylation, acetylation, or adenylation. A variety of bacterial enzymes make these modifications to prevent kanamycin B from attaching to the small ribosomal subunit.

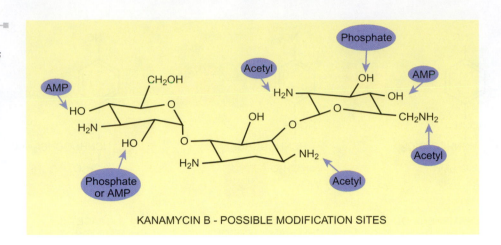

KANAMYCIN B - POSSIBLE MODIFICATION SITES

chloramphenicol and aminoglycosides. Rather than inactivating tetracycline by modification, R-plasmids produce proteins that pump the antibiotic out of the bacteria. Tetracycline actually binds to both prokaryotic and eukaryotic ribosomes. Bacteria are more sensitive than animal cells because tetracyclines are actively taken up by bacterial cells, but not by eukaryotic cells. In fact, eukaryotic cells naturally actively export tetracyclines. In tetracycline-resistant bacteria, the antibiotic is actively taken into the cell, but then pumped out again. As there is no similarity between tetracycline and any known transportable nutrients, the purpose of the bacterial transport system that takes up tetracycline and its mechanism of operation are still baffling. However, the Tet-resistance protein is part of a large family of sugar transporter proteins, and may have evolved from recognizing sugar to recognizing tetracycline.

> Tetracycline resistance is due to energy-driven export of the antibiotic.

Plasmid-encoded tetracycline resistance is typically two-fold. A basal constitutive level of resistance protects bacteria by 5–10 fold relative to sensitive bacteria. In addition, exposure to tetracycline induces a second higher resistance level. Both resistance levels are due to production of proteins that are found in the cytoplasmic membrane and actively expel tetracycline from the cell. Tetracycline enters the cell as the protonated form by an active transport system. Inside the cell it binds Mg^{2+}. The Tet-resistance protein uses energy to expel the Tet-Mg^{2+} complex by proton antiport (Fig. 20.15).

4.6. Resistance to Sulfonamides and Trimethoprim

Both sulfonamides and trimethoprim are antagonists of the vitamin folic acid. The reduced form of folate, tetrahydrofolate, is used as a co-factor by enzymes that synthesize methionine, adenine, thymine, and other metabolites whose synthesis involves adding a one-carbon fragment. **Sulfonamides** are completely-synthetic antibiotics and are analogs of *p*-aminobenzoic acid (Fig. 20.16), a precursor of the vitamin folic acid. Sulfonamides inhibit dihydropteroate synthetase, an enzyme in the synthetic pathway for folate. **Trimethoprim** is an analog of the pterin ring portion of tetrahydrofolate. It inhibits dihydrofolate reductase, the bacterial enzyme that converts dihydrofolate to tetrahydrofolate. Animal cells rely on folate in their food and so these antibiotics are effective against bacteria that normally manufacture their own tetrahydofolate without having any effect on animals or humans.

Plasmid-mediated resistance to both sulfonamides and trimethoprim involves synthesis of folic acid biosynthetic enzymes that no longer bind the antibiotic.

---

sulfonamides   Synthetic antibiotics that are analogs of *p*-aminobenzoic acid, a precursor of the vitamin folic acid. Sulfonamides inhibit dihydropteroate synthetase

trimethoprim   Antibiotic that is an analog of the pterin ring portion of the folate co-factor. It inhibits dihydrofolate reductase

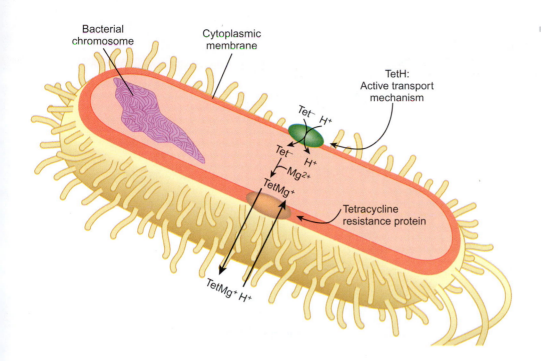

■ **FIGURE 20.15**
Expulsion of Tetracycline from Resistant Bacteria

The bacterial chromosome contains the gene for TetH, a protein that takes tetracycline from the environment and actively pumps the antibiotic and a proton into the cell. Once inside the cell, tetracycline complexes with Mg^{2+}, and may bind to the ribosome. In bacterial cells with an R-plasmid for tetracycline, another transport protein, called the tetracycline resistance protein, is manufactured. This protein allows a proton to enter the cell to produce energy for export of the Tet-Mg^+ complex.

R-plasmid-encoded dihydropteroate synthetase has the same affinity for *p*-aminobenzoic acid as the chromosomal enzyme but is resistant to sulfonamides. Similarly, R-plasmid-encoded dihydrofolate reductase is resistant to trimethoprim. Sulfonamides plus trimethoprim are often used in combination for double blockade of the folate pathway. As a result, sulfonamide and trimethoprim resistance are often found together on the same R-plasmid.

> Resistance to trimethoprim and sulfonamides is due to replacement of the target enzyme.

5. Plasmids May Provide Aggressive Characters

The first plasmids drew attention because they provided their host bacteria with resistance to antibiotics. Other plasmids protect bacteria against heavy metal toxicity. However, many plasmids are known that confer aggressive, rather than defensive properties. These may be subdivided into two broad groups. Bacteriocin plasmids encode toxic proteins used by certain strains of bacteria to kill related bacteria. **Virulence plasmids** carry genes for a variety of characters deployed by bacteria that infect higher organisms, both plants and animals, including humans.

Generally speaking, bacteria are most likely to attack their close relatives. The reason is that the more closely related they are, the more likely two strains of bacteria will compete for the same resources. Proteins made by bacteria to kill their relatives are known generally as **bacteriocins**. Particular bacteriocins are named after the species that makes them. So, for example, many strains of *E. coli* deploy a wide variety of **colicins**, intended to kill other strains of the same species. (Since most work has been done on *E. coli*, bacteriocins from other bacteria are often referred to as colicins, although this is not strictly correct.) Surveys suggest that 10–15% of enteric bacteria make bacteriocins.

On several occasions, *Yersinia pestis*, the bacterium that causes Black Death (bubonic plague), has wiped out a third of the human population of Europe, and

---

bacteriocin Toxic protein made by bacteria to kill closely-related bacteria
colicin Toxic protein or bacteriocin made by *Escherichia coli* to kill closely-related bacteria
virulence plasmid Plasmid that carries genes for virulence factors that play a role in bacterial infection

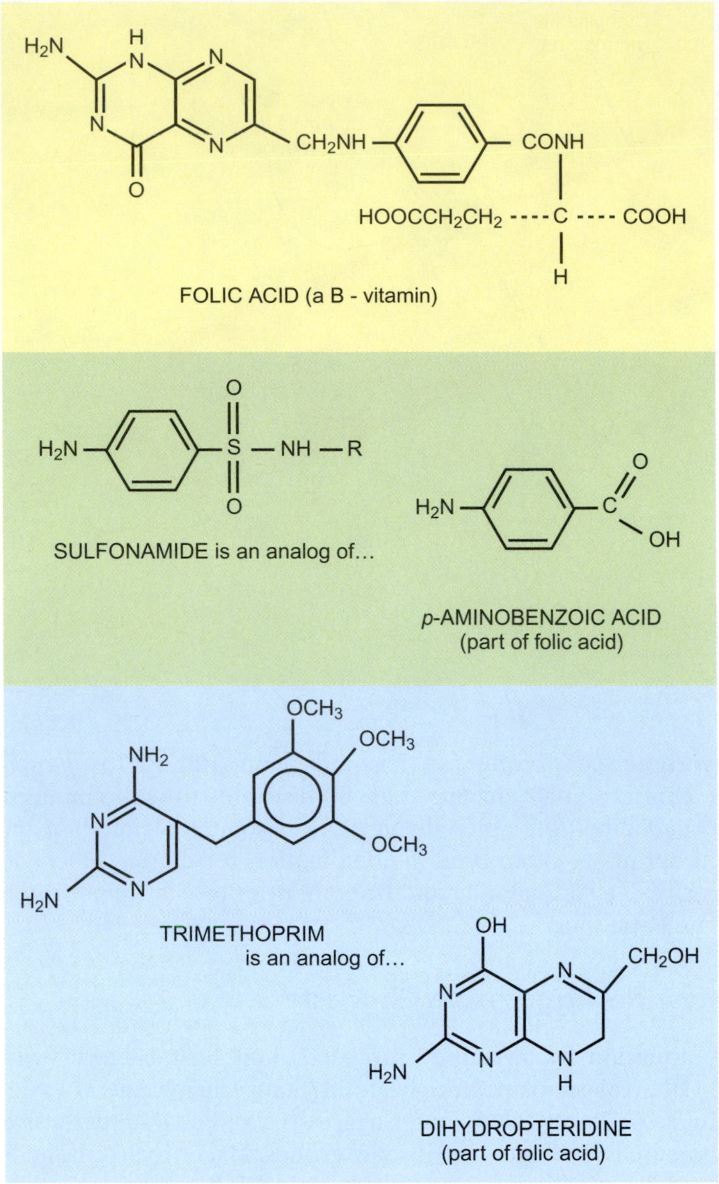

FIGURE 20.16
Trimethoprim, Sulfonamides, and the Folate Co-factor

Bacterial cells make folic acid, whereas animal cells do not. The antibiotic sulfonamide is an analog of the *p*-aminobenzoic acid portion of folic acid. Trimethoprim is an analog of the dihydropteridine portion of folic acid. Both trimethoprim and sulfonamide bind to the biosynthetic enzymes and prevent synthesis of folic acid from its precursors.

FOLIC ACID (a B - vitamin)

SULFONAMIDE is an analog of...

p-AMINOBENZOIC ACID (part of folic acid)

TRIMETHOPRIM is an analog of...

DIHYDROPTERIDINE (part of folic acid)

Many bacteria make toxic proteins—bacteriocins—to kill closely-related bacteria that compete for the same resources.

Bacteriocins are usually encoded on plasmids. These provide the starting point for many genetic-engineering vectors.

probably most of Africa and Asia. The virulence factors required for infection are carried on a series of plasmids (see below). As if this was not enough, *Yersinia pestis* also makes bacteriocins, called pesticins in this case, which are designed to kill competing strains of its own species.

The ability to make bacteriocins is usually due to the presence of a plasmid in the producer cell. The best known examples are the three related **ColE plasmids** of *E. coli*, ColE1 (Fig. 20.17), ColE2, and ColE3. These are small plasmids that exist in 50 or more copies per cell and have been used to derive many of the cloning vectors used in genetic engineering (see Ch. 7). These cloning vectors have the actual colicin genes removed. A variety of other colicin plasmids also occur, including the ColI and ColV plasmids. These are large single-copy plasmids and are usually transferable from one strain of *E. coli* to another. Many ColI and ColV plasmids also carry genes for antibiotic resistance.

ColE plasmid Small multicopy plasmid that carries genes for colicins of the E group. Used as the basis of many widely-used cloning vectors

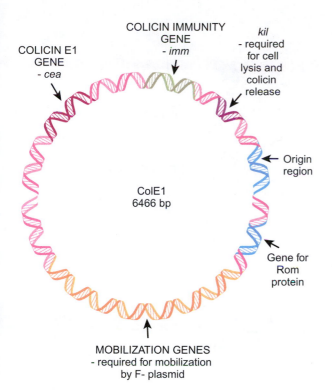

COLICIN E1 GENE - *cea*

COLICIN IMMUNITY GENE - *imm*

kil - required for cell lysis and colicin release

ColE1 6466 bp

→ Origin region

→ Gene for Rom protein

MOBILIZATION GENES - required for mobilization by F- plasmid

■ **FIGURE 20.17**
ColE1 is an Example of a Colicin Plasmid

The ColE1 plasmid of *E. coli* carries genes for colicin E1 (*cea*), immunity to colicin E1 (*imm*), and the *kil* gene, required for liberation of colicin from the producer cell. The Rom gene is involved in copy-number control as discussed above. ColE1 is the basis for many plasmids used in genetic engineering. The mobilization genes allow ColE1 to be transferred from cell-to-cell during conjugation mediated by the F-plasmid.

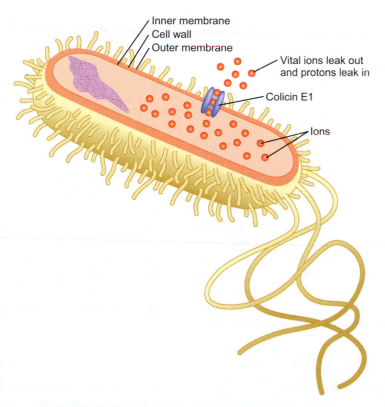

Inner membrane
Cell wall
Outer membrane

Vital ions leak out and protons leak in

Colicin E1

Ions

■ **FIGURE 20.18**
Some Colicins Damage the Cell Membrane

When colicin E1 protein attacks a bacterial cell, it punctures a hole through the outer membrane, cell wall, and inner membrane. The hole allows protons to leak into the bacteria and vital ions to leak out. A single channel abolishes energy generation.

5.1. Most Colicins Kill by One of Two Different Mechanisms

The Col plasmids allow the strains of *E. coli* that possess them to kill other related bacteria. There are two basic approaches to this. The first is to damage the victim's cell membrane. A gene on the ColE1 plasmid encodes the colicin E1 protein that inserts itself through the membrane of the target cell and creates a channel allowing vital cell contents, including essential ions, to leak out and protons to flood into the cell (Fig. 20.18). The influx of protons collapses the proton motive force. The energy

FIGURE 20.19
Colicin Immunity System

In order to protect itself, a colicin-making cell also produces an immunity protein (right). This protein is also encoded by the colicin plasmid. It blocks the active site of the colicin thus preventing the cell from killing itself. The immunity protein is specific and only inhibits one type of colicin. If a cell lacks immunity protein, the colicin is able to kill the cell (left).

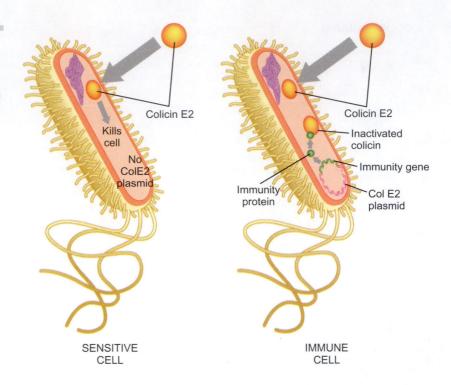

SENSITIVE CELL

IMMUNE CELL

The two most popular modes of action for bacteriocins are: a) damaging the cell membrane or b) destroying nucleic acids.

derived from the proton motive force drives the production of ATP and the uptake of many nutrients, without which the bacteria quickly die. A single molecule of colicin E1 that penetrates the membrane is enough to kill the target cell. Colicin I and colicin V operate by a similar mechanism.

Colicin M and Pesticin A1122 destroy the peptidoglycan of the cell wall rather than puncturing the cytoplasmic membrane. These colicins need to penetrate only as far as the outer surface of the cytoplasmic membrane (i.e., the site of peptidoglycan assembly). Without the peptidoglycan, the bacterial cell loses shape and eventually bursts. Pesticin A1122 is made by *Yersinia pestis* and kills *Y. pseudotuberculosis*, *Y. enterocolitica*, plasmid-free *Y. pestis*, and many strains of *E. coli* (although curiously not *E. coli* K12).

The second approach is to degrade the nucleic acids of the victim. The ColE2 and ColE3 plasmids both encode nucleases, enzymes that degrade nucleic acids. The colicin E2 and E3 proteins are very similar over their N-terminal region and as a result they share the same receptor on the surface of sensitive bacteria. They differ in the C-terminus and have different nucleic acid targets. Colicin E2 is a deoxyribonuclease that cuts up the chromosome of the target cell. Colicin E3 is a ribonuclease that snips the 16S rRNA of the small ribosomal subunit at a specific sequence, releasing a fragment of 49 nucleotides from the 3' end. This abolishes protein synthesis and though much more specific than colicin E2, is just as lethal. Again, a single colicin molecule that enters the victim is enough to kill the target cell.

5.2. Bacteria are Immune to Their Own Colicins

Those bacterial cells producing a particular colicin are immune to their own type, but not to other types of colicin. Immunity is due to specific **immunity proteins** that bind to the corresponding colicin proteins and cover their active sites (Fig. 20.19). For example, the ColE2 plasmid carries genes for both colicin E2 and a soluble immunity protein that binds colicin E2. This immunity protein does not protect against any

immunity protein Protein that provides immunity. In particular bacteriocin immunity proteins bind to the corresponding bacteriocins and render them harmless

other colicin, including the closely-related colicin E3. Immunity to membrane active colicins is due to a plasmid-encoded inner membrane protein that blocks the colicin from forming a pore in the host cell. For example, the Ia immunity protein protects membranes against colicin Ia but not against the closely-related colicin Ib even though colicins Ia and Ib share the same receptor, have the same mode of action, and have extensive sequence homology. Although the immune systems of animals are much more complex, the concept of immunity is based on the ability of immune system proteins to recognize and neutralize specific alien or hostile molecules.

> Bacteria that make bacteriocins also make immunity proteins to protect themselves.

5.3. Colicin Synthesis and Release

In a population of ColE plasmid-carrying bacteria, most cells do not produce colicin. Every now and then an occasional cell goes into production and manufactures large amounts of colicin. It then bursts and releases the colicin into the medium. Note that it is the burst and release mechanism that kills the producer cell, not the colicin. All sensitive bacteria in the area are wiped out, but those with the ColE plasmid have immunity protein and survive.

> Bacteriocin production is often a suicidal process.

About 1 in 10,000 cells actually produce colicin in each generation. Thus, release of colicin E is a communal action in the sense that a small minority of producer cells sacrifice themselves so that their relatives carrying the same ColE plasmid can take over the habitat. Colicin-E production involves expression of two plasmid genes, *cea* (colicin protein) and *kil* (lysis protein). LexA, the repressor of the SOS DNA repair system (Ch. 23), normally represses these genes. Thus, colicin production is induced by DNA damage and those cells that sacrifice themselves were probably injured anyway. (Note that many lysogenic bacteriophage are also induced by DNA damage monitored via the SOS system (see Ch. 21).)

Not all colicins are produced by the suicidal mechanism. Many colicins made by large single-copy plasmids (e.g., colicin V, colicin I) are apparently made continuously in smaller amounts. These colicins tend to remain attached to the surface of the producer cell rather than being released as freely-soluble proteins, like the E colicins. When the producer bumps into a sensitive bacterium the colicin may be transferred, with lethal results.

5.4. Virulence Plasmids

Virulence plasmids help bacteria infect humans, animals, or even plants, by a variety of mechanisms. Some **virulence factors** are toxins that damage or kill animal cells, others help bacteria to attach to and invade animal cells (Fig. 20.20), whereas yet others protect bacteria against retaliation by the immune system.

> Many pathogenic bacteria carry genes for virulence on plasmids or other mobile genetic elements.

Although most strains of *E. coli* are harmless, occasional rogue strains cause disease. These pathogenic *E. coli* generally rely on plasmid-borne virulence factors. A variety of toxins are found in different pathogenic *E. coli* strains, including heat-labile **enterotoxin** (resembles **choleratoxin**), heat-stable enterotoxin, **hemolysin** (lyses red blood cells), and Shiga-like toxin (similar to the toxin of dysentery-causing *Shigella*). There is a similar variety of **adhesins** or "**colonization factors**," proteins that enable bacteria to stick to the surface of animal cells. Adhesins form filaments that vary in length and thickness, but generally resemble pili. Consequently, the symptoms and severity of infection by *E. coli* vary greatly.

---

adhesin   Protein that enables bacteria to attach themselves to the surface of animal cells. Same as colonization factor
choleratoxin   Type of toxin made by *Vibrio cholerae* the cholera bacterium
colonization factor   Protein that enables bacteria to attach themselves to the surface of animal cells. Same as adhesin
enterotoxins   Types of toxin made by enteric bacteria including some pathogenic strains of *E. coli*
hemolysin   Type of toxin that lyses red blood cells
virulence factors   Proteins that promote virulence in infectious bacteria. Include toxins, adhesins, and proteins protecting bacteria from the immune system

FIGURE 20.20
Toxins and Adhesins

Bacteria are able to attack animal cells by attaching to the cellular membrane and releasing toxins. The bacteria contain plasmids that encode adhesins, which are protein filaments able to recognize and attach to cell-surface receptors found on animal cells. Once attached, the bacteria secrete toxins, which can penetrate the animal cell membrane and kill the cell.

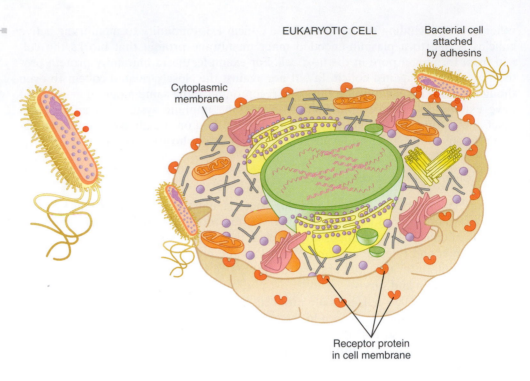

EUKARYOTIC CELL

Bacterial cell attached by adhesins

Cytoplasmic membrane

Receptor protein in cell membrane

Other enteric bacteria, such as *Salmonella typhi* (typhoid) and *Yersinia pestis* (bubonic plague), cause severe infections. They also carry virulence plasmids. In *Salmonella* the majority of the virulence genes are on the chromosome, but there are also ones that are plasmid-borne. In contrast, in *Yersinia* several plasmids carry the bulk of the virulence genes. In addition to toxins and adhesins, these "professional" pathogens possess more sophisticated virulence factors that protect against host defenses. Although plasmids have been investigated most intensively in enteric bacteria, it is clear that virulence in many other bacteria often depends on at least some plasmid-borne genes.

6. Ti Plasmids Are Transferred from Bacteria to Plants

Although the F-plasmid of *E. coli* is limited in its host range to a few enteric bacteria, it can actually promote DNA transfer between *E. coli* and yeast! Similarly, broad host range plasmids of the IncP, IncQ, and IncW incompatibility groups can mobilize DNA from gram-negative bacteria into both gram-positive bacteria and yeast. In both cases, the range of species in which the plasmid can survive and replicate is much smaller than the range of species to which DNA may be transferred. Therefore, many plasmids are degraded or destroyed after they are transferred to an incompatible cell. Some DNA mobilized in this manner may survive if it is recombined with the host chromosome or resident plasmids.

> The Ti plasmids can mediate transfer of DNA from bacteria to plant cells.

The greatest versatility in plasmid transfer is shown by the highly-specialized **Ti plasmids** (Ti = tumor-inducing) that allow certain bacteria to insert DNA into the nucleus of plant cells. The Ti plasmid is carried by soil bacteria of the *Agrobacterium* group, in particular *A. tumefaciens*, and confers the ability to infect plants and produce tumors, inside which the bacteria grow and divide happily. This results in tumor-like swellings on the stems of infected plants, a condition known as "**crown gall disease**." The related Ri plasmid is carried by *Agrobacterium rhizogenes*, which infects roots and causes hairy root disease.

---

crown gall Type of tumor formed on plants due to infection by *Agrobacterium* carrying a Ti plasmid
Ti plasmid Tumor-inducing plasmid. Plasmid that is carried by soil bacteria of the *Agrobacterium* group and confers the ability to infect plants and produce tumors

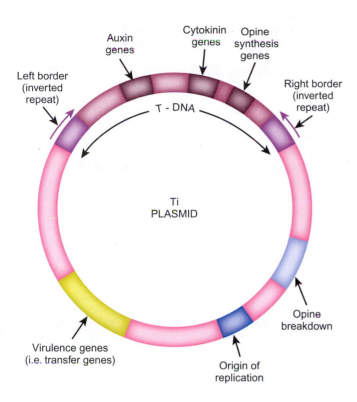

Left border (inverted repeat)

Auxin genes

Cytokinin genes

Opine synthesis genes

Right border (inverted repeat)

T - DNA

Ti PLASMID

Virulence genes (i.e. transfer genes)

Origin of replication

Opine breakdown

■ FIGURE 20.21
Structure of the Ti Plasmid

The Ti plasmid of *Agrobacterium* has several regions. The T-DNA region is flanked by two inverted repeats and contains genes for auxin and cytokinin, which induce the plant cells to grow, and genes for opine synthesis, a carbon source for *Agrobacterium*. This region is transferred into the plant cell by the expression of the transfer genes found on the other part of the Ti plasmid. The Ti plasmid also has an origin of replication and genes for opine breakdown.

Agrobacterium is attracted by chemicals, such as acetosyringone, which are released by wounded plants. It then enters via the wound and transfers a portion of the Ti plasmid into the plant cell by a mechanism similar to bacterial conjugation. A slight abrasion that is trivial to the health of the plant is of course sufficient for the entry of a microorganism. The result is a crown gall tumor that provides a home for the *Agrobacterium* at the expense of the plant.

The Ti plasmid consists of several regions (Fig. 20.21), but only one segment, the **T-DNA** (tumor-DNA), is actually transferred into the plant cell, where it enters the nucleus. The T-DNA is flanked by 25 bp inverted repeats. Any DNA included within these repeats will be transferred into the plant cell. Consequently, Ti plasmids have been widely used in the genetic engineering of plants. The virulence genes on the plasmid are responsible for cell-to-cell contact and transfer of the T-DNA but do not themselves enter the plant cells.

Acetosyringone, which attracts the bacteria to the wounded plant, also induces the virulence genes, thus facilitating the transfer of the T-DNA region (Fig. 20.22). Acetosyringone binds to VirA protein in the *Agrobacterium* membrane. This activates VirG, which in turn switches on the other *vir* genes, including *virD* and *virE2*. VirD makes a single-stranded nick in the Ti plasmid at the left border of the T-DNA and the T-DNA unwinds from the cut site. The single-stranded T-DNA is bound by VirE2 protein and unwinding stops at the right border. The Ti plasmid then replicates by a rolling circle mechanism as the single-stranded T-DNA region enters the plant cell. The mechanism resembles bacterial conjugation and the "virulence" genes of the Ti-plasmid are equivalent to the *tra* genes of other plasmids. The T-DNA is covered by the VirE2 protein made by *Agrobacterium* and also, when it enters the plant cell, by the plant VIP1 protein. The VirF protein enters the plant nucleus along with the T-DNA and removes these two proteins from the DNA (see Focus on Relevant Research). The T-DNA then integrates at random into one of the plant chromosomes. Overall, this results in DNA transfer from the bacteria into the plant cells.

Bacteria-carrying Ti plasmids infect plants and cause the formation of tumors.

Only part of the Ti plasmid enters the plant cell, where it integrates into the plant chromosomes.

T-DNA (tumor-DNA) Region of the Ti plasmid that is transferred into the plant cell nucleus

FIGURE 20.22
Formation of Tumor by Agrobacterium

Agrobacterium are attracted to an injured region of a plant by sensing molecules of acetosyringone. The bacteria enter the plant through the open wound, and begin colonizing the area. The plant cells are stimulated to divide and a tumor forms around the bacteria.

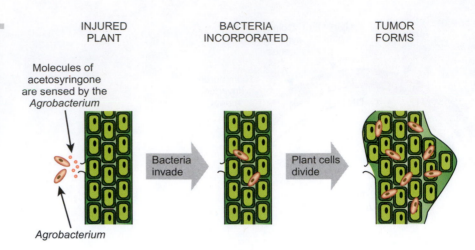

INJURED PLANT

Molecules of acetosyringone are sensed by the *Agrobacterium*

Agrobacterium

Bacteria invade

BACTERIA INCORPORATED

Plant cells divide

TUMOR FORMS

FOCUS ON RELEVANT RESEARCH

Zaltsman A, Krichevsky A, Loyter A, and Citovsky V (2010) *Agrobacterium* induces expression of a host F-box protein required for tumorigenicity. Cell Host Microbe 7: 197–209.

When the T-DNA from the *Agrobacterium* Ti plasmid enters the plant cell it is protected by both a bacterial protein, VirE2, and also a plant protein, VIP1. These must be removed when the T-DNA enters the nucleus in order to allow integration of the T-DNA into the plant genome. This is accomplished by the F-box protein VirF, which assembles a complex of host proteins that degrade VirE2 and VIP1.

This paper shows that VirF is not needed in all plant species. In some plants *Agrobacterium* induces expression of a plant F-box protein (VBF) that replaces VirF. The *vbf* gene can be inserted into *Agrobacterium* that is deleted for *virF* and VBF can functionally replace the bacterial VirF protein. VBF is normally induced by bacterial infection of plants. The authors suggest that *Agrobacterium* has taken over a plant-defense response.

Once inserted, the genes in the T-DNA are switched on. The enzymes they encode synthesize two plant hormones, **auxin** and **cytokinin**. Auxin makes plant cells grow bigger and cytokinin makes them divide. When this happens rapidly in the absence of normal cell differentiation, the result is a tumor (Figs. 20.23 and 20.24).

The T-DNA also carries genes that subvert the plant cell into making opines. These are unusual nutrient molecules that are made at the expense of the plant cell but can only be used by bacteria that carry special genes for opine breakdown. The genes for opine degradation are found on the part of the Ti plasmid that does not enter the plant cell. So the *Agrobacterium* can grow by using opines but the plant cannot use them. Other bacteria that might infect the plant are also excluded as they do not have opine breakdown genes either.

Modified Ti plasmids are widely used in genetic engineering of plants.

Modified Ti plasmids are widely used in the genetic engineering of plants. The genes for plant hormones and opine synthesis are removed and the genes to be transferred into the plant are inserted in their place. In practice, *Agrobacterium* carrying an engineered Ti plasmid is used to transfer genes of interest into plants using plant tissue culture.

In addition to inserting external genes into plants the Ti plasmid system may be used for analysis of plant gene function. Insertion of T-DNA into the plant chromosome may disrupt a plant gene if insertion occurs into the coding sequence (or essential regulatory sequences). The model plant, *Arabidopsis thaliana*, has been used to

auxin   Plant hormone that induces plant cells to grow bigger
cytokinin   Plant hormone that induces plant cells to divide

■ **FIGURE 20.23**
Crown Gall Tumor Caused by Agrobacterium

A crown gall tumor formed by *Agrobacterium* is shown on a tree trunk. *(Credit: E.R. Degginger, Photo Researchers Inc.)*

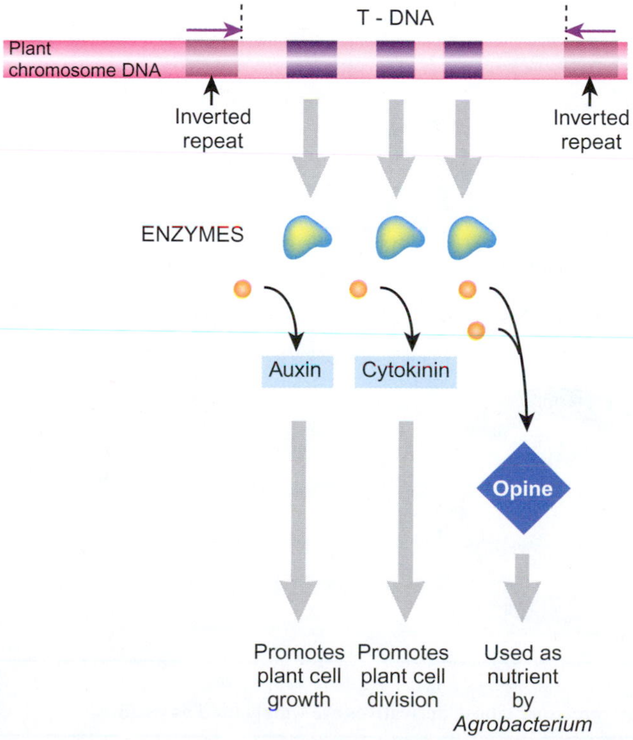

FIGURE 20.24 diagram

T - DNA

Plant chromosome DNA

Inverted repeat · · · · Inverted repeat

ENZYMES

Auxin   Cytokinin

Opine

Promotes plant cell growth · Promotes plant cell division · Used as nutrient by *Agrobacterium*

■ **FIGURE 20.24**
Expression of Genes on T-DNA

Survival of *Agrobacteria* in the plant requires space for them to grow and a carbon source to provide energy. The genes of the T-DNA region trick the plant cell into providing these factors. The genes for auxin and cytokinin are growth factors that induce the plant cells to grow at the site of infection, providing the space. Opine is a carbon source for the bacteria, providing a constant food supply.

generate a set of gene knockouts by random insertion of T-DNA. The locations of several 100,000 such insertions have been determined. These include insertions into almost all of the estimated 27,000 genes of *Arabidopsis*. These insertions may be used to investigate the functions of the inactivated genes by comparing the knockout mutants with the parental wild-type plant.

> ### Box 20.3 The Ti Plasmid Can Enter Animal and Yeast Cells
>
> Although Ti plasmids are normally transferred from *Agrobacterium* to plant cells, the Ti plasmid is capable of entering the cells of other eukaryotes, at least in the laboratory. Yeast, some filamentous fungi, and the cultivated mushroom *Agaricus* have all successfully received the Ti plasmid by conjugation from *Agrobacterium*. The T-DNA appears to integrate at random into the fungal chromosomes as it does in plants. Although no actual animals have yet been tested, the Ti plasmid entered cultured Human HeLa cells and the T-DNA integrated into the human chromosomes.
>
> Whether the Ti plasmid can be transferred from *Agrobacterium* to eukaryotes other than plants in the natural environment is unknown. Laboratory data suggest that if this does happen it will be at much lower frequency than to the "natural" plant hosts.

7. The 2µ Plasmid of Yeast

Plasmids are found in higher organisms, although they are less common than in bacteria. The yeast *Saccharomyces cerevisiae* has been used as a model organism for the investigation of eukaryotic molecular biology. Most strains of yeast harbor a plasmid known as the **2µ circle** or **2µ plasmid**. This is a circular molecule consisting of 6318 bp of double-stranded DNA. It is present at 50–100 copies per haploid genome and is located in the nucleus of the yeast cell, where it is bound by histones and forms nucleosomes like chromosomal DNA. The 2µ plasmid has been widely used in genetic engineering as the basis for multicopy eukaryotic cloning vectors. Similar plasmids are found in other species of yeast.

The 2µ plasmid contains two perfect inverted repeats of 599 bp that separate the plasmid into two regions of 2774 and 2346 bp, respectively (Fig. 20.25). The

FIGURE 20.25
The 2µ Plasmid of Yeast

Two alternate forms of the 2µ plasmid are inter-converted by recombination. The plasmid has two inverted repeats (IVR1 and IVR2), which can align. The enzyme, Flp recombinase, recognizes the FRT sites (flip recombination target) and makes a crossover that inverts one half of the plasmid relative to the other. Notice the top plasmid has origin (*ori*) close to the Rep2 sequence, whereas the bottom plasmid has the origin on the other side (close to the *FLP* gene). The Rep1 and Rep2 proteins regulate both the *FLP* gene and the replication of the plasmid itself.

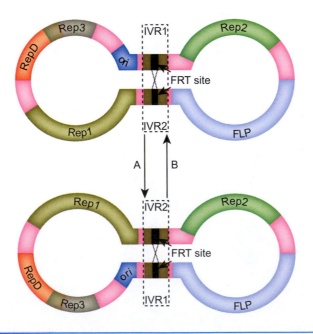

2µ circle   Same as 2µ plasmid
2µ plasmid (or 2µ circle)   A multicopy plasmid found in the yeast *Saccharomyces cerevisiae*, whose derivatives are widely used as vectors

plasmid-encoded Flp protein (**Flp recombinase or "flippase"**) catalyzes recombination between the inverted repeats. Flp recognizes a 48 bp target site (**Flp recombination target, or FRT site**) located within the inverted repeats. The result is the inversion of one half of the plasmid relative to the other. The two forms of the plasmid are found in roughly equal proportions. The Rep1 and Rep2 proteins regulate the expression of the *FLP* gene and also bind to the origin of replication (*ori*) and the *REP3* DNA sequence.

The Flp recombinase is used in genetic engineering to control the expression of a variety of genes by inverting segments of DNA. Flp is functional in bacteria, plants, and animals provided the correct recognition sites are present. In addition to the inversion reaction, Flp recombinase will promote site-specific insertion and deletion of segments flanked by FRT sites. The Flp/FRT system is similar to the widely used Cre/*loxP* recombinase system of bacterial virus P1.

> Many yeast strains contain a small multicopy plasmid, the "2 micron circle."

> Flippase catalyses inversion of the DNA located between its recognition sites.

8. Certain DNA Molecules May Behave as Viruses or Plasmids

There are several similarities between the behavior of plasmids and viruses. In fact, some circles of DNA can choose to live either as a plasmid or as a virus. The bacterial virus P1 is a good example. It can indeed behave as a virus, in which case it destroys

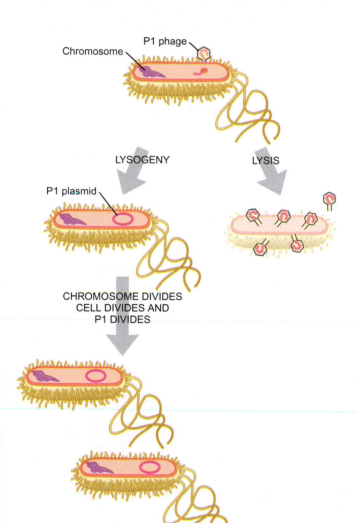

FIGURE 20.26
Lysis versus Lysogeny

Some plasmids, such as the P1 plasmid of bacteria, have a dual personality. P1 can exist in a lysogenic state as a plasmid, using bidirectional replication to divide when the host cell divides. P1 can also grow as a virus and destroy the cell. During such lytic growth, P1 divides by the rolling circle mechanism, creating a large number of copies. It then packages genome-sized units into new virus particles and lyses the bacterial cell.

Flp recombinase (or flippase)   Enzyme encoded by the 2μ plasmid of yeast that catalyzes recombination between inverted repeats (FRT sites)
Flp recombination target (or FRT site)   Recognition site for Flp recombinase
FRT site   Flp recombination target, the recognition site for Flp recombinase

the bacterial cell, replicates by rolling circle mode, and manufactures large numbers of virus particles to infect more bacterial cells. This is known as **lytic growth** since the host cells are "lysed" (derived from the Greek word for broken).

Alternatively, P1 can choose to live as a plasmid and divide in step with the host cell. In this case, the circular P1 DNA uses bidirectional replication like a typical plasmid (Fig. 20.26). Each descendant of the infected bacterial cell gets a single copy of P1 DNA. The cell is unharmed and no virus particles are made. This state is known as **lysogeny** and a host cell containing such a virus in its plasmid mode is called a **lysogen**.

Changing conditions may stimulate a lysogenic virus to return to destructive virus mode. This tends to happen if the host cell is injured, in particular if there is severe damage to the host cell DNA. The virus decides to make as many virus particles as possible before the cell dies. If, on the other hand, the host cell is growing and dividing in a healthy manner, the virus will most likely decide to lie dormant and divide in step with its host. Further aspects of virus behavior are covered in Chapter 21.

Key Concepts

- Plasmids are "extra" self-replicating genetic elements found in cells. They are distinct from chromosomes in being non-essential.
- Replicons are DNA or RNA molecules with their own origin of replication.
- Plasmids are classified into families based on incompatibility.
- Most plasmids consist of circular DNA but occasional plasmids are linear or made of RNA.
- Some plasmids' DNA undergoes bidirectional replication like miniature bacterial chromosomes.
- Other plasmids replicate by the rolling circle mechanism, as do many viruses.
- The copy number varies from one plasmid to another and may be controlled by antisense RNA.
- Many larger plasmids make toxins that kill the cell if the plasmid is lost, a phenomenon referred to as plasmid addiction.
- Plasmids often provide functions useful to the host cell.
- R-plasmids confer antibiotic resistance on the host cell.
- The mechanism of antibiotic resistance often involves inactivation of the antibiotic by plasmid-encoded enzymes.
- Colicin plasmids carry genes for toxic proteins that bacteria use to kill related bacteria.
- Most colicins kill bacteria either by damaging the cell membrane or by degrading nucleic acids.
- Bacteria are immune to the colicins they produce themselves due to synthesis of a specific immunity protein.
- Virulence plasmids carry genes that enable bacteria to damage the host animal or plant cell.
- Ti plasmids are transferred from bacteria to plants and hence are used for the genetic engineering of plants.
- The best-known plasmid of yeast is the 2μ plasmid that is widely used in genetic engineering.
- Some molecules of DNA may behave as viruses under certain conditions and as plasmids under others.

lysogeny   State in which a virus replicates its genome in step with the host cell without making virus particles or destroying the host cell. Same as latency, but generally used to describe bacterial viruses
lysogen   Host cell containing a lysogenic virus
lytic growth   Growth of virus resulting in death of cell and release of many virus particles

Review Questions

1. Why are plasmids not regarded as part of the host cell genome?
2. What are the most important properties of plasmids?
3. How does copy number affect other plasmid-borne characteristics?
4. What is the term used to describe the ability of a plasmid to move from one cell to another?
5. What does it mean for two plasmids to be incompatible within a host cell? Why are plasmids incompatible?
6. What are the two structures for the ends of linear plasmids? Why are these structures necessary for linear plasmids?
7. What is the difference between linear plasmids in bacteria and linear chromosomes?
8. What are the two methods for plasmid replication? Describe the steps in each.
9. What are some regulatory mechanisms for controlling plasmid copy number?
10. What type of molecule is ribonuclease H specific for? What happens if ribonuclease H fails to cleave the pre-primer RNAII? What situation would prevent ribonuclease H from cleaving its target?
11. What is plasmid addiction? Give an example of this type of plasmid.
12. What are some examples of plasmid-encoded genes that are beneficial to a host cell?
13. What families do antibiotic-resistance plasmids belong to? Would these plasmids be incompatible with an F-plasmid? Why or why not?
14. What is the usual difference between antibiotic resistance due to chromosomal mutations and plasmid-borne resistance to antibiotics?
15. To what family of antibiotics do penicillins and cephalosporins belong? What is the mode of action for these antibiotics? Why are human cells not affected the same way as bacterial cells?
16. What enzyme confers resistance to penicillins and cephalosporins? What is its mode of action? What compounds can inhibit this enzyme?
17. What is the mode of action for chloramphenicol? What enzyme inhibits this action? How does this inhibition occur?
18. What is the mode of action for aminoglycosides? In what ways can this family of antibiotics be inactivated?
19. What is the mode of action for tetracycline? How does the mechanism of tetracycline resistance differ from chloramphenicol and aminoglycosides? Why does tetracycline not affect human cells even though it is able to inhibit human cell ribosomes?
20. What is the mode of action for sulfonamides and trimethoprim? What do these antibiotics antagonize? What is the resistance mechanism to these antibiotics? Why do these antibiotics not affect human cells?
21. Why would bacteria want to kill their close relatives? What is the name of toxic proteins used to kill other bacteria?
22. What are the two major mechanisms used by colicins to kill bacteria? Describe the general properties of the Col plasmids.
23. How do colicin immunity proteins function? In what molecule are the genes encoded for the immunity proteins? What would happen to the bacterial cell if the genes for immunity proteins were not present?
24. Why is bacteriocin production often a suicidal process? What is the significance of colicin release?
25. Describe the regulation of colicin E production. What circumstances induce colicin production and release?
26. What other process besides the suicidal process releases colicin? Describe this process.
27. What are virulence plasmids?
28. What advantage do virulence plasmids confer on bacterial cells? What toxins might be involved in infectivity?
29. What is a Ti plasmid? What advantage does this plasmid confer on *Agrobacterium tumefaciens*?
30. What compound attracts *Agrobacterium tumefaciens* to a wounded plant and also induces virulence gene expression?
31. What is T-DNA? How is it transferred? What gene products are encoded on T-DNA?
32. Why is the relationship between *Agrobacterium tumefaciens* and the infected plant not considered a symbiotic relationship? Which organism (bacteria or plant) benefits most from this relationship and how?

33. How are Ti plasmids used in genetic engineering?
34. What is the 2μ plasmid? Which organism harbors this plasmid?
35. What are the characteristics of the 2μ plasmid? What does Flp recombinase or "flippase" do? What is the target site called for this enzyme?
36. How is Flp recombinase used in genetic engineering?
37. Describe the ways in which the bacteriophage P1 can live as either a virus or a plasmid.
38. What is the difference between lytic growth and lysogeny?
39. What conditions can stimulate a switch between lysogeny and lytic growth?

Conceptual Questions

1. Fill in the following table:

Antibiotic	How it kills bacteria	Bacterial-resistance proteins	Mechanism of bacterial resistance proteins
tetracycline			
kanamycin			
sulfonamide			
chloramphenicol			
neomycin			

2. For each of the above antibiotics, describe why it does not harm humans.
3. Many different naturally-occurring antibiotic resistance genes are used as genetic markers in laboratory experiments. In order to study promoter elements, promoters are fused to genes encoding different enzymes so when the promoter is active the enzyme is made. One marker used to assess promoter function is CAT, or chloramphenicol acetyl transferase. Using your knowledge of its function in resistant bacteria, discuss how you would use the *cat* gene to make promoter fusions.
4. A researcher has two different genes (ZIP and ZAP) that he wants to express in bacteria. He decided to clone each gene and put it into a plasmid so that he can express it in bacteria. In his experiment, he clones ZIP into a ColE1 plasmid with the *bla* gene and he clones ZAP into ColE3 which has CAT. He takes all the bacteria he transformed and plates them onto ampicillin and chloramphenicol. He has no colonies on his plates. Why?
5. The following two plasmids were created by the lab down the hall from where you work. They give you the following map, but do not give you any information of how to grow the bacteria after you transform the plasmid into *E. coli*. Based on these two maps, what do you need to include for selection of *E. coli* that have these plasmids? What do you need to add to express the gene of interest?

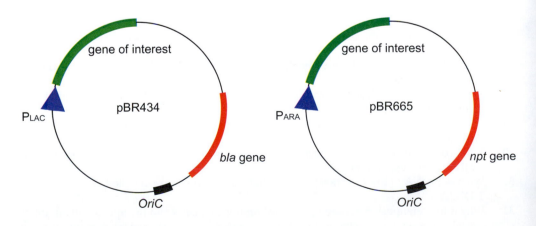

Viruses

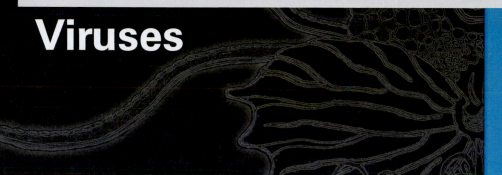

Are viruses alive or not? For many years this question has received a great deal of attention but no satisfactory answer. Nonetheless, the question illustrates several important points about the nature of life and the role of genetic information. Viruses possess their own genetic information yet need a host cell in which to replicate. Thus, they are genetic entities but not living cells. Some scientists have argued that a virus-infected cell can be regarded as a living organism. This viewpoint emphasizes the way in which the virus takes control of the cell and uses its resources to assemble more virus particles, rather than the damage that virus infection causes to the host. On the other hand, the virus particle that travels between host cells is clearly inert. Viruses show immense diversity in genome structure and size. Unlike living cells, some viruses have RNA genomes, illustrating that RNA can be used as the primary genetic material. Even more minimalist than viruses are the viroids, genetic entities with tiny RNA genomes that have no genes and do not encode any proteins.

FIGURE 21.01
Viruses Consist of Protein plus DNA or RNA

A simple virus contains a genome of RNA (blue) or DNA (pink) surrounded by a capsid made of proteins.

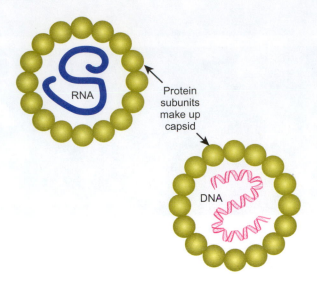

Protein subunits make up capsid

1. Viruses Are Infectious Packages of Genetic Information

Viruses are packages of genes inside protective protein shells. Viruses cannot grow or divide by themselves. In order to replicate, a virus must first infect a host cell. Only then are the virus genes expressed and the virus components manufactured using the host cell machinery. Viruses are not merely pieces of nucleic acid like plasmids or transposons and neither are they true living cells with the ability to generate energy and make protein. They lie in the gray area between. Viruses cannot make their own proteins or generate their own energy. They can only multiply when they have entered a suitable host cell and taken over the cellular machinery. Despite this a virus is certainly not inert; it does replicate if it can subvert a host cell.

Virus particles contain proteins plus genetic information in the form of DNA or RNA (Fig. 21.01). The virus particle, or **virion**, consists of a protein shell, known as a **capsid**, surrounding a length of nucleic acid, either RNA or DNA, which carries the virus genes and is often referred to as the **viral genome**. Many simple viruses have only these two components.

Trying to define precisely what is living and what is non-living can be quite confusing. Here, we will sidestep the issue of defining life by noting that being alive and being a **living cell** are not necessarily the same. To qualify as a genuine living cell, a structure must send genetic messages (RNA) from its genes (DNA) to its own ribosomes to make its own proteins. A living cell generates the energy to produce these proteins and maintain cellular integrity (Fig. 21.02).

In contrast to self-sufficient living cells, viruses rely upon their host for many functions. Living cells store information as DNA and make their messages out of RNA, whereas viruses can store their genetic information as either RNA or DNA and rely upon the host cell to make their messenger RNA (mRNA). Genuine cells possess ribosomes that are capable of making proteins. Viruses are parasitic and rely on the host cell to provide the ribosomes for translating virus mRNA into protein. Cells transport nutrients and metabolize them to generate energy and make a variety of metabolic intermediates. Viruses rely on host cell metabolism for energy and precursors.

> Viruses are subcellular parasites that rely on a cell to provide energy and raw material.

> Virus particles contain DNA or RNA protected by a shell of protein.

> A key property of a living cell is that it possesses its own ribosomes for making proteins. Viruses do not contain ribosomes but use the host cell ribosomes to make proteins.

capsid   Shell or protective layer that surrounds the DNA or RNA of a virus particle
living cell   A unit of life that possesses a genome made of DNA and sends genetic messages (RNA) from its genes (DNA) to its own ribosomes to make its own proteins with energy it generates itself
virion   Virus particle
viral genome   Molecule of DNA or RNA that carries the genes of a virus
virus   Infectious agent, consisting of DNA or RNA inside a protective shell of protein, that must infect a host cell in order to replicate

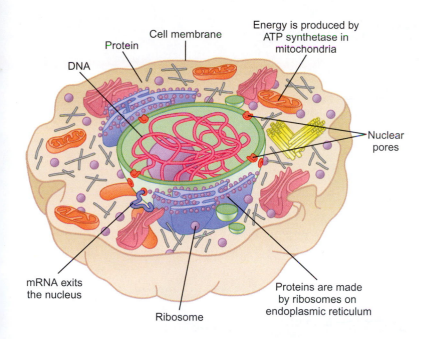

DNA

Protein

Cell membrane

Energy is produced by
ATP synthetase in
mitochondria

Nuclear
pores

mRNA exits
the nucleus

Ribosome

Proteins are made
by ribosomes on
endoplasmic reticulum

■■ **FIGURE 21.02**

***Characteristics of a Living
Cell***

This simplified cell shows all the
essential characteristics of a living
cell: an energy source to provide
ATP (ATP synthetase), genetic
information (chromosomal DNA
and messenger RNA), ribosomes
to convert genetic information
into proteins, and a biological
membrane that maintains cellular
integrity.

Living cells are surrounded by metabolically active cell membranes. Most simple viruses have only a protein shell and no true membrane, although some complex virus particles have a membrane stolen from the previous host. However, the membranes around virus particles are not active metabolically either in energy generation or nutrient transport. Nonetheless, virus particles do have an outer covering and can survive on their own outside their host cells (admittedly without multiplying).

Viruses are all **parasites** that cannot multiply without a host cell. Furthermore, viruses are **intracellular parasites**; that is to say that they must actually enter the cells of the host organism to replicate. Note that not all intracellular parasites are viruses. Certain disease-causing bacteria and protozoans may enter the cells of higher organisms and live inside them as parasites. However, these parasites are nonetheless living cells themselves and contain their own ribosomes to make their own proteins. This chapter does not attempt to cover the realm of virology systematically. Rather, examples are given to illustrate novel aspects of molecular biology found among the viruses.

> Living cells have membranes, whereas most viruses do not.

1.1. Lifecycle of a Virus

A virus alternates between two forms, an inert virus particle, the virion, which survives outside the host cell, and an active intracellular stage. The lifecycle of a typical virus goes through the following stages (Fig. 21.03):

a. Attachment of virion to the correct host cell

b. Entry of the virus genome

c. Replication of the virus genome

d. Manufacture of the virus proteins

e. Assembly of new virus particles (virions)

f. Release of new virions from the host cell

> Virus genes subvert the host cell into manufacturing more virus particles.

Attachment of a virus requires a protein on the virus particle to recognize a molecule on the surface of the target cell. Sometimes this receptor is another protein;

intracellular parasite   Parasite that lives inside the cells of its host organism
parasite   An organism or infectious agent that uses the resources of another organism in order to grow and multiply

FIGURE 21.03
Virus Lifecycle

The lifecycle of a virus starts when the viral DNA or RNA enters the host cell. Once inside, the virus uses the host cell to manufacture more copies of the virus genome and to make the protein coats for assembly of virus particles. Once multiple copies of the virus have been assembled, the host cell is burst open to allow the viral progeny to escape and find new host cells to infect.

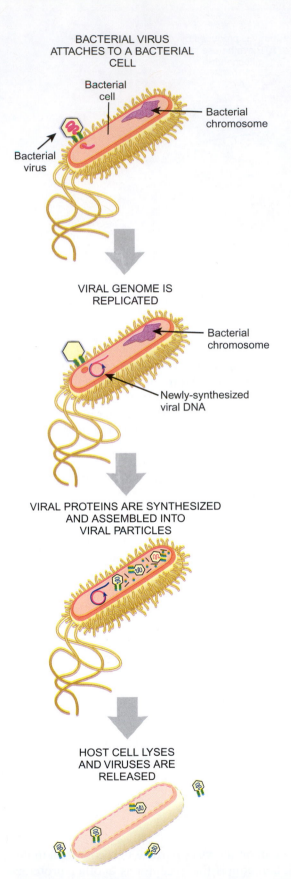

BACTERIAL VIRUS ATTACHES TO A BACTERIAL CELL

Bacterial cell

Bacterial chromosome

Bacterial virus

VIRAL GENOME IS REPLICATED

Bacterial chromosome

Newly-synthesized viral DNA

VIRAL PROTEINS ARE SYNTHESIZED AND ASSEMBLED INTO VIRAL PARTICLES

HOST CELL LYSES AND VIRUSES ARE RELEASED

sometimes it is a carbohydrate. Often it is a glycoprotein; that is, a protein with carbohydrate groups attached. On some virus particles, the recognition proteins form spikes or prongs sticking out from the surface. Most bacterial and plant viruses abandon their protein coat when they infect a new host cell. Only the genetic material

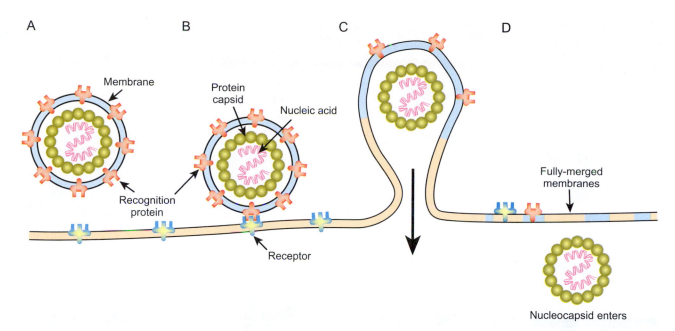

FIGURE 21.04
Enveloped Viruses Merge with the Animal Cell Membrane

When the virus particle left the previous host cell, it surrounded itself with a layer of the host cell membrane. This outer layer contains viral recognition proteins previously inserted into the host cell membrane during virus infection. The recognition proteins bind to the cell membrane receptors of another animal cell. The protein complex triggers the animal cell to take in the particle by fusing the two membranes. The nucleocapsid structure enters the animal cell.

(DNA or RNA) enters the cell. Animal viruses vary in regard to when exactly they disassemble their protein coat.

Many animal viruses have an extra envelope outside the protein shell. This is made of membrane stolen from the previous host cell into which virus proteins have been inserted. These virus-encoded proteins detect and bind to receptors on the next target cell. When an enveloped virus enters a new animal cell, its envelope layer merges with the cell membrane and the inner protein shell containing the nucleic acid (the "**nucleocapsid**") enters (Fig. 21.04). Once inside, the protein shell disassembles, exposing the genome.

Once inside the host cell, the virus genome has two major functions. First, it must replicate to produce more virus genomes. Second, it must subvert the cell to manufacture lots of virus proteins for the assembly of new virus particles. Note that viruses do not divide like cells. They are assembled from components manufactured by the host cell using genetic information in the virus genome (Fig. 21.05).

The genes of viruses are often divided into "**early genes**" and "**late genes**." The early genes have promoters that resemble those of the host cell and encode for proteins responsible for replicating the virus genome. Consequently, they are transcribed by host cell RNA polymerase and are expressed immediately after infection. In very small viruses, host enzymes are largely responsible for replicating the virus genome so there may be very few "early genes" involved in replication. Conversely, in viruses that have large numbers of genes, such as bacterial virus T4 or the poxviruses of animals, regulation is obviously more complex and there may be several subcategories of genes such as "immediate early," "delayed early," etc.

The late genes have promoters that are not recognized by host polymerase alone. These genes are expressed late in infection and encode the structural proteins of the

> Before entering the host cell, a virus must bind to a receptor on the cell surface.

> Viruses do not divide. Instead, their genes code for components that are assembled into new virus particles.

---

early genes   Genes expressed early during virus infection and that mainly encode enzymes involved in virus DNA (or RNA) replication
late genes   Genes expressed later in virus infection and that mainly encode enzymes involved in virus particle assembly
nucleocapsid   Inner protein shell of a virus particle that contains the nucleic acid

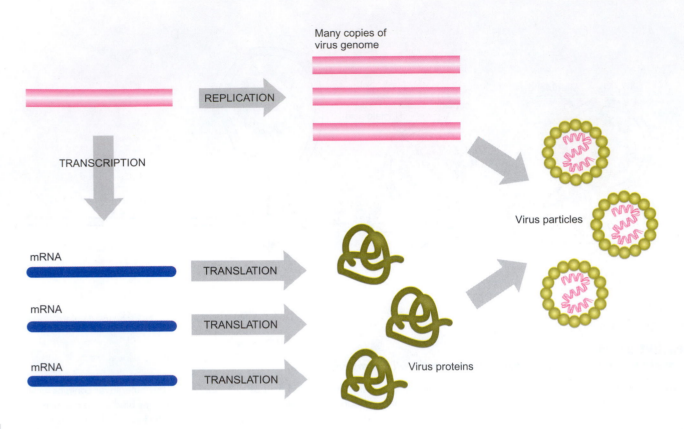

FIGURE 21.05
Synthesis and Assembly of Virus Components

The viral genome (in pink) directs the host cell to replicate many copies of the virus genome. The viral genome is also transcribed and the mRNA is translated, giving viral proteins. The viral genome carries all the genes needed for making the protein coat. Finally, the coat proteins and the viral genomes are assembled to give new virus particles.

virus particle together with proteins involved in the assembly and packaging processes and in lysing the host cell. Some viruses, such as bacterial virus T7, encode their own RNA polymerase in order to express late genes; others, such as bacterial virus T4, modify the host RNA polymerase. For example, T4 gene 55 encodes an alternative sigma factor that recognizes the promoters of T4 late genes.

1.2. Bacterial Viruses are Known as Bacteriophage

Viruses that infect bacteria are often called **bacteriophage** or **phage** for short. Phage is derived from the Greek word for "eat" and refers to the way in which bacterial viruses eat holes or "**plaques**" in a lawn of bacteria growing on the surface of agar (Fig. 21.06). Bacterial viruses were heavily used in the early days of molecular biology to investigate the nature of the gene because viruses only contain DNA or RNA surrounded by a protein coat and because bacteriophage infect the simplest of all cells, bacteria.

Bacteria have a cell wall protecting their cell membrane and so bacterial viruses cannot simply merge with the membrane, as do animal viruses. Therefore, bacterial viruses do not bother with an outer envelope layer. They just have a protein shell surrounding the DNA or RNA. After binding to the cell surface, they inject their nucleic acid into the bacterial cell and the outer protein coat of the virus particle is left

bacteriophage (phage)   Virus that infects bacteria
phage   Short for bacteriophage, a virus that infects bacteria
plaque   (When referring to viruses) A clear zone caused by virus killing and lysing bacteria that are growing as a lawn on a surface of an agar plate

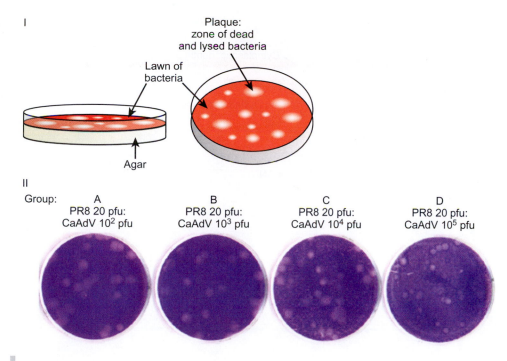

FIGURE 21.06
Formation of Plaques in Lawn of Bacteria

I) Bacteriophage are viruses that infect bacteria. To isolate individual types of bacteriophage, plaques are made. A mixture of bacteriophage is added to a large number of bacteria and poured onto a nutrient agar plate. The bacteria grow quickly, covering the agar with a cloudy layer of bacteria, known as a lawn (red). Wherever a bacteriophage infects a bacterial cell, it destroys the cell and produces many more bacteriophage. These spread out to infect neighboring bacteria, forming a clear zone in the lawn that is called a plaque. Each plaque contains descendents of the single original bacteriophage that landed in that region of the lawn. If needed, purified lines of bacteriophage can be isolated from individual plaques. II) Real photo of the plaques that form when influenza virus is mixed with cultured human cells. The larger clear plaques seen in A–D are made by influenza strain PR8 and the small plaques (clearest in C) are due to influenza strain CaAdV. Notice that C contains both large and small plaques. *(Credit: Murata, et al., (2011) Plaque purification as a method to mitigate the risk of adventitious-agent contamination in influenza vaccine virus seeds. Vaccine 29(17): 3155–3161.)*

behind (Fig. 21.07). Many well-known bacterial viruses have a complex capsid that resembles a miniature moon-lander. The capsid has an icosahedral head, a tail, and six landing legs with attachment proteins at the tips. The tail contracts and injects the DNA through the bacterial cell envelope.

In 1952 Hershey and Chase performed a classic experiment using bacteriophage T4 to demonstrate that only the DNA entered the host cell, *Escherichia coli* (Fig. 21.08). The protein and the DNA of the virus particles were radioactively labeled with two different isotopes. The labeled viruses were added to bacterial cells and the fate of the two radioactive labels was followed. The protein, labeled with ^{35}S, was left outside and the DNA, labeled with ^{32}P, entered the cells. Moreover, some of the ^{32}P-labeled DNA was found in the new generation of virus particles liberated when the infected cells burst. Since only the virus DNA enters the host cell and the other components are abandoned outside, this provides further evidence that nucleic acids rather than proteins carry genetic information. Historically, the Hershey and Chase experiment demonstrated for the first time that DNA alone was the carrier of the genetic information and that the associated protein was not required. These findings prompted other researchers to investigate the structure of DNA and its role as the genetic material.

> Bacterial viruses leave their protein shell behind and only their genome enters the host cell.

FIGURE 21.07 ────────
Bacterial Viruses Inject their Nucleic Acid

To enter a bacterial cell, bacterial viruses must get their genomes through three layers, the bacterial outer membrane, the cell wall, and the inner membrane. The bacterial wall structure prevents the virus from simply merging membranes, as in animal cells. To overcome the defenses, the virus punches a hole through the three layers (1) and injects its DNA (or RNA) (2) into the cytoplasm. The empty shell (3) never enters the bacterial cell.

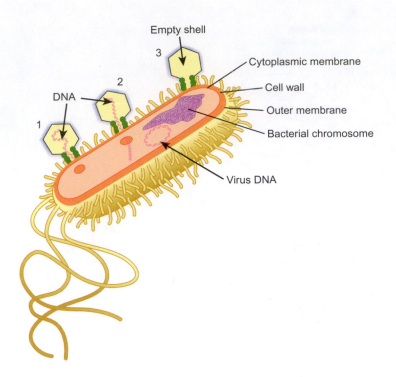

Box 21.1 Phages are the Most Numerous Life Form

It is likely that there are more bacteriophages on our planet than any other life form. There are an estimated 10^{30} bacteria on the planet, and it is estimated that there are probably about 10 phages for every living bacterial cell, which would give a total of around 10^{31} phage. Virus particles, including bacteriophage, are ubiquitous on Earth. Examination of seawater under the electron microscope has shown typical counts of 50×10^6 virus particles per milliliter. It has been estimated that phages destroy up to 40% of the bacteria in the ocean every day. Remnants of these lysed bacteria add significant amounts of organic matter to the ocean water and may affect global carbon cycling. A colossal amount of novel genetic material is present in the vast number of phages present in natural habitats. Preliminary surveys have indicated that around 75% of the genes carried by phages are unrelated to anything presently in the DNA data banks.

Many of the virus particles in the environment are probably orphaned in the sense that susceptible host cells are no longer available in their habitats. In some cases the host cell may be extinct or perhaps only mutants resistant to the virus have survived. Conversely, many virus particles are inherently defective and are incapable of successfully infecting host cells, even if available. This is especially true of RNA viruses where the mutation rate is extremely high. The benefit of a high mutation rate is that the virus constantly changes and so evades recognition by the host defense systems. The downside is that most mutations are deleterious and a high percentage of defective virus genomes are made. Indeed, for some RNA viruses the majority of virus particles released are defective mutants and only a minority of the population has infectious particles.

1.3. Lysogeny or Latency by Integration

When an infecting virus generates many virus particles and destroys the cell, this is known as **lytic growth** because the cell is burst or lysed. When instead, the virus divides in step with the host chromosome, this is known as **lysogeny,** and a cell

---

lysogeny   Type of virus infection in which the virus becomes largely quiescent, makes no new virus particles, and duplicates its genome in step with the host cell. Same as latency but used of bacterial viruses
lytic growth   Type of infection in which a virus generates many virus particles and destroys the host cell to release them

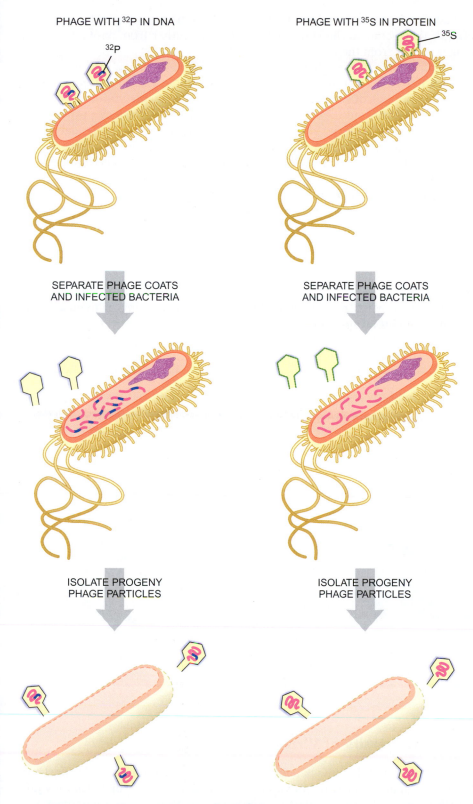

PHAGE WITH ³²P IN DNA

PHAGE WITH ³⁵S IN PROTEIN

³²P

³⁵S

SEPARATE PHAGE COATS
AND INFECTED BACTERIA

SEPARATE PHAGE COATS
AND INFECTED BACTERIA

ISOLATE PROGENY
PHAGE PARTICLES

ISOLATE PROGENY
PHAGE PARTICLES

■ **FIGURE 21.08**
*Only the DNA of
Bacteriophage T4 Enters
the Cell*

In a classic experiment by Hershey
and Chase, the viral DNA was
found to enter the bacterial cell
but the protein coat did not.
To trace the DNA and protein,
bacteriophage T4 was first grown
with radioactive precursors
(³²P-dATP and ³⁵S-methionine,
respectively). The radioactively-
labeled T4 were isolated and
purified. Phage labeled with ³²P
(blue DNA) were added to new
host *E. coli*, and the *E. coli* were
separated from remaining virus
particles (left panel). After viral
replication, the progeny viruses
were isolated. The ³²P-labeled
DNA was found in these viral
particles, implying that the original
virus DNA entered the host *E. coli*,
and was eventually packaged in
new particles. The same experiment
was performed with ³⁵S-methionine
labeled T4 (green coats), but the
final viral particles did not contain
any ³⁵S-methionine (right panel).
This suggested that none of the
original protein coat entered the
host *E. coli*.

containing such a virus is a **lysogen**. The term **latency** means the same as lysogeny but
is usually used when referring to animal cells. In Chapter 20 we discussed the close
relationships between plasmids and viruses. Some gene creatures can choose to live
either as a plasmid or as a virus. Some plasmids are probably derived from viruses

---

latency   Type of virus infection in which the virus becomes largely quiescent, makes no new virus particles, and duplicates its genome in step with
the host cell. Same as lysogeny but used of animal viruses
lysogen   A cell containing a lysogenic virus

Viruses may replicate aggressively, killing the host cell. Alternatively, they may limit themselves to duplicating their genome in step with cell division.

that have lost the ability to grow lytically. Conversely, some viruses may have evolved from plasmids that obtained the genes for lytic growth, either from another virus or, over a longer period, from the host cell.

Lysogeny or latency means that the virus has decided to divide in step with the host cell instead of killing it. It does not necessarily mean the virus has decided to live as a plasmid. Many cases of lysogeny or latency are caused by integration of the virus DNA into a host cell chromosome. Such an integrated virus is known as a **provirus** (or **prophage** in the case of bacterial viruses). The virus DNA becomes a physical part of the chromosome and is replicated when the chromosome divides. Integrated viruses are found even in the symbiotic bacteria that are necessary for the survival of certain insects and nematodes (see Focus on Relevant Research).

The bacterial virus **lambda (λ)**, which infects the bacterium *E. coli*, recognizes and integrates into a special sequence of DNA on the chromosome of its host cell, known as *att*λ (λ **attachment site**). Integration occurs by site-specific recombination as described in Chapter 24. This allows lambda to occasionally pick up and carry bacterial genes as described in the chapter on bacterial genetics (Ch. 25). Some animal viruses also insert themselves into the chromosomes of their host cells. Some have special recognition sites, while others insert at random. Retroviruses, which carry their genes as RNA in the virus particle, must first make a DNA copy of themselves to insert into the host chromosome (see below).

Kent BN and Bordenstein SR (2010) Phage WO of *Wolbachia*: lambda of the endosymbiont world. Trends Microbiol 18: 173–182.

Many insects and nematode worms contain bacterial endosymbionts, which are symbiotic bacteria that reside inside the cells of the animal. Most of these bacteria provide essential amino acids and occasional vitamins and co-factors that the insects cannot synthesize themselves. In some cases the endosymbiont is essential for fertility. Conversely, the bacteria depend on nutrients supplied by the

FOCUS ON RELEVANT RESEARCH

host and most are incapable of living on their own. Bacterial endosymbionts are important because many co-exist with human parasites. If scientists can understand

their relationship, perhaps we can kill the parasite by killing its endosymbiont. A parasite without its supply of extra nutrients will die. Endosymbiotic bacteria themselves are also host to unusually high levels of mobile elements including both transposons and viruses that integrate into the bacterial chromosome (i.e., lysogenic bacteriophages).

The authors focus on the bacteriophage WO that infects the widespread bacterial symbiont *Wolbachia*. Study of this virus has revealed unusually high levels of genomic change in *Wolbachia*. The bacterial symbionts may be transmitted horizontally between insect species and the WO phage may be transmitted horizontally between bacterial strains. The WO phages, in turn, carry a wide range of insertion sequences that are often involved in bacterial chromosome rearrangements.

The WO phage also plays a role in the symbiotic relationship between bacterium and animal cells. A variety of genes encoding what are probably virulence factors and toxins are present on different strains of phage WO. How these affect the interaction of *Wolbachia* with its animal host cells is a matter for ongoing investigation. As for parasitic insect hosts of *Wolbachia*, perhaps the parasite can be eradicated by killing the bacteriophage WO rather than by killing the bacteria or the parasite itself.

---

2. The Great Diversity of Viruses

Viruses have been found that attack animal cells, plant cells, and bacterial cells. There is colossal variation in the structure of viruses and the detailed way in which they take over the cells they invade. The smallest viruses have only three genes; the largest have two or three hundred and carry out some extremely complicated genetic maneuvers

---

λ attachment site (*att*λ) Recognition sequence on the chromosome of *Escherichia coli* where bacteriophage lambda integrates
lambda (λ) Virus that infects the *Escherichia coli* and may integrate into a special sequence of DNA on the bacterial chromosome
prophage Bacteriophage genome that is integrated into the DNA of the bacterial host cell
provirus Virus genome that is integrated into the host cell DNA

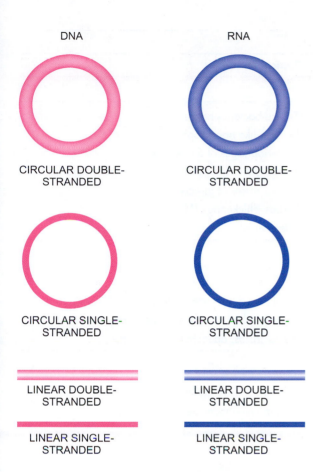

DNA

CIRCULAR DOUBLE-STRANDED

CIRCULAR SINGLE-STRANDED

LINEAR DOUBLE-STRANDED

LINEAR SINGLE-STRANDED

RNA

CIRCULAR DOUBLE-STRANDED

CIRCULAR SINGLE-STRANDED

LINEAR DOUBLE-STRANDED

LINEAR SINGLE-STRANDED

FIGURE 21.09
Variety of Virus Genomes

Viral genomes come in all shapes and sizes. They may be comprised of DNA or RNA. They may be circular or linear, and they may be double-stranded or single-stranded.

to outwit their host cells. Mimivirus is presently the largest known virus. It is roughly icosahedral with a diameter of 0.75 microns. It has a 1.2 Mbp genome encoding over 900 genes. The largest known bacterial virus genome is that of Bacteriophage G, which infects *Bacillus megaterium*. It has nearly 700 genes, more than some bacteria. Some viruses have DNA genomes while others have RNA. Furthermore, the nucleic acid may be either single- or double-stranded and either linear or circular. All of these possibilities exist (Fig. 21.09; Table 21.01), though some are more common than others. Some viruses even have segmented genomes made up of several pieces of DNA or RNA. We shall survey a range of viruses, partly to illustrate their genetic diversity. We will also include some viruses that are widely used in molecular biology or are especially notorious for causing disease.

The details of virus replication also vary considerably. Nonetheless, many viruses use versions of the rolling circle scheme for replication, which was discussed in Chapter 20. Viruses whose particles contain linear DNA or RNA will often circularize after infection in order to perform rolling circle replication. Viruses with single-stranded DNA or RNA synthesize a second strand upon infection, making a double-stranded circular version of the genome known as the **replicative form (RF)**. DNA viruses may or may not rely on host enzymes for replication, but RNA viruses must provide a special RNA polymerase to replicate their RNA genomes, called an **RNA replicase**.

Viruses may also be classified according to the structure of the virus particle, or virion. The three major shapes seen are spherical, filamentous, and complex. Spherical viruses are not truly spherical but have 20 triangular faces and are thus icosahedrons

Different viruses have from three to just over 1000 genes.

Many viruses replicate their genomes by some form of rolling circle mechanism.

Virus particles come in a wide range of sizes and shapes.

replicative form (RF)   Circular double-stranded version of a virus genome used for rolling circle replication
RNA replicase   Special RNA polymerase used by RNA viruses to replicate their RNA genomes

TABLE 21.01	Families of Viruses
Virus Family	**Typical Examples**

Double-Stranded DNA Viruses

Virus Family	Typical Examples
Myoviridae	T4-like bacteriophages
Siphoviridae	lambda-like bacteriophages
Fuselloviridae	bacteriophages of *Sulfolobus*
Poxviridae	cowpox, vaccinia, smallpox (= *Variola*), ectromelia
Baculoviridae	baculoviruses of insects
Herpesviridae	herpes, chickenpox (*Varicellavirus*)
Adenoviridae	adenovirus

Single-Stranded DNA Viruses

Virus Family	Typical Examples
Inoviridae	small filamentous bacteriophage (e.g., fd)
Microviridae	small spherical bacteriophage (e.g., ΦX174)
Geminiviridae	assorted plant viruses (e.g., beet curly top virus)
Parvoviridae	parvoviruses, adeno-associated virus

Reverse Transcribing Viruses

Virus Family	Typical Examples
Hepadnaviridae	hepatitis B virus, cauliflower mosaic virus
Retroviridae	human immunodeficiency virus (HIV)

Double-Stranded RNA Viruses

Virus Family	Typical Examples
Reoviridae	reoviruses, bluetongue virus of sheep, rotaviruses
Birnaviridae	*Drosophila* X virus
Totiviridae	*Saccharomyces cerevisiae* virus L-A

Negative Single-Stranded RNA Viruses

Virus Family	Typical Examples
Paramyxoviridae	parainfluenza, measles, mumps
Rhabdoviridae	vesicular stomatitis virus, rabies
Filoviridae	Marburg virus, Ebola virus
Orthomyxoviridae	influenza
Bunyaviridae	hantavirus, tomato spotted wilt virus

Positive Single-Stranded RNA Viruses

Virus Family	Typical Examples
Leviviridae	bacteriophages MS2 and Qβ
Picornaviridae	polio, common cold (*Rhinovirus*), hepatitis A, foot-and-mouth virus
Tombusviridae	tobacco necrosis virus (TNV)
Flaviviridae	yellow fever, hepatitis C
Togaviridae	rubella (German measles), tobacco mosaic virus (TMV)

(Fig. 21.10). Cross-sections through an icosahedron may be five- or six-sided depending on where the cut is made. As with most other biological filaments, filamentous viruses consist of helically-arranged protein subunits forming cylindrical shells. These may be either open or closed off at the ends. Complex viruses are large viruses with multiple structural components that do not fit neatly into the spherical versus filamentous classification. In addition, outer envelopes, partly derived from the cytoplasmic membrane of the previous host cell, surround some virus particles (Fig. 21.11).

RNA

Protein

HELICAL
(part of tobacco
mosaic virus)

ICOSAHEDRAL
(adenovirus)

■- **FIGURE 21.10**
***Filamentous and Spherical
Virus Structures***

Filamentous viruses are actually
built from helical arrays of protein.
The proteins of tobacco mosaic
virus coat the helical RNA molecule
to form a cylinder. Spherical
viruses, such as adenovirus, are
actually icosahedral. They have
a 20-sided form that may or may
not have protein spikes sticking
out from the protein coat at the
vertexes. These protruding knobs
carry the proteins that recognize
virus receptors on the host cell
surface.

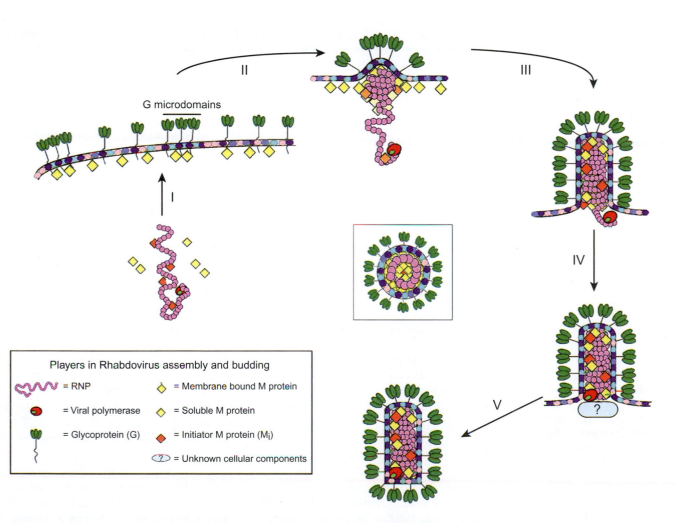

Players in Rhabdovirus assembly and budding

∿∿ = RNP

● = Viral polymerase

🌿 = Glycoprotein (G)

◇ = Membrane bound M protein

◇ = Soluble M protein

◆ = Initiator M protein (M$_i$)

⬭? = Unknown cellular components

FIGURE 21.11
Rhabdovirus Is an Enveloped Virus

Rhabdoviruses are examples of enveloped RNA viruses. The group includes vesicular stomatitis, rabies, and related viruses. The envelope is generated
by budding of the virus from the cytoplasmic membrane of the infected host cell, as shown. (I) The viral ribonucleoprotein (RNP) associates with the
inside of the membrane via M protein. (II) RNP condensation begins and bud sites form. Budding occurs preferentially where the viral glycoprotein
(G) has clustered. (III) Assembly and extrusion of the virion. (IV) Cellular proteins of unknown identity (?) complete the budding and promote release of
virions (V) via membrane fission. The inset shows a virion in cross-section with its central M protein scaffold and glycoprotein spikes protruding from the
viral envelope. *(Credit: Jayakar et al, (2004) Rhabdovirus assembly and budding. Virus Research 106:117–132.)*

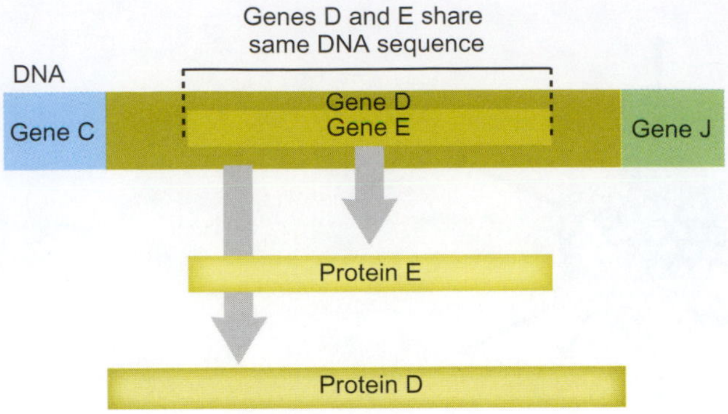

FIGURE 21.12
Bacteriophage ΦX174 has Overlapping Genes

Since many viral genomes are very small, the same region of the genome sometimes encodes two different genes. In the ΦX174 genome, the gene for E protein is entirely embedded within the gene for D protein. The two gene sequences are not identical, even where they overlap, since D uses a different reading frame than E.

Genes D and E share same DNA sequence

DNA

Gene C | Gene D / Gene E | Gene J

Protein E

Protein D

2.1. Small Single-Stranded DNA Viruses of Bacteria

Bacteriophage ΦX174 is a small simple spherical virus that contains 5,386 bases of circular single-stranded DNA. At each of the 12 vertices of the virion is a spike made of two different proteins involved in recognizing the host cell. Overall, ΦX174 looks rather like a World War II naval mine. The most interesting property of ΦX174 is that it has so little DNA that five of its 11 genes overlap others. For example, gene E is completely inside the DNA for gene D. Genes D and E are read in two different reading frames so they produce two totally different proteins (Fig. 21.12). D-protein helps assemble the virus capsid, though it does not form part of the final structure, and E-protein destroys the cell wall of the host bacterium to allow the newly made virus particles out. A mutation in the DNA for gene E will also alter gene D. Although overlapping saves on DNA, the two genes are no longer free to evolve separately. Although overlapping genes are quite often found in small viruses, they are only found in real cells under exceptional circumstances.

Small viruses quite often have overlapping genes.

Some other small bacterial DNA viruses are filamentous rather than spherical. For example, **bacteriophage M13** has single-stranded circular DNA like ΦX174, but the virus particle is a long thin filament. M13 is **"male-specific,"** which means that it only infects bacteria carrying the F-plasmid; that is, "male" bacteria (Ch. 25). M13 does not kill its host when progeny virus particles are released; therefore, scientists use this bacteriophage to make single-stranded DNA for sequencing (see Ch. 8). Since the host is not killed more DNA can be isolated if needed.

2.2. Complex Bacterial Viruses with Double-Stranded DNA

There is a large family of bacterial viruses that all have a complex form made up of a head, tail, and tail fibers (Fig. 21.13). The head of the virus particle contains a large molecule of linear double-stranded DNA. They include bacteriophages T4, Lambda, P1, and Mu, which are all used in bacterial genetics (see Ch. 25) and molecular biology. The number of genes ranges from Mu with approximately 40 genes to T4 with nearly 200. T4 and its close relatives are some of the most complex types of virus known.

Complex bacterial viruses have their genome in a head structure and a tail that contracts for injecting their genome into the host.

The head of the Mu virus particle is more or less spherical (i.e., it is a symmetrical icosahedron). However, the head of T4 is relatively elongated in order to accommodate the much greater amount of DNA. Attached to the head is a tail with tail fibers that act as landing legs. These viruses bind to bacterial cells by means of recognition proteins on the end of their tail fibers. After setting down like lunar landers, their tails contract and they inject their DNA like miniature hypodermic syringes.

bacteriophage ΦX174   A small spherical virus that contains circular single-stranded DNA and infects *Escherichia coli*
bacteriophage M13   A small male-specific filamentous virus that contains circular single-stranded DNA and infects *Escherichia coli*
male-specific phage   Virus that only infects "male" bacteria (i.e., those bacteria carrying the F-plasmid)

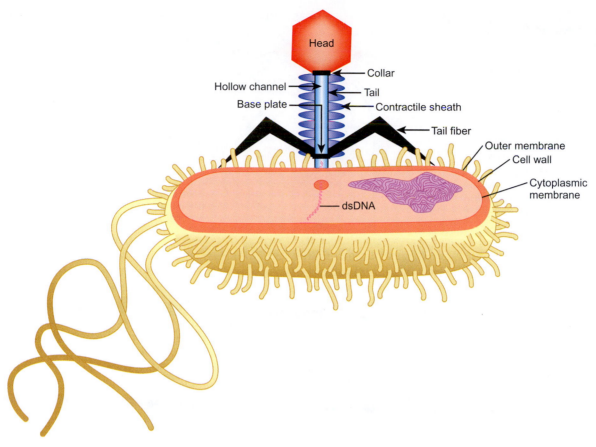

FIGURE 21.13
Bacteriophage with Heads and Tails

Complex bacterial viruses such as Mu and T4 have a head region that holds the double-stranded DNA, a tail region that injects the DNA into the bacteria, and tail fibers that recognize specific proteins on the bacterial outer membrane and so facilitate attachment. The contractile sheath helps inject the DNA from the virus head into the inside of the bacterium.

2.3. DNA Viruses of Higher Organisms

Most DNA viruses of animals contain double-stranded DNA. For example, **Simian virus 40 (SV40)** is a smallish, spherical virus that causes cancer in monkeys by inserting its DNA into the host chromosome. Another double-stranded DNA virus, **herpesvirus**, is spherical with an extra outer envelope of material stolen from the nuclear membrane of the host cell (Fig. 21.14). The internal nucleic acid with its protein shell is referred to as the nucleocapsid. This family includes viruses that cause cold sores and genital herpes as well as chickenpox and infectious mononucleosis. The herpesviruses are difficult to cure completely as they are capable of remaining in a latent state where they cause no damage but merely replicate in step with the host cell. Latent herpesviruses are found in the nucleus where they replicate as circular extrachromosomal plasmids. Active infections may then break out again after a long period of quiescence, due to stress or other factors.

> Herpesviruses can remain latent for long periods of time.

   Poxviruses are the most complex animal viruses and are so large they may just be seen with a light microscope (Fig. 21.14). They are approximately 0.4 by 0.2 microns, compared to 1.0 by 0.5 microns for bacteria like *E. coli*. Unlike other animal DNA viruses which all replicate inside the cell nucleus, poxviruses replicate their

---

herpesviruses   A family of spherical animal DNA viruses with an outer envelope of material stolen from the nuclear membrane of the host cell
poxviruses   A family of large and complex double-stranded DNA animal viruses with 150–200 genes
simian virus 40 (SV40)   A small, spherical double-stranded DNA virus that causes cancer in monkeys by inserting its DNA into the host chromosome

FIGURE 21.14
Herpesviruses and Poxviruses

Many animal viruses use double-stranded DNA for their genomes. Herpesvirus is a simple virus that has a protein coat and outer envelope surrounding the double-stranded DNA genome. Poxvirus has two envelope layers. A protein layer, known as the palisade, is embedded within the core envelope. In addition, pre-made viral enzymes are also packaged with the genome to allow replication immediately upon infection. These viruses infect animals and in both cases the outermost viral membrane is derived from the membrane of the previous host cell.

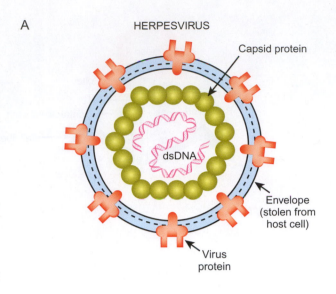

A HERPESVIRUS

Capsid protein

dsDNA

Envelope (stolen from host cell)

Virus protein

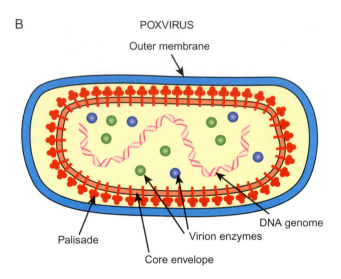

B POXVIRUS

Outer membrane

Palisade

Core envelope

Virion enzymes

DNA genome

Poxviruses are the largest animal viruses.

Strong promoters from cauliflower mosaic virus are used to express genes in the genetic engineering of plants.

double-stranded DNA in the cytoplasm of the host cell. Virus particles are manufactured inside subcellular factories known as inclusion bodies. Poxviruses have 150–200 genes, about the same number as the T4 family of complex bacterial viruses.

Plant viruses containing DNA are relatively rare. One example is **cauliflower mosaic virus (CMV)**, which has circular DNA inside a small spherical shell and kills cauliflower and its relatives such as cabbages and Brussels sprouts. This virus is of note because the promoters from some of its genes are extremely strong and have been used in plant genetic engineering to express insect-killing toxins or other transgenes. Most plant viruses have RNA genomes, as will be discussed below.

2.4. Mimivirus and other Giant Viruses

The largest viruses known, both in physical size and in genome size, are mimivirus and its relatives. As seems fitting for the largest known virus, mimivirus has been analyzed by the world's most powerful X-ray laser, the Linac Coherent Light Source (LCLS) at the SLAC National Accelerator Laboratory in Stanford, California, which is a billion times brighter than previous X-ray sources. Consequently, it is

cauliflower mosaic virus (CMV)   A small spherical virus of plants with circular DNA. Some of its promoters are used in plant genetic engineering

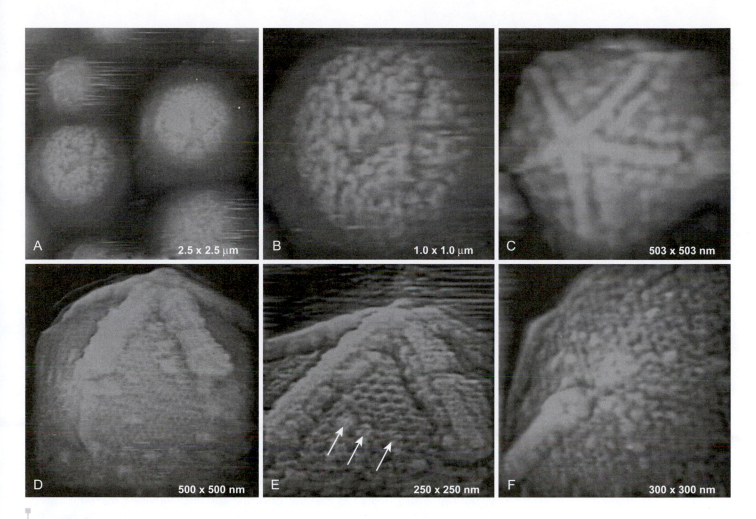

FIGURE 21.15
Atomic Force Microscopy of Mimivirus Particles

Mimivirus particles are coated with surface fibers except for a single vertex where a star-shaped depression is present (A and B). When these surface materials are removed by enzyme treatment, the star-shaped structure that attaches the virus particle to the membrane of its host cell becomes visible (C through F). *(Credit: Kuznetsov, et al., (2010) Atomic force microscopy investigation of the giant Mimivirus. Virology 104: 127–137.)*

no longer necessary to crystallize samples. Instead, untreated single virus particles are sprayed into the beam. They are vaporized into plasma at over 100,000 degrees, but not before an X-ray diffraction image is recorded. Mimivirus possess an outer protein shell around an inner capsid of protein lined with phospholipid membranes. The viral particle has a total diameter of about 0.75 μm and consists of an icosahedral capsid of about 0.5 μm diameter surrounded by filaments of 125 nm (Fig. 21.15). It contains 1.2 Mbp of double-stranded DNA. This was originally estimated to carry just over 900 genes, however, more recent data have increased the number of genes to 1018, of which 979 encode proteins, 6 encode tRNAs, and 33 are other noncoding RNAs. Thus, the mimivirus genome is larger than that of several bacteria, such as the mycoplasmas. Of the mimivirus proteins a mere 300 have functions predictable by homology; nonetheless, there is evidence that many of the genes encoding unique proteins are expressed and not merely "genetic junk."

Mimivirus belongs to a group of large viruses with large genomes known as the **nucleocytoplasmic large DNA viruses** (NCLDV). These include the poxviruses, which

nucleocytoplasmic large DNA viruses (NCLDV)   A grouping of different families of virus including mimivirus and poxvirus that are large in size and have large genomes and that typically infect eukaryotes

infect animals, and the phycodnaviruses, which infect algae. Mimivirus itself infects the amoeba *Acanthamoeba polyphaga*. As for the poxviruses, mimivirus particles are manufactured inside subcellular factories in the host cytoplasm. The NCLDV family of viruses share a set of homologous genes, mostly involved in DNA replication. DNA sequences found in environmental samples suggest that viruses of the mimivirus group are common in nature.

The relatively large amount of unique genetic information carried by giant viruses has prompted rethinking of ideas on the origins of viruses. The traditional view of virus origins holds that viruses originated as rogue elements of DNA or RNA that somehow escaped from cells. However, the recent massive increase in sequence data has shown that most virus proteins have no known homologs among present-day cellular proteins. This implies that much viral genetic information originated either during the replication of viruses themselves or that it derives from cellular lineages that are now extinct. Some specifically viral proteins, including capsid proteins, are shared among viruses that infect host cells from all three domains of life. This argues that some viruses are extremely ancient and that their ancestors predated the Last Universal Common Ancestor of cellular life forms (see Ch. 26 for details on cellular origins). The origin of viruses and, in particular, the invention of the virus particle as a means for moving between host cells is still a mystery.

Box 21.2 Sputnik the Virophage Infects Mimiviruses

Mimiviruses are themselves infected by parasites known as virophages, the first of which was named "Sputnik." Technically, virophages are satellite viruses; that is, they are defective viruses that need a helper virus to provide missing functions (see Section 5.1). Typically, satellite viruses infect and hence damage the same host cell as their helper virus. However, virophages differ in a major respect. Instead of replicating separately in the host cell they invade and infect the virus assembly compartment where mimivirus replicates. Consequently, they cripple the replication of mimivirus. This, paradoxically, benefits the host cell since the virophages are tiny relative to the colossal mimivirus (50 nm versus 750 nm; 20 genes versus 1000). Virophages also infect the phycodnaviruses that infect algae. They reduce algal death from phycodnaviruses and overall promote growth of the algae in the marine environment.

3. Viruses with RNA Genomes Have Very Few Genes

The smallest genomes are those of RNA-containing viruses. As explained in Chapter 23, "Mutations and Repair," this is related to their high mutation rates. RNA-based genomes have rates of mutation that are 1000-fold higher than DNA. Since each mutant gene product must continue to interact with other virus components, this limits the number of genes to no more than a dozen or so. The advantage of a high mutation rate to these viruses is that it allows them to change their proteins rapidly so evading recognition by host defense systems. The overall success of this strategy is shown by the fact that the majority of the best-known viral diseases, such as flu, are due to RNA viruses.

Given an RNA genome, there are three alternatives: double-stranded RNA, single-stranded positive RNA, and single-stranded negative RNA. The terms positive and negative refer to the coding strand and the non-coding strand, respectively. Remember that when mRNA is made only one of the two DNA strands is used as the template (see Fig. 21.16). The mRNA will be complementary in sequence to the template strand and identical to the non-transcribed strand of the DNA (except for using U instead of T in RNA). The non-transcribed strand of DNA and the mRNA

RNA viruses have small genomes and very high mutation rates.

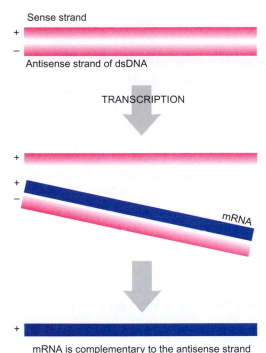

Sense strand

+

−

Antisense strand of dsDNA

TRANSCRIPTION

+

+

−

mRNA

+

mRNA is complementary to the antisense strand
and has the same sequence as the sense strand

■ **FIGURE 21.16**
Plus and Minus Strands

Double-stranded DNA has two strands, the sense strand and the antisense strand. The sense strand can also be referred to as the plus strand and the antisense strand as the minus strand. The sequence of mRNA is by definition the sense or coding sequence and is identical to the plus strand and complementary to the minus strand.

are both coding strands and are therefore **positive** or **"plus" strands,** whereas the template strand is the **negative** or **"minus" strand**.

If the virus contains the plus strand it can use this RNA directly as an mRNA to make proteins immediately upon entering the host cell. Such viruses are called positive-strand RNA viruses. Conversely, if the virus particle contains the minus strand, it must first make the complementary, positive strand before moving on to manufacture proteins. Despite this apparent disadvantage, it is the negative-stranded RNA viruses that appear to be the most successful and widespread.

Positive-strand RNA viruses contain RNA that can be directly translated.

Box 21.3 The Puzzling Origin of Vaccinia Virus

The story of how Jenner invented the technique of **vaccination** against infectious diseases is well known. He observed that milkmaids who worked in close contact with cattle often caught cowpox, a mild disease. After recovering, they became immune to the much more severe disease, smallpox. So Jenner then tested this connection experimentally in 1796. After deliberately infecting patients with cowpox he found that they had indeed become immune to smallpox. Jenner called the material he used "vaccine" after the cows where it originated (vacca = cow in Latin) and the novel technique was named vaccination.

For a long time it was believed that cowpox virus and the vaccine virus were identical. However, in 1939 it was found that cowpox virus (re-isolated from cows) and the virus cultures maintained for use in vaccination were actually quite distinct. The vaccine cultures were then named vaccinia to distinguish them from cowpox. It seems clear that Jenner himself did indeed vaccinate his patients with cowpox. Since viruses could not be cultured or purified in the laboratory until much later, those who followed Jenner constantly re-isolated new strains of what they thought was cowpox virus from cows and horses. At some time the original cowpox was replaced by a different virus, the present vaccinia virus. No virus presently circulating in the wild corresponds to vaccinia so its precise origin is unknown. Furthermore, the differences between vaccinia and cowpox are too large for mutation to be responsible for cowpox evolving into vaccinia. Presumably another poxvirus was circulating among cattle and horses during the nineteenth century and was eventually kept because it is even milder than cowpox.

---

negative or **"minus" strand**   The non-coding strand of RNA or DNA
positive or **"plus" strand**   The coding strand of RNA or DNA
vaccination   Immunization of a patient by introducing a milder or inactivated form of the disease-causing agent

3.1. Bacterial RNA Viruses

Some of the smallest virus genomes are found among the RNA-containing bacteriophages. A minimalist virus can get by with only three genes (Fig. 21.17). These encode the protein coat, the RNA replicase that replicates the viral genome, and the lysis protein that destroys the wall of the host cell so the newly-made virus particles can get out. An example is bacteriophage Qβ that has approximately 3,500 bases of single-stranded RNA and infects the bacterium *E. coli*. Qβ and its relatives, such as MS2, are small spherical viruses with single-stranded RNA and only three or four genes. They are "male-specific" as they only infect bacteria carrying the F-plasmid (i.e., "male" bacteria) because they attach to the sex pilus, which is only found on the surface of F^+ cells (see Ch. 25). Very few bacterial viruses have double-stranded RNA.

3.2. Double-Stranded RNA Viruses of Animals

Double-stranded RNA viruses are relatively uncommon. Among animal viruses, the reoviruses are the best-known double-stranded RNA viruses. They are spherical, with two concentric protein shells but no envelope and contain a dozen or so separate double-stranded RNA molecules, each coding for a single virus protein. Their most famous member is the rotavirus (Fig. 21.18), which causes infant diarrhea, a disease infecting vast numbers of babies and small children.

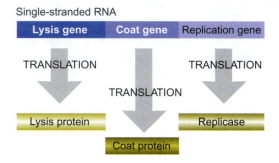

FIGURE 21.17
Genome of Bacteriophage Qβ

Qβ is a small spherical virus that contains a short single-stranded RNA genome. The virus only makes three proteins: a replicase to replicate its RNA, a coat protein, and lysis protein to lyse the infected bacterial cell. Although the genome is very small, the three proteins are all that the virus needs to successfully infect bacteria and replicate itself.

FIGURE 21.18
Rotaviruses Have Multiple Double-Stranded RNA Molecules

Rotavirus particles have two protein coats that encapsulate about a dozen double-stranded RNA molecules. Each of the double-stranded RNA pieces encodes one of the proteins necessary for viral survival and propagation.

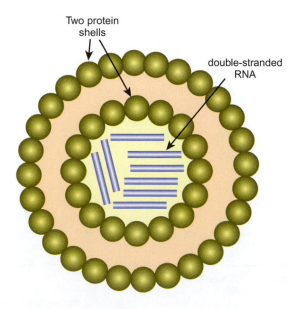

3.3. Positive-Stranded RNA Viruses Make Polyproteins

Picornaviruses are small, spherical single-stranded RNA viruses. They include Polio, the common cold, Hepatitis A, and Foot and Mouth Disease. Their genome is long enough for about a dozen genes. Since they are positive-strand RNA viruses, their RNA can be directly used as mRNA. The viral RNA does have a poly(A) tail attached to its 3' end but has no cap. Instead a protein, the Vpg protein, is attached covalently to the 5' end of the virus RNA. This allows the virus to use an internal ribosome entry site (see Ch. 13, Box 13.2). Not only does the virus avoid using cap-dependent translation itself, but it shuts it down, thus preventing host protein synthesis (see Focus on Relevant Research).

FOCUS ON RELEVANT RESEARCH

Ho BC, Yu SL, Chen JJ, Chang SY, Yan BS, Hong QS, Singh S, Kao CL, Chen HY, Su KY, Li KC, Cheng HW, Lee JY, Lee CN, and Yang P (2010) Enterovirus-induced miR-141 contributes to shutoff of host protein translation by targeting the translation initiation factor eIF4E. Cell Host Microbe 9: 58–69.

Because picornaviruses use an internal ribosome entry site for viral protein synthesis they can shut down host-cell protein synthesis by suppressing cap dependent translation. This is done by dual mechanisms. In the previously known mechanism, viral proteases cleave the host translation initiation factors eIF4GI and eIF4GII.

The newly-discovered second mechanism reported here depends on microRNA (see Ch. 18, Section 2.2.). The miR-141 microRNA is induced upon viral infection and targets another host translation initiation factor, eIF4E. The authors showed that an antisense RNA that was complementary to miR-141 prevented this and rescued eIF4E production. Note that miR-141 is a host encoded microRNA that is induced upon viral infection. This contrasts with the herpesviruses, which encode their own microRNAs.

However, there is another technical problem. Unlike bacteria where a single mRNA molecule may code for several proteins (an operon; see Ch. 11), in higher organisms each molecule of mRNA only encodes a single protein. Indeed, eukaryotic ribosomes will only translate the first reading frame on an RNA message, even if it carries several. Picornaviruses do indeed use their positive single-stranded RNA molecule directly as an mRNA. They avoid the problem by using the RNA to code for a single giant polypeptide that uses all of their genetic information (Fig. 21.19). This "**polyprotein**" is then chopped up into 10–20 smaller proteins.

> Polyproteins are made from a single giant gene and then cut up to give several final proteins.

3.4. Strategy of Negative-Strand RNA Viruses

The negative-strand RNA viruses are divided into several families and include the agents of well-known diseases such as rabies, mumps, measles, and influenza as well as more exotic emerging pathogens such as Ebola virus. In all of these, the single-stranded RNA in the virus particle is complementary to the mRNA and is therefore the minus strand. These viruses vary in shape and structure but are similar in having an outer envelope derived from the membrane of the host cell where they were assembled.

After infiltrating the cell, the first mission of a negative-strand RNA virus is to make its RNA double-stranded by synthesizing the corresponding positive RNA strand. Once it becomes double-stranded, it uses them both as templates. The plus strand is used as a template to manufacture more negative strands for the next generation of virus particles. The minus strand is used as a template to manufacture multiple positive strands that act as mRNA molecules (Fig. 21.20). This strategy is not only an effective division of labor but also avoids the problem of translating multiple reading frames on a single incoming virus RNA molecule.

> The RNA in negative-strand RNA viruses is the antisense strand.

---

polyprotein   A long polypeptide that is cut up to generate several smaller proteins

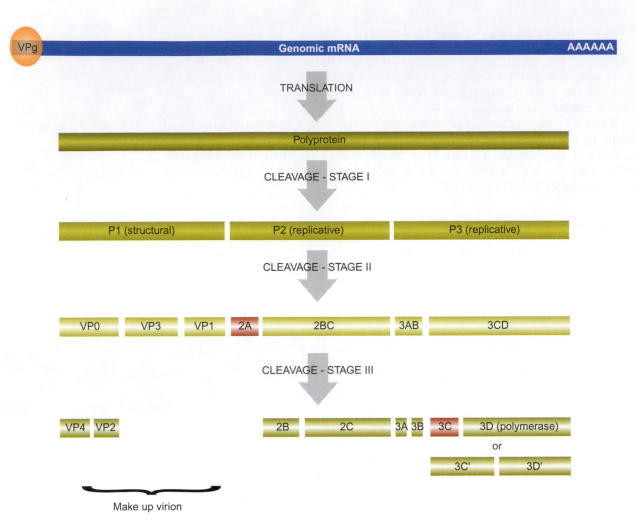

FIGURE 21.19
Polyprotein Strategy of Plus-Strand RNA Viruses

Rather than having many positive single-stranded RNA molecules, the picornaviruses have one large positive single-stranded RNA molecule (top). Translation of the single-stranded RNA produces a large polyprotein that is cleaved into successively smaller pieces. In Stage I the structural proteins are cleaved from the replicative proteins. In Stage II the segments are cleaved into separate proteins. For example, the P1 structural region is cleaved into VP0, VP3, and VP1 and then VP0 is cleaved into VP4 and VP2. The four proteins VP1, VP2, VP3, and VP4 combine to form the coat of the virion. The proteins shown in red are proteases. The Vpg protein is attached to the 5' end of the virus RNA.

3.5. Plant RNA Viruses

Most plant viruses are small with just a few genes and contain single-stranded RNA. Some, like cucumber mosaic virus, are spherical, while others are rod-shaped, like **tobacco mosaic virus (TMV)**, the most widespread plant virus. TMV attacks lots of plants including vegetables like the tomato, pepper, beet, and turnip as well as tobacco. The name "mosaic" refers to the yellowish blotches on the leaves of infected plants (Fig. 21.21). The virus coat consists of 2,130 identical protein molecules arranged in a helix with the RNA in the center (Fig. 21.22). TMV and its relatives have around 10,000 bases of RNA, enough for about 10 genes.

TMV is notable for its use in early structural investigations. X-ray crystallography and other physical methods were used to show that cylindrical viruses are in fact made of helically-arranged protein subunits. In addition, the capsid of TMV shows

tobacco mosaic virus   A filamentous single-stranded RNA virus that infects a wide range of plants

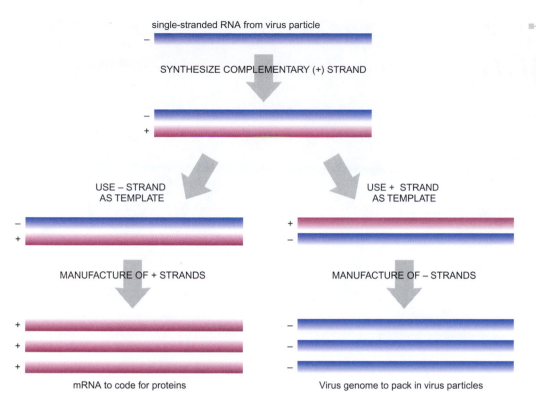

single-stranded RNA from virus particle

SYNTHESIZE COMPLEMENTARY (+) STRAND

USE – STRAND AS TEMPLATE

USE + STRAND AS TEMPLATE

MANUFACTURE OF + STRANDS

MANUFACTURE OF – STRANDS

mRNA to code for proteins

Virus genome to pack in virus particles

▶ **FIGURE 21.20**
Strategy of Negative-Strand RNA Virus

The negative strand of RNA has a sequence complementary to the coding strand. Therefore, viruses that use this type of genome must synthesize the complementary plus strand upon entry into the host cell. The plus RNA strand can then be used as a template to manufacture more viral genomes (right side). The negative RNA strand is then free to manufacture more copies of the plus strand (left side). These act as mRNA and direct viral protein synthesis.

▶ **FIGURE 21.21**
Plants with Tobacco Mosaic Virus

Tobacco plant infected with tobacco mosaic virus, Lafayette County, Florida. *(Credit: Norm Thomas, Photo Researchers, Inc.)*

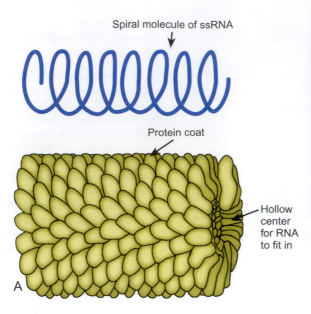

Spiral molecule of ssRNA

Protein coat

Hollow center for RNA to fit in

A

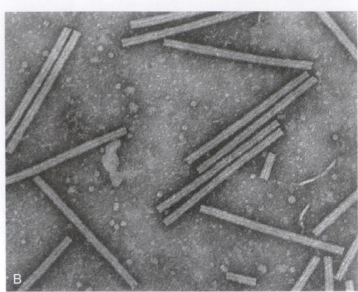

B

FIGURE 21.22
Tobacco Mosaic Virus Structure

Tobacco mosaic virus has a simple cylindrical structure. A) Its single-stranded RNA genome is packaged inside a shell of proteins that are arranged as a helix. B) Electron micrograph of tobacco mosaic virus rod-shaped particles. *(Credit: Rothamsted Experimental Station.)*

self-assembly. Purified capsid proteins plus virus RNA will spontaneously form virus particles when mixed. The virus may be disassembled by cooling and reassembled by gentle warming. This illustrates that the capsid proteins are held together largely by the hydrophobic force (see Ch. 14), which (unlike other forms of chemical bonding) grows weaker at lower temperatures.

4. Retroviruses Use Both RNA and DNA

The **retroviruses** infect animals and include **human immunodeficiency virus (HIV)** the notorious virus that causes **acquired immunodeficiency syndrome (AIDS)**. They are unique in having both RNA and DNA versions of their genome. The virus particle contains single-stranded RNA (Fig. 21.23) inside two protein shells surrounded by an outer envelope layer. The retrovirus envelope is made from the cell membrane of its previous victim. This membrane layer has retrovirus proteins both inserted through it and covering its surface (Fig. 21.24).

> Retroviruses carry RNA in the particle but make a DNA copy after entering the host cell.

What makes a retrovirus "retro" is that upon entering a host cell it reverses the normal flow of genetic information by making a DNA copy of its RNA genome. Upon infecting an animal cell the retrovirus converts the single-stranded RNA into a double-stranded DNA copy by using the enzyme **reverse transcriptase** (Fig. 21.25). The retrovirus DNA is then inserted into the host cell DNA. Once integrated, the retrovirus DNA is never excised but remains permanently inserted in the host genome. Consequently, retroviruses are impossible to get rid of completely after infection and integration, at least using current medications and procedures.

> Reverse transcriptase synthesizes DNA from an RNA template.

Retrovirus particles carry single-stranded RNA of the plus conformation. Although this RNA has a sequence identical to an mRNA, it is not used as an mRNA.

---

aquired immunodeficiency syndrome (AIDS)   Disease caused by human immunodeficiency virus (HIV) that damages the immune system
human immunodeficiency virus (HIV)   The retrovirus that causes AIDS
retroviruses   A family of animal viruses with single-stranded RNA inside two protein shells surrounded by an outer envelope. They possess reverse transcriptase, which is used to convert the single-stranded RNA version of the genome to a double-stranded DNA copy
reverse transcriptase   An enzyme that uses single-stranded RNA as a template for making double-stranded DNA
self-assembly   The spontaneous assembly of a biological structure from its subunits

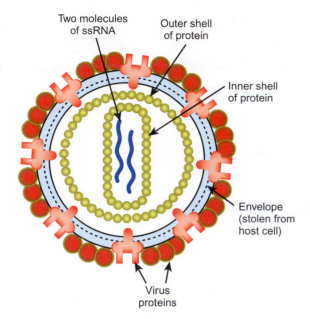

Two molecules of ssRNA
Outer shell of protein
Inner shell of protein
Envelope (stolen from host cell)
Virus proteins

■ **FIGURE 21.23**
Retrovirus Particle

Retroviral particles such as HIV have two single-stranded RNA (ssRNA) molecules surrounded by a double-protein shell. Like many other animal viruses, the retrovirus has an outer membrane derived from the host cell. This envelope contains viral proteins that coat the entire surface of the virus particle. Some of the proteins are embedded in the outer envelope while others sit upon the surface.

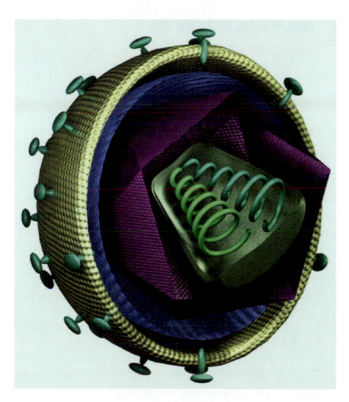

■ **FIGURE 21.24**
Structure of Retrovirus Particle

Three-dimensional cut-away view of a retrovirus particle. The two coils at the center (green and pale blue) are the copies of the virus genomic RNA. The purple and dark blue layers surrounding the RNA are protein shells. The outermost fawn layer is the viral envelope, with protruding proteins. *(Credit: The Universal Virus Database of the International Committee on Taxonomy of Viruses.)*

Instead, it is used as a template for reverse transcriptase to make a DNA copy of the retrovirus genome (Fig. 21.25). Reverse transcriptase first uses single-stranded RNA to make a complementary DNA strand. It then degrades the RNA and uses the first DNA strand as a template to make a second DNA strand. The process of making a double-stranded DNA copy from an RNA sequence is known as **reverse transcription** and its discovery forced the first major revision to the Central Dogma of Molecular Biology. Previously it was believed that information flowed only from DNA to RNA to protein, never in reverse.

reverse transcription   The process in which single-stranded RNA is used as a template for making double-stranded DNA

FIGURE 21.25
Reverse Transcriptase

During normal transcription, mRNA is made using the template strand of DNA (left). Retroviruses convert their single-stranded RNA (ssRNA) to a double-stranded DNA (dsDNA) molecule using an enzyme called reverse transcriptase. Making double-stranded DNA from single-stranded RNA is a two-step process. First, a complementary strand of DNA is made forming an RNA–DNA hybrid molecule. The original RNA is then degraded and a second DNA strand is made.

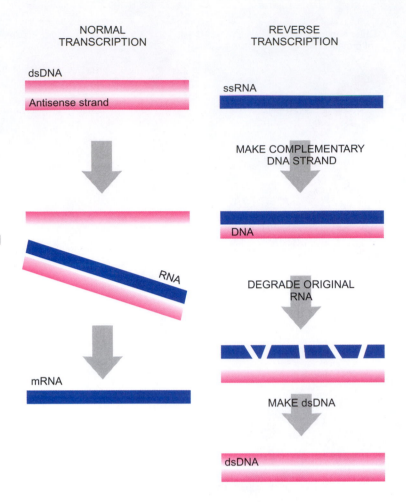

NORMAL TRANSCRIPTION

dsDNA

Antisense strand

RNA

mRNA

REVERSE TRANSCRIPTION

ssRNA

MAKE COMPLEMENTARY DNA STRAND

DNA

DEGRADE ORIGINAL RNA

MAKE dsDNA

dsDNA

The DNA version of a retrovirus integrates into the chromosomes of the host cell.

Although a retrovirus genome consists of single-stranded RNA, the virus particle actually contains two identical single-stranded RNA molecules (Fig. 21.26). These are bound together by base-pairing with two molecules of transfer RNA (tRNA) stolen from the previous host cell. In addition to binding the two virus RNA molecules together in the virion, the tRNA has another role. It is used as a primer by the reverse transcriptase when starting a new strand of DNA.

When a retrovirus enters a new cell the outer envelope merges with the host cell membrane and the core particle or nucleocapsid is released into the cytoplasm and disassembles, liberating the single-stranded RNA. One of the two single-stranded RNA molecules is then used by the reverse transcriptase to make the double-stranded DNA form of the retrovirus. This double-stranded DNA now enters the nucleus of the host cell. The retrovirus double-stranded DNA has repeated sequences at each end, the **long terminal repeats (LTRs)**. These are direct, not inverted, repeats and are required for integration of the retrovirus DNA into the host cell DNA (Fig. 21.27A). The site of integration is more or less random and once integrated, the retrovirus DNA is there to stay. It has become a permanent part of the host-cell chromosome.

The integrated retrovirus DNA is transcribed to give mRNA molecules that are capped and tailed just like the mRNA of a typical eukaryotic cell (Fig. 21.27B; also see Ch. 12, Section 1 for mRNA processing). The retrovirus RNA molecules exit the nucleus to the cytoplasm. Some are translated to produce viral proteins and others

long terminal repeats (LTRs)   Direct repeats found at the ends of the retrovirus genome which are required for integration of the retrovirus DNA into the host cell DNA

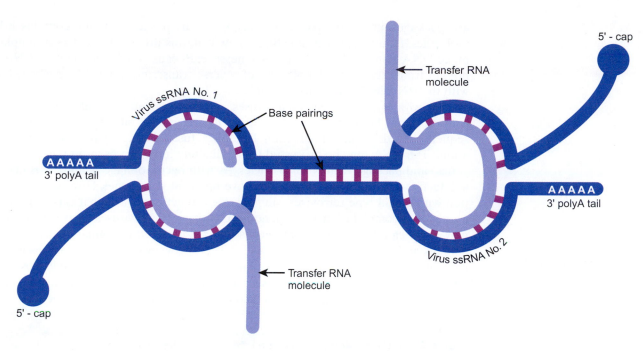

FIGURE 21.26
Retrovirus Particle Contains Two Single-Stranded RNA Molecules

The genome of retroviruses has a unique structure. Two molecules of single-stranded RNA (ssRNA) are held together by base-pairing. In addition, two transfer RNA molecules from the previous host are also base-paired with the two single-stranded RNA molecules. The tRNA acts as a primer for reverse transcriptase.

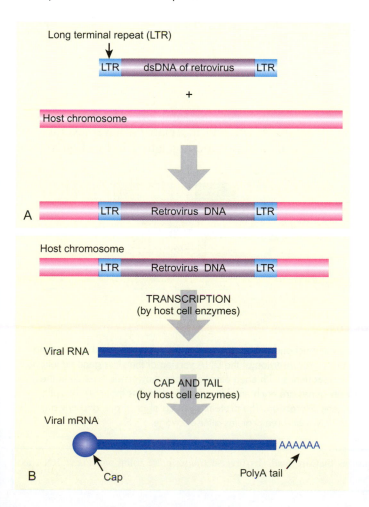

FIGURE 21.27
Integration and Transcription of Retrovirus DNA

A) After double-stranded retrovirus DNA is made by reverse transcriptase, the double-stranded DNA (dsDNA) integrates into the host chromosome. The DNA is flanked by two long terminal repeats (LTR) that facilitate the insertion. B) Once integrated into the host chromosome, the retrovirus DNA is transcribed and expressed as any other gene in the host cell. The retrovirus RNA is processed by addition of a 7-methyl-G-cap and a poly(A) tail.

are packaged into virus particles. Thus, the virus particle actually contains mRNA molecules. However, when infecting a new cell, this mRNA is used as a template to make DNA instead of being used as a message. Because the incoming viral RNA does not get translated, several molecules of reverse transcriptase must be packaged along with the RNA in the retrovirus particle.

As with integrated bacterial viruses, retroviruses can transduce host cell genes. Chromosomal DNA close to the retrovirus integration site may be fused to virus sequences by deletion of intervening DNA. The fused DNA may then be transcribed as a unit. The RNA may be processed (removing any introns present in the original host genes) and packaged into virus particles. As with bacteriophage lambda, this first gives a defective virus, in which host genes have replaced virus genes. However, recombination with a wild-type retrovirus can generate a functional virus that carries host genes as well as a complete retrovirus genome. Occasional retroviruses carry oncogenes and may cause cancer. Oncogenes are genes that are involved in regulating cell division in animals and when mutated cause cancer. The oncogenes carried by viruses are originally of animal origin and were picked up by the virus from some previous host animal. Although these cancer-causing viruses have attracted most notice, in principle any host gene close to the site of integration could be moved by retrovirus transduction.

> Retroviruses sometimes pick up host genes and carry them to another animal.

Box 21.4 Using Reverse Transcriptase to Make cDNA

Reverse transcriptase is now widely used in genetic engineering for making DNA copies of RNA (see Ch. 7 for details). It is especially useful in obtaining copies of eukaryotic genes that lack the non-coding intervening sequences. Many genes from higher animals have more non-coding DNA than coding sequence and for many purposes the coding sequence alone is easier to handle. Such **complementary DNA (cDNA)** copies of eukaryotic genes are made using the mRNA (which lacks introns due to processing) as a template. Although the cDNA lacks the introns found in the original natural DNA gene it still encodes the same protein.

4.1. Genome of the Retrovirus

A typical retrovirus has three major genes: *gag, pol,* and *env.* The human immunodeficiency virus (HIV), which causes AIDS, and its relatives, have several extra small genes (Fig. 21.28). Two of these, *tat* and *rev*, are regulatory in function and the

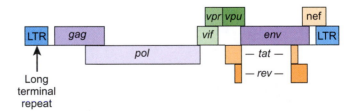

FIGURE 21.28
Genome of AIDS Retrovirus

The genome of a retrovirus is very compact and only encodes nine genes. It is flanked by two long terminal repeats (LTR) that are required for insertion of the DNA version of the virus genome into host DNA. Many of the genes overlap in sequence with each other and are shown side-by-side in the diagram. The regulatory *tat* and *rev* genes are each encoded as two segments that must be spliced together at the RNA level during gene expression. The protein products of *gag, pol,* and *env* are actually polyproteins that give rise to two or three proteins after cleavage.

> **complementary DNA (cDNA)** Copies of eukaryotic genes lacking introns that are made by reverse transcriptase using messenger RNA as template

others are involved in various aspects of virus maturation and/or modulating host cell metabolism. The products of the three major genes, *gag*, *pol*, and *env*, are polyproteins that are cleaved to generate several shorter proteins (Table 21.02).

TABLE 21.02	Retrovirus Gene Products and Proteins	
Gene Product	**Functions**	
Major Proteins	**Cleavage Products**	
Gag	MA	matrix protein (between nucleocapsid and envelope)
	CA	capsid protein (major structural protein of virion)
	NC	nucleocapsid protein (protects RNA)
Pol	PR	protease (cleaves precursor proteins)
	RT	reverse transcriptase (makes DNA copy of viral genome)
	IN	integrase (integrates virus DNA into host genome)
Env	SU	surface protein (forms spikes on virus surface that recognize host cell receptors, in particular **CD4 protein**)
	TM	transmembrane protein (fuses virus envelope with host cell membrane)
Regulatory Proteins		
Tat	Increases transcription of virus mRNA—binds to TAR (trans-activator response element).	
Rev	Regulates alternative splicing of virus RNA and export of RNA from nucleus into cytoplasm—binds to RRE (Rev response element)	
Accessory Proteins		
Nef	Decreases surface expression of the CD4 protein and has other effects on host cell immune system.	
Vif	a) Prevents premature cleavage of Gag and Pol polyproteins.	
	b) Attaches ubiquitin to APOBEC3G, which is then degraded by the proteasome.	
	(APOBEC3G is a host anti-viral protein that deaminates C to U in retrovirus cDNA.)	
Vpr	a) Prevents host cell entering mitosis.	
	b) Promotes entry of viral genome into nucleus.	
Vpu	a) Forms an ion channel and is needed for maturation and release of virions.	
	b) Newly-made host CD4 protein that has not yet reached the cell surface can bind Env protein and block virus release. Vpu binds to and promotes degradation of this CD4.	

CD4 protein   A protein found on the surface of T-cells that is one of the receptors recognized by HIV during infection

Box 21.5 Defunct Retroviruses Make Up 7% of the Human Genome

Endogenous retroviruses are the remains of integrated retroviruses that are no longer in circulation as virus particles. About 7% of human DNA is derived from these defunct sequences. The vast majority of such sequences are defective and many have large internal deletions. The DNA versions of the genomes of about 20 groups of endogenous retroviruses have been found located in our chromosomes. From one to several thousand copies of the genomes of each type of these retroviruses may be present. About 10 times as many single LTR elements have been found as endogenous retroviruses. These are created when homologous recombination between two LTR sequences deletes out all the internal retroviral genes, leaving behind a single LTR.

Based on sequence comparison, most endogenous retroviruses occupied their present sites before divergence of humans and other primates. A few have entered the human genome since the human/chimpanzee split (approximately 5 million years ago). These are all in the same place in different human groups, implying that new additions happen very infrequently and that once integrated the endogenous retroviruses rarely move from one position to another.

Whether these endogenous retroviruses are a health hazard remains disputed. For example, virus particles related to endogenous retrovirus sequences have been isolated from patients with multiple sclerosis, but whether these contribute to the disease or are a side effect of tissue damage is unknown. There is also evidence that some human promoters, enhancers, and alternative splice sites may have been recruited from the remains of endogenous retroviruses. For example, alternative splicing of the receptor for the hormone leptin, which controls fat metabolism, depends on sequences within a defunct retrovirus LTR.

5. Subviral Infectious Agents

A variety of **subviral agents** are found, most of which have RNA genomes. These may be self-replicating but lack a protein coat or they may be defective and depend on a helper virus to provide replication genes and/or a protein coat. These entities are listed in Table 21.03. We will discuss only the **satellite viruses** and the **viroids** in more detail.

5.1. Satellite Viruses

Some viruses are parasitic on other viruses. Such satellite viruses are only capable of replication and/or packaging when another virus, the **helper virus**, is present to provide the necessary gene products. Note that satellite viruses are not merely deleted variants of the helper virus. Deletion mutants of viruses do of course exist and can sometimes replicate in the presence of wild-type virus as helper. Satellite viruses are distinct entities that rely on a helper virus for a variety of functions. For example, satellite tobacco necrosis virus (STNV) has 1,221 nucleotides of single-stranded RNA and encodes a single protein—the virus coat protein. For replication it relies on tobacco necrosis virus (TNV), which contains 3,759 nucleotides of single-stranded RNA and encodes six proteins. The two viral RNAs share no homology and have quite distinct coat proteins. Replication of STNV requires only the RNA polymerase of TNV; the other TNV gene products are not used by the satellite.

Satellite viruses are incomplete and rely on a helper virus to provide essential genes.

---

helper virus   A virus that provides essential functions for defective viruses, satellite viruses, and satellite RNA
subviral agent   Infectious agents that are more primitive than viruses and encode fewer of their own functions
satellite virus   A defective virus that needs an unrelated helper virus to infect the same host cell in order to provide essential functions
viroid   Naked single-stranded circular RNA that forms a stable highly base-paired rod-like structure and replicates inside infected plant cells. Viroids do not encode any proteins but possess self-cleaving ribozyme activity

TABLE 21.03	Subviral Infectious Agents

Satellite Virus. A defective DNA or RNA virus that needs an unrelated helper virus to infect the same host cell. The helper virus provides the essential functions that it lacks, such as supplying a replicase, capsid proteins, or other functions that allow the satellite virus to survive inside the host cell.

Virusoid. An RNA molecule that does not encode any proteins and depends on a helper virus for replication and capsid formation. The virusoid genome resembles a viroid and consists of circular ssRNA with self-cleaving ribozyme activity.

Satellite RNA. Any small RNA molecule that requires a helper virus for replication and capsid formation. Satellite RNAs range from 200–1700 nucleotides and the larger ones may encode a protein. Virusoids are sometimes regarded as a subtype of satellite RNA.

Defective Interfering RNA (DI-RNA). Shorter RNA molecule derived from viral RNA by deletions that remove essential functions. DI-RNA depends on the original parental virus for replication. The presence of DI-RNA usually reduces the yield of the parental virus.

Viroid. Self-replicating pathogen of plants that consists solely of naked single-stranded circular RNA. The viroid RNA forms an extremely stable highly base-paired rod-like structure. Viroids do not code for any proteins. Most viroid RNAs have self-cleaving ribozyme activity.

Two other satellite viruses are hepatitis delta virus (HDV) and adeno-associated virus (AAV). HDV is a small single-stranded RNA satellite of Hepatitis B virus. Despite replicating in human cells, HDV resembles the smaller plant **satellite RNA** viruses. Adeno-associated virus is a satellite of adenovirus that is being used as a eukaryotic cloning vector in gene therapy.

The level of degeneracy varies from one satellite virus to another. For example, bacteriophage P4 of *E. coli* has 11.6 kb of double-stranded DNA encoding several genes. Alone, P4 can replicate as a plasmid or integrate into the host chromosome but cannot form virus particles. P4 relies on phage P2 as a helper for structural components and assembly of the phage particle. To switch on the P2 genes it needs, P4 deploys an anti-repressor protein (E protein). This prevents the repressor protein of P2 switching off the genes of the P2 virus genome. The structural genes of P2 are therefore expressed and the proteins they code for can be used by P4. The genome of P4 is much shorter than that of its helper P2. Therefore, P4 makes smaller capsids although it uses P2 components. The P4 *sid* gene product controls capsid *size* determination. When *sid* is expressed by P4, the capsids become too small for the P2 genome and P2 can no longer even be packaged by its own capsid protein!

5.2. Viroids are Naked Molecules of Infectious RNA

Viroids are infectious agents that consist only of naked RNA without any protective layer such as a protein coat. Viroids infect plants (but no other forms of life) and are replicated at the expense of the host cell. Viroid genomes are small single-stranded circles of RNA that are only 250–400 bases long. For example, the coconut cadang-cadang viroid has only 246 bases of RNA (Fig. 21.29).

Although viroid RNA is single-stranded, base-pairing occurs between bases on opposite halves of the circle to produce a rod-like structure (Fig. 21.30). Because viroids have no protein coat, they lack attachment proteins and cannot recognize and penetrate healthy cells as can a true virus. Viroids can infiltrate a plant cell only when its surrounding membrane is already damaged. They often take advantage of damage done to plant tissue by insects. Once inside, viroids may be passed from one plant cell to another via cellular junctions.

Viroids have no protective coat, just single-stranded RNA.

satellite RNA   Parasitic RNA molecule that requires a helper virus for replication and capsid formation

CUGGGGAAAU	CUACAGGGCA	CCCCAAAAAC	CACUGCAGGA	GAGGCCGCUU
GAGGGAUCCC	CGGGGAAACG	UCAAGCGAAU	CUGGGAAGGG	AGCGUACCUG
GGUCGAUCGU	GCGCGUUGGA	GGAGACUCCU	UCGUAGCUUC	GACGCCCGGC
CGCCCCUCCU	CGACCGCUUG	GGAGACUACC	CGGUGGAUAC	AACUCACGCG
GCUCUUACCU	GUUGUUAGUA	AAAAAAGGUG	UCCCUUUGUA	GCCCCU

FIGURE 21.29
Coconut Cadang-Cadang Viroid, Variant CCCVd.1

Complete sequence (246 bases) of coconut cadang-cadang viroid.

FIGURE 21.30
Viroid RNA Forms a Rod-Like Structure

Viroids are naked pieces of RNA, which can only infiltrate an already-damaged plant cell. The viroid is a single-stranded piece of circular RNA that has an unusual structure due to complementary base-pairing. Some form a simple rod-like structure, whereas other viroids have a complex branched structure.

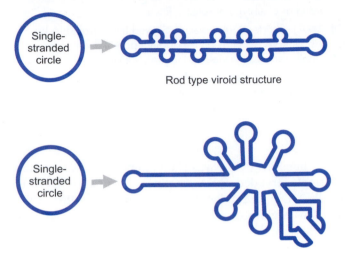

Rod type viroid structure

Branched viroid structure

Viroids have no protein coding genes, but the viroid RNA itself acts as a ribozyme.

Viruses all encode at least one protein needed for replication of the virus genome. However, viroid RNA does not contain any genes that encode proteins; it merely carries signals for its own replication by the host machinery. Although the viroid encodes no protein enzymes, the viroid RNA itself acts as a **ribozyme**; that is, the RNA catalyzes an enzymatic reaction. Whether or not a viroid has any genes depends on whether we count the sequence of RNA that possesses ribozyme activity as a gene.

Viroids replicate by a rolling circle mechanism (Fig. 21.31). The viroids own ribozyme activity is used for self-cleavage of the multimeric RNA generated during replication. Host enzymes provide all other functions. First, the circular plus strand is copied by host RNA polymerase to form a multimeric minus strand. Site-specific cleavage of this strand by the viroid ribozyme gives monomers that are circularized by a host RNA ligase. The minus-stranded circles are the templates for a second round of rolling circle replication by RNA polymerase. The resulting multimeric plus strand undergoes ribozyme cleavage to create monomers. These are circularized to produce the progeny viroids (circular, positive single-stranded RNA).

Most viroids replicate in the plant cell nucleus and rely on RNA polymerase II for RNA synthesis. A smaller group of viroids (e.g., chrysanthemum chlorotic mottle viroid) have a highly-branched structure, rather than a rod with bulges, and replicate in the chloroplast.

Some viroids cause no detectable symptoms in their host plants, whereas others cause massive damage. How viroids cause major damage to plants is still mysterious. Recently, it has been proposed that the viroid and/or its replicative intermediates trigger RNA interference or interfere with the related microRNA regulation (see Ch. 18).

ribozyme   An RNA enzyme; that is, an RNA molecule with catalytic activity

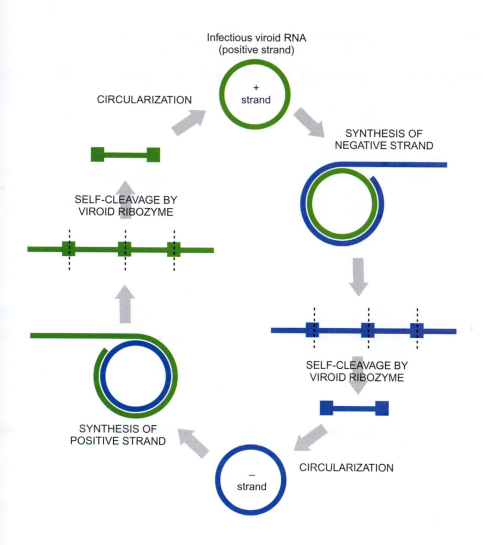

Infectious viroid RNA
(positive strand)

CIRCULARIZATION

+
strand

SYNTHESIS OF
NEGATIVE STRAND

SELF-CLEAVAGE BY
VIROID RIBOZYME

SELF-CLEAVAGE BY
VIROID RIBOZYME

SYNTHESIS OF
POSITIVE STRAND

CIRCULARIZATION

–
strand

► FIGURE 21.31
***Viroids Replicate by a
Rolling Circle Mechanism***

Two rounds of rolling circle
replication are used by viroids to
replicate themselves. Upon entry
into a plant cell, the circular,
positive single-stranded RNA uses
the plant RNA polymerase to make
a minus strand. The polymerase
continues to make multiple copies
using the rolling circle mechanism.
The linear, negative single-stranded
RNA uses its own catalytic activity
to cut itself into genome-sized units
that are circularized. The circular,
negative single-stranded RNA then
undergoes another round of rolling
circle replication and self-cleavage
to produce multiple copies of the
linear plus strand. Finally, these are
circularized to give the infectious
circular, positive single-stranded
RNA form.

5.3. Prions are Infectious Proteins

Until recently, all infectious diseases were thought to be caused by germs that carry at
least some of their own genes. Some diseases are due to living cells like bacteria, while
others are due to viruses or viroids, but all contain their own genetic information in
the form of DNA or RNA. However, several bizarre diseases of the nervous system
are caused by infectious agents containing absolutely no nucleic acid. These diseases
are due to rogue proteins known as **prions**. The first to be described was a disease of
sheep known as **scrapie**. The name refers to the tendency of diseased sheep to scrape
themselves against trees, fences, etc., which often creates bare patches. Records of
scrapie go back to the early 1700s in Europe.

The **prion protein (PrP)** is actually encoded by a gene (*Prnp*) belonging to the
victim. This gene is transcribed and translated normally and produces a protein found
attached to the outside surface of nerve cells, especially in the brain. The prion protein is a glycoprotein, with one or two attached carbohydrate groups, and is attached
to the cell membrane by a covalently-attached phospholipid molecule (phosphatidyl
inositol). Its proper function in the brain is still uncertain, although it binds copper

> A few diseases of the nervous
> system are caused by infectious
> proteins known as prions.

---

prion   A protein that can misfold into an alternative pathological form that then promotes its own formation auto-catalytically. Improperly folded
  prion proteins are responsible for the neurodegenerative diseases known as spongiform encephalopathies that include scrapie, kuru, and BSE
prion protein (PrP)   The prion protein found in the nervous tissue of mammals and whose improperly folded form is responsible for prion
  diseases
scrapie   An infectious disease of sheep that causes degeneration of the brain and is caused by misfolded prion proteins

FIGURE 21.32
Re-Folding of Normal Prion is Triggered by Rogue Prion

Normal prions can be induced to change their conformation by contact with a misfolded prion. When the pathological improperly folded form PrP^{sc} comes in contact with the normal prion, PrP^{c}, the normal protein changes into the abnormal form. The alternate conformation has a tendency to aggregate, forming clumps that damage nerve cells.

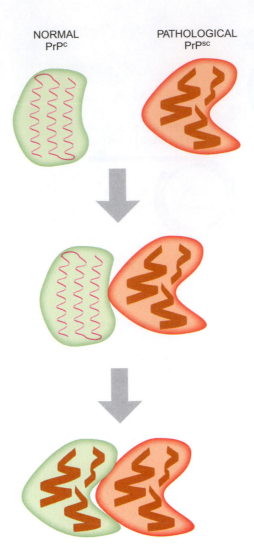

NORMAL
PrP^{c}

PATHOLOGICAL
PrP^{sc}

ions and may be involved in protection against oxidative stress. More recently, evidence has emerged suggesting that the normal form of the prion protein may protect brain cells by dampening the response to N-Methyl-D-Aspartate (NMDA) receptors.

The critical property of the prion protein is that it can fold into two alternative conformations. Occasionally, the normal, properly-folded form (**cellular PrP** or **PrP^{C}**) rearranges to produce the pathological form of the protein (**scrapie PrP** or **PrP^{Sc}**), which then polymerizes to form fibrillar aggregates. The prion protein is not chemically altered; it merely changes shape. Healthy prions consist largely of α-helical segments, whereas rogue prions have less α-helix and more β-sheet regions instead. PrP^{C} is estimated to have 42% α-helix and 3% β-sheet, whereas PrP^{Sc} has 30% α-helix and 43% β-sheet (see Ch. 14 for protein structures). The presence of the improperly folded PrP^{Sc} form induces the normal PrP^{C} proteins to also change conformation (Fig. 21.32). Thus, once a few molecules of PrP^{Sc} are present, they propagate themselves by catalyzing the conversion of PrP^{C} to more of the PrP^{Sc} isoform. Precisely how these changes in protein conformation and aggregation damage nerve cells is still obscure. Nonetheless, once the change to the scrapie agent has been initiated, slow nervous degeneration and eventual death are inevitable.

Prion proteins exist in two alternative forms. The pathogenic form provokes the normal form to change conformation so making more of the pathogenic version.

cellular PrP (PrP^{C})   The healthy, normal form of the prion protein
scrapie PrP (PrP^{Sc})   The pathological form of the prion protein, sometimes known as the scrapie agent

Prion disease occurs by three mechanisms, all of which lead to a similar final result. In sporadic prion disease, a spontaneous change in conformation occurs in a PrP molecule within a brain cell of a normal, uninfected individual. This change is propagated as described above, and over many years the pathological form of the prion protein gradually accumulates until symptoms appear, usually in old age. This happens to about one person in a million.

In the inherited form of prion disease a mutation in the prion gene results in a mutant prion protein that changes more often into the disease-causing form. This scenario is little different than a typical inherited disease. Several different mutations are known in the *Prnp* gene and the symptoms of the resulting disease vary slightly. Three main subcategories are known: Creutzfeldt-Jacob disease (CJD), Gerstmann-Straussler-Scheinker syndrome, and Fatal Familial Insomnia. These account for about 15% of the human cases of prion disease.

In infectious prion disease, also known as **transmissable spongiform encephalopathy (TSE)**, the pathological form of the prion protein is passed from one individual to another. Such an infection can be passed from one cell to another within the same organism, or one animal to another. The two individuals may be of the same or different species. Infection of a new victim by prions is relatively difficult. It requires uptake of rogue prion proteins from infected nervous tissue, especially brain, but the details of infection remain obscure. An epidemic of **mad cow disease,** officially known as **bovine spongiform encephalopathy (BSE)**, began in England in 1986. Although originally blamed on scrapie, it is due to transfer of pathogenic prions from diseased cattle.

What happens if an incoming PrP^{sc} protein finds no healthy prions to alter? The answer is that no disease results. This may seem surprising, but is logical given the mechanism of prion action. Mice that have been engineered with both copies of the *Prnp* gene disrupted and which thus do not produce PrP^c are resistant to infection with pathogenic prions. Interestingly, such mice also live for a normal time and do not show any obvious behavioral abnormalities.

It was originally believed that prions were unique to mammals. They have not been found in any other group of animals, although prions have been found in yeast. Instead of causing disease the prions adapt the yeast to alternate environmental conditions (see Focus on Relevant Research).

> Prion disease may be infectious, inherited, or spontaneous.

Halfmann R, Alberti S, and Lindquist S (2010) Prions, protein homeostasis, and phenotypic diversity. Trends in Cell Biology Vol. 20: 125–134.

Although the prions responsible for mammalian brain disease are the most famous, prions with completely different effects have also been found in other unrelated organisms. Several fungi possess prions and these have been most intensively investigated in the yeast *Saccharomyces cerevisiae*.

The authors argue that the fundamental property of prions is that they form protein aggregates that can be inherited epigenetically. In other words, they provide molecular memories made of protein rather than nucleic acids. The two alternative prion states in yeast are not "healthy" versus "pathogenic" as in mammals.

On the contrary, they adapt the yeast to different environmental conditions by affecting the genetic regulation of alternative sets of genes. Examples include the use of alternative nitrogen or carbon sources. Such prions may be regarded as "bet-hedging" mechanisms that allow the yeast to adapt rapidly to changing environments.

FOCUS ON RELEVANT RESEARCH

---

bovine spongiform encephalopathy (BSE)   Same as mad cow disease
mad cow disease   Infectious prion disease transmitted from cattle to humans
transmissable spongiform encephalopathy (TSE)   Technical name for infectious prion disease

Key Concepts

- Virus particles consist of genetic information (the virus genome) enclosed in protein shells (the capsid).
- Viruses must enter and take over a host cell in order to replicate.
- Viruses that infect bacteria are known as bacteriophage.
- Viruses may use either DNA or RNA as their genetic material.
- Some viruses integrate into the chromosomes of their host cells and are maintained in a largely inactive state.
- Viruses are extremely diverse in the structures of both their particles and their genomes as well as in their modes of replication.
- Viruses may have anywhere from three to just over a thousand genes.
- Complex bacterial viruses have a head and tail structure and contain double-stranded DNA.
- Pox viruses are the most complex animal viruses. Unlike most viruses, which replicate in the nucleus, pox viruses replicate in the cytoplasm inside subcellular factories.
- Mimivirus and its relatives are the largest viruses known and some may have 1000 genes. They infect primitive eukaryotes.
- Viruses with RNA genomes mutate rapidly and possess relatively few genes.
- Positive-stranded RNA viruses make giant polyproteins that are cleaved into individual proteins.
- Negative-strand RNA viruses make the complementary (sense) strand upon entering their host cell.
- Retroviruses use both RNA and DNA for their genome at different stages of their lifecycle.
- Subviral infectious agents include various defective viruses as well as viroids and prions.
- Satellite viruses are defective viruses that rely on other, intact viruses, to provide missing functions.
- Viroids are short molecules of infectious RNA that lack protein coats and have no protein coding sequences.
- Prions are infectious proteins that exist in two alternative conformations and lack nucleic acids in their infectious form.

Review Questions

1. What are viruses? What is needed in order for viruses to multiply?
2. What is a key property of a living cell that viruses do not have?
3. Describe the six phases of a typical virus lifecycle.
4. What molecules do viruses recognize on host cells?
5. Once inside the host cell, what are the two main functions of the virus genome?
6. What are the differences between "early genes" and "late genes"?
7. Why can bacteriophage not just "merge" with a bacterial cell?
8. What important experiment demonstrated that genetic information was carried on DNA? What scientists performed this experiment? Describe the experiment in detail.
9. What is the term used to describe an integrated bacterial virus? How is the viral genome replicated in this situation?
10. How does integration of lambda DNA occur? Into which site does this integration happen?
11. What are retroviruses? What must first happen before a retroviral genome can be integrated into the host genome?
12. Upon entry into a host, what is the first thing that single-stranded DNA or RNA viruses must do? What is the molecule it makes first called?

13. What is the name of the enzyme required by RNA viruses?
14. What is a property of very small viruses, such as ΦX174 (phiX174), which allows more genes to be encoded on a small amount of genetic material?
15. What is meant by "male-specific" for bacteriophage M13? How is M13 useful in molecular biology?
16. What are the components of a complex bacterial virus? Name some examples of complex bacterial viruses.
17. Where do poxviruses replicate their genetic material?
18. Compare viral RNA genomes with viral DNA genomes with regards to size and mutation rates.
19. What are the three types of viral RNA genomes?
20. What is meant by single-stranded positive RNA and single-stranded negative RNA? Name one advantage to having a positive-strand RNA genome.
21. What is the minimum number of genes needed in the smallest bacterial RNA virus? What do these genes encode? Give two examples of this type of virus and name which bacteria they infect.
22. Give an example of a double-stranded RNA virus of animals.
23. What are some examples of single-stranded, positive RNA viruses? What are the characteristic structural features of their RNA?
24. What is a polyprotein? Why can't a single-stranded RNA virus genome be translated into individual proteins like a bacterial operon?
25. Give examples of negative-strand RNA viruses. Upon entry into a host cell, what is the first process performed by these viruses?
26. What advantage do negative-strand RNA viruses have over positive-strand RNA viruses?
27. What is "self-assembly" as it refers to viruses? Which types of interaction are involved in this process? What can hinder or help this process?
28. Why are retroviruses unique?
29. What type of genome does HIV have?
30. What does reverse transcriptase do? What happens to the single-stranded RNA template?
31. How did the discovery of reverse transcriptase affect the Central Dogma of Molecular Biology?
32. Where do retroviral genomes integrate into the host genome? Can retroviruses ever be excised from the host genome once integrated?
33. What is the role of tRNA in the retroviral lifecycle?
34. What enzyme must be packaged inside a new retroviral particle? Why?
35. What are oncogenes? Why can retroviruses transfer them?
36. What are the three major genes of HIV? What proteins do these genes encode?
37. What are satellite viruses? What do helper viruses provide?
38. What are viroids? How do they replicate? What is a ribozyme? What is the function of a ribozyme in viroid replication?
39. What is a prion? What are the two different prion forms or conformations?
40. What are the three mechanisms by which prion disease occurs?

Conceptual Questions

1. Lambda bacteriophage is able to grow both in a lytic cycle and a lysogenic cycle. A researcher found a mutant strain of lambda, λLytonly, that cannot enter a lysogenic cycle. The researcher finds this mutant genome is simply missing one short region, and no other defects are found. Explain what λ genomic sequence is missing and why it prevents the virus from entering lysogeny.
2. Why are plasmids and viruses considered gene creatures, and usually not classified as living organisms? Some scientists do regard viruses as alive. Give your opinion on this issue.
3. Discuss the consequences of a frameshift mutation in the gene for protein E of bacteriophage ΦX174. What is the consequence of a conservative substitution in this gene?
4. Why do you think that retroviruses are so prone to mistakes during replication of their genome? Think about how DNA polymerase prevents errors.
5. Adeno-associated virus (AAV) is used as a cloning vector in gene therapy for humans. Why do you think that researchers chose this virus as a way to add a gene into human cells?

Chapter 22

Mobile DNA

When geneticists talk about "**mobile DNA**" they are not referring to the DNA of chromosomes or plasmids being transferred between cells. Mobile DNA is a special class of DNA that can move between different locations within chromosomes, plasmids, and virus genomes. In other words, we have here the idea of a host DNA molecule providing an environment with another, smaller segment of DNA acting as an inhabitant or parasite. Most mobile DNA consists of transposable elements of one kind or another, although the much rarer integrons and homing introns are also included here. Movement of mobile DNA requires the action of proteins that bind recognition sequences on the mobile DNA and enzymes that catalyze the movement. These functions may be carried out by the same protein, the transposase in the case of simpler forms of transposition. In other cases, the mechanism may be more complicated and multiple proteins may be involved.

1. Subcellular Genetic Elements as Gene Creatures

In addition to the chromosomes of living cells, a whole variety of **genetic elements** are found in nature. These range in complexity from viruses that possess both a genome plus a protective protein shell to genetic elements such as plasmids and viroids that are merely single molecules

---

mobile DNA   Segment of DNA that moves from site-to-site within or between other molecules of DNA

genetic element   Any molecule or segment of DNA or RNA that carries genetic information and acts as a heritable unit

---

Molecular Biology.

of nucleic acid (see Chs. 20 and 21). These subcellular elements are all parasitic in the sense that they rely on a host cell to provide energy and raw materials. In some cases, as with many viruses, the host cell may be destroyed. In other cases, as with most plasmids, the host cell is unharmed and may, in fact, benefit by genes carried on the plasmid.

Since all of these elements possess their own genetic information they may be regarded as life forms of a sort. We will sometimes refer to them as **gene creatures** since they lack their own cells but carry genetic information. From the perspective of a cell, gene creatures are parasites. From the perspective of a gene creature, the cell it inhabits is simply its environment. The latter view may seem strange, but it often helps to understand genetic mechanisms if we view them from the perspective of the genetic element, rather than the cell or organism.

Some of these genetic elements do not even exist as independent genomes but are found as lengths of DNA integrated into other DNA molecules. Some of these have alter egos as viruses or plasmids, whereas others are only ever found as segments of integrated DNA. Some of these integrated DNA segments are able to move around from site-to-site within host DNA molecules and are known as **transposable elements** or **transposons**. Other stretches of parasitic DNA are stuck permanently where they are and are probably the remains of once mobile gene creatures. They have degenerated into what was originally called junk DNA. Much of the large human genome is comprised of these types of DNA that are no longer active.

> Living cells can be viewed as providing habitats for viruses and other genetic elements.

2. Most Mobile DNA Consists of Transposable Elements

Although the DNA of certain viruses can insert itself into the chromosomes of the host cell, most mobile DNA consists of genetic elements known as transposons or transposable elements. They are also sometimes called "**jumping genes**" because they may hop around from place to place among the chromosomes. The process of jumping from one site to another is called **transposition**. Transposons are not merely dependent on a host cell like plasmids and viruses; they are dependent on a host DNA molecule! Transposons are always inserted into other DNA molecules so they are never free as separate molecules (Fig. 22.01).

As noted in Chapter 10, any molecule of DNA that possesses its own origin of replication is known as a **replicon**. Chromosomes, plasmids, and virus genomes are

> Certain segments of mobile DNA possess the ability to move from place to place on the same or different DNA molecules.

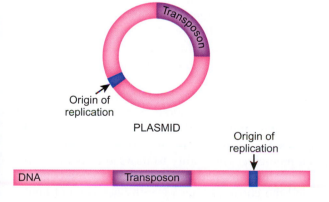

Origin of replication

PLASMID

Origin of replication

DNA   Transposon

CHROMOSOME

FIGURE 22.01
Transposable Elements are Never Free

Transposable elements are stretches of DNA able to move from one position to another, but are always found within a DNA molecule such as a bacterial plasmid (top) or a eukaryotic chromosome (bottom). Transposons do not contain their own origin of replication, but rely on the host DNA to provide this feature.

---

gene creature   Genetic entity that consists primarily of genetic information, sometimes with a protective covering, but without its own machinery to generate energy or replicate macromolecules

jumping gene   Popular name for a transposable element

transposable element   A mobile segment of DNA that is always inserted in another host molecule of DNA. It has no origin of replication of its own and relies on the host DNA for replication. Includes both DNA-based transposons and retrotransposons

transposition   The process by which a transposon moves from one host DNA molecule to another

transposon   Same as transposable element, although the term is usually restricted to DNA-based elements that do not use reverse transcriptase

replicon   A molecule of DNA or RNA that is self-replicating; that is, it has its own origin of replication

replicons and may be regarded as self-replicating. In contrast, transposons lack a replication origin of their own and are not replicons. They can only be replicated by integrating themselves into other molecules of DNA, such as chromosomes, plasmids, or virus DNA. As long as the DNA molecule of which the transposon is part gets replicated, the transposon will also be replicated. If the transposon inserts itself into a DNA molecule with no future, the transposon "dies" with it.

Transposable elements are classified based on their mechanism of movement. The major division is between those that move via an RNA intermediate and need reverse transcriptase to generate this, and those with a DNA-based mechanism. The DNA-based transposons are subdivided into two main groups according to whether a new copy is generated during transposition (complex or replicative transposition) or whether the original copy moves, leaving a gap in the DNA in its previous location (conservative or "cut-and-paste" transposition). Table 22.01 summarizes the main groups of transposable elements whose properties will be discussed in more detail below.

2.1. The Essential Parts of a DNA Transposon

All DNA-based transposons possess two essential features. First, they have **inverted repeats** at either end. Second, transposons must have at least one gene that encodes the **transposase**, the enzyme needed for movement (Fig. 22.02). The transposase recognizes the inverted repeats and moves the segment of DNA bounded by them from one site to another. The frequency of transposition varies from one transposon to another. Typically, it ranges from 1 in 1,000 to 1 in 10,000 per transposon per cell generation.

In fact, the transposase recognizes two different DNA sequences. First, it recognizes the inverted repeats at the transposon ends, and this tells it which piece of DNA must be moved. In addition, the transposase must also recognize a specific sequence on the DNA molecule it has chosen as its future home. This is known as the **target sequence** and is usually from three to nine base pairs long. Most often it is an odd number of base pairs, with nine base pairs being the most common length. Transposases will often accept a target site with a sequence that is a near match to the preferred target sequence. In some cases, recognition of the target sequence is so lax that it may be difficult to derive a consensus sequence. Due to the short length and low specificity, multiple copies of the target sequence will be found on most DNA molecules of reasonable length and insertion will often be almost random. Whenever a transposon moves, the target sequence is duplicated due to the mechanism of transposition (see below). The result is that two identical copies of the target sequence are found, one on each side of the transposon.

Many larger transposons carry a variety of genes unrelated to transposition itself. Antibiotic resistance genes, virulence genes, metabolic genes, and others may be located within transposons and become mobile within the genome. Some of the first transposons identified have genes for antibiotic resistance, which protect the host bacteria from attack by human medicine. The ability to move blocks of genetic material from one DNA molecule to another has made transposons very important in the evolution of plasmids, viruses, and the chromosomes of both bacteria and higher organisms.

> Chromosomes, plasmids, and virus genomes are capable of self-replication; transposons are not.

> The enzymes responsible for transposition must recognize repeated sequences at each end of the transposon.

> Although transposons are inserted at target sequences, the specificity is low and transposition appears more or less random.

inverted repeats   Sequences of DNA that are the same but in opposite orientations
transposase   Enzyme responsible for moving a transposon
target sequence   Sequence on host DNA molecule into which a transposon inserts itself

TABLE 22.01	Transposons and Related Elements			
Type of Element	**Length (approximate)**	**Terminal Repeats**	**Mechanism of Mobility**	**Mechanism to Exit Cell**
DNA-based elements				
Insertion sequence	750–1,500	Inverted	cut-and-paste transposition	No
Simple transposon	1,300–5,000	Inverted	cut-and-paste transposition	No
Composite transposon	2,500–10,000	IS sequences	cut-and-paste transposition	No
Complex transposon	5,000	Inverted	replicative transposition	No
Bacteriophage Mu	37 kb	Inverted	replicative transposition	Virus particle
Conjugative transposon	30 kb–150 kb	None	transfer plus integration	Conjugation
Integron & cassettes	1,500	None	accumulation of cassettes	No
Helitrons	5,000–10,000	None	replicase plus helicase	No
Retroelements (possess reverse transcriptase)				
Retrovirus	7,000–10,000	Direct (LTR)	via RNA intermediate	Virus particle
Endogenous retrovirus	7,000–10,000	Direct (LTR)	via RNA intermediate	Defective virus particle
Retrovirus-like element	7,000–10,000	Direct (LTR)	via RNA intermediate	No
Retrotransposon	6,000	Direct	via RNA intermediate	No
LINE	6,500	None	via RNA intermediate	No
Retron	1,300–2,000	None	unknown	No
Retrointron	2,000–3,000	None	via RNA intermediate	No
Retroderived elements (or "Retrotranscripts")				
Processed pseudogene	1,000–3,000	None	immobile	No
SINE	300	None	transcription & reintegration	No

Notes: None of these elements have their own origin of replication, except for conjugative transposons, which have an origin of transfer (but <u>not</u> of vegetative replication).

Length refers to "typical" elements. Since many transposable elements may acquire extra internal DNA by a variety of mechanisms, longer elements carrying a variety of extra genes may be found. In addition, shorter defective elements are frequently found in many classes of transposable element. The lengths given refer to intact functional elements.

The length of Mu DNA inside a virus particle is 39 kb and consists of 37 kb of Mu DNA plus approximately 2,000 bases of chromosomal DNA attached to the Mu ends. Note that when Mu first enters a new host cell the original integration event uses cut-and-paste transposition, not replicative transposition.

All intermediates between fully-functional retroviruses and sequences presumed to be derived from defective retroviruses are found in eukaryotic genomes, as well as individual LTR elements with no internal sequences. Consequently, the classification of retroelements varies somewhat among different authors. LINES and related elements are sometimes known as non-LTR-retrotransposons to distinguish them from "true" retrotransposons, which have terminal repeats equivalent to the LTRs of retroviruses.

Retrointrons are mobile group I or II introns that encode reverse transcriptase.

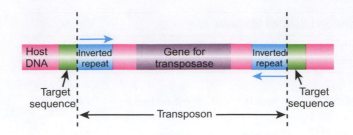

FIGURE 22.02
Essential Components of a Transposon

Essential features of a transposon are two inverted repeats at the ends and a gene for transposase. Other, non-essential genes may be found between the inverted repeats. During transposition, the transposon duplicates the host DNA target sequence and therefore one copy flanks each side of the transposon.

FIGURE 22.03
Outline of Conservative Transposition

The transposon moves from one DNA molecule to another. It inserts into the target sequence on the recipient DNA molecule and leaves behind a double-stranded break in its original location.

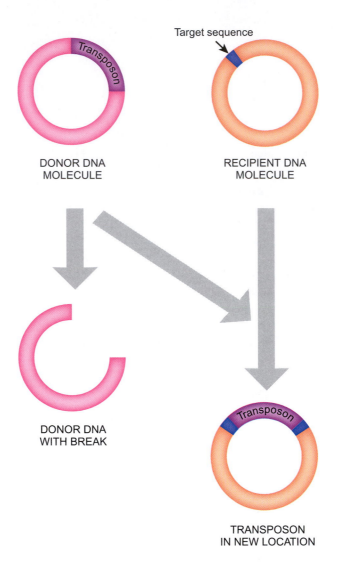

2.2. Movement by Conservative Transposition

The simplest transposons move by a mechanism known as **conservative** or **"cut-and-paste" transposition** (Fig. 22.03). This requires three elements: the transposase, the inverted repeats at the ends of the transposon, and a suitable target sequence on another segment of DNA somewhere else in the cell. The transposon may move to

conservative transposition Same as cut-and-paste transposition
cut-and-paste transposition Type of transposition in which a transposon is completely excised from its original location and moves as a whole unit to another site

another site in the same molecule of DNA or to a separate molecule of DNA. The DNA molecule into which a transposon jumps can be a plasmid, a virus, or a chromosome; any DNA molecule will do, as long as it has a reasonable target sequence.

The transposase starts by binding the inverted repeats at the transposon ends. It then cuts the transposon loose from its original site (Fig. 22.04A). Next, the transposase finds a suitable target sequence in the molecule of DNA that will be the transposon's new home. It makes a staggered cut that opens the target sequence to give overhanging ends (Fig. 22.04B). Finally, it inserts the transposon into the gap.

This leaves a structure with two short regions of single-stranded DNA. These are recognized by the bacterial host, which synthesizes the second strand. The net result is that the transposon has moved and the target sequence has been duplicated in the process. This type of transposition process is known as conservative transposition because the DNA of the transposon is not altered during the move.

Where the transposon cut itself out of its original home, it leaves a double-stranded break in the DNA. There is a significant likelihood that this damaged DNA molecule will not be repaired and is doomed. Clearly, high-frequency transposition would severely damage the host chromosomal DNA. Even if the break is sealed, the duplicated target sequence that is left behind may cause mutations that alter gene expression. Consequently, as remarked above, transposition must be tightly regulated.

> Conservative transposition leaves behind a double-stranded break in the DNA.

2.3. Complex Transposons Move by Replicative Transposition

Transposition does not always leave behind damaged DNA with double-stranded breaks. Some transposons are capable of **replicative transposition**, during which the transposon creates a second copy of itself (Fig. 22.05). Consequently, both the original home site and the newly-selected target location end up with a copy of the transposon. The original home DNA molecule is not abandoned or damaged. Transposons using this mechanism are known as **complex transposons** because the process is more complex than the simple cut-and-paste mechanism described above.

Complex transposons have a transposase that recognizes their inverted repeats and the host target sequence just like other kinds of transposons. In addition, replicative transposons need an extra enzyme, **resolvase**, and an extra DNA sequence, the **internal resolution site (IRS)** that is recognized by the resolvase (Fig. 22.06). Although complex transposons are replicated while moving, they are not replicons, as they have no origin of replication. The transposon does not make a new copy of itself that is liberated as a free intermediate to find a new home. Instead, complex transposons trick the host cell into duplicating their DNA during the transposition process.

Replicative transposition proceeds as follows. The transposase starts by making single-stranded nicks at the ends of both the transposon and the target sequence. Next, it joins the free ends to create a **cointegrate** in which both DNA molecules are linked together via single strands of transposon DNA (Fig. 22.07). The presence of single-stranded DNA alerts the host cell repair systems, which synthesize the complementary strands, thus duplicating the transposon. This leaves a cointegrate of two double-stranded DNA molecules linked by two transposons. Note that each copy of the transposon consists of one old and one new strand of DNA.

Resolvase untwists the cointegrate and separates the two DNA molecules. It does this by recognizing the two IRS sequences, in the middle of the two copies of the transposon, and carrying out recombination between them (Fig. 22.07). This generates

> Replicative transposition does not cause damage to the original DNA host molecule.

---

complex transposon   A transposon that moves by replicative transposition
cointegrate   A temporary structure formed by linking the strands of two molecules of DNA during transposition, recombination, or similar processes
internal resolution site (IRS)   Site within a complex transposon where resolvase cuts the DNA to release two separate molecules of DNA from the cointegrate during replicative transposition
replicative transposition   Type of transposition in which two copies of the transposon are generated; one in the original site and another at a new location
resolvase   An enzyme that cuts apart a cointegrate releasing two separate molecules of DNA

FIGURE 22.04
Movement by Conservative Transposition

A) Transposase, produced by the transposon, recognizes and cuts the inverted repeats, freeing the transposon from the chromosome. The chromosome is left with a double-stranded break that needs to be repaired. B) The transposon bound by transposase identifies a target sequence, and transposase directs a staggered cut into the target DNA. The single-stranded DNA ends of the target sequence are joined to the inverted repeats of the transposon. The resulting single-stranded regions are filled in by the host cell, thus duplicating the target sequence.

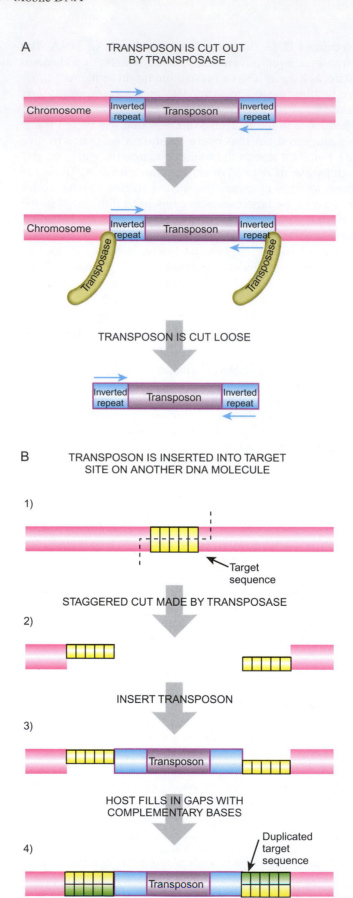

A   TRANSPOSON IS CUT OUT BY TRANSPOSASE

TRANSPOSON IS CUT LOOSE

B   TRANSPOSON IS INSERTED INTO TARGET SITE ON ANOTHER DNA MOLECULE

1)

Target sequence

STAGGERED CUT MADE BY TRANSPOSASE

2)

INSERT TRANSPOSON

3)

HOST FILLS IN GAPS WITH COMPLEMENTARY BASES

4)

Duplicated target sequence

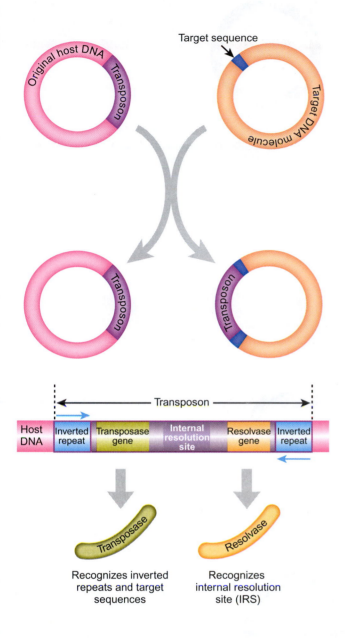

► **FIGURE 22.05**
Outline of Replicative Transposition

The transposon is duplicated as it moves from one DNA molecule to another. It inserts into the target sequence on the recipient DNA molecule and leaves behind a copy of the transposon in the original location.

► **FIGURE 22.06**
Components of a Complex Transposon

Complex (replicative) transposons have a gene for resolvase and an internal resolution site, in addition to the gene for transposase and two flanking inverted repeats.

two free DNA molecules, each carrying a single copy of the transposon. Note that resolution further scrambles the two copies of the transposon as shown in Figure 22.07.

Most of the known complex transposons carry other genes in addition to those involved in transposition and resolution. For example, Tn1 and Tn3 are complex transposons that carry resistance to antibiotics of the penicillin family and are found in both the plasmids and chromosomes of many bacteria. Movement of these complex transposons is traced by observing the expression of antibiotic resistance.

Replicative transposons need an extra enzyme, resolvase, and a site in the DNA for it to act on.

2.4. Replicative and Conservative Transposition are Related

Although replicative and conservative transpositions seem quite different, the actual mechanisms of the transposase steps are closely related. In both cases the target sequence is opened by a staggered cut (Fig. 22.08). In both cases, the transposase cuts at the junction between the ends of the transposon and the host DNA. However, in conservative transposition, both strands are cut, whereas in replicative transposition only one strand is cut. In either case, the free 3′ ends of the transposon are joined to the 5′ ends of the opened target sequence. This sequence of events moves a

FIGURE 22.07
Replicative Transposition Forms a Cointegrate

Single-stranded cuts are made flanking the transposon in the donor molecule and a staggered cut is made in the target site on the recipient molecule. The ends are joined as shown, which causes the transposon to separate, resulting in two single-stranded copies of the transposon. The single-stranded DNA alerts the host to repair the defect thus making both transposons double-stranded. The recipient DNA is now joined to the donor DNA via duplicated transposons and forms what is called a cointegrate. Resolvase, produced by the transposon, then resolves the cointegrate at the two IRS sequences and releases the donor and recipient molecules. Notice that a copy of the transposon is now located on each molecule of DNA.

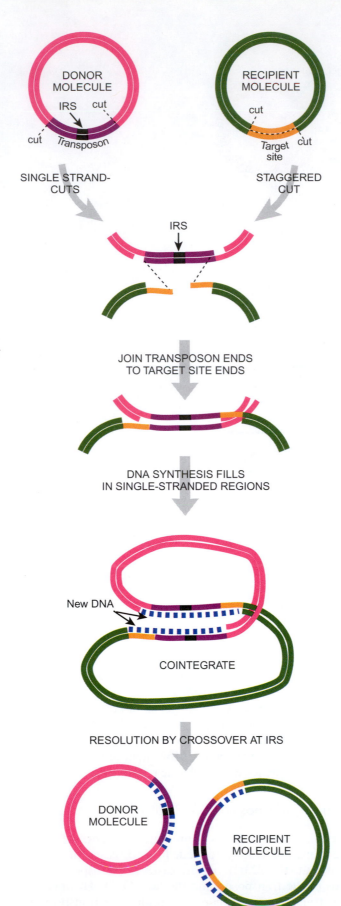

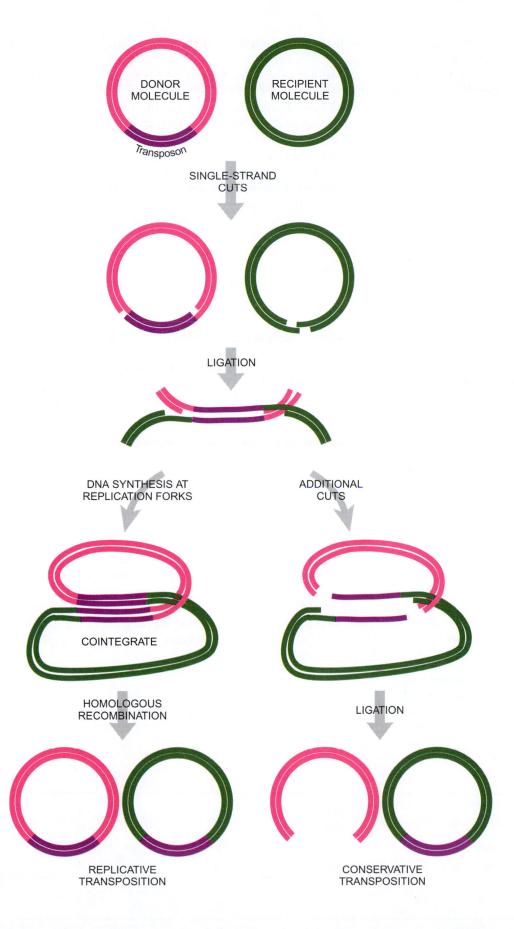

FIGURE 22.08
Replicative and Conservative Transposition are Related

Both conservative and replicative transpositions begin with single-stranded cuts and the joining of ends. In conservative transposition (right side) a second cut completely releases the transposon from the donor molecule after the donor and recipient molecule have joined. In contrast, in replicative transposition, the transposon separates into two single-stranded copies at this stage. Next, the host cell fills in the single-stranded gaps, which is just the short target sequence in conservative replication, but the entire transposon in replicative transposition.

conservative transposon to its new position while it creates a cointegrate in the case of the complex transposon.

The next step is again very similar. The host cell enzymes fill in the single-stranded regions using the free 3' ends of the opened target sequence as primers. In conservative transposition the new DNA is merely a handful of nucleotides and this step just duplicates the target sequence. In the case of replicative transposition, the single-stranded regions are longer and this step duplicates the transposon itself.

This similarity is illustrated by the transposon Tn7, which normally operates by the conservative mechanism. Tn7 is unusual in having a transposase consisting of two proteins. TnsA makes single-stranded nicks at the 5' ends of Tn7, and TnsB carries out the nicking and joining at the 3' ends of Tn7, therefore, TnsA and TnsB create a double-stranded cut when both are expressed. Mutants of Tn7 exist that have a defective TnsA protein and no longer cut the 5' strand. However, TnsB continues to cut and rejoin the 3' strand, forming cointegrates as in replicative transposition. Therefore, TnsB resembles the transposase of complex transposons. TnsA protein, which cuts Tn7 free of its original site, has a structure similar to a type II restriction endonuclease (see Ch. 5).

2.5. Insertion Sequences—the Simplest Transposons

The simplest and shortest transposons, known as **insertion sequences**, were first found in bacteria. They are designated IS1, IS2, etc. Typical insertion sequences are 750–1,500 base pairs (bp) long with terminal inverted repeats of 10–40 bp (Table 22.02). Often, the inverted repeats at the ends of insertion sequences are not quite exact repeats. For example, the inverted repeats of IS1 match in 20 out of 23 positions.

Insertion sequences only encode a single enzyme, the transposase, the enzyme needed for movement. Between the inverted repeats is a region that actually contains two open reading frames, *orfA* and *orfB*. The transposase itself is derived from both open reading frames by a frameshift that occurs during translation (Fig. 22.09). Since the frameshift is a rare event, transposase is produced in very low amounts, which ensures that the transposons do not move very often. Unregulated transposition would lead to massive damage to the host chromosome. When no frameshift occurs during translation, only the first open reading frame, *orfA*, is converted into a protein. This gene product is a transcriptional regulator that controls the production of transposase and of itself.

Insertion sequences are found in the chromosomes of bacteria and also in the DNA of their plasmids and viruses. For example, several copies each of the insertion sequences IS1, IS2, and IS3 are found in the chromosome of *E. coli*. The F-plasmid

TABLE 22.02	Some Insertion Sequences		
Insertion sequence	**Total length**	**Inverted repeat**	**Target length**
IS1	768	23	9
IS2	1327	41	5
IS3	1258	40	3
IS4	1426	18	11–13
IS5	1195	16	4
IS10	1329	22	9
IS50	1531	9	9
IS903	1057	18	9

insertion sequence   A simple transposon consisting only of inverted repeats surrounding a gene-encoding transposase

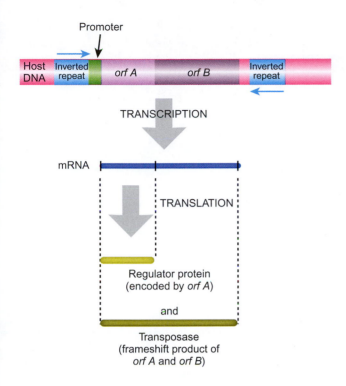

TRANSCRIPTION

mRNA

TRANSLATION

Regulator protein
(encoded by *orf A*)

and

Transposase
(frameshift product of
orf A and *orf B*)

■ **FIGURE 22.09**
*Structure of an Insertion
Sequence*

Insertion sequences, like all
transposons, are flanked by
inverted repeats. Two open
reading frames within the insertion
sequence encode the transposase
gene. When a frameshift occurs
during translation, transposase
is produced and the insertion
sequence "jumps" to a new
location. If the frameshift does not
occur, then only the *orfA* gene
product is expressed. This protein
regulates transposition by turning
off transcription of *orfA* and *orfB* at
the promoter.

has zero copies of IS1, one copy of IS2, and two copies of IS3. When the F-plasmid and chromosome possess identical IS sequences, the plasmid can insert itself into the host chromosome. This, in turn, allows transfer of chromosomal genes by the F-plasmid as explained in Chapter 25.

Insertion sequences contain no genes that provide a convenient phenotype. Originally their presence was recognized because movement of the insertion sequence inactivated genes with a noticeable phenotype. Such insertion mutations usually abolish gene function completely and are often polar to downstream genes of the operon. In addition, they repair themselves only at very low frequencies.

2.6. Composite Transposons

A **composite transposon** consists of two inverted repeats from two separate transposons moving together as one unit and carrying the DNA between them (Fig. 22.10). For example, consider a segment of DNA flanked at both ends by two identical insertion sequences. The transposase will move any segment of DNA surrounded by a pair of the inverted repeats that it recognizes. Consequently, when transposition occurs there are several possibilities. First, each of the insertion sequences may move independently. Second, the whole structure between the two outermost inverted repeats may move as a unit, that is as a composite transposon.

Many of the well-known bacterial transposons that carry genes for antibiotic resistance or other useful properties are composite transposons. Three of the best known are Tn5 (kanamycin resistance), Tn9 (chloramphenicol resistance), and Tn10 (tetracycline resistance). Usually the pair of insertion sequences at the ends of the transposon is inverted relative to each other, as in Tn5 and Tn10. The IS elements of Tn5 and Tn10 are named IS50 and IS10, respectively. The IS10 element has been found alone in bacterial chromosomes, often in multiple copies, however IS50 has not. Less often the two IS elements face in the same direction, as in Tn9, which is flanked by direct repeats of IS1.

> Transposable elements may be built up in a modular fashion from simpler transposons plus entrapped DNA.

composite transposon   A transposon that consists of two insertion sequences surrounding a central block of genes

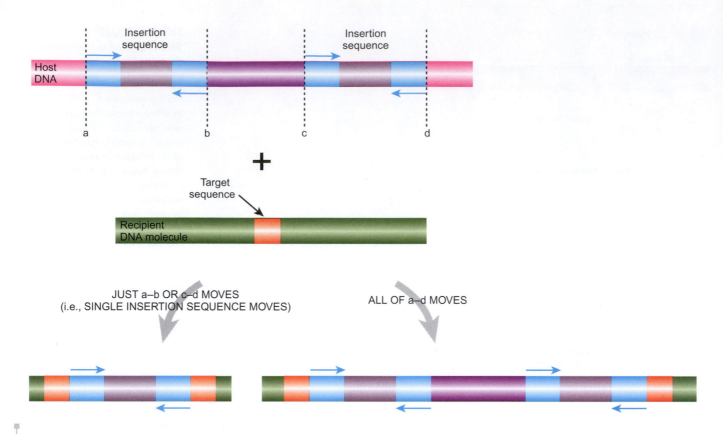

FIGURE 22.10
Principle of the Composite Transposon

Two identical insertion sequences can move as two independent transposons or as one composite transposon. Composite transposition (region a–d) moves both insertion sequences as well as any intermediate DNA to the new location.

Once a useful composite transposon has assembled by chance, natural selection will act to keep the parts together. Mutations accumulate that inactivate the inner-most pair of inverted repeats, which prevents the insertion sequences from jumping independently. Often, one of the two transposase genes is also lost. The result is that the two ends and the middle are now permanently associated and always move as a unit (Fig. 22.11). In practice, all stages from newly-formed to fully-fused composite transposons are found in bacteria. For that matter, novel composite transposons can be assembled in the laboratory by genetic manipulation.

2.7. Transposition May Rearrange Host DNA

Insertions, deletions, and inversions of host cell DNA may result from anomalous or failed attempts at movement by a composite transposon. As mentioned above, transposase will move any segment of DNA surrounded by a pair of correct inverted repeats. Composite transposons have four such terminal repeats. Two of these are "inside ends," relative to the transposon, and two are "outside ends" (Fig. 22.10 above). Any pair that face in opposite directions may be used together. Transposition of a single IS element involves one "inside end" and one "outside end." Normal movement of the whole composite transposon uses the two "outside ends."

But suppose that the two "inside ends" are used for transposition. The whole of the DNA molecule outside the transposon will be moved. This is easiest to see if we consider a small circular DNA molecule, such as a plasmid (Fig. 22.12), that carries a composite transposon. Using the "outside ends" moves the transposon, using the "inside ends" moves the rest of the DNA molecule. (Notice that the "inside ends" and the "outside ends" face in different directions.) The result, in the case illustrated, is the insertion of a segment of plasmid DNA into the host chromosome. Note that the segment that

> Movement of transposons, or their subcomponents, may cause rearrangements of the host DNA molecule.

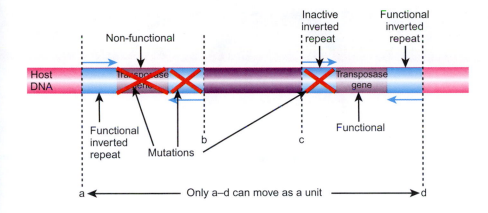

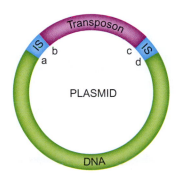

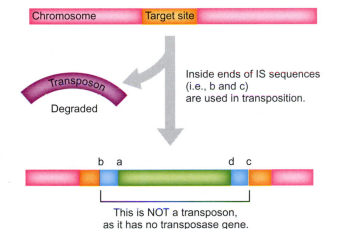

■ **FIGURE 22.11**
Evolution of a Composite Transposon

Since composite transposons no longer require two separate transposase genes and two of the four inverted repeats, mutations accumulate in the non-essential regions. These mutations generate a transposon that only moves as a composite unit. This is especially important for bacterial evolution if the transposon carries internal genes that enhance the survival of the host cell.

■ **FIGURE 22.12**
Insertion Created by Using Inside Ends to Transpose

Plasmid DNA can integrate into a chromosome if the "inside ends" (b and c) of the composite transposon are moved by transposase. Inverted repeats b and c point outwards from the transposon, but the transposase is still capable of moving the DNA between them—that is, the DNA making up most of the plasmid. If the transposase gene is located inside the transposon (in the purple segment), the DNA that jumped will not be able to move itself again (i.e., it is not itself a genuine transposon).

moved may not be able to move again if the gene for the transposase was left behind during this maneuver.

If both of the "inside ends" are used in jumping to another site on the same host DNA molecule the result will be the insertion of the host DNA into itself (Fig. 22.13). Depending on which pairs of ends are re-joined after breakage, the host DNA will suffer either a deletion or an inversion. The inside region of the composite transposon is lost during this process. This process is sometimes known as abortive transposition.

2.8. Transposons in Higher Life Forms

Transposons are scattered throughout the DNA of all forms of life. Although we have used examples from bacteria to illustrate how they work, Barbara McClintock

FIGURE 22.13
Deletions and Inversions Made by Abortive Transposition

"Inside ends" (b and c) are incorrectly used by transposase to move the host DNA rather than the transposon. If the target sequence (p–q) is found on the same DNA molecule, transposase also cuts this site, creating a double-stranded break. There are two alternative ways of rejoining the ends. Separate molecules may be formed, thus creating a deletion in the host DNA molecule. Alternately, the small piece (p–b) may rejoin the larger fragment such that region p joins IS sequence c–d and region q joins IS sequence a–b, thus inverting the DNA.

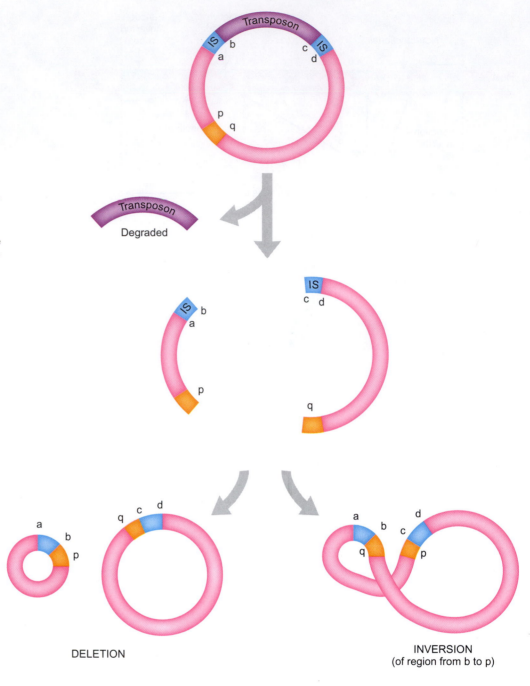

DELETION

INVERSION
(of region from b to p)

observed the first jumping genes during genetic crosses in maize (corn) plants. She named them the Activator (Ac) and Dissociation (Ds) elements. She worked in the 1940s before the DNA double helix was even discovered, but nonetheless realized that segments of the plant genetic material must be moving around. When she presented her conclusions to the public in 1951, no one believed her. Later investigations in bacteria revealed the molecular details of transposition and confirmed what Barbara McClintock had observed in the 1940s. Barbara McClintock received her Nobel Prize in 1983 when the significance of her work was more fully realized.

The Ac/Ds family of transposons in corn is simple and conservative. Family members leave behind double-stranded gaps in the DNA when they move. They have inverted terminal repeats of 11 base pairs and insert at an 8 bp target sequence. The

Plants frequently contain simple transposons, such as the Ac/Ds elements of corn.

Ac element

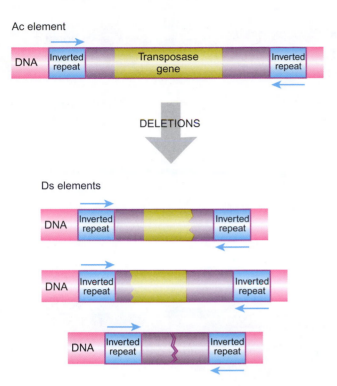

DELETIONS

Ds elements

■ **FIGURE 22.14**
*Ac/Ds Family of
Transposons in Maize*

The simple Ac element of maize
contains two inverted repeats and
a functional transposase gene. Ds
elements are derived from Ac by
deletion of the transposase gene,
either completely or partially.

Ac element is 4,500 bp long and is a fully-functional transposon with the ability to move itself. The **Ds elements** vary in size and are defective. They are derived from Ac by deletion of all or part of the transposase gene and they cannot move by themselves. If a cell contains an Ac element anywhere in its DNA, then the transposase enzyme made by Ac can also move any Ds element. Therefore, to remain mobile, the Ds elements must keep the inverted repeats; otherwise, the Ac transposase will not recognize them (Fig. 22.14). The Ac and Ds elements do not need to be on the same chromosome for transposition to occur.

Barbara McClintock traced the development of corn kernel color to ascertain that there is movement of genetic elements. She looked at the pattern of the kernel color and compared this to the chromosomal structure of the corn. If a Ds element was inserted into the gene for purple kernels of corn, the gene no longer made purple pigment and the kernels were colorless. If there is no Ac element in any of the cells of the entire corn kernel the white color is stably inherited. If an Ac element is also present in some of the cells of the corn kernel, it may move the Ds element. The cells where Ac transposase moves the Ds element return to the original purple color since the gene for purple pigment is no longer disrupted. In addition, the chromosomes for these cells show alterations in chromosome structure from the loss of the Ds element. All the daughter cells inherit this defect and will also be purple. Eventually, a patch of purple will appear on the kernel as it grows. If the transposition occurs when the kernel is just beginning to develop, a large patch of purple will appear, and if the transposition occurs when the kernel is almost fully developed the patch of purple will be very small. This produces a mottled kernel of corn (Fig. 22.15).

Animal and plant cells frequently contain multiple transposons of the same kind of which many are defective. These may vary greatly in their overall lengths. Not only are there defective members that need help to move because they lack transposase, but we also find totally inactive transposons. These have suffered mutations in their

Ac element   Intact and active version of a transposon found in maize
Ds elements   Defective version of the Ac transposon of maize; it cannot move alone but needs the Ac element to provide transposase

FIGURE 22.15
Movement of Ds Element Gives Mottled Corn

A) The gene for purple color in maize produces a purple kernel of corn. B) If the gene for purple color is disrupted by a Ds element, the kernel is white since the gene cannot direct the synthesis of purple pigment. C) If both the Ac and Ds element are present in the same cell, transposase from the Ac element moves the Ds element from the purple gene, restoring its ability to direct pigment synthesis, and then returning the cell to the original purple color. As the cell divides, the restored purple pigment gene is inherited in the daughter cells, and a patch of purple forms on the white kernel. Random transposition events in several such cells results in a mottled kernel. D) Photograph of mottled corn kernels showing reddish-purple streaks due to transposition of Ac/Ds. The patterns caused by transposition may be blotches, dots, irregular lines, or streaks. *(Credit: Ds tagging logo courtesy of Tom Brutnell, Jon Duvick and Erik Vollbrecht.)*

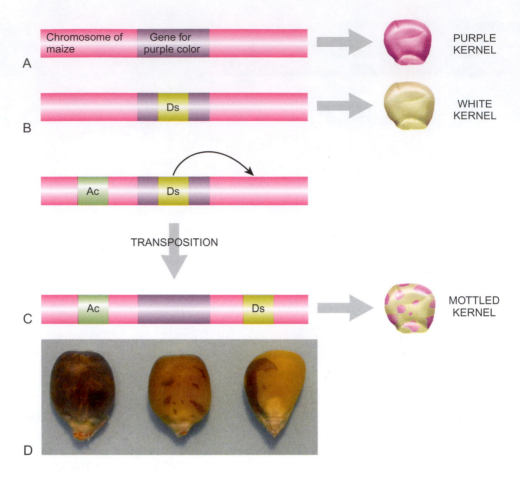

terminal repeats, which make them unrecognizable by transposase and therefore immobile. Plants in particular contain large numbers of miniature inverted-repeat transposable elements (MITEs). These very short transposons lack transposase and cannot move by themselves. In some cases, corresponding autonomous transposons are known; in other cases, the presumed autonomous elements have not been identified and may indeed no longer exist. Overall, transposable elements comprise the majority of the DNA of higher plant genomes and have had a major effect on the evolution of plants (see Focus on Relevant Research).

Multiple defective copies of transposons are often found in higher organisms. They rely on an intact copy to help them move.

FOCUS ON RELEVANT RESEARCH

Tenaillon MI, Hollister JD, and Gaut BS (2010) A triptych of the evolution of plant transposable elements. Trends Plant Sci 15: 471–478.

The majority of the DNA in the genome of higher plants is derived from transposable elements. Many different families of both DNA transposons and retroelements are found. Some of these are well characterized, whereas others remain somewhat mysterious. This review discusses the effect that this mobile DNA has had on plant evolution. Transposition generates more copies of transposable elements. However, the host cell tries to limit the increased copy number of transposable elements by a variety of mechanisms. Furthermore, deletion of substantial lengths of genomic DNA may occur by recombination or other mechanisms and will eliminate considerable numbers of transposable elements. Finally, population genetics affects the survival of both host and associated transposable elements.

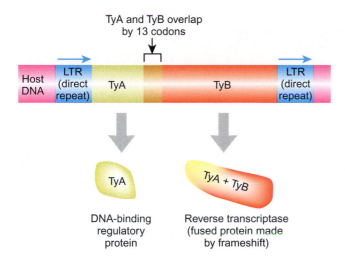

FIGURE 22.16
Structure of Ty-1 Retrotransposon

Ty-1 is flanked by two direct repeats (LTRs) and contains the genes for reverse transcriptase and a DNA-binding regulatory protein. Reverse transcriptase is only produced if a frameshift occurs during translation. If no frameshift occurs, the truncated gene product binds to DNA and acts as a regulatory protein.

Transposons are divided into families based on the evolutionary relationships of their transposase enzymes. The Ac/Ds elements of maize are categorized in the *hAT* family of transposons (named after its early members, hobo from flies and Ac and Tam3 from plants). Members of the *hAT* transposon family have been found in most groups of eukaryotes, including plants, fungi, animals, and man.

The most widely distributed DNA-based transposons in higher organisms are those of the Tc1/mariner family. The first members to be found were **Tc1**, from the nematode *Caenorhabditis*, and the **mariner** transposon of *Drosophila*. Members of this family are found in fungi, plants, animals (including humans), and protozoans. They range from roughly 1300–2500 bp in length, contain a single transposase gene, and are flanked by inverted repeats. They move using the conservative cut-and-paste mechanism. The breaks they leave behind are often mended by the eukaryotic double-stranded break repair systems (see Ch. 23).

> Conservative transposons of the mariner family are widespread in eukaryotes.

3. Retroelements Make an RNA Copy

The transposons we have discussed so far move as segments of DNA. In addition, there is a vast array of mobile elements that move by using an intermediate RNA form. These range from retroviruses (see Ch. 21), which can integrate their DNA form into the host cell chromosome, to transposon-like elements. These may all be classed as **retroelements** since they all use **reverse transcriptase** to convert an RNA transcript back into DNA. Whenever retroelements insert into host DNA the target sequence is duplicated as for DNA-based transposons. Retroelements are more common in higher organisms, whereas DNA-based transposons predominate in bacteria.

> Some transposons move via an RNA intermediate and rely on reverse transcriptase.

Retrotransposons or (for short) **retroposons** are transposable elements that rely on reverse transcriptase for movement. They are found most often in eukaryotes, especially animals and higher plants. The **Ty1** (Transposon of yeast 1) retrotransposon of yeast is around 6,000 bp long, contains a gene-encoding reverse transcriptase, and is flanked by direct repeats of 334 bp that correspond to the **long terminal repeats (LTRs)** of a retrovirus (Fig. 22.16).

---

long terminal repeats (LTRs)   Direct repeats of several hundred base pairs found at the ends of retroviruses and some other retroelements
mariner elements   A widespread family of conservative DNA-based transposons first found in *Drosophila*
retroelement   A genetic element that uses reverse transcriptase to convert the RNA form of its genome to a DNA copy
reverse transcriptase   Enzyme that synthesizes a DNA copy from an RNA template
retrotransposon   A transposable element that uses reverse transcriptase to convert the RNA form of its genome to a DNA copy
retroposon   Short for retrotransposon
Tc1 element   Transposon *Caenorhabditis* 1. A transposon of the mariner family found in the nematode *Caenorhabditis*
Ty1 element   Transposon yeast 1. A retrotransposon of yeast that moves via an RNA intermediate

FIGURE 22.17 ————
Movement of Ty-1 Retrotransposon

Ty-1 elements use host cell transcription to create a single-stranded RNA. Reverse transcriptase uses the RNA as a template to synthesize a strand of DNA. The DNA/RNA hybrid is next converted into double-stranded DNA. The double-stranded DNA form of the retrotransposon can be inserted into a new target site within the host DNA.

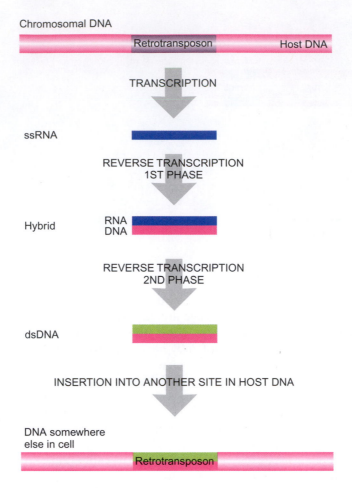

When moving, the first step is to make a single-stranded RNA copy (Fig. 22.17). The host provides the RNA polymerase II that transcribes the Ty-1 element starting in the left-hand LTR. Next, reverse transcriptase makes a double-stranded DNA copy. First, it makes a single strand of DNA, so creating an RNA/DNA hybrid. A second round of synthesis replaces the RNA with DNA, forming double-stranded DNA. Finally, the DNA is inserted into a new site within the host cell DNA. There are 30–40 copies of Ty1 per yeast cell.

Box 22.1 Human Genetic Defects Due to Retroposon Insertion

Since non-coding DNA vastly outnumbers coding DNA in higher organisms, it is hardly surprising that most insertions of transposable elements into the genome occur in the noncoding regions. Nonetheless, occasional examples are known where insertion of a retrotransposon inactivates a gene so causing a hereditary defect. The first human case to be identified is a form of muscular dystrophy known as Fukuyama-type congenital muscular dystrophy (FCMD), which is particularly common in Japan although rare elsewhere. FCMD is one of the most common autosomal recessive disorders in Japan, occurring at a frequency of approximately 0.7–1.2 per 10,000 births, with a carrier frequency estimated to be as high as 1 in 80.

This condition is caused by the insertion of a retrotransposon into the 3'-untranslated region (3' UTR) of the *FCMD* gene. As a result, little to no detectable FCMD messenger RNA (mRNA) is seen in cells harboring the retrotransposon. Since the retrotransposon disrupts the 3' UTR, the defect is most likely due to disrupting mRNA stability rather than

Continued

Box 22.1 Continued

actually affecting protein translation directly. The *FCMD* gene codes for a 461-amino-acid protein that is normally expressed in brain, skeletal muscle, and heart and is involved in muscle function.

Genetic analysis indicates that the retrotransposon insertion in the *FCMD* gene could have been derived from a single ancestor who lived 2,000–2,500 years ago. This was about the time that the Yayoi people migrated to Japan from Korea and China, giving rise to the possibility, as yet unproven, that these immigrants brought the defective *FCMD* allele into the Japanese population. (It is thought that humans migrated from the Asian continent to Japan in two waves. The first wave brought hunter-gatherers of the Jomon culture over 10,000 years ago. The second wave brought the Yayoi people from the Korean Peninsula about 2,300 years ago.)

Typical retroposons have only one or two genes and lack the ability to make virus particles and infect other cells. Nonetheless, more complex retroelements that are intermediate between retroposons and retroviruses do exist. Some retroelements pack their RNA into defective virus-like particles. However, these particles are not released from the cell where they were assembled and therefore cannot infect other cells. These elements are sometimes called endogenous retroviruses. Defective versions of these, known as retrovirus-like elements, are common in animal cells where they often exist in multiple copies.

> Some retroelements are closely related to retroviruses.

3.1. Repetitive DNA of Mammals

A substantial portion of the DNA of both animals and plants consists of repeated sequences as discussed in Chapter 4. Many of these were probably derived by retrotransposition events. In mammals, there are two major classes of moderate-to-highly-repetitive DNA, **short interspersed elements**, or **SINEs**, and **long interspersed elements**, or **LINEs**. Although these were originally defined in terms of their length, the SINEs are derived from host DNA (see below), whereas the LINEs are derived from retrotransposons.

> The highly-repetitive sequences known as LINEs are related to retrotransposons.

The **LINE-1 (or L1) element** of mammals is a retroposon that moves much as the Ty1 element of yeast. However, it lacks LTR sequences and at its 3′ end is a run of AT base pairs that are derived from the poly-A tail typical of eukaryotic mRNA (Fig. 22.18). In humans the LINE-1 element is present in over 100,000 copies and makes up 5% or more of the total DNA! The vast majority of the LINE-1 repeats are defective.

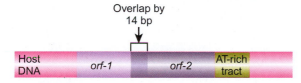

FIGURE 22.18
Structure of LINE-1 Element

LINE-1 elements contain two open reading frames that encode reverse transcriptase followed by an A/T-rich region. Unlike other retrotransposons, LINE-1 elements are not flanked by inverted repeats.

LINE-1 (L1) element   A particular LINE found in many copies in the genome of humans and other mammals
long interspersed element (LINE)   Long repeated sequence that makes up a significant fraction of the moderately- or highly-repetitive DNA of mammals
short interspersed element (SINE)   Short repeated sequence that makes up a significant fraction of the moderately- or highly-repetitive DNA of mammals

The complete LINE-1 sequence is 6,500 bp long and contains a gene for reverse transcriptase. However, only about 3,000 of the LINE-1 sequences are full length and most of these are crippled by point mutations.

Very rarely LINE-1 makes a new copy of itself and may insert it somewhere else in the DNA. This movement causes one type of hemophilia, an inherited condition caused by a defect in blood-clotting factors. A few very rare cases are due to the insertion of LINE-1 into the gene for blood clotting factor VIII on the X-chromosome. The intact LINE-1 that jumped came from chromosome 22. This still active copy of LINE-1 is found in the same location in the DNA of the gorilla, implying that it has been lurking in the same place for millions of years of primate evolution.

3.2. Retro-Insertion of Host-Derived DNA

A pseudogene is a duplicate, but functionless, copy of a gene (see Ch. 4). Pseudogenes may be generated by DNA duplications and so often lie close to the original genes. Pseudogenes made in this way will contain both the introns and exons of the original gene.

Sometimes the reverse transcriptase of a retroelement may work by accident on mRNA that derives from a host gene. Since mRNA has had the introns removed by splicing, reverse transcriptase generates a DNA copy missing the introns; that is, **cDNA (complementary DNA)**. Occasionally, this cDNA copy may be re-integrated into the host genome (Fig. 22.19). This will give a pseudogene that lacks the introns

FIGURE 22.19
Creation of a Processed Pseudogene

When a host gene is expressed, the primary RNA transcript is processed so the introns are removed and a poly-A tail is added. Occasionally, reverse transcriptase will convert host mRNA into a cDNA copy. The cDNA may be inserted into the host DNA (at another site) creating a processed pseudogene. This pseudogene will not be expressed as there is no promoter at its new insertion site.

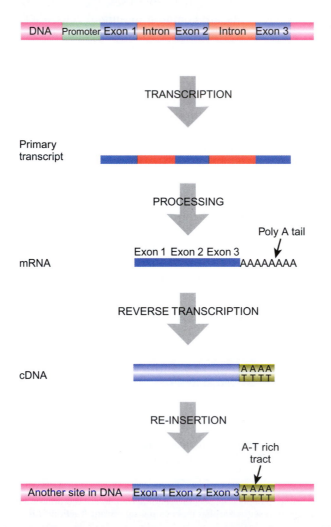

cDNA (complementary DNA)   DNA copy made by reverse transcription from mRNA and therefore lacking introns

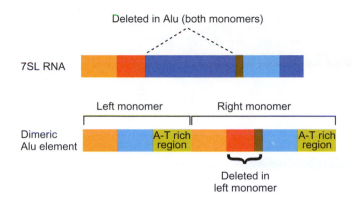

7SL RNA

Deleted in Alu (both monomers)

Left monomer | Right monomer

Dimeric Alu element

A-T rich region

A-T rich region

Deleted in left monomer

►**FIGURE 22.20**
Origin of the Alu Element from 7SL RNA

7SL RNA has given rise to the Alu element by a complex process involving reverse transcription and re-integration. The sequence of 7SL RNA is shown at the top. The Alu element below is actually derived from two fused 7SL RNA sequences that have suffered several deletions.

and consists solely of the coding sequence and is therefore known as a **processed pseudogene** or a **retro-pseudogene**.

The SINEs are a special class of processed pseudogenes that were originally derived from host DNA sequences. The most common SINE is the **Alu element**, which is derived from the **7SL RNA** gene (Fig. 22.20). This non-coding RNA is part of the machinery for protein export across membranes. It is normally transcribed from an internal promoter by RNA polymerase III (see Ch. 11). Standard pseudogenes lack a promoter (unless by chance they were integrated just next to one). Consequently, they cannot be transcribed and no additional copies are produced. In contrast, each copy of the Alu element carries an internal promoter. As long as this promoter remains intact, each Alu element can serve as the source for another round of transcription, reverse transcription, and re-integration. As a result the number of Alu elements has mushroomed until there are now around a million copies in the human genome. Although most insertions are harmless, animals have developed mechanisms to control transposable element proliferation, especially in germ line cells where insertion into an actual gene might be fatal to the offspring (see Focus on Relevant Research).

> Some pseudogenes originated via reverse transcriptase acting on mRNA.

> The highly-repetitive Alu element is a pseudogene that contains an internal promoter.

Saito K and Siomi MC (2010) Small RNA-mediated quiescence of transposable elements in animals. Developmental Cell 19: 687–697.

Transposable elements form a major component of the DNA in the genome of higher organisms. Although most transposable elements are located in intergenic regions, their unrestrained proliferation would eventually cause serious problems. Consequently, many organisms possess specialized genetic systems to monitor and inhibit the replication and movement of transposable elements. The Argonaute (AGO) family of proteins that is involved in RNA interference (see Ch. 18, section 2.4) plays a central role in these defense mechanisms. Two kinds of small RNA are also vital components, piRNA in germline cells and endogenous siRNA in somatic cells (see Ch. 18, section 2.2).

Most animals possess multiple AGO proteins. Some are specialized for RNA interference; others for defense against transposable elements. Some of these are usually expressed specifically in the reproductive cells where they bind piRNAs and

suppress transposition primarily by cleaving the transcripts that come from the transposable elements. The piRNAs mostly originate from transposable elements or related

FOCUS ON
RELEVANT RESEARCH

sequences and are generated without the help of Dicer (which makes siRNA). Somatic cells also suppress transposable elements, but less intensively than the germline. They use endogenous siRNA that also originate from transposable elements or related sequences and bind to the AGO proteins of the RNA interference system.

Alu element   The most common short interspersed element (or SINE) in the highly-repetitive DNA of mammalian cells
processed pseudogene   Pseudogene lacking introns because it was reverse transcribed from messenger RNA by reverse transcriptase
retro-pseudogene   Another name for a processed pseudogene
7SL RNA   Non-coding RNA that forms part of the machinery for protein export across intracellular membranes in eukaryotic cells

FIGURE 22.21
Structure of a Retron and its Gene Products

A retron consists of a single gene for reverse transcriptase that is preceded by a long non-coding region that gives rise to untranslated RNA.

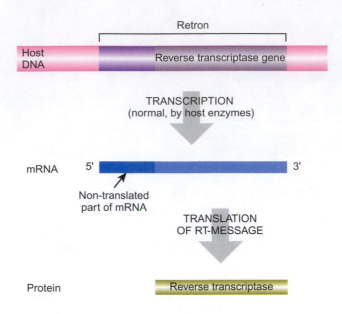

Retrons are retroelements found in bacteria that generate bizarre DNA/RNA hybrid structures.

3.3. Retrons Encode Bacterial Reverse Transcriptase

Retrons, found in bacteria, are shorter, stranger relatives of the retroposons of higher cells. They have just one gene that is transcribed into an RNA molecule consisting of a sizeable untranslated region followed by the coding region for reverse transcriptase (Fig. 22.21).

The reverse transcriptase uses the non-coding portion of the RNA molecule as both a template and a primer to make multiple copies of a bizarre branched molecule that is part RNA and part DNA (Fig. 22.22). Typically, the RNA region of the retron product is around 100 bases long and the DNA 50–75 bases. The DNA is joined to the RNA by the 2′-OH of a specific G residue. In some retrons the single-stranded DNA is cleaved from the RNA; in other cases, the two stay attached. The retron DNA has not been observed to reintegrate into the chromosome. However, the chromosomes of some strains of myxobacteria have repeated sequences related to retron DNA.

It is assumed that retrons move around, but it is unknown how. Perhaps the copies observed so far are defective and the mobile master copy has not yet been found. Retrons are found in relatively few bacteria and only a single copy of the retron is usually found in the bacterial chromosome. Retrons are often inserted into the DNA of bacterial viruses that have in turn inserted into the bacterial chromosome. How and where the virus acquired the retron remains obscure.

4. The Multitude of Transposable Elements

We have discussed a wide range of transposable elements that vary in their mechanism of movement, and whether or not they have an RNA phase. Other transposons exist whose detailed mechanisms of action are variations on the above themes. Some larger and more complicated elements possess the ability to use multiple mechanisms under different conditions. In others, a variety of extra genes oversee the frequency of transposition or the specificity of insertion. For example, Tn7 normally inserts at only a single specific sequence on the *Escherichia coli* chromosome. However, mutants exist that have inactive specificity proteins. These insert more or less at random like more typical transposons.

retron   Genetic element found in bacteria that encodes reverse transcriptase and uses it to make a bizarre RNA/DNA hybrid molecule

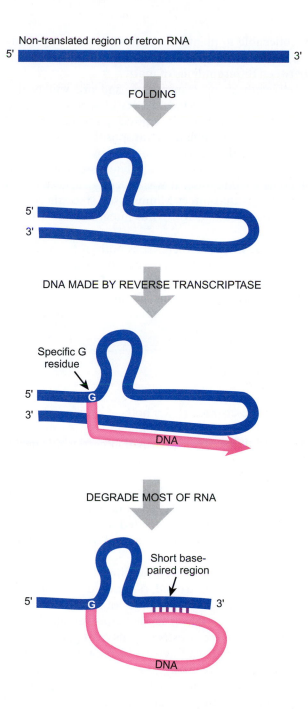

Non-translated region of retron RNA

FOLDING

DNA MADE BY REVERSE TRANSCRIPTASE

Specific G residue

DNA

DEGRADE MOST OF RNA

Short base-paired region

DNA

■– **FIGURE 22.22**
Retron RNA and RNA/DNA Hybrid

Retron RNA only contains the gene for reverse transcriptase and a large 5′ UTR. This folds into a hairpin structure so that the 5′ end meets the 3′ end. Reverse transcriptase recognizes a specific guanine near the front of the hairpin and begins DNA synthesis here. The RNA is partly degraded, resulting in a strange RNA/DNA hybrid structure.

Recently discovered transposable elements whose mechanism of movement is not completely understood include the **helitrons** and the **polintons** (both DNA-based). Helitrons, sometimes known as "rolling circle" transposons, lack terminal inverted repeats and do not generate duplications of their target sites upon insertion. They lack transposase and rely on two enzyme activities, rolling circle replication initiator (Rep) and helicase (Hel), to move. Both activities are generally present on a single large protein. The Rep domain is homologous to the rolling circle initiator proteins of plasmids and viruses. Helitrons promote genomic alterations by moving

helitron   A transposable element found in eukaryotes that transposes via rolling circle replication
polintons   Self-synthesizing element that transposes from one location to another via excision of the original copy, synthesis of another DNA copy by Polinton-encoded DNA polymerase and then reintegration using Polinton-encoded integrates

to new locations. This is especially noticeable in plants where helitrons are especially numerous. Helitrons account for 2% of the maize genome and their movement is responsible for massive variation between different lines of maize.

Polintons are also known as self-synthesizing transposons and are widely distributed in eukaryotes. Polintons have terminal-inverted repeats of several hundred nucleotides and generate 6-bp target site duplications. They possess a protein-primed DNA polymerase and a retroviral-type integrase. Polinton transposition may involve excision of a Polinton DNA molecule from the genome followed by synthesis of a double-stranded DNA copy by the Polinton-encoded DNA polymerase. The copy is then inserted back into the host genome by integrase. It was recently suggested that polintons have evolved from the virophages that infect Mimiviruses, since these also possess both protein-primed DNA polymerase and retroviral integrase (see Ch. 21, Box 21.2).

Other genetic elements exist that possess the characteristics both of transposable elements and some other life form, such as a plasmid or virus. For example, **conjugative transposons** both transpose and promote conjugation like fertility plasmids. Again, bacteriophage Mu is both a virus and a transposon. Plasmids and viruses are discussed in Chapters 20 and 21, but the transposon-like aspects of these hybrid elements are discussed below.

> Hybrid elements that combine the properties of virus and transposon are known.

4.1. Bacteriophage Mu Is a Transposon

Hybrid gene creatures exist that possess the characteristics both of transposable elements and some other genetic element. For example, **bacteriophage Mu** is both a virus and a transposon. (Note that we are not talking about a virus that carries a transposon inserted within its DNA—a frequent occurrence—but about a genetic element that <u>behaves</u> as both a virus and a transposon simultaneously.) When Mu DNA enters *E. coli*, its bacterial host, it integrates at random into the host chromosome by transposition (Fig. 22.23). In other words, the whole of the Mu genome is a transposon. If Mu inserts into the middle of a host gene this will be inactivated. Early investigators noticed that infection with this virus caused frequent mutations and therefore named it Mu for "mutator" phage.

Like many bacterial viruses, Mu may lie dormant as a prophage or go lytic (see Ch. 21). When Mu replicates it does so by uncontrolled replicative transposition, not by replicating as a typical virus. This creates multiple copies of Mu inserted into the host DNA and destroys so many host genes that it is inevitably lethal. Despite this, Mu manages to avoid inserting into and hence destroying its own genome. This transposition immunity relies on the differential binding of Mu B protein to DNA (see Focus on Relevant Research). Unlike other viruses, the DNA of Mu is never found replicating as an independent molecule, free of the chromosome. The segments of DNA packaged into the virus particle contain a whole Mu genome plus small stretches of host DNA at the ends. Thus, even inside the virus particle, Mu DNA is still inserted into host DNA! When the virus DNA infects a new cell, the Mu genome transposes out of the tiny fragment of host DNA it brought with it. Thus, Mu is a true transposon and its DNA is never free.

Note that many viruses may integrate into the host chromosome. This includes bacterial viruses such as lambda and animal viruses such as the retroviruses (see Ch. 21). Nonetheless, these viruses do not replicate by transposition and are not surrounded by host DNA when packaged into their virus particles. Thus, neither lambda nor retroviruses are transposons.

---

bacteriophage Mu A bacterial virus that replicates by transposition and causes mutations by insertion into host cell genes
conjugative transposon A transposon that is also capable of moving from one bacterial cell to another by conjugation

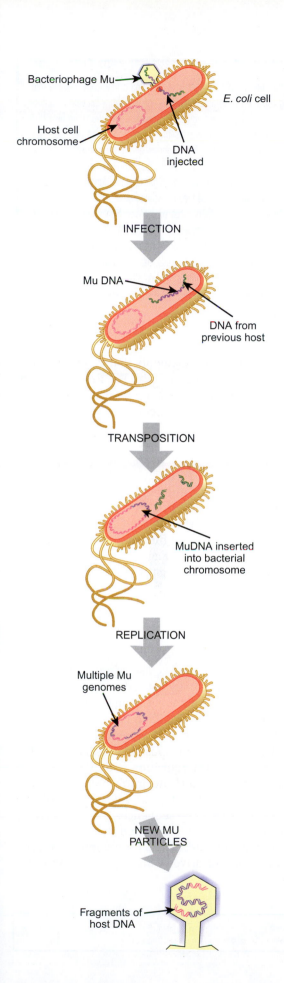

Bacteriophage Mu

Host cell
chromosome

E. coli cell

DNA
injected

INFECTION

Mu DNA

DNA from
previous host

TRANSPOSITION

MuDNA inserted
into bacterial
chromosome

REPLICATION

Multiple Mu
genomes

NEW MU
PARTICLES

Fragments of
host DNA

▬ **FIGURE 22.23** ────────
*Bacteriophage Mu Is a
Transposon*

Bacteriophage Mu attaches to the
E. coli cell and injects its DNA
into the cytoplasm. Once inside,
the Mu DNA inserts into the host
chromosome via transposition.
Notice that the flanking DNA from
the previous host is not inserted.
Once integrated, Mu DNA may
undergo so many transpositions
that cellular functions are
destroyed. When the Mu DNA is
packaged into virus particles, short
lengths of host chromosome are
also packaged attached to the Mu
ends; therefore, Mu DNA is always
integrated into host DNA.

Han YW and Mizuuchi K (2010) Phage Mu transposition immunity: protein pattern formation along DNA by a diffusion-ratchet mechanism. Mol. Cell 39: 48–58.

FOCUS ON RELEVANT RESEARCH

Several transposable elements are known that possess transposition immunity (also known as target DNA immunity). This has two related functions.

Firstly, it prevents the transposable element from inserting a new copy into the middle of a pre-existing copy of the same element. Secondly, it may prevent or reduce insertion of another copy of the same transposable element into the same host DNA molecule. The first of these obviously protects the transposable element itself. The second reduces damage to the host cell. Both operate by the same mechanism, and the host protection is distance dependent; that is, the closer to the first copy the lower the likelihood of inserting a second copy. Consequently, larger genomes are susceptible to multiple insertions provided they are spread out.

In this paper the authors focus on transposition immunity in bacteriophage Mu. This is due to the interaction of Mu A and Mu B protein. DNA with the Mu B protein bound is a potential target for insertion of another copy of the Mu genome. However, Mu A protein binds to Mu ends and also displaces Mu B protein from DNA. Consequently, the closer to Mu ends the less stably Mu B binds to DNA.

4.2. Conjugative Transposons

> Conjugative transposons combine the ability to move by transposition and to move from one bacterial cell to another.

Conjugative transposons, found in bacteria, are hybrid elements that can both transpose and can move from cell-to-cell, like a transmissible plasmid. The first of these to be discovered, Tn916, confers tetracycline resistance and was found in the bacterium *Enterococcus faecalis*. Tn916 carries several genes needed for conjugative transfer and is therefore much larger than most transposons.

Tn916 jumps by the cut-and-paste mechanism. However, it differs in two respects from typical conservative transposons. First, the target sequence is not duplicated when Tn916 inserts itself. Second, it can excise itself precisely, leaving the host cell DNA intact. When moving from one bacterial cell to another, Tn916 is thought to excise itself temporarily from the DNA of the original cell. It then transfers itself into the recipient and, once inside, it transposes into the DNA of the new host cell (Fig. 22.24). Tn916 and related elements can enter many different groups of bacteria, both gram-positive and gram-negative. Because the host range of conjugative transposons is so broad they are partly to blame in the spread of antibiotic resistance genes among diverse groups of bacteria.

4.3. Integrons Collect Genes for Transposons

Many bacteria that carry multiple drug resistance have emerged since antibiotic use has become widespread. Antibiotic resistance genes are usually carried on plasmids, many of which may be transferred between bacteria. In many cases the antibiotic resistance genes are actually carried within transposons that are inserted into the plasmids.

> Integrons accumulate genes by integration of DNA modules flanked by recognition sequences.

Novel antibiotic resistance genes often appear first in transposons of the Tn21 family found in gram-negative bacteria. This group of transposons possesses an internal element known as an **integron** that acts as a gene acquisition and expression system. An integron consists of a recognition region, the *attI* site, into which a variety of gene cassettes may be integrated, plus a gene encoding the enzyme responsible for insertion, the **integrase**. The *attI* site is flanked by two 7 bp sequences that act as recognition sites for the integrase (Fig. 22.25). Two promoters, facing in different directions, are situated between the integrase gene and the *attI* site. One is for the integrase gene; the other faces into the gene collection region and drives transcription of whatever gene has been integrated.

integron   Genetic element consisting of an integration site plus a gene encoding an integrase
integrase   Enzyme that integrates one segment of DNA into another DNA molecule at a specific site

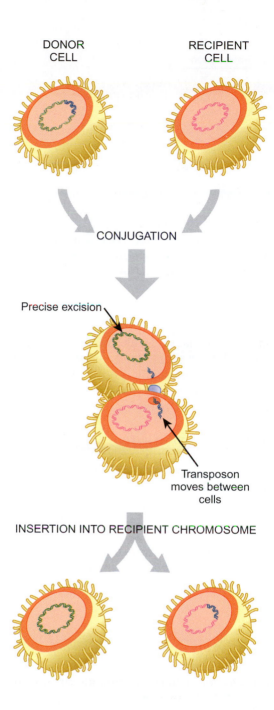

DONOR CELL

RECIPIENT CELL

CONJUGATION

Precise excision

Transposon moves between cells

INSERTION INTO RECIPIENT CHROMOSOME

FIGURE 22.24
Conjugative Transposon

Conjugative transposons move from donor cell to recipient cell during bacterial conjugation. The transposon moves in a precise manner, leaving the donor DNA intact and integrating into the recipient DNA without duplicating the target sequence.

Gene cassettes suitable for integration consist of a structural gene lacking its own promoter plus an integrase recognition sequence, the *attC* site. The *attC* sites are rather variable, except for the conserved 7 bp sequences at the ends. (The *attC* sites were originally referred to as "59 base elements" because the first examples discovered were 59 bp long. However, later examples were found that varied in length and internal sequence.) Integration of cassettes occurs by a novel version of site-specific recombination in which a single-stranded version of the cassette is inserted. One strand of the incoming cassette is folded so that its *attC* region is double-stranded due to complementary base-pairing between strands. After insertion, the second strand of the cassette is synthesized.

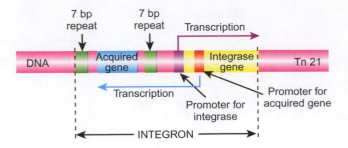

FIGURE 22.25
Integrons Collect Antibiotic Resistance Genes

The structure of the integron shows two regions. The upstream region has two 7 bp repeat sequences that are recognition sites for integrase. The downstream region contains the gene for integrase. Expression of the integrase gene causes the capture of various other genes, most noticeably antibiotic resistance genes. Once integrated, these captured genes are expressed from a promoter within the integrase gene.

Gene cassettes may exist temporarily as free circular molecules incapable of replication and gene expression or else integrated into the *attI* site of an integron. The ultimate source of the genes on the gene cassettes is presently obscure. The reason why most known integron cassettes carry genes for antibiotic resistance is presumably due to observer bias—antibiotic resistance is clinically important and therefore noticed more often.

Transposons of the Tn21 family are widespread and frequently trade antibiotic resistance genes. The Tn2501 transposon carries no antibiotic resistance genes and appears to have an "empty" integron. In other Tn21 family members, the integron has multiple genes, each flanked by 7 bp boxes. Other similar integrons are found on various plasmids and transposons of other families.

Although most integrons are located on transposons and/or plasmids, a few are found in the chromosomes of gram-negative bacteria. Such chromosomal integrons may collect several hundred genes and are then known as super-integrons. The best known example is on the second chromosome of *Vibrio cholerae* (causative agent of cholera) and has collected approximately 200 genes, mostly of unknown function and unknown origin.

4.4. Homing Introns

Mobile DNA does not only consist of transposable elements. **Homing introns** are a strange and relatively-rare type of mobile DNA. As their name indicates, homing introns are intervening sequences that are inserted into genes between two exons. Each homing intron is located in one unique position within one particular gene of the cell it inhabits. This target gene can exist in two versions, with or without the homing intron inserted.

> Homing introns are mobile DNA segments that can occupy only a single site within a host cell gene.

Movement of homing introns is very restricted as they can only occupy this one specific site. Mobilization can only happen when a cell contains two copies of the target gene, one with and the other without the homing intron. The homing intron will then insert itself into the target gene that lacks a copy of the intron. In eukaryotes, this situation may occur after mating when chromosomes from two different parental cells come together in the zygote. In bacteria, it may occur when DNA enters a recipient cell via any of a variety of mechanisms (see Ch. 25).

homing intron   A mobile intron that encodes a protein enabling it to insert itself into a recognition sequence within a target gene

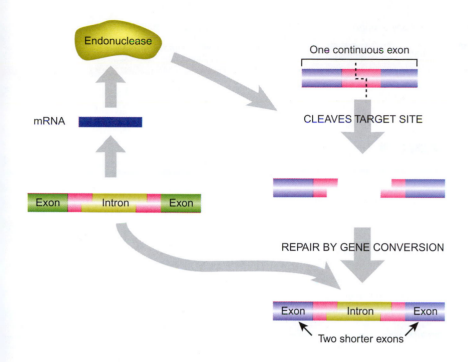

► **FIGURE 22.26**
Homing Intron Inserts in a Unique Location

A homing intron contains a single gene for an endonuclease. This enzyme cleaves a very specific target site, which is only found in a copy of the target gene not containing the homing intron. This situation only occurs when two copies of the target gene are present in a cell; one with the homing intron and one without. The double-stranded break tricks the host cell into repairing the break by gene conversion, thus duplicating the homing intron.

Homing introns encode an endonuclease that is responsible for their movement. The endonuclease cleaves a recognition sequence within the target gene and generates short overhanging ends. This double-stranded break triggers a gene conversion event (see Ch. 23) in which the intact version of the gene is copied and used to repair the break (Fig. 22.26). Thus, the homing intron merely encodes an enzyme to cut the DNA and leaves the host cell to repair the damage. Insertion of the homing intron disrupts the recognition sequence. Thus, the endonuclease only cuts the target gene if the intron is absent. The recognition sequences of homing introns may be as long as 18–20 bp and are the longest and most specific known for any nuclease. This ensures that the intron inserts only into a single unique site in the genome of each cell.

Homing introns of group I use simple endonucleases as described above. These are found in various bacteria and lower eukaryotes. Homing introns of group II are more complex. They are found in bacteria and the organelles of lower eukaryotes. Group II homing introns are retroelements that use an RNA intermediate. The protein they encode has both endonuclease and reverse transcriptase activity. As before, the endonuclease makes a double-stranded break in the middle of the recognition sequence. Next, the reverse transcriptase generates a DNA copy of the intron for inserting into the break. For a template, it uses the primary transcript from the copy of the target gene that contains the intron (Fig. 22.27). The free 3'-OH generated by the endonuclease cleavage is used as primer for starting DNA synthesis. Consequently, the DNA copy is made already attached to the edge of the double-stranded break.

Box 22.2 Junk Philosophy—A Rant from your Author

The situation in the eukaryotic genome reminds me of modern society. In the human genome a small proportion of coding DNA is outnumbered by repetitive and/or noncoding DNA, much of it junk. Similarly, most of the e-mail I get is spam and most of my regular mail is junk mail. Like selfish DNA, most of "my" mail is promoting someone else's advertising agenda. Adding to the melee, my mail also contains urban legends, distortions generated by the media, propaganda put around by politicians, new religious cults, and alternative medicine. Perhaps it is inevitable that in any complex system, whether a cellular genome or a human society, most of the information is valueless, false, or parasitic.

FIGURE 22.27 —————
*Homing Retro-Intron Inserts
via RNA Intermediate*

Group II homing introns integrate
themselves via reverse transcriptase
rather than gene conversion. The
homing intron expresses an enzyme
with both reverse transcriptase and
endonuclease activity. The
endonuclease generates a double-
stranded break at the specific
target site. The reverse transcriptase
generates a DNA copy of the intron
using the 3'-OH of the double-
stranded break as a primer for
synthesis and the mRNA of the intron
as a template. These two functions
are shown separated in the figure
for clarity. However, in real life, the
dsDNA copy of the intron made by
reverse transcriptase is attached to
the 3' end of the double-stranded
break while it is being made.

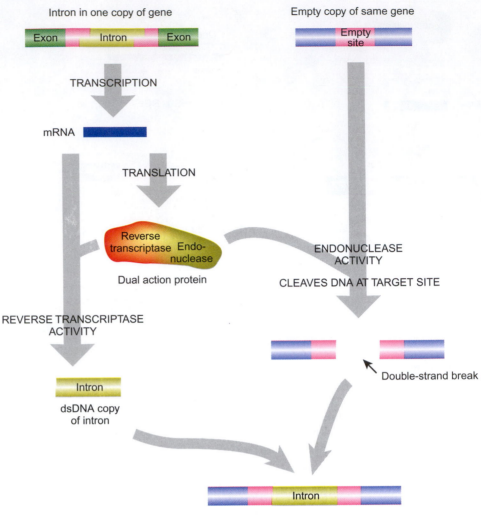

5. Junk DNA and Selfish DNA

DNA sequences that perform no useful function but merely "inhabit" the chromosomes of other organisms may be regarded as genetic parasites of a very degenerate
kind. They are not merely subcellular but submolecular parasites as they are always
found as constituents of other DNA molecules. These include the assorted transposable elements discussed above, both DNA-based and retroelements.

This general type of DNA has been named "**selfish DNA**" because it behaves as if
motivated by its own interests, not those of the host DNA. If the host DNA is degraded,
the parasite dies with it. Thus, the spread of selfish DNA is limited by the need to avoid
destroying the host DNA molecule or inactivating too many host cell genes. Apart from
such considerations, the selfish DNA multiplies inside its host DNA molecule just like a
virus replicating inside a cell, or an infectious bacterium multiplying inside a patient. In
most higher organisms, a substantial proportion of the DNA consists of multiple copies
of such selfish DNA. Why doesn't the host cell purge its chromosome of these parasites? In small, efficient, fast-growing cells, like bacteria or even yeasts, there is very little selfish DNA. Presumably cells with a significant burden of extra DNA divide more
slowly and are weeded out by competition. In large, slow-growing cells the parasites
replicate faster relative to the host DNA and gradually increase in numbers.

> Junk DNA is mostly derived
> from mobile selfish DNA that
> has degenerated.

---

selfish DNA Any segment of DNA that replicates but which is of no use to the host cell it inhabits

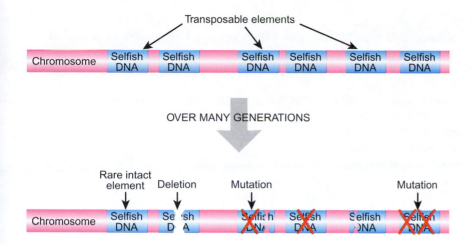

Transposable elements are often termed selfish DNA because they are parasitic DNA sequences that inhabit a host genome. Over time, many copies of selfish DNA are inactivated by mutations and deletions, leaving DNA remnants called junk DNA.

Most selfish DNA is probably the remains of viruses or transposons that inserted into the chromosome long ago. Over long periods of time the copies diverge both by single base mutations and by deletions. Eventually most of the copies become defective and lose the ability to form virus particles or to move around; they degenerate into mere "**junk DNA**" (Fig. 22.28). The genomes of eukaryotic cells generally contain large amounts of junk DNA; indeed, the great majority of mammalian DNA consists of non-coding sequences of one kind or another.

Key Concepts

- Some subcellular genetic elements may be regarded as gene creatures that inhabit cells as their environment.
- Mobile DNA refers to segments of DNA that move as discrete units between different locations on other DNA molecules.
- Transposable elements or transposons are the most common form of mobile DNA. They are always found inserted into other DNA molecules.
- DNA transposons have inverted repeats at the ends and encode an enzyme known as transposase responsible for their movement.
- Conservative transposons move by a cut-and-paste mechanism that may damage host DNA.
- Complex transposons move by replicative transposition and require a second enzyme, resolvase.
- Movement of transposons may sometimes cause rearrangements, such as deletions or inversions, of the host DNA molecule.
- Transposons are found in all forms of life. In higher organisms they are often present in multiple copies, many of which are incomplete.
- Retroelements, including retrotransposons, move by using reverse transcriptase to make an RNA copy of their genome.
- Much of the repetitive DNA of animals and plants is derived from retroelements and/or retrotransposition of host sequences.
- Retrons are retroelements found in bacteria that produce unusual DNA-RNA hybrid molecules.

junk DNA Defective selfish DNA that is of no use to its host cell and can no longer either move or express its genes

- Some genetic elements, for example bacteriophage Mu, combine the properties of both viruses and transposons.
- Conjugative transposons combine the properties of transposons and transmissible plasmids.
- Integrons possess recognition sites for integration plus an integrase enzyme and accumulate mobile gene cassettes.
- Homing introns are mobile DNA segments that are inserted into specific genes between two exons.
- Junk DNA is functionless DNA that has accumulated by the degeneration of mobile DNA.

Review Questions

1. Where can "mobile DNA" be inserted?
2. What are examples of self-replicating genomes? Why are transposons not considered replicons?
3. Where must transposons integrate in order to "survive"?
4. Describe the two main classes of transposable elements. What is the major difference between the two classes?
5. What are the two DNA-based mechanisms of transposition?
6. What are the two essential features of transposons?
7. Besides the gene for transposase, what other genes may be found on transposons?
8. Draw a schematic of an insertion sequence. What does *orfA* encode?
9. What is the importance of frame-shifting in the regulation of transposase production by insertion sequences?
10. What are the three elements required for conservative or "cut-and-paste" transposition?
11. What potentially harmful effects are produced on the original DNA molecule during conservative transposition?
12. Compare and contrast conservative and replicative transposition.
13. What extra enzyme is needed in replicative transposition (compared to conservative transposition)? What is the name of the DNA sequence that this enzyme recognizes?
14. What is the difference between replicons and complex transposons?
15. During which type of transposition is a cointegrate formed?
16. What is the classic example of genes carried on composite transposons? Name three of the best known.
17. How many terminal repeats do composite transposons contain?
18. What is the term sometimes used to describe the process of moving a composite transposon using the "inside ends"? What can happen to the host DNA during this transposition?
19. Compare the Ac and Ds elements of plants. Why can't Ds elements move themselves? What is required for Ds element movement?
20. What is the most widely-distributed type of transposable element in higher organisms?
21. What enzyme is used to convert an RNA transcript back into DNA?
22. What is different about the repeat sequences in retrotransposons than those in DNA-based transposons?
23. What are the differences between SINEs and LINEs?
24. What are two examples of genetic defects causing disease due to transposable elements?
25. What is an Alu element? How is it different from other pseudogenes? Why are there a million copies of the Alu element in the human genome?
26. What is the origin of the Alu element?
27. What enzyme do retrons encode? What nucleic acid product do retrons generate?
28. Name one genetic element that possesses both the characteristics of a transposable element and a virus? What is the host organism?
29. How does the replication of bacteriophage Mu affect host DNA?
30. What are conjugative transposons?
31. What is the role of transposons in antibiotic resistance in bacteria?

32. What is an integron?
33. What is the function of integrase (in an integron)? Where is the gene encoding this enzyme found? What sites does integrase recognize?
34. What is the best-known example of a super-integron? How does a super-integron form?
35. Why is it difficult to get rid of "selfish DNA" from a genome?
36. Where are homing introns inserted? Are homing introns specific for their insertion site?
37. What do homing introns encode? How does this enzyme "know" not to cleave DNA if a homing intron is already present?
38. What is the difference between group I and group II homing introns?

Conceptual Questions

1. The inverted or direct repeats that flank transposable elements are sometimes involved in a recombination event. What would happen to the following transposable element if there was a recombination event between the inverted repeats? Do you think this would affect the function of the transposable element?

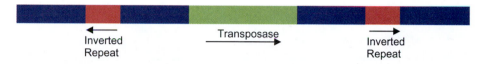

2. A researcher studying the lactose permease gene discovered a very interesting phenomenon. When a particular strain of *E. coli* containing the gene for lactose permease was plated on MacConkey agar containing lactose, some colonies were red and others were white. MacConkey agar is a nutrient agar with an indicator dye. When the bacteria are able to take up lactose via the lactose permease, the colonies turn red; when the bacteria cannot take up lactose, the colonies appear white. The color of the colony is therefore an indicator of whether or not the lactose permease gene is being expressed as a functional protein. The researcher isolated the gene for lactose permease from one red colony and one white colony and determined that the gene from the white colony was about 500 base pairs longer than the gene from the red colony. The researcher repeatedly took one white colony (which is derived from one single bacterium) and plated that colony on new lactose MacConkey plates. Every plate that started from one white colony was able to make a mixture of white and red colonies. Every time the researcher analyzed the genes, the lactose permease gene from the white colony contained the 500 bp insert, whereas the lactose permease gene from the red colony did not contain the extra sequence. Assuming that each colony derived from one single bacterial cell, suggest a possible reason for the white colony being able to produce offspring bacterium with red colonies. (This is a true story!)

3. Describe the mechanism of transposition for both retroposons and complex transposons highlighting the differences between the two categories.

4. Transposons are used extensively in the laboratory to help researchers identify new genes. Transposon tagging is a technique used to identify the phenotype of new genes and then isolate and clone that gene. The *copia* element of *Drosophila* is used often to study and clone new genes. In a recent experiment, a *Drosophila melanogaster* strain that had an active *copia* element was grown in the lab. Each offspring was looked at under a microscope to see if there was any change in phenotype. After looking at 1000 flies, the researcher found a male and female fly that both had white eyes instead of the normal red. If you mated these two flies and looked at the offspring, would you expect all the flies to have white eyes?

5. A researcher was studying two different strains of *E. coli*; one called K8380 was resistant to the antibiotic ampicillin and the other called M765 was resistant to tetracycline. She wondered what would happen if the two were cultured together in liquid media and then plated onto agar containing both tetracycline and ampicillin. What do you think would happen? Would any of the bacteria grow?

UNIT 6

Changing the DNA Blueprint

Mutations and Repair

Although our discussions of DNA replication, transcription, and translation present a picture of well-defined and elegant processes, nature is not perfect and mistakes happen. An error in a cell's genetic material is known as a **mutation**. As might be expected, many mutations are detrimental. However, detrimental mutations tend to be overestimated because they are more noticeable. Very often the negative effect of a mutation is minimal, and in fact, the majority of mutations have little or no significant effect on the survival of an organism—they are essentially neutral. Furthermore, occasional mutations may turn out to be beneficial to the survival and reproduction of the organism. The accumulation of such beneficial mutations allows the organism to evolve in response to changing environmental conditions (see Ch. 26).

1. Mutations Alter the DNA Sequence

At the molecular level, mutations are alterations in the DNA molecule. Consequently, when a DNA molecule replicates, any changes due to mutation of the original DNA base sequence will be duplicated and passed on to the next generation of cells. In single-celled organisms, mutations are passed from one generation to the next when the organism divides. Among multicellular organisms, the situation is more

---

mutation An alteration in the DNA (or RNA) that comprises the genetic information

Mutations are alterations in the genetic material of any organism or genetic element.

complicated. Mutations are inherited by the next generation of organisms only if they occur in the cells of the germ line and are passed on during sexual reproduction. When humans carry a mutation in their reproductive cells that leads to an observable defect in their children, the resulting condition is referred to as **inherited disease**. On the other hand, mutations that occur in somatic cells will only be passed on to the descendents of those cells. Such mutant cell lines will be restricted to the original multicellular organism where the mutation occurred.

Since DNA is used as a template in transcription, a mutation in the DNA sequence will be passed on to the messenger RNA (mRNA) molecule, and in turn to the protein during translation. Defective proteins may alter the cell's function. Somatic mutations that result in unregulated cell growth are responsible for the emergence of cancers. Other somatic mutations merely result in particular cell lines or organs being genetically different from the rest of the body. (Errors may also occur during transcription and translation, but do not affect the DNA. These result in occasional defective RNA or protein molecules. These are not regarded as mutations as they are not passed on to the descendents of the cell where they occurred.)

Most mutations cause little observable harm. A few unusually severe mutations are responsible for inherited disease.

There are many different types of mutation and most have only minor effects; in fact, many cause no noticeable defect at all. One reason for the majority of mutations having no effect is that higher organisms have two copies of each gene. This means that if one copy is damaged by mutation, there is a back-up copy that can be used to produce the correct protein. This often suppresses the potential defect, unless the mutation is dominant.

It has been estimated that a typical human carries enough harmful mutations to total approximately eight lethal equivalents per genome. Put another way, if humans were haploid, with only a single copy of each gene, the average person would be dead eight times over. These silent mutations accumulate throughout the generations and provide the basis for human evolution. Recent estimates from the 1000 Genomes Project suggest that each person contains on average 75–100 potentially damaging alleles, although almost all are present as only one of the two copies of each gene we possess. In addition, we each contain around 20,000 changes that are largely or totally harmless.

2. The Major Types of Mutation

A single mutation results from a single event and a multiple mutation is due to several events. A single mutational event, however large or complex in effect, is regarded as a single mutation. A **null mutation** totally inactivates a gene; the expression "null mutation" is a genotypic term. Complete absence of a gene product may or may not cause a detectable phenotype. A **tight mutation** is one whose phenotype is clear-cut. The complete loss of a particular enzyme may result in no product in a particular biochemical pathway. For example, a tight mutation would prevent a bacterium from growing unless there was an essential nutrient in the growth medium. A **leaky mutation** is one where partial activity remains. For example, 10% residual enzyme activity might allow a bacterium to still grow, albeit very slowly.

A **silent mutation** is an alteration in the DNA sequence that has no effect on the operation of the cell and is therefore not so much silent as invisible from the outside. By definition, silent mutations have no effect on the phenotype. Silent mutations can occur in the non-coding DNA between genes, in introns (except for the splice site recognition sequences), or in the third base of a codon. The third base is often

---

inherited disease   Disease due to a genetic defect that is passed on from one generation to the next
leaky mutation   Mutation where partial activity remains
null mutation   Mutation that totally inactivates a gene
silent mutation   An alteration in the DNA sequence that has no effect on the phenotype
tight mutation   Mutation whose phenotype is clear-cut due to the complete loss of function of a particular gene product

degenerate, and changing it does not alter the amino acid encoded because of the wobble rules (see Ch. 13). This is known as **third base redundancy** or **codon degeneracy**, and in such cases mutation of the third base does not alter the protein sequence. The caveat to this type of silent mutation is that there are sometimes different amounts of available transfer RNAs (tRNAs) for different codons for the same amino acid. Changing the codon from a commonly used codon to a codon that does not have as much tRNA can result in slower protein production albeit of normal proteins.

The sequence of a DNA molecule may be altered in many different ways. Such mutations have a variety of outcomes that depend in part on the nature of the change and in part on the role of the DNA sequence that was altered. The major types of sequence alteration are as follows and will be discussed separately below:

Base substitution: one base is replaced by another base.

Insertion: one or more bases are inserted into the DNA sequence.

Deletion: one or more bases are deleted from the DNA sequence.

Inversion: a segment of DNA is inverted, but remains at the same overall location.

Duplication: a segment of DNA is duplicated; the second copy usually remains near the same location as the original.

Translocation: a segment of DNA is transferred from its original location to another position either on the same DNA molecule or on a different DNA molecule.

Much of the discussion below considers what happens when mutations occur within genes that encode proteins. However, it is important to realize that mutations may also occur within those genes whose products are tRNA, ribosomal RNA (rRNA), or other non-translated RNA molecules. Alterations in these molecules may have drastic effects on ribosome function, splicing, or other vital processes. Furthermore, mutations may also fall within promoter sequences or other regulatory sites that do not actually encode any gene product. Nonetheless, such regulatory sites are important for gene expression and altering them may have major effects.

2.1. Base Substitution Mutations

A mutation that involves only a single base is known as a **point mutation**. If one base is replaced by another, this is a base substitution mutation. These may be subdivided into **transitions** and **transversions**. In a transition, a pyrimidine is replaced by another pyrimidine (i.e., T is replaced by C or vice versa) or a purine is replaced by another purine (i.e., A is replaced by G or vice versa). A transversion occurs when one base is replaced by another of a different type (i.e., a pyrimidine is replaced by a purine or vice versa).

DNA molecules are double-stranded. If a mutation occurs and a single base is replaced with another, the DNA molecule will temporarily contain a pair of mismatched bases (Fig. 23.01). When the DNA molecule replicates, complementary bases

---

base substitution   Mutation in which one base is replaced by another
codon degeneracy   Situation where a set of codons all code for the same amino acid and the identity of the third codon base makes no difference during translation
deletion   Mutation in which one or more bases is lost from the DNA sequence
duplication   Mutation in which a segment of DNA is duplicated
insertion   Mutation in which one or more extra bases are inserted into the DNA sequence
inversion   Mutation in which a segment of DNA has its orientation reversed, but remains at the same location
point mutation   Mutation that affects a single base pair
third base redundancy   Situation where a set of codons all code for the same amino acid and thus the identity of the third codon base makes no difference during translation
transition   Mutation in which a pyrimidine is replaced by another pyrimidine or a purine is replaced by another purine
transversion   Mutation in which a pyrimidine is replaced by a purine or vice versa
translocation   Mutation in which a segment of DNA is transferred from its original location to another site on the same or a different DNA molecule

FIGURE 23.01
Segregation of Base Alterations in DNA

A) A mutation has occurred causing a C to be replaced by a T. B) During replication, one DNA molecule (in yellow) matches the original base and the other strand (green) matches the mutated base. C) The strands segregate into the progeny, giving one wild-type and one mutant.

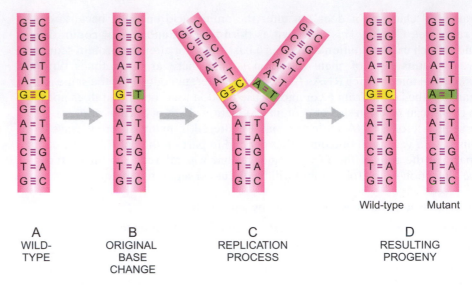

A WILD- TYPE	B ORIGINAL BASE CHANGE	C REPLICATION PROCESS	D RESULTING PROGENY

will be incorporated into the new strands opposite the bases making up the mismatch. The result is one wild-type daughter molecule and one mutant DNA molecule.

When mutations are induced by experimental treatment, it is necessary to allow the cells time to divide after treatment before imposing any selection. This allows the original DNA strands to separate and the cell to make new DNA molecules that are either fully wild-type or fully mutant. This process is sometimes referred to as **segregation**, as the originally mutated cell segregates the mutation and the wild-type into separate daughter cells upon cell division.

2.2. Missense Mutations May Have Major or Minor Effects

Overall, the location of a single nucleotide change largely determines the effect of the mutation. A **missense mutation** occurs when a change in the base sequence alters a codon so that one amino acid in a protein is replaced with a different amino acid. The severity of a missense mutation depends on the location and the nature of the amino acid that is substituted.

Mutant versions of genes are numbered as they are discovered. Thus, *genX123* refers to the 123rd mutation isolated in the gene *genX*. A mutation that results in a codon for one amino acid being replaced by another may be written as *genX123* (Arg185Leu), or in one-letter code (R185L). This indicates that arginine at position 185 has been replaced by leucine.

Proteins must assume their correct 3D structure in order to function properly. Moreover, most proteins, especially enzymes, contain an active site whose role is critical. This region contains relatively few of the many amino acids that make up a typical protein, and these may be separated from each other in the primary sequence. Sequence comparison of the same protein from different organisms usually shows that only the amino acids in a few positions are invariant or nearly so. These highly-conserved amino acid residues generally include those in the active site(s) plus others that are critical for correct folding of the protein. Thus, mutations that alter active site residues will usually have major effects. Mutations that alter residues important for structure will also have a major impact (although these changes may be more forgiving). However, mutations affecting less vital parts of the protein will often have minor effects and substantial activity may remain.

> Mutations in amino acids that are invariant among related species tends to affect the active site or amino acids important for 3D structure of the protein.

missense mutation   Mutation in which a single codon is altered so that one amino acid in a protein is replaced with a different amino acid
segregation   Replication of a hybrid DNA molecule (whose two strands differ in sequence) to give two separate DNA molecules, each with a different sequence

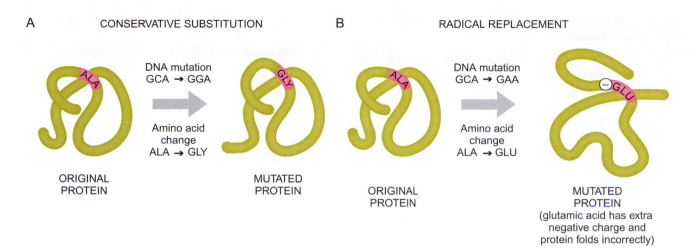

FIGURE 23.02
Conservative Substitution and Radical Replacement

A) A mutation resulting in DNA change from GCA to GGA will result in the conservative substitution of an alanine for a glycine. Since both amino acids have similar properties, it is likely that the mutant protein will fold similarly to the wild-type. B) A mutation resulting in the substitution of a glutamate for an alanine is a radical replacement as the glutamate has an extra negative charge that will probably cause the protein to fold quite differently from the wild-type.

The chemical properties of the original amino acid and the one replacing it in the mutant are also important. Suppose the codon UCU, which codes for serine, is changed to ACU, which codes for threonine. Both serine and threonine are small, hydrophilic amino acids with hydroxyl groups. Replacing one amino acid with another that has similar chemical and physical properties is known as a **conservative substitution** or **neutral mutation** (Fig. 23.02A). Swapping serine for threonine in the less critical regions of a protein will probably not alter its structure very much and the protein may still work, at least partially. In rare instances, the protein may actually work better. On the other hand, if the exchange is made in a critical region of the protein, such as the active site, even a conservative substitution may completely destroy activity.

> Replacing an amino acid with a chemically similar one often has little effect on a protein.

Replacing one amino acid with another that has different chemical and physical properties is known as a **radical replacement** (Fig. 23.02B). If the codon GUA, which codes for valine, is changed to GAA, this yields a glutamic acid. This replaces a bulky hydrophobic residue with a smaller, hydrophilic residue that carries a strong negative charge. Under most circumstances, replacing valine with glutamic acid will seriously cripple or totally incapacitate most proteins. If the residue in question is on the surface of the protein, it is sometimes possible to get away with a radical replacement, provided that the change does not affect a critical binding site or alter the solubility of the protein too drastically.

> Replacing an amino acid with one that has very different properties often causes significant damage to the protein.

Mutations whose effects vary depending on the environment are known as **conditional mutations.** An interesting and sometimes useful type of missense mutation is the **temperature sensitive (ts) mutation** (Fig. 23.03). As its name indicates, the mutant protein folds properly at low temperature (the "permissive" temperature) but is unstable at higher temperatures and unfolds. Consequently, the protein is inactive at the higher or "restrictive" temperature. If a protein is essential, a missense mutation will often be lethal to the cell. However, a temperature sensitive mutant can be grown and used for genetic experiments at the lower permissive temperature, where

---

conservative substitution   Replacement of an amino acid with another that has similar chemical and physical properties
conditional mutation   Mutation whose phenotypic effects depend on environmental conditions such as temperature or pH
neutral mutation   Replacement of an amino acid with another that has similar chemical and physical properties
radical replacement   Replacement of an amino acid with another that has different chemical and physical properties
temperature sensitive (ts) mutation   Mutation whose phenotypic effects depend on temperature

FIGURE 23.03
Temperature Sensitive Mutation

The wild-type gene encodes a protein that folds similarly at high and low temperature. The mutant protein folds normally at low temperature but unfolds at high temperature, and consequently, no longer works properly.

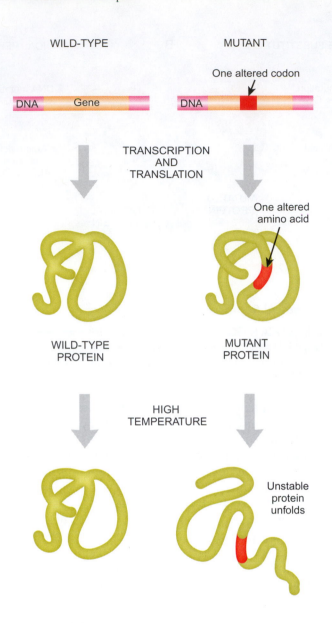

WILD-TYPE   MUTANT

One altered codon

DNA   Gene   DNA

TRANSCRIPTION AND TRANSLATION

One altered amino acid

WILD-TYPE PROTEIN   MUTANT PROTEIN

HIGH TEMPERATURE

Unstable protein unfolds

Mutant proteins may sometimes be defective only under certain conditions, such as high temperature.

Mutations whose effects vary depending on a variety of environmental conditions are well known.

it remains alive. To analyze the damage caused by the mutation, the temperature is then shifted upward to the restrictive temperature at which the protein is inactivated and the organism may eventually die. An example in the fruit fly, *Drosophila*, is the *para(ts)* mutation. This affects a protein that forms sodium channels necessary for transmitting nerve impulses. At high temperatures the mutant protein is inactive and the flies are paralyzed. At lower temperatures, they are capable of normal flight.

Naturally-occurring temperature sensitive mutations have given rise to the patterns of fur coloration in some animals. Many light colored animals have black tips to their paws, tails, ears, and noses. This is due to a temperature-sensitive mutation in the enzyme that synthesizes melanin, the black skin pigment of mammals. In these cases, the mutant enzyme is inactive at normal mammalian body temperature, but active at the lower temperatures found at the extremities. Consequently, melanin is made only in the cooler outlying parts of the body (Fig. 23.04).

Cold-sensitive mutations do occur but are much rarer than high temperature sensitive mutations. One possibility is when multisubunit proteins are held together by hydrophobic patches on the surfaces of the subunits (see Ch. 14). The hydrophobic interaction is weaker at lower temperatures, and therefore it is possible to get altered proteins that fail to assemble at low temperatures but are normal at higher temperatures. For example, microtubule proteins are temperature-dependent. Microtubules are cylinders made from the helical assembly of the monomer tubulin.

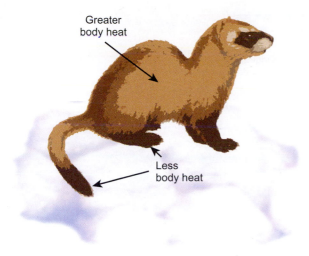

Greater body heat

Less body heat

■ **FIGURE 23.04**
Temperature-Sensitive Fur Coloration

Some animals carry a mutation in the gene for melanin, which makes the enzyme more active at lower temperatures. As a result, the cooler body parts are darker in color than the warmer parts of the body.

In *Saccharomyces cerevisiae*, mutations that caused cold sensitivity were concentrated in amino acids at the interfaces between adjacent α-tubulin subunits. Other cold-sensitive mutations are found that respond to the osmotic pressure or ionic strength of the medium.

2.3. Nonsense Mutations Cause Premature Polypeptide Chain Termination

Not all codons encode amino acids. Three (UAA, UAG, and UGA) are stop codons that signal the end of a polypeptide chain. A **nonsense mutation** occurs when the codon for an amino acid is mutated to give a stop codon. Suppose that the codon UCG for serine is changed by replacing the middle base, C, with A. This gives the stop codon UAG. When the ribosome translates the mRNA, it comes to the mutant codon that used to be serine. But this is now a stop codon, so the ribosome stops and the rest of the protein does not get made. Release factor recognizes the premature stop codon and releases the partly-made polypeptide. Hence, nonsense mutations are sometimes called **chain termination mutations**. Usually, the shortened polypeptide chain cannot fold properly (Fig. 23.05). Such misfolded proteins are detected and degraded by the cell. The result, in practice, is normally the total absence of this particular protein. Nonsense mutations are often lethal if they affect important proteins. In eukaryotes, nonsense mutations that generate shortened mRNA transcripts are not translated. Instead, the mRNA is degraded by nonsense mediated decay (See Chapter 12, section 11.1).

2.4. Deletion Mutations Result in Shortened or Absent Proteins

Mutations that remove one or more bases are known as deletions and those that add extra bases are known as insertions. Clearly, the effect of a deletion (or insertion) depends greatly on how many bases are removed (or inserted). In particular, we should distinguish between point mutations where one or a very few bases are affected and gross deletions and insertions that affect long segments of DNA. Point deletions and insertions may have major effects due to disruption of the reading frame if they fall within the coding region of a gene—see Section 2.6.

Larger deletions are indicated by the symbol Δ or by DE. Thus, Δ(*argF-lacZ*) or DE(*argF-lacZ*) indicates a deletion of the region (of the *Escherichia coli* chromosome in this case) from the *argF* to the *lacZ* gene. Obviously, deletion of the DNA sequence

---

chain termination mutation   Same as nonsense mutation
nonsense mutation   Mutation due to changing the codon for an amino acid to a stop codon

FIGURE 23.05
Nonsense Mutation

A mutation in the DNA has produced a new stop codon that causes premature termination of the protein. The shortened protein remains unfolded and is usually detected by the cell and degraded.

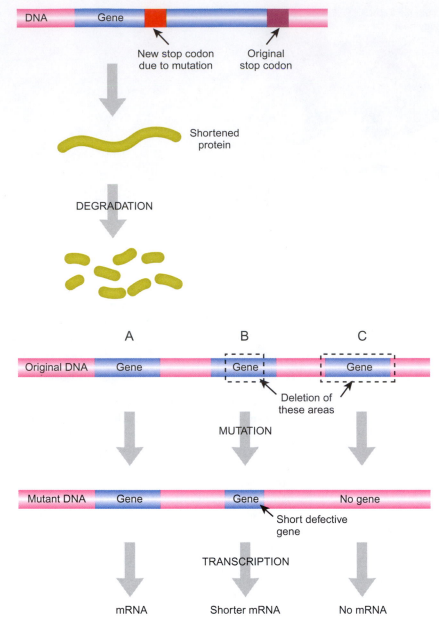

FIGURE 23.06
Effects of Deletion Mutations

A) The wild-type gene produces a normal mRNA and a normal protein. B) A large deletion causes a shorter mRNA and a short unstable protein. C) Deletion of an entire gene results in no mRNA and no protein.

Deletions may remove critical segments of DNA or largely functionless regions of DNA.

for a whole gene means that no mRNA and no protein will be made (Fig. 23.06). If the protein is essential, then the deletion will be lethal (in a haploid organism). Large deletions may remove part of a gene, an entire gene, or several genes. Deletions may also remove all or part of the regulatory region for a gene. Depending on the precise region removed, gene expression may be decreased or increased. For example, a deletion that removes the binding site for a repressor may result in a large increase in activity of the gene in question. Thus, loss of DNA may result in elevated activity. Again, deletions may remove largely functionless DNA, such as the non-coding sequences between genes or the introns found within genes. In this case, the effects may be small or marginal.

DNA INSERTION OF TRANSPOSON

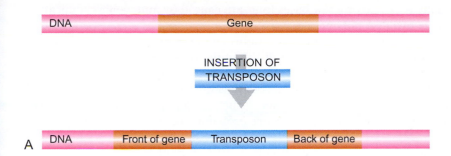

A

POLAR EFFECT IN BACTERIA

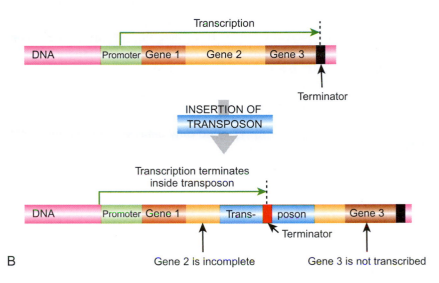

B

■ **FIGURE 23.07**
Effects of Insertion Mutations

A) Insertion of a transposon into the middle of a gene interrupts the coding sequence. B) Insertion of a transposon into the second gene of a bacterial operon with three genes. Gene 1 is the only gene correctly transcribed since the transposon disrupts gene 2 and causes premature termination. Gene 3 will not be transcribed, although its coding sequence is still intact.

Deletion mutations are surprisingly frequent. About 5% of spontaneous mutations in bacteria such as *E. coli* are deletions. Although bacteria lack introns and the intergenic spacer regions are very short, it is still possible to generate nonlethal deletions of considerable size. The main reason is that many genes are needed only under certain limited environmental conditions. Thus, deletions of the entire *lac* operon in *E. coli* prevent the organism from using lactose as a source of carbon, yet have no other deleterious effects.

2.5. Insertion Mutations Commonly Disrupt Existing Genes

Genes may also be inactivated by insertions of DNA. If a foreign segment of DNA is inserted into the coding region, then the gene is said to be disrupted. Usually the gene will be completely inactivated; however, the precise result will vary depending on the length and sequence of the inserted DNA as well as on its precise location (Fig. 23.07). Thus, if an insertion occurs very close to the 3′ end of a gene, most of the coding sequence will remain intact, and sometimes a more or less functional protein may still be made. The cause of insertion mutations may be divided into two distinct categories. Some of these mutations are the result of **mobile genetic elements**, usually

---

mobile genetic element A discrete segment of DNA that is able to change its location within larger DNA molecules by transposition or integration and excision

Most naturally-occurring large insertions are due to mobile elements, including transposons, retroposons, and certain viruses.

Large insertions within operons often prevent transcription of genes downstream from the insertion site.

thousands of bases long, inserting themselves into a gene (see Ch. 22). These include **transposons** and **retrotransposons** (see Ch. 22) or viruses (see Ch. 21). Other insertion mutations, usually only one or a few bases long, are caused by mutagenic chemicals or by mistakes made by DNA polymerase.

Insertions are indicated by the symbol :: between the target gene and the inserted element. Thus, *lacZ*::Tn10 indicates the insertion of the transposon Tn10 into the *lacZ* gene. Insertion of such large genetic elements disrupts the target gene and completely disrupts its proper function. The presence of transposons greatly increases the frequency of various other DNA rearrangements, such as deletions and inversions. The precise mechanisms are uncertain, but they are probably due to abortive transposition attempts (see Ch. 22).

Transposable elements and viruses usually contain multiple transcriptional terminators. Consequently, RNA polymerase cannot transcribe through them. Therefore, their presence blocks transcription of any other downstream genes that share the same promoter as the gene that suffered the insertion event. This effect is referred to as **polarity**. Since bacterial genes are often found clustered in operons and are co-transcribed onto the same mRNA (see Ch. 11), they are much more likely than eukaryotic genes to show polarity effects due to insertions.

Occasionally, insertions may activate genes. If an insertion occurs in the recognition site for a repressor, binding of the repressor will be prevented and activation of the gene may result. In addition, a few transposons are known to have promoters close to their ends, facing outwards (Fig. 23.08). Insertion of these may activate a previously silent gene. Examples are known of "cryptic" genes with potentially functional gene products that cannot be expressed due to defective promoters. Thus, the *bgl* operon of *E. coli* is inactive in the wild-type and only expressed in mutants when a transposon carrying an outward-facing promoter is inserted just upstream of the operon and reactivates it.

FIGURE 23.08
Unusual Activating Effects of Insertion Mutations

A) The gene shown is under the control of its own promoter. B) A transposon is inserted between the normal promoter and the structural gene. The gene is now expressed under control of a promoter carried by the transposon.

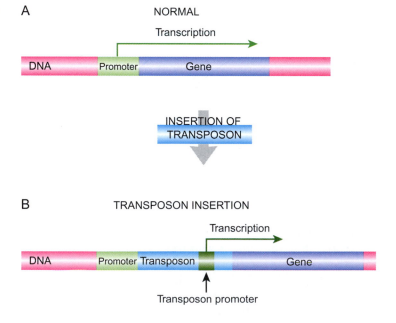

polarity   When the insertion of a segment of DNA affects the expression of downstream genes, usually by preventing their transcription
retrotransposon or retroposon   A transposable element that uses reverse transcriptase to convert the RNA form of its genome to a DNA copy
transposon   Same as transposable element, although the term is usually restricted to DNA-based elements that do not use reverse transcriptase

2.6. Frameshift Mutations Sometimes Produce Abnormal Proteins

Bases are read as codons, which are groups of three nucleotides, when translated into amino acids during protein synthesis (Ch. 13). The introduction or removal of one or two bases can have drastic effects since the alteration changes the reading frame of the afflicted gene. If a single base of a coding sequence is inserted or removed, the reading frame for all codons following the insertion or deletion will be changed (Fig. 23.09). The result will be a completely garbled protein sequence. Such **frameshift mutations** usually completely destroy the function of a protein, unless they occur extremely close to the far end. Insertion or deletion of two bases also changes the reading frame and alters protein function.

However, insertion or deletion of three bases adds or removes a whole codon and the reading frame is retained. Apart from the single amino acid that is gained or lost, the rest of the protein is unchanged. If the deleted (or inserted) amino acid is in a relatively less vital region of the protein, a functional protein may be made. Adding or deleting more than three bases will give a similar result as long as the number is a multiple of three. In other words, a whole number of codons must be added or subtracted to avoid the consequences of changing the reading frame.

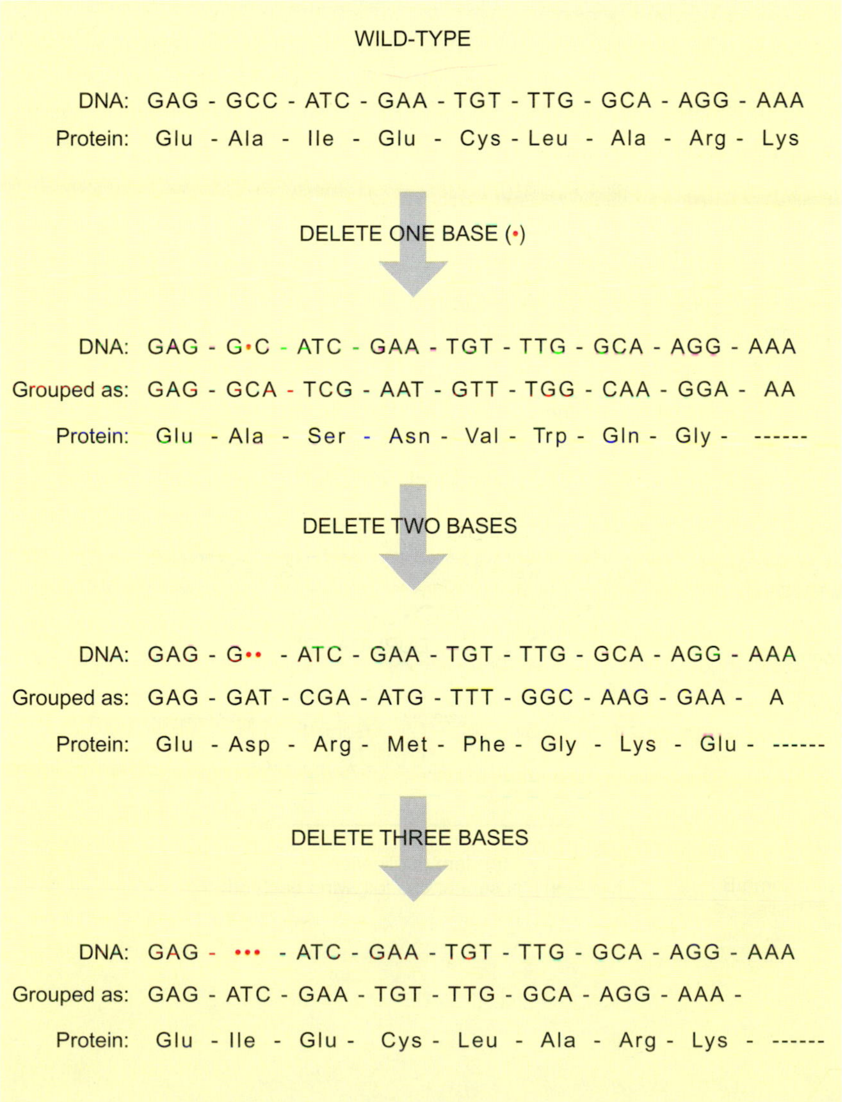

> A mutation that alters the reading frame usually disrupts protein function completely.

► FIGURE 23.09
Frameshift Mutations

The effect of deleting one, two, or three bases on the reading frame and the protein produced is illustrated. Note that by deleting three consecutive bases of a reading frame, the amino acid sequence stays the same although there is one amino acid missing.

WILD-TYPE

DNA: GAG - GCC - ATC - GAA - TGT - TTG - GCA - AGG - AAA
Protein: Glu - Ala - Ile - Glu - Cys - Leu - Ala - Arg - Lys

DELETE ONE BASE (•)

DNA: GAG - G•C - ATC - GAA - TGT - TTG - GCA - AGG - AAA
Grouped as: GAG - GCA - TCG - AAT - GTT - TGG - CAA - GGA - AA
Protein: Glu - Ala - Ser - Asn - Val - Trp - Gln - Gly - ------

DELETE TWO BASES

DNA: GAG - G•• - ATC - GAA - TGT - TTG - GCA - AGG - AAA
Grouped as: GAG - GAT - CGA - ATG - TTT - GGC - AAG - GAA - A
Protein: Glu - Asp - Arg - Met - Phe - Gly - Lys - Glu - ------

DELETE THREE BASES

DNA: GAG - ••• - ATC - GAA - TGT - TTG - GCA - AGG - AAA
Grouped as: GAG - ATC - GAA - TGT - TTG - GCA - AGG - AAA -
Protein: Glu - Ile - Glu - Cys - Leu - Ala - Arg - Lys - ------

frameshift Mutation in which the reading frame of a structural gene is altered by insertion or deletion of one or a few bases

2.7. DNA Rearrangements Include Inversions, Translocations, and Duplications

An inversion is just what its name implies, an inverted segment of DNA (Fig. 23.10A). Transcribing an inverted piece of DNA completely changes the coding sequence. Inversions within genes are usually highly detrimental. On the other hand, inversions do not always disrupt genes. If the endpoints of an inversion are in intergenic DNA, then inversion of a DNA segment carrying one or more intact genes, together with their promoters, may have only mild effects. In this case the orientation of the gene(s) will be reversed relative to the rest of the chromosome and it will be transcribed in the opposite direction.

A translocation is the removal of a section of DNA from its original position and its insertion in another location, either on the same chromosome or on a completely different chromosome (Fig. 23.10B). If an intact gene is merely moved from one place to another, it may still work and little damage may result. On the other hand, if, for example, half of one gene is moved and inserted into the middle of another gene, the results will be doubly chaotic.

A duplication occurs when a segment of DNA is duplicated and both copies are retained. In most cases, the duplicate is located just following the original copy (Fig. 23.10C);

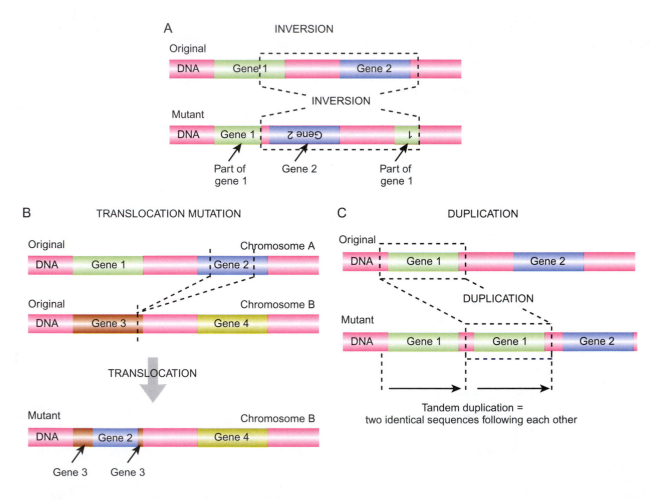

FIGURE 23.10
Inversions, Translocations, and Duplications

A) The inversion shown encompasses gene 2 and part of gene 1. The inverted regions are indicated by the backward spelling. B) Two chromosomes are shown with four genes. Part of gene 2 is moved from its location on chromosome A to chromosome B to a position that splits gene 3. C) Gene 1, along with some non-coding DNA, is duplicated in a tandem duplication.

in other words, it is a **tandem duplication**. Duplication within a gene will seriously disrupt the gene product, whereas duplication of a large segment of DNA may generate extra copies of genes. Such duplication followed by sequence divergence is thought to be a major source of new genes over the course of evolution (see Ch. 26).

3. Chemical Mutagens Damage DNA

Mutations caused by agents that damage DNA are known as **induced mutations**. Agents that mutate DNA are called **mutagens** and are of three main types: mutagenic chemicals, radiation, and heat. Even if there are no dangerous chemicals or radiation around, mutations still occur, though less frequently. These are **spontaneous mutations**. Some of these are due to errors in DNA replication. The enzymes of DNA replication are not perfect and occasionally make mistakes. In addition, DNA undergoes certain spontaneous chemical reactions (alterations) at a low but detectable rate, and this rate goes up with increasing temperature.

> DNA may be damaged by a variety of chemicals and by radiation.

The most common mutagens are toxic chemicals that react with DNA and alter the chemical structure of the bases. For example, EMS (ethyl methane sulfonate) is an alkylating agent widely used by molecular biologists to mutagenize growing cells. It adds an ethyl group to bases in DNA and so changes their shape and their base-pairing properties (Fig. 23.11A). Nitrite converts amino groups to hydroxyl groups and so converts the base cytosine to uracil (Fig. 23.11B). Nitrite is used experimentally to mutate purified DNA, such as a cloned gene carried on a plasmid, while the plasmid is in the test tube. The mutagenized DNA is then transferred back into a cell to identify the mutations that were generated. During replication, DNA polymerase misidentifies these altered bases and puts in the wrong bases in the new complementary strand of DNA it is making.

> Base analogs are mistaken by the cell for the natural nucleic acid bases.

Base analogs are chemical mutagens that mimic the bases found in natural DNA. For example, bromouracil resembles thymine in shape. It is converted by the cell to the DNA precursor bromouridine triphosphate, which DNA polymerase inserts where thymine should go. Unfortunately, bromouracil can flip-flop between two alternative shapes (Fig. 23.12). In its alternate form, bromouracil resembles cytosine and pairs with guanine. If bromouracil is in its misleading form when DNA polymerase arrives, a G will be put into the new strand opposite the bromouracil instead of A.

> Intercalating agents result in the insertion of an extra base pair during DNA replication.

Some mutagens imitate the structure of a base pair rather than a single base. For example, **acridine orange** has three rings and is about the size and shape of a base pair. Acridine orange is not chemically incorporated into the DNA. Instead, it squeezes in between the base pairs in the DNA (Fig. 23.13), a process referred to as **intercalation**. During DNA replication, DNA polymerase mistakes the intercalating agent for a base pair and puts in an extra base when making the new strand. As discussed above, insertion of an extra base will change the reading frame of the protein encoded by a gene. Since this will completely destroy the function of the protein, intercalating agents are highly hazardous mutagens.

A **teratogen** is an agent that causes abnormal development of the embryo, which results in gross structural defects. Teratogens may or may not cause mutations. The most famous example is thalidomide, which caused the birth of malformed children with missing limbs. Thalidomide interferes with the development of embryos as opposed to causing mutations. Although the mechanism responsible for the malformations remains uncertain, it is known that thalidomide prevents blood vessels from

---

acridine orange   A mutagenic agent that acts by intercalation
base analog   Chemical mutagen that mimics a DNA base
induced mutation   Mutation caused by external agents such as mutagenic chemicals or radiation
intercalation   Insertion of a flat chemical molecule between the bases of DNA, often leading to mutagenesis
mutagen   Any agent, including chemicals and radiation, that can cause mutations
spontaneous mutation   Mutation that occurs "naturally" without the help of mutagenic chemicals or radiation
tandem duplication   Mutation in which a segment of DNA is duplicated and the second copy remains next to the first
teratogen   An agent that causes abnormal embryo development leading to gross structural defects

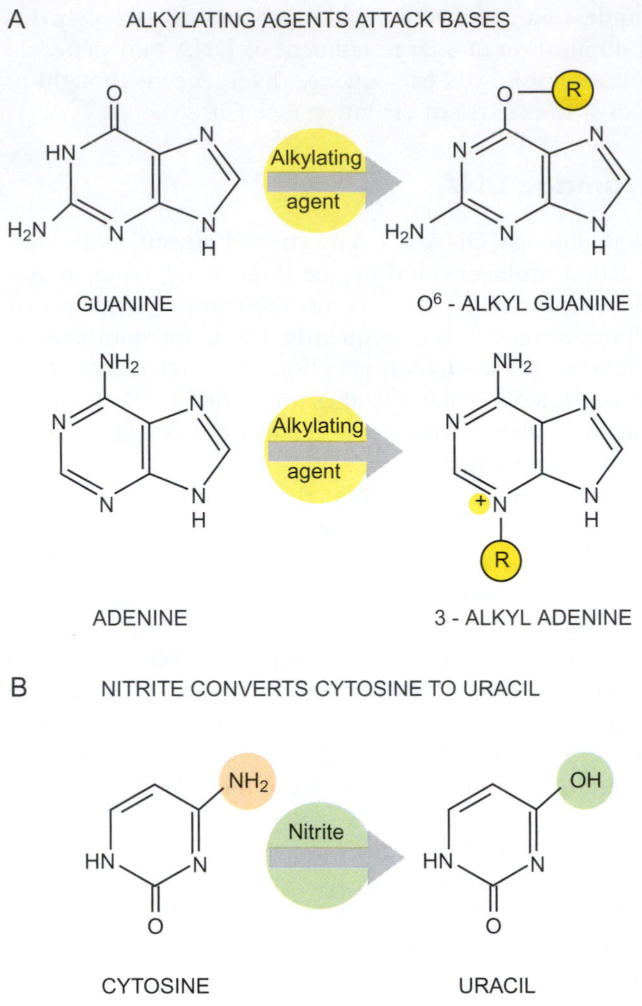

A   ALKYLATING AGENTS ATTACK BASES

GUANINE → O^6 - ALKYL GUANINE

ADENINE → 3 - ALKYL ADENINE

B   NITRITE CONVERTS CYTOSINE TO URACIL

CYTOSINE → URACIL

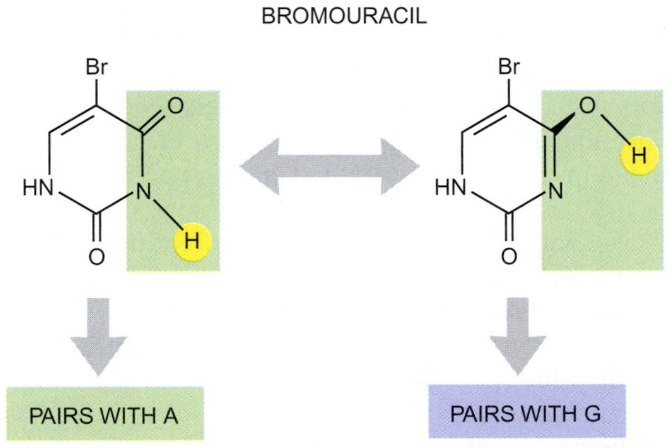

BROMOURACIL

PAIRS WITH A   PAIRS WITH G

forming, which may partly explain the drug's ability to cause birth defects. Curiously, thalidomide has recently come back into clinical use to treat cancers such as multiple myeloma.

3.1. Radiation Causes Mutations

Some types of radiation cause mutations. High-frequency electromagnetic radiation, including ultraviolet radiation (UV light) and X-rays, and also gamma rays (γ-rays),

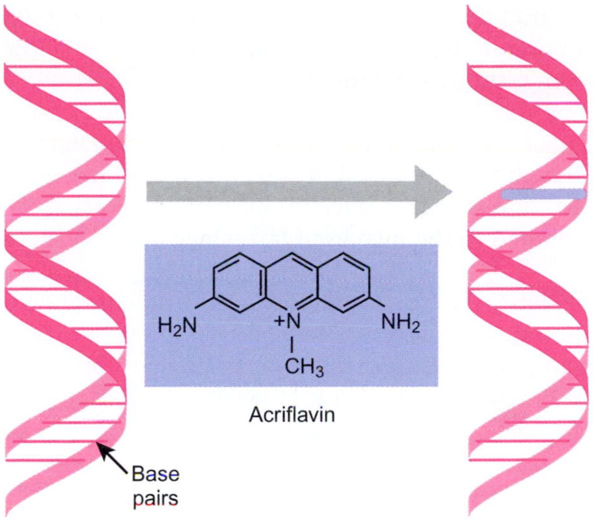

◗ **FIGURE 23.13**
Intercalating Agents

An intercalating agent, such as acriflavin, can insert itself between base pairs and mimic a whole extra base pair. When replication occurs, the intercalating agent causes an extra base pair to be inserted into the new DNA. Commercial acriflavin is actually a mixture of the structure shown plus the derivative without the N-methyl group.

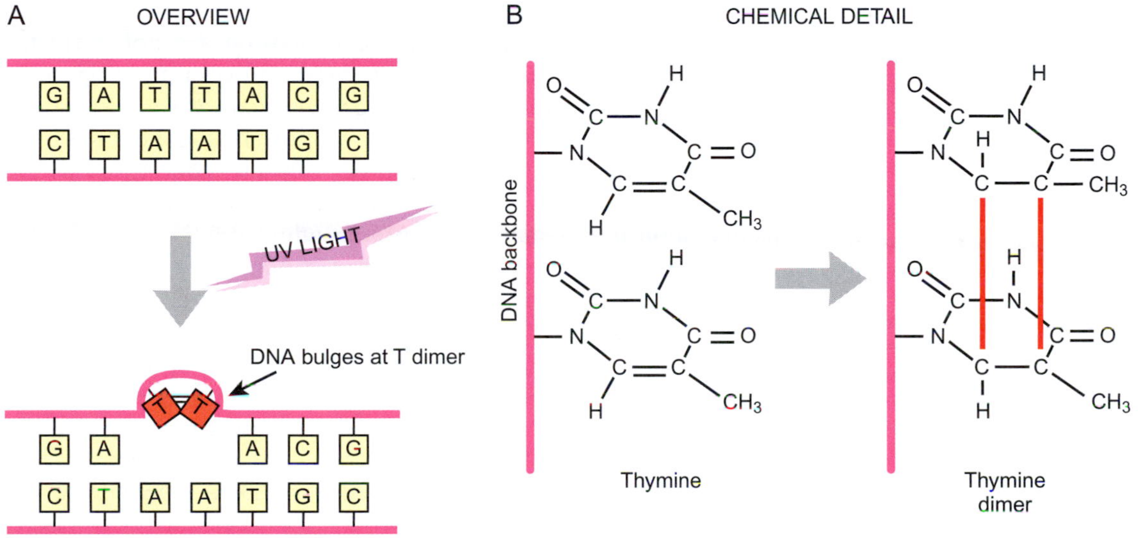

FIGURE 23.14
Thymine Dimers

A) Ultraviolet light (UV) sometimes results in the formation of a thymine dimer (red). B) The detailed chemical structure of the thymine dimer is shown.

directly damage DNA. X-rays and γ-rays are **ionizing radiation**; that is, they react with water and other molecules to generate ions and free radicals, notably hydroxyl radicals. Ionizing radiation is responsible for about 70% of the radiation damage to DNA. The other 30% of the radiation damage is due to direct interaction of X-rays and γ-rays with DNA itself. In the early days of molecular biology, X-rays were often used to generate mutations in the laboratory. X-rays tend to produce multiple mutations and often yield rearrangements of the DNA, such as deletions, inversions, and translocations.

Ultraviolet radiation is electromagnetic radiation with wavelengths from 100–400 nm. It is non-ionizing and acts directly on the DNA. The bases of DNA show an absorption peak at around 254 nm, and UV close to this wavelength is absorbed very efficiently by DNA. In particular, UV causes two neighboring pyrimidine bases to cross-react with each other to give dimers. Thymine dimers are especially frequent (Fig. 23.14). Although DNA polymerase can proceed by skipping over thymine

High-energy radiation damages DNA.

Ultraviolet radiation promotes formation of thymine dimers.

ionizing radiation   Radiation that ionizes molecules that it strikes

dimers, this leaves a single-stranded region that needs repairing. The repair process in turn causes the insertion of incorrect bases in the newly-synthesized strand (see below for details on error-prone repair), which results in mutations.

Ultraviolet radiation is emitted by the sun. Most of it is absorbed by the ozone layer in the upper atmosphere, so it does not reach the surface of the Earth. Damage to the ozone layer by the chlorinated hydrocarbons used in aerosol sprays and refrigerants has allowed more UV radiation to reach the surface of this planet, especially in certain areas. This has probably contributed to the increased frequency of skin cancer noted in recent years.

In addition to electromagnetic radiation, there are other forms of radiation, such as the α-particles and β-particles emitted by radioactive materials along with γ-rays. Most α-particles are too weak even to penetrate skin, but β-particles may cause significant damage to DNA and other biological molecules. However, α-emitters can be mutagenic if they have entered the body, for example, by being breathed in or swallowed.

3.2. Spontaneous Mutations Can Be Caused by DNA Polymerase Errors

The enzymes that replicate DNA during cell division are not perfect. They make errors at a rate that is low, but nonetheless significant over a long period of time. As discussed in Chapter 10, DNA polymerases carry out **proofreading** and check recently inserted nucleotides for mistakes before moving on. In some cases, the proofreading ability is part of the polymerase itself. In other cases, it is due to an accessory protein such as the DnaQ protein associated with *E. coli* DNA polymerase III. Cells carrying mutations that abolish or damage these proofreading abilities show much higher rates of spontaneous mutation. Genes that give rise to altered mutation rates when they themselves are mutated are known as **mutator genes**. Hence, *E. coli dnaQ* mutants were originally named *mutD* (for mutator D).

> **DNA polymerase makes spontaneous mistakes that result in mutations.**

The error rate for DNA replication in *E. coli* is approximately one base in 10 million. About 20 times as many errors occur in the lagging strand as in the leading strand. This probably results from DNA polymerase I having a less effective proofreading capability than DNA polymerase III. (The lagging strand is made discontinuously (see Ch. 10) and the gaps are filled in by DNA polymerase I, whereas the leading strand is all made by Pol III.)

> **DNA polymerase may slip when replicating short sequence repeats.**

In addition to putting in an occasional wrong base, DNA polymerase may very rarely omit bases or insert extra bases. This is due to strand slippage. If a run of several identical bases occurs, the template strand and newly-synthesized strand of DNA may become misaligned (Fig. 23.15). Depending on which strand slips, a base may be inserted or omitted during replication.

Slippage may also occur in regions of DNA where there are multiple repeats of a short sequence, perhaps two or three bases (Fig. 23.16). In this case, a whole repeat unit of several bases will be added or deleted. Well-known cases occur in the human trinucleotide repeat expansion diseases, such as fragile X syndrome and Huntington's disease. Here, copies of a three-base repeat are added or lost due to slippage.

3.3. Mutations Can Result from Mispairing and Recombination

Recombination may occur between closely-related sequences of DNA, such as two alleles of the same gene. Many DNA rearrangements, including deletions, inversions,

---

mutator gene   Gene whose mutation alters the mutation frequency of the organism, usually because it codes for a protein involved in DNA synthesis or repair
proofreading   Process that checks whether the correct nucleotide has been inserted into new DNA. Usually refers to DNA polymerase checking whether it has inserted the correct base

Template strand of DNA

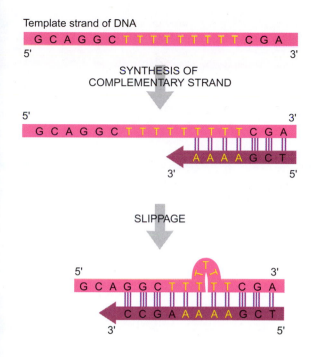

FIGURE 23.15
Strand Slippage Creates Small Insertions or Deletions

The template strand of DNA shown contains numerous thymines (T) in a row (shown in yellow). When replication occurs thymine pairs with adenine (A). However, a long tract of identical bases may cause confusion and some thymines may slip and pair out of register. The extra T residues of the template strand do not pair and form a bulge. In the case shown a small deletion of three bases has occurred in the newly synthesized strand.

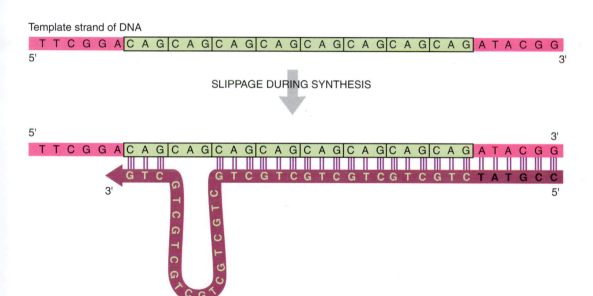

FIGURE 23.16
Strand Slippage of Trinucleotide Repeat

Multiple trinucleotide repeats, such as CAG, may cause strand slippage during DNA replication. In the case illustrated, looping out has occurred in the newly-synthesized strand of DNA. The result will be an insertion of six trinucleotide repeats.

translocations, and duplications may result from mistaken pairing of similar sequences followed by recombination. The mechanism of recombination is dealt with in Chapter 24; here, the overall result of mispairing will be considered. If the similar sequences are in the same orientation, mispairing followed by crossing over will generate duplications on one molecule of DNA and a corresponding deletion on the other (Fig. 23.17).

If two copies of a sequence are on the same DNA molecule but face each other (i.e., are in opposite orientations), mispairing followed by crossing over will generate an inversion (Fig. 23.18). For example, the chromosome of *E. coli* contains seven copies of the genes for ribosomal RNA (rRNA). Strains of *E. coli* are known in which the whole segment of the bacterial chromosome between two of these rRNA operons has been inverted. Such strains grow slightly slower, but nonetheless are viable.

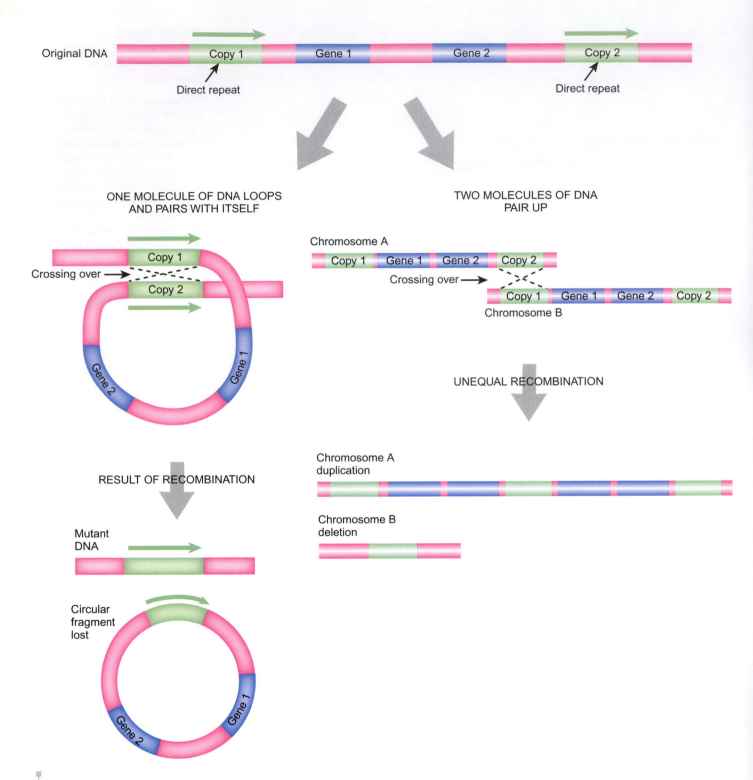

FIGURE 23.17
Mispairing of Direct Repeats Generates Deletions and Duplications

Direct repeats in a DNA molecule may undergo two fates. On the left, the two repeats in a single DNA molecule pair up and recombine. This yields two products; the original DNA molecule suffers a deletion of DNA between the two repeats and a separate circular molecule of DNA is released. On the right, a repeat on chromosome A pairs with another repeat on chromosome B. The result after recombination is a duplication on chromosome A and a deletion of the corresponding region from chromosome B.

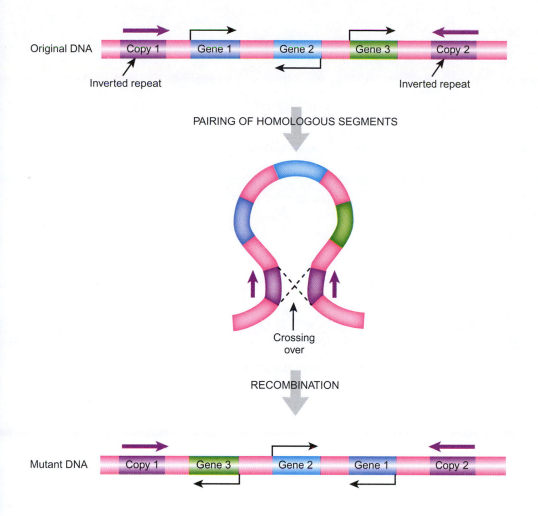

■ **FIGURE 23.18**
Inversion of DNA by Mispairing of Inverted Repeats

The DNA molecule shown has two copies of a sequence that are inverted relative to each other. Three intervening genes (genes 1, 2, and 3) with their directions of transcription (arrows) are also shown. The duplicate sequences may pair up, forming a stem and loop, and undergo recombination. The result is an inversion of the region between the duplicate sequences. This reverses the direction of transcription of the three enclosed genes with respect to the DNA molecule.

3.4. Spontaneous Mutation May Result from Tautomerization

However sophisticated DNA polymerase may be, there are chemical limits on the accuracy of DNA replication. Even if DNA polymerase inserts the correct base, errors may still occur. This is due to the **tautomerization** of the bases that constitute DNA. Each base may exist as two possible alternative structures that interconvert. Such structural isomers that exist in dynamic equilibrium are known as tautomers. In each case, one isomer is much more stable, and the vast majority of the base is found in this form. However, the less stable alternative tautomer will appear very rarely. If this happens just as the replication fork is passing, the rare tautomer may cause incorrect base-pairing.

Thymine has keto and enol tautomers (Fig. 23.19). The common, keto form pairs with adenine, but the rare enol tautomer forms a base pair with guanine. Guanine also has keto and enol tautomers. In this case the rare *enol*-guanine base pairs with thymine rather than cytosine. Similarly, adenine equilibrates between common amino and rare imino tautomers. The rare *imino*-adenine forms a base pair with cytosine instead of thymine. Cytosine alone does not have the potential to introduce mismatches. Although it does have amino and imino tautomers, both pair with guanine. As the temperature increases, the probability that a base is in the incorrect tautomeric state also increases and so, therefore does the mutation frequency.

tautomerization   Alternation of a molecule, in particular a base of a nucleic acid, between two different isomeric structures

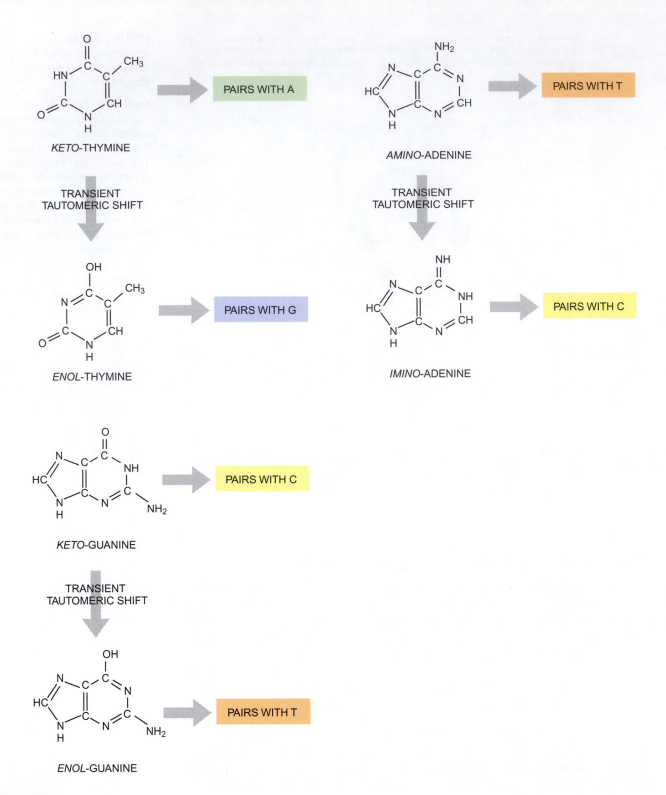

FIGURE 23.19
Base Tautomerization May Cause Mismatches

The tautomers of thymine, adenine, and guanine are shown. In each instance the lower, short-lived tautomer pairs with the inappropriate base.

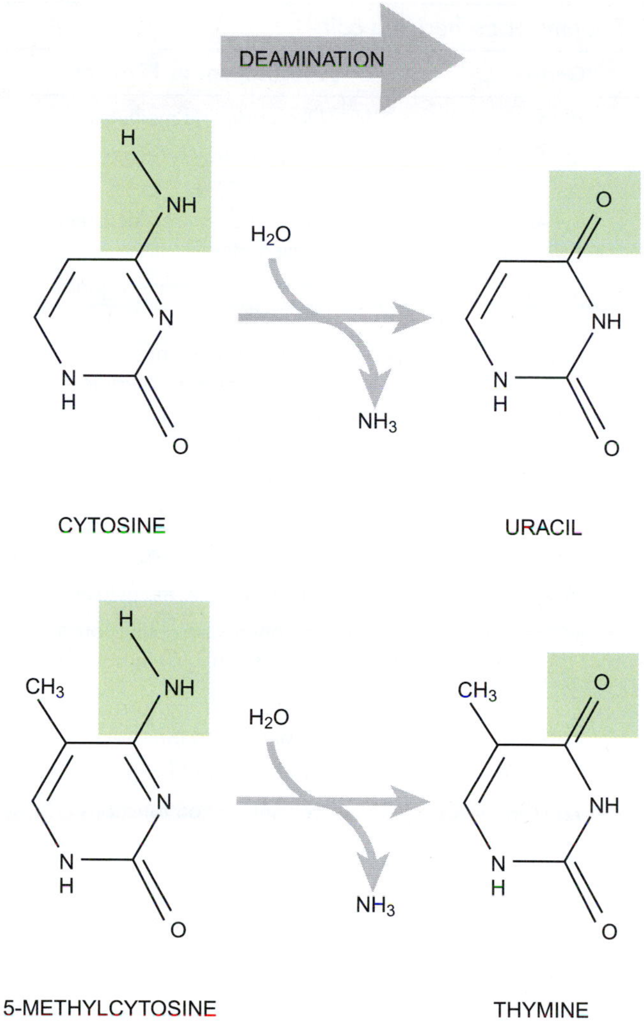

FIGURE 23.20
***Deamination of Cytosine
and 5-Methylcytosine***

The deamination of cytosine yields
uracil and the deamination of the
methylated form of cytosine yields
thymine—both inappropriate
bases.

DEAMINATION

CYTOSINE

URACIL

5-METHYLCYTOSINE

THYMINE

3.5. Spontaneous Mutation Can Be Caused by Inherent Chemical Instability

Although DNA is relatively stable, some of its components do show a low level of spontaneous chemical reaction. Several bases undergo slow but measurable loss of their amino group (i.e., **deamination**). Adenine, guanine, and cytosine may all spontaneously deaminate, but by far the most frequent is the deamination of cytosine to give uracil (Fig. 23.20). In addition, the modified base 5-methylcytosine is especially prone to deamination, so giving "methyl-uracil;" in other words, thymine. The result, in both cases, is the replacement of C by T. Deamination of A (to hypoxanthine) and G (to xanthine) occurs at only 2–3% of the rate for cytosine. Both hypoxanthine and xanthine usually (but not exclusively) base pair with C, so mutations may be introduced in some cases.

Oxidative damage to DNA is also significant. Hydroxyl and superoxide radicals derived from molecular oxygen will attack several bases. The most common target is guanine, which is oxidized to 8-hydroxyguanine, which pairs preferentially with A. Hence, a G/C base pair may be mutated into a T/A pair.

Non-enzymatic methylation of bases occurs at a low frequency. The methyl donor, S-adenosyl methionine, is normally used by enzymes that attach methyl groups to their substrates. However, it is sufficiently reactive to attack several bases spontaneously at a very low rate. The major problem is the formation of 3-methyladenine, which tends to block DNA elongation.

deamination   Loss of an amino group

TABLE 23.01	DNA Repair Systems of *Escherichia coli*	
Repair System	**Genes**	**Mechanism or Function**
Mismatch repair	*dam*	DNA adenine methylase
	mutSHL	base mismatch recognition and excision
Nucleotide excision or "cut and patch" repair	*uvrABCD*	finds and excises incorrect nucleotides
Guanine oxidation repair	*mutMYT*	removal of oxidized guanine derivatives
Alkylated base repair	*ada*	alkyl removal and transcriptional activator
	alkA	alkylpurine removal (glycosylase)
Uracil removal	*ung*	uracil-N-glycosylase removes uracil from DNA
Base excision	*xthA, nfo*	AP endonucleases
Very short patch repair	*dcm*	DNA cytosine methylase
	vsr	endonuclease cutting on 5' side of T in T/G mismatch
Photoreactivation	*phr*	photolyase
Recombination repair	*recA*	single-strand binding
	recBCD	double-strand break repair
	recFOR	recombination functions
SOS repair system or "error-prone repair"	*recA, lexA*	regulation of SOS system
	umuDC	DNA polymerase V
	dinB	DNA polymerase IV

Occasionally, the bonds linking the bases of DNA to deoxyribose may spontaneously hydrolyze. This occurs more often with purines than with pyrimidines, generating empty, apurinic sites. Such missing bases tend to block DNA replication and are also an invitation to DNA polymerase to insert an incorrect base.

4. Overview of DNA Repair

Even if the DNA has been damaged, all is not lost. Most cells contain a variety of damage control systems and some of these can repair damaged DNA. There are several different DNA repair systems designed to deal with different problems. Some repair systems act in a general manner by looking for overall distortions of DNA structure, whereas others focus on specific chemical defects. In some cases, the DNA repair process itself may lead to mutations.

The details of DNA repair are derived mostly from investigation of the bacterium *E. coli*. However, most organisms repair their DNA and possess a variety of repair systems that are probably similar in many respects to those of *E. coli*. Defects in certain DNA repair systems of humans lead to a higher rate of mutation in both somatic and germline cells. This results in a higher incidence both of heritable defects and of cancer. Table 23.01 lists the known DNA repair systems of *E. coli* together with some of the genes involved. (The synthetic enzymes involved in replacing damaged DNA are not included in Table 23.01 if they are also used in normal DNA replication.) Selected repair systems will be discussed in more detail below.

> All organisms contain a variety of systems that repair damage to the DNA.

4.1. DNA Mismatch Repair System

Some repair systems monitor the DNA double helix for structural defects, rather than looking for any specific chemical error. In *E. coli* there are two of these, the **mismatch repair system** and the **excision repair system** (see below). Both detect structural distortions of the DNA double helix, but the mismatch system is more sensitive than the excision repair system. A variety of alterations result in base pairs that don't pair properly. If two opposite bases do not match (e.g., G/A) and therefore do not hydrogen bond correctly, a slight bulge will form in the DNA helix. Mismatches of genuine DNA bases, the presence of base analogs, chemically-altered bases, and frameshift mutations all cause distortions that alert the mismatch repair system.

The mismatch repair system cuts out part of the DNA strand containing the wrong base. The gap is then filled in by DNA polymerase III ("Pol III"), which hopefully inserts the correct bases to give correctly matched base pairs (Fig. 23.21). The involvement of Pol III is unusual as most repair systems use DNA polymerase I to replace short damaged regions of DNA.

> Mismatch repair corrects mispaired bases.

But how does the cell know which of the two mispaired bases was the wrong one? Since most mismatches arise during DNA replication, the repair systems need to know which strand came from the mother cell and which strand was the recently synthesized (and error-carrying) daughter strand. In *E. coli*, the chromosome is methylated by two systems that serve to distinguish the new and old strands of DNA. **DNA adenine methylase (Dam)** (product of the *dam* gene) converts adenine in the sequence GATC to 6-methyladenine. **DNA cytosine methylase (Dcm)** converts cytosine in the sequences CCAGG and CCTGG to 5-methylcytosine. Note that all these recognition sequences are palindromic so that the DNA will be methylated equally on both strands. These methylated bases do not perturb base pairing, as 6-methyladenine and 5-methylcytosine form correct base pairs with T and G, respectively (Fig. 23.22). Recall that thymine, which is actually 5-methyluracil, and uracil both base pair identically with adenine.

Immediately after DNA replication, the old strands will be methylated, but the new strands of DNA will not. The Dam and Dcm enzymes take just a couple of minutes to methylate the new strands. Until this is done, there is a brief period in which the DNA is said to be **hemi-methylated** (Fig. 23.23). During this period, several repair systems check the DNA, looking for mismatches where the wrong base was inserted. The difference in methylation tells them which strand is original and which is the newly-made strand. This delay in fully methylating new DNA also helps control the initiation of chromosome replication in bacteria, in a manner not fully understood (see Ch. 10). Different groups of bacteria use different recognition sequences, but the principle of distinguishing old and new strands by methylation remains the same.

> In bacteria, the parental strand of DNA is identified by methylation.

> Newly-made DNA is not methylated and is checked for errors before methylation occurs.

The major mismatch repair system of *E. coli*, the MutSHL system, uses the methylation of GATC sequences by DNA adenine methylase to monitor which strand is newly made (Fig. 23.24). Many of the genes involved in DNA repair in *E. coli* are named *mut* for "mutator" since defects in these genes result in a higher mutation rate. The mismatch repair system consists of three genes: *mutS, mutH,* and *mutL*. The MutS protein recognizes the distortion caused by mismatched base pairs. The MutH protein finds the nearest GATC site and nicks the non-methylated strand. The MutL protein apparently holds the complex together. The nearest GATC site may be some distance from the mismatch and so it is thought that the DNA is pulled through the

---

DNA adenine methylase (Dam)   A bacterial enzyme that methylates adenine in the sequence GATC
DNA cytosine methylase (Dcm)   A bacterial enzyme that methylates cytosine in the sequences CCAGG and CCTGG
excision repair system   Also known as "cut and patch" repair. A DNA repair system that recognizes bulges in the DNA double helix, removes the damaged strand, and replaces it
hemi-methylated   Methylated on only one strand
mismatch repair system   DNA repair system that recognizes mispaired bases and cuts out part of the DNA strand containing the wrong base

FIGURE 23.21
Principle of Mismatch Repair

Base pairs with incorrect hydrogen bonding cause distortions in the double helix. The mismatch repair system identifies and corrects these distortions.

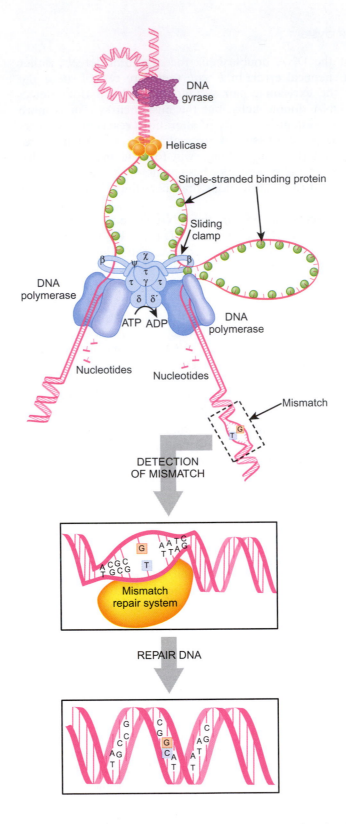

MutSHL complex forming a loop until a GATC site is reached (Fig. 23.24). Pol III then attaches to the DNA and repairs the gap created by the MutSHL system.

4.2. General Excision Repair System

The most widely-distributed system for dealing with damaged DNA is excision repair, often referred to as "cut and patch" repair. This system recognizes bulges in the

Cut and patch repair recognizes distortions of the DNA double helix.

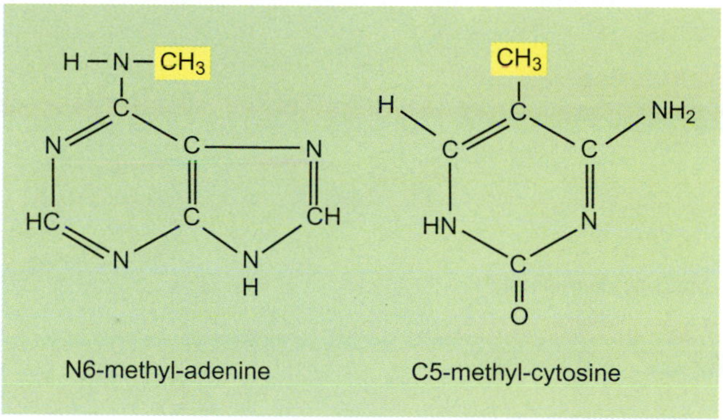

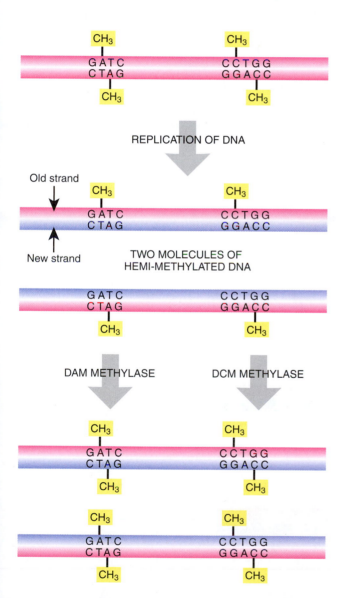

■ FIGURE 23.22
Methylated Bases—Chemical Structure

Sites for methylation of adenine and cytosine are highlighted in yellow.

■ FIGURE 23.23
Hemi-methylated DNA: Old Strands Versus New

When DNA is replicated, the old strand is methylated, but there is a delay in methylating the new strand and thus the DNA double helix, as a whole, is "hemi-methylated." Dam methylase and Dcm methylase methylate the new strand at their specific recognition sequences as described in the text shortly after mismatch repair has a chance to scan the new DNA for errors.

DNA double helix, but is not as sensitive as the mismatch system. The excision repair system does not detect mismatches, base analogs, or certain methylated bases that cause only slight distortions. It does repair most damage due to UV radiation such as thymine dimers and other cross-linked products. Defects in the genes for excision repair result in lowered resistance to UV light and are therefore named *uvr* (UV resistance).

The UvrAB excision repair system is similar to the mismatch system. The UvrAB complex cruises the DNA looking for bulges. When it finds a defect, UvrA departs

FIGURE 23.24
The MutSHL Mismatch Repair System

MutS recognizes a mismatch shortly after DNA replication. MutS recruits MutL and two MutH proteins to the mismatch. MutH locates the nearest GATC of the new strand by identifying the methyl group attached to the "mother" strand. MutH cleaves the non-methylated strand and the DNA between the cut and the mismatch is degraded. The region is replaced and the mismatch is corrected.

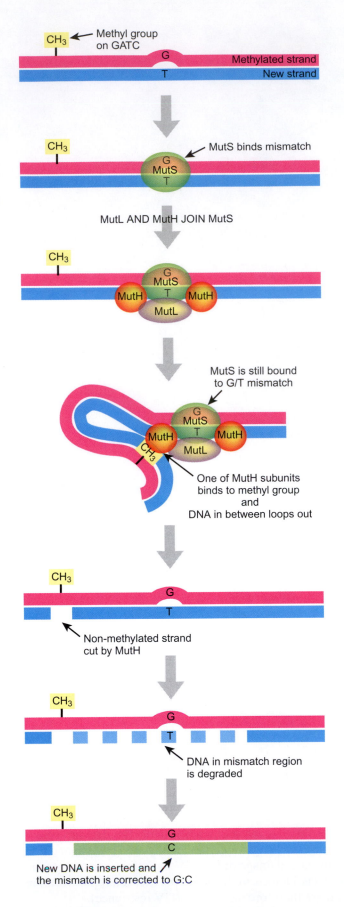

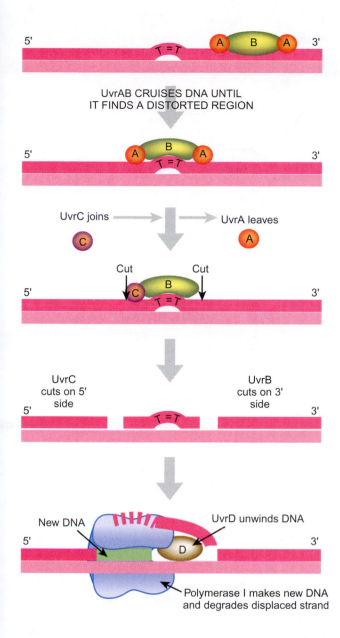

UvrAB CRUISES DNA UNTIL
IT FINDS A DISTORTED REGION

UvrC joins → → UvrA leaves

Cut   Cut

UvrC
cuts on 5'
side

UvrB
cuts on 3'
side

New DNA   UvrD unwinds DNA

Polymerase I makes new DNA
and degrades displaced strand

► **FIGURE 23.25**
***The UvrABC Excision
Repair System***

A cross-linked thymine in the top
(dark pink) strand is recognized by
the UvrAB complex. UvrA protein
is replaced with UvrC, which cuts
the defective strand just before the
thymine dimer (to the 5' side). UvrB
nicks the strand downstream of
the thymine dimer (3' side). UvrD
unwinds the damaged strand,
which is degraded and replaced by
DNA polymerase I.

and is replaced by UvrC protein. UvrB nicks the DNA to the 3' side of the damage
and UvrC nicks to the 5' side (Fig. 23.25). UvrD is a helicase that unwinds the single-
stranded region that has been cut out by UvrBC. Next, DNA polymerase I (Pol I) fills
in the gap with a new strand of DNA. Note that Pol I has a 5'-exonuclease activity
that allows it to remove the old strand of DNA in front of it. Thus, as it moves along
synthesizing the new strand, Pol I also nibbles away at the old strand. Finally, the nicks
are closed by DNA ligase.

4.3. DNA Repair by Excision of Specific Bases

In contrast to those that recognize distortions in the DNA double helix, there are
several repair systems that recognize specific chemical changes in DNA. These are
the **base excision repair** systems. In particular, methylation and deamination create

---

base excision repair   Repair systems that recognize mutations in DNA that do not cause distortions in the helix

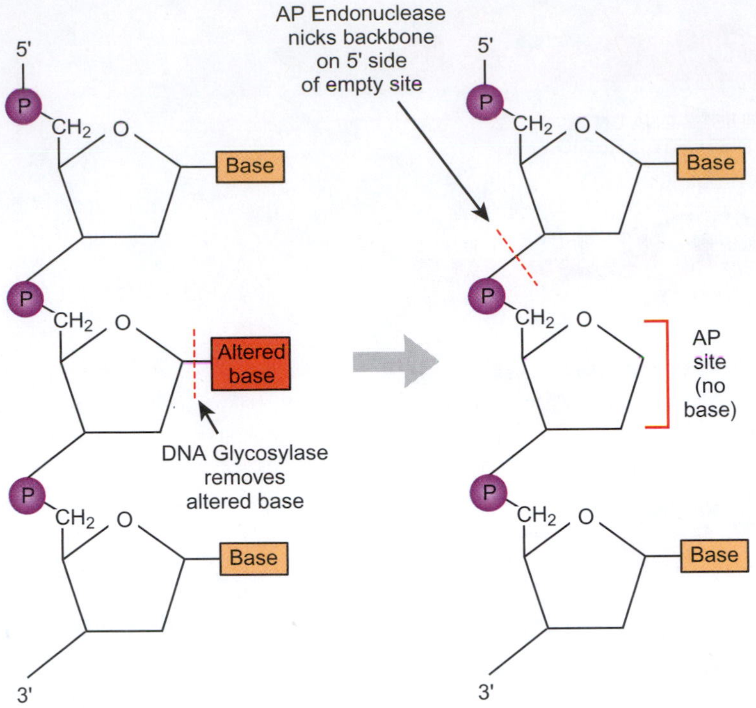

FIGURE 23.26
Removal of Unnatural Bases

An altered base (such as hypoxanthine) is removed by DNA glycosylase leaving an empty AP site. The AP site is recognized by AP endonuclease, which nicks at the 5' side of the sugar backbone. This leaves a free 3'-OH on the base pair just upstream of (above) the AP site. PolI recognizes the 3'-OH and replaces a stretch of single-stranded DNA that includes the AP site.

> Bases that do not occur naturally are removed from DNA by a variety of specific repair systems.

unnatural bases that are not normally found in DNA. In such cases, there is no question which member of a mismatched base pair is wrong. Obviously, the non-DNA base should be removed and a variety of enzymes exist that do just this.

Deamination generates hypoxanthine (from adenine), xanthine (from guanine), and uracil from cytosine. These three bases are all removed by **DNA glycosylases** that break the bond between the base and the deoxyribose sugar of the DNA backbone. Specific DNA glycosylases exist for each unnatural base. Thus, **uracil-N-glycosylase**, **Ung protein**, removes uracil from DNA. Some methylated derivatives, such as 3-methyladenine and 3-methylguanine, are removed in the same manner.

Removal of bases leaves an empty space in the DNA known as an **AP site**. AP stands either for apurinic or apyrimidinic depending on which type of base was removed to create the AP site (Fig. 23.26). Next, an **AP endonuclease** cuts the backbone of the DNA next to the missing base leaving a free 3'-OH group. DNA polymerase I (Pol I) then makes a short new piece of DNA starting with the free 3'-OH group. As Pol I moves along it degrades the single strand in front of it and as usual, the final nick is sealed by DNA ligase. Curiously, over-activity of DNA glycosylases may increase the mutation frequency and this underlies certain human cancers (see Focus on Relevant Research).

AP endonuclease   Endonuclease that nicks DNA next to an AP site
AP site   A site in DNA where a base is missing (AP site = apurinic site or apyrimidinic site depending on the nature of the missing base)
DNA glycosylase   Enzyme that breaks the bond between a base and the deoxyribose of the DNA backbone
uracil-N-glycosylase   Enzyme that removes uracil from DNA
Ung protein   Same as uracil-N-glycosylase

Klapacz J, Lingaraju GM, Guo HH, Shah D, Moar-Shoshani A, Loeb LA, and Samson LD (2010) Frameshift mutagenesis and microsatellite instability induced by human alkyladenine DNA glycosylase. Mol. Cell 37: 843–853.

In humans, alkyladenine DNA glycosylase (hAAG; encoded by the *hAAG* gene) recognizes a variety of mutated bases. These include N3- and N7-alkylated purines, hypoxanthine, xanthine, and cyclic etheno adducts of adenine and guanine (Fig. 23.27). Alkyladenine DNA glycosylase removes these aberrant bases, leaving an AP site that is repaired by AP endonuclease, AP lyase, and DNA polymerase.

When the *hAAG* gene is overexpressed it increases levels of spontaneous frameshift mutations and alters microsatellite repeat numbers. This can disrupt normal gene expression. If cell cycle control genes are affected, the cells may become cancerous. The *hAAG* gene is overexpressed in patients with ulcerative colitis, an inflammation of the intestines known to have a major hereditary component. These patients also show microsatellite instability and have a greater risk of developing colorectal cancer.

FOCUS ON RELEVANT RESEARCH

The authors used site-directed mutagenesis to alter the amino acids that line the active site of hAAG and expressed the altered proteins in *E. coli*. They then monitored the frequency of mutations generating antibiotic resistance. They found that *hAAG* with the double mutation Y127I/H136L increased the mutation rate. In yeast, the *hAAG* double mutant greatly increased frameshift mutations. In cultured human cells, the *hAAG* mutation increased the number of microsatellite repeats. The double mutant was unable to excise defective bases like wild-type *hAAG*, but could still bind the altered bases. This binding obstructs other repair systems and so causes frameshifts and microsatellite alterations.

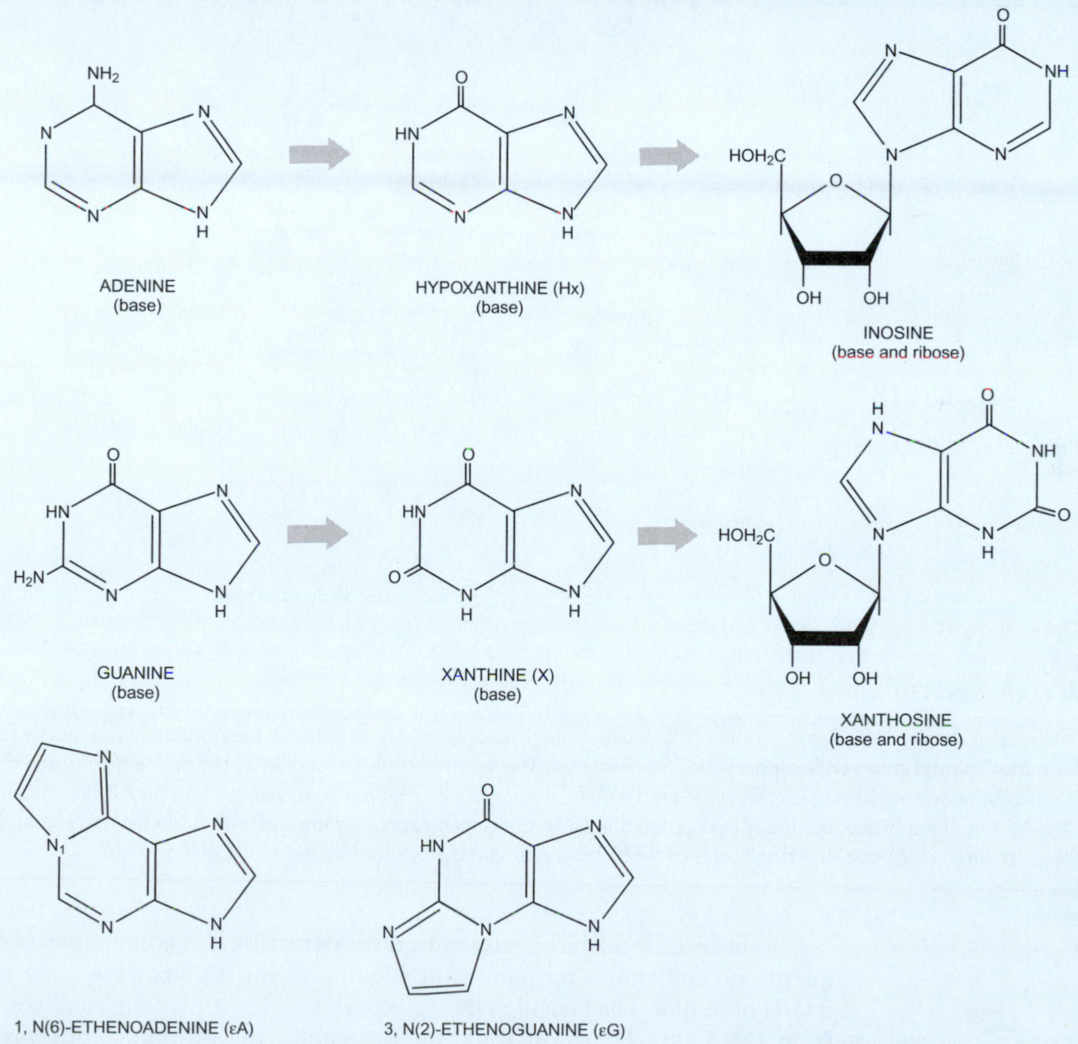

FIGURE 23.27
Structure of Aberrant Bases

Normally guanine and adenine have the structure on the left. In some cases, chemical modifications can alter the base structure so that the DNA sequence will change from adenine to inosine and from guanine to xanthosine. Other chemical changes in the bases, adenine, and guanine create cyclic adducts such as ethenoadenine and ethenoguanine.

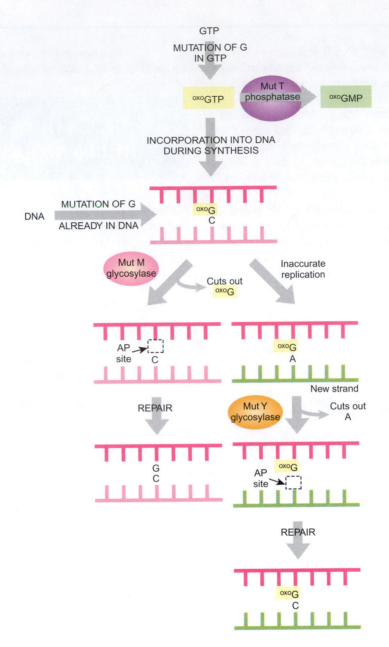

FIGURE 23.28
Dealing with Oxidized Guanine

Three different methods of dealing with 8-oxoguanine (^{oxo}G) exist. If the 8-oxoguanine is detected in the nucleotide precursors for DNA synthesis (oxoGTP), MutT phosphatase dephosphorylates the precursor. The mono-phosphate (oxoGMP) cannot be incorporated into DNA. If 8-oxoguanine is already incorporated and is correctly paired with C, it is cut out by MutM glycosylase leaving an AP site. AP endonuclease and PolI then repair the AP site. If the 8-oxoguanine is incorporated in DNA but is incorrectly mismatched with A, then MutY glycosylase removes the A and this AP site is repaired as above. The resulting 8-oxoG/C base pair can be repaired by the MutM glycosylase.

Some unnatural bases are caused by oxidation. 8-Oxoguanine is especially frequent and sometimes mispairs with adenine giving an 8oxoG/A base pair instead of a G/C base pair. The specific DNA glycosylase, MutM protein, removes 8-oxoguanine from DNA (see Focus on Relevant Research). In addition, MutY protein removes adenine when (and only when) it is found opposite 8-oxoguanine (Fig. 23.28). In both cases removal generates an AP site that is processed as described above. Finally, 8-oxoguanine may arise in the precursors used to synthesize DNA. To avoid incorporation of the defective nucleotide, MutT protein finds the 8-oxoguanine derivatives of GTP and cleaves off two phosphate groups.

Jiricny J (2010) DNA repair: how MutM finds the needle in a haystack. Curr. Biol 20:R145–147.

MutM removes 8-oxoguanine as well as certain other oxidized bases. It is known that when such DNA glycosylases find a wrong base they flip it out of the DNA double helix into their recognition pockets. This also weakens the bond between base and sugar backbone, which is then cut if the base is indeed incorrect.

But how do repair enzymes such as MutM find an occasional wrong base among millions of correct bases in the first place? It appears that they cruise the DNA, bending it slightly as they travel along the double helix. This bending causes oxidized bases to push against the sugar phosphate backbone of DNA and distort the sugar. This then alerts MutM to the presence of an incorrect base.

**FOCUS ON
RELEVANT RESEARCH**

4.4. Specialized DNA Repair Mechanisms

Several more exotic repair systems exist for special cases. Deamination of 5-methylcytosine (which pairs with G) produces thymine causing a T·G mismatch. Since thymine is a naturally-occurring DNA base, the mismatch repair system would only catch a T·G mismatch if generated by a replication error and found in newly-made DNA. However, deamination occurs spontaneously at any time during the cell cycle. Consequently, deamination of 5-methylcytosine often goes unrepaired, and the presence of 5-methylcytosine leads to mutation hotspots as discussed below. Unlike bacteria, animals possess a thymine DNA glycosylase that can deal with such T·G mismatches generated after replication.

> Thymine derived from methylcytosine is removed by the "very short patch repair" system.

Nonetheless, most of the 5-methylcytosine in *E. coli* is made by Dcm methylase and is found in the sequences CCAGG and CCTGG. Whenever T is found replacing C and therefore paired with G in one of these sequences, it is removed. A specific endonuclease nicks the DNA next to the T of the T·G mismatch. This system is sometimes known as **"very short patch repair"** and the nicking enzyme as **Vsr endonuclease**. DNA polymerase I then removes a short length of the strand with the incorrect T and replaces it with a new piece of DNA. Although methylating cytosine in CCAGG/CCTGG is unique to *E. coli*, other organisms also make 5-methylcytosine and presumably contain analogous repair systems with different sequence specificities.

Although some methylated bases are removed by DNA glycosylases, those with the methyl group attached to oxygen get special treatment. The methyl groups of O^6-methylguanine or O^4-methylthymine are removed by suicidal proteins that transfer the methyl group from the base to themselves (Fig. 23.29). This leaves behind the correct base, with no need for synthesizing a new stretch of DNA, but inactivates the repair protein, which is then degraded. Such single-use proteins are not true enzymes, as they do not catalyze a reaction that occurs multiple times.

> Ada protein not only removes methyl groups from both the bases and phosphate backbone of DNA, but also acts as a transcriptional activator.

The Ada ("adaptation to alkylation") protein of *E. coli* is especially interesting since it has both C-terminal and N-terminal active sites, both used to remove methyl groups from DNA. When methyl groups from methylated bases are attached near the C-terminus of Ada, it is inactivated. However, alkylating agents attack the phosphate groups of the DNA backbone as well as bases. Ada protein will also remove methyl groups from the phosphate backbone. In this case, the methyl group is transferred to a site near the front (N-terminus) of Ada protein. This turns Ada into a transcriptional activator that increases expression of several genes that combat DNA damage by alkylation (Fig. 23.30).

4.5. Photoreactivation Cleaves Thymine Dimers

So far we have considered repair systems that are specific for a single altered base. UV light generates pyrimidine dimers, in particular, thymine dimers as well as some

> Light energy can be used to split apart thymine dimers.

"very short patch repair"   System that removes a short length of single-stranded DNA around a T/G mismatched base pair within the Dcm methylase recognition sequences CCAGG or CCTGG

Vsr endonuclease   Enzyme that nicks or cuts the DNA backbone next to a T/G mismatched base pair in the "very short patch repair" system

FIGURE 23.29
Suicide Demethylase for O-Methyl Bases

Demethylase protein recognizes CH₃ attached via oxygen (O) to either guanine or thymine. The demethylase transfers the CH₃ to itself then degrades itself. The correct base structure is restored by removal of the methyl group without any other modification.

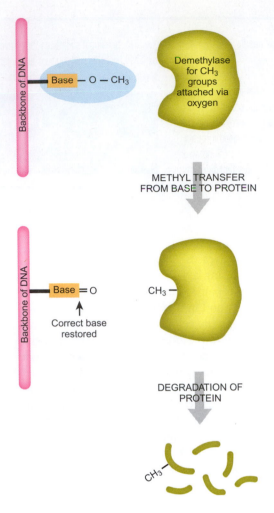

FIGURE 23.30
Ada Plays a Dual Role in Removing Alkyl Groups

Ada protein has two roles in DNA repair. Ada is a classic demethylase that accepts CH₃ from an altered base then degrades itself. Ada can also accept CH₃ from the phosphate backbone of the DNA. This CH₃ is attached to the N-terminal domain of Ada and transforms Ada into a transcriptional activator. This form of Ada turns on genes used to combat DNA alkylation.

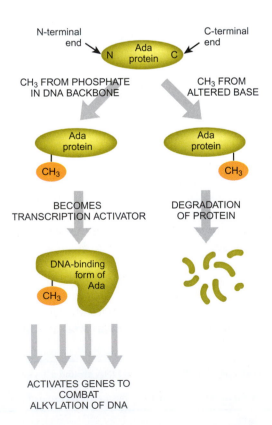

other types of damage. Although the Uvr excision repair system will remove pyrimidine dimers due to their distortion of the DNA double helix, a specific repair system also exists. The photoreactivation repair system specifically identifies pyrimidine dimers and cleaves them, thus regenerating the original bases directly (Fig. 23.31). No DNA synthesis is involved. Photoreactivation only occurs when visible light is present. A single protein is responsible, an enzyme known as photolyase, which absorbs blue light (350–500 nm) and uses the energy to drive the cleavage reaction.

Curiously, photolyase even helps remove pyrimidine dimers in the dark! Under these conditions it binds to the dimer but lacks the energy to remove the crosslink. Nonetheless, binding of photolyase alerts the excision repair system to replace the damaged DNA. Photoreactivation was the first DNA repair system discovered and is present in almost all organisms—except placental mammals, including humans.

4.6. Repair by Recombination

Although repairing all damaged DNA would be ideal, in practice cells often have to settle for less. Some cases of DNA damage prevent replication and if left unattended would kill the cell. In such a predicament, avoiding mutations is a secondary issue. For example, thymine dimers cause DNA polymerase to stall. DNA replication then restarts beyond the blockage. However, this leaves a single-stranded region that was not replicated. If left unfilled, this gap would cause a break in the chromosome next time a replication fork passed through. Although a gap is left in one strand, the replication enzymes did synthesize a new strand of DNA opposite the undamaged mother strand. Thus, there are two copies of the damaged region; one complete and the other single-stranded with the blockage still in position.

To fill this gap, the bacterial RecA protein carries out recombination between the single-stranded region and the complete double-stranded copy. RecA protein binds the single strand and then forms a triplex structure with the corresponding double-stranded region. Recombination fills the gap in the defective strand, at the cost of leaving a single-stranded gap in the other, undamaged double-stranded version of this region. However, since this gap is opposite an undamaged strand of DNA, it can be filled by DNA polymerase (Fig. 23.32).

The net result of this indirect process is that replication has occurred. Although the old damaged DNA strand remains un-repaired, the newly-made DNA molecule is correct. This process to circumvent a blockage may occur during multiple rounds of replication and eventually only one descendent cell out of thousands will have a defective strand of DNA. Higher organisms contain analogous systems and can even use recombination to repair double-stranded breaks. In the case of damage that is difficult or impossible to repair, such damage limitation systems are extremely useful.

4.7. SOS Error-Prone Repair in Bacteria

DNA damage induces expression of several genes whose products help to minimize the effects of the damage. Some of these are repair enzymes and others delay cell division until the damage has been repaired. Yet others provide a pathway of last resort—they allow DNA replication to precede through severely-damaged zones, even at the cost of introducing mutations, a process known as **error-prone repair**.

The **SOS system** of *E. coli* was so named because it responds to severe and potentially lethal DNA damage. Many single-stranded regions of DNA are generated by damage, partly due to excision repair and to stalled replication reinitiating beyond the damage. Single-stranded DNA activates RecA. Activated RecA induces the SOS system by activating LexA, a transcriptional repressor of the SOS genes, by inducing

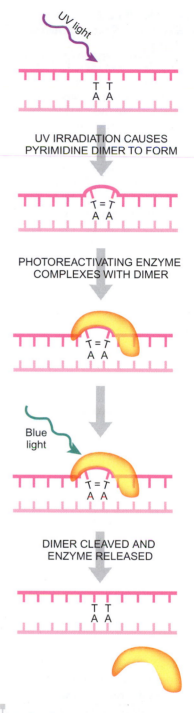

FIGURE 23.31
Photoreactivation Cleaves Pyrimidine Dimers

UV light crosslinks two thymines found next to each other on the DNA molecule. The distortion in the helix is recognized by photolyase. Light activates photolyase to remove the crosslink, restoring the thymines to their original conformation.

---

error-prone repair Type of DNA repair process that fixes severely-damaged DNA even if the repair induces mutations
SOS system An error-prone repair system of bacteria that responds to severe DNA damage

FIGURE 23.32
RecA and Recombination Repair

Since DNA damage stalls the replication enzymes, thymine dimers create large single-stranded gaps (green strand). The other strand (light pink) can be replicated as usual. To repair the gap, RecA directs crossovers on both sides of the single-stranded region. This transfers the gap to another DNA molecule where the correct complementary strand is present. The gap can now be filled in by DNA polymerase. Notice that the thymine dimer is not repaired during this process.

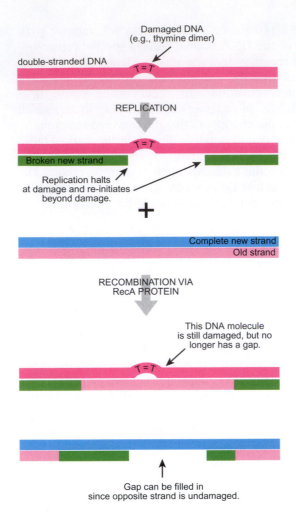

Damaged DNA
(e.g., thymine dimer)

double-stranded DNA

T = T

REPLICATION

T = T

Broken new strand

Replication halts at damage and re-initiates beyond damage.

+

Complete new strand
Old strand

RECOMBINATION VIA RecA PROTEIN

This DNA molecule is still damaged, but no longer has a gap.

T = T

Gap can be filled in since opposite strand is undamaged.

The SOS repair system tackles severely damaged DNA, especially regions of single-stranded DNA.

Error-prone repair fills dangerous gaps in the DNA at the cost of generating base changes (i.e., point mutations).

its self-cleavage. Once cleaved, LexA no longer blocks transcription of SOS genes. The SOS proteins combat the DNA damage (Fig. 23.33).

Among the genes induced by the SOS response are *umuC* and *umuD* (umu = ultraviolet mutagenesis), which encode **DNA polymerase V** (PolV). This polymerase lacks a proofreading subunit so it can replicate past pyrimidine dimers and missing bases (i.e., AP sites). PolV makes mistakes when passing damaged DNA and, for example, tends to put in GA (rather than the correct AA) opposite a thymine dimer. Normal DNA polymerase (PolIII) cannot replicate past such damage because its proofreading subunit stops it from proceeding until a correct base pair has been inserted. If this is impossible due to damage, PolIII grinds to a halt and PolV takes over. The PolV subunits, UmuC and UmuD, form a complex of two UmuD plus one UmuC protein. However, when first made, UmuD$_2$C does not act as a polymerase but delays normal DNA replication in order to allow time for repair. Activated RecA then induces UmuD to cleave itself giving UmuD'. This converts the complex to UmuD'$_2$C, which is the actual error-prone polymerase, PolV (Fig. 23.34).

Like *E. coli*, yeast, flies, and humans all have error-prone DNA polymerases that respond to DNA damage and can replicate past damaged regions. In higher organisms these repair enzymes appear to be more specialized and less error-prone. For example, when human **Polymerase Eta** passes a symmetrical thymine dimer it usually puts in AA. However, although more accurate, it cannot pass other types of pyrimidine dimers like *E. coli* PolV. Although error-prone DNA polymerases generate mutations, this is better than failure to replicate the DNA.

DNA polymerase V   A repair polymerase in bacteria that can replicate past pyrimidine dimers and AP sites
Polymerase Eta   A repair DNA polymerase in animals that can replicate past thymine dimers

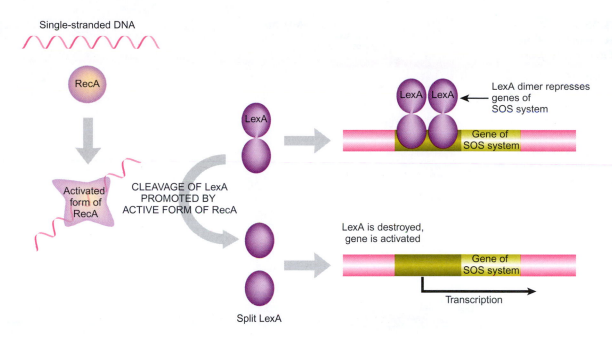

Single-stranded DNA

RecA

Activated form of RecA

CLEAVAGE OF LexA PROMOTED BY ACTIVE FORM OF RecA

LexA

LexA dimer represses genes of SOS system

LexA   LexA

Gene of SOS system

Split LexA

LexA is destroyed, gene is activated

Gene of SOS system

Transcription

FIGURE 23.33
RecA and LexA Control the SOS System

Binding of RecA to single-stranded DNA activates RecA so that it cleaves LexA protein. Cleaved LexA protein can no longer bind to DNA and is released. The genes of the SOS system are no longer blocked from transcription and the gene products combat DNA damage.

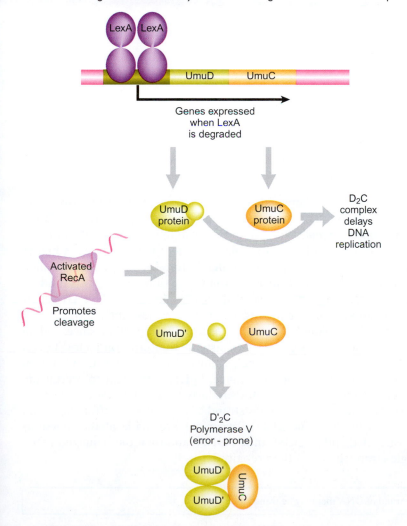

LexA   LexA

UmuD   UmuC

Genes expressed when LexA is degraded

Activated RecA

Promotes cleavage

UmuD protein

UmuC protein

D_2C complex delays DNA replication

UmuD'   UmuC

D'_2C
Polymerase V
(error - prone)

UmuD'
UmuD'
UmuC

FIGURE 23.34
DNA Polymerase V is Part of the SOS System

Activated RecA cleaves LexA dimers, allowing expression of UmuC and UmuD. Two UmuD combine with one UmuC protein and this complex slows replication. DNA repair mechanisms then have time to repair some damage. If the damage is extensive, activated RecA (bound to single-stranded DNA) cleaves UmuD to form UmuD'. When two UmuD' and one UmuC proteins combine, PolV is formed and replicates past any unrepaired damage.

FIGURE 23.35
Eukaryotic Transcription-Coupled Excision Repair

Distortion of eukaryotic DNA is recognized by TFIIH as it unwinds the DNA during transcription. The excision repair system recognizes TFIIH and nicks the unwound region of DNA, releasing the damaged strand. DNA polymerase repairs the gap and transcription can resume.

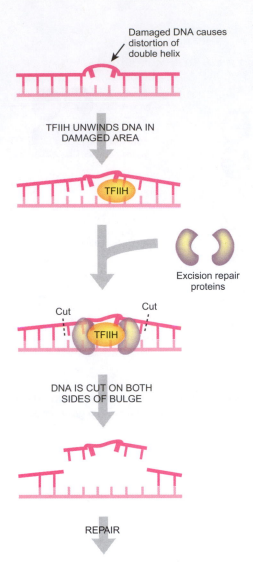

Damaged DNA causes distortion of double helix

TFIIH UNWINDS DNA IN DAMAGED AREA

TFIIH

Excision repair proteins

Cut   Cut

TFIIH

DNA IS CUT ON BOTH SIDES OF BULGE

REPAIR

4.8. Transcriptional Coupling of Repair

If the template strand of DNA is damaged in a gene that is being transcribed the RNA polymerase may grind to a halt. Such damage is repaired preferentially. This is known as **transcription-coupled repair**. If the mutation is in the non-template strand it is less likely to be repaired. This implies that repairing the template strand is of higher priority. In bacteria the transcription-repair coupling factor, TRCF, detects a stalled RNA polymerase and directs the UvrAB proteins to the site of the block. It is thought that TRCF detaches the RNA polymerase so it does not hinder the UvrAB repair proteins. The excision repair system then mends the damage as described above.

In eukaryotes, transcription-coupled repair is especially important since coding regions are relatively few and far between. The eukaryotic transcription factor, TFIIH (see Ch. 11), possesses helicase activity and helps unwind DNA when transcription begins. If the DNA is distorted by chemical damage, TFIIH is thought to recruit the eukaryotic excision repair system. This operates similarly to its bacterial counterpart, except that in eukaryotes the damaged single strand is cut out only after the DNA is unwound to produce a bubble (Fig. 23.35). Two nicks are made at the junctions between the double-stranded and single-stranded DNA, and then the damaged DNA is removed. The single-stranded area is then repaired.

| **transcription-coupled repair** | Preferential repair of the template strand of DNA that may be transcribed |

4.9. Repair in Eukaryotes

Many of the repair systems described for bacteria have counterparts in animals. However, these are usually less well characterized and much of our knowledge comes from comparing the properties of the eukaryotic enzymes to bacterial DNA repair enzymes. Defects in human DNA repair systems cause assorted health problems. In particular, the higher mutation rates that occur in the absence of DNA repair cause a higher frequency of various forms of cancer.

For example, the human *hMSH2* (human MutS homolog 2) gene encodes a protein very similar to the MutS protein of *E. coli*. Defects in this gene greatly increase the likelihood of several types of cancer. Such patients have a relatively high frequency of short deletions and insertions, although mostly in non-coding microsatellite regions. These would normally be corrected by the mismatch repair system that repairs errors that occur during DNA replication (similar to MutSHL in *E. coli*). Curiously, when the human *hMSH2* gene is cloned and expressed in *E. coli* it increases the bacterial mutation frequency! This is apparently due to interference between hMSH2 and MutS.

Patients with a defective *BRCA1* (breast cancer A1) gene are more susceptible to breast and ovarian cancer. The BRCA1 protein affects both the mending of double-strand breaks and transcription-coupled excision repair. These processes are defective when the *BRCA1* gene is damaged. The role of mutation and DNA repair in cancer is extremely complex and many details are still obscure.

The recessive hereditary disorder xeroderma pigmentosum is due to a failure of the excision repair system that removes thymine dimers and other bulky base adducts. Defects in any of approximately 10 genes involved in excision repair will give xeroderma pigmentosum. The result is hypersensitivity of the skin to sunlight and ultraviolet radiation (see Focus on Relevant Research).

Hendriks G, Calleja F, Besaratinia A, Vreiling H, Pfeifer GP, Mullenders LHF, Jansen JG, and de Wind N (2010) Transcription-dependent cytosine deamination is a novel mechanism in ultraviolet light-induced mutagenesis. Curr. Biol. 20: 170–175.

Exposure to ultraviolet (UV) sunlight is one cause of human skin cancer. As discussed in the text, UV is well-known to form pyrimidine dimers also called photolesions, and various repair systems work to ensure these mutations are fixed. Previous work has found that UV-induced mutations are prevalent in actively transcribed genes of skin stem cells. These cells continually divide to replenish the layers of skin, and to accomplish this growth, many genes associated with the cell cycle are actively transcribed. When a cell cycle gene is mutated, then the cells grow uncontrollably, creating a tumor.

This associated article provides evidence that UV exposure in skin stem cells deaminates cytosine, creating uracil within the photolesions. Although these lesions are prevalent on both strands of DNA, the transcription-associated mutations are more prevalent in an actively transcribed reporter gene. In addition, the paper provides evidence that transcription of the damaged DNA strand creates intragenic deletions. The mechanism for these mutations seems to be independent of the cytosine deamination mechanism of mutation. Finally, the paper provides evidence

FOCUS ON RELEVANT RESEARCH

for the transcription-coupled nucleotide excision repair system in repairing both types of mutations, which is a newly-discovered role for this repair system. These findings provide insights into the formation of mutations that may lead to skin cancer.

4.10. Double-Strand Repair in Eukaryotes

Double-strand breaks may be caused by ionizing radiation and certain chemical mutagens. They may also be left behind when some transposable elements excise themselves and move (see Ch. 22). Both yeast and mammalian cells have a similar

FIGURE 23.36
Non-Homologous End Joining in Mammals

Double-stranded breaks are recognized by the Ku proteins, which bind one to each end. The two Ku proteins recruit DNA-PK to the complex. DNA-PK phosphorylates XRCC4 protein, which then recruits DNA ligase IV to join the two broken ends.

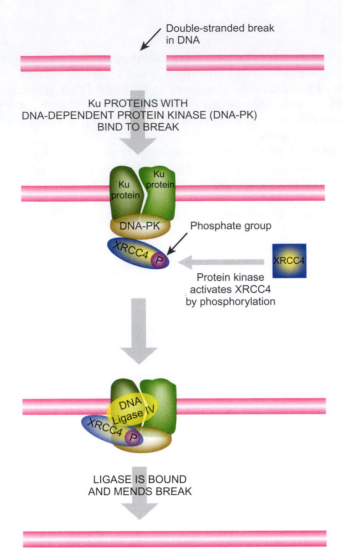

Double-stranded break in DNA

Ku PROTEINS WITH DNA-DEPENDENT PROTEIN KINASE (DNA-PK) BIND TO BREAK

Ku protein
Ku protein
DNA-PK
XRCC4 P
Phosphate group
XRCC4
Protein kinase activates XRCC4 by phosphorylation

DNA Ligase IV
XRCC4 P

LIGASE IS BOUND AND MENDS BREAK

Eukaryotic cells can mend double-stranded breaks in the DNA.

system for repairing double-strand breaks by a process known as **non-homologous end joining** (Fig. 23.36). First, two Ku proteins bind, one on either side of the break. DNA-dependent protein kinase (DNA-PK), which is attached to the Ku complex, then activates the XRCC4 protein, which in turn directs DNA ligase IV to repair the break.

Occasionally, the non-homologous end joining system joins together lengths of DNA that were never previously attached. This can be viewed as a form of non-homologous recombination and generates new combinations of genetic material, including rarities such as chromosomal rearrangements.

In addition, double-strand breaks may be repaired by homologous recombination. This makes use of the presence of pairs of chromosomes in eukaryotic cells. Here, the corresponding sequence on the sister chromosome is used to repair its damaged partner. This process involves a complex that includes Rad51 and other Rad proteins.

5. Mutations Occur More Frequently at Hotspots

If the same gene is mutated thousands of times, are the mutations all different and are they distributed at random throughout the DNA sequence of that gene? Many of them are, but here and there in the DNA sequence are locations where mutations

non-homologous end joining   DNA repair system found in eukaryotes that mends double-stranded breaks

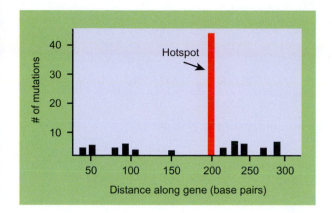

► **FIGURE 23.37**
Hotspots in Distribution of Mutations

After comparing the same gene from many different organisms, the number of mutations at each specific nucleotide was graphed. One nucleotide at around position 200 is mutated in more copies of the gene than any other location. This is considered a mutational hotspot.

happen many times more often than average (Fig. 23.37). All the mutations occurring at such a site will usually be identical. These sites are called **hotspots**.

Most hotspots are due to the presence of occasional methylcytosine bases in the DNA (see Ch. 4 for DNA methylation). These are made from cytosine after DNA synthesis and they pair correctly with guanine, just like normal cytosine. However, every now and then methylcytosine spontaneously deaminates to give thymine (= methyluracil). This base pairs with adenine, not guanine, and so when the DNA is replicated next, an error results.

Hotspots also occur for deletions, insertions, and other major DNA rearrangements. Depending on the mechanism of mutation, certain sequence motifs will favor particular genetic events. As noted above, many rearrangements are due to illegitimate recombination between two nearby regions of DNA with similar sequences.

6. Reversions Are Genetic Alterations That Change the Phenotype Back to Wild-Type

Obviously, it is possible for DNA that already carries one mutation to be mutated again. There is a small chance that the second mutation will reverse the effect of the first. This process is called **reversion** and refers to the observable outward characteristics of an organism. Reversion thus refers to the phenotype.

What is the chance that a single previously-mutated base will mutate again and revert to the wild-type? Obviously, the likelihood that precisely this one base out of millions will be the very one to mutate again is extremely low. Those rarities where the original base sequence is exactly restored are known as **true revertants**. More often, revertants actually contain a second-base change that cancels out the effect of the first one. These are known as **second-site revertants** and the second mutation is known as a **suppressor mutation**.

Not surprisingly, mutations that involve more than a single base change are much less likely to revert. Since part of the original DNA sequence has been completely lost, deletions are completely non-revertible; at least as far as restoring the original DNA sequence is concerned. Precise reversal of insertions, inversions, and

A reversion that precisely restores the original DNA sequence is highly unlikely.

Most revertants have changes in their DNA that cancel out the effects of the previous mutation.

---

hotspots Site in DNA or RNA where mutations are unusually frequent
reversion Alteration of DNA that reverses the effects of a prior mutation
second-site revertant Revertant in which the change in the DNA, which suppresses the effect of the mutation, is at a different site to the original mutation
suppressor mutation A mutation that restores function to a defective gene by suppressing the effect of a previous mutation
true revertant Revertant in which the original base sequence is exactly restored

FIGURE 23.38
Second-Site Reversion of Frameshift Mutation

The DNA sequence and the encoded amino acids are shown for wild-type, an original frameshift mutant and a second-site revertant. In the original mutant a single base deletion alters the reading frame. The second-site revertant has an extra base inserted, which reverses the original frameshift. Although the DNA sequence is not identical to the wild-type, the amino acid sequence of the protein has been restored.

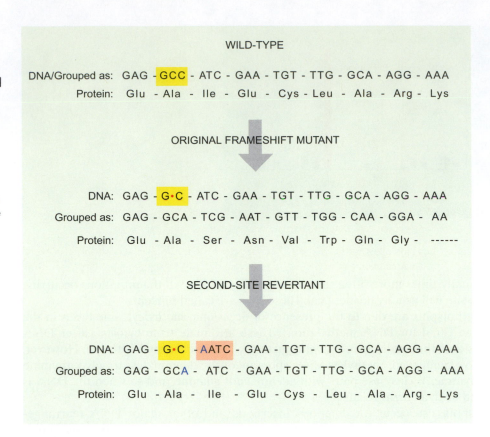

translocations is also extremely rare, though not theoretically impossible. Nonetheless, in such cases, reversion is almost always due to compensatory changes in another gene(s).

A second-site reversion can occur if the original mutation was a frameshift caused by the deletion or insertion of a single base. The frameshift mutation alters the reading frame and garbles the protein sequence, as shown in Figure 23.38. But suppose an extra base is inserted a little way farther along the sequence. This second-site insertion will restore the original reading frame. Although the DNA sequence is not identical to its original state, because of third-base redundancy the protein may be exactly restored. Similarly, an insertion mutation can be corrected by a second-site deletion.

A less obvious but more frequent case of reversion occurs where the original mutation was a base change. Again, the key to successful reversion is to restore activity to the protein. The precise restoration of the original DNA sequence is less important. Consider a protein whose correct 3D structure depends on the attraction between a positively-charged amino acid at, say, position 25 and a negatively-charged one at position 50. Suppose the original mutation changes codon 50 from GAA for glutamic acid (negatively-charged) to AAA, encoding lysine, a positively-charged amino acid. The folding of the protein is now disrupted due to charge repulsion. A true revertant could be made by replacing AAA with GAA. However, the attraction between residues 25 and 50 could also be restored by mutating codon 25 to give a negatively-charged amino acid. This yields a negative charge at position 25 and a positive charge at position 50. Because the attraction between these two regions has been restored, the protein may fold properly again (Fig. 23.39). Will such a revertant protein work correctly? Sometimes yes, sometimes no; it depends on a variety of other factors, such as whether folding is completely restored and whether the alterations damage the active site.

6.1. Reversion Can Occur by Compensatory Changes in Other Genes

When selecting revertants of a particular gene based on phenotypic differences, a mixture of new mutations will be found. Experimentally this is most easily done

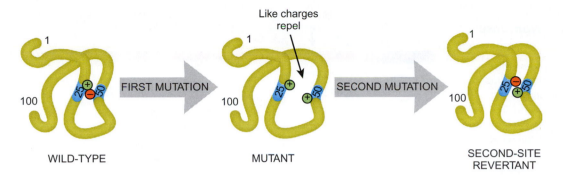

FIGURE 23.39
Second-Site Reversion of Base Change Mutation

A) The original mutation alters amino acid #50 from negatively- to positively-charged. This causes a change in conformation due to charge repulsion. B) A second mutation alters amino acid #25 from positively- to negatively-charged. This restores the attraction between position #25 and #50 and the protein reverts to its original conformation.

using cultures of bacteria or yeast, but the principle applies to higher organisms, too. Occasional true revertants will regain the original DNA sequence. Various possible second-site revertants will occur that restore at least some activity to the protein. These may lie within the gene that was originally mutated or in other genes. Restoring activity to a protein by a second mutation within the same gene is sometimes known as **intragenic suppression**. This contrasts with **extragenic suppression**, where the effects of a mutation are suppressed by a compensatory mutation in a second, quite separate gene.

Extragenic suppression is also confusingly called intergenic suppression to indicate that the two genes involved interact in some way. A whole variety of possible mechanisms exists for such effects. Since an alteration in one gene is making up for a defect in another, extragenic suppression rarely restores function completely.

Two examples of extragenic expression are presented: one due to metabolic compensation and the other due to altered tRNA (see below). Many organisms have multiple genes that code for closely-related proteins. For example, the fumarate reductase and the succinate dehydrogenase of *E. coli* both catalyze the same reaction, the interconversion of fumarate with succinate. The fumarate reductase (FRD) is used to make succinate from fumarate during anaerobic growth, whereas the succinate dehydrogenase (SDH) functions in the Krebs cycle to convert succinate to fumarate during aerobic growth. Although there are slight differences in the DNA and protein sequence, the major difference is in the mode of regulation; the *frd* genes are normally expressed only anaerobically and the *sdh* genes only in the presence of oxygen. Consequently, a mutation in *frd* may be suppressed by a regulatory mutation that allows expression of succinate dehydrogenase anaerobically. Conversely, a mutation in *sdh* may be suppressed by a regulatory mutation that turns on fumarate reductase in the presence of oxygen.

> Sometimes reversion is due to compensatory changes in a completely different gene.

6.2. Altered Decoding by tRNA May Cause Suppression

Nonsense mutations can be suppressed by alterations in tRNA. As noted above, a nonsense mutation occurs when a codon for an amino acid is changed to a stop codon. This results in a truncated and usually non-functional protein. Such a defect may be suppressed, at least partially, by changing the anticodon sequence of a tRNA molecule so that it recognizes the stop codon instead. Consider the stop codon UAG.

extragenic suppression   Reversion of a mutation by a second change that is within another distinct gene
intragenic suppression   Reversion of a mutation by a second change at a different site but within the same gene

FIGURE 23.40
Mechanism of Nonsense Suppression

A gene containing a nonsense codon suffers a premature stop during translation and a short defective protein is made. However, a tRNA whose anticodon is mutated (from GUC to AUC) can recognize the stop codon and insert an amino acid (glutamine in this case). A full-length protein will be made that has only one amino acid different from the original wild-type.

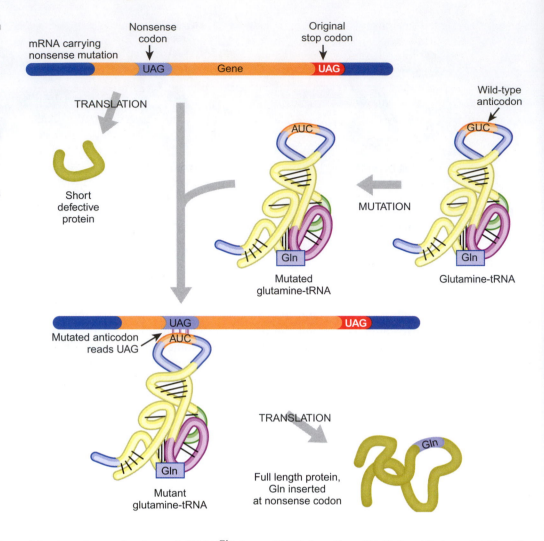

Mutant tRNA molecules are known that read stop codons and insert amino acids. This suppresses nonsense mutations.

Altering the anticodon of tRNA^{Gln} from GUC (reading CAG for Gln) to AUC will make it recognize UAG instead. Such an altered tRNA will insert glutamine wherever it finds a UAG stop codon (Fig. 23.40).

Such altered tRNA molecules are known as **suppressor tRNAs**. The UAG stop codon is known as *amber* and the UAA stop codon as *ochre*. The UGA stop codon has no universally accepted name, but is sometimes called *opal*. Amber suppressors are mutant tRNAs that read UAG instead of their original codon. Ochre suppressor tRNAs read both UAA and UAG due to wobble. Opal suppressors are rare.

Suppressor tRNA mutations can only occur if a cell has more than one tRNA that reads a particular codon. One tRNA may be mutated while the other must carry out the original function; otherwise, the loss of the original tRNA would be lethal. In practice, cells often have multiple tRNA genes and so suppressor mutations are reasonably common, at least in microorganisms. Bacterial suppressor mutations have been found in tRNAs for glutamine, leucine, serine, tyrosine, and tryptophan. The amino acid inserted by the suppressor tRNA may be identical to the original amino acid whose codon mutated to give the stop codon. In this case, the protein made will be fully restored. Alternatively, a different amino acid may be inserted and a partially active protein may be produced.

Remember that stop codons are normally recognized by release factor and have no cognate tRNAs. Since suppressor tRNA competes with release factor, suppression is never complete and typically ranges from 10–40%. This may provide enough

suppressor tRNA   A mutant tRNA that recognizes a stop codon and can insert an amino acid instead of release factor terminating translation of the suppressed protein for the cells to survive. However, the suppressor tRNA will also suppress other stop codons in the same cell and so generate longer (and incorrect) versions of many proteins whose genes were never mutated. Not surprisingly, cells with suppressor mutations grow more slowly. Only bacteria and lower eukaryotes (e.g., yeasts, roundworms) can tolerate suppressor mutations. In both insects and mammals, suppressor mutations are lethal.

Frameshift suppressor tRNAs are also occasionally found among bacteria. These mutant tRNA molecules have an enlarged anticodon loop and a four-base anticodon. This enables them to insert a single amino acid by reading four bases in the mRNA. They can suppress the effects of frameshift mutations caused by the insertion of a single extra base. Frameshift suppressor tRNAs with five-base anticodons have been made artificially, but have not been isolated naturally.

Box 23.1 Amber, Ochre, and Opal

The stop codons were originally identified by mutations in bacteriophage T4. The first one identified was UAG, the amber codon, which received its name in a curiously convoluted manner. The laboratory of Seymour Benzer at Caltech was looking for a mutation that would allow a certain kind of bacteriophage mutant to grow. Benzer said that whoever identified the mutation would have it named after him. The mutation was eventually isolated by a student named Harris Bernstein. Since "Bernstein" is German for "amber" UAG was named the amber codon. The second stop codon to be found (UAA) was called "ochre" to keep the color theme. The third stop codon (UGA) is less common and so the use of "opal" or less often "umber" is less frequent and not fully settled.

---

7. Site-Directed Mutagenesis

DNA may be manipulated in the test tube and the altered DNA construct may then be inserted into the target organism. The simplest form of such *in vitro* mutagenesis is to treat purified DNA with mutagenic chemicals or radiation. However, a variety of more sophisticated techniques have been used to deliberately construct mutations utilizing genetic engineering technology. These techniques are usually known as **directed mutagenesis** (or, sometimes, **site-directed mutagenesis**, when the site of mutation is carefully controlled). Some techniques introduce changes in one or a few bases, whereas others involve more drastic alterations. Some *in vitro* techniques generate semi-random base changes, whereas others are extremely specific. Polymerase chain reaction or PCR(see Ch. 6) is now widely used to create most mutations. These applications are summarized for reference in Table 23.02.

> DNA alterations are often constructed by a variety of genetic engineering techniques.

TABLE 23.02 Techniques Used for *In Vitro* Mutagenesis

Chemical mutagenesis of cloned DNA

The gene to be mutagenized is cloned onto a suitable vector, usually a plasmid. DNA carrying the target gene is extracted and purified and treated with a chemical mutagen *in vitro*. The altered DNA is then transformed back into the original organism and screening is carried out to identify organisms that received a mutant version of the gene.

Gene disruption by restriction and ligation (see Ch. 5)

A DNA cassette, often carrying a gene for resistance to some antibiotic to allow selection, is inserted into the target gene by using restriction enzymes and DNA ligase. This approach is often used if convenient restriction sites are available. If not, then PCR-based introduction of extra DNA is a good alternative.

(Continued)

---

directed mutagenesis   Deliberate alteration of the DNA sequence of a gene by any of a variety of artificial techniques
site-directed mutagenesis   Deliberate alteration of a specific DNA sequence by any artificial technique

TABLE 23.02	Continued

In vitro DNA synthesis (see Ch. 5)

Single-stranded DNA is sometimes generated for sequencing by using M13 vectors. *In vitro* DNA synthesis may be performed using such single-stranded DNA as template using T7 polymerase and a supply of nucleoside triphosphates. DNA polymerization may be initiated using artificially-synthesized primers whose sequence has been altered by a few bases. This will generate a mutagenized product that incorporates these changes. This technique has largely been replaced by PCR based methods.

PCR based techniques (see Ch. 6)

a) Introduction of specific base changes

Using PCR primers whose sequence has been altered will generate a PCR product that incorporates these changes.

b) Localized random mutagenesis

Manganese ions cause errors in PCR reactions. Hence, random mutations may be introduced into the segment of DNA being amplified.

c) Generation of insertion or deletion by PCR

Using PCR primers that include sequences homologous to the target location allows replacement of a region of chromosome with a segment of DNA generated by PCR.

Transgenic technology

Transgenic technology creates genetically-modified organisms. It may therefore be regarded as a form of mutagenesis. Extra DNA sequences may be introduced from other organisms by a variety of techniques.

Key Concepts

- Mutations are inheritable alterations in the DNA sequence.
- Most mutations have little noticeable effect on the phenotype.
- In base substitution mutations one base is replaced by another.
- Missense mutations occur when one amino acid in a protein is replaced by another.
- Nonsense mutations create new stop codons and cause premature termination of polypeptide chains.
- Deletion mutations may eliminate all or part of a gene or they may be largely harmless if they remove inessential non-coding DNA.
- Insertion mutations may disrupt coding sequences or regulatory regions.
- Frameshift mutations alter the reading frame and generate garbled protein sequences.
- Rearrangements of DNA include duplications, inversions, and translocations.
- Some mutations are induced by chemical mutagens; others are caused by radiation; in particular, UV generates pyrimidine dimers.
- Spontaneous mutations may result from errors by DNA polymerase during replication or from inherent chemical instability of DNA bases.
- Most cells contain several DNA repair systems to deal with different types of DNA damage.
- The DNA mismatch repair system and the general excision repair system monitor DNA for structural defects.
- Some DNA repair systems cut out specific incorrect bases.
- Photoreactivation splits the thymine dimers generated by UV light.
- Error-prone repair actually generates mutations but allows replication of severely-damaged segments of DNA.

- Eukaryotes possess the ability to repair double-stranded DNA breaks by non-homologous end joining.
- Hotspots are sites on the DNA where mutations are unusually frequent.
- Reversions are genetic alterations that reverse the effect of mutations.
- Some revertants are due to compensatory changes in genes different from the one with the original mutation.
- Suppressor tRNAs insert amino acids at stop codons and can thus suppress certain nonsense mutations.

Review Questions

1. What is a mutation? Why are errors in transcription or translation not considered mutations?
2. What is the difference between a tight mutation and a leaky mutation?
3. What is the difference between a transition and a transversion?
4. Compare and contrast conservative substitution with radical replacement. What is the likely effect of both mutations if they occur (a) at a location far from the active site or (b) at the active site?
5. Describe how temperature sensitive mutations affect protein folding and function.
6. What are the consequences of a nonsense mutation? What is the fate of the resulting protein?
7. The repressor protein encoded by *lacI* prevents expression of the *lac* operon. Predict the consequences of deleting *lacI*. Would this be a lethal mutation? Why or why not?
8. What is polarity and why are polar effects more common in prokaryotes than in eukaryotes?
9. Why is inserting or deleting three bases less detrimental to a gene than inserting or deleting only one or two bases?
10. What are DNA inversions, translocations, and duplications?
11. Define and provide examples of silent mutations.
12. What are the three main types of mutagens?
13. What would be the final outcome (i.e., mutated sequence) if the following double-stranded DNA sequence was treated with bromouracil: GAC/CTG?
14. What kind of mutation would an intercalating agent, such as acridine orange, potentially cause in a DNA sequence?
15. Why is too much sunbathing or "tanning" not a good idea?
16. What effect does UV light have on DNA? How does this result in mutations?
17. Describe how DNA strand slippage can result in mutation. The slippage of which strand (template or daughter) can result in nucleotide insertion? The slippage of which strand can result in nucleotide deletion?
18. Summarize how deletions, duplications, and inversions can result from DNA mispairing and recombination.
19. What is a tautomer? What role does tautomerization play in spontaneous mutation?
20. What compound results from cytosine deamination?
21. What compound is formed by oxidation of guanine? If this results in a GC→TA mutation, is this an example of a transition or transversion? Explain.
22. What are DNA hotspots? What is the most common cause of a hotspot?
23. What is meant by reversion? Which is more common: true reversion or second-site reversion? Why?
24. What is the difference between intragenic and extragenic suppression?
25. What part of a suppressor tRNA has mutated to allow suppression to occur? Describe the mechanism of suppression by suppressor tRNAs.
26. What is site-directed mutagenesis?
27. What are the consequences of not repairing DNA damage or of the replication mechanism proceeding through damaged areas?
28. Which repair mechanisms monitor the DNA double helix for damage rather than individual errors? Which mechanism is more sensitive?

29.   How does the mismatch repair system know which base is the wrong one?
30.   What kind of sequences do Dam and Dcm methylase recognize? What is special about these sequences?
31.   What is hemi-methylation and when does it take place?
32.   The major mismatch repair system in *E. coli* is made up of three proteins; what are they and what is their function?
33.   What kind of damage does the excision repair system fix?
34.   In the UvrAB excision repair system which proteins find the mismatches? Which proteins cut out the damage?
35.   Describe the removal of unnatural bases. What are DNA glycosylases?
36.   What is "very short patch repair," and what enzymes are involved?
37.   What is photoreactivation and what kind of damage does it repair?
38.   How does the transcription-coupled excision repair system work?
39.   How does DNA polymerase V differ from other DNA polymerases?
40.   What proteins are involved in the SOS response?
41.   What is the difference between *E. coli* polymerase V and human Polymerase Eta?
42.   What happens when defects in repair systems occur in humans? Give examples.
43.   What is non-homologous end joining?
44.   What is another way (apart from non-homologous end joining) to repair double-stranded breaks?

Conceptual Questions

1.   The genome of a diploid human cell has about 3.2×10^9 base pairs. The spontaneous rate of depurinations is 3×10^{-9} per purine nucleotide per minute. How many spontaneous depurinations occur in 24 hours?
2.   Nitrous acid is a chemical mutagen that causes oxidative deamination of DNA bases. Deamination of adenine creates hypoxanthine, which causes an A/T base pair to convert to a G/C base pair. What effect would oxidative deamination of adenine have on a histidine triplet codon? If the mutant codon was within an open reading frame, would the mutation affect the protein coding sequence? How would you categorize the nucleotide substitution and the amino acid substitutions (if there are any)?
3.   In the human inherited disorder called Nijmegan breakage syndrome, there is a defect in the *NBS1* gene on chromosome 8 that causes chromosome instability, microcephaly (small cranium), and a higher propensity for developing cancer. What are some possible functions for this gene? The sequence for *NBS1* has similarity to the Ku protein, and based on this information, what could be the molecular mechanism of *NBS1* in humans?
4.   A researcher wants to create a genetic mutant that cannot metabolize lactose, so she takes the wild-type *E. coli* and exposes it to large doses of UV light. She grows her mutagenized cells on plates that only contain lactose sugar and does not see any mutant colonies. Why didn't this experiment work? What could she do to increase her odds of identifying a mutant?
5.   A genetic mutant was isolated for the Ada protein from *E. coli* that was unable to activate the transcription of genes that remove alkyl groups from DNA. What part of the Ada protein has the mutation?

Recombination

Alterations in the DNA sequence may occur due to mutation, as discussed in the previous chapter. Another mechanism is recombination. In this case, novel DNA sequences are generated by recombining two sequences that already exist. Recombination is especially important in promoting genetic diversity in higher organisms that are diploid. In such organisms each cell has two copies of the genetic information available to recombine with each other. Recombination may also occur in bacteria and viruses under a variety of circumstances.

1. Overview of Recombination

The exchange of genetic information between chromosomes occurs in a variety of organisms and under a variety of circumstances. At the molecular level, this involves exchanging segments of DNA molecules by a mechanism known as **recombination**. During sexual reproduction in eukaryotes, the process of meiosis allows the exchange of segments of DNA between homologous chromosomes. This generates greater genetic diversity among the offspring, which in turn allows much greater opportunity for evolutionary selection. Although prokaryotes do not practice sexual reproduction in the same manner as higher organisms, they have several mechanisms to promote genetic exchange. Among bacteria, fragments of DNA may be recombined into the chromosome after entering the cell as a result of transformation, transduction, or conjugation (see Ch. 25 for a description of DNA transfer mechanisms among prokaryotes). Recombination therefore

---

recombination Exchange of genetic information between chromosomes or other molecules of DNA

FIGURE 24.01
Two Crossovers Result in Recombination

Two paired double-stranded DNA molecules align related regions (A with a; B with b, etc.). Breakage occurs in each paired DNA molecule followed by crossing over and rejoining of the ends. This exchanges part of one DNA strand with another.

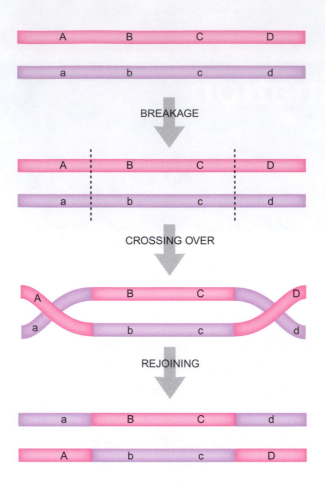

Different DNA molecules may swap segments, usually of related sequence.

Two crossovers between DNA molecules are needed for recombination to occur.

Recombination between unrelated DNA sequences can occur due to the involvement of specific recognition proteins.

has a major impact on bacterial evolution (see Focus on Relevant Research). Even virus genomes may undergo recombination under certain circumstances.

In all cases of recombination, two DNA molecules are broken and rejoined to each other forming a **crossover** (Fig. 24.01). A single crossover usually forms short-lived hybrid DNA molecules. If two crossovers occur, the segment of DNA between them will be transferred from one DNA molecule to the other. This is recombination. (In fact, a single crossover can promote recombination by exchanging the ends of a pair of linear chromosomes. However, a single crossover cannot cause recombination between two circular molecules of DNA.)

Recombination may be divided into **homologous** and **non-homologous recombination**. For homologous recombination to occur, the DNA sequences at the crossover region must be sufficiently similar to base pair. In practice, homologous recombination normally occurs between two copies of the same chromosome (as in meiosis) or between two copies of closely related DNA. Crossovers due to base homology may occur in DNA as short as 20–30 bases, however, 50–100 bases is needed for reasonable crossover frequencies. Non-homologous recombination is much rarer and involves specific proteins that recognize particular sequences and supervise the formation of crossovers between them (Fig. 24.02). In both cases, the molecular details come mostly from bacteria, especially *Escherichia coli,* and the details in higher organisms are much more complicated and poorly understood.

crossover Structure formed when the strands of two DNA molecules are broken and joined to each other
homologous recombination Recombination between two lengths of DNA that are identical, or nearly so, in sequence
non-homologous recombination Recombination between two lengths of DNA that are largely unrelated. It involves specific proteins that recognize particular sequences and form crossovers between them. Same as site-specific recombination

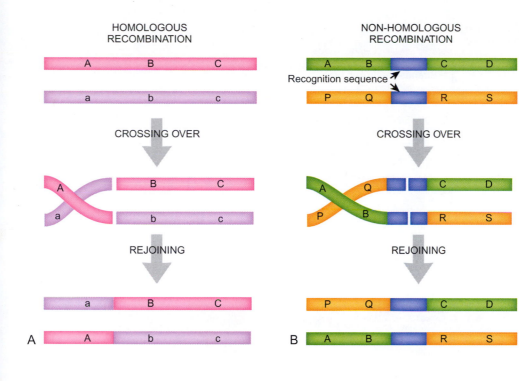

Didelot X and Maiden MCJ (2010) Impact of recombination on bacterial evolution. Trends Microbiology 18: 315–322.

FIGURE 24.02
Homologous versus Non-homologous Recombination

A) In homologous recombination, two DNA molecules have similar sequences such that the pink (top) strand can align with the purple (bottom) strand. If a double-stranded break occurs within the aligned regions, a crossover event will exchange regions of the DNA. B) In non-homologous recombination, related protein recognition sequences lie within two unrelated regions of DNA. Proteins bind to the recognition sequences and carry out recombination. The proteins direct double-stranded breakage and crossing over. Genetic exchange can thus occur between two unrelated DNA molecules. (This event could also theoretically be classified as a translocation.)

Since bacteria divide by binary fission, that is, they reproduce "asexually," it was once believed that little genetic exchange occurred in most bacterial populations. Today we know that this is not true and that genetic exchange, including recombination, plays a major role in bacterial evolution. Gene transfer in bacteria is due to three mechanisms: conjugation, transduction, and transformation (discussed in Ch. 25). However, once DNA from an external source has entered a bacterial cell, its continued long-term existence usually depends on recombination into the host genome.

Recombination thus plays a major role in the evolution of most bacteria. This review discusses the evidence for such recombination in bacterial populations and its impact. The frequency and extent of recombination vary greatly among different bacteria, even between different sublineages of the same species.

FOCUS ON RELEVANT RESEARCH

Most recombination events seen in the wild are no larger than a few thousand base pairs. Although much larger segments of DNA can easily be recombined in the laboratory, it seems likely that in the wild such large alterations to the genome reduce fitness and are eliminated.

2. Molecular Basis of Homologous Recombination

During homologous recombination two double-stranded DNA molecules recognize each other and form a crossover. This involves breaking one strand of each DNA duplex, exchanging strands, and rejoining the ends (Fig. 24.03). This results in the formation of a **Holliday junction**, named after Robin Holliday who proposed this model in 1964. The Holliday junction contains two **heteroduplex** regions where single strands from the two separate DNA molecules have paired up. (A heteroduplex is any region of double-stranded nucleic acid, DNA or RNA, where the two strands come from two different original molecules.)

heteroduplex A DNA double helix composed of single strands from two different DNA molecules
Holliday junction DNA structure formed during recombination and found at the crossover point where the two molecules of DNA are joined

FIGURE 24.03
Formation of a Crossover

Two homologous molecules of DNA align in regions of similar sequence. A single-stranded break occurs in the backbone of each molecule. The two ends switch with each other creating a crossover of single-stranded DNA.

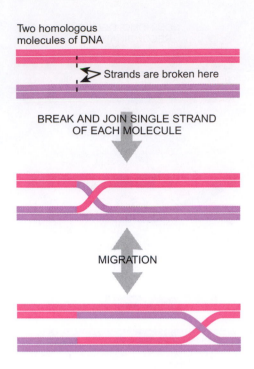

Two homologous
molecules of DNA

Strands are broken here

BREAK AND JOIN SINGLE STRAND
OF EACH MOLECULE

MIGRATION

The Holliday junction can twist around and rearrange itself. The two inter-convertible forms shown in Figure 24.04 do not require any change in bonding or base-pairing; they are simply alternative conformations. The important issue is that two genuinely different products can be formed from the breakdown or **resolution** of the Holliday junction. Which product is obtained depends on which conformation the junction is in when it is resolved. One possible result is the regeneration of the two original DNA molecules. In fact, they are not absolutely the same as before and are sometimes known as "**patch recombinants**" as a short patch of heteroduplex remains in each molecule. The alternative is the formation of two hybrid DNA molecules by crossing-over. Resolution of the Holliday junction to give two separate DNA molecules requires an enzyme known as a **resolvase**. In *E. coli*, the RuvC and RecG proteins both act as resolvases and can substitute for each other. Resolvases cut and rejoin the second (previously unbroken) strands at the junction, generating the complete, double-stranded crossover.

Another interesting property of the Holliday junction is that it can migrate along the DNA (Fig. 24.05), a process called "branch migration." This involves breaking and re-forming hydrogen bonds. This should, in theory, require no overall energy input, as an equal number of bonds are broken as re-formed. However, in practice, spontaneous migration is extremely slow and energy-dependent enzymes are needed to speed up the process. In *E. coli*, the RuvA protein binds to the junction and RuvB drives migration.

2.1. Single-Strand Invasion and Chi Sites

A major question is how the homologous sequences of DNA actually find and recognize each other. Remember that we are not talking about two single strands of DNA base-pairing, but about the merger of two separate double helixes of DNA in which the bases are turned inwards. The approach seems to differ somewhat between bacteria and higher organisms.

> The mechanism of crossover formation involves a temporary triple helix and specific recognition sequences, the chi sites.

---

patch recombinant DNA double helix with a short patch of heteroduplex due to transient formation of a crossover
resolution Cleavage of the junction where two DNA molecules are fused together so releasing two separate DNA molecules. Refers to the breakdown both of crossovers formed during recombination and of co-integrates formed by transposition
resolvase Enzyme that carries out resolution of DNA

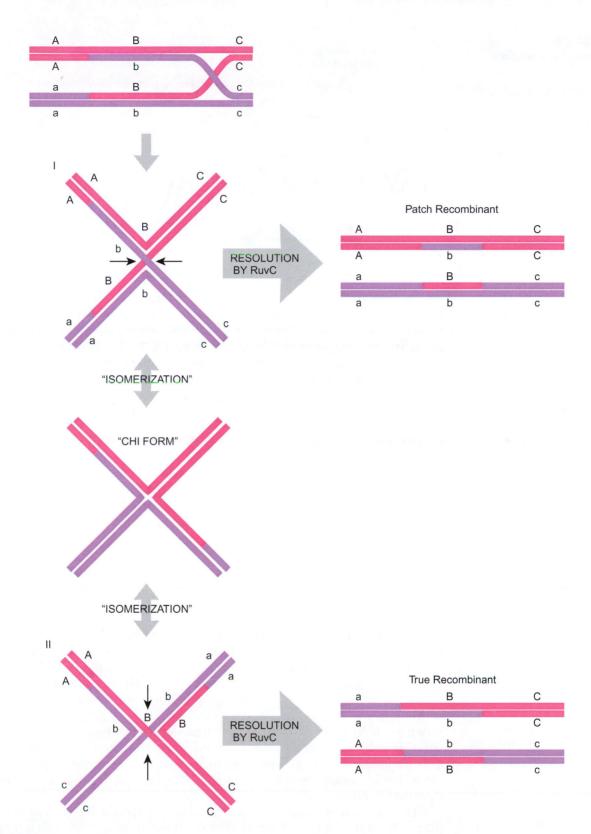

FIGURE 24.04
Rearrangement and Resolution of a Holliday Junction

The Holliday junction can isomerize or change into two alternate conformations. Form I isomerizes into the chi form by rotation of the lower arms at the junction. Another isomerization of the two right arms of the chi form creates form II. Resolution of the crossover can occur at any time. Resolution of form I exchanges gene "B" from the pink molecule with gene "b" from the purple molecule resulting in a patch recombinant. The other strands of the two DNA molecules are unaffected, thus only a small region of heteroduplex DNA results. For form II, resolution by RuvC hybridizes both DNA strands of the double helix rather than just one, and both strands are exchanged.

FIGURE 24.05
Migration of a Holliday Junction

A complex of four RuvA and six RuvB proteins is able to break and re-form hydrogen bonds between base pairs, thus allowing the crossover to migrate along the DNA helix.

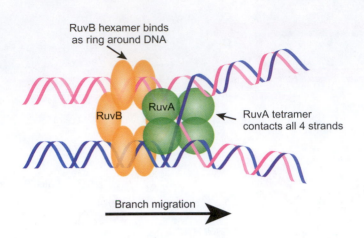

RuvB hexamer binds as ring around DNA

RuvB

RuvA

RuvA tetramer contacts all 4 strands

Branch migration

In bacteria, the major proposed mechanism is single-strand invasion. This requires a single-stranded region to form in one of the DNA double helixes. This single strand then intrudes into the second DNA double helix to give a triple-stranded helix. The two stages in this process depend on the RecBCD and RecA proteins, respectively. Originally it was thought that crossovers could form between any two homologous sequences, but specific sequences called **chi sites** are also needed. However, since chi sequences (5′-GCTGGTGG-3′) are very common, crossovers do form more or less at random in any sufficient length of bacterial DNA. The chi site (chi = crossover hotspot initiator) was named because of the resemblance of a crossover to the Greek letter chi, χ.

The first step in the process of single-strand invasion is RecBCD binding to DNA at double-strand breaks (Fig. 24.06). The helicase complex, RecBCD, is then driven along the DNA by two molecular motors, the RecB and RecD proteins. When RecBCD reaches a chi site, it pauses. The RecB and RecD motor proteins move at different speeds. RecD, which is faster, is replaced as lead motor by the slower RecB when the complex pauses at the chi site. Consequently, the complex slows down after it passes the chi site. At the chi site the RecD endonuclease also clips one of the strands to the 3′ side of the chi sequence before it dissociates from the DNA. The RecBC helicase continues unwinding the DNA generating a single strand. As it passes chi sites, the RecBCD complex also becomes capable of recruiting RecA, which binds single-stranded DNA.

Note that the RecBCD complex also plays another role. It is used by bacteria to degrade foreign DNA after this has been cleaved by restriction enzymes. Under these conditions, there is no need to slow down and so, during the degradation of foreign DNA, the complex moves permanently in top gear, driven by RecD.

In the next step of single-strand invasion, **RecA protein** binds to the single strand with the free 3′ end and inserts it into another DNA double helix to give a temporary triple helix (Fig. 24.07). RecA stabilizes the single-stranded region of DNA. This strand invasion causes displacement of one of the strands of the second double helix. This displaced strand will eventually pair with the remaining single strand of the DNA that was originally unwound by RecBCD. The resulting crossover is resolved as described in Figure 24.04.

The question remains—how did the double-stranded break appear in the first place? Bacteria avoid generating double-stranded breaks in their chromosomes. Furthermore, bacteria are haploid and therefore do not contain pairs of homologous chromosomes that recombine during sexual reproduction. In practice, recombination

---

chi sites   Specific sequences on the DNA of prokaryotes where crossovers form
RecA protein   Protein involved in recombination and repair of DNA in *E. coli* that binds single-stranded DNA

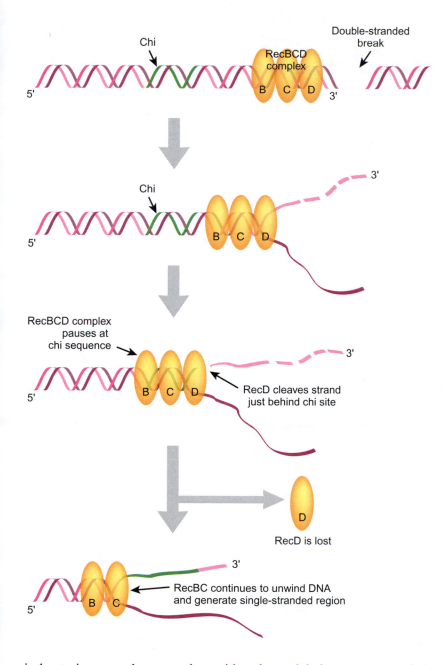

■ **FIGURE 24.06**
RecBCD Recognizes Chi Sites

The complex of RecB, RecC, and RecD proteins recognizes the ends of a double-stranded break and travels along the DNA until reaching the closest chi site. RecD cleaves the backbone of one strand and dissociates from the complex. RecBC continue to unwind the DNA beyond the chi site creating a stretch of single-stranded DNA.

in bacteria occurs between the resident bacterial chromosome and shorter fragments of incoming DNA. These fragments enter the bacterial cell by a variety of processes (see Ch. 25 for details). They may be taken up as free DNA from the outside medium (transformation), carried inside a virus particle (transduction), or received from another cell during mating (conjugation). In most cases, the incoming DNA will consist of relatively short linear fragments that provide the ends necessary for recognition by RecBCD. These mechanisms provide genetic diversity to haploid, non-sexual bacteria, allowing them to adapt to changing environments.

3. Site-Specific Recombination

Recombination can occur between two molecules of DNA that have little sequence similarity. This is known as non-homologous or **site-specific recombination**. Instead of

---

site-specific recombination Recombination between two lengths of DNA that are largely unrelated. It involves specific proteins that recognize particular sequences and form crossovers between them. Same as non-homologous recombination

FIGURE 24.07
RecA Promotes Strand Invasion

RecA binding stabilizes the unwound single-stranded DNA from Figure 24.06. The stabilized strand is able to invade the homologous double-stranded DNA forming a triple helix.

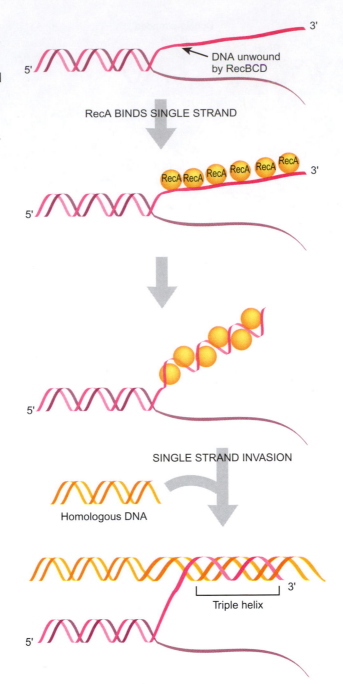

aligning regions of DNA by sequence homology, the DNA contains a short recognition sequence for a specific protein. The protein then initiates the recombination event.

The classic case is the integration of the DNA of bacteriophage lambda (λ) into the chromosome of *Escherichia coli*. Each of these contains a λ **attachment site (*att*λ)**. These are designated *attBOB'* (on the bacterial chromosome) and *attPOP'* (on the lambda genome). The central core of 15 bases (designated O) of these is identical, but the outermost regions (B and P) differ in size and sequence between host and phage. The core region of both attachment sites is recognized by lambda **integrase** or **Int protein**. This makes a staggered double-strand cut in each core sequence. The ends are

Int protein   Same as integrase
integrase   Enzyme that inserts a segment of dsDNA into another DNA molecule at a specific recognition sequence. In particular, lambda integrase inserts lambda DNA into the chromosome of *E. coli*
lambda attachment site (*att*λ)   Recognition site on DNA used during integration of lambda DNA into *E. coli* chromosome

PHAGE
DNA

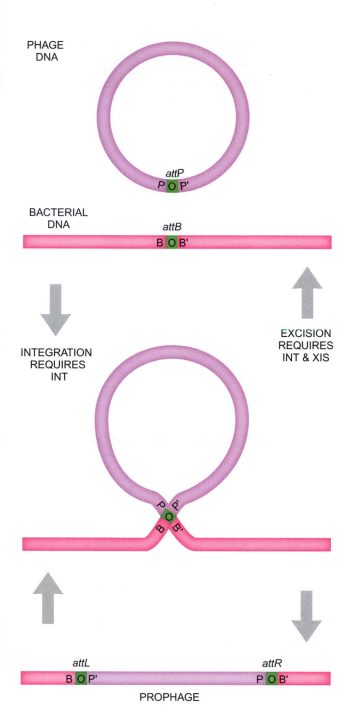

attP
P O P'

BACTERIAL
DNA

attB
B O B'

INTEGRATION
REQUIRES
INT

EXCISION
REQUIRES
INT & XIS

P
P
O
B
B'

attL
B O P'

attR
P O B'

PROPHAGE

■ **FIGURE 24.08**
*Integration of Lambda
DNA—Overview*

Bacterial DNA and λ-phage DNA
are aligned at the "O" region of
the attachment sites. Int protein
induces two double-stranded
breaks that are resolved, giving a
crossover. Since the two recognition
sites are altered in their flanking
regions, λ cannot be excised by Int
alone but needs another protein,
known as excisionase or Xis, in
addition.

joined, and the result is that the circle of lambda DNA is inserted into the bacterial chromosome (Fig. 24.08).

In fact, the strands are cut and joined one at a time. The first round of cutting and joining gives a Holliday junction and the second round resolves it leading to **integration**. The arrangement at the crossover is shown in Figure 24.09. Note that a single cross-over event is sufficient for integration of a circular molecule of DNA into another molecule. (Although lambda DNA is linear inside the virus particle, it circularizes upon entering the bacterial cell and before integration (see Ch. 21 for lifecycle of lambda).) After integration, lambda DNA is flanked by two hybrid *att*λ sites, *attBOP'*

Integration of lambda into
the chromosome of *E. coli*
depends on recognition of the
attachment site by a specific
integrase protein.

---

integration   Insertion of a segment of dsDNA into another DNA molecule at a specific recognition sequence

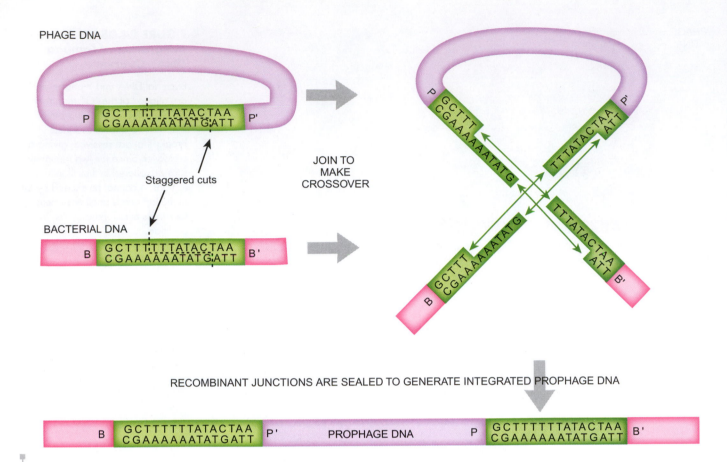

PHAGE DNA

BACTERIAL DNA

Staggered cuts

JOIN TO
MAKE
CROSSOVER

RECOMBINANT JUNCTIONS ARE SEALED TO GENERATE INTEGRATED PROPHAGE DNA

FIGURE 24.09
Integration of Lambda DNA—Detail of Crossover

The "O" region (green) is cut so that an overhang is generated. Rejoining the cut ends of λ with the bacterial chromosome allows λ to integrate its DNA into the host cell.

and *attPOB'*. Int protein cannot carry out recombination between these hybrid sites and cannot therefore reverse the integration event. Excision of lambda requires **Xis protein ("excisionase")** in addition to Int. The control of Xis and Int activity determines whether or not λ stays latent in the bacterial chromosome or emerges and replicates. This decision plays a large part in controlling the lifecycle of the virus.

4. Recombination in Higher Organisms

Recombination in eukaryotes occurs mostly during the early stages of meiosis. For crossing over to occur between the pairs of homologous chromosomes, double-stranded breaks must be introduced into them—a hazardous procedure. Double-stranded breaks appear in eukaryotic chromosomes during the first stage of meiosis (prophase I), known as leptotene, and the paired chromosomes are joined together during the next stage (zygotene) to form the hybrid junction structures needed for recombination (Fig. 24.10). It is assumed that these resemble the Holliday junction of bacteria, but the details are obscure. Resolution of the crossovers then occurs during the third stage of meiosis (pachytene). Finally, the crossovers dissociate, releasing recombinant chromosomes in the final stage of meiosis (diplotene). During mitosis,

excisionase Enzyme that reverses DNA integration by removing a segment of dsDNA and resealing the gap. In particular, lambda excisionase removes integrated lambda DNA
Xis protein Enzyme that reverses DNA integration by removing a segment of dsDNA and resealing the gap leaving behind an intact recognition sequence. Same as excisionase. Not to be confused with Xist RNA involved in X chromosome silencing

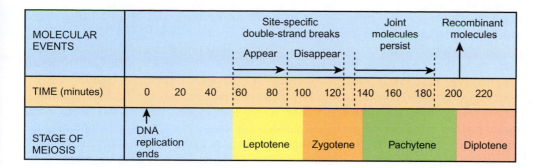

MOLECULAR EVENTS					Site-specific double-strand breaks				Joint molecules persist			Recombinant molecules	
					Appear		Disappear						
TIME (minutes)	0	20	40	60	80	100	120	140	160	180	200	220	
STAGE OF MEIOSIS	DNA replication ends			Leptotene		Zygotene		Pachytene			Diplotene		

FIGURE 24.10
Timeline of Eukaryotic Recombination in Yeast

Site-specific, double-stranded breaks in DNA appear 60–90 minutes after DNA replication. The breaks disappear as hybrid molecules are made during the zygotene phase of meiosis. Resolution of the hybrids then occurs during pachytene. Recombinant molecules appear approximately 120 minutes after the appearance of double-stranded breaks. Therefore, eukaryotic recombination occurs in a span of approximately 2 hours.

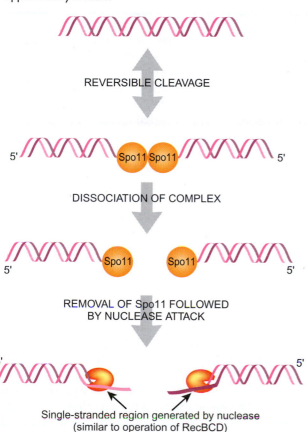

REVERSIBLE CLEAVAGE

5' Spo11 Spo11 5'

DISSOCIATION OF COMPLEX

5' Spo11   Spo11 5'

REMOVAL OF Spo11 FOLLOWED
BY NUCLEASE ATTACK

5'   5'
Single-stranded region generated by nuclease
(similar to operation of RecBCD)

FIGURE 24.11
Spo11 Promotes Double-Strand Breaks

Spo11 binds and cleaves double-stranded DNA in yeast. After cleaving the DNA, Spo11 is replaced by a nuclease that generates the single-stranded region. The two DNA fragments are held together by other proteins that are not shown here.

recombination mostly serves to repair double-strand breaks or single-strand gaps (see Focus on Relevant Research).

A complex of a dozen or more proteins, many poorly characterized, is needed to generate the double-stranded breaks. In yeast, the Spo11 protein is probably responsible for making the double-stranded breaks (Fig. 24.11). Other proteins then displace it, but keep hold of the loose ends of the DNA. Single strands with free 3' ends are then generated. As in bacteria, this single strand is then used to invade another DNA double helix. In yeast the **Rad proteins** oversee this process. In particular, the Rad51

During meiosis in eukaryotic cells, frequent recombination occurs between pairs of homologous chromosomes.

Rad proteins   Group of proteins involved in recombination and repair of DNA damage in yeast and animal cells. Rad51 corresponds to the prokaryotic RecA protein

recombinase of yeast is homologous to RecA of bacteria and binds to free single strands of DNA forming helical, protein-covered filaments (see Fig. 24.07).

Single-strand invasion requires several other Rad proteins, in addition to Rad51. Defects in the *RAD* genes of yeast cause radiation sensitivity as many of them are also involved in repair of radiation damage to DNA (see Chapter 23, Section 4.10)—hence their name. Recombination during mitosis depends on the Rad system alone. However, recombination during meiosis is 100-fold more frequent and needs additional factors. The RAD52 protein is needed for crossing over between homologous chromosomes. RAD52 forms rings that were originally thought to surround single-stranded DNA and help it to invade double-stranded DNA. However, recent data suggest that the DNA is wrapped around the RAD52 ring.

Another homolog of RecA and Rad51 is the Dmc1 recombinase. This protein is found in almost all eukaryotes and is meiosis specific. Rad51 is essential for recombination during both mitosis and meiosis. Thus, Dmc1 appears to play a specialized, but as yet undefined role in meiosis. It has been suggested that alterations in Dcm1 may be responsible for some cases of human infertility.

FOCUS ON RELEVANT RESEARCH

Ho CK, Mazon G, Lam AF, and Symington LS (2010) Mus81 and Yen1 Promote Reciprocal Exchange during Mitotic Recombination to Maintain Genome Integrity in Budding Yeast. Mol. Cell 40: 988–1000.

Formation of crossovers requires resolution of the Holliday junction by resolvase enzymes. This paper describes the characterization of two resolvases in budding yeast, Mus81 and Yen1. Mus81 was originally found via its association with the Rad proteins. Mutants in *mus81* accumulate meiotic intermediates, although the extent varies in different types of yeast. Budding yeast, where the effect of *mus81* mutations is relatively mild, also possesses the Yen1 resolvase, whereas fission yeast where *mus81* mutations are severe does not. The Yen1 enzyme serves as a back up when Mus81 is not available and when the DNA tracts are short. When both are defective, there is a major increase in the faulty segregation of chromosomes.

During mitosis, homologous recombination is used to repair double-strand breaks. When homologous recombination is defective the frequency of mutations, including chromosome rearrangements, increases.

---

5. Gene Conversion

Generally, crossing over is expected to be symmetrical and different alleles from two different parents will be inherited according to Mendel's laws (see Ch. 1). In other words, when two parents reproduce sexually, different alleles from each parent should appear with equal frequency in the offspring. Occasional exceptions to this occur by a mechanism known as **gene conversion**. The name refers to the fact that one allele is converted to the other. The mechanism involves the mismatch repair system operating upon the intermediate structures generated by recombination.

Under normal circumstances, the two strands of a DNA molecule are complementary in sequence and therefore represent the same genetic information. Consequently, any particular double-helical DNA molecule carries a single allele of a gene. However, if the two strands of a DNA molecule do not base pair completely then, strictly speaking, each strand represents a different allele. Such heteroduplex DNA only exists transiently and will soon be corrected by the mismatch repair system; nonetheless, its existence provides the opportunity for gene conversion.

> Sometimes one allele of a gene is converted into another by mismatch repair.

---

gene conversion Recombination and repair of DNA during meiosis that leads to replacement of one allele by another. This may result in a non-Mendelian ratio among the progeny of a genetic cross

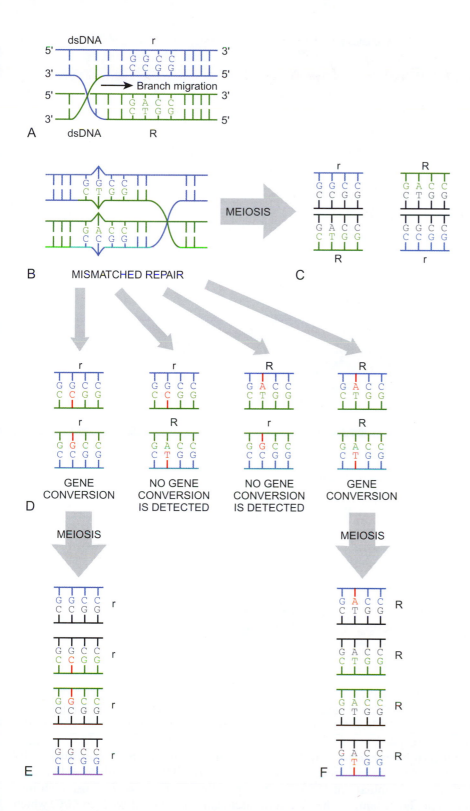

FIGURE 24.12
***Gene Conversion
Following Crossing Over***

A) Crossing over between two double-stranded DNA molecules (dsDNA) can occur in a region where two different alleles of the same gene have a single-nucleotide change. In the blue strand of DNA the "r" allele has the sequence GGCC. In the green strand of DNA, the "R" allele has the sequence GACC. B) As branch migration of the Holliday junction moves past the nucleotide difference, each DNA strand now has a mismatched base pair. C) If meiosis occurs before mismatch repair, then the four haploid spores will have two "R" alleles and two "r" alleles. D) If mismatch repair fixes the mistakes, then one of the mismatched bases will be removed and replaced with the correct complementary base (shown in red). This figure shows four possible ways to fix the strands. E) When mismatch repair fixed the G/T and A/C mismatch, T was changed to C and A was changed to G. After meiosis of the corrected dsDNAs, four "r" haploid spores are created. F) When mismatch repair fixed the G/T and A/C mismatched base pairs, the G was changed to A and the C was changed to T. After meiosis, the four haploid spores have only the "R" allele.

During recombination short heteroduplex regions are created in the DNA next to the crossover point, as discussed above. (This is true whether crossing over generates hybrid DNA molecules or whether the "original" DNA molecules are regenerated from the Holliday junction as shown in Fig. 24.04.) Consider a crossover between two DNA molecules carrying different alleles of the same gene. The "r" allele has the sequence 5′ GGCC 3′, and the "R" allele has the sequence 5′ GACC 3′ (Fig. 24.12A). If the crossover occurs within the coding sequence of the gene of interest, then heteroduplexes will be formed in which there is a mismatched base (Fig. 24.12B). Two

FIGURE 24.13 ———
Mendelian Ratios in Ascospore Formation

In certain fungi a structure known as an ascus keeps together groups of spores ("ascospores") derived from a single zygote. This allows the observation of Mendelian ratios from a single occurrence of meiosis.

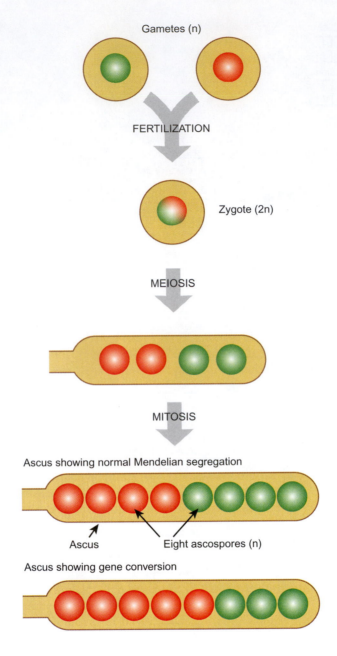

Gametes (n)

FERTILIZATION

Zygote (2n)

MEIOSIS

MITOSIS

Ascus showing normal Mendelian segregation

Ascus   Eight ascospores (n)

Ascus showing gene conversion

Gene conversion may be widespread, but is difficult to detect under normal conditions.

alternative possibilities now exist. Replication and meiosis may occur immediately, in which case the four haploid genomes of DNA produced will have two r alleles and two R alleles (Fig. 24.12C). Alternatively, the mismatch repair system may correct the mismatched base pairs in the heteroduplex region before replication (Fig. 24.12D). One of four possible scenarios could occur to repair the mismatched bases. Two of the possibilities convert the incorrect base and then the two DNA strands look like they did before the recombination. The other two possibilities convert the mismatch into the other allele. For example, the G/T mismatch can be repaired back to G/C, which is the sequence for the "r" allele. For the other strand though, the A/C mismatch is also fixed to be a G/C. This is also an "r" allele. When meiosis occurs here, the four haploid spores all have the "r" allele, and "R" is not passed onto the next generation (Fig. 24.12E). In a similar manner, the other possibility exists where the G/T mismatch is repaired to an A/T, and the A/C mismatch is repaired to A/T also. After meioisis, all four haploid genomes now contain the "R" allele, and "r" is not passed onto the next generation (Fig. 24.12F).

Such occasional deviations are difficult to detect since gene conversion is equally likely in either direction. In Figure 24.12D, the mismatch repair system can fix the mismatch in four different ways. Two of these four possible repairs look just like the parent and are not detectable after meiosis. Only the two events on the far left and far right actually skew the Mendelian segregation of the parental genotype. Gene conversion is thought to occur in all or most organisms, but is only detectable under special circumstances. In practice, it is seen most easily in fungi of the **Ascomycete** group (yeasts, *Neurospora*, etc.) because of their developmental pattern. Sexual reproduction results in the formation of a zygote from the fusion of egg and sperm. Meiosis in these fungi then produces a cluster of four spores all derived from the same zygote (Fig. 24.13). Sometimes mitosis immediately follows meiosis and results in eight final spores. These four (or eight) "ascospores" all stay together inside a special bag-like structure, the **ascus**. It is thus possible to count the Mendelian ratio separately for each group of four (or eight) offspring derived from the same individual zygote.

Key Concepts

- Genetic recombination is the exchange of segments between two different molecules of DNA.
- Recombination proceeds via the formation of crossovers between the two participating DNA molecules.
- In homologous recombination the DNA sequences at the crossover region must be sufficiently similar to base pair.
- In bacteria crossover formation starts by single-strand invasion of DNA at recognition sequences known as chi sites.
- Site-specific recombination occurs when two molecules of DNA have little sequence similarity. Specific proteins that bind to recognition sites are involved.
- Recombination in higher organisms occurs mostly during the early stages of meiosis by a process similar to that in bacteria.
- Gene conversion is when one allele is converted to another during recombination.

Review Questions

1. What is the process by which DNA can be exchanged between chromosomes?
2. What is a crossover?
3. What are the differences between homologous and non-homologous recombination?
4. What is a Holliday junction? What is a heteroduplex?
5. What are the possible results from the resolution of a Holliday junction?
6. What is a resolvase?
7. What occurs during single-strand invasion?
8. What is needed to complete single-strand invasion?
9. Which kind of recombination takes place when there is little sequence similarity?
10. How does lambda integrate into the *E. coli* chromosome?
11. What is the function of the attachment sites on the bacterial chromosome and the lambda genome?
12. How does RecA promote single-strand invasion?
13. What is needed for excision of lambda from a bacterial chromosome?
14. When in the eukaryotic cell cycle does recombination take place? Why is it confined to this stage?
15. How is the yeast Rad51 protein similar to the RecA protein of bacteria?

Ascomycete   Type of fungus that produces four (or sometimes eight) spores in a structure known as an ascus
ascus   Specialized spore forming structure of ascomycete fungus

Conceptual Questions

1. Create a list of steps that occur during creation of Holliday junctions in *E. coli*. Include the following terms in your step-by-step instructions: *chi* sites, RecA, and RecBCD.

2. Explain the mechanism of integration of bacteriophage lambda into the chromosome of *E. coli*. Explain why lambda excision is not merely the opposite of lambda integration. Describe the roles of *att* sites, integrase, and excisionase.

3. Two genetically different yeast parents are mated. The one parent has two identical alleles of the dominant gene *Sss*, and the other parent is heterozygous such that it has one dominant allele *Sss*, and one recessive allele designated *sss*. After meiosis, the majority of the resulting asci have four haploid spores with three containing *Sss*, and one spore with *sss*. One ascus was found to have two spores with *Sss* and two spores with *sss*. What genetic event(s) could have created this mutant spore?

Bacterial Genetics

Chapter 25

In many ways, bacterial genetics underlies molecular biology. The discovery of gene transfer in bacteria, and the involvement of plasmids in the mechanism for this, provided the foundations for molecular cloning. The genetics of bacteria is very different from that of higher organisms. Firstly, bacteria are generally haploid, with one copy of each gene on a single circular chromosome (unlike eukaryotes, which are diploid with multiple linear chromosomes). Secondly, gene transfer in bacteria is normally unidirectional; that is, a donor cell transfers genes to a recipient cell rather than two cells sharing genetic information to generate progeny as seen in the more familiar forms of reproduction in higher organisms. Gene transfer in bacteria occurs by one of three major mechanisms, which form the main topics of this chapter.

1. Reproduction versus Gene Transfer

Sex and reproduction are not at all the same thing in all organisms. In animals, reproduction normally involves sex, but in bacteria, and in many lower eukaryotes, these are two distinct processes. Bacteria divide by **binary fission**. First, they replicate their single chromosome and then the cell elongates and divides down the middle. No resorting

---

binary fission Simple form of cell division in which the cell replicates its DNA, elongates, and divides down the middle

---

In bacteria, cell division and the reassortment of genetic information are completely separate processes.

of the genes between two individuals (that is, no sex) is involved and so this is known as **asexual** or **vegetative reproduction**.

From a biological perspective, **sexual reproduction** serves the purpose of reshuffling genetic information. This will sometimes produce offspring with combinations of genes superior to those of either parent (and, of course, sometimes worse!). Although bacteria normally grow and divide asexually, gene transfer may occur between bacterial cells. During sexual reproduction in higher organisms, germline cells from two parents fuse to form a zygote that contains equal amounts of genetic information from each parent. In contrast, in bacteria gene transfer is normally unidirectional and cell fusion does not occur. Genes from one bacterial cell are donated to another. We thus have a **donor cell** that donates DNA and a **recipient cell** that receives the DNA.

Gene transfer between bacteria may involve uptake of naked DNA, transport of DNA via virus particles, or transfer of DNA via a specialized cell-to-cell connection.

The transfer of genes between bacteria fulfils a similar evolutionary purpose to the mingling of genes during sexual reproduction in higher organisms. However, mechanistically it is very different. Consequently, some scientists regard bacterial gene transfer as a primitive or aberrant form of sex, whereas others believe that it is quite distinct, and that use of the same terminology is misleading.

Molecular biologists use bacteria together with their plasmids and viruses to carry most cloned genes, whether they are originally from cabbages or cockroaches. Consequently, an understanding of bacterial gene transfer is needed to understand the genetic engineering of plants and animals. Gene transfer in bacteria occurs by three basic mechanisms. Only in **conjugation** are genes transferred by cell-to-cell contact. In **transduction**, genes are transferred via virus particles, and in **transformation**, free molecules of DNA are taken up by a bacterial cell. Before considering these three mechanisms in detail, we will discuss what happens to the DNA after uptake, as similar considerations apply in all three cases.

2. Fate of the Incoming DNA after Uptake

Irrespective of its mode of entry, DNA that enters a bacterial cell has one of three possible fates. It may survive as an independent DNA molecule, it may be completely degraded, or part may survive by integration or recombination with the host chromosome before the rest is degraded.

Linear fragments of DNA will be destroyed by cells that receive them unless they are a replicon.

For incoming DNA to survive inside a bacterial cell as a self-replicating DNA molecule, it must be a replicon. In other words, it must have its own origin of replication and lack exposed ends. For survival in the vast majority of bacteria, this means that it must be circular. In those few bacteria, such as *Borrelia* and *Streptomyces* (see Ch. 4) with linear replicons, the ends must be properly protected. In eukaryotes, long-term survival of a linear DNA molecule requires a replication origin, a centromere sequence, and telomeres to protect the ends (see Ch. 4).

A linear fragment of double-stranded DNA that enters a bacterial cell will normally be broken down by exonucleases that attack the exposed ends. For any of its genes to survive, they must be incorporated into the chromosome of the recipient cell by the process of recombination (see Ch. 24). For recombination to occur, crossovers must form between regions of DNA of similar sequence—that is, homologous sequences. DNA between two crossover points will be swapped by the two DNA molecules (Fig. 25.01). Consequently, if genes from incoming DNA are incorporated, the corresponding original genes of the recipient cell are lost.

Incoming fragments of DNA may be preserved from destruction by recombination onto the host chromosome.

Such homologous recombination normally only occurs between closely related molecules of DNA—for example, DNA from two strains of the same bacterial species.

---

asexual or vegetative reproduction   Form of reproduction in which there is no reshuffling of the genes between two individuals
conjugation   Process in which genes are transferred from one bacterium to another by cell-to-cell contact
donor cell   Cell that donates DNA to another cell
recipient cell   Cell that receives DNA from another cell
sexual reproduction   Form of reproduction that involves reshuffling of the genes between two individuals
transduction   Process in which genes are transferred from one bacterium to another via virus particles
transformation   (As used in bacterial genetics) Process in which genes are transferred into a cell as free molecules of DNA

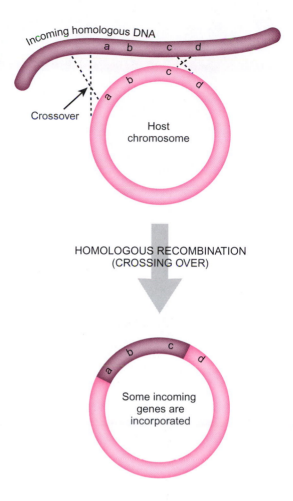

■ **FIGURE 25.01**
*Recombination Allows
Survival of Transformed
DNA*

In most cases, incoming linear
DNA molecules are degraded
by the host cell exonucleases. If
there are homologous regions
between incoming DNA and the
host chromosome, crossing over
may replace regions of the host
chromosome with part of the
incoming DNA.

Unrelated DNA may be incorporated by recombination provided it is surrounded by sequences that are related (Fig. 25.02). Another possibility is that the incoming DNA contains a transposon that can function in the recipient cell. If so, then the transposon may survive by abandoning the incoming DNA molecule and jumping into the chromosome of the new host cell.

If the incoming DNA is a plasmid that can replicate on its own, recombination into the chromosome is not necessary for survival. For genetic engineering purposes, it is usually more convenient to avoid adding genes into the bacterial chromosome via recombination. Consequently, molecular biologists often put the genes they are working with onto plasmids (see Ch. 20).

In addition to exonuclease attack, incoming DNA is often susceptible to restriction. This is a protective mechanism designed to destroy incoming foreign DNA. Most bacteria assume that foreign DNA is more likely to come from an enemy, such as a virus, than from a harmless relative, and they cut it into small fragments with restriction enzymes. This applies to both linear and circular DNA, since the degradative enzymes are endonucleases that cut DNA molecules in the middle (see Ch. 5 for details). Only DNA that has been modified by methylating the appropriate recognition sequences is accepted as friendly. In genetic engineering, restriction negative host strains are used to surmount this obstacle.

Incoming circular DNA with its own origin of replication can survive without recombination.

Restriction enzymes degrade unmethylated foreign DNA, whether linear or circular.

3. Transformation Is Gene Transfer by Naked DNA

The simplest way to transfer genetic information is for one cell to release DNA into the medium and for another cell to import it. The transfer of "pure" or "naked" DNA from the external medium into the bacterium is known as transformation (Fig. 25.03). By "naked," we mean no other biological macromolecules, such as protein, are present to enclose or protect the DNA. No actual cell-to-cell contact happens during

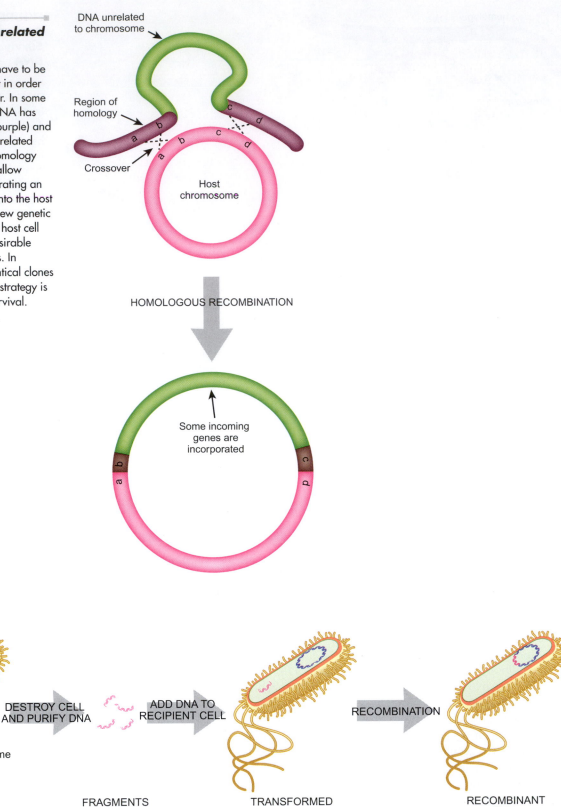

FIGURE 25.02
Incorporation of Unrelated DNA

Incoming DNA does not have to be entirely related to the host in order for recombination to occur. In some instances, the incoming DNA has regions that are related (purple) and regions that are totally unrelated (green). The regions of homology may be large enough to allow recombination, thus integrating an unrelated piece of DNA into the host chromosome. Receiving new genetic material may provide the host cell with a new trait that is desirable to changing environments. In organisms that make identical clones during reproduction, this strategy is critical to evolutionary survival.

DNA unrelated to chromosome

Region of homology

Crossover

Host chromosome

HOMOLOGOUS RECOMBINATION

Some incoming genes are incorporated

DESTROY CELL AND PURIFY DNA

ADD DNA TO RECIPIENT CELL

RECOMBINATION

Chromosome

ORIGINAL BACTERIAL CELL

FRAGMENTS OF DNA

TRANSFORMED CELL

RECOMBINANT CELL

FIGURE 25.03
Gene Transfer by Transformation

Under the right conditions, bacteria can take up pieces of naked DNA from the external environment. The fragment of DNA may pass through the outer cell layers without the aid of a protein or virus. Once inside the bacteria, the fragment of DNA must recombine with the chromosome to prevent degradation by exonucleases or restriction enzymes.

transformation, nor is the DNA packaged inside a virus particle. Bacterial cells can often take up naked DNA molecules and may incorporate the genetic information they carry.

In practice, transformation is mostly a laboratory technique. The DNA is extracted from one organism by the experimenter and offered to other cells in culture. Cells able to take up DNA are said to be "**competent**." Some species of bacteria readily take up external DNA without any pretreatment. Probably they use this ability to take up DNA under natural conditions. From time to time, bacteria in natural habitats die and disintegrate, which releases DNA that nearby cells may import.

Other bacteria must first be chemically treated to make them competent. Since these rarely undergo natural transformation, there are two different laboratory techniques used to make bacteria competent for transformation. One method is to chill the bacterial cells in the presence of metal ions, especially high concentrations of Ca^{2+}, that damage their cell walls and then to heat shock them briefly. This loosens the structure of the cell walls and allows DNA to enter. Another method is electroshock treatment. Bacteria are placed in an "**electroporator**" and zapped with a high-voltage discharge that opens the cell wall and allows the DNA to get into the cell. Laboratory transformation techniques are an essential tool in genetic engineering. After genes or other useful segments of DNA have been cloned in the test tube, it is almost always necessary to put them into some bacterial cell for analysis or manipulation. *Escherichia coli* is normally treated by some variant of the Ca^{2+}/cold-shock treatment and does not require electroshock. Yeast cells may also be transformed. Since yeast has a very thick cell wall, electroshock is used. Conversely, animal cells, which lack cell walls, often take up DNA readily, and only require a mild chemical treatment.

> Cells that have cell walls usually need some sort of treatment before they can take up DNA.

3.1. Transformation as Proof that DNA Is the Genetic Material

Transformation was first observed by Oswald Avery in 1944 and provided the earliest strong evidence that purified DNA carries genetic information and, therefore, that genes are made of DNA. *Pneumococcus pneumoniae* (now renamed *Streptococcus pneumoniae*) has two variants; one forms smooth colonies when grown on nutrient agar, the other has a rough appearance. The smooth variant has a capsule that surrounds the bacterial cell wall, whereas the bacteria in the rough colonies lack the capsule. The ability to make a capsule affects both colony shape and virulence as the capsule protects bacteria from the animal immune system. Thus, if smooth isolates of *S. pneumoniae* are injected into a live mouse, it dies of bacterial pneumonia. In contrast, rough strains are non-virulent. Avery exploited this difference to prove that DNA from one strain could "transform" or change the other strain. Avery used DNA extracted from virulent strains of *S. pneumonia*. He purified the DNA and added it to harmless strains of the same bacterial species. Some of the harmless bacteria took up the DNA and were transformed into virulent strains. Hence, Avery named this process transformation (Fig. 25.04). (Strictly speaking, Avery's transforming DNA could have interacted in some unknown way with the host chromosome to promote a genetic change. His experiment was therefore not absolute proof that DNA is the genetic material. Nonetheless, this is the most obvious interpretation and this observation convinced many scientists that genes were very likely made of DNA.)

> The transfer of inherited characteristics due to the uptake of pure DNA was part of the original proof that DNA was the genetic information.

The use of viruses to transfer DNA into a bacterium provided more evidence that DNA was the genetic material that passed from one generation to the next. Special terminology is used when scientists use naked viral DNA during transformation. In a viral *infection*, the virus punctures a hole in the bacterial cell wall and injects DNA from the viral particle into the cytoplasm. The viral DNA induces the host to manufacture new viral particles. When viruses infect cells naturally, they often leave their protein coats behind and only the viral genome enters (see Ch. 21). The term **transfection** (a hybrid of

---

competent   A state in which a bacterium is able to take free DNA from the surrounding medium into its cytoplasm
electroporator   Device that uses a high-voltage discharge to make cells competent to take up DNA
transfection   Process in which purified viral DNA enters a cell by transformation. Often used to refer to entry of any DNA, even if not of viral origin, into an animal cell

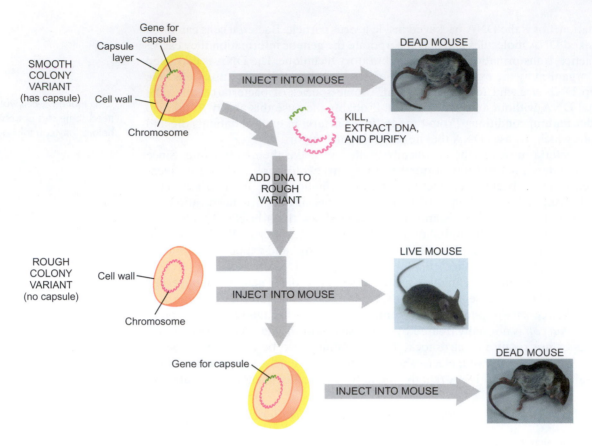

FIGURE 25.04
Avery's Experiment

Avery isolated DNA from the virulent variant and added it to the rough variant of *S. pneumonia*. He noticed that the virulent DNA "transformed" or changed the rough variant into a smooth variant. To confirm that the bacteria were truly transformed, he exposed mice to the newly created smooth variants, and the mice died. Thus, the transformed bacteria had gained both the smooth appearance and virulence by taking up DNA from the original virulent strains.

transformation with infection) refers to the use of purified viral DNA in transformation. In this case the experimenter purifies the viral genome from the virus particle and offers it to **competent cell**s (Fig. 25.05). If taken up, purified viral DNA induces the cell to synthesize virus, illustrating that the virus coat is only necessary to protect the viral DNA outside the host cell and does not carry any of the virus genetic information.

Transformation and transfection also have two other meanings. Cancer specialists use the term "**transformation**" to refer to the changing of a normal cell into a cancer cell, even though in most cases no extra DNA enters the cell. (Note that alterations in the DNA are indeed involved in creating cancer cells, but as a result of mutation.) Supposedly to avoid ambiguity, researchers who use animal cells often use the term "transfection" to refer to the uptake of DNA (by transformation!) whether it is of viral origin or not.

3.2. Transformation in Nature

More detailed investigation of *Streptococcus pneumoniae* and other gram-positive bacteria, including *Bacillus*, shows that they develop natural competence in dense cultures. Competence is induced by competence **pheromones**. (A pheromone is a hormone that travels between organisms, rather than circulating within the same organism.) Competence pheromones are short peptides that are secreted into the culture medium

---

competent cell   Cell that is capable of taking up DNA from the surrounding medium
transformation   (As used of cancer) Changing a normal cell into a cancer cell, even if no extra DNA enters the cell
pheromone   Hormone or messenger molecule that travels between organisms, rather than circulating within the same organism

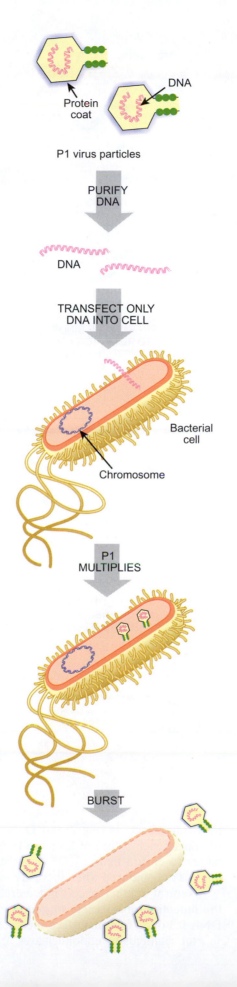

P1 virus particles

PURIFY
DNA

DNA

TRANSFECT ONLY
DNA INTO CELL

Bacterial
cell

Chromosome

P1
MULTIPLIES

BURST

■ **FIGURE 25.05**
Transfection

During viral transfection, an experimenter first isolates pure viral DNA from virus particles. In this diagram, DNA is isolated from P1 virus. Next, the bacterial cell wall is made competent to take up naked DNA (usually by treating with calcium ions or by electroshock). The isolated DNA and the competent bacteria are mixed. If the bacteria take up the P1 DNA, the bacteria will start producing viral particles and burst to release the viral progeny. Thus, viral DNA alone can give the same end result as infection with whole virus particles.

FIGURE 25.06
Competence Pheromones

Dense cultures of *Streptococcus pneumoniae* start producing competence pheromones that induce nearby cells to take up DNA. First, certain cells of the culture produce polypeptide precursors, which are digested into a small peptide, or competence pheromone. The small peptide is secreted from the producer cell and binds to a receptor on a nearby cell. The receptor then signals that cell to make proteins used in DNA uptake.

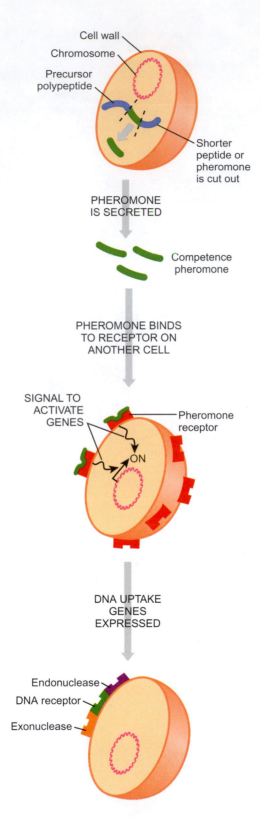

Cell wall
Chromosome
Precursor polypeptide
Shorter peptide or pheromone is cut out

PHEROMONE IS SECRETED

Competence pheromone

PHEROMONE BINDS TO RECEPTOR ON ANOTHER CELL

SIGNAL TO ACTIVATE GENES
Pheromone receptor
ON

DNA UPTAKE GENES EXPRESSED

Endonuclease
DNA receptor
Exonuclease

Transformation occurs among certain bacteria in the natural environment.

by dividing bacteria (Fig. 25.06). Only when the density of bacteria is high will the pheromones reach sufficient levels to trigger competence. This mechanism is presumably meant to ensure that any DNA taken up will come from related bacteria as competence is only induced when there are many nearby cells of the same species.

Natural competence is not merely due to random entry of DNA, but involves the induction of a variety of genes whose products take part in DNA uptake. First, DNA is bound by cell-surface receptors (Fig. 25.07). Then the bound DNA is cut into shorter segments by endonucleases, and one of the strands is completely degraded by

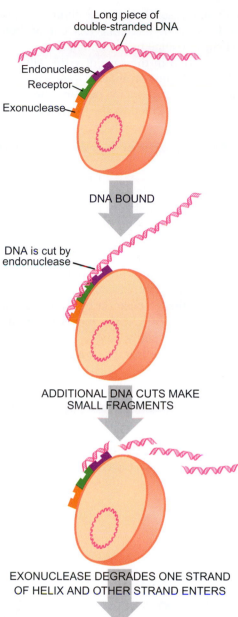

Long piece of
double-stranded DNA

Endonuclease

Receptor

Exonuclease

DNA BOUND

DNA is cut by
endonuclease

ADDITIONAL DNA CUTS MAKE
SMALL FRAGMENTS

EXONUCLEASE DEGRADES ONE STRAND
OF HELIX AND OTHER STRAND ENTERS

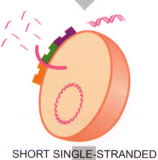

SHORT SINGLE-STRANDED
DNA IS TAKEN INTO CELL

◾▪ FIGURE 25.07
*Mechanism of Natural
Competence*

A cell that is naturally competent
takes DNA into its cytoplasm by
a protein-mediated process. First,
the long molecule of double-
stranded DNA is recognized by
a receptor on the surface of the
competent cell. A cell-surface
endonuclease digests the DNA into
small fragments. An exonuclease
then degrades one strand of
the DNA. The remaining single-
stranded fragment is taken into the
cytoplasm of the bacterium.

an exonuclease. Only the resulting short single-stranded segments of DNA enter the cell. Part of the incoming DNA may then displace the corresponding region of the host chromosome by recombination.

Note that in the case of artificially-induced competence, the mechanism is quite different. Double-stranded DNA enters the cell through a cell wall that is seriously damaged. Indeed, many, perhaps the majority, of the cells that are made artificially competent are killed by the treatment. It is the few survivors who take up the DNA.

4. Gene Transfer by Virus—Transduction

When a virus succeeds in infecting a bacterial cell it manufactures more virus particles, each of which should contain a new copy of the virus genome. Occasionally, viruses make mistakes in packaging DNA, and fragments of bacterial DNA get packaged into the virus particle. From the viewpoint of the virus, this results in a defective particle. Nonetheless, such a virus particle, carrying bacterial DNA, may infect another bacterial cell. If so, instead of injecting viral genes, it injects DNA from the previous bacterial victim. This mode of gene transfer is known as transduction.

Bacterial geneticists routinely carry out gene transfer between different but related strains of bacteria by transduction using bacterial viruses, or bacteriophages (phages for short). If the bacterial strains are closely related the incoming DNA is accepted as "friendly" and is not destroyed by restriction. In practice, transduction is the simplest way to replace a few genes of one bacterial strain with those of a close relative.

To perform transduction, a bacteriophage is grown on a culture of the donor bacterial strain. These bacteria are destroyed by the phage, leaving behind only DNA fragments that carry some of their genes and that are packaged inside phage particles. If required, this phage sample can be stored in the fridge for weeks or months before use. Later, the phage is mixed with a recipient bacterial strain and the virus infects the bacteria, injecting their DNA. Most recipients get genuine phage DNA and are killed. However, others get donor bacterial DNA and are successfully transduced (Fig. 25.08).

> Transduction is when viruses pick up fragments of current host DNA and carry them to another host cell.

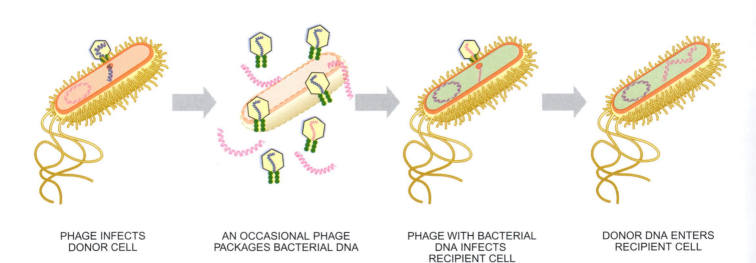

| PHAGE INFECTS DONOR CELL | AN OCCASIONAL PHAGE PACKAGES BACTERIAL DNA | PHAGE WITH BACTERIAL DNA INFECTS RECIPIENT CELL | DONOR DNA ENTERS RECIPIENT CELL |

FIGURE 25.08
Principle of Transduction

Occasionally, when a phage infects a bacterium, one of the virus coats will be packaged with host bacterial DNA (pink). The defective phage particle still infects a nearby cell where it injects the bacterial DNA. This cell will survive since it is not injected with viral DNA. The incoming DNA may recombine with the host chromosome, thus this cell may gain new genetic information.

4.1. Generalized Transduction

There are two distinct types of transduction. In **generalized transduction** fragments of bacterial DNA are packaged more or less at random in the phage particles. This is the case for bacteriophage P1 as described above (Fig. 25.08). Consequently, all genes have roughly the same chance of being transferred. In **specialized transduction** certain regions of the bacterial DNA are carried preferentially—see below.

For a bacterial virus to transduce, several conditions must be met. In particular, the phage must not degrade the bacterial DNA. For example, phage T4 normally destroys the DNA of *E. coli* after infection. However, mutants of T4 that have lost the ability to degrade host DNA work well as transducing phages. The packaging mechanism is also critical. Some phages, such as lambda, use specific recognition sequences when packaging their DNA into the virus particle and so will not package random fragments of DNA (see Section 4.2, below). In other cases, packaging depends on the amount of DNA the head of the virus particle can hold. Such "**headful packaging**" is essential for generalized transduction.

Two examples of generalized transducing phages are **P1**, which works on *E. coli*, and **P22**, which infects *Salmonella*. The ratio of transducing particles to live virus is about 1:100 in both cases; that is, for every 100 virus particles made, one will contain bacterial host DNA. The likelihood of the transduced DNA recombining into the recipient chromosome is roughly 1–2 in 100. P1 can package approximately 2% of the *E. coli* chromosome (about 90 kb of DNA), whereas P22 is smaller and can carry only 1% of the *Salmonella* chromosome. Taken all together, about 1 in 500,000 P1 particles will successfully transduce any particular gene on the *E. coli* chromosome. This may seem a low probability, but as both typical bacterial cultures and preparations of P1 contain about 10^9 per ml, transduction happens at useful frequencies in practice. P1 can also transduce DNA from *E. coli* into certain other gram-negative bacteria, such as *Klebsiella*.

4.2. Specialized Transduction

During specialized transduction, certain specific regions of the bacterial chromosome are favored. This is due to integration of the bacteriophage into the host chromosome (see Ch. 21). If the virus enters its lytic cycle and manufactures virus particles, those bacterial genes nearest the virus integration site are most likely to be incorrectly packaged into the viral particles. As discussed in Chapter 21, when bacteriophage **lambda** (or λ) infects *E. coli*, it sometimes inserts its DNA into the bacterial chromosome (Fig. 25.09). This occurs at a single specific location, known as the **lambda attachment site (***att***λ)**, which lies between the *gal* and *bio* genes. The integrated virus DNA is referred to as a **prophage**.

When lambda is induced, it excises its DNA from the chromosome and goes into lytic mode. The original donor cell is destroyed and several hundred virus particles containing lambda DNA are produced. Just like generalized transducing phages, a small fraction of lambda virus particles contain bacterial DNA. There are, however, two major differences. First, only chromosomal genes next to the lambda attachment

> Some viruses can carry fragments of host DNA. In generalized transduction, random pieces of the host DNA are packaged and injected into a different host.

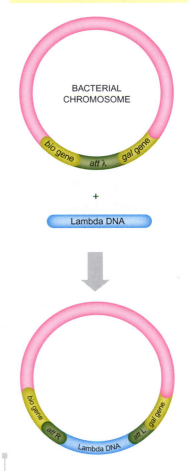

FIGURE 25.09
Integration of Lambda into the **E. coli** *Chromosome*

When bacteriophage lambda infects a host *E. coli* cell, it can integrate its phage DNA into the chromosome. The phage DNA will only integrate at a site called *att*λ, which is found between the *bio* gene and *gal* gene of the chromosome. Once integrated, the phage is referred to as a prophage.

---

headful packaging   Type of virus packaging mechanism that depends on the amount of DNA the head of the virus particle can hold (as opposed to using specific recognition sequences)

generalized transduction   Type of transduction where fragments of bacterial DNA are packaged at random and all genes have roughly the same chance of being transferred

lambda attachment site (*att***λ)**   Site where lambda inserts its DNA into the bacterial chromosome

lambda (or λ)   Specialized transducing phage of *Escherichia coli* that may insert its DNA into the bacterial chromosome

P1   Generalized transducing phage of *Escherichia coli*

P22   Generalized transducing phage of *Salmonella*

prophage   Virus DNA that is integrated into the host chromosome

specialized transduction   Type of transduction where certain regions of the bacterial DNA are carried preferentially

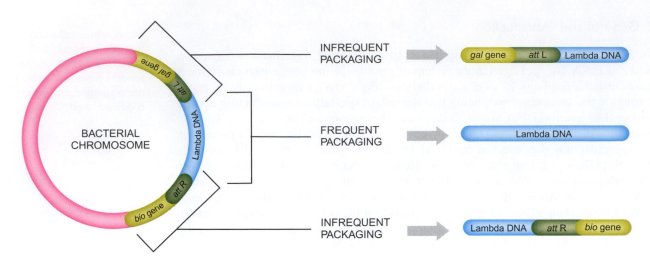

FIGURE 25.10
Packaging of Host DNA During Transduction by Lambda

When lambda phage enters its lytic cycle and makes phage particles, it usually packages the lambda DNA between the *att*L and *att*R sites. Occasionally, a mistake will occur, and part of the bacterial chromosome DNA will be packaged. Since lambda DNA normally integrates between the *gal* and *bio* genes of the *E. coli* chromosome, the defective lambda particles will most likely contain one or other of these genes.

Specialized transduction occurs in viruses that integrate into host DNA at specific sites in the host chromosome. Only adjacent genes to the integration site are transduced.

site are transduced. Second, the specialized transducing particles contain a hybrid-DNA molecule comprising both lambda and chromosomal DNA (Fig. 25.10). This hybrid molecule results from mistakes during excision of the lambda prophage. Chromosomal DNA to the right or to the left of the prophage, but not both, may be included in the transducing phage. In practice this means that either the *gal* or *bio* genes are picked up.

Mistakes in excision of lambda only occur at a rate of 1 in a million relative to correct excision. Furthermore, the defective excision must generate a segment of DNA approximately the same length as the lambda genome in order for it to fit into the phage head. Consequently, specialized transducing particles arise only at extremely low frequency. However, once a lambda transducing phage has been created, it may re-integrate its DNA into the chromosome of another host cell. This may occur either in the *att*λ site or into the chromosomal copy of the gene (usually *gal* or *bio*) carried by the lambda-transducing phage. Inducing this defective prophage DNA will give a second generation of transducing phage particles at a much higher frequency.

The properties of lambda transducing phages depend on which lambda genes were lost in exchange for chromosomal DNA. The λd*gal*-transducing phages lack lambda genes needed for making head and tail components and instead, the virus contains the *E. coli gal* gene. These are therefore "defective" (hence, the "d" in λd*gal*). **Defective phage** may be grown together with a wild-type lambda as a **helper phage**, which provides the missing functions. In the case of λd*gal*, helper phage would make the head and tail components. Conversely, the λp*bio*-transducing phages lack the lambda *int* gene, which integrates the phage DNA into the *att*λ site, and instead contains the *bio* gene from *E. coli*. Since the phage cannot integrate, λp*bio* must enter the lytic phase and are thus obligate plaque formers (hence, the "p" in λp*bio*). If wild-type helper phage is added, the *int* function is restored, and the phage forms lysogens. Cloning vectors derived from lambda are widely used in genetic engineering (see Ch. 7).

defective phage   Mutant phage that lacks genes for making virus particles
helper phage   Phage that provides the necessary genes so allowing a defective phage to make virus particles

Since the cloned DNA replaces many of the lambda genes, such vectors need to be grown in the presence of helper phages.

5. Transfer of Plasmids between Bacteria

Transferability is the ability of certain plasmids to move from one bacterial cell to another. Many medium-sized plasmids, such as the F-type and P-type plasmids, are able to move and are referred to as **Tra**$^+$ (transfer-positive). For transfer to occur, the two bacterial cells must make physical contact and transfer the DNA by a process known as bacterial conjugation. DNA moves in one direction only, from the plasmid-carrying donor to the recipient (Fig. 25.11). The donor cell manufactures a **sex pilus** that binds to a suitable recipient and draws the two cells together. Next, a **conjugation bridge** forms between the two cells and provides a channel for DNA to move from donor to recipient. In real life, mating bacteria actually tend to cluster together in groups of five to ten (Fig. 25.12).

The genes for formation of the sex pilus and conjugation bridge and for overseeing the DNA transfer process are known as *tra genes* and are all found on the plasmid itself. Since plasmid transfer requires over 30 genes, only medium or large plasmids possess this ability. Very small plasmids, such as the ColE plasmids, do not have enough DNA to accommodate the genes needed. Plasmids that enable a cell to donate DNA are called **fertility plasmids** and the most famous of these is the **F-plasmid** of *E. coli*, which is approximately 100 kbp long. Donor cells are sometimes known as F$^+$ or "male" and recipient cells as F$^-$ or "female" and conjugation is sometimes referred to as bacterial mating. Note, however, that the "sex" of a bacterial cell is determined by the presence or absence of a plasmid and that DNA transfer is unidirectional, from donor to recipient. When a recipient cell has received the F-plasmid it becomes F$^+$. From a human perspective it has been transmuted from "female" into a "male"! Thus, bacterial mating is not at all equivalent to sexual reproduction among higher organisms.

> Transferable plasmids move from one cell to another via the conjugation bridge.

Plasmid DNA transfer involves replication by the rolling circle mechanism (Fig. 25.13). First, one of the two strands of the double-stranded DNA of the F-plasmid opens up at the origin of transfer. This linearized single strand of DNA moves through the conjugation bridge from the donor into the recipient cell. An unbroken single-stranded circle of F-plasmid DNA remains inside the donor cell. This is used as a template for the synthesis of a new second strand to replace the one that just left. As the linear single strand of F-plasmid DNA enters the female cell, a new complementary strand of DNA is made using the incoming strand as template. Thus, only one strand of F-plasmid DNA is transferred from the donor to the recipient.

> A single strand of newly-made DNA is transferred from the donor to the recipient cell during conjugation.

The detailed physical mechanism of DNA transfer via the conjugation bridge was only solved relatively recently. The earliest proposals were that DNA traveled through the central channel of the sex pilus itself. Although this is incorrect, the DNA does in fact travel through the central channel of the basal structure on which the pilus is built. When the sex pilus is assembled, its protein subunits travel through the channel in the basal structure (also known as the transfer apparatus). After donor and recipient have made contact, the pilus is retracted and the pilus subunits return through the same channel. This brings the two cells into contact and leaves the basal structure bridging the inner and outer membranes of the donor and in contact with

---

conjugation bridge   Junction that forms between two cells and provides a channel for DNA to move from donor to recipient during conjugation
F-plasmid   Fertility plasmid that allows *E. coli* to donate DNA by conjugation
fertility plasmid   Plasmid that enables a cell to donate DNA by conjugation
sex pilus   Protein filament made by donor bacteria that binds to a suitable recipient and draws the two cells together
transferability   Ability of certain plasmids to move themselves from one bacterial cell to another
Tra$^+$ Transfer-positive (refers to a plasmid capable of self-transfer)
tra genes   Genes needed for plasmid transfer

FIGURE 25.11
Bacterial Conjugation

Certain plasmids, called Tra$^+$ or transfer-positive, are able to move a copy of their DNA into a different cell through a mechanism called bacterial conjugation. First, the cell containing a Tra$^+$ plasmid manufactures a rod-like extension on the surface of the outer membrane called a sex pilus. The sex pilus binds to a nearby cell and pulls the two cells together by retracting. Once the cells are in contact, the basal structure of the pilus makes a connection between the two cells known as the conjugation bridge. This connects the cytoplasm of the two cells, so the plasmid can transfer a copy of itself to the recipient cell.

FORMATION OF MATING PAIRS

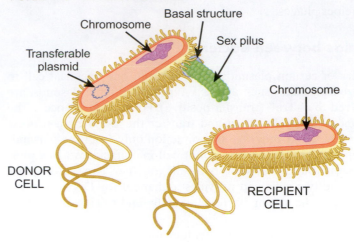

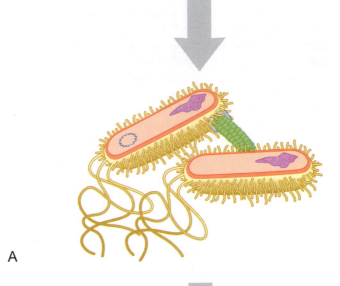

A

FORMATION OF A CONJUGATION BRIDGE

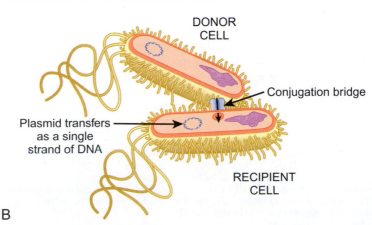

B

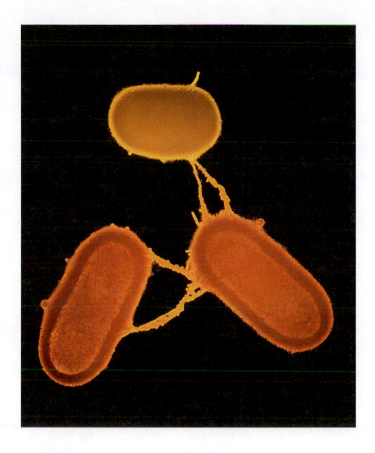

■ **FIGURE 25.12**
Conjugating Cells of **E. coli**

False-color transmission electron
micrograph (TEM) of a male
E. coli bacterium (bottom-right)
conjugating with two females.
This male has attached two F-pili
to each of the females. The tiny
bodies covering the F-pili are
bacteriophage MS2, a virus
that attacks only male bacteria
and binds specifically to F-pili.
Magnification: ×11,250. *(Credit:
Dr. L. Caro, Photo Researchers,
Inc.)*

REPLICATION

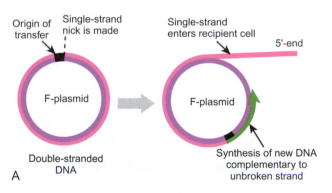

TRANSFER

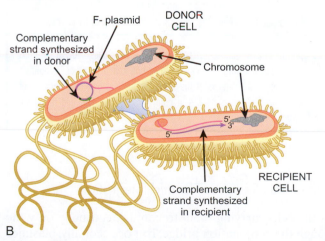

■ **FIGURE 25.13**
*Plasmid Transfer Involving
Rolling Circle Replication*

A) During bacterial conjugation,
the F-plasmid of *E. coli* is
transferred to a new cell by rolling
circle replication. First, one strand
of the F-plasmid is nicked at the
origin of transfer. The two strands
start to separate and synthesis of
a new strand starts at the origin
(green strand). B) The single-
strand of F-plasmid DNA that is
displaced (pink strand) crosses the
conjugation bridge and enters the
recipient cell. The second strand
of the F-plasmid is synthesized
inside the recipient cell. Once
the complete plasmid has been
transferred, it is re-ligated to form
a circle once again.

FIGURE 25.14
Basal Structure of the Sex Pilus

The basal structure of the sex pilus resembles a type IV secretion system. It crosses both the inner and outer membranes and its central channel is large enough for the transit of proteins or DNA. The details of individual components vary somewhat between organisms, depending on the specific role of the system. This diagram is a simplified version showing the common core structures.

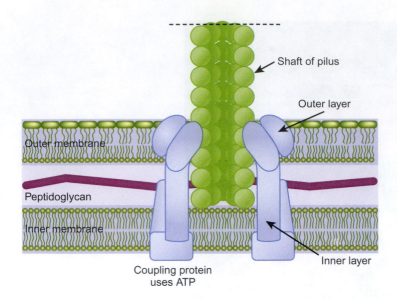

the recipient. DNA then moves through the channel of the basal structure into the recipient. The basal structure belongs to the family of type IV secretion systems. These are used by a variety of bacteria for protein secretion as well as DNA uptake and DNA transfer (Fig. 25.14).

Although ColE and other small plasmids are not self-transferable, they are often mobilizable (Mob⁺). A transferable plasmid, such as the F-plasmid, can mobilize the ColE plasmid if they both inhabit the same cell. The F-plasmid oversees conjugation and forms the conjugation bridge and the ColE-plasmid is transferred through this. The *mob* (mobilization) genes of the ColE-plasmid are responsible for making a single-stranded nick at the origin of transfer of ColE and for unwinding the strand to be transferred. Recently, interconnections between bacterial cells in biofilms have been discovered. These allow plasmids that lack the ability to transfer themselves to move between cells under these conditions (see Focus on Relevant Research).

> Plasmids unable to transfer themselves may be able to hitchhike using the transfer systems of other plasmids.

FOCUS ON RELEVANT RESEARCH

Dubey GP and Sigal Ben-Yehuda S (2011) Intercellular nanotubes mediate bacterial communication. Cell 144:590–600.

It has been known for some time that the cells of multicellular higher organisms are often connected by thin tubes. The best known are the plasmodesmata of plants, which are cytoplasmic tubes that allow transfer of nutrients, proteins, and other macromolecules between cells. Recently, membrane nanotubes have been found linking mammalian cells. These allow movement of macromolecules and even viruses and organelles.

In this paper the authors report the discovery of nanotubes between bacterial cells growing in biofilms. The nanotubes were visualized by electron microscopy. Proteins and nucleic acids could be transferred between cells. Transfer was observed using molecules with fluorescent labels. The nanotubes allowed plasmids that are typically non-transferable to move from one cell to another. Nanotubes formed not only between members of the same species, but between members of different genera, even between *Bacillus*, a gram-positive bacterium and the gram-negative *E. coli*.

5.1. Transfer of Chromosomal Genes Requires Plasmid Integration

Although many plasmids allow the cells carrying them to conjugate, usually only the plasmid itself is transferred through the conjugation bridge. But occasionally, plasmids

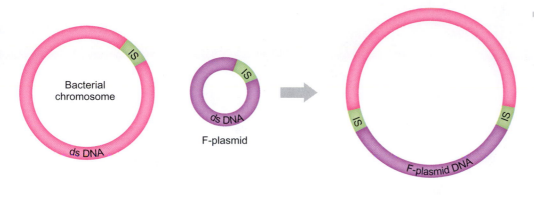

Integration of F-Plasmid into Chromosome

If recombination occurs between two insertion sequences, one on the F-plasmid and one on the host bacterial chromosome, the entire F-plasmid becomes integrated into the chromosome.

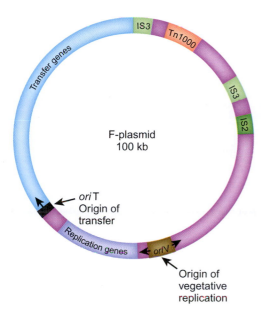

Insertion Sequences on F-Plasmid and Chromosome

Insertion sequences are scattered throughout the F-plasmid and chromosome of *E. coli*. The F-plasmid has two IS3 elements and one IS2 element. Even more copies of IS2 and IS3 are found on the chromosome (not shown). A recombination event between any of the chromosomal IS2 or IS3 elements and the corresponding element on the F-plasmid will integrate the entire F-plasmid into the chromosome. Tn1000 (also known as γδ) is another insertion sequence, although not generally involved in F-plasmid integration in *E. coli*.

> Transferable plasmids sometimes move chromosomal DNA from one cell to another.

mediate transfer of the host chromosome when they move from one bacterial cell to another. In order to transfer chromosomal genes, a plasmid must first physically integrate itself into the chromosome of the bacterium. This event involves pairs of identical (or nearly identical) DNA sequences, one on the plasmid and the other on the chromosome. In practice, **insertion sequences** (see Ch. 22) are used for integration of the F-plasmid into the chromosome of *E. coli* (Fig. 25.15).

A variety of different insertion sequences are found on the chromosome of *E. coli* and in its plasmids and viruses. The F-plasmid has three insertion sequences (Fig. 25.16): two copies of IS3 and a single copy of IS2. The chromosome of *E. coli* has 13 copies of IS2 and six copies of IS3 scattered around more or less at random. Integration of the F-plasmid may occur in either orientation at any of these 19 sites.

When an F-plasmid that is integrated into the chromosome transfers itself by conjugation, it drags along the chromosomal genes to which it is attached (Fig. 25.17). Just like the unintegrated F-plasmid, only a single strand of the DNA moves and the recipient cell has to make the complementary strand itself. Bacterial strains with an F-plasmid integrated into the chromosome are known as **Hfr-strains** because they transfer chromosomal genes at <u>h</u>igh <u>fr</u>equency. A prolonged mating of 90 minutes or so is needed to transfer the whole chromosome of *E. coli*. More often, bacteria break off after a shorter period of, say, 15–30 minutes, and only part of the chromosome is

---

Hfr-strain   Bacterial strain that transfers chromosomal genes at high frequency due to an integrated fertility plasmid (F-plasmid)
insertion sequence   A simple transposon consisting only of inverted repeats surrounding a gene that encodes transposase

FIGURE 25.17
Transfer of Chromosomal Genes by F-Plasmid

An integrated F-plasmid can still induce bacterial conjugation and rolling circle transfer of DNA into another bacterial cell. Since rolling circle replication does not stop until the entire circle is replicated, the attached chromosome is also transferred into the recipient cell. First, a single-stranded nick is made at the *ori*T, or transfer origin of the integrated plasmid. The free 5′ end (black triangle) enters the recipient cell through the conjugation bridge. Notice that the transfer of the single-stranded DNA does not end with the F-plasmid DNA and continues into the chromosomal DNA. Genes closest to the site of plasmid integration are transferred first (in the order a, b, c, d, e, f, in this example). The amount of chromosomal DNA that is transferred depends on how long the two bacteria remain attached by the conjugation bridge.

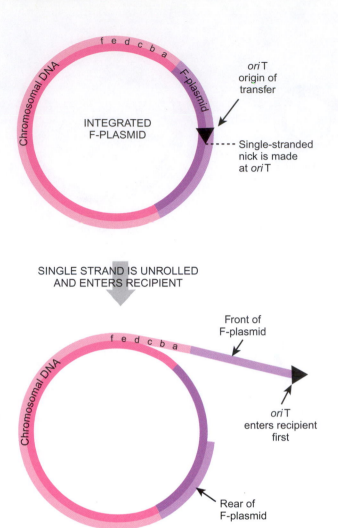

In order to mobilize chromosomal DNA, the plasmid must first integrate into the chromosome.

transferred. Since different Hfr-strains have their F-plasmids inserted at different sites on the bacterial chromosome, transfer of chromosomal genes begins at different points. In addition, the F-plasmid may be inserted in either orientation. Consequently, gene transfer may be either clockwise or counterclockwise for any particular Hfr strain.

Hfr strains were used in earlier times to identify the order of genes on the *E. coli* chromosome. To monitor whether the recipient has received the gene in question, the donor and recipient strains must have different alleles of this gene that can be distinguished phenotypically, usually by their growth properties. For example, the recipient may have a mutation that makes it unable to grow with lactose as carbon source. The donor Hfr strain would have the allele that restores the ability to grow on lactose. Using this method, genetic maps were constructed by two major approaches. First, the **cotransfer frequency** of two genes was measured. Thus, if genes a and b are close to each other, the donor Hfr strain would transfer them together at high frequency. Conversely, if genes a and b were on opposite sides of the chromosome, the Hfr strain would usually only transfer gene a, and the co-transfer frequency would be low.

Secondly, time of entry measurements were made to determine gene order around the bacterial chromosome. Hfr strains transfer genes starting where the F-plasmid is integrated and proceeding sequentially around the circular chromosome

cotransfer frequency Frequency with which two genes remain associated during transfer of DNA between bacterial cells

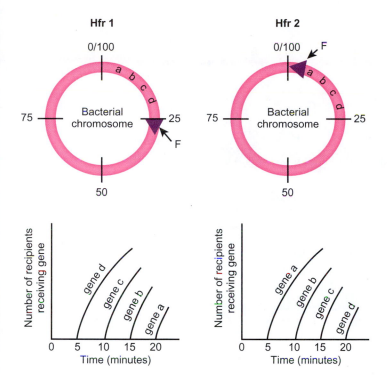

To determine the time of entry by conjugation, the Hfr strain is mixed with a recipient strain carrying a defective copy of a particular gene, "a." After conjugation has proceeded for a specific time, a sample of the mixture is removed. This is plated on agar, which prevents growth of the Hfr and only allows growth of strains carrying the wild-type version of gene "a." Survivors are derivatives of the recipient that have gained the wild-type version of gene "a" from the Hfr. This is repeated for several time points. The whole procedure is then repeated for the other genes. In strain Hfr 1 (left panel), the integrated F-plasmid is closest to gene "d" and only begins transferring gene "a" after about 20 minutes. In strain Hfr 2 (right panel), the F-plasmid is integrated closer to gene "a," which therefore begins to appear in the recipient as early as five minutes after transfer begins.

(Fig. 25.18). The length of time it takes for a gene to enter the recipient gives an estimate of its relative distance from the origin of transfer of the Hfr strain used. For time of entry mapping the site and orientation of the F-plasmid must be known. In addition, mutations in the genes being studied (a, b, c, and d) must give recognizable phenotypes. Finally, the recipient must be resistant to some antibiotic (e.g., streptomycin) so that it can be selected on medium that prevents growth of the Hfr strain. Different Hfr strains will transfer the same genes in different orders and at different times, depending on their location relative to the integration site of the F-plasmid.

F-plasmids can excise themselves from the chromosome by reversing the integration process. Sometimes they excise carrying pieces of chromosomal DNA, which creates **F′- or F-prime plasmids**. This typically occurs by recombination between a different pair of IS sequences than used during integration. Such F′-plasmids may be transferred to F-minus recipients, carrying with them the chromosomal segment from their previous host. If the chromosomal segment is homologous, the F′ can reintegrate via homologous recombination. Historically, F-primes were used to carry part of the *lacZ* gene in the **alpha-complementation** method for screening recombinant plasmids (see Ch. 7).

6. Gene Transfer among Gram-Positive Bacteria

Traditionally, the bacteria are divided into two major groups: the **gram-negative** and the **gram-positive bacteria**. This division was originally based on their response to the gram stain. The differences in staining reflect differences in the chemical composition and structure of the cell envelope. The envelope of gram-negative bacteria consists of the following layers (from inside to outside): cytoplasmic membrane, cell wall (peptidoglycan), and **outer membrane** (Fig. 25.19). The envelope of gram-positive bacteria

---

alpha-complementation   Assembly of functional beta-galactosidase from N-terminal alpha fragment plus the remaining part of the protein
F′ or F-prime plasmid   Fertility plasmid of *E. coli* that has excised itself from its host's chromosomal DNA and may contain segments of the host DNA in addition to the regular plasmid DNA
gram-negative bacteria   Major division of Eubacteria that possess an extra outer membrane lying outside the cell wall
gram-positive bacteria   Major division of Eubacteria that lack an extra outer membrane lying outside the cell wall
outer membrane   Extra membrane lying outside the cell wall in gram-negative but not gram-positive bacteria

FIGURE 25.19
Differences in Envelopes of Gram-negative and Gram-positive Bacteria

The outer surfaces of gram-positive and gram-negative bacteria have different structures. A) In gram-negative bacteria, such as *E. coli*, there are three surface layers. The outermost layer, called the outer membrane, is a lipid-bilayer that contains various proteins embedded within the lipids, and an outer coating of lipopolysaccharide. Next, within the periplasmic space, the cell wall contains a single layer of peptidoglycan. Lipoproteins connect this cell wall to the outer membrane. The layer closest to the cytoplasm, called the inner membrane, is a lipid bilayer embedded with various proteins. B) The outer surface of gram-positive bacteria only has two layers, a thick wall of peptidoglycan plus teichoic acid surrounding the cell membrane.

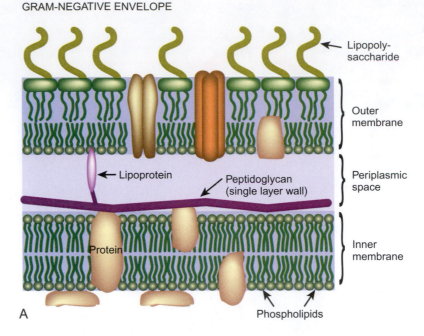

GRAM-NEGATIVE ENVELOPE

A

GRAM-POSITIVE ENVELOPE

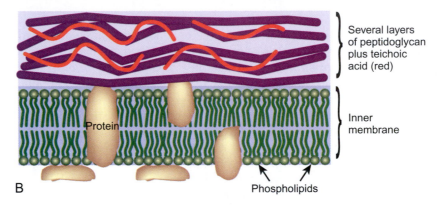

B

Gram-negative bacteria, including *E. coli*, have an extra outer membrane.

Transfer of plasmids between gram-positive bacteria is often promoted by pheromones.

is simpler and lacks the outer membrane. Both kinds of bacteria sometimes have an extra protective layer, the capsule, on the very outside.

The gram-negative bacterium *E. coli is* widely used as a host for cloning and expressing genes from a variety of other organisms. The synthesis of large amounts of a purified recombinant protein is often desirable. Secretion of recombinant protein into the culture medium would be very convenient since this avoids purifying it away from all the other proteins inside the bacterial cell. However, the complex envelope of gram-negative bacteria is a major hindrance in the export of proteins into the culture medium. In contrast, secretion across the simpler gram-positive envelope is easier. Indeed, many gram-positive bacteria, such as *Bacillus*, excrete proteins into the culture medium naturally. As a result, there is considerable interest in using gram-positive bacteria as hosts in genetic engineering. Unfortunately, the genetics of gram-positive bacteria is far behind that of the intensively studied *E. coli* and its relatives. Nonetheless, mechanisms of gene transfer are available in gram-positive bacteria.

Self-transmissible plasmids are widespread among gram-positive bacteria and many of these plasmids are rather promiscuous. Since the cell envelope is simpler in gram-positive bacteria, plasmid transfer is also simpler and a sex pilus is not needed.

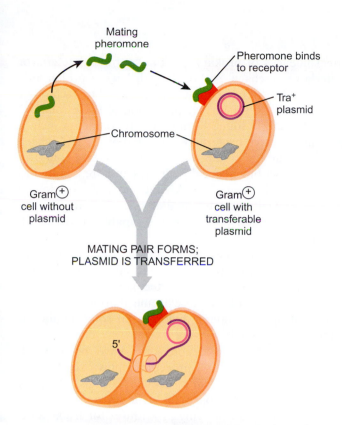

■ **FIGURE 25.20**
Pheromones Induce Mating in Gram-positive Bacteria

In gram-positive bacteria such as *Enterococcus*, cells without plasmids secrete pheromones to attract bacteria with transferable (Tra⁺) plasmids. Mating pheromones bind to receptors on the surface of cells containing Tra⁺-plasmids. Binding the receptor activates the transfer genes to form a conjugation bridge and transfer the plasmid by rolling circle replication. Each pheromone is specific and only attracts bacteria with certain plasmids.

Apparently, only half a dozen genes are required to encode the transfer functions. Some gram-positive bacteria, such as *Enterococcus*, secrete mating pheromones into the culture medium (see Focus on Relevant Research). These are short peptides that induce the *tra* genes of plasmids in neighboring bacteria. This results in aggregation and plasmid transfer (Fig. 25.20). Different pheromones are specific for different plasmids. Only bacteria that lack a particular plasmid secrete the corresponding pheromone. Furthermore, the plasmid only expresses its transfer genes when a suitable recipient is nearby.

Dunny GM and Johnson CM (2011) Regulatory circuits controlling enterococcal conjugation: lessons for functional genomics. Current Opinion Microbiology 14: 174–180.

Plasmid transfer in the gram-positive bacterium *Enterococcus* is relatively well studied. Transfer is regulated by peptide pheromones that are bound by a pheromone receptor protein. Two different peptide signal molecules, the pheromone and the inhibitor peptide, compete for binding to the same site on the same receptor protein, PrgX. When the inhibitor peptide binds to PrgX it acts as a repressor and prevents transcription from the promoter of prgQ. Conversely, when the pheromone binds to PrgX it dissociates from the promoter and the *prgQ* gene is de-repressed. This ultimately results in activation of the plasmid-transfer system.

FOCUS ON RELEVANT RESEARCH

Gram-positive bacteria also harbor **conjugative transposons** (e.g., Tn916 of *Enterococcus*). These can transfer themselves from one bacterial cell to another (see Ch. 22). These elements excise themselves temporarily from the chromosome of the donor cell before conjugation. Once inside the recipient, they reinsert themselves into the bacterial chromosome.

conjugative transposon Transposons that can transfer themselves from one bacterial cell to another by conjugation

7. Archaeal Genetics

There are two genetically-distinct lineages of prokaryotes, the "normal" bacteria or **Eubacteria** and the **Archaebacteria** or **Archaea**. Although both have a prokaryotic cell without a nucleus, the Eubacteria and Archaea are no more related to each other than either is to the eukaryotes. (See Ch. 26 for further discussion of these relationships.) The Eubacteria include most bacteria found in normal environments, including both the gram-negative and gram-positive bacteria discussed above. The Archaea include the methane bacteria and a variety of less well-known bacteria found in extreme environments. Many have strange biochemical pathways and are adapted to extremes of temperature, pH, or salinity. This makes the Archaea an attractive source of novel enzymes or proteins with unusual properties and/or resistance to extreme conditions. There are many possible industrial uses for enzymes capable of withstanding extreme temperatures, for example.

Although many complete genome sequences are available for members of the Archaea, development of systems for gene transfer has lagged way behind the Eubacteria. There are many practical problems, including the need to grow many Archaea under extreme conditions. For example, some extreme thermophiles grow at temperatures high enough to melt agar. Obtaining colonies on solid media has required the development of alternative materials.

Plasmids have been found in several Archaea and some have been developed into cloning vectors (Fig. 25.21). Transformation procedures now exist for getting DNA into several Archaea. They rely on removal of divalent cations, especially Mg^{2+}, which results in the disassembly of the glycoprotein layer surrounding many archaeal cells. (Note the contrast with the corresponding procedures for Eubacteria, which involve cold-shock in the <u>presence</u> of divalent cations!) It has been possible to express the *lacZ* reporter gene in methane bacteria under control of an archaeal promoter. However, staining of β-galactosidase with Xgal requires exposure to air, which kills methane bacteria! Consequently, colonies must first be replicated and one set sacrificed for analysis.

A major problem is choice of a selectable marker. Most standard antibiotics do not affect Archaea due to their unusual biochemistry. For example, Archaea do not have cell walls made of peptidoglycan and are therefore not susceptible to penicillins. In addition, many resistance proteins from normal organisms are denatured at the extremes of temperature, salinity, or pH under which many Archaea grow. Novobiocin (a DNA gyrase inhibitor—see Ch. 10) and mevinolin (an inhibitor of the isoprenoid pathway) have been used to inhibit halophiles, and puromycin and neomycin (both protein synthesis inhibitors—see Ch. 13) will inhibit methane bacteria.

Viruses have been discovered that infect many Archaea. So far, only one, the ΨM1 phage of *Methanobacterium thermoautotrophicum*, has been shown to transduce genes of its host bacterium. Unfortunately, this is of no practical use because of the low burst size—about six phage particles are liberated per cell after infection. The SSV1 phage of *Sulfolobus solfataricus* integrates into the bacterial chromosome and may be of future use.

Conjugation in Archaea is of two types, and different kinds of surface structures are used (see Focus on Relevant Research). Self-transferable plasmids that promote conjugation are found in *Sulfolobus*. In contrast, some halobacteria form conjugation bridges without the participation of fertility plasmids. Moreover, in these cases DNA transfer is bidirectional. Neither of these phenomena has so far been developed into routine gene-transfer systems.

Archaebacteria (or Archaea) Type of bacteria forming a genetically-distinct domain of life. Includes many bacteria growing under extreme conditions

Eubacteria Bacteria that are more familiar to us, and have peptidoglycan in their cell walls. These are a separate domain from Archeae

---

Sidebar notes:

Archaea are genetically distinct and often live under unusual or extreme conditions.

Gene transfer in Archaea is widespread but still poorly understood.

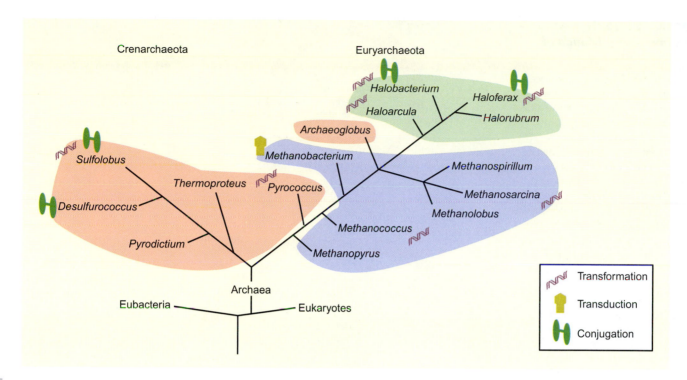

FIGURE 25.21
Groups of Archaea and their Gene-Transfer Mechanisms

Phylogenetic tree of the Archaea lineage illustrating that different types of gene transfer can occur. The green zone contains salt tolerant organisms, the blue zone indicates methane producers and the red zone contains Archaea that grow at extremely high temperatures. Some Archaea use transformation whereas others use conjugation. Rare cases of viral transduction also occur. The modes of gene transfer seen within each family do not correlate well with either lifestyle or evolutionary relationships. The Crenarchaeota and the Euryarchaeota are the two major branches of the Archaea.

Pohlschroder M, Ghosh A, Tripepi M and Albers S-V (2011) Archaeal type IV pilus-like structures—evolutionarily conserved prokaryotic surface organelles. Current Opinion Microbiology 14: 1–7.

Several different types of pili are found on the surface of bacterial cells. Some of these are also found in the Archaea, including the type IV pilus and related structures. A key characteristic of type IV structures is the filament consisting of three helical protein strands. In addition to the core components that are found in all of these related structures there are other components that vary from species to species. Consequently, the detailed structure and the biological role may vary considerably.

In bacteria type IV pili are largely used for attachment. In contrast to bacteria, multiple type IV structures with distinct functions are often found in the same Archaeal cell. Thus, in Archaea type IV-related structures are not only used as pili for attachment but as flagella for swimming. In addition, it seems likely that

FOCUS ON RELEVANT RESEARCH

some Archaea, such as *Sulfolobus*, use type IV-related structures for transfer of DNA between cells. However, other Archaea, such as the Halobacteria, use other structures unrelated to type IV pili.

8. Whole-Genome Sequencing

The techniques for gene transfer described in this chapter have allowed the construction of detailed genetic maps for *E. coli* and a few other well-investigated bacteria. However, for the vast majority of microorganisms, no "classical" genetics exists. Nowadays these are largely being investigated by more modern techniques, such as gene cloning and DNA sequencing.

Since the development of rapid automated techniques for sequencing DNA (see Ch. 8) many whole genomes have been totally sequenced. The first genome sequence

The bacterium *Hemophilus influenzae* was the first organism to have its DNA completely sequenced.

FIGURE 25.22 ————
Pathogenicity Islands of Salmonella

Comparison of the *E. coli* genome with its close relative, *Salmonella*, reveals large regions of DNA that have no homology (orange). The remaining regions have similar genes that are in identical order. For example, *Salmonella* genes d through J are clustered together in the exact same order as *E. coli* genes d through J. Since *Salmonella* is pathogenic and *E. coli* is not, the regions of no homology probably encode the genes required for pathogenicity; therefore, they are termed pathogenicity islands. The islands are flanked by inverted repeats, suggesting the DNA may have been acquired through transposition. (*Note: This figure is not drawn to scale; the pathogenicity islands are greatly exaggerated relative to the rest of the chromosome for purposes of illustration.*)

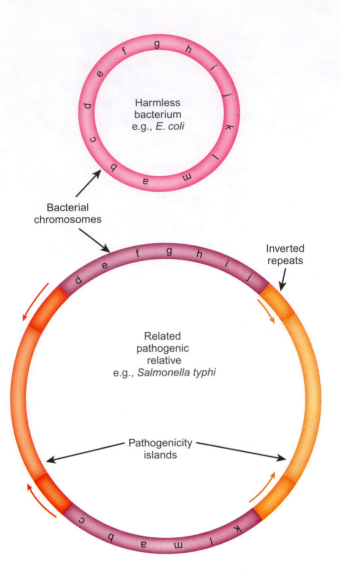

Harmless bacterium e.g., *E. coli*

Bacterial chromosomes

Inverted repeats

Related pathogenic relative e.g., *Salmonella typhi*

Pathogenicity islands

Virulence genes are often clustered together forming "islands."

to be finished was from the bacterium *Hemophilus influenzae* in 1995. Since then, hundreds of bacterial genomes have been sequenced. Sequence comparison with genes of well-investigated organisms allows provisional identification of many genes. However, even in *E. coli*, the function of about a third of the genes remains uncertain.

Whole-genome sequencing of pathogenic bacteria and comparison with their harmless relatives often reveals extra blocks of genes responsible for causing disease. Many virulence genes are carried on plasmids as discussed in Chapter 20. Others are found clustered together in regions of the chromosome known as "**pathogenicity islands**." Most genes of *Salmonella*, as well as their order around the chromosome, correspond to those of its close relative *E. coli*, as would be expected. However, extra segments of DNA are found in *Salmonella* that are lacking in *E. coli*. Some of these are pathogenicity islands (Fig. 25.22). Such extra regions are often flanked by inverted repeats, implying that the whole region was inserted into the chromosome by transposition at some period in the evolutionary past. In agreement with this idea, such islands are often found in some strains of a particular species but not others. In addition, these islands tend to have different G/C to A/T ratios and/or codon usage

pathogenicity island   Region of bacterial chromosome containing clustered genes for virulence

frequencies from the rest of the chromosome, suggesting their origin in some other organism. Conversely, *E. coli* possesses a few DNA segments missing in *Salmonella*. Interestingly, one of these is the area including the *lac* operon and a few surrounding genes. Thus, the classic *lac* operon, the most-studied "typical" gene of the "standard organism" is probably a relatively recent intruder into the *E. coli* genome!

Pathogenicity islands are simply the best-known case of "specialization islands." These are blocks of contiguous genes, presumed to have a "foreign" origin, which contribute to some specialized function that is not needed for simple survival. Not surprisingly, medical relevance has drawn most human interest. Other examples include genes encoding pathways for the biodegradation of aromatic hydrocarbons, herbicides, and other products of human industry and pollution.

Movement of genes "sideways" is designated **lateral** or **horizontal gene transfer** in distinction to the "vertical" transfer of genes from ancestors to their direct descendents. Horizontal gene transfer can occur by conjugation, natural transformation, viral transduction, or transposon jumping. Horizontal gene transfer may occur between closely-related organisms or those far apart taxonomically. Estimates suggest that in typical bacteria around 5% of the genes have been obtained by lateral gene transfer, and in rare cases up to 25%. *Thermotoga* is a Eubacterium adapted to life at very high temperatures and which consequently shares its habitat with several Archaea. *Thermotoga* has apparently gained around 25% of its genes by transfer from thermophilic Archaea such as *Archaeoglobus* and *Pyrococcus*. When we remember that the F-plasmid of *E. coli* can mediate DNA transfer into yeast (see Ch. 20), these results are perhaps not so surprising.

> Differences in G/C and A/T ratios reveal segments of chromosomes with foreign origins.

> Horizontal transfer of genes is especially significant in bacteria.

8.1. Bacterial Genome Assembly and Transplantation

The Venter Institute has performed an intriguing set of genetic manipulations intended to pave the way for the synthesis of artificial life. They first showed that it is possible to transform a whole bacterial genome into a suitable recipient cell. For this they used the bacteria of the genus *Mycoplasma*, which has one of the smallest bacterial genomes (just under 600,000 nucleotides in length). The genome from one species of *Mycoplasma* was purified and then transformed into a cell of another *Mycoplasma* species. The incoming chromosome was selected by antibiotic resistance and displaced the resident chromosome. Technically, this "genome transplantation" converted one species of the genus *Mycoplasma* to another.

The next step was to synthesize the whole genome of *Mycoplasma* chemically and then insert it into a cell (Figure 25.23). This was done in several hierarchical stages. First, about 100 segments of DNA of 5000–7000 bases long were chemically synthesized. These had overlapping sequences at their ends that allowed them to be joined together by recombination in *E. coli*. Assembly proceeded via units of 24 kb, 72 kb, and 144 kb (quarter genomes) all carried on bacterial artificial chromosomes. Final assembly of the four quarters into a complete genome was performed in yeast. The genome was then transplanted into a *Mycoplasma* host cell and selected as before. Artificial "watermark" sequences were included in the artificially-assembled genome to verify its presence. Several variants of this procedure have been carried out. One of the most recent versions contains the rather larger genome (1.08 megabases) of *Mycoplasma mycoides* chemically synthesized and modified to contain "watermark" sequences including the authors' names and famous quotes, one being *"What I cannot build, I cannot understand,"* by the physicist Richard Feynman. This cell has been nicknamed Synthia Venter!

---

horizontal gene transfer   Movement of genes sideways between unrelated organisms. Same as lateral gene transfer
lateral gene transfer   Movement of genes sideways between unrelated organisms. Same as horizontal gene transfer

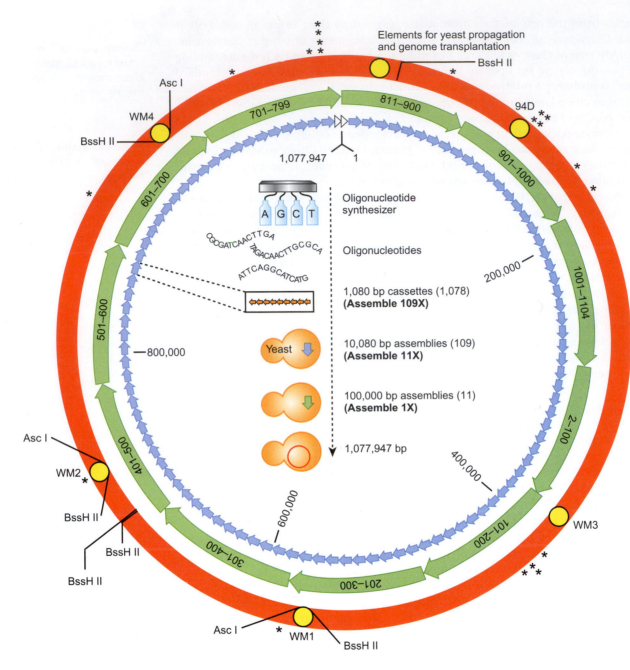

FIGURE 25.23
Assembly of Synthia Venter

An assembly of a whole genome for a synthetic organism called *M. mycoides* was assembled in yeast. First, segments of DNA about 1000 basepairs in length were chemically synthesized using an oligonucleotide synthesizer. These were combined into groups of 10 to create the 10 kB fragments (blue arrows). The 10 kB pieces were combined into groups of 10 to create 100 kB fragments (green arrows). Finally, the 11 fragments were combined into one genome (red circle). The pieces were assembled using homologous recombination in yeast. (*Credit: Gibson, D. G. et al. (2010). Creation of a bacterial cell controlled by a chemically synthesized genome. Science, 329:52–56.*)

Key Concepts

- In bacteria reproduction and sex are two distinct processes.
- Gene transfer between bacteria may occur by uptake of unprotected DNA, movement of DNA via virus particles, or specialized cell-to-cell DNA transfer.

- DNA that enters a bacterial cell may survive on its own if it is a complete replicon. Otherwise, it will be degraded unless it is recombined into the host chromosome.
- Gene transfer by the uptake of unprotected or "naked" DNA is known as transformation.
- The transfer of inherited characters by transformation was part of the original proof that DNA (not protein) is the genetic material.
- Transformation occurs in certain bacteria under natural conditions.
- Gene transfer between bacteria by DNA in virus particles is known as transduction.
- In generalized transduction, random fragments of bacterial DNA are carried by virus particles.
- In specialized transduction, specific regions of the bacterial chromosome are preferentially packaged in virus particles.
- Many plasmids can transfer themselves between bacterial cells by a process known as conjugation.
- Transfer of chromosomal genes by plasmids requires integration of the plasmid into the bacterial chromosome.
- Plasmid transfer between gram-positive bacteria is often regulated by mating pheromones secreted into the culture medium.
- Conjugative transposons can both transpose and transfer themselves between gram-positive bacteria by conjugation.
- Gene transfer in Archaea is common but still poorly investigated. Both plasmids and viruses exist that can transfer genes in these organisms.
- For most bacteria, genetic information has been gathered by sequencing the whole genome.
- Genome specialization islands are blocks of contiguous genes usually with a "foreign" origin that perform some specialized function, such as virulence or biodegradation.
- Whole bacterial genomes have been chemically synthesized and successfully inserted into bacterial cells.

Review Questions

1. What are the three mechanisms of bacterial gene transfer? Briefly define each type of transfer.
2. What are the three possible fates of incoming DNA fragments during gene transfer?
3. What is required for an incoming piece of DNA to survive without a recombination event? What types of DNA molecules can do this?
4. What type of DNA must be incorporated into the chromosome in order to survive?
5. What are "competent" cells?
6. Describe two ways to make cells "competent."
7. What is the term used to describe the uptake of naked viral DNA? What is the viral DNA alone able to do? What is the purpose of viral coats (capsids)?
8. What does the word "transformation" mean to a cancer specialist?
9. What is the role of competence pheromones? How do they work?
10. Compare and contrast natural competence and artificially-induced competence.
11. Describe transduction. How is transduction performed in a laboratory?
12. What are the two types of transduction? Describe each type. Give examples of bacteriophage that perform each type of transduction.
13. What are the conditions required for transduction?
14. What is meant by "headful packaging"? For which type of transduction is "headful packaging" essential?
15. What is *att*λ (*att*-lambda)? Between which genes and in which organism is this site located?
16. What are the two major differences between generalized and specialized transduction with regard to the bacterial DNA carried by the bacteriophages?

17. Why do specialized transducing particles arise at extremely low frequencies?
18. Give examples of defective lambda phage. Why are they defective and what is usually required to restore the defects?
19. What is the term used to describe the ability of certain plasmids to move themselves from one bacterial cell to another? What families of plasmids have this ability?
20. Describe the process of conjugation. What is a sex pilus and conjugation bridge? How is the plasmid replicated during transfer?
21. What are "*tra* genes" and where are they located?
22. Why are very small plasmids not able to transfer themselves?
23. Under what circumstances could a ColE plasmid be transferred via a conjugation bridge? What does mobilizable mean?
24. What determines the "sex" of a bacterial cell?
25. How can plasmids mediate the transfer of chromosomal genes by conjugation?
26. What are Hfr-strains? Why are they useful?
27. How long does it take to transfer the entire *E. coli* chromosome during conjugation?
28. Why is it possible for gene transfer by conjugation to be either clockwise or counterclockwise?
29. What two methods may be used to construct genetic maps using Hfr-strains? What three things are necessary for measuring time of entry?
30. What are F-prime- (or F'-) plasmids?
31. What are the major differences between the cell envelopes of gram-negative and gram-positive bacteria?
32. What usually promotes the transfer of plasmids between gram-positive bacteria?
33. What are conjugative transposons and how are they transferred?
34. What types of environments do Archaea often thrive in?
35. Contrast the transfer of genetic material in Archaea and Eubacteria.
36. What was the first organism to have its DNA completely sequenced?
37. What are "pathogenicity islands"? What usually flanks these regions? What do these flanking regions indicate?
38. What do differences in G/C to A/T ratios within "pathogenicity islands" indicate? What else gives a similar indication?
39. Give an example of an extra segment of DNA that *E. coli* has that is missing in *Salmonella*.
40. Why are "pathogenicity islands" described as "specialization islands"? What are three properties that describe them?
41. What is horizontal or lateral gene transfer? By which mechanisms may this transfer occur?

Conceptual Questions

1. Why do λ-phage heads preferentially package the *E. coli* genes *gal* and *bio*?
2. Describe Oswald Avery's experiment to prove that DNA is the genetic material. Why was his experiment not <u>absolute</u> proof that DNA is the genetic material?
3. What are some of the problems associated with the development of genetic systems within Archaea?
4. Based on the co-transfer frequencies for the following genes, determine the order in which these genes occur on the chromosome and their relative distances:

upy and *dny*	30%
sdw and *dny*	10%
sdw and *nmt*	0%
upy and *sdw*	3%
upy and *nmt*	10%
dny and *nmt*	2%

5. A researcher was doing a transformation experiment and wanted to create a plasmid based on the ColE plasmids. The researcher isolated the gene he was studying and added it to a unique Bam*HI*- restriction site on the ColE plasmid (below). He isolated his clone, made competent *E. coli*, and plated the bacteria on nutrient plates without any antibiotics. His results show a lawn of bacteria with no distinct colonies. Based on your knowledge of plasmids, what went wrong?

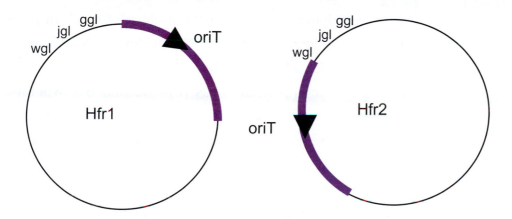

6. Compare the two different Hfr *E. coli* strains below. Which Hfr-strain is going to transfer gene *wgl* first in a time of entry conjugation experiment? Which strain will transfer *ggl* first?

7. Discuss the similarities and differences between conjugation in gram-negative bacteria and conjugation in gram-positive bacteria.

Chapter 26

Molecular Evolution

Molecular evolution deals with the mechanisms underlying evolution at the molecular level. In particular, it is concerned with the changes in DNA sequence that affect the selectable characters of organisms. The recent genomics revolution has provided an overwhelming amount of sequence information that is still being digested. Consequently, we can now largely chart the history of life in terms of genetic information and DNA sequences, rather than by comparing the structure of modern-day organisms, together with their fossil record. In particular, the classification and evolution of bacteria and, even more, of viruses, is highly dependent on sequence information. We start here by following the emergence of organic molecules capable of carrying genetic information. This leads into the RNA world scenario, and then, ultimately to our modern living world, largely based on DNA sequences.

1. Getting Started—Formation of the Earth

The big bang is estimated to have happened about 20,000,000,000 years ago. About 15,000,000,000 years later, a cloud of interstellar dust and gas coalesced and condensed due to gravity into a large ball of gas that we call the sun, orbited by smaller spherical bodies of variable composition, called the planets. The universe consists mostly of the light molecular weight gases hydrogen and helium, which account for most of the material of the stars. The heavier elements together comprise only about 0.1% of the total and form the planets (Table 26.01).

TABLE 26.01	Elemental Compositions in Atoms per 100,000			
Element	**Universe**	**Earth**	**Crust**	**Life**
H	92,700	120	2,900	60,600
He	7,200	<0.1	<0.1	0
O	50	48,900	60,400	26,700
Ne	20	<0.1	<0.1	0
N	15	0.3	7	2,400
C	8	99	55	10,700
Si	2.3	14,000	20,500	<1
Mg	2.1	12,500	1,800	11
Fe	1.4	18,900	1,900	<1

As the Earth formed, heat was released by the collapse due to gravity and also by the radioactivity of elements present in the original dust. During its first few hundred million years the Earth was too hot for water to liquefy and so H_2O was only present as steam. Later, as the Earth cooled off, the steam condensed to form oceans and lakes. The "classic theory" of the origin of life suggests that life originated by means of chemical reactions occurring in the atmosphere followed by further reactions in the primeval oceans and lakes (the hydrosphere). More recent suggestions have emphasized subsurface reactions, especially those involving sulfur and hydrogen. We will discuss the "classic theory" first.

1.1. The Early Atmosphere

The Earth's original atmosphere, the **primary atmosphere**, consisted mostly of hydrogen and helium, but the Earth is too small a planet to hold such light gases and they floated away into space. The Earth then accumulated a **secondary atmosphere**, mostly by volcanic out-gassing. Volcanic activity was much greater on the hotter primitive Earth. Volcanic gas consists mostly of steam (95%) and variable amounts of CO_2, N_2, SO_2, H_2S, HCl, B_2O_3, elemental sulfur, and smaller quantities of H_2, CH_4, SO_3, NH_3, and HF but no O_2. Of all these, the concentration of CO_2 was the second highest in amount (about 4%). In addition, water vapor reacted with primeval minerals such as nitrides to give ammonia, with carbides to give methane and with sulfides to give hydrogen sulfide. There was no free oxygen.

> Life evolved under a reducing atmosphere lacking any free oxygen.

Our present atmosphere, the **tertiary atmosphere**, is of biological origin. The methane, ammonia, and other reduced gases have been consumed and the inert components (nitrogen, traces of argon, xenon, etc.) have remained largely unchanged. Substantial amounts of oxygen have been produced by photosynthesis. This could not occur until the cyanobacteria, the first oxygen-releasing photosynthetic organisms, had evolved about 2.5 thousand million years ago (see Table 26.02 for time scale). As more and more photosynthetic organisms appeared, the oxygen content of the atmosphere increased. The oxygen content of the atmosphere reached 1% about 800 million years ago and 10% about 400 million years ago. Today it is about 20%. Although there has been an overall upward trend, the increase in oxygen has not been smooth and there have been temporary decreases.

> Oxygen in today's atmosphere is the result of photosynthesis.

---

primary atmosphere   The original atmosphere of the Earth consisting mostly of hydrogen and helium
secondary atmosphere   The atmosphere of the Earth after the light gases were lost and resulting mostly from volcanic out-gassing. It contained reduced gases but no oxygen
tertiary atmosphere   The present atmosphere of the Earth resulting from biological activity

TABLE 26.02	Approximate Evolutionary Time Scale
Millions of Years Ago	**Major Events**
20,000	Big Bang
5,000	Origin of planets and sun
3,500	Origin of life
3,000	Primitive bacteria start using solar energy
2,500	Advanced photosynthesis releases oxygen
1,500	First eukaryotic cells
1,000	Multicellular organisms
600	First skeletons give nice fossils
1.8	First true humans—*Homo erectus*
0.1	African Eve gives birth to modern Man
0.01	Domestication of crops and animals begins
0.002	Roman Empire
0.0001	Darwin's theory of evolution
0.00006	Double helix discovered by Watson and Crick
0.000010	Human genome sequenced

Evidence for the increase in O_2 content of the atmosphere comes partly from the finding that rocks of different ages are oxidized to different extents. Thus, rocks of age 1,800–2,500 million years sometimes contain UO_2, FeS, ZnS, PbS, and FeO, all of which are unstable in the presence of even small amounts of gaseous O_2. Later rocks contain mostly Fe^{3+} rather than Fe^{2+} and more oxidized ores of U, Zn, and Pb.

2. Oparin's Theory of the Origin of Life

Ultraviolet radiation from the sun, together with lightning discharges, caused the gases in the primeval atmosphere to react, forming simple organic compounds. These dissolved in the primeval oceans and continued to react, forming what is sometimes referred to as the "**primitive soup**" or "**primordial soup**." The primitive soup contained amino acids, sugars, and nucleic acid bases, among other randomly synthesized molecules (Fig. 26.01). Further reactions formed polymers and these associated, eventually forming globules. Ultimately, these evolved into the first primitive cells. This theory of the origin of life was put forward by the Russian biochemist Alexander Oparin in the 1920s. Charles Darwin himself had actually proposed that life might have started in a warm little pond provided with ammonia and other necessary chemicals. However, it was Oparin who outlined all the necessary steps and realized the critical point: life evolved before there was any oxygen in the air. Oxygen is highly reactive and would have reacted with the organic precursor molecules formed in the atmosphere, oxidizing them back to water and carbon dioxide.

Reactions in the primeval atmosphere led to the accumulation of organic material in oceans and lakes—the primitive soup.

2.1. The Miller Experiment

In the early 1950s, the biochemist Stanley Miller mimicked the reactions proposed to occur in the primitive atmosphere. An imitation atmosphere containing methane, ammonia, and water vapor was subjected to a high-voltage discharge (to simulate lightning) or to ultraviolet light (Fig. 26.02). The gases were circulated around the

primitive soup Mixture of random molecules, including amino acids, sugars, and nucleic acid bases, found in solution on the primeval Earth
primordial soup Same as primitive soup

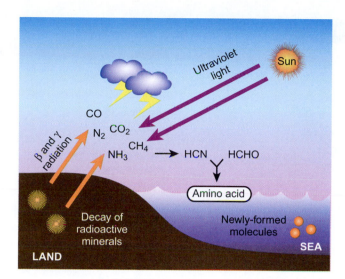

FIGURE 26.01
Formation of Primitive Soup

According to Oparin's theory of the origin of life, conditions on planet Earth were sufficient for forming early biological molecules. The atmosphere at this point contained CO, CO_2, CH_4, N_2, and NH_3. When energy was supplied by electrical discharge from lightning, ultraviolet radiation from the sun, and/or β and γ radiation from the Earth, early organic compounds such as HCN and $HCHO$ would form. These compounds would combine in the vapor phase and in the water to form amino acids. Dissolving in water protects the precursor molecules from being degraded again by the energy sources that triggered their formation.

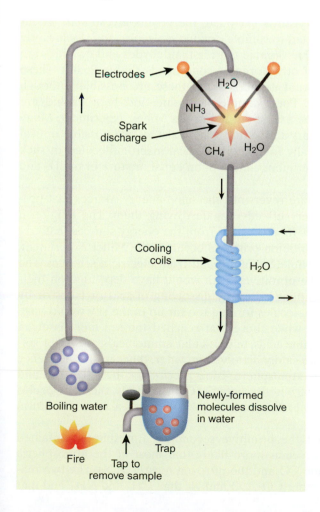

FIGURE 26.02
Miller's Experiment

Miller's experiment used a closed system to simulate the primeval conditions on early planet Earth. Water was boiled to make steam (lower left), which then mixed with NH_3 and CH_4 in another chamber. This upper chamber simulates Earth's early atmosphere and was subjected to either electrical discharge (shown) or ultraviolet radiation (not shown). The resulting products were cooled and condensed as they passed coils filled with cold water. The condensed products dissolved in a flask of water, which simulated the early oceans. The newly-formed molecules were analyzed at various times during the experiment.

TABLE 26.03	Typical Products from Miller's Experiment	
Molecule	**Name**	**Relative Yield**
H-COOH	formic acid	1000
H_2N-CH_2-COOH	glycine	275
HO-CH_2-COOH	glycolic acid	240
H_2N-CH(CH_3)-COOH	alanine	150
HO-CH(CH_3)-COOH	lactic acid	135
H_2N-CH_2CH_2-COOH	beta-alanine	65
CH_3-COOH	acetic acid	65
CH_3-CH_2-COOH	propionic acid	55
CH_3-NH-CH_2-COOH	sarcosine	20
HOOC-CH_2CH_2-COOH	succinic acid	17
H_2N-CO-NH_2	urea	9
HOOC-CH_2CH_2CH(NH_2)-COOH	glutamic acid	2.5
HOOC-CH_2CH(NH_2)-COOH	aspartic acid	1.7

apparatus so that any organic compounds formed in the artificial atmosphere could dissolve in a flask of water, intended to represent the primeval ocean. These compounds could continue to react with each other in the water.

There are many variants of this experiment (different gas mixtures, different energy sources, etc.). As long as oxygen is excluded, the results are similar. About 10–20% of the gas mixture is converted to soluble organic molecules and significantly more is converted to a non-analyzable organic tar. First, aldehydes and cyanides are formed, and then a large variety of other organic compounds (Table 26.03). These experiments produce multiple different isomers where these are possible; although not all are biologically significant. For example, sarcosine and beta-alanine are both isomers of alanine and all three are generated in the Miller experiment. Many organic molecules, in particular sugars and amino acids, exist as two possible optical isomers, only one of which is normally found in biological macromolecules. In such cases the products of the Miller experiment consist of an equal mixture of the D- and L-isomers.

Since many chemical reactions are reversible, the same energy sources that produce organic molecules are also very effective at destroying them. The long-term build-up of organic material requires its protection from the energy sources that created it. This is the function of the imitation primeval ocean in the Miller experiment. Water shields molecules from ultraviolet radiation and from electric discharges. The survival of organic molecules on the primitive Earth would have depended on their escape from UV radiation and lightning either by dissolving in seas or lakes or by sticking to minerals. Most organic molecules formed too far up in the sky would have been destroyed again very quickly, while those that reached the sea and dissolved would have survived. Note that organic acids, in particular amino acids, are water soluble and non-volatile. Once they are safely dissolved in water, there is little tendency for such molecules to return to the atmosphere. Their precursors, the aldehydes and cyanides, are not only highly reactive but also volatile. Consequently, these molecules do not survive for long. Thus, even at this early stage, there was a form of natural selection between molecules.

Originally it was thought that the primitive secondary atmosphere contained mostly NH_3 and CH_4. However, it seems more likely that most of the atmospheric carbon was CO_2 with perhaps some CO and the nitrogen mostly as N_2. The two reasons for this are: volcanic gas has more CO_2, CO and N_2 than CH_4 and NH_3, and that

Experiments to mimic reactions in the primeval atmosphere have generated many of the metabolites and monomers found in modern cells.

Water protects organic molecules from destruction by UV radiation or lightning.

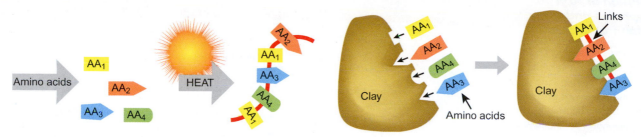

FIGURE 26.03
Formation of Proteinoids by Mild Heat or Clay Catalysis

A) A mixture of separate amino acids will form artificial polypeptide chains or "proteinoids" when subjected to heat in the absence of water for a few hours. B) Amino acids can also form bonds when they bind to certain types of clay. The clay has binding sites for amino acids in close proximity, therefore, once bound the amino acids condense into a proteinoid.

UV radiation destroys NH_3 and CH_4, therefore, these molecules would have been short-lived. The UV destruction of CH_4 occurs in two steps. First, UV light photolyses H_2O to $H^\bullet$ and $^\bullet OH$ radicals. These then attack methane, eventually giving CO_2 and H_2, which would be lost into space. In Miller's experiment, gas mixtures containing CO, CO_2, N_2, etc., give much the same products as those containing CH_4 and NH_3 so long as there is no O_2. Since CO, CO_2, and N_2 do not supply H atoms, these come mostly from water vapor photolysis. In fact, in order to generate aromatic amino acids under primitive Earth conditions it is necessary to use less hydrogen-rich gaseous mixtures. Most of the natural amino acids, hydroxy-acids, purines, pyrimidines, and sugars have been produced in variants of the Miller experiment.

2.2. Polymerization of Monomers to Give Macromolecules

Polymerization of monomers to give biological macromolecules usually requires the removal of H_2O. Clearly, water is in excess in the oceans and removal of H_2O from dissolved molecules is therefore unfavorable. Consequently, the assembly of macromolecules such as proteins and nucleic acids needs energy to form the linkages and/or remove the water. Before the high-energy phosphates used in modern cells were available, some other form of energy was needed.

> Formation of biological polymers requires removal of water.

Imitation protein polymers, containing randomly-linked amino acids, are known as "**proteinoids**." They can be formed by heating dry amino acid mixtures at around 150°C for a few hours (Fig. 26.03A). Whereas biological proteins are bonded using only the α-NH_2 and α-COOH groups of α-amino acids, these "primeval polypeptides" contain substantial numbers of bonds involving side chain residues. They contain up to 250 amino acids and can sometimes perform primitive enzymatic activities. Such dry heat could have occurred near volcanoes or when pools left behind by a changing coastline evaporated. Much of the early work on proteinoids was done by Sydney Fox who proposed their thermal origin. However, another way to randomly polymerize amino acids is by using clay minerals with special binding properties (Fig. 26.03B). Binding of small molecules to the surface of catalytic minerals can promote many reactions. For example, certain clays, such as Montmorillonite, will condense amino acids to form polypeptides up to 200 residues long.

> Water can be removed by moderate heating, by certain clay minerals, or by chemical condensing agents.

Polymerization of amino acids may have also occurred in solution, but another component, a condensing agent, is required to withdraw water. Several possible primeval condensing agents have been proposed, including reactive cyanide derivatives and, more biologically relevant, **polyphosphates**. Inorganic polyphosphates

---

polyphosphate   Compound consisting of multiple phosphate groups linked by high-energy phosphate bonds
proteinoid   Artificially-synthesized polypeptide containing randomly-linked amino acids

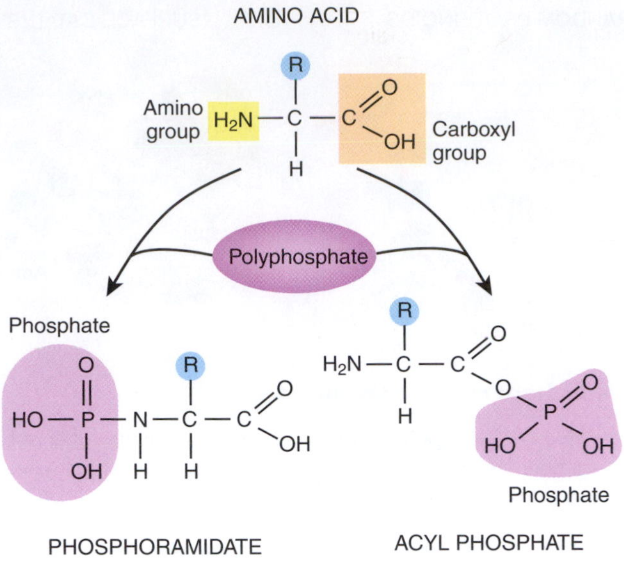

FIGURE 26.04
Formation of Acyl Phosphates and Phosphoramidates

Condensing single amino acids into polypeptide chains could have occurred in solution as long as a condensing agent was present. One possible condensing agent was a polyphosphate that would react with the amino group or carboxyl group of individual amino acids. The two possible products, phosphoramidates and acyl phosphates, can form polypeptide chains by heating in solution.

would have been present in primeval times (formed by volcanic heat from phosphates, for example). Polyphosphates can react with many organic molecules to give organic phosphates. Amino acids give two possible products (Fig. 26.04). **Acyl phosphates** have the phosphate group attached to the carboxyl group of the amino acid ($NH_2CHRCO\text{-}OPO_3H_2$) and **phosphoramidates** have the phosphate attached to the amino group of the amino acid ($H_2O_3P\text{-}NH\text{-}CHR\text{-}COOH$). Gentle heating or irradiation of such derivatives will give polypeptides. Modern life uses acyl-phosphate derivatives during protein synthesis, although from a chemical viewpoint, either derivative would work. In fact, laboratory synthesis of DNA does use phosphoramidates. Analogous reactions can produce AMP from adenine plus polyphosphate and polynucleotides can then form by polymerization (see below).

2.3. Enzyme Activities of Random Proteinoids

Interestingly, random proteinoids stewed up in modern laboratories under fake primeval Earth conditions will carry out some simple enzyme reactions (Table 26.04). They are far slower and less accurate than enzymes made by real cells, but nonetheless they can perform recognizable enzymatic reactions. For example, random proteinoids can often remove carbon dioxide from molecules like pyruvate or oxaloacetate and split organic esters. About 50% of all modern enzymes contain metal ions as co-factors and the addition of metal ions greatly extends the enzyme activities of random proteinoids. The presence of traces of copper promotes reactions involving amino groups and iron mediates oxidation-reduction reactions. Incorporation of zinc allows the breakdown of ATP, which is used by modern cells, both as a precursor of nucleic acids and as an energy carrier. Most modern enzymes that process nucleic acids possess a zinc atom as co-factor.

Artificial random proteinoids show inefficient but detectable enzyme activities.

3. Origin of Informational Macromolecules

Biological information is passed on by template-specific polymerization of nucleotides. A mixture of polyphosphate, purines, and pyrimidines will produce random

acyl phosphate Phosphate derivative in which the phosphate is attached to a carboxyl group
phosphoramidate Phosphate derivative in which the phosphate group is attached to an amino group

TABLE 26.04	Enzyme Activities of Random Proteinoids	
Reaction	**Requirements**	**Substrate**
esterase	histidine	p-nitrophenyl-phosphate
ATPase	Zn^{2+}	ATP
amination and deamination	Cu^{2+}	α-ketoglutarate $\leftrightarrow$ glutamate
peroxidase and catalase	heme, basic proteinoids	H_2O_2 and H-donors e.g., hydroquinone, NADH
decarboxylation	basic proteinoids, or acidic proteinoids	oxaloacetate, pyruvate

nucleic acid chains if ribose or deoxyribose is included. One problem, not yet solved, is that life uses 3′–5′ linked nucleic acids, whereas primeval syntheses give RNA molecules with a mixture of linkages, but mostly 2′–5′. In contrast, deoxyribose has no 2′-OH and so cannot give 2′–5′ links. However, it is generally thought that RNA probably provided the first informational molecule and that DNA is a later invention designed to store information in a more stable and accurate form.

When an RNA template is incubated with a mixture of nucleotides, plus a primeval condensing agent, a complementary piece of RNA is synthesized. This non-enzymatic reaction is catalyzed by lead ions, with an error rate of about 1 wrong base in 10. With zinc ions, a great improvement is seen and lengths of up to 40 bases are produced, with an error rate of about 1 in 200. All modern-day RNA and DNA polymerases contain zinc. If a 3′–5′ linked RNA template is used about 75% of the newly–formed RNA is 3′–5′ linked. However, this does not surmount the problem that the original formation of random RNA type polymers favors the non-biological 2′–5′ linkage very heavily.

If a mixture of nucleoside triphosphates (or nucleotides plus polyphosphate) is incubated under primeval conditions, using Zn as a catalyst, a molecule of single-stranded RNA with a random sequence will form. This original polymerization step is very slow. However, once an initial RNA polymer is present, it will act as a template for the assembly of a complementary strand. Template-directed synthesis is much more rapid, even in the absence of any enzyme. The complementary strand will in turn act as a template to generate more of the original RNA molecule. The net result is that once the first random sequence emerges, it will multiply rapidly and take over the incubation mixture (Fig. 26.05). The result will be a set of sequences with frequent mistakes but which are nonetheless clearly related (a molecular "**quasi-species**"). If a series of similar incubations are carried out, each individual sample will yield a "quasi-species" of related sequences. However, the sequence that takes over will be different for each incubation mixture.

3.1. Ribozymes and the RNA World

Which came first the chicken or the egg?—Protein or nucleic acid? Since it is possible for random RNA molecules to be assembled and duplicate themselves under primeval conditions, it is generally held that nucleic acids appeared first. Furthermore, although most modern-day enzymes are indeed proteins, examples of **ribozymes** (i.e., RNA acting as an enzyme) do exist. This scenario suggests that the primeval nucleic acid replicated alone and that proteins emerged later.

One rather extreme viewpoint is the idea that the earliest organisms had both genes and enzymes made of RNA and formed a so-called "**RNA world.**" This idea

Primeval synthesis of proteins or nucleic acids gives polymers with a mixture of natural and unnatural linkages.

Zn or Pb ions can catalyze non-enzymatic synthesis of RNA.

The first strands of RNA to form act as templates for the assembly of later molecules.

The ability of RNA to act as an enzyme as well as encode genetic information suggests that RNA came first.

---

quasi-species   A set of closely-related sequences whose individual members vary from consensus by frequent errors or mutations
ribozyme   RNA molecule that is enzymatically active
RNA world   The hypothetical stage of early life in which RNA encoded genetic information and carried out enzyme reactions without the need for either DNA or protein

FIGURE 26.05
Assembly and Duplication of Random RNA

A mixture of nucleosides and polyphosphates can form random stretches of RNA in the presence of zinc ions. The first strand of RNA forms very slowly. However, once the first strand is assembled, it may be used as a template to assemble complementary strands of RNA. Such template-driven assembly of nucleotides is much faster than formation of the original random RNA strand. However, the non-enzymatic synthesis of RNA incorporates many wrongly-paired bases.

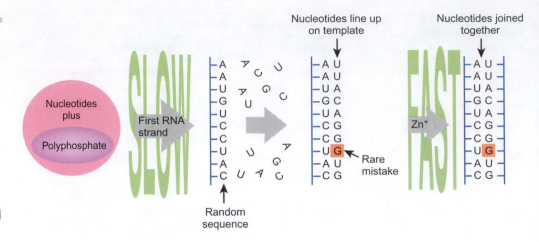

RNA molecules with enzyme activity include ribosomal RNA, ribonuclease P, self-splicing introns, and viroids.

Newly-made nucleic acid molecules all start with a stretch of RNA.

was proposed by Walter Gilbert in 1986 and seeks to avoid the paradoxical problem that nucleic acids are needed to encode proteins, but that enzymes made of protein are needed to replicate nucleic acids. During the RNA world stage, RNA supposedly carried out both functions. Later, proteins infiltrated and took over the role of enzymes and DNA appeared to store the genetic information, leaving RNA as a mere intermediate between genes and enzymes.

Several examples illustrate the ability of RNA to perform enzymatic reactions and regulatory roles as well as encode genetic information. The following cases favor the primacy of RNA:

1. Ribozymes are RNA molecules that are enzymatically active. There is a growing list of known and suspected ribozymes. The most important is the ribosomal RNA of the large subunit, which is directly involved in the reactions of protein synthesis (see Ch. 13). One of the best-known ribozymes is **ribonuclease P**. This enzyme has both RNA and protein components and processes certain transfer RNA molecules. It is the RNA part of ribonuclease P that carries out the reaction. The protein serves only to hold together the ribozyme and the tRNA it operates on. In concentrated solution, the protein is not even necessary, and the RNA component will work on its own.

2. Self-splicing introns are an example of catalytic RNA. The genes of eukaryotes are often interrupted by non-coding regions (introns), which must be removed from the mRNA before translation into protein. Normally, this is done by a spliceosome made up of several proteins and small RNA molecules. Occasionally, the intron RNA splices itself out without help from any protein. Such self-splicing is found in a few nuclear genes of some protozoans, in the mitochondria of fungal cells, and the chloroplasts of plant cells (see Ch. 12 for details). Note that genuine enzymes process large numbers of other molecules and are not permanently altered by the reaction. Therefore, self-splicing RNA is not a true enzyme as it works only once.

3. Viroids are infectious RNA molecules that infect plants. As noted in Chapter 21, viroid RNA carries out a self-cleavage reaction during replication (i.e., viroids act as ribozymes).

4. DNA polymerase cannot initiate new strands but can only elongate pre-existing strands (see Ch. 10). Primers made of RNA must be used whenever new strands of DNA are started. RNA polymerase is capable both of initiation and

ribonuclease P   A ribozyme found in many bacteria that processes certain transfer RNA molecules

elongation. This suggests that RNA polymerase and RNA may have evolved before DNA polymerase and DNA.

5. The precursors to DNA, deoxyribonucleotides containing deoxyribose, are not assembled from their subcomponents. Instead they are made by the reduction of ribonucleotides by ribonucleotide reductase. Thus, deoxyribose is never made as a separate molecule, whereas ribose is. Similarly, thymine (only in DNA) is made by methylation of uracil (found in RNA).

6. The ribose versions of nucleotides are used as energy currency (ATP, GTP), in modifying proteins (AMP groups, ADP-ribose), and as parts of many co-factors (NAD, FAD, coenzyme A, etc.).

7. Small guide molecules of RNA are used in a variety of processes. These include the removal of introns, the modification and editing of messenger RNA (see Ch. 12 for details), and the extension of the ends of eukaryotic chromosomes by telomerase.

8. Riboswitches are binding motifs in messenger RNA that directly bind small molecules and change conformation. This enables them to control gene expression in the absence of regulatory proteins (see Ch. 18 for details).

In a way, the critical question is whether RNA can copy itself without the need for DNA or help from protein enzymes. Although no RNA polymerases that are ribozymes still exist, it has proven possible to generate them artificially. Altered RNA molecules can be selected by a form of Darwinian evolution at the molecular level. Pre-existing ribozymes can be used as starting materials. Alternatively some experimenters have used random pools of artificially-generated RNA sequences. In one experiment, RNA molecules showing primitive RNA ligase activity were selected from a pool of random RNA sequences. Such artificial ribozymes can link together two chains of RNA by a typical ligase reaction just like protein enzymes in modern cells. The best of these RNA ligase ribozymes was then subjected to further rounds of mutation and selection. The result was a ribozyme of 189 bases that uses an RNA template to synthesize a complementary strand of RNA with about 96–99% accuracy. This ribozyme adds single nucleotides, one at a time, to an RNA primer using nucleoside triphosphates as substrates (Fig. 26.06). However, it is very slow and can only extend chains by around 14 nucleotides because it is not "processive." In other words, the ribozyme dissociates from the template after adding each nucleotide, whereas true polymerases remain attached and proceed along the template adding nucleotides in quick succession.

One problem with the "RNA world" concept is that RNA is more reactive than DNA. Although RNA would form more easily than DNA under primeval conditions, it would also be less stable. Thus, DNA, though slower to form initially, might tend to accumulate under such conditions. Moreover, the primordial soup would contain a mixture of the subcomponents of both types of nucleic acid as well as proteins, lipids, and carbohydrates. So it seems perhaps more likely that an ill-defined mixture, perhaps even hybrid nucleic acid molecules with both RNA and DNA components, emerged first.

> Artificial ribozymes can be isolated by screening pools of random RNA sequences for particular enzymatic reactions.

3.2. The First Cells

Forming biologically-relevant molecules in the primitive Earth would have been the first step on the road to forming the first primitive cells. Possibly random proteins and greasy lipid molecules collected around the primeval RNA (or DNA), so forming a microscopic membrane-covered organic blob. Eventually this proto-cell learned how to use RNA to code for its protein sequences. The lipids formed a membrane around the outside to keep the other components together. Early on, protein and RNA shared the enzymatic functions. Later, RNA lost most of its enzymatic roles as the more versatile proteins took these over (Fig. 26.07). It is generally thought that RNA was the first information-storage molecule and that DNA was a later invention.

FIGURE 26.06
Artificial Evolution of Ribozyme RNA Polymerase

A random pool of RNA was generated and screened for the ability to seal a nick in one strand of a broken piece of double-stranded RNA. Occasional random molecules of RNA that had this enzymatic activity were found and isolated. Through successive rounds of mutation and selection, this primitive ribozyme was altered just enough to actually catalyze the elongation of the RNA using a single-stranded template with a primer. Unlike true protein RNA polymerases, the RNA polymerizing ribozyme only added one nucleotide at a time, dissociating after each addition.

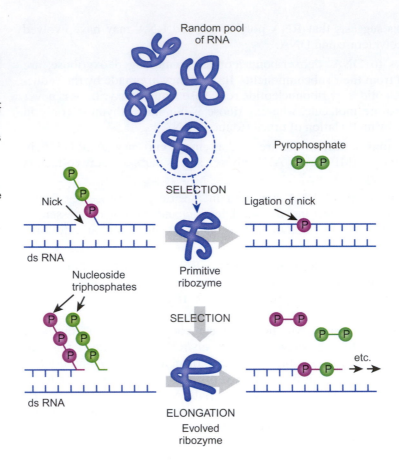

FIGURE 26.07
Emergence of the Proto-Cell

An early primitive cell may have contained RNA molecules as the information-coding material. RNA molecules would have acted as ribozymes to make copies of the RNA genetic material. Over time, the RNA would have evolved the ability to synthesize proteins to do much of the enzymatic work.

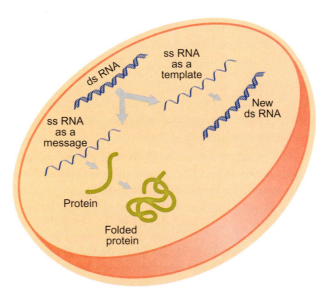

The first cells probably used RNA in multiple roles, some of which were later taken over by proteins or DNA.

Because DNA is more stable than RNA, it would store and transmit information with fewer errors.

This primitive cell vaguely resembles primitive bacteria and lived off the organic compounds in the primitive soup. Eventually the supply of pre-made organic molecules was consumed. The proto-cell was forced to find a new source of energy and it turned to the sun (Fig. 26.08A). The earliest forms of photosynthesis probably used solar energy coupled to the use of sulfur compounds to provide reducing power. Later, more advanced photosynthesis used water instead of sulfur compounds. The water was split, releasing oxygen into the atmosphere.

A   PHOTOSYNTHETIC APPARATUS

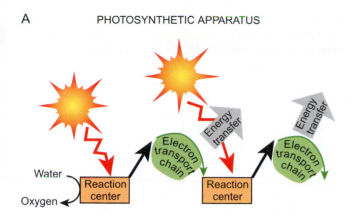

B   RESPIRATION RELIES ON OXYGEN FROM PHOTOSYNTHESIS

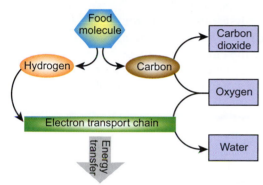

► **FIGURE 26.08**
Development of Photosynthesis Sets the Stage for Respiration

A) When early primitive cells ran out of pre-made organic molecules to supply energy, the cells started using the energy of the sun. The light energy was harvested by reaction centers that used water to provide the reducing power and released molecular oxygen into the atmosphere. Electrons carrying energy derived from the sunlight passed down electron transport chains. This allowed the conversion of the energy to a form used by the cell. B) The accumulation of oxygen in the atmosphere allowed some of the proto-cells to change their biochemistry. Instead of using water and sunlight, these cells started to use oxygen to oxidize organic matter. This provided the cell with energy and released carbon dioxide and water into the environment.

Before this, the atmosphere had been void of oxygen. The addition of oxygen completely altered the primitive Earth. Once oxygen became available, respiration could be developed. Cells reorganized components from the photosynthetic machinery to release energy by oxidizing food molecules with oxygen (Fig. 26.08B). Photosynthesis emits oxygen and consumes carbon dioxide, whereas respiration does the reverse. The overall result is an ecosystem where plants and animals complement each other biochemically.

Primitive photosynthesis probably used sulfur compounds as a source of electrons, more advanced photosynthesis resulted in the release of oxygen from water.

4. The Autotrophic Theory of the Origin of Metabolism

There is an alternative theory for the chemical origin of life. According to this view, the first proto-cells were not heterotrophic scavengers of organic molecules but were autotrophic and fixed carbon dioxide into organic matter themselves. An autotroph is defined as any organism that uses an inorganic source of carbon and makes its own organic matter as opposed to a heterotroph, which uses pre-made organic matter. The most familiar autotrophs are plants that use energy from sunlight to convert carbon dioxide into sugar derivatives. However, a variety of bacteria exist that fix carbon dioxide without light but instead rely on other sources of energy. Furthermore, the pathways of carbon dioxide fixation vary. In particular, some autotrophic bacteria incorporate carbon dioxide into carboxylic acids rather than generating sugar derivatives like plants.

The autotrophic theory of the origin of life postulates the chemical oxidation of readily available iron compounds as the primeval energy source. In particular, the conversion of ferrous sulfide (FeS) to pyrite (FeS_2) by hydrogen sulfide (H_2S) releases energy and provides H atoms to reduce carbon dioxide to organic matter. (Anaerobic bacteria are found today that generate energy by the oxidation of iron

An alternative theory suggests that energy released by the reaction of iron and sulfur compounds powered the earliest life forms.

Fe²⁺ compounds to Fe³⁺, as well as others that generate energy by oxidizing sulfur compounds. Thus, a primeval metabolism based on iron and sulfur seems reasonable.)

Several possible schemes have been suggested for the first carbon dioxide fixation reactions. One scheme involves iron-catalyzed insertion of CO_2 into sulfur derivatives of those carboxylic acids still found today as metabolic intermediates (e.g., acetic acid, pyruvic acid, succinic acid, etc.). These early reactions would have occurred on the surface of iron sulfide minerals buried underground, rather than in a primordial soup. This leaves open the question of where such organic acids came from originally. One possibility is that they resulted from a Miller type synthesis, as described above.

More radical is the suggestion that the first organic molecules were derived directly from carbon monoxide plus hydrogen sulfide. It has been demonstrated that a mixed FeS/NiS catalyst can convert carbon monoxide (CO) plus methane thiol (CH_3SH) into a thioester ($CH_3\text{-}CO\text{-}SCH_3$), which then hydrolyses into acetic acid. Inclusion of catalytic amounts of selenium allows conversion of CO plus H_2S alone to CH_3SH (and then to the thioester and acetic acid). Recently it was shown that carbon monoxide (CO) activated by the same mixed FeS/NiS catalyst can also drive the formation of peptide bonds between alpha-amino acids in hot aqueous solution. Not surprisingly, this system will also hydrolyze polypeptides.

The use of such transition metal catalysts offers a way round the chicken-and-egg paradox of the origin of enzyme activity: How could the monomers (amino acids or nucleotides) have been available before the enzymes (proteins or ribozymes) that catalyze their synthesis? A variety of transition metal elements (e.g., Cu, Fe, Ni, Mo) and small organic molecules to complex with would have been available in the subsurface, especially the hydrothermal vents in the ocean.

5. Evolution of DNA, RNA, and Protein Sequences

Consider the genes of an ancient ancestral organism. Over millions of years, mutations will occur in the DNA sequences of its genes at a slow but steady rate (see Ch. 23). Most mutations will be selected against because they are detrimental, but some will survive. Most mutations that are incorporated permanently into the genes will be neutral mutations with no harmful or beneficial effects on the organism. Occasionally, mutations that improve the function of a gene and/or the protein encoded by it will occur, although these are relatively rare. Sometimes a mutation that was originally harmful may turn out to be beneficial under new environmental conditions.

The actual function of the protein matters the most, not the exact sequence of the gene. If the protein can still operate normally, a mutation in the gene may be acceptable. Many of the amino acids making up a protein chain can be varied, within reasonable limits, without damaging the function of the protein too much. Replacement of one amino acid by a similar one (i.e., a conservative substitution—see Ch. 23) will rarely abolish the function of a protein. If we compare the sequences of the same protein taken from many different modern-day organisms we will find that the sequences can be lined up and are very similar. For example, the α chain of hemoglobin is identical in humans and chimpanzees. Yet, 13% of the hemoglobin amino acids are different in pigs compared with humans, 25% are different in chickens, and 50% are different in fish. Nonetheless, all of these proteins are functional. This divergence in sequence correlates with other estimates of evolutionary relatedness. Figure 26.09 shows an example of one of these alignments. It shows a highly-conserved, iron-binding site found in a related family of enzymes—a group of alcohol dehydrogenases found in microorganisms—that use iron in their active site mechanism.

It is possible, then, to construct an evolutionary tree using a set of sequences for a protein as long as it is found in all the creatures being compared. The α chain of hemoglobin is only found in our blood relatives. In contrast, cytochrome c is a protein involved in energy generation in all higher organisms, including plants and fungi. It even has recognizable relatives in many bacteria. A cytochrome c tree is shown in Fig. 26.10. Humans and fish differ in amino acid sequence by only 18% for cytochrome c. From humans to either plants or fungi gives about 45% divergence.

Margin notes:

Organic matter can be generated from simple gas molecules using metallic catalysts.

Due to mutation, the sequences of DNA and the encoded proteins will gradually change over long periods of time.

Related organisms contain genes and proteins with related sequences. These may be used to construct evolutionary trees.

```
ECO    NPVARERVHS    AATIAGIAFA    NAFLGVCHSM    AHKLGSQFHI    PHGLANALLI    CHNVIRYNAND
SALM   NPVARERLHS    AAYIAPIAFA    NAFLGVCHWM    AHKLGSQFHI    PHGPFNARYR    HSVRR    AQS
CLOE   NEKAREKMAH    ASTMAGMASA    NAFLGLCHSM    AIKLSSEHNI    PSGIANALLI    EEVIRKFNAVD
CLOB   E  AREQMHY    AQCLAGMAFS    NALLGICHSM    AHKTGAVFHI    PHGCANAIYL    PYVIKFNSKT
ZYMMO  DMPAREAMAY    AQFLAGMAFN    NASLGYVHAM    AHQLGGYYNL    PHGVCNAVLL    PHVLAYNASV
ENTH   DLEAREKMHN    AATIAGMAFA    SAFLGMDHSM    AHKVGAAFHL    PHGRCVAVLL    PYHVIRYNGQ
YEAST  DKKARTDMCY    AEYLAGMAFN    NASLGYVHAL    AHQLGGFYHL    PHGVCNAVLL    PHVQ    EANM
PRD    D  AGEEMAL    GQYVAGMGFS    NVGLGLVHGM    AHPLGAFYNT    PHGVANAILL    PHVMRYNADF
GLD    EKCEQTFKY     GK            LAYESVK       AKVVTPALE     AVVEANTL      LSGLGFESGG
```

IRON-BINDING SITE

ECO = *E. coli* alcohol dehydrogenase
SALM = *Salmonella typhimurium* alcohol dehydrogenase
CLOB = *Clostridium acetobutylicum* butanol dehydrogenase
CLOE = *C. acetobutylicum* alcohol dehydrogenase
ZYMMO = *Zymomonas mobilis* alcohol dehydrogenase II
ENTH = *Entamoeba histolytica* alcohol dehydrogenase
YEAST = Yeast alcohol dehydrogenase IV
PRD = *E. coli* propanediol dehydrogenase
GLD = *Bacillus* glycerol dehydrogenase (a zinc enzyme)

■ **FIGURE 26.09**
Alignment of Related Sequences

Amino-acid sequences of related polypeptides are given in one letter code. The conserved iron-binding site is shown in bold. In particular, the iron atom is bound by the two conserved histidine (H) residues. The glycerol dyhydrogenase from *Bacillus* is related to the other members of this protein family but no longer uses iron and, as can be seen, the iron-binding sequence has diverged and both histidines have been replaced by other amino acids.

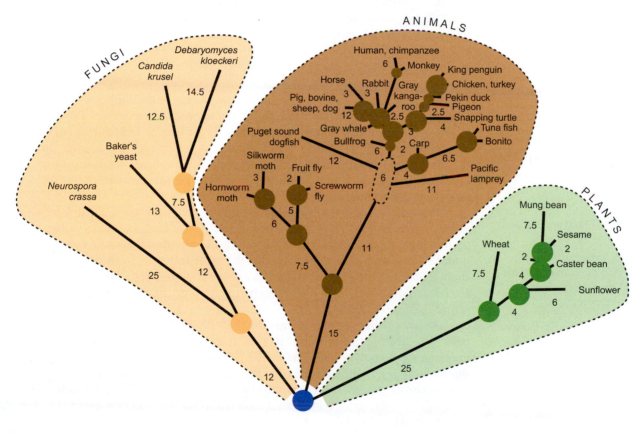

FIGURE 26.10
Evolutionary Tree based on Cytochrome c

All three kingdoms, fungi, animals, and plants, have the gene for cytochrome c, which is involved in production of energy. This phylogenetic tree was constructed by comparing the amino-acid sequence of cytochrome c from each of these organisms. The closer the branches are together, the more similar the sequences. The numbers along the branches represent the number of amino-acid differences.

However, plants and fungi also differ by 45%, which tells us that, by this measure, plants have diverged as far from fungi as animals have from plants.

Individual mutations may revert and restore the ancestral sequence of a gene or protein at a particular location. However, genes almost never mutate backwards to resemble the ancestors they diverged from many mutations ago. This is essentially a

FIGURE 26.11
Duplication Creates New Genes

During evolution, the entire ancestral globin gene was duplicated. The two copies were then able to diverge independently of each other. The first gene developed into hemoglobin, which is only found in red blood cells. The second copy developed into myoglobin, which is found in muscle tissue. Both proteins still carry oxygen, but they have tissue specificity.

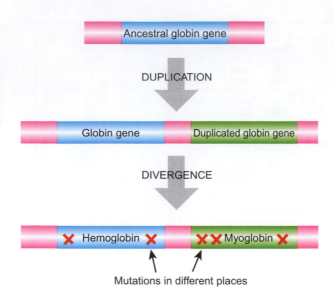

Mutations in different places

matter of probability. There is nothing forbidding any particular mutation to revert to the original sequence, but if this change reduces the function of the protein, it will be lost due to selection pressure.

5.1. Creating New Genes by Duplication

How do new genes arise? The standard way is via gene duplication. As discussed in Chapter 23, mutations may cause the duplication of a segment of DNA that carries a whole gene or several genes. The original copy must be kept for its original function but the extra copy is free to mutate and may be extensively altered. In most cases the mutations that accumulate will inactivate the duplicate copy. Less often, the extra copy will remain active and be altered so as to perform a related but different function from the original copy.

Duplication followed by sequence divergence may result in a family of related genes that carry out related functions. One of the best examples is the **globin** family of genes. Hemoglobin carries oxygen in the blood, whereas myoglobin carries it in muscle. These two proteins have much the same function, have similar 3D shapes, and their sequences are related. After the ancestral globin gene duplicated, the two genes for hemoglobin and myoglobin slowly diverged as they specialized to operate in different tissues (Fig. 26.11).

The actual hemoglobin of mammalian blood has two alpha (α)-globin and two beta (β)-globin chains forming an $\alpha2/\beta2$ tetramer, unlike myoglobin, which is a monomer of a single polypeptide chain. The α-globin and β-globin were derived by further duplication of the ancestral hemoglobin gene. In addition, the ancestral α-globin gene split again to give modern α-globin and zeta (ζ)-globin. The ancestral β-globin gene split again, twice, to give modern β-globin and the gamma (γ)-, delta (δ)-, and epsilon (ϵ)-globins (Fig. 20.12).

These globin variants are used during different stages of development. At each stage, the hemoglobin tetramer consists of two α-type and two β-type chains. The ζ-globin and ϵ-globin chains appear in early embryos, which possess $\zeta2/\epsilon2$ hemoglobin. In the fetus, the ϵ-chain is replaced by the γ-chain and the ζ-chain is replaced by the α-chain, so giving $\alpha2/\gamma2$ hemoglobin. A fetus needs to attract oxygen away from the mother's blood, so the $\alpha2/\gamma2$ hemoglobin binds oxygen better than the adult $\alpha2/\beta2$ hemoglobin (Fig. 26.12).

Duplication of segments of DNA creates a supply of new genes.

The steady accumulation of mutations in duplicated genes results in families of genes with related sequences.

globins   Family of related proteins, including hemoglobin and myoglobin, that carry oxygen in the blood and tissues of animals

GLOBIN FAMILY TREE

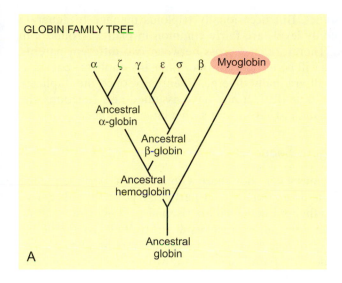

A

FETAL HEMOGLOBIN IS BETTER

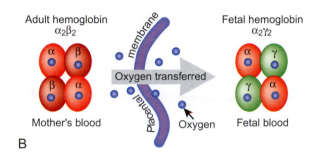

B

A) Over the course of evolution, a variety of gene duplication and divergence events gave rise to a family of closely-related genes. The first ancestral globin gene was duplicated giving hemoglobin and myoglobin. After another duplication, the hemoglobin gene diverged into the ancestral α-globin and ancestral β-globin genes. Continued duplication and divergence created the entire family of globin genes. B) The different members of the hemoglobin family are adapted for specific functions during development. Thus, the fetus uses two α chains and two γ chains to form its hemoglobin tetramer. This form is able to extract the oxygen from the mother's blood because it has a higher affinity for oxygen than the adult form of hemoglobin.

The globin genes are an example of a **gene family**, a group of closely-related genes that arose by successive duplication. The individual members are obviously related in their sequences and carry out similar roles. During evolution, continued gene duplication may give rise to multiple new genes whose functions steadily diverge until their ancestry may be difficult to recognize; this gives a **gene superfamily**. The genes of the immune system provide good examples of gene families and superfamilies.

In eukaryotes, retro-elements that encode reverse transcriptase are relatively common (see Ch. 22). Consequently, occasional reverse transcription of cellular mRNA molecules may occur. This gives a complementary DNA copy that may be integrated into the genome. This results in a duplicate copy of the gene, although this lacks the introns and promoter of the original gene. Such inactive copies are known as pseudogenes and usually accumulate mutations that inactivate the coding sequence. Rarely, a pseudogene may end up next to a functional promoter and be expressed. This gives a duplicate functional copy of the original gene that may be altered by mutation as already discussed.

Rare mistakes during cell division may result in the whole genome being duplicated. In particular, errors in meiosis may give diploid gametes. Fusion of two diploid gametes would give a tetraploid zygote and hence a tetraploid individual. More likely is the formation of a triploid individual by fusion of one mutant diploid gamete plus one normal haploid gamete. Most triploids are sterile, as they give gametes with

Reverse transcriptase may generate duplicate genes that lack introns and promoters and are located far away from the original copy.

Occasionally whole genomes may be duplicated.

gene family Group of closely-related genes that arose by successive duplication and perform similar roles
gene superfamily Group of related genes that arose by several stages of successive duplication. Members of a superfamily have often diverged so far that their ancestry may be difficult to recognize

incorrect numbers of chromosomes. But occasionally triploids manage to generate tetraploid progeny. Aberrant ploidy levels are fairly common in plants. Around 5 in 1000 plant gametes are diploid. Therefore, in a cross between two different parents, approximately 2.5 in 10^{-5} zygotes will be tetraploid. Over time, the duplicate copies of genes in a tetraploid organism will gradually diverge. Eventually, once the duplicate copies have diverged far enough to be distinct and have assumed new functions, the organism will effectively become "diploid" again, but with a larger genome.

5.2. Paralogous and Orthologous Sequences

Sequences are said to be **homologous** when they share a common ancestral sequence. If several organisms each contain single copies of a particular gene that all derive from the same common ancestor then sequence comparison should give an accurate evolutionary tree. However, gene duplication may result in multiple copies of the same gene within a single organism. These alternatives are illustrated in Figure 26.13. **Orthologous** genes are those found in separate species and which diverged when the organisms containing them diverged. **Paralogous** genes are multiple copies due originally to gene duplication within a single organism. Later, divergence of organisms

FIGURE 26.13
Paralogous and Orthologous Sequences

Homologous genes are genes that derive from one ancestral gene found in an ancestral organism. If that gene duplicates within the ancestral organism and one evolves into a new function, these two genes (A and B) are now called paralogs. If after the duplication event, the ancestral organism speciated into human and chimp, then gene A of chimps and gene A of humans are considered orthologs, and gene B of chimp and gene B of human are also orthologs. In this evolutionary tree, another duplication of gene B occurred in chimps and humans. The B' and B'' genes of chimp or human are within the species, and therefore, these are also paralogs.

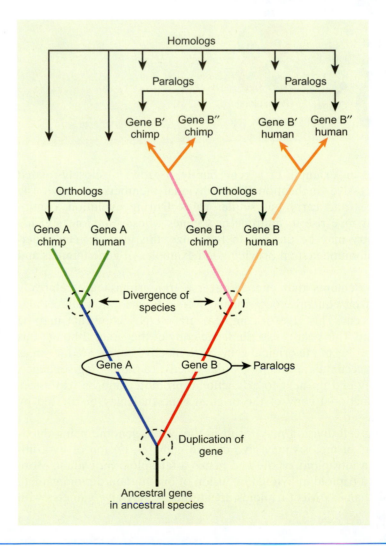

homologous   Related in sequence to an extent that implies common genetic ancestry
orthologous genes   Homologous genes that are found in separate species and which diverged when the organisms containing them diverged
paralogous genes   Homologous genes that are located within the same organism due to gene duplication

containing paralogs will generate many, somewhat confusing, pairs of paralogs as shown in Figure 26.13.

To generate an accurate evolutionary tree, orthologous genes must be compared. For example, we must compare the sequences of α-globin from one animal with the orthologous α-globin from another and not with the paralogous β-globin. Since paralogous sets of genes have similar sequences, these may cause confusion unless their correct ancestry is known. In particular, we need to know whether or not an organism contains multiple sequences derived from the same ancestor. Suppose that, due to partial information, we only knew of α-globin from pigs and β-globin from dogs. If we were unaware of the presence of other members of the globin family in these organisms we might compare these two sequences as if they were orthologous. This would lead to an incorrect relationship.

> Orthologs are homologous genes in two different species that are derived from the same ancestral gene in the genome of the common ancestor of the two species.

> Paralogs are copies originally derived from duplication of an ancestral gene within a single species.

5.3. Creating New Genes by Shuffling

Another way to create new genes is by using pre-made modules. Segments from two or more genes may be fused together by DNA rearrangements so generating a novel gene consisting of regions derived from several sources (Fig. 26.14A). An example of the formation of a new gene from several diverse components is the LDL receptor (Fig. 26.14B). LDL is low-density lipoprotein that carries cholesterol around in the blood. The LDL receptor is found on the surface of cells that take up LDL. The gene for the LDL receptor consists of several regions, two of which are derived from other genes. Towards the front, there are seven repeats of a sequence also appearing in the C9 factor of complement, an immune system protein. Farther along is a segment related to part of epidermal growth factor (a hormone). When such a gene mosaic is transcribed and translated, we get a patchwork protein consisting of several different domains that provides a new function for the organism. In addition to duplication and re-shuffling new proteins may form both by minor gene modifications and, occasionally, by the emergence of a completely novel sequence (see Focus on Relevant Research).

> New genes may be made by shuffling of modular segments of DNA.

PRINCIPLE OF MODULAR EVOLUTION

A

LDL RECEPTOR - AN EXAMPLE OF MODULAR EVOLUTION

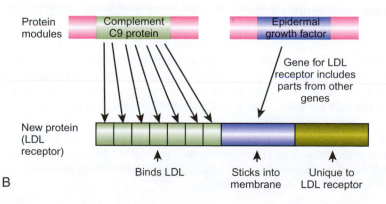

B

FIGURE 26.14

Novel Gene Derived from Pre-Existing Modules

A) Modular evolution of a new gene may involve the fusion of separate gene modules or functional units. For example, the purple module of gene 1 may provide a function that complements the blue module of gene 2. If these two domains fuse they might then form a novel but functional gene. B) The LDL receptor has domains found in several other proteins. The first part of the LDL receptor gene has seven repeated modules or functional units also found in the C9 complement factor of the immune system. The next module allows the protein to bind the cell membrane. This module is very similar to a portion of the epidermal growth factor receptor. Finally, the LDL receptor has a module that is unique.

Bornberg-Bauer E, Huylmans AK, and Sikosek T (2010) How do new proteins arise? Current Opinion Struct Biol 20:390–396.

FOCUS ON RELEVANT RESEARCH

The emergence of novel proteins is a complex matter. The recent availability of large amounts of comparative sequence data has shed new light on this issue. Gene duplications and re-shuffling can create new genes, and consequently new proteins. However, in addition new proteins often arise by constant minor modifications to the genes that encode them, such as extension by the recruitment and modification of short segments of nearby DNA.

Sometimes, protein creation is more radical and some genes have emerged from random sequences of non-coding DNA. Surveys of novel genes in *Drosophila* suggest that about 50% are due to duplications, 30% from re-shuffling and/or extension, about 10% from transposition events (especially retrotransposition), and about 10% from non-coding sequences. Cryptic signals for transcription start sites and sequences close to consensus for other regulatory roles are often found in non-coding regions. These can occasionally be activated and the result is a novel transcript. This may originally be an RNA gene with no protein product. Coding sequences may be generated later by a series of minor alterations.

6. Different Proteins Evolve at Very Different Rates

> The sequences of different genes and proteins evolve at different rates. Less critical sequences are free to evolve faster.

Obviously, we should not rely on a single protein to build an evolutionary tree. If we make trees for several proteins, we often get rather similar evolutionary relationships. However, different proteins evolve at different speeds. As noted above, humans and fish differ by 50% in the α chain of hemoglobin but by less than 20% in their cytochrome c. If we plot the number of amino acid changes versus the evolutionary time scale (Fig. 26.15), we can see this easily for cytochrome c (slow), hemoglobin (both α and β chains evolve at medium speed), and fibrinopeptides A and B (rapid evolution).

Table 26.05 gives the evolutionary rates for an assortment of proteins. Fibrinopeptides are involved in the blood-clotting process. They need an arginine at the end and must be mildly acidic overall. Apart from this they can vary widely as there are so few constraints on what is needed. In contrast, histones bind to DNA and are responsible for its correct folding. Almost all changes to a histone would be lethal for the cell, so they evolve extremely slowly.

Cytochrome c is an enzyme whose function depends most critically on a few amino acid residues at the active site, which bind to its heme co-factor. Consequently, these active site residues rarely vary, even though amino acids around them change. Of 104 residues, only Cys-17, His-18, and Met-80 are totally invariant. In other places variation is low; large, non-polar, amino-acid residues always fill positions 35 and 36. Several cytochrome c molecules have been examined by X-ray crystallography and all have the same 3D structure. Although cytochrome c molecules may vary by as many as 88% of their residues, they retain the same 3D conformation. Thus, little variation is seen with the amino acids that are essential to the function or structure of cytochrome c.

Insulin is a hormone that evolves at much the same rate as cytochrome c. Insulin consists of two protein chains (A and B) encoded by a single insulin gene. During protein synthesis, a long pro-insulin molecule is made. The central C-peptide is cut out and discarded. Disulfide bonds hold the A and B chains together. Since the C chain is not part of the final hormone, it is free to evolve much faster and it changes at almost 10 times the rate of the A and B chains. Notice that all these proteins maintain their critical residues throughout evolution. It is important to note that mutations are random. A mutation is just as likely to occur in the A, B, and C chains of the insulin gene. Mutations that do occur in the A and B chains are most likely detrimental to the organism; consequently, these mutations are rarely passed on to successive generations. On the other hand, mutations in the C chain do not alter insulin function and are much more likely to get passed on to the progeny.

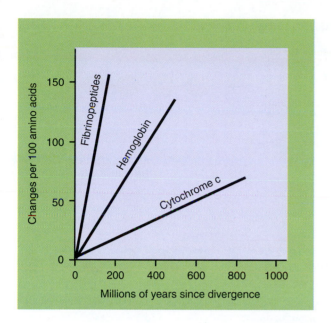

■ **FIGURE 26.15**
Rates of Protein Evolution

During the course of evolution, some proteins accumulate more mutations than others. The cytochrome c gene is very stable, and only 50 changes/100 amino acids have occurred in 800 million years. Fibrinopeptides A and B, on the other hand, have accumulated 50 changes/100 amino acids in less than 100 million years.

TABLE 26.05	Rates of Evolution for Different Proteins
Protein	**Rate of Evolution**
Neurotoxins	110–125
Immunoglobulins	100–140
Fibrinopeptide B	91
Fibrinopeptide A	59
Insulin C peptide	53
Lysozyme	40
Hemoglobin α chain	27
Hemoglobin β chain	30
Somatotropin	25
Insulin	7.1
Cytochrome c	6.7
Histone H2	1.7
Histone H4	0.25

The rate of evolution is given as the number of mutations per 100 amino acid residues per 100 million years.

6.1. Molecular Clocks to Track Evolution

A rapidly-evolving protein will eventually become so altered in sequence between diverging organisms that the relationship will no longer be recognizable. Conversely, a protein that evolves very slowly will show little or no difference between two different organisms. Therefore, we need to use slowly-changing sequences to work out distant evolutionary relationships and fast-evolving sequences for closely-related organisms.

Most human proteins have identical sequences to those of the closely-related chimpanzee. Even if we examine the rapidly-evolving fibrinopeptides, humans and chimpanzees end up on the same branch of the evolutionary tree. So how can we tell people apart from chimps? Mutations that do not affect the sequence of proteins accumulate much faster during evolution, since they have little or no detrimental

Rapidly-evolving sequences reveal the relationships between closely-related organisms. Slowly-changing sequences are needed to compare distantly-related organisms.

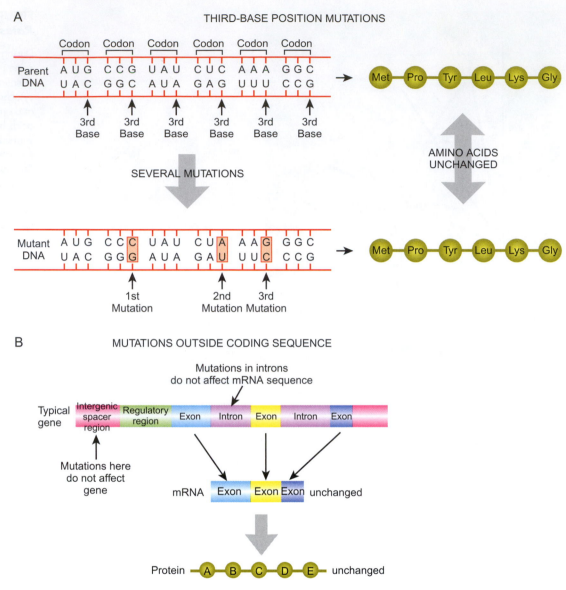

FIGURE 26.16
Non-Coding DNA Evolves Faster

(A) During evolution mutations in the third position of the triplet codon rarely alter the amino acid sequence of the protein; therefore, they are rarely deleterious. B) Non-coding DNA regions often have no obvious function and so many mutations accumulate within these regions. Mutations in intragenic spacer regions or in introns have no effect on protein sequence or function.

effect. So if instead of protein sequences we look at the DNA sequences of very closely-related organisms we find many more differences. These are located mainly in non-coding regions or in the third codon position. As discussed in Chapter 13, changing the third base of most codons does not alter the encoded amino acid and so leaves the encoded protein unaltered (Fig. 26.16).

Introns are non-coding sequences that are spliced out of the primary transcript and so do not appear in the mRNA (see Ch. 12). Intron sequences are therefore not represented in the final protein. Apart from the intron boundaries and splice-recognition sites, the DNA sequence of an intron is free to mutate extensively. Other non-coding sequences exist between genes and, if not involved in regulation, they are also relatively free to mutate.

The early data on cytochrome c, hemoglobin, etc., were obtained by direct sequencing of proteins. Since DNA sequencing is much easier and more accurate, most recent protein sequences are deduced from the DNA sequence. We now have a vast amount of DNA information on closely-related animals. This helps fine-tune the evolutionary relationship between animals such as humans and chimpanzees.

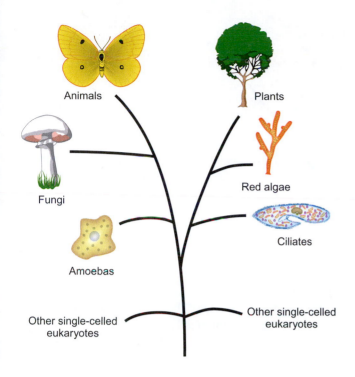

■ **FIGURE 26.17**

***Eukaryote Kingdoms
Based on Ribosomal RNA
Sequences***

Comparing the ribosomal RNA sequences has allowed scientists to deduce how closely the major divisions of eukaryotic organisms are related. A variety of single-celled eukaryotes were the earliest groups to branch from the ancestral eukaryotic lineage. The relative positions of many of these groups is still undecided. As can be seen, the fungi are more closely related to animals than to plants. The branch leading to higher plants includes several groups of algae, including the red algae, as shown here.

6.2. Ribosomal RNA—A Slowly Ticking Clock

A major problem is how to construct an evolutionary tree that includes all living organisms and shows the relationships between all the main groups of organisms. To achieve this, first we need a molecule that is present in all organisms. Second, the chosen molecule must evolve slowly so as to still be recognizable in all the major groups of living things.

Although histones evolve very slowly, only eukaryotic cells possess them; they are missing from bacteria. The solution is to use ribosomal RNA. In practice, the DNA of the genes that encode the RNA of the small ribosomal subunit (16S or 18S rRNA) is sequenced and the rRNA sequence is then deduced. All living organisms have to make proteins and they all have ribosomes. Furthermore, since protein synthesis is so vital, ribosomal components are highly constrained and evolve slowly. The only group excluded is the viruses, which have no ribosomes. (Whether viruses are truly alive is debatable and their evolutionary origins are still controversial; see Ch. 21.)

Use of relationships based on ribosomal RNA has allowed the creation of large-scale evolutionary trees encompassing all the major groups of organisms. Higher organisms traditionally consist of three main groups—plants, animals, and fungi (Fig. 26.17). Analysis of rRNA indicates that the ancestral fungus was never photosynthetic but split off from the plant ancestor before the capture of the chloroplast. Despite traditionally being studied by botanists, fungi are actually more closely related to animals than plants. Varieties of single-celled organisms sprout off the eukaryotic tree near the bottom and do not fall into any of the three major kingdoms.

Most eukaryotic cells contain mitochondria and, in addition, plant cells contain chloroplasts. These organelles are derived from symbiotic bacteria and contain their own ribosomes (see below). The rRNA sequences of mitochondria and chloroplasts reveal their relationship to the bacteria. Relationships among eukaryotes, like those shown in Figure 26.17, are therefore made by using the rRNA of the ribosomes found in the cytoplasm of eukaryotic cells. These ribosomes have their rRNA encoded by genes in the cell nucleus.

When an rRNA-based tree is made that includes both prokaryotes and eukaryotes it turns out that life on Earth consists of three lineages (Fig. 26.18). These three

Ribosomal RNA sequences have been used to construct a global evolutionary tree.

Sequence analysis indicates that fungi are closer to being immobile animals than non-photosynthetic plants.

Sequencing of rRNA reveals that chloroplasts and mitochondria are related to the bacteria.

FIGURE 26.18
The Three Domains of Life

All of today's organisms belong to one of three main divisions based on relationships among ribosomal RNA: the Eubacteria, the Archaea (or Archaebacteria), and the eukaryotes. Mitochondria and chloroplasts have rRNA that most closely resembles Eubacteria.

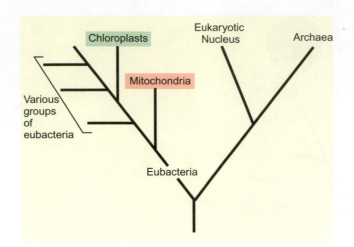

Life consists of three domains— the Eubacteria, the Archaea, and the eukaryotes.

domains of life are the **Eubacteria** ("true" bacteria, including the mitochondria and chloroplasts), the **Archaea** or **Archaebacteria** ("ancient" bacteria), and the eukaryotes. There is as much difference between the two genetically-distinct types of prokaryote as between prokaryotes and eukaryotes. Sequencing of organelle rRNA indicates that mitochondria and chloroplasts belong to the eubacterial lineage.

One bizarre aspect of classifying life forms by rRNA is that the organism itself is not needed. A sample of DNA containing the genes for 16S rRNA is sufficient. Although many microorganisms present in the sea or in soil have never been successfully cultured, DNA can be extracted from soil or seawater directly. Using PCR (see Ch. 6) it is possible to amplify this DNA and get enough of the 16S rRNA gene to obtain a sequence. Several new groups of Eubacteria and Archaea have been discovered by this method, despite never being cultured.

6.3. The Archaea versus the Eubacteria

Genetically, the Archaea are somewhat closer to the eukaryotes than to the Eubacteria.

Both Eubacteria and Archaea have microscopic cells without a nucleus. They both have single circular chromosomes and divide in two by simple binary fission. In short, they both conform to the definition of a prokaryotic cell. Until sequence analysis of ribosomal RNA was performed, there was no obvious reason to suspect from their physical structure that they were so radically different.

Of the two groups of prokaryotes, the Archaea are more closely-related genetically to eukaryotes (i.e., to their nuclear genome). In many Archaea, DNA is packaged by histone-like proteins that show sequence homology to the true histones of higher organisms. In addition, the details of protein synthesis and the translation factors of Archaea resemble those of eukaryotes, rather than Eubacteria. Such similarities suggest that the primeval eukaryote evolved from an archaeal-type ancestor.

Archaea differ biochemically from Eubacteria in several other major respects. Archaea have no peptidoglycan and their cytoplasmic membrane contains unusual lipids, which are made up from C5 isoprenoid units rather than C2 units as with normal fatty acids (Fig. 26.19). Moreover, the isoprenoid chains are attached to glycerol by ether linkages instead of esters. Some double-length isoprenoid hydrocarbon chains stretch across the whole membrane.

Archaea tend to be found in bizarre environments and many of them are adapted to extreme conditions. Most Archaea fall into two major divisions, the Crenarchaeota, which are more basal and share rather more characteristics with the eukaryotes, and

Archaea   New name for Archaebacteria, one of the three domains of life
Archaebacteria   One of the three domains of life comprising the "ancient" bacteria
Eubacteria   One of the three domains of life comprising the "true" bacteria, including the organelles

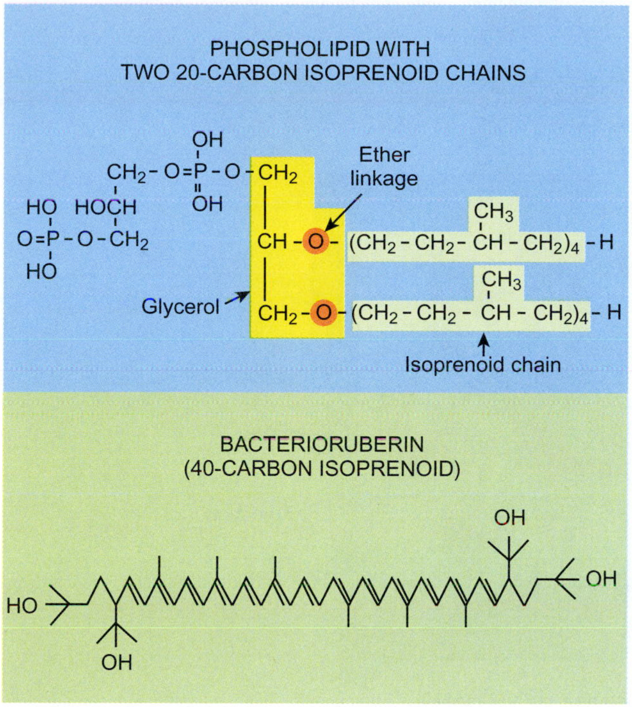

■ **FIGURE 26.19**
Unusual Lipids of Archaea

The Archaea have lipid chains made of 5-carbon isoprenoid units rather than 2-carbon units as seen in Eubacteria. The isoprenoid chains are linked to a glycerol via an ether link rather than an ester link. In some instances the isoprenoid lipid chains may contain 40 carbons (bacterioruberin, for example). These longer lipids span the whole membrane of the Archaea.

the Euryarchaeota. Most Crenarchaeota are thermophilic and/or possess a sulfur-based metabolism. An example is *Sulfolobus,* which lives in geothermal springs, grows best at a pH of 2–3 and a temperature of 70–80°C, and oxidizes sulfur to sulfuric acid. The Euryarchaeota are more varied and include two distinct thermophilic lineages, two groups of methane-generating bacteria, and the Halobacteria. The latter are found in the super salty Dead Sea and Great Salt Lake. They are extremely salt-tolerant and grow in up to 5 M NaCl but will not grow below 2.5 M NaCl (sea water is only 0.6 M). They respire normally but many also trap energy from sunlight using bacterio-rhodopsin, which is related to the rhodopsin used as a photo-detector in animal eyes.

7. Symbiotic Origin of Eukaryotic Cells

Unlike the cells of prokaryotes, eukaryotic cells are divided into compartments by membranes (Fig. 26.20). The most important compartment is the nucleus, where the chromosomes reside. Eukaryotes are defined by the possession of a nucleus that typically contains several linear chromosomes. Higher eukaryotes are normally diploid and have pairs of homologous chromosomes, although this is not always the case for less advanced eukaryotes.

In addition to a nucleus, almost all eukaryotic cells (animal, plant, or fungus) contain mitochondria. Plant cells contain chloroplasts as well as mitochondria. Both of these organelles provide the majority of energy for all cellular processes. These organelles contain their own genomes and encode at least a few of their own proteins. The organelle genome is prokaryotic in nature. It consists of a circular DNA molecule that is not bound by histones. Mitochondria and chloroplasts synthesize their own ribosomes, which are more closely related to those of bacteria than to those of the eukaryotic cytoplasm. Both mitochondria and chloroplasts are roughly the same size and shape as bacterial cells, and the organelles grow and divide in the same manner as bacteria (Fig. 26.21). When a eukaryotic cell divides, each daughter cell inherits some of its parent's mitochondria and chloroplasts. If either organelle is lost, it cannot be reconstructed because the nucleus does not have all of the genetic information needed to assemble the entire organelle.

Eukaryotic cells are divided into compartments, including the nucleus, by membranes.

Eukaryotes are derived from the merger of two or more ancestral organisms.

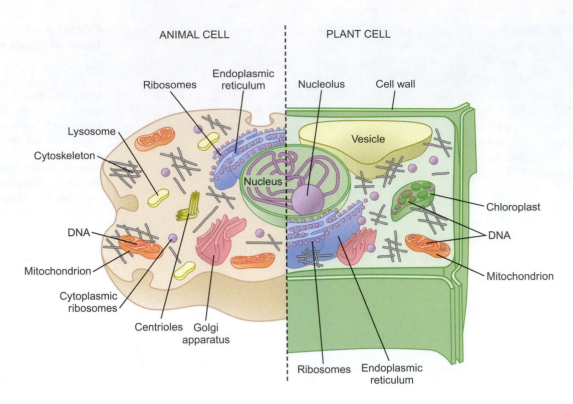

ANIMAL CELL   PLANT CELL

FIGURE 26.20
Defining Features of Eukaryotic Cells

Eukaryotic cells have membrane bound compartments that are not found in prokaryotes. In a typical animal cell, these compartments include the nucleus, mitochondria, endoplasmic reticulum, golgi apparatus, and lysosomes. In addition a typical plant cell also has chloroplasts, which harvest light energy and convert it into ATP. Animal cells maintain their 3D shape with an internal cytoskeleton composed of microtubules and microfilaments. In contrast, plant cells maintain their shape by a rigid cell wall surrounding the cytoplasm.

FIGURE 26.21
Chloroplasts arise by Division

Transmission electron micrograph of a dividing chloroplast in a bean seedling. Chloroplasts and mitochondria divide independently of the eukaryotic cell in which they reside. The organelles divide by binary fission in a manner reminiscent of prokaryotic cells. The plastid shown here is technically an "etioplast," a precursor chloroplast that has not yet developed any green pigment. (Credit: Biochemistry and Molecular Biology of Plants by Buchanan, Gruissem and Jones, 2000, American Society of Plant Physiologists.)

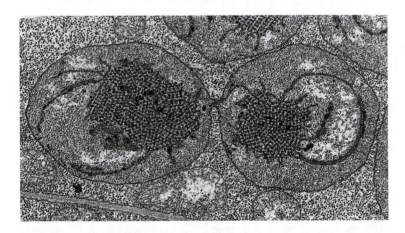

The **symbiotic theory** proposes that the complex eukaryotic cell arose by a series of symbiotic events in which organisms of different lineages merged. The cells of higher organisms are thus symbiotic associations. The word "**symbiosis**" is from the Greek word "sym," meaning together, and "bios," meaning life. The nuclear genes of eukaryotic cells are sometimes referred to as derived from the "**urkaryote**."

symbiotic theory   Theory that the organelles of eukaryotic cells are derived from symbiotic prokaryotes
symbiosis   Association of two living organisms that interact
urkaryote   Hypothetical ancestor that provided the genetic information of the eukaryotic nucleus

The urkaryote is the hypothetical ancestor that provided the genetic information found in the present-day eukaryotic nucleus.

According to the symbiotic theory, mitochondria are descended from bacteria that colonized the ancestors of modern eukaryotic cells. These bacteria received shelter and nutrients, and in return, devoted themselves to generating energy by respiration. During the following eons, these bacteria became narrowly specialized for energy production, lost the ability to survive on their own, and evolved into mitochondria. The term **endosymbiosis** is sometimes used for symbiotic associations where one partner is physically inside the other ("endo" is from the Greek for inside), as in the present case. Similarly, chloroplasts are descended from photosynthetic bacteria that took up residence in the ancestors of modern-day plants. The term **plastid** refers to all organelles that are genetically equivalent to chloroplasts, whether functional or not. Since fungi do not contain chlorophyll, the green light-absorbing pigment of plants, fungi were once thought to be degenerate plants that had lost their chlorophyll during evolution. However, fungi contain no trace of a plastid genome, meaning they were never photosynthetic, confirming the evolutionary tree based on rRNA sequences (see Fig. 26.17).

> Mitochondria are derived from ancestral bacteria that specialized in respiration, whereas chloroplasts are descended from ancestral photosynthetic bacteria.

7.1. The Genomes of Mitochondria and Chloroplasts

Both mitochondria and chloroplasts contain a genome consisting of a circular DNA molecule, presumably derived from the ancestral bacterial chromosome. Over evolutionary time, these organelle genomes lost many genes that were unnecessary for life as an organelle inside a host cell. In addition, many genes that are still necessary have been transferred to the chromosomes in the nucleus. Consequently, the mitochondria of animals have very little DNA left. For example, human mitochondrial DNA has only 13 protein-encoding genes, together with genes for several rRNA and tRNA molecules (Fig. 26.22). However, about 400 different proteins are present in mitochondria. The genes for most of these reside in the nucleus and these polypeptides are imported into the organelle after synthesis on the ribosomes of the eukaryotic cytoplasm.

> Chloroplasts and mitochondria possess small circular genomes.

The chloroplasts of higher plants retain rather more DNA than mitochondria—approximately enough for a hundred genes—but this is still much less than their bacterial ancestors. It is estimated that 1000 or more genes from the ancestral photosynthetic bacterium have been transferred to the plant cell nucleus.

> Defects due to mutations in mitochondrial genes are inherited maternally.

During sexual reproduction, mitochondria and chloroplasts are inherited maternally. When a sperm fertilizes an egg cell to create a zygote, the organelles of the sperm are lost. The new individual retains the organelles from the egg cell (i.e., those from the female parent only). Certain inherited defects of humans are due to mutations in the mitochondrial DNA. These affect the generation of energy by respiration and affect the function of muscle and nerve cells in particular. These defects are passed on through the maternal line because all children with the same mother inherit the same mitochondria.

Partial exceptions to the rule of maternal inheritance for organelles occur in a few single-celled eukaryotes. *Chlamydomonas* is a single-celled green alga whose cells contain a single chloroplast. During mating, about 5% of the zygotes receive two chloroplasts (one from each parent) rather than one. In these cells recombination can occur between the two different chloroplast genomes. Division of the zygote gives cells with only a single chloroplast each. These may be examined to determine the outcome of the genetic crosses.

---

endosymbiosis Form of symbiosis where one organism lives inside the other
plastid Any organelle that is genetically equivalent to a chloroplast, whether functional in photosynthesis or not

FIGURE 26.22
Genetic Map of Human Mitochondrial DNA

A) During evolution, the mitochondrial genome has been streamlined. Many of the genes necessary for mitochondrial function have moved to the nucleus, causing the mitochondrial genome to shrink in size. B) The mitochondrial DNA of humans contains the genes for ribosomal RNA, transfer RNA, and some proteins of the electron transport chain.

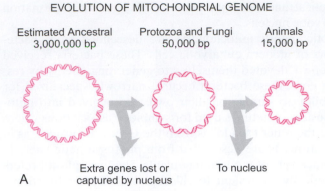

EVOLUTION OF MITOCHONDRIAL GENOME

Estimated Ancestral 3,000,000 bp   Protozoa and Fungi 50,000 bp   Animals 15,000 bp

Extra genes lost or captured by nucleus   To nucleus

A

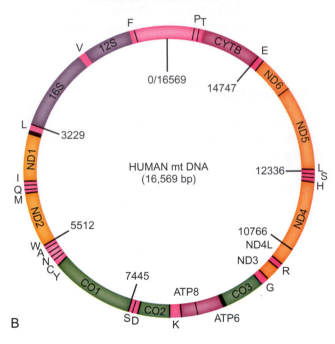

HUMAN MITOCHONDRIAL DNA

HUMAN mt DNA (16,569 bp)

B

7.2. Primary and Secondary Endosymbiosis

A symbiotic relationship where one organism lives inside the other is known as endosymbiosis. **Primary endosymbiosis** refers to the original internalization of prokaryotes by an ancestral eukaryotic cell, resulting in the formation of the mitochondria and chloroplasts. Two membranes surround mitochondria and chloroplasts. The inner one is derived from the bacterial ancestor and the outer "mitochondrial" or "chloroplast" membrane is actually derived from the host-cell membrane. However, several lineages of protozoans appear to have engulfed other single-celled eukaryotes, in particular algae. Several groups of algae therefore have chloroplasts acquired at second-hand by what is termed **secondary endosymbiosis**.

In contrast to the typical two membranes of primary organelles, four membranes surround chloroplasts obtained by secondary endosymbiosis. In most cases, the nucleus of the engulfed eukaryotic alga has disappeared without trace. Occasionally,

primary endosymbiosis   Original uptake of prokaryotes by the ancestral eukaryotic cell, giving rise to mitochondria and chloroplasts
secondary endosymbiosis   Uptake by an ancestral eukaryotic cell of another single-celled eukaryote, usually an alga, thus providing chloroplasts at second-hand

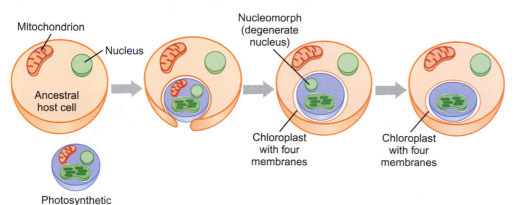

PRIMARY ENDOSYMBIOSIS

SECONDARY ENDOSYMBIOSIS

▸─ **FIGURE 26.23** ────────
Primary versus Secondary Endosymbiosis

Primary endosymbiosis yields organelles with two membranes. In this example, the original independent cyanobacterium has a cytoplasmic membrane, which is retained, and an outer membrane, which is lost during symbiosis. When the two cells associate, the host-cell cytoplasmic membrane surrounds the cyanobacterium, which is therefore left surrounded by two membranes. In contrast to primary endosymbiosis, secondary endosymbiosis occurs when an ancestral host cell engulfs a photosynthetic eukaryotic alga. The alga already has a chloroplast with two membranes as well as a nucleus and other organelles. Since the host cell only needs the energy from the chloroplast, the other captured organelles degenerate and eventually disappear. However, the membranes often remain and the chloroplast is left with four membranes, rather than two.

the remains of this nucleus are still to be found lying between the two pairs of membranes (Fig. 26.23). This structure is termed a **nucleomorph** and can be seen in cryptomonad algae where it represents the remains of the nucleus of a red alga that was swallowed by an amoeba-like ancestor. The nucleomorph contains three vestigial linear chromosomes totaling 550 kb of DNA. These carry genes for rRNA that is incorporated into a few eukaryotic type ribosomes that are also located in the space between the two pairs of membranes.

Cells resulting from secondary endosymbiosis are composites of four or five original genomes. These include the primary ancestral eukaryote nucleus and its mitochondrion, plus the nucleus, mitochondrion, and chloroplast from the secondary endosymbiont. Many genes from the subordinate genomes have been lost during evolution and no trace has ever been found of the secondary mitochondrion. Some genes from the secondary endosymbiont nucleus have been transferred to the primary eukaryotic nucleus. The protein products of about 30 of these are made on ribosomes belonging to the primary nucleus and shipped from the primary eukaryotic cytoplasm back into the nucleomorph compartment. In turn, the nucleomorph contains genes for proteins that are made on the 80S ribosomes in the nucleomorph compartment and transported across the inner two membranes into the chloroplast. Finally, there are proteins now encoded by the primary nucleus that must be translocated across both sets of double membranes from the primary cytoplasm into the chloroplast!

nucleomorph   Degenerate remains of the nucleus of a symbiotic eukaryote that was incorporated by secondary endosymbiosis into another eukaryotic cell

Box 26.1 Is Malaria Really a Plant?

Malaria is a disease that affects many millions of people world wide and is responsible for two or three million deaths each year, mostly in Africa. Malaria is caused by the single-celled eukaryote **Plasmodium**. The malaria parasite and other related single-celled eukaryotes are members of the phylum **Apicomplexa**. Although these parasites live inside humans and mosquitoes, far from the sunlight, they possess plastids as well as mitochondria. These plastids are degenerate, non-photosynthetic chloroplasts with a circular genome. In *Plasmodium* the plastid DNA is 35 kb and encodes rRNA, tRNA, and a few proteins, mostly involved in translation (Fig. 26.24).

The malarial plastid or "**apicoplast**" is thought to derive from secondary endosymbiosis. The ancestor of the Apicomplexa appears to have swallowed a single-celled eukaryotic alga that possessed a chloroplast. The algal nucleus has been completely lost, but the plastid

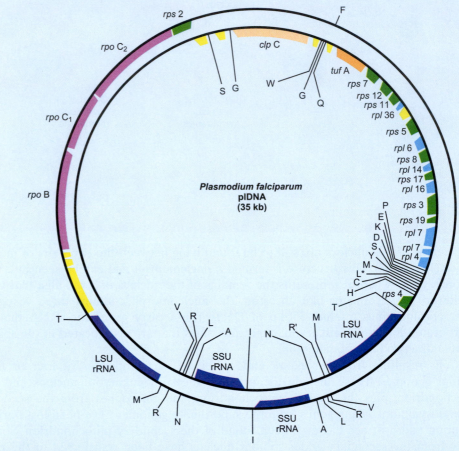

FIGURE 26.24
Plastid Genome of Plasmodium

The circular genome of the plastid of *Plasmodium* has genes for rRNA, tRNA, and protein synthesis. The tRNA genes are denoted by the single-letter amino-acid code, for example, S for the tRNA for serine.

Continued

Apicomplexa Phylum of parasitic single-celled eukaryotes, including malaria, which contain both mitochondria and degenerate non-photosynthetic chloroplasts
apicoplast Degenerate non-photosynthetic chloroplast found in members of the Apicomplexa
Plasmodium The malaria parasite, a protozoan belonging to the Apicomplexa

Box 26.1 Continued

was kept and is surrounded by four membranes. Sequence comparisons suggest the malarial apicoplast is most closely related to the chloroplast of red algae.

Although it does not convert light into energy, the apicoplast is essential for the survival of *Plasmodium*. The apicoplast plays a vital role in lipid metabolism. Several enzymes of fatty-acid synthesis are encoded in the nucleus but translocated into the apicoplast where fatty-acid synthesis occurs. As a result, certain herbicides that prevent fatty-acid synthesis in the chloroplasts of green plants are effective against *Plasmodium* and other pathogenic apicomplexans such as *Toxoplasma* and *Cryptosporidium*. For example, clodinafop targets the acetyl-CoA carboxylase and triclosan inhibits the enoyl ACP reductase of plants and bacteria. These herbicides have no effect on fatty-acid synthesis in animals or fungi. In addition, the herbicide fosmidomycin inhibits the isoprenoid pathway of plants and bacteria, which differs from that of animals. Fosmidomycin inhibits growth of *Plasmodium* and cures malaria-infected mice. *Plasmodium* and its relatives are also inhibited by chloramphenicol, rifamycin, macrolides, and quinolones, all of which are antibacterial antibiotics. These are also thought to act via the apicoplast.

8. DNA Sequencing and Biological Classification

Before the sequencing of DNA became routine, animals and plants were classified reasonably well, fungi and other primitive eukaryotes were classified poorly, and bacterial classification was a lost cause due to lack of observable characters. Using gene sequences for classification was developed for bacteria and has since spread to other types of organism. Nowadays ancestries may be traced by comparing the sequences of DNA, RNA, or proteins that are more representative of fundamental genetic relationships than are many superficial characteristics. Furthermore, in situations where division into species, genera, families, etc., is arbitrary, sequence data can provide quantitative measurements of genetic relatedness. Even if we cannot unambiguously define a species, we can be consistent in how much sequence divergence is needed to allocate organisms to different species or families.

Originally rRNA sequences were used for molecular classification. However as ever more sequence data is obtained, including whole genomes, an increasing number of other genes can be taken into account. Computer programs exist for calculating the relative divergence of the sequences and can generate trees such as that in Figure 26.25. Here, we have four bacteria, all in different genera but belonging to the same family, the Enterobacteria. To root such a tree correctly we also need the sequence from an organism in an "out-group;" in this case, the bacterium *Pseudomonas*, which is only distantly related to enteric bacteria. The nodes in Figure 26.25 represent the deduced common ancestors. The branch lengths are often scaled to represent the number of mutations needed and the numbers indicate how many base changes are needed to convert the sequence at each branch point into the next. (The total length of the 16s rRNA of Enteric bacteria is 1542 bases.)

Parasites are highly adapted to living at the expense of their hosts and have often lost many ancestral characteristics that they no longer need. Consequently, it is often very difficult to deduce phylogenetic relationships among parasites based on external traits. Fortunately, gene sequences can often be used to trace the ancestry of parasites or other aberrant life forms. For example, moles, which have adapted to living underground, have lost their eyes since these organs are no longer useful. Sometimes vestigial remnants remain even though the animal has no use for the structure. Whales have the atrophied remains of hind limbs, which imply that whales are not fish, but mammals that have become fish-like in general form by adapting to ocean life. Until gene sequencing emerged, it remained unknown what mammals are the whale's closest relatives. It now appears that whales are related to the artiodactyls, hoofed mammals such as hippos, giraffes, pigs, and camels.

Evolutionary trees are often constructed that show the number of sequence differences in rRNA.

Sequences are especially useful in classifying aberrant organisms that have lost structural characters normally used for comparison.

The ancestries of sequences can be confirmed if they share major insertions or deletions.

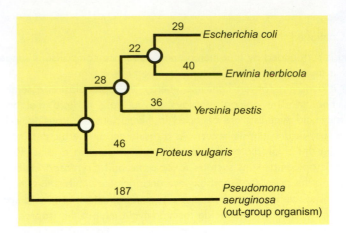

FIGURE 26.25
Phylogenetic Tree for Enteric Bacteria

The phylogenetic relationship between bacteria can be deduced by comparing ribosomal RNA sequences. Here, the sequences of the 16S rRNA genes are compared for the four enteric bacteria, *E. coli, Erwinia herbicola, Yersinia pestis,* and *Proteus vulgaris.* The relatively unrelated bacterium *Pseudomona aeruginaosa* is used as an out-group organism to provide the base or root of the tree. From these comparisons, it can be deduced that *P. vulgaris* was the first to branch from the primitive ancestor and that *E. coli* and *E. herbicola* were the latest.

One major problem with sequence comparison is that base changes can revert. Although statistical comparison of multiple sequences with many altered sites is often sufficient to establish a lineage, ambiguity sometimes remains. A useful way to help resolve ambiguities is by using conserved insertions or deletions—known as signature sequences or "indels." Although a single base insertion or deletion might possibly revert, the likelihood that an insertion or deletion of several bases might revert so as to exactly restore the original length and sequence is vanishingly small. Consequently, if a subgroup of a family of related sequences all contain an indel of defined length and sequence at the same location, they must all have been derived from the same ancestral sequence.

8.1. Mitochondrial DNA—A Rapidly Ticking Clock

Although mitochondria contain circular molecules of DNA reminiscent of bacterial chromosomes, the mitochondrial genome is much smaller. mtDNA of animals accumulates mutations much faster than the nuclear genes. In particular, mutations accumulate rapidly in the third codon position of structural genes and even faster in the intergenic regulatory regions. This means that mtDNA can be used to study the relationships of closely-related species or of races within the same species. Most of the variability in human mtDNA occurs within the D-loop segment of the regulatory region. Sequencing this segment allows us to distinguish between people of different racial groups.

One apparent drawback to using mitochondrial DNA is that mitochondria are all inherited from the mother. Although sperm cells do contain mitochondria, these are not released during fertilization of the egg cell and are not passed on to the descendants. On the other hand, analysis of mitochondria gives an unambiguous female ancestry, as complications due to recombination may be ignored. Furthermore, a eukaryotic cell contains only one nucleus but has many mitochondria so there are usually thousands of copies of the mitochondrial DNA. This makes extraction and sequencing of mtDNA easier from a technical viewpoint.

Mitochondrial DNA can sometimes be obtained from museum samples or extinct animals. Mitochondrial DNA extracted from frozen mammoths found in Siberia differed in four to five bases out of 350 from both Indian elephants and

Mitochondrial DNA changes fast enough to be used to classify subgroups within the same species.

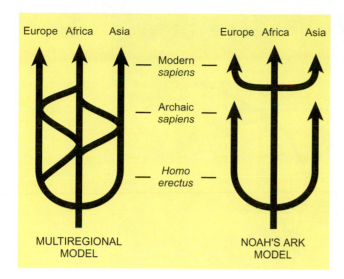

FIGURE 26.26
Multiregional and Noah's Ark Models of Human Evolution

The multiregional model of human evolution (left) suggests that *Homo sapiens* developed from multiple interactions between several ancestral lines. The early *Homo erectus* ancestor branched and migrated from Africa to Asia and Europe. Traits developing in each branch were transmitted to the other branches implying genetic exchanges between the three branches, even though many thousands of miles separated the early ancestral groups. The Noah's Ark model (right) seems more plausible based on genetic analysis. The model suggests that modern *Homo sapiens* developed from one ancestral group in Africa. Other branches of archaic *sapiens* did develop and inhabited different regions in Europe and Asia for a while before dying out. The modern *sapiens* branch has then evolved into several branches from a relatively-recent African ancestor.

African elephants. This DNA analysis supports the three-way split proposed based on anatomical relationships. The quagga is an extinct animal, similar to the zebra. It grazed the plains of Southern Africa only a little over a hundred years ago. A pelt preserved in a German museum has yielded muscle fragments from which DNA has been extracted and sequenced. The two gene fragments used were from the quagga mtDNA. The quagga DNA differed in about 5% of its bases from the modern zebra. The quagga and mountain zebra are estimated from this to have had a common ancestor about three million years ago.

8.2. The African Eve Hypothesis

Attempts to sort out human evolution from skulls and other bones led to two alternative schemes. The multiregional model proposes that *Homo erectus* evolved gradually into *Homo sapiens* simultaneously throughout Africa, Asia, and Europe. The Noah's Ark model proposes that most branches of the human family became extinct and were replaced, relatively recently, by descendants from only one local sub-group (Fig. 26.26). Although anthropologists take both theories seriously, few geneticists regard the multiregional model as plausible. This model implies continuous genetic exchange between widespread and relatively-isolated tribes over a long period of prehistory. Not surprisingly, recent molecular analysis supports the Noah's Ark model.

Although mitochondria evolve fast, the overall variation among people of different races is surprisingly small. Calculations based on the observed divergence and the estimated rates suggest that our common ancestor lived in Africa between 100,000 and 200,000 years ago. Since mitochondria are inherited maternally, this ancestor has been named "**African Eve.**" This African origin is supported by the deeper "genetic

Mitochondrial sequence analysis suggests all modern humans are derived from a small group of ancestors who lived in Africa around 100,000 years ago.

African Eve   Hypothetical female human ancestor thought to have lived in Africa around 100,000–200,000 years ago

FIGURE 26.27
African Eve Hypothesis I—DNA

This phylogenetic relationship was deduced by comparing mitochondrial DNA sequences from living humans. Numbers shown are estimated years before the present (BP). According to the African Eve theory, early humans developed in Africa about 150,000 years BP and diverged into many different tribal groups, most of which remained in Africa. The European and Asian races are derived from those relatively few groups of African ancestors who emigrated into Eurasia via the Middle East.

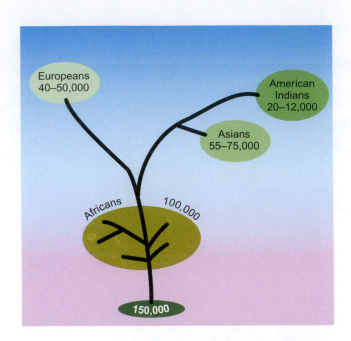

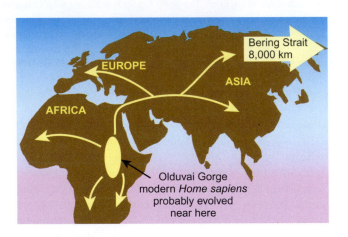

FIGURE 26.28
African Eve Hypothesis II—Migrations

The divergence of the African ancestor into the modern African, European, and Asian races included migration into different parts of the world. Scientists believe that modern *Homo sapiens* evolved in eastern Africa, around the Olduvai Gorge. Descendents of these early ancestors migrated to Europe and Asia as well as other areas in Africa. Descendents of some Asian groups crossed the Bering Strait to inhabit the American continent. Once isolated, these various groups evolved independently.

roots" of modern-day African populations. In other words, different subgroups of Africans branched off from each other before the other races branched off from the Africans as a whole (Fig. 26.27). The ancestors of today's Europeans split off from their Euro-Asian forebears and wandered into Europe via the Middle East around 40,000–50,000 years ago (Fig. 26.28). American Indians appear to derive from mainland Asian populations who migrated from 13,000–16,000 years ago when the Bering Strait was frozen over. The colonization of parts of Oceania is even more recent and still controversial (see Focus on Relevant Research).

Besides using mitochondrial DNA, sequences of microsatellite regions of the chromosomes have been compared among the different races. The phylogenetic results are very similar. They also give a primary African—non-African split, and if anything, they suggest an even more recent date for the common ancestor, nearer 100,000 years ago.

But what about Adam, or "**Y-guy**" as he is sometimes called by molecular biologists? The shorter human Y-chromosome does not recombine with its longer partner, the X-chromosome over most of its length. This allows us to follow the male lineage without complications due to recombination. For example, the *ZFY* gene on the Y-chromosome is handed on from father to son and is involved in sperm maturation. The sequence data for *ZFY* suggest a split between humans and chimps about 5 million years ago and a common male ancestor for modern mankind about 250,000 years ago. However, recent data from a much larger number of genetic markers on the Y-chromosome dates Y-guy to somewhat less than 100,000 years ago. Recent analyses of clusters of mutations on the Y-chromosome are incompatible with the multiregional model and confirm the recent African origin of modern humans.

> Y-chromosome sequences confirm the recent African origin of humans.

FOCUS ON RELEVANT RESEARCH

Soares P, Rito T, Trejaut J, Mormina M, Hill C, Tinkler-Hundal E, Braid M, Clarke DJ, Loo JH, Thomson N, Denham T, Donohue M, Macaulay V, Lin M, Oppenheimer S, and Richards MB (2011) Ancient voyaging and Polynesian origins. Am J Hum Genet 88:239–247.

The human colonization of Oceania occurred in two major stages. The occupation of "Near Oceania" (including Borneo, New Guinea, and Australia) occurred around 27,000 years ago. This was followed more recently by the settlement of "Remote Oceania" (including Fiji and Polynesia).

The dating and details of this more recent migration have been controversial. One theory proposes a rapid dispersal of maternal lineages from Taiwan approximately 4000 years ago. The Taiwan theory is supported by linguistic and cultural data as well as some DNA evidence. Another suggestion is an origin in East Indonesia. However, based on more detailed analysis of mitochondrial DNA, the authors suggests that this migration may be substantially older and originated from the population of "Near Oceania" rather than Taiwan or Indonesia. The authors argue that later movements of relatively-small numbers of people were responsible for the transmission of language and culture.

Box 26.2 Our Cousins—Neanderthals, Denisovans, and Hobbits

Neanderthal man survived to live alongside the modern races of *Homo sapiens* in Europe and the Middle East until relatively recently (approximately 30,000 years ago). Early DNA sequence analysis with limited material suggested that the Neanderthals came to a dead end. However, more recently the whole Neanderthal genome has become available. Improved analysis suggests that the Neanderthals probably did interbreed to some small extent with modern Man. In particular, Neanderthals show more sequence similarities with modern humans in Eurasia than those from Africa. This implies cross-breeding in those regions where Neanderthals and modern Man co-existed, rather than inheritance from shared ancestry.

Recently a new human relative was discovered from sequences of DNA extracted from bones in a cave at Denisova in the Altai Mountains in Siberia. The site dates to 30,000–50,000 years ago. What is especially fascinating is that both Neanderthals and modern humans were also present in the area at the same time! This implies that three species of humans shared this region for several thousand years. The Denisovans differ from modern humans more than Neanderthals do. Sequence data suggest that the common ancestor of the Denisovans and Neanderthals split off from the ancestral modern human lineage around 800,000 years ago. The Denisovans and Neanderthals then separated around 650,000 years ago. Apparently, the Denisovans interbred to some extent with the ancestors of modern Melanesians (inhabitants of Oceania only distantly related to typical Polynesians).

It has been suggested, although it is still controversial, that the miniature humans (nicknamed "hobbits") found on the island of Flores in Indonesia also constitute a separate human species, *Homo floresiensis*. The "hobbits" survived until around 12,00 years ago. Unfortunately, no DNA sequences are available and the dispute over the status of the "hobbits" is based on skeletal remains.

Y-guy   Hypothetical male human ancestor thought to have lived in Africa around 100,000–200,000 years ago

8.3. Ancient DNA from Extinct Animals

Apart from the occasional mummy or mammoth, DNA sequences from still living creatures are normally used to construct evolutionary schemes. However, ancient DNA extracted from the fossilized remains of extinct creatures can provide a valuable check on estimated evolutionary rates. To date, the oldest identified animal DNA is approximately 50,000 years old and comes from mammoths preserved in the permafrost of Siberia. The permafrost has also yielded identifiable plant DNA from grasses and shrubs around 300,000–400,000 years old.

DNA has also been successfully extracted from Egyptian mummies. DNA sequences have been obtained from mummies that are several thousand years old, although the amounts of DNA obtained are only 5% or so of those from fresh, modern, human tissue. Fragments of known human genes (amelogenin and beta-actin) have been identified. In addition, the mummy DNA contained repetitive Alu elements that are characteristic of human DNA.

Earlier reports of much older ancient DNA are now largely discredited. These include claims of dinosaur DNA and of insect DNA from samples trapped in amber that were supposedly millions of years old. Although amber acts as a preservative and the internal structure of individual cells from trapped insects can still be seen with an electron microscope, the DNA has long since been degraded.

> Small stretches of DNA sequence have been rescued from extinct life forms.

Box 26.3 Genghis Khan's Y-chromosome

Large-scale surveys have shown that about 1 in 12 men in Asia carry a variant of the Y-chromosome that originated in Mongolia roughly 1000 years ago. Around 30 natural genetic markers were surveyed in several thousand men. The markers included deletions and insertions, sequence polymorphisms, and repetitive sequences. Most men carry Y-chromosomes with more or less unique combinations of such DNA markers. However, about 8% of Asian males carry Y-chromosomes with the same (or almost the same) combination of genetic markers. This phenomenon was not seen among men from other continents.

Furthermore, the Asian men with the special "Mongol cluster" of genetic markers were found only among those populations who formed part of the Mongol Empire of Genghis Khan. For example, the "Mongol cluster" was absent from Japan and southern China, which were not incorporated into the Mongolian Empire, but was present in 15 different populations throughout the area of Mongolian domination (Fig. 26.29). In addition,

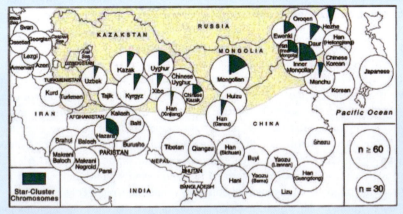

FIGURE 26.29
Genghis Khan's Empire and the Y-chromosome

The relative proportion of Mongol cluster chromosomes at various geographical locations is represented by the green segments in the circles. The size of the circles indicates sample size. (Credit: Zerjal and Tyler-Smith, The genetic legacy of the Mongols, *American Journal of Human Genetics* (2003) 72:717–721.)

Continued

Box 26.3 Continued

although very few Pakistanis have the "Mongol cluster," about 30% of a small tribal group known as the Hazara does possess it. The Hazara are known to be of Mongolian origin and claim to be direct descendents of Genghis Khan. Since present-day Pakistan is outside Genghis Khan's area of conquest they presumably migrated to their present location later.

This particular variant has therefore been proposed to be the Y-chromosome of Genghis Khan the great Mongolian conqueror. About 800 years ago the warlord Temujin united the Mongols and in 1206 assumed the title of Genghis Khan ("Lord of Lords"). The Mongols massacred many of the males and impregnated many of the women in areas they conquered. The present-day distribution of Y-chromosomes apparently reflects these practices. Whether this special variant of the Y-chromosome was present in Genghis Khan himself or just frequent among his Mongol warriors cannot be known for certain. Nonetheless, it is more likely than not that Genghis Khan himself had this Y-chromosome, as all the warriors in such tribes were usually closely related.

9. Evolving Sideways: Horizontal Gene Transfer

Standard Darwinian evolution involves alterations in genetic information passed on from one generation to its descendants. However, it is also possible for genetic information to be passed "sideways" from one organism to another that is not one of its descendents or even a near relative. The term **vertical gene transfer** refers to gene transmission from the parental generation to its direct descendants. Vertical transmission thus includes gene transmission by all forms of cell division and reproduction that create a new copy of the genome, whether sexual or not. **Horizontal gene transfer** (also known as "**lateral gene transfer**") happens when genetic information is passed sideways, from a donor organism to an unrelated organism.

For example, when antibiotic resistance genes are carried on plasmids they can be passed between unrelated types of bacteria (see Ch. 20). Since genes carried on plasmids are sometimes incorporated into the chromosome, a gene can move from the genome of one organism to that of an unrelated one in a couple of steps. The complete genomes of many bacteria have now been fully sequenced. Estimates using this data suggest that about 5–6% of the genes in an average bacterial genome have been acquired by horizontal transfer. The effects of horizontal transfer are especially noticeable in a clinical context. Both virulence factors and antibiotic resistance are commonly carried on transmissible bacterial plasmids.

Horizontal transfer may occur between members of the same species (e.g., the transfer of a plasmid between two closely-related strains of *Escherichia coli*) or over major taxonomic distances (e.g., the transfer of a Ti-plasmid from bacteria to plant cells). Horizontal gene transfer over long distances depends on carriers that cross the boundaries from one species to another (see Focus on Relevant Research). Viruses, plasmids, and transposons are all involved in such sideways movement of genes and have been discussed in their own chapters (see Chs. 20–22).

An extreme example among the bacteria is *Thermotoga*, which shares extremely hot environments with thermophilic Archaea. *Thermotoga* has around 10% archaeal genes. Perhaps more curious is that around 40% of their genes are closely related to those of thermophilic clostridia. This conflicts with their 16S rRNA sequences that place *Thermotoga* in the bacterial phylum Aquificales, which is only very distantly related to the clostridia.

> Genetic information may be passed "vertically" from an organism to its direct descendents or "horizontally" to other organisms that are not descendents.

> Horizontal gene transfer usually involves viruses, plasmids, or transposons.

---

horizontal gene transfer Transfer of genetic information "sideways" from one organism to another that is not directly related
lateral gene transfer Movement of genes sideways between unrelated organisms. Same as horizontal gene transfer
vertical gene transfer Transfer of genetic information from an organism to its descendents

FIGURE 26.30
Horizontal Transfer of Type-C Virogene in Mammals

The type-C virogene was present during evolution of Old World monkeys from their common ancestor. Surprisingly, a version of this gene closely related to the one in baboons was identified in North African and European cats. Since baboons and cats are not closely related, the gene must have moved from one group to another via horizontal transfer. Further supporting the idea of horizontal transfer, the gene is not found in cats like the lion or cheetah, which developed before the North African and European cats branched off.

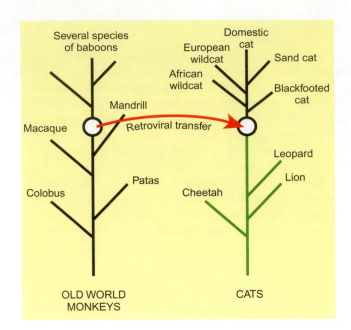

Retroviruses are noteworthy for inserting themselves into the chromosomes of animals, picking up genes, and moving them into another animal species. One such example of horizontal transfer in animals concerns the type-C virogene shared by baboons and all other Old World monkeys. This gene was present in the common ancestor of these monkeys, about 30 million years ago, and since then has diverged in sequence like any other normal monkey gene. Related sequences are also found in a few species of cats. Only the smaller cats of North Africa and Europe possess the baboon type-C virogene. American, Asian, and Sub-Saharan African cats all lack this sequence. Therefore, the original cat ancestor did not have this gene. Furthermore, the sequence in North African cats resembles that of baboons more closely than the sequences in monkeys closer to the ancestral stem (see Fig. 26.30). This suggests that about 5–10 million years ago a retrovirus carried the type-C virogene horizontally from the ancestor of modern baboons to the ancestor of small North African cats. Thus, the domestic pussycat carries the type-C virogene. However, other cats that diverged more than 10 million years ago lack these sequences.

FOCUS ON RELEVANT RESEARCH

Dunning Hotopp JC (2011) Horizontal gene transfer between bacteria and animals. Trends in Genetics 27:157–163.

Examples of gene transfer from bacteria to higher organisms are relatively frequent. The most common cases occur between bacterial symbionts, which live permanently in insects and nematodes, and their hosts. Many insects contain bacterial symbionts, such as *Wolbachia* or *Buchnera*, within their cells. These provide the insects with vital amino acids and vitamins. Most of these symbiotic bacteria are no longer capable of independent growth. As many as 70% of insect hosts have DNA from *Wolbachia* in their chromosomes.

A scatttering of other invertebrates that lack bacterial symbionts have also been found to have inserted bacterial DNA. These include *Hydra* and rotifers. DNA from other organisms tends to be concentrated close to the telomeres and most is undoubtedly non-functional. Nonetheless, some foreign genes are transcribed.

Horizontal gene transfer from animals to bacteria is also known, although with fewer examples. One of the most impressive is the occurrence of over 20 eukaryotic proteins in the bacterium *Legionella*, the cause of Legionnaires' disease. These bacteria often inhabit amoebas in the wild. Many other possible cases of animal to bacteria transfer exist, but few have been fully investigated.

9.1. Problems in Estimating Horizontal Gene Transfer

When the human genome was sequenced, several hundred human genes were at first attributed to horizontal transfer from bacteria. However, later analyses indicated that very few of these were genuine cases of horizontal transfer (see Ch. 9). Several factors have contributed to such over-estimates of horizontal transfer, both for the human genome and in other cases:

a. Sampling Bias. Relatively few eukaryotic genomes have been fully sequenced, whereas hundreds of bacterial genomes have been sequenced. Thus, the absence of sequences homologous to a human gene from a handful of other eukaryotes is insufficient evidence for an external (bacterial) origin. As more eukaryotic sequence data has become available many genes supposedly of "bacterial" origin have been found in other eukaryotes.

b. The loss of homologs in related lineages may suggest that a gene originated externally to the group of organisms that retain it. As in the related case (a) above, the solution to this artifact is the collection of more sequence data from many related lineages.

c. Gene duplication followed by rapid divergence may give rise to apparently novel genes that are missing from the direct vertical ancestor of a group of organisms.

d. Intense evolutionary selection for a particular gene may result in a greatly increased rate of sequence alteration. Rapidly-evolving genes will tend to be misplaced when sequence comparison is used to construct evolutionary trees.

e. The ease of horizontal transfer of genetic information by plasmids, viruses, and transposons under laboratory conditions is misleading. Under natural conditions there are major barriers to such movements. Furthermore, the results of horizontal transfer are often only temporary. Newly-acquired genes, especially those on plasmids, transposons, etc., are easily lost. Such genes tend to be acquired in response to selection such as antibiotic resistance and, conversely, they will be lost when the original selective conditions disappear.

f. Experimental problems such as DNA contamination. Bacterial and viral parasites are associated with essentially all higher organisms and completely purifying the eukaryotic DNA is not always easy.

Many of the originally proposed examples of widespread horizontal gene transfer have been severely compromised by the above factors. However, some examples do seem to be valid. One of the most interesting is the recent finding of relatively-frequent horizontal gene transfer between the mitochondrial genomes of flowering plants. The genes for certain mitochondrial ribosomal proteins have apparently been transferred from an early monocotyledonous lineage to several different dicotyledonous lineages. Examples include transfer of the *rps2* gene to kiwifruit (*Actinidia*) and the *rps11* gene to bloodroot (*Sanguinaria*).

Another interesting example involves two major eukaryotic lineages. This is the transfer of genes for carotenoid synthesis from fungi to pea aphids. Carotenoids are not normally made by animals and this is the only known case of carotenoid genes in any animal genome. The carotenoid pigments provide the red, yellow, and green colors typical of these insects. Red and green versions of the aphids are found that are recognized and eaten by different predators. The green variants have a defect in carotenoid desaturase and consequently lack the red version of the pigment.

Key Concepts

- The Earth's primary atmosphere of hydrogen plus helium was lost as the Earth is too small to hold such light gases.
- The Earth's secondary atmosphere was largely generated by volcanic outgassing. It contained no free oxygen.
- The chemical theory of the origin of life was put forward by the Russian biochemist Alexander Oparin in the 1920s.
- Miller's experiment mimics the formation of organic molecules in the primeval atmosphere.
- Polymerization of monomers into biological macromolecules requires the removal of water. This may be done by mild heat, clay minerals, or chemical-condensing agents.
- Random proteinoids show some slow and simple enzyme activities.
- Primeval synthesis can generate strands of RNA around 50 nucleotides long but with both 2′–5′ and 3′–5′ linkages.
- Ribozymes are enzymes made of RNA rather than protein. Although few in number they do exist in modern living cells.
- The RNA World scenario proposes that the first cells had both genes and enzymes made of RNA.
- The autotrophic theory of origins argues that the earliest life forms used energy released by the reaction of iron and sulfur compounds.
- The sequences of DNA and its encoded proteins will gradually change over long periods of time due to the accumulation of mutations.
- The amino-acid sequences of different proteins evolve at very different rates.
- Duplication is a major mechanism for creating new genes. New genes may also be created by mixing and shuffling segments of pre-existing genes.
- Although the *Archaea* share prokaryotic cell structure with the *Bacteria*, they are more closely related to Eukaryotes genetically.
- The mitochondria and chloroplasts of eukaryotic cells are derived from symbiotic bacteria that gradually lost their independence.
- Both mitochondria and chloroplasts still possess their own small genomes.
- Several protozoan lineages have arisen by engulfing other single-celled algae and thus have chloroplasts acquired by what is known as secondary endosymbiosis.
- The ancestries and relationships of groups of organisms may be derived by comparison of DNA sequences.
- Slowly-changing sequences, such as ribosomal RNA, are needed to compare distantly-related organisms.
- Rapidly-changing sequences, such as mitochondrial DNA, are used to compare closely-related organisms.
- Mitochondrial DNA analysis implies that modern humans originated in Africa about 100,000 years ago.
- DNA may be extracted from dead or extinct organisms and used to reveal their relationships.
- Horizontal (or lateral) gene transfer occurs when genetic information is passed "sideways" to a relatively unrelated organism (as opposed to a direct descendent).
- The extent of horizontal gene transfer is difficult to measure accurately and has often been over-estimated.

Review Questions

1. How long ago is it estimated that the Big Bang occurred? How long ago was the origin of life?
2. Which two chemical elements make up the majority of the universe?
3. What are the characteristics of Earth's primary, secondary, and tertiary atmospheres? What is unique about the tertiary atmosphere?
4. Why is the absence of oxygen vital to the evolution of life?
5. Summarize the Miller Experiment. Why is water vital to the production of biological molecules?
6. Most (but not all) amino acids, sugars, and other basic biological monomers common to life have been produced by the Miller Experiment. Do you view this as a problem? Why or why not?
7. Why does excess water present a problem when monomers attempt to polymerize? What alternatives may have allowed polymerization of amino acids on the early Earth?
8. What are acyl phosphates and phosphoramidates?
9. What types of catalytic activity do random proteinoids exhibit?
10. Briefly describe how primeval RNA could have replicated.
11. Define quasi-species and describe how it relates to RNA molecules.
12. Why is it thought that RNA was the first genetic macromolecule to evolve?
13. Describe the "RNA World" hypothesis. What is a major problem with this hypothesis?
14. List six lines of evidence that support the primacy of RNA.
15. Why is DNA preferred over RNA as the repository of genetic information?
16. How did the emergence of an oxygenated atmosphere affect the evolution of life?
17. What is an autotroph? Summarize the autotrophic theory of the origin of metabolism.
18. How can protein sequences be used to determine evolutionary relatedness?
19. Describe an example of how gene duplication can lead to the formation of a gene family.
20. Describe a process by which many new genes can evolve that occurs mostly in plants.
21. What is meant by the term "homologous sequence"? What is the difference between paralogous and orthologous sequences? When making evolutionary comparisons, which type of sequences must be compared? Why?
22. How can new genes be made by shuffling?
23. Why do proteins evolve at varying rates? What could you infer about a protein that evolves very quickly and whose sequence is highly divergent among various species?
24. What molecules are most appropriate to determine evolutionary relatedness between two closely-related species? Why?
25. What molecule has been used to determine a global evolutionary tree? Why?
26. What are the three domains of life? What molecule was used to determine this?
27. Explain how it is possible to sequence a gene without actually isolating or culturing the organism.
28. Give examples detailing how Archaea are very different from Eubacteria. In what kinds of environments are Archaea typically found?
29. What impact has DNA sequencing had on the science of biological classification? What is a major problem associated with using DNA sequences to classify organisms?
30. Mitochondrial DNA can be used to determine the relationships between what kinds of organisms?
31. Why does mitochondrial DNA only provide information about female ancestry?
32. Compare and contrast the multiregional and Noah's ark models of human evolution. Which model does the genetic evidence support?
33. Who is "African Eve"? Who is "Y-Guy"?
34. How is amber useful for studying ancient DNA? Does this lend credibility to "science fiction" scenarios such as Michael Crichton's *Jurassic Park*?
35. How have scientists been able to recover ancient bacteria?
36. Why is it thought that approximately 1 in 12 Asian men is a descendent of Genghis Khan?
37. What is the difference between vertical and horizontal gene transfer? Provide one example of each.
38. How do scientists account for the presence of the type C virogene in both baboons and certain species of cats?
39. List and describe six problems typically encountered when estimating the extent of horizontal gene transfer.
40. Do scientists believe horizontal gene transfer is less common, as common, or more common than was thought previously?

Conceptual Questions

1. Name three different regions of a nuclear genome that accumulate mutations faster.

2. Phylogenetic trees are drawn to understand the relationship between organisms. There are rooted trees where the base represents the basic sequence on which the other organisms are compared, and there are unrooted trees that only show the relationship between the sequences or organisms. Each branch of the tree represents one of the sequences or organisms, and the bifurcation point represents a common ancestor between the two branches. The first step to making a phylogenetic tree is to align the sequences to make comparisons, determine how similar the sequences are in a pairwise manner (comparing two at a time), and group the sequences based on similarity. Here are four different sequences for the original gene (top). Below are lines that represent homologous sequences found in different organisms. A [] symbol marks a deletion of the original gene, a G/C or A/T represents single-nucleotide changes, and the TE represents an insertion of a transposable element. Based on the sequence information, place each sequence on the rooted tree drawn below.

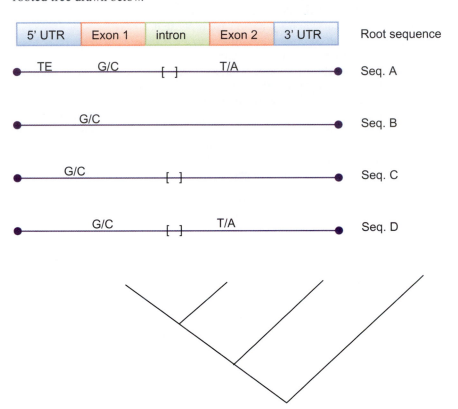

3. Understanding molecular evolution is actually very complex, and the use of computers is necessary to analyze large numbers of sequence homologies. At the National Center for Biotechnology Information (NCBI) (http://www.ncbi.nlm.nih.gov), there are listings of every piece of sequenced DNA from every species and organism. New sequences are deposited into the sequence database called GenBank (http://www.ncbi.nlm.nih.gov/Genbank/) and this can be searched for certain genes. Go to the GenBank website and search for accession number NM_019955 using "Nucleotide." List the gene name and the most recent article that describes the gene. Then using the BLAST option under Sequence Analysis Tools, list the other organism (not mouse) that has a nucleotide sequence most similar to this gene. How similar are these two genes?

4. Label the orthologous genes and paralogous genes for the following hypothetical gene family:

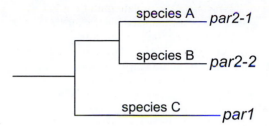

2-micron plasmid: See 2μ plasmid

2μ circle: Same as 2μ plasmid

2μ plasmid (or 2μ circle): A multicopy plasmid found in the yeast *Saccharomyces cerevisiae*, whose derivatives are widely used as vectors

30 nanometer fiber: Chain of nucleosomes that is arranged helically, approximately 30nm in diameter

30S initiation complex: Initiation complex for translation that contains only the small subunit of the bacterial ribosome

30S subunit: Small subunit of a 70S ribosome

3′ splice site: Recognition site for splicing at the downstream or 3′ end of the intron

3′-exonuclease: An enzyme that degrades nucleic acids from the 3′ end

3′-untranslated region (3′-UTR): Sequence at the 3′ end of mRNA, downstream of the final stop codon, which is not translated into protein

40S subunit: Small subunit of an 80S ribosome

454 sequencing: Second-generation method of sequencing that uses pyrosequencing to determine what nucleotide is added by DNA polymerase

4-methylumbelliferyl phosphate: An artificial substrate that is cleaved by alkaline phosphatase, releasing a fluorescent molecule

50S subunit: Large subunit of a 70S ribosome

5′ splice site: Recognition site for splicing at the upstream or 5′ end of the intron

5′-terminal oligopyrimidine tract (5′-TOP): Long pyrimidine-rich tracts located between the 5′ end of mRNA and the start codon

5′-untranslated region (5′-UTR): Region of an mRNA between the 5′ end and the translation start site

60S subunit: Large subunit of an 80S ribosome

70S initiation complex: Initiation complex for translation that contains both subunits of the bacterial ribosome

70S ribosome: Type of ribosome found in bacterial cells

7SL RNA: Non-coding RNA that forms part of the machinery for protein export across intracellular membranes in eukaryotic cells

80S ribosome: Type of ribosome found in cytoplasm of eukaryotic cells

−10 region: Region of bacterial promoter 10 bases back from the start of transcription that is recognized by RNA polymerase

−35 region: Region of bacterial promoter 35 bases back from the start of transcription that is recognized by RNA polymerase

α- (alpha) helix: A helical secondary structure found in proteins

β- (beta) sheet: A flat sheet-like secondary structure found in proteins

β-galactosidase: Enzyme that splits lactose and other compounds of galactose

β-lactamase: Enzyme that destroys antibiotics of the β-lactam class, which includes penicillins and cephalosporins

β-mercaptoethanol (BME): A small molecule with free sulfhydryl groups often used to break disulfide bonds in proteins

λ attachment site (*att* λ): Recognition sequence on the chromosome of *Escherichia coli* where bacteriophage lambda integrates

A (acceptor) site: Binding site on the ribosome for the tRNA that brings in the next amino acid

Ac element: Intact and active version of a transposon found in maize

acceptor stem: Base paired stem of tRNA to which the amino acid is attached

acetylation: Addition of an acetyl (CH3CO) group

aconitase: An enzyme of the Krebs cycle that, in animals, also acts as an iron-regulatory protein

acridine orange: A mutagenic agent that acts by intercalation

actin: A long filament of small subunits that is a component of cellular cytoskeleton

activation domain: The part of a transcription factor that interacts with the transcription apparatus

activator protein: Protein that switches a gene on

active site: Special site or pocket on a protein where other molecules are bound and the chemical reaction occurs

acyl phosphate: Phosphate derivative in which the phosphate is attached to a carboxyl group

adenine (A): A purine base that pairs with thymine, found in DNA or RNA

adenosine monophosphate (AMP): The nucleotide consisting of adenine, (deoxy)ribose, and one phosphate

adhesin: Protein that enables bacteria to attach themselves to the surface of animal cells. Same as colonization factor

A-DNA: A rare alternative form of double-stranded helical DNA

A-form: An alternative form of the double helix, with 11 base pairs per turn, often found for double-stranded RNA, but rarely for DNA

African Eve: Hypothetical female human ancestor thought to have lived in Africa around 100,000–200,000 years ago

agarose: A polysaccharide from seaweed that is used to form gels for separating nucleic acids by electrophoresis

agarose gel electrophoresis: Technique for separation of nucleic acid molecules by passing an electric current through a gel made of agarose

alkaline phosphatase: An enzyme that cleaves phosphate groups from a wide range of molecules

allele: One particular version of a gene, or more broadly, a particular version of any locus on a molecule of DNA

allo-lactose: An isomer of lactose that is the true inducer of the *lac* operon

allosteric enzyme: Enzyme that changes shape when it binds a small molecule

allosteric protein: Protein that changes shape when it binds a small molecule

alpha- (α-) carbon: Central carbon atom of an amino acid that carries both the amino group and the carboxyl group

alpha complementation: Assembly of functional β-galactosidase from N-terminal alpha fragment plus rest of protein

alpha DNA: Tandem DNA repeats found around the centromere in human DNA

alpha fragment: N-terminal fragment of β-galactosidase

alternative sigma factor: A non-standard sigma factor needed to recognize a specialized subset of genes

alternative splicing: Alternative ways to make two or more different final mRNA molecules by using different segments from the same original gene

Alu element: An example of a SINE, a particular short DNA sequence found in many copies on the chromosomes of humans and other primates

amino acid: Monomer from which the polypeptide chains of proteins are built

amino- or N-terminus: The end of a polypeptide chain that is made first and that has a free amino group

amino-acyl tRNA synthetase: Enzyme that attaches an amino acid to tRNA

aminoglycosides: Class of antibiotics that inhibits protein synthesis; includes streptomycin, neomycin, kanamycin, amikacin, and gentamycin

amp gene: Gene conveying resistance to ampicillin and related antibiotics and encoding beta-lactamase. Same as *bla* gene

ampicillin: A widely-used antibiotic of the penicillin group

anaerobic respiration: Respiration using other oxidizing agents (e.g., nitrate) instead of oxygen

analog: A chemical substance that mimics another well enough to be mistaken for it by biological macromolecules, in particular enzymes, receptor proteins, or regulatory proteins

anchor sequence: Sequence added to primers or probes that may be used for binding to a support or may incorporate convenient restriction sites, primer binding sites for future manipulations, or primer binding sites for subsequent PCR reactions

aneuploid: Having irregular numbers of different chromosomes

annealing: The re-pairing of separated single strands of DNA to form a double helix

anti-anti-sigma factor: Protein that binds to an anti-sigma factor and so prevents the anti-sigma factor from binding to and inhibiting a sigma factor

antibiotics: Chemical substances that inhibit specific biochemical processes and thereby stop bacterial growth selectively; that is, without killing the patient too

antibody: Protein made by the immune system to recognize and bind to foreign proteins or other macromolecules

anticodon: Group of three complementary bases on tRNA that recognizes and binds to a codon on the mRNA

anticodon loop: Loop of tRNA molecule that contains the anticodon

antifreeze protein: Protein that prevents freezing of blood, tissue fluids, or cells of organisms living at subzero temperatures

antiparallel: Parallel, but running in opposite directions

antisense RNA: An RNA molecule that is complementary to messenger RNA or another functional RNA molecule

antisense strand: Strand of DNA used as a guide for synthesizing a new strand by complementary base-pairing

anti-Shine-Dalgarno sequence: Sequence on 16S rRNA that is complementary to the Shine-Dalgarno sequence of mRNA

anti-sigma factor: Protein that binds to a sigma factor and blocks its role in the initiation of transcription

anti-termination factor: Protein that allows transcription to continue through a transcription terminator

anti-terminator protein: Protein that allows transcription to continue through a transcription terminator

AP endonuclease: Endonuclease that nicks DNA next to an AP site

AP site: A site in DNA where a base is missing (AP site = apurinic site or apyrimidinic site depending on the nature of the missing base)

Apicomplexa: Phylum of parasitic single-celled eukaryotes, including malaria, which contain both mitochondria and degenerate non-photosynthetic chloroplasts

apicoplast: Degenerate non-photosynthetic chloroplast found in members of the Apicomplexa

apoprotein: That portion of a protein consisting only of the polypeptide chains without any extra co-factors or prosthetic groups

apoptosis: Programmed suicide of unwanted cells

aquired immunodeficiency syndrome (AIDS): Disease caused by human immunodeficiency virus (HIV) that damages the immune system

araBAD operon: Operon that encodes proteins involved in metabolism of the sugar arabinose

arabinose: A 5-carbon sugar often found in plant cell wall material that can be used as a carbon source by many bacteria

archaea: New name for archaebacteria, one of the three domains of life

Archaebacteria (or Archaea): Type of bacteria forming a genetically-distinct domain of life. Includes many bacteria growing under extreme conditions

archaebacteria: One of the three domains of life comprising the "ancient" bacteria

Argonaut (AGO) family: Enzymes found within the RISC complex that degrade any cellular RNAs that are complementary to the guide strand of an siRNA

ascomycete: Type of fungus that produces four (or sometimes eight) spores in a structure known as an ascus

ascus: Specialized spore forming structure of ascomycete fungus

asexual or vegetative reproduction: Form of reproduction in which there is no re-shuffling of the genes between two individuals

asymmetric center: Carbon atom with four different groups attached. This results in optical isomerism

attenuation protein: Regulatory protein involved in attenuation and that binds to the leader region of mRNA

attenuation: Type of transcriptional regulation that works by premature termination and depends on alternative stem and loop structures in the leader region of the mRNA

attλ: Lambda-attachment site—site where lambda inserts its DNA into the bacterial chromosome

autogenous regulation: Self regulation; that is, when a DNA-binding protein regulates the expression of its own gene

autoradiography: Laying a piece of photographic film on top of a gel in order to identify the exact location of the radioactive DNA

auxin: Plant hormone that induces plant cells to grow bigger

avidin: A protein from egg white that binds biotin very tightly

B1 element: An example of a SINE found in mice; the precursor sequence from which the human Alu element evolved

bacteria: Primitive, relatively simple, single-celled organisms that lack a cell nucleus

bacterial (70S) ribosome: Type of ribosome found in bacterial cells

bacterial artificial chromosome (BAC): Single copy vector based on the F-plasmid of E. coli that can carry very long inserts of DNA. Widely used in the human genome project

bacteriocin: Toxic protein made by bacteria to kill closely-related bacteria

bacterioferritin: The bacterial analog of ferritin, an iron-storage protein

bacteriophage (phage): Virus that infects bacteria

bacteriophage lambda: Virus of E. coli with both lytic and lysogenic alternatives to its life cycle, which is widely used as a cloning vector

bacteriophage M13: A small male-specific filamentous virus that contains circular single-stranded DNA and infects Escherichia coli

bacteriophage Mu: A bacterial virus that replicates by transposition and causes mutations by insertion into host cell genes

bacteriophage T7: A bacteriophage that infects E. coli whose promoters are only recognized by its own RNA polymerase

bacteriophage ΦX174: A small spherical virus that contains circular single-stranded DNA and infects Escherichia coli

bait: The fusion between the DNA-binding domain of a transcriptional activator protein and another protein as used in two-hybrid screening

bandshift assay: Method for testing binding of a protein to DNA by measuring the change in mobility of DNA during gel electrophoresis. Same as gel retardation or mobility shift assay

barcode sequence: A unique 20 base pair sequence added to a cassette to mark the gene with a tag for subsequent analysis

Barr body: Inactive and highly condensed X-chromosome as seen in the light microscope

base: Alkaline chemical substance, in molecular biology especially refers to the cyclic nitrogen compounds found in DNA and RNA

base analog: Chemical mutagen that mimics a DNA base

base excision repair: Repair systems that recognize mutations in DNA that do not cause distortions in the helix

base pair: A pair of two complementary bases (A with T or G with C) held together by hydrogen bonds

base substitution: Mutation in which one base is replaced by another

bent DNA: Double-helical DNA that is bent due to several runs of As

beta-galactosidase (β-galactosidase): Enzyme that cleaves lactose and other β-galactosides, so releasing galactose

beta-lactamase (β-lactamase): Enzyme that degrades beta-lactam antibiotics, including penicillins and cephalosporins

beta-lactams or β-lactams: Family of antibiotics that inhibit crosslinking of the peptidoglycan of the bacterial cell wall; includes penicillins and cephalosporins

B-form or B-DNA: The normal form of the DNA double helix, as originally described by Watson and Crick

bi-directional replication: Replication that proceeds in two directions from a common origin

binary fission: Simple form of cell division in which the cell elongates and divides down the middle after replication of the DNA

binding protein: Protein whose role is to bind another molecule

bioinformatics: The computerized analysis of large amounts of biological sequence data

biopanning: Method of screening a phage display library for a desired displayed protein by binding to a bait molecule attached to a solid support

biotin: One of the B family of vitamins that is also widely used for chemical labeling of DNA molecules

***bla* gene:** Gene conveying resistance to ampicillin and related antibiotics and encoding beta-lactamase. Same as *amp* gene

blue/white screening: Screening procedure based on insertional inactivation of the gene for β-galactosidase

blunt ends: Ends of a double-stranded DNA molecule that are fully base paired and have no unpaired single-stranded overhang

bovine spongiform encephalopathy (BSE): Same as mad cow disease

bp: Abbreviation for base pair(s)

branch site: Site in the middle of an intron where branching occurs during splicing

budding: Type of cell division seen in yeasts in which a new cell forms as a bulge on the mother cell, enlarges, and finally separates

CAAT box: A sequence often found in the upstream region of eukaryotic promoters and which binds transcription factors

calmodulin: A small calcium-binding protein of animal cells

cap: Structure at the 5′ end of eukaryotic mRNA consisting of a methylated guanosine attached in reverse orientation

capsid: Shell or protective layer that surrounds the DNA or RNA of a virus particle

carboxy- or C-terminus: The end of a polypeptide chain that is made last and has a free carboxy group

carboxy-terminal domain (CTD): Repetitive region at the C-terminus of RNA polymerase II that may be phosphorylated

carrier protein: Protein that carries other molecules around the body or within the cell

CAT: Chloramphenicol acetyl transferase

catenane: Structure in which two or more circles of DNA are interlocked

cauliflower mosaic virus (CMV): A small spherical virus of plants with circular DNA. Some of its promoters are used in plant genetic engineering

CD4 protein: A protein found on the surface of T-cells that acts as a receptor during the immune response

cDNA (complementary DNA): DNA copy made by reverse transcription from mRNA and therefore lacking introns

cDNA library: Collection of genes in their cDNA form, lacking introns

cell cycle: Series of stages that a cell goes through from one cell division to the next

cell: The cell is the basic unit of life. Each cell is surrounded by a membrane and usually has a full set of genes that provide it with the genetic information necessary to operate

cellular PrP (PrPC): The healthy, normal form of the prion protein

Cen sequence: See centromere sequence

central dogma: Basic plan of genetic information flow in living cells which relates genes (DNA), message (RNA), and proteins

centrifugation: Process in which samples are spun at high speed and the centrifugal force causes the larger or heavier components to sediment to the bottom

centromere (Cen) sequence: Sequence at centromere of eukaryotic chromosome that is needed for correct partition of chromosomes during cell division

centromere: Region of eukaryotic chromosome, usually more or less central, where the microtubules attach during mitosis and meiosis

centromere sequence (CEN): A recognition sequence found at the centromere and needed for attachment of the spindle fibers

cephalosporins: Group of antibiotics of the β-lactam type that inhibit crosslinking of the peptidoglycan of the bacterial cell wall

CG-islands: Region of DNA in eukaryotes that contains many clustered CG sequences that are used as targets for cytosine methylation

chain termination mutation: Same as nonsense mutation

chain termination sequencing: Method of sequencing DNA by using dideoxynucleotides to terminate synthesis of DNA chains. Same as dideoxy sequencing

chaotropic agent: Chemical compound that disrupts water structure and so helps hydrophobic groups to dissolve

chaperone: Sometimes "molecular chaperone"; same as chaperonin

chaperonin: A protein that helps other proteins fold correctly

Chargaff's rule: For each strand of DNA the ratio of purines to pyrimidines is always 1:1 because A always pairs with T and G always pairs with C.

charged tRNA: tRNA with an amino acid attached

chemiluminescence: Production of light by a chemical reaction

chi sites: Specific sequences on the DNA of prokaryotes where crossovers form

chiasma (pl. chiasmata): point at which two homologous chromosomes break and rejoin opposite strands

chimera: Hybrid molecule of DNA that has DNA from more than one source or organism

chiral center: Same as asymmetric center

chloramphenicol: An antibiotic that inhibits bacterial protein synthesis

chloramphenicol acetyl transferase (CAT): Enzyme that inactivates chloramphenicol by adding acetyl groups

chlorophyll: Green pigment that absorbs light during photosynthesis

choleratoxin: Type of toxin made by *Vibrio cholerae*, the cholera bacterium

chromatid: Single double-helical DNA molecule making up whole or half of a chromosome; a chromatid also contains histones and other DNA-associated proteins

chromatin: Complex of DNA plus protein which constitutes eukaryotic chromosomes

chromatin immunoprecipitation (ChIP): Technique that identifies the DNA binding site for a particular transcription factor by crosslinking the DNA to the transcription factor, and then immunoprecipitating the transcription factor

chromatin remodeling complex: A protein assembly that rearranges the histones of chromatin in order to allow transcription

chromid: A term that combines chromosome and plasmid used to describe bacterial genetic material with characteristics of a chromosome and a plasmid

chromID: The brand name of a chromogenic culture medium sold by bioMérieux

chromogenic substrate: Colorless or pale substrate that is converted to a strongly colored product by an enzyme

chromosome banding technique: Visualization of chromosome bands by using specific stains that emphasize regions lacking genes

chromosome: Structure containing the genes of a cell and made of a single molecule of DNA

chromosome walking: Method for cloning neighboring regions of a chromosome by successive cycles of hybridization using overlapping probes

cI **gene:** Gene encoding the lambda repressor or cI protein

cI **protein:** Lambda repressor protein responsible for maintaining bacteriophage lambda in the lysogenic state

ciprofloxacin: A fluoroquinolone antibiotic that inhibits DNA gyrase

cistron: Segment of DNA (or RNA) that encodes a single polypeptide chain

clamp-loading complex: Group of proteins that loads the sliding clamp of DNA polymerase onto the DNA

clavulanic acid: And its derivatives bind to β-lactamases and react forming a covalent bond to the protein that kills the enzyme

cloning vector: Any molecule of DNA that can replicate itself inside a cell and is used for carrying cloned genes or segments of DNA. Usually a small multicopy plasmid or a modified virus

cloverleaf structure: 2-D structure showing base-pairing in a tRNA molecule

coding strand: The strand of DNA equivalent in sequence to the mRNA (same as plus strand)

co-dominance: when two different alleles both contribute to the observed properties

codon degeneracy: Situation where a set of four codons all code for the same amino acid and the identity of the third codon base makes no difference during translation

codon: Group of three RNA or DNA bases that encodes a single amino acid

co-factor: Extra chemical group bound (often temporarily) to a protein but which is not part of the polypeptide chain

cofactor: Extra chemical group non-covalently attached to a protein that is not part of the polypeptide chain

co-immunoprecipitation: Method of identifying protein-protein interaction by using antibodies to one of the proteins

cointegrate: A temporary structure formed by linking the strands of two molecules of DNA during transposition, recombination, or similar processes

"cold": Slang for non-radioactive

ColE plasmid: Small multicopy plasmid that carries genes for colicins of the E group. Used as the basis of many widely-used cloning vectors

colicin: Toxic protein or bacteriocin made by *Escherichia coli* to kill closely-related bacteria

colonization factor: Protein that enables bacteria to attach themselves to the surface of animal cells. Same as adhesin

community profiling: Assessing the abundance and diversity of bacteria in an environment using PCR

competent cell: Cell that is capable of taking up DNA from the surrounding medium

competitive inhibitor: Chemical substance that inhibits an enzyme by mimicking the true substrate well enough to be mistaken for it

complementary DNA (cDNA): Copies of eukaryotic genes lacking introns that are made by reverse transcriptase using messenger RNA as template

complementary sequences: Two nucleic acid sequences whose bases pair with each other because A, T, G, C in one sequence correspond to T, A, C, G, respectively, in the other

complex transposon: A transposon that moves by replicative transposition

conditional mutation: Mutation whose phenotypic effects depend on environmental conditions such as temperature or pH

conjugated protein: Complex of protein plus another molecule

conjugation bridge: Junction that forms between two cells and provides a channel for DNA to move from donor to recipient during conjugation

conjugation: Process in which genes are transferred by cell-to-cell contact

conjugative transposon: A transposon that is also capable of moving from one bacterial cell to another by conjugation

consensus sequence: Idealized base sequence consisting of the bases most often found at each position

conservative substitution: Replacement of an amino acid with another that has similar chemical and physical properties

conservative transposition: Same as cut-and-paste transposition

constitutive gene: Gene that is expressed all the time

contig: A stretch of known DNA sequence that is contiguous and lacks gaps

controlled pore glass (CPG): Glass with pores of uniform sizes that is used as a solid support for chemical reactions such as artificial DNA synthesis

Coomassie Blue: A blue dye used to stain proteins

copy number: the number of copies of a gene that are present

copy number variation (CNVs): A form of structural variation in which one genome will have either an insertion or deletion relative to another genome from a different individual

core enzyme: Bacterial RNA polymerase without the sigma (recognition) subunit

co-repressor: In prokaryotes—a small signal molecule needed for some repressor proteins to bind to DNA; in eukaryotes—an accessory protein, often a histone deacetylase, involved in gene repression

***cos* sequences (lambda cohesive ends):** Complementary 12bp long overhangs found at each end of the linear form of the lambda genome

cosmid: Small multicopy plasmid that carries lambda *cos* sites and can carry around 45kb of cloned DNA

co-transfer frequency: Frequency with which two genes remain associated during transfer of DNA between cells

cotranslational export: Export of a protein across a membrane while it is still being synthesized by a ribosome

coupled transcription-translation: When ribosomes of bacteria start translating an mRNA molecule that is still being transcribed from the DNA

covalently closed circular DNA (cccDNA): Circular DNA with no nicks in either strand

cPABP (chloroplast polyadenylate-binding protein): A translational activator protein that controls expression of chloroplast mRNA

CRISPR locus: Region of the bacterial genome that has a series of small sequence elements acquired from invading nucleic acids of viruses

CRISPR system: A bacterial defense system that uses an enzyme complex associated with a short single-stranded RNA derived from the CRISPR locus to find and destroy any RNA with complementary sequences

crista (plural cristae): Infolding of the respiratory membranes of mitochondria

crossing over: when two different strands of DNA are broken and are then joined to one another

crossover resolvase: Bacterial enzyme that separates covalently fused chromosomes

crossover: Structure formed when the strands of two DNA molecules are broken and joined to each other

crown gall: Type of tumor formed on plants due to infection by *Agrobacterium* carrying a Ti plasmid

CRP (cyclic AMP receptor protein): Bacterial protein that binds cyclic AMP and then binds to DNA

cruciform structure: Cross-shaped structure in double-stranded DNA (or RNA) formed from an inverted repeat

cryptic plasmid: A plasmid that confers no identified characteristics or phenotypic properties

cut-and-paste transposition: Type of transposition in which a transposon is completely excised from its original location and moves as a whole unit to another site

cycloheximide: An antibiotic that inhibits eukaryotic protein synthesis

cytokinesis: Cell division

cytokinin: Plant hormone that induces plant cells to divide

cytoplasm: The portion of a cell that is inside the cell membrane but outside the nucleus

cytosine (C): One of the pyrimidine bases found in DNA or RNA and which pairs with guanine

cytoskeleton: Internal structural elements in eukaryotic cells that keep the cellular shape and provide structures to move intracellular materials and organelles from one location to another

data mining: The use of computer analysis to find useful information by filtering or sifting through large amounts of data

***de novo* methylase:** An enzyme that adds methyl groups to wholly non-methylated sites

deaminase: An enzyme that removes an amino group

deamination: Loss of an amino group

defective phage: Mutant phage that lacks genes for making virus particles

degenerate primer: Primer with several alternative bases at certain positions

deletion: Mutation in which one or more bases is lost from the DNA sequence

demethylase: An enzyme that removes methyl groups

denaturant: Chemical compound that destroys the 3D structure of proteins, especially by breaking hydrogen bonds

denaturation: In reference to DNA, the breaking apart of a double strand of DNA into two single strands; when used of proteins, refers to the loss of correct 3D structure

deoxynucleoside: A nucleoside containing deoxyribose as the sugar

deoxynucleotide: A nucleotide containing deoxyribose as the sugar

deoxyribonuclease (DNase): Enzyme that cuts or degrades DNA

deoxyribonuclease I (DNase I): Non-specific nuclease that cuts DNA between any two nucleotides. Often used in footprint analysis

deoxyribonucleic acid (DNA): Nucleic acid polymer of which the genes are made

deoxyribonucleoside 5′-triphosphate (deoxyNTP): Precursor for DNA synthesis consisting of a base, deoxyribose, and three phosphate groups

deoxyribose: The sugar with five carbon atoms that is found in DNA

detergent: Molecule that is hydrophobic at one end and highly hydrophilic at the other and which is used to dissolve lipids or grease

Dicer: Ribonuclease that cleaves double-stranded RNA into segments of 21–23bp

dideoxy sequencing: Method of sequencing DNA by using dideoxynucleotides to terminate synthesis of DNA chains. Same as chain termination sequencing

dideoxynucleotide: Nucleotide whose sugar is dideoxyribose instead of ribose or deoxyribose

dideoxyribose: Derivative of ribose that lacks the oxygen of both the 2′ and the 3′ hydroxyl groups

dif **site:** Site on bacterial chromosome used by crossover resolvase to separate covalently-fused chromosomes

differential display PCR: Variant of RT-PCR that specifically amplifies messenger RNA from eukaryotic cells using oligo(dT) primers

differentiation: Progressive changes in the structure and gene expression of cells belonging to a single organism that leads to the formation of different types of cell

digoxigenin: A steroid from foxglove plant widely used for chemical labeling of DNA molecules

dihydrofolate (DHF): Co-factor with a variety of roles including making precursors for DNA and RNA synthesis

dihydrofolate reductase: Enzyme that converts dihydrofolate back to tetrahydrofolate

dimethoxytrityl (DMT) group: Group used for blocking the 5′-hydroxyl of nucleotides during artificial DNA synthesis

diphthamide: Modified amino acid found only in eukaryotic elongation factor eEF2 that is the target for diphtheria toxin

diploid: Possessing two copies of each gene

dipolar ion: Same as zwitterion; a molecule with both a positive and a negative charge

directed mutagenesis: Deliberate alteration of the DNA sequence of a gene by any of a variety of artificial techniques

disulfide bond: A sulfur to sulfur bond formed between two sulfhydryl groups, in particular between those of cysteine, and which binds together two protein chains

Dmd **gene:** Gene responsible for Duchenne muscular dystrophy

DNA: Deoxyribonucleic acid; nucleic acid polymer of which the genes are made

DNA adenine methylase (Dam): A bacterial enzyme that methylates adenine in the sequence GATC

DNA binding domain: The part of a transcription factor that binds to DNA

DNA chip: Chip used to simultaneously detect and identify many short DNA fragments by DNA-DNA hybridization. Also known as DNA array or oligonucleotide array detector

DNA cytosine methylase (Dcm): A bacterial enzyme that methylates cytosine in the sequences CCAGG and CCTGG

DNA fingerprint: Individually unique pattern due to multiple bands of DNA produced using restriction enzymes, separated by electrophoresis and usually visualized by Southern blotting

DNA glycosylase: Enzyme that breaks the bond between a base and the deoxyribose of the DNA backbone

DNA gyrase: An enzyme that introduces negative supercoils into DNA, a member of the type II topoisomerase family

DNA helicase: Enzyme that unwinds double-helical DNA

DNA library: Collection of cloned segments of DNA that is big enough to contain at least one copy of every gene from a particular organism. Same as gene library

DNA ligase: Enzyme that joins DNA fragments covalently, end to end

DNA microarray or DNA array: Chip-carrying array of DNA segments used to simultaneously detect and identify many short RNA or DNA fragments by hybridization. Also known as DNA chip or oligonucleotide array

DNA polymerase: Enzyme that synthesizes DNA

DNA polymerase I (Pol I): Bacterial enzyme that makes small stretches of DNA to fill in gaps between Okazaki fragments or during repair of damaged DNA

DNA polymerase III (Pol III): Enzyme that makes most of the DNA when bacterial chromosomes are replicated

DNA polymerase V: A repair polymerase in bacteria that can replicate past pyrimidine dimers and AP sites

DNA polymerase α: Enzyme that makes short segment of initiator DNA during replication of animal chromosomes

DNA polymerase δ: Enzyme that makes most of the lagging strand DNA when animal chromosomes are replicated

DNA polymerase ε: Enzyme that makes most of the leading strand DNA when animal chromosomes are replicated

DNA virus: A virus whose genome consists of DNA

DnaA protein: Protein that binds to the origin of bacterial chromosomes and helps initiate replication

domain (of life):   Highest ranking group into which living creatures are divided, based on the most fundamental genetic properties

domain (of protein):   A region of a polypeptide chain that folds up more or less independently to give a local 3D structure

dominant allele: allele whose properties are expressed in the phenotype whether present as a single or double copy

donor cell: Cell that donates DNA to another cell

double helix: Structure formed by twisting two strands of DNA spirally around each other

Ds elements: Defective version of the Ac transposon of maize; it cannot move alone but needs the Ac element to provide transposase

Duchenne muscular dystrophy: One of several inherited diseases affecting muscle function

duplication: Mutation in which a segment of DNA is duplicated

E (exit) site: Site on the ribosome that a tRNA occupies just before leaving the ribosome

early genes: Genes expressed early during virus infection and that mainly encode enzymes involved in virus DNA (or RNA) replication

eco-trawling: Isolating useful genes from the environment by PCR

EDTA (ethylene diamine tetraacetate):   A widely-used chelating agent that binds di-positive ions such as Ca^{2+} and Mg^{2+}

electrophoresis: Movement of charged molecules due to an electric field; used to separate and purify nucleic acids and proteins

electroporation: Inducing small pores or openings in a cellular membrane with an electrical current; used for bacteria, mammalian cells, yeast, and other small organisms

electroporator: Device that uses a high-voltage discharge to make cells competent to take up DNA

electrospray ionization (ESI):   Type of mass spectrometry in which gas-phase ions are generated from ions in solution

elongation factors: Proteins that are required for the elongation of a growing polypeptide chain

enantiomers: A pair of mirror-image optical isomers (i.e., D- and L-isomers)

endonuclease: A nuclease that cuts a nucleic acid in the middle

endoplasmic reticulum: Internal system of membranes found in eukaryotic cells

endosymbiosis: Form of symbiosis where one organism lives inside the other

enhancer: Regulatory sequence outside, and often far away from, the promoter region that binds transcription factors

***Entamoeba*:**   A very primitive single-celled eukaryote that lacks mitochondria

enterotoxins: Types of toxin made by enteric bacteria including some pathogenic strains of *E. coli*

enzyme: A protein or RNA molecule that catalyzes a chemical reaction

epigenetics: Inheritance of phenotypic differences that occur without alteration of the nucleotide DNA sequences; often refers to the methylation pattern of DNA or the post-translational modification pattern on histones

epistasis: When a mutation in one gene masks the effect of alterations in another gene

error-prone repair: Type of DNA repair process that introduces mutations

erythromycin: An antibiotic that inhibits bacterial protein synthesis

***Escherichia coli*:**   A bacterium commonly used in molecular biology

ethidium bromide: A stain that specifically binds to DNA or RNA and appears orange if viewed under ultraviolet light

Eubacteria: Bacteria of the normal kind as opposed to the genetically-distinct *Archaebacteria*

euchromatin: Normal chromatin, as opposed to heterochromatin

eukaryote: Higher organism with advanced cells, which have more than one chromosome within a compartment called the nucleus

eukaryotic (80S) ribosome: Type of ribosome found in cytoplasm of eukaryotic cell and encoded by genes in the nucleus

excision repair system: Also known as "cut and patch" repair. A DNA repair system that recognizes bulges in the DNA double helix, removes the damaged strand, and replaces it

excisionase: Enzyme that reverses DNA integration by removing a segment of dsDNA and resealing the gap. In particular, lambda excisionase removes integrated lambda DNA

exon cassette selection: Type of alternative splicing that makes different mRNA molecules by choosing different selections of exons from the primary transcript

exon: Segment of a gene that codes for protein and that is still present in the messenger RNA after processing is complete

exon trapping: Experimental procedure for isolating exons by using their flanking splice recognition sites

exonuclease: A nuclease that cuts a nucleic acid at the end

expressed sequence tag (EST):   A special type of STS derived from a region of DNA that is expressed by transcription into mRNA

expression vector: Vector specifically designed to place a cloned gene under control of a plasmid-borne promoter

extein: A segment of a protein that remains after the splicing out of any inteins

extragenic suppression: Reversion of a mutation by a second change that is within another distinct gene

Fe$_4$S$_4$ cluster: A group of inorganic iron and sulfur atoms found as a co-factor in several proteins

ferritin: An iron storage protein

fertility plasmid: Plasmid that enables a cell to donate DNA by conjugation

filial generations: successive generations of descendants from a genetic cross, which are numbered F1, F2, F3, etc., to keep track of them

FISH: See Fluorescence in Situ Hybridization

FLAG tag: A short peptide tag (AspTyrLysAspAsp AspAspLys) that is bound by a specific anti-FLAG antibody that may be attached to a resin for use in column purification of proteins

Flp recombinase (or flippase): Enzyme encoded by the 2μ plasmid of yeast that catalyzes recombination between inverted repeats (FRT sites)

Flp recombination target (or FRT site): Recognition site for Flp recombinase

fluorescence activated cell sorter (FACS): Instrument that sorts cells (or chromosomes) based on fluorescent labeling

fluorescence in situ hybridization (FISH): Using a fluorescent probe to visualize a molecule of DNA or RNA in its natural location

fluorescence: Process in which a molecule absorbs light of one wavelength and then emits light of another, longer, lower energy wavelength

fluorescence resonance energy transfer (FRET): Transfer of energy from short-wavelength fluorophore to long-wavelength fluorophore so quenching the short wave emission

fluorophore: Fluorescent group

folate: Co-factor involved in carrying one carbon group in DNA synthesis

footprint: Method for testing binding of a protein to DNA by its protection of DNA from chemical degradation

F-plasmid: A particular plasmid that confers ability to mate on its bacterial host, *Escherichia coli*

frameshift: Alteration in the reading frame during polypeptide synthesis

FRT site: Flp recombination target, the recognition site for Flp recombinase

functional genomics: The study of the whole genome and its expression

Fur (ferric uptake regulator): Global regulatory protein that senses iron levels in bacteria

fusidic acid: An antibiotic that inhibits protein synthesis

G1 phase: Stage of the eukaryotic cell cycle following cell division; cell growth occurs here

G2 phase: Stage of the eukaryotic cell cycle between DNA synthesis and mitosis: preparation for division

galactoside: Compound of galactose, such as lactose, ONPG, or X-gal

gametes: Cells specialized for sexual reproduction that are haploid (have one set of genes)

gametophyte: Haploid phase of a plant, especially of lower plants such as mosses and liverworts, where it forms a distinct multicellular body

gap: A break in a strand of DNA or RNA where bases are missing

GC ratio: The amount of G plus C divided by the total of all four bases in a sample of DNA. The GC ratio is usually expressed as a percentage

gel electrophoresis: Electrophoresis of charged molecules through a gel meshwork in order to sort them by size

gel retardation: Method for testing binding of a protein to DNA by measuring the change in mobility of DNA during gel electrophoresis. Same as bandshift assay or mobility shift assay

gene: A unit of genetic information

gene cassette: Deliberately designed segment of DNA that is flanked by convenient restriction sites and usually carries a gene for resistance to an antibiotic or some other easily observed characteristic

gene conversion: Recombination and repair of DNA during meiosis that leads to replacement of one allele by another. This may result in a non-Mendelian ratio among the progeny of a genetic cross

gene creature: Genetic entity that consists primarily of genetic information, sometimes with a protective covering, but without its own machinery to generate energy or replicate macromolecules

gene family: Group of closely-related genes that arose by successive duplication and perform similar roles

gene fusion: Structure in which parts of two genes are joined together, in particular when the regulatory region of one gene is joined to the coding region of a reporter gene

gene library: Collection of cloned segments of DNA that is big enough to contain at least one copy of every gene from a particular organism. Same as DNA library

gene product: End product of gene expression; usually a protein but includes various untranslated RNAs such as rRNA, tRNA, and snRNA

gene superfamily: Group of related genes that arose by several stages of successive duplication. Members of a superfamily have often diverged so far that their ancestry may be difficult to recognize

GeneChip array: The first brand of DNA chip, made by Affymetrix Corporation

general transcription factor: Transcription factor required for expression of most eukaryotic genes

generalized transduction: Type of transduction where fragments of bacterial DNA are packaged at random and all genes have roughly the same chance of being transferred

generation time: The time from the start of one cell division to the start of the next

genetic code: System for encoding amino acids as groups of three bases (codons) of DNA or RNA

genetic element: Any molecule or segment of DNA or RNA that carries genetic information and acts as a heritable unit

genome mining: The use of computer analysis to find useful information by filtering or sifting through large amounts of biological sequence data

genome: The entire genetic information from an individual

genomics: Study of genomes as a whole rather than one gene at a time

genotype: The genetic make-up of an organism

genus: A group of closely-related species

germ cells: Cells specialized to pass genetic information to the next generation of organisms; see gametes

germline cells: Reproductive cells producing eggs or sperm that take part in forming the next generation

gigabase pair (Gbp): 109 base pairs

global regulation: Regulation of a large group of genes in response to the same stimulus

global regulator: A regulator that controls a large group of genes, generally in response to some stimulus or developmental stage

globins: Family of related proteins, including hemoglobin and myoglobin, that carry oxygen in the blood and tissues of animals

glutathione-S-transferase (GST): Enzyme that binds to the tripeptide, glutathione. GST is often used in making fusion proteins

glycine: The simplest amino acid

glycogen: Storage carbohydrate found both in bacteria and in the livers of animals

glycoprotein: Complex of protein plus carbohydrate

Golgi apparatus: A membrane-bound organelle that takes part in export of materials from eukaryotic cells

gram-negative bacteria: Major division of Eubacteria that possess an extra outer membrane lying outside the cell wall

gram-negative bacterium: Type of bacterium that has both an inner (cytoplasmic) membrane plus an outer membrane that is located outside the cell wall

gram-positive bacteria: Major division of Eubacteria that lack an extra outer membrane lying outside the cell wall

gram-positive bacterium: Type of bacterium that has only an inner (cytoplasmic) membrane and lacks an outer membrane

gratuitous inducer: A molecule (usually artificial) that induces a gene but is not metabolized like the natural substrate; the best known example is the induction of the *lac* operon by IPTG

green fluorescent protein (GFP): A jellyfish protein that emits green fluorescence and is widely used in genetic analysis

Gregor Mendel: Discovered the basic laws of genetics by crossing pea plants

guanidine: Non-ionized form of guanidinium

guanidinium chloride: A widely-used denaturant of proteins

guanine (G): A purine base found in DNA or RNA that pairs with cytosine

guide RNA (gRNA): Small RNA used to locate sequences on a longer mRNA during RNA editing

hairpin: A double-stranded base-paired structure formed by folding a single strand of DNA or RNA back upon itself

haploid genome: A complete set containing a single copy of all the genes (generally used to describe organisms that have two or more sets of each gene)

haploid: Possessing only a single copy of each gene

haplotypes: A combination of alleles or genetic markers that are inherited as one unit during meiosis

H-DNA: A form of DNA consisting of a triple helix; its formation is promoted by acid conditions and by runs of purine bases

headful packaging: Type of virus-packaging mechanism that depends on the amount of DNA the head of the virus particle can hold (as opposed to using specific recognition sequences)

heat shock protein (HSP): Protein induced in response to high temperature. Many heat shock proteins are chaperonins

heat shock proteins: A set of proteins that protect the cell against damage caused by high temperatures

heat shock response: Response to high temperature by expressing a set of genes that encode heat shock proteins

helitron: A transposable element found in eukaryotes that transposes via rolling circle replication

helix-loop-helix (HLH): One type of DNA-binding motif common in proteins

helix-turn-helix (HTH): One type of DNA-binding motif common in proteins

helper phage: Phage that provides the necessary genes so allowing a defective phage to make virus particles

helper virus: A virus that provides essential functions for defective viruses, satellite viruses, and satellite RNA

hemi-methylated: Methylated on only one strand

hemolysin: Type of toxin that lyses red blood cells

herpesvirus: A family of spherical animal DNA viruses with an outer envelope of material stolen from the nuclear membrane of the host cell

heterochromatin: A highly condensed form of chromatin that cannot be transcribed because it cannot be accessed by RNA polymerase

heterodimer: Dimer composed of two different subunits

heteroduplex: A DNA double helix composed of single strands from two different DNA molecules

heterozygous: Having two different alleles of the same gene

Hfr-strain: Bacterial strain that transfers chromosomal genes at high frequency due to an integrated fertility plasmid

highly repetitive DNA: DNA sequences that exist in hundreds of thousands of copies

His tag: Six tandem histidine residues that are fused to proteins, so allowing purification by binding to nickel ions that are attached to a solid support. Also known as polyhistidine tag

histone: Special positively-charged protein that binds to DNA and helps to maintain the structure of chromosomes in eukaryotes

histone acetyl transferase (HAT): Enzyme that adds acetyl groups to histones

histone deacetylase (HDAC): Enzyme that removes acetyl groups from histones

histone-like protein: Bacterial protein that binds non-specifically to DNA and participates in maintaining the structure of the nucleoid; they do not actually have much in common with true histones

H-NS protein (histone-like nucleoid structuring protein): A bacterial protein that binds non-specifically to DNA and helps maintain the higher level structure of the nucleoid

Holliday junction: DNA structure formed during recombination and found at the crossover point where the two molecules of DNA are joined

holoenzyme: An active enzyme complex consisting of multiple functional subunits that are made of multiple proteins

holoprotein: Complete protein consisting of the polypeptide chains plus any extra metal ions, co-factors, or prosthetic groups

homing intron: A mobile intron that encodes a protein enabling it to insert itself into a recognition sequence within a target gene

homologous: Related in sequence to an extent that implies common genetic ancestry

homologous chromosomes: Two chromosomes are homologous when they carry the same sequence of genes in the same linear order

homologous recombination: Recombination between two lengths of DNA that are identical, or nearly so, in sequence

Hoogsteen base pair: A type of non-standard base pair found in triplex DNA, in which a pyrimidine is bound sideways onto a purine

horizontal gene transfer: Movement of genes sideways between unrelated organisms. Same as lateral gene transfer

"hot": Slang for radioactive

hot start PCR: PCR in which *Taq* polymerase is sequestered by antibodies or blocking proteins from the remaining ingredients until the template DNA is fully denatured

hotspots: Site in DNA or RNA where mutations are unusually frequent

housekeeping genes: Genes that are switched on all the time because they are needed for essential life functions

HU protein (heat-unstable nucleoid protein): A bacterial protein that binds to DNA with low specificity and is involved in bending of DNA

Human Genome Project: Program to sequence the entire human genome

human immunodeficiency virus (HIV): The retrovirus that causes AIDS

hybrid DNA: Artificial double-stranded DNA molecule made by pairing two single strands from two different sources

hybridization: Formation of double-stranded DNA molecule by annealing of two single strands from two different sources

hydrogen bond: Bond resulting from the attraction of a positive hydrogen atom to both of two other atoms with negative charges

hydrophilic: Water-loving; readily dissolves in water

ice nucleation factor: Protein found on surface of certain bacteria that promotes the formation of ice crystals

IHF (integration host factor): A bacterial protein that bends DNA, so helping the initiation of transcription of certain genes; named after its role in helping the integration of bacteriophage lambda into the chromosome of *E. coli*

Illumina/Solexa sequencing: Second-generation method of sequencing DNA that uses reversible dye terminators to identify the nucleotide that is added by DNA polymerase

immunity protein: Protein that provides immunity. In particular bacteriocin immunity proteins bind to the corresponding bacteriocins and render them harmless

immunization: Process of preparing the immune system for future infection by treating the patient with weak or killed versions of an infectious agent

immunological screening: Screening procedure that relies on the specific binding of antibodies to the target protein

imprinting: When the expression of a particular allele depends on whether it originally came from the father or the mother (imprinting is a rare exception to the normal rules of genetic dominance)

***in vitro* packaging:** Procedure in which virus proteins are mixed with DNA *in vitro* to assemble infectious virus particles. Often used for packaging recombinant DNA into bacteriophage lambda

incompatibility: The inability of two plasmids of the same family to co-exist in the same host cell

induced fit: When the binding of the substrate induces a change in enzyme conformation so that the two fit together better

induced mutation: Mutation caused by external agents such as mutagenic chemicals or radiation

inducer: A signal molecule that turns on a gene by binding to a regulatory protein

inherited disease: Disease due to a genetic defect that is passed on from one generation to the next

initiation complex (for replication): Assemblage of proteins that binds to the origin and initiates replication of DNA

initiation factors: Proteins that are required for the initiation of a new polypeptide chain

initiator box: Sequence at the start of transcription of a eukaryotic gene

initiator DNA (iDNA): Short segment of DNA made just after the RNA primer during replication of animal chromosomes

initiator tRNA: The tRNA that brings the first amino acid to the ribosome when starting a new polypeptide chain

inosine: A purine nucleoside, found most often in transfer RNA, that contains the unusual base hypoxanthine

insertion: Mutation in which one or more extra bases are inserted into the DNA sequence

insertion sequence: A simple transposon consisting only of inverted repeats surrounding a gene-encoding transposase

insertional inactivation: Inactivation of a gene by inserting a foreign segment of DNA into the middle of the coding sequence

insulator: A DNA sequence that shields promoters from the action of enhancers and also prevents the spread of heterochromatin

insulator-binding protein (IBP): Protein that binds to insulator sequence and is necessary for the insulator to function

Int protein: Same as integrase

integrase: Enzyme that inserts a segment of dsDNA into another DNA molecule at a specific recognition sequence. In particular, lambda integrase inserts lambda DNA into the chromosome of *E. coli*

integration: Insertion of a segment of dsDNA into another DNA molecule at a specific recognition sequence

integron: Genetic element consisting of an integration site plus a gene encoding an integrase

intein: An intervening sequence in a protein—a segment of a protein that can splice itself out

intercalation: Insertion of a flat chemical molecule between the bases of DNA, often leading to mutagenesis

intergenic DNA: Non-coding DNA that lies between genes

intergenic region: DNA sequence between genes

internal resolution site (IRS): Site within a complex transposon where resolvase cuts the DNA to release two separate molecules of DNA from the cointegrate during replicative transposition

interphase: Part of the eukaryotic cell cycle between two cell divisions and consisting of G1, S, and G2 phases

intervening sequence: An alternative name for an intron

intracellular parasite: Parasite that lives inside the cells of its host organism

intragenic suppression: Reversion of a mutation by a second change at a different site but within the same gene

intron: Segment of a gene that does not code for protein but is transcribed and forms part of the primary transcript

inverse PCR: Method for using PCR to amplify unknown sequences by circularizing the template molecule

inversion: Mutation in which a segment of DNA has its orientation reversed, but remains at the same location

inverted repeat: Sequence of DNA that is the same when read forwards as when read backwards, but on the other complementary strand; one type of palindrome

ionizing radiation: Radiation that ionizes molecules that it strikes

IPTG (*iso*-propyl-thiogalactoside): A gratuitous inducer of the *lac* operon

iron sulfur cluster: Group of iron and sulfur atoms found in proteins and involved in oxidation/reduction reactions

iron-regulatory protein (IRP): Translational regulator that controls expression of mRNA in animals in response to the level of iron

iron-responsive element (IRE): Site on mRNA where the IRP binds

irreversible inhibition: Type of inhibition in which an enzyme is permanently inactivated by a chemical change

isoelectric focusing: Technique for separating proteins according to their charge by means of electrophoresis through a pH gradient

isoschizomers: Restriction enzymes from different species that share the same recognition sequence

ISWI ("imitation switch") complex: Smaller type of chromatin remodeling complex

jumping gene: Popular name for a transposable element

junk DNA: Defective selfish DNA that is of no use to its host cell and can no longer either move or express its genes

kanamycin: Antibiotic of the aminoglycoside family that inhibits protein synthesis

karyotype: The complete set of chromosomes found in the cells of a particular individual

kilobase ladder: Standard set of DNA fragments with known length used to compare DNA fragments of unknown size in gel electrophoresis

kinase: Enzyme that attaches a phosphate group to another molecule

kinetic proofreading: Proofreading of DNA that occurs during the process of DNA synthesis

kinetochore: Protein structure that attaches to the DNA of the centromere during cell division and also binds the microtubules

kingdom: Major subdivision of eukaryotic organisms; in particular the plant, fungus, and animal kingdoms

Klenow polymerase: DNA polymerase I from *E. coli* that lacks the 5′ to 3′ exonuclease domain

Km: See Michaelis constant

L- and D-forms: The two isomeric forms of an optically active substance; also called L- and D-isomers

LacI protein: Repressor that controls the *lac* operon

LacI: The lactose repressor protein

lactose permease (LacY): The transport protein for lactose

lacZ gene: Gene encoding β-galactosidase; widely used as a reporter gene

lagging strand: The new strand of DNA that is synthesized in short pieces during replication and then joined later

lambda (or λ): Specialized transducing phage of *Escherichia coli* that may insert its DNA into the bacterial chromosome

lambda attachment site (*attλ*): Recognition site on DNA used during integration of lambda DNA into *E. coli* chromosome

lambda left promoter (*P_L*): One of the promoters repressed by binding of the lambda repressor or cI protein

lambda repressor (cI protein): Repressor protein responsible for maintaining bacteriophage lambda in the lysogenic state

large subunit: The larger of the two ribosomal subunits, 50S in bacteria, 60S in eukaryotes

lariat structure: Branched, lariat-shaped segment of RNA generated by splicing out an intron

late genes: Genes expressed later in virus infection and that mainly encode enzymes involved in virus particle assembly

latency: Type of virus infection in which the virus becomes largely quiescent, makes no new virus particles, and duplicates its genome in step with the host cell. Same as lysogeny but used of animal viruses

lateral gene transfer: Movement of genes sideways between unrelated organisms. Same as horizontal gene transfer

leader peptidase: Enzyme that removes the leader sequence after protein export

leader peptide: The short protein produced in attenuated genes that has codons that correspond to the amino acid in which the operon synthesizes

leader region: The region of an mRNA molecule in front of the structural genes, especially when involved in regulation by the attenuation mechanism

leading strand: The new strand of DNA that is synthesized continuously during replication

leaky mutation: Mutation where partial activity remains

leucine zipper: One type of DNA-binding motif common in proteins

LINE: Long interspersed element

LINE-1 (L1) element: A particular LINE found in many copies in the genome of humans and other mammals

linkage group: a group of alleles carried on the same DNA molecule (that is, on the same chromosome)

linkage: two alleles are linked when they are inherited together more often than would be expected by chance, usually this is because they reside on the same DNA molecule (that is, on the same chromosome)

linking number (L): The sum of the superhelical turns (the writhe, W) plus the double helical turns (the twist, T)

lipoprotein: Complex of protein plus lipid

L-isomer: That one of a pair of optical isomers that rotates light in an anticlockwise direction

living cell: A unit of life that possesses a genome made of DNA and sends genetic messages (RNA) from its genes (DNA) to its own ribosomes to make its own proteins with energy it generates itself

lock and key model: Model of enzyme action in which the active site of an enzyme fits the substrate precisely

locus (plural, loci): A place or location on a chromosome; it may be a genuine gene or just any site with variations in the DNA sequence that can be detected, like RFLPs or VNTRs

LOD score (logarithm of the odds) (Z): Statistical estimate of whether two loci are found near each other along a chromosome

long interspersed element (LINE): Long repeated sequence that makes up a significant fraction of the moderately- or highly-repetitive DNA of mammals

long non-coding RNA (lncRNA): Longer regulatory RNA molecules (>200 bases) of eukaryotic cells

long PCR: PCR reaction used specifically to amplify longer target sequences than standard PCR

long terminal repeats (LTRs): Direct repeats found at the ends of the retrovirus genome which are required for integration of the retrovirus DNA into the host cell DNA

luc gene: Gene-encoding luciferase from eukaryotes

luciferase: Enzyme that consumes energy and generates light

luciferin: Chemical substrate used by luciferase to emit light

lumi-phos: Substrate for alkaline phosphatase that releases light upon cleavage

***lux* gene:** Gene-encoding luciferase from bacteria

Lyme disease: Infection caused by *Borrelia burgdorferii* and transmitted by ticks

lysogen: A cell containing a lysogenic virus

lysogeny: State in which a virus replicates its genome in step with the host cell without making virus particles or destroying the host cell. Same as latency, but generally used to describe bacterial viruses

lysosome: A membrane-bound organelle of eukaryotic cells that contains degradative enzymes

lytic growth: Growth of virus resulting in death of cell and release of many virus particles

M13: Rod-shaped bacteriophage that infects *E. coli*, contains a circle of single-stranded DNA, and is used to manufacture DNA for sequencing

macromolecule: Large polymeric molecule in living cells, especially DNA, RNA, protein, or polysaccharide

mad cow disease: Infectious prion disease transmitted from cattle to humans

maintenance methylase: Enzyme that adds a second methyl group to the other DNA strand of half-methylated sites

MALDI: see Matrix-assisted laser desorption-ionization

male-specific phage: Virus that only infects "male" bacteria (i.e., those bacteria carrying the F-plasmid)

maltose-binding protein (MBP): Protein of E. coli that binds maltose during transport. MBP is often used in making fusion proteins

Mariner elements: A widespread family of conservative DNA-based transposons first found in *Drosophila*

mass spectrometry: Technique for measuring the mass of molecular ions derived from volatilized molecules

matrix attachment region (MAR): Site on eukaryotic DNA that binds to proteins of the nuclear matrix or of the chromosomal scaffold—same as SAR sites

matrix-assisted laser desorption-ionization (MALDI): Type of mass spectrometry in which gas-phase ions are generated from a solid sample by a pulsed laser

maximum velocity (Vm or Vmax): Velocity reached when all the active sites of an enzyme are filled with substrate

Mbp: Megabase pairs or million base pairs

mechanical protein: Protein that uses chemical energy to perform physical work

mediator: A protein complex that transmits the signal from transcription factors to the RNA polymerase in eukaryotic cells

meiosis: Formation of haploid gametes from diploid parent cells

melting temperature (Tm): The temperature at which the two strands of a DNA molecule are half unpaired

melting: When used of DNA, refers to its separation into two strands as a result of heating

membrane: A thin flexible structural layer made of protein and phospholipid that is found surrounding all living cells

membrane-bound organelles: Organelles that are separated from the rest of the cytoplasm by membranes

Mendelian character: Trait that is clear cut and discrete and can be unambiguously assigned to one category or another

Mendelian ratios: Whole number ratios of inherited characters found as a result of a genetic cross

messenger RNA (mRNA): The class of RNA molecule that carries genetic information from the genes to the rest of the cell

metabolism: The processes by which nutrient molecules are transported and transformed within the cell to release energy and to provide new cell material

metabolome: The total complement of small molecules and metabolic intermediates of a cell or organism

metagenomic library: Collection of cloned segments of DNA that has genes from multiple organisms found in a particular environment

metagenomics: The genome level study of whole biological communities

metallothionein: Protein that protects animal cells by binding toxic metals

methotrexate (or amethopterin): Anti-cancer drug that inhibits dihydrofolate reductase of animals

methylcytosine-binding protein (MeCP): Proteins in eukaryotes that recognize methylated CG islands

Michaelis constant (Km): The substrate concentration that gives half maximal velocity in an enzyme reaction. It is an inverse measure of the affinity of the substrate for the active site

Michaelis-Menten equation: Equation describing relationship between substrate concentration and the rate of an enzyme reaction

micro RNA (miRNA): Small regulatory RNA molecules of eukaryotic cells

microsatellite: Another term for a VNTR (variable number tandem repeats) with repeats around 13 base pairs in length

minisatellite: Another term for a VNTR (variable number tandem repeats) with repeats around 25 base pairs in length

mirror-like palindrome: Sequence of DNA that is the same when read forwards and backwards on the same strand; one type of palindrome

mismatch repair: DNA repair system that recognizes and corrects wrongly-paired bases

mismatch repair system: DNA repair system that recognizes mispaired bases and cuts out part of the DNA strand containing the wrong base

mismatch: Wrong pairing of two bases in a double helix of DNA

missense mutation: Mutation in which a single codon is altered so that one amino acid in a protein is replaced with a different amino acid

mistranslation: Errors made during translation

mitochondrion: Membrane-bound organelle found in eukaryotic cells that produces energy by respiration

mitosis: Division of eukaryotic cell into two daughter cells with identical sets of chromosomes

mobile DNA: Segment of DNA that moves from site-to-site within or between other molecules of DNA

mobile genetic element: A discrete segment of DNA that is able to change its location within larger DNA molecules by transposition or integration and excision

mobility shift assay: Method for testing binding of a protein to DNA by measuring the change in mobility of DNA during gel electrophoresis. Same as bandshift assay or gel retardation

mobilizability: Ability of a non-transferable plasmid to be moved from one host cell to another by a transferable plasmid

moderately repetitive sequence: DNA sequences that exist in thousands of copies (but less than a hundred thousand)

modification enzyme: Enzyme that binds to the DNA at the same recognition site as the corresponding restriction enzyme but methylates the DNA

modified base: Nucleic acid base that is chemically altered after the nucleic acid has been synthesized

modifier gene: Gene that modifies the expression of another gene

molecular beacon: A fluorescent probe molecule that contains both a fluorophore and a quenching group and that fluoresces only when it binds to a specific DNA target sequence

molecular sewing: Creation of a hybrid gene by joining segments from multiple sources using PCR

monocistronic mRNA: mRNA carrying the information of a single cistron, which is a coding sequence for only a single protein

multimeric: Formed of multiple subunits

multiple cloning site (MCS): A stretch of artificially synthesized DNA that contains cut sites for seven or eight widely used restriction enzymes. Same as polylinker

multiplex PCR: Using different fluorescent dyes on different probes in a quantitative or real-time PCR reaction in order to assess the amplification of more than one target sequence

mutagen: Any agent, including chemicals and radiation, that can cause mutations

mutation: An alteration in the DNA (or RNA) that comprises the genetic information

mutator gene: Gene whose mutation alters the mutation frequency of the organism, usually because it codes for a protein involved in DNA synthesis or repair

MyoD: A eukaryotic transcription factor that takes part in muscle cell differentiation

nalidixic acid: A quinolone antibiotic that inhibits DNA gyrase

nanopore detector: Detector that allows a single strand of DNA through a molecular pore and records its characteristics as it passes through

negative control or regulation: Regulatory mode in which a repressor keeps a gene switched off until it is removed

negative elongation factor (NELF): A complex of proteins that inhibits RNA polymerase elongation in eukaryotes

negative feedback: Form of negative regulation where the final product of a pathway inhibits the first enzyme in the pathway

negative or "minus" strand: The non-coding strand of RNA or DNA

negative regulation: Control by a repressor that prevents expression of a gene unless it is somehow removed

negative supercoiling: Supercoiling with a left-handed or counterclockwise twist

neomycin: Antibiotic of the aminoglycoside family that inhibits protein synthesis

neomycin phosphotransferase: Enzyme that inactivates the antibiotics kanamycin and neomycin by adding a phosphate group

neutral buoyancy: Point at which the density of the substance is the same as the solution in which it is floating

neutral mutation: Replacement of an amino acid with another that has similar chemical and physical properties

N-formyl-methionine or fMet: Modified methionine used as the first amino acid during protein synthesis in bacteria

nick: A break in the backbone of a DNA or RNA molecule (but where no bases are missing)

nick translation: The removal of a short stretch of DNA or RNA, starting from a nick, and its replacement by newly-made DNA

NMR spectroscopy: Technique to determine protein structure that uses alternating magnetic fields to change the spin of electrons within the sample

non-coding DNA: DNA sequences that do not code for proteins or functional RNA molecules

non-coding regulatory RNA: An RNA molecule that sequesters regulatory proteins from functioning as translational repressors (e.g., CsrB and CsrC)

non-coding RNA: Any RNA molecule that is not translated to give protein

non-homologous end joining: DNA repair system found in eukaryotes that mends double-stranded breaks

non-homologous recombination: Recombination between two lengths of DNA that are largely unrelated. It involves specific proteins that recognize particular sequences and form crossovers between them. Same as site-specific recombination

nonsense mutation: Mutation due to changing the codon for an amino acid to a stop codon

nonsense-mediated decay (NMD): Mechanism in eukaryotes used to destroy mRNAs that contain a premature stop codon

norfloxacin: A fluoroquinolone antibiotic that inhibits DNA gyrase

Northern blotting: Hybridization technique in which a DNA probe binds to an RNA target molecule

novobiocin: An antibiotic that inhibits type II topoisomerases, especially DNA gyrase, by binding to the B-subunit

npt **gene:** Gene for neomycin phosphotransferase. Provides resistance against the antibiotics kanamycin and neomycin

nuclear envelope: Envelope consisting of two concentric membranes that surrounds the nucleus of eukaryotic cells

nuclear matrix: A mesh of filamentous proteins found on the inside of the nuclear membrane and used in anchoring DNA

nuclear pore: Pore in nuclear membrane that allows proteins and RNA into and out of the nucleus

nuclease: Enzyme that cuts or degrades nucleic acids

nucleic acid: Class of polymer molecule consisting of nucleotides that carries genetic information

nucleocapsid: Inner protein shell of a virus particle that contains the nucleic acid

nucleocytoplasmic large DNA viruses (NCLDV): A grouping of different families of virus including mimivirus and poxvirus that are large in size and have large genomes and that typically infect eukaryotes

nucleoid: Area within a bacterial cell in which the chromosome is usually found; not surrounded by membranes

nucleolar organizer: Chromosomal region associated with the nucleolus; actually, a cluster of rRNA genes

nucleolus: Region of the nucleus where ribosomal RNA is made and processed

nucleomorph: Degenerate remains of the nucleus of a symbiotic eukaryote that was incorporated by secondary endosymbiosis into another eukaryotic cell

nucleoprotein: Complex of protein plus nucleic acid

nucleoside: The union of a purine or pyrimidine base with a pentose sugar

nucleosome: Subunit of a eukaryotic chromosome consisting of DNA coiled around histone proteins

nucleotide: Monomer or subunit of a nucleic acid, consisting of a pentose sugar plus a base plus a phosphate group

nucleus: An internal compartment surrounded by the nuclear membrane and containing the chromosomes. Only the cells of higher organisms have nuclei

null allele: Mutant version of a gene that completely lacks any activity

null mutation: Mutation that totally inactivates a gene

Nus proteins: A family of bacterial proteins involved in termination of transcription and/or in anti-termination

NusA protein: A bacterial protein involved in termination of transcription

oil drop model: Model of protein structure in which the hydrophobic groups cluster together on the inside away from the water

Okazaki fragments: The short pieces of DNA that make up the lagging strand

oligo(dT): DNA strand consisting only of thymidine

oligo(U): Stretch of single-stranded RNA consisting solely of U or uridine residues

oligonucleotide array: DNA array used to simultaneously detect and identify many short RNA or DNA fragments by hybridization. Also known as DNA array or DNA chip

oligonucleotide array detector: Chip used to simultaneously detect and identify many short DNA fragments by DNA-DNA hybridization. Also known as DNA array or DNA chip

o-**nitrophenyl galactoside: (ONPG)** Artificial substrate that is split by β-galactosidase, releasing yellow *o*-nitrophenol

o-**nitrophenyl phosphate:** Artificial substrate that is split by alkaline phosphatase, releasing yellow *o*-nitrophenol

ONPG (*o*-nitrophenyl galactoside): Artificial substrate that is split by β-galactosidase, releasing yellow *o*-nitrophenol

open circle: Circular DNA with one strand nicked and hence with no supercoiling

open reading frame (ORF): Sequence of bases (either in DNA or RNA) that can be translated (at least in theory) to give a protein

operator: Site on DNA to which a repressor protein binds

operon: A cluster of prokaryotic genes that are transcribed together to give a single mRNA (i.e., polycistronic mRNA)

optical isomers: Isomers where the molecules differ only in their 3D arrangement and consequently affect the rotation of polarized light

organelle: Subcellular structure that carries out a specific task. Membrane-bound organelles are separated from the rest of the cytoplasm by membranes, but other organelles such as the ribosome are not.

origin of replication: Site on a DNA molecule where replication begins

orthologous genes: Homologous genes that are found in separate species and which diverged when the organisms containing them diverged

ortho-**nitrophenyl galactoside (ONPG):** Artificial substrate for β-galactosidase that yields a yellow color upon cleavage

outer membrane: Extra membrane lying outside the cell wall in gram-negative but not gram-positive bacteria

overlap primer: PCR primer that matches small regions of two different gene segments and is used in joining segments of DNA from different sources

P (peptide) site: Binding site on the ribosome for the tRNA that is holding the growing polypeptide chain

P1 artificial chromosome (PAC): Single copy vector based on the P1-phage/plasmid of *E. coli* that can carry very long inserts of DNA

P1: Generalized transducing phage of *Escherichia coli*

P22: Generalized transducing phage of *Salmonella*

palindrome: A sequence that reads the same backwards as forwards

paralogous genes: Homologous genes that are located within the same organism due to gene duplication

parasite: An organism or genetic entity that replicates at the expense of another creature

partial dominance: When a functional allele only partly masks a defective allele

partitioning: Movement of each replicated copy of the chromosome into each respective daughter cell during cell division

patch recombinant: DNA double helix with a short patch of heteroduplex due to transient formation of a crossover

pathogenic: Disease-causing

pathogenicity island: Region of bacterial chromosome containing clustered genes for virulence

PCNA protein: The sliding clamp for the DNA polymerase of eukaryotic cells (PCNA = proliferating cell nuclear antigen)

PCR machine: See thermocycler

PCR primers: Short pieces of single-stranded DNA that match the sequences at the ends of the target DNA segment, which are needed to initiate DNA synthesis in PCR

penetrance: Variability in the phenotypic expression of an allele

penicillin: An antibiotic made by a mold called *Penicillium*, which grows on bread producing a blue layer of fungus

penicillins: Group of antibiotics of the β-lactam type that inhibit crosslinking of the peptidoglycan of the bacterial cell wall

pentose: A five-carbon sugar, such as ribose or deoxyribose

peptide bond: Type of chemical linkage holding amino acids together in a protein molecule

peptide nucleic acid (PNA): Artificial analog of nucleic acids with a polypeptide backbone

peptidoglycan: Mixed polymer of carbohydrate and amino acids that comprises the structural layer of bacterial cell walls

peptidyl transferase: Enzyme activity on the ribosome that makes peptide bonds; actually 23S rRNA (bacterial) or 28S rRNA (eukaryotic)

permease: A protein that transports nutrients or other molecules across a membrane

phage display: Fusion of a protein or peptide to the coat protein of a bacteriophage whose genome also carries the cloned gene encoding the protein. The protein is displayed on the outside of the virus particle and the corresponding gene is carried on the inside

phage display library: Collection of a large number of modified phages displaying different peptide or protein sequences

phage: Short for bacteriophage, a virus that infects bacteria

pharmacogenetics: Studying the particular genes that affect how a person reacts to drugs

pharmacogenomics: A field of research that correlates individual genotypes relative to the person's reaction to a pharmaceutical agent

phenol extraction: Technique for removing protein from nucleic acids by dissolving the protein in phenol

phenotype: The visible or measurable effect of the genotype

pheromone: Hormone or messenger molecule that travels between organisms, rather than circulating within the same organism

phoA **gene:** Gene-encoding alkaline phosphatase; widely used as a reporter gene

phosphatase: An enzyme that removes phosphate groups

phosphate group: Group of four oxygen atoms surrounding a central phosphorus atom found in the backbone of DNA and RNA

phosphodiester: The linkage between nucleotides in a nucleic acid that consists of a central phosphate group esterified to sugar hydroxyl groups on either side

phospholipid: A hydrophobic molecule found making up cell membranes and consisting of a soluble head group and two fatty acids both linked to glycerol phosphate

phosphoramidate: Phosphate derivative in which the phosphate group is attached to an amino group

phosphoramidite method: Method for artificial synthesis of DNA that utilizes the reactive phosphoramidite group to make linkages between nucleotides

phosphorothioate: A phosphate group in which one of the four oxygen atoms around the central phosphorus is replaced by sulfur

phylum (plural phyla): Major groups into which animals are divided, roughly equivalent in rank to the divisions of plants or bacteria

Piwi-interacting RNA (piRNA): Small RNA molecules slightly longer than siRNA that are derived from genomic tandem repeats clustered at the centromere and prevent the spread of transposons through an Argonaut-like protein

plaque: (When referring to viruses) A clear zone caused by virus destruction in a layer of cultured cells or a lawn of bacteria

plasmid: Accessory molecule of nucleic acid capable of self-replication. Does not normally carry genes needed for existence of host cell. Usually consists of double-stranded circular DNA but occasional plasmids that are linear or made of RNA exist

Plasmodium: The malaria parasite, a protozoan belonging to the Apicomplexa

plastid: Any organelle that is genetically equivalent to a chloroplast, whether functional in photosynthesis or not

ploidy: The number of sets of chromosomes possessed by an organism

PNA clamp: Two identical PNA strands that are joined by a flexible linker and are intended to form a triple helix with a complementary strand of DNA or RNA

point mutation: Mutation that affects a single base pair

polarity: When the insertion of a segment of DNA affects the expression of downstream genes, usually by preventing their transcription

polinton: Self-synthesizing element that transposes from one location to another via excision of the original copy, synthesis of another DNA copy by Polinton-encoded DNA polymerase, and then reintegration using Polinton-encoded integrates

poly(A) polymerase: Enzyme that adds the poly(A) tail to the end of mRNA

poly(A) tail: A stretch of multiple adenosine residues found at the 3′ end of mRNA

poly(A)-binding protein (PABP): Protein that binds to mRNA via its poly(A) tail

polyacrylamide: Polymer used in separation of proteins or very small nucleic acid molecules by gel electrophoresis

polyacrylamide gel electrophoresis (PAGE): Technique for separating proteins by electrophoresis on a gel made from polyacrylamide

polyadenylation complex: Protein complex that adds the poly(A) tail to eukaryotic mRNA

polycistronic mRNA: mRNA carrying multiple coding sequences that may be translated to give several different protein molecules; only found in prokaryotic (bacterial) cells

Polycomb group (PcG) proteins: A large protein complex that controls developmental expression of genes important for growth by methylating histones

polyhistidine tag (His tag): Six tandem histidine residues that are fused to proteins, so allowing purification by binding to nickel ions that are attached to a solid support

polylinker: A stretch of artificially synthesized DNA that contains cut sites for seven or eight widely used restriction enzymes. Same as multiple cloning site (MCS)

polymerase chain reaction (PCR): Amplification of a DNA sequence by repeated cycles of strand separation and replication

polymerase: Enzyme that synthesizes nucleic acids

polymerase eta: A repair DNA polymerase in animals that can replicate past thymine dimers

polymorphism: A difference in DNA sequence between two related individual organisms

polypeptide chain: A polymer that consists of amino acids

polyphosphate: Compound consisting of multiple phosphate groups linked by high-energy phosphate bonds

polyploidy: Possessing more than two copies of each gene

polyprotein: A long polypeptide that is cut up to generate several smaller proteins

polysome: Group of ribosomes bound to and translating the same mRNA

positive control or regulation: Control by an activator that promotes gene expression when it binds

positive or "plus" strand: The coding strand of RNA or DNA

positive regulation: Control by an activator that promotes gene expression when it binds

post-translational modification: Modification of a protein or its constituent amino acids after translation is finished

potential intrastrand triplex (PIT): Stretch of DNA that might be expected from its sequence to form H-type triplex DNA

poxviruses: A family of large and complex dsDNA animal viruses with 150–200 genes

pre-loading complex (pre-LC): Complex of proteins that forms prior to binding to the origin of replication, but is essential to promoting the correct association of the pre-RC

pre-replicative complex (pre-RC): A complex of enzymes (ORC, Cdc6, Cdt1, and MCM) that assemble at the origin of replication during eukaryotic DNA replication

prey: The fusion between the activator domain of a transcriptional activator protein and another protein as used in two-hybrid screening

PriA: Protein of the primosome that helps primase bind

Pribnow box: Another name for the −10 region of the bacterial promoter

primary atmosphere: The original atmosphere of the Earth consisting mostly of hydrogen and helium

primary endosymbiosis: Original uptake of prokaryotes by the ancestral eukaryotic cell, giving rise to mitochondria and chloroplasts

primary structure: The linear order in which the subunits of a polymer are arranged

primary transcript: RNA molecule produced by transcription before it has been processed in any way

primase: Enzyme that starts a new strand of DNA by making an RNA primer

primer extension: Method to locate the 5′ start site of transcription by using reverse transcriptase to extend a primer bound to mRNA so locating the 5′ end of the transcript

primer walking: Approach to sequencing a long cloned DNA molecule by using successive primers located at stages along the molecule

primitive soup: Mixture of random molecules, including amino acids, sugars, and nucleic acid bases, found in solution on the primeval Earth

primosome: Cluster of proteins (including PriA and primase) that synthesizes a new RNA primer during DNA replication

prion: A protein that can mis-fold into an alternative pathological form that then promotes its own formation auto-catalytically. Improperly folded prion proteins are responsible for the neurodegenerative diseases known as spongiform encephalopathies that include scrapie, kuru, and BSE

prion protein (PrP): The prion protein found in the nervous tissue of mammals and whose improperly folded form is responsible for prion diseases

probe molecule: Molecule that is tagged in some way (usually radioactive or fluorescent) and is used to bind to and detect another molecule

probe: Short for probe molecule

processed pseudogene: Pseudogene lacking introns because it was reverse transcribed from messenger RNA by reverse transcriptase

prokaryote: Lower organism, such as a bacterium, with a primitive type of cell containing a single chromosome and having no nucleus

promoter: Region of DNA in front of a gene that binds RNA polymerase and so promotes gene expression

proofreading: Process that checks whether the correct nucleotide has been inserted into new DNA. Usually refers to DNA polymerase checking whether it has inserted the correct base

prophage: Bacteriophage genome that is integrated into the DNA of the bacterial host cell

prosthetic group: Extra chemical group bound (often covalently) to a protein but which is not part of the polypeptide chain

protease: Same as proteinase; an enzyme that degrades proteins

proteasome: Protein assembly found in eukaryotic cells that degrades proteins

protein: Polymer made from amino acids that does most of the work in the cell

protein A: Antibody-binding protein from *Staphylococcus* that is often used in making fusion proteins

protein interactome: The total of all the protein-protein interactions in a particular cell or organism

protein kinase: An enzyme that adds phosphate groups to another protein

protein microarray: Microarray of immobilized proteins used for proteome analysis and normally screened by fluorescent or radioactive labeling

protein primer: Protein used instead of RNA as a primer for DNA synthesis in some bacteria and viruses

proteinoid: Artificially-synthesized polypeptide containing randomly-linked amino acids

proteome: The total set of proteins encoded by a genome or the total protein complement of an organism

protomer: A single polymer chain that is itself a subunit for a higher level of assembly

provirus: Virus genome that is integrated into the host cell DNA

pseudogene: Defective copy of a genuine gene

pseudouridine: An isomer of uridine that is introduced into some RNA molecules by post-transcriptional modification

purine: Type of nitrogenous base with a double ring found in DNA and RNA

pyrimidine: Type of nitrogenous base with a single ring found in DNA and RNA

pyrosequencing: Sequencing method based on the generation of light pulses when a base is added onto a growing nucleotide chain by DNA polymerase

pyrrolysine (Pyl): 22nd genetically-encoded amino acid, derived from lysine

quasi-species: A set of closely-related sequences whose individual members vary from consensus by frequent errors or mutations

quaternary structure: Aggregation of more than one polymer chain in the final structure

quenching group: Molecule that prevents fluorescence by binding to the fluorophore and absorbing its activation energy

quinolone antibiotics: A family of antibiotics, including nalidixic acid, norfloxacin, and ciprofloxacin that inhibit DNA gyrase and other type II topoisomerases by binding to the A-subunit

RACE: See rapid amplification of cDNA ends

Rad proteins: Group of proteins involved in recombination and repair of DNA damage in yeast and animal cells. Rad51 corresponds to the prokaryotic RecA protein

radiation hybrid: A cell (usually from a rodent) that contains fragments of chromosomes (generated by irradiation) from another species

radical replacement: Replacement of an amino acid with another that has different chemical and physical properties

radioisotope: Radioactive form of an element

random coil: Region of polypeptide chain lacking secondary structure

randomly amplified polymorphic DNA (RAPD): Method for testing genetic relatedness using PCR to amplify arbitrarily chosen sequences

rapid amplification of cDNA ends (RACE): RT-PCR-based technique that generates the complete 5′ or 3′ end of a cDNA sequence starting from a partial sequence

reading frame: One of three alternative ways of dividing up a sequence of bases in DNA or RNA into codons

RecA protein: Protein involved in recombination and repair of DNA in *E. coli* that binds single-stranded DNA

recessive allele: the allele whose properties are not observed because they are masked by the dominant allele

recipient cell: Cell that receives DNA from another cell

recombinants: gametes in which genetic recombination occurred

recombination: Exchange of genetic information between chromosomes or other molecules of DNA

recombineering: Technique that uses homologous recombination to insert a piece of DNA into a vector

regulatory nucleotide: A nucleic acid base that is modified and used as a signaling molecule

regulatory protein: A protein that regulates the expression of a gene or the activity of another protein

regulatory region: DNA sequence in front of a gene, used for regulation rather than to encode a protein

regulon: A set of genes or operons that are regulated by the same regulatory protein even though they are at different locations on the chromosome

release factor: Protein that recognizes a stop codon and brings about the release of a finished polypeptide chain from the ribosome

renaturation: Re-annealing of single-stranded DNA or refolding of a denatured protein to give the original natural 3D structure

repeated sequences: DNA sequences that exist in multiple copies

repetitive sequences: Same as repeated sequences

replication bubble (replication eye): Bulge where DNA is in the process of replication

replication: Duplication of DNA prior to cell division

replication factor C (RFC): Eukaryotic protein that binds to initiator DNA and loads DNA polymerase δ plus its sliding clamp onto the DNA

replication fork: Region where the enzymes replicating a DNA molecule are bound to untwisted, single-stranded DNA

replicative form (RF): Circular double-stranded version of a virus genome used for rolling circle replication

replicative transposition: Type of transposition in which two copies of the transposon are generated; one in the original site and another at a new location

replicon: A molecule of DNA or RNA that is self-replicating; that is, it has its own origin of replication

replisome: Assemblage of proteins (including primase, DNA polymerase, helicase, SSB protein) that replicates DNA

reporter gene: Gene that is used in genetic analysis because its product is convenient to assay or easy to detect

reporter protein: A protein that is easy to detect and gives a signal that can be used to reveal its location and/or indicate levels of gene expression

repressor: Regulatory protein that prevents a gene from being transcribed

resolution: Cleavage of the junction where two DNA molecules are fused together so releasing two separate DNA molecules. Refers to the breakdown both of crossovers formed during recombination and of cointegrates formed by transposition

resolvase: An enzyme that cuts apart a cointegrate releasing two separate molecules of DNA

restriction enzyme: Type of endonuclease that cuts double-stranded DNA at a specific sequence of bases, the recognition site

restriction fragment length polymorphism (RFLP): A difference in restriction enzyme sites between two related DNA molecules that results in production of restriction fragments of different lengths

restriction map: A diagram showing the location of restriction enzyme cut sites on a segment of DNA

retroelement: A genetic element that uses reverse transcriptase to convert the RNA form of its genome to a DNA copy

retron: Genetic element found in bacteria that encodes reverse transcriptase and uses it to make a bizarre RNA/DNA hybrid molecule

retroposon: Short for retrotransposon

retro-pseudogene: Another name for a processed pseudogene

retrotransposon: A transposable element that uses reverse transcriptase to convert the RNA form of its genome to a DNA copy

retrovirus: Type of virus that has its genes as RNA in the virus particle but converts this to a DNA copy inside the host cell by using reverse transcriptase

reverse transcriptase: An enzyme that uses single-stranded RNA as a template for making double-stranded DNA

reverse transcriptase PCR (RT-PCR): Variant of PCR that allows genes to be amplified and cloned as intron-free DNA copies by starting with mRNA and using reverse transcriptase

reverse transcription: The process in which single-stranded RNA is used as a template for making double-stranded DNA

reverse turn: Region of polypeptide chain that turns around and goes back in the same direction

reversion: Alteration of DNA that reverses the effects of a prior mutation

R-group: Any unspecified chemical group; in particular, the side chain of an amino acid

Rho (ρ) protein: Protein factor needed for successful termination at certain transcriptional terminators

Rho-dependent terminator: Transcriptional terminator that depends on Rho protein

Rho-independent terminator: Transcriptional terminator that does not need Rho protein

ribonuclease (RNase): Enzyme that cuts or degrades RNA

ribonuclease: A nuclease that cuts RNA

ribonuclease H (RNase H): Enzyme that degrades the RNA strand of DNA:RNA hybrid double helixes. In bacteria it removes the major portion of RNA primers used to initiate DNA synthesis

ribonuclease H: A ribonuclease of bacterial cells that is specific for RNA-DNA hybrids

ribonuclease III: A ribonuclease of bacteria whose main function is processing rRNA and tRNA precursors

ribonuclease P: A ribonuclease involved in processing tRNA in bacteria that consists of an RNA ribozyme plus an accessory protein

ribonucleic acid (RNA): Nucleic acid that differs from DNA in having ribose in place of deoxyribose and having uracil in place of thymine

ribonucleoside: A nucleoside whose sugar is ribose (not deoxyribose)

ribonucleotide reductase: Enzyme that reduces ribonucleotides to deoxyribonucleotides

ribose: The five-carbon sugar found in RNA

ribosomal RNA (rRNA): Class of RNA molecule that makes up part of the structure of a ribosome

ribosome binding site (RBS): Same as Shine-Dalgarno sequence; sequence close to the front of mRNA that is recognized by the ribosome; only found in prokaryotic cells

ribosome modulation factor (RMF): Protein that inactivates surplus ribosomes during slow growth or stationary phase in bacteria

ribosome recycling factor (RRF): Protein that dissociates the ribosomal subunits after a polypeptide chain has been finished and released

ribosome: The cell's machinery for making proteins

riboswitch: Domain of messenger RNA that directly senses a signal and controls translation by alternating between two structures

ribozyme: An RNA enzyme; that is, an RNA molecule with catalytic activity

rickettsia: Type of degenerate bacterium that is an obligate parasite and infects the cells of higher organisms

right-handed double helix: In a right-handed helix, as the observer looks down the helix axis (in either direction), each strand turns clockwise as it moves away from the observer

R-loop analysis: Hybridization of the DNA copy of a gene to the corresponding mRNA that results in the appearance of loops that represent the intervening sequences or introns in the DNA

RNA editing: Changing the coding sequence of an RNA molecule after transcription by altering, adding, or removing bases

RNA interference (RNAi): Response that is triggered by the presence of double-stranded RNA and results in the degradation of mRNA or other RNA transcripts homologous to the inducing dsRNA

RNA polymerase: Enzyme that synthesizes RNA

RNA polymerase I: Eukaryotic RNA polymerase that transcribes the genes for the large ribosomal RNAs

RNA polymerase II: Eukaryotic RNA polymerase that transcribes the genes encoding proteins

RNA polymerase III: Eukaryotic RNA polymerase that transcribes the genes for 5S ribosomal RNA and transfer RNA

RNA primer: Short segment of RNA used to initiate synthesis of a new strand of DNA during replication

RNA replicase: Special RNA polymerase used by RNA viruses to replicate their RNA genomes

RNA thermosensor: A specialized riboswitch that responds to temperature to control mRNA translation

RNA virus: A virus whose genome consists of RNA

RNA world: The hypothetical stage of early life in which RNA encoded genetic information and carried out enzyme reactions without the need for either DNA or protein

RNA-dependent RNA polymerase (RdRP): RNA polymerase that uses RNA as a template and is involved in the amplification of the RNAi response

RNA-induced silencing complex (RISC): Protein complex induced by siRNA that degrades single-stranded RNA corresponding in sequence to the siRNA

RNA-seq: the use of high-throughput cDNA sequencing to characterize an RNA sample

rolling circle replication: Mechanism of replicating double-stranded circular DNA that starts by nicking and unrolling one strand and using the other, still circular, strand as a template for DNA synthesis. Used by some plasmids and viruses

R-plasmid: Plasmid that carries genes for antibiotic resistance

Rubisco (ribulose bisphosphate carboxylase): A critical enzyme in the fixation of carbon dioxide during photosynthesis

S1 nuclease: Endonuclease from *Aspergillus oryzae* that cleaves single-stranded RNA or DNA but does not cut double-stranded nucleic acids

S1 nuclease mapping: Method using S1 nuclease to locate the 5′ end or 3′ end of a transcript

satellite DNA: Highly repetitive DNA of eukaryotic cells that is found as long clusters of tandem repeats and is permanently coiled tightly into heterochromatin

satellite RNA: Parasitic RNA molecule that requires a helper virus for replication and capsid formation

satellite virus: A defective virus that needs an unrelated helper virus to infect the same host cell in order to provide essential functions

saturated: (Referring to enzymes) When all the active sites are filled with substrate and the enzyme cannot work any faster

scaffold attachment region (SAR): Site on eukaryotic DNA that binds to proteins of the chromosomal scaffold or of the nuclear matrix—same as MAR sites

scintillant: Molecule that emits pulses of light when hit by a particle of radioactivity

scintillation counter: Machine that detects and counts pulses of light

scintillation counting: Detection and counting of individual microscopic pulses of light

Scorpion primer: DNA primer joined to a molecular beacon by an inert linker. When the probe sequence binds target DNA, the quencher and fluorophore are separated allowing fluorescence

scrapie: An infectious disease of sheep that causes degeneration of the brain and is caused by mis-folded prion proteins

scrapie PrP (PrPSc): The pathological form of the prion protein, sometimes known as the scrapie agent

secondary atmosphere: The atmosphere of the Earth after the light gases were lost and resulting mostly from volcanic out-gassing. It contained reduced gases but no oxygen

secondary endosymbiosis: Uptake by an ancestral eukaryotic cell of another single-celled eukaryote, usually an alga, thus providing chloroplasts at second hand

secondary structure: Initial folding up of a polymer due to hydrogen bonding

second-site revertant: Revertant in which the change in the DNA, which suppresses the effect of the mutation, is at a different site to the original mutation

segregation: Replication of a hybrid DNA molecule (whose two strands differ in sequence) to give two separate DNA molecules, each with a different sequence

selenocysteine (Sec): Amino acid resembling cysteine but containing selenium instead of sulfur

selenocysteine insertion sequence (SECIS element): Recognition sequence that signals for insertion of selenocysteine at a UGA stop codon

self-assembly: Automatic assembly of protein subunits without need of any outside assistance

selfish DNA: Any segment of DNA that replicates but is of no use to the host cell it inhabits

self-splicing: Splicing out of an intron by the ribozyme activity of the RNA molecule itself without the requirement for a separate protein enzyme

semi-conservative replication: Mode of DNA replication in which each daughter molecule gets one of the two original strands and one new complementary strand

sense RNA: Normal RNA that has been produced from the non-coding strand of DNA

sense strand: The strand of DNA equivalent in sequence to the mRNA (same as plus strand)

sensor kinase: A protein that phosphorylates itself when it senses a specific signal (often an environmental stimulus, but sometimes an internal signal)

septum: Cross-wall that separates two new bacterial cells after division

Sequenase®: Genetically-modified DNA polymerase from bacteriophage T7 used for sequencing DNA

sequence tagged site (STS): A short sequence (usually 100–500bp) that is unique within the genome and can be easily detected, usually by PCR

sequestration protein (SeqA): Protein that binds the origin of replication, thereby delaying its methylation

serial analysis of gene expression (SAGE): Method to monitor level of multiple mRNA molecules by sequencing a DNA concatemer that contains many serially-linked sequence tags derived from the mRNAs

sex pilus: Protein filament made by donor bacteria that binds to a suitable recipient and draws the two cells together

sex-linked: A gene is sex-linked when it is carried on one of the sex chromosomes

sexual reproduction: Form of reproduction that involves re-shuffling of the genes between two individuals

Shine-Dalgarno (S-D) sequence: Same as RBS; sequence close to the front of mRNA that is recognized by the ribosome; only found in prokaryotic cells

short interfering RNA (siRNA): Double-stranded RNA molecules of 21–23 nucleotides involved in triggering RNA interference in eukaryotes

short interspersed element (SINE): Short repeated sequence that makes up a significant fraction of the moderately- or highly-repetitive DNA of mammals

shotgun sequencing: Approach in which the genome is broken into many random short fragments for sequencing. The complete genome sequence is then assembled by computerized searching for overlaps between individual sequences

shuttle vector: A vector that can survive in and be moved between more than one type of host cell

sigma subunit: Subunit of bacterial RNA polymerase that recognizes and binds to the promoter sequence

signal molecule: A small molecule that triggers a regulatory response by binding to a regulatory protein

signal sequence: Short, largely hydrophobic sequence of amino acids at the front of a protein that labels it for export

silencing: In genetic terminology, refers to switching off genes in a relatively non-specific manner

silent mutation: An alteration in the DNA sequence that has no effect on the phenotype

simian virus 40 (SV40): A small, spherical dsDNA virus that causes cancer in monkeys by inserting its DNA into the host chromosome

simple sequence length polymorphism (SSLP): Any DNA region consisting of tandem repeats that vary in number from individual to individual, including VNTRs, microsatellites, and other tandem repeats

SINE: Short interspersed element

single nucleotide polymorphism (SNP): A difference in DNA sequence of a single base change between two individuals

single strand binding protein (SSB protein): A protein that keeps separated strands of DNA apart

site-directed mutagenesis: Deliberate alteration of a specific DNA sequence by any artificial technique

site-specific recombination: Recombination between two lengths of DNA that are largely unrelated. It involves specific proteins that recognize particular sequences and form crossovers between them. Same as non-homologous recombination

Slicer: Ribonuclease activity of the RISC complex

sliding clamp: Subunit of DNA polymerase that encircles the DNA, thereby holding the core enzyme onto the DNA

small cytoplasmic RNA (scRNA): Small RNA molecules of varied function found in the cytoplasm of eukaryotic cells

small nuclear ribonucleoprotein (snRNP): Complex of snRNA plus protein

small nuclear RNA (snRNA): Small RNA molecules that are involved in RNA splicing in the nucleus of eukaryotic cells

small nucleolar RNA (snoRNA): Small RNA molecules that are involved in ribosomal RNA base modification in the nucleolus of eukaryotic cells

small RNA (sRNA): An RNA molecule that sequesters regulatory proteins from functioning as translational repressors (e.g., CsrB and CsrC)

small subunit: The smaller of the two ribosomal subunits, 30S in bacteria, 40S in eukaryotes

SMRT sequencing: (for single-molecule real-time) Third-generation sequencing method that identifies the nucleotide added by DNA polymerase onto a growing strand of DNA using fluorescently labeled pyrophosphate

snurp: snRNP or small nuclear ribonucleoprotein

sodium dodecyl sulfate (SDS): A detergent widely used to denature and solubilize proteins before separation by electrophoresis

somatic cell: Cell making up the body, as opposed to the germline

SOS system: An error-prone repair system of bacteria that responds to severe DNA damage

Southern blotting: A method to detect single-stranded DNA that has been transferred to nylon paper by using a probe that binds DNA

specialized transduction: Type of transduction where certain regions of the bacterial DNA are carried preferentially

species: A group of closely-related organisms with a relatively recent common ancestor. Among animals, species are populations that breed among themselves but not with individuals of other populations. No satisfactory definition exists for bacteria or other organisms that do not practice sexual reproduction

specific regulation: Regulation that applies to a single gene or operon or to a very small number of related genes

specific transcription factor: Transcription factor needed for expression of certain specific genes under specific conditions

S-phase: Stage in the eukaryotic cell cycle in which chromosomes are duplicated

spliceosome: Complex of proteins and small nuclear RNA molecules that removes introns during the processing of messenger RNA

splicing: Removal of intervening sequences and re-joining the ends of a molecule; usually refers to removal of introns from RNA

spontaneous mutation: Mutation that occurs "naturally" without the help of mutagenic chemicals or radiation

spore: A cell specialized for survival under adverse conditions and/or designed for distribution

star activity: Imprecise or random cleavage of DNA by restriction enzymes that only occurs under specific reaction conditions

start codon: The special AUG codon that signals the start of a protein

stem and loop: Structure made by folding an inverted repeat sequence

steroid receptor: Protein that binds steroid hormones

sticky ends: Ends of a double-stranded DNA molecule that have unpaired single-stranded overhangs, generated by a staggered cut

stop codon: Codon that signals the end of a protein

streptavidin: Protein from Streptococcus that binds biotin very tightly

streptomycin: An antibiotic of the aminoglycoside family that inhibits protein synthesis

stringent response: Decreasing transcription of non-essential genes when nutrients are in limited supply

structural gene: Sequence of DNA (or RNA) that codes for a protein or for an untranslated RNA molecule

structural protein: A protein that forms part of a cellular structure

substrate: Molecule that binds to an enzyme and is the target of enzyme action

subtractive hybridization: Technique used to remove unwanted DNA or RNA by hybridization, so leaving behind the DNA or RNA molecule of interest

subviral agent: Infectious agents that are more primitive than viruses and encode fewer of their own functions

sulfhydryl group: -SH; Chemical group of sulfur and hydrogen

sulfonamide: Antibiotic that inhibits the synthesis of the folate cofactor

sulfonamides: Synthetic antibiotics that are analogs of *p*-aminobenzoic acid, a precursor of the vitamin folic acid. Sulfonamides inhibit dihydropteroate synthetase

supercoiling: Higher level coiling of DNA that is already a double helix

suppressor mutation: A mutation that restores function to a defective gene by suppressing the effect of a previous mutation

suppressor tRNA: A mutant tRNA that recognizes a stop codon and can insert an amino acid when it reads a stop codon on the mRNA

S-value: The sedimentation coefficient is the velocity of sedimentation divided by the centrifugal field. It is dependent on mass and is measured in Svedberg units

Swi/Snf ("switch sniff") complex: Larger type of chromatin remodeling complex

SYBR Green I: A DNA-binding fluorescent dye that binds only to double-stranded DNA and becomes fluorescent only when bound

symbiosis: Association of two living organisms that interact

symbiotic theory: Theory that the organelles of eukaryotic cells are derived from symbiotic prokaryotes

synapsis: the process in which homologous paternal and maternal chromosomes align so each gene is in the same location

systems biology: A term that refers to the integration of many different types of research on an organism with the goal of defining the biological state of an organism within a certain environment

T4 ligase: Type of DNA ligase from bacteriophage T4 and which is capable of ligating blunt ends

TA cloning: Procedure that uses *Taq* polymerase to generate single 3'-A overhangs on the ends of DNA segments that are used to clone DNA into a vector with matching 3'-T overhangs

TA cloning vector: Vector with single 3'-T overhangs (in its linearized form) that is used to clone DNA

segments with single 3'-A overhangs generated by Taq polymerase

tail specific protease: Enzyme that destroys mis-made proteins by degrading them tail-first; that is, from the carboxyl end

tandem duplication: Mutation in which a segment of DNA is duplicated and the second copy remains next to the first

tandem mass spectrometry (MS/MS): Two successive rounds of mass spectrometry in which a parent ion is first isolated and then fragmented into daughter ions for more detailed analysis

tandem repeats: Repeated sequences of DNA (or RNA) that lie next to each other

Taq **polymerase:** Heat-resistant DNA polymerase from *Thermus aquaticus* that is used for PCR

TaqMan probe: Fluorescent probe consisting of two fluorophores linked by a DNA probe sequence. Fluorescence increases only after the fluorophores are separated by degradation of the linking DNA

target DNA: DNA that is the target for binding by a probe during hybridization or the target for amplification by PCR

target sequence: Sequence on host DNA molecule into which a transposon inserts itself

TATA binding protein (TBP): Transcription factor that recognizes the TATA box

TATA box: Binding site for a transcription factor that guides RNA polymerase II to the promoter in eukaryotes

TATA box factor: Another name for TATA-binding protein

tautomerization: Alternation of a molecule, in particular a base of a nucleic acid, between two different isomeric structures

Tc1 element: Transposon *Caenorhabditis* 1. A transposon of the mariner family found in the nematode *Caenorhabditis*

T-DNA (tumor-DNA): Region of the Ti plasmid that is transferred into the plant cell nucleus

telomerase: Enzyme that adds DNA to the end, or telomere, of a chromosome

telomere: Specific repetitive sequence of DNA found at the end of linear eukaryotic chromosomes

temperature-sensitive (ts) mutation: Mutation whose phenotypic effects depend on temperature

template strand: Strand of DNA used as a guide for synthesizing a new strand by complementary base-pairing

Ter **site:** Site in the terminus region that blocks movement of a replication fork

teratogen: An agent that causes abnormal embryo development leading to gross structural defects or monstrosities

terminator: DNA sequence at end of a gene that tells RNA polymerase to stop transcribing

terminus: Region on a chromosome where replication finishes

tertiary atmosphere: The present atmosphere of the Earth resulting from biological activity

tertiary structure: Final 3D folding of a polymer chain

tet **operon:** Bacterial genes that produce proteins that confer resistance to the antibiotic tetracycline

tetracycline: Antibiotic that binds to 16S ribosomal RNA and inhibits protein synthesis

tetracyclines: Family of antibiotics that inhibit protein synthesis

tetrad: A structure found in prophase I of meiosis in which the two duplicated sister chromatids align to create a complex of four homologous chromosomes

tetrahydrofolate (THF): Reduced form of dihydrofolate cofactor that is needed for making precursors for DNA and RNA synthesis

tetraploid: Having four copies of each gene

thermocycler: Machine used to rapidly shift samples between several temperatures in a pre-set order (for PCR)

***Thermus aquaticus*:** Thermophilic bacterium found in hot springs and used as a source of thermostable DNA polymerase

theta-replication: Mode of replication in which two replication forks go in opposite directions around a circular molecule of DNA

third-base redundancy: Situation where a set of four codons all code for the same amino acid and thus the identity of the third codon base makes no difference during translation

thymidylate synthetase: Enzyme that adds a methyl group, so converting the uracil of dUMP to thymine

thymine (T): A pyrimidine base found in DNA that pairs with adenine

Ti plasmid: Tumor-inducing plasmid. Plasmid that is carried by soil bacteria of the *Agrobacterium* group and confers the ability to infect plants and produce tumors

tight mutation: Mutation whose phenotype is clear-cut due to the complete loss of function of a particular gene product

tiling array: Type of microarray that consists of probes covering the whole genome, not just coding sequences

time-of-flight (TOF): Type of mass spectrometry detector that measures the time for an ion to fly from the ion source to the detector

tmRNA: Specialized RNA used to terminate protein synthesis when a ribosome is stalled by a damaged mRNA

tobacco mosaic virus: A filamentous single-stranded RNA virus that infects a wide range of plants

topoisomerase: Enzyme that alters the level of supercoiling or catenation of DNA (i.e., changes the topological conformation)

topoisomerase IV: A particular topoisomerase involved in DNA replication in bacteria

topoisomers: Isomeric forms that differ in topology (i.e., their level of supercoiling or catenation)

totipotent: Capable of giving rise to a complete multicellular organism

tra **genes:** Genes needed for plasmid transfer

transcription: Conversion of information from DNA into its RNA equivalent

transcription bubble: Region where DNA double helix is temporarily opened up so allowing transcription to occur

transcription factor: Protein that regulates gene expression by binding to DNA in the control region of the gene

transcription-coupled repair: Preferential repair of the template strand of DNA that may be transcribed

transcriptome: The total sum of the RNA transcripts found in a cell, under any particular set of conditions

transduction: Process in which genes are transferred inside virus particles

transfection: Process in which purified viral DNA enters a cell by transformation. Often used to refer to entry of any DNA, even if not of viral origin, into an animal cell

transfer RNA (tRNA): RNA molecules that carry amino acids to a ribosome

transferability: Ability of a plasmid to move itself from one host cell to another

transformation: (As used of cancer) Changing a normal cell into a cancer cell, even if no extra DNA enters the cell

transition: Mutation in which a pyrimidine is replaced by another pyrimidine or a purine is replaced by another purine

transition state analog: Enzyme inhibitor that mimics the reaction intermediate or transition state, rather than the substrate

transition state: Another term for the activated intermediate in a chemical reaction

transition state energy: Energy difference between the reactants and the activated reaction intermediate or transition state

translation: Making a protein using the information provided by messenger RNA

translational activator: A protein that binds to mRNA and promotes its translation

translational repression: Form of control in which the translation of a messenger RNA is prevented

translational repressor: A protein that binds to mRNA and prevents its translation

translatome: The total set of proteins that have actually been translated and are present in a cell under any particular set of conditions

translocase: Enzyme complex that transports proteins across membranes

translocation: a) Transport of a newly-made protein across a membrane by means of a translocase; b) sideways movement of the ribosome on mRNA during translation; and c) removal of a segment of DNA from a chromosome and its reinsertion in a different place

transmissable spongiform encephalopathy (TSE): Technical name for infectious prion disease

transport protein: A protein that carries other molecules across membranes or around the body

transposable element: A mobile segment of DNA that is always inserted in another host molecule of DNA. It has no origin of replication of its own and relies on the host DNA for replication. Includes both DNA-based transposons and retrotransposons

transposable element or transposon: Segment of DNA that can move as a unit from one location to another, but which always remains part of another DNA molecule

transposase: Enzyme responsible for moving a transposon

transposition: The process by which a transposon moves from one host DNA molecule to another

transposon insertion profiling (TIP)-chip: A method to determine the location of scattered repeated elements within a genome such as LINEs or SINEs

transposon: Same as transposable element, although the term is usually restricted to DNA-based elements that do not use reverse transcriptase

trans-splicing: Splicing of a segment from one RNA molecule into another distinct RNA molecule

transversion: Mutation in which a pyrimidine is replaced by a purine or vice versa

Tra+: Transfer-positive (refers to a plasmid capable of self-transfer)

trimethoprim: Antibiotic that inhibits dihydrofolate reductase of bacteria

triploid: Having three copies of each gene

trisomy: Having three copies of a particular chromosome

true revertant: Revertant in which the original base sequence is exactly restored

trypanosome: Type of single-celled eukaryotic microorganism that lives as a parasite in higher animals and causes diseases such as sleeping sickness

Tus protein: Bacterial protein that binds to *Ter* sites and blocks movement of replication forks

twist, T: The number of double helical turns in a molecule of DNA (or double-stranded RNA)

two-component regulatory system: A regulatory system consisting of two proteins, a sensor kinase and a DNA-binding regulator

two-hybrid system: Method of screening for protein-protein interactions that uses fusions of the proteins being investigated to the two separate domains of a transcriptional activator protein

Ty1 element: Transposon yeast 1. A retrotransposon of yeast that moves via an RNA intermediate

type I restriction enzyme: Type of restriction enzyme that cuts the DNA a thousand or more base pairs away from the recognition site

type I topoisomerase: Topoisomerase that cuts a single strand of DNA and therefore changes the linking number by one

type II restriction enzyme: Type of restriction enzyme that cuts the DNA in the middle of the recognition site

type II topoisomerase: Topoisomerase that cuts both strands of DNA and therefore changes the linking number by two

U1: Snurp (snRNP) that recognizes the upstream splice site

U2: Snurp (snRNP) that recognizes the branch site

U2AF (U2 accessory factor): Protein involved in splicing of introns that recognizes the downstream splice site

ubiquitin: Small protein attached to other proteins as a signal that they should be degraded; used by eukaryotic cells, not bacteria

uncharged tRNA: tRNA without an amino acid attached

unequal crossing over: Crossing over in which the two segments that cross over are of different lengths; often due to misalignment during pairing of DNA strands

Ung protein: Same as uracil-N-glycosylase

universal genetic code: Version of the genetic code used by almost all organisms

upstream element: DNA sequence upstream of the TATA box in eukaryotic promoters that is recognized by specific proteins

upstream region: Region of DNA in front (i.e., beyond the 5′ end) of a structural gene; its bases are numbered negatively counting backwards from the start of transcription

uracil (U): A pyrimidine base found in RNA that may pair with adenine

uracil-N-glycosylase: Enzyme that removes uracil from DNA

urea: A nitrogen waste product of animals; also widely used as a denaturant of proteins

urkaryote: Hypothetical ancestor that provided the genetic information of the eukaryotic nucleus

vaccination: Artificial induction of the immune response by injecting foreign proteins or other antigens

variable number tandem repeats (VNTR): Cluster of tandemly-repeated sequences in the DNA whose number of repeats differs from one individual to another

vector: (a) In molecular biology a vector is a molecule of DNA that can replicate and is used to carry cloned genes or DNA fragments; (b) in general biology a vector is an organism (such as a mosquito) that carries and distributes disease-causing microorganisms (such as yellow fever or malaria)

vertical gene transfer: Transfer of genetic information from an organism to its descendents

vertical gene transmission: Transfer of genetic information from an organism to its descendents

"very short patch repair": System that removes a short length of single-stranded DNA around a T/G mismatched base pair within the Dcm methylase recognition sequences CCAGG or CCTGG

viral genome: Molecule of DNA or RNA that carries the genes of a virus

virion: A virus particle

viroid: Naked single-stranded circular RNA that forms a stable highly base-paired rod-like structure and replicates inside infected plant cells. Viroids do not encode any proteins but possess self-cleaving ribozyme activity

virulence factors: Proteins that promote virulence in infectious bacteria. Include toxins, adhesins, and proteins protecting bacteria from the immune system

virulence plasmid: Plasmid that carries genes for virulence factors that play a role in bacterial infection

virus: Infectious agent, consisting of DNA or RNA inside a protective shell of protein, that must infect a host cell in order to replicate

VNTR: See variable number tandem repeats

Vsr endonuclease: Enzyme that nicks or cuts the DNA backbone next to a T/G mismatched base pair in the "very short patch repair" system

Western blot: Detection method in which an antibody is used to identify a specific protein

Western blotting: Detection technique in which a probe, usually an antibody, binds to a protein target molecule

wild-type: The original or "natural" version of a gene or organism

wobble rules: Rules allowing less rigid base-pairing but only for codon/anticodon pairing

writhe: Same as writhing number, W

writhing number, W: The number of supercoils in a molecule of DNA (or double-stranded RNA)

X-chromosome: female sex chromosome; possession of two X-chromosomes creates a female gender in mammals

X-gal (5-bromo-4-chloro-3-indolyl β-D-galactoside): Artificial substrate that is split by β-galactosidase, releasing a blue dye

X-inactivation: The condensation and complete shutting down of gene expression of one of the two X-chromosomes in cells of female mammals

Xis protein: Enzyme that reverses DNA integration by removing a segment of dsDNA and resealing the gap leaving behind an intact recognition sequence. Same as excisionase. Not to be confused with Xist RNA involved in X chromosome silencing

***Xist* gene:** A gene that causes the inactivation of the X-chromosome that carries it

X-phos: 5-bromo-4-chloro-3-indolyl phosphate, an artificial substrate that is split by alkaline phosphatase, releasing a blue dye

Y-chromosome: Male sex chromosome; possession of a Y-chromosome plus an X-chromosome results in male gender in mammals

yeast artificial chromosome (YAC): Single copy vector based on yeast chromosome that can carry very long inserts of DNA. Widely used in the human genome project

Y-guy: Hypothetical male human ancestor thought to have lived in Africa around 100,000–200,000 years ago

Z-DNA: An alternative form of DNA double helix with left-handed turns and 12 base pairs per turn

zero-mode waveguides: Small nanosized metal cylindrical wells that reduce background light so that only a very small portion of the cylinder can be visualized for fluorescent light flashes

Z-form: An alternative form of double helix with left-handed turns and 12 base pairs per turn. Both DNA and dsRNA may be found in the Z-form

zinc finger: One type of DNA-binding motif common in proteins

zipcode sequence: See barcode sequence; a unique 20 base pair sequence added to a cassette to mark the gene with a tag for subsequent analysis

zoo blotting: Comparative Southern blotting using DNA target molecules from several different animals to test whether the probe DNA is from a coding region

zwitterion: Same as dipolar ion; a molecule with both a positive and a negative charge

zygote: Cell formed by union of sperm and egg which develops into a new individual

Index

In this index a *b* following a page number denotes information found in a box, an *f* denotes a figure, and a *t* denotes a table